Universitext

Universitext

Universitext is a series of textbooks that presents material from a wide variety of mathematical disciplines at master's level and beyond. The books, often well class-tested by their author, may have an informal, personal even experimental approach to their subject matter. Some of the most successful and established books in the series have evolved through several editions, always following the evolution of teaching curricula, to very polished texts.

Thus as research topics trickle down into graduate-level teaching, first textbooks written for new, cutting-edge courses may make their way into *Universitext.*

More information about this series at http://www.springer.com/series/223

Jean-Paul Penot

Analysis

From Concepts to Applications

Jean-Paul Penot
Université Pierre et Marie Curie
Paris, France

ISSN 0172-5939 ISSN 2191-6675 (electronic)
Universitext
ISBN 978-3-319-32409-8 ISBN 978-3-319-32411-1 (eBook)
DOI 10.1007/978-3-319-32411-1

Library of Congress Control Number: 2016960582

Mathematics Subject Classification (2010): 03-01, 06-AXX, 26XX, 28AXX, 34XX, 35AXX, 40-00, 46-01, 47-01, 49-00, 49-01, 54-00, 54-01, 90-01

Printed on acid-free paper

This Springer imprint is published by Springer Nature
The registered company is Springer International Publishing AG Switzerland

To the memory of Hélène

Preface

> Nothing is lost, nothing is created out of nothing, everything is transformed.
> *Lavoisier*

> Mathematics has been created by human beings for their needs. The teaching mathematician must remain a teacher of action.
> *Henri Lebesgue*

> I hear, I forget; I see, I remember; I do, I understand!
> *Chinese proverb*

The purpose of this book is to present some basic results in analysis that can be used to solve various problems, so it may serve as a kind of toolbox. We do not intend to give a complete panorama of the field. In particular, we concentrate on real analysis and leave complex analysis aside; however we consider complex functions now and then. Although we cannot claim that this volume is devoted to applied analysis, our choice of topics is driven by our wish to present results that can be applied to concrete problems. It is in this sense that this book could be called "Motivated Analysis". This expression is due to J.-P. Aubin and means that we shall gather some results of analysis that may be useful in applications, even if the nature of these applications is not the focus of the text. Thus, our route is not the choice of some important problems that occur in applications, as in [22, 62, 87, 88, 101, 129, 130, 140, 178, 191, 236, 241, 245], but neither is it the panorama of deep advances in mathematical concepts presented in [100] and [101].

A leading thread throughout this book is a strategy often advised by mathematicians, given in many instances: if a problem seems to be difficult, try to change it into a more tractable problem. This can be done in various ways. A first approach consists in reformulating the question, keeping its crucial features and dropping inessential details as much as possible (in a sense, mathematics can be considered as an art of strip tease). In doing so, one is often led to a more general problem that is no more difficult to solve. On the contrary, because its framework is rather bare and simple, its solution is often easier to reach. Of course, some knowledge of general mathematics may help.

Another means of solving a problem consists in using a transformation. This approach can been combined with the simplification just mentioned. Quite often, a hidden or rather intricate operation is interchanged with a simple operation via the transformation. For instance, a convolution may be transformed into an addition or a product. Then the problem may appear as much more tractable. Many transformations have been proposed by mathematicians. Among them are those due to Cole-Hopf, Fenchel, Fourier, Hilbert, Laplace, Legendre, Mellin, Radon and Vaslov. For this reason, we focus our attention on some useful transforms, even if we just give an introduction to them. Their applications to economics, engineering, mathematics, medicine and physics are numerous and important. The passage from monotone operators to convex functions (and the inverse passage) is another example of such a transformation that is not yet classical, but it is fruitful and it has attracted a number of mathematicians recently. We give some attention to it in Chap. 9.

The different chapters of this book may serve as separate courses. However, we consider it is important not to neglect the interdependence of the subjects. This is revealed in several instances throughout the book. Integration can be taught without a knowledge of abstract measure theory, but the latter can serve as an important foundation and it is the basis of probability theory. Nemytskii operators play a crucial role in nonlinear partial differential equations. Functional analysis permeates almost all of the topics considered in the book. Elementary differential calculus can be set in Euclidean spaces rather than in normed vector spaces. However the main lines may be hidden by a heavy use of partial derivatives and components and some applications would be lost if we confined differential calculus to such a restricted framework.

The balance between generality and simplicity is not easy to reach. Take the notion of convergence for example. It can be considered as the keystone of analysis. For that reason we present the main lines of the concept, but it could be expounded more thoroughly. In fact, we encourage the reader to prune rather than to develop what is presented here, since we have given more than the essentials of what is needed in practice. Also, the theory of distributions could be given a more prominent role than the one we offer it in our presentation of Sobolev spaces; incidentally the convergence approach is certainly simpler than the topological approach to distributions considered as elements of a dual space. However, we prefer to give a rather direct treatment, even if the concept of transposition is central to the understanding of generalized derivatives and generalized solutions.

The difficulty in choosing the presentation of a subject is well reflected in the following quotation from Julian Barnes, *Staring at the Sun*, (Jonathan Cape, London, 1986) mentioned by I. Smith, in Bulletin of the American Mathematical Society 52 (3) p. 415 (2015):

> ...everything you wanted to say required a context. If you gave the full context, people thought you a rambling old fool. If you didn't give the context, people thought you a laconic old fool.

In his remarkable book [117] L.C. Evans writes "notation is a nightmare". It certainly presents difficulties, in particular when one tries to conciliate various uses.

However, this difficulty is no greater than the challenge posed by the choice of topics and the writing of clear and concise proofs. Our experience with texts that were written one or more lifetimes ago showed us that the need of rigour and precision has increased and is likely to increase more in the future. Thus, we avoid some common abuses such as confusing a function f or $f(\cdot)$ with its value $f(x)$, a sequence (x_n) with its general term x_n or its set of values $\{x_n\}$, and a space with its dual. We distinguish the adjoint A^* of a continuous linear map A and its transpose $A^\intercal$ and we distinguish the derivative Df of a function and its gradient ∇f. If you are not convinced by such distinctions, you are challenged to give at once the higher derivatives of a composite function. Also, we refrain from using the notation f_{x_1} to denote the partial derivative of a function f with respect to its first variable x_1 because f_{x_1} may denote the function $f(x_1, \cdot)$ of the variables other than x_1, the variable x_1 being "frozen". On the other hand, our position is not rigid, as the next two examples show. We do not follow the choice of C. Zalinescu in [265] who denotes by $\alpha^{\#}$ the especial conjugate of a function α defined on $\mathbb{R}_+$ rather than $\mathbb{R}$; albeit his choice is wise, we prefer the reader to see at first glance the link with convex conjugacy, even if there is a risk of confusion (but this risk is limited since we only use nonnegative values of the arguments). Also, we retain the classical terminology concerning "locally uniformly convex" norms despite the fact that the notion is a pointwise notion rather than a local one.

Our general choice of discarding ambiguity may lead to unusual expressions or notations. We hope the reader will not be disturbed by such novelties and that authors will support them. Most mathematicians are unaware that the notation they use was once considered as shocking. Rarely is credit given to their inventors or promoters, such as Oresme (for coordinates and exponents), Recorde (for the sign $=$), Leibniz (for derivatives, products, quotients and $\int$, as the first letter of the Latin *summa* or sum was denoted at that time), and Peano (for inclusion...). In this respect we must say that the notation $\mathbb{P}$ used here for the set of positive numbers (more present in analysis than the set $\mathbb{Q}$ of rational numbers) is, to our knowledge, due to J.M. Borwein. The scalar product of two vectors u, v is denoted here by the compromise $\langle u \mid v \rangle$ between the mathematicians' and the physicists' uses; but, from time to time, we use the dot notation $u \cdot v$ in view of its simplicity. We avoid the notation (u, v) which may mean either a pair or an open interval. In general, we try to avoid ambiguity since mathematics aims at being a clear language. For that reason, we sometimes depart from the most common terminology in order to avoid confusion. But, consciously or unconsciously, we also use some abuses of notation that tradition has made acceptable (until they are rejected?).

In preparing this book, we have benefited from the teaching of our masters, some of whom were actors in the clarification or the setting up of the subject. We also benefited from several excellent books. The readers will easily detect these sources. Our choices were dictated by our wish to present the most elegant proofs. The word "elegant" may seem a strange choice for a mathematical book. However, mathematicians are often sensitive to aesthetics when considering proofs. Under this term they often appreciate bold, clear, concise proofs. A frequent drawback is the effort required from the reader. But the reward is a better understanding of the nature of the result and of its possible applications. For that reason, we encourage the reader to approach proofs with a paper and a pen in order to devise variants or developments or to mark the decisive steps. Experiencing proofs is the best apprenticeship for a field. We are so convinced of the value of proofs that sometimes we give two proofs of the same result; on the other hand we skip some proofs, either because they are too involved or because they are outside the scope of the book. Exercises are often augmented with hints, but not complete solutions. The most difficult exercises are marked with an asterisk; they are included as complements rather than for training. An asterisk also marks results or sections that can be omitted on a first reading.

It is a great pleasure to thank my colleagues and friends who kindly read and criticized parts of the successive versions of this book; in particular, the contributions of Luc Barbet, Marc Dambrine, Marc Durand, Emmanuel Giner, Dena Kazerani, Alexander Kruger, Khadra Nachi, Steve Robinson, Lionel Thibault, Constantin Zălinescu, Nadia Zlateva have been precious for reducing the number of mistakes or misprints and for bringing me some encouragement.

I hope the reader will find this toolbox useful and will enjoy the rich legacy of our predecessors. Comments and criticisms will be welcome at penotj@ljll.math.upmc.fr.

Paris, France
July 2016

Jean-Paul Penot

Gallery

The list of mathematicians that follows is neither a pantheon nor an academy. It is directly connected to the results expounded in this book and is intended to give the reader an idea of the historical appearance of the ideas involved here. The brief indications that follow are by no means biographies. They just give some elements sketching an historical perspective: we endeavor to devote just one line to each author. It is a pity such elements are so brief and in some sense unfair for works whose extent is often large and varied (in particular, for Euler, Gauss, Cauchy, Poincaré and von Neuman). In some cases a biography would be rather short (for instance, Gateaux was killed in 1914 shortly after leaving the Ecole Normale Supérieure), but in many cases mathematicians have enjoyed a rather long life. Also, one may regret that some anecdotes about people who were living persons with various talents and weaknesses could not be reproduced here. For instance, many mathematicians were concerned with philosophical or logical questions, others were involved in the study of physical phenomena. Hausdorff published literary and philosophical works under the pseudonym of Paul Mongré; Vandermonde was one of the founders of Conservatoire des Arts et Métiers as he was involved in metallurgy; Green was a miller and he hardly attended school; W.H. Young worked on so many topics with his wife Grace Chisholm-Young that it is not clear whether the results attributed to him are joint results or not.

Because nationalism has brought so much terrible suffering to human beings, instead of mentioning nationalities, and because present nationalities are not the same as those of the past, in general we pick the names of a few places where the discoveries were elaborated rather than the nationalities of authors. This is also a credit to their scientific environments. When the authors have conducted research in many places, we simply state the name of the country. Our choices suffer from outrageous simplifications, not just of locations, but also for what concern achievements. Happily, many sources can complete the glimpse we give hereafter, among which are: [28, 39, 49, 87, 105, 110, 148, 156] and the web sites

en.wikipedia.org
www.-history.mcs.st-andrews.ac.uk/history/BiogIndex.html
www.storyofmathematics.com

Abel, Niels Henrik, 1802–1829, Oslo: algebraic equations, elliptic functions, integrals, series.
Aleksandrov, Pavel Sergeïlevich, 1896–1982, Moscow: general and algebraic topology.
d'Alembert Le Rond, Jean, 1717–1783, Paris: analysis, complex numbers, statistics
Archimedes, circa 287–212 BC, Greek inventor and mathematician: π, curves.
Arzela, Cesare, 1847–1912: functional analysis, integration.
Ascoli, Giulio, 1843–1896, Milano: sets whose points are functions.
Baire, René, 1874–1932, Montpellier, Dijon: theory of functions, general topology.
Banach, Stefan, 1892–1945, Lwow: functional analysis.
Bernoulli, Jacob, 1654–1705, Basel: curves, polar coordinates, probabilities.
Bernstein, Felix, 1878–1956, Göttingen and USA: set theory
Bessel, Friedrich, 1784–1846, Koënigsberg, astronomer: special functions.
Bochner, Salomon, 1899–1982, Munich, Princeton: integration theory.
du Bois-Raymond, 1831–1889, Heidelberg, Berlin, Freiburg: analysis, Fourier series.
Bolzano, Bernhard, 1781–1848, Prague: set theory, logic, topology.
Boole, George, 1815–1864, Britain, Cork: logic, order, set theory.
Borel, Emile, 1871–1956, Paris: measure theory, probabilities, game theory.
Brouwer, Luitzen Egbertus Jan, 1881–1966, Amsterdam: intuitionism, fixed points.
Brunn, Herman Karl, 1862–1939, Germany: measure and geometry.
Bunyakovsky, Viktor, 1804–1889, Saint Petersburg: analysis, probabilities.
Cantor, Georg Ferdinand Ludwig, 1845–1918, Göttingen, Halle: cardinality, logic.
Carleson, Lennart, born 1928, Stockholm, Los Angeles: harmonic analysis.
Cauchy, Augustin-Louis, 1789–1857, Paris: real and complex analysis.
Cartan, Henri, 1904–2008, Paris: algebraic topology, homology, complex variables.
Caratheodory, Constantin, 1873–1950, Greece, Germany: PDE, calculus of variations.
Cavalieri, Francesco Bonaventura, 1598–1647 Bologna: precursor of integration theory.
Cesàro, Ernesto, 1859–1906, Naples: divergent series and trigonometric series.
Chasles, Michel, 1793–1880, Paris: geometry, in particular projective geometry.
Chebyshev, Pafnuty, 1821–1894, Saint Petersburg: probability, number theory, mechanics.
Dini, Ulisse, 1845–1918, Pisa: analysis, derivatives, geometry.
Dirac, Paul Adrien Maurice, 1902–1984, British physicist using generalized functions.
Dirichlet, Gustav Lejeune, 1805–1859, Göttingen: number theory, series, mechanics.
Egorov, Dmitri Fyodorovich, 1869–1931, Moscow: measure theory.
Euler, Leonhard, 1707–1783, Basel, Saint Petersburg, Berlin: analysis and more.
Fatou, Pierre, 1878–1929, Paris: astronomy, integral calculus, complex analysis.
Fejér, Lipót, 1880–1959, Budapest: Fourier series.
de Fermat, Pierre, 1601–1665, Toulouse: number theory, geometry, optimization.
Fischer, Ernst Sigismund, 1875–1954, Erlangen, Köln: functional analysis.
Fourier, Joseph Jean-Baptiste, 1768–1830, Auxerre, Paris, Grenoble: series, analysis.
Fréchet, Maurice, 1878–1973, Poitiers, Strasbourg, Paris: analysis, metric spaces.
Fresnel, Augustin-Jean, 1788–1827, Paris physicist and engineer: light, integrals.
Friedrichs, Kurt Otto, 1901–1982, Aachen, Braunschweig, New York: Sobolev spaces.
Fubini, Guido, 1879–1943, Pisa, Torino, Princeton: differential equations, integration.
Gagliardo, Emilio, 1930–2008, Pavia: analysis.
Gårding, Lars, 1919–2014, Lund, Sweden: partial differential equations, bird songs
Gateaux, René, 1889–1914, Paris, Roma: functional analysis, differential calculus.
Gauss, Carl Friedrich, 1777–1855, Göttingen: geometry, number theory, analysis.
Gram, Jørgen Pedersen, 1850–1916, Copenhagen: approximation, number theory determinants.
Green, George, 1793–1841, English mathematician and physicist: a formula.

Guldin, Paul, 1577–1643, Vienna, Graz: area, geometry, physics.
Hadamard, Jacques, 1865–1963, Paris: P.D.E., differentiability, well-posedness.
Hahn, Hans, 1879–1934, Vienna, Bonn: measure theory, functional analysis.
Hausdorff, Felix, 1868–1942, Greisswald, Berlin: topology, measure.
Hermite, Charles, 1822–1901, Paris: polynomial functions, elliptic functions, algebra.
Hilbert, David, 1862–1943, Königsberg, Göttingen, foundations, functional analysis.
Hille, Einar Carl, 1894–1980, Princeton, Yale: semigroups, evolution equations.
Hölder, Ludwig Otto, 1859–1937, Göttingen, Tubingen: analysis, group theory, logic.
Hörmander, Lars, 1931–2012, Stockholm: convexity, analysis, P.D.E.
Jacobi, Carl Gustav Jacob, 1804–1851, Berlin, elliptic functions, determinants.
Jensen, Johan Ludwig W.V., 1859–1925, Engineer of the Copenhagen Telephone Co
John, Fritz, 1910–1994 American mathematician of German origin, analysis
Jordan, Camille, 1838–1922, Paris: curves, equations, measure theory.
Kantorovich, Leonid 1912–1986, Petrograd: linear programming, economics, lattices.
Komolgorov, Andrei Nikolaevich, 1903–1987, Moscow: probability, complexity.
Lagrange, Joseph Louis, 1736–1813 Torino, Berlin, Paris: optimization, analysis.
Laguerre, Edmond Nicolas, 1834–1886, Paris: geometry, algebraic equations.
Landau, Edmund Georg Hermann, 1877–1938, Berlin, Jerusalem: numbers, complex analysis
Laplace, Pierre-Simon, 1749–1827, Paris: astronomy, probability, analysis, P.D.E.
Lax, Peter David, Hungarian-born, New York: analysis, P.D.E.
Lebesgue, Henri, Léon, 1875–1941, Rennes, Poitiers, Paris: integration, topology.
Legendre, Adrien-Marie, 1752–1833, Paris: polynomial functions, number theory.
Leibniz, Gottfried, 1646–1716, Leipzig, Berlin, Hanover: philosophy, analysis.
Levi, Beppo, 1875–1961, Torino, Rosario: integration, quantum mechanics.
Lindelöf, Lorentz, 1870–1946, Helsinski: functional analysis, topology.
Lions, Jacques-Louis, 1928–2001, Paris, numerical analysis, modeling, control, P.D.E.
Lipschitz, Rudolf Otto, 1832–1903, Berlin, Breslau, Bonn: Brown, Cornell, Princeton differential equations, geometry.
Mazur, Stanisław, 1905–1981, Lwów, Lodz: functional analysis, convexity.
Meyers, Norman G., Minneapolis: partial differential equations
Minkowski, Hermann, 1864–1909, Zurich, Göttingen: geometry, arithmetic, relativity.
Monge, Gaspard, 1746–1818, Paris: geometry, P.D.E., optimization.
Moreau, Jean-Jacques, 1923–2014, Montpellier: mechanics, convexity, algorithms.
Morrey, Charles B. Jr, 1907–1980, Berkeley, Brown, Cornell, Princeton: calculus of variations, P.D.E.
Morse, Marston, 1892–1977: differential topology.
von Neuman, János (John), 1903–1957, Berlin, Princeton: algebras, games, economics.
Newton, Isaac, 1642–1727, Cambridge: mechanics, analysis, differentials, physics.
Nikodým, Otton Marcyn, 1887–1974, Poland, U.S.A.: measure theory, topology.
Nirenberg, Louis, born 1925, New York, P.D.E., nonlinear analysis.
Ostrogradski, Mikhail, 1801–1861, St Petersburg, math. physics, celestial mechanics.
Parseval, Marc Antoine, 1755–1836, Paris: geometry, trigonometric series.
Peano, Giuseppe, 1858–1932, Torino: axiomatic of sets, analysis (famous curve), area.
Picard, Emile, 1856–1941, Paris: geometry, differential equations, analytic functions, groups.
Plancherel, Michel, 1885–1967, Friburg, Zürich: integration theory, Fourier transform.
Plateau, Joseph, 1801–1883, Belgium, Liège, Gand: minimal surfaces, vision.
Poincaré, Henri,1854–1912, Paris, differential equations, chaos, topology, physics.
Poisson Siméon Denis, 1741–1840, Paris probability, P.D.E., mathematical physics.
Pythagoras, circa 570–500 BC Greece, philosopher: geometry, irrational numbers.
Rademacher, Hans 1892–1969, Breslau, Pennsylvany: analysis, number theory.

Radon, Johann, 1887–1956, Vienna, measure theory, integration.
Riemann, Bernhard, 1826–1866, Göttingen, geometry, complex numbers, analysis.
Riesz, Frigyes, 1880–1956, Cluj-Napoca, Szeged, Budapest: functional analysis.
Rockafellar, Tyrrell, USA, convex analysis, variational analysis, measurability.
Rolle, Michel, 1652–1719, Paris: algebra, analysis.
Schauder, Juliusz Pavel, 1896–1943, Lwów: functional analysis, fixed points, P.D.E.
Schmidt, Erhard, 1876–1959, Berlin: integral equations, functional analysis.
Schwarz, Karl Hermann Amandus, 1843–1921: analysis, differential geometry.
Schwartz, Laurent, 1915–2002, Nancy, Paris, distributions, functional analysis, measure.
Serrin, James B., 1926–2012, Minneapolis: continuum mechanics, nonlinear analysis, P.D.E.
Šmulian, Vitold, 1914–1944 Russia: functional analysis.
Sobolev, Sergueï, 1908–1989, Moscow, Novosibirsk, math. physics, functional analysis.
Steinhaus, Hugo Dyonis, 1887–1967, Lwow, Wroclaw: functional analysis.
Stieltjes, Thomas, 1856–1894, Toulouse: analysis, integration.
Stokes, George Gabriel, 1819–1903, London, Cambridge, Analysis.
Stone, Marshall, 1903–1989, U.S.A., topology, functional analysis, Boolean algebra.
Tietze, Heinrich Franz Friedrich, 1880–1964, Vienna, Brno, Munich: topology.
Tonelli, Leonido, 1885–1946, Parma, Pisa, Roma: integration, calculus of variations.
Tychonov, Andrey, 1906–1993, Moscow: topology, fixed point theory, ill-posedness.
Urysohn, Pavel Samuilovich, 1898–1924, Moscow: topology.
de La Vallée Poussin, Charles, 1866–1962, Louvain: analysis.
Vandermonde, Alexandre, 1735–1796, Paris, determinants, substitution groups.
Vitali, Giuseppe, 1875–1932, Padova, Bologna, functional analysis, measure theory.
Volterra, Vito, 1860–1940, Roma: dynamics of populations, integral equations, finance.
Wallis, John, 1616–1703, Cambridge, Oxford: infinitesimal calculus, integration.
Weierstrass, Karl, 1815–1897, Berlin: series, analytic functions, elliptic functions.
Yosida, Kôsaku, 1909–1990, Tokyo: functional analysis, semigroups.
Young, William Henry, 1863–1942, Calcutta, Liverpool: differential calculus.

Notation

i.e.	That is to say
f.i.	For instance
$:=$	Equality by definition
$=$	Equality
$\leq$	Less than or equal to
$\geq$	Greater than or equal to
$<$	Less than
$>$	Greater than
$\forall$	For all
$\forall_\mu$	μ-almost everywhere
$\exists$	There exists
$\in$	Belongs to
$\subset$	Included in
$\cap$ or $\bigcap$	Intersection
$\cup$ or $\bigcup$	Union
$\sqcup$ or $\coprod$	Union of mutually disjoint sets
$\vee$ or $\bigvee$	Supremum
$\wedge$ or $\bigwedge$	Infimum
$\sum$	Sum
$\int$	Integral
λ, λ_d	The Lebesgue measure on $\mathbb{R}^d$
$\to$	Converges
$\mapsto$	Is sent to
$\to_S$	Converges while remaining in S
$\to r_+$	Converges to r while remaining greater than r
$\stackrel{*}{\to}$	Converges in the weak* topology
$\varnothing$	The empty set
∞	Infinity
$\mathbb{C}$	The set of complex numbers
$\mathbb{N}$	The set of natural integers

$\mathbb{N}_k$	The set $\{1, \ldots, k\}$
$\mathbb{P}$	The set of positive real numbers
$\mathbb{Q}$	The set of rational numbers
$\mathbb{R}$	The set of real numbers
$\mathbb{R}_+$	The set of nonnegative real numbers
$\mathbb{R}_-$	The set of nonpositive real numbers
$\overline{\mathbb{R}}_+$	The set $\mathbb{R}_+ \cup \{+\infty\}$ of nonnegative extended real numbers
$\overline{\mathbb{R}}$	The set $\mathbb{R} \cup \{-\infty, +\infty\}$ of extended real numbers
$\mathbb{R}_\infty$	The set $\mathbb{R} \cup \{+\infty\}$
$\mathbb{Z}$	The set of integers
Δ_m	The canonical m-simplex $\{(t_1, \ldots, t_m) \in \mathbb{R}^m_+ : t_1 + \ldots + t_m = 1\}$
$A \backslash B$	The set of $a \in A$ that do not belong to B
$A \Delta B$	The symmetric difference of A and B: $(A\backslash B) \cup (B \backslash A)$
(a, b)	The pair formed by a and b
$]a, b[$	The open interval with end points a and b
$[a, b]$	The closed interval with end points a and b
$[a, b[$	The semi-closed interval $[a, b] \backslash \{b\}$
$]a, b]$	The semi-closed interval $[a, b] \backslash \{a\}$
$\langle \cdot, \cdot \rangle$	The coupling function between a normed space and its dual
$\cdot$	An unspecified variable
$(\cdot \mid \cdot)$	The scalar product
$\Vert \cdot \Vert$	Or $\Vert \cdot \Vert_E$ the norm of a normed space (n.v.s.) E
$\Vert f \Vert_p$	The norm of a function f in $\mathcal{L}_p(S, \mu, E)$ for $p \in [1, \infty[$
$\Vert f \Vert_\infty$	$\sup_x \Vert f(x) \Vert$ or $\operatorname{esssup} \Vert f \Vert$
$\operatorname{supp} f$	The support of a function f
I_S	Or just I, the identity map from S to S
F^{-1}	The inverse of a multimap (or multifunction) F
$A^\intercal$	The transpose of the continuous linear map A
A^*	The adjoint of the continuous linear map A between Hilbert spaces
1_S	The characteristic function of the subset S: 1 on S, 0 elsewhere
ι_S	The indicator function of the subset S: 0 on S, $+\infty$ elsewhere
σ_S or h_S	The support function of the subset S
d_S, $d(\cdot, S)$	Distance to a subset S of a metric space
$\operatorname{card} S$	Cardinal of a set S
$\operatorname{int} S$	The interior of the subset S of a topological space
$\operatorname{cl} S$	The closure of the subset S of a topological space
$\operatorname{bdry} S$	The boundary of the subset S of a topological space
$\operatorname{diam} S$	The diameter of a subset S of a metric space (X, d)
$\dim X$	The dimension of a vector space
$\operatorname{co} S$	The convex hull of the subset S of a linear space
$\overline{\operatorname{co}}\, S$	The closed convex hull of the subset S of a n.v.s.
$\overline{\operatorname{co}}^*\, S$	The weak* closed convex hull of a subset S of a dual space
$\operatorname{span} S$	The smallest linear subspace containing S
S^0	The polar set $\{x^* \in X^* : \forall x \in S\ \langle x^*, x \rangle \leq 1\}$ of S
$B(x, r)$	The open ball with center x and radius r in a metric space

$B[x, r]$	The closed ball with center x and radius r in a metric space
B_X	The closed unit ball in a n.v.s. (normed space) X
$L(X, Y)$	The set of continuous linear maps from X to another n.v.s. Y
X^*	The topological dual space $L(X, \mathbb{R})$ of a n.v.s. X
$\Gamma(X)$	The set of closed proper convex functions on a n.v.s. X
$C(X)$	The space of continuous functions on a topological space X
$C_b(X)$	The space of bounded continuous functions X
$C(X, Y)$	The set of continuous maps from X to another space Y
$C^k(W)$	The set of functions of class C^k on an open subset W of a n.v.s.
$C_b^k(W)$	The space of functions in $C^k(W) \cap C_b(W)$ with derivatives in $C_b(W)$
$C_c^k(W)$	The set of functions of class C^k with compact support on W
$C^k(W, Y)$	The set of maps of class C^k from W into a n.v.s. Y
T	An interval of $\mathbb{R}$
$R(T, E)$	The space of regulated functions from T into a n.v.s. E
$\mathcal{L}_p(S, \mu, E)$	The space of p-integrable maps from S into E
$\mathcal{L}_p(S, \mu)$	The space $\mathcal{L}_p(S, \mu, E)$ for $E := \mathbb{R}$ or $\mathcal{L}_p(\mu)$, $\mathcal{L}_p(S)$
$L_p(S, \mu, E)$	The space of classes for equality a.e. of elements of $\mathcal{L}_p(S, \mu, E)$
$\mathcal{P}(X)$	The set of subsets of a set X
$\mathcal{O}$	The family of open subsets of a topological space X
$\mathcal{N}(x)$	The family of neighborhoods of x in a topological space
$\mathcal{G}_\delta$	The family of countable intersections of open subsets
$\mathcal{S}$	A ring or a σ-algebra on a set S
f'	The derivative of a function f or a map f or Df
$D_i f$	The partial derivative of f with respect to the i-th variable or $\frac{\partial}{\partial x_i} f$
$D^\alpha f$	The partial derivative of f for the multi-index α or $\frac{\partial^{\lvert\alpha\rvert}}{\partial x_1^{\alpha_1} \ldots \partial x_d^{\alpha_d}} f$
∇f	The gradient of a function f
$\partial f(x)$	The subdifferential of a function f at x
Δf	The Laplacian of the function f

Contents

Chapter 1
Sets, Orders, Relations and Measures

Measure what is measurable, and make measurable what is not so. (Misura ciò che è misurabile, e rendi misurabile ciò che non lo è)

Galileo Galilei (1564–1642)

Abstract This chapter is devoted to some preliminary subjects and techniques. Sets and orders are briefly considered, in particular for defining nets and sequences which are used throughout the book. Basic facts about countability are reviewed. Since a practice in set theory is needed for topology and analysis, measure spaces are chosen for training. The classical extension results are presented and the product measure is treated without using integration theory. The Lebesgue and Stieltjes measures are introduced as examples.

A knowledge of basic set theory is desirable for a reading of the present book, as in various branches of mathematics. However, it is not our purpose to enter into the subtleties of logic arguments involved in such foundations. We just assume the reader is familiar with the usual operations $\cup$, $\cap$, $\times$, $\subset$ of set theory and has a basic knowledge of the fundamental sets $\mathbb{N}$, $\mathbb{Z}$, $\mathbb{R}$, $\mathbb{C}$. We refer to the list of notations for the definitions of these sets and further information.

We devote this opening chapter to some preliminary material dealing with sets, orders, correspondences and particular families of subsets of a set. We gather this material because we believe the reader must have a certain familiarity with these basic techniques before using more specific concepts. In some cases the reader will have to jump to the next chapter when some topological notions are intertwined with measure spaces.

The concepts of measurable space and of measure offer an opportunity to train in the practice of sets and maps. For this reason, we present them in this first chapter. Such a choice has a drawback: in some places we have to anticipate some topological notions. We suggest the reader skip these passages or briefly look at the next chapter to fetch the required information. We hope the benefit of the gained familiarity with the basic tools of analysis will compensate any inconvenience. Integration will be presented much later (Chap. 7) because it requires many more

J.-P. Penot, *Analysis*, Universitext, DOI 10.1007/978-3-319-32411-1_1

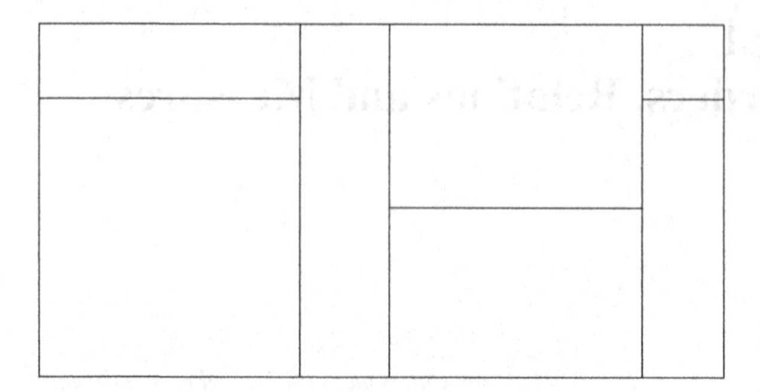

Fig. 1.1 A paved yard

concepts of analysis such as normed vector spaces and completeness. Introducing measure spaces at this stage and not later may also be useful to readers beginning a study of probability.

In order to demonstrate the need for the training we mentioned, let us consider the following problem.

Problem A yard is paved with rectangular stones. Knowing that the area of a rectangle is the product of its length by its width, prove that the area of the yard is the sum of the areas of the stones (Fig. 1.1).

1.1 Sets and Orders

Many facts in mathematics (and real life) can be formulated in terms of relations. A *relation* R between two sets X, Y is a subset R of $X \times Y$. Given $x \in X$ one sets $R(x) := \{y \in Y : (x, y) \in R\}$. Viewed in this way a relation can be considered as a map from X into the set $\mathcal{P}(Y)$ of subsets of Y. We shall return to this viewpoint in Sect. 1.3.

Order is an important topic in mathematics (and in real life). However not all orders are total orders for which two elements are always comparable as in $\mathbb{N}$, $\mathbb{Z}$, $\mathbb{R}$. Authority in a modern family provides an example of order that is not total: neither of the two parents is above the other, even if both are above any child. Using a sheet of paper and the order induced by the height on the page, one can make drawings of various ordered situations (Fig. 1.2).

In precise terms, a *preorder* or *partial preorder* or preference relation on a set X is a relation A between elements of X, often denoted by $\leq$, with $A(x) := \{y \in X : x \leq y\}$, that is *reflexive* ($x \leq x$ or $x \in A(x)$ for all $x \in X$) and *transitive* ($A \circ A \subset A$, i.e. $x \leq y,\ y \leq z \Rightarrow x \leq z$ for $x, y, z \in X$). One also writes $y \geq x$ instead of $x \leq y$ or $y \in A(x)$ and one reads: y is above x or y is preferred to x. The preorder is an *order* whenever it is *antisymmetric* in the sense that for any $x, y \in X$ one has $x = y$ whenever $x \leq y$ and $y \leq x$. Two elements x, y of a preordered set $(X, \leq)$ are

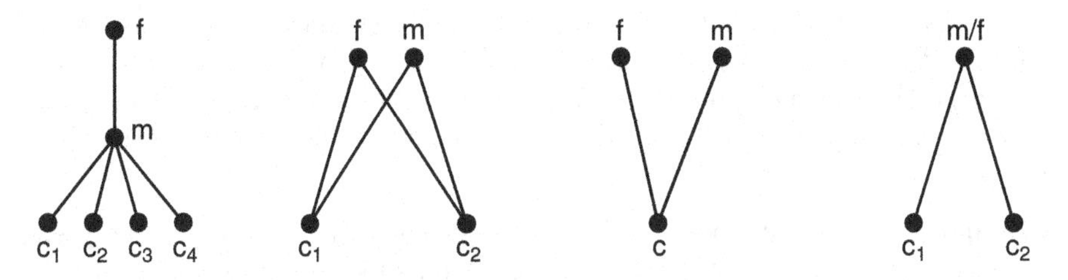

Fig. 1.2 Authority in various families

said to be *comparable* if either $x \leq y$ or $y \leq x$. If this is the case for all pairs of X, one says that $(X, \leq)$ is *totally ordered*. As mentioned above, this is not always the case (think of the set $X := \mathcal{P}(S)$ of subsets of a set S with inclusion when S has at least two elements). Given a subset S of $(X, \leq)$ an element m of X is called an *upper bound* (resp. a *lower bound*) of S if one has $s \leq m$ (resp. $m \leq s$ for all $s \in S$). An upper bound of S that belongs to S is a greatest element of S; but it may happen that S has upper bounds but no greatest element. For instance, 1 is a least upper bound of $[0, 1[:= \{r \in \mathbb{R} : 0 \leq r < 1\}$ in $\mathbb{R}$ but not a greatest element of this set. An element m of X is a *least upper bound* of S if for any upper bound m' of S one has $m \leq m'$. If $\leq$ is an order, such an element is unique and one writes $m = \sup S$ or $m = \bigvee S$. The definition of a greatest lower bound of S can be deduced from the definition of a least upper bound by reversing the preorder, i.e. by introducing the preorder $\leq'$ given by $x \leq' y$ if $y \leq x$. An ordered set $(X, \leq)$ is called a *lattice* if for any $(x, x') \in X \times X$ the set $\{x, x'\}$ has a least upper bound denoted by $x \vee x'$ and a greatest lower bound denoted by $x \wedge x'$. If $\sup S$ (resp. $\inf S$) exist for any nonempty subset S of X that has upper (resp. lower) bounds, $(X, \leq)$ is called a *complete lattice*.

A map $f : X \to X'$ between two preordered sets $(X, \leq)$ and $(X', \leq')$ will be called *homotone* or *order-preserving* or *increasing* if for any $x_1 \leq x_2$ in X one has $f(x_1) \leq' f(x_2)$. It is *isotone* if it is bijective, homotone and if f^{-1} is homotone. If f is homotone with respect to the reverse order on X' one says that f is *antitone* or *order reversing* or *decreasing*. Since the term "monotone" is often used for one-variable real-valued functions that are either order-preserving or order-reversing, we prefer to avoid it. When the order in X' is not total we also avoid the term "nondecreasing" which is ambiguous. If f is such that $f(x_1) < f(x_2)$ whenever $x_1 < x_2$, i.e. $x_1 \leq x_2$ and $x_1 \neq x_2$, we say that f is *strictly increasing*.

The preceding notions are illustrated in the following fixed point theorem.

Theorem 1.1 (Knaster-Tarski) *Let $(L, \leq)$ be an upper complete lattice, i.e. an ordered set in which any nonempty subset has a least upper bound. Suppose $f : L \to L$ is order-preserving and there exists some $z \in L$ such that $z \leq f(z)$. Then the set $F := \{x \in L : f(x) = x\}$ of fixed points of f is nonempty and F has a greatest element.*

Proof Let $D := \{x \in L : x \leq f(x)\}$. It is a nonempty subset of L (as $z \in D$), hence it has a least upper bound u. For all $x \in D$ one has $x \leq u$, hence $f(x) \leq f(u)$ and, by transitivity, $x \leq f(u)$. Thus $f(u)$ is an upper bound of D. Since u is the least upper bound of D, we have $u \leq f(u)$. Thus $u \in D$. Since f is homotone, we observe that $f(u) \leq f(f(u))$, so that $f(u) \in D$. Then, by definition of u, we get $f(u) \leq u$, hence $f(u) = u$ and $u \in F$. Finally, since F is contained in D, for any $v \in F$ one has $v \leq u$. □

Example Let L be the *power set* $\mathcal{P}(X)$ (also denoted by 2^X) of a nonempty set X, i.e. the set of subsets of X. With respect to the order given by inclusion L is a complete lattice, the least upper bound (resp. greatest lower bound) of a family $(S_i)_{i \in I}$ of subsets of X being $\bigcup_{i \in I} S_i$ (resp. $\bigcap_{i \in I} S_i$). Since $(L, \subset)$ has the empty set $\varnothing$ as a least element, if $F : \mathcal{P}(X) \to \mathcal{P}(X)$ is an order-preserving map, there exists some $M \in \mathcal{P}(X)$ such that $F(M) = M$.

Theorem 1.2 (Cantor-Bernstein-Schröder) *If there exist injective maps $f : X \to Y$ and $g : Y \to X$ between the sets X and Y, then there exists a bijective map $h : X \to Y$.*

Proof As in the preceding example, let $L := \mathcal{P}(X)$ be the power set of X. Let us define $F : \mathcal{P}(X) \to \mathcal{P}(X)$ by

$$\forall A \in \mathcal{P}(X) \qquad F(A) := X \backslash g(Y \backslash f(A))$$

where for two subsets B, C of a set Z one writes $B \backslash C := \{b \in B : b \notin C\}$. It is easy to show that F is homotone. Using the preceding example, we conclude that there exists some subset M of X such that $F(M) = M$. This relation means that $g(Y \backslash f(M)) = X \backslash M$. Let us define $h : X \to Y$ by $h(x) := f(x)$ if $x \in M$, $h(x) := g^{-1}(x)$ if $x \in X \backslash M$. Given $x \neq x'$ in X we can show that it is impossible that $h(x) = h(x')$ by considering separately the cases $x, x' \in M$, $x, x' \in X \backslash M$ and $x \in M$, $x' \in X \backslash M$ (in the latter case we have $h(x) \in f(M)$ and $h(x') \in Y \backslash f(M)$). On the other hand, for all $y \in Y$, we have either $y \in f(M) = h(M)$ or $y \in Y \backslash f(M)$ and then $y = h(x)$ for $x := g(y)$. Thus h is a bijection of X onto Y. □

The preceding theorem opens the way to comparison of the "size" of sets. One says that two sets X and Y are *equipotent* or have the same *cardinality* and one writes card X = card Y if there exists a bijection from X onto Y. This defines an equivalence relation between sets. If X and Y are finite sets, equipotence means that X and Y have the same number of elements. For infinite subsets, equipotence may be more mysterious. For instance the set $\mathbb{Q}$ of rational numbers is equipotent to the set $\mathbb{N}$ of natural numbers and to the set $\mathbb{Z}$ of integers. More surprising is the fact that $\mathbb{R}$ and $\mathbb{R}^2$ are equipotent. Still some rules are more familiar and easy to prove. For instance, if X and X' are equipotent sets and if Y and Y' are equipotent sets, then $X \times Y$ and $X' \times Y'$ are equipotent. Also, the disjoint union of X and Y is equipotent to the disjoint union of X' and Y'. However, the following comparison theorem, whose last assertion is a rephrasing of the Cantor-Bernstein-Schröder theorem is not an obvious result.

Theorem 1.3 *For any two sets X and Y at least one of the following two assertions holds:*

(a) X is equipotent to a subset of Y;
(b) Y is equipotent to a subset of X.

Moreover, if these assertions hold simultaneously, then X and Y are equipotent.

If X is equipotent to a subset of Y but is not equipotent to Y itself, one writes card $X <$ card Y. A set X is said to be *infinite* if there exists a subset X' of X equipotent to X and different from X. A set X is said to be *countable* or *denumerable* if it is equipotent to $\mathbb{N}$ or a subset of $\mathbb{N}$. It is *uncountable* if it is not countable. The existence of different sorts of infinite sets is revealed by the following result.

Theorem 1.4 (Cantor) *For any set X one has* card $\{x\} <$ card $\mathcal{P}(X)$.

Proof Since the map that assigns to x the singleton x is injective, it suffices to prove that there is no surjective map $f : X \to \mathcal{P}(X)$. Otherwise, setting $A := \{x \in X : x \notin f(x)\}$, we cannot find $x \in X$ such that $f(x) = A$ since if $x \in A$ we have $x \notin f(x)=A$, and if $x \in X\backslash A$ we have $x \in f(x)=A$, a contradiction in both cases. □

Let us note that the Knaster-Tarski Theorem does not use the *axiom of choice* or its equivalent statements. This axiom asserts that for any set X there exists a map $f : \mathcal{P}(X) \to X$ such that for any nonempty subset A of X one has $f(A) \in A$. Such an assertion seems to be plausible, as is the assertion that a product of nonempty subsets is nonempty. However, such assertions are equivalent to other statements such as Zermelo's axiom and Zorn's axiom which are not obvious. Zermelo's axiom (or theorem) asserts that on any set X one can introduce an order such that any nonempty subset of X has a least element. Most mathematicians work in a framework accepting such assertions. Since Zorn's axiom (or lemma) is the most useful statement for our purposes, let us give an account of it with the corresponding terminology.

A subset C of $(X, \leq)$ that is totally ordered with respect to the induced preorder is called a *chain*. A preorder on X is said to be *upper inductive* (resp. *lower inductive*) if any chain C has an upper bound (resp. a lower bound). Recall that an element $\overline{x}$ of a preordered space $(X, \leq)$ is said to be *maximal* if for any $x \in X$ such that $\overline{x} \leq x$ one has $x \leq \overline{x}$; it is called *minimal* if it is maximal for the reverse preorder. Zorn's Lemma can be stated as follows.

Theorem 1.5 (Zorn's Lemma or Zorn's Axiom) *Any preordered set whose preorder is upper (resp. lower) inductive has at least one maximal (resp. minimal) element.*

As mentioned above, this statement is equivalent to a number of other axioms such as the Zermelo's axiom or *well-ordering principle* (a non-intuitive assertion) or the *axiom of choice*, which seems to be very natural. We shall not deal with such aspects of the foundations of mathematics.

Corollary 1.1 *Let $(X, \leq)$ be a preordered set whose preorder is upper (resp. lower) inductive. Then, for any $x_0 \in X$ there exists a maximal (resp. minimal) element $\overline{x}$ satisfying $x_0 \leq \overline{x}$ (resp. $\overline{x} \leq x_0$).*

Proof Suppose $\leq$ is lower inductive, the other case being settled by considering the opposite order. The set $X_0 := \{x \in X : x \leq x_0\}$ is lower inductive, any chain C in X_0 being a chain in X and any lower bound of C in X_0 being a lower bound of C in X. Thus X_0 has a minimal element $\overline{x}$. Since any minimal element of X_0 is a minimal element of X, the corollary is established. □

We need some other concepts, in particular for convergence questions. A map $f : H \to I$ between two preordered spaces is said to be *filtering* if for all $i \in I$ there exists an $h \in H$ such that $f(k) \geq i$ whenever $k \in H$ satisfies $k \geq h$. A preordered set $(I, \leq)$ is said to be *directed* if any finite subset F of I has an upper bound. This occurs as soon as any pair of elements in I has an upper bound. A subset J of a preordered set $(I, \leq)$ is said to be *cofinal* if for all $i \in I$ there exists some $j \in J$ such that $j \geq i$.

Exercises

1. Show that a subset C of a preordered space $(X, \leq)$ is a chain iff (if and only if) $C \times C \subset A \cup A^{-1}$, where $A := \{(x, y) : x \leq y\}$, $A^{-1} := \{(x, y) : (y, x) \in A\}$.
2. Let $(I, \leq)$ be a directed set. Show that if $J \subset I$ is not cofinal, then $I \backslash J$ is cofinal.
3. Let $(X, \leq)$ be a preordered space. Verify that the relation $<$ defined by $x < y$ if $x \leq y$ and not $y \leq x$ is transitive.
4. Show that if a map $f : H \to I$ between two preordered spaces is a homotone bijection, if $(H, \leq)$ is totally ordered and if $(I, \leq)$ is ordered, then f is isotone.
5. Show that a homotone map $f : H \to I$ between two preordered spaces is filtering iff $f(H)$ is cofinal.
6. Give an example of a subset J of a preordered space $(I, \leq)$ having more than one supremum. Verify that when $\leq$ is an order, a subset of I has at most one supremum.
7. Show that when a subset J of a preordered space $(I, \leq)$ has a greatest element k then k is a supremum of J and for any supremum s of J one has $k \leq s$ and $s \leq k$. Note that when a supremum s of J belongs to J, then s is a greatest element of J.
8. Let $(I, \leq)$ and $(J, \leq)$ be two ordered sets. Verify that the relation $(i, j) \leq (i', j')$ if $i < i'$ (i.e. $i \leq i'$ and $i \neq i'$) or if $i = i'$ and $j \leq j'$ is an order relation (the *lexicographic order*) on $I \times J$. Show that this order is total if the orders on I and J are total.
9. Show that the union of a countable family of countable sets is countable. Deduce from this that the set $\mathcal{P}_f(\mathbb{N})$ of finite subsets of $\mathbb{N}$ is countable.
10. Let (f_n) be a sequence of maps from $\mathbb{N}$ into $\mathbb{N}$. Let $f : \mathbb{N} \to \mathbb{N}$ be defined by $f(n) = f_n(n) + 1$ for $n \in \mathbb{N}$. Show that there is no $k \in \mathbb{N}$ such that $f = f_k$. Deduce from this that the set of maps from $\mathbb{N}$ into $\mathbb{N}$ is uncountable.
11. Show that the set $\mathbb{Q}$ of rational numbers is countable.
12. Let E be an infinite set and let F be a finite subset of E. Show that E and $E \backslash F$ are equipotent. Prove the same conclusion when F is countable and $E \backslash F$ is infinite, admitting that any infinite set contains an infinite countable subset.

13. Deduce from the preceding exercise that the set $\mathbb{R}$ of real numbers is equipotent to $[0, 1]$.

14. Let X be a set and let $\mathcal{S} \subset \mathcal{P}(X)$. Consider $f : X \times \mathcal{P}(X) \to \mathbb{R}$ such that $f(x, S) = \min_{i \in I} f(x, S_i)$ for all $x \in X$ and all families $(S_i)_{i \in I}$ in $\mathcal{S}$ whose union is $S \in \mathcal{S}$. Here $\min_{i \in I} f(x, S_i)$ means that there exists some $j \in I$ such that $f(x, S_j) \leq f(x, S_i)$ for all $i \in I$. Given $S \in \mathcal{P}(X)$ let

$$W_S(\mathcal{S}) := \{x \in X : f(x, S) \leq f(x, S') \ \forall S' \in \mathcal{S}\}.$$

Show that for any family $(S_i)_{i \in I}$ in $\mathcal{S}$ whose union is $S \in \mathcal{S}$ one has

$$W_S(\mathcal{S}) = \bigcup_{i \in I} W_{S_i}(\mathcal{S}).$$

Given a family $(\mathcal{S}_\alpha)_{\alpha \in A}$ of subsets of $\mathcal{P}(X)$ and $\mathcal{S} := \bigcup_{\alpha \in A} \mathcal{S}_\alpha$, verify that

$$W_S(\mathcal{S}) = \bigcap_{\alpha \in A} W_S(\mathcal{S}_\alpha).$$

Give an interpretation when $f(x, S) := \min\{d(x, y) : y \in S\}$ where $d : X \times X \to \mathbb{R}_+$ is some function. If d is the distance on the surface of the globe and $\mathcal{S}$ is the family of states, $W(S)$ can be considered as the territorial waters of S.

15. Given ordered sets $\mathcal{S}, \mathcal{T}$, a map $D : \mathcal{S} \to \mathcal{T}$ is called a *duality* if for any family $(S_i)_{i \in I}$ in $\mathcal{S}$ having an infimum $\bigwedge_{i \in I} S_i$ the family $(D(S_i))_{i \in I}$ has a supremum $\bigvee_{i \in I} D(S_i)$ and if

$$D(\bigwedge_{i \in I} S_i) = \bigvee_{i \in I} D(S_i).$$

Suppose $\mathcal{S}$ is a complete inf-lattice in the sense that any nonempty family in $\mathcal{S}$ has an infimum. Show that for any antitone map $D : \mathcal{S} \to \mathcal{T}$ there exists a greatest antitone map $D' : \mathcal{T} \to \mathcal{S}$ such that $D'(D(S)) \leq S$ for all $S \in \mathcal{S}$ and that D' is given by

$$D'(T) := \bigwedge \{S \in \mathcal{S} : D(S) \leq T\}.$$

Show that D' is a duality when D is a duality and that $D(D'(T)) \leq T$ for all $T \in \mathcal{T}$.

16. Given sets $X, Y, \mathcal{S} \subset \mathcal{P}(X), \mathcal{T} \subset \mathcal{P}(Y)$ endowed with the order induced by the inclusion and such that $\mathcal{S}$ is a complete sup-lattice, a map $P : \mathcal{S} \to \mathcal{T}$ is called a *polarity* if for any family $(S_i)_{i \in I}$ in $\mathcal{S}$ having a supremum in $\mathcal{S}$ one has

$$P(\bigvee_{i \in I} S_i) = \bigcap_{i \in I} P(S_i).$$

Show that there exists a smallest antitone map $P' : \mathcal{T} \to \mathcal{S}$ such that $P'(P(S)) \geq S$ for all $S \in \mathcal{S}$ and that P' is given by

$$P'(T) := \bigvee \{S \in \mathcal{S} : T \subset P(S)\}.$$

Show that when $\mathcal{S} = \mathcal{P}(X)$ one has

$$P'(T) := \{x \in X : T \subset P(\{x\})\}.$$

17. Given sets X, Y, a point s of Y and $f : X \times Y \to \mathbb{R}_+$, for a subset T of Y let

$$V_s(T) := \{x \in X : f(x,s) \leq f(x,t) \ \forall t \in T\}$$

with $V_s(\varnothing) := X$. The set $V_s(T)$ is called the *Voronoi cell* associated with s and T. Show that $V_s : \mathcal{P}(Y) \to \mathcal{P}(X)$ is a *polarity* on the set $\mathcal{P}(Y)$ of subsets of Y, i.e. for any family $(T_i)_{i \in I}$ of $\mathcal{P}(Y)$ one has

$$V_s(\bigcup_{i \in I} T_i) = \bigcap_{i \in I} V_s(T_i).$$

Assume that $\mathcal{C}$ is a subset of $\mathcal{P}(Y)$ containing Y and the singletons of Y and that $\mathcal{C}$ is stable under intersections. Prove that for all $B \in \mathcal{P}(Y)$ there exists a smallest subset $\mathrm{clo}(B)$ in $\mathcal{C}$ containing B. Let $\mathcal{V}$ be a subset of $\mathcal{P}(X)$ such that $V_s(C) \in \mathcal{V}$ for all $C \in \mathcal{C}$. Verify that the restriction of V_s to $\mathcal{C}$ and $\mathcal{V}$ is still a polarity from $\mathcal{C}$ into $\mathcal{V}$ in the sense that

$$V_s(C) = \bigcap_{i \in I} V_s(C_i) \text{ for } C := \bigvee_{i \in I} C_i := \mathrm{clo}(\bigcup_{i \in I} C_i).$$

Describe the polarity $V'_s : \mathcal{V} \to \mathcal{C}$ associated with V_s as in the preceding exercise.

1.2 Convergence and Summability in $\mathbb{R}$

We assume the reader is familiar with the usual properties of $\mathbb{R}$. On the other hand, it may be useful to review some convergence properties generalizing the convergence of sequences and series. A general approach will be presented later on.

A *net* of real numbers comprises the data of a directed set $(I, \leq)$ and of a family $(x_i)_{i \in I}$ of elements of $\mathbb{R}$. One says that a net $(x_i)_{i \in I}$ *converges* to $x \in \mathbb{R}$, and one writes $(x_i)_{i \in I} \to x$, if for every $\varepsilon > 0$ there exists some $i_\varepsilon \in I$ such that $x - \varepsilon < x_i < x + \varepsilon$ for all $i \geq i_\varepsilon$. One also writes $x = \lim_{i \in I} x_i$ or just $x = \lim_i x_i$. The limit is unique: if $(x_i)_{i \in I} \to x$ and $(x_i)_{i \in I} \to y$ with $x < y$, taking $\varepsilon := (y - x)/2$, since I

is directed we can find some $k \in I$ such that for $i \geq k$ we have both $x_i < x + \varepsilon$ and $x_i > y - \varepsilon$, an impossibility. It is easy to see that if $(x_i)_{i\in I} \to x$ and $(y_i)_{i\in I} \to y$, then one has $(x_i + y_i)_{i\in I} \to x + y$. Also, for all $r \in \mathbb{R}$ one has $(rx_i)_{i\in I} \to rx$.

The next sufficient condition for convergence is similar to a well known condition for sequences. Taking nets instead of sequences is useful to obtain a sound understanding of some matters such as the Riemann integral. Here we say that $(x_i)_{i\in I}$ is an increasing net if for $i \leq j$ in I we have $x_i \leq x_j$.

Theorem 1.6 *Let $(x_i)_{i\in I}$ be an increasing net of real numbers that is bounded above. Then $(x_i)_{i\in I} \to x := \sup_{i\in I} x_i$.*

Proof Given $\varepsilon > 0$ we can find some $h \in I$ such that $x_h > x - \varepsilon$. Then, for $i \geq h$ we have $x_i \geq x_h > x - \varepsilon$ since $(x_i)_{i\in I}$ is increasing, and of course, $x_i < x + \varepsilon$. □

Consequently, a bounded below decreasing net $(x_i)_{i\in I}$ converges to $\inf_{i\in I} x_i$.

A general convergence criterion is the *Cauchy criterion*. It concerns *Cauchy nets*, i.e. nets $(x_i)_{i\in I}$ having the property that for every $\varepsilon > 0$ there exists some $h \in I$ such that $\left|x_i - x_j\right| \leq \varepsilon$ whenever $i, j \in I$ satisfy $i \geq h, j \geq h$. A convergent net is a Cauchy net. The converse is interesting because it enables us to assert that a net converges without knowing its limit.

Theorem 1.7 (Cauchy Criterion) *Any Cauchy net in $\mathbb{R}$ converges.*

Proof Let $(x_i)_{i\in I}$ be a Cauchy net. For $i \in I$ let $a_i := \sup_{j\geq i} x_j$, $b_i := \inf_{j\geq i} x_j$; they are finite for i large enough. Then $(a_i)_{i\in I}$ is decreasing and $(b_i)_{i\in I}$ is increasing. Moreover, for all $\varepsilon > 0$ there exists some $h_\varepsilon \in I$ such that $|a_i - b_i| \leq \varepsilon$ for all $i \in I$ satisfying $i \geq h_\varepsilon$. Then, for $i \geq h_1$ in I we have $a_i \geq b_{h_1} - 1$ and $b_i \leq a_{h_1} + 1$. Thus the nets $(a_i)_{i\in I}$ and $(b_i)_{i\in I}$ are convergent. Their respective limits a and b satisfy $a \geq b$ since $a_i \geq b_i$ for all $i \in I$ and for every $\varepsilon > 0$ we can find $h_\varepsilon \in I$ such that for $i \geq h_\varepsilon$ we have $a \leq a_i \leq b_i + \varepsilon \leq b + \varepsilon$. Thus $a = b$ and for $i \geq h_\varepsilon$ we have $x_i \leq a_i \leq b + \varepsilon$ and $x_i \geq b_i \geq a - \varepsilon$, so that $(x_i) \to \ell := a = b$. □

Let us turn to summability questions. Let $(r_t)_{\in T}$ be an arbitrary family of real numbers, T being an arbitrary set. For J in the family $\mathcal{J}$ of finite subsets of T let $s_J := \Sigma_{j\in J} r_j$. The set $\mathcal{J}$ ordered by inclusion is directed (for J', $J'' \in \mathcal{J}$ the set $J := J' \cup J''$ is greater than or equal to J' and J''). This observation gives a meaning to the following definition.

Definition 1.1 A family $(r_t)_{t\in T}$ of real numbers is said to be summable if the net $(s_J)_{J\in\mathcal{J}}$ of finite sums converges to some $s \in \mathbb{R}$ called the sum of the family $(r_t)_{t\in T}$. One writes $s = \Sigma_{t\in T} r_t$.

Clearly, if all but a finite number of members r_t of the family $(r_t)_{t\in T}$ are null, then the family $(r_t)_{t\in T}$ is summable. Theorem 1.6 ensures that if $(r_t)_{t\in T}$ is a family of nonnegative numbers and if there is some $c \in \mathbb{R}$ such that $s_J \leq c$ for all $J \in \mathcal{J}$, then the family $(r_t)_{t\in T}$ is summable and $\Sigma_{t\in T} r_t = \sup_{J\in\mathcal{J}} s_J$. Another criterion is as follows.

Proposition 1.1 (Cauchy Summability Criterion) *A family $(r_t)_{t\in T}$ of real numbers is summable whenever it satisfies the condition: for all $\varepsilon > 0$ there exists a finite subset H_ε of T such that for any finite subset F of T contained in $T\backslash H_\varepsilon$ one has $|s_F| \leq \varepsilon$.*

Proof This follows from the fact that the net $(s_J)_{J\in\mathcal{J}}$ satisfies the Cauchy criterion if the family $(r_t)_{t\in T}$ satisfies the Cauchy summability criterion: given $\varepsilon > 0$, let $H_\varepsilon \in \mathcal{J}$ be such that $|s_F| \leq \varepsilon$ for any $F \in \mathcal{J}$ contained in $T\backslash H_\varepsilon$; then, for J, $K \in \mathcal{J}$ containing H_ε, since $|s_J - s_K| \leq |s_{J\backslash K}| + |s_{K\backslash J}| \leq 2\varepsilon$ since $J\backslash K$ and $K\backslash J$ are contained in $t\backslash H_\varepsilon$. □

Exercise Prove the converse: if the family $(r_t)_{t\in T}$ is summable, then it satisfies the Cauchy summability criterion.

Corollary 1.2 *Let $(r_t)_{t\in T}$ be a summable family of real numbers. Then, for any subset T' of T the family $(r_t)_{t\in T'}$ is summable.*

Corollary 1.3 *If $(r_t)_{t\in T}$ is a summable family of real numbers, then the set T' of $t \in T$ such that $r_t \neq 0$ is countable.*

Proof The converse of the Cauchy summability criterion ensures that for all $n \in \mathbb{N}$ there exists a finite subset H_n of T such that for all $t \in T\backslash H_n$ one has $|r_t| \leq 1/(n+1)$ (take $F := \{t\}$). Thus $T' = \cup_n H_n$ is at most countable. □

Let us say that a family $(r_t)_{t\in T}$ of real numbers is *absolutely summable* if the family $(|r_t|)_{t\in T}$ is summable. We have the following surprising result as it differs from a well known fact for series.

Proposition 1.2 *A family $(r_t)_{t\in T}$ of real numbers is absolutely summable if and only if it is summable.*

Proof If $(r_t)_{t\in T}$ is absolutely summable, for every $\varepsilon > 0$ we can find a finite subset H_ε of T such that for any finite subset F of $T\backslash H_\varepsilon$ one has $\Sigma_{t\in F} |r_t| < \varepsilon$. Then $|\Sigma_{t\in F} r_t| < \varepsilon$, and $(r_t)_{t\in T}$ satisfies the Cauchy summability condition, hence is summable.

Conversely, suppose $(r_t)_{t\in T}$ is summable. Let $T_+ := \{t \in T : r_t \geq 0\}$, $T_- := T\backslash T_+$. Setting $r_t^+ := \max(r_t, 0)$, $r_t^- := \max(-r_t, 0)$ we see that the family $(r_t^+)_{t\in T}$ is summable and its sum is the sum of the family $(r_t)_{t\in T_+}$. Similarly, the family $(r_t^-)_{t\in T}$ is summable and its sum is the sum of the family $(-r_t)_{t\in T_-}$. Then the family $(|r_t|)_t = (r_t^+ + r_t^-)_{t\in T}$ is summable and its sum is the sum of $\Sigma_{t\in T} r_t^+$ and of $\Sigma_{t\in T} r_t^-$. □

Corollary 1.4 *A countable family $(r_n)_{n\in\mathbb{N}}$ of real numbers is summable iff (if and only if) the series Σr_n is absolutely convergent.*

Proof Saying that the series Σr_n is absolutely convergent means that the partial sums $\Sigma_{0\leq k\leq n} |r_k|$ converge as $n \to +\infty$. Then the finite sums $\Sigma_{t\in J} |r_t|$ are bounded above. Thus the family $(r_t)_{t\in T}$ is absolutely summable, hence summable. □

In general, since there is no order on the set T one can say that the summability of a family $(r_t)_{t\in T}$ is a commutative property in the sense that if $f : T' \to T$ is a bijection, and if $r'_{t'} := r_{f(t')}$, the family $(r_t)_{t\in T}$ is summable if and only if the family $(r'_{t'})_{t'\in T'}$ is summable.

Let us present some properties.

Proposition 1.3 *If $c \in \mathbb{R}$ and if $(r_t)_{t\in T}$ and $(r'_t)_{t\in T}$ are summable families of real numbers, then $(cr_t + r'_t)_{t\in T}$ is summable and its sum is $c\Sigma_{t\in T}r_t + \Sigma_{t\in T}r'_t$.*

Proof The result stems from the fact that for any finite subset J of T one has $\Sigma_{t\in J}(cr_t + r'_t) = c\Sigma_{t\in J}r_t + \Sigma_{t\in J}r'_t$. □

We dispose of an associativity property that enables us to sum by gathering bunches.

Theorem 1.8 *Let $(T_a)_{a\in A}$ be a partition of T in the sense that the subsets T_a of T are mutually disjoint and such that $T = \cup_{a\in A}T_a$. If $(r_t)_{t\in T}$ is a summable family of real numbers, then for all $a \in A$ the family $(r_t)_{t\in T_a}$ is summable and if s_a denotes its sum the family $(s_a)_{a\in A}$ is summable and its sum is the sum s of the family $(r_t)_{t\in T}$:*

$$\sum_{a\in A}\sum_{t\in T_a} r_t = \sum_{t\in T} r_t.$$

Proof We already know that for all $a \in A$ the family $(r_t)_{t\in T_a}$ is summable. Let us prove the other two assertions. For $a \in A$, let us denote by $\mathcal{J}_a$ (resp. $\mathcal{J}$) the family of finite subsets of T_a (resp. T). Given $\varepsilon > 0$ let $K \in \mathcal{J}$ be such that for all $J \in \mathcal{J}$ containing K one has $|s_J - s| \le \varepsilon$. Let

$$C := \{a \in A : \ K \cap T_a \neq \varnothing\},$$

so that $K = \cup_{c\in C}K \cap T_c$ and C is finite. Let B be a finite subset of A containing C. Let n be the number of elements of B. For each $b \in B$ we can find some $J_b \in \mathcal{J}_b$ containing $K \cap T_b$ such that $|s_{J_b} - s_b| \le \varepsilon/n$. Then the set $J := \cup_{b\in B}J_b$ contains K and since the sets J_b ($b \in B$) are disjoint, we have $\Sigma_{b\in B}s_{J_b} = \Sigma_{j\in J}r_j$,

$$\left|\sum_{b\in B} s_{J_b} - s\right| = \left|\sum_{j\in J} r_j - s\right| \le \varepsilon, \qquad \left|\sum_{b\in B} s_{J_b} - \sum_{b\in B} s_b\right| \le \varepsilon$$

hence

$$\left|\sum_{b\in B} s_b - s\right| \le 2\varepsilon.$$

Since $\varepsilon > 0$ is arbitrarily small, this shows that the family $(s_a)_{a\in A}$ is summable with sum s. □

Exercises

1. Let $(T_a)_{a \in A}$ be a partition of a set T. For each $a \in A$ let $(r_t)_{t \in T_a}$ be a summable family of nonnegative real numbers, with sum s_a. Suppose that the family $(s_a)_{a \in A}$ is summable with sum s. Show that the family $(r_t)_{t \in T}$ is summable and its sum is s.
2. Let I and J be two sets and let $(a_i)_{i \in I}$, $(b_j)_{j \in J}$ be two (absolutely) summable families of real numbers. Show that the family $(a_i b_j)_{(i,j) \in I \times J}$ is summable and its sum is $(\Sigma_{i \in I} a_i)(\Sigma_{j \in J} b_j)$.
3. Let b be an integer greater than 1.
   (a) Verify that for any sequence (x_n) of nonnegative integers less than b the series $\Sigma_{n \geq 0} x_n b^{-n}$ converges to some $s \in [0, 1]$.
   (b) Conversely, show that any $s \in [0, 1]$ is the sum of such a series and that this series is unique provided s does not belong to the set of numbers of the form kb^{-m} for some $k, m \in \mathbb{N}$. In the latter case there are exactly two such series with sum s.
   (c) Taking $b = 2$ in what precedes, prove that $[0, 1]$ is equipotent to $\mathcal{P}(\mathbb{N})$.
   (d) Using Theorem 1.4 and Exercise 13 of Sect. 1.1 prove that $\mathbb{R}$ is uncountable.

1.3 Maps and Multimaps (Relations)

This section forms an introduction to multivalued analysis, which assumes an increasingly more important position in analysis.

Mathematics is usually viewed as a precise field in which there is no ambiguity. That is not the case. All mathematicians use some abuses of notation or abuses of terminology. This is not a severe weakness as long as these abuses are well recognized and mastered. However, it is preferable to limit their uses to specific cases in which a more precise notation or terminology would be too heavy or cumbersome. For instance, given a map $f : X \to Y$ between two sets and $y \in Y$, one often writes $f^{-1}(y)$ instead of $f^{-1}(\{y\})$, where, for $B \subset Y$ one sets

$$f^{-1}(B) := \{x \in X : f(x) \in B\}.$$

Such an abuse can hardly lead to mistakes. However, some common abuses may lead to misunderstandings or mistakes and we encourage the reader to avoid them. For instance, we recommend to denote by f or $f(\cdot)$ a map $f : X \to Y$ between two sets rather than $f(x)$. If f cannot be identified with one of its values, it cannot be assimilated to its image, i.e. the set $f(X)$ of its values: two different maps f, $g : X \to Y$ may have the same image. For this reason, we avoid the notation $\{x_n\}$ for a sequence in a set X, i.e. a map $s : \mathbb{N} \to X$, with $x_n := s(n)$. Let us recall that a *sequence* in a set X is a map s from $\mathbb{N}$ to X, hence is an element of $X^{\mathbb{N}}$. While the notations $(x_n)_{n \in \mathbb{N}}$, $(x_n)_{n \geq 0}$ or just (x_n) are unambiguous, the notations x_n, $\{x_n : n \in \mathbb{N}\}$ or $\{x_n\}$ should be avoided.

It may be of interest to recall that the correspondence $y \mapsto f^{-1}(y)$ (called the inverse image) is not a map from Y into X but a *multimap* or *multifunction*, i.e. a map from Y into the set $\mathcal{P}(X)$ of subsets of X. It enjoys nice properties: for any family $(B_i)_{i\in I}$ of subsets of Y one has

$$f^{-1}(\bigcup_{i\in I} B_i) = \bigcup_{i\in I} f^{-1}(B_i), \qquad f^{-1}(\bigcap_{i\in I} B_i) = \bigcap_{i\in I} f^{-1}(B_i)$$

and, if $B \subset B' \in \mathcal{P}(Y), f^{-1}(B) \subset f^{-1}(B'), f^{-1}(B'\backslash B) = f^{-1}(B')\backslash f^{-1}(B)$, where $B'\backslash B := \{y \in B' : y \notin B\}$ is the complement of B in B'. For direct images, given a family $(A_i)_{i\in I}$ of subsets of X one just has

$$f(\bigcup_{i\in I} B_i) = \bigcup_{i\in I} f(B_i), \qquad f(\bigcap_{i\in I} B_i) \subset \bigcap_{i\in I} f(B_i)$$

as $f(A) \subset f(A')$ when $A \subset A'$.

If $F : X \to \mathcal{P}(Y)$ is a multimap, one defines direct and inverse images of subsets by

$$F(A) := \bigcup_{a\in A} F(a), \qquad F^{-1}(B) := \{x \in X : F(x) \cap B \neq \varnothing\}$$

for $A \in \mathcal{P}(X)$, $B \in \mathcal{P}(Y)$. Note that F^{-1} appears as a multimap $F^{-1} : Y \to \mathcal{P}(X)$ given, for $y \in Y$, by $F^{-1}(y) := F^{-1}(\{y\}) = \{x \in X : y \in F(x)\}$. This notation is compatible with the one we used for inverse images, considering a map $f : X \to Y$ as a multimap F whose values are the singletons $F(x) := \{f(x)\}$. In order to underline the analogy with maps, a multimap $F : X \to \mathcal{P}(Y)$ is often denoted by $F : X \rightrightarrows Y$. Like maps, multimaps can be composed: given multimaps $F : X \rightrightarrows Y,\ G : Y \rightrightarrows Z$, the composition of F and G is the multimap $G \circ F : X \rightrightarrows Z$ given by

$$(G \circ F)(x) := G(F(x)),$$

where $G(B)$, for $B := F(x)$, is defined as above. Then one has the associativity rule

$$H \circ (G \circ F) = (H \circ G) \circ F.$$

It is often convenient to associate to a multimap $F : X \rightrightarrows Y$ its *graph*

$$\mathrm{gph}(F) := \{(x, y) \in X \times Y : y \in F(x)\}\,.$$

This subset of $X \times Y$ (also denoted by $G(F)$ when no confusion may arise) characterizes F since

$$F(x) = \{y \in Y : (x, y) \in G(F)\}\,. \tag{1.1}$$

Conversely, to any subset G of $X \times Y$ one can associate a multimap $F : X \rightrightarrows Y$ by setting

$$F(x) = \{y \in Y : (x, y) \in G\},$$

so that G is the graph of F. Moreover, when G is the graph $G(M)$ of some multimap M, one gets $F = M$ via this reverse process. Thus, there is a one-to-one correspondence between subsets of $X \times Y$ and multimaps from X into Y. This correspondence is simpler than the correspondence between maps and their graphs, since in the latter correspondence one has to consider only subsets G whose vertical slices $G \cap (\{x\} \times Y)$ (for $x \in X$) are singletons. In view of this one-to-one correspondence between a multimap and its graph, it is often convenient to identify a multimap with its graph and to say that a multimap has a property $\mathcal{P}$ if its graph has this property (such as closedness, convexity...). This viewpoint is often fruitful and without any important risk of confusion; however, when X and Y are endowed with some operation $\circledast$ one has to be aware that $F \circledast F'$ usually denotes the multimap $x \mapsto F(x) \circledast F'(x)$ and not the multimap whose graph is $G(F) \circledast G(F')$. Moreover, one has to be careful with the order of the terms in the product $X \times Y$ since they determine the direction of the multimap. When X is a product $X = X_1 \times X_2$ one has to be precise when one associates to a subset G of $X \times Y$ a multimap, as a partial multimap also can be defined in this way.

Note that if $F : X \rightrightarrows Y$ is a multimap and if A (resp. B) is a subset of X (resp. Y) one has

$$F(A) = p_Y(\mathrm{gph}(F) \cap (A \times Y)), \qquad F^{-1}(B) = p_X(\mathrm{gph}(F) \cap X \times B),$$

where $p_X : X \times Y \to X$ and $p_Y : X \times Y \to Y$ are the so-called *canonical projections* defined by $p_X(x, y) := x$ and $p_Y(x, y) := y$. We also observe that

$$\mathrm{gph}(F^{-1}) = (\mathrm{gph}(F))^{-1} := \{(y, x) \in Y \times X : (x, y) \in \mathrm{gph}(F)\},$$

where for $G \subset X \times Y$ one sets $G^{-1} := \{(y, x) \in Y \times X : (x, y) \in G\}$.

Let us note that here we do not leave the realm of multimaps, whereas the inverse of a mapping is in general a multimap, not a map (and this fact is a source of many mistakes for beginners in mathematics). The *domain* $\mathrm{dom}\, F$ or $D(F)$ of a multimap $F : X \rightrightarrows Y$ is given by

$$\mathrm{dom}\, F := D(F) := \{x \in X : F(x) \neq \varnothing\}.$$

It is also the *range* or *image* $R(F^{-1}) := \mathrm{Im}\, F^{-1} := F^{-1}(Y)$ of F^{-1}. Conversely, $\mathrm{dom}\, F^{-1}$ is the image of F : the roles of F and F^{-1} are fully symmetric and $F = (F^{-1})^{-1}$.

For any subsets A, A' of X (resp. B, B' of Y) one has $F(A \cup A') = F(A) \cup F(A')$ and $F^{-1}(B \cup B') = F^{-1}(B) \cup F^{-1}(B')$. Let us observe that, contrary to what occurs for maps, in general one has

$$F^{-1}(B \cap B') \neq F^{-1}(B) \cap F^{-1}(B'). \tag{1.2}$$

Note that since F^{-1} may be an arbitrary multimap from Y to X and since for a multimap $M : Y \rightrightarrows X$ one has $M(A \cap B) \neq M(A) \cap M(B)$ in general, taking $F = M^{-1}$, so that $F^{-1} = M$, we obtain (1.2).

The following proposition can be considered as a preparation for the use of multimaps. It will be useful later when considering monotone multimaps (also called monotone operators); see Sect. 9.4.3 for the meaning of the terms used here.

Proposition 1.4 *Let X be a vector space, let $r \in \mathbb{R}\backslash\{0\}$ and let $M : X \rightrightarrows X$ be a multimap. Then the resolvants of M are related to the Yosida regularizations of M by*

$$r^{-1}[I_X - (I_X + rM)^{-1}] = (rI_X + M^{-1})^{-1}. \tag{1.3}$$

In particular, for $r = 1$,

$$I_X - (I_X + M)^{-1} = (I_X + M^{-1})^{-1}.$$

Here and elsewhere I_X (resp. I_Y) denotes the *identity map* of X (resp. Y).

Proof Relation (1.3) is a consequence in the following equivalences:

$$\begin{aligned} y \in r^{-1}[I_X - (I_X + rM)^{-1}](x) &\Leftrightarrow ry - x \in -(I_X + rM)^{-1}(x) \\ &\Leftrightarrow x \in (I_X + rM)(x - ry) \\ &\Leftrightarrow x - (x - ry) \in rM(x - ry) \\ &\Leftrightarrow y \in M(x - ry) \Leftrightarrow x - ry \in M^{-1}(y) \\ &\Leftrightarrow x \in (rI_X + M^{-1})(y) \Leftrightarrow y \in (rI_X + M^{-1})^{-1}(x). \end{aligned}$$

□

Exercises

1. Give an example showing that, for a multimap $F : X \rightrightarrows Y$, one may have

$$F^{-1} \circ F \neq I_X \qquad F \circ F^{-1} \neq I_Y.$$

2. Show that $I_X \subset F^{-1} \circ F$ (the inclusion being the inclusion of graphs or images) if and only if $\mathrm{dom}(F) = X$. Also show that $I_Y \subset F \circ F^{-1}$ if and only if $F(X) = Y$.
3. Given multimaps $F : X \rightrightarrows Y$, $G : Y \rightrightarrows Z$ and $H := G \circ F$, give a sufficient condition in order to have $F \subset G^{-1} \circ H$. Show that this inclusion may not hold.
4. Give an example showing that for a multimap $F : X \rightrightarrows Y$ and subsets B, B' of Y in general one has $F^{-1}(B \cap B') \neq F^{-1}(B) \cap F^{-1}(B')$.
5. Given multimaps $F : X \rightrightarrows Y$, $G : Y \rightrightarrows Z$ show that

$$\mathrm{gph}(G \circ F) = (I_X \times G)(\mathrm{gph}F) = (F \times I_Z)^{-1}(\mathrm{gph}G).$$

6. Considering an order relation $\preceq$ on a set X as a multimap $F : X \rightrightarrows X$ whose graph is the set $\{(x, y) \in X \times X : x \preceq y\}$, write the properties of $\preceq$ as properties of F. Do the same with an equivalence relation.

1.4 Measurable Spaces

The family $\mathcal{P}(X)$ of all the subsets of a set X has a rich algebraic structure inherited from the operations $\cap$, $\cup$ and from the inclusion $\subset$. The aim of the present section is a study of some remarkable subclasses of $\mathcal{P}(X)$. The inclusion $\subset$ endows $\mathcal{P}(X)$ with a lattice structure. It is even a *complete lattice* in the sense that for any family $(A_i)_{i\in I}$ of subsets of X, the intersection (resp. union) of the family $(A_i)_{i\in I}$ is the greatest lower bound (resp. smallest upper bound) of $(A_i)_{i\in I}$ with respect to the order defined by $\subset$. Moreover, $\varnothing$ (resp. X) is the least element (resp. greatest element) of $\mathcal{P}(X)$ and $\mathcal{P}(X)$ is *complemented*, i.e. every element A of $\mathcal{P}(X)$ has a complement A^c such that $A \cap A^c = \varnothing$ and $A \cup A^c = X$; obviously $A^c = X \backslash A := \{x \in X : x \notin A\}$. Furthermore, $\mathcal{P}(X)$ is *distributive* in the sense that for any family $(A_i)_{i\in I}$ of subsets of X and any $B \in \mathcal{P}(X)$ one has

$$(\bigcup_{i\in I} A_i) \cap B = \bigcup_{i\in I}(A_i \cap B),$$

$$(\bigcap_{i\in I} A_i) \cup B = \bigcap_{i\in I}(A_i \cup B).$$

One says that $\mathcal{P}(X)$ is a Boolean lattice or a *Boolean algebra.*

Besides union and intersection, $\mathcal{P}(X)$ is endowed with the operation Δ called the *symmetric difference* given by

$$\forall A,\ B \in \mathcal{P}(X) \qquad A\Delta B := (A\backslash B) \cup (B\backslash A) = (A \cup B)\backslash(A \cap B),$$

where for $C,\ D \in \mathcal{P}(X)$ one sets $C\backslash D := \{x \in C :\ x \notin D\}$. The operation Δ is commutative: $A\Delta B = B\Delta A$ and associative: for all $A,\ B,\ C \in \mathcal{P}(X)$ one has

$$(A\Delta B)\Delta C = A\Delta(B\Delta C)$$

(exercise).

It can be convenient to embed $\mathcal{P}(X)$ into the set $\mathcal{F}(X, \mathbb{R})$ of real-valued functions on X by using the *characteristic map*

$$\theta : \mathcal{P}(X) \to \mathbb{R}^X := \mathcal{F}(X, \mathbb{R})$$

given by $\theta(A) = 1_A$, where $1_A \in \mathcal{F}(X, \mathbb{R})$ is the *characteristic function* of $A \in \mathcal{P}(X)$ (sometimes called the indicator function, but we prefer to keep this term for another function) defined by

$$1_A(x) = 1 \text{ if } x \in A, \qquad 1_A(x) := 0 \text{ if } x \in A^c := X \backslash A.$$

It is such that $\theta(A \cap B) = \theta(A).\theta(B)$, the product of the two functions $\theta(A)$ and $\theta(B)$ on X. It is still more illuminating to consider θ as taking its values in the set $\mathcal{F}(X, \mathbb{Z})$ of integer-valued functions and to compose θ with the map $p : \mathcal{F}(X, \mathbb{Z}) \to \mathcal{F}(X, \mathbb{Z}_2)$ induced by the quotient map $q : \mathbb{Z} \to \mathbb{Z}_2 := \mathbb{Z}/2\mathbb{Z}$ by setting $p(f) := q \circ f$. Then one gets the characteristic map $\chi := p \circ \theta : \mathcal{P}(X) \to \mathcal{F}(X, \mathbb{Z}_2)$. Identifying $\mathbb{Z}_2$ with the set $\{0, 1\}$, the operations on the ring $\mathbb{Z}_2$ are carried into operations on $\{0, 1\}$ corresponding to evenness, since $q(n) = 0$ if n is even, $q(n) = 1$ if n is odd. Then, for all $A, B \in \mathcal{P}(X)$ one has

$$\chi(A \cap B) = \chi(A).\chi(B), \qquad \chi(A \Delta B) = \chi(A) + \chi(B).$$

Then one realizes that $\mathcal{P}(X)$ is given a ring structure (in the usual algebraic sense) with the two operations Δ (for the addition) and $\cap$ (for the product). In order to check the usual rules for rings, it suffices to observe that χ is injective (since $A = B$ whenever $1_A = 1_B$); in fact χ is a bijection whose inverse is the map $g \mapsto g^{-1}(1)$. Then χ becomes a ring isomorphism from $\mathcal{P}(X)$ onto $\mathcal{F}(X, \mathbb{Z}_2)$. Moreover, the empty set $\varnothing$ is the neutral element for Δ and X is the unit for $\cap$. These observations justify the following terminology.

Definition 1.2 A nonempty subclass $\mathcal{A}$ of $\mathcal{P}(X)$ is called a *ring* if for all $A, B \in \mathcal{A}$ one has $A \Delta B \in \mathcal{A}$ and $A \cap B \in \mathcal{A}$.

If, moreover, $X \in \mathcal{A}$ one says that $\mathcal{A}$ is a (Boolean) *algebra.*

A subclass $\mathcal{S}$ of $\mathcal{P}(X)$ is called a *σ-algebra* (resp. a *σ-ring*) if it is an algebra (resp. a ring) and if the union of a countable family of elements of $\mathcal{S}$ is in $\mathcal{S}$.

The following criterion may appear as more convenient.

Proposition 1.5 *A nonempty subclass $\mathcal{A}$ of $\mathcal{P}(X)$ is a ring if and only if it is such that for all $A, B \in \mathcal{A}$ one has $A \cup B \in \mathcal{A}$ and $A \backslash B \in \mathcal{A}$.*

A nonempty subclass $\mathcal{A}$ of $\mathcal{P}(X)$ is a σ-algebra if and only if it is such that $X \in \mathcal{A}$, $\cup_n A_n \in \mathcal{A}$ whenever $A_n \in \mathcal{A}$ for all $n \in \mathbb{N}$, and $A^c := X \backslash A \in \mathcal{A}$ for all $A \in \mathcal{A}$.

Proof For the characterization of rings, the only if assertion follows from the relations

$$A \cup B = A \Delta B \Delta (A \cap B), \qquad A \backslash B = A \Delta (A \cap B).$$

The converse is a consequence in the relations

$$A \Delta B = (A \backslash B) \cup (B \backslash A), \qquad A \cap B = (A \cup B) \Delta (A \Delta B).$$

The only if assertion of the characterization of σ-algebras is obvious. The if assertion follows from the preceding characterization since $A \backslash B = A \cap B^c$ and $A \cap B = X \backslash (A^c \cup B^c)$. □

Note that when $\mathcal{A}$ is closed under finite unions, since $A\backslash B = (A \cup B)\backslash B$ the requirement $A\backslash B \in \mathcal{A}$ for all $A, B \in \mathcal{A}$ is satisfied whenever $\mathcal{A}$ is *relatively complemented* in the sense that $A\backslash B \in \mathcal{A}$ for all $A, B \in \mathcal{A}$ satisfying $B \subset A$.

Note also that a nonempty ring contains the empty set $\varnothing$ since for any $A \in \mathcal{A}$ one has $A\Delta A = \varnothing$. Among the preceding notions, the notion of a σ-algebra is the most important one, as the next definition and the sequel show.

Definition 1.3 A *measurable space* is a pair $(X, \mathcal{S})$ where $\mathcal{S}$ is a σ-algebra of subsets of X.

A map f between two measurable spaces $(X, \mathcal{S})$, $(Y, \mathcal{T})$ is said to be *measurable* if for all $B \in \mathcal{T}$ one has $f^{-1}(B) \in \mathcal{S}$.

Clearly, given measurable spaces $(X, \mathcal{S})$, $(Y, \mathcal{T})$, $(Z, \mathcal{U})$, and measurable maps $f : (X, \mathcal{S}) \to (Y, \mathcal{T})$, $g : (Y, \mathcal{T}) \to (Z, \mathcal{U})$, the map $g \circ f : (X, \mathcal{S}) \to (Z, \mathcal{U})$ is measurable.

If $(X, \mathcal{S})$ is a measurable space, if W is a set, and if $j : W \to X$ is a map, the family $\mathcal{S}_W$ of subsets of W of the form $j^{-1}(B)$ with $B \in \mathcal{S}$ is a σ-algebra called the *inverse image* of $\mathcal{S}$ by j. In particular, if $(X, \mathcal{S})$ is a measurable space and if W is a subset of X, taking for j the canonical injection, we see that the family $\mathcal{S}_W$ of subsets A of W such that there is some $B \in \mathcal{S}$ satisfying $B \cap W = A$ is a σ-algebra called the σ-algebra *induced* by $\mathcal{S}$ on W. Note that when $W \in \mathcal{S}$, for $A \in \mathcal{P}(W)$ one has $A \in \mathcal{S}_W$ if and only if $A \in \mathcal{S}$.

The intersection of a family $(\mathcal{S}_i)_{i \in I}$ of rings (resp. σ-algebras) of X is a ring (resp. a σ-algebra). As a consequence, for any subset $\mathcal{G}$ of $\mathcal{P}(X)$ there is a smallest ring $\mathcal{S}$ (resp. σ-algebra) containing $\mathcal{G}$: $\mathcal{S}$ is the intersection of the family of rings (resp. σ-algebras) containing $\mathcal{G}$ (since $\mathcal{P}(X)$ is itself a σ-algebra, this family is nonempty). One says that $\mathcal{S}$ is the ring (resp. σ-algebra) *generated* by $\mathcal{G}$. A similar observation holds for algebras and σ-rings.

Let us note the following useful observation.

Lemma 1.1 *Let $j : W \to X$ be a map and let $\mathcal{B}$ be the σ-algebra on X generated by some subset $\mathcal{G}$ of $\mathcal{P}(X)$. Then the σ-algebra $\mathcal{A} := \{j^{-1}(B) : B \in \mathcal{B}\}$ on W is generated by $\mathcal{F} := \{j^{-1}(G) : G \in \mathcal{G}\}$.*

In particular, if W is a subset of a set X and if $\mathcal{B}$ is the σ-algebra generated by some subset $\mathcal{G}$ of $\mathcal{P}(X)$, then the induced σ-algebra $\mathcal{A} := \{B \cap W : B \in \mathcal{B}\}$ on W is generated by $\mathcal{F} := \{G \cap W : G \in \mathcal{G}\}$.

Proof Let $\mathcal{A}'$ be a σ-algebra on W containing $\mathcal{F}$. Set

$$\mathcal{B}' := \{B' \in \mathcal{P}(X) : j^{-1}(B') \in \mathcal{A}'\}.$$

Since $\mathcal{B}'$ contains $\mathcal{G}$ and is a σ-algebra, $\mathcal{B}'$ contains $\mathcal{B}$. Thus, for all $B \in \mathcal{B}$ we have $j^{-1}(B) \in \mathcal{A}'$. In other words we have $\mathcal{A} \subset \mathcal{A}'$. Since $\mathcal{A}$ is clearly a σ-algebra, this shows that $\mathcal{A}$ is the smallest σ-algebra on W containing $\mathcal{F}$. □

The preceding construction can be generalized and applied to products by taking for g_i below the canonical projections.

Proposition 1.6 *Let X be a set and for $i \in I$ let $g_i : X \to X_i$ be a map. Given σ-algebras $\mathcal{S}_i$ on X_i there is a smallest σ-algebra $\mathcal{S}$ on X such that for all $i \in I$ the map $g_i : (X, \mathcal{S}) \to (X_i, \mathcal{S}_i)$ is measurable. Moreover, if $\mathcal{S}_i$ is generated by $\mathcal{G}_i \subset \mathcal{S}_i$, then $\mathcal{S}$ is generated by the collection $\mathcal{G}$ of finite intersections $\cap_{j \in J} g_j^{-1}(G_j)$ for J a finite subset of I and $G_j \in \mathcal{G}_j$.*

If $(W, \mathcal{R})$ is a measurable space, a map $f : (W, \mathcal{R}) \to (X, \mathcal{S})$ is measurable if and only if for all $i \in I$ the map $g_i \circ f$ is measurable.

Proof Given $\mathcal{G}_i \subset \mathcal{S}_i$ generating $\mathcal{S}_i$ we denote by $\mathcal{S}$ the σ-algebra generated by the union $\mathcal{F}$ of the families $\mathcal{F}_i := \{g_i^{-1}(G_i) : G_i \in \mathcal{G}_i\}$ for $i \in I$. Then, for all $i \in I$, the class $\mathcal{A}_i := \{A \in \mathcal{S}_i : g_i^{-1}(A_i) \in \mathcal{S}\}$ is a σ-algebra containing $\mathcal{G}_i$, so that $\mathcal{A}_i = \mathcal{S}_i$ and g_i is measurable. Clearly $\mathcal{S}$ is the smallest σ-algebra satisfying this property and $\mathcal{S}$ is generated by the class $\mathcal{G}$ of the statement.

If $f : (W, \mathcal{R}) \to (X, \mathcal{S})$ is measurable, then for all $i \in I$ the map $g_i \circ f$ is measurable. Conversely, if for all $i \in I$ the map $g_i \circ f$ is measurable, then for all $F \in \mathcal{F} := \cup_{i \in I} \mathcal{F}_i$ one has $f^{-1}(F) \in \mathcal{R}$ so that f is measurable. □

Let us describe an inverse construction consisting in endowing the image of a map with a σ-algebra. Given a map $f : X \to Y$ between two sets and a σ-algebra $\mathcal{S}$ in X, the family $\mathcal{T}$ of subsets B of Y such that $f^{-1}(B) \in \mathcal{S}$ is a σ-algebra. The σ-algebra $\mathcal{T}$ is called the *direct image* of $\mathcal{S}$ by f and is denoted by $f_*(\mathcal{S})$.

More generally, one can glue together measurable spaces.

Lemma 1.2 *Let X be a set and let $(X_i)_{i \in I}$ be a family of subsets of X. If for all $i \in I$, $\mathcal{A}_i$ is a ring (resp. a σ-ring) of X_i then*

$$\mathcal{A} := \{A \in \mathcal{P}(X) : \forall i \in I \;\; A \cap X_i \in \mathcal{A}_i\}$$

is a ring (resp. σ-ring) of X. If $\mathcal{A}_i$ is an algebra (resp. σ-algebra) of X_i then $\mathcal{A}$ is an algebra (resp. σ-algebra).

Proof The first assertion stems from the relations

$$(A \cap B) \cap X_i = (A \cap X_i) \cap (B \cap X_i), \quad (A \backslash B) \cap X_i = (A \cap X_i) \backslash (B \cap X_i),$$

$$(\bigcup_n A_n) \cap X_i = \bigcup_n (A_n \cap X_i).$$

The second one is a consequence in the relation $A^c \cap X_i = X_i \backslash A$. □

Anticipating the notion of a topological space, let us mention that if $(X, \mathcal{O})$ is a topological space, the σ-algebra $\mathcal{B}$ generated by the family $\mathcal{O}$ of open subsets of X is called the *Borel algebra* of $(X, \mathcal{O})$ and its members are called the *Borel subsets* of $(X, \mathcal{O})$. One often denotes $\mathcal{B}$ by $\mathcal{B}(X)$ rather than $\mathcal{B}(X, \mathcal{O})$.

Proposition 1.7 *Given measurable spaces $(X, \mathcal{S})$, $(Y, \mathcal{T})$, and a family of subsets $\mathcal{G}$ generating $\mathcal{T}$, a map $f : X \to Y$ is measurable with respect to $\mathcal{S}$ and $\mathcal{T}$ if and only if for all $G \in \mathcal{G}$ one has $f^{-1}(G) \in \mathcal{S}$.*

In particular, if $(Y, \mathcal{O})$ is a topological space and if $\mathcal{T}$ is the Borel σ-algebra of $(Y, \mathcal{O})$, a map $f : X \to Y$ is measurable if and only if for all $O \in \mathcal{O}$ one has $f^{-1}(O) \in \mathcal{S}$.

Equivalently, in this particular case, $f : X \to Y$ is measurable if and only if, for any closed subset C of Y, the set $f^{-1}(C)$ is in $\mathcal{S}$. The second assertion of the proposition entails that any continuous map between two topological spaces is measurable for the associated Borel σ-algebras.

Proof It suffices to show that f is measurable whenever $f^{-1}(G) \in \mathcal{S}$ for all $G \in \mathcal{G}$. This follows from the fact that $f_*(\mathcal{S}) := \{H \in \mathcal{P}(Y) : f^{-1}(H) \in \mathcal{S}\}$ is a σ-algebra containing $\mathcal{G}$, so that $f_*(\mathcal{S})$ contains $\mathcal{T}$. □

It is of interest to consider functions with values in $\overline{\mathbb{R}} := \mathbb{R} \cup \{-\infty, +\infty\}$ rather than in $\mathbb{R}$ because $\overline{\mathbb{R}}$ is compact and every family in $\overline{\mathbb{R}}$ has a supremum and an infimum. We use the fact that there exists an increasing bijection $h : \overline{\mathbb{R}} \to [-1, +1]$ such as the one given by $h(r) := r/(|r|+1)$ for $r \in \mathbb{R}$, $h(-\infty) := -1$, $h(+\infty) := 1$ that can serve to define a topology on $\overline{\mathbb{R}}$. Moreover, the topology of $\mathbb{R}$ is the induced topology of $\overline{\mathbb{R}}$ and the Borel algebra $\mathcal{B}(\mathbb{R})$ of $\mathbb{R}$ is the σ-algebra induced by the Borel algebra $\mathcal{B}(\overline{\mathbb{R}})$ of $\overline{\mathbb{R}}$. Thus a function $f : X \to \mathbb{R}$ is measurable if and only if it is measurable as a function from X to $\overline{\mathbb{R}}$.

Corollary 1.5 *Given a measurable space $(X, \mathcal{S})$ and a function $f : X \to \mathbb{R}$ (resp. $f : X \to \overline{\mathbb{R}}$), endowing $\mathbb{R}$ (resp. $\overline{\mathbb{R}}$) with its Borel σ-algebra, f is measurable if and only if for all $r \in \mathbb{R}$ or all $r \in \mathbb{Q}$ the set $\{f > r\} := f^{-1}(]r, +\infty[)$ is in $\mathcal{S}$.*

Proof This is a consequence in the fact that $\mathcal{B}(\mathbb{R})$ (resp $\mathcal{B}(\overline{\mathbb{R}})$) is generated by the family of intervals $]r, +\infty[$ (resp. $]r, +\infty]$) for $r \in \mathbb{Q}$. □

Proposition 1.8 *If $f, g : X \to \overline{\mathbb{R}}$ are measurable, then $f \vee g := \sup(f, g)$ and $f \wedge g := \inf(f, g)$ are measurable. If f and g are finitely valued, then $(f, g) : X \to \mathbb{R}^2$ is measurable for the Borel σ-algebra of $\mathbb{R}^2$ and $f+g$, $f-g$, and $f.g$ are measurable.*

Proof The first assertion follows from the relations

$$(f \vee g)^{-1}(]r, +\infty]) = f^{-1}(]r, +\infty]) \cup g^{-1}(]r, +\infty]),$$

$$(f \wedge g)^{-1}(]r, +\infty]) = f^{-1}(]r, +\infty]) \cap g^{-1}(]r, +\infty]).$$

To show that $h := (f, g)$ is measurable when f and g are measurable it suffices to observe that for open subsets U, V of $\mathbb{R}$ the set $h^{-1}(U \times V) = f^{-1}(U) \cap g^{-1}(V)$ is in $\mathcal{S}$ and that the Borel σ-algebra of $\mathbb{R}^2$ is generated by the family of products of open subsets. Using the continuous functions $(r, s) \mapsto r + s$, $(r, s) \mapsto r - s$, $(r, s) \mapsto rs$ and taking their compositions with h, we get the last assertion. □

Let us note the following stability result.

Lemma 1.3 *Let $(X, \mathcal{S})$ be a measurable space, let (Y, d) be a metric space and let (f_n) be a sequence of measurable maps from X into Y which converges pointwise in the sense that for all $x \in X$ one has $(f_n(x)) \to f(x)$. Then the limit f of (f_n) is measurable.*

Proof This stems from the fact that, for every nonempty closed subset C of Y, for $C_k := \{y \in Y : d(y, C) \leq 2^{-k}\}$ one has

$$f^{-1}(C) = \cap_{k\in\mathbb{N}} \cup_{m\in\mathbb{N}} \cap_{n\geq m} f_n^{-1}(C_k).$$

□

In an important case, the construction of the ring generated by a subset of $\mathcal{P}(X)$ can be made explicit. It will be convenient to use the following notion.

Definition 1.4 A subclass $\mathcal{C}$ of the class $\mathcal{P}(X)$ of all the subsets of a set X is a *semi-ring* if for all A, B in $\mathcal{C}$ one has $A \cap B \in \mathcal{C}$ and if $A \backslash B$ is the union of a finite family of disjoint elements of $\mathcal{C}$.

Let us note that if a semi-ring $\mathcal{C}$ is nonempty, then it contains the empty set $\varnothing$ since otherwise, given $C \in \mathcal{C}$, one cannot obtain $C \backslash C$ as a finite union of nonempty sets. In $\mathbb{R}$ the family $\mathcal{C}$ of intervals of the form $[a, b[:= \{r \in \mathbb{R} : a \leq r < b\}$ with a, $b \in \mathbb{R}$, $a \leq b$, is an important example of a semi-ring.

In the sequel if A and B are two disjoint subsets of X (in the sense that $A \cap B = \varnothing$) we write $A \sqcup B$ for $A \cup B$. It will be convenient to say that a class $\mathcal{F}$ of subsets of a set X is *disjoint* if distinct members of $\mathcal{F}$ are disjoint. If $(A_i)_{i\in I}$ is a disjoint family of subsets of X, we write $\sqcup_{i\in I} A_i$ for $\cup_{i\in I} A_i$. We say that $(A_i)_{i\in I}$ is a *partition* of $A \subset X$ if $A = \sqcup_{i\in I} A_i$.

Lemma 1.4 *The ring generated by a semi-ring $\mathcal{C}$ on X is the set $\mathcal{A}$ formed by the unions of disjoint finite subfamilies of $\mathcal{C}$. Moreover, $\mathcal{A}$ is also the set of unions of finite subfamilies of $\mathcal{C}$ and for C, $C_1, \ldots, C_n$ in $\mathcal{C}$ one can find finite families $(D_j)_{j\in J}$, $(D_{i,j})_{j\in J_i}$ ($i \in \mathbb{N}_n$) of members of $\mathcal{C}$ such that*

$$C \backslash (\bigcup_{i=1}^{n} C_i) = \coprod_{j\in J} D_j, \tag{1.4}$$

$$\bigcup_{i=1}^{n} C_i = \coprod_{i=1}^{n} \coprod_{j\in J_i} D_{i,j} \qquad \text{with } D_{i,j} \subset C_i,\ D_{i,j} \in \mathcal{C} \text{ for } i \in \mathbb{N}_n,\ j \in J_i. \tag{1.5}$$

Proof Given A, $B \in \mathcal{A}$ let us first show that $A \cap B$ belongs to $\mathcal{A}$. Let I, J be finite sets and let $(C_i)_{i\in I}$, $(D_j)_{j\in J}$ be two families of disjoints elements of $\mathcal{C}$ such that $A = \cup_{i\in I} C_i$ and $B = \cup_{j\in J} D_j$. Then $A \cap B = \cup_{(i,j)\in I\times J} C_i \cap D_j$ and the family $(C_i \cap D_j)_{(i,j)\in I\times J}$ is a family of disjoint elements of $\mathcal{C}$, so that $A \cap B \in \mathcal{A}$.

Now $A \backslash B = \cup_{i\in I}(C_i \backslash B) = \cup_{i\in I} \cap_{j\in J} (C_i \backslash D_j)$ and since $C_i \backslash D_j \in \mathcal{A}$ we have $C_i' := \cap_{j\in J}(C_i \backslash D_j) \in \mathcal{A}$ by what precedes and an induction. Since the family $(C_i')_{i\in I}$ is formed of disjoint subsets in $\mathcal{C}$ we have $A \backslash B \in \mathcal{A}$ and $\mathcal{A}$ is a ring by Proposition 1.5. Since any ring containing $\mathcal{C}$ contains the elements of $\mathcal{A}$, the ring generated by $\mathcal{C}$ is $\mathcal{A}$.

Relation (1.4) can be established by induction on n since we know the result for $n = 1$ and $C \backslash (\cup_{1\leq i\leq n} C_i) = (C \backslash (\cup_{1\leq i\leq n-1} C_i)) \backslash C_n$. Relation (1.5) is a consequence in the following relation

$$\bigcup_{i=1}^{n} C_i = \coprod_{i=1}^{n} C_i' \qquad \text{with } C_i' := C_i \backslash (\bigcup_{h=1}^{i-1} C_h) \text{ for } i \geq 2,\ C_1' := C_1,$$

the property $C_i' \cap C_j' = \varnothing$ for $i < j$ stemming from the inclusion $C_i \subset \cup_{1 \leq h \leq j-1} C_h$. Finally, relation (1.5) ensures that the family $\mathcal{A}'$ of unions of elements of $\mathcal{C}$ coincides with $\mathcal{A}$. □

The next two results take into account the fact that $\mathcal{P}(X)$ has a natural order given by inclusion. They are closely related and have important consequences. Let us say that a class $\mathcal{D}$ of subsets of a set X is an *increasing class* if it includes the union of any increasing sequence of members of $\mathcal{D}$; here a sequence (A_n) of $\mathcal{P}(X)$ is said to be *increasing* (resp. *decreasing*) if $A_n \subset A_{n+1}$ (resp. $A_{n+1} \subset A_n$) for all $n \in \mathbb{N}$. A class $\mathcal{D}$ of subsets of a set X is a *decreasing class* if it includes the intersection of any decreasing sequence of members of $\mathcal{D}$. A class $\mathcal{M}$ of subsets of X is called a *monotone class* if it is an increasing class and a decreasing class. Let us recall that a class $\mathcal{D}$ of subsets of a set X is *relatively complemented* if for all $A, B \in \mathcal{D}$ such that $B \subset A$ one has $A \backslash B \in \mathcal{D}$. Given $\mathcal{C} \subset \mathcal{P}(X)$ there is a smallest relatively complemented increasing class (resp. monotone class) containing $\mathcal{C}$. It is called the relatively complemented increasing class (resp. the monotone class) generated by $\mathcal{C}$.

Theorem 1.9 (Increasing Class Theorem) *Let X be a set and let $\mathcal{C} \subset \mathcal{P}(X)$ be closed under the formation of finite intersections. Then the relatively complemented increasing class $\mathcal{D}$ generated by $\mathcal{C}$ is the σ-ring $\mathcal{R}$ generated by $\mathcal{C}$.*

If X is the union of a countable subfamily of $\mathcal{C}$, then $\mathcal{D}$ is the σ-algebra generated by $\mathcal{C}$.

The proof below shows that instead of assuming that $\mathcal{C}$ is closed under finite intersections one may assume that for all $C, C' \in \mathcal{C}$ one has $C \cap C' \in \mathcal{D}$. However, in the applications we have in view, the class $\mathcal{C}$ is closed under intersections.

Proof Since a σ-ring is a relatively complemented increasing class, one has $\mathcal{D} \subset \mathcal{R}$. In order to prove the reverse inclusion we begin by showing that $\mathcal{D}$ is closed under (finite) intersections. Let

$$\mathcal{D}' := \{A \in \mathcal{D} : A \cap C \in \mathcal{D}\ \forall C \in \mathcal{C}\}.$$

By assumption, $\mathcal{C}$ is contained in $\mathcal{D}'$. The relation $(A \backslash B) \cap C = (A \cap C) \backslash (B \backslash C)$ shows that $\mathcal{D}'$ is relatively complemented. Moreover, $\mathcal{D}'$ is clearly an increasing class as is $\mathcal{D}$. Thus $\mathcal{D}' = \mathcal{D}$. Now let

$$\mathcal{D}'' := \{A \in \mathcal{D} : A \cap D \in \mathcal{D}\ \forall D \in \mathcal{D}\}.$$

The relation $\mathcal{D}' = \mathcal{D}$ implies that $\mathcal{C} \subset \mathcal{D}''$. Moreover, the same arguments show that $\mathcal{D}''$ is a relatively complemented increasing class, hence that $\mathcal{D}'' = \mathcal{D}$. Therefore $\mathcal{D}$

is closed under finite intersections, hence is a σ-ring. It follows that $\mathcal{R} \subset \mathcal{D}$ and that $\mathcal{R} = \mathcal{D}$.

If X is the union of a countable subfamily of $\mathcal{C}$, then $X \in \mathcal{R} = \mathcal{D}$: $\mathcal{D}$ is a σ-algebra. □

Theorem 1.10 (Monotone Class Theorem) *Let $\mathcal{G}$ be a class of subsets of a set X. Suppose the monotone class $\mathcal{M}$ generated by $\mathcal{G}$ contains the complements of the members of $\mathcal{G}$ and the finite intersections of members of $\mathcal{G}$. Then $\mathcal{M}$ is the σ-algebra $\mathcal{S}$ generated by $\mathcal{G}$.*

The same conclusion holds if $\mathcal{M}$ contains the finite unions of members of $\mathcal{G}$ and the complements of members of $\mathcal{G}$.

In particular, if $\mathcal{G}$ is an algebra, then $\mathcal{M}$ is the σ-algebra generated by $\mathcal{G}$.

Proof The class $\mathcal{M}'$ of sets M in $\mathcal{M}$ such that $M^c := X \backslash M$ belongs to $\mathcal{M}$ is a monotone class and by assumption it contains $\mathcal{G}$. Thus $\mathcal{M}' = \mathcal{M}$ and $\mathcal{M}$ is closed by taking complements. Let $\mathcal{C}$ be the class of finite intersections of elements of $\mathcal{G}$, so that $\mathcal{C}$ is closed under finite intersections. Assuming $\mathcal{C}$ is contained in $\mathcal{M}$ means that $\mathcal{M}$ is the monotone class generated by $\mathcal{C}$. Theorem 1.9 shows that $\mathcal{M}$ is the σ-ring generated by $\mathcal{C}$. Thus $\mathcal{M}$ is also the σ-ring generated by $\mathcal{G}$. Since $\mathcal{M}$ is closed under complementation, $\mathcal{M}$ is the σ-algebra generated by $\mathcal{G}$.

The case $\mathcal{M}$ contains the finite unions of members of $\mathcal{G}$ and the complements of members of $\mathcal{G}$ can be proved by taking complements. □

Exercises

1. Given a set X and two nonempty subsets A, B of X, describe the rings, algebras, σ-rings, and σ-algebras generated by the families $\mathcal{G} := \{A\}, \mathcal{G}' := \{A, B\}$.
2. Show that the family $\mathcal{C}_d$ of boxes of $\mathbb{R}^d$ of the form $[a_1, b_1[\times \ldots \times [a_d, b_d[$ is a semi-ring.
3. Prove that the Borel family of $\mathbb{R}^d$ is generated by the semi-ring $\mathcal{C}_d$ of Exercise 2.
4. Let (A_n) be a sequence in a σ-ring $\mathcal{A}$ of subsets of a set X. Prove that the intersection of the family $\{A_n : n \in \mathbb{N}\}$ is in $\mathcal{A}$ as are the sets

$$\bigcap_m \bigcup_{n \geq m} A_n \qquad \bigcap_m \bigcup_{n \geq m} A_n,$$

often denoted by $\limsup_n A_n$ and $\liminf_n A_n$, respectively.
5. Let X be a set and let $((X_i, \mathcal{S}_i))_{i \in I}$ be a family of measurable spaces. Given maps $f_i : X \to X_i$ show that there exists a unique σ-algebra $\mathcal{S}$ in X such that for any measurable space $(W, \mathcal{U})$ a map $g : (W, \mathcal{U}) \to (X, \mathcal{S})$ is measurable if and only if for all $i \in I$ the map $f_i \circ g : (W, \mathcal{U}) \to (X_i, \mathcal{S}_i)$ is measurable. Verify that $\mathcal{S}$ is the smallest σ-algebra on X for which the maps f_i are measurable.

6. Deduce from the preceding exercise that the product of a family of measurable spaces can be endowed with the structure of a measurable space in a canonical way making the projections measurable.
7. Given measurable functions $f, g : X \to \mathbb{R}$ for some σ-algebra $\mathcal{S}$ in X, check that the measurability of $f + g$ is a consequence in the fact that $\{f + g > r\}$ is the union over $q \in \mathbb{Q}$ of the sets $\{f > q\} \cap \{g > r - q\}$.
8*. (**Stone's Theorem**) Show that any Boolean lattice $(S, \leq)$ (in the algebraic sense) is isomorphic to a ring $\mathcal{S}$ of subsets of a set X. [Hint: consider the set $X \subset \mathbb{Z}_2^S$ of homomorphisms $x : S \to \mathbb{Z}_2 := \{0, 1\}$ and for $s \in S$ set $f(s) := \{x \in X : x(s) = 1\}$; to show that $f : S \to \mathcal{P}(X)$ is injective, consider first the case when S is finite and given $\overline{s} \in S$ use a compactness argument involving the family of finite subrings of S containing $\overline{s}$ to construct some $x \in f(\overline{s})$.]
9. Verify that a relatively complemented increasing class of subsets of a set X is a monotone class. [Hint: given a decreasing sequence (A_n) of subsets of X note that $\cap_n A_n = A_0 \backslash (\cup B_n)$ for $B_n := A_0 \backslash A_n$.] From this observation deduce that Theorem 1.9 is a consequence in Theorem 1.10 (and thus that the two results are equivalent).

1.5 Measures

The concept of measure space is a fundamental notion linked with some additivity properties.

Definition 1.5 Given a set X and a subset $\mathcal{C}$ of $\mathcal{P}(X)$, a function $\mu : \mathcal{C} \to \overline{\mathbb{R}}_\infty := \mathbb{R} \cup \{+\infty\}$ is said to be *additive* (resp. *subadditive*) if for every finite sequence $(C_1, \ldots, C_n)$ of disjoint elements of $\mathcal{C}$ whose union C is in $\mathcal{C}$ one has $\mu(C) = \mu(C_1) + \ldots + \mu(C_n)$ (resp. $\mu(C) \leq \mu(C_1) + \ldots + \mu(C_n)$).

The function μ is said to be *countably additive* or *σ-additive* (resp. *countably subadditive* or σ-subadditive) if for any sequence (C_n) of disjoint elements of $\mathcal{C}$ whose union C is in $\mathcal{C}$ one has $\mu(C) = \sum_n \mu(C_n)$ (resp. $\mu(C) \leq \sum_n \mu(C_n)$).

The function μ is said to be finite if it takes its values in $\mathbb{R}$. It is said to be σ-finite if there exists a sequence (A_n) of sets in $\mathcal{C}$ whose union is X such that $\mu(A_n) < +\infty$ for all $n \in \mathbb{N}$.

Definition 1.6 A measure on $(X, \mathcal{C})$ with $\varnothing \in \mathcal{C}$ is a countably additive function $\mu : \mathcal{C} \to \overline{\mathbb{R}}_+ := [0, +\infty]$ that satisfies $\mu(\varnothing) = 0$.

For the sake of simplicity we adopt this (usual) terminology although countably additive functions with values in $\mathbb{R}$ are of interest; they will be called signed measures (see Chap. 8 and the exercises).

When $\mathcal{C}$ is a σ-algebra $\mathcal{S}$ and $\mu : \mathcal{S} \to \overline{\mathbb{R}}_+$ is a measure, the triple $(X, \mathcal{S}, \mu)$ is called a *measure space*. If moreover $\mu(X) = 1$ one says that μ is a *probability* on X.

Example The *counting measure* on a set X is defined on $\mathcal{P}(X)$ by $\mu(A) := 0$ if A is the empty set, $\mu(A) := n$ if A has n distinct points, and $\mu(A) := +\infty$ if A is infinite.

Example Given a function $\omega : X \to \mathbb{R}_+$ on a set X, the *discrete measure* associated with the weights $(\omega_x)_{x\in X}$ with $\omega_x := \omega(x)$ is defined by $\mu(A) := \Sigma_{x\in A}\omega_x$. If $\omega_x = 1$ for all x we get the counting measure on X.

We observe that any additive function μ on a ring $\mathcal{A}$ is nondecreasing since for $A, B \in \mathcal{A}$ with $A \subset B$ one has $\mu(B) = \mu(A) + \mu(B\backslash A) \geq \mu(A)$. This observation justifies the existence of the limits in the next lemma.

Let us stress the difference between additivity and σ-additivity by pointing out the following "continuity" property.

Lemma 1.5 *For an additive function $\mu : \mathcal{A} \to \overline{\mathbb{R}}_+$ on a ring $\mathcal{A}$ of subsets of X, μ is σ-additive if and only if for any increasing sequence (A_n) of elements of $\mathcal{A}$ whose union is in $\mathcal{A}$ one has $\mu(A) = \lim_n \mu(A_n)$.*

Moreover, if $\mu : \mathcal{A} \to \overline{\mathbb{R}}_+$ is σ-additive, if (B_n) is a decreasing sequence of elements of $\mathcal{A}$ whose intersection B is in $\mathcal{A}$, and if for some $m \in \mathbb{N}$, $\mu(B_m)$ is finite, then one has $\mu(B) = \lim_n \mu(B_n)$.

If μ is finite and additive and if for every decreasing sequence (B_n) of elements of $\mathcal{A}$ whose intersection is empty one has $\lim_n(\mu(B_n)) \to 0$, then μ is σ-additive.

Proof Let μ be σ-additive and let (A_n) be an increasing sequence in $\mathcal{A}$ whose union A is in $\mathcal{A}$. Setting $B_0 := A_0$, $B_n := A_n\backslash A_{n-1}$ for $n \in \mathbb{N}\backslash\{0\}$, we get a sequence in disjoint elements of $\mathcal{A}$ whose union is A (since $B_0 \cup \ldots \cup B_n = A_n$) so that we get

$$\mu(A) = \sum_{n=0}^{\infty} \mu(B_n) = \lim_{n\to\infty} \sum_{k=0}^{n} \mu(B_k) = \lim_{n\to\infty} \mu(B_0 \cup \ldots \cup B_n) = \lim_{n\to\infty} \mu(A_n).$$

Conversely, suppose μ is additive and satisfies the above property. Let us show that μ is σ-additive. Given a sequence (A_n) of disjoint elements of $\mathcal{A}$ whose union A is in $\mathcal{A}$, let us set $B_n := A_0 \cup \ldots \cup A_n$. Then $B_n \in \mathcal{A}$, (B_n) is increasing and its union is $A \in \mathcal{A}$, so that

$$\mu(A) = \lim_{n\to\infty} \mu(B_n) = \lim_{n\to\infty} \sum_{k=0}^{n} \mu(A_k) = \sum_{k=0}^{\infty} \mu(A_k).$$

The second assertion is obtained by considering the sequence $(B_m\backslash B_n)_{n\geq m}$ whose union is $B_m\backslash B$ and in using a passage to the limit and the fact that $\mu(B) = \mu(B_m) - \mu(B_m\backslash B)$ and $\mu(B_n) = \mu(B_m) - \mu(B_m\backslash B_n)$ for $n \geq m$.

Finally, let us prove the last assertion. Let $A \in \mathcal{A}$ and let (A_n) be a partition of A by elements of $\mathcal{A}$. Let $B_n := A\backslash \cup_{p\geq n} A_p$, so that $B_n = \cap_{p\geq n}(A\backslash A_p)$. Since $B_{n+1} \subset B_n$ and $\cap_n B_n = \varnothing$ we have $(\mu(B_n)) \to 0$. Moreover, $A = A_0 \sqcup \ldots \sqcup A_{n-1} \sqcup B_n$, so that $\mu(A) = \mu(A_0) + \ldots + \mu(A_{n-1}) + \mu(B_n)$. This means that $\mu(A) = \Sigma_n \mu(A_n)$. □

Theorem 1.9 yields uniqueness results for extensions.

Theorem 1.11 *Let $\mathcal{C}$ be a family of subsets of a set X closed under the formation of finite intersections and let $\mu : \mathcal{C} \to \overline{\mathbb{R}}_+$ be σ-additive. If two measures ν, ν' on the σ-ring $\mathcal{S}$ generated by $\mathcal{C}$ extend μ, then they coincide provided one of the following two conditions is satisfied:*

(a) ν is finite;
(b) μ is σ-finite.

Proof Let us first suppose ν is finite on $\mathcal{S}$. Let

$$\mathcal{D} := \{D \in \mathcal{S} : \nu(D) = \nu'(D)\}.$$

Since $\mathcal{D}$ contains $\mathcal{C}$, if we prove that $\mathcal{D}$ is a relatively complemented increasing class we get the result since, by Theorem 1.9, $\mathcal{S}$ is the relatively complemented increasing class generated by $\mathcal{C}$, so that $\mathcal{S} \subset \mathcal{D}$. For all $D, E \in \mathcal{D}$ with $E \subset D$ we have $D \backslash E \in \mathcal{D}$ since $D \backslash E \in \mathcal{S}$ and

$$\nu(D \backslash E) = \nu(D) - \nu(E) = \nu'(D) - \nu'(E) = \nu'(D \backslash E).$$

Let (D_n) be an increasing sequence in $\mathcal{D}$ and let $D := \cup D_n \in \mathcal{S}$. Setting $E_0 := D_0$, $E_1 := D_1 \backslash D_0, \ldots, E_n := D_n \backslash D_{n-1}, \ldots$ we get a sequence of disjoint elements of $\mathcal{D}$ such that $E_0 \sqcup \ldots \sqcup E_n = D_0 \cup \ldots \cup D_n$. By σ-additivity we get

$$\nu(D) = \nu(\bigcup_n E_n) = \sum_n \nu(E_n) = \sum_n \nu'(E_n) = \nu'(\bigcup_n E_n) = \nu'(D)$$

so that $D \in \mathcal{D}$. Thus $\mathcal{D}$ is a relatively complemented increasing class and $\mathcal{D} = \mathcal{S}$.

Now let us suppose X is the union of a countable family (A_n) of $\mathcal{C}$, each A_n having a finite measure; note that then $\mathcal{S}$ is the σ-algebra generated by $\mathcal{C}$. Let

$$\mathcal{E} := \{E \in \mathcal{S} : \nu(E \cap A_k) = \nu'(E \cap A_k)\ \forall k \in \mathbb{N}\}.$$

Since $\mathcal{C}$ is closed under intersections, $\mathcal{C}$ is contained in $\mathcal{E}$. Using Lemma 1.5, we see that if (E_n) is an increasing sequence in $\mathcal{E}$, then $\cup_n E_n \in \mathcal{E}$. Now if $E, F \in \mathcal{E}$ are such that $F \subset E$, since $(E \backslash F) \cap A_k = (E \cap A_k) \backslash (F \cap A_k)$ we see as above that $E \backslash F \in \mathcal{E}$. Again, Theorem 1.9 ensures that $\mathcal{E} = \mathcal{S}$. Let us define an increasing sequence (B_n) in $\mathcal{S}$ by setting $B_0 := A_0$, $B_1 = B_0 \cup (A_1 \backslash B_0), \ldots, B_n := B_{n-1} \cup (A_n \backslash B_{n-1})$. Since $A_n \in \mathcal{C}$ and $B_{n-1} \in \mathcal{S} = \mathcal{E}$, for all $n \geq 1$ we have

$$\nu'(A_n \backslash B_{n-1}) = \nu'(A_n) - \nu'(A_n \cap B_{n-1}) = \nu(A_n) - \nu(A_n \cap B_{n-1}) = \nu(A_n \backslash B_{n-1}).$$

By induction we show that for $n \geq 1$ we have $\nu'(B_{n-1}) = \nu(B_{n-1})$. For $n = 1$ this equality holds since $B_0 = A_0$. Now if $\nu'(B_{n-1}) = \nu(B_{n-1})$, by the preceding relations and by additivity we get $\nu'(B_n) = \nu(B_n)$. Thus $\nu'(B_n) = \nu(B_n)$ for all $n \in \mathbb{N}$ and $X = \cup_n B_n$.

Let us prove by induction that for all $S \in \mathcal{S}$ and all $n \in \mathbb{N}$ we have

$$\nu'(S \cap B_n) = \nu(S \cap B_n). \tag{1.6}$$

This relation is satisfied for $n = 0$ since $B_0 = A_0$ and $\mathcal{S} = \mathcal{E}$. Suppose it holds for n. Since

$$S \cap B_{n+1} = (S \cap B_n) \cup (A_{n+1} \cap (S \backslash B_n))$$

with $S \backslash B_n \in \mathcal{S} = \mathcal{E}$, so that $\nu'(A_{n+1} \cap (S \backslash B_n)) = \nu(A_{n+1} \cap (S \backslash B_n))$, by additivity we get (1.6) with n changed into $n+1$. Thus, for all $S \in \mathcal{S}$ and all $n \in \mathbb{N}$ (1.6) holds. Then, for all $S \in \mathcal{S}$ we get

$$\nu'(S) = \lim_n \nu'(S \cap B_n) = \lim_n \nu(S \cap B_n) = \nu(S).$$

□

Let us give a first extension result.

Lemma 1.6 *Let $\mu : \mathcal{C} \to \overline{\mathbb{R}}_+$ be an additive function on a semi-ring $\mathcal{C}$ of subsets of X. Then there exists a unique additive function λ on the ring $\mathcal{A}$ generated by $\mathcal{C}$ whose restriction to $\mathcal{C}$ is μ.*

Proof Lemma 1.4 ensures that for any $A \in \mathcal{A}$ there exists a finite family $(C_1, \ldots, C_n)$ of disjoint elements of $\mathcal{C}$ such that $A = C_1 \cup \ldots \cup C_n$. Then, by the additivity requirement, $\lambda(A)$ cannot be anything else than

$$\lambda(A) = \mu(C_1) + \ldots + \mu(C_n).$$

Let us show that this value is independent of the decomposition of A. Given another family $(D_1, \ldots, D_p)$ of disjoint elements of $\mathcal{C}$ whose union is A we observe that the family $(E_{i,j}) := (C_i \cap D_j)$ is formed with disjoint elements of $\mathcal{C}$ and its union is A. Since for $j \in \mathbb{N}_p := \{1, \ldots, p\}$ we have $D_j = E_{1,j} \cup \ldots \cup E_{n,j} \in \mathcal{C}$, we get $\mu(D_j) = \mu(E_{1,j}) + \ldots + \mu(E_{n,j})$ and $\Sigma_{j=1}^p \mu(D_j) = \Sigma_{j=1}^p \Sigma_{i=1}^n \mu(E_{i,j})$ and similarly $\Sigma_{i=1}^n \mu(C_i) = \Sigma_{i=1}^n \Sigma_{j=1}^p \mu(E_{i,j})$. Thus $\lambda(A)$ is unambiguously defined and any additive function λ' on $\mathcal{A}$ extending μ gives to A the same value. In particular, if $A \in \mathcal{C}$ we have $\lambda(A) = \mu(A)$.

Given $A \in \mathcal{A}$ and $B \in \mathcal{C}$ with $A \cap B = \varnothing$ our construction shows that $\lambda(A \cup B) = \lambda(A) + \mu(B) = \lambda(A) + \lambda(B)$. Given $A, B \in \mathcal{A}$ with $A \cap B = \varnothing$, writing $B := C_1 \cup \ldots \cup C_n$ with $C_i \in \mathcal{C}$ and $C_i \cap C_j = \varnothing$ for $i \neq j$ we can prove by induction on n that $\lambda(A \cup B) = \lambda(A) + \mu(C_1) + \ldots + \mu(C_n) = \lambda(A) + \lambda(B)$. Thus λ is additive on $\mathcal{A}$. □

A refinement of the preceding lemma can be given. It will serve to define a product measure; but since another means can be used, this lemma can be skipped in a first reading.

Lemma 1.7 *Let $\mu : \mathcal{C} \to \overline{\mathbb{R}}_+$ be a σ-additive function on a semi-ring $\mathcal{C}$ of subsets of X. Then the additive function λ on the ring $\mathcal{A}$ generated by $\mathcal{C}$ extending μ is σ-additive.*

Proof Let $A \in \mathcal{A}$ be the union of a sequence (A_n) of disjoint members of $\mathcal{A}$. Since $A \in \mathcal{A}$ we can find a finite family $(C_1, \ldots, C_m)$ of disjoint elements of $\mathcal{C}$ whose union is A. Then, for all $i \in \mathbb{N}_m$, we have $C_i = \cup_{n \geq 0}(A_n \cap C_i)$. Since for all $(i, n) \in \mathbb{N}_m \times \mathbb{N}$ the set $A_{i,n} := A_n \cap C_i$ belongs to $\mathcal{A}$, we can find a finite family $(C_{1,i,n}, \ldots, C_{p(i,n),i,n})$ of disjoint elements of $\mathcal{C}$ whose union is $A_{i,n}$. By σ-additivity of μ and associativity of sums we have

$$\mu(C_i) = \sum_{n=0}^{\infty} \sum_{j=1}^{p(i,n)} \mu(C_{j,i,n}) = \sum_{n=0}^{\infty} \lambda(A_n \cap C_i).$$

Thus, by definition of λ, we get

$$\lambda(A) = \sum_{i=1}^{m} \mu(C_i) = \sum_{i=1}^{m} \sum_{n=0}^{\infty} \lambda(A_n \cap C_i).$$

Since the terms of this series are nonnegative, since λ is additive, and since $(A_n \cap C_1) \cup \ldots \cup (A_n \cap C_m) = A_n \cap A = A_n$ and $(A_n \cap C_i) \cap (A_n \cap C_{i'}) = \varnothing$ for $i \neq i'$, we get

$$\lambda(A) = \sum_{n=0}^{\infty} \sum_{i=1}^{m} \lambda(A_n \cap C_i) = \sum_{n=0}^{\infty} \lambda(A_n).$$

Thus λ is σ-additive. □

A general means to get a measure uses the notion of an outer measure, a notion that is less exacting than the concept of measure. An *outer measure* on a set X is an increasing, σ-subadditive function $\omega : \mathcal{P}(X) \to \overline{\mathbb{R}}_+$ satisfying $\omega(\varnothing) = 0$.

Outer measures are easily obtained.

Proposition 1.9 *Let X be a set, let $\mathcal{C} \subset \mathcal{P}(X)$ with $\varnothing \in \mathcal{C}$, and let $\mu : \mathcal{C} \to \overline{\mathbb{R}}_+$ be such that $\mu(\varnothing) = 0$. For any subset S of X let $\mathcal{C}(S)$ be the collection of sequences (C_n) of elements of $\mathcal{C}$ whose unions contain S and let*

$$\omega(S) := \inf\{\sum_{n} \mu(C_n) : (C_n) \in \mathcal{C}(S)\}. \tag{1.7}$$

Then ω is an outer measure on X such that $\omega(C) \leq \mu(C)$ for all $C \in \mathcal{C}$.

If $\mathcal{C}$ is a semi-ring and if μ is σ-subadditive, then the restriction of ω to $\mathcal{C}$ is μ.

In this definition we use the familiar convention that the infimum of the empty set in $\overline{\mathbb{R}}$ is $+\infty$.

Proof Clearly $\omega(\varnothing) = 0$. Let $S \subset T \subset X$. Since $\mathcal{C}(T) \subset \mathcal{C}(S)$ we get $\omega(S) \leq \omega(T)$.

Let us show that ω is σ-subadditive. Let (S_k) be a sequence in $\mathcal{P}(X)$ and let S be the union of the S_k's. If for some $k \in \mathbb{N}$ one has $\mathcal{C}(S_k) = \varnothing$, then $\mathcal{C}(S) = \varnothing$ and the relation

$$\omega(S) \leq \sum_k \omega(S_k) \tag{1.8}$$

holds, each side being $+\infty$. If $\mathcal{C}(S_k) \neq \varnothing$ for all k, given $\varepsilon > 0$, we pick sequences $(C_{k,n})_n \in \mathcal{C}(S_k)$ such that

$$\sum_n \mu(C_{k,n}) \leq \omega(S_k) + 2^{-k}\varepsilon.$$

Then the family $(C_{k,n})_{k,n}$ can be organized in a sequence $(B_j)_{j\geq 0}$ by using a bijection $h : \mathbb{N} \to \mathbb{N}^2$ and setting $B_j := C_{h(j)}$ so that (B_j) covers S and

$$\omega(S) \leq \sum_j \mu(B_j) = \sum_{k,n} \mu(C_{k,n}) \leq \sum_k (\omega(S_k) + 2^{-k}\varepsilon) = \sum_k \omega(S_k) + \varepsilon.$$

Since $\varepsilon > 0$ is arbitrary, relation (1.8) is established and ω is an outer measure.

Given $C \in \mathcal{C}$, setting $C_0 := C$, $C_n = \varnothing$ for $n > 0$, we see that $\omega(C) \leq \mu(C)$.

Finally, let us show that when $\mathcal{C}$ is a semi-ring and μ is countably subadditive the restriction of ω to $\mathcal{C}$ is μ. Given $C \in \mathcal{C}$, for any $(C_n) \in \mathcal{C}(S)$ we may suppose the union of the sets C_n is C (replacing C_n with $C_n \cap C$) and that the sets C_n are disjoint (replacing C_n with a finite family of disjoint sets $D_{i,n} \in \mathcal{C}$ whose union is $C_n \setminus C_{n-1} \setminus \ldots \setminus C_0$ and relabelling them). Then we have

$$\mu(C) \leq \sum_n \mu(C_n),$$

hence, taking the infimum over $\mathcal{C}(S)$, $\mu(C) \leq \omega(C)$ and equality holds. □

Example The trivial outer measure on a set X is given by $\omega(\varnothing) = 0$ and $\omega(S) = 1$ for all nonempty $S \in \mathcal{P}(X)$.

Example The counting measure on a set X is a measure on $\mathcal{P}(X)$, hence is an outer measure.

Example Let X be an infinite set and let ω be given by $\omega(S) = 0$ if S is countable, $\omega(S) := 1$ if S is uncountable. Then ω is an outer measure.

Example Let $\mathcal{C}$ be the family of intervals of $\mathbb{R}$ of the form $[a, b[$ with $a \leq b$ and let $\mu : \mathcal{C} \to \mathbb{R}_+$ be given by $\mu([a, b[) := b - a$, the length of $[a, b[$. We shall show in Sect. 1.7 that one can define an outer measure ω extending μ called the *Lebesgue outer measure* on $\mathbb{R}$.

Example Let $\mathcal{C}$ be the family of rectangles of $\mathbb{R}^d$ of the form $[a_1, b_1[\times \ldots \times [a_d, b_d[$ with $a_i \leq b_i$ for $i = 1, \ldots, d$, and let $\mu : \mathcal{C} \to \mathbb{R}_+$ be given by $\mu([a_1, b_1[\times \ldots \times [a_d, b_d[) := (b_1 - a_1) \ldots (b_d - a_d)$. We shall show in the Sect. 1.7 that one can define an outer measure ω extending μ called the *Lebesgue outer measure* on $\mathbb{R}^d$. □

Now let us show how to get a measure from an outer measure ω on a set X. Let us say that a subset M of X is *ω-measurable* if

$$\forall S \in \mathcal{P}(X) \qquad \omega(S) \geq \omega(M \cap S) + \omega(M^c \cap S), \tag{1.9}$$

with $M^c := X \setminus M$. In fact, since ω is subadditive, this relation is an equality.

Theorem 1.12 (Caratheodory) *Let X be a set and let $\omega : \mathcal{P}(X) \to \overline{\mathbb{R}}_+$ be an outer measure on the family $\mathcal{P}(X)$ of all subsets of X. Then*

(a) the collection $\mathcal{M}$ of all ω-measurable subsets of X is a σ-algebra;
(b) the restriction of ω to $\mathcal{M}$ is a measure on $\mathcal{M}$.

Proof Since relation (1.9) is symmetric in M and M^c, one has $M^c \in \mathcal{M}$ whenever $M \in \mathcal{M}$. Given A, B in $\mathcal{M}$, let us show that $A \cap B \in \mathcal{M}$. For any $S \in \mathcal{P}(X)$, since B is ω-measurable we have

$$\omega(A \cap S) = \omega(A \cap S \cap B) + \omega(A \cap S \cap B^c).$$

Adding $\omega(A^c \cap S)$ to both sides of this relation, on the left we obtain $\omega(S)$ since $A \in \mathcal{M}$. Thus, to prove that $A \cap B \in \mathcal{M}$ it suffices to show that

$$\omega((A \cap B)^c \cap S) = \omega(A \cap S \cap B^c) + \omega(A^c \cap S).$$

This is seen by using the fact that $A \in \mathcal{M}$ and by replacing M and S with A and $(A \cap B)^c \cap S$ respectively in (1.9):

$$\omega((A \cap B)^c \cap S) = \omega(A \cap S \cap (A \cap B)^c) + \omega(A^c \cap S \cap (A \cap B)^c)$$

since $A \cap S \cap (A \cap B)^c = A \cap S \cap B^c$ and $A^c \cap S \cap (A \cap B)^c = A^c \cap S$. Thus $A \cap B \in \mathcal{M}$.

Now, let A and B be disjoint elements of $\mathcal{M}$. Since $A \cap B \in \mathcal{M}$ we have $A \cup B = (A^c \cap B^c)^c \in \mathcal{M}$. Moreover, given an arbitrary element S of $\mathcal{P}(X)$, using relation (1.9) with A instead of M and $(A \cup B) \cap S$ instead of S, we get

$$\begin{aligned} \omega((A \cup B) \cap S) &= \omega(A \cap (A \cup B) \cap S) + \omega(A^c \cap (A \cup B) \cap S) \\ &= \omega(A \cap S) + \omega(B \cap S). \end{aligned}$$

In particular, taking $S = X$ we see that ω is additive on $\mathcal{M}$.

Now, let (A_n) be a sequence in disjoint ω-measurable subsets of X, let A be its union, and let $S \in \mathcal{P}(X)$. For any $k \in \mathbb{N}\backslash\{0\}$ we have

$$\omega((A_0 \cup \ldots \cup A_k) \cap S) = \sum_{i=0}^{k} \omega(A_i \cap S)$$

as an induction shows, the case $k = 1$ being already established. Using relation (1.9) with $A_0 \cup \ldots \cup A_k \in \mathcal{M}$ instead of M, we deduce from the preceding relation that

$$\begin{aligned}\omega(S) &= \omega((A_0 \cup \ldots \cup A_k) \cap S) + \omega((A_0 \cup \ldots \cup A_k)^c \cap S) \\ &\geq \sum_{i=0}^{k} \omega(A_i \cap S) + \omega(A^c \cap S),\end{aligned}$$

by the inclusion $A^c \cap S \subset (A_0 \cup \ldots \cup A_k)^c \cap S$ and the fact that ω is increasing. Whence, since k is arbitrarily large and since ω is σ-subadditive,

$$\omega(S) \geq \sum_{i=0}^{\infty} \omega(A_i \cap S) + \omega(A^c \cap S) \geq \omega(A \cap S) + \omega(A^c \cap S).$$

This proves that $A \in \mathcal{M}$. Thus $\mathcal{M}$ is a σ-algebra and since the preceding relations are equalities when $S = A$ we see that ω is σ-additive on $\mathcal{M}$. □

Combining this theorem with Proposition 1.9, we get an extension result.

Proposition 1.10 (Hahn) *Let $\mu : \mathcal{A} \to \overline{\mathbb{R}}_+$ be a measure on a semi-ring or a ring $\mathcal{A}$ of subsets of a set X. Then there exists a measure $\nu : \mathcal{A}_\sigma \to \overline{\mathbb{R}}_+$ on the σ-algebra $\mathcal{A}_\sigma$ generated by $\mathcal{A}$ whose restriction to $\mathcal{A}$ is μ.*

If moreover μ is σ-finite, then the extension ν of μ is unique.

Proof By Lemma 1.6 there is no loss of generality in assuming that $\mathcal{A}$ is a ring. Let ω be the outer measure associated with μ given by Proposition 1.9 and let $\mathcal{M}$ be the σ-algebra of ω-measurable subsets of X.

In order to show that $\mathcal{A}$ is contained in $\mathcal{M}$ let us introduce the family $\mathcal{M}'$ formed by the subsets $M \in \mathcal{P}(X)$ such that

$$\forall A \in \mathcal{A} \qquad \omega(A) \geq \omega(A \cap M) + \omega(A \cap M^c). \tag{1.10}$$

Clearly, $\mathcal{M} \subset \mathcal{M}'$ and since by Proposition 1.9 the restriction of ω to $\mathcal{A}$ is additive as it is the measure μ, we have $\mathcal{A} \subset \mathcal{M}'$. Let us show that $\mathcal{M}' = \mathcal{M}$. This will imply that $\mathcal{A} \subset \mathcal{M}$, hence that the σ-algebra $\mathcal{A}_\sigma$ generated by $\mathcal{A}$ is contained in $\mathcal{M}$ and Theorem 1.12 will ensure that the restriction ν of ω to $\mathcal{A}_\sigma$ is a measure. Then the restriction of ν to $\mathcal{A}$ being also the restriction of ω to $\mathcal{A}$ is μ.

We have to show that every $M \in \mathcal{M}'$ belongs to $\mathcal{M}$ or that

$$\forall S \in \mathcal{P}(X) \qquad \omega(S) \geq \omega(S \cap M) + \omega(S \cap M^c). \tag{1.11}$$

This relation is trivially satisfied if $\omega(S) = +\infty$. If $\omega(S)$ is finite, for every $\varepsilon > 0$ one can find a sequence (A_n) of $\mathcal{A}$ such that $S \subset \cup_n A_n$ and $\sum_n \mu(A_n) < \omega(S) + \varepsilon$. Since for all n we have $A_n \in \mathcal{A}$, hence $\mu(A_n) = \omega(A_n) \geq \omega(A_n \cap M) + \omega(A_n \cap M^c)$, by (1.10), summing we get

$$\begin{aligned} \omega(S) + \varepsilon &\geq \sum_n \omega(A_n \cap M) + \sum_n \omega(A_n \cap M^c) \\ &\geq \omega(S \cap M) + \omega(S \cap M^c) \end{aligned}$$

since ω is σ-subadditive and increasing. Since ε is arbitrarily small, relation (1.11) is satisfied and $M \in \mathcal{M}$.

The uniqueness assertion in the case when μ is σ-finite is given by Theorem 1.11.

□

Exercises

1. Let $(X, \mathcal{M}, \mu)$ be a measure space and let $A \in \mathcal{M}$. Verify that any partition of M into sets of positive measure is at most countable.
2. Let $(X, \mathcal{M}, \mu)$ be a measure space and let $A_1, \ldots, A_n \in \mathcal{M}$ with finite measures. Show that $\mu(A_1 \cup A_2) = \mu(A_1) + \mu(A_2) - \mu(A_1 \cap A_2)$ and
   $\mu(A_1 \cup A_2 \cup A_3) = \Sigma_{i=1}^3 \mu(A_i) + \mu(\cap_{i=1}^3 A_i) - \mu(A_1 \cap A_2) - \mu(A_2 \cap A_3) - \mu(A_3 \cap A_1)$.
   Generalize the preceding relations to the case of four or more subsets.
3. Let $(X, \mathcal{M}, \mu)$ be a finite measure space and let $A_1, \ldots, A_n \in \mathcal{M}$. Show that if $\Sigma_{i=1}^n \mu(A_i) > (n-1)\mu(X)$ then $\cap_{i=1}^n A_i$ is nonempty.
4. Verify that an additive function $\mu : \mathcal{C} \to \overline{\mathbb{R}}_+$ on a semi-ring is increasing.
5. Give an example of an additive function on a σ-algebra that is not σ-additive. [Hint: consider an infinite set X and $\mu : \mathcal{P}(X) \to \overline{\mathbb{R}}_+$ given by $\mu(S) = 0$ if S is finite, $\mu(S) = +\infty$ if S is infinite.]
6. Verify that there exists a decreasing sequence (B_n) of Borel subsets of $\mathbb{R}$ such that $\lambda(\cap_n B_n) \neq \lim_n \lambda(B_n)$, λ being the Lebesgue measure. [Hint: take $B_n := \{x \in \mathbb{R} : |x| > n\}$.]
7. Let X be a set, let $\mathcal{C} \subset \mathcal{P}(X)$ be closed under finite intersections, with $\varnothing \in \mathcal{C}$. Let $\mu : \mathcal{C} \to \overline{\mathbb{R}}_+$ be such that $\mu(\varnothing) = 0$ and such that for any sequence (C_n) of $\mathcal{C}$ whose union C is in $\mathcal{C}$ one has $\mu(C) \leq \Sigma_n \mu(C_n)$. Show that the restriction to $\mathcal{C}$ of the outer measure ω deduced from μ coincides with μ. [Hint: if $C \in \mathcal{C}$ and if $(C_n)_n \in \mathcal{C}(C)$ consider the sequence $(C_n \cap C)_n$].

8. Let X be a set, let $\mu : \mathcal{A} \to \mathbb{R}$ be a measure on a ring of subsets of X, and let ω be the outer measure associated with μ. Show that for every increasing sequence (S_n) of $\mathcal{P}(X)$ one has $\omega(\cup_n S_n) = \lim_n \omega(S_n)$. [Hint: for any $\varepsilon > 0$ and any $n \in \mathbb{N}$ pick B_n in the σ-algebra $\mathcal{A}_\sigma$ generated by $\mathcal{A}$ such that $S_n \subset B_n$ and $\omega(B_n) < \omega(S_n) + \varepsilon/2^n$ and let $C_n := \cap_{p=n}^{\infty} B_p$ so that (C_n) is increasing, $S_n \subset C_n \subset B_n$ and $\lim_n \omega(S_n) = \lim_n \omega(C_n) = \omega(\cup_n C_n) \geq \omega(\cup_n S_n)$ and use the fact that ω is increasing.]
9. Let (X, d) be a metric space (see Sect. 2.3). An outer measure $\omega : \mathcal{P}(X) \to \overline{\mathbb{R}}_+$ is called a *metric exterior measure* if for any $A, B \in \mathcal{P}(X)$ one has $\omega(A \cup B) = \omega(A) + \omega(B)$ whenever $\mathrm{gap}(A, B) > 0$, where $\mathrm{gap}(A, B) := \inf\{d(a, b) : a \in A,\ b \in B\}$. Show that if ω is a metric exterior measure, then the Borelian subsets of X are measurable and the restriction of ω to the Borel σ-algebra is a measure. [See [56, p. 214], [240, p. 267].]
10. Let (X, d) be a metric space and let $\alpha > 0$. For $\varepsilon > 0$ and a subset A of X, let $\mathcal{R}_\varepsilon(A)$ be the set of countable coverings of A by balls whose diameter is at most ε. Set

$$\mu_{\alpha,\varepsilon}(A) := \inf\{\sum_n (\mathrm{diam} B_n)^\alpha \ :\ (B_n) \in \mathcal{R}_\varepsilon(A)\}.$$

Show that the function $\varepsilon \mapsto \mu_{\alpha,\varepsilon}(A)$ is nonincreasing on $\mathbb{P} :=]0, \infty[$. Verify that is an outer measure on X. Deduce from this that $\omega_\alpha := \lim_{\varepsilon \to 0_+} \mu_{\alpha,\varepsilon}$ is an outer measure on X. It is called the *α-Hausdorff measure*.
11. Keeping the data and the notation of the preceding exercise with $X = \mathbb{R}^d$, verify that for all $A \subset X$ and $\varepsilon \in]0, 1]$ the functions $\alpha \mapsto \varepsilon^{-\alpha}\mu_{\alpha,\varepsilon}(A)$ and $\alpha \mapsto \omega_\alpha(A)$ are nonincreasing on $\mathbb{P}$.

 Given $\alpha \geq d$ and $\varepsilon > 0$, show that there exists some $c_d > 0$ such that for every rectangle R of $\mathbb{R}^d$ one has $\mu_{\alpha,\varepsilon}(R) \leq c_d \varepsilon^{\alpha - d}\lambda_d(R)$, where λ_d is the measure on $\mathbb{R}^d$ associated with the Lebesgue outer measure. Deduce from this that for all $\alpha > d$ and all bounded subset A of $\mathbb{R}^d$ one has $\omega_\alpha(A) = 0$.

 The *Hausdorff dimension* $h(A)$ of a bounded subset A of $\mathbb{R}^d$ is defined by

$$h(A) := \inf\{\alpha > 0 : \omega_\alpha(A) = 0\}.$$

 Show that $\omega_\alpha(A) = 0$ for $\alpha > h(A)$ and that $\omega_\alpha(A) = +\infty$ for $\alpha < h(A)$. Show that if A is a nonempty rectangle of $\mathbb{R}^d$ then $h(A) = d$.

1.6 Completion of a Measure

Given a measure space $(X, \mathcal{S}, \mu)$ it is often necessary to consider properties which hold almost everywhere, i.e. outside a negligible set. A precise definition is as follows.

Definition 1.7 Given a measure μ on a ring $\mathcal{S}$ of subsets of a set X, a subset N of X is called a *null set* or a *negligible set* if for every $\varepsilon > 0$ there exists some $S \in \mathcal{S}$ such that $N \subset S$ and $\mu(S) \leq \varepsilon$.

If every null set belongs to $\mathcal{S}$ one says that μ is *complete* .

If some ambiguity may arise one says that N is μ-null. Clearly, any subset of a null set is a null set. If $\mathcal{S}$ is a σ-ring, any countable union N of null sets N_n is a null set since for any $\varepsilon > 0$ one can find some $S_n \in \mathcal{S}$ such that $N_n \subset S_n$ and $\mu(S_n) \leq \varepsilon/2^n$, so that one has $N \subset \cup_{n\geq 1} S_n \in \mathcal{S}$ and $\mu(\cup_{n\geq 1} S_n) \leq \varepsilon$. If $\mathcal{S}$ is stable under countable intersections, $N \in \mathcal{P}(X)$ is a null set if and only if N is contained in some $S \in \mathcal{S}$ with $\mu(S) = 0$. Also, if μ is complete, a subset N of X is a null set if and only if $N \in \mathcal{S}$ and $\mu(N) = 0$.

Proposition 1.11 *Given a measure μ on a ring (resp. σ-ring, resp. σ-algebra) $\mathcal{S}$ of subsets of a set X, the family $\mathcal{S}_\mu$ of sets of the form $S \cup N$, where $S \in \mathcal{S}$ and N is a null set is a ring (resp. σ-ring, resp. σ-algebra) called the completion of $\mathcal{S}$. Setting $\overline{\mu}(S \cup N) = \mu(S)$ for $S \in \mathcal{S}$, N null, one gets a measure $\overline{\mu}$ on $\mathcal{S}_\mu$ extending μ and $\overline{\mu}$ is complete. Moreover, the null sets for $\overline{\mu}$ are the null sets for μ.*

Proof Let $T := S \cup N$ with N null and $S \in \mathcal{S}$. Replacing N with $N \backslash S$ (which is still a null set) we may suppose S and N are disjoint, so that $T = S \Delta N$. Since the symmetric difference is an associative and commutative operation, for $T' := S' \Delta N'$ with $S' \in \mathcal{S}$ and N' null we have

$$T \Delta T' = (S \Delta S') \Delta (N \Delta N') \in \mathcal{S}_\mu.$$

Since $(S \cup N) \cap (S' \cup N') = (S \cap S') \cup N''$ with $N'' := (S \cap N') \cup (S' \cap N) \cup (N \cap N')$ we see that $\mathcal{S}_\mu$ is stable under finite intersections. Moreover, if $\mathcal{S}$ is a σ-ring, one easily checks that the union of a countable family of elements of $\mathcal{S}_\mu$ is in $\mathcal{S}_\mu$. Thus $\mathcal{S}_\mu$ is a σ-ring (and a σ-algebra if $\mathcal{S}$ is a σ-algebra).

Now let us observe that the definition of $\overline{\mu}$ is coherent: if $S \cup N = S' \cup N'$ with N, N' null sets and $S, S' \in \mathcal{S}$, we have $S \Delta S' \subset N \cup N'$, so that $S \Delta S'$ is a null set, and $\mu(S \Delta S') = 0$, $\mu(S \backslash S') = 0 = \mu(S' \backslash S)$,

$$\mu(S) = \mu(S \cap S') + \mu(S \backslash S') = \mu(S \cap S')$$

and similarly $\mu(S') = \mu(S \cap S')$ so that $\mu(S') = \mu(S)$. Obviously $\overline{\mu}(\varnothing) = 0$. Let (T_n) be a sequence in disjoint members of $\mathcal{S}_\mu$, with $T_n := S_n \cup N_n$, $S_n \in \mathcal{S}$, N_n null. Since the union N of the family (N_n) is a null set, the union T of the family (T_n) is such that $T = S \cup N$, where S is the union of (S_n), so that $\overline{\mu}(T) = \mu(S) = \sum_n \mu(S_n) = \sum_n \mu(T_n)$. Clearly $\overline{\mu}$ extends μ. If $Y \in \mathcal{P}(X)$ is a $\overline{\mu}$-null set, for every $\varepsilon > 0$ we can find some $T := S \cup N \in \mathcal{S}_\mu$ containing Y with $\overline{\mu}(T) = \mu(S) \leq \varepsilon/2$, $S \in \mathcal{S}$, $S' \in \mathcal{S}$ with $N \subset S'$, $\mu(S') \leq \varepsilon/2$, so that we have $Y \subset S \cup S'$ with $\mu(S \cup S') \leq \varepsilon$ and Y is a null set, so that $Y = \varnothing \cup Y \in \mathcal{S}_\mu$. Thus $(X, \mathcal{S}_\mu, \overline{\mu})$ is complete. □

Proposition 1.12 *Let $\mu : \mathcal{S} \to \overline{\mathbb{R}}_+$ be a measure on a σ-algebra $\mathcal{S}$ of subsets of a set X and let $(X, \mathcal{S}_\mu, \overline{\mu})$ be the completion of $(X, \mathcal{S}, \mu)$. Let $\omega : \mathcal{P}(X) \to \overline{\mathbb{R}}_+$ be the outer measure deduced from μ. Then every $T \in \mathcal{S}_\mu$ belongs to the class $\mathcal{M}$ of ω-measurable subsets and one has $\omega(T) = \overline{\mu}(T)$.*

If μ is σ-finite, then $\mathcal{S}_\mu = \mathcal{M}$.

Proof Let us first show that every element T of $\mathcal{S}_\mu$ belongs to $\mathcal{M}$. We start with a null set N. By the observations following Definition 1.7 there exists an $N' \in \mathcal{S}$ containing N that satisfies $\mu(N') = 0$; thus by (1.7) $\omega(N) = 0$. Moreover, for every $Y \in \mathcal{P}(X)$ we have $\omega(N \cap Y) \leq \omega(N) = 0$, hence

$$\omega(Y) \geq \omega(N^c \cap Y) \geq \omega(N \cap Y) + \omega(N^c \cap Y),$$

so that $N \in \mathcal{M}$. Now for every $T := S \cup N \in \mathcal{S}_\mu$ with N null, $S \in \mathcal{S}$, and $S \cap N = \varnothing$ we have $T \in \mathcal{M}$ since $\mathcal{M}$ is a σ-algebra containing $\mathcal{S}$ and $N \in \mathcal{M}$. Moreover, since ω is subadditive and increasing,

$$\omega(T) \leq \omega(S) + \omega(N) = \omega(S) \leq \omega(T).$$

Since $\omega(S) = \mu(S) = \overline{\mu}(T)$, we get $\omega(T) = \overline{\mu}(T)$.

It remains to show that $\mathcal{M} \subset \mathcal{S}_\mu$ when μ is σ-finite. Let us first suppose $M \in \mathcal{M}$ with $\omega(M) < +\infty$ and construct some $A \in \mathcal{S}$ such that $M \subset A$ and $\mu(A) = \omega(M)$. By definition of ω, for every $k \in \mathbb{N}\backslash\{0\}$ we can find a covering $(A_{k,n})_n$ of M by members of the σ-ring $\mathcal{S}$ such that

$$\sum_n \mu(A_{k,n}) \leq \omega(M) + \frac{1}{k}.$$

For $A_k := \cup_n A_{k,n}$ we have $A_k \in \mathcal{S}$, $M \subset A_k$, and $\mu(A_k) \leq \omega(M) + 1/k$. Let $A := \cap_k A_k$, so that $A \in \mathcal{S}$, $M \subset A$ and $\mu(A) \leq \inf_k \mu(A_k) \leq \omega(M)$. Since $\omega(M) \leq \omega(A) = \mu(A)$ we get $\mu(A) = \omega(M)$. Thus $M = A\Delta(A\backslash M)$ with $\omega(A\backslash M) = 0$ and $A\backslash M \in \mathcal{M}$ (but not $A\backslash M \in \mathcal{S}$!). Replacing M with $A\backslash M$ in what precedes, we find some $B \in \mathcal{S}$ such that $A\backslash M \subset B$ and $\mu(B) = \omega(A\backslash M) = 0$. This shows that $A\backslash M$ is a null set and that $M = A\Delta(A\backslash M) \in \mathcal{S}_\mu$.

In the general case, taking a sequence (S_n) in $\mathcal{S}$ whose union is S and is such that $\mu(S_n) < +\infty$ for all n we reduce the question to the case of $M_n := M \cap S_n$, which is of finite measure. □

Exercises

1. Find two Borel subsets A, B of $\mathbb{R}$ whose Lebesgue measure is 0 and whose sum $A + B$ is $\mathbb{R}$.
2. With the notation of Proposition 1.11 show that $\mathcal{S}_\mu$ is the smallest σ-ring containing $\mathcal{S}$ on which a complete measure extending μ can be defined.

3. Let $(X, \mathcal{S}, \mu)$ be a measure space, the values of μ on $\mathcal{S}$ being finite. For $A, B \in \mathcal{S}$ set $d_0(A, B) = \mu(A \Delta B)$. Verify that d_0 is a semimetric: for any $A, B, C \in \mathcal{S}$

$$d_0(A, A) = 0, \quad d_0(B, A) = d_0(A, B), \quad d_0(A, C) \leq d_0(A, B) + d_0(B, C).$$

Let $\widehat{\mathcal{S}}$ be the quotient space of $\mathcal{S}$ with respect to the equivalence relation $A \sim B$ if $d_0(A, B) = 0$ and let $p : \mathcal{S} \to \widehat{\mathcal{S}}$ be the quotient map. Verify that the function $d : \widehat{\mathcal{S}} \times \widehat{\mathcal{S}} \to \mathbb{R}$ characterized by $d(p(A), p(B)) := d_0(A, B)$ for all $A, B \in \mathcal{S}$ is a metric on $\widehat{\mathcal{S}}$ and prove that $(\widehat{\mathcal{S}}, d)$ is complete. [Hint: using the relation $A_0 \Delta A_n \subset \cup_{i=0}^{n-1} A_i \Delta A_{i+1}$ show that for any sequence (A_n) of $\mathcal{S}$ satisfying $d_0(A_n, A_{n+1}) \leq 2^{-n}$ one has $(d_0(A_n, B)) \to 0$, for $B := \cup_n \cap_{p \geq n} A_p$, $(d_0(A_n, C)) \to 0$ for $C := \cap_n \cup_{p \geq n} A_p$.]

Prove that if $\mathcal{S}$ is the σ-algebra generated by a subalgebra $\mathcal{A}$ of $\mathcal{S}$ then the image of $\mathcal{A}$ by p is dense in $\widehat{\mathcal{S}}$.

4. (**Poincaré's Recurrence Theorem**) Let $(X, \mathcal{S}, \mu)$ be a finite measure space and let $T : X \to X$ be a measure-preserving map, i.e. $\mu(T(A)) = \mu(A)$ for all $A \in \mathcal{S}$. Show that for all $A \in \mathcal{S}$ there exists a null set N of A such that for all $x \in A \backslash N$ one has $T^{(n)}(x) \in A$ for infinitely many n. [Hint: first prove that there exists some null set N_0 of A such that for all $x \in A \backslash N_0$ one has $T^n(x) \in A$ for at least one $n \in \mathbb{N} \backslash \{0\}$; then apply this preliminary result to $T^{(2)} := T \circ T$, $T^{(3)}, \ldots$ and conclude.]

1.7 Lebesgue and Stieltjes Measures

Anticipating on Sect. 2.2, let us observe that if $\mathcal{B}$ is the σ-algebra generated by a family $\mathcal{O}$ of subsets of a set X and if there exists a countable subfamily $\mathcal{C}$ of $\mathcal{B}$ such that each element of $\mathcal{O}$ is the union of a subfamily of $\mathcal{C}$, then $\mathcal{B}$ is the σ-algebra generated by $\mathcal{C}$. In particular, if $(X, \mathcal{O})$ is a topological space, if $\mathcal{B}$ is the Borel σ-algebra generated by $\mathcal{O}$, and if every open set is the union of a family contained in a countable subfamily $\mathcal{C}$ of $\mathcal{B}$, then $\mathcal{B}$ is also generated by $\mathcal{C}$. That happens when $\mathcal{O}$ has a countable base. For $X = \mathbb{R}^d$ one can take for $\mathcal{C}$ the family of open balls whose radii are rational and whose centers have rational coordinates. One can also take closed such balls. Since any open interval $]a, b[$ of $\mathbb{R}$ is the countable union of a family of semi-closed intervals of the form $[a_n, b_n[$, one can also take for $\mathcal{C}$ the family $\mathcal{R}_d$ of semi-closed rectangles we call boxes $C := [a_1, b_1[\times \ldots \times [a_d, b_d[$ with $a_i, b_i \in \mathbb{R}$, with the convention $[a_i, b_i[= \varnothing$ if $a_i \geq b_i$. Such a family is a semi-ring. For the case $d = 1$ this follows from the relations

$$[a, b[\cap [a', b'[= [\sup(a, a'), \inf(b, b')[,$$

$$[a, b[\backslash [a', b'[= [a, \inf(a', b)[\cup [\sup(a, b'), b[,$$

the two intervals of the last right-hand side being disjoint. The general case follows by induction and the observation made above. Let us note that it is the stability of $\mathcal{R}_d$ under intersection which justifies our choice of semi-closed intervals, and, in dimension $d > 1$ the choice of boxes for $\mathcal{C}$. Moreover, $\mathcal{C}$ is a semi-ring. The preceding observations and the extension results of the Sect. 1.5 entail that there is a unique measure λ_d on the Borel σ-algebra $\mathcal{B}_d := \mathcal{B}(\mathbb{R}^d)$ of $\mathbb{R}^d$ whose restriction to $\mathcal{R}_d$ is the volume: $\lambda_d(C) := (b_1 - a_1) \ldots (b_d - a_d)$ for $C := [a_1, b_1[\times \ldots \times [a_d, b_d[$, provided one shows that λ_d is σ-additive on $\mathcal{R}_d$. That will be shown for the larger class of Stieltjes measures. The completion of $\mathcal{B}_d$ for the measure λ_d is called the *Lebesgue σ-algebra*. The measure on this algebra is still denoted by λ_d rather than $\overline{\lambda_d}$.

We are ready to study *Stieltjes measures* μ on the Borel σ-algebra $\mathcal{B}$ of $\mathbb{R}$ (or on the Lebesgue σ-algebra of $\mathbb{R}$). They are the measures that are finite on any interval bounded above. First, we associate a Stieltjes measure to an increasing, left-continuous function $g : \mathbb{R} \to \mathbb{R}$. We start with the Stieltjes outer measure $\omega := \omega_g$ associated with g. Let us denote by $\mathcal{C}$ the collection of bounded intervals of the form $C := [a, b[$ with $a, b \in \mathbb{R}$ with $a \leq b$ and for $C := [a, b[$ let us set $\mu(C) := g(b) - g(a)$. For a subset S of $\mathbb{R}$, ω is defined by

$$\omega(S) = \inf\{\sum_n \mu(C_n) : (C_n) \in \mathcal{C}(S)\}$$

where $\mathcal{C}(S)$ is the collection of sequences (C_n) of $\mathcal{C}$ whose unions contain S. In order to prove that the restriction of ω to $\mathcal{C}$ coincides with μ, by Proposition 1.9 it suffices to show that μ is σ-additive. We do that by using the notion of compactness of Sect. 2.2.4.

Let $(C_n)_{n\geq 0} := ([a_n, b_n[)_{n\geq 0}$ be a sequence of disjoint elements of $\mathcal{C}$ whose union $C := [a, b[$ belongs to $\mathcal{C}$. Since by Lemma 1.6 μ can be extended to an additive function also denoted by μ on the ring $\mathcal{A}$ generated by $\mathcal{C}$, for all $p \in \mathbb{N}\backslash\{0\}$ we have

$$\sum_{n=0}^{p} \mu(C_n) = \mu(C_0 \cup \ldots \cup C_p) \leq \mu(C),$$

hence $\sum_{n=0}^{\infty} \mu(C_n) \leq \mu(C)$. Let us prove the reverse inequality. Given $\varepsilon > 0$ let $b' < b$ be such that $g(b') > g(b) - \varepsilon/2$ and for all $n \in \mathbb{N}\backslash\{0\}$ let $a'_n < a_n$ be such that $g(a'_n) > g(a_n) - \varepsilon/2^{n+2}$. By compactness of $[a, b']$ we can find some $p \in \mathbb{N}$ such that

$$[a, b'[\subset [a, b'] \subset \bigcup_{n=0}^{p}]a'_n, b_n[\subset \bigcup_{n=0}^{p} [a'_n, b_n[$$

so that

$$g(b) - g(a) < g(b') - g(a) + \frac{\varepsilon}{2} = \mu([a, b'[) + \frac{\varepsilon}{2} \leq \sum_{n=0}^{p} \mu([a'_n, b_n[) + \frac{\varepsilon}{2}$$

$$\leq \sum_{n=0}^{p} (g(b_n) - g(a'_n)) + \frac{\varepsilon}{2} \leq \sum_{n=0}^{p} (g(b_n) - g(a_n)) + \sum_{n=0}^{p} \frac{\varepsilon}{2^{n+2}} + \frac{\varepsilon}{2},$$

$$\mu(C) \leq \sum_{n=0}^{p} \mu(C_n) + \varepsilon.$$

Since $\varepsilon > 0$ is arbitrarily small, the required inequality is established and μ is σ-additive on $\mathcal{C}$ and its extension to $\mathcal{A}$ is also σ-additive. Moreover, the restriction of ω to $\mathcal{C}$ coincides with μ. Since $\mathbb{R}$ is the union of the intervals $[n, n+1[$ with finite measure, Proposition 1.10 ensures that the restriction of ω to the σ-algebra $\mathcal{A}_\sigma$ generated by $\mathcal{A}$ is an extension of μ. Since $\mathcal{A}_\sigma = \mathcal{B}$ we have proved the following result.

Proposition 1.13 *Given a nondecreasing, left continuous function $g : \mathbb{R} \to \mathbb{R}$, there exists a unique σ-finite measure $\mu := \mu_g$ on the Borel σ-algebra $\mathcal{B}$ of $\mathbb{R}$ such that $\mu([a, b[) = g(b) - g(a)$ for all $a, b \in \mathbb{R}$ with $a \leq b$.*

If g is bounded below, μ_g is called the Stieltjes measure associated with g.

Since any bounded closed interval $T := [a, b]$ is the intersection of a decreasing sequence (C_n) of intervals C_n of the form $[a, b_n[$, Lemma 1.5 ensures that $\mu(T) = \lim_n g(b_n) - g(a) = g(b_+) - g(a)$ with $g(b_+) := \lim_{b'(>b) \to b} g(b')$. Similarly, writing $]a, b[$ as an increasing sequence of intervals in $\mathcal{C}$, we get $\mu(]a, b[) = g(b) - g(a_+)$.

The measure μ_h associated with the function $h := g + c$, where c is a constant, obviously coincides with μ_g. Thus, when g is bounded below we can assume that $\inf g(\mathbb{R}) = 0$ and then, for all $r \in \mathbb{R}$, we have $\mu(]-\infty, r]) = g(r)$. If g is bounded μ_g is finite. In particular, if $\inf g(\mathbb{R}) = 0$ and $\sup g(\mathbb{R}) = 1$ we get a probability on $\mathcal{B}$. Now we show that any Stieltjes measure on the Borel σ-algebra $\mathcal{B}$ of $\mathbb{R}$ is obtained in the preceding manner.

Proposition 1.14 *Let $\mu : \mathcal{B} \to \mathbb{R}$ be a measure on the Borel σ-algebra of $\mathbb{R}$ such that $\mu(]-\infty, r[)$ is finite for all $r \in \mathbb{R}$. Then, the function $g : \mathbb{R} \to \mathbb{R}$ given by $g(r) := \mu(]-\infty, r[)$ is nondecreasing, left continuous and* $\inf g(\mathbb{R}) = 0$. *Moreover, the measure μ_g associated with g coincides with μ.*

Proof Given μ and g as in the statement, for $r \leq s$ in $\mathbb{R}$ we get $g(r) \leq g(s)$ since $]-\infty, r[\subset]-\infty, s[$. Let (r_n) be an increasing sequence in $]-\infty, r[$ with limit r (we write $(r_n) \to r_-$). Since $]-\infty, r[= \cup_n]-\infty, r_n[$ Lemma 1.5 ensures that $(g(r_n)) \to g(r)$. We easily deduce from this the left continuity of g. The relation $\inf g(\mathbb{R}) = 0$ similarly follows from Lemma 1.5. Since for $a \leq b$ in $\mathbb{R}$ one has $[a, b[=]-\infty, b[\setminus]-\infty, a[$ we have $\mu([a, b[) = g(b) - g(a)$. The uniqueness assertion of the preceding proposition ensures that the measure μ_g coincides with μ. □

The function g associated with a Stieltjes measure μ on $\mathcal{B}$ is called the *distribution function* of μ. In particular, the study of probabilities on $\mathcal{B}$ can be somewhat reduced to the study of functions.

The *Lebesgue measure* λ is obtained as the special case of Proposition 1.13 obtained by taking for g the identity function of $\mathbb{R}$. Then for any bounded interval T of $\mathbb{R}$ the measure $\lambda(T)$ of T is its length. Let us note that the proof of Proposition 1.10 shows that λ is a measure on the σ-algebra $\mathcal{M}$ of measurable subsets of $\mathbb{R}$ for the outer measure ω associated with the length. Countable subsets of $\mathbb{R}$ have Lebesgue measure 0, a singleton $\{a\}$ being the interval $[a, a]$ with length 0. However there are uncountable subsets S of $\mathbb{R}$ such that $\lambda(S) = 0$, for example the *Cantor set*.

Example (The Cantor Set) Consider the sequence (C_n) of subsets of $[0, 1]$ obtained by taking $C_0 := [0, 1]$, $C_1 := [0, 1/3] \cup [2/3, 1]$, C_{n+1} being obtained from C_n by removing the open middle third of each interval forming C_n. The Cantor set is the set $C := \cap_n C_n$. Its Lebesgue measure is 0 since $\lambda(C_n) = (2/3)^n$. It can be shown that C is the image of the set $\{0, 1\}^{\mathbb{N}}$ of sequences $(k_n)_n$ with $k_n = 0$ or 1 under the map $(k_n) \mapsto \Sigma_n 2k_n/3^{n+1}$. Since this map is injective, C is uncountable (it has the cardinality of the continuum).

Proposition 1.15 *The Borel σ-algebra $\mathcal{B}$ and the Lebesgue σ-algebra $\mathcal{M}$ on $\mathbb{R}$ are invariant under all translations $t_c : x \mapsto x - c$ and dilations $h_r : x \mapsto rx$ for $c \in \mathbb{R}$ and $r > 0$. Moreover, $\lambda(t_c(S)) = \lambda(S)$ and $\lambda(h_r(S)) = r\lambda(S)$ for all $S \in \mathcal{M}$.*

Proof The class of sets $B \in \mathcal{B}$ (resp. $B \in \mathcal{M}$) such that $t_c(B) \in \mathcal{B}$ (resp. $h_r(B) \in \mathcal{M}$) is a σ-algebra, as is easily seen. Since it contains the class of intervals, it coincides with $\mathcal{B}$ (resp. $\mathcal{M}$). Moreover, λ, $\lambda \circ t_c$, and $r^{-1}\lambda \circ h_r$ coincide on the class $\mathcal{C}$ of intervals, hence coincide on $\mathcal{M}$. □

Exercises

1. Let $\mathcal{C}$ be the semi-ring of bounded semi-closed intervals of $\mathbb{R}$ and let λ be the length measure on $\mathcal{C}$. Let $C \in \mathcal{C}$ and $C_1, \ldots, C_m$ be disjoint elements of $\mathcal{C}$ such that $C_1 \cup \ldots \cup C_m \subset C$. Prove that $\lambda(C_1) + \ldots + \lambda(C_m) \leq \lambda(C)$.

Let C and $C_1, \ldots, C_m$ be elements of $\mathcal{C}$ such that $C \subset C_1 \cup \ldots \cup C_m$. Prove that $\lambda(C) \leq \lambda(C_1) + \ldots + \lambda(C_m)$.

Deduce from these facts that there is a unique (countably additive) measure λ on the σ-ring generated by $\mathcal{C}$.

2*. Let $X := [0, 1]$ and let $I := X/\sim$ be the quotient of X by the equivalence relation defined by $x \sim y$ if $x - y \in \mathbb{Q}$. Using the axiom of choice, define a map $s : I \to X$ such that $s(i) \in p^{-1}(i)$ for all $i \in I$, where $p : X \to X/\sim$ is the quotient map. Show that $A := s(I)$ does not belong to the σ-algebra of Lebesgue measurable subsets of X. [See [240, pp. 24–25].]

3. Prove that the Lebesgue σ-algebra $\mathcal{L}$ of $\mathbb{R}$ is equipotent to $\mathcal{P}(\mathbb{R})$. [Using the fact that the Cantor set C is equipotent to $\mathbb{R}$ and belongs to the class $\mathcal{N}$ of null sets of $\mathbb{R}$, note the inclusions $\mathcal{P}(C) \subset \mathcal{N} \subset \mathcal{L} \subset \mathcal{P}(\mathbb{R})$.]
4*. (*A Lebesgue measurable set that is not Borel*) Let $h : [0, 1] \to [0, 1]$ be given by $h(1) := 1$, $h(x) := \Sigma_{n\geq 1} 2x_n/3^n$ if $x := \Sigma_{n\geq 1} x_n/2^n$ with $x_n \in \{0, 1\}$ and such that for all $k \in \mathbb{N}$ there exists an $n \geq k$ with $x_n \neq 1$. Verify that h is strictly increasing, hence Borel measurable and that $h([0, 1]) = C$, the Cantor set. Let A be the set obtained in Exercise 2. Show that $S := h(A)$ is Lebesgue measurable but not a Borel set. [Hint: S is a null set as $S \subset C$, but since $h^{-1}(S) = A$ is not a Borel set, S cannot be a Borel set.]
5*. (*A non-Borel function that is Riemann-integrable*) Let C_n be the set described in the construction of the Cantor set C and let $f_n := 1_{C_n}$ be its characteristic function. Verify that $\int_0^1 f_n = (2/3)^n \to 0$. Deduce from this that if f is the characteristic function of the subset S of C obtained in Exercise 4, then f is not a Borel function but is Riemann-integrable with a null integral. [Hint: note that $0 \leq 1_S \leq 1_C \leq 1_{C_n}$.]

1.8 * Product Measures

Because the content of this section is somewhat involved and can be obtained more easily by using integration theory, we suggest that this section should be skipped on first reading. On the other hand, it has its place in the present chapter and it can be considered as a good training in set theory.

Given σ-finite measure spaces $(X, \mathcal{M}, \mu)$ and $(Y, \mathcal{N}, \nu)$ can one endow the product space $X \times Y$ with a canonical measure? It is the purpose of the present section to give a positive answer to this question. We first consider product rings and σ-algebras. Given sets X, Y and rings $\mathcal{A}$, $\mathcal{B}$ on X and Y respectively, we denote by $\mathcal{C}$ the collection of rectangles, i.e. sets of the form $C := A \times B$ with $A \in \mathcal{A}$, $B \in \mathcal{B}$ and by $\mathcal{A} \boxtimes \mathcal{B}$ the collection of unions of finite families of disjoint elements of $\mathcal{C}$. The relations

$$(A \times B) \cap (A' \times B') = (A \cap A') \times (B \cap B'),$$
$$(A \times B)\backslash(A' \times B') = [(A\backslash A') \times B] \cup [(A \cap A') \times (B\backslash B')],$$
$$(A \times B)^c = (A^c \times Y) \cup (A \times B^c)$$

show that $\mathcal{C}$ is a semi-ring and Lemma 1.4 entails that $\mathcal{A} \boxtimes \mathcal{B}$ is the ring generated by $\mathcal{C}$. Moreover, $\mathcal{A} \boxtimes \mathcal{B}$ is an algebra if $\mathcal{A}$ and $\mathcal{B}$ are algebras. The following proposition shows that $\mathcal{A} \boxtimes \mathcal{B}$ is a σ-algebra if $\mathcal{A}$ and $\mathcal{B}$ are σ-algebras. Given a class $\mathcal{G}$ of subsets of a set Z, we denote by $\mathcal{G}_\sigma$ the σ-algebra generated by $\mathcal{G}$.

Proposition 1.16 *Given rings $\mathcal{A}$, $\mathcal{B}$ on the sets X and Y respectively, the σ-algebra $\mathcal{A} \otimes \mathcal{B} := (\mathcal{A} \boxtimes \mathcal{B})_\sigma$ generated by the semi-ring $\mathcal{C}$ or by the ring $\mathcal{A} \boxtimes \mathcal{B}$ coincides with the σ-algebra $\mathcal{A}_\sigma \otimes \mathcal{B}_\sigma := (\mathcal{A}_\sigma \boxtimes \mathcal{B}_\sigma)_\sigma$ generated by the ring $\mathcal{A}_\sigma \boxtimes \mathcal{B}_\sigma$.*

Proof Since $\mathcal{D} := \mathcal{A} \boxtimes \mathcal{B} \subset \mathcal{A}_\sigma \otimes \mathcal{B}_\sigma$ it follows that $\mathcal{A} \otimes \mathcal{B} := \mathcal{D}_\sigma \subset \mathcal{A}_\sigma \otimes \mathcal{B}_\sigma$. Given $B \in \mathcal{B}$, the set $\mathcal{A}_B := \{M \in \mathcal{P}(X) : M \times B \in \mathcal{D}_\sigma\}$ contains $\mathcal{A}$ and is closed under finite intersections, relative complementation since $(M \backslash N) \times B = (M \times B) \backslash (N \times B)$, and symmetric differences since $(M \times B) \Delta (M' \times B) = (M \Delta M') \times B$. Moreover, $\mathcal{A}_B$ is closed under countable unions. Thus $\mathcal{A}_B$ contains $\mathcal{A}_\sigma$. This means that for all $M \in \mathcal{A}_\sigma$ and all $B \in \mathcal{B}$ we have $M \times B \in \mathcal{D}_\sigma$. Thus the set $\mathcal{B}_M := \{B \in \mathcal{P}(Y) : M \times B \in \mathcal{D}_\sigma\}$ contains $\mathcal{B}$. As above one can show it is a σ-algebra. Thus $\mathcal{B}_M$ contains $\mathcal{B}_\sigma$, so that for all $M \in \mathcal{A}_\sigma$ and all $N \in \mathcal{B}_\sigma$ we have $M \times N \in \mathcal{D}_\sigma$. Thus $\mathcal{A}_\sigma \otimes \mathcal{B}_\sigma \subset \mathcal{D}_\sigma$ and equality holds. □

We denote by $p_X : X \times Y \to X$ and $p_Y : X \times Y \to Y$ the canonical projections. If $f : X \times Y \to Z$ is a map with values in a measurable space $(Z, \mathcal{S})$ and if $P \in \mathcal{P}(X \times Y)$, for $x \in X$ we denote by $f_x : Y \to Z$ the partial map of f and by $P_x \in \mathcal{P}(Y)$ the slice of P defined by

$$f_x(y) := f(x, y), \ y \in Y \qquad P_x := \{y \in Y : (x, y) \in P\}.$$

Lemma 1.8 *Let X, Y be sets and let $\mathcal{A}$, $\mathcal{B}$ be rings (resp. σ-algebras) on X and Y respectively. Then, for all $P \in \mathcal{A} \boxtimes \mathcal{B}$ (resp. $\mathcal{A} \otimes \mathcal{B}$) and for all $x \in X$ one has $P_x \in \mathcal{B}$.*

For any measurable map $f : (X \times Y, \mathcal{A} \otimes \mathcal{B}) \to (Z, \mathcal{S})$ and any $x \in X$, f_x is measurable.

If $(W, \mathcal{W})$ is a measurable space, a map $g : W \to X \times Y$ is measurable with respect to $\mathcal{W}$ and $\mathcal{A} \otimes \mathcal{B}$ if and only if $p_X \circ g$ and $p_Y \circ g$ are measurable.

Proof Let $\mathcal{Q}$ be the collection of sets $Q \in \mathcal{A} \boxtimes \mathcal{B}$ (resp. $\mathcal{A} \otimes \mathcal{B}$) such that $Q_x \in \mathcal{B}$ for all $x \in X$. We want to show that $\mathcal{Q} = \mathcal{A} \boxtimes \mathcal{B}$ (resp. $\mathcal{A} \otimes \mathcal{B}$). Since $(X \times Y)_x = Y$ and since $\mathcal{Q}$ contains the family $\mathcal{C}$ of rectangles of the form $A \times B$ with $A \in \mathcal{A}$, $B \in \mathcal{B}$, it suffices to show that $\mathcal{Q}$ is a ring (resp. a σ-algebra). This follows from the relations

$$(P \backslash Q)_x = P_x \backslash Q_x, \quad (P \cup Q)_x = P_x \cup Q_x, \quad (\bigcup_n P_n)_x = \bigcup_n (P_n)_x$$

for all $P, Q, P_n \in \mathcal{P}(X \times Y)$ and all $x \in X$.

Let $f : (X \times Y, \mathcal{A} \otimes \mathcal{B}) \to (Z, \mathcal{S})$ be measurable and let $x \in X$. For every $S \in \mathcal{S}$ one has

$$f_x^{-1}(S) = (f^{-1}(S))_x \in \mathcal{B}$$

by the first part of the lemma. Thus f_x is measurable.

Let $g : W \to X \times Y$. If g is measurable with respect to $\mathcal{W}$ and $\mathcal{A} \otimes \mathcal{B}$, since p_X and p_Y are measurable, $p_X \circ g$ and $p_Y \circ g$ are measurable. Conversely, if $p_X \circ g$ and $p_Y \circ g$ are measurable, for all $A \in \mathcal{A}$, $B \in \mathcal{B}$ one has $g^{-1}(A \times B) = (p_X \circ g)^{-1}(A) \cap (p_Y \circ g)^{-1}(B) \in \mathcal{W}$. Since $\mathcal{C}$ generates the σ-algebra $\mathcal{A} \otimes \mathcal{B}$, g is measurable. □

For the remaining part of this section, $(X, \mathcal{M}, \mu)$ and $(Y, \mathcal{N}, \nu)$ are σ-finite measure spaces. We want to construct a natural measure $\pi := \mu \otimes \nu$ on $(X \times Y, \mathcal{M} \otimes \mathcal{N})$. We denote by $\mathcal{A}$ (resp. $\mathcal{B}$) the ring formed by those $A \in \mathcal{M}$ (resp. $B \in \mathcal{N}$) such that $\mu(A) < +\infty$ (resp. $\nu(B) < +\infty$). In the sequel we assume μ and ν are σ-finite, hence that $\mathcal{M} = \mathcal{A}_\sigma$ and $\mathcal{N} = \mathcal{B}_\sigma$. Then, Proposition 1.16 ensures that

$$\mathcal{M} \otimes \mathcal{N} = (\mathcal{A} \boxtimes \mathcal{B})_\sigma = \mathcal{C}_\sigma.$$

It is natural to define $\pi := \mu \otimes \nu$ on $\mathcal{C}$ by setting

$$(\mu \otimes \nu)(A \times B) := \mu(A)\nu(B) \qquad A \in \mathcal{A},\ B \in \mathcal{B}. \tag{1.12}$$

Let us first show that π is additive on $\mathcal{C}$. We need a refinement of Lemma 1.4.

Lemma 1.9 *Given a finite family $(C_i)_{i\in I}$ of elements of a semi-ring $\mathcal{C}$ one can find a finite partition $(E_j)_{j\in J}$ of $C := \cup_{i\in I} C_i$ by elements of $\mathcal{C}$ such that for all $i \in I$ one has $C_i = \cup_{j\in J_i} E_j$ for some subset J_i of J.*

Proof We prove this assertion by induction on the number n of elements of I. For $n = 2, I := \{1, 2\}$ we take $J := \{1, 2, 3\}, J_1 = \{1, 3\}, J_2 := \{2, 3\}$, and we write

$$C_1 \cup C_2 = E_1 \sqcup E_2 \sqcup E_3 \text{ with } E_1 := C_1 \backslash C_2, E_2 := C_2 \backslash C_1, E_3 := C_1 \cap C_2.$$

In order to pass from $n - 1$ to $n \geq 3$, given a family $(C_i)_{i\in I}$, with $I := \mathbb{N}_n$, we write $C_1 \cup \ldots \cup C_{n-1} = \sqcup_{h\in H} E_h$, $C_i := \sqcup_{h\in H_i} E_h$ with $H_i \subset H \subset \mathbb{N}$ by our induction assumption. Then, replacing (C_1, C_2) with $(\sqcup_{h\in H} E_h, C_n)$ in the preceding relation, we get the decomposition

$$C_1 \cup \ldots \cup C_{n-1} \cup C_n = (\sqcup_{h\in H}(E_h \cap C_n)) \sqcup (C_n \backslash (\sqcup_{h\in H} E_h)) \sqcup (\sqcup_{h\in H}(E_h \backslash C_n)).$$

For $h \in H$ let $F_h := E_h \cap C_n \in \mathcal{C}$. Since $C_n \backslash (\sqcup_{h\in H} E_h) = \cap_{h\in H}(C_n \backslash E_h)$ belongs to the ring $\mathcal{A}$ generated by $\mathcal{C}$, we can find a finite subset K of $\mathbb{N} \backslash H$ and a partition $(F_k)_{k\in K}$ of $C_n \backslash (\sqcup_{h\in H} E_h)$ by members of $\mathcal{C}$. Similarly, we can find a finite subset L of $\mathbb{N} \backslash (H \cup K)$, a partition $(L_h)_{h\in H}$ of L, and partitions $(F_\ell)_{\ell\in L_h}$ of $E_h \backslash C_{n+1}$ by elements of $\mathcal{C}$. Then, for $J := H \cup K \cup L$, the family $(F_j)_{j\in J}$ is a partition of $C := C_1 \cup \ldots \cup C_n$. For $i \in \mathbb{N}_{n-1}$ we have $C_i = \sqcup_{j\in J_i} F_j$ for $J_i := H_i \cup (\cup_{h\in H_i} L_h)$ and for $i := n$ we have $C_n = \sqcup_{j\in J_n} F_j$ for $J_n := H \cup K$ so that the induction step is established. □

Lemma 1.10 *The function $\mu \otimes \nu$ is additive on $\mathcal{C} := \{A \times B : A \in \mathcal{A},\ B \in \mathcal{B}\}$.*

Proof Let us consider a finite partition $(C_i)_{i\in I}$ of $C := A \times B \in \mathcal{C}$ by elements $C_i := A_i \times B_i$ of $\mathcal{C}$. The families $(A_i)_{i\in I}$ and $(B_i)_{i\in I}$ are not partitions in general. However, by the preceding lemma, we can find finite sets J, K, partitions $(E_j)_{j\in J}$, $(F_k)_{k\in K}$ of A and B by nonempty members of $\mathcal{A}$ and $\mathcal{B}$ respectively such that for all $i \in I$, there exist some subsets J_i, K_i of J and K respectively, such that $(E_j)_{j\in J_i}$ and $(F_k)_{k\in K_i}$ are partitions of A_i and B_i respectively.

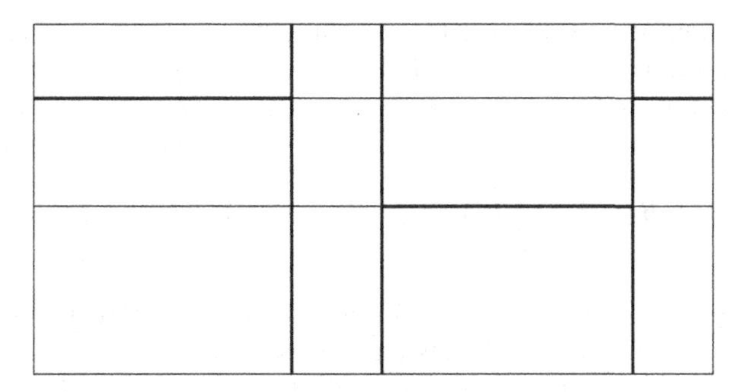

Fig. 1.3 Refined pavement of the yard of Fig. 1.1

Then $(E_j \times F_k)_{(j,k)\in J\times K}$ is a partition of $A \times B$: for all $(j,k) \in J \times K$, picking some $(x,y) \in E_j \times F_k$ we can find some $i \in I$ such that $(x,y) \in C_i := A_i \times B_i$, so that $(j,k) \in J_i \times K_i$; on the other hand, when $(E_j \times F_k) \cap (E_{j'} \times F_{k'}) \neq \varnothing$ we have $(j,k) = (j',k')$. Then

$$(\mu \otimes \nu)(C) = \mu(A)\nu(B) = (\sum_{j\in J} \mu(E_j))(\sum_{k\in K} \nu(F_k)) = \sum_{(j,k)\in J\times K} \mu(E_j)\nu(F_k).$$

Similarly, since $(E_j \times F_k)_{(j,k)\in J_i\times K_i}$ is a partition of $A_i \times B_i$,

$$(\mu \otimes \nu)(C_i) = \mu(A_i)\nu(B_i) = (\sum_{j\in J_i} \mu(E_j))(\sum_{k\in K_i} \nu(F_k)) = \sum_{(j,k)\in J_i\times K_i} \mu(E_j)\nu(F_k).$$

Now, let us note that $(J_i \times K_i)_{i\in I}$ is a partition of $J \times K$ since for $h \neq i$ in I and $(j,k) \in (J_h \times K_h) \cap (J_i \times K_i)$ we have $E_j \subset A_h \cap A_i$, $F_k \subset B_h \cap B_i$ hence $E_j \times F_k \subset (A_h \times B_h) \cap (A_i \times B_i) = \varnothing$. Thus, by associativity of summation,

$$\sum_{(j,k)\in J\times K} \mu(E_j)\nu(F_k) = \sum_{i\in I} \sum_{(j,k)\in J_i\times K_i} \mu(E_j)\nu(F_k) = \sum_{i\in I} (\mu \otimes \nu)(C_i).$$

It follows that $(\mu \otimes \nu)(C) = \Sigma_{i\in I}(\mu \otimes \nu)(C_i)$. □

Then, by Lemma 1.6, $\mu \otimes \nu$ can be uniquely extended into an additive function on the ring $\mathcal{A} \boxtimes \mathcal{B}$. In order to prove that $\mu \otimes \nu$ is σ-additive we need an estimate (Fig. 1.3).

Lemma 1.11 *Let $c \in \mathbb{R}_+$ and let $P \in \mathcal{A} \boxtimes \mathcal{B}$ be such that $\nu(P_x) \leq c$ for all $x \in X$. Then one has $(\mu \otimes \nu)(P) \leq c\mu(p_X(P))$.*

Proof Every element of $\mathcal{A} \boxtimes \mathcal{B}$ is a finite union of a disjoint family of rectangles in $\mathcal{C}$. We carry out the proof by an induction on the number n of such rectangles. For $n = 1$, $P := A \times B$ and since $P_x = B$ if $x \in A$ and $P_x = \varnothing$ if $x \in X \backslash A$, we have $(\mu \otimes \nu)(P) = \mu(A)\nu(B) \leq c\mu(A)$ if $\nu(B) \leq c$.

Now we assume the result is established when P is the disjoint union of at most $n-1$ rectangles and we prove it when $P = C_1 \sqcup \ldots \sqcup C_n$ with $C_i := A_i \times B_i$, where $A_i \in \mathcal{A}$, $B_i \in \mathcal{B}$ for $i \in \mathbb{N}_n$. Let $A := A_1 \cup \ldots \cup A_{n-1}$ and let

$$D = (A_n \backslash A) \times B_n \quad E = \coprod_{i=1}^{n-1} (A_i \backslash A_n) \times B_i \quad F = \bigcup_{i=1}^{n-1} (A_i \cap A_n) \times (B_i \cup B_n).$$

These sets are disjoint because their projections on X are disjoint and we easily see that $P = D \cup E \cup F$. Since these sets are unions of at most $n-1$ rectangles, the induction assumption and the inclusions $D_x \subset P_x$, $E_x \subset P_x$, $F_x \subset P_x$ for all $x \in X$ yield

$$\begin{aligned}\pi(P) &= \pi(D) + \pi(E) + \pi(F) \leq c\mu(p_X(D)) + c\mu(p_X(E)) + c\mu(p_X(F)) \\ &\leq c(\mu(A_n \backslash A) + \mu(A \backslash A_n) + \mu(A \cap A_n)) = c\mu(A \cup A_n) = c\mu(p_X(P)).\end{aligned}$$

□

Theorem 1.13 *The function $\pi := \mu \otimes \nu$ has a unique extension as a measure on $\mathcal{M} \otimes \mathcal{N}$.*

Proof Since $\mathcal{M} \otimes \mathcal{N}$ is the σ-algebra generated by $\mathcal{A} \boxtimes \mathcal{B}$, by Theorem 1.10 it suffices to show that π is σ-additive on $\mathcal{A} \boxtimes \mathcal{B}$.

We first assume that μ and ν are finite measures; let $a := \mu(X) > 0, b := \nu(Y) > 0$ (the cases $a = 0$ and $b = 0$ are trivial). Since π is finite and additive on $\mathcal{A} \boxtimes \mathcal{B}$, Lemma 1.5 ensures that it suffices to show that for every decreasing sequence (P_n) of elements of $\mathcal{A} \boxtimes \mathcal{B}$ whose intersection is empty one has $\lim_n \pi(P_n) = 0$. Assume on the contrary that there exist $r > 0$ and a decreasing sequence (P_n) such that $\cap_n P_n = \varnothing$ and $\lim_n \pi(P_n) > r$.

Since for all $P \in \mathcal{A} \boxtimes \mathcal{B}$ and all $x \in X$ the set $P_x \in \mathcal{A}$ and $x \mapsto \nu(P_x)$ is a sum of characteristic functions, the sets

$$A_n := \{x \in X : \nu((P_n)_x) > r/2a\}$$

are measurable. Let $Q_n := P_n \backslash (A_n \times Y)$, so that for all $x \in p_X(Q_n) \subset X \backslash A_n$ we have $\nu((Q_n)_x) = \nu((P_n)_x) \leq r/2a$. The preceding lemma entails that $\pi(Q_n) \leq (r/2a)\mu(p_X(P)) \leq r/2$. Since

$$r < \pi(P_n) = \pi(P_n \backslash Q_n) + \pi(Q_n) \leq \pi(A_n \times Y) + r/2,$$

we have $b\mu(A_n) > r/2$. Since the sequence (A_n) is decreasing we cannot have $\cap_n A_n = \varnothing$. Let $\overline{x} \in \cap_n A_n$. By definition of A_n we have $\nu((P_n)_{\overline{x}}) > r/2a$. Since the sequence $((P_n)_{\overline{x}})_n$ is decreasing, its intersection must be nonempty. Let $\overline{y} \in \cap_n (P_n)_{\overline{x}}$. Then we have $(\overline{x}, \overline{y}) \in P_n$ for all $n \in \mathbb{N}$, contradicting our assumption that $\cap_n P_n = \varnothing$.

When $(X, \mathcal{M}, \mu)$ and $(Y, \mathcal{N}, \nu)$ are σ-finite measure spaces, the function $\mu \otimes \nu$ is σ-finite. Given $C := A \times B \in \mathcal{C}$ and a countable partition (C_n) of C by elements of $\mathcal{C}$, replacing the measures μ and ν by the measures $\mu_A : M \mapsto \mu(M \cap A)$ and $\nu_B : N \mapsto \nu(N \cap B)$, what precedes shows that $(\mu \otimes \nu)(C) = \Sigma_n(\mu \otimes \nu)(C_n)$. Then, Hahn's extension theorem (Theorem 1.10) shows that $\mu \otimes \nu$ can be extended to an additive function on $\mathcal{M} \otimes \mathcal{N}$. □

In general, a product of complete measure spaces is not complete. To see this, let $\pi := \mu \otimes \nu$ be the product measure on $X \times Y$ of two complete measure spaces $(X, \mathcal{A}, \mu)$ and $(Y, \mathcal{B}, \nu)$. We denote by $\mathcal{N}_\lambda$ the set of null sets with respect to a measure λ with $\lambda = \mu, \nu, \pi$. Then we observe that (with the usual convention $0 \times (+\infty) = 0$)

$$(\mathcal{N}_\mu \times \mathcal{P}(Y)) \cup (\mathcal{P}(X) \times \mathcal{N}_\nu) \subset \mathcal{N}_\pi. \tag{1.13}$$

It follows that if $A \in \mathcal{N}_\mu \backslash \{\varnothing\}$ and $B \in \mathcal{P}(Y) \backslash \mathcal{B}$, then $P := A \times B \notin \mathcal{A} \otimes \mathcal{B}$ since otherwise, by Lemma 1.8, for $x \in A$ we would have $B = P_x \in \mathcal{B}$.

In particular, since the Lebesgue σ-algebra $\mathcal{L}$ on $\mathbb{R}$ obtained by completing the Borel σ-algebra $\mathcal{B}$ of $\mathbb{R}$ is different from $\mathcal{P}(\mathbb{R})$ and since the family $\mathcal{N}$ of its null sets is not reduced to $\varnothing$, we get that $(\mathbb{R}^2, \mathcal{L} \otimes \mathcal{L}, \lambda \otimes \lambda)$ is not complete.

For $d \in \mathbb{N} \backslash \{0\}$ let $\mathcal{L}_d$ be the σ-algebra on $\mathbb{R}^d$ obtained by completing the Borel σ-algebra $\mathcal{B}_d$ of $\mathbb{R}^d$ and let $\mathcal{C}_d$ be the semi-ring on $\mathbb{R}^d$ formed by the semi-closed rectangles $C := \prod_{i=1}^d [a_i, b_i[$ with $a_i \le b_i$ for all $i \in \mathbb{N}_d$. Since any open subset of $\mathbb{R}^d$ is a countable union of such rectangles, $\mathcal{B}_d$ is generated by $\mathcal{C}_d$.

Let us make precise the links between completed measures and product measures.

Proposition 1.17 *Let $(X, \mathcal{A}, \mu)$ and $(Y, \mathcal{B}, \nu)$ be two σ-finite measure spaces and let $(X, \mathcal{A}_\mu, \overline{\mu})$ and $(Y, \mathcal{B}_\nu, \overline{\nu})$ be the completed measure spaces. Then one has*

$$(\mathcal{A}_\mu \otimes \mathcal{B}_\nu)_{\overline{\mu} \otimes \overline{\nu}} = (\mathcal{A} \otimes \mathcal{B})_{\mu \otimes \nu}, \qquad \mathcal{N}_{\overline{\mu} \otimes \overline{\nu}} = \mathcal{N}_{\mu \otimes \nu}, \qquad \overline{\overline{\mu} \otimes \overline{\nu}} = \overline{\mu \otimes \nu}.$$

Proof The inclusions $\mathcal{A} \otimes \mathcal{B} \subset \mathcal{A}_\mu \otimes \mathcal{B}_\nu \subset (\mathcal{A}_\mu \otimes \mathcal{B}_\nu)_{\overline{\mu} \otimes \overline{\nu}}$ are obvious. Given $A_\mu := A \cup M \in \mathcal{A}_\mu$, $B_\nu := B \cup N \in \mathcal{B}_\nu$ with $A \in \mathcal{A}$, $B \in \mathcal{B}$, $M \in \mathcal{N}_\mu$, $N \in \mathcal{N}_\nu$, setting $(\mathcal{A} \otimes \mathcal{B}) \overline{\cup} \mathcal{N}_{\mu \otimes \nu} := \{D \cup N : D \in \mathcal{A} \boxtimes \mathcal{B}, N \in \mathcal{N}_{\mu \otimes \nu}\}$, we have

$$A_\mu \times B_\nu = (A \times B) \cup (M \times B) \cup (A \times N) \cup (M \times N) \in (\mathcal{A} \boxtimes \mathcal{B}) \overline{\cup} \mathcal{N}_{\mu \otimes \nu}$$

in view of relation (1.13). Thus

$$\mathcal{A}_\mu \boxtimes \mathcal{B}_\nu \subset (\mathcal{A} \boxtimes \mathcal{B}) \overline{\cup} \mathcal{N}_{\mu \otimes \nu} \subset (\mathcal{A} \otimes \mathcal{B})_{\mu \otimes \nu}.$$

Since the restriction $(\overline{\mu} \otimes \overline{\nu}) \mid_{\mathcal{A} \otimes \mathcal{B}}$ of $\overline{\mu} \otimes \overline{\nu}$ to $\mathcal{A} \otimes \mathcal{B}$ coincides with $\mu \otimes \nu$ as both measures coincide on the ring $\mathcal{A} \boxtimes \mathcal{B}$ that generates $\mathcal{A} \otimes \mathcal{B}$, every $\mu \otimes \nu$-null set P in $\mathcal{A} \otimes \mathcal{B}$ is $\overline{\mu} \otimes \overline{\nu}$-null: $\mathcal{N}_{\mu \otimes \nu} \subset \mathcal{N}_{\overline{\mu} \otimes \overline{\nu}}$. Now, since $\mathcal{A}_\mu \otimes \mathcal{B}_\nu$ is a σ-algebra, given $P \in \mathcal{N}_{\overline{\mu}\otimes\overline{\nu}}$ we can find some $Q \in \mathcal{A}_\mu \otimes \mathcal{B}_\nu \subset (\mathcal{A} \otimes \mathcal{B})_{\mu\otimes\nu}$ such that $P \subset Q$ and $(\mu \otimes \nu)(Q) = 0$. The characterization of null sets in $(\mathcal{A} \otimes \mathcal{B})_{\mu\otimes\nu}$ ensures that there exists some $R \in \mathcal{A} \otimes \mathcal{B}$ such that $Q \subset R$ and $(\mu \otimes \nu)(R) = 0$. Thus $P \subset R$ with $(\mu \otimes \nu)(R) = 0$, so that $P \in \mathcal{N}_{\mu\otimes\nu}$. Therefore $\mathcal{N}_{\overline{\mu}\otimes\overline{\nu}} = \mathcal{N}_{\mu\otimes\nu}$. For every $S \in (\mathcal{A}_\mu \otimes \mathcal{B}_\nu)_{\overline{\mu}\otimes\overline{\nu}}$ we can find $T \in \mathcal{A}_\mu \otimes \mathcal{B}_\nu$ and $P \in \mathcal{N}_{\overline{\mu}\otimes\overline{\nu}}$ such that $S = T \cup P$. Then, by what precedes, we have $T \in (\mathcal{A} \otimes \mathcal{B}) \cup \mathcal{N}_{\mu\otimes\nu}$ and $P \in \mathcal{N}_{\mu\otimes\nu}$, so that $S = (T' \cup P') \cup P = T' \cup (P \cup P')$ with $T' \in \mathcal{A} \otimes \mathcal{B}$ and $P \cup P' \in \mathcal{N}_{\mu\otimes\nu}$. Thus $S \in (\mathcal{A} \otimes \mathcal{B})_{\mu\otimes\nu}$ and we conclude that $(\mathcal{A}_\mu \otimes \mathcal{B}_\nu)_{\overline{\mu}\otimes\overline{\nu}} \subset (\mathcal{A} \otimes \mathcal{B})_{\mu\otimes\nu}$. The reverse inclusion is obvious since for $R \in \mathcal{A} \otimes \mathcal{B}$ and $N \in \mathcal{N}_{\mu\otimes\nu} = \mathcal{N}_{\overline{\mu}\otimes\overline{\nu}}$ we have $R \in \mathcal{A}_\mu \otimes \mathcal{B}_\nu$ and $N \in \mathcal{N}_{\overline{\mu}\otimes\overline{\nu}}$, hence $R \cup N \in (\mathcal{A}_\mu \otimes \mathcal{B}_\nu)_{\overline{\mu}\otimes\overline{\nu}}$. Moreover, $(\overline{\overline{\mu} \otimes \overline{\nu}})(R \cup N) = (\overline{\mu} \otimes \overline{\nu})(R) = (\mu \otimes \nu)(R) = (\overline{\mu \otimes \nu})(R \cup N)$. Thus $\overline{\overline{\mu} \otimes \overline{\nu}} = \overline{\mu \otimes \nu}$. □

Exercises

1. Let $(X, \mathcal{M}, \mu)$ be a σ-finite complete measure space. Show that a nonnegative function f on X is measurable if and only if its positive hypograph $H_f := \{(x, r) \in X \times \mathbb{R}_+ : r \leq f(x)\}$ is measurable. Give properties of the map $f \mapsto (\mu\otimes\lambda)(H_f)$.
2. Let $((X_n, \mathcal{M}_n, \mu_n))$ be a sequence of probability spaces (this means that $\mu_n(X_n) = 1$ for all n). Let $\mathcal{M}$ be the σ-algebra on $X := \prod_n X_n$ generated by the families of sets of the form $p_n^{-1}(A_n)$ with $A_n \in \mathcal{M}_n$, where $p_n : X \to X_n$ is the canonical projection. Show that there exists a probability μ on $\mathcal{M}$ satisfying $\mu(A \times \prod_{k>n} X_k) = (\mu_1 \otimes \ldots \otimes \mu_n)(A)$ for all $n \in \mathbb{N}$ and all $A \in \mathcal{M}_1 \otimes \ldots \otimes \mathcal{M}_n$.
3. (*Brunn-Minkowski inequality*) Prove that for two compact subsets A, B of $\mathbb{R}^d$ the following inequality holds, with $A + B := \{a + b : a \in A,\ b \in B\}$, λ being the Lebesgue measure of $\mathbb{R}^d$:

$$\lambda^{1/d}(A + B) \geq \lambda^{1/d}(A) + \lambda^{1/d}(B).$$

   [See: [190, p. 88].]

1.9 * Regular Measures on Metric Spaces

In this section we anticipate the notion of a metric space. The reader may either obtain the required knowledge from Chap. 2 or suppose the metric space (X, d) we consider is $\mathbb{R}^d$ endowed with one of its usual norms. Note that the Borel σ-algebra $\mathcal{B}$ of (X, d), i.e. the σ-algebra generated by the family $\mathcal{O}$ of open subsets of X, is not easy to describe. Thus, our aim is to obtain estimates of the measures of *Borelian subsets*, i.e. sets in $\mathcal{B}$. We denote by $\mathcal{O}$ (resp. $\mathcal{F}$, resp. $\mathcal{K}$) the family of open (resp. closed, resp. compact) subsets of (X, d).

Definition 1.8 A measure μ on $(X, \mathcal{B})$ is said to be *outer regular* (resp. *inner regular*) if for all $B \in \mathcal{B}$ one has

$$\mu(B) = \inf\{\mu(G) : G \in \mathcal{O},\ B \subset G\}$$
$$(\text{resp.} \quad \mu(B) = \sup\{\mu(K) : K \in \mathcal{K},\ K \subset B\} \quad).$$

The measure μ is said to be *regular* if it is both outer regular and inner regular.

Proposition 1.18 *Let μ be a finite measure on $(X, \mathcal{B})$. Then for every $B \in \mathcal{B}$ and every $\varepsilon > 0$ there exist $G \in \mathcal{O}$ and $F \in \mathcal{F}$ such that $F \subset B \subset G$ and $\mu(G \backslash F) < \varepsilon$.*

Proof We want to show that the family $\mathcal{A}$ of elements A of $\mathcal{B}$ such that for all $\varepsilon > 0$ there exist $G \in \mathcal{O}$ and $F \in \mathcal{F}$ such that $F \subset B \subset G$ and $\mu(G \backslash F) < \varepsilon$ is a σ-algebra containing $\mathcal{O}$, whence $\mathcal{B}$. Let $A \in \mathcal{O}$; we take $G = A$ and we observe that for $(r_n) \to 0_+$ and $F_n := \{x \in X : d(x, X \backslash A) \geq r_n\}$ we have $F_n \in \mathcal{F}$, $F_n \subset A$, and $\cup_n F_n = A$, so that by Lemma 1.5 $\mu(A) = \lim_n \mu(F_n)$. Thus, we can take $F := F_n$ with n large enough and we get $\mathcal{O} \subset \mathcal{A}$.

Now using Proposition 1.5, let us show $\mathcal{A}$ is a σ-algebra. Let (A_n) be a sequence in $\mathcal{A}$ and let $A := \cup_n A_n$. Given $\varepsilon > 0$, for all $n \in \mathbb{N}$ we pick $G_n \in \mathcal{O}$ and $F_n \in \mathcal{F}$ such that

$$F_n \subset A_n \subset G_n \qquad \text{and} \qquad \mu(G_n \backslash F_n) < \varepsilon/2^{n+2}.$$

Let $G := \cup_n G_n \in \mathcal{O}$, $E := \cup_n F_n$ so that $E \subset A \subset G$ and $G \backslash E \subset \cup_n (G_n \backslash F_n)$. By σ-subadditivity we get $\mu(G \backslash E) \leq \Sigma_n \mu(G_n \backslash F_n) \leq \varepsilon/2$. On the other hand, since $\mu(E) = \lim_n \mu(\cup_{k=0}^n F_k)$, we can find $m \in \mathbb{N}$ such that $\mu(E \backslash F) < \varepsilon/2$ for $F := \cup_{k=0}^m F_k$. Then F is closed and $\mu(G \backslash F) \leq \mu(G \backslash E) + \mu(E \backslash F) < \varepsilon$. Thus $A \in \mathcal{A}$.

Finally, if $A \in \mathcal{A}$, then $A^c := X \backslash A \in \mathcal{A}$ since if $(G, F) \in \mathcal{O} \times \mathcal{F}$ is such that $F \subset A \subset G$, for $G^c := X \backslash G \in \mathcal{F}$ and $F^c := X \backslash F \in \mathcal{O}$ we have $G^c \subset A^c \subset F^c$ and $F^c \backslash G^c = G \backslash F$, hence $\mu(F^c \backslash G^c) = \mu(G \backslash F) < \varepsilon$. □

The conclusion of the proposition is equivalent to the following assertion: for all $B \in \mathcal{B}$ one has

$$\mu(B) = \sup\{\mu(F) : F \in \mathcal{F},\ F \subset B\} = \inf\{\mu(G) : G \in \mathcal{O},\ B \subset G\}.$$

Let us give a generalization to the case when μ is σ-finite.

Theorem 1.14 *If μ is a σ-finite measure on $(X, \mathcal{B})$, then for all $B \in \mathcal{B}$ one has*

$$\mu(B) = \sup\{\mu(F) : F \in \mathcal{F},\ F \subset B\}.$$

If X is the countable union of a sequence (X_n) of open subsets with finite measures, then μ is outer regular.

If X is the countable union of a sequence (X_n) of compact subsets with finite measures, then μ is inner regular.

Proof Let (X_n) be a sequence in $\mathcal{B}$ with union X such that $\mu(X_n) < +\infty$ for all n. We may suppose (X_n) is increasing. Then, for all $B \in \mathcal{B}$ we have $\mu(B) = \lim_n \mu(B \cap X_n)$. Given $r < \mu(B)$ there exists an $m \in \mathbb{N}$ such that $\mu(B \cap X_m) > r$. The measure μ_m given by $\mu_m(A) := \mu(A \cap X_m)$ being finite, the preceding proposition yields some $F \in \mathcal{F}$ such that $F \subset B$ and $\mu_m(F) > r$. Thus $\mu(F) \geq \mu(F \cap X_m) = \mu_m(F) > r$. This proves the first assertion.

Now suppose X_n is open (and with finite measure) for all $n \in \mathbb{N}$. The measure μ_n on $\mathcal{B}$ given by $\mu_n(A) := \mu(A \cap X_n)$ being finite, the preceding proposition yields some $G_n \in \mathcal{O}$ such that $B \subset G_n$ and $\mu_n(G_n) < \mu_n(B) + \varepsilon/2^{n+1}$ i.e.

$$\mu(G_n \cap X_n) < \mu(B \cap X_n) + 2^{-n-1}\varepsilon. \tag{1.14}$$

Let $G := \cup_n G_n \in \mathcal{O}$ and let $H_n := \cup_{k=0}^n G_k \cap X_k \in \mathcal{O}$. Let us show by induction on n that

$$\mu(H_n) \leq \mu(B \cap X_n) + \varepsilon - 2^{-n-1}\varepsilon. \tag{1.15}$$

For $n = 0$, this relation is satisfied by our choice of G_n. Let us assume it is satisfied for $n - 1$. Since $H_n \backslash H_{n-1} \subset (G_n \cap X_n) \backslash H_{n-1}$ we have

$$\mu(H_n \backslash H_{n-1}) \leq \mu((G_n \cap X_n) \backslash H_{n-1}) = \mu(G_n \cap X_n) - \mu((G_n \cap X_n) \cap H_{n-1}).$$

Moreover, since

$$B \cap X_{n-1} \subset G_{n-1} \cap X_{n-1} \subset H_{n-1}, \qquad B \cap X_{n-1} \subset B \cap X_n \subset G_n \cap X_n,$$

hence $B \cap X_{n-1} \subset G_n \cap X_n \cap H_{n-1}$, $\mu(B \cap X_{n-1}) \leq \mu(G_n \cap X_n \cap H_{n-1}) < \infty$, we deduce from (1.14) and the preceding inequalities that

$$\begin{aligned}
\mu(H_n) &\leq \mu(H_n \backslash H_{n-1}) + \mu(H_{n-1}) \\
&\leq \mu(G_n \cap X_n) - \mu(B \cap X_{n-1}) + \mu(H_{n-1}) \\
&\leq \mu(B \cap X_n) + 2^{-n-1}\varepsilon + \varepsilon - 2^{-n}\varepsilon = \mu(B \cap X_n) + \varepsilon - 2^{-n-1}\varepsilon
\end{aligned}$$

and relation (1.15) holds for all $n \in \mathbb{N}$. Passing to the limit in relation (1.15), for $H := \cup_{k \geq 0} H_k$ we get $\mu(H) \leq \mu(B) + \varepsilon$ and $B = \cup_n (B \cap X_n) \subset \cup_n (G_n \cap X_n) = H$.

For the last assertion, one just observes that $F \cap X_n \in \mathcal{K}$ and $\mu(F) = \lim \mu(F \cap X_n)$. □

Corollary 1.6 *Let (X, d) be a separable locally compact metric space and let μ be a measure on the Borel σ-algebra $\mathcal{B}$ of X that is finite on compact subsets of X. Then μ is regular. In particular, any measure on the Borel σ-algebra $\mathcal{B}_d$ of $\mathbb{R}^d$ that is finite on compact subsets of X is regular.*

Proof Let $\{x_n : n \in \mathbb{N}\}$ be a countable dense subset of X and let I be the set of pairs $(n, r) \in \mathbb{N} \times \mathbb{Q}$ such that $B[x_n, r]$ is compact. Since I is countable, we can find a sequence $(I_k)_{k\geq 0}$ of finite subsets whose union is I. Let $X_k := \cup_{(n,r)\in I_k} B[x_n, r] \in \mathcal{K}$. For each $x \in X$ there exists a compact neighborhood K_x of x. Taking $(n, r) \in \mathbb{N} \times \mathbb{Q}$ such that $x \in B(x_n, r)$ and $B[x_n, r] \subset K_x$, we get that $x \in X_k$ for some $k \in \mathbb{N}$. Thus $X = \cup_k X_k$ and $X = \cup_k \mathrm{int}(X_k)$. The regularity of μ follows from the theorem. □

Exercises

1. Show that any finite measure μ on the Borel σ-algebra of a Polish space (X, d), i.e. a complete and separable metric space, is regular in the sense of Definition 1.8. Prove moreover that there is a sequence (K_n) of compact subsets of X such that $\lim_n \mu(X \backslash K_n) = 0$.
2. Show that two measures μ, ν on the Borel σ-algebra of a metric space (X, d) coincide whenever one of the following conditions is satisfied.

   (a) μ and ν coincide on the family $\mathcal{O}$ of open subsets of X and X is the union of a countable family of open subsets of finite measures.
   (b) μ and ν are finite and coincide on the family $\mathcal{K}$ of compact subsets of X and X is separable and locally compact.
   (c) μ and ν coincide on the family $\mathcal{K}$ of compact subsets of X and X is separable and complete with $\mu(X) = \nu(X) < +\infty$.

3. Let (X, d) be a separable metric space and let μ be a measure on the Borel σ-algebra of X. The *essential support* of a measurable function $f : X \to \mathbb{R}$ is defined as the intersection $S_\mu(f)$ of the family of closed subsets F of X such that $f = 0$ on $X \backslash F$.

   (a) Show that when f is continuous one has $S_\mu(f) \subset \mathrm{supp}(f) := \mathrm{cl}(\{x \in X : f(x) \neq 0\})$.
   (b) Show that $S_\mu(f) = \mathrm{supp}(f)$ when f is continuous and $\mu(O) > 0$ for all $O \in \mathcal{O} \backslash \{\varnothing\}$.
   (c) Show that for any measurable function f on X one has $f = 0$ a.e. on $X \backslash S_\mu(f)$.
   (d) Verify that for two measurable functions f, g on X satisfying $|f| \leq |g|$ a.e. one has $S_\mu(f) \subset S_\mu(g)$. Deduce from this that $S_\mu(f) = S_\mu(g)$ whenever $f = g$ a.e.
   (e) Let (f_n) be an increasing sequence in nonnegative measurable functions with limit f. Show that $S_\mu(f) = \cup_n S_\mu(f_n)$.

4. Let (X, d) be a separable metric space and let μ be a measure on the Borel σ-algebra of X. Let $\mathcal{O}_\mu := \{O \in \mathcal{O} : \mu(O) = 0\}$. Show that $\mathcal{O}_\mu$ is nonempty and has a greatest element O_μ. The *support of the measure* μ is defined to be $X \backslash O_\mu$. Determine the support of a Dirac measure δ_a, of the counting measure μ_c on a finite set X, and of the Lebesgue measure λ_d on $X := \mathbb{R}^d$.

5. Let (X,d) be a separable metric space, let μ be a measure on the Borel σ-algebra of X, and let $f : X \to \mathbb{R}_+$ be a measurable function. Show that the support of the measure $\mu_f := f\mu$ coincides with the essential support $S_\mu(f)$ of f defined in Exercise 3.

Notes, Remarks, and Additional Reading

A basic knowledge of elementary analysis is required for the reading of the present book. Among the numerous monographs devoted to such a topic one is referred to [10, 30, 34, 59, 63, 84, 125, 146, 169, 176, 177, 183, 184, 213, 223, 224, 270].

The Cantor-Bernstein Theorem was stated by Georg Cantor during the period 1895–1897 and proved by his student Felix Bernstein (aged 18!) in 1896. It was published in the book "Leçons sur la théorie des fonctions" by Emile Borel in 1898 under a proposal of G. Cantor, who was facing harsh criticisms, in particular from Leopold Kronecker. Ernst Schröder produced an (imperfect) proof the same year. David Hilbert gave his strong approval of the advances of G. Cantor in the analysis of the various forms of infinity but some mathematicians are still reluctant to adopt the axiom of choice.

Historical views on the subjects treated in this book can be found in the references [28, 39, 49, 87, 104, 105, 110, 148, 156, 166].

More on set-valued analysis is contained in the books [12, 13, 93, 208, 221].

More on measure theory can be found in the following references: [38, 39, 41, 42, 49, 54, 64, 80, 99, 100, 105, 118, 126, 127, 132, 150, 156, 168, 182, 183, 185, 190, 193, 224, 226, 247].

The notion of metric spaces is due to Maurice Fréchet. It will be expounded in the next chapter.

Chapter 2
Encounters With Limits

Mathematics is not an arid land in the scientific universe. It is simultaneously the queen, maid and daughter of the observational sciences.

La mathématique ne constitue pas une terre aride dans l'univers scientifique. Elle est à la fois reine, servante et fille des sciences de l'observation.

Gustave Choquet (1915–2006)

Abstract The notion of limit is central in analysis. Thus the concept of convergence is presented in a general framework and then in the classes of topological spaces and metric spaces. Compactness, connectedness, completeness are studied in detail. Baire's Theorem is included as well as Ekeland's Variational Principle. The contraction theorem is proved and as an application an existence result for ordinary differential equations is presented.

The use of limits is the central theme of analysis. A detailed account of the history of limits and infinitesimals may be found in [28] in the context of nonstandard analysis. The rules in $\mathbb{R}$, $\mathbb{R}_\infty := \mathbb{R} \cup \{+\infty\}$, $\overline{\mathbb{R}} := \mathbb{R} \cup \{-\infty, +\infty\}$, $\mathbb{C}$, $\mathbb{R}^d$ serve as models for more general spaces. There are several ways to reach such generalizations. A first approach consists in selecting some rules for convergence; we evoke it briefly. A second approach uses some real-valued functions such as metrics, norms or seminorms in order to reduce convergence in an abstract set to convergence in $\mathbb{R}$. A third approach consists in introducing topologies, a set theoretical approach that is particularly versatile; it focusses attention on closed sets, i.e. sets that are stable when taking limits. However, it is usually formulated in terms of open sets, i.e. sets whose complements are closed.

The present chapter is devoted to a first glimpse of such approaches. In later chapters a more detailed study will be undertaken when special structures are at hand.

As an illustration let us consider the intuitive fact that if we write the letter Q with a segment that is shorter and shorter this letter "converges" to the letter O: Since the addresses we write on envelopes are read by machines and since these machines have a limited resolution, there is some risk that our letter will go to a wrong address if we are lazier and lazier in drawing the little slanted segment (Fig. 2.1).

J.-P. Penot, *Analysis*, Universitext, DOI 10.1007/978-3-319-32411-1_2

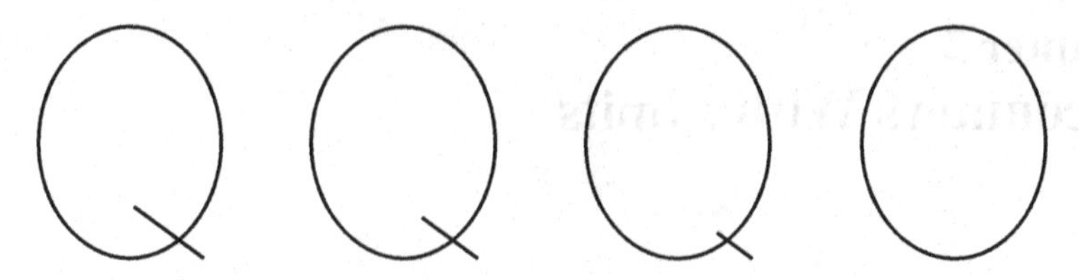

Fig. 2.1 Deformed letters

2.1 Convergences

The first approach to convergences appeared after the first quarter of the twentieth century. It is not often used nowadays, even if in some cases it is simpler than a topological approach (this is the case for pointwise convergence and for the second example given below). Since the selected rules are prevalent in practice, it is worth stating them in a formal definition. Let us first recall that a *sequence* in a set X is a map s from $\mathbb{N}$ to X, hence an element of $X^{\mathbb{N}}$. Setting $x_n := s(n)$ for $n \in \mathbb{N}$ we often write $(x_n)_{n\in\mathbb{N}}$, $(x_n)_{n\geq 0}$ or just (x_n) instead of s. A *subsequence* of $s := (x_n)$ is a sequence $s' := (x'_n)$ of X such that there exists a strictly increasing map $k : \mathbb{N} \to \mathbb{N}$ satisfying $s' = s \circ k$, i.e. $x'_n = x_{k(n)}$. Thus one obtains s' by deleting some terms of s and by reindexing the remaining terms.

Definition 2.1 A *space with limits* is a set X such that a relation denoted by $\to$ is defined between the sets $X^{\mathbb{N}}$ and X and read as (x_n) *converges* to x or x is a limit of (x_n), the relation $\to$ being required to satisfy the following properties:

(L1) any constant sequence with value x converges to x;
(L2) if $(x_n) \to x$, then any subsequence (x'_n) of (x_n) converges to x;
(L3) if $x \in X$ and $(x_n) \in X^{\mathbb{N}}$ are such that any subsequence (x'_n) of (x_n) has a subsequence (x''_n) converging to x, then $(x_n) \to x$.

For some needs, it is useful to add a uniqueness condition:

(U) if $(x_n) \to x$ and $(x_n) \to x'$ then $x = x'$.

Clearly, the convergences in $\mathbb{R}$, $\mathbb{R}_\infty$, $\overline{\mathbb{R}} := \mathbb{R}$, $\mathbb{C}$, $\mathbb{R}^d$ satisfy these four conditions.

Example If X is the set of real-valued functions on a set S, or, in other terms, if $X = \mathbb{R}^S$, then *pointwise convergence* on X is defined by $(x_n) \to x$ if and only if for all $s \in S$ one has $(x_n(s))_n \to x(s)$ as $n \to +\infty$. Here the element $x := (x_s)_{s\in S}$ of $\mathbb{R}^S$ is identified with the function $f : S \to \mathbb{R}$ defined by $f(s) := x_s$ and is also denoted by $(x(s))_{s\in S}$.

Example *Uniform convergence* on the set X of real-valued functions on S is defined in the following way: $(f_n) \xrightarrow{u} f$ if and only if for all $\varepsilon > 0$ one can find some $n(\varepsilon) \in \mathbb{N}$ such that $|f_n(s) - f(s)| \leq \varepsilon$ for all $n \geq n(\varepsilon)$ and all $s \in S$.

Example Let X be the set of continuous functions on $\mathbb{R}^d$ that vanish outside a bounded subset. Declare that a sequence $(f_n) \to f$ if there exists a bounded subset B of $\mathbb{R}^d$ such that the functions f_n and f are null on $\mathbb{R}^d \backslash B$ and if the net of restrictions to B uniformly converges to the restriction of f: $(f_n \mid B) \xrightarrow{u} f \mid B$. Variants of such a convergence are used in the theory of distributions.

Sometimes sequences are inadequate and one must replace them with generalized sequences, also called nets (see Exercise 5). A *net* in a set X is a map s from a directed set $(I, \leq)$ into X. Setting $x_i := s(i)$, one also denotes it by $(x_i)_{i \in I}$. A *subnet* is a net $s' := (x'_j)_{j \in J}$ such that there exists a filtering map $h : J \to I$ such that $x'_j = x_{h(j)}$ for all $j \in J$. Note that, in contrast to what occurs for subsequences, one takes for J a directed set that may differ from I. It is often of the form $J := I \times K$, where K is another directed set, or a subset of $I \times K$, h being the first projection or a restriction of it. In some simple cases one can take for J a cofinal subset of I and for h the injection of J into I, but that it not always possible. The axioms we select are the analogues of those of the preceding definition.

Definition 2.2 A *convergence space* is a set X such that for every directed set I there is a relation denoted by $\to$ between the set X^I of nets of X indexed by I and X itself in such a way that the following conditions are satisfied:

(C1) for every $x \in X$ the constant net with value x converges to x;
(C2) if $(x_i)_{i \in I} \to x$ and if $(x'_j)_{j \in J}$ is a subnet of $(x_i)_{i \in I}$, then $(x'_j)_{j \in J} \to x$;
(C3) if $x \in X$ and $(x_i)_{i \in I}$ is a net in X such that for every subnet $(x'_j)_{j \in J}$ of $(x_i)_{i \in I}$ there exists a subnet $(x''_k)_{k \in K}$ of $(x'_j)_{j \in J}$ that converges to x, then $(x_i)_{i \in I} \to x$.

The preceding examples can be adapted to convergence spaces.

A map $f : X \to Y$ between two convergent spaces is said to be *continuous* at $x \in X$ if for any net $(x_i)_{i \in I}$ converging to x the net $(f(x_i))_{i \in I}$ converges to $f(x)$. It is continuous on some subset A of X if for all $x \in A$ it is continuous at x.

If $(W, \to)$ is a convergence space and if X is a subset of W, the induced convergence on X is defined by $(x_i)_{i \in I} \to_X x$ if $(x_i)_{\in I} \to x$ in W. It is easy to verify that the conditions (C1)–(C3) are satisfied and that a map $f : V \to X$ from a convergence space $(V, \to_V)$ into $(X, \to_X)$ is continuous at $\overline{v} \in V$ if and only if f is continuous at $\overline{v}$ when considered as a map from V into W.

If $(X_a)_{a \in A}$ is a family of convergence spaces, the product $X := \prod_{a \in A} X_a$ is made a convergence space by requiring that a net $(x_i)_{i \in I} \to x$ if for all $a \in A$ the a-component $(p_a(x_i)) \to p_a(x)$. Then, a map $f : W \to X$ from a convergence space W to X is continuous at $\overline{w} \in W$ if and only if for all $a \in A$ the a-component $f_a := p_a \circ f$ is continuous at $\overline{w}$.

The *lower limit* (resp. *upper limit*) of a net $(r_i)_{i \in I}$ of real numbers is defined by

$$\liminf_{i \in I} r_i := \sup_{h \in I} \inf_{i \in I,\, i \geq h} r_i \qquad (\text{resp. } \limsup_{i \in I} r_i := \inf_{h \in I} \sup_{i \in I,\, i \geq h} r_i)$$

These substitutes for the limit always exist in $\overline{\mathbb{R}} := \mathbb{R} \cup \{-\infty, +\infty\}$.

Whereas it is useful to evoke the use of limits, the two concepts of limit spaces and of convergence spaces are not of great use in the present state of analysis. So, in the next sections we turn to other means of dealing with limits.

Exercises

1. Given a net $(r_i)_{i\in I}$ of real numbers, show that there exist subnets $(s_j)_{j\in J}$ and $(t_k)_{k\in K}$ such that $\limsup_{i\in I} r_i = \lim_{j\in J} s_j$ and $\liminf_{i\in I} r_i = \lim_{k\in K} t_k$.

2. Given a net $(r_i)_{i\in I}$ of real numbers, show that for any subnet $(q_h)_{h\in H}$ of $(r_i)_{i\in I}$ that converges one has $\liminf_{i\in I} r_i \le \lim_{h\in H} q_h \le \limsup_{i\in I} r_i$.

3. Define pointwise convergence of nets for functions from a set S into a convergence space $(X, \to)$ and verify conditions (C1)–(C3). Do the same for uniform convergence of functions with values in $\mathbb{R}^d$.

4. Let X be the set of real-valued continuous functions on $\mathbb{R}^d$ that vanish outside a bounded subset. Declare that a net $(f_i) \to f$ if there exist a bounded subset B of $\mathbb{R}^d$ and some $\bar{i} \in I$ such that for all $i \ge \bar{i}$ in I the functions f_i and f are null on $\mathbb{R}^d \backslash B$ and if the net of restrictions to B uniformly converges to the restriction of f: $(f_i \mid B) \overset{u}{\to} f \mid B$. Verify conditions (C1)–(C3).

5. (**Sequences do not suffice**). Let S be an infinite uncountable set and let X be the set of real-valued functions on S equipped with pointwise convergence. Let Y be the subset of X formed by those $f \in X$ that are null off a finite subset. Show that Y is *dense* in X in the sense that every $f \in X$ is the limit of a net $(f_i)_{i\in I}$ of Y. Verify that if $f \in X$ is the limit of a sequence $(f_n)_{n\in\mathbb{N}}$ of Y, then the set $S_f := \{s \in S : f(s) \ne 0\}$ is countable. Deduce from this the fact that the *sequential closure* of Y, i.e. the set of limits of sequences in Y, is different from its closure X.

6. Using the notion of topology displayed in the next section, define a convergence $\to$ on a topological space $(X, \mathcal{O})$ by setting $(x_i)_{i\in I} \to x$ if for any $O \in \mathcal{O}$ there exists some $h \in I$ such that $x_i \in O$ for all $i \in I$ satisfying $i \ge h$.

7. Conversely, associate to any convergence space $(X, \to)$ a topology $\mathcal{O}$ by taking for $\mathcal{O}$ the set of subsets O of X such that for any $x \in O$ and any net $(x_i)_{i\in I} \to x$ one has $x_i \in O$ whenever $i \ge h$ for some $h \in I$. Verify the assumptions (O1), (O2) of the next definition. Prove that the convergence $\to$ is stronger than (i.e. implies) the convergence $\to_{\mathcal{O}}$ associated with the topology $\mathcal{O}$. Show that for any map $f : X \to Y$ with values in a topological space $(Y, \mathcal{O}_Y)$ the map f is continuous from $(X, \mathcal{O})$ into $(Y, \mathcal{O}_Y)$ if and only if f is continuous for the convergence $\to$ on X and the convergence associated with the topology of Y.

8*. Find an additional condition on a convergence in order that it is the convergence associated with a topology. [See [174].]

2.2 Topologies

The success of topology is due to two features: first, convergences are defined through an intuitive notion of neighborhoods for each point; second, the formalism and the rules of set theory can be used efficiently.

2.2.1 General Facts About Topologies

A topology on a set X is obtained by selecting a family of subsets called the family of closed subsets having some stability property. Equivalently, one usually introduces topologies by considering the family of complements of closed sets. These sets are called open sets.

Definition 2.3 A *topology* on a set X is the data comprising a family $\mathcal{O}$ of so-called *open* subsets that satisfies the following two requirements:

(O1) the union of any subfamily of $\mathcal{O}$ belongs to $\mathcal{O}$;
(O2) the intersection of any finite subfamily of $\mathcal{O}$ belongs to $\mathcal{O}$.

By convention, we admit that these two conditions include the requirements that X and the empty set $\varnothing$ belong to $\mathcal{O}$. A topological space $(X, \mathcal{O})$ is also denoted by X if the choice of the topology $\mathcal{O}$ is unambiguous. A subset F of X is declared to be *closed* if $X \backslash F$ belongs to $\mathcal{O}$.

Exercise Give conditions characterizing the family of closed (resp. open) subsets of a topological space in terms of nets for the convergence associated to $\mathcal{O}$ defined in Exercise 6 of the preceding section and reminded a few lines below.

A subset V of a topological space $(X, \mathcal{O})$ is a *neighborhood* of some $\overline{x} \in X$ if there exists some $U \in \mathcal{O}$ such that $\overline{x} \in U \subset V$. For $x \in X$ we denote by $\mathcal{N}(x)$ the family of neighborhoods of x. The topology $\mathcal{O}$ is determined by $(\mathcal{N}(x))_{x \in X}$, as shown by the exercises: O is open iff for all $x \in O$ one has $O \in \mathcal{N}(x)$.

To any topology on X one can associate a convergence $\to$ by setting:

$$((x_i)_{i \in I} \to x) \iff (\forall\ V \in \mathcal{N}(x)\ \exists\ i_V \in I\ :\ i \in I,\ i \geq i_V \Rightarrow x_i \in V)\,.$$

When a limit is unique one also writes $x = \lim_{i \in I} x_i$.

Exercise Verify that the conditions (C1), (C2), (C3) are satisfied. Moreover, a subset C of X is closed if and only if for any net $(x_i)_{i \in I}$ of C and any $x \in X$ satisfying $(x_i)_{i \in I} \to x$ one has $x \in C$.

Definition 2.4 A map $f\ :\ (X, \mathcal{O}) \to (X', \mathcal{O}')$ between two topological spaces is said to be *continuous* at $\overline{x} \in X$ if for any $V' \in \mathcal{N}(f(\overline{x}))$ there exists some $V \in \mathcal{N}(\overline{x})$ such that $f(V) \subset V'$. The map f is said to be continuous on some subset A of X if it is continuous at all $\overline{x} \in A$.

The definition of continuity of f at $\bar{x}$ is natural: in order that $f(x)$ be close enough to $f(\bar{x})$ it suffices to take x close enough to $\bar{x}$. However, one has to remember that in this condition, the neighborhood V' of $f(\bar{x})$ should be prescribed first. Continuity of f at $\bar{x}$ can be expressed by requiring that for any $V' \in \mathcal{N}(f(\bar{x}))$ one has $f^{-1}(V') \in \mathcal{N}(\bar{x})$.

Exercise Show that $f : X \to X'$ is continuous (on X) if and only if for all $O' \in \mathcal{O}'$ its inverse image $f^{-1}(O') := \{x \in X : f(x) \in O'\}$ belongs to $\mathcal{O}$.

The composition of two continuous maps is clearly continuous.

A bijection that is continuous and whose inverse is continuous is called a *homeomorphism*. Topology is the study of properties that are preserved under homeomorphisms.

A topology $\mathcal{O}'$ on X is said to be *weaker* than a topology $\mathcal{O}$ if the *identity map* $I_X : (X, \mathcal{O}) \to (X, \mathcal{O}')$ is continuous, i.e. if any member of $\mathcal{O}'$ is in $\mathcal{O}$, i.e. if $\mathcal{O}' \subset \mathcal{O}$. One also says that $\mathcal{O}$ is *finer* or stronger than $\mathcal{O}'$.

Example On a set X there is a topology that is weaker than any other one, the *rough topology*: its family of open sets is $\mathcal{O}_R := \{\varnothing, X\}$. There is a topology that is finer than any other one, the *discrete topology*, for which any subset is open: $\mathcal{O}_D := \mathcal{P}(X)$.

Given a family $\mathcal{G}$ of subsets of a set X, there is a topology $\mathcal{O}$ on X that is the weakest among those containing $\mathcal{G}$. It is obtained as the intersection of the family of topologies $\mathcal{O}_i$ satisfying $\mathcal{G} \subset \mathcal{O}_i$. Then one says that $\mathcal{G}$ *generates* $\mathcal{O}$. If $\mathcal{B} \subset \mathcal{O}$ is such that any element of $\mathcal{O}$ is a union of elements of $\mathcal{B}$, one says that $\mathcal{B}$ is a *base* of $\mathcal{O}$. It is easy to verify that when $\mathcal{G}$ generates $\mathcal{O}$, the family $\mathcal{B}$ of finite intersections of elements of $\mathcal{G}$ is a base of $\mathcal{O}$. A family $\mathcal{U}$ of subsets of X is a *base of neighborhoods* of $\bar{x}$ if $\mathcal{U}$ is contained in the family $\mathcal{N}(\bar{x})$ of neighborhoods of $\bar{x}$ and if for any $V \in \mathcal{N}(\bar{x})$ there exists some $U \in \mathcal{U}$ such that $U \subset V$. Given $\mathcal{B} \subset \mathcal{O}$, we see that $\mathcal{B}$ is a base of $\mathcal{O}$ if and only if for all $x \in X$, $\mathcal{B}(x) := \{U \in \mathcal{B} : x \in U\}$ is a base of neighborhoods of x.

The notion of continuity can be localized by using neighborhood bases. A map $f : (X, \mathcal{O}) \to (X', \mathcal{O}')$ is continuous at $\bar{x} \in X$ if for any neighborhood W in some neighborhood base of $f(\bar{x})$ in $(X', \mathcal{O}')$ there exists some $V \in \mathcal{N}(\bar{x})$ such that $f(V) \subset W$.

Given a set X, a family $(X_a, \mathcal{O}_a)_{a \in A}$ of topological spaces, and a family $(g_a)_{a \in A}$ of maps $g_a : X \to X_a$, among all the topologies on X for which all g_a are continuous, there is one that is weaker than any other. It is the topology $\mathcal{O}_X$ generated by the sets $g_a^{-1}(G_a)$ for $a \in A$ and $G_a \in \mathcal{O}_a$. It is easy to verify that a map $f : W \to X$ from a topological space $(W, \mathcal{O}_W)$ into X is continuous with respect to $\mathcal{O}_X$ if and only if $g_a \circ f : W \to X_a$ is continuous for all $a \in A$. When X is the product $\Pi_{a \in A} X_a$ and g_a is the canonical projection, the topology $\mathcal{O}_X$ is called the *product topology*. When X is a subset of a topological space $(Y, \mathcal{O}_Y)$ and one considers for the family $(g_a)_{a \in A}$ the sole canonical injection $j : X \to Y$ one says that $\mathcal{O}_X$ is the *induced topology*. Then $O \subset X$ belongs to $\mathcal{O}_X$ if and only if there exists some $G \in \mathcal{O}_Y$ such that $O = G \cap X$. It is easy to verify that the associated convergence to $\mathcal{O}_X$ is the induced convergence.

Besides the notion of limit associated with a topology $\mathcal{O}$ on X, one disposes of a weaker notion. One says that a net $(x_i)_{i\in I}$ of X has a *cluster point* $\overline{x} \in X$ if for any $V \in \mathcal{N}(\overline{x})$ and any $i \in I$ one can find some $j \in I$ such that $j \geq i$ and $x_j \in V$.

One can show that $\overline{x} \in X$ is a cluster point of $(x_i)_{i\in I}$ if and only if there exists a subnet of $(x_i)_{i\in I}$ that converges to $\overline{x}$. The if condition is immediate. For the necessary condition one can take $J := \{(i, V) \in I\times\mathcal{N}(\overline{x}) : x_i \in V\}$, a cofinal subset of $I\times\mathcal{N}(\overline{x})$ for the product order, and define $h : J \to I$ by $h(i, V) := i$.

A topology $\mathcal{O}$ on a set X is said to be *Hausdorff* if for every pair (x, x') of distinct points of X one can find neighborhoods $V \in \mathcal{N}(x)$, $V' \in \mathcal{N}(x')$ that are disjoint. This property is equivalent to uniqueness of limits of nets.

Proposition 2.1 *A topology $\mathcal{O}$ on a set X is* Hausdorff *if and only if all nets in X have at most one limit. In fact, if $\mathcal{O}$ is Hausdorff and if a net $(x_i)_{i\in I}$ of X has a limit $\overline{x}$, it cannot have a different cluster point.*

Proof Suppose $\mathcal{O}$ is Hausdorff and a net $(x_i)_{i\in I}$ of X has a limit $\overline{x}$ and a cluster point $\overline{y} \neq \overline{x}$. Let $V \in \mathcal{N}(\overline{x})$, $W \in \mathcal{N}(\overline{y})$ be such that $V \cap W = \varnothing$. By definition, there exists an $i_V \in I$ such that $x_i \in V$ for all $i \geq i_V$. Thus one cannot find some $j \geq i_V$ such that $x_j \in W$, contradicting the assumption that $\overline{y}$ is a cluster point of $(x_i)_{i\in I}$.

Now suppose $\mathcal{O}$ is not Hausdorff: there exists a pair $(\overline{x}, \overline{y})$ of distinct points of X such that for any $V \in \mathcal{N}(\overline{x})$, $W \in \mathcal{N}(\overline{y})$ one has $V\cap W \neq \varnothing$. Denoting $\mathcal{N}(\overline{x})\times\mathcal{N}(\overline{y})$ by I and giving to I the order opposite to inclusion, for $i := (V, W) \in I$ one can pick $x_i \in V \cap W$. Then the net $(x_i)_{i\in I}$ converges to $\overline{x}$ and to $\overline{y}$. □

Corollary 2.1 *In a Hausdorff topological space $(X, \mathcal{O})$ finite subsets, in particular singletons, are closed.*

Proof It suffices to show that a singleton $S := \{\overline{x}\}$ is closed. If $\overline{y} \in X\backslash S$ one can find $V \in \mathcal{N}(\overline{x})$ and $W \in \mathcal{N}(\overline{y})$ that are disjoint. Then W is contained in $X\backslash S$. This proves that $X\backslash S$ is open and S is closed. □

A topology $\mathcal{O}$ on X is uniquely determined by its associated convergence for nets: a subset C of X is closed if and only if it contains the limits of its convergent nets. In general $\mathcal{O}$ is not determined by the convergence of sequences. The following proposition shows that nets may be convenient. It also shows that continuity in topological spaces coincides with continuity for the induced convergence spaces.

Proposition 2.2 *A map $f : X \to Y$ between two topological spaces is continuous at $\overline{x} \in X$ if and only if for any net $(x_i)_{i\in I}$ of X converging to $\overline{x}$, the net $(f(x_i))_{i\in I}$ converges to $f(\overline{x})$.*

When $\mathcal{N}(\overline{x})$ has a countable base sequences can be used instead of nets.

Proof Necessity is immediate. Let us show sufficiency. Suppose f is not continuous at $\overline{x}$. Then, there exists a $V \in \mathcal{N}(f(\overline{x}))$ such that for all $U \in \mathcal{N}(\overline{x})$ there exists some $x_U \in U$ with $f(x_U) \notin V$. Then, the net $(x_U)_{U\in\mathcal{N}(\overline{x})} \to \overline{x}$ but $(f(x_U))_{U\in\mathcal{N}(\overline{x})}$ does not converge to $f(\overline{x})$. □

The *closure* $\mathrm{cl}(S)$ of a subset S of a topological space $(X, \mathcal{O})$ is the intersection of the family of all closed subsets of X containing S. It is clearly the smallest closed subset of $(X, \mathcal{O})$ containing S. The *interior* $\text{int}(S)$ of S is the union of all the open subsets of $(X, \mathcal{O})$ contained in S. It is the largest open subset of X contained in S. Thus $\text{int}(S) = X \backslash \text{cl}(X \backslash S)$. The *boundary* or frontier of S is $\text{bdry}(S) := \text{cl}(S) \backslash \text{int}(S)$.

Exercise Prove that the closure $\text{cl}(S)$ of a subset S of a topological space $(X, \mathcal{O})$ is the set of limits of nets of S that converge in X.

Proposition 2.3 *The closure* $\text{cl}(S)$ *of a subset* S *of a topological space* $(X, \mathcal{O})$ *is the set of points* $x \in X$ *such that for any neighborhood* V *of* x *one has* $S \cap V \neq \varnothing$.

Proof If $\overline{x} \in X \backslash \text{cl}(S)$ there exists some closed set C containing S such that $\overline{x} \in X \backslash C$. Then $V := X \backslash C$ is an open neighborhood of $\overline{x}$ and $S \cap V = \varnothing$. Conversely, if for some $V \in \mathcal{N}(\overline{x})$ one has $S \cap V = \varnothing$, taking some open subset U satisfying $\overline{x} \in U \subset V$ we see that $C := X \backslash U$ is a closed subset containing S (since $U \subset V$ and $S \cap V = \varnothing$) and $\overline{x} \notin C$, hence $\overline{x} \notin \text{cl}(S)$. □

Corollary 2.2 *A point* $\overline{x}$ *of a topological space* $(X, \mathcal{O})$ *is a cluster point of a net* $(x_i)_{i \in I}$ *of* X *if* $\overline{x} \in \cap_{i \in I} C_i$, *where* $C_i := \text{cl}(\{x_j : j \in I, j \geq i\})$.

Proof This follows from the fact that $\overline{x}$ is a cluster point of $(x_i)_{i \in I}$ if and only if for all $i \in I$ and all $V \in \mathcal{N}(\overline{x})$ one has $V \cap \{x_j : j \in I, j \geq i\} \neq \varnothing$. □

A subset D of $(X, \mathcal{O})$ is said to be *dense* in a subset E of X if $D \subset E$ and if E is contained in the closure of D. A topological space is said to be *separable* if it contains a countable dense subset.

Example Given a directed set $(I, \leq)$, let $I_\infty := I \cup \{\infty\}$, where ∞ is an additional element satisfying $i \leq \infty$ for all $i \in I$. Then one can endow I_∞ with the topology $\mathcal{O}$ defined by $G \in \mathcal{O}$ if either G is contained in I, else if there exists some $h \in I$ such that $i \in G$ for all $i \in I_\infty$ such that $i \geq h$. Thus I is dense in I_∞. Given a topological space $(X, \mathcal{O})$, $x \in X$ and a net $(x_i)_{i \in I}$ of X, one easily checks that $(x_i)_{i \in I} \to x$ if and only if the map $f : I_\infty \to X$ given by $f(i) := x_i, f(\infty) := x$ is continuous at ∞.

Definition 2.5 Given two topological spaces $(W, \mathcal{O})$, $(X', \mathcal{O}')$, a subset X of W and $w \in \text{cl}(X)$, one says that $f : X \to X'$ has a limit $\overline{x}'$ as $x \to_X w$ (i.e. $x \to w$ with $x \in X$), or that f converges to $\overline{x}'$ as $x \to_X w$ and one writes $\overline{x}' = \lim_{x \to_X w} f(x)$, if for any $V' \in \mathcal{N}(\overline{x}')$ there exists some $V \in \mathcal{N}(w)$ such that $f(V \cap X) \subset V'$.

Thus f converges to $\overline{x}'$ as $x \to_X w$ iff for any net $(x_i)_{i \in I}$ of X satisfying $(x_i)_{i \in I} \to w$ in W one has $(f(x_i))_{i \in I} \to \overline{x}'$ in X'. If $X = W$, one just writes $\overline{x}' = \lim_{x \to w} f(x)$. Thus, f is continuous at w if and only if f has the limit $f(w)$ as $x \to w$. We invite the reader to verify that the notion $(x_n) \to 0_+$ in $\mathbb{R}$ (i.e. $(x_n) \to 0$ and $x_n > 0$ for all $n \in \mathbb{N}$) corresponds to the case when $W := \mathbb{R}$, $X := \mathbb{P}$, the set of positive real numbers.

The preceding definition is a special case of a more general concept. Given another map $g : X \to Y$ with values in another topological space $(Y, \mathcal{G})$ and some $\overline{y} \in Y$, one says that f has a limit $\overline{x}'$ as $g(x) \to \overline{y}$, or that f converges to $\overline{x}'$ as $g(x) \to \overline{y}$ and one writes $\overline{x}' = \lim_{g(x) \to \overline{y}} f(x)$, if for any $V' \in \mathcal{N}(\overline{x}')$ there exists a $W \in \mathcal{N}(\overline{y})$

such that $f(x) \in V'$ for all $x \in g^{-1}(W)$. Taking for g the canonical injection of X into $(Y, \mathcal{G}) := (W, \mathcal{O})$, one recovers the preceding notion of limit.

The next result is often used for uniqueness purposes.

Proposition 2.4 *Let $(W, \mathcal{O})$, $(X', \mathcal{O}')$ be two topological spaces, let X be a dense subset of W and let $f, g : W \to X'$ be two continuous maps. If the restrictions of f and g to X coincide, then f and g coincide.*

Proof The set $Z := \{w \in W : f(w) = g(w)\}$ is a closed subset of W containing X. Since X is dense in W, we have $Z = W$ since $W = \mathrm{cl}(X) \subset Z$. □

Exercises

1. Let $f : \mathbb{R} \to \mathbb{R}$ be a map such that $f(r + s) = f(r) + f(s)$ for all $r, s \in \mathbb{R}$. Show that $f(q) = qf(1)$ for all $q \in \mathbb{Q}$. Prove that f is linear over $\mathbb{R}$ when moreover f is continuous or monotone.
2. Write the alphabet with capital letters and decide which letters are mutually homeomorphic.
3. Let $(X, \mathcal{O})$ be a Hausdorff topological space and let $f : X \to X$ be a continuous map. Show that the set $F := \{x \in X : f(x) = x\}$ is closed in X.
4. Let $(X, \mathcal{O}_X)$, $(Y, \mathcal{O}_X)$ be topological spaces and let $f : X \to Y$ be a continuous map. Show that the set $G := \{(x, y) \in X \times Y : y = f(x)\}$ is homeomrphic to X.
5. Show that a topological space $(X, \mathcal{O})$ is Hausdorff if and only if the diagonal $\Delta_X := \{(x, x') \in X^2 : x = x'\}$ is closed in X^2.
6. Let X and Y be topological spaces, let $f : X \to Y$ and let $(X_i)_{i \in I}$ be a covering of X, i.e. a family of subsets of X whose union is X. Suppose the restriction f_i of f to X_i is continuous. Show that if every X_i is open, then f is continuous.
7. With the notation of the preceding exercise, suppose that I is finite and that every X_i is closed. Show that f is continuous if every f_i is continuous. Give an example showing that the assumption that I is finite cannot be dropped. Give an example showing that the assumption that every X_i is closed cannot be dropped.
8. Let X and Y be topological spaces, let $s \in X$, and let $f : X \times Y \to \mathbb{R}$ be separately continuous (i.e. f is continuous in each of its two variables). For a subset T of Y let
$$V_s(T) := \{x \in X : f(x, s) \leq f(x, t)\ \forall t \in T\}$$
with $V_s(\varnothing) := X$ be the Voronoi cell associated with s and T as in Exercise 17 of Sect. 1.1. Show that $V_s(T)$ is closed and that $V_s(\mathrm{cl}(T)) = V_s(T)$ for each $T \in \mathcal{P}(Y)$.

2.2.2 Connectedness

It is sometimes useful to know that a space is not made of several pieces: this can be used to globalize some properties or for uniqueness results. For example, if X is an open subset of $\mathbb{R}$ and if $f : X \to \mathbb{R}$ is a differentiable function whose derivative is 0 we cannot conclude that f is constant because X can be the union of disjoint open intervals. We must give a precise definition.

Definition 2.6 A topological space $(X, \mathcal{O})$ is said to be *connected* if $\varnothing$ and X are the only subsets of X that are both open and closed.

The space $(X, \mathcal{O})$ is *arcwise connected* if any two points x_0, x_1 of X can be joined by a continuous arc, i.e. if for any x_0, x_1 there is a continuous map $c : [0,1] \to X$ such that $c(0) = x_0$, $c(1) = x_1$.

Connectedness is a more important notion than arcwise connectedness, but the latter is more intuitive and the proofs of several propositions below for arcwise connectedness are very easy. The reader is invited to verify this assertion. The definition of connectedness can be rephrased by saying that X is connected if any partition of X into two open subsets is improper, i.e. if one of the subsets is empty and the other one is the whole space, a *partition* of X being a covering by disjoint subsets. The reader must be aware that there are topological spaces that are not connected; in such a space, showing that a subset is not closed does not prove that it is open (a frequent mistake). For instance, if X is the union of two open disjoint intervals A, B of $\mathbb{R}$, A and B are also closed in X since their complements are open.

Example A bounded, closed interval $X := [a, b]$ of $\mathbb{R}$ is connected (with respect to the induced topology). To prove this, let us consider a nonempty subset C of X that is both open and closed and let us show that $C = X$. We may suppose $a \in C$ (otherwise we consider $C' := X \setminus C$). Let $s := \sup T$ with $T := \{t \in X : [a, t] \subset C\}$. Since C is open in X we have $s > a$. Moreover, $s \in T$ since there exists a sequence (t_n) of $T \subset C$ converging to s so that $[a, s[= \cup_n [a, t_n[\subset C$ and since C is closed we have $[a, s] \subset C$. Let us show that assuming $s < b$ leads to a contradiction. Since C is open in X and $s \in C$, we can find $\varepsilon > 0$ such that $s + \varepsilon \leq b$ and $[s, s + \varepsilon] \subset C$. Then since $s \in T$ we get $[a, s + \varepsilon] = [a, s] \cup [s, s + \varepsilon] \subset C$; this means that $s + \varepsilon \in T$, contradicting the definition of s.

It follows from Proposition 2.7 below that any interval of $\mathbb{R}$ is connected. □

The use of connectedness for existence results is illustrated by the following properties. The second one is often called the Intermediate Value Theorem.

Proposition 2.5 (Customs Lemma) *Let C be a connected subset of a topological space $(X, \mathcal{O})$ and let S be a subset of X. If $C \cap \mathrm{int}(S)$ and $C \cap (X \setminus \mathrm{cl}(S))$ are nonempty, then C contains some point of the boundary of S (Fig. 2.2).*

Fig. 2.2 The customs lemma

Fig. 2.3 The Daisy property

Proof If, on the contrary, C does not meet the boundary of S then C is the union of the two disjoint sets $C \cap \text{int}(S)$ and $C \cap (X \backslash \text{cl}(S))$, contradicting the connectedness of C. □

Proposition 2.6 (Bolzano) *Let $(X, \mathcal{O})$ be a connected topological space and let $f : X \to \mathbb{R}$ be a continuous map. Let $a < b < c$ in $\mathbb{R}$ be such that $a \in f(X)$, $c \in f(X)$. Then there exists some $x \in X$ such that $f(x) = b$.*

Proof If $f^{-1}(b) = \varnothing$, the sets $f^{-1}(]-\infty, b[)$ and $f^{-1}(]b, +\infty[)$ form a partition of X into two open subsets, an impossibility if X is connected. □

The following property can be used as a convenient criterion (Fig. 2.3).

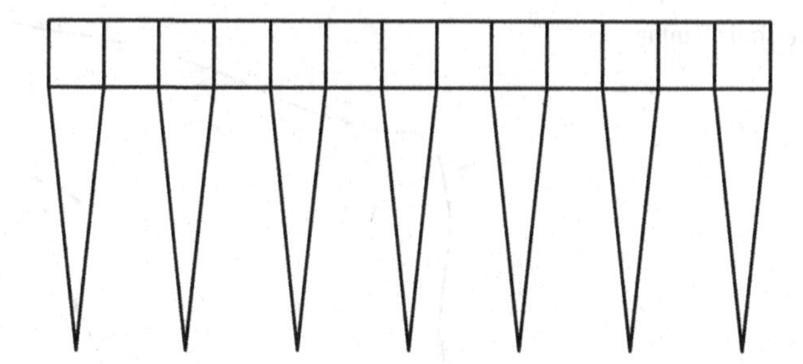

Fig. 2.4 The Comb or Rake property

Proposition 2.7 (The Daisy Property) *Let $(X, \mathcal{O})$ be a topological space that is the union of a family $(X_i)_{i \in I}$ of connected subspaces. If $\cap_{i \in I} X_i$ is nonempty, then X is connected.*

Proof Let $\bar{x} \in \cap_{i \in I} X_i$. Let C be a subset of X that is closed and open. Changing C into $X \backslash C$ we may suppose $\bar{x} \in C$. Then $C \cap X_i$ is open and closed in X_i and contains $\bar{x}$. Since X_i is connected, we have $C \cap X_i = X_i$. Thus $C \supseteq X_i$ for all $i \in I$, hence $C = X$. □

A slight refinement can be given (Fig. 2.4).

Corollary 2.3 (The Comb or Rake Property) *Let $(X, \mathcal{O})$ be a topological space that is the union of a family $(X_i)_{i \in I}$ of connected subspaces. If there is some nonempty connected subspace Y of X such that $X_i \cap Y \neq \varnothing$ for all $i \in I$, X is connected.*

Proof Set $Y_i := X_i \cup Y$. Then, by the preceding proposition, Y_i is connected. Since $X = \cup_{i \in I} Y_i$ and $\cap_{i \in I} Y_i$ contains $Y \neq \varnothing$, then X is connected. □

Proposition 2.8 *A product of two connected spaces is connected.*

Proof Let $Z := X \times Y$, the spaces X, Y being connected and nonempty. Let $x_0 \in X$. Then $Y_0 := \{x_0\} \times Y$ is homeomorphic to Y, hence is connected and for all $y \in Y$ the subspace $X_y := X \times \{y\}$ is connected and meets Y_0 as $(x_0, y) \in X_y \cap Y_0$. Since $Z = \cup_{y \in Y} X_y$, Z is connected by the rake property (2.3).

Now let us consider the general case of an arbitrary product $X := \prod_{i \in I} X_i$ of connected spaces. Given two nonempty open subsets, A, B of X satisfying $A \cup B = X$, the construction of the product topology ensures that there exist a finite subset J of I and open subsets A_J, B_J of $X_J := \prod_{j \in J} X_j$ such that $A = A_J \times X_{I \backslash J}$ and $B = B_J \times X_{I \backslash J}$. Then $A_J \cup B_J = X_J$. Since X_J is connected in view of the first part of the proof and of an induction, we have $A_J \cap B_J \neq \varnothing$. Thus $A \cap B \neq \varnothing$ and X is connected. □

The preceding proof used the obvious fact that when two topological spaces are homeomorphic, both are connected when one of them is connected. This fact is also a consequence in a more general property that is obviously valid for arcwise connectedness.

Proposition 2.9 *Let $(X, \mathcal{O}_X)$, $(Y, \mathcal{O}_Y)$ be two topological spaces and let $f : X \to Y$ be continuous. If X is connected, then $f(X)$ is connected with respect to the induced topology.*

Proof There is no loss of generality in assuming $f(X) = Y$. Then, if $G \in \mathcal{O}_Y$ is nonempty, open and closed, then so is $f^{-1}(G)$. Since X is connected we have $f^{-1}(G) = X$, hence $G = f(f^{-1}(G)) = f(X) = Y$. □

Corollary 2.4 *An arcwise connected space is connected.*

Proof Let $(X, \mathcal{O})$ be an arcwise connected space and let $x_0 \in X$. By definition, for all $x \in X$ there exists a continuous map $f_x : [0, 1] \to X$ such that $f_x(0) = x_0$ and $f_x(1) = x_1$. Since $C_x := f_x([0, 1])$ is connected and since $X = \cup_{x \in X} C_x$ with $x_0 \in C_x$ for all x, X is connected. □

Given a topological space $(X, \mathcal{O})$ and $x \in X$, Proposition 2.7 implies that the union $C(x)$ of all connected subsets of X containing x is connected. It is clearly the largest connected subset containing x. Moreover, X can be split into a partition of connected subsets called the *connected components* of X by taking those $C(x)$ that are disjoint (note that if $C(x) \cap C(x') \neq \emptyset$ then $C(x) = C(x')$). It follows from Exercise 1 below that the connected components of X are closed subsets.

A topological space $(X, \mathcal{O})$ is said to be *locally connected* if every point of X has a base of neighborhoods formed by connected sets. Clearly $\mathbb{R}$ is locally connected, but $\mathbb{Q}$ and $\mathbb{R} \backslash \mathbb{Q}$ are not locally connected. It is easy to show that a space is locally connected if and only if the connected components of any open subset are open.

Exercises

1. Let A and B be two subsets of a topological space $(X, \mathcal{O})$ such that $A \subset B \subset \mathrm{cl}(A)$. Show that B is connected whenever A is connected. Deduce from this that $\mathrm{cl}(A)$ is connected and that any connected component of X is closed.
2. Let $(X, \mathcal{O})$ be a topological space such that X is the union of a sequence $(X_n)_n$ of connected subsets satisfying $X_n \cap X_{n+1} \neq \emptyset$ for all $n \in \mathbb{N}$. Show that X is connected.
3. Show that the connected subsets of $\mathbb{R}$ are the intervals.
4. Prove that any open subset of $\mathbb{R}$ is the union of a finite or countable family of disjoint open intervals.
5. Verify that a topological space $(X, \mathcal{O})$ is connected if and only if any continuous map $f : X \to \mathbb{Z}$ is constant.
6. Let A and B be two nonempty closed subsets of a topological space. Show that if $A \cap B$ and $A \cup B$ are connected, then A and B are connected. Show by an example of two subsets of $\mathbb{R}$ that the assumption that A and B are closed cannot be omitted.
7. Let $G := \{(r, \sin(1/r)) : r \in]0, 1]\}$ and let X be its closure in $\mathbb{R}^2$. Prove that X is connected but that X is not arcwise connected and not locally connected.

2.2.3 Lower Semicontinuity

In order to deal with minimization problems, one may use a one-sided weakening of continuity when a continuity assumption is not realistic. A precise definition is as follows.

Definition 2.7 A function $f : X \to \overline{\mathbb{R}} := \mathbb{R} \cup \{-\infty, +\infty\}$ on a topological space X is said to be *lower semicontinuous* (l.s.c.) at some $\overline{x} \in X$ if for every real number $r < f(\overline{x})$ there exists some member V of the family $\mathcal{N}(\overline{x})$ of neighborhoods of $\overline{x}$ such that $r < f(v)$ for all $v \in V$. A function f is upper semicontinuous (u.s.c.) at $\overline{x}$ whenever $-f$ is l.s.c. at $\overline{x}$.

The function f is said to be lower semicontinuous (l.s.c.) on some subset S of X if f is lower semicontinuous at each point of S.

We observe f is automatically l.s.c. at $\overline{x}$ when $f(\overline{x}) = -\infty$; when $f(\overline{x}) = +\infty$ the lower semicontinuity of f means that the values of f can be as large as required provided one remains in a small enough neighborhood of $\overline{x}$. When $f(\overline{x})$ is finite the definition amounts to assigning to any $\varepsilon > 0$ a neighborhood V_ε of $\overline{x}$ such that $f(v) > f(\overline{x}) - \varepsilon$ for each $v \in V_\varepsilon$. Thus, lower semicontinuity allows sudden upward changes of the value of f but excludes sudden downward changes. Obviously, f is continuous at $\overline{x}$ iff it is both l.s.c. and u.s.c. at $\overline{x}$.

Example The function $f : \mathbb{R} \to \mathbb{R}$ given by $f(x) = 1$ for $x < 0$, $f(x) = 0$ for $x \in \mathbb{R}_+$ is l.s.c. but not continuous at 0.

Example The *indicator function* ι_A of a subset A of X, defined by $\iota_A(x) = 0$ for $x \in A$, $\iota_A(x) = +\infty$ for $x \in X \backslash A$ is l.s.c. if and only if, A is closed, as is easily seen. Such a function is of great use in optimization theory and nonsmooth analysis.

Example The *characteristic function* 1_A of a subset A of X, defined by $1_A(x) = 1$ for $x \in A$, $1_A(x) = 0$ for $x \in X \backslash A$ is l.s.c. if and only if, A is open. Such a function is of primary importance in integration theory.

Example *The Length Function* Given a metric space (M, d), let $X := C(T, M)$ be the space of continuous maps (curves or arcs of M) from $T := [0, 1]$ to M. Given a subdivision $\sigma := \{t_0 = 0 < t_1 < \ldots < t_n = 1\}$ of T, let us set for $x \in X$

$$\ell_\sigma(x) := \sum_{i=1}^{n} d(x(t_{i-1}), x(t_i)),$$

and let $\ell(x)$ be the supremum of $\ell_\sigma(x)$ as σ varies in the set of finite subdivisions of T. The properties devised below yield that ℓ is l.s.c. when X is endowed with the topology of uniform convergence (and even when X is endowed with the topology of pointwise convergence). However ℓ is not continuous: one can increase ℓ by following a nearby curve which makes many small changes (a fact any dog knows, when tied with a lease). Details are given in Exercise 3 below.

The following characterizations are global ones.

Proposition 2.10 *For a function $f : X \to \overline{\mathbb{R}}$ the following assertions are equivalent:*

(a) f is l.s.c.;
(b) the epigraph $E := \{(x,r) \in X \times \mathbb{R} : r \geq f(x)\}$ of f is closed;
(c) for each $s \in \mathbb{R}$ the sublevel set $S(s) := \{x \in X : f(x) \leq s\}$ is closed.

Proof (a)⇒(b) It suffices to prove that $(X \times \mathbb{R}) \setminus E$ is open when f is l.s.c. Given $(\overline{x}, \overline{r}) \in (X \times \mathbb{R}) \setminus E$, i.e. such that $\overline{r} < f(\overline{x})$, for any $r \in]\overline{r}, f(\overline{x})[$ one can find a neighborhood V of $\overline{x}$ such that $r < f(v)$ for all $v \in V$. Then $V \times] - \infty, r[$ is a neighborhood of $(\overline{x}, \overline{r})$ in $X \times \mathbb{R}$ which does not meet E. Hence $(X \times \mathbb{R}) \setminus E$ is open.

(b)⇒(c) It suffices to observe that for each $s \in \mathbb{R}$ one has $S(s) \times \{s\} = E \cap (X \times \{s\})$.

(c)⇒(a) Given $\overline{x} \in X$ and $r \in \mathbb{R}$ such that $r < f(\overline{x})$ one has $\overline{x} \in X \setminus S(r)$ which is open and for all $v \in V := X \setminus S(r)$ one has $r < f(v)$. □

The notion of lower semicontinuity is intimately tied to the concept of lower limit (denoted by liminf), which is a one-sided concept of limit which can be used even when there is no limit. In the following definition we assume X is a subspace of a larger space W, $w \in \mathrm{cl}(X)$ (a situation which will be encountered later, for instance when $X = \mathbb{P} :=]0, \infty[$, $W = \mathbb{R}$ and $w = 0$) and we denote by $\mathcal{N}(w)$ the family of neighborhoods of the point w in W.

Definition 2.8 Given a topological space W, a subspace X and a point w in the closure of X, the lower limit of a function $f : X \to \overline{\mathbb{R}}$ at w is the extended real number

$$\liminf_{x \to_X w} f(x) := \sup_{V \in \mathcal{N}(w)} \inf_{v \in V \cap X} f(v).$$

Setting $m_V := \inf f(V \cap X)$, the supremum over $V \in \mathcal{N}(w)$ of the family $(m_V)_V$ can also be considered as the limit of the net $(m_V)_V$; this explains the terminology. One can show that $\sup_V m_V$ is also the least cluster point of $f(x)$ as $x \to w$ in X. When W is metrizable, one can replace the family $\mathcal{N}(w)$ by the family of balls centered at w, so that $\liminf_{x \to w} f(x) = \sup_{r>0} m_r$, with $m_r := \inf f(B(w,r) \cap X)$, is the limit of a sequence.

Exercise Deduce from the preceding that the lower limit of a function $f : X \to \overline{\mathbb{R}}$ at w is the least of the limits of the converging nets $(f(x_i))_{i \in I}$ where $(x_i)_{i \in I}$ is a net in X converging to w.

Lower semicontinuity can be characterized using the notion of lower limit (here $W = X$).

Lemma 2.1 *A function $f : X \to \overline{\mathbb{R}}$ on a topological space X is l.s.c. at some $w \in X$ iff one has $f(w) \leq \liminf_{x \to w} f(x)$.*

Proof Clearly, when f is l.s.c. at w, one has $f(w) \leq \liminf_{x \to w} f(x)$. Conversely, when this inequality holds, for any $r < f(w)$, by the definition of the supremum over $\mathcal{N}(w)$, one can find $V \in \mathcal{N}(w)$ such that $r < \inf_{v \in V} f(v)$, so that f is l.s.c. at w. □

One can also use nets for such a characterization.

Lemma 2.2 *A function $f : X \to \overline{\mathbb{R}}$ on a topological space X is l.s.c. at some $\overline{x} \in X$ iff for any net $(x_i)_{i \in I}$ in X converging to $\overline{x}$ one has $f(\overline{x}) \leq \liminf_{i \in I} f(x_i)$. When $\overline{x}$ has a countable base of neighborhoods, one can replace nets by sequences in that characterization.*

Proof The condition is necessary: if a net $(x_i)_{i \in I}$ in X converges to $\overline{x}$, for any $r < f(\overline{x})$ there exists some $V \in \mathcal{N}(\overline{x})$ such that $f(v) > r$ for all $v \in V$, and there exists some $h \in I$ such that $x_i \in V$ for $i \geq h$, so that $\inf_{i \geq h} f(x_i) \geq r$, hence $\liminf_{i \in I} f(x_i) \geq \inf_{i \geq h} f(x_i) \geq r$.

Conversely, suppose f is not l.s.c. at $\overline{x}$ and let $(V_i)_{i \in I}$ be a base of neighborhoods of $\overline{x}$: there exists some $r < f(\overline{x})$ such that for any $i \in I$ there exists some $x_i \in V_i$ such that $f(x_i) < r$. Ordering I by $j \geq i$ if $V_j \subset V_i$, we get a net $(x_i)_{i \in I}$ which converges to $\overline{x}$ and is such that $\liminf_{i \in I} f(x_i) \leq r$.

The second assertion follows from the fact that when $\overline{x}$ has a countable base of neighborhoods, one can take a decreasing sequence of neighborhoods for a base. □

Let us give some useful calculus rules (with the convention $0r = 0$ for all $r \in \overline{\mathbb{R}}$).

Exercise For any $\alpha \in \mathbb{R}_+$ and $f : X \to \overline{\mathbb{R}}$ one has $\liminf_{x \to \overline{x}} \alpha f(x) = \alpha \liminf_{x \to \overline{x}} f(x)$.

Exercise If $f, g : X \to \overline{\mathbb{R}}$ are such that $\{\liminf_{x \to \overline{x}} f(x), \liminf_{x \to \overline{x}} g(x)\} \neq \{-\infty, +\infty\}$, then

$$\liminf_{x \to \overline{x}} (f + g)(x) \geq \liminf_{x \to \overline{x}} f(x) + \liminf_{x \to \overline{x}} g(x).$$

The family of lower semicontinuous functions enjoys stability properties.

Proposition 2.11 *If $(f_i)_{i \in I}$ is a family of functions which are l.s.c. at $\overline{x}$, then the function $f := \sup_{i \in I} f_i$ is l.s.c. at $\overline{x}$.*

For any $\alpha \in \mathbb{R}_+$ and f, g which are l.s.c. at $\overline{x}$, the functions $\inf(f, g)$ and αf are l.s.c. at $\overline{x}$; the same is true for $f + g$ provided $\{f(\overline{x}), g(\overline{x})\} \neq \{-\infty, +\infty\}$. If moreover f and g are nonnegative, then fg is l.s.c. at $\overline{x}$.

If $f : X \to \mathbb{R}$ is finite and l.s.c. at $\overline{x}$ and if $g : \overline{\mathbb{R}} \to \overline{\mathbb{R}}$ is nondecreasing and l.s.c. at $f(\overline{x})$, then $g \circ f$ is l.s.c. at $\overline{x}$.

One may observe that in the first assertion one cannot replace lower semicontinuity by continuity, as shown by the above example of arc length.

Proof Let $r \in \mathbb{R}$ be such that $r < f(\overline{x})$. There exists some $j \in I$ such that $r < f_j(\overline{x})$, hence one can find some $V \in \mathcal{N}(\overline{x})$ such that $r < f_j(v) \leq f(v)$ for all $v \in V$. The proofs of the other assertions are also straightforward or follow from the preceding lemma. □

Proposition 2.12 *For any function $f : X \to \overline{\mathbb{R}}$ on a topological space X, the family of l.s.c. functions majorized by f has a greatest element $\bar{f}$ called the lower semicontinuous hull of f (in short, the l.s.c. hull of f). Its epigraph is the intersection with $X \times \mathbb{R}$ of the closure* $\mathrm{cl}(\mathrm{epi} f)$ *of the epigraph of f in $X \times \overline{\mathbb{R}}$. The function $\bar{f}$ is given by*

$$\bar{f}(x) = \liminf_{u \to x} f(u) = \sup_{U \in \mathcal{N}(x)} \inf_{u \in U} f(u).$$

Proof The first assertion is a direct consequence in Proposition 2.11. The second assertion easily stems from the fact that setting $g(x) = \min\{r : (x, r) \in \mathrm{cl}(\mathrm{epi} f)\}$ one defines a lower semicontinuous function that is the greatest lower semicontinuous function majorized by f. The proof of the explicit expression of $\bar{f}$ is left as an exercise. □

The treatment of the following example requires the knowledge of some material in integration theory to be found in Sects. 7.4 and 8.5. Its importance justifies its presence here.

Example Let E be a Banach space, let $(S, \mathcal{S}, \mu)$ be a measure space and let $L : E \times S \to \mathbb{R}$ be a measurable function such that for a null set N in S the function $L_s : e \mapsto L(e, s)$ is lower semicontinuous whenever $s \in S \backslash N$. Suppose that for some $p \in [1, \infty[$ and some $a \in \mathbb{R}$, $b \in \mathcal{L}_1(S)$ one has

$$L(e, s) \geq b(s) - a \|e\|^p \qquad \forall (e, s) \in E \times (S \backslash N).$$

Then the function $j : L_p(S, E) \to \overline{\mathbb{R}}$ given by $j(x) := \int_S L(x(s), s) d\mu(s)$ is lower semicontinuous.

We first prove this assertion in the case $a = 0$, $b = 0$. Let (x_n) be a sequence in $L_p(S, E)$ converging to some $x \in L_p(S, E)$. Taking a subsequence if necessary we may suppose $(j(x_n))$ converges to $\liminf_n j(x_n)$. Taking a further subsequence we may assume that $(x_n(s)) \to x(s)$ a.e. in S. Our lower semicontinuity assumption on L ensures that

$$\liminf_n L(x_n(s), s) \geq L(x(s), s) \qquad \text{a.e.},$$

so that, by Fatou's lemma, we obtain

$$\liminf_n \int_S L(x_n(s), s) d\mu(s) \geq \int_S \liminf_n L(x_n(s), s) d\mu(s) \geq \int_S L(x(s), s) d\mu(s),$$

or $\liminf_n j(x_n) \geq j(x)$, the required lower semicontinuity property.

In the general case we set $M(e, s) := L(e, s) - b(s) + a \|e\|^p \in \mathbb{R}_+$ and $m(x) := \int_S M(x(s), s) d\mu(s)$ for $x \in L_p(S, E)$. Given $x \in L_p(S, E)$ and a sequence $(x_n) \to x$ in

$L_p(S,E)$, we have

$$\liminf_n j(x_n) = \liminf_n (m(x_n) - a\int_S \|x_n(s)\|^p \, d\mu(s) + \int_S b(s)d\mu(s)$$
$$\geq m(x) - a\,\|x\|_p^p + \int_S b(s)d\mu(s) = j(x).$$

That shows that j is lower semicontinuous. □

Exercises

1. Using the relation $E = \bigcap_{i\in I} E_i$, where E_i is the epigraph of a function f_i and E is the epigraph of $f := \sup_{i\in I} f_i$, show that f is l.s.c. on X when each f_i is l.s.c. on X. Use a similar argument with sublevel sets.
2. Suppose X is a metric space. Show that f is l.s.c. at $\overline{x}$ iff for any sequence (x_n) converging to $\overline{x}$ one has $f(\overline{x}) \leq \liminf_n f(x_n)$.
3. Let (M,d) be a metric space and let $X := C(T,M)$, where $T := [0,1]$. Given some $x \in X$ and some element s of the set S of nondecreasing sequences $s := (s_n)_{n\geq 0}$ satisfying $s_0 = 0$, $s_n = 1$ for n large, let

$$\ell_s(x) := \sum_{n\geq 0} d(x(s_n), x(s_{n+1}))$$

(observe that the preceding sum contains only a finite number of non-zero terms). Define the length of a curve $x \in X$ by $\ell(x) := \sup_{s\in S} \ell_s(x)$. Show that $\ell_s : X \to \mathbb{R}$ is continuous when X is endowed with the metric of uniform convergence (and even when X is provided with the topology of pointwise convergence). Conclude that the length ℓ is a l.s.c. function on X.

   Show that ℓ is not continuous by taking $M := \mathbb{R}^2$, $\overline{x}$ given by $\overline{x}(t) := (t,0)$ and by showing that there is some $x_n \in X$ such that $d(x_n, \overline{x}) \to 0$ and $\ell(x_n) \geq \sqrt{2}$ [Hint: for $n > 0$ define $x_n(t) = t - \frac{2k}{2n}$ for $t \in [\frac{2k}{2n}, \frac{2k+1}{2n}]$, $k \leq n$ and $x_n(t) = -t + \frac{2k+2}{2n}$ for $t \in [\frac{2k+1}{2n}, \frac{2k+2}{2n}]$, $k \leq n-1$].
4. Show that the infimum of an infinite family of l.s.c. functions is not necessarily l.s.c. [Hint: observe that any function f on X is the infimum of the family $(f_a)_{a\in X}$ of functions given by $f_a(x) = f(a)$ if $x = a$, $+\infty$ else].
5. Let $f : X \to \mathbb{R}_\infty$ be a l.s.c. function on a topological space X and let A be a nonempty subset of X. Show that $\inf f(A) = \inf f(\mathrm{cl}A)$. Can one replace inf by sup?
6. Show that the supremum of a family of continuous functions is not necessarily continuous.
7. (**Ritz's method**) Let $f : X \to \mathbb{R}$ be an upper semicontinuous function on a topological space X and let $(X_n)_{n\geq 0}$ be a sequence in subspaces such that for

all $x \in X$ there exists a sequence $(x_n) \to x$ satisfying $x_n \in X_n$ for all n. Let $m := \inf f(X)$, $m_n := \inf f(X_n)$. Show that $m = \lim_n m_n$.

2.2.4 *Compactness*

The existence of limits being a frequent aim, the following definition is of interest.

Definition 2.9 A topological space $(X, \mathcal{O})$ is said to be *compact* if it is Hausdorff and if any net in X has a convergent subnet.

Equivalently, since a cluster point of net is the limit of some subnet, a topological space $(X, \mathcal{O})$ is compact if every net in X has a cluster point. Moreover, if a net $(x_i)_{i\in I}$ of a compact space $(X, \mathcal{O})$ has at most one cluster point $\overline{x}$, then it converges to $\overline{x}$: if this were not the case, one could find an open neighborhood V of $\overline{x}$ and a cofinal subset J of I such that $x_j \notin V$ for all $j \in J$ and then $(x_j)_{j\in J}$ would have a convergent subnet whose limit would be in $X \backslash V$, contradicting the uniqueness of the cluster point of $(x_i)_{i\in I}$.

The property of the definition can be characterized in different ways. The usual one deals with open coverings of X, i.e. families $(O_i)_{i\in I}$ of open subsets whose union is X. Another one deals with families $(C_i)_{i\in I}$ satisfying the *finite intersection property*, i.e. such that for any finite subset J of I one has $\cap_{j\in J} C_j \neq \varnothing$.

Theorem 2.1 *A Hausdorff topological space $(X, \mathcal{O})$ is compact if and only if every open covering $(O_i)_{i\in I}$ of X has a finite covering, if and only if any family $(C_i)_{i\in I}$ of closed subsets with the finite intersection property has a nonempty intersection.*

Proof Setting $O_i := X \backslash C_i$ (and conversely $C_i := X \backslash O_i$) one sees that the last two properties are equivalent, since for all $J \subset I$ the family $(O_j)_{j\in J}$ is a covering of X if and only if $\cap_{j\in J} C_j$ is empty. Let us assume the last property is satisfied and let $(x_i)_{i\in I}$ be a net in X. Setting $C_i := \mathrm{cl}\{x_j : j \geq i\}$, we see that for every finite subset J of I and for any $k \in I$ such that $k \geq j$ for all $j \in J$ (such a k exists since I is directed) one has $\cap_{j\in J} C_j \supset C_k \neq \varnothing$, so that $\cap_{i\in I} C_i$ is nonempty. Since by Proposition 2.3 $\cap_{i\in I} C_i$ is the set of cluster points of $(x_i)_{i\in I}$, we get a cluster point of $(x_i)_{i\in I}$.

Now let $(C_i)_{i\in I}$ be a family of closed subsets satisfying the finite intersection property. Let $\mathcal{J}$ be the family of finite subsets of I and for $J \in \mathcal{J}$ let $x_J \in C_J := \cap_{j\in J} C_j$. Since $\mathcal{J}$ is directed with respect to inclusion, we get a net $(x_J)_{J\in\mathcal{J}}$ in X. Let us show that if $\cap_{i\in I} C_i = \varnothing$ the net $(x_J)_{J\in\mathcal{J}}$ cannot have a cluster point. Suppose $\overline{x}$ is a cluster point of $(x_J)_{J\in\mathcal{J}}$. Since $\overline{x} \notin \cap_{i\in I} C_i$ there exists some $k \in I$ such that $\overline{x} \in V := X \backslash C_k$. Then, for $J \in \mathcal{J}$ satisfying $J \supset K := \{k\}$ we cannot have $x_J \in V$ since $x_J \in C_J \subset C_k$. Thus $\overline{x}$ cannot be a cluster point of $(x_J)_{J\in\mathcal{J}}$. □

Corollary 2.5 *Let X be a subset of a Hausdorff topological space $(W, \mathcal{O}_W)$ endowed with the induced topology $\mathcal{O}_X := \{O \cap X : O \in \mathcal{O}_W\}$. Then $(X, \mathcal{O}_X)$ is compact if and only if every covering $(O_i)_{i\in I}$ of X by members of $\mathcal{O}_W$ has a finite subcovering.*

Here a family $(W_i)_{i\in I}$ of subsets of W is called a *covering* of X if $X \subset \cup_{i\in I} W_i$ and a subcovering is a subfamily $(W_j)_{j\in J}$ of $(W_i)_{i\in I}$ that is still a covering of X.

Proof If $(X, \mathcal{O}_X)$ is compact, for every covering $(O_{i\in I})$ of X by open subsets of W, the family $(O_i \cap X)_{i\in I}$ being a covering of X for the induced topology $\mathcal{O}_X$, one can find a finite subset J of I such that $(O_j \cap X)_{j\in J}$ is a covering of X. Then $(O_j)_{j\in J}$ is a finite subcovering of X.

Conversely, suppose every covering $(O_i)_{i\in I}$ of X by members of $\mathcal{O}_W$ has a finite subcovering $(O_j)_{j\in J}$. Then, if $(G_i)_{i\in I}$ is an open covering of $(X, \mathcal{O}_X)$, picking $O_i \in \mathcal{O}_W$ such that $G_i = O_i \cap X$, we can find a finite subset J of I such that $X \subset \cup_{j\in J} O_j$ and then $(G_j)_{j\in J}$ is a finite subcovering of $(G_i)_{i\in I}$. Thus $(X, \mathcal{O}_X)$ is compact. □

Example A discrete space, i.e. a set endowed with the discrete topology, is compact if and only if it is finite.

Example If (x_n) is a convergent sequence in a Hausdorff topological space $(W, \mathcal{O})$ and if $\overline{x} := \lim_n x_n$, then the set $X := \{x_n : n \in \mathbb{N}\} \cup \{\overline{x}\}$ is compact with respect to the induced topology. In fact, given a covering $(O_i)_{i\in I}$ of X by open subsets of W we can find some $k \in I$ such that $\overline{x} \in O_k$. Since $(x_n) \to \overline{x}$ there exists some $m \in \mathbb{N}$ such that $x_n \in O_k$ for $n > m$. Then, taking for all $j \in \mathbb{N}, j \leq m$ some $i(j) \in I$ such that $x_j \in O_{i(j)}$, we obtain a finite subcover of X by taking the family $\{O_{i(j)} : 0 \leq j \leq m\} \cup \{O_k\}$. □

Another important example is given by the following theorem.

Theorem 2.2 (Heine-Borel-Lebesgue) *Every closed bounded interval of* $\mathbb{R}$ *is compact.*

Proof Let $X := [a, b]$ with $a \leq b$ in $\mathbb{R}$ and let $(O_i)_{i\in I}$ be an open covering of X by open subsets of $\mathbb{R}$. Let A be the set of $x \in X$ such that $[a, x]$ is covered by a finite number of members of $(O_i)_{\in I}$. Then A is nonempty (since $a \in A$), hence it has a least upper bound $c \leq b$. Let $h \in I$ be such that $c \in O_h$. Suppose $c < b$. Given $\varepsilon > 0$ such that $c + \varepsilon \leq b$ and $[c - \varepsilon, c + \varepsilon] \subset O_h$, one can find some $x \in A$ such that $c - \varepsilon < x \leq c$. By definition of A one can find a finite subset J of I such that $[a, x] \subset \cup_{j\in J} O_j$. Then $c \in A$ and $[a, c + \varepsilon] \subset \cup_{k\in K} O_k$ for $K := J \cup \{h\}$, so that $c + \varepsilon \in A$, a contradiction. Thus $c = b$ and $b \in A$: $[a, b]$ can be covered by a finite subfamily of $(O_i)_{i\in I}$. □

Corollary 2.6 *The space* $\overline{\mathbb{R}} := \mathbb{R} \cup \{-\infty, +\infty\}$ *is compact.*

Proof This follows from the fact that there exists a homeomorphism h from $\overline{\mathbb{R}}$ onto $[-1, 1]$, for instance $h : r \mapsto r/(1 + |r|)$ for $r \in \mathbb{R}$, $h(-\infty) := -1$, $h(+\infty) := 1$. □

Let us give some permanence properties.

Proposition 2.13 *Let* X *be a closed subset of a compact topological space* $(W, \mathcal{O}_W)$. *Then, denoting by* $\mathcal{O}_X := \{O \cap X : O \in \mathcal{O}_W\}$ *the induced topology,* $(X, \mathcal{O}_X)$ *is compact.*

Proof If $(x_i)_{i\in I}$ is a net in X, it has a subnet $(x_j)_{j\in J}$ that converges in W. But since X is closed in W, the limit $\overline{x}$ of $(x_j)_{j\in J}$ belongs to X and $(x_j)_{j\in J}$ converges in $(X, \mathcal{O}_X)$. □

Proposition 2.14 *Let X be a subset of a Hausdorff topological space $(W, \mathcal{O}_W)$. If X is compact with respect to the induced topology $\mathcal{O}_X$, then X is closed in $(W, \mathcal{O}_W)$.*

Proof Let $\overline{w}$ be an element of the closure of X. Then, there exists a net $(x_i)_{i\in i}$ of X that converges to $\overline{w}$. But since $(X, \mathcal{O}_X)$ is compact, $(x_i)_{i\in I}$ has a cluster point $\overline{x} \in X$. Then Proposition 2.1 ensures that $\overline{x} = \overline{w}$, so that $\overline{w} \in X$. □

Corollary 2.7 *The compact subsets of $\mathbb{R}$ (with respect to the induced topology) are the closed bounded subsets of $\mathbb{R}$.*

Proof Since $\mathbb{R}$ is Hausdorff, the preceding proposition shows that a compact subset X of $\mathbb{R}$ is closed. It is bounded since otherwise we could find a sequence (x_n) in X satisfying $|x_n| > n$; such a sequence cannot have a cluster point.

Conversely, let X be a closed bounded subset of $\mathbb{R}$. Then there exists a closed bounded interval $W := [a, b]$ containing X. Since W is compact and X is closed in W, X is compact by Proposition 2.13. □

Exercise Show that in a Hausdorff topological space the union of a finite family of compact subsets is compact and the intersection of a family of compact subsets is compact.

Theorem 2.3 *Let $(X, \mathcal{O}_X)$ and $(Y, \mathcal{O}_Y)$ be two Hausdorff topological spaces, and let $f : X \to Y$ be a continuous map. If X is compact, then $Z := f(X)$ is compact.*

Proof Let $(z_i)_{i\in I}$ be a net in Z. If $x_i \in X$ is such that $z_i = f(x_i)$, then $(x_i)_{i\in I}$ has a convergent subnet $(x_j)_{j\in J}$. Then $(z_j)_{j\in J}$ is a subnet of $(z_i)_{i\in I}$ that converges to $f(\lim_j x_j)$. □

Corollary 2.8 *Let $(X, \mathcal{O}_X)$ and $(Y, \mathcal{O}_Y)$ be two Hausdorff topological spaces, and let $f : X \to Y$ be a continuous injective map. If X is compact, then f is a homeomorphism from X onto $f(X)$.*

Proof If C is a closed subset of X, C is compact and $f(C)$ is compact too, so that $f(C)$ is closed in $f(X)$. □

The proof of the following theorem requires Zorn's Lemma when the product has an infinite number of factors.

Theorem 2.4 (Tykhonov) *The product of a family of compact topological spaces is compact.*

Proof We admit the result when the product has an infinite number of factors. The case of a finite number of factors is reduced to the case of two factors by an induction. Let $Z := X \times Y$ be the product of two compact spaces and let $(z_i)_{i\in I} := ((x_i, y_i))_{i\in I}$ be a net in Z. A subnet $(x_j)_{j\in J}$ of $(x_i)_{i\in I}$ converges. In turn, a subnet $(y_k)_{k\in K}$ of $(y_j)_{j\in J}$ converges. Then $(z_k)_{k\in K}$ is a convergent subnet of $(z_i)_{i\in I}$. □

Corollary 2.9 *The compact subsets of* $\mathbb{R}^d$ *(with respect to the induced topology) are the closed bounded subsets of* $\mathbb{R}^d$.

Here a subset of $\mathbb{R}^d$ is said to be *bounded* if its projections are bounded.

Proof If S is a compact subset of $\mathbb{R}^d$, it is closed and its projections are compact, hence bounded. Conversely, if S is a closed bounded subsets of $\mathbb{R}^d$ then S is contained in a product of closed bounded intervals. Thus S is a closed subset of a compact space, hence S is compact. □

A subset X of a topological space $(W, \mathcal{O}_W)$ is said to be *relatively compact* if its closure is compact. Thus, any subset of a relatively compact subset is relatively compact.

Since fixed point results enable us to solve equations, the interest of the notion of compactness is illustrated by the following theorem. It has been given many proofs; an elementary one can be found in [197, 222] and in the appendix. Recall that a subset C of a vector space is said to be *convex* if for all $x_0, x_1 \in C$ and $t \in [0, 1]$ one has $(1-t)x_0 + tx_1 \in C$.

Theorem 2.5 (Brouwer) *Let* X *be a compact convex subset of* $\mathbb{R}^d$ *(or of a finite dimensional normed vector space) and let* $f : X \to X$ *be a continuous map. Then there exists some* $\overline{x} \in X$ *such that* $f(\overline{x}) = \overline{x}$.

Corollary 2.10 (Hairy Ball Theorem) *Let* X *be a (finite dimensional) Euclidean space with scalar product* $\langle \cdot \mid \cdot \rangle$, *unit closed ball* B_X, *unit sphere* S_X, *let* $r > 0$, *and let* $g : rB_X \to X$ *be continuous and pointing inside* rB_X, *i.e. such that*

$$\langle g(x) \mid x \rangle \leq 0 \qquad \forall x \in rS_X.$$

Then there exists some $\overline{z} \in rB_X$ *such that* $g(\overline{z}) = 0$.

Proof Suppose on the contrary that $g(x) \neq 0$ for all $x \in rB_X$. Then $h : rB_X \to X$ given by

$$h(x) := \frac{r}{\|g(x)\|} g(x) \qquad x \in rB_X$$

is continuous, takes its values in $rS_X \subset rB_X$, hence has a fixed point $\overline{x} \in rB_X$ by Brouwer's Theorem. Then we get the contradiction

$$r^2 = \|h(\overline{x})\|^2 = \langle h(\overline{x}) \mid \overline{x} \rangle = \frac{r}{\|g(\overline{x})\|} \langle g(\overline{x}) \mid \overline{x} \rangle \leq 0.$$

This contradiction proves that g has a zero in rB_X. □

Corollary 2.11 *Let* X *be a (finite dimensional) Euclidean space, let* $b, r \in \mathbb{R}_+$, *and let* $f : X \to X$ *be continuous and such that* $\langle f(x) \mid x \rangle \geq b \|x\|$ *for all* $x \in rS_X$. *Then for all* $y \in bB_X$ *the equation* $f(x) = y$ *has a solution* $x \in rB_X$.

If $\langle f(x) \mid x \rangle \geq c(\|x\|)\,\|x\|$ *for some function* $c(\cdot)$ *such that* $c(r) \to \infty$ *as* $r \to \infty$, *then* $f(X) = X$. *Moreover, there exists a function* $k : \mathbb{R}_+ \to \mathbb{R}_+$ *such that for all* x, $y \in X$ *satisfying* $f(x) = y$ *one has* $\|x\| \leq k(\|y\|)$.

Proof Given $y \in bB_X$ let us set $g(\cdot) := y - f(\cdot)$, so that g is continuous and for $x \in rS_X$

$$\langle g(x) \mid x \rangle = \langle y \mid x \rangle - \langle f(x) \mid x \rangle \leq b\,\|x\| - b\,\|x\| \leq 0.$$

Thus, there exists some $x \in rB_X$ such that $g(x) = 0$ or $f(x) = y$.

Assuming that $\langle f(x) \mid x \rangle \geq c(\|x\|)\,\|x\|$ for $c(\cdot)$ satisfying $\lim_{r\to\infty} c(r) = \infty$, setting

$$k(s) := \sup\{r \in \mathbb{R}_+ : c(r) \leq s\} \qquad s \in \mathbb{R}_+$$

we see that k takes finite values and that whenever x is such that $f(x) = y$ we have $c(\|x\|).\,\|x\| \leq \langle f(x) \mid x \rangle = \langle y \mid x \rangle \leq \|y\|\,.\,\|x\|$, hence $\|x\| \leq k(\|y\|)$. □

Some topological spaces of interest have a local behavior involving compactness.

Definition 2.10 A topological space $(X, \mathcal{O}_X)$ is said to be locally compact if it is Hausdorff and if each point of X has a compact neighborhood.

A compact space is obviously locally compact. A discrete space (i.e. a space in which each subset is open) is locally compact, but it is not compact if it is infinite. The space $\mathbb{R}$ with its usual topology is locally compact but not compact. An open subset of a compact space or locally compact space is locally compact, as the following proposition shows.

Proposition 2.15 *In a compact space, or more generally in a locally compact space, each point has a base of neighborhoods formed by compact sets.*

Proof Let $(X, \mathcal{O}_X)$ be a compact space, let $\overline{x} \in X$ and let $U \in \mathcal{N}(\overline{x})$. We want to show that there exists some closed $V \in \mathcal{N}(\overline{x})$ contained in U. Without loss of generality we may assume U is open. If no such V exists, since $\mathcal{N}(\overline{x})$ is stable under finite intersections, the family $\{V \backslash U : V \in \mathcal{N}(\overline{x}),\ \mathrm{cl}(V) = V\}$ of closed subsets of $X \backslash U$ has the finite intersection property. Since $X \backslash U$ is compact by Proposition 2.13, one can find some $\overline{y} \in X \backslash U$ belonging to any closed neighborhood V of $\overline{x}$. Then we have $\overline{y} \neq \overline{x}$ and since X is Hausdorff we can find some $V \in \mathcal{N}(\overline{x})$, $W \in \mathcal{N}(\overline{y})$ with $V \cap W = \varnothing$. Since we may assume W is open, replacing V with its closure we may suppose V is closed. Then $\overline{y} \in V$. Since $\overline{y} \in W$ and $V \cap W = \varnothing$ we get a contradiction.

Now suppose X is locally compact. Let $\overline{x} \in X$ and let $U \in \mathcal{N}(\overline{x})$. By assumption there is some $W \in \mathcal{N}(\overline{x})$ that is compact. The preceding yields some compact neighborhood V of $\overline{x}$ in W contained in $U \cap W$. Then V is a neighborhood of $\overline{x}$ in X and is contained in U. □

Proposition 2.16 *Every open or closed subset X of a locally compact space $(W, \mathcal{O}_W)$ is locally compact.*

Proof If X is open, for every $\overline{x} \in X$ one has $X \in \mathcal{N}(\overline{x})$, so that there is some compact $V \in \mathcal{N}(\overline{x})$ contained in X and V is a neighborhood of $\overline{x}$ with respect to the induced topology on X.

If X is closed in W and if $\overline{x} \in X$, taking a neighborhood V of $\overline{x}$ in W that is compact, we see that $W \cap X$ is compact and is a neighborhood of $\overline{x}$ in X with respect to the induced topology. Thus X is locally compact. □

Proposition 2.17 *The product of a finite family of locally compact spaces is locally compact. In particular, $\mathbb{R}^d$ is locally compact.*

Proof It suffices to prove that the product $Z := X \times Y$ of two locally compact spaces is locally compact. Given $\overline{z} := (\overline{x}, \overline{y}) \in Z$ we pick $U \in \mathcal{N}(\overline{x})$, $V \in \mathcal{N}(\overline{y})$ that are compact. Then $U \times V$ is a compact neighborhood of $\overline{z}$. □

Given a topological space $(X, \mathcal{O})$ there are different ways of embedding it into a compact space. When $(X, \mathcal{O})$ is locally compact the simplest approach consists in adding a point $\overline{w}$ to X and declaring that a subset O of $W := X \cup \{\overline{w}\}$ is open if either it belongs to $\mathcal{O}$ or if $\overline{w} \in O$ and $X \backslash O$ is compact. It is easy to see that the resulting topological space is compact. It is called the *Alexandroff compactification* of X or one point compactification of X.

Exercise Show that the Alexandroff compactification of $\mathbb{R}^d$ is homeomorphic to the unit sphere $\mathbb{S}^d$ of $\mathbb{R}^{d+1}$. [Hint: use the stereographic projection $p : \mathbb{S}^d \setminus \{n\} \to \mathbb{R}^d$, where n is the "north pole" $n := (0, \ldots, 0, 1)$ of $\mathbb{S}^d$ defined as follows: for $x \in \mathbb{S}^d \setminus \{n\}$, $p(x)$ is the point of $\mathbb{R}^d \times \{0\}$ that belongs to the half-line $x + \mathbb{R}_+(x - n)$.]

The following theorem is the main existence result in optimization theory.

Theorem 2.6 (Weierstrass) *Let $f : X \to \overline{\mathbb{R}}$ be a lower semicontinuous function on a (nonempty) compact topological space X. Then the set $M := \{w \in X : f(w) \leq f(x)\ \forall x \in X\}$ of minimizers of f is nonempty.*

Proof We may suppose $m := \inf f(X) < \infty$ for otherwise f is constant with value ∞. Setting $S_f(r) := \{x \in X : f(x) \leq r\}$ for $r \in \mathbb{R}$, the family $\{S_f(r) : r > m\}$ is formed of nonempty closed subsets and any finite subfamily has a nonempty intersection: $\cap_{1 \leq i \leq k} S_f(r_i) = S_f(r_j)$ where $r_j := \min_{1 \leq i \leq k} r_i$. Therefore $M = \cap_{r > m} S_f(r)$ is nonempty. □

Given a topological space X, one may try to weaken its topology in order to enlarge the family of compact subsets. Then, a continuous function on X may not remain continuous. There are interesting cases, for instance making use of convexity assumptions, for which the function still remains lower semicontinuous, so that the preceding generalization of the classical existence of a minimizer under a continuity assumption is of interest.

Let us give a criterion for the lower semicontinuity of a function obtained by minimization.

Proposition 2.18 *Let W, X be topological spaces, let $\overline{w} \in W$ and let $f : W \times X \to \overline{\mathbb{R}}$ be a function which is lower semicontinuous at $(\overline{w}, \overline{x})$ for every $\overline{x} \in X$. If the following compactness assumption is satisfied, then the performance function $p : W \to \overline{\mathbb{R}}$ given by $p(w) := \inf_{x \in X} f(w, x)$ is lower semicontinuous at $\overline{w}$:*

(C) for any net $(w_i)_{i\in I} \to \overline{w}$ there exist a subnet $(w_j)_{j\in J}$, a convergent net $(x_j)_{j\in J}$ in X and $(\varepsilon_j)_{j\in J} \to 0_+$ such that $f(w_j, x_j) \leq p(w_j) + \varepsilon_j$ for all $j \in J$.

Proof Given a net $(w_i)_{i\in I} \to \overline{w}$ such that $(p(w_i))_{i\in I}$ converges, let $(w_j)_{j\in J}$ be a subnet of $(w_i)_{i\in I}$, and let $(x_j)_{j\in J}$ and $(\varepsilon_j) \to 0_+$ be as in (C). Then, if $\overline{x}$ is the limit of $(x_j)_{j\in J}$, one has

$$p(\overline{w}) \leq f(\overline{w}, \overline{x}) \leq \liminf_{j\in J} f(w_j, x_j) \leq \liminf_{j\in J} \left(p(w_j) + \varepsilon_j\right) = \lim_{i\in I} p(w_i).$$

Since $\liminf_{w\to\overline{w}} p(w)$ is the limit of $(p(w_i))_{i\in I}$ for some net $(w_i)_{i\in I} \to \overline{w}$ such that $(p(w_i))_{i\in I}$ converges, these inequalities show that $p(\overline{w}) \leq \liminf_{w\to\overline{w}} p(w)$. □

Corollary 2.12 *Let W and X be topological spaces, X being compact, and let $f : W \times X \to \mathbb{R}_\infty := \mathbb{R} \cup \{+\infty\}$ be lower semicontinuous. Then the performance function p defined as above is lower semicontinuous.*

Proof We give a proof in the case when f is just lower semicontinuous at each point of $\{\overline{w}\} \times X$; when f is lower semicontinuous on $W \times X$, a simpler proof can be given using the Weierstrass' Theorem. Condition (C) is clearly satisfied when X is compact since for any net $(w_i)_{i\in I} \to \overline{w}$ and for any sequence $(\alpha_n) \to 0_+$ one can take $H := I \times \mathbb{N}$, $w_h := w_i$, $\varepsilon_h := \alpha_n$ for $h := (i, n)$ and pick $x_h \in X$ satisfying $f(w_h, x_h) \leq p(w_h) + \varepsilon_h$, and take a subnet $(x_j)_{j\in J}$ of $(x_h)_{h\in H}$ which converges in X.

□

Exercises

1. For every net $(r_i)_{i\in I}$ in the compact space $\overline{\mathbb{R}}$ show that $\liminf_{i\in I} r_i$ is the least cluster point of $(r_i)_{i\in I}$ and $\limsup_{i\in I} r_i$ is the greatest cluster point of $(r_i)_{i\in I}$.
2. Let X and Y be topological spaces and let Z be a closed subset of $X \times Y$. Show that if Y is compact, then the projection $p_X(Z)$ of Z on X is closed. Give an example showing that one cannot drop the assumption that Y is compact.
3. Let X and Y be topological spaces, Y being Hausdorff and let $f : X \to Y$ be a map. Show that if f is continuous then the graph $G := \{(x, f(x)) : x \in X\}$ of f is closed in $X \times Y$. Give an example with $X = Y = \mathbb{R}$ showing that the converse is not true. Prove that if Y is compact, then the converse is true.
4. Let $(K_n)_n$ be a decreasing sequence in nonempty compact subsets of a topological space X. Show that $K := \cap_n K_n$ is nonempty and that for any open subset G of X containing K there exists some $m \in \mathbb{N}$ such that $K_m \subset G$. Give a generalization to a filtered family $(K_i)_{i\in I}$ of nonempty compact subsets.

5. Let X be a Hausdorff topological space and let A and B be two disjoint compact subsets of X. Show that there exist open subsets U, V of X such that $A \subset U$, $B \subset V$ and $U \cap V = \varnothing$. [Hint: start with the case when B is a singleton.]
6. Let $X := [0,1] \times [0,1]$ be endowed with the lexicographic order $\preceq$. Let $\mathcal{O}$ be the topology generated by the open intervals $T_{w,z} := \{x \in X : w \prec x \prec z\}$ of X and the intervals $\{x \in X : x \prec z\}$, $\{x \in X : w \prec x\}$, where $x \prec x'$ means that $x \preceq x'$ and $x \neq x'$. Show that X is compact with respect to this topology.
7. Show that for any compact subset K of a locally compact space X and any neighborhood U of K there exists some compact neighborhood V of K contained in U.
8. Show that the intersection of two locally compact subsets of a topological space is locally compact.
9. Prove that the union of two locally compact subsets of a topological space is not always locally compact. [Hint: in $\mathbb{R}$ take $A := \{a\}$ where a is the limit of a sequence (a_n) of $\mathbb{R}\backslash\{a\}$ and $B := \mathbb{R}\backslash(\{a_n : n \in \mathbb{N}\} \cup \{a\})$ or in $\mathbb{R}^2$ take $A := \mathbb{R} \times \mathbb{P}$, $B := \{(0,0)\}$.]
10. Show by an example that the image of a locally compact space X under a continuous map $f : X \to Y$ is not necessarily locally compact. [Hint: take a bijection f from $\mathbb{N}$ onto $\mathbb{Q}$.]
11*. Prove the odd **Hairy Ball Theorem**: for d odd there is no continuous vector field on the unit sphere $\mathbb{S}^{d-1} := S_{\mathbb{R}^d}$ tangent to $S_{\mathbb{R}^d}$ [Hint: use the appendix.]
12. (**Beals**) Let B be the closed unit ball of the space $X := c_0$ of sequences $x := (x_n)_{n \geq 0}$ with limit 0 endowed with the supremum norm. Prove that the map $f : B \to B$ given by $f(x) = (1, x_0, x_1, \ldots)$ for $x := (x_0, x_1, \ldots) \in B$ is nonexpansive, i.e. does not increase distances, and has no fixed point.
13. (**Stone**) A Boolean ring is a ring A such that $a^2 = a$ for all $a \in A$. A topological space $(X, \mathcal{O})$ is said to be *totally disconnected* if the family $\mathcal{B}$ of subsets of X which are simultaneously open and closed forms a base of its topology. Show that $\mathcal{B}$ forms a Boolean ring with unit when the product is $\cap$ and the addition is Δ as in Definition 1.2. Conversely, show that any Boolean ring with unit is isomorphic to the ring of subsets of a compact topological space which are simultaneously open and closed. [See [106, p. 41].]

2.3 Metric Spaces

A usual means of studying convergence on a set X is to reduce the question to the case of convergence in $\mathbb{R}$. That can be done if one disposes of functions from $X \times X$ to $\mathbb{R}$ that allow such a transfert. Metrics (also called distances) and semimetrics are such means.

2.3.1 General Facts About Metric Spaces

Definition 2.11 A *semimetric* on a set X is a function $d : X \times X \to \mathbb{R}_+ := [0, +\infty[$ satisfying the properties:

(SM1) for all $x \in X$ one has $d(x, x) = 0$;
(SM2) for all $x, x' \in X$ one has $d(x, x') = d(x', x)$;
(SM3) for all $x, x', x'' \in X$, one has $d(x, x'') \leq d(x, x') + d(x', x'')$.

A *metric* is a semimetric d such that for all $x, x' \in X$, $d(x, x') = 0 \Rightarrow x = x'$.
A *pseudo-metric* on X is a function $d : X \times X \to \overline{\mathbb{R}}_+ := [0, +\infty]$ satisfying (SM1)–(SM3).

The relation in (SM3) is called the *triangle inequality*.

A *metric space* (resp. *semimetric space*) is a pair (X, d) formed by a set X and a metric (resp. semimetric) d on X. When there is no ambiguity on d, we just write X. A *uniform space* is a pair $(X, (d_a)_{a \in A})$, where $(d_a)_{a \in A}$ is a family of pseudo-metrics on X. In a metric space (X, d) (resp. a uniform space $(X, (d_a)_{a \in A})$) one can introduce a convergence by setting

$$(x_i)_{i \in I} \to x \iff (d(x_i, x))_{i \in I} \to 0$$

$$(\text{resp. } (x_i)_{i \in I} \to x \iff \forall a \in A \quad (d_a(x_i, x))_{i \in I} \to 0).$$

Exercise Verify the axioms (C1), (C2), (C3) of convergence spaces (Definition 2.2)

In fact, a semimetric d induces a topology $\mathcal{O}$ on X defined by: $G \in \mathcal{O}$ iff for all $x \in G$ there exists some $r > 0$ such that the *open ball*

$$B(x, r) := \{x' \in X : d(x, x') < r\}$$

is contained in G. Thus $\mathcal{O}$ is the topology generated by the family of open balls. This family is a base of $\mathcal{O}$ and for all $\overline{x} \in X$, the family of open balls centered at $\overline{x}$ is a base of neighborhoods of $\overline{x}$. In the sequel, the *closed ball* with center x and radius $r \in \mathbb{R}_+$ is the set

$$B[x, r] := \{x' \in X : d(x, x') \leq r\}.$$

The family of closed balls centered at x with positive radius is also a base of neighborhoods of x. A topology can also be associated with a uniform space $(X, (d_a)_{a \in A})$: it is the topology generated by the balls $B_a(x, r) := d_a(x, \cdot)^{-1}([0, r[)$ for $a \in A$, $x \in X$, $r > 0$. It is easy to show that the convergence on (X, d) or $(X, (d_a)_{a \in A})$ described above is the convergence associated with the topology we just defined. Moreover, when d is a metric, the associated topology is Hausdorff: given $x, x' \in X$ such that $x \neq x'$, for $r \in]0, d(x, x')/2[$ the balls $B(x, r)$ and $B(x', r)$ are disjoint in view of the triangle inequality. Hence convergent nets or sequences have a unique limit by Proposition 2.1. The existence of a metric d on X implies a noticeable property of accumulation points. We propose it in the next exercise. A point a is called an *accumulation point* of a subset S of a topological space X if every neighborhood V of a contains some point $x \in S, x \neq a$.

Exercise Show that every neighborhood V of an accumulation point a of a subset S of a metric space (X, d) contains an infinite family of points of S. [Hint: by induction define a sequence (x_n) of $S \backslash \{a\}$ such that $d(x_{n+1}, a) < d(x_n, a)$.]

Given a subset S of a metric space (X, d) the *diameter* of S is $\text{diam}(S) := \sup\{d(x, y) : x, y \in S\}$ and the *distance to* S is the function $d_S : X \to \mathbb{R}_+$ given by

$$d_S(x) := \inf\{d(x, y) : y \in S\}, \qquad x \in X.$$

This notion is often convenient. If A, B are two subsets of X, their *gap* $g(A, B)$ is

$$\text{gap}(A, B) := \inf\{d(a, b) : a \in A, \ b \in B\}.$$

This number is sometimes called the distance between A and B but this terminology is improper as $(A, B) \mapsto \text{gap}(A, B)$ is not a metric on the set $\mathcal{P}(X)$ of subsets of X or on the set of closed subsets of X.

In metric spaces continuity can be expressed with the help of ε's and δ's: a map $f : (X, d) \to (X', d')$ is continuous at $\overline{x} \in X$ if and only if for all $\varepsilon > 0$ one can find some $\delta > 0$ such that $d'(f(x), f(\overline{x})) < \varepsilon$ for all $x \in X$ satisfying $d(x, \overline{x}) < \delta$. In metric spaces, one can avoid nets and just use sequences in a convenient way:

Proposition 2.19 *For a map $f : (X, d) \to (X', d')$ between two metric spaces and $\overline{x} \in X$ the following assertions are equivalent:*

(a) f is continuous at $\overline{x} \in X$;
(b) for every sequence $(x_n) \to \overline{x}$ one has $(f(x_n)) \to f(\overline{x})$;
(c) for every sequence $(x_n) \to \overline{x}$ one can find a subsequence $(x_{k(n)})$ such that $(f(x_{k(n)})) \to f(\overline{x})$.

Proof (a)$\Rightarrow$(b) Sequences being particular nets, the implication stems from Proposition 2.2. (b)$\Rightarrow$(c) being obvious, it remains to show that (c)$\Rightarrow$(a).

(c)$\Rightarrow$(a) If f is not continuous at $\overline{x}$ there exists some $\varepsilon > 0$ such that for all $\delta > 0$ one has $f(B(\overline{x}, \delta)) \not\subseteq B(f(\overline{x}), \varepsilon)$. Taking a sequence $(\delta_n) \to 0_+$ we can find some $x_n \in B(\overline{x}, \delta_n)$ such that $f(x_n) \notin B(f(\overline{x}), \varepsilon)$. Then we have $(x_n) \to \overline{x}$ but for any subsequence $(x_{k(n)})$ the sequence $(f(x_{k(n)}))$ does not converge to $f(\overline{x})$. □

Proposition 2.20 *In a metric space (X, d) the closure $\text{cl}(S)$ of a subset S is the set of limits of sequences in S.*

Proof If x is the limit of a sequence (x_n) of S, then clearly $x \in \text{cl}(S)$. Conversely, if $x \in \text{cl}(S)$, for any $n \geq 1$ the set $S \cap B(x, 1/n)$ is nonempty. Picking $x_n \in S \cap B(x, 1/n)$ we get a sequence (x_n) converging to x. □

Proposition 2.21 *In a metric space (X, d) any cluster point of a sequence in X is the limit of a subsequence.*

Proof Let $\overline{x}$ be a cluster point of a sequence (x_n) of X. Given a sequence $(r_n) \to 0_+$ an induction on n gives a sequence $(k(n))$ of $\mathbb{N}$ such that $k(n+1) > k(n)$ and $x_{k(n)} \in B(\overline{x}, r_n)$ for all n. Then $(x_{k(n)})_n$ is a subsequence in (x_n) that converges to $\overline{x}$. □

A map $f : (X, d) \to (X', d')$ is said to be *uniformly continuous* if for all $\varepsilon > 0$ one can find some $\delta > 0$ such that $d'(f(w), f(x)) < \varepsilon$ for all $w, x \in X$ satisfying $d(w, x) < \delta$. Such a map is clearly continuous at each $\overline{x} \in X$ and one sees that δ does not depend on $\overline{x}$. If there exists a *modulus* μ, i.e. a function $\mu : \mathbb{R}_+ \to \overline{\mathbb{R}}_+ := [0, +\infty]$ satisfying $\mu(t) \to 0$ as $t \to 0$, such that $d'(f(w), f(x)) \leq \mu(d(w, x))$ for all $w, x \in X$, then f is uniformly continuous. The converse is true and we invite the reader to produce a modulus satisfying the preceding property (and even the least such modulus μ_f called the modulus of uniform continuity of f). The case when μ is linear deserves some attention, but there are other cases of interest, for example the case $\mu(t) = ct^\alpha$ with $c \in \mathbb{R}_+, \alpha > 0$ (then f is said to be *Hölderian*).

A map $f : (X, d) \to (X', d')$ is said to be *Lipschitzian* if there exists some $c \in \mathbb{R}_+$ such that $d'(f(x_1), f(x_2)) \leq cd(x_1, x_2)$ for all $x_1, x_2 \in X$. The constant c is called a Lipschitz constant (or rate, or rank). The least such constant is called the (exact) *Lipschitz rate* of f. If this rate is 1, f is said to be *nonexpansive*. If this rate is less than 1 one says that f is contractive or a *contraction*. If f is a bijection and if for all $x_1, x_2 \in X$ one has $d'(f(x_1), f(x_2)) = d(x_1, x_2)$ one says that f is an *isometry*; then f^{-1} is also an isometry. For $x \in X$ the *Lipschitz rate* of f *at* x is the infimum of the Lipschitz rates of the restrictions of f to the neighborhoods of x (and $+\infty$ if there is no neighborhood of x on which f is Lipschitzian). If for all $x \in X$ there is a neighborhood V of x such that the restriction $f \mid V$ is Lipschitzian, f is said to be *locally Lipschitzian*.

If $(X, (d_a)_{a \in A})$ and $(Y, (d_b)_{b \in B})$ are two uniform spaces, a map $f : X \to Y$ is said to be *uniformly continuous* if for all $b \in B$ and all $\varepsilon \in \mathbb{P}$ one can find a finite subset $A(b, \varepsilon)$ of A and $\delta > 0$ such that $d_b(f(x), f(x')) \leq \varepsilon$ whenever $x, x' \in X$ satisfy $d_a(x, x') \leq \delta$ for all $a \in A(b, \varepsilon)$.

Exercise Show that the composition of two uniformly continuous maps is uniformly continuous.

Different equivalence properties can be defined on the set of metrics on a set X. Two metrics d, d' are said to be *topologically equivalent* if the topologies they define coincide. They are said to be *uniformly equivalent* (resp. *metrically equivalent*) if the identity map from (X, d) into (X, d') and its inverse are uniformly continuous (resp. Lipschitzian).

Proposition 2.22 *A metric space (X, d) is separable if and only if its topology has a countable base.*

Proof If the topology of (X, d) has a countable base $\mathcal{B} := \{B_n : n \in \mathbb{N}\}$, picking an arbitrary point $a_n \in B_n$ for all n we get a dense subset $A := \{a_n : n \in \mathbb{N}\}$ since for all $x \in X$ and any open subset U containing x there exists some $n \in \mathbb{N}$ such that $B_n \subset U$.

Conversely, suppose X contains a countable dense subset $A := \{a_n : n \in \mathbb{N}\}$. Then we claim that the family $\mathcal{B} := \{B(a_n, q) : n \in \mathbb{N},\ q \in \mathbb{Q},\ q > 0\}$ is a base of the topology of (X, d). In fact, given $\overline{x} \in X$ and $r > 0$ we can find $q \in \mathbb{Q}$ such that $0 < q < r/2$ and if $n \in \mathbb{N}$ is such that $a_n \in B(\overline{x}, q)$, for all $x \in B(a_n, q)$ we have $x \in B(\overline{x}, r)$ by the triangle inequality. □

Example The set $\mathbb{R}$ of real numbers endowed with its usual metric has a countable base since the set $\mathbb{Q}$ of rational numbers is dense in $\mathbb{R}$.

Corollary 2.13 *A subspace of a separable metric space is separable.*

Proof This follows from the proposition and from the fact that if $\mathcal{B}$ is a countable base of (X, d), then for a subspace Y, the family $\mathcal{B}_Y := \{B \cap Y : B \in \mathcal{B}\}$ is a base for the induced topology on Y. □

On the product $Z := X \times Y$ of two metric spaces (X, d_X), (Y, d_Y), a metric d is called a *product metric* if the canonical projections $p_X : Z \to X$, $p_Y : Z \to Y$ and the insertions $j_b : x \mapsto (x, b)$, $j_a : y \mapsto (a, y)$ are nonexpansive.

Exercise Show that a metric d on $Z := X \times Y$ is a product metric if and only if for all (u, v), $(x, y) \in X \times Y$ one has

$$\max(d_X(u, x), d_Y(v, y)) \leq d((u, v), (x, y)) \leq d_X(u, x) + d_Y(v, y).$$

The left (resp. right) side defines a convenient metric usually denoted by d_∞ (resp. d_1). Describe its balls.

Whereas the product X of a family of metric spaces (X_s, d_s) ($s \in S$, an arbitrary set) cannot be provided with a (sensible, i.e. inducing the product topology) metric in general, we have seen that a product of topological spaces $(X_s, \mathcal{O}_s)$ ($s \in S$) can always be endowed with a topology $\mathcal{O}$ that makes the projections $p_s : X \to X_s$ continuous and that is as weak as possible, namely it is the topology generated by the sets $p_s^{-1}(O_s)$ for $s \in S$, $O_s \in \mathcal{O}_s$. Its associated convergence is componentwise convergence. When $(X_s, \mathcal{O}_s) := (Y, \mathcal{O}_Y)$ for all $s \in S$, identifying the product X with the set Y^S of maps from S to Y, the convergence associated with the product topology $\mathcal{O}$ on X coincides with *pointwise convergence*: $(f_i)_{i \in I} \to f$ in Y^S if, for all $s \in S$, $(f_i(s))_{i \in I} \to f(s)$. When $\mathcal{O}_Y$ is the topology associated with a metric d_Y on Y, a stronger convergence can be defined on Y^S: it is the so-called *uniform convergence* for which $(f_i)_{i \in I} \to f$ iff $(d_\infty(f_i, f)) := (\sup_{s \in S} d_Y(f_i(s), f(s))) \to 0$. This convergence is adapted to bounded functions, but it can be considered for any set of maps from a set S into a metric space Y. It enjoys better preservation properties, such as the following one.

Theorem 2.7 *Let X be a topological space and let (Y, d) be a metric space. Let $(f_i)_{i \in I}$ be a net (or a sequence) of continuous functions from X into Y that converges uniformly to some map $f : X \to Y$. Then f is continuous.*

Proof Let $\overline{x} \in X$ and let $\varepsilon > 0$ be given. We can find $k \in I$ such that for $i \geq k$ we have $\sup_{x \in X} d(f_i(x), f(x)) \leq \varepsilon/3$. Since f_k is continuous there exists a neighborhood V of $\overline{x}$ such that $d(f_k(x), f_k(\overline{x})) \leq \varepsilon/3$ for all $x \in V$. Then, for $x \in V$ we have

$$d(f(x), f(\overline{x})) \leq d(f(x), f_k(x)) + d(f_k(x), f_k(\overline{x})) + d(f_k(\overline{x}), f(\overline{x})) \leq \varepsilon.$$

This proves that f is continuous at $\overline{x}$. □

Some constructions can be obtained by using metrics.

Lemma 2.3 (Urysohn's Lemma) *Let A and B be two disjoint nonempty closed subsets of a metric space (X, d). Then there exists inuous function $h : X \to [-1, 1]$ such that $h(x) = -1$ for all $x \in A$ and $h(x) = 1$ for all $x \in B$.*

If C is a closed subset of X and if $f : C \to [-1, 1]$ is a continuous function, there exists a continuous function $g : X \to [-1/3, 1/3]$ such that $|f(x) - g(x)| \leq 2/3$ for all $x \in C$.

Proof Since A and B are disjoint, for all $x \in X$ we have $g(x) := d(x, A) + d(x, B) > 0$. Setting $h(x) := (d(x, A) - d(x, B))/g(x)$ we obtain the required function.

Let us set $A := \{x \in C : f(x) \leq -1/3\}$ and $B := \{x \in C : f(x) \geq 1/3\}$. If A is empty we take $g := 1/3$; if B is empty we take $g := -1/3$. If A and B are both nonempty we take $g := (1/3)h$ where h is as in the first assertion. We obtain the relation $|f(x) - g(x)| \leq 2/3$ for all $x \in C$ by considering the three cases $x \in A$, $x \in B$, $x \in C \backslash (A \cup B)$. □

Theorem 2.8 (Tietze-Urysohn) *Let C be a closed subset of a metric space (X, d) and let $f : C \to \mathbb{R}$ a bounded continuous function. Then there exists a continuous function $g : X \to \mathbb{R}$ such that $g \mid_C = f$, $\inf g(X) = \inf f(C)$, and $\sup g(X) = \sup f(C)$.*

Proof The result is obvious when f is constant. Changing f into $af + b$, where a and b are appropriate real numbers, we may assume $\inf f(C) = -1$ and $\sup f(C) = 1$.

The second assertion of the lemma yields a continuous function g_0 with values in $[-1/3, 1/3]$ such that $|f(x) - g_0(x)| \leq 2/3$ for all $x \in C$.

Let us suppose that inductively we have defined for $n \in \mathbb{N}_k := \{1, \ldots, k\}$ a continuous function g_n such that

$$|g_n| \leq 1 - (\frac{2}{3})^{n+1} \qquad |g_n \mid_C - f| \leq (\frac{2}{3})^{n+1} \tag{2.1}$$

Applying the second assertion of the lemma to the function $(\frac{3}{2})^{k+1}(g_k \mid_C - f)$, we get a continuous function h_{k+1} with values in $[-2^{k+1}/3^{k+2}, 2^{k+1}/3^{k+2}]$ such that for all $x \in C$

$$|f(x) - g_k(x) - h_{k+1}(x)| \leq (\frac{2}{3})^{k+2}. \tag{2.2}$$

Setting $g_{k+1} := g_k + h_{k+1}$ we obtain the two inequalities of relation (2.1) for $n = k + 1$. Since for $x \in X$ we have $|g_{k+1}(x) - g_k(x)| \leq 2^{k+1}/3^{k+2}$ for all $k \in \mathbb{N}$, the sequence (g_k) converges uniformly on X to a function g that is continuous by Theorem 2.7. Moreover, for $x \in C$, passing to the limit in relation (2.2) we get $g(x) = f(x)$. □

Metrics can be used in order to tackle optimization problems with constraints: a natural idea consists in introducing some penalty terms, replacing the objective f by a penalized objective. If one has to minimize f on an admissible subset A of a metric space (X, d) one may consider the minimization of

$$f_s := f + sd_A(\cdot)$$

on the whole space X, expecting that for a large penalization rate s the effect will be similar. Here $d_A(x) := d(x, A) := \inf\{d(x, w) : w \in A\}$ for $x \in X$. In general one has to replace s by an infinite sequence $(s_n) \to \infty$, so that one has to solve a sequence of unconstrained problems. In some favorable cases a single penalized problem suffices as in the simple situation presented in the following result.

Proposition 2.23 (Exact Penalization) *Let X be a metric space and let $f : X \to \mathbb{R}$ be a Lipschitzian function with rate r. Then, for any $s \geq r$ and any nonempty subset A of X*

$$\inf_{x \in A} f(x) = \inf_{x \in X} (f(x) + sd_A(x)) . \tag{2.3}$$

Moreover, $\overline{x} \in A$ is a minimizer of f on A if and only if, $\overline{x}$ is a minimizer of $f_s := f + sd_A$ on X. If A is closed and if $s > r$, any minimizer z of f_s belongs to A.

A local version can be given by replacing X with a neighborhood of $\overline{x}$.

Proof Since $f_s = f$ on A, we have $m := \inf f(A) \geq \inf f_s(X)$. If we had strict inequality we could find $x \in X$ such that $f_s(x) < m$. Then we would have $sd_A(x) < m - f(x)$, so that we could pick $x' \in A$ such that $sd(x, x') < m - f(x)$. Since f is Lipschitzian with rate $r \leq s$ we would get $f(x') \leq f(x) + sd(x, x') < m$, a contradiction. The second assertion follows from (2.3).

Suppose now that A is closed and that for some $s > r$ a minimizer z of f_s does not belong to A. Then $d_A(z)$ is positive, so that, by the relations $\inf f(A) = \inf f_s(X) = f(z) + sd_A(z)$,

$$rd_A(z) < sd_A(z) = \inf f(A) - f(z)$$

one can find $a \in A$ such that $rd(a, z) < \inf f(A) - f(z)$, contradicting the relations $\inf f(A) = \inf f_r(X) \leq f(z) + rd(a, z)$. □

When the admissible set A is defined by equalities or inequalities, it is sensible to take these relations into account. In the case when the admissible set A is defined as $A := g^{-1}(C)$, where $g : X \to W$ is a map with values in another metric space (W, d') and C is a closed subset of W such that for some $c \in \mathbb{P}$ one has $d_A(x) \leq cd'(g(x), C)$

for all $x \in X$, one can replace f_s with $f + scd'(g(x), C)$. Since d_C is often easier to compute than the distance to the implicitly defined set $A := g^{-1}(C)$, as is the case when $W := \mathbb{R}^m$, $C := \mathbb{R}^m_-$, such a penalized problem is often more tractable.

Exercises

1. A function $h : \mathbb{R}_+ \to \mathbb{R}_+$ is said to be *subadditive* if it satisfies $h(r + s) \leq h(r) + h(s)$ for all $r, s \in \mathbb{R}_+$. Let H be the set of subadditive functions $h : \mathbb{R}_+ \to \mathbb{R}_+$ satisfying $h(0) = 0$ and $h(r) > 0$ for all $r > 0$. Verify that for $h, k \in H$ one has $h + k \in H$, $h \vee k \in H$, $h \circ k \in H$ and that H is stable under pointwise limits and suprema. Prove that if $h : \mathbb{R}_+ \to \mathbb{R}_+$ is concave, increasing and such that $h(0) = 0$, then $h \in H$. Show that for every metric d on a set X, the function $h \circ d$ is a metric on X. Using the functions $r \mapsto r/(1 + r)$ and $r \mapsto \min(r, 1)$ show that any metric d on X is uniformly equivalent to a bounded metric.
2. Give examples of disjoint closed subsets A, B of a metric space (X, d) such that $\text{gap}(A, B) = 0$. Show that the triangle inequality $\text{gap}(A, C) \leq \text{gap}(A, B) + \text{gap}(B, C)$ is not valid.
3. (**Hausdorff-Pompeiu metric**) Let (X, d) be a metric space and let $\mathcal{B}_0$ be the set of nonempty bounded subsets of X. For $A, B \in \mathcal{B}_0$ let $\mathrm{e}(A, B)$ be the *excess* of A over B defined by $\mathrm{e}(A, B) := \sup\{d(a, B) : a \in A\}$. Verify that $\mathrm{e}(A, B) = 0$ if and only if $A \subset \text{cl}(B)$. Verify that $d_H : (A, B) \mapsto \max(\mathrm{e}(A, B), \mathrm{e}(B, A))$ is a semimetric on $\mathcal{B}_0$ and a metric on the set $\mathcal{F}_b$ of nonempty closed bounded subsets of X.
4. Let (X, d) a metric space and let $(Y, \mathcal{O}_Y)$ be a topological space. Show that a map $f : X \to Y$ is continuous at $\overline{x} \in X$ if and only if for any sequence $(x_n) \to \overline{x}$ one has $(f(x_n)) \to f(\overline{x})$.
5. Prove that if $f : \mathbb{R}^d \to \mathbb{R}$ is uniformly continuous, then there exist $a, b \in \mathbb{R}_+$ such that $f(x) \leq ad(x, 0) + b$ for all $x \in \mathbb{R}^d$.
6. Let S be a subset of a metric space (X, d). For $r > 0$ let $B(S, r) := \{x \in X : d(x, S) < r\}$. Verify that $B(S, r)$ is the union over $x \in S$ of the balls $B(x, r)$ and that $\cap_{r>0} B(S, r) = \text{cl}(S)$. Examine whether similar conclusions hold for $B[S, r] := \{x \in X : d(x, S) \leq r\}$.
7. Let (M, d) be a metric space in which closed balls are compact. Suppose X is arcwise connected. Show that any pair of points x_0, x_1 in X can be joined by a curve with least length (a so-called *geodesic*) [see [75]]. Identify such a curve when M is the unit sphere $\mathbb{S}^{d-1}$ of $\mathbb{R}^d$ and when M is a circular cylinder in $\mathbb{R}^3$. Such curves have prompted the development of differential geometry.
8. Given a metric space (M, d), let X be the set of subsets of M. For a net $(S_i)_{i \in I}$ in X define

$$\liminf_{i \in I} S_i := \{x \in M : \lim_{i \in I} d(x, S_i) = 0\},$$

$$\limsup_{i \in I} S_i := \{x \in M : \liminf_{i \in I} d(x, S_i) = 0\}$$

and write $(S_i)_{i\in I} \to S$ if $\liminf_{i\in I} S_i = \limsup_{i\in I} S_i = S$. Verify the axioms of convergence spaces.

9. Devise another proof of the Tietze-Urysohn extension theorem that gives an explicit expression for the extension g of f in the case $\inf f(C) = 1$, $\sup f(C) = 2$:

$$g(x) := \frac{1}{d(x,C)} \inf_{w\in C} f(w)d(w,x) \qquad x \in X\backslash C$$

and $g(x) = f(x)$ for $x \in C$. Show that g is well defined on X, satisfies $\inf g(X) = \inf f(C)$, $\sup g(X) = \sup f(C)$, and is continuous. [Hint: note that $x \mapsto \inf_{w\in C} f(w)d(w,x)$ is Lipschitzian with rate 2 and that for $x \in X\backslash C$ and $D(x) := C \cap B(x, 2d(x,C))$ one has $g(x) = (1/d(x,C)) \inf_{w\in D(x)} f(w)d(w,x)$.]

10*. Prove that any continuous function f on a metric space (X,d) can be uniformly approximated by a locally Lipschitz function [see [133].]

11. Verify the following counterexample showing that a continuous function f on a metric space (X,d) cannot always be uniformly approximated by a Lipschitz function: take $X := \mathbb{R}$ with its usual distance and f continuous such that $f(n) = 0$ for $n \in \mathbb{N}$, $f(n + r_n) = 1$ for $n \in \mathbb{N}$, where (r_n) is a sequence in $]0,1[)$ with limit 0. This counterexample and the preceding reference have been communicated to the author by G. Beer.

12. Let (X,d) be a metric space, let $s \in X$ and for a subset T of X let

$$V_s(T) := \{x \in X : d(x,s) \leq d(x,t)\ \forall t \in T\}$$

be the Voronoi cell associated with s and T (with $V_s(\varnothing) := X$ and $f := d$). Verify that $s \in \text{int}V_s(T)$ if and only if $s \in X\backslash \text{cl}(T)$.

13. Let (X,d) be a metric space such that for all $w, x \in X$ and $r > 0$ such that $r < d(w,x)$ one has $d(x, B[w,r]) < d(x,w)$. Using the notation of the preceding exercise, show that when $s \in X\backslash\text{cl}(T)$ one has $V_s(T) = V_s(\text{bdry}(T))$.

When $s \in \text{cl}(T)$ show that the relation $V_s(T) = V_s(\text{bdry}(T))$ may or may not hold. [Hint: for $X := \mathbb{R}^2$, $s = (0,0)$, $T = \mathbb{R}^2_+$ one has $V_s(T) = V_s(\text{bdry}(T)) = \mathbb{R}^2_-$ whereas for $T := \mathbb{R} \times \mathbb{R}_+$ one has $V_s(T) = \{0\} \times \mathbb{R}_-$, $V_s(\text{bdry}(T)) = \{0\} \times \mathbb{R}$.]

Show that the assumption on (X,d) is satisfied whenever for any $w, x \in X$ with $w \neq x$ there exists a connected set S containing w and x such that $d(w,z) + d(z,x) = d(w,x)$ for all $z \in S$.

14. (**McShane**, **Whitney**, 1934) Let W be a nonempty subset of a metric space (X,d) and let $f : W \to \mathbb{R}$ be a Lipschitz function with rate r. Show that the functions $f^\flat, f^\# : X \to \mathbb{R}$ defined by

$$f^\flat(x) := \inf_{w\in W}(f(w) - rd(w,x)), \qquad f^\#(x) := \sup_{w\in W}(f(w) + rd(w,x))$$

are Lipschitzian with rate r and extend f. Prove that any Lipschitzian extension g of f with rate r satisfies $f^\flat \leq g \leq f^\#$.

2.3.2 Complete Metric Spaces

The structure of metric space is richer than the structure of topological space. In particular, one disposes of the notion of a Cauchy sequence: a sequence (x_n) of (X, d) is called a *Cauchy sequence* if $(d(x_n, x_p)) \to 0$ as $n, p \to +\infty$. A metric space is said to be *complete* if its Cauchy sequences are convergent. The interest of such a notion is the fact that one can assert the convergence of such a sequence without knowing its limit. The following result is worth noting; its proof is immediate.

Proposition 2.24 *If a Cauchy sequence (x_n) in a metric space (X, d) has a converging subsequence, then (x_n) is converging.*

It is often convenient to replace Cauchy sequences with more special sequences. We call a sequence (x_n) in a metric space (X, d) an *Abel sequence* if there exists some $c \in \mathbb{R}_+$ and some $r \in]0, 1[$ such that $d(x_n, x_{n+1}) \leq cr^n$ for all $n \in \mathbb{N}$. We observe that Abel sequences can be a substitute to Cauchy sequences in view of the following lemma, the easy proof of which is left to the reader.

Lemma 2.4 *Any Abel sequence in a metric space is a Cauchy sequence.*
Any Cauchy sequence has a subsequence that is an Abel sequence.

Corollary 2.14 *A metric space (X, d) is complete if and only if any Abel sequence in X is convergent.*

Let us give some permanence properties.

Proposition 2.25 *If (X, d_X) is a subspace of a metric space (W, d_W) with the induced metric and if (X, d_X) is complete, then X is closed in W.*
Conversely, any closed subset X of a complete metric space (W, d_W) is complete with respect to the induced metric.

Proof Let us show that any point $\overline{w}$ in the closure $\mathrm{cl}(X)$ of X belongs to X. Proposition 2.20 ensures that some sequence (x_n) of X converges to $\overline{w}$. Such a sequence is a Cauchy sequence, hence has a limit $\overline{x} \in X$. Uniqueness of limits in metric spaces implies that $\overline{w} = \overline{x} \in X$.

For the converse, let X be a closed subset of a complete metric space (W, d_W). Since a Cauchy sequence in X with respect to the induced metric d_X is a Cauchy sequence in (W, d_W), it converges in W, and in fact in X since X is closed. Thus X is complete with respect to d_X. □

Other permanence properties concern function spaces. It is extremely useful to consider metric spaces formed with functions or maps.

Proposition 2.26 *Let S be a set and let (M, d) be a metric space. On the set $B(S, M)$ of $f : S \to M$ that are bounded, i.e. such that $f(S)$ is bounded in (M, d), one can* *define a metric by setting for* $f, g \in B(S, M)$

$$d_\infty(f, g) := \sup_{s \in S} d(f(s), g(s)).$$

If (M, d) *is complete,* $(B(S, M), d_\infty)$ *is complete.*

Proof It is easy to see that d_∞ is a metric on $B(S, M)$. Let $(f_n)_n$ be a Cauchy sequence in $(B(S, M), d_\infty)$. Since for every $x \in S$ the evaluation map $f \mapsto f(x)$ is nonexpansive, the sequence $(f_n(x))_n$ is a Cauchy sequence in (M, d). When (M, d) is complete, this sequence converges. Let $f(x)$ be its limit. If $\varepsilon \mapsto k(\varepsilon)$ is such that $d_\infty(f_m, f_n) \leq \varepsilon$ for $n \geq m \geq k(\varepsilon)$, passing to the limit as $n \to \infty$ we see that $d(f_m(x), f(x)) \leq \varepsilon$ for $m \geq k(\varepsilon)$ and all $x \in S$. Equivalently we have $d_\infty(f_m, f) \leq \varepsilon$ for $m \geq k(\varepsilon)$, so that $(f_n) \to f$ in $(B(S, M), d_\infty)$ since f is bounded as $\sup_{x,x' \in S} d(f(x), f(x')) \leq \sup_{x,x' \in S} d(f_m(x), f_m(x')) + 2\varepsilon < +\infty$. □

When a sequence (f_n) converges to f for d_∞ one says that (f_n) *converges uniformly* to f. This property is stronger than *pointwise convergence*, which means that for all $x \in S$ one has $(f_n(x))_n \to f(x)$.

Proposition 2.27 *Let X be a topological space and let* (M, d) *be a metric space. The subspace* $C_b(X, M)$ *of bounded continuous functions from X to M is closed in* $B(X, M)$ *with respect to the metric* d_∞*. Thus, if* (M, d) *is complete,* $(C_b(X, M), d_\infty)$ *is complete.*

Proof We have to prove that the uniform limit f of a sequence (f_n) in $C_b(X, M)$ belongs to $C_b(X, M)$. We know from the preceding proof that f is bounded. The continuity of f is established in Theorem 2.7. □

Completeness can be used with great success for extension results.

Theorem 2.9 *Let* (W, d_W) *and* (Y, d_Y) *be complete metric spaces and let* $f : X \to Y$ *be uniformly continuous, where X is a dense subset of W. Then f can be extended uniquely to a uniformly continuous map* $\overline{f} : W \to Y$.

Proof Uniqueness of the extension stems from Proposition 2.4. Let us prove the existence of a uniformly continuous extension. Let $m : \mathbb{R}_+ \to \overline{\mathbb{R}}_+$ be a modulus of uniform continuity of f in the sense that $d_Y(f(x), f(x')) \leq m(d_W(x, x'))$ for all $x, x' \in X$. Given $w \in W$, let (x_n) be a sequence in X with limit w. Since (x_n) is a Cauchy sequence in X, $(f(x_n))$ is a Cauchy sequence in (Y, d_Y). Since (Y, d_Y) is complete, $(f(x_n))$ has a limit $y \in Y$. This limit does not depend on the choice of the sequence (x_n) since given two such sequences (x_n), (x'_n) one has $d_Y(f(x_n), f(x'_n)) \leq m(d_W(x_n, x'_n))$, so that, passing to the limit, we get $(f(x'_n)) \to y$. Denoting this limit by $\overline{f}(w)$ we define a map $\overline{f} : W \to Y$. For all $t > 1$ this map satisfies $d_Y(\overline{f}(w), \overline{f}(w')) \leq m(td_W(w, w'))$ since for any sequences $(x_n) \to w$, $(x'_n) \to w'$ in X one has $d_W(x_n, x'_n) \leq td_W(w, w')$ for n large enough, so that $d_Y(\overline{f}(w), \overline{f}(w')) = \lim_n d_Y(f(x_n), f(x'_n)) \leq m(td_W(w, w'))$. Let us observe that when m is continuous (and not just continuous at 0), m is also a modulus of uniform continuity of the map $\overline{f}$. □

The notion of a complete metric space enables us to present a powerful fixed point theorem. Here we deal with *contraction maps* , i.e. Lipschitzian maps whose Lipschitz rate is less than 1 and the process is called the method of *successive approximations*.

Theorem 2.10 (Contraction Theorem, Picard, Banach) *Let (X, d) be a complete metric space and let $f : X \to X$ be a contraction. Then f has a fixed point $\overline{x}$. Moreover, $\overline{x}$ is unique and for every $x_0 \in X$ one has $(f^{(n)}(x_0)) \to \overline{x}$ for $f^{(n)} := f \circ f^{(n-1)}$.*

Proof Let $c \in [0, 1[$ be the Lipschitz rate of f. Uniqueness of the fixed point stems from the contraction property: if $\overline{x}$ and $\overline{y}$ satisfy $f(\overline{x}) = \overline{x}, f(\overline{y}) = \overline{y}$ we have $d(\overline{x}, \overline{y}) = d(f(\overline{x}), f(\overline{y})) \leq cd(\overline{x}, \overline{y})$ and as $c \in [0, 1[$ this is possible only if $d(\overline{x}, \overline{y}) = 0$, i.e. $\overline{x} = \overline{y}$.

Given $x_0 \in X$ let us define inductively $x_{n+1} = f(x_n)$ for $n \in \mathbb{N}$. Since for $n \geq 1$ we have $x_n = f(x_{n-1})$ and $x_{n+1} = f(x_n)$, the contraction property yields

$$d(x_{n+1}, x_n) \leq cd(x_n, x_{n-1}).$$

By induction, this relation entails that $d(x_{n+1}, x_n) \leq c^n d(x_1, x_0)$ for all $n \in \mathbb{N}$. The sequence (x_n) is thus an Abel sequence (hence a Cauchy sequence), hence has a limit $\overline{x} \in X$. Passing to the limit in the relation $d(f(x_n), x_n) \leq c^n d(x_1, x_0)$ we get $d(f(\overline{x}), \overline{x}) = 0$, hence $f(\overline{x}) = \overline{x}$. □

Let us observe that the convergence of (x_n) to $\overline{x}$ is rapid enough since

$$d(x_{n+p}, x_n) \leq \sum_{k=n+1}^{n+p} d(x_k, x_{k-1}) \leq d(x_1, x_0) \sum_{k=n}^{\infty} c^k = \frac{d(x_1, x_0)}{1-c} c^n \tag{2.4}$$

and, passing to the limit as $p \to +\infty$, $d(\overline{x}, x_n) \leq (1-c)^{-1} c^n d(f(x_0), x_0)$.

Corollary 2.15 *Let W be a topological space, let (X, d) be a complete metric space and let $f : W \times X \to X$ be a continuous map such that for some $c \in [0, 1[$ and all $w \in W$ the partial map $f_w : x \mapsto f(w, x)$ is Lipschitzian with rate c. Then there exists a unique continuous map $g : W \to X$ such that $g(w) = f(w, g(w))$ for all $w \in W$.*

Proof For $w \in W$, let $g(w)$ be the unique fixed point of f_w. We have to show that g is continuous. Let $z \in W$ and given $\varepsilon > 0$ let $V \in \mathcal{N}(z)$ be such that $d(f(w, g(z)), f(z, g(z))) \leq \varepsilon$ for all $w \in V$. The triangle inequality yields for $w \in V$

$$\begin{aligned} d(g(w), g(z)) &= d(f_w(g(w)), f_z(g(z))) \\ &\leq d(f_w(g(w)), f_w(g(z))) + d(f_w(g(z)), f_z(g(z))) \\ &\leq cd(g(w), g(z)) + \varepsilon, \end{aligned}$$

so that $d(g(w), g(z)) \leq (1-c)^{-1}\varepsilon$. This shows that g is continuous at z. □

A remarkable property of complete metric spaces is the Baire property.

Theorem 2.11 (Baire) *Let (X, d) be a complete metric space.*

If (G_n) is a sequence in open dense subsets of X, then $\cap_n G_n$ is dense.

If X is the union of a countable family (F_n) of closed subsets, then one of them has a nonempty interior.

Proof Let (G_n) be a sequence of open dense subsets of X. Let us show that $G := \cap_n G_n$ is dense. Let (s_n) be a sequence of positive numbers with limit 0. Given a nonempty open subset U of X, the set $G_n \cap U$ is nonempty and open; in particular, $G_0 \cap U$ contains some closed ball $B[x_0, r_0]$ with $r_0 \in]0, s_0]$. Assume by induction that we have constructed open balls $B(x_k, r_k)$ with $r_k \leq s_k$, $B[x_k, r_k] \subset B(x_{k-1}, r_{k-1}) \cap G_k$ for $k = 1, \ldots, n$. Since $B(x_n, r_n)$ meets G_{n+1}, we can find a closed ball $B[x_{n+1}, r_{n+1}] \subset G_{n+1} \cap B(x_n, r_n)$ with $r_{n+1} \leq s_{n+1}$. The sequence (x_n) obtained in this way is a Cauchy sequence (since $d(x_{n+p}, x_n) \leq s_n$ for all n, p). Its limit $\overline{x}$ belongs to $B[x_m, r_m] \subset G_m$ for all $m \in \mathbb{N}$ and in particular $\overline{x} \in B[x_0, r_0] \subset U$ and $\overline{x} \in G$, so that $G \cap U$ is nonempty: G is dense.

Now suppose $X = \cup_n F_n$, where each F_n is closed. Then $G_n := X \backslash F_n$ is open and if F_n has an empty interior then G_n is dense. If this happens for all $n \in \mathbb{N}$, then $\cap_n G_n$ is dense, an impossibility since $\cap_n G_n = \varnothing$. Thus, at least one F_n has a nonempty interior. □

The preceding result can be expressed in terms of genericity. A subset G of some topological space T is *generic* if it contains the intersection of a countable family of open subsets of T (a so-called $\mathcal{G}_\delta$ set, the notation being a reminder of the German term "Gebiete", while the notation F stems from the French "fermé" for closed) that are dense in T; other terminologies are that G is *residual* or that the complement of G is *meager* or a *set of first category*. It is convenient to say that a property involving a point is generic if it holds on a generic subset. The main feature of this notion is that the intersection of a finite (or countable) family of generic subsets is still generic, a property that does not hold for dense subsets (consider the set of rational numbers and the set of irrational numbers in $\mathbb{R}$). The (equivalent) properties of Theorem 2.11 can be phrased as follows: in a complete metric space any generic subset is dense. A topological space satisfying this property is called a Baire space. Locally compact topological spaces are also Baire spaces.

Complete metric spaces can be characterized by an approximate minimization principle that is extremely useful. Here, given $\varepsilon > 0$, we say that a point $\overline{x}$ of a set X is an *ε-minimizer* of a function $f : X \to \mathbb{R}_\infty$ if $f(\overline{x}) \leq \inf f(X) + \varepsilon$.

Theorem 2.12 (Ekeland) *Let (X, d) be a complete metric space and let $f : X \to \mathbb{R}_\infty$ be a bounded below lower semicontinuous function taking at least one finite value. Given $\varepsilon > 0$, an ε-minimizer $\overline{x}$ of f, and given $c, r > 0$ satisfying $cr \geq \varepsilon$, one can find $u \in B[\overline{x}, r]$ such that $f(u) + cd(u, \overline{x}) \leq f(\overline{x})$ and*

$$f(u) < f(x) + cd(u, x) \quad \textit{for all } x \in X \backslash \{u\}. \tag{2.5}$$

Thus, not too far from $\overline{x}$, we can find a strict mimimizer u of the modified function $f_{c,u} := f + cd(u, \cdot)$. Replacing the metric d with the metrics $d' := d/(1 + d)$ or $d'' := \min(d, 1)$ and c with the general term of a sequence $(c_n) \to 0_+$, we can ensure that the approximate function $f_{c,u}$ is as close to f as required with respect to the uniform metric. However, there is a trade off between the accuracies of the two approximating elements u, $f_{c,u}$: one cannot expect to get arbitrarily accurate approximations of f and of $\overline{x}$ at the same time.

Proof We associate to f and c an order on X defined by $w \preceq x$ if $f(w) + cd(w, x) \leq f(x)$. Let

$$B(x) := \{w \in X : f(w) + cd(w, x) \leq f(x)\} \qquad x \in X$$

be the set of elements below x. We have $x \in B(x)$ for all $x \in X$ and the relations $y \in B(x)$, $x \in B(y)$ imply $d(x, y) = 0$ or $x = y$. Let us verify that the relation B satisfies the transitivity property $B(y) \subset B(x)$ for all $x \in X$, $y \in B(x)$. We may assume $x \in \mathrm{dom} f := f^{-1}(\mathbb{R})$, so that $f(y) < \infty$. Then, for all $z \in B(y)$ we also have $f(z) < \infty$ and $cd(y, z) \leq f(y) - f(z)$. Since $y \in B(x)$, we also have $cd(x, y) \leq f(x) - f(y)$. Adding the respective sides of these two inequalities, and using the triangle inequality, we get $cd(x, z) \leq f(x) - f(z)$, or $z \in B(x)$. Thus B defines an order; we shall construct a minimal element.

Given $\overline{x} \in \mathrm{dom} f$, we can define inductively a sequence starting from $x_0 := \overline{x}$ by picking $x_{n+1} \in B(x_n)$ satisfying

$$f(x_{n+1}) \leq \frac{1}{2} f(x_n) + \frac{1}{2} \inf f(B(x_n)). \tag{2.6}$$

Such a choice is possible: it suffices to use the definition of an infimum when $\inf f(B(x_n)) < f(x_n)$ and to take $x_{n+1} = x_n$ when $\inf f(B(x_n)) = f(x_n)$. Since $x_n \in B(x_n)$, (2.6) ensures that the sequence $(f(x_n))$ is nonincreasing, hence is convergent as f is bounded below. Let $\ell := \lim_n f(x_n)$.

Since $x_{n+1} \in B(x_n)$ we have $cd(x_n, x_{n+1}) \leq f(x_n) - f(x_{n+1})$ and by induction

$$cd(x_n, x_{n+p}) \leq f(x_n) - f(x_{n+p}) \tag{2.7}$$

for all $n, p \geq 0$. Thus (x_n) is a Cauchy sequence, hence has a limit we denote by u.

Because f is lower semicontinuous, for each $n \in \mathbb{N}$ the set $B(x_n)$ is closed. Since relation (2.7) says that $x_{n+p} \in B(x_n)$ for all $p \geq 0$, we get $u \in B(x_n)$. In particular, taking $n = 0$ and remembering that $x_0 = \overline{x}$, we get

$$f(u) + cd(\overline{x}, u) \leq f(\overline{x}).$$

Moreover, by the transitivity property of relation B, for all $v \in B(u)$ and all $n \in \mathbb{N}$, we have $v \in B(x_n)$. Thus $\inf f(B(x_n)) + cd(x_n, v) \leq f(v) + cd(x_n, v) \leq f(x_n)$ and relation (2.6) yields

$$cd(x_n, v) \leq f(x_n) - \inf f(B(x_n)) \leq 2\,(f(x_n) - f(x_{n+1})) \to 0,$$

hence $d(v, u) = 0$. It follows that $B(u) = \{u\}$. This relation means that (2.5) is satisfied.

If $\overline{x}$ is such that $f(\overline{x}) \leq \inf f(X) + \varepsilon$, and $\varepsilon \leq cr$, we have

$$\inf f(X) + cd(u, \overline{x}) \leq f(u) + cd(u, \overline{x}) \leq f(\overline{x}) \leq \inf f(X) + cr,$$

so that $d(u, \overline{x}) \leq r$. □

Given a metric space (X, d) one may look for a complete metric space $(\widehat{X}, \hat{d})$ that is close enough to (X, d). A precise answer can be given, even for semimetric spaces.

Theorem 2.13 *Given a semimetric space (X, d) there is a complete semimetric space $(\widehat{X}, \hat{d})$ and an isometry j from X onto a dense subset of $(\widehat{X}, \hat{d})$. Moreover, if (Y, d_Y) is a complete metric space and if $f : X \to Y$ is a uniformly continuous map, there is a unique uniformly continuous map $\hat{f} : \widehat{X} \to Y$ such that $f = \hat{f} \circ j$.*

Proof This last property ensures that the pair $(\widehat{X}, j)$ is unique up to an isometry. In fact, if $(\widehat{X'}, j')$ is another pair, there is a uniformly continuous map $\widehat{j'} : \widehat{X} \to \widehat{X'}$ such that $j' = \widehat{j'} \circ j$. Similarly, there is a uniformly continuous map $\hat{j} : \widehat{X'} \to \widehat{X}$ such that $j = \hat{j} \circ j'$. Then $j = \hat{j} \circ (\widehat{j'} \circ j) = I_{\widehat{X}} \circ j$, so that the two maps $\hat{j} \circ \widehat{j'}$ and $I_{\widehat{X}}$ coincide by the uniqueness requirement (or the density of $j(X)$ in $\widehat{X}$). Similarly, one shows that $\widehat{j'} \circ \hat{j}$ coincides with the identity map $I_{\widehat{X'}}$ on $\widehat{X'}$.

Several specific constructions can be given for $(\widehat{X}, j)$. One consists in taking for $\widehat{X}$ the set of equivalence classes of Cauchy sequences for the relation $(x_n) \sim (x'_n)$ if $\lim_n d(x_n, x'_n) = 0$. We invite the reader to complete the construction by defining $\hat{d}$ and taking for $j(x)$ the class of the constant sequence with value x.

When d is a metric that is bounded above on X^2 we can also take an embedding j into the space $C_b(X) := C_b(X, \mathbb{R})$ endowed with the metric d_∞ defined by $d_\infty(f, g) = \sup_{x \in X} |f(x) - g(x)|$. Given $x \in X$ we define $j(x)$ as the function $w \mapsto d(w, x)$ on X. Then, for x, $y \in X$, the triangle inequality yields $d_\infty(j(x), j(y)) \leq d(x, y)$. In fact this inequality is an equality since $|j(x)(y) - j(y)(y)| = d(x, y)$. Taking for $\widehat{X}$ the closure of $j(X)$ in $C_b(X)$ for $\hat{d} := d_\infty$ we get a complete space in which $j(X)$ is dense. The last assertion of the statement is a consequence in Theorem 2.4. When d is an arbitrary metric, turning the requirement that j be isometric into the requirement that j be nonexpansive, i.e. Lipschitzian with rate 1, we can define j by setting $j(x)(w) := \min(d(w, x), 1)$. □

If $(d_a)_{a \in A}$ is a family of semimetrics on X, one can consider on X the topology generated by the balls $B_a(x, r) := \{w \in X : d_a(w, x) < r\}$. However, X has a structure richer than a topology. It is called a uniformity and $(X, (d_a)_{a \in A}$ is called a *uniform space*. A sequence (x_n) of X is called a *Cauchy sequence* if for all $a \in A$ one has $(d_a(x_n, x_p)) \to 0$ as $n, p \to +\infty$. A uniform space is said to be complete if any Cauchy sequence is convergent.

2.3.3 Application to Ordinary Differential Equations

We intend to give a simple existence theorem for ordinary differential equations. One of its proofs relies on a generalization of the Contraction Theorem.

Proposition 2.28 *Let (X, d) be a complete metric space and let $f : X \to X$ be a continuous map such that for some $k \in \mathbb{N}\backslash\{0\}$ the k times iterated map $f^{(k)} := f \circ \ldots \circ f$ is a contraction. Then f has a unique fixed point.*

Proof If $\overline{x}$ and $\overline{y}$ are fixed points of f, then they are fixed points of $f^{(k)}$, hence $\overline{x} = \overline{y}$. Let $\overline{x}$ be the fixed point of $f^{(k)}$. We know that for any $x_0 \in X$ we have $(f^{(kn)}(x_0))_n \to \overline{x}$. In particular, taking $x_0 := f(\overline{x})$ we get that $(f^{(kn+1)}(\overline{x}))_n \to \overline{x}$. On the other hand, since $f^{(kn)}(\overline{x}) = \overline{x}$, we have $f(f^{(kn)}(\overline{x})) = f(\overline{x})$. By uniqueness of limits, we get $f(\overline{x}) = \overline{x}$. □

For this existence result, we anticipate to some notions of derivation and integration to be found later. The reader may suppose $E := \mathbb{R}$ for the sake of simplicity.

Theorem 2.14 *Let T be a bounded interval of $\mathbb{R}$, let E be a Banach space and let $f : T \times E \to E$ be a continuous map such that for some $c \in \mathbb{R}_+$ one has $\|f(t, e) - f(t, e')\| \leq c\,\|e - e'\|$ for all $t \in T$, e, $e' \in E$. Then, given $t_0 \in T$, $e_0 \in E$ there exists a unique solution $x \in C(T, E)$ with a continuous derivative x' to the equation*

$$x'(t) = f(t, x(t)) \quad t \in T, \qquad x(t_0) = e_0. \tag{2.8}$$

Proof We admit that $x \in C(T, E)$ satisfies (2.8) if and only if x satisfies the integral equation

$$x(t) = e_0 + \int_{t_0}^{t} f(s, x(s))ds. \tag{2.9}$$

Let us endow the space $X := C_b(T, E)$ of bounded continuous maps from T to E with the norm $\|\cdot\|_\infty$ and let us consider the map $F : X \to X$ given by

$$F(x)(t) := e_0 + \int_{t_0}^{t} f(s, x(s))ds.$$

For $x, y \in C_b(T, E)$ we have

$$\begin{aligned} \|F(x) - F(y)\|_\infty &= \sup_{t \in T} \left\| \int_{t_0}^{t} (f(s, x(s)) - f(s, y(s)))ds \right\| \\ &\leq \sup_{t \in T} \left| \int_{t_0}^{t} c\,\|x(s) - y(s)\|\, ds \right| \leq \ell(T)c\,\|x - y\|_\infty \end{aligned}$$

where $\ell(T)$ is the length of T. Thus, for $\ell(T)c < 1$, F is a contraction and F has a fixed point that is a solution to (2.9). In the general case it can be shown by induction that for all $k \in \mathbb{N}\setminus\{0\}$, $x, y \in C_b(T, E)$ one has

$$\left\| F^{(k)}(x) - F^{(k)}(y) \right\|_\infty \leq \frac{c^k \ell(T)^k}{k!} \|x - y\|_\infty .$$

For k large enough one has $c^k \ell(T)^k / k! < 1$ and we can apply the preceding proposition to find a fixed point of F, hence a solution to (2.8). □

Exercises

1. Observe that the proof we gave of Corollary 2.15 only uses the continuity of the partial map $w \mapsto f(w, g(z))$. Prove that such an assumption is equivalent to the continuity of f at $(z, g(z))$ in view of the hypothesis that for all $w \in W$ the map f_w is Lipschitzian with rate c.
2. Let (X, d) be a compact metric space and let $f : X \to X$ be a map such that for $x, x' \in X$ with $x \neq x'$ one has $d(f(x), f(x')) < d(x, x')$. Show that f has a unique fixed point that can be found by the method of successive approximations.
3. Let (X, d) be a complete metric space, let $c > 1$, and let $f : X \to X$ be a continuous map such that for $x, x' \in X$ one has $d(f(x), f(x')) \geq cd(x, x')$. Show that f has a unique fixed point.
4. Let $T := [a, b]$, let $X := L_2(T)$, let $k \in L_2(T \times T)$ be such that $\int_{T^2} k^2(s, t)dsdt < 1$, and let $f : \mathbb{R}\times T^2 \to \mathbb{R}$ be such that for all $(r, r', s, t) \in \mathbb{R}^2 \times T^2$ one has $|f(r, s, t) - f(r', s, t)| \leq k(s, t) |r - r'|$ and such that for all $x \in X$ the function $t \mapsto \int_a^b f(x(s), s, t)ds$ belongs to X. Prove that for every $y \in X$ the following *integral equation* has a unique solution

$$x(t) = y(t) + \int_a^b f(x(s), s, t)ds.$$

5. Let X be an open subset of a complete metric space (W, d), with $X \neq W$. For x, $x' \in X$, let

$$d_X(x, x') := \left| 1/d(x, W\setminus X) - 1/d(x', W\setminus X) \right| .$$

Show that d_X is a metric on X whose associated topology is the induced topology. Prove that (X, d_X) is complete. Does this contradicts the fact that in general X is not closed in W?
6. Show that on $\mathbb{R}$ the function $d' : \mathbb{R} \times \mathbb{R} \to \mathbb{R}$ given by $d'(w, x) = \left| w^3 - x^3 \right|$ is a metric topologically equivalent to the usual metric d given by $d(w, x) := |w - x|$, but not uniformly equivalent to d. Verify that the Cauchy sequences for d and d' are the same.

7. Let (X, d) be a complete metric space and let $f, g : X \to X$ be two contractions with rate $c \in]0, 1[$. Let $\overline{x}$ (resp. $\overline{y}$) be the fixed point of f (resp. g). Prove that $d(\overline{x}, \overline{y}) \leq (1 - c)^{-1} d_\infty(f, g)$ where $d_\infty(f, g) := \sup_{x \in X} d(f(x), g(x))$. [Hint: In relation (2.4) take $n = 0, x_0 := \overline{y}$, pass to the limit on p and note that $d(f(\overline{y}), \overline{y}) \leq d_\infty(f, g)$.]

2.3.4 Compact Metric Spaces

Some additional properties of compact spaces can be obtained when they are *metrizable*, i.e. when their topologies can be associated with metrics. They are even valid for sequentially compact spaces, a topological space X being called *sequentially compact* if every sequence in X has a convergent subsequence. First we note a uniformity property related to coverings.

Lemma 2.5 (Lebesgue) *Let (W, d) be a metric space and let X be a subset of W that is sequentially compact. Then for any family $(U_i)_{i \in I}$ of open subsets of W whose union contains X there exists some $r > 0$ such that for all $x \in X$ the ball $B(x, r)$ is contained in some U_i.*

Proof If no such r exists, for all $n \in \mathbb{N} \backslash \{0\}$ one can find some $x_n \in X$ such that for all $i \in I$ the ball $B(x_n, 1/n)$ is not contained in U_i. Let $\overline{x} \in X$ be the limit of a subsequence $(x_{k(n)})_n$ of (x_n). Let $r_n := d(x_{k(n)}, \overline{x})$ and let $j \in I$ be such that $\overline{x} \in U_j$. Let $s > 0$ be such that $B(\overline{x}, s) \subset U_j$. For n large enough we have $r_n + 1/k(n) < s$ so that $B(x_{k(n)}, 1/k(n))$ is contained in $B(\overline{x}, s)$, hence in U_j. This contradicts the choice of the balls $B(x_n, 1/n)$. □

The second property we consider captures the idea that such a space can be approximated by a finite set.

Definition 2.12 A metric space (X, d) is said to be *precompact* if for any $\varepsilon > 0$ there exists a finite subset F_ε of X such that for all $x \in X$ one has $d(x, F_\varepsilon) < \varepsilon$.

This means that for all $\varepsilon > 0$ there is a covering of X by a finite number of balls of radius ε. Such a space is clearly bounded and separable. But we have more.

Theorem 2.15 *For a metric space (X, d) the following properties are equivalent:*

(a) every sequence (x_n) of X has a cluster point;
(b) X is sequentially compact;
(c) X is precompact and complete;
(d) every infinite subset S of X has an accumulation point;
(e) X is compact.

Proof The equivalence (a)⇔(b) stems from Proposition 2.21 and the implication (e)⇒(a) is obvious. Let us prove (b)⇒(c). If X is sequentially compact then X is complete since every Cauchy sequence in X has a convergent subsequence, hence is convergent. If X is not precompact there exists some $\varepsilon > 0$ such that X is not covered by a finite number of balls of radius ε. Thus, by induction we build a sequence (x_n) such that, for all $n \in \mathbb{N}$, x_{n+1} is not contained in $B(x_0, \varepsilon) \cup \ldots \cup B(x_n, \varepsilon)$. Such a sequence cannot have a convergent subsequence. Now let us show that (c)$\Rightarrow$(d). For every $n \in \mathbb{N}$ there exists a finite subset F_n of X such that the balls with center in F_n and radius 2^{-n} cover X, hence covers S. One of these balls contains an infinite number of points of S. Let us denote by $x_n \in F_n$ its center. Similarly there exists some $x_{n+1} \in F_{n+1}$ such that $B(x_{n+1}, 2^{-n-1})$ contains an infinite number of points of $S \cap B(x_n, 2^{-n})$. By the triangle inequality we have $d(x_{n+1}, x_n) < 2^{-n+1}$. Since X is complete, the sequence (x_n) built in this way converges to some $\overline{x} \in X$ since it is a Cauchy sequence (in fact an Abel sequence). Again, the triangle inequality shows that $\overline{x}$ is an accumulation point of S. The implication (d)$\Rightarrow$(a) is immediate: given a sequence (x_n) of X either $S := \{x_n : n \in \mathbb{N}\}$ is finite and then a subsequence of (x_n) is constant, hence convergent, else S is infinite and any accumulation point a of S is a cluster point of (x_n) since we know that any neighborhood V of a contains an infinite number of points of S. It remains to show that (a)$\Rightarrow$(e). Let $(U_i)_{i \in I}$ be an open covering of X. The preceding lemma yields some $r > 0$ such that for all $x \in X$ there exists some $i(x) \in I$ such that $B(x, r) \subset U_{i(x)}$. On the other hand, since (a)$\Rightarrow$(c), X is precompact and there exists a finite subset F of X such that $\{B(x, r) : x \in F\}$ is a covering of X. Then $\{U_{i(x)} : x \in F\}$ is a finite covering of X: X is compact. □

Corollary 2.16 *A compact metric space is separable.*

Proof Let X be a compact metric space. Since X is precompact, for any sequence $(r_n) \to 0_+$ one can find a finite subset F_n of X such that $\{B(x, r_n) : x \in F_n\}$ is a covering of X. Then $D := \cup_n F_n$ is a dense countable subset of X. □

Another important consequence of compactness and metrizability follows.

Theorem 2.16 *Let (W, d_W) be a metric space and let X be a relatively compact subset of W. If $f : W \to Y$ is a continuous map with values in another metric space (Y, d_Y), then f is uniformly continuous around X in the following sense: for every $\varepsilon > 0$ there exists some $\delta > 0$ such that for all $w \in W$, $x \in X$ satisfying $d_W(w, x) < \delta$ one has $d_Y(f(w), f(x)) < \varepsilon$.*

Of course, if $X = W$ the conclusion is just usual uniform continuity of f.

Proof Given $\varepsilon > 0$, since f is continuous, for all $z \in \mathrm{cl}(X)$ there exists some open neighborhood U_z of z such that $f(U_z) \subset B(f(z), \varepsilon/2)$. Lemma 2.5 yields some $r > 0$ such that for all x in the compact set $\mathrm{cl}(X)$ one has $B(x, r) \subset U_{z(x)}$ for some $z(x) \in \mathrm{cl}(X)$. Then, for $w \in B(x, r)$ one has $d_Y(f(w), f(x)) \leq d_Y(f(w), f(z(x))) + d_Y(f(z(x)), f(x)) < \varepsilon$ since $w, x \in B(x, r) \subset U_{z(x)}$. Thus we can take $\delta = r$. □

Exercise Give a proof by contradiction using sequences and subsequences.

Exercise Give a proof using sequences and cluster points.

Exercise Show that the result is valid if X is a semi-metric space and Y is a uniform space.

The compactness assumption in Theorem 2.6 can be relaxed by using a notion of coercivity. Let us say that a function $f : X \to \overline{\mathbb{R}}$ on a metric space X is *coercive* if for all $r \in \mathbb{R}$ the sublevel set $S_f(r) := f^{-1}(]-\infty, r])$ is bounded, or, equivalently, if $f(x) \to \infty$ as $d(x, x_0) \to \infty$ (x_0 being an arbitrary point of X). This notion is essentially used in the case when X is a normed space and $f(x) \to \infty$ as $\|x\| \to \infty$. We will say that f is *compactly coercive* if for all $r \in \mathbb{R}$ the sublevel set $S_f(r)$ is compact. When f is lower semicontinuous and the closed balls of X are compact, both notions coincide. For such a function, the existence of minimizers is ensured.

Lemma 2.6 *Let $f : X \to \mathbb{R}_\infty$ be a compactly coercive function on a metric space X. Then f attains its minimum. In particular, when the closed balls of X are compact, any coercive lower semicontinuous function on X attains its minimum.*

Proof The result being obvious when f only takes the value $+\infty$, let us take $r \in \mathbb{R}$ such that $S_f(r) := f^{-1}(]-\infty, r])$ is nonempty. By assumption, $S_f(r)$ is compact. By the Weierstrass' Theorem, f attains its infimum on $S_f(r)$. Since $\inf f(X) = \inf f(S_f(r))$, any minimizer of the restriction $f \mid S_f(r)$ of f is also a minimizer of f. □

Proposition 2.29 *Let C be a nonempty closed convex subset of a Euclidean space X and let $a \in X$. Then there exists some $p \in C$ called the best approximation of a in C such that $\|p - a\| \leq \|x - a\|$ for all $x \in C$. Moreover, p is characterized by the inequality*

$$\forall x \in C \qquad \langle a - p \mid x - p \rangle \leq 0. \tag{2.10}$$

Proof The function $x \mapsto \|x - a\|$ is continuous and coercive and the closed balls of X are compact since they are dilations of smaller balls and since X is locally compact. Setting $f(x) = \|x - a\|$ if $x \in C$ and $f(x) := +\infty$ if $x \in X \backslash C$, we get a coercive lower semicontinuous function on X. It attains its infimum at some point p of C, so that $\|p - a\| \leq \|x - a\|$ for all $x \in C$. Given $x \in C$, for $t \in]0, 1]$ we have $x_t := p + t(x - p)$ in C by convexity, hence

$$\|p - a\|^2 \leq \|x_t - a\|^2 = \|p - a\|^2 + 2t\langle p - a \mid x - p \rangle + t^2 \|x - p\|^2.$$

Simplifying both sides and dividing by t and then passing to the limit we get (2.10). □

Exercises

1. Let X be a closed subset of $\mathbb{R}^d$ and let $f : X \to \overline{\mathbb{R}}$ be lower semicontinuous and *pseudo-coercive* in the sense that there exists some $x_0 \in X$ such that $f(x_0) < \liminf_{\|x\| \to \infty,\, x \in X} f(x)$. Show that f attains its infimum.

2. Let X be a closed subset of $\mathbb{R}^d$ and let $f : X \to \overline{\mathbb{R}}$. Assume f is *finitely minimizable* in the sense that there exists an $r \in \mathbb{R}_+$ such that, for any $t > m := \inf f(X)$, there exists some $x \in X$ with $\|x\| \leq r, f(x) < t$. Show that any pseudo-coercive function is finitely minimizable and that any finitely minimizable lower semicontinuous function on X attains its infimum at some point of $X \cap B[0, r]$, where r is the radius of essential minimization, i.e. the infimum of the real numbers r for which the above definition is satisfied.
3. Prove Weierstrass' Theorem in the case when X is a compact metric space by using a *minimizing sequence* of f, i.e. a sequence (x_n) of X such that $(f(x_n)) \to \inf f(X)$.
4. Show that any l.s.c. function f on $[0, 1]$ (or on a separable metric space X) is the supremum of the family of continuous functions majorized by f.
5. Prove Corollary 2.12 by using open subsets.
6. Show that among all cylindrical barrels of a given area s there is one with greatest volume.
7*. (**d'Alembert-Gauss Theorem**) Prove that any polynomial P with complex coefficients has at least one root in $\mathbb{C}$. [Hint: verify that $|P(\cdot)|$ is coercive and show that if z_0 is such that $|P(z_0)| = \inf\{|P(z)| : z \in \mathbb{C}\}$, then $P(z_0) = 0$.]
8. Show that the following properties of a locally compact metric space (X, d) are equivalent:

   (a) X is the union of an increasing sequence (X_n) of open relatively compact subsets of X such that $\mathrm{cl}(X_n) \subset X_{n+1}$ for all $n \in \mathbb{N}$;
   (b) X is the union of a countable family of compact subsets;
   (c) X is separable.

Additional Reading

[59, 159, 163, 169, 174, 176, 177, 183, 225, 255]

Chapter 3
Elements of Functional Analysis

Everything should be made as simple as possible, but not simpler.

Albert Einstein.

Abstract In this central chapter, the fundamental elements of functional analysis are presented. Although topological vector spaces are considered, essentially for the use of weak topologies, the focus is on normed spaces. Normed spaces in duality or metric duality form a convenient framework. The main pillars of functional analysis are presented: separation properties, the uniform boundedness theorem, the open mapping theorem, the closed graph theorem... Some special properties such as reflexivity, separability, and uniform convexity are given some attention. An account of spectral theory for linear operators is presented and the case of compact operators is considered. The chapter ends with a presentation of regulated functions and functions of bounded variation. This enables us to dispose of an elementary integration theory that is often sufficient for simple purposes.

Besides subsets of Euclidean spaces, spaces of functions are the most common examples of topological spaces. Historically, they prompted the study of abstract metric spaces and topological spaces: the notion of a metric space appeared as a transposition of the usual notion in $\mathbb{R}^d$. Then, it was realized that the concept of a norm is even a simplification of the concept of a metric and that the framework of linear spaces equipped with a norm is a very convenient framework to tackle and solve a number of problems. This chapter is devoted to such a tool. We focus on linearity, leaving aside the powerful tools of nonlinear functional analysis. Most of the results we expound have been obtained during the twentieth century. They have applications in various problems and thus are important.

Since many problems can be solved by using a fixed point theorem, let us quote one such result in a simple form:

(**Brouwer's Fixed Point Theorem**) For any continuous map $f : B_X \to B_X$ from the closed unit ball B_X of a Euclidean space X into itself there exists some $\overline{x} \in B_X$ such that $f(\overline{x}) = \overline{x}$.

J.-P. Penot, *Analysis*, Universitext, DOI 10.1007/978-3-319-32411-1_3

A proof of this result is presented in the appendix; it involves several tools from various chapters of the book. Whereas a generalization due to J. Schauder of this result has to be used to solve problems in functional spaces (see Theorem 3.28 below), we first present a proof for the simple one-dimensional case for which $B_X = [-1, 1]$. Then, setting $g(x) = f(x) - x$, we see that $g(-1) \geq 0$, $g(1) \leq 0$ and the existence of some $\overline{x} \in [-1, 1]$ such that $g(\overline{x}) = 0$ stems from the intermediate value theorem.

3.1 Normed Spaces

When a set is endowed with a topology (or a metric) and has an algebraic structure, it is interesting to study the case when these two structures are compatible in the sense that the operations are continuous. If this is not the case, one might face difficulties.

3.1.1 General Properties of Normed Spaces

Given a linear space X, it is natural to select metrics or semimetrics that are compatible with the operations, so that one may expect simplifications. Let us say that a semimetric d on a linear space is *compatible* (with the linear structure) if for all $x, y, v \in X$ one has $d(x + v, y + v) = d(x, y)$ (invariance by translations) and if for all $\lambda \in \mathbb{R}$ (or $\lambda \in \mathbb{C}$ if X is a linear space over $\mathbb{C}$) one has $d(\lambda x, \lambda y) = |\lambda|\, d(x, y)$. The following result is immediate.

Lemma 3.1 *If a semimetric d on a linear space X is compatible, then the function $p : X \to \mathbb{R}_+$ given by $p(x) := d(x, 0)$ is a seminorm, i.e. a function $p : X \to \mathbb{R}_+$ satisfying the following properties:*

(SN1) $p(\lambda x) = |\lambda|\, p(x)$ *for all* $\lambda \in \mathbb{R}$ *(or* $\lambda \in \mathbb{C}$*) and all* $x \in X$*;*
(SN2) $p(x + y) \leq p(x) + p(y)$ *for all* $x, y \in X$.

This last relation is called the *triangle inequality*. It stems from the relations

$$d(x + y, 0) \leq d(x + y, y) + d(y, 0) = d(x, 0) + d(y, 0).$$

Conversely, given a seminorm p on X, one gets a compatible semimetric d by setting $d(x, y) := p(x - y)$. For $x, y, z \in X$ one has

$$d(x, z) = p(x - z) = p((x - y) + (y - z)) \leq p(x - y) + p(y - z) = d(x, y) + d(y, z).$$

Since $p(-v) = p(v)$ for all $v \in X$, one also has $d(y, x) = d(x, y)$. Moreover, from (SN1) one deduces that $p(0) = 0$, hence $d(x, x) = 0$ for all $x \in X$. The compatibility conditions are immediate consequences of the definition of d and of (SN1). A seminorm satisfying the condition

(N0) $p(x) = 0$ implies $x = 0$

is called a *norm*. The associated semimetric is then a metric and the converse is true. A norm is often denoted by $x \mapsto \|x\|$. A *normed space* is a linear space equipped with a norm. A *Banach space* is a complete normed space.

Example On $\mathbb{R}^d$ familiar norms are defined as follows for $x = (x_1, \ldots, x_d)$:

$$\|x\|_1 = |x_1| + \ldots + |x_d|,$$
$$\|x\|_\infty = \max(|x_1|, \ldots, |x_d|),$$
$$\|x\|_p = (|x_1|^p + \ldots + |x_d|^p)^{1/p} \qquad (p \in [1, \infty[).$$

The norm $\|\cdot\|_2$ corresponding to the choice $p = 2$ is called the *Euclidean norm.* It has interesting properties due to the fact that it is associated with a scalar product but it is not always as convenient as $\|\cdot\|_1$ or $\|\cdot\|_\infty$.

Example If S is an arbitrary set and if $(E, \|\cdot\|)$ is a normed space, it is easy to show that on the space $B(S,E)$ (resp. $B(S)$) of bounded maps from S into E (resp. of bounded real-valued functions on S) one gets a norm $\|\cdot\|_\infty$ by setting $\|f\|_\infty := \sup_{s \in S} \|f(s)\|$ (resp. $\|f\|_\infty := \sup_{s \in S} |f(s)|$). This norm is called the norm of uniform convergence or the sup norm.

It is convenient to denote by B_X the closed ball $B[0, 1]$ centered at 0 with radius 1, called the (closed) *unit ball* of X. A subset B of X is seen to be bounded if and only if there exists some $r \in \mathbb{R}_+$ such that $B \subset rB_X$. The *unit sphere* of X is the set S_X of $u \in X$ such that $\|u\| = 1$.

Two norms $\|\cdot\|$ and $\|\cdot\|'$ are said to be *equivalent* if there exist two positive constants c, c' such that $(1/c)\|\cdot\| \leq \|\cdot\|' \leq c'\|\cdot\|$. Such a relation defines an equivalence relation among norms. Moreover, if the norms $\|\cdot\|$ and $\|\cdot\|'$ are equivalent then the associated metrics d and d' are metrically (hence uniformly and topologically) equivalent.

The following result can be considered as a training in the use of norms (Fig. 3.1).

Proposition 3.1 (Riesz) *Let Y be a closed linear subspace of a normed space $(X, \|\cdot\|)$ with $Y \neq X$. Then, for every $\varepsilon > 0$ one can find some $x \in X$ such that $\|x\| = 1$ and $d(x, Y) \geq 1 - \varepsilon$, for $d(x, Y) := \inf\{d(x, y) : y \in Y\}$.*

Proof Let $z \in X \backslash Y$ and let $\varepsilon > 0$ be given. We can find some $y_\varepsilon \in Y$ such that

$$r := \|z - y_\varepsilon\| \leq (1 + \varepsilon) d(z, Y).$$

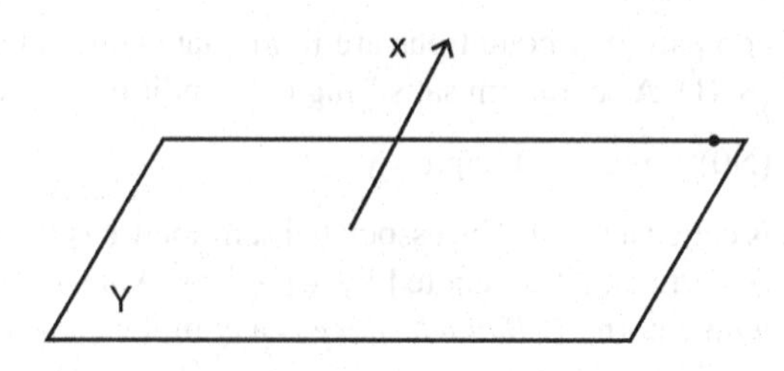

Fig. 3.1 Approximate orthogonality

Let $x = r^{-1}(z - y_\varepsilon)$. If $y \in Y$ we have $y_\varepsilon + ry \in Y$, hence $\|z - y_\varepsilon - ry\| \geq d(z, Y)$,

$$\|x - y\| = \left\|r^{-1}(z - y_\varepsilon) - y\right\| = r^{-1}\|z - y_\varepsilon - ry\| \geq r^{-1}d(z, Y) \geq (1+\varepsilon)^{-1}.$$

Taking the infimum over $y \in Y$, it follows that $d(x, Y) \geq (1+\varepsilon)^{-1} \geq 1 - \varepsilon$. □

Exercises

1. Show that in a normed space $(X, \|\cdot\|)$ the closure of the open ball $B(a, r)$ is the closed ball $B[a, r]$ and that the interior of $B[a, r]$ is the open ball $B(a, r)$. Give an example showing that this property is not always satisfied in a general metric space.
2. Let A be a closed subset of a normed space $(X, \|\cdot\|)$ and let B be a compact subset of X. Show that $A + B := \{x + y : x \in A,\ y \in B\}$ is closed. Give an example showing that the sum of two closed subsets of X is not always closed. [Hint: take $X := \mathbb{R}^2$, $A := \mathbb{R}\times\{0\}$, $B := \{(x, y) \in \mathbb{R}_+ \times \mathbb{R}_+ : xy = 1\}$.]
3. Let X (often denoted by c_{00}) be the space of sequences $x := (x_n)$ of real numbers such that x_n is null for n large enough. Show that the norms given by $\|x\| := \sup_n |x_n|$ and $\|x\|_1 = \Sigma_n |x_n|$ are not equivalent.
4. Show that two norms $\|\cdot\|$ and $\|\cdot\|'$ on a linear space X are equivalent if and only if the associated metrics d and d' are metrically equivalent (resp. uniformly equivalent, resp. topologically equivalent).
5. Let $(X, \|\cdot\|)$ be a normed vector space, let $x, y \in X$ and let $r \in \mathbb{R}_+$. Show that $r\|x\| = \lim_{n\to\infty}(\|(r+n)x + y\| - \|nx + y\|)$.
6. Let X be the space of functions of class C^1 on some compact interval T of $\mathbb{R}$ equipped with the norm defined by $\|x\| := \sup_{t\in T}|x(t)| + \sup_{t\in T}|x'(t)|$. Show that setting $\|x\|' := \sup_{t\in T}(|x(t)| + |x'(t)|)$ one defines an equivalent norm.

7*. Let $(X, \|\cdot\|)$ be a normed vector space and let $x, y \in X\backslash\{0\}$. Show that

$$\|x - y\| \geq \frac{1}{2}\max(\|x\|, \|y\|). \left\| \frac{x}{\|x\|} - \frac{y}{\|y\|} \right\|,$$

$$\|x - y\| \geq \frac{1}{4}(\|x\| + \|y\|). \left\| \frac{x}{\|x\|} - \frac{y}{\|y\|} \right\|.$$

Show that the constants $\frac{1}{2}$ and $\frac{1}{4}$ cannot be replaced with greater constants.

3.1.2 Continuity of Linear and Multilinear Maps

Continuity of linear maps can be handled in several convenient ways.

Proposition 3.2 *A linear map $\ell : X \to Y$ between two normed spaces $(X, \|\cdot\|_X)$, $(Y, \|\cdot\|_Y)$ is continuous (and in fact Lipschitzian) if and only if there exists some $c \in \mathbb{R}_+$ such that $\|\ell(x)\|_Y \leq c\,\|x\|_X$ for all $x \in X$. The least such constant is called the norm of ℓ.*

Proof Let us prove more, namely that the following assertions about a linear map $\ell : X \to Y$ are equivalent:

(a) ℓ is *bounding*, i.e. for any bounded subset B of X, $\ell(B)$ is bounded;
(b) $\ell(B_X)$ is bounded;
(c) there exists some $c \in \mathbb{R}_+$ such that $\|\ell(x)\|_Y \leq c\,\|x\|_X$ for all $x \in X$;
(d) ℓ is Lipschitzian;
(e) ℓ is continuous;
(f) ℓ is continuous at 0.

The implications (a)⇒(b), (d)⇒(e)⇒(f) are obvious.

The implication (b)⇒(c) is a consequence in the homogeneity of ℓ: if $\ell(B_X)$ is contained in cB_Y, then for all $x \in X\backslash\{0\}$, setting $r := \|x\|_X$, we have $\|\ell(x/r)\|_Y \leq c$, hence $\|\ell(x)\|_Y \leq cr = c\,\|x\|_X$ and $\|\ell(0)\|_Y = 0$.

The implication (c)⇒(d) is a consequence in the additivity of ℓ: for all $x, x' \in X$ we have $\|\ell(x) - \ell(x')\|_Y = \|\ell(x - x')\|_Y \leq c\,\|x - x'\|_X$.

Finally, let us prove that (f)⇒(a). When ℓ is continuous at 0 there exists some $\delta > 0$ such that $\ell(\delta B_X) \subset B_Y$. Then if B is a subset of rB_X for some $r \in \mathbb{R}_+$ one has $\ell(B) \subset (r/\delta)B_Y$: $\ell(B)$ is bounded. □

For a *linear form*, i.e. a linear map from X into $\mathbb{R}$, a simple characterization of continuity is available. In the sequel we say that a subset H of a linear space X is a *hyperplane* if there exist $c \in \mathbb{R}$ and a linear form $h \neq 0$ on X such that $H = h^{-1}(c)$.

Corollary 3.1 *Let X be a (real) normed vector space, let $c \in \mathbb{R}$, and let h be a non-null linear form on X. The hyperplane $H := h^{-1}(c)$ is closed if and only if h is continuous.*

Proof Obviously, when h is continuous, H is closed, $\{c\}$ being closed in $\mathbb{R}$. Conversely, suppose that H is closed. Since h is non null, $X\backslash H$ is nonempty. Let $x_0 \in X\backslash H$, so that there exists some $r > 0$ for which $B(x_0, r) \subset X\backslash H$. Assuming $h(x_0) < c$ (the other possibility is dealt with changing h, c into $-h, -c$), let us note that $h(x) < c$ for all $x \in B(x_0, r)$: otherwise, if there exists an $x_1 \in B(x_0, r)$ such that $h(x_1) > c$, for $t := (h(x_1) - c)(h(x_1) - h(x_0))^{-1}$ one has $h((1-t)x_0 + tx_1) = c$ and $(1-t)x_0 + tx_1 \in B(x_0, r)$, contradicting $B(x_0, r) \subset X\backslash H$. Then, for all $u \in B(0, 1)$ we have $h(x_0 + ru) < c$ or $h(u) < r^{-1}(c - h(x_0))$ and h is continuous. □

Corollary 3.2 *Let X be a dense linear subspace of a normed space W and let Y be a Banach space. If $f : X \to Y$ is a continuous linear map, then there exists a unique continuous linear map $g : W \to Y$ whose restriction to X is f.*

Proof Since f is Lipschitzian with rate $\|f\|$ as we have seen, f has a unique extension g to W as a Lipschitzian map. The linearity of g follows from a passage to the limit in the relation $f(x_n + ry_n) = f(x_n) + rf(y_n)$ with $(x_n) \to x$, $(y_n) \to y$. □

The space $L(X, Y)$ of continuous linear maps from X to Y can be turned into a normed space.

Proposition 3.3 *The map $u \mapsto \|u\| := \sup\{\|u(x)\|_Y : x \in B_X\}$ on the space $L(X, Y)$ of continuous linear maps from X into Y is a norm on $L(X, Y)$. In particular, the (topological) dual space $X^* := L(X, \mathbb{R})$ of a normed space X is a normed space.*

Proof The fact that this map is a seminorm is an easy consequence in the definitions. It is a norm since when $\|u\| = 0$ one has $u(x) = 0$ for all $x \in B_X$, hence for all $x \in X$ by homogeneity. □

Example In general the supremum in the definition of $\|u\|$ is not attained, even for $Y = \mathbb{R}$. Taking $T := [0, 1]$ and $X := \{x \in C(T) : x(0) = 0\}$ with $\|x\| = \|x\|_\infty := \sup_{t \in T} |x(t)|$ for $x \in X$, one can see that for $u \in X^*$ given by $u(x) = \int_0^1 x(t)dt$ one has $\|u\| = 1$ since for $x_n(t) = t^{1/n}$ one has $u(x_n) = n(n+1)^{-1}$. However, if $u(\overline{x}) = 1$ for some $\overline{x} \in B_X$ we have $\int_0^1 (1 - \overline{x}(t))dt = 0$, hence $\overline{x}(t) = 1$ for all $t \in T$, contradicting $x(0) = 0$.

Exercise Show that for $u \in L(X, Y)$, $\|u\|$ is the least $c \in \mathbb{R}_+$ such that $\|u(x)\|_Y \leq c \|x\|_X$ for all $x \in X$. Moreover, $\|u\| = \sup\{\|u(x)\|_Y : x \in S_X\} = \sup\{\|u(x)\|_Y / \|x\|_X : x \in X\backslash\{0\}\}$.

Exercise Show that if X, Y, Z are normed spaces and if $u \in L(X, Y)$, $v \in L(Y, Z)$, then $\|v \circ u\| \leq \|v\| . \|u\|$.

Proposition 3.4 *If X and Y are normed spaces and if Y is complete, then $L(X, Y)$ is complete.*

Proof Let (u_n) be a Cauchy sequence in $L(X, Y)$. For each $x \in X$, since $\|u_n(x) - u_m(x)\| \leq \|u_n - u_m\| . \|x\|$, the sequence $(u_n(x))$ is a Cauchy sequence in Y, hence it converges to some $y_x \in Y$ which we denote by $u(x)$. The rules for limits in Y show that u is a linear map. For $x \in B_X$, passing to the limit on n in the relation

$$\|u_n(x)\| \leq \|u_n(x) - u_m(x)\| + \|u_m(x)\|$$

we see that $u(B_X)$ is bounded. Another passage to the limit in the relation $\sup_{x \in B_X} \|u_n(x) - u_m(x)\| \leq \|u_n - u_m\|$ shows that $(\|u - u_m\|)_m \to 0$, so that $L(X, Y)$ is complete. □

Corollary 3.3 *If X is a normed space, then its (topological) dual $X^* := L(X, \mathbb{R})$ is complete.*

A norm $\|\cdot\|$ on the product $X \times Y$ of two normed spaces is a *product norm* if its associated metric is a product metric. This amounts to the following inequalities for all $(x, y) \in X \times Y$:

$$\|(x, y)\|_\infty := \max(\|x\|_X, \|y\|_Y) \leq \|(x, y)\| \leq \|(x, y)\|_1 := \|x\|_X + \|y\|_Y.$$

Corollary 3.4 *If X is a normed space, there exists a complete normed space $\widehat{X}$ and an isometry j from X onto a dense subspace of $\widehat{X}$.*

Proof It can be checked directly that the completion $\widehat{X}$ of X as described above as the set of equivalence classes of Cauchy sequences of X can be given a linear structure and a norm inducing on X the original norm. However, we prefer to invoke a result below ensuring that the canonical embedding $e_X : X \to X^{**} := (X^*)^*$ given by $\langle e_X(x), x^* \rangle := \langle x^*, x \rangle$ is an isometry onto its image $e_X(X)$. Then we can take for $\widehat{X}$ the closure of $e_X(X)$ in X^{**}. □

Now let us turn to the continuity of multilinear maps.

Proposition 3.5 *Let $X_1, \ldots X_m$, Y be normed vector spaces and let $u : X_1 \times \ldots \times X_m \to Y$ be an m-linear map, i.e. a map that is linear with respect to each of its m variables. Then u is continuous if and only if there exists some $c \in \mathbb{R}_+$ such that for all $(x_1, \ldots, x_m) \in X := X_1 \times \ldots \times X_m$ one has*

$$\|u(x_1, \ldots, x_m)\| \leq c \|x_1\| \ldots \|x_m\|. \tag{3.1}$$

Proof For the sake of simplicity, we give the proof for $m = 2$ only. The general case is similar, but is more laborious to write. Let us endow X with the supremum norm. Condition (3.1) is clearly sufficient to prove the continuity of u at $(0, 0)$. The continuity of u at arbitrary $\overline{x} := (\overline{x}_1, \overline{x}_2) \in X$ ensues since for $(x_1, x_2) \in B[\overline{x}, \delta]$ with $\delta \in]0, 1]$ one has $\|x_1\| \leq \|\overline{x}\| + 1$, hence

$$\|u(x_1, x_2) - u(\overline{x}_1, \overline{x}_2)\| = \|u(x_1, x_2 - \overline{x}_2) + u(x_1 - \overline{x}_1, \overline{x}_2)\| \leq c\delta(2\|\overline{x}\| + 1).$$

Now let us prove the necessity of (3.1). Suppose u is continuous at $(0, 0)$. Then, there exists some $r > 0$ such that for all $x := (x_1, x_2) \in B[0, r]$ one has $u(x_1, x_2) \in B[0, 1]$.

Given $x := (x_1, x_2) \in X$, let $w := (w_1, w_2) \in B[0, r]$ be such that $r^{-1} \|x_i\| w_i = x_i$ for $i = 1, 2$ (if $x_i \neq 0$ we take $w_i := rx_i / \|x_i\|$ and if $x_i = 0$ we take $w_i := 0$). Then we have $u(x_1, x_2) = r^{-2} \|x_1\| \|x_2\| u(w_1, w_2)$, hence $\|u(x_1, x_2)\| \leq r^{-2} \|x_1\| \|x_2\|$.

□

We can turn the space $L_m(X_1, \ldots, X_m; Y)$ of continuous m-linear maps from $X_1 \times \ldots \times X_m$ into Y into a normed space by setting

$$\|u\| := \sup\{\|u(x_1, \ldots, x_m)\| : \sup(\|x_1\|, \ldots, \|x_m\|) \leq 1\}$$

for $u \in L_m(X_1, \ldots, X_m; Y)$. It can be shown that $\|u\|$ is the least constant c such that (3.1) holds for all $(x_1, \ldots, x_m) \in X := X_1 \times \ldots \times X_m$.

It can be shown that the space $L_m(X_1, X_2, \ldots, X_m; Y)$ is isometric to the space $L(X_1, L(X_2, \ldots, L(X_m, Y) \ldots))$. In particular it is complete if Y is complete. We just state and prove the case $m = 2$; the general case follows by induction.

Proposition 3.6 *Given normed spaces X, Y, Z, and $u \in L_2(X, Y; Z)$, for $x \in X$ let $u_x \in L(Y, Z)$ be given by $u_x(y) := u(x, y)$. Then $\tilde{u} : x \mapsto u_x$ is a linear continuous map from X into $L(Y, Z)$ and the map $u \mapsto \tilde{u}$ is a linear isometry from $L_2(X, Y; Z)$ onto $L(X, L(Y, Z))$.*

In particular, $L_2(X, Y; \mathbb{R})$ is isometric to $L(X, Y^*)$.

Proof Since for all $(x, y) \in X \times Y$ we have $\|u_x(y)\| = \|u(x, y)\| \leq \|u\| \|x\| \|y\|$, we see that u_x is linear and continuous and that

$$\sup_{x \in B_X} \|u_x\| = \sup_{x \in B_X} \sup_{y \in B_Y} \|u(x, y)\| = \sup_{(x,y) \in B_X \times B_Y} \|u(x, y)\| = \|u\|.$$

This shows that the map $u \mapsto \tilde{u}$ (which is obviously linear) from $L_2(X, Y; Z)$ into $L(X, L(Y, Z))$ is an isometry onto its image. Let us prove this map is surjective. Given $v \in L(X, L(Y, Z))$, setting $u(x, y) := v(x)(y)$ we clearly define a bilinear map that is continuous since $\|u(x, y)\| \leq \|v(x)\| \|y\| \leq \|v\| \|x\| \|y\|$ and $\tilde{u} = v$. □

Exercises

1. Let f be a map from a normed space X into another one Y that is bounded on the unit ball of X and additive, i.e. such that $f(x + x') = f(x) + f(x')$ for all $x, x' \in X$. Prove that f is linear.
2. Let $h \neq 0$ be a continuous linear form on a normed space X, and let $H := h^{-1}(0)$. Show that for all $x \in X$ the distance $d(x, H)$ from x to the hyperplane H is given by $d(x, H) = |h(x)| / \|h\|$.
3. Let $X := c_0$ be the space of real sequences (x_n) with limit 0 endowed with the norm given by $\|x\|_\infty := \sup_n |x_n|$ for $x := (x_n)$. Let $h \in X^*$ be given by $h(x) := \Sigma_n 2^{-n} x_n$ for $x := (x_n) \in X$. Compute $\|h\|$. Show that there is no $x \in X$

such that $\|x\| = 1$ and $h(x) = \|h\|$. Setting $H := h^{-1}(0)$, show that for every $x \in X\backslash H$ there is no $v \in H$ such that $d(x, H) = \|x - v\|$.

4. A function $q : X \to \mathbb{R}$ on a normed space is said to be *quadratic* if there exists a symmetric bilinear map $b : X \times X \to \mathbb{R}$ such that $q(x) = b(x, x)$ for all $x \in X$. Verify that such a bilinear map is unique. Prove that b is continuous if and only if q is continuous. [Hint: define b by $b(x, y) := \frac{1}{2}[q(x + y) - q(x) - q(y)]$ for $x, y \in X$.]
5. Let X (often denoted by c_0) be the space of sequences $x := (x_n)$ of real numbers such that $\lim_n x_n = 0$. Let Y (often denoted by ℓ_1) be the space of sequences $y := (y_n)$ of real numbers such that $\Sigma_n |y_n| < +\infty$. Given $y \in Y$, show that $f_y : x \mapsto \Sigma_n x_n y_n$ is a continuous linear form on X. Prove that $y \mapsto f_y$ is a linear isometry from Y onto the dual X^* of X.
6. Let Z (often denoted by ℓ_∞) be the space of bounded sequences $z := (z_n)$ of real numbers. Given $z \in Z$, show that $g_z : y \mapsto \Sigma_n z_n y_n$ is a continuous linear form on ℓ_1. Prove that $z \mapsto g_z$ is a linear isometry from Z onto the dual Y^* of Y. Verify that the canonical injection $j : X \to Z$ is compatible with the injection $X \to X^{**}$ via the preceding identifications.
7. Prove that on any infinite dimensional normed space X there is a linear form f that is not continuous. [Hint: by Zorn's Lemma there exists a maximal family $\{e_i : i \in I\}$ of linearly independent elements of the unit sphere S_X. It is an algebraic basis of X and since X is infinite dimensional, one can find a subset N of I and a bijection $j : N \to \mathbb{N}$. Setting $f(e_n) = j(n)$ for $n \in N, f(e_i) = 0$ for $i \in I\backslash N$ one gets an unbounded linear form.]

3.1.3 Finite Dimensional Normed Spaces

Finite dimensional normed spaces have interesting properties. We start with $\mathbb{R}^d$ endowed with the product topology and the norm $\|\cdot\|_\infty$.

Proposition 3.7 *A subset S of $\mathbb{R}^d$ is compact with respect to the induced topology if and only if it is closed and bounded.*

Proof If S is compact, by Proposition 2.14 it is closed in $\mathbb{R}^d$. It is bounded since the norm $\|\cdot\|_\infty$ on $\mathbb{R}^d$ is continuous. Conversely, if S is bounded it is contained in some ball $B[0, r]$ with respect to the norm $\|\cdot\|_\infty$ which is a product of intervals $[-r, r]$, hence is compact. If moreover S is closed, it is compact as a closed subset of a compact space. □

Proposition 3.8 *All norms on a finite dimensional vector space X are equivalent.*

Proof Taking an isomorphism from $\mathbb{R}^d$ onto X, where d is the dimension of X, we may suppose $X = \mathbb{R}^d$. Thus it suffices to prove that any norm $\|\cdot\|$ on $\mathbb{R}^d$ is equivalent to the norm $\|\cdot\|_\infty$. Setting $c := \|e_1\| + \ldots + \|e_d\|$, where $(e_1, \ldots, e_d)$ is the canonical basis of $\mathbb{R}^d$, for all $x = (x_1, \ldots, x_d) \in \mathbb{R}^d$ we have

$$\|x\| \leq |x_1| \, \|e_1\| + \ldots + |x_d| \, \|e_d\| \leq c \, \|x\|_\infty .$$

Moreover, $\|\cdot\|$ is continuous and even Lipschitzian since for all $x, y \in \mathbb{R}^d$ we have

$$|\|x\| - \|y\|| \leq \|x - y\| \leq c \|x - y\|_\infty .$$

Since the unit sphere S of $\mathbb{R}^d$ with respect to the norm $\|\cdot\|_\infty$ is compact, the function $\|\cdot\|$ attains its infimum m on S. Since $0 \notin S$ this infimum cannot be 0. Thus m is positive and for all $x \in S$ we have $\|x\| \geq m \|x\|_\infty$. By homogeneity this inequality is valid for all $x \in \mathbb{R}^d$ and completes the first inequality. □

Corollary 3.5 *If X is a finite dimensional vector subspace of a normed vector space W, then X is closed in W.*

Proof Taking a basis of X we get an isometry of X onto $\mathbb{R}^d$ equipped with a certain norm. Since this norm is equivalent to $\|\cdot\|_\infty$, X is complete, hence closed in W. □

Theorem 3.1 (Riesz) *A normed vector space is locally compact if and only if it is finite dimensional.*

Proof If X is a finite dimensional vector space, taking a base of X we get an isomorphism from $\mathbb{R}^d$ onto X. The norm of X being transformed to a norm on $\mathbb{R}^d$ that is equivalent to the norm $\|\cdot\|_\infty$, we see that X is locally compact.

Conversely, suppose X is locally compact. Taking an equivalent norm if necessary, we may assume the unit ball B_X of X is compact. Thus it can be covered by a finite number of balls $B(x_i, 1/2)$, $i \in \mathbb{N}_m$. Let Y be the subspace generated by $x_1, \ldots, x_m$. By the preceding corollary, Y is closed in X. Let us show that assuming $Y \neq X$ leads to a contradiction. Taking $\varepsilon = 1/2$ in Proposition 3.1 we get some $x \in X$ satisfying $\|x\| = 1$ and $d(x, Y) > 1/2$. This contradicts the fact that there exists some $i \in \mathbb{N}_m$ such that $x \in B(x_i, 1/2)$. Thus $Y = X$ and X is finite dimensional. □

Exercises

1. Show that if X is finite dimensional, every linear map from X into a normed vector space Y is continuous. [Hint: use an isomorphism of $\mathbb{R}^d$ onto X.]
2. Show that if X is a finite dimensional subspace of a normed vector space W and if Y is a closed subspace of W, then $X + Y$ is closed in W.
3. Show that if X is a finite dimensional supplement of a vector subspace Y of a normed space W, then X is a topological supplement of Y in the sense that W is the topological direct sum of X and Y, i.e. that the projections of W onto X and Y are continuous.
4. Prove that a normed vector space whose unit sphere S_X is compact is finite dimensional. [Hint: use the fact that the unit ball B_X is the image of $[0, 1] \times S_X$ under the map $(u, r) \mapsto ru$.]
5. Show that an infinite dimensional Banach space cannot have a countable algebraic basis $\{e_n : n \in \mathbb{N}\}$. [Hint: let X_n be the linear subspace generated

by $\{e_0, \dots, e_n\}$. Using Baire's Theorem show that $\mathrm{int}(X_k) \neq \varnothing$ for some $k \in \mathbb{N}$, hence $X_k = X$, a contradiction.]

3.1.4 Series and Summable Families

The study of summable families of real numbers we made can easily be extended to families of elements of normed spaces. We slightly change the notation of Sect. 1.2.

Definition 3.1 A family $(x_i)_{i \in I}$ of elements of a normed vector space $(X, \|\cdot\|)$ is said to be *summable* with sum s if the family $(s_J)_{J \in \mathcal{J}}$ of finite sums $s_J := \Sigma_{j \in J} x_j$ converges to s.

Here $\mathcal{J}$ denotes the directed set of finite subsets of I with the order given by inclusion. The next properties are immediate consequences of the definition.

Proposition 3.9 *If $r \in \mathbb{R}$ and if $(x_i)_{i \in I}$, $(x'_i)_{i \in I}$ are summable families of X, then $(rx_i)_{i \in I}$ and $(x_i + x'_i)_{i \in I}$ are summable families.*

Proposition 3.10 *If X and Y are normed spaces, if $\ell : X \to Y$ is a continuous linear map and if $(x_i)_{i \in I}$ is a summable family of X, then $(\ell(x_i))_{i \in I}$ is a summable family of Y with sum $\ell(s)$.*

Proof This follows from the fact that the net $(\Sigma_{j \in J} \ell(x_j))_{J \in \mathcal{J}}$ converges to $\ell(s)$, where s is the sum $\Sigma_{i \in I} x_i$ since $\Sigma_{j \in J} \ell(x_j) = \ell(s_J)$ with $s_J := \Sigma_{j \in J} x_j$ and since $(s_J)_{J \in \mathcal{J}}$ converges to s. □

Proposition 3.11 (Cauchy Summability Criterion) *A family $(x_i)_{i \in I}$ of elements of a complete normed space X is summable if and only if it satisfies the condition: for all $\varepsilon > 0$ there exists a finite subset H_ε of I such that for any finite subset F of I contained in $I \backslash H_\varepsilon$ one has $\|s_F\| \leq \varepsilon$.*

Proof The condition is sufficient since the net $(s_J)_{J \in \mathcal{J}}$ satisfies the Cauchy criterion whenever the family $(x_i)_{i \in I}$ satisfies the Cauchy summability criterion: given $\varepsilon > 0$, let $H_\varepsilon \in \mathcal{J}$ be such that $\|s_F\| \leq \varepsilon$ for any $F \in \mathcal{J}$ contained in $I \backslash H_\varepsilon$; then, for J, $K \in \mathcal{J}$ containing H_ε, since $\|s_J - s_K\| = \|s_F\|$ for $F := J \Delta K \subset I \backslash H_\varepsilon$ we have $\|s_J - s_K\| \leq \varepsilon$.

Conversely, let $(x_i)_{i \in I}$ be a summable family of X with sum s. Given $\varepsilon > 0$ we can find $H \in \mathcal{J}$ such that for $J \in \mathcal{J}$ containing H we have $\|s_J - s\| \leq \varepsilon/2$. Then, for any $F \in \mathcal{J}$ contained in $I \backslash H$, setting $J := H \cup F$ we have $\|s_F\| = \|s_J - s_H\| \leq \|s_J - s\| + \|s - s_H\| \leq \varepsilon$. □

Corollary 3.6 *A subfamily of a summable family of a Banach space X is summable.*

A simple means of showing that a family of elements of a Banach space is summable consists in reducing the question to the summability of a family of nonnegative real numbers. This relies on the next proposition and on the notion of an *absolutely summable* family: a family $(x_i)_{i\in I}$ is said to be absolutely summable if the family $(\|x_i\|)_{i\in I}$ is summable.

Proposition 3.12 *Any absolutely summable family of a Banach space is summable.*

Proof It suffices to show that an absolutely summable family $(x_i)_{i\in I}$ satisfies the Cauchy summability criterion. This follows from the Cauchy summability criterion for $(\|x\|_i)_{i\in I}$ and the fact that for any finite subset F of I one has $\|s_F\| \leq \Sigma_{i\in F} \|x_i\|$. □

It can be shown that in any infinite dimensional Banach space there exist summable families that are not absolutely summable. That does not occur in finite dimensional spaces.

Proposition 3.13 *In a finite dimensional normed space a family is summable if and only if it is absolutely summable.*

Proof Since any finite dimensional normed space is isomorphic to $\mathbb{R}^d$ for some $d \geq 1$, it suffices to show that a summable family $(x_i)_{i\in I}$ of $\mathbb{R}^d$ is absolutely summable. Proposition 3.10 ensures that for all $k \in \mathbb{N}_d$ the family $(x_i^k)_{i\in I}$ of the k-components of $(x_i)_{i\in I}$ is summable since $x_i^k = p^k(x_i)$, where $p^k : \mathbb{R}^d \to \mathbb{R}$ is the k-th projection. Then $(x_i^k)_{i\in I}$ is absolutely summable, so that $(\|x_i\|_1)_{i\in I} = (|x_i^1| + \ldots + |x_i^d|)_{i\in I}$ is summable. □

Let us turn to series in a normed space $(X, \|\cdot\|)$. A *series* in X is a pair of sequences (x_n), (s_n) of X such that $s_n = x_0 + \ldots + x_n$ for all $n \in \mathbb{N}$; x_n is called the n-th term and s_n is called the n-th partial sum of the series. The pair (x_n), (s_n) is often denoted by $\Sigma_n x_n$, a formal notation. One says that the series $\Sigma_n x_n$ *converges* if the sequence (s_n) converges. The limit s of (s_n) is then called the sum of the series. Then, the remainder $r_n := \Sigma_{p\geq n} x_p$ converges to 0.

For series of nonnegative general terms, convergence can be simply characterized.

Lemma 3.2 *Let $(x_n)_n$ be a sequence in nonnegative real numbers. Then the following assertions are equivalent:*

(a) the series with general term x_n is convergent;
(b) the partial sums s_n are bounded above;
(c) the family $(x_n)_{n\in\mathbb{N}}$ is summable.

Proof (a)⇒(b) Since the sequence (s_n) is nondecreasing, if it converges one has $s_n \leq \lim_n s_n$.

(b)⇒(c) Let $c \geq s_n$ for all $n \in \mathbb{N}$. For any finite subset J of $\mathbb{N}$ one can find n such that $J \subset [0, n]$, so that $s_J \leq c$ and the nondecreasing net $(x_J)_{J\in\mathcal{J}}$ is convergent.

(c)⇒(a) The implication is a general fact, as shown below. In fact, if $(x_n)_{n\in\mathbb{N}}$ is summable with sum s, for every $\varepsilon > 0$ one can find a finite subset H of $\mathbb{N}$ such that for $n \geq \sup H$ and $J := \{0, \ldots, n\}$ one has $s_n = s_J$, hence $|s - s_n| = |s - s_J| < \varepsilon$: $(s_n) \to s$. □

The rules for convergence of sequences yield rules for convergence of series (sums, images) as above. Cauchy criterion can be adapted as follows.

Proposition 3.14 *A series with general term x_n in a Banach space is convergent if and only if for all $\varepsilon > 0$ there exists an $k_\varepsilon \in \mathbb{N}$ such that for $p > n \geq k_\varepsilon$ one has $\left\| \Sigma_{k=n+1}^{k=p} x_k \right\| \leq \varepsilon$.*

Proof This statement is merely Cauchy criterion for the sequence $(s_n)_n$. □

It follows that the general term of a convergent series tends to 0. It is well known that this property is far from being sufficient.

Let us compare convergence of series with summability. We start with a simple observation.

Lemma 3.3 *Let $(x_n)_{n\geq 0}$ be a sequence in a normed space X. If the family $(x_n)_{n\in\mathbb{N}}$ satisfies the Cauchy summability criterion and if the series with general term x_n is convergent with sum s, then the family $(x_n)_{n\in\mathbb{N}}$ is summable and its sum is s.*

Proof Given $\varepsilon > 0$ we can find a finite subset H of $\mathbb{N}$ such that $\|s_F\| < \varepsilon/2$ for any finite subset F of $\mathbb{N}\backslash H$ and we can find $k \in \mathbb{N}$ such that $\|s_n - s\| < \varepsilon/2$ for all $n \geq k$. Without loss of generality we may suppose $k \geq \max H$. Then, for $K := \{0, 1, \ldots, k\}$ we have $s_k = s_K$ and for any finite subset J of $\mathbb{N}$ containing K, setting $F := J\backslash K$ we have

$$\|s_J - s\| \leq \|s_K + s_F - s\| \leq \|s_K - s\| + \|s_F\| < \varepsilon.$$

Thus the family (x_n) is summable with sum s. □

Theorem 3.2 *A series with general term x_n in a normed vector space is commutatively convergent (in the sense that for any permutation π of $\mathbb{N}$ the series with general term $x_{\pi(n)}$ is convergent) if and only if the family $(x_n)_{n\in\mathbb{N}}$ is summable.*

Proof Suppose the family $(x_n)_{n\in\mathbb{N}}$ is summable, with sum s. Let π be a permutation of $\mathbb{N}$ and let $y_n := x_{\pi(n)}$, $t_n := y_0 + \ldots + y_n$, $s_J := \Sigma_{j\in J} x_j$ for J a finite subset of $\mathbb{N}$. For any $\varepsilon > 0$ there exists some finite subset H of $\mathbb{N}$ such that for any finite subset J of $\mathbb{N}$ containing H one has $\|s_J - s\| \leq \varepsilon$. Since H is finite there exists a $k \in \mathbb{N}$ such that $H \subset \pi(\{0, \ldots, k\})$. Then, for $n \geq k$ one has $J := \pi(\{0, \ldots, n\}) \supset H$, hence $\|t_n - s\| = \|s_J - s\| \leq \varepsilon$: the series $\Sigma_n y_n$ converges to s.

Conversely, suppose the series with general term x_n is commutatively convergent. The lemma asserts that if the family $(x_n)_{n\in\mathbb{N}}$ satisfies the Cauchy summability criterion, then it is summable. It remains to show that if the Cauchy summability criterion is not satisfied we get a contradiction. In such a case there exists an $\varepsilon > 0$ such that for any $H \in \mathcal{J}$ one can find $F(H) \in \mathcal{J}$ contained in $\mathbb{N}\backslash H$ such that $\left\|s_{F(H)}\right\| > \varepsilon$. Taking $H_0 := \{0\}$ and inductively setting $H_{n+1} := H_n \cup F(H_n)$, we get a sequence $(F(H_n))$ of disjoint finite subsets of $\mathbb{N}$ such that $\left\|s_{F(H_n)}\right\| > \varepsilon$. Keeping the order on $F(H_n)$ and ranking in a consecutive order the sets $F(H_n)$, we get a strictly increasing sequence $(k(n))_n$ and a bijective map $\pi : \mathbb{N} \to \mathbb{N}$ such that $F(H_n) = \{\pi(j) : k(n) \leq j < k(n+1)\}$. The series with general term $y_n := x_{\pi(n)}$ cannot be convergent since for all $n \in \mathbb{N}$ we have $\left\| \Sigma_{j=k(n)}^{j=k(n+1)-1} y_j \right\| = \left\|s_{F(H_n)}\right\| > \varepsilon$. This is the required contradiction. □

A series with general term x_n in a normed space is said to be *absolutely convergent* if the series with general term $\|x_n\|$ is convergent. Lemma 3.2 shows that the series Σx_n is absolutely convergent if and only if the family $(x_n)_{n\in\mathbb{N}}$ is summable. Combining this observation with the preceding theorem we get the next statement.

Corollary 3.7 *In a Banach space, any absolutely convergent series is commutatively convergent.*

Adding Proposition 3.13 to the preceding implications, we get nice equivalences.

Proposition 3.15 *For a sequence (x_n) in a finite dimensional space the following assertions are equivalent:*

(a) the series Σx_n is commutatively convergent;
(b) the series Σx_n is absolutely convergent;
(c) the family $(x_n)_{n\in\mathbb{N}}$ is absolutely summable;
(d) the family $(x_n)_{n\in\mathbb{N}}$ is summable.

When X is a space of functions from a set T to a Banach space E, besides pointwise convergence, one may consider uniform convergence. Taking for X the space $B(T,E)$ of bounded functions from T to E with the supremum norm $\|\cdot\|_\infty$, it may be convenient to find a convergent series $\Sigma_n r_n$ such that $\|x_n(t)\| \leq r_n$ for all $t \in T$. Then one says that the series $\Sigma_n x_n$ is *normally convergent*; then it is absolutely convergent in $X := B(T,E)$.

Let us provide some means for studying series that are convergent but not necessarily absolutely convergent. The *Abel transformation* is such a means. It presents some similarity with integration by parts.

Lemma 3.4 *Let (x_n) be a sequence in a normed space X such that $c_n := \sup_{k\geq 0}\|x_n + x_{n+1} + \ldots + x_{n+k}\| < \infty$ for all $n \in \mathbb{N}$ and let (r_n) be a nonincreasing sequence of nonnegative numbers. Then for all n, $k \in \mathbb{N}$ one has*

$$\|r_n x_n + r_{n+1}x_{n+1} + \ldots + r_{n+k}x_{n+k}\| \leq r_n c_n. \tag{3.2}$$

Proof For $n \in \mathbb{N}$ let $y_{n,0} = 0$, and for $k \in \mathbb{N}\backslash\{0\}$ let $y_{n,k} := x_n + x_{n+1} + \ldots + x_{n+k}$. Since $x_{n+j} = y_{n,j} - y_{n,j-1}$ for $j = 1, \ldots, k$, we have

$$\sum_{j=0}^{k} r_{n+j}x_{n+j} = r_n y_{n,0} + \sum_{j=1}^{k} r_{n+j}(y_{n,j} - y_{n,j-1})$$

$$= (r_n - r_{n+1})y_{n,0} + \ldots + (r_{n+k-1} - r_{n+k})y_{n,k-1} + r_{n+k}y_{n,k},$$

$$\|\sum_{j=0}^{k} r_{n+j}x_{n+j}\| \leq c_n((r_n - r_{n+1}) + \ldots + (r_{n+k-1} - r_{n+k}) + r_{n+k}) = c_n r_n.$$

□

It follows from this estimate that when $(c_n r_n) \to 0$ the series with general term $r_n x_n$ satisfies the Cauchy criterion.

Proposition 3.16 (Abel) *Let (x_n) be a sequence in a complete normed space X and let (r_n) be a nonincreasing sequence of nonnegative real numbers. If either the series $\Sigma_n x_n$ converges or $(r_n) \to 0$ and the sequence (s_n) of partial sums $s_n := x_0 + \ldots + x_n$ is bounded in X, then the series $\Sigma r_n x_n$ converges.*

Proof Since X is complete, in view of the preceding lemma it suffices to show that $(c_n r_n) \to 0$ in each of these two cases. In the first case, the convergence of $\Sigma_n x_n$ implies that $(c_n) \to 0$. Since $r_n \leq r_0$ we have $c_n r_n \leq c_n r_0$ and we get $(c_n r_n) \to 0$. In the second case we have $y_{n,k} := x_n + \ldots + x_{n+k} = s_{n+k} - s_{n-1}$, so that $c_n := \sup_{k \geq 0} \|y_{n,k}\|$ is bounded above by some c. Then $c_n r_n \leq c r_n$ and $(c_n r_n) \to 0$. □

Example An alternate series is a series whose general term is $(-1)^n r_n$ where (r_n) is a nonincreasing sequence of nonnegative numbers. Since s_n is either 1 or 0, if $(r_n) \to 0$ then the series $\Sigma_n (-1)^n r_n$ converges.

Example The case $x_n = z^n$ with $z \in \mathbb{C}$, $|z| = 1$, $z \neq 1$ also pertains to the second case of the proposition. Since $s_n = (1 - z^{n+1})(1 - z)^{-1}$ one has $|s_n| \leq 2\,|1 - z|^{-1}$, so that if $(r_n) \to 0$ and (r_n) is nonincreasing, the series $\Sigma_n r_n z^n$ is convergent. Setting $z := e^{it}$ we see that for $t \notin 2\pi\mathbb{Z}$ the series with general terms $r_n \cos nt$ and $r_n \sin nt$ are convergent.

As an application of series in Banach spaces, let us consider the question of invertibility of linear maps. A similar study could be made in Banach algebras (i.e. Banach spaces endowed with a continuous bilinear and associative product). When X and Y are finite dimensional normed spaces and f is a linear isomorphism, we know that any linear map g that is close enough to f with respect to some norm on the space $L(X, Y)$ of linear continuous maps from X into Y is still an isomorphism: taking bases in X and Y we see that if g is close enough to f its determinant will remain different from 0. A similar result holds in infinite dimensional Banach spaces.

Proposition 3.17 *Let f be a linear isomorphism between two Banach spaces X and Y. Then any $g \in L(X, Y)$ such that $\|f - g\| < \|f^{-1}\|^{-1}$ is an isomorphism. Thus, the set of linear continuous maps that are isomorphisms is open in the space $L(X, Y)$.*

Proof Let us first consider the case $X = Y$, $f = I_X$. Let $u := I_X - g$, so that $u \in L(X, X)$ satisfies $\|u\| < 1$. Since the map $(v, w) \mapsto w \circ v$ is continuous, since

$$I_X - u^{n+1} = (I_X - u) \circ \left(\sum_{k=0}^{n} u^k \right) = \left(\sum_{k=0}^{n} u^k \right) \circ (I_X - u),$$

and since the series $\sum_{k=0}^{\infty} u^k$ is absolutely convergent (as $\left\| u^k \right\| \leq \|u\|^k$), we get that its sum is a right and left inverse of $I_X - u$. Thus $I_X - u$ is invertible.

The general case can be deduced from this special case. Given $g \in L(X, Y)$ such that $\|f - g\| < r := \|f^{-1}\|^{-1}$, setting $u := I_X - f^{-1} \circ g$, we observe that $\|u\| \leq \|f^{-1} \circ (f - g)\| \leq \|f^{-1}\| \cdot \|f - g\| < 1$. Therefore, by what precedes, $f^{-1} \circ g = I_X - u$ is invertible. It follows that g is invertible, with inverse $(I_X - u)^{-1} \circ f^{-1}$. □

Exercises

1. Prove that if $(x_i)_{i \in I}$ is a summable family of a Banach space X the set $D := \{i \in I : x_i \neq 0\}$ is countable.
2. Recall that a *(vector) basis* of a vector space E is a family $(e_i)_{i \in I}$ of elements of E such that any element x of E can be written in a unique way as a linear combination of a finite subfamily of $(e_i)_{i \in I}$.
   Prove that an infinite dimensional Banach space cannot have a countable basis. [Hint: given a sequence $(e_n)_n$ of linearly independent vectors of norm 1, let X_n be the space generated by $\{e_0, \ldots, e_n\}$; define inductively a sequence (r_n) of nonnegative numbers by $r_0 := 0$, $r_1 := 0$, $r_{n+1} = (1/3)d(r_n e_n, X_{n-1})$ for $n \geq 1$ and show that the series $\Sigma_n r_n e_n$ is absolutely convergent but that its sum does not belong to any of the subspaces X_n.]
3. Let $T := [0, 1]$, $X := C(T)$, let $f \in X$ be given by $f(0) = 0, f(t) := t \sin^2(\pi/t)$ for $t \in T \backslash \{0\}$ and let $f_n \in X$ be defined by $f_n := f 1_{[1/(n+1), 1/n]}$. Show that the family (f_n) is summable but not absolutely summable.
4. In the space ℓ_∞ of bounded sequences of real numbers provided with the norm given by $\|x\| := \sup_n |x_n|$ for $x := (x_n)$, let a_n be the element of X whose components are all 0 except for the nth one which is $1/(n+1)$. Show that the family (a_n) is not absolutely summable but is summable.
5. Let X, Y, Z be normed spaces and let $b : X \times Y \to Z$ be normed spaces. If $(x_i)_{i \in I}$ (resp. $(y_j)_{j \in J}$) is an absolutely summable family of X (resp. Y), prove that $(b(x_i, y_j))_{(i,j) \in I \times J}$ is absolutely summable with sum $b(\Sigma_i x_i, \Sigma_j y_j)$ when Z is complete.
6. Let $(k(n))_n$ be a strictly increasing sequence of integers with $k(0) = 0$ and let $\Sigma_n x_n$ be a convergent series in a normed space X. For $n \in \mathbb{N}$ let $y_n := x_{k(n)} + \ldots + x_{k(n+1)-1}$. Show that the series $\Sigma_n y_n$ converges to the sum s of the series $\Sigma_n x_n$.
7. Let X and Y be Banach spaces and let $S \subset L(X, Y)$ be the set of $A \in L(X, Y)$ such that there exists a $B \in L(Y, X)$ (called a left inverse of A) satisfying $B \circ A = I_X$. Show that S is open in $L(X, Y)$. [Hint: use Proposition 3.17.]

3.1.5 Spaces of Continuous Functions

The question arises of characterizing subsets of a space of functions that are compact or complete. We already considered the question of completeness. Let us give a criterion for compactness. It uses the following definition.

Definition 3.2 Given a topological space X and a metric space Y, a set F of maps from X into Y is said to be *equicontinuous* at $x \in X$ if for every $\varepsilon > 0$ one can find some $V \in \mathcal{N}(x)$ such that for all $f \in F$ and all $v \in V$ one has $d(f(v), f(x)) \leq \varepsilon$.

The set F is said to be equicontinuous if it is equicontinuous at all $x \in X$.

Clearly, any finite set of continuous maps is equicontinuous. If X is a metric space and if there exist a neighborhood U of x, $c > 0$ and $\alpha > 0$ such that $d(f(u), f(x)) \leq cd(u, x)^\alpha$ for all $u \in U$ and $f \in F$, then F is equicontinuous at x. In particular, if the functions in F are Lipschitzian with the same rate c, then F is equicontinuous.

Theorem 3.3 (Ascoli–Arzela) *Let X be a compact topological space, let Y be a complete metric space, and let F be a subset of the space $C(X, Y)$ of continuous maps from X into Y. Then, endowing $C(X, Y)$ with the metric d_∞ of uniform convergence, F is relatively compact in $C(X, Y)$ if and only if F is equicontinuous and for all $x \in X$ the set $F(x) := \{f(x) : f \in F\}$ is relatively compact in Y.*

Proof If the closure $\overline{F}$ of F in $C(X, Y)$ is compact, since for every $x \in X$ the evaluation $e_x : f \mapsto f(x)$ is continuous, $F(x)$ is contained in the compact set $e_x(\overline{F})$, hence is relatively compact in Y. Since $\overline{F}$ is precompact by Theorem 2.15, given $\varepsilon > 0$ one can find some finite subset $\{f_i : i \in \mathbb{N}_m\}$ of $\overline{F}$ such that the balls $B(f_i, \varepsilon/3)$ $(i \in \mathbb{N}_m)$ cover $\overline{F}$. Let $x \in X$ and let $V \in \mathcal{N}(x)$ be such that $d(f_i(x), f_i(v)) \leq \varepsilon/3$ for all $i \in \mathbb{N}_m$ and all $v \in B(x, \delta)$. Then, for all $f \in F$ and all $v \in V$, picking $i \in \mathbb{N}_m$ such that $f \in B(f_i, \varepsilon/3)$, we have

$$d(f(x), f(v)) \leq d(f(x), f_i(x)) + d(f_i(x), f_i(v)) + d(f_i(v), f(v)) \leq \varepsilon :$$

F is equicontinuous at x.

For the converse, since $(C(X, Y), d_\infty)$ is complete, it suffices to prove that F is precompact when F is equicontinuous and when $F(x)$ is relatively compact for all $x \in X$. Given $\varepsilon > 0$, by equicontinuity, for all $x \in X$ we can find some $V_x \in \mathcal{N}(x)$ such that $d(f(v), f(x)) \leq \varepsilon/4$ for all $v \in V_x$ and all $f \in F$. Let A be a finite subset of X such that $\{V_a : a \in A\}$ is a cover of X. Since $F(a)$ is relatively compact in Y for all $a \in A$, so is $F(A) := \cup_{a \in A} F(a)$. Let $b_1, \ldots, b_m$ be points of $F(A)$ such that $\{B(b_i, \varepsilon/4) : i \in \mathbb{N}_m\}$ is a cover of $F(A)$. Let K be the finite set of maps k from A into $\mathbb{N}_m$. For all $f \in F$ there exists some $k \in K$ such that $f(a) \in B(b_{k(a)}, \varepsilon/4)$ for all $a \in A$. Thus F is covered by the sets

$$F_k := \{g \in F : \forall a \in A \; g(a) \in B(b_{k(a)}, \varepsilon/4)\} \qquad k \in K.$$

Since for $g, h \in F_k$ and $x \in X$ one can find some $a \in A$ such that $x \in V_a$ one has

$$d(g(x), h(x)) \leq d(g(x), g(a)) + d(g(a), b_{k(a)}) + d(b_{k(a)}, h(a)) + d(h(a), h(x)) \leq \varepsilon,$$

the diameters of the sets F_k are at most ε and F is precompact. □

Under a compactness assumption and a monotonicity assumption one can pass from pointwise convergence to uniform convergence.

Theorem 3.4 (Dini) *Let X be a compact topological space and let (f_n) be a sequence of continuous real-valued functions that pointwise converges to some continuous function f. If (f_n) is increasing (in the sense that $f_n \leq f_{n+1}$ for all $n \in \mathbb{N}$) then $(f_n) \to f$ uniformly.*

Proof Given $\varepsilon > 0$, for $n \in \mathbb{N}$ let $U_n := \{x \in X : f(x) - f_n(x) < \varepsilon\}$. Since f and f_n are continuous, U_n is open (we note that it would suffice to suppose f is upper semicontinuous and f_n is lower semicontinuous). For all $x \in X$, since $(f_n(x)) \to f(x)$, we have $x \in U_n$ for n large enough. Since X is compact we can select a finite subcovering $\{U_k : k \in \mathbb{N}_m\}$ from the covering $\{U_n : n \in \mathbb{N}\}$. The sequence (f_n) being increasing, we have $U_k \subset U_m$ for $k \in \mathbb{N}_m$. Thus $U_m = X$ and for $n \geq m$ we have $|f(x) - f_n(x)| = f(x) - f_n(x) < \varepsilon$. □

It is often useful to approximate a continuous function by simple functions. A prototype of such a process is the following result in which $r \mapsto \sqrt{r}$ could be replaced by any continuous function, as we shall show later.

Lemma 3.5 (Weierstrass) *There exists a sequence (p_n) of polynomials which converges uniformly on $[0, 1]$ to $r \mapsto \sqrt{r}$ and is increasing on this interval.*

Proof We define p_n by induction, setting $p_0 = 0$ and

$$p_{n+1}(r) := p_n(r) + (1/2)(r - p_n^2(r)) \qquad n \geq 1,\ r \in \mathbb{R}. \tag{3.3}$$

We show by induction that $p_n(r) \leq \sqrt{r}$ for $r \in [0, 1]$ thanks to the relation

$$\sqrt{r} - p_{n+1}(r) = (\sqrt{r} - p_n(r))[1 - (1/2)(\sqrt{r} + p_n(r))] \geq 0$$

since $(1/2)(\sqrt{r} + p_n(r)) \leq \sqrt{r} \leq 1$. It follows that $p_{n+1}(r) \geq p_n(r)$ for $n \in \mathbb{N}$ and $r \in [0, 1]$. The increasing sequence $(p_n(r))$ of $[0, 1]$ converges to some $q(r) \in [0, 1]$. Passing to the limit in relation (3.3) we get $r - q^2(r) = 0$, hence $q(r) = \sqrt{r}$. Since $[0, 1]$ is compact and (p_n) is increasing on $[0, 1]$, Dini's Theorem ensures that the convergence is uniform. □

In the sequel S is a compact topological space and we say that a subset A of the space $C(S)$ of continuous real-valued functions on S *separates the points* of S if for any pair (x, y) of distinct points of S there exists some $f \in A$ such that $f(x) \neq f(y)$. Let us consider a subalgebra A of $C(S)$, i.e. a vector subspace of $C(S)$ such that $fg \in A$ whenever $f, g \in A$.

Lemma 3.6 *If A is a subalgebra of $C(S)$, for any $f \in A$ one has $|f| \in \overline{A} := \mathrm{cl}(A)$. Moreover, $\overline{A}$ is a sublattice of $C(S)$: for all $f, g \in \overline{A}$ one has $f \wedge g, f \vee g \in \overline{A}$.*

Proof Let $f \in A$. We may suppose $r := \|f\|_\infty > 0$. Then, setting $f_n(s) := p_n(f^2(s)/r^2)$, where (p_n) is as in the preceding lemma, we get that $f_n \in A$ as A is an algebra and $(f_n) \to |f|/r$ for $\|\cdot\|_\infty$. Thus $|f|/r \in \overline{A}$. Since $\overline{A}$ is a subalgebra of $C(S)$, as is easily seen, for all $f \in \overline{A}$ we have $|f| \in \overline{A}$.

For all $f, g \in \overline{A}$ one has $f \vee g = (1/2)(f + g + |f - g|) \in \overline{A}$ and $f \wedge g = (1/2)(f + g - |f - g|) \in \overline{A}$. □

Lemma 3.7 *If A is a subalgebra of $C(S)$ that separates the points of S and contains the constant functions, then for every pair x, y of distinct points of S and any pair r, s of real numbers there is some $f \in A$ such that $f(x) = r$ and $f(y) = s$.*

Proof Given $x \neq y$ in S, by assumption there is some $g \in A$ such that $g(x) \neq g(y)$. Setting $t := (r - s)/(g(x) - g(y))$, since A contains the constant functions, we get that $f := r + t(g - g(x)) \in A, f(x) = r$, and $f(y) = s$. □

Lemma 3.8 *If A is a subalgebra of $C(S)$ that separates the points of S and contains the constant functions, then for all $f \in C(S)$, $\varepsilon > 0$ and all $x \in S$ there is some $g \in \overline{A}$ such that $g(x) = f(x)$ and $g \leq f + \varepsilon$.*

Proof Let $f \in C(S)$, $\varepsilon > 0$ and $x \in S$ be given. For all $y \in S \backslash \{x\}$, by the preceding lemma, there exists some $g_y \in A$ such that $g_y(x) = f(x)$ and $g_y(y) < f(y) + \varepsilon$. Let V_y be an open neighborhood of y such that $g_y(v) < f(v) + \varepsilon$ for all $v \in V_y$. Since S is compact, there exists a finite subset Y of S such that $\{V_y : y \in Y\}$ is a covering of S. Then, by Lemma 3.6 $g := \inf_{y \in Y} g_y \in \overline{A}$, $g(x) = f(x)$ and for all $w \in S$ we can find $y \in Y$ such that $w \in V_y$, so that $g(w) \leq g_y(w) < f(w) + \varepsilon$. □

Theorem 3.5 (Stone-Weierstrass) *Let S be a compact topological space and let A be a subalgebra of $C(S)$ that separates the points of S and contains the constant functions. Then A is dense in $(C(S), \|\cdot\|_\infty)$.*

Proof Let $f \in C(S)$ and let $\varepsilon > 0$ be given. By the preceding lemma, for all $x \in S$ there exists some $g_x \in \overline{A}$ such that $g_x(x) = f(x)$ and $g_x \leq f + \varepsilon$. Since f and g_x are continuous, there exists an open neighborhood U_x of x such that $g_x(u) \geq f(u) - \varepsilon$ for all $u \in U_x$. Let X be a finite subset of S such that $\{U_x : x \in X\}$ covers S. Then $h := \sup_{x \in X} g_x \in \overline{A}$ by Lemma 3.6 and satisfies $h \leq f + \varepsilon$, $h \geq f - \varepsilon$ since every $z \in S$ belongs to some U_x with $x \in X$. Thus $\|h - f\|_\infty \leq \varepsilon$ and $f \in \mathrm{cl}(\overline{A}) = \overline{A}$. □

The corresponding conclusion with $C(S)$ replaced by the space $C(S, \mathbb{C})$ of complex-valued continuous functions on S is not true. However, one can get a similar conclusion by adding an assumption.

Corollary 3.8 *Let S be a compact topological space and let A be a subalgebra of $C(S, \mathbb{C})$ that separates the points of S, contains the constant functions and is such that $\overline{f} \in A$ for all $f \in A$. Then A is dense in $(C(S, \mathbb{C}), \|\cdot\|_\infty)$.*

Proof Let $B := \{f \in A : \overline{f} = f\}$. Then B is a real subalgebra of $C(S)$ that contains the constant functions. Moreover, since for all $f \in A$ we have $(1/2)(f + \overline{f}) \in B$ and $(1/2i)(f - \overline{f}) \in B$, B separates the points of S. Thus B is dense in $C(S)$, hence $A := B + iB$ is dense in $C(S, \mathbb{C}) = C(S) + iC(S)$. □

Since the polynomial functions on $\mathbb{R}^d$ separate the points and form an algebra, we get the following announced consequence.

Corollary 3.9 *For any compact subset S of $\mathbb{R}^d$ and any continuous function f on S there exists a sequence (p_n) of polynomials on $\mathbb{R}^d$ that converges uniformly to f on S.*

Corollary 3.10 *If S is a compact metric space, the spaces $C(S)$ and $C(S, \mathbb{C})$ are separable.*

Proof Since $C(S, \mathbb{C})$ is the topological direct sum of $C(S)$ and $iC(S)$, it suffices to prove the result for $C(S)$. Now S is separable by Corollary 2.16, hence there exists a countable base $\{G_n : n \in \mathbb{N}\}$ of the topology of S. Set $g_n := d(\cdot, S\backslash G_n)$. Given a pair x, y of distinct points of S there exists some $n \in \mathbb{N}$ such that $x \in G_n$ and $y \in X\backslash G_n$, so that $g_n(x) > 0$ and $g_n(y) = 0$. Thus the algebra A generated by $\{g_n : n \in \mathbb{N}\}$ and the constant functions separates the points of A, hence it is dense in $C(S)$. Since A is formed by the monomials $g_1^{\alpha_1} \ldots g_n^{\alpha_n}$ with $\alpha_1, \ldots, \alpha_n \in \mathbb{N}$, A is countable and the Stone-Weierstrass Theorem ensures that A is dense in $C(S)$. □

The following result shows a link between algebraic properties and topological properties. It will be used for the spectral analysis of operators on a Hilbert space. Let us recall that a subset J of a ring or an algebra R is an *ideal* if it is a subring (or a subgroup) of R such that $fg \in J$ whenever $f \in R$ and $g \in J$.

Theorem 3.6 (Ideal Theorem) *Let S be a compact space and let J be a closed ideal of $R := C(S)$ endowed with the sup norm. Let $Z := \{x \in S : \forall f \in J, f(x) = 0\}$. Then J coincides with the set of $f \in R$ such that $f(z) = 0$ for all $z \in Z$.*

In other words, the set $\mathcal{C}$ of closed subsets of S is in bijection with the set $\mathcal{J}$ of ideals of $C(S)$ via the map $T \mapsto J_T := \{f \in C(S) : f(z) = 0 \ \forall z \in T\}$.

Proof Let $f \in R$ be such that $f(z) = 0$ for all $z \in Z$. Given $\varepsilon > 0$, we will find some $f_\varepsilon \in J$ such that $\|f_\varepsilon - f\|_\infty \leq \varepsilon$. Since J is closed, this will prove that $f \in J$. Let $Z_\varepsilon := \{x \in S : |f(x)| < \varepsilon\}$, $S_\varepsilon := S\backslash Z_\varepsilon$. Since Z is contained in Z_ε, for all $y \in S_\varepsilon$ there exists some $g_y \in J$ such that $g_y(y) \neq 0$ and some open neighborhood V_y of y such that $g_y(v) \neq 0$ for all $v \in V_y$. Let $y_1, \ldots, y_k$ be such that $(V_{y_1}, \ldots, V_{y_k})$ is a finite covering of the compact set S_ε. Let

$$g := g_{y_1}^2 + \ldots + g_{y_k}^2,$$

so that $g \in J$, g takes nonnegative values and is positive on S_ε. Then, for all $n \in \mathbb{N}$ the function $g_n := n(1+ng)^{-1}g$ is in J since $1+ng \geq 1$ and $n(1+ng)^{-1} \in C(S)$. Since g is bounded below by some $\alpha > 0$ on the compact set S_ε, we see that $(g_n) \to 1$ and $(g_n f) \to f$ uniformly on S_ε. On the other hand, since $0 \leq n(1+ng)^{-1}g \leq 1$ we have $|g_n(x)f(x) - f(x)| < \varepsilon$ for all $x \in Z_\varepsilon$. Thus, taking n large enough, for $f_\varepsilon := g_n f$, we have $\|f_\varepsilon - f\|_\infty \leq \varepsilon$ and $g_n f \in J$. □

Exercises

1. Verify that the union of a finite family of equicontinuous sets of maps from a topological space X into a metric space Y is equicontinuous. Deduce from this result that in particular any finite set of such maps is equicontinuous.
2. Let X be a topological space, let Y be a metric space and let $(f_i)_{i\in I}$ be a net of maps from X into Y that converges pointwise to some $f : X \to Y$. Show that if the family $\{f_i : i \in I\}$ is equicontinuous at $x \in X$, then f is continuous at x.
3. With the data of Exercise 2 show that the closure in $(B(X, Y), d_\infty)$ of an equicontinuous subset F is equicontinuous.
4. Let X be a compact topological space, let Y be a metric space and let (f_n) be a sequence in $C(X, Y)$ that is equicontinuous and converges pointwise to some $f : X \to Y$. Prove that $(f_n) \to f$ uniformly.
5. Let X be a topological space, let Y be a metric space and let $\{f_n : n \in \mathbb{N}\}$ be a family of maps from X into Y that is equicontinuous at $\overline{x} \in X$. Show that if $(f_n(\overline{x}))_n$ converges to some $\overline{y} \in Y$ and if $(x_n) \to \overline{x}$, then $(f_n(x_n)) \to \overline{y}$.
6. Let S and T be two compact topological spaces and let $A := C(S) \otimes C(T)$ be the set of finite sums of separable functions, i.e. functions of the form $(s, t) \mapsto f(s)g(t)$ with $f \in C(S)$, $g \in C(T)$. Show that A is dense in $C(S \times T)$.
7. Let $S = T = [0, 1]$ and let $\{r_i : i \in \mathbb{N}_m\}$ be a finite family of distinct points of S. Prove that the functions $r \mapsto |r - r_i|$ $(i \in \mathbb{N}_m)$ are linearly independent. Deduce from this observation that the function $h : (s, t) \mapsto |s - t|$ cannot be an element of $C(S) \otimes C(T)$.
8. Prove that the additional condition $\overline{f} \in A$ for all $f \in A$ in Corollary 3.8 cannot be omitted. [Hint: let $S := \{z \in \mathbb{C} : |z| \leq 1\}$, let A be the set of restrictions to S of the complex polynomial functions on $\mathbb{C}$. Note that for all $f \in A$, hence for all $f \in \overline{A}$, one has $f(0) = (1/2\pi) \int_0^{2\pi} f(e^{it})dt$, a relation that is not satisfied by all $f \in C(S)$.]
9. Let S be a compact subset of a metric space (T, d). Deduce from Theorem 3.5 that any $f \in C(S)$ is the restriction to S of some $g \in C(T)$. [Hint: Take $A := \{g\,|_S : g \in C(T)\}$ and note that for all $f \in A$ one can find a $g \in C(T)$ such that $\sup g = \sup f$ and $\inf g = \inf f$. Given a sequence $(\varepsilon_n) \to 0_+$ find a sequence (g_n) of $C(T)$ such that $\|f - \Sigma_{0\leq k\leq n} g_k\,|_S\| \leq \varepsilon_n$, $\sup_{t\in T} |g_n(t)| \leq \varepsilon_{n-1}$ and take $g := \Sigma_n g_n$.]

3.2 Topological Vector Spaces. Weak Topologies

In infinite dimensional normed spaces, it appears that compact subsets are scarce. A natural means to get a richer family of compact subsets on a normed space $(X, \|\cdot\|)$ is to weaken the topology: then there will be more convergent nets and, since open covers will be not as rich, finding finite subfamilies will be easier. The drawbacks are that continuity of maps issued from X will be lost in general and that no norm will be available to define the weakened topology if X is infinite dimensional. A partial remedy for the first inconvenience will be proposed in the next subsection. Now, the lack of a norm will not be too dramatic if one realizes that the structure of *topological linear space* is preserved. This means that the two operations $(x, y) \mapsto x + y$ and $(\lambda, x) \mapsto \lambda x$ will be continuous with respect to the new topology. One will even dispose of a family of seminorms defining the topology. Recall that a *seminorm* on a linear space X being a function $p : X \to \mathbb{R}_+$ that is *subadditive* (i.e. such that $p(x+y) \leq p(x)+p(y)$ for all $(x, y) \in X \times X$) and *absolutely homogeneous* (i.e. such that $p(\lambda x) = |\lambda| p(x)$ for all $(\lambda, x) \in \mathbb{R} \times X$) or equivalently subadditive, *positively homogeneous* (i.e., such that $p(\lambda x) = \lambda p(x)$ for all $(\lambda, x) \in \mathbb{R}_+ \times X$) and *even* (i.e., such that $p(-x) = p(x)$ for all $x \in X$). Note that a seminorm p is a *norm* iff $p^{-1}(0) = \{0\}$. The topology associated with a family $(p_i)_{i \in I}$ of seminorms on X is the topology generated by the family of semi-balls $B_i(a, r) := \{x \in X : p_i(x - a) < r\}$ for all $a \in X$, $r \in \mathbb{P}$, $i \in I$. Such a topology is clearly compatible with the operations on X, so that X becomes a topological linear space. It is even a *locally convex topological linear space* in the sense that each point has a base of neighborhoods that are convex. One can show that this property is equivalent to the existence of a family of seminorms defining the topology.

On the (topological) *dual space* X^* of a topological linear space X, i.e. on the space of continuous linear forms on X, a natural family of seminorms is the family $(p_x)_{x \in X}$ given by $p_x(f) := |f(x)|$ or, adopting a notation we will use frequently, $p_x(x^*) := |\langle x^*, x\rangle|$ for $x^* \in X^*$. Then a net $(f_i)_{i \in I}$ of X^* converges to some $f \in X^*$ if and only if for all $x \in X$, $(f_i(x))_{i \in I} \to f(x)$; then we write $(f_i)_{i \in I} \overset{*}{\to} f$. Thus, the obtained topology on X^*, denoted by $\sigma^* := \sigma(X^*, X)$ and called the *weak** *topology*, is just the topology induced by pointwise convergence. It is the weakest topology on X^* for which the evaluations $f \mapsto f(x)$ are continuous, for all $x \in X$. Although this topology is poor, it preserves some continuity properties. In particular, if X and Y are normed spaces and if $A \in L(X, Y)$, its *transpose* map (often called the adjoint) $A^\intercal : Y^* \to X^*$ defined by $A^\intercal(y^*) := y^* \circ A$ for $y^* \in Y^*$ or

$$\langle A^\intercal(y^*), x\rangle = \langle y^*, A(x)\rangle \qquad (x, y^*) \in X \times Y^*$$

is not just continuous with respect to the topologies induced by the dual norms (the so-called *strong topologies*) since $\|A^\intercal(y^*)\| = \|y^* \circ A\| \leq \|A\| . \|y^*\|$ for all $y^* \in Y^*$; it is also continuous with respect to the weak* topologies: when $(y_i^*)_{i \in I} \overset{*}{\to} y^*$ one has $(A^\intercal(y_i^*))_{i \in I} \overset{*}{\to} A^\intercal(y^*)$ since for all $x \in X$ one has $(A^\intercal(y_i^*)(x))_{i \in I} = (y_i^*(A(x)))_{i \in I} \to y^*(A(x))$. Note that when X and Y are *Hilbert spaces* (i.e., Banach spaces whose norms derive from scalar products), so that they can be identified with their dual spaces, $A^\intercal$ corresponds to the *adjoint* $A^* : Y \to X$ of A characterized by $\langle A^*(y) \mid x\rangle_X = \langle y \mid A(x)\rangle_Y$ for all $x \in X$, $y \in Y$, $\langle \cdot \mid \cdot\rangle_X$ (resp. $\langle \cdot \mid \cdot\rangle_Y$) denoting the scalar product in X (resp. Y).

Let us show there are sufficiently many linear forms on X^* that are continuous with respect to the weak* topology.

Proposition 3.18 *The set of continuous linear forms on X^* endowed with the weak* topology can be identified with X.*

Proof By definition, for all $x \in X$, the linear form $e_x : x^* \mapsto \langle x^*, x\rangle$ on X^* is continuous with respect to the weak* topology σ^*. Let us show that any continuous linear form f on (X^*, σ^*) coincides with some e_x. We can find $\delta > 0$ and a finite family $(a_1, \ldots, a_m)$ in X such that $|f(x^*)| < 1$ for all $x^* \in X^*$ satisfying $p_{a_i}(x^*) := |\langle x^*, a_i\rangle| < \delta$ for $i \in \mathbb{N}_m := \{1, \ldots, m\}$. Setting $x_i := a_i/\delta$, we get $|f(x^*)| \leq \max_{1\leq i\leq m} |\langle x^*, x_i\rangle|$ since otherwise, by homogeneity, we could find $x^* \in X^*$ such that $|f(x^*)| = 1$ and $\max_{1\leq i\leq m} |\langle x^*, x_i\rangle| < 1$, contradicting the choice of a_i and x_i. Changing the indexing if necessary, we may suppose that, for some $k \in \mathbb{N}_m$, $x_1, \ldots, x_k$ form a basis of the linear space spanned by $x_1, \ldots, x_m$. Let

$$A := (e_{x_1}, \ldots, e_{x_k}) : X^* \to \mathbb{R}^k.$$

Then, denoting by N the kernel of A and by $p : X^* \to X^*/N$ the canonical projection, f can be factorized into $f = g \circ p$ for some linear form g on X^*/N. Since A is surjective, there is also an isomorphism $B : X^*/N \to \mathbb{R}^k$ such that A is factorized into $A = B \circ p$. Then, $p = B^{-1} \circ A$ and $f = g \circ B^{-1} \circ A$, hence $f(x^*) = c_1x^*(x_1) + \ldots + c_kx^*(x_k)$ for all $x^* \in X^*$, where $c_1, \ldots, c_k$ are the components of $g \circ B^{-1}$ in $(\mathbb{R}^k)^*$. Thus $f = e_x$ for $x := c_1x_1 + \ldots + c_kx_k \in X$. □

The following result shows that in introducing the weak* topology we have attained our aim of getting sufficiently many compact subsets.

Theorem 3.7 (Alaoglu-Bourbaki) *Every weak* closed, bounded subset of the dual space X^* of X is weak* compact, i.e. is compact with respect to the weak* topology.*

Proof It suffices to show that the closed unit ball $B^* := B_{X^*}$ of X^* is weak* compact. To do so, let us denote by S the closed unit sphere S_X of X, by H the space of positively homogeneous functions on X and by H_S the space of all the restrictions to S of the elements of H. The restriction operator $r : H \to H_S$ is then a bijection, with inverse given by $r^{-1}(h)(x) = th(t^{-1}x)$ for $x \in X\backslash\{0\}$, $t := \|x\|$, $r^{-1}(h)(0) = 0$. Then r and r^{-1} are continuous with respect to the pointwise convergence topologies on H and H_S, and for this topology H_S is homeomorphic to the product space $\mathbb{R}^S$. The subset B^* of H is easily seen to be closed with respect to the pointwise convergence topology on H. Moreover, $r(B^*)$ is contained in $[-1, 1]^S$, which is compact, by the Tychonov Theorem. Thus, $r(B^*)$ and B^* are compact in H_S and H, respectively. It follows that B^* is compact in X^* endowed with the weak* topology. □

The weak topology on X is the topology $\sigma(X, X^*)$ associated with the semi-norms $p_{x^*} : x \mapsto |x^*(x)|$. It will be shown later that this topology is the topology on X induced by $\sigma(X^{**}, X^*)$ when X is considered as a subspace of its bidual space $X^{**} := (X^*)^*$.

Exercises

1. Prove that the interior with respect to the weak* topology of the unit ball of the dual of an infinite dimensional Banach space is empty. [Hint: any neighborhood of 0 with respect to the weak* topology contains an unbounded subset, in fact a non-trivial linear space.]
2. A *cone* of a linear space is a subset stable under the homotheties $h_t : x \mapsto tx$ for all $t > 0$. Show that a weakly* closed cone Q of the dual of a normed space X is weakly* *locally compact* (i.e., for each point $\overline{x}^*$ of Q there exists a weak* neighborhood V of $\overline{x}^*$ such that $Q \cap V$ is weakly* compact) if and only if there exists a neighborhood U of 0 such that $Q \cap U$ is weakly* compact. [Hint: let $u_1, \ldots, u_n \in X$ be such that $U_1 \cap \ldots \cap U_n \subset U$ for $U_i := \{x^* \in X^* : x^*(u_i) \leq 1\}$. Given $\overline{x}^* \in Q$, let $t > \max(1, \overline{x}^*(u_1), \ldots, \overline{x}^*(u_n))$. Let $V_i := f_i^{-1}((-\infty, t])$ for $i = 1, \ldots, n$. Then $V := V_1 \cap \ldots \cap V_n$ is a weak* neighborhood of $\overline{x}^*$ and $Q \cap V = t(Q \cap t^{-1}V) \subset t(Q \cap U)$ which is weakly* compact.]
3. Let $p \in]1, \infty[$ and let ℓ_p be the space of sequences $x := (x_n)$ such that $\|x\|_p := (\Sigma_n |x_n|^p)^{1/p} < \infty$. Show that the dual of ℓ_p is ℓ_q for $q := (1 - 1/p)^{-1}$. Given $x := (x_n) \in \ell_q$ and a sequence $(x^{(k)})_{k \geq 0}$ in ℓ_q, show that $(x^{(k)})$ converges weakly* to x if and only if it is bounded and for each $n \in \mathbb{N}$ one has $(x_n^{(k)})_k \to x_n$.
4. Let ℓ_1 be the space of sequences $x := (x_n)$ such that $\|x\|_1 := \Sigma_n |x_n| < \infty$. Show that the dual of ℓ_1 is the space ℓ_∞ of bounded sequences with the norm $\|\cdot\|_\infty$ given by $\|x\|_\infty := \sup_n |x_n|$. Prove that in ℓ_1 a sequence converges weakly if and only if it converges strongly.

   For $k \in \mathbb{N}$, let $x^{(k)} := (x_n^{(k)})_n$ be given by $x_n^{(k)} := \delta_n^k := 1$ if $n = k$, 0 otherwise. Deduce from what precedes that the bounded sequence $(x^{(k)})_k$ has no subsequence that converges weakly.
5. Let C be the subset of the space ℓ_∞ defined in the preceding exercise that consists in those $x := (x_n)$ such that $\lim_n x_n = 1$ and $x_n \in [0, 1]$ for all $n \in \mathbb{N}$. Show that C is a closed, convex, bounded subset of ℓ_∞, hence is weakly closed, but that C is not weak* closed i.e. closed for $\sigma(\ell_\infty, \ell_1)$.
6. Let $A \in L(X, Y)$ be a continuous linear map between two normed spaces. Show that the transpose map $A^\intercal : Y^* \to X^*$ is continuous and that $\|A^\intercal\| = \|A\|$.
7. **Komolgorov's normability criterion**. Prove that a locally convex topological vector space X is normable i.e. its topology is associated with a norm, if and only if some neighborhood V of 0 is bounded.

3.3 Separation and Extension. Polarity

This section is devoted to one of the major tools of functional analysis, the possibility of making use of linear continuous functionals on a normed space. In particular, we show that this family is rich enough and enables us to separate disjoint convex sets under some additional conditions. We first review some properties of convex subsets.

3.3.1 Convex Sets and Convex Functions

Let us recall that a subset C of a linear space X is said to be *convex* if a segment whose extremities are in C is entirely contained in C: for all $x_0, x_1 \in C$, $t \in [0,1]$, one has $x_t := (1-t)x_0 + tx_1 \in C$. Among convex subsets, the simplest ones are *affine subspaces* obtained by translating linear subspaces, *half-spaces* (subsets D such that there exist a linear form ℓ on X and $r \in \mathbb{R}$ for which $D = \ell^{-1}(]-\infty, r[)$ or $D = \ell^{-1}(]-\infty, r])$ and *convex cones*. The latter are the subsets that are stable under addition and positive homotheties $h_r : x \mapsto rx$ (with $r \in \mathbb{P} :=]0, \infty[$ fixed), as easily checked. From antiquity to the present days, *polyhedral subsets*, i.e. finite intersections of closed half-spaces, have played a special role among convex subsets as they enjoy particular properties not shared by all convex sets.

A function f from a linear space X to $\overline{\mathbb{R}} := \mathbb{R} \cup \{-\infty, \infty\}$ is said to be *convex* if its *epigraph*

$$E_f := \operatorname{epi} f := \{(x,r) \in X \times \mathbb{R} : r \geq f(x)\}$$

is convex or, equivalently, if for any $t \in [0,1]$, $x_0, x_1 \in X$

$$f((1-t)x_0 + tx_1) \leq (1-t)f(x_0) + tf(x_1)$$

(with the convention that $(-\infty) + (+\infty) = +\infty$ and $0.(+\infty) = +\infty$, $0.(-\infty) = -\infty$ we adopt in the sequel). It is easy to show that f is convex if and only if its *strict epigraph*

$$E'_f := \operatorname{epi}_s f := \{(x,r) \in X \times \mathbb{R} : r > f(x)\}$$

is convex. A function f is *concave* if $-f$ is convex. A function $s : X \to \overline{\mathbb{R}}$ is said to be *sublinear* if its epigraph is a convex cone, i.e. if it is *subadditive* ($s(x + x') \leq s(x) + s(x')$ for all $x, x' \in X$) and *positively homogeneous* ($s(tx) = ts(x)$ for all $t \in \mathbb{P}$, $x \in X$). A sublinear function p with nonnegative values is called a *gauge*; if moreover p is finite and *even* i.e., if $p(-x) = p(x)$ for every $x \in X$, then p is a *semi-norm*.

Example Let $g : X \to \mathbb{R}_\infty$ be a convex function. The associated (positively) homogeneous function is the function $h : X \times \mathbb{R} \to \mathbb{R}_\infty$ given by $h(x,r) := rg(x/r)$ for $(x,r) \in X \times \mathbb{P}$, $h(0,0) := 0$, $h(x,r) = +\infty$ otherwise. Then h is sublinear since it is clearly positively homogeneous and convex since $(x,r,s) \in \operatorname{epi} h$ if and only if $(x,s,r) \in \mathbb{P}(\operatorname{epi} g \times \{1\}) \cup \{(0,0,0)\}$.

Example The *support function* of a subset S of a normed space is the function σ_S or $h_S : X^* \to \overline{\mathbb{R}}$ given by

$$\sigma_S(x^*) := h_S(x^*) := \sup\{\langle x^*, x\rangle : x \in S\} \qquad x^* \in X^*. \tag{3.4}$$

Here we use the fact that the supremum of a family of convex functions is convex since the intersection of a family of convex subsets of a vector space is convex.

A convex function taking the value $-\infty$ is very special (on any straight line there are at most two points at which it takes a finite value and if the function is lower semicontinuous no such point exists); therefore we will usually discard them and only consider functions with values in $\mathbb{R}_\infty := \mathbb{R} \cup \{\infty\}$. In contrast, it is useful to admit functions taking the value $+\infty$. Among them is the *indicator function* ι_C of a subset C of X: let us recall it is given by $\iota_C(x) = 0$ for $x \in C$, $\iota_C(x) = +\infty$ for $x \in X\backslash C$. For instance, in a minimization problem one can take a constraint C into account by replacing an objective function f by $f_C := f + \iota_C$. One calls a function *proper* if it does not take the value $-\infty$ and takes at least one finite value. The expression nonimproper would be less ambiguous, but the risk of confusion with the topological concept is limited, so that we retain the usual terminology. Moreover, the epigraph of a function $f : X \to \mathbb{R}_\infty$ is a proper subset (nonempty and not the whole space) of $X \times \mathbb{R}$ if and only if f is proper. We denote by D_f or $\mathrm{dom} f$ the *domain* of f, i.e. the projection on X of $E_f := \mathrm{epi} f$:

$$D_f := \mathrm{dom} f := \{x \in X : f(x) < +\infty\}.$$

The following statement will be used repeatedly; it relies on the obvious fact that the image of a convex set under a linear map is convex.

Lemma 3.9 *Let W and X be linear spaces and let $f : W \times X \to \overline{\mathbb{R}}$ be convex. Then the performance function $p : W \to \overline{\mathbb{R}}$ defined as follows is convex*

$$p(w) := \inf_{x \in X} f(w, x).$$

Proof The result follows from the fact that the strict epigraph of p is the projection on $W \times \mathbb{R}$ of the strict epigraph of f. □

Let us add that if f is positively homogeneous in the variable w, so is p.

Example If C is a convex subset (resp. a convex cone) of a normed space, then the associated *distance function* $d_C : w \mapsto \inf_{x \in C} \|w - x\|$ is convex (resp. sublinear).

Example Given $f, g : X \to \overline{\mathbb{R}}$, their *infimal convolution* $f \square g : X \to \overline{\mathbb{R}}$ defined by

$$(f \square g)(w) := \inf\{f(u) + g(v) : u, v \in X,\ u + v = w\} = \inf_{x \in X}(f(w - x) + g(x))$$

is convex whenever f and g are convex. If f and g are sublinear then $f \square g$ is sublinear. The preceding example is the case corresponding to $f := \|\cdot\|$, $g := \iota_C$.

Besides the indicator function, the support function, and the distance function, another function associated with a convex set plays a noteworthy role. If C is a subset of X containing the origin, the *gauge function* (or Minkowski gauge) μ_C of C is defined by

$$\mu_C(x) := \inf\{r \in \mathbb{R}_+ : x \in rC\} \qquad\qquad x \in X.$$

Clearly, μ_C is positively homogeneous and one has $C \subset \mu_C^{-1}([0,1])$. If C is *starshaped*, i.e. if for all $x \in C$, $t \in [0,1]$ one has $tx \in C$, then $\mu_C^{-1}([0,1[) \subset C$. If moreover C is algebraically closed in the sense that its intersection with every ray $L_u := \mathbb{R}_+ u$, $u \in X \backslash \{0\}$ is closed in L_u, then $C = \mu_C^{-1}([0,1])$. In particular, the gauge function of the closed unit ball B_X of a normed space $(X, \|\cdot\|)$ is just $\|\cdot\|$. We leave the proof of the next lemma as an exercise using the fact that $\mu_C(x) = \inf_r h(x,r)$ where $h(x,r) := rg(x/r)$ with $g := \iota_C + 1$. Hereafter, a subset C of a linear space X is said to be *absorbing* if for all $x \in X$ there exists some $r > 0$ such that $x \in rC$.

Lemma 3.10 *The gauge μ_C of a convex subset C of X is sublinear.*
A subset C of X is absorbing if and only if μ_C is finitely valued.

Since the intersection of a family of convex subsets is convex, any nonempty subset A of a linear space X is contained in a convex set C that is the smallest in the family $\mathcal{C}_A$ of convex sets containing A. It is denoted by $\mathrm{co}(A)$ and called the convex set generated by A or the *convex hull* of A. It is obtained as the intersection of the family $\mathcal{C}_A$. It is easy to see that $\mathrm{co}(A)$ is the set of convex combinations of elements of A, i.e. $\mathrm{co}(A)$ is the set of $x \in X$ that can be written as

$$t_1 a_1 + \ldots + t_n a_n$$

with $n \in \mathbb{N} \backslash \{0\}$, $a_i \in A$, $t := (t_1, \ldots, t_n)$ being an element of the *canonical simplex* Δ_n, i.e. the set of $t := (t_1, \ldots, t_n) \in \mathbb{R}_+^n$ satisfying $t_1 + \ldots + t_n = 1$. The *convex hull* $\mathrm{co}(f)$ of a function $f : X \to \mathbb{R}_\infty$ is the greatest convex function g bounded above by f. Its epigraph is almost the convex hull of the epigraph E_f of f. In fact, it is the vertical closure of $\mathrm{co}(E_f)$ in the sense that one has $\mathrm{epi}_s\, g \subset \mathrm{co}(E_f) \subset \mathrm{epi}\, g$. Thus

$$g(x) := \inf_{m \geq 1} \inf\{\sum_{i=1}^{m} t_i f(x_i) : (t_1, \ldots, t_m) \in \Delta_m,\ x_i \in X,\ t_1 x_1 + \ldots + t_m x_m = x\}.$$

Exercise Show that for $g := \mathrm{co}(f)$ the inclusions $\mathrm{epi}_s\, g \subset \mathrm{co}(\mathrm{epi} f) \subset \mathrm{epi}\, g$ may be strict. [Hint: consider $f : \mathbb{R} \to \mathbb{R}$ given by $f(0) := 1$ and $f(x) := |x|$ for $x \in \mathbb{R} \backslash \{0\}$.]

Note that in general, the union of a family (C_p) of convex subsets is no longer convex; but when (C_p) is an increasing sequence (with respect to inclusion), the union is convex. Similarly, the infimum of a countable family (k_p) of convex functions is convex when the sequence (k_p) is decreasing; but that is not the case if the sequence (k_p) does not satisfy this property.

When X is a normed space, any subset S is contained in a smallest closed convex subset, its *closed convex hull* $\overline{\mathrm{co}}(S)$. Using the following elementary result, it is easy to check that this set is just the closure of $\mathrm{co}(S)$. In fact, the lemma and the preceding assertion are valid in any topological linear space. In the sequel, a number of results given for normed spaces are valid for topological linear spaces. We leave the proofs of the next two results as exercises.

Lemma 3.11 *The closure* $\mathrm{cl}(C)$ *and the interior* $\mathrm{int}(C)$ *of a convex subset* C *of a normed space are convex. Moreover, if* $t \in [0, 1[$, $x_0 \in \mathrm{int}(C)$ *and* $x_1 \in C$ *then* $(1-t)x_0 + tx_1 \in \mathrm{int}(C)$.

Lemma 3.12 *If the interior of a convex subset* C *of a normed space is nonempty, then one has* $\mathrm{cl}(C) = \mathrm{cl}(\mathrm{int}(C))$ *and* $\mathrm{int}(\mathrm{cl}(C)) = \mathrm{int}(C)$.

Lemma 3.13 *If* C *is a nonempty convex subset of a finite dimensional space, then* C *has a nonempty interior (called the* relative interior *and denoted by* $\mathrm{ri}(C)$*) in the affine subspace* A *it generates.*

Proof By definition, A is the smallest affine subspace containing C. Using a translation, we may suppose $0 \in C$, so that A is the linear subspace generated by C. Let n be the dimension of A and let m be the greatest integer k such that there exists a linearly independent family $\{e_1, \ldots, e_k\}$ in C satisfying

$$\mathrm{co}\{e_1, \ldots, e_k\} = \{t_1e_1 + \ldots + t_ke_k : (t_1, \ldots, t_k) \in \Delta_k\} \subset C.$$

Let $\{e_1, \ldots, e_m\}$ be such a family and let L be the linear space it generates. Then C is contained in L: otherwise, we could find some $e \in C\backslash L$ and the family $\{e_1, \ldots, e_m, e\}$ would satisfy the above conditions and be strictly larger than $\{e_1, \ldots, e_m\}$. Thus $L = A$ and the set $\mathrm{co}\{e_1, \ldots, e_k\}$ has a nonempty interior in A with respect to the unique Hausdorff linear topology on A obtained by transporting the topology of $\mathbb{R}^m$ by the isomorphism defined by the base $\{e_1, \ldots, e_m\}$. □

Exercises

1. Let A and B be convex subsets of a normed space $(X, \|\cdot\|)$. Show that the set $C := \{(1-t)a + tb : a \in A,\ b \in B,\ t \in [0, 1]\}$ is the convex hull $\mathrm{co}(A \cup B)$ of $A \cup B$, i.e. the smallest convex subset of X containing $A \cup B$.
2. Let A and B be compact convex subsets of a normed space $(X, \|\cdot\|)$. Show that the sets $C := \mathrm{co}(A \cup B)$ and $S := A + B$ are compact. Give an example showing that the convex hull of the sum of two closed convex subsets of $\mathbb{R}^2$ is not always closed. [Hint: take $A := \mathbb{R}\times\{1\}$, $B := \{(x, y) \in \mathbb{R}_+ \times \mathbb{R}_+ : xy \geq 1\}$.]
3. (**Hermite-Hadamard**) Given a continuous convex function $f : [a, b] \to \mathbb{R}$ prove the inequalities

$$f(\frac{a+b}{2}) \leq \frac{1}{b-a}\int_a^b f(x)dx \leq \frac{f(a)+f(b)}{2}.$$

[Hint: for the right inequality use the relation $f(x) \leq f(a) + (f(b) - f(a))(b - a)^{-1}(x - a)$ and integrate; for the left inequality, split the interval $[a, b]$ into $[a, (a + b)/2] \cup [(a + b)/2, b]$.]

4. Given a sequence $(E_n)_{n\geq 1}$ of nonempty subsets of a linear space Z, show that the convex hull C of the union E of the E_n's is the union over $p \in \mathbb{N}\backslash\{0\}$ of the convex hulls C_p of $E_1 \cup \cdots \cup E_p$:

$$C := \mathrm{co}(E) = \bigcup_p C_p \quad \text{where } C_p := \mathrm{co}\left(\bigcup_{1\leq n\leq p} E_n\right).$$

For $m, p \in \mathbb{N}\backslash\{0\}$, setting $\mathbb{N}_m := \{1, \ldots, m\}$ and denoting by $J_{m,p}$ the set of maps $j : \mathbb{N}_m \to \mathbb{N}_p$, show that the set C_p is given by

$$C_p := \bigcup_{m\geq 1} \bigcup_{j\in J_{m,p}} \{\sum_{i=1}^{m} t_i x_i : t := (t_1, \ldots, t_m) \in \Delta_m,\ x_i \in E_{j(i)}\}.$$

5. Given a sequence (h_n) of functions on a linear space X, show that the convex hull k of the function $h := \inf_n h_n$ is the infimum over $p \in \mathbb{N}\backslash\{0\}$ of the convex hulls $k_p := \mathrm{co}\left(h_1, \ldots, h_p\right)$ of the functions $h_1, \ldots, h_p$. The function k_p is given by

$$k_p(x) = \inf_{m\geq 1} \inf_{j\in J_{m,p}} \inf\{\sum_{i=1}^{m} t_i h_{j(i)}(x_i) : (t_1, \ldots, t_m) \in \Delta_m,\ x_i \in X,\ \sum_{i=1}^{m} t_i x_i = x\}.$$

6. Let C be a closed subset of a normed space that is *midconvex* in the sense that for any $x, y \in C$ one has $\frac{1}{2}x + \frac{1}{2}y \in C$. Show that C is convex.
7. Let $f : X \to \mathbb{R}_\infty$ be a *midconvex function* in the sense that for any $x, y \in X$ one has $f(\frac{1}{2}x + \frac{1}{2}y) \leq \frac{1}{2}f(x) + \frac{1}{2}f(y)$. Prove that f is convex if f is lower semicontinuous.
8. Let X be a normed space, let $A \in L(X, X^*)$, $b \in X^*$, $c \in \mathbb{R}$, and let $f : X \to \mathbb{R}$ be given by $f(x) := \frac{1}{2}\langle Ax, x\rangle + \langle b, x\rangle + c$. Show that f is convex if A is positive semidefinite in the sense that $\langle Ax, x\rangle \geq 0$ for all $x \in X$.

3.3.2 Separation and Extension Theorems

Looking at both the analytical face and the geometrical face of the results of this section is fruitful. In fact, the following extension and separation theorems are closely intertwined. We start with a finite dimensional separation property.

Theorem 3.8 (Finite Dimensional Separation Property) *Let C be a nonempty convex subset of a finite dimensional vector space X and let $a \in X\backslash C$. Then, there* *exists some* $f \in X^*\backslash\{0\}$ *such that* $f(a) \geq \sup f(C)$. *If moreover* C *is closed, one can request that* $f(a) > \sup f(C)$.

Proof Let us first consider the case when C is closed. Since X is finite dimensional, we may endow X with the norm associated with a scalar product $\langle \cdot \mid \cdot \rangle$. Then, by Corollary 2.29, the point a has a best approximation p in C characterized by

$$\forall z \in C \qquad \langle z - p \mid a - p \rangle \leq 0.$$

For $f \in X^*$ defined by $f(x) := \langle x \mid a - p \rangle$, for every $z \in C$ we have $f(p) \geq f(z)$ and the second conclusion is established since $f(a) - f(p) = \|a - p\|^2 > 0$, as $a \notin C$.

Now let us consider the general case in which C is not assumed to be closed. Let S_{X^*} be the unit sphere of X^* and for $x \in C$ let

$$S_x := \{u^* \in S_{X^*} : \langle u^*, x \rangle \leq \langle u^*, a \rangle\},$$

so that $\ell \in S_{X^*}$ is such that $\ell(a) \geq \sup \ell(C)$ if and only if $\ell \in \cap_{x \in C} S_x$. Since X is finite dimensional, S_{X^*} is compact, hence this intersection is nonempty provided the family of closed subsets $(S_x)_{x \in C}$ has the finite intersection property. Thus, we have to show that for any finite subset $F := \{x_1, \ldots, x_n\}$ of C one has $S_{x_1} \cap \ldots \cap S_{x_n} \neq \varnothing$. Let

$$\Delta_n := \{(t_1, \ldots, t_n) \in \mathbb{R}^n_+ : t_1 + \ldots + t_n = 1\},$$
$$h : (t_1, \ldots, t_n) \mapsto t_1 x_1 + \ldots + t_n x_n,$$

$E := \mathrm{co}(F) := h(\Delta_n)$. Since the canonical simplex Δ_n is compact and h is continuous, E is compact, hence closed, and contained in C. The first part of the proof yields some ℓ in $X^*\backslash\{0\}$ satisfying $\ell(a) > \ell(z)$ for all $z \in E$, in particular for $z \in F$. Without loss of generality we may suppose $\|\ell\| = 1$. Thus $\ell \in S_x$ for all $x \in F$: the family $(S_x)_{x \in C}$ has the finite intersection property. □

Corollary 3.11 *Let* A *and* B *be two disjoint nonempty convex subsets of a finite dimensional space* X. *Then there exists some* $f \in X^*\backslash\{0\}$ *such that*

$$\forall a \in A, \ \forall b \in B \qquad f(a) \geq f(b).$$

Proof Since $C := A - B$ is convex and since A and B are disjoint, one has $0 \notin C$ and it suffices to take the linear form f provided by Theorem 3.8. □

Now let us deal with the possibly infinite dimensional case, for which one has to use the axiom of choice in the form of Zorn's Lemma. The analytical versions are intimately linked with the geometrical versions. In the latter case one is led to detect the special place of half-spaces among convex subsets; in the analytical versions, one sheds light on the special place of linear forms among sublinear forms.

We first observe that a sublinear form s on a linear space X is linear if (and only if) it is odd: for any $x, y \in X$, $r \in \mathbb{R}$, $r < 0$, one has

$$s(x+y) \le s(x) + s(y) = -s(-x) - s(-y) \le -s(-x-y) = s(x+y),$$
$$s(rx) = s(-|r|\, x) = -|r|\, s(x) = rs(x).$$

Proposition 3.19 *The space $S(X)$ of finite sublinear functions on the vector space X, ordered by the pointwise order, is (lower) inductive, hence has minimal elements. Each such element is a linear form.*

Proof We have to show that any totally ordered subset C of $S(X)$ has a lower bound. Let s_0 be a fixed element of C. For every $s \in C$, $x \in X$, we have

$$s(x) \ge \inf(s_0(x), -s_0(-x)),$$

since we have either $s \ge s_0$ or $s(x) \ge -s(-x) \ge -s_0(-x)$ if $s \le s_0$. It follows that $p : X \to \mathbb{R}$ given by $p(x) := \inf\{s(x) : s \in C\}$ is finite, and, as easily checked, it is sublinear. Thus, by Zorn's Lemma, $S(X)$ has minimal elements.

The second assertion is a consequence in the next lemma. This lemma is motivated by the observation preceding the statement which incites us to look for sublinear forms that are odd on some linear subspaces.

Lemma 3.14 *Let $s \in S(X)$ and let $u \in X$. Then the function s_u given by*

$$s_u(x) := \inf\{s(x - tu) - s(-tu) : t \in \mathbb{R}_+\}$$

is sublinear and such that $s_u \le s$, $s_u(u) = -s_u(-u)$.

Thus, when s is minimal in $S(X)$, one has $s_u = s$ and $s(u) = -s(-u)$ for all $u \in X$ and the proof of the proposition will be reduced to the following proof.

Proof We first observe that the infimum in the definition of $s_u(x)$ is finite since for all $t \in \mathbb{R}_+$ we have $s(-tu) \le s(x - tu) + s(-x)$, hence

$$\forall t \in \mathbb{R}_+ \qquad -s(-x) \le s(x - tu) - s(-tu).$$

Moreover, the inequality $s_u(x) \le s(x)$ stems from the choice $t = 0$ in the definition of s_u. It is easy to see that s_u is sublinear. Taking $t = 1$ in the definition of $s_u(x)$, we get $s_u(u) \le -s(-u)$. But since $0 \le s_u(u) + s_u(-u)$ and $s_u \le s$, we obtain

$$-s(-u) \le -s_u(-u) \le s_u(u) \le -s(-u),$$

hence $-s_u(-u) = s_u(u)$. □

Corollary 1.1 of Zorn's Lemma yields the following consequence.

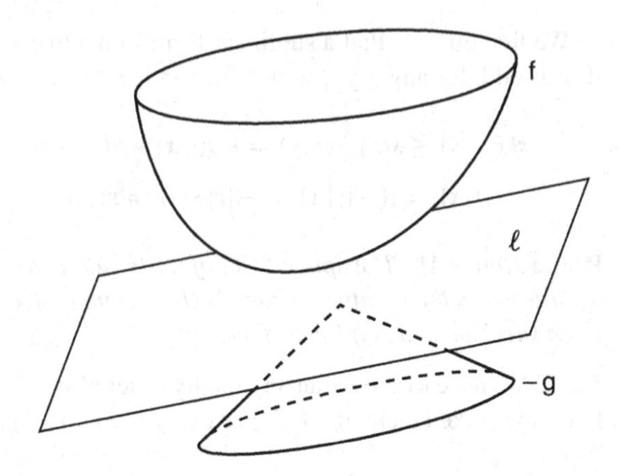

Fig. 3.2 The Sandwich Theorem

Corollary 3.12 *For every $s \in S(X)$ there exists some linear form ℓ on X such that $\ell \leq s$.*

The next statement is suggestive. It holds under more general assumptions (Exercise 1; Fig. 3.2). In fact, if $f : X \to \mathbb{R}$ is convex, g is as in the statement and $-g \leq f$ there exists a linear form ℓ on X such that $-g \leq \ell \leq f$. This follows from Theorem 3.9 by considering h given by $h(x) := inf_{t>0}(1/t)(f(tx) - f(0))$ that is sublinear and such that $-g \leq h \leq f$.

Theorem 3.9 (Sandwich Theorem) *Let $g : X \to \mathbb{R}_\infty := \mathbb{R} \cup \{+\infty\}$ and $h : X \to \mathbb{R}$ be sublinear functions on a linear space X. If $-g \leq h$ there exists a linear form ℓ on X such that $-g \leq \ell \leq h$.*

Proof Let $s : X \to \overline{\mathbb{R}}$ be defined by

$$s(x) := \inf\{h(x+y) + g(y) : y \in X\}.$$

Since $h(y) \leq h(x+y) + h(-x)$ and since $h(y) \geq -g(y)$ for all $y \in X$, we have

$$h(x+y) + g(y) \geq h(y) - h(-x) + g(y) \geq -h(-x),$$

so that $s(x) > -\infty$ for all $x \in X$, and, of course, $s(x) \leq h(x) < +\infty$. We easily verify that s is sublinear (in fact, s is the infimal convolution of h and $k : X \to \mathbb{R}_\infty$ given by $k(x) = g(-x)$ for $x \in X$) and that $s \leq h$, $s \leq k$. Thus, taking a linear form $\ell \leq s$, as in the preceding corollary, we have $\ell \leq h$, $\ell \leq k$, hence, for $x \in X$, $\ell(x) = -\ell(-x) \geq -k(-x) = -g(x)$. □

Theorem 3.10 (Hahn-Banach) *Let X_0 be a vector subspace of a real vector space X, let ℓ_0 be a linear form on X_0 and let $h : X \to \mathbb{R}$ be a sublinear functional such that $\ell_0(x) \leq h(x)$ for every $x \in X_0$. Then there exists a linear form ℓ on X extending ℓ_0 such that $\ell \leq h$.*

Proof Let $g : X \to \mathbb{R}_\infty$ be given by $g(x) = -\ell_0(x)$ for $x \in X_0$, $g(x) := +\infty$ for $x \in X \backslash X_0$. It is easy to see that g is sublinear and that $-g \leq h$. Taking ℓ such that $-g \leq \ell \leq h$, for $x \in X_0$ we get $\ell_0(x) = -g(x) \leq \ell(x)$ and similarly $\ell_0(-x) \leq \ell(-x)$, so that $\ell(x) = \ell_0(x)$ for all $x \in X_0$. □

Now let us turn to the case when X is endowed with a topology.

Corollary 3.13 *Let X be a topological vector space and let $h : X \to \mathbb{R}$ be a continuous sublinear functional. Then there exists a continuous linear form ℓ on X such that $\ell \leq h$.*

Proof By the preceding corollary there exists a linear form ℓ on X such that $\ell \leq h$. Let us prove that ℓ is continuous. Given $\varepsilon > 0$ we take a symmetric neighborhood V of 0 such that $h(x) \leq \varepsilon$ for all $x \in V$. Then for $x \in V$ we have $\ell(-x) \leq h(-x) \leq \varepsilon$, so that $|\ell(x)| \leq \varepsilon$ for all $x \in V$. Thus ℓ is continuous. □

In particular, if $p : X \to \mathbb{R}$ is a seminorm on X one can find a linear form ℓ on X such that $\ell \leq p$. Such an assertion can be made more precise. We just give a version with a norm.

Corollary 3.14 *Let X be a normed vector space and let $\overline{x} \in X$. Then there exists a continuous linear form ℓ on X such that $\|\ell\| = 1$ and $\ell(\overline{x}) = \|\overline{x}\|$.*

Proof Let $X_0 := \mathbb{R}\overline{x}$ and let ℓ_0 be the linear form on X_0 given by $\ell_0(r\overline{x}) = r\,\|\overline{x}\|$ for $r \in \mathbb{R}$. Thus, for every $x \in X_0$, one has $\ell_0(x) \leq h(x) := \|x\|$. The Hahn-Banach Theorem yields some linear form ℓ on X extending ℓ_0 such that $\ell \leq h$. Then one has $\|\ell\| \leq 1$ and $\ell(\overline{x}) = \|\overline{x}\|$, hence $\|\ell\| = 1$. □

The preceding corollary can be rephrased by saying that the *duality map* $J : X \to \mathcal{P}(X^*)$ defined as follows has nonempty values:

$$J(x) := \{x^* \in X^* : \langle x^*, x \rangle = \|x\|^2,\ \|x^*\| = \|x\|\}.$$

This (multi)map or set-valued map is a useful tool, in particular for the geometry of normed spaces and for the study of dissipative operators. Explicit expressions for some spaces are to be found in the exercises below.

Another tool is given in the next corollary. On a special class of normed spaces called inner product spaces or Hilbert spaces we shall dispose of a better tool satisfying a bilinear property.

Corollary 3.15 (Lumer) *For any normed space $(X, \|\cdot\|)$ there exists a* semi-scalar product, *i.e. a function $[\cdot,\cdot] : X \times X \to \mathbb{R}$ such that for all $x, y, z \in X$, $r \in \mathbb{R}$ one has*

$$[x, y + z] = [x, y] + [x, z], \quad [x, ry] = r[x, y],$$

$$|[x, y]| \leq \|x\| \cdot \|y\|, \qquad [x, x] = \|x\|^2.$$

Proof It suffices to take a selection j of J, i.e. a map $j : X \to X^*$ such that $j(x) \in J(x)$ for all $x \in X$ and to set $[x, y] = \langle j(x), y \rangle$. Note that we can even require that $[tx, y] = t[x, y]$ for all $(t, x, y) \in \mathbb{R}_+ \times X \times X$. □

Corollary 3.14 is a special case of the next corollary.

Corollary 3.16 *Let X be a normed vector space and let Y be a vector subspace of X. Then any continuous linear form y^* on Y has a linear continuous extension x^* to X such that $\|x^*\| = \|y^*\|$.*

Proof Let $c := \|y^*\|$. Theorem 3.10 yields some linear form ℓ on X extending y^* and satisfying $\ell \leq c \|\cdot\|$. Then $x^* := \ell$ is continuous and $\|x^*\| = c$. □

Corollary 3.17 *Let Y be a closed linear subspace of a normed space X. If $Y \neq X$ there exists a non-null continuous linear form f on X that is null on Y.*

Proof Let $p : X \to X/Y$ be the quotient map. Since $Y \neq X$ one can find some non-null $z \in X/Y$. Then Corollary 3.14 yields some ℓ in the dual of X/Y such that $\ell(z) \neq 0$. Then $f = \ell \circ p$ is non null on X and null on Y. □

Corollary 3.18 *Let Y be a closed vector subspace of a Banach space X. Then Y^* is isometric to $X^*/Y^\perp$, where $Y^\perp := \{x^* \in X^* : x^*(y) = 0 \ \forall y \in Y\}$.*

Proof Let $r : X^* \to Y^*$ be the restriction map given by $r(x^*) := x^* \mid_Y$. Corollary 3.16 ensures that r is onto. The kernel of r being precisely $Y^\perp$, one can factorize r as $r = q \circ p$, where $p : X^* \to X^*/Y^\perp$ is the canonical projection and $q : X^*/Y^\perp \to Y^*$ is bijective and continuous. Giving to $X^*/Y^\perp$ the quotient norm defined by $\|z\| := \inf\{\|x^*\| : x^* \in p^{-1}(z)\}$, Corollary 3.16 can serve to prove that q is isometric. □

Corollary 3.19 *Given normed spaces X, Y, the transpose $A^\intercal$ of $A \in L(X, Y)$ satisfies $\|A^\intercal\| = \|A\|$.*

Proof We have seen that $\|A^\intercal\| \leq \|A\|$. Given $\overline{x} \in X$, Corollary 3.14 yields some $\overline{y}^* \in Y^*$ such that $\|\overline{y}^*\| = 1$ and $\langle \overline{y}^*, A\overline{x} \rangle = \|A\overline{x}\|$. Then $\|A\overline{x}\| = \langle A^\intercal \overline{y}^*, \overline{x} \rangle \leq \|A^\intercal \overline{y}^*\| . \|\overline{x}\| \leq \|A^\intercal\| . \|\overline{x}\|$. Thus $\|A\| \leq \|A^\intercal\|$ and equality holds. □

Now let us turn to geometric forms of the Hahn-Banach Theorem. We first consider an algebraic version. We recall that a subset C of a vector space X is said to be *absorbing* if for all $x \in X$ there exists some $r > 0$ such that $rx \in C$. The *core* of a convex subset C of X is the set of points $a \in X$ such that $C - a$ is absorbing.

Proposition 3.20 *Let C be an absorbing convex subset of a vector space X and let $e \in X \backslash \operatorname{core} C$. Then there exists a hyperplane H of X such that $e \in H$ and $H \cap \operatorname{core} C = \varnothing$. Moreover, C is contained in one of the strict half-spaces determined by H.*

Proof Let $j := \mu_C$ be the Minkowski gauge of C:

$$j(x) := \inf\{t > 0 : x \in tC\}.$$

Since C is absorbing and convex, j is finite on X and sublinear. For all $x \in \operatorname{core} C$ one has $j(x) < 1$ since there exists some $r > 0$ such that $rx \in C - x$, hence $j(x) \leq (1 + r)^{-1}$. Conversely, if $j(x) < 1$ then $x \in \operatorname{core} C$ since for all $u \in X$ and for $\varepsilon > 0$ such that $\varepsilon \max(j(u), j(-u)) < 1 - j(x)$ one has, for all $r \in [-\varepsilon, \varepsilon]$, $j(x + ru) \leq j(x) + j(ru) < 1$, hence $x + ru \in tC \subset C$ for some $t \in]0, 1[$. Since $e \in X \backslash \operatorname{core} C$, we have $j(e) \geq 1$. Let $X_0 := \mathbb{R}e$, and let $\ell_0 : X_0 \to \mathbb{R}$ be given by $\ell_0(re) := rj(e)$. Then, since $rj(e) \leq 0 \leq j(re)$ for $r \leq 0$, we have $\ell_0 \leq j \mid X_0$, so that there exists some linear form h on X extending ℓ_0 with $h \leq j$. Let $H := \{x \in X : h(x) = j(e)\}$. Then $e \in H$ and for $x \in \operatorname{core} C$ we have $h(x) \leq j(x) < 1 \leq j(e)$ hence $x \notin H$ and $\operatorname{core} C \subset h^{-1}(] - \infty, j(e)[)$. □

A topological version is eased by the following observation.

Lemma 3.15 *Let X be a normed vector space, let $c \in \mathbb{R}$, and let h be a non-null linear form on X. The hyperplane $H := h^{-1}(c)$ is closed if and only if h is continuous.*

Proof Obviously, when h is continuous, H is closed, $\{c\}$ being closed in $\mathbb{R}$. Conversely, suppose that H is closed. Since h is non-null, $X \backslash H$ is nonempty. Let $x_0 \in X \backslash H$, so that there exists some $r > 0$ for which $B(x_0, r) \subset X \backslash H$. Assuming $h(x_0) < c$ (the other possibility is dealt with by changing h, c into $-h$, $-c$), let us note that $h(x) < c$ for all $x \in B(x_0, r)$: otherwise, if there exists an $x_1 \in B(x_0, r)$ such that $h(x_1) > c$, for $t := (f(x_0 - c)(f(x_0) - f(x_1))^{-1}$ one has $h((1-t)x_0 + tx_1) = c$ and $(1 - t)x_0 + tx_1 \in B(x_0, r)$, contradicting $B(x_0, r) \subset X \backslash H$. Then, for all $u \in B(0, 1)$ we have $h(x_0 + ru) < c$ or $h(u) < r^{-1}(c - h(x_0))$ and h is continuous. □

Theorem 3.11 (Eidelheit) *Let A and B be two disjoint nonempty convex subsets of a topological vector space X. If A is open, then there exists a closed hyperplane H separating A and B: for some $f \in X^* \backslash \{0\}$, $r \in \mathbb{R}$ one has*

$$\forall a \in A,\ \forall b \in B \qquad f(a) > r \geq f(b).$$

Proof Let $D := A - B := \{a - b : a \in A,\ b \in B\}$. It is a convex subset of X which is open as the union over $b \in B$ of the translated sets $A - b$, and $0 \notin D$. Taking $e \in D$ and setting $C := e - D$, we see that $e \notin C$, $0 \in C$ and C is absorbing. Thus, there exist some $s > 0$ and some linear form f on X such that $f(e) = s$ and $f(x) < s$ for all $x \in C$. Since f is bounded above on the neighborhood C of 0, f is continuous. Moreover, for $a \in A$, $b \in B$ one has $f(e - a + b) < s = f(e)$, hence $f(a) \geq \sup f(B)$. In fact, since A is open and $f \neq 0$, one must have $f(a) > r := \sup f(B)$. □

Theorem 3.12 (Hahn-Banach Strong Separation Theorem) *Let A and B be two disjoint nonempty convex subsets of a normed space (or a locally convex topological vector space) X. If A is compact and B is closed, then there exists some $f \in X^* \backslash \{0\}$ and some $r \in \mathbb{R}$, $\delta > 0$ such that*

$$\forall a \in A,\ \forall b \in B \qquad f(a) > r + \delta > r > f(b).$$

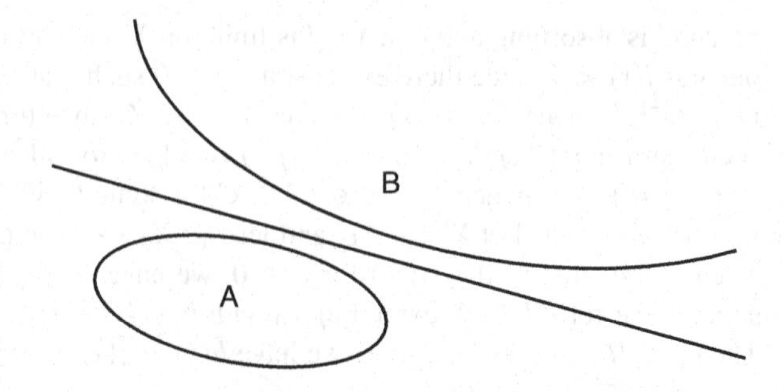

Fig. 3.3 Separation of two convex subsets

Proof For every $a \in A$ there exists a symmetric open convex neighborhood V_a of 0 in X such that $(a + 2V_a) \cap B = \varnothing$. Let F be a finite subset of A such that the family $(a + V_a)_{a \in F}$ forms a finite covering of A. Then, if V is the intersection of the family $(V_a)_{a \in F}$, V is an open neighborhood of 0 and $A' \cap B = \varnothing$ for $A' := A + V$. The Edelheit Theorem yields $f \in X^* \setminus \{0\}$ and $s \in \mathbb{R}$ such that $f(a) > s \geq f(b)$ for all $a \in A'$, $b \in B$. The compactness of A ensures that there exists a $\delta > 0$ such that $f(a) > s + 2\delta$ for all $a \in A$. Setting $r := s + \delta$, we get the result (Fig. 3.3). □

Example The compactness assumption on A cannot be omitted, as shown by the example of $X = \mathbb{R}^2, A := \{(r,s) \in \mathbb{R}^2_+ : rs \geq 1\}, B := \mathbb{R} \times]-\infty, 0]$.

The following application to approximate solutions to linear systems will be used later on.

Lemma 3.16 (Helly) *Let X be a normed space, let $f_1, \ldots, f_n$ in the dual X^* of X and let $a_1, \ldots, a_n$ be real numbers. The following assertions are equivalent:*

(a) for any $\varepsilon > 0$ there exists some $x \in B_X$ such that $|f_i(x) - a_i| \leq \varepsilon$ for all $i \in \mathbb{N}_n$;
(b) for all $(r_1, \ldots, r_n) \in \mathbb{R}^n$ one has $\left|\sum_{i=1}^n r_i a_i\right| \leq \left\|\sum_{i=1}^n r_i f_i\right\|$.

Proof (a)⇒(b) Given $(r_1, \ldots, r_n) \in \mathbb{R}^n$, let $s := |r_1| + \ldots + |r_n|$, so that by (a), given $\varepsilon > 0$ one can find $x_\varepsilon \in B_X$ such that

$$\left|\sum_{i=1}^n r_i f_i(x_\varepsilon) - \sum_{i=1}^n r_i a_i\right| \leq \sum_{i=1}^n |r_i| \, |f_i(x_\varepsilon) - a_i| \leq \varepsilon s.$$

Thus, since $\|x_\varepsilon\| \leq 1$,

$$\left|\sum_{i=1}^n r_i a_i\right| \leq \left|(\sum_{i=1}^n r_i f_i)(x_\varepsilon)\right| + \varepsilon s \leq \left\|\sum_{i=1}^n r_i f_i\right\| + \varepsilon s.$$

Since $\varepsilon > 0$ is arbitrarily small, we get the inequality of assertion (b).

(b)⇒(a) Let us consider the map $f : X \to \mathbb{R}^n$ with components $f_1, \ldots, f_n$ and let $a := (a_1, \ldots, a_n)$. Assertion (a) means that a belongs to the closure $\mathrm{cl}(f(B_X))$ of the image of B_X under f. If that does not hold, by compactness of $f(B_X)$, one can find $x^* = (r_1, \ldots, r_n) \in \mathbb{R}^n$ and $c \in \mathbb{R}$ such that

$$\forall x \in B_X \qquad x^*.f(x) < c < x^*.a.$$

Then (b) does not hold since these relations imply that

$$\left\| \sum_{i=1}^{n} r_i f_i \right\| = \sup_{x \in B_X} \left| (\sum_{i=1}^{n} r_i f_i)(x) \right| = \sup_{x \in B_X} |x^*.f(x)| \leq c < \sum_{i=1}^{n} r_i a_i.$$

□

A special case of the Fenchel transform we will study later on is the passage from closed convex subsets (or their indicator functions) to their support functions. Recall that the *support function* h_C or σ_C of a subset C of a normed space X is the function $h_C : X^* \to \overline{\mathbb{R}}$ given by

$$h_C(x^*) := \sup\{\langle x^*, x\rangle : x \in C\}.$$

Corollary 3.20 (Hörmander) *The map $h : C \mapsto h_C$ is an injective lattice morphism from the set $\mathcal{C}(X)$ of nonempty closed convex subsets of the normed space X into the space $\mathcal{H}(X)$ of positively homogeneous functions on X null at* 0. *Moreover, $h_{\lambda C} = \lambda h_C$ for all $\lambda \in \mathbb{R}_+$, $C \in \mathcal{C}(X)$ and $h_{\mathrm{cl}(A+B)} = h_A + h_B$ for all $A, B \in \mathcal{C}(X)$.*

Proof We just prove the injectivity of h, leaving the other assertions as exercises. It suffices to prove that for $C, D \in \mathcal{C}(X)$ satisfying $h_C \geq h_D$ one has $C \supset D$ since the roles of C and D can be interchanged. Given $b \in X \setminus C$ we can find $x^* \in X$ such that $\langle x^*, b\rangle > \sup_{x \in C}\langle x^*, x\rangle$. Then we cannot have $b \in D$ since otherwise we would have $h_D(x^*) \geq \langle x^*, b\rangle > \sup_{x \in C}\langle x^*, x\rangle = h_C(x^*)$. □

Exercises

1. Let X and Y be finite dimensional spaces, let $A \in L(X, Y)$ and let $f : X \to \mathbb{R}_\infty$, $g : Y \to \mathbb{R}_\infty$ be convex functions such that $\mathbb{R}_+(\mathrm{dom} g - A(\mathrm{dom} f)) = Y$ and $f \geq -g \circ A$. Prove that there exist some $\ell \in X^*$ and $c \in \mathbb{R}$ satisfying $f \geq \ell - c \geq -g \circ A$.
2. Prove the **Mazur-Orlicz Theorem**: Let $h : X \to \mathbb{R}$ be a sublinear functional on some vector space X and let C be a nonempty convex subset of X. Then there exists a linear form ℓ on X such that $\ell \leq h$ and $\inf \ell(C) = \inf h(C)$ [See [232, p.13].]
3. Prove the **Mazur-Bourgin Theorem**: Let C be a convex subset with nonempty interior in a topological vector space X and let A be an affine subspace of X such

that $A \cap \text{int}C = \varnothing$. Prove that there exists a hyperplane H of X containing A which does not meet int$C$ [See [164, p. 5].]

4. Prove the **Mazur Theorem**: Let (x_n) be a sequence in a normed space X that weakly converges to some $x \in X$. Then there exists a sequence (y_n) strongly converging to x such that, for all $k \in \mathbb{N}$, y_k is a convex combination of the x_n's. [Hint: Consider the closed convex hull of $\{x_n : n \in \mathbb{N}\}$.]
5. Prove the Sandwich Theorem using the Eidelheit's Theorem.
6. Prove the **Stone's Theorem**: Let A and B be disjoint convex subsets of a normed space X. Show that there exists a pair (C, D) of disjoint convex subsets satisfying $A \subset C$ and $B \subset D$ which is maximal with respect to the order induced by inclusion. Show that when A is open one can take for C and D opposite half spaces, C being open.
7. Verify that for $p \in]1, \infty[$, $q := (1 - 1/p)^{-1}$ and the usual norm, the *duality map* $J : L_p(S, \mu) \to L_q(S, \mu)$ is single-valued and is given by $J(x)(s) = \|x\|_p^{2-p} |x(s)|^{p-2} x(s)$ for $s \in S$ such that $x(s) \neq 0$, $J(x)(s) = 0$ for $s \in S$ such that $x(s) = 0$.
8. Verify that the *duality (multi)map* $J : L_1(S, \mu) \to (L_1(S, \mu))^*$ is given by $J(x)(s) = \{\|x\|_1 x(s)/ |x(s)|\}$ for $s \in S$ such that $x(s) \neq 0$, $J(x)(s) = \{y(s) : y \in L_\infty(S, \mu),\ \|y\|_\infty \leq \|x\|_\infty\}$ for $s \in x^{-1}(0)$, $(L_1(S, \mu))^*$ being identified with $L_\infty(S, \mu)$.

3.3.3 *Polarity and Orthogonality*

Let us give a short account of *polarity*, a passage from a subset of a normed space X to a subset of the dual X^* of X (or the reverse, or, more generally from a subset of X to a subset of a space Y paired with X by a bilinear coupling function). This correspondence is a geometric analogue of a correspondence for functions, the Fenchel conjugacy, we will study in Chap. 6.

The *polar set* S^0 of a subset S of X is the set defined as follows with the help of its support function

$$S^0 := h_S^{-1}(] - \infty, 1]) := \{x^* \in X^* : \forall x \in S\ \langle x^*, x\rangle \leq 1\}.$$

Clearly, S^0 is a weak* closed convex subset of X^* containing 0. If S is a cone, then S^0 is a convex cone and $S^0 := \{x^* \in X^* : \forall x \in S\ \langle x^*, x\rangle \leq 0\}$; if S is a linear subspace, then S^0 is the linear subspace

$$S^\perp := \{x^* \in X^* : \forall x \in S\ \langle x^*, x\rangle = 0\},$$

also called the *orthogonal* of S. It is also easy to show that

$$(S \cup T)^0 = S^0 \cap T^0.$$

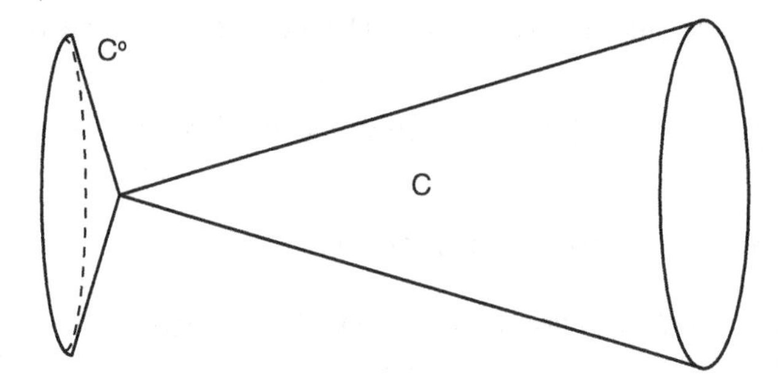

Fig. 3.4 Polar cones

A base of neighborhoods of 0 for the weak* topology is formed by the polar sets of finite subsets. On the other hand, one has the following classical theorem (Fig. 3.4).

Theorem 3.13 (Alaoglu-Bourbaki) *Let X be a normed space and let S be a neighborhood of 0. Then S^0 is weak* compact.*

Proof Since $S^0 \subset T^0$ when $T \subset S$ and since S^0 is weak* closed, it suffices to prove the result when S is a ball centered at 0. Since $(rS)^0 = r^{-1}S^0$ for $r > 0$, we may suppose $S = B_X$. Then $S^0 = B_{X^*}$ and the result has been shown in Theorem 3.7 in that case. □

The polar S_*^0 of a subset S_* of X^* is defined similarly by

$$S_*^0 := \{x \in X : \forall x^* \in S_* \ \langle x^*, x\rangle \leq 1\}.$$

If S is a subset of X, then its *bipolar* is the set $S^{00} := (S^0)^0$.

Corollary 3.21 (Bipolar Theorem) *For every nonempty subset S of a normed space X, its bipolar is the closed convex hull of $S \cup \{0\}$: $S^{00} := \overline{\text{co}}(S \cup \{0\})$. In particular, if S is a convex subset of X containing 0, then S^{00} is the closure* $\text{cl}(S)$ *of S.*

Proof Let $C := \overline{\text{co}}(S \cup \{0\})$. Since one has $S \subset S^{00}$, and since S^{00} is closed convex and contains 0, one has $C \subset S^{00}$. Given $b \in X\backslash C$, Theorem 3.12 yields $x^* \in X^*$ and $r \in \mathbb{R}$ such that $\langle x^*, b\rangle > r > \langle x^*, a\rangle$ for all $a \in C$. Since $0 \in C$, one has $r > 0$ and $r^{-1}x^* \in C^0 \subset S^0$, hence $b \notin S^{00}$. Therefore $S^{00} = C$. □

Remark One must be careful with the use of polar sets. One cannot apply the preceding result with S a subset of the dual space of X, unless one replaces the closure with the weak* closure $\overline{\text{co}}^*(S \cup \{0\})$ of $\text{co}(S \cup \{0\})$, using Proposition 3.18.

Let us give some calculus rules. They will be completed by a more refined result (Theorem 3.25).

Proposition 3.21 *Let G and H be cones (resp. linear subspaces) of a normed space E. Then*

$$G^0 \cap H^0 = (G+H)^0 \qquad (\text{resp. } G^\perp \cap H^\perp = (G+H)^\perp\ ), \tag{3.5}$$

so that when G and H are convex cones one has $(G^0 \cap H^0)^0 = \mathrm{cl}(G+H)$.

If G and H are closed, convex cones, then

$$G \cap H = (G^0 + H^0)^0 \qquad (\text{resp. } G \cap H = (G^\perp + H^\perp)^\perp). \tag{3.6}$$

Proof Since $M^0 \subset L^0$ when $L \subset M$ and since $L^0 = (L \cup \{0\})^0$, we have $(G + H)^0 \subset G^0 \cap H^0$. Conversely, for any $f \in G^0 \cap H^0$ and any $x \in G$, $y \in H$ we have $\langle f, x+y\rangle = \langle f, x\rangle + \langle f, y\rangle \leq 0$, hence $f \in (G+H)^0$.

If G and H are closed, convex cones, replacing G and H with their polar cones in relation (3.5), we get $G \cap H = G^{00} \cap H^{00} = (G^0 + H^0)^0$. □

The following is a prototype of a duality result for minimization problems.

Proposition 3.22 *Given a convex cone C of a normed space X and an element* $\overline{x}$ *of X one has*

$$d(\overline{x}, C) = \max\{\langle x^*, \overline{x}\rangle : x^* \in C^0,\ \|x^*\| \leq 1\}.$$

If D is a weak closed cone of* X^**, then for all* $\overline{x}^* \in X^*$ *one has*

$$d(\overline{x}^*, D) = \sup\left\{\langle \overline{x}^*, x\rangle : x \in D^0 \cap B_X\right\}.$$

Proof For every $x^* \in C^0 \cap B_{X^*}$ and any $x \in C$ we have

$$\langle x^*, \overline{x}\rangle \leq \langle x^*, \overline{x} - x\rangle \leq \|x^*\| . \|\overline{x} - x\| \leq \|\overline{x} - x\|\,, \tag{3.7}$$

hence, taking the infimum over $x \in C$, $\langle x^*, \overline{x}\rangle \leq \delta := d(\overline{x}, C)$. Now, since C and the open ball $B(\overline{x}, \delta)$ are disjoint, we can separate them by a hyperplane $H := \{x \in X : \langle \overline{x}^*, x\rangle = c\}$ with $\|\overline{x}^*\| = 1$, $c \in \mathbb{R}$:

$$\forall x \in C,\ y \in B(\overline{x}, \delta) \qquad \langle \overline{x}^*, x\rangle \leq c < \langle \overline{x}^*, y\rangle.$$

Replacing x by tx with $t \to \infty$ and $t \to 0_+$, we see that $\overline{x}^* \in C^0$ and $c \geq 0$. Taking the infimum over $y \in B(\overline{x}, \delta) = \overline{x} + B(0, \delta)$ we get $c \leq \langle \overline{x}^*, \overline{x}\rangle - \delta$, hence $\delta \leq \langle \overline{x}^*, \overline{x}\rangle$. With (3.7) this shows that the supremum of $\{\langle x^*, \overline{x}\rangle : x^* \in C^0 \cap B_{X^*}\}$ is attained for $x^* = \overline{x}^*$ and its value is δ.

Now let D be a weak* closed cone of X^* and let $\overline{x}^* \in X^*$. Inequalities similar to those in (3.7) show that $\sup\{\langle \overline{x}^*, x\rangle : x \in D^0 \cap B_X\} \leq \delta := d(\overline{x}^*, D)$. For all $r \in]0, \delta[$

the closed ball $B[\overline{x}^*, r]$ is weak* compact and disjoint from the weak* closed convex set D. Theorem 3.12 yields some $c_r \in \mathbb{R}$ and $x_r \in X$, the dual of $(X^*, \sigma(X^*, X))$, such that $\|x_r\| = 1$ and

$$\forall x^* \in D,\ y^* \in B[\overline{x}^*, r] \qquad \langle x^*, x_r\rangle \leq c_r < \langle y^*, x_r\rangle .$$

By homogeneity we see that $x_r \in D^0$. Since $0 \in D$ we have $c_r \geq 0$ and taking the infimum over $y^* \in B[\overline{x}^*, r]$ we get $c_r \leq \langle \overline{x}^*, x_r\rangle - r$. Thus, $r \leq \langle \overline{x}^*, x_r\rangle \leq \sup\{\langle \overline{x}^*, x\rangle : x \in D^0 \cap B_X\}$ and since r is arbitrarily close to δ we get the second equality of the statement. □

Remark Moreover, if the infimum over $w \in C$ of the distances $\|\overline{x} - w\|$ is attained at $\overline{w} \in C$, then one has $\langle \overline{x}^*, \overline{x} - \overline{w}\rangle = \|\overline{x} - \overline{w}\|$ since the following inequalities are equalities:

$$\|\overline{x} - \overline{w}\| = \langle \overline{x}^*, \overline{x}\rangle \leq \langle \overline{x}^*, \overline{x} - \overline{w}\rangle \leq \|\overline{x} - \overline{w}\| .$$

□

The following lemma will be used to establish a minimax theorem. In it, we denote by $\vee$ and $\wedge$ the operations given by

$$r_1 \vee \ldots \vee r_k = \max(r_1, \ldots, r_k), \qquad r_1 \wedge \ldots \wedge r_k = \min(r_1, \ldots, r_k)$$

for $r_i \in \mathbb{R}$, $i = 1, \ldots, k$ and, as above, Δ_k stands for the canonical simplex of $\mathbb{R}^k$: $\Delta_k := \{(s_1, \ldots, s_k) \in \mathbb{R}^k_+ : s_1 + \ldots + s_k = 1\}$. As usual, max (resp. min) means that one has attainment of the supremum (resp. infimum) when it is finite.

Lemma 3.17 *Let $f_1, \ldots, f_k$ be convex functions on a convex subset C of a vector space X. Then*

$$\inf_C (f_1 \vee \ldots \vee f_k) = \max\{\inf_C (s_1 f_1 + \ldots + s_k f_k) : s := (s_1, \ldots, s_k) \in \Delta_k\}.$$

If $g_1, \ldots, g_k$ are concave functions on C then

$$\sup_C (g_1 \wedge \ldots \wedge g_k) = \min\{\sup_C (s_1 g_1 + \ldots + s_k g_k) : s := (s_1, \ldots, s_k) \in \Delta_k\}.$$

Proof Let $h := f_1 \vee \ldots \vee f_k$. Then for each $s := (s_1, \ldots, s_k) \in \Delta_k$ we have $h \geq h_s := s_1 f_1 + \ldots + s_k f_k$, hence $\inf_C h \geq \inf_C h_s$ and $\inf_C h \geq \sup\{\inf_C (s_1 f_1 + \ldots + s_k f_k) : s := (s_1, \ldots, s_k) \in \Delta_k\}$, with equality if $\inf_C h = -\infty$. Now let

$$A := \{r = (r_1, \ldots, r_k) \in \mathbb{R}^k : \exists x \in C,\ r_i > f_i(x)\ i = 1, \ldots, k\},$$

which is convex. For $t \leq \inf_C h$ one has $b := (t, \ldots, t) \notin A$. The finite dimensional separation theorem yields some $\bar{s} = (\bar{s}_1, \ldots, \bar{s}_k) \in \mathbb{R}^k \backslash \{0\}$ such that

$$\bar{s}_1 r_1 + \ldots + \bar{s}_k r_k \geq \bar{s}_1 t + \ldots + \bar{s}_k t \qquad \forall r = (r_1, \ldots, r_k) \in A.$$

We have $\bar{s}_i \geq 0$ for $i = 1, \ldots, k$ since r_i can be arbitrarily large. Since $\bar{s} \neq 0$, by homogeneity, we may suppose $\bar{s}_1 + \ldots + \bar{s}_k = 1$, i.e. $\bar{s} \in \Delta_k$. Then, for each $x \in C$, since r_i can be arbitrarily close to $f_i(x)$ we get

$$\bar{s}_1 f_1(x) + \ldots + \bar{s}_k f_k(x) \geq \bar{s}_1 t + \ldots + \bar{s}_k t = t.$$

Therefore $\inf_C(\bar{s}_1 f_1 + \ldots + \bar{s}_k f_k) \geq t$ and since t can be arbitrarily close to $\inf_C h$, we get $\sup_{s \in \Delta_k} \inf_C(s_1 f_1 + \ldots + s_k f_k) \geq \inf_C h$, so that equality holds. When $\inf_C h$ is finite we can take $t = \inf_C h$ and the inequality $\inf_C(\bar{s}_1 f_1 + \ldots + \bar{s}_k f_k) \geq t$ shows that we have attainment for this $\bar{s} \in \Delta_k$.

The second assertion is obtained by setting $f_i := -g_i$. □

Theorem 3.14 (Infimax Theorem) *Let A and B be nonempty convex subsets of vector spaces X and Y respectively, and let $\ell : A \times B \to \mathbb{R}$ be a function that is convex in its first variable and concave in its second variable. Then, if B is compact with respect some topology on Y and if ℓ is upper semicontinuous in its second variable, one has*

$$\inf_{x \in A} \max_{y \in B} \ell(x, y) = \max_{y \in B} \inf_{x \in A} \ell(x, y).$$

Proof The inequality $\alpha := \inf_{x \in A} \sup_{y \in B} \ell(x, y) \geq \sup_{y \in B} \inf_{x \in A} \ell(x, y) := \beta$ is valid without any assumption. Here we can write max instead of sup since B is compact and $\ell(x, \cdot)$ is u.s.c. for each $x \in A$ as is $\inf_{x \in A} \ell(x, \cdot)$. Given $k \in \mathbb{N} \backslash \{0\}$ and $a_1, \ldots, a_k \in A$, applying the preceding lemma with $C = B$, $g_i = \ell(a_i, \cdot)$, we can find $s \in \Delta_k$ such that

$$\sup_{b \in B}(\ell(a_1, b) \wedge \ldots \wedge \ell(a_k, b)) = \sup_{b \in B}(s_1 \ell(a_1, b) + \ldots + s_k \ell(a_k, b)).$$

Since $\ell(\cdot, b)$ is convex for each $b \in B$, we get

$$\sup_{b \in B}(\ell(a_1, b) \wedge \ldots \wedge \ell(a_k, b)) \geq \sup_{b \in B}(\ell(s_1 a_1 + \ldots + s_k a_k, b)) \geq \alpha.$$

Introducing for $a \in A$ the closed subset $B_a := \{b \in B : \ell(a, b) \geq \alpha\}$, which is nonempty by the Weierstrass' Theorem, we deduce from these inequalities that $B_{a_1} \cap \ldots \cap B_{a_k}$ is nonempty. The finite intersection property of the compact space B ensures that $\bigcap_{a \in A} B_a$ is nonempty. This means that there exists some $\bar{b} \in B$ such that $\inf_{a \in A} \ell(a, \bar{b}) \geq \alpha$. Thus $\beta \geq \alpha$ and equality holds. □

Exercises

1. Let $j : Y \to X$ be the canonical injection of a vector subspace Y of a normed space X into X and let $j^{\intercal} : X^* \to Y^*$ be its transpose map given by $j^{\intercal}(x^*) := x^* \circ j$ for $x^* \in X^*$. Rephrase Corollary 3.16 as: $j^{\intercal}$ is surjective. Show that the kernel of $j^{\intercal}$ is the polar Y^0 of Y and that Y^* can be isometrically identified with X^*/Y^0.
2. Let $A : W \to X$ be a continuous linear operator with transpose map $A^{\intercal} : X^* \to W^*$ given by $A^{\intercal}(x^*) = x^* \circ A$. Show that for $D := A(C)$ one has $D^0 = (A^{\intercal})^{-1}(C^0)$.
3. Let $A : W \to X$ be a continuous linear operator with transpose map $A^{\intercal} : X^* \to W^*$ and let D be a closed convex subset of X containing the origin. Prove that $(A^{-1}(D))^0 = \mathrm{cl}^*(A^{\intercal}(D^0))$.
4. Let G and H be closed convex cones of a normed space X. Prove that $(G \cap H)^0 = \mathrm{cl}^*(G^0 + H^0)$ and that one can omit the weak* closure if $G \cap \mathrm{int}H \neq \varnothing$.
5. Verify that the sets $G := \{(x, y, z) : x^2 + y^2 \leq z^2,\ z \geq 0\}$ and $H := \{(x, y, z) : y = z\}$ are closed convex cones of $\mathbb{R}^3$. Describe the polar cones G^0, H^0, $(G \cap H)^0$. Show that $(1, 1, -1) \in (G \cap H)^0 \backslash (G^0 + H^0)$ and deduce that the sum of two closed convex cones is not necessarily closed.
6. (**Ky Fan's Infimax Theorem**) Prove the conclusion of Theorem 3.14 with the assumptions that A is a nonempty set, that B is a nonempty compact topological space, and that f is upper semicontinuous in its second variable and is *convex-concave-like* in the following sense: for any $t \in [0, 1]$ and any $x_1,\ x_2 \in A$, $y_1, y_2 \in B$ there exist some $x_3 \in A$, $y_3 \in B$ such that

$$\ell(x_3, y) \leq (1 - t)\ell(x_1, y) + t\ell(x_2, y) \qquad \forall y \in B,$$
$$\ell(x, y_3) \geq (1 - t)\ell(x, y_1) + t\ell(x, y_2) \qquad \forall x \in A.$$

   [Hint: adapt the proof of the Infimax Theorem.]
7. (**Sion's Infimax Theorem**) Prove the conclusion of Theorem 3.14 with the assumption that f is convex-concave changed into the assumption that f is quasiconvex-quasiconcave, i.e. for all $r \in \mathbb{R}$, $x \in A$, $y \in B$ the sets $\{a \in A : f(a, y) \leq r\}$ and $\{b \in B : f(x, b) \geq r\}$ are convex.

3.4 Couplings and Reflexivity

3.4.1 Couplings

It may be useful to consider weak topologies in a symmetric way. For such a purpose, given normed vector spaces X and Y, we consider a *coupling* $c : X \times Y \to \mathbb{R}$ (or $c : X \times Y \to \mathbb{C}$ if X and Y are complex spaces, but here we consider only real vector spaces), i.e. a continuous bilinear function such that the maps $c_X : X \to Y^*$ and $c_Y : Y \to X^*$ given by

$$c_X(x) := c(x, \cdot) \quad x \in X,$$
$$c_Y(y) := c(\cdot, y) \quad y \in Y$$

are injective. This means that if $x \in X$ is such that $c(x, y) = 0$ for all $y \in Y$ then $x = 0$ and symmetrically if $y \in Y$ is such that $c(x, y) = 0$ for all $x \in X$ then $y = 0$. We say that c is a *metric coupling* if c_X and c_Y are *monometries*, i.e. isometries onto their images, or, in other terms, preserve the norms.

For any coupling $c : X \times Y \to \mathbb{R}$ the map $c_Y : Y \to X^*$ allows us to identify Y with a subspace of X^* endowed with a stronger norm (and symmetrically c_X allows us to identify X with a subspace of Y^*). When c is a metric coupling the norm of Y (resp. X) is the induced norm.

The main example of a coupling is the evaluation map $e : X \times X^* \to \mathbb{R}$ given by $e(x, x^*) := x^*(x)$ for $(x, x^*) \in X \times X^*$. The assertion that it is a coupling relies on Corollary 3.14. Let us note that it is a metric coupling. Since e_{X^*} is just the identity map I_{X^*} of X^*, we have $\|e_{X^*}(\cdot)\|_{X^*} = \|\cdot\|_{X^*}$. On the other hand, Corollary 3.14 asserts that for all $x \in X$ one can find some $\ell_x \in X^*$ such that $\|\ell_x\| = 1$ and $\ell_x(x) = \|x\|$; since $|x^*(x)| \leq \|x\|$ for all x^* in the closed unit ball B_{X^*} of X^*, this means that

$$\|e_X(x)\|_{X^{**}} := \sup_{x^* \in B_{X^*}} e_X(x)(x^*) = \sup_{x^* \in B_{X^*}} x^*(x) = \ell_x(x) = \|x\| .$$

The map $e_X : X \to X^{**}$ is called the *canonical embedding*. When it is surjective the space X is said to be *reflexive*.

3.4.2 Reflexivity and Weak Topologies

Not all Banach spaces are reflexive (see the exercises). We shall soon give examples of reflexive Banach spaces.

Proposition 3.23 *If $c : X \times Y \to \mathbb{R}$ is a metric coupling between Banach spaces and if X is reflexive, then Y can be identified with the dual X^* of X via the map c_Y.*

Proof Let $Z := c_Y(Y)$ in X^*. Since c_Y is an isometry from Y onto Z, Z is a closed subspace of X^*. In order to prove that $Z = X^*$ it suffices to show that any $x^{**} \in X^{**}$ such that $\langle x^{**}, z\rangle = 0$ for all $z \in Z$ is null. Since X is reflexive, there exists an $x \in X$ such that $x^{**} = e_X(x)$. Then, for all $y \in Y$, setting $x^* := c_Y(y) \in X^*$ we have

$$c(x, y) = \langle c_Y(y), x\rangle = \langle x^*, x\rangle = \langle x^{**}, x^*\rangle = \langle x^{**}, c_Y(y)\rangle = 0.$$

Since c is a coupling, these equalities show that $x = 0$, hence $x^{**} = 0$. □

Given a coupling $c : X \times Y \to \mathbb{R}$, since Y can be identified with its image $c_Y(Y)$ in X^*, one can endow Y with the topology $\sigma(Y, X)$ induced by the weak* topology $\sigma(X^*, X)$. Similarly, identifying X with the subspace $c_X(X)$ of Y^* one can endow X with the topology $\sigma(X, Y)$ induced by $\sigma(Y^*, Y)$. An adaptation of the proof of Proposition 3.18 proves the following result.

Proposition 3.24 *Let $c : X \times Y \to \mathbb{R}$ be a coupling. Then the (topological) dual of $(Y, \sigma(Y, X))$ is X.*

Taking $Y := X^*$ and taking for c the evaluation e, the topology on X induced by the canonical embedding e_X of X into X^{**} endowed with its weak* topology $\sigma(X^{**}, X^*)$ is called the *weak topology* on X and is denoted by $\sigma(X, X^*)$. Thus, $\sigma := \sigma(X, X^*)$ is the topology induced by the family $(p_f)_{f \in X^*}$ of seminorms on X given by $p_f(x) := |f(x)|$ for $x \in X, f \in X^*$ and one has

$$(x_i)_{i \in I} \overset{\sigma}{\to} x \Longleftrightarrow \forall x^* \in X^* \quad (x^*(x_i))_{i \in I} \to x^*(x).$$

Thus, the weak topology of X is indeed weaker than the topology associated with the norm; the latter is often called the *strong topology*. In the sequel we often write $(x_i)_{i \in I} \overset{*}{\to} x$ for weak or weak* convergence and $\langle x, y \rangle$ instead of $c(x, y)$.

Remark If $A : X \to Y$ is a continuous linear map between two normed spaces, then A is continuous for the weak topologies on X and Y: if $(x_i)_{i \in I} \overset{*}{\to} x$, then, for all $y^* \in Y^*$ one has

$$(\langle Ax_i, y^* \rangle)_{i \in I} = (\langle x_i, A^\intercal(y^*) \rangle)_{i \in I} \to \langle x, A^\intercal(y^*) \rangle = \langle Ax, y^* \rangle,$$

so that $(Ax_i)_{i \in I} \overset{*}{\to} Ax$ and A is weakly continuous. □

When X is finite dimensional, the weak* topology on X^* coincides with the strong topology (and, similarly, the weak topology of X coincides with the topology associated with the norm). In fact, a net $(f_i)_{i \in I}$ in X^* converges to some $f \in X^*$ if and only if for every element b of a base of X the net $(f_i(b))_{i \in I}$ converges to $f(b)$, and this is enough to imply the convergence for the dual norm. If X is infinite dimensional, the weak (resp. weak*) topology never coincides with the strong topology (the topology induced by the norm or dual norm). This stems from the fact that no neighborhood V of 0 in the weak or weak* topology is bounded since it contains the intersection of the kernels of a finite family of linear forms. Although $\sigma(X, X^*)$ does not coincide with the strong topology, for the class of convex subsets closedness for these two topologies is the same. One has to be warned that this fact is not valid for convex subsets of a dual space in the weak* topology.

Proposition 3.25 (Mazur) *Closed convex subsets of a Banach space X are weakly closed.*

Proof Let C be a nonempty closed convex subset of X. If $C = X$, C is obviously weakly closed. Suppose $C \neq X$ and let $a \in X \setminus C$. Taking $A := \{a\}$ and $B := C$ in

Theorem 3.12, we get some $f \in X^* \backslash \{0\}$ and some $r \in \mathbb{R}$ such that $f(a) > r > f(b)$ for all $b \in C$. Thus $W := f^{-1}(]r, +\infty[)$ is an open neighborhood of a with respect to the weak topology contained in $X \backslash C$: $X \backslash C$ is open in the weak topology. □

In general, the weak topology does not provide compact subsets as easily as does the weak* topology. However, when X is reflexive, since then the weak topology coincides with the weak* topology obtained by considering X as the dual of X^*, we do get a rich family of compact subsets. We state this fact in the following corollary.

Corollary 3.22 *Every bounded weakly closed subset of a reflexive Banach space X is weakly compact.*

In particular, every bounded, closed, convex subset of X is weakly compact.

In order to show that this property characterizes reflexivity, let us prove a noteworthy lemma.

Lemma 3.18 (Goldstine) *The image $e_X(B_X)$ of the unit ball B_X of a normed space X via the canonical embedding $e_X : X \to X^{**}$ is dense in the closed unit ball $B_{X^{**}}$ of X^{**} for the $\sigma(X^{**}, X^*)$ topology.*

Proof The conclusion means that for any $\overline{x}^{**} \in B_{X^{**}}$ and any $\sigma(X^{**}, X^*)$-neighborhood V of $\overline{x}^{**}$ we have $e_X(B_X) \cap V \neq \varnothing$. By construction of $\sigma(X^{**}, X^*)$ we can find $\varepsilon > 0$ and a finite set $F := \{f_1, \ldots, f_n\}$ in X^* such that

$$W := \{x^{**} \in X^{**} : |\langle x^{**} - \overline{x}^{**}, f_i \rangle| \leq \varepsilon, \ i = 1, \ldots, n\} \subset V.$$

Since $\|\overline{x}^{**}\| \leq 1$, setting $a_i := \langle \overline{x}^{**}, f_i \rangle$, for all $(r_1, \ldots, r_n) \in \mathbb{R}^n$ we have

$$\left| \sum_{i=1}^{n} r_i a_i \right| = \left| \langle \overline{x}^{**}, \sum_{i=1}^{n} r_i f_i \rangle \right| \leq \left\| \sum_{i=1}^{n} r_i f_i \right\|.$$

Then we deduce from Lemma 3.16 that there exists some $x \in B_X$ such that $|f_i(x) - a_i| \leq \varepsilon$, i.e. $|\langle e_X(x) - \overline{x}^{**}, f_i \rangle| \leq \varepsilon$ or $e_X(x) \in W \subset V$. □

Theorem 3.15 *A normed space X is reflexive if and only if its closed unit ball B_X is weakly compact.*

Proof If X is reflexive the canonical embedding $e_X : X \to X^{**}$ is an isomorphism (and even an isometry). Its inverse e_X^{-1} is continuous, hence, by the preceding remark, it is continuous with respect to the weak topologies $\sigma(X^{**}, X^{***}) = \sigma(X^{**}, X^*)$ on X^{**} and $\sigma(X, X^*)$ on X. Since the unit ball $B_{X^{**}}$ is $\sigma(X^{**}, X^*)$-compact, B_X is $\sigma(X, X^*)$-compact.

Conversely, let us suppose B_X is $\sigma(X, X^*)$-compact. Since e_X is continuous, hence continuous with respect to the weak topologies $\sigma(X, X^*)$ on X and $\sigma(X^{**}, X^{***})$ on X^{**} and a fortiori with respect to the topology $\sigma(X^{**}, X^*)$ on X^{**} we get that $e_X(B_X)$ is $\sigma(X^{**}, X^*)$-compact, hence $\sigma(X^{**}, X^*)$-closed. Since $e_X(B_X)$ is dense in $B_{X^{**}}$ in the topology $\sigma(X^{**}, X^*)$, we get that $e_X(B_X) = B_{X^{**}}$ and by homogeneity, $e_X(X) = X^{**}$. □

Remark In the preceding proof some care is required concerning the topologies on X^{**}. In fact, if X is a Banach space, $e_X(B_X)$ is closed in $B_{X^{**}}$ in the norm topology. But if X is not reflexive, then $e_X(B_X)$ is not closed in $B_{X^{**}}$ in the topology $\sigma(X^{**}, X^*)$ since it is dense in $B_{X^{**}}$ with respect to this topology but distinct from this ball. □

Let us give some elementary permanence properties of reflexive spaces.

Proposition 3.26 *A closed subspace Y of a reflexive Banach space X is reflexive.*

Proof Let us first observe that the weak topology $\sigma(Y, Y^*)$ on Y coincides with the topology induced by $\sigma(X, X^*)$ on Y. In fact, by definition, a net $(y_i)_{i\in I}$ in Y converges to $y \in Y$ for $\sigma(Y, Y^*)$ if and only if for every $g \in Y^*$ one has $(g(y_i))_{i\in I} \to g(y)$. But since g is the restriction to Y of some $f \in X^*$, if $(y_i)_{i\in I} \to y$ in the topology induced by $\sigma(X, X^*)$ then one has $(g(y_i))_{i\in I} = (f(y_i))_{i\in I} \to f(y) = g(y)$. Conversely, if $(y_i)_{i\in I} \to y$ in the $\sigma(Y, Y^*)$ topology, given $f \in X^*$, setting $g := f \mid_Y$ we get $(f(y_i))_{i\in I} = (g(y_i))_{i\in I} \to g(y) = f(y)$, so that $(y_i)_{i\in I} \to y$ in the topology induced by $\sigma(X, X^*)$.

Now $B_Y = B_X \cap Y$ and Y is weakly closed (by Proposition 3.25). Thus B_Y is weakly closed in B_X, hence is weakly compact, so that Y is reflexive. □

Proposition 3.27 *A Banach space X is reflexive if and only if its dual X^* is reflexive.*

Proof When X is reflexive one has $\sigma(X^*, X^{**}) = \sigma(X^*, X)$ and since B_{X^*} is compact in the $\sigma(X^*, X)$ topology, it is compact in the $\sigma(X^*, X^{**})$ topology, so that X^* is reflexive by Theorem 3.15.

When X^* is reflexive, X^{**} is reflexive by the preceding. Since $e_X(X)$ is a closed subspace of X^{**}, $e_X(X)$ is reflexive by the preceding proposition. Since X and $e_X(X)$ are isometric, X is reflexive. □

The following results are of interest but they are outside the scope of our purposes, although they have some bearing on our study in the reflexive case. We refer to [106, 109, 119, 165] for the proofs.

Theorem 3.16 (Eberlein-Šmulian) *For a subset S of a Banach space X the following assertions are equivalent:*

(a) the weak closure $\mathrm{w}-\mathrm{cl}(S)$ of S is weakly compact;
(b) every sequence in S has a weakly convergent subsequence;
(c) every sequence in S has a weak cluster point.

Theorem 3.17 (Banach, Dieudonné, Krein, Šmulian) *Let C be a convex subset of the dual X^* of a Banach space X. If for all $r \in \mathbb{R}_+$ the set $C \cap rB_{X^*}$ is closed for $\sigma(X^*, X)$, then C is closed for $\sigma(X^*, X)$.*

Theorem 3.18 (James) *Let A be a bounded and weakly closed subset of a Banach space X. If every continuous linear form on X attains its supremum on A then A is weakly compact.*

In particular, if every continuous linear form on X attains its supremum on B_X then X is reflexive.

Exercises

1. Prove that the quotient X/Y of a reflexive space by a closed subspace Y is reflexive.
2. Prove that if the quotient X/Y of a Banach space by a closed reflexive subspace Y is reflexive then X is reflexive. [See [165, p. 126].]
3. Show that a weakly lower semicontinuous function f on a reflexive space X attains it infimum if it is *coercive* in the sense that $f(x) \to \infty$ when $\|x\| \to \infty$.
4. The purpose of this exercise is to show that weak continuity of continuous maps cannot be expected in general in an infinite dimensional Banach space X.

   (a) Prove that the unit sphere S_X of X is dense in the closed unit ball B_X endowed with the topology induced by the weak topology σ.
   (b) Verify the continuity of the retraction $r : X \to B_X$ given by $r(x) := x/\max(\|x\|, 1)$.
   (c) Given $x \in X$ such that $\|x\| = 1/2$, let $(x_i)_{i\in I}$ be a net in S_X weakly converging to x. Observe that $(r(2x_i))_{i\in I} = (x_i)_{i\in I}$ weakly converges to x and not to $r(2x) = 2x$.

5*. Show that the weak topology of a Banach space X need not be *sequential*. Here a topology $\mathcal{T}$ on X is said to be sequential if the closure cl(S) for $\mathcal{T}$ of any subset S of X is the set of limits of convergent sequences in S. [Hint: in a separable Hilbert space X with orthonormal base (e_n) show that 0 is in the weak closure of the set $S := \{e_m + me_n : m, n \in \mathbb{N}, m < n\}$ but no sequence in S weakly converges to 0.]
6*. (**Šmulian's Theorem**). Prove that any sequence in a weakly compact subset of a Banach space has a weakly convergent subsequence.
7*. Let I be an infinite set and let $X := \ell_\infty(I)$ be the space of bounded functions on I with the supremum norm. Show that the unit ball B_{X^*} of X^* contains a weak* compact subset that has no weak* convergent sequence besides the ones that are eventually constant.
8. Show that the class $\mathcal{W}$ of Banach spaces having weak* sequentially compact dual balls is stable under the following operations (see [97, p. 227]): (a) taking dense continuous linear images; (b) taking quotients; (c) taking subspaces.
9. Prove a result similar to the one of exercise 5 of Sect. 3.2 for a weakly closed cone of a normed space endowed with the weak topology.
10. (a) Show that the *polar set* $P^0 := \{x^* \in X^* : \langle x^*, x\rangle \leq 1 \ \forall x \in P\}$ of a cone P of a normed space X is a cone and is given by $P^0 = \{x^* \in X^* : \langle x^*, x\rangle \leq 0 \ \forall x \in P\}$.

(**b**) A *base* of a convex cone Q is a convex subset C of Q such that $0 \notin C$ and $Q = \mathbb{R}_+ C$. Show that a closed convex cone P of a Banach space has a nonempty interior if and only if its polar cone $Q := P^0$ has a weak* compact base.

11. (**a**) Verify that the polar cone Q of the cone $P := \{0\} \times \mathbb{R}_+ \subset \mathbb{R}^2$ is locally compact but does not have a compact base.

(**b**) Prove that if Q is a weak* closed convex cone of the dual of a Banach space, Q has a weak* compact base if and only if it is locally compact. [See [109].]

12*. (**Davis-Figiel-Johnson-Pelczynski Theorem**) Let Q be a weakly compact symmetric convex subset of a Banach space X. Show that there exists a weakly compact symmetric convex subset P of X containing Q such that the linear span Y of P endowed with the gauge of P is a reflexive space.

3.4.3 Uniform Convexity

In this subsection we give a useful criterion for reflexivity. Let us call a *gage* or a *forcing function* a function $\gamma : \mathbb{R}_+ \to \mathbb{R}_+$ that is nondecreasing and such that $\gamma(r) = 0$ if and only if $r = 0$. Let us say that a subset A of a normed space $(X, \|\cdot\|)$ is *uniformly rotund* or *uniformly convex* if there is a gage γ such that for any $x, y \in A$ one has $\frac{1}{2}(x+y) + \gamma(\|\frac{1}{2}(x-y)\|)B_X \subset A$. The space $(X, \|\cdot\|)$ is said to be *uniformly rotund* or *uniformly convex* if its unit ball B_X is uniformly rotund. Thus $(X, \|\cdot\|)$ is uniformly convex if and only if the following property holds:

(UC) for every $\varepsilon > 0$ there exists some $\delta > 0$ such that

$$x, y \in B_X, \|x - y\| \geq \varepsilon \Rightarrow \left\| \frac{1}{2}(x + y) \right\| \leq 1 - \delta.$$

Such a property is of metric character and is not preserved under isomorphisms as one can see by taking different usual norms on $\mathbb{R}^d$. It is easy to see that X is uniformly convex if and only if there exists a *modulus* μ (i.e. a nondecreasing function $\mu : \mathbb{R}_+ \to \overline{\mathbb{R}}_+$ such that $\mu(0) = 0$ and μ is continuous at 0) satisfying $\|x - y\| \leq \mu(1 - \frac{1}{2}\|x + y\|)$ for all $x, y \in B_X$ (Fig. 3.5).

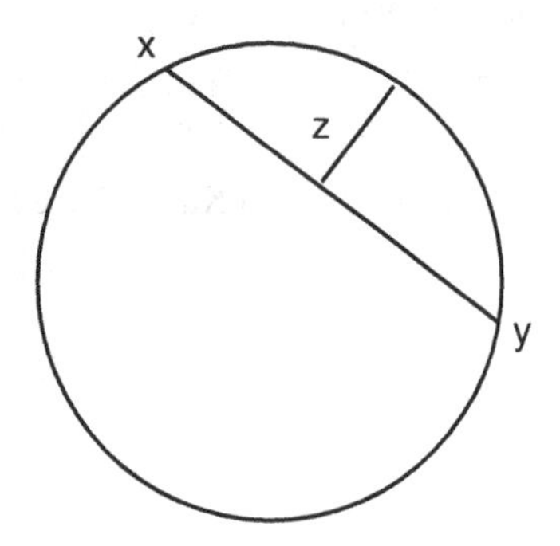

Fig. 3.5 Uniform convexity

Theorem 3.19 (Milman-Pettis) *Any uniformly convex Banach space is reflexive.*

Proof Let $e_X : X \to X^{**}$ be the canonical embedding. Since $e_X(X)$ is closed in X^{**} in the norm topology (since it is complete), it suffices to show that any $\overline{x}^{**} \in X^{**}$ belongs to the closure of $e_X(X)$. Without loss of generality we may suppose $\|\overline{x}^{**}\|_{**} = 1$. Given $\varepsilon > 0$ we want to find some $x \in X$ such that $\|e_X(x) - \overline{x}^{**}\|_{**} \leq \varepsilon$. Let $\delta > 0$ be associated with ε as in (UC). Since $\|\overline{x}^{**}\|_{**} = 1$ we can find $x^* \in S_{X^*}$ such that $\langle \overline{x}^{**}, x^* \rangle > 1 - \delta/2$. Let

$$V := \{x^{**} \in X^{**} : |\langle x^{**} - \overline{x}^{**}, x^* \rangle| < \delta/2\}.$$

Since V is open in the $\sigma(X^{**}, X^*)$-topology and since $e_X(B_X)$ is $\sigma(X^{**}, X^*)$-dense in $B_{X^{**}}$ by Lemma 3.18, we can find some $x \in B_X$ such that $e_X(x) \in V$. Let us show that assuming that $\|e_X(x) - \overline{x}^{**}\|_{**} > \varepsilon$ leads to a contradiction. This inequality means that $\overline{x}^{**} \in W := X^{**} \backslash (e_X(x) + \varepsilon B_{X^{**}})$, a $\sigma(X^{**}, X^*)$-open subset of X^{**} since $\varepsilon B_{X^{**}}$ is $\sigma(X^{**}, X^*)$-compact. Applying again Lemma 3.18, we can find some $y \in B_X$ such that $e_X(y) \in V \cap W$: $\|e_X(y) - e_X(x)\| > \varepsilon$ and $|\langle e_X(y) - \overline{x}^{**}, x^* \rangle| < \delta/2$. Since $e_X(x) \in V$ by our choice of x, we also have $|\langle e_X(x) - \overline{x}^{**}, x^* \rangle| < \delta/2$ and we get $|\langle e_X(x+y) - 2\overline{x}^{**}, x^* \rangle| < \delta$ hence $\langle e_X(x+y), x^* \rangle > 2\langle \overline{x}^{**}, x^* \rangle - \delta > 2 - 2\delta$,

$$\|x + y\| = \|e_X(x+y)\| = \|e_X(x+y)\| \,.\, \|x^*\| > 2 - 2\delta.$$

Since $\frac{1}{2}\|x+y\| > 1 - \delta$, (UC) ensures that $\|x - y\| \leq \varepsilon$, contradicting the relation $\|e_X(y) - e_X(x)\| > \varepsilon$ and the fact that e_X is isometric. □

The preceding criterion is just a sufficient condition, not a necessary condition. But it enables us to prove that some usual Banach spaces such as Hilbert spaces and Lebesgue spaces $L_p(S)$ for $p \in]1, +\infty[$ are reflexive. Also, uniform convexity implies a useful property relating weak and strong topologies:

Proposition 3.28 (Kadec-Klee) *Let $(x_i)_{i \in I}$ be a net (or a sequence) in a uniformly convex Banach space X which weakly converges to $x \in X$ and is such that $(\|x_i\|)_{i \in I} \to \|x\|$ (or just $\limsup_{i \in I} \|x_i\| \leq \|x\|$). Then $(\|x_i - x\|)_{i \in I} \to 0$.*

Proof Since the result is obvious if $x = 0$, we may suppose $x \neq 0$. Let $r_i := \max(\|x_i\|, \|x\|)$, so that $(r_i)_{i \in I} \to r := \|x\|$. Let $u_i := x_i/r_i$, $u := x/\|x\|$, so that u_i, $u \in B_X$ and $(u_i)_{i \in I} \to x/r = u$ in the $\sigma(X, X^*)$ topology. Then $(\frac{1}{2}(u_i + u)) \to u$ in the $\sigma(X, X^*)$ topology, so that $\|u\| \leq \liminf_{i \in I}(\frac{1}{2}\|u_i + u\|)$. But since $\|u\| = 1$ and $\|u + u_i\| \leq \|u\| + \|u_i\| \leq 2$, we get $(\frac{1}{2}\|u_i + u\|) \to 1$, hence, by condition (UC), $(\|u_i - u\|)_{i \in I} \to 0$. We conclude that (x_i) converges to x in the norm topology. □

In turn, the dual Kadec-Klee Property implies that the duality map is continuous.

Proposition 3.29 *Let $(X, \|\cdot\|)$ be a normed space such that the dual norm has the sequential dual Kadec-Klee Property: $(x_n^*) \to x^*$ whenever $(x_n^*) \to x^*$ in the weak* topology and $(\|x_n^*\|_*) \to \|x^*\|_*$. Then, if the duality map J is single-valued, it is continuous from X to X^* with respect to the topologies associated with the norms.*

Proof For simplicity, we assume that B_{X^*} is sequentially compact. Let $(x_n) \to x$ in $(X, \|\cdot\|)$, so that $(\|x_n\|) \to \|x\|$ and $(\|J(x_n)\|_*) \to \|J(x)\|_*$. From the bounded sequence $(J(x_n))$ one can extract a subsequence $(J(x_{k(n)}))$ that weak* converges to some $x^* \in X^*$. Then $\langle x^*, x\rangle = \lim_n \langle J(x_{k(n)}), x_{k(n)}\rangle = \lim_n \left\|x_{k(n)}\right\|^2 = \|x\|^2$, so that $\|x^*\|_* \geq \|x\|$. But since $\|\cdot\|_*$ is weak* lower semicontinuous, one has $\|x^*\|_* \leq \liminf_n \left\|J(x_{k(n)})\right\|_* = \|x\|$. Thus $\|x^*\|_* = \|x\|$ and $\langle x^*, x\rangle = \|x\|^2$, hence $x^* = J(x)$. Since $(J(x_{k(n)}))$ can be extracted from any subsequence in $(J(x_n))$, one has $(J(x_n)) \to J(x)$, $(\|J(x_n)\|_*) \overset{*}{\to} \|J(x)\|_*$ and, by the sequential dual Kadec-Klee Property, $(\|J(x_n) - x^*\|_*) \to 0$. □

Exercises

1. Let $(X, \|\cdot\|)$ be a normed vector space. Show that the following properties are equivalent:

   (a) $\forall u, v \in S_X, u \neq v \Longrightarrow \|u + v\| < 1$;
   (b) $\forall u, v \in X, u \neq 0, \|u + v\| = \|u\| + \|v\| \Longrightarrow \exists r \in \mathbb{R}_+\colon v = ru$.

   When the norm satisfies these properties, it is said to be *strictly convex* and the space is said to be *strictly convex.*
2. Show that when the dual norm $\|\cdot\|_*$ on the dual of a normed vector space $(X, \|\cdot\|)$ is strictly convex, the duality map J is single-valued.
3. Let $(X, \|\cdot\|)$ be a normed vector space such that there exists on X an equivalent norm $\|\cdot\|'$ that is strictly convex. Show that for all $c \in]1, \infty[$ there exists some $b \in \mathbb{P}$ such that the norm $\|\cdot\|_b := \|\cdot\| + b\,\|\cdot\|'$ satisfies $\|\cdot\| \leq \|\cdot\|_b \leq c\,\|\cdot\|$ and is strictly convex.
4. Let $(X, \|\cdot\|)$ be a uniformly convex normed vector space and let $p \in]1, \infty[$. Prove that for every $\delta > 0$ there exists an $\varepsilon > 0$ such that for all $x, y \in B_X$ one has

$$\|x - y\| \geq \delta \Longrightarrow \left\|\frac{1}{2}(x + y)\right\|^p \leq \frac{1}{2}\|x\|^p + \frac{1}{2}\|y\|^p - \varepsilon.$$

5. Let $(X, \|\cdot\|)$ be a normed vector space such that there exists on X an equivalent norm $\|\cdot\|'$ that is uniformly convex. Using the preceding exercise, show that for all $c \in]1, \infty[$ there exists some $b \in \mathbb{P}$ such that the norm $\|\cdot\|_b := \|\cdot\| + b\,\|\cdot\|'$ satisfies $\|\cdot\| \leq \|\cdot\|_b \leq c\,\|\cdot\|$ and is uniformly convex.
6. Prove that any uniformly convex normed space is strictly convex.

7. Let C be a nonempty closed convex subset of a strictly convex Banach space X and let $w \in X$. Show that there is at most one point $a \in C$ such that $\|w - a\| = d_C(w) := \inf_{x \in C} \|x - w\|$.
8. Let C be a nonempty closed convex subset of a reflexive Banach space. Show that for every $w \in W$ there exists some point $a \in C$ such that $\|w - a\| = d_C(w)$.
9. Let C be a nonempty closed convex subset of a strictly convex reflexive Banach space X. Show that there exists a unique point $a := P_C(w) \in C$ such that $\|w - a\| = d_C(w)$.
10. Let C be a nonempty closed convex subset of a uniformly convex Banach space X. Show that for every $w \in X$, any sequence (x_n) in C satisfying $(\|x_n - w\|)_n \to d_C(w)$ is a Cauchy sequence. Conclude that there exists a unique point $a := P_C(w) \in C$ such that $\|w - a\| = d_C(w)$. Prove that the map $P_C : X \to C$ is uniformly continuous on bounded subsets of X.

11*. For $p \in]1, \infty[$ let $h : \mathbb{R} \to \mathbb{R}$ be given by $h(t) := |t|^p + 1 - 2^{1-2p} |t + 1|$. After a study of h, prove that there exists some $c_p \in \mathbb{P}$ such that for all $r, s \in \mathbb{R}$ one has

$$|r - s|^p \leq c_p(|r|^p + |s|^p)(|r|^p + |s|^p - 2^{1-p} |r + s|)^{p/2}.$$

Deduce from this the fact that for $p \in]1, 2]$ the space $L_p(S, \mu)$ is uniformly convex. We shall return to this question in Sect. 8.5.

3.4.4 Separability

It is often important to know whether one can use sequences when dealing with weak topologies. Such a question is related to the metrizability of the topology induced by a weak topology on a bounded subset. We recall that a topological space is said to be *metrizable* if its topology can be associated with a metric.

Theorem 3.20 *Let X and Y be normed spaces in metric duality. If X is separable, then the topology induced by $\sigma(Y, X)$ on any bounded subset of Y is metrizable.*

If Y is the dual of X or if X is the dual of Y, the converse is true.

Proof If X is separable, then B_X is separable by Corollary 2.13. Let $\{a_n : n \in \mathbb{N}\}$ be a countable dense subset of B_X. It is easy to see that the function

$$y \mapsto \|y\|_\sigma := \sum_{n \geq 1} 2^{-n} |\langle a_n, y \rangle|$$

is a norm on Y. Let us show that the topology induced by $\sigma(Y, X)$ on B_Y coincides with the topology associated with the norm $\|\cdot\|_\sigma$. Then the same will hold on any bounded subset.

Given $\overline{y} \in B_Y$ and $r > 0$, let us find a neighborhood V of $\overline{y}$ in $(B_Y, \sigma(Y,X))$ such that $V \subset U := B_\sigma(\overline{y}, r) := \{y \in Y : \|y - \overline{y}\|_\sigma < r\}$. We take V of the form

$$V := \{y \in B_Y : |\langle a_i, y - \overline{y}\rangle| < \varepsilon,\ i \in \mathbb{N}_k\}$$

for some $\varepsilon \in]0, r[$ and $k \in \mathbb{N}\backslash\{0\}$ such that $\varepsilon + 2^{1-k} < r$. Then, for $y \in V$, since $\|y - \overline{y}\| \leq 2$ we have $y \in U$ since

$$\|y - \overline{y}\|_\sigma \leq \sum_{n=1}^{k} 2^{-n} |\langle a_n, y - \overline{y}\rangle| + \sum_{n=k+1}^{\infty} 2^{-n} |\langle a_n, y - \overline{y}\rangle| < \varepsilon + \sum_{n=k+1}^{\infty} 2^{-n} 2 < r.$$

Conversely, given a neighborhood W of $\overline{y}$ in $(B_Y, \sigma(Y,X))$, let us find $r > 0$ such that $B_\sigma(\overline{y}, r) \cap B_Y \subset W$. We may assume that there is a finite subset $\{x_1, \ldots, x_m\}$ of X and $\varepsilon > 0$ such that

$$W := \{y \in Y : |\langle x_i, y - \overline{y}\rangle| < \varepsilon,\ i \in \mathbb{N}_m\}.$$

Without loss of generality we may assume $x_i \in B_X$ for all $i \in \mathbb{N}_m$. Since $\{a_n : n \in \mathbb{N}\}$ is dense in B_X, for each $i \in \mathbb{N}_m$ we can pick some $n(i) \in \mathbb{N}$ such that $\left\|x_i - a_{n(i)}\right\| < \varepsilon/4$. Let us pick $r > 0$ such that $2^{n(i)} r < \varepsilon/2$ for all $i \in \mathbb{N}_m$ and let us show that $B_\sigma(\overline{y}, r) \cap B_Y \subset W$. Given $y \in B_\sigma(\overline{y}, r) \cap B_Y$, for $i \in \mathbb{N}_m$ we have $2^{-n(i)} \left|\langle a_{n(i)}, y - \overline{y}\rangle\right| < r$, so that $y \in W$ since

$$|\langle x_i, y - \overline{y}\rangle| \leq \left|\langle x_i - a_{n(i)}, y - \overline{y}\rangle\right| + \left|\langle a_{n(i)}, y - \overline{y}\rangle\right| < 2\varepsilon/4 + 2^{n(i)} r < \varepsilon.$$

Finally, let us prove the converse in the case $Y = X^*$: assuming that the topology induced by $\sigma(Y,X)$ on B_Y is defined by a metric d, let us show that X is separable. For $n \in \mathbb{N}\backslash\{0\}$ let $U_n := \{y \in B_Y : d(y, 0) < 1/n\}$ and let V_n be a neighborhood of 0 in $(B_Y, \sigma(Y,X))$ such that $V_n \subset U_n$. Shrinking V_n if necessary we can find $\varepsilon_n > 0$ and a finite subset F_n of X such that

$$V_n := \{y \in B_Y : \forall x \in F_n \quad |\langle x, y\rangle| < \varepsilon_n\}.$$

Then $\cap_n V_n \subset \cap_n U_n = \{0\}$ so that if $y \in Y$ is such that $\langle x, y\rangle = 0$ for all $x \in F := \cup_n F_n$ one has $y = 0$. Therefore, the linear space generated by F is dense in X. Since F is countable, this means that X is separable (consider the set of linear combinations of elements of F with rational coefficients).

The case $X = Y^*$ is more delicate and we refer to monographs on functional analysis [106, p. 426] f.i. □

From the relative weak* compactness and the metrizability of bounded subsets of the dual of a separable Banach space one gets the following useful statement.

Corollary 3.23 *Let (f_n) be a bounded sequence in the dual of a separable Banach space X. Then (f_n) has a subsequence that converges in the weak* topology $\sigma(X^*, X)$.*

The following fact related to the preceding duality relationship is noteworthy.

Proposition 3.30 *If the dual X^* of a Banach space X is separable, then X is separable.*

Proof Let $\{x_n^* : n \in \mathbb{N}\}$ be a countable dense subset of S_{X^*}. For all $n \in \mathbb{N}$ let $x_n \in S_X$ be such that $\langle x_n^*, x_n \rangle > 1/2$. Let Y be the smallest closed linear subspace containing $\{x_n : n \in \mathbb{N}\}$. It suffices to prove that $Y = X$. If $Y \neq X$ one can find some $x^* \in S_{X^*}$ such that $\langle x^*, y \rangle = 0$ for all $y \in Y$. Let $k \in \mathbb{N}$ be such that $\|x_k^* - x^*\| < 1/2$. Then one gets the contradiction:

$$0 = \langle x^*, x_k \rangle = \langle x_k^*, x_k \rangle - \langle x_k^* - x^*, x_k \rangle \geq \langle x_k^*, x_k \rangle - \|x_k^* - x^*\| \cdot \|x_k\| > 0.$$

□

Corollary 3.24 *A Banach space X is reflexive and separable if and only if its dual X^* is reflexive and separable.*

Proof If X^* is reflexive and separable then X is reflexive by Proposition 3.27 and it is separable by the preceding proposition. Conversely, if X is reflexive and separable then its dual X^* is reflexive and it is separable since $(X^*)^* = X$ is separable. □

Theorem 3.21 *Any bounded sequence in a reflexive Banach space has a subsequence that converges in the weak topology.*

Proof Let (x_n) be a bounded sequence in a reflexive Banach space X. The closed linear subspace Y generated by $\{x_n\}$ is separable and reflexive by Proposition 3.26. Then Y^* is separable by the preceding corollary. Corollary 3.23 ensures that (x_n) has a convergent subsequence with respect to the topology $\sigma(Y, Y^*)$. The proof of Proposition 3.26 showed that $\sigma(Y, Y^*)$ is the topology induced by $\sigma(X, X^*)$ on Y. The conclusion ensues. □

Exercises

1. Show that if X is an infinite dimensional Banach space the weak topology $\sigma(X, X^*)$ is not metrizable.
2. Prove that if X is a reflexive infinite dimensional normed space there exists in the unit sphere S_X of X a sequence that converges to 0 in the $\sigma(X, X^*)$ topology.
3. Prove the same conclusion if X has a separable dual and is infinite dimensional.
4. Let $X := \ell_1$ be the space of sequences $x := (x_n)_{n \in \mathbb{N}}$ of real numbers such that $\|x\|_1 := \Sigma_{n \geq 0} |x_n| < +\infty$, endowed with the norm $\|\cdot\|_1$. Prove that any weakly convergent sequence in X converges in the norm topology.

5. Show that the space ℓ_1 is separable.
6. Show that the space c_0 of sequences of real numbers with limits 0 endowed with the supremum norm is separable.
7. Prove that the space ℓ_∞ of bounded real sequences endowed with the supremum norm is not separable. [Hint: Considering a sequence as a function on $\mathbb{N}$ and denoting by 1_A the characteristic function of a subset A of $\mathbb{N}$ defined by $1_A(n) = 1$ if $n \in A$, $1_A(n) = 0$ else, note that for distinct subsets A, B of $\mathbb{N}$ one has $\|1_A - 1_B\| = 1$. Deduce from Cantor's Theorem that the space ℓ_∞ is not separable as it cannot have a countable base of open sets.]
8. Using James' Theorem, give another proof of the fact that a uniformly convex space X is reflexive. [Hint: for $x^* \in X^*$ with norm $r > 0$ show that a sequence (x_n) of B_X satisfying $\lim_n x^*(x_n) = r$ is a Cauchy sequence.]

3.5 Some Key Results of Functional Analysis

We devote this section to some classical theorems of functional analysis and to some applications and refinements. Most of them are consequences of Baire's Theorem. With the Hahn-Banach Theorem, they form the pillars of linear functional analysis.

3.5.1 Some Classical Theorems

The Uniform Boundedness Theorem, the Open Mapping Theorem and the Closed Range Theorem that are on our agenda are among the cornerstones of linear functional analysis. They have many consequences and applications. We start with a side result that enables us to reduce an interiority condition to an algebraic condition.

Recall that the *core* (or *algebraic interior*) of a convex subset C of a vector space X is the set of points $a \in X$ such that for all $v \in X$ there exists some $\alpha > 0$ for which $a + [-\alpha, \alpha]v \subset C$. One has the following characterizations.

Lemma 3.19 *For a nonempty convex subset C of a vector space X and $a \in X$, the following assertions are equivalent:*

(a) $a \in \operatorname{core} C$;
(b) $C - a$ *is absorbing: for all* $x \in X$ *there exists a* $t > 0$ *such that* $tx \in C - a$;
(c) $X = \mathbb{R}_+(C - a) := \{r(c - a) : r \in \mathbb{R}_+,\ c \in C\}$;
(d) $a \in C$ *and for all* $v \in X$ *there exists an* $\varepsilon > 0$ *such that* $a + [0, \varepsilon]v \subset C$.

Proof Since a subset D of X is absorbing if and only if, for every $x \in X$ there exists some $r > 0$ such that $x \in rD$, the implications (a)⇒(b)⇒(c) are obvious. Assertion (d) is clearly satisfied when $X = \{0\}$. Let us show that (c)⇒(d) when $X \neq \{0\}$. Taking $v \neq 0$ in X, we can write $v = r(c - a)$, $-v = r'(c' - a)$, for some $c, c' \in C$ and $r, r' > 0$ (as $v \neq 0$), so that $0 = r(r + r')^{-1}c + r'(r + r')^{-1}c' - a \in C - a$

by convexity, hence $a \in C$. Moreover, setting $\varepsilon := 1/r$, for $s \in [0, \varepsilon]$, we have $a + sv = (1 - sr)a + src \in C$. Assuming (d) holds, given $v \in X$, taking $\varepsilon, \varepsilon' > 0$ such that $a + [0, \varepsilon]v \subset C$, $a + [0, \varepsilon'](-v) \subset C$ and setting $\alpha := \min(\varepsilon, \varepsilon')$ we see that $a + [-\alpha, \alpha]v \subset C$ so that $a \in \operatorname{core} C$. □

Let us give comparisons between the core and the interior of a convex subset of a normed vector space. The inclusion

$$\operatorname{int} C \subset \operatorname{core} C$$

always holds since for all $a \in \operatorname{int} C$ and all $v \in X$ the map $f : t \mapsto a + tv$ is continuous and $f(0) \in \operatorname{int} C$, so that $f(t) \in C$ for $t > 0$ small enough. We leave the first two comparisons that follow as exercises.

Exercise If C is a convex subset of a normed vector space and if the interior of C is nonempty, then $\operatorname{int} C = \operatorname{core} C$. [Hint: for $\overline{x} \in \operatorname{core} C$ and $a \in \operatorname{int} C$ let $\varepsilon > 0$ be such that $z := \overline{x} + \varepsilon(\overline{x} - a) \in C$. Then note that $h : x \mapsto z + \varepsilon(1 + \varepsilon)^{-1}(x - z)$ is a homeomorphism of X onto X that maps C onto a neighborhood of $\overline{x}$ contained in C.]

Exercise If C is a convex subset of a finite dimensional vector space, then $\operatorname{int} C = \operatorname{core} C$. [Hint: use the preceding exercise and the fact that C has a nonempty interior in the affine subspace it generates, which is the whole space if $\operatorname{core} C$ is nonempty.]

The next relationship requires a proof.

Proposition 3.31 *The core of a closed convex subset C of a Banach space X coincides with its interior.*

Proof We may suppose $\operatorname{core} C \neq \varnothing$ and, using a translation if necessary, that $0 \in \operatorname{core} C$. Then X is the union of the closed subsets nC for $n \in \mathbb{N}\backslash\{0\}$. Since X is a Baire space, one of these sets has a nonempty interior. Thus C has a nonempty interior and the result follows by the first exercise above. □

The preceding proposition can be considered as a special case of the next one when in it one takes $X := \{0\}$, $F := \{0\} \times C$. Note that the projection of a closed subset is not necessarily closed, so that new arguments are required.

Proposition 3.32 (Robinson) *Let $C := p_Y(F)$ be the projection on Y of a closed convex subset F of the product $X \times Y$ of two Banach spaces. Then* $\operatorname{core} C = \operatorname{int} C$.

Proof in the case X is reflexive In such a case a simple proof using a compactness argument can be given. Again, we may suppose $\operatorname{core} C \neq \varnothing$ and, using translations if necessary, that $0 \in \operatorname{core} C$ and $(0, 0) \in F$. Then, for all $y \in Y$ there exist some $(u, v) \in F$ and some $k \in \mathbb{N}\backslash\{0\}$ such that $y = kv$. Taking $m \in \mathbb{N}\backslash\{0\}$ such that $u \in mB_X$, where B_X is the closed unit ball of X, we see that $y = kmp_Y(u/m, v/m) \in kmF(B_X)$, where $F(B_X) := p_Y(F \cap (B_X \times Y))$, F being considered as a multimap from X into Y. Thus $Y = \cup_{n \geq 1} nF(B_X)$ and $0 \in \operatorname{core} F(B_X)$. It remains to prove that $F(B_X)$ is (weakly) closed and convex, so that, by the preceding proposition,

$0 \in \operatorname{int} F(B_X) \subset \operatorname{int} C$. Convexity is obvious and closedness follows from Exercise 2 of Sect. 2.2.4. Let us adapt the argument to the present case. Let y be the (weak) limit of a net $(y_i)_{i\in I}$ in $F(B_X)$. For all $i \in I$ we pick $x_i \in B_X$ such that $y_i \in F(x_i)$, i.e. $(x_i, y_i) \in F$. Since B_X is weakly compact, a subnet $(x_j)_{j\in J}$ of $(x_i)_{i\in I}$ has a weak limit x in B_X. Since F is weakly closed, we have $(x, y) \in F$ and $y \in F(B_X)$. □

Proof of Proposition 3.32 *in the general case* Again, by translations and homotheties we can reduce the task to showing that if $0 \in \operatorname{core} C$ and $(0, 0) \in F$, then, for some $s > 0$, C contains the ball sB_Y. As above, considering F as a multimap from X into Y we have $Y = \cup_{n\geq 1} nF(B_X)$.

The Baire category theorem asserts that for some $n \geq 1$ the set $\operatorname{cl} nF(B_X)$ has a nonempty interior. Thus $D := \operatorname{cl} F(B_X)$ has a nonempty interior. Let $\overline{y} \in \operatorname{int}(D)$. Since $0 \in \operatorname{core} C$ there exists some $s > 0$ such that $-s\overline{y} \in C$. If $u \in X$ is such that $-s\overline{y} \in F(u)$, taking $t \in]0, s]$ such that $ts^{-1}u \in B_X$, by convexity we see that $-t\overline{y} \in F(B_X) \subset D := \operatorname{cl} F(B_X)$. Then, since D is convex, by Lemma 3.11 one has

$$0 = \frac{t}{1+t}\overline{y} + \frac{1}{1+t}(-t\overline{y}) \subset \frac{t}{1+t}\operatorname{int}(D) + \frac{1}{1+t}D \subset \operatorname{int}(D).$$

Thus there exists an $r > 0$ such that $rB_Y \subset D = \operatorname{cl} F(B_X)$. The rest of the proof is devoted to showing that $(1-q)rB_Y \subset F(B_X)$ for all $q \in]0, 1[$ (the reader can just take $q := 1/2$ but since $\operatorname{int}(rB_Y)$ is the union of the balls $r'B_Y$ for $r' \in [0, r[$, our proof shows more: $\operatorname{int}(rB_Y) \subset F(B_X)$).

Let $y \in (1-q)rB_Y$. Since by convexity, for $t \in [0, 1]$, $trB_Y \subset \operatorname{cl} F(tB_X)$, given $w \in trB_Y$ and $\varepsilon > 0$ we can find $u \in tB_X$, $v \in F(u)$ such that $\|v - w\| < \varepsilon$. In particular, taking $t := 1 - q$, $w := y$ we can find $x_1 \in (1-q)B_X$, $y_1 \in F(x_1)$ such that $\|y_1 - y\| < \varepsilon_0 := \frac{1}{2}(1-q)q^2 r$. Then, setting $x_0 := 0$, we show by induction on $n \in \mathbb{N}\backslash\{0\}$ that there exists some $(x_n, y_n) \in F$ such that

$$\|x_n - x_{n-1}\| \leq q^{n-1}(1-q), \qquad \|y_n - y\| \leq \frac{1}{2}(1-q)q^{n+1}r. \tag{3.8}$$

These relations imply $\|x_n\| = \left\|x_0 + \Sigma_{j=1}^n (x_j - x_{j-1})\right\| \leq \Sigma_{j=1}^n q^{j-1}(1-q) \leq 1 - q^n$ and they are satisfied for $n = 1$. Assuming $x_0, x_1, \ldots, x_k$ are constructed in such a way that (3.8) is satisfied for $n \leq k$, taking $w_k := y + t_k(y - y_k)$, with $t_k := 2q^{-k}(1-q)^{-1}$, so that $\|w_k - y\| = t_k \|y_k - y\| \leq qr$ and $\|w_k\| \leq qr + \|y\| \leq r$, or $w_k \in rB_Y$, we pick $u_k \in B_X$, $v_k \in F(u_k)$ such that $\|v_k - w_k\| < \varepsilon_k := (1-q)q^{k+2}r/2$ and we set

$$(x_{k+1}, y_{k+1}) := \frac{t_k}{1+t_k}(x_k, y_k) + \frac{1}{1+t_k}(u_k, v_k) \in F$$

since F is convex. Moreover, using the relation $1/(1+t_k) \leq 1/t_k$, we get

$$\|x_{k+1} - x_k\| \leq \frac{1}{1+t_k}(\|x_k\| + \|u_k\|) \leq \frac{q^k(1-q)}{2}(2 - q^k) \leq q^k(1-q).$$

By our choice of w_k

$$y_{k+1} - y = \frac{t_k}{1+t_k}(y_k - y) + \frac{1}{1+t_k}[(w_k - y) + (v_k - w_k)] = \frac{1}{1+t_k}(v_k - w_k),$$

so that we get $\|y_{k+1} - y\| \leq (1+t_k)^{-1} \|v_k - w_k\| \leq \varepsilon_k := (1-q)q^{k+2}r/2$ and relation (3.8) is established. The Cauchy sequence (x_n) has a limit x and $(x, y) = \lim_n (x_n, y_n) \in F$ since F is closed and $x \in B_X$ since $x_n \in B_X$ for all $n \in \mathbb{N}$. Thus $y \in F(B_X)$. □

Remark Interchanging the roles of X and Y, we get that for any multimap $F : X \rightrightarrows Y$ with closed convex graph with domain $D := p_X(F)$ one has $\mathrm{core}\, D = \mathrm{int}\, D$.

Taking for F the epigraph of a lower semicontinuous convex function f, we get the next corollary.

Corollary 3.25 *Let W be a Banach space and let $f : W \to \overline{\mathbb{R}}$ be a lower semicontinuous convex function. Then* core (dom$f$) = int (dom$f$).

The following generalization of the Open Mapping Theorem is a versatile tool.

Theorem 3.22 (Robinson-Ursescu) *Let X, Y be Banach spaces, let $F : X \rightrightarrows Y$ be a multimap with closed convex graph. Then, for any $(\overline{x}, \overline{y})$ in (the graph of) F such that $\overline{y} \in \mathrm{core}\, F(X)$, the multimap F is open at $(\overline{x}, \overline{y})$ in the sense that for any $U \in \mathcal{N}(\overline{x})$ one has $F(U) \in \mathcal{N}(\overline{y})$.*

In fact, F is open at $(\overline{x}, \overline{y})$ with a linear rate: there exists some $c > 0$ such that

$$\forall t \in]0, 1] \qquad B(\overline{y}, tc) \subset F(B(\overline{x}, t)).$$

Proof Without loss of generality, we may suppose $(\overline{x}, \overline{y}) = (0, 0)$, so that $F(X)$ is absorbing. Let B be the closed ball with center $\overline{x} = 0$ and radius r in X and let $C := F(B) = p_Y((B \times Y) \cap \mathrm{gph} F)$. The beginning of the first proof of Proposition 3.32 shows that C is absorbing, but let us show that again, which will prove that $0 \in \mathrm{int} C$ by Proposition 3.32 in which we replace F with $(B \times Y) \cap \mathrm{gph} F$. Let y be an arbitrary point of Y; since $F(X)$ is absorbing, there exist some $s > 0$ and $x \in X$ such that $sy \in F(x)$. If $x \in B$, then $sy \in C$. If $x \in X \backslash B$, let $t := r \|x\|^{-1} \in]0, 1[$, so that $tx \in B$. Since F is convex we have

$$sty = tsy + (1-t)0 \in tF(x) + (1-t)F(0) \subset F(tx + (1-t)0) \subset F(B) = C.$$

Thus, C is absorbing and any neighborhood of $\overline{x}$ is mapped by F onto a neighborhood of $\overline{y}$: F is open at $(\overline{x}, \overline{y})$.

The last assertion stems from the convexity of F: if $B(\overline{y}, c)$ is contained in $F(B(\overline{x}, 1))$, then, for $t \in]0, 1]$, we have $tF(B(\overline{x}, 1)) + (1-t)F(\overline{x}) \subset F(tB(\overline{x}, 1) + (1-t)\overline{x})$, hence

$$B(\overline{y}, tc) = tB(\overline{y}, c) + (1-t)\overline{y} \subset tF(B(\overline{x}, 1)) + (1-t)F(\overline{x}) \subset F(B(\overline{x}, t)).$$

□

Taking for F a surjective linear map with closed graph, we get a classical result.

Theorem 3.23 (Banach-Schauder Open Mapping Theorem) *Let X and Y be Banach spaces and let $A : X \to Y$ be a continuous linear mapping such that $A(X) = Y$. Then there exists some $c > 0$ such that for all $y \in Y$ one can find $x \in X$ satisfying $A(x) = y$ and $\|x\| \leq c \|y\|$ and A is open, i.e. for any open subset U of X, its image $V := A(U)$ is an open subset of Y.*

We immediately point out some important consequences. The first is obtained by replacing Y with $A(X)$, which is a Banach space when it is closed in Y.

Corollary 3.26 *Let X and Y be Banach spaces and let $A \in L(X, Y)$ be such that $R(A) := A(X)$ is closed. Then there exists some $c > 0$ such that for all $y \in R(A)$ one can find $x \in X$ satisfying $A(x) = y$ and $\|x\| \leq c \|y\|$.*

When A is bijective, the conclusion means that A^{-1} is continuous, a remarkable fact.

Theorem 3.24 (Banach Isomorphism Theorem) *If A is a linear continuous bijection between two Banach spaces, then A is an isomorphism.*

Corollary 3.27 (Closed Graph Theorem) *Any linear map with closed graph between two Banach spaces is continuous.*

Proof Let $B : Y \to Z$ be such a map. The graph X of B, being a closed linear subspace of $Y \times Z$, is a Banach space and $A : (y, By) \mapsto y$ is a continuous bijection from X onto Y. Then its inverse $y \mapsto (y, By)$ is continuous and B is continuous. □

Exercise Let X and Y be Banach spaces and let $A : X \to Y$ be a linear map such that, for every $y^* \in Y^*$, the linear form $y^* \circ A$ is continuous on X. Show that A is continuous. [Hint: note that the graph G of A is closed as $G = \{(x, y) \in X \times Y : \forall y^* \in Y^* \; y^*(A(x) - y) = 0\}$.]

A factorization result is often helpful.

Lemma 3.20 *Let X, Y be Banach spaces, let $A \in L(X, Y)$ be such that $A(X) = Y$ and let $\ell \in X^*$ be such that $\ell(x) = 0$ for all $x \in N := \ker A$. Then there exists some y^* in the dual Y^* of Y such that $\ell = y^* \circ A$.*

Proof Since A is onto and since for any $x, x' \in X$ satisfying $A(x) = A(x')$ one has $\ell(x) = \ell(x')$, there exists a map $k : Y \to \mathbb{R}$ such that $\ell = k \circ A$. It is easy to see that k is linear. Now, the Banach Open Mapping Theorem asserts that there exists some $c > 0$ such that for all $y \in Y$ one can find some $x \in A^{-1}(y)$ satisfying $\|x\| \leq c \|y\|$. It follows that for all $y \in Y$ one has $k(y) = k(A(x)) = \ell(x) \leq \|\ell\| \, c \, \|y\|$. Thus k is continuous and we can take $y^* = k$. □

Remark Instead of using the Banach Open Mapping Theorem, one can introduce the canonical projection $p : X \to X/N$, observe that ℓ can be factorized as $\ell = \overline{\ell} \circ p$ for some $\overline{\ell}$ in the dual of X/N and use the Banach Isomorphism Theorem to get that the map $\overline{A} : X/N \to Y$ induced by A is an isomorphism, so that one has $\ell = y^* \circ A$ with $y^* := \overline{\ell} \circ \overline{A}^{-1}$. □

Corollary 3.28 *Let G and H be two closed subspaces of a Banach space E such that $G+H$ is closed. Then there exists some $c>0$ such that every $z\in G+H$ can be decomposed into $z=x+y$ with $x\in G$, $y\in H$, $\|x\|\le c\,\|z\|$ and $\|y\|\le c\,\|z\|$.*

Proof This is a special case of Corollary 3.26 obtained by taking $X:=G\times H$ and $A:(x,y)\mapsto x+y$, so that $R(A)=G+H$. □

Corollary 3.29 *Let X and Y be closed subspaces of a Banach space Z such that $Z=X+Y$ and $X\cap Y=\{0\}$. Then Z is the direct* topological sum *of X and Y in the sense that the map $S:(x,y)\mapsto x+y$ is an isomorphism from $X\times Y$ onto Z.*

Then the projectors p_X and p_Y of Z onto X and Y are continuous since $(p_X,p_Y)=S^{-1}$. One says that X (and Y) are (topologically) *complemented*. In Hilbert spaces, every closed subspace is complemented. This is not the case in general in Banach spaces. For example, the closed subspace c_0 of ℓ_∞ is not complemented (see Exercises 5 and 6 of Sect. 3.3.3 for the definitions of these spaces).

Corollary 3.30 *Let X and Y be closed subspaces of a Banach space W such that $Z:=X+Y$ is closed. Then there exists a $c>0$ such that, for all $w\in W$,*

$$d(w,X\cap Y)\le cd(w,X)+cd(w,Y).$$

Proof Given $t>1$ and $w\in W$ we can find $x\in X$, $y\in Y$ such that

$$\|w-x\|\le td(w,X),\qquad \|w-y\|\le td(w,Y).$$

Taking $z:=x-y$ in Corollary 3.28, we obtain some $x'\in X$, $y'\in Y$ such that

$$x-y=x'+y',\quad \|x'\|+\|y'\|\le c\,\|x-y\|\,.$$

Then $x-x'=y'+y\in X\cap Y$ and since $\|x'-y'\|\le\|x'\|+\|y'\|\le c\,\|x-y\|$ we get

$$\begin{aligned}
d(w,X\cap Y)&\le\frac{1}{2}\left\|2w-(x-x')-(y+y')\right\|\\
&\le\frac{1}{2}(\|w-x\|+\|w-y\|+\|x'-y'\|)\\
&\le\frac{1}{2}(\|w-x\|+\|w-y\|+c\,\|x-w\|+c\,\|w-y\|)\\
&\le\frac{1}{2}t(c+1)(d(w,X)+d(w,Y)).
\end{aligned}$$

Since t is arbitrarily close to 1, the constant of the statement can be estimated as $\frac{1}{2}(c+1)$, where c is as in Corollary 3.28. □

Exercise Let A be a linear continuous map from a Banach space X onto a Banach space Y. Show that A has a continuous right inverse B (i.e. a map $B\in L(Y,X)$ such that $A \circ B = I_Y$) if and only if the kernel N of A has a topological complement M, i.e. if X is the direct topological sum of M and N).

Exercise Let A be an injective linear continuous map from a Banach space X into a Banach space Y. Show that A has a continuous left inverse B (i.e. a map $B \in L(Y, X)$ such that $B \circ A = I_X$) if and only if the image $R(A)$ of A has a topological complement Q.

The next theorem is a result of independent interest completing Proposition 3.21.

Theorem 3.25 *Let G and H be closed linear subspaces of a Banach space E. The following assertions are equivalent:*

$$G + H \text{ is closed in } E \tag{3.9}$$

$$G^{\perp} + H^{\perp} \text{ is closed in } E^* \tag{3.10}$$

$$G + H = (G^{\perp} + H^{\perp})^{\perp} \tag{3.11}$$

$$G^{\perp} + H^{\perp} = (G \cap H)^{\perp}. \tag{3.12}$$

Proof (3.11)⇒(3.9) and (3.12)⇒(3.10) are immediate implications. The reverse implication (3.9)⇒(3.11) is a consequence in Proposition 3.21.

(3.9)⇒(3.12) Since $G^{\perp} \subset (G \cap H)^{\perp}$ and $H^{\perp} \subset (G \cap H)^{\perp}$, hence $G^{\perp} + H^{\perp} \subset (G \cap H)^{\perp}$, it suffices to prove that $(G \cap H)^{\perp} \subset G^{\perp} + H^{\perp}$. Let $f \in (G \cap H)^{\perp}$. Given two decompositions $z = x + y = x' + y'$ of $z \in G + H$ with $x, x' \in G$, y, $y' \in H$ we have $f(x) = f(x')$ since $x - x' = y' - y \in G \cap H$. Thus one can define $\ell : F := G + H \to \mathbb{R}$ by $\ell(z) := f(x)$ for $z = x + y$ with $x \in G$, $y \in H$ and ℓ is seen to be linear. Moreover, ℓ is continuous since by Corollary 3.28 there exists a $c > 0$ such that for all $z \in F$ we can find $x \in G$, $y \in H$ such that $z = x + y$, $\|x\| \le c \|z\|$, hence $|\ell(z)| \le c \|f\| \|z\|$. The Hahn-Banach Theorem allows us to consider ℓ as the restriction to F of a continuous linear form on E still denoted by ℓ. Then we can write f as $f = (f - \ell) + \ell$ with $f - \ell \in G^{\perp}$ and $\ell \in H^{\perp}$.

(3.10)⇒(3.9) Corollary 3.30 asserts that there exists a $c > 0$ such that

$$\forall z^* \in E^* \qquad d(z^*, G^{\perp} \cap H^{\perp}) \le cd(z^*, G^{\perp}) + cd(z^*, H^{\perp}). \tag{3.13}$$

Now by the second part of Proposition 3.22, for $D := G^{\perp}, H^{\perp}, G^{\perp} \cap H^{\perp}$ we have

$$\forall z^* \in E^* \qquad d(z^*, D) = \sup\{\langle z^*, z\rangle : z \in B_E \cap D^{\perp}\} = h_{B_{D^{\perp}}}(z^*).$$

By relation (3.5) we have $(G^{\perp} \cap H^{\perp})^{\perp} = (G + H)^{\perp\perp} := F$ where F is the closure of $G + H$. Now, by Corollary 3.20 about support functions, (3.13) implies that

$$c^{-1} B_F \subset \text{cl}(B_G + B_H).$$

The proof of the Open Mapping Theorem applied to the map $A : (x, y) \to x + y$ from $G \times H$ to F shows that $\text{int}(c^{-1}B_F) \subset A(B_{G\times H}) = B_G + B_H$ when $G \times H$ is endowed with the sup norm. This shows that A is surjective or that $F = G + H$: $G + H$ is closed. □

The following classical result is known as the *Closed Range Theorem.*

Theorem 3.26 (Closed Range Theorem) *Let X and Y be Banach spaces and let $A^\intercal$ be the transpose map of $A \in L(X, Y)$. Then the range $R(A^\intercal)$ of $A^\intercal$ is closed if and only if the range $R(A)$ of A is closed.*

Proof Suppose $R(A)$ is closed. Corollary 3.26 yields some $c > 0$ such that for all $y \in R(A)$ one can find $x \in X$ satisfying $A(x) = y$ and $\|x\| \leq c\,\|y\|$. Then, given $y^* \in Y^*$, for all $y \in R(A)$ we have

$$|\langle y^*, y\rangle| = |\langle y^*, Ax\rangle| = |\langle A^\intercal y^*, x\rangle| \leq \|A^\intercal y^*\| \,.\, \|x\| \leq c\,\|A^\intercal y^*\| \,.\, \|y\|\,,$$

so that $\|y^*\| \leq c\,\|A^\intercal y^*\|$. This inequality shows that any Cauchy sequence in $R(A^\intercal)$ is the image under $A^\intercal$ of a Cauchy sequence in Y^*. Since $A^\intercal$ is continuous, we see that any Cauchy sequence in $R(A^\intercal)$ converges. Thus $R(A^\intercal)$ is closed in X^*.

Conversely, suppose $R(A^\intercal)$ is closed. Then $R(A^{\intercal\intercal})$ is closed in Y^{**}. Considering X (resp. Y) as a subspace of X^{**} (resp. Y^{**}), for all $x \in X$, $y^* \in Y^*$ we have

$$\langle y^*, A^{\intercal\intercal}x\rangle = \langle A^\intercal y^*, x\rangle = \langle y^*, Ax\rangle,$$

so that $A^{\intercal\intercal}x = Ax$. As above we can find some $c > 0$ such that $\|x^{**}\| \leq c\,\|A^{\intercal\intercal}x^{**}\|$ for all $x^{**} \in X^{**}$. Given $y^{**} \in \text{cl}(A(X)) \subset \text{cl}(A^{\intercal\intercal}(X))$ and a sequence (x_n) in X such that $(A^{\intercal\intercal}(x_n)) \to y^{**}$ we get that (x_n) is a Cauchy sequence in X, hence has a limit x. Since $A^{\intercal\intercal}$ is continuous we have $y^{**} = A^{\intercal\intercal}(x) = A(x)$, hence $y^{**} \in A(X)$. This shows that $R(A) := A(X)$ is closed in Y^{**}, hence in Y. □

The next classical theorem is often useful.

Theorem 3.27 (Banach-Steinhaus or Uniform Boundedness Theorem) *Let X, Y be normed spaces, X being complete, and let F be a subset of the space $L(X, Y)$ of continuous linear maps from X to Y. If for all $x \in X$, the set $F(x) := \{f(x) : f \in F\}$ is bounded in Y, then F is bounded in $L(X, Y)$ with respect to the usual norm.*

Proof Denoting by B_Y the unit ball of Y, for $n \in \mathbb{N}$ consider the closed set

$$X_n := \{x \in X : \forall f \in F \;\; \|f(x)\| \leq n\} = \bigcap_{f\in F} f^{-1}(nB_Y).$$

By assumption, X is the union of the family (X_n). By Baire's Theorem 2.11, there is some $k \in \mathbb{N}$ such that X_k has a nonempty interior. If $a \in \text{int}X_k$, we also have $-a \in \text{int}X_k$ since X_k is symmetric with respect to 0. It follows that $0 \in \text{int}X_k$. Thus, if $r > 0$ is such that $rB_X \subset X_k$, we have $\|f\| \leq r^{-1}k$ for all $f \in F$. □

Corollary 3.31 *A weakly* bounded subset of the dual* X^* *of a Banach space* X *is bounded. A weakly bounded subset of* X *is bounded. In particular, weak* convergent sequences of* X^* *and weakly convergent sequences of* X *are bounded.*

Here a subset F of X^* (resp. $S \subset X$) is said to be *weakly** (resp. *weakly*) *bounded* if for all $x \in X$ (resp. $x^* \in X^*$) the set $\{f(x) : f \in F\}$ (resp. $\{x^*(x) : x \in S\}$) is bounded in $\mathbb{R}$.

Proof The first assertion is the special case of the theorem corresponding to $Y := \mathbb{R}$. The second stems from the fact that the canonical embedding of X into X^{**} is isometric. □

We end this subsection by deriving from Brouwer's Theorem a fixed point theorem that may be used to solve equations in functional spaces. The proof relies on two useful tools: an approximation method and partitions of unity.

Theorem 3.28 (Schauder) *Let* X *be a normed space, let* C *be a nonempty closed convex subset of* X *and let* $f : C \to C$ *be a continuous map such that* $f(C)$ *is contained in a compact subset* K *of* C*. Then* f *has a fixed point.*

Proof Given $r > 0$ let $(B(y_i, r/2))_{i \in I}$ be a finite covering of K by open balls with radius $r/2$, with $y_i \in K$ for all $i \in I$. Let us set

$$p_i(x) := \max(r - \|f(x) - y_i\|, 0), \qquad p(x) := \sum_{i \in I} p_i(x).$$

Then p_i and p are continuous and for all $x \in C$ we can find some $i \in I$ such that $f(x) \in B(y_i, r/2)$, so that $p(x) \geq p_i(x) \geq r/2$. Thus, the map f_r defined by

$$f_r(x) := \frac{1}{p(x)} \sum_{i \in I} p_i(x) y_i$$

is continuous and $\|f_r(x) - f(x)\| \leq r$ for all $x \in C$: indeed, setting $I(x) := \{i \in I : p_i(x) > 0\}$, we have $\|f(x) - y_i\| < r$ for all $i \in I(x)$, hence

$$\begin{aligned} \|f_r(x) - f(x)\| &= \left\| \frac{1}{p(x)} \sum_{i \in I(x)} p_i(x) y_i - \frac{1}{p(x)} \sum_{i \in I(x)} p_i(x) f(x) \right\| \\ &\leq \frac{1}{p(x)} \sum_{i \in I(x)} p_i(x) \|f(x) - y_i\| \leq r. \end{aligned}$$

Let C_r be the convex hull of $\{y_i : i \in I\}$. It is a compact convex subset of the linear span of $\{y_i : i \in I\}$. Since $f_r(C_r)$ is contained in C_r, the map $f_r \mid_{C_r}$ has a fixed point $x_r \in C_r$ by Theorem 2.5. Since $\|x_r - f(x_r)\| = \|f_r(x_r) - f(x_r)\| \leq r$ and $f(x_r) \in K$, we can find a sequence $(r(n))$ with limit 0 such that $(f(x_{r(n)}))_n$ and $(x_{r(n)})_n$ converge to the same limit $\overline{x}$. Then, by continuity of f, we have $f(\overline{x}) = \overline{x}$. □

Exercises

1. Show that if $(e_i)_{i \in I}$ is a basis of a Banach space X, then I is uncountable. [Hint: if $I = \mathbb{N}$, for $n \in \mathbb{N}$ let X_n be the linear span of $\{e_i : i \in \mathbb{N}_n\}$, so that X is the union of the closed subspaces X_n. Using Baire's Theorem, obtain a contradiction.]
2. Let X and Y be Banach spaces and let $b : X \times Y \to \mathbb{R}$ be a bilinear map. Assuming that b is separately continuous in each of its two variables, show that b is continuous. Is the conclusion valid if b is not bilinear? [Hint: define $B : X \to Y^*$ by $B(x) := b(x, \cdot)$ and use the Uniform Boundedness Theorem.]
3. Let X and Y be Banach spaces and let (A_n) be a sequence in $L(X, Y)$ such that for all $x \in X$ the sequence $(A_n x)$ has a limit denoted by Ax. Show that $A \in L(X, Y)$ and that for any convergent sequence $(x_n) \to x$ in X one has $(A_n x_n) \to Ax$.
4. Let X be a reflexive Banach space and let (x_n) be a sequence in X such that for all $f \in X^*$ the sequence $(f(x_n))$ has a limit. Deduce from the preceding exercise that there exists some $x \in X$ such that $(x_n) \to x$ weakly.

   Exhibit a sequence in a (non-reflexive) Banach space X such that the preceding property does not hold. [Hint: take for X the space of sequences $v := (v_n)$ of real numbers such that $(v_n) \to 0$ and take $x_n := 1_{[0,n]}$ with $x_{n,k} := 1$ for $k \leq n$, 0 for $k > n$.]
5. Let X be a Banach space and let $A : X \to X^*$ be a linear operator such that $\langle Ax, x \rangle \geq 0$ for all $x \in X$. Show that A is continuous. [Hint: use the closed graph theorem.]
6. Let X and Y be Banach spaces and let $A \in L(X, Y)$ be surjective. Show that the kernel $N(A)$ of A has a complement if and only if A has a right inverse $B \in L(Y, X)$ (meaning that $A \circ B = I_Y$).
7. Let X and Y be Banach spaces and let $A \in L(X, Y)$ be injective. Show that $R(A)$ is closed and has a complement if and only if A has a left inverse $B \in L(Y, X)$ (that means that $B \circ A = I_X$).
8. (**Robinson**) Given a Banach space X, a normed space Y, and a closed convex subset F of $X \times Y$ such that $p_X(F)$ is bounded, show that $\mathrm{int}(p_Y(F)) = \mathrm{int}(\mathrm{cl}(p_Y(F)))$. [Hint. Adapt the proof of Proposition 3.32.]
9. (**Metric regularity property**) Given $c \in \mathbb{P}$ and a multimap $F : X \rightrightarrows Y$ whose graph is convex satisfying $B(\overline{y}, c) \subset F(B(\overline{x}, 1))$ for some $(\overline{x}, \overline{y}) \in F$, show that for all $(x, y) \in X \times B(\overline{y}, c)$ one has

$$d(x, F^{-1}(y)) \leq \frac{1 + \|x - \overline{x}\|}{c - \|y - \overline{y}\|} d(y, F(x)).$$

3.5.2 Densely Defined Operators and Transposition

In many problems involving derivatives or partial derivatives one is faced with operators that are not defined everywhere, such as $A : x \mapsto x'$ on $C([0, 1])$ which is defined on the subspace $C^1([0, 1])$ of continuously differentiable functions on

$[0, 1]$, or $\Delta : C(\Omega) \to C(\Omega)$ given by $\Delta u = \Sigma_{i=1}^{d} D_i^2 u$ for $u \in C^2(\Omega)$, where Ω is an open subset of $\mathbb{R}^d$ and $C^2(\Omega)$ is the space of functions having continuous derivatives of order one and two on Ω. Thus, it is of interest to perform a general study of so-called unbounded operators, i.e. operators whose domain is a subspace. Applications to quantum mechanics can be found in [67, 74, 185] for instance.

In the rest of this subsection X and Y are Banach spaces and $A : D(A) \to Y$ is a linear map whose domain $D(A)$ is a linear subspace of X. We first show that an elementary rule for integration and the notion of transposition can be extended to closed densely defined operators, i.e. operators whose graphs are closed and whose domains are dense.

Lemma 3.21 *Let $A : D(A) \to Y$ be a closed linear map, let $T := [a, b]$ be a compact interval of $\mathbb{R}$, and let $f \in C(T, X)$ be such that $f(T) \subset D(A)$ and $A \circ f \in C(T, Y)$. Then $\int_T f \in D(A)$ and $A(\int_T f) = \int_T A \circ f$.*

Proof By assumption, $g : t \mapsto (f(t), A(f(t)))$ is continuous from T into the graph $G(A)$ of A endowed with the norm induced by $X \times Y$. Since $G(A)$ is closed, hence complete, $\int_T g \in G(A)$. Since $\int_T g = (\int_T f, \int_T A \circ f)$, we get $\int_T A \circ f = A(\int_T f)$. □

Definition 3.3 Let $A : D(A) \to Y$ be a linear map with dense domain $D(A)$. Let $D(A^\mathsf{T})$ be the set of $y^* \in Y^*$ such that $y^* \circ A$ is continuous on $D(A)$ with respect to the induced topology. Then there exists a unique linear map $A^\mathsf{T} : D(A^\mathsf{T}) \to X^*$ satisfying the relation

$$\langle A^\mathsf{T} y^*, x \rangle = \langle y^*, Ax \rangle \qquad \forall y^* \in D(A^\mathsf{T}),\ \forall x \in D(A).$$

It is called the *transpose* (or conjugate or dual operator) of A.

The fact that $A^\mathsf{T} y^* \in X^*$ is well defined when $y^* \in D(A^\mathsf{T})$ stems from the property that the linear continuous map $y^* \circ A$ has a unique (linear) continuous extension to X since $D(A)$ is dense in X and $y^* \circ A$ is uniformly continuous. A routine argument shows that $D(A^\mathsf{T})$ is a linear subspace and that A^T is linear.

The transpose of A is often called the *adjoint* of A, but we prefer to keep this term (and the notation A^*) for a related notion specific to Hilbert spaces. Namely, if $J_X : X \to X^*$ and $J_Y : Y \to Y^*$ are the duality maps of Hilbert spaces X and Y one sets

$$A^* := J_X^{-1} \circ A^\mathsf{T} \circ J_Y,$$

using the fact that J_X and J_Y are isomorphisms. In other terms, A^* is characterized by

$$\forall x \in D(A),\ y \in J_Y^{-1}(D(A^\mathsf{T})) \qquad (A^* y \mid x) = (y \mid Ax),$$

where $(\cdot \mid \cdot)$ is the scalar product. If X and Y are finite dimensional Euclidean spaces, both notions can be identified.

It can be shown that if $A : D(A) \to Y$ and $B : D(B) \to Y$ are such that $D(A) \cap D(B)$ is dense in X, then $(A + B)^\mathsf{T} = A^\mathsf{T} + B^\mathsf{T}$ on $D(A^\mathsf{T}) \cap D(B^\mathsf{T})$. Let us note that when $D(A) = X$ and A is continuous, then $D(A^\mathsf{T}) = Y$ since for all $y^* \in Y^*$ the linear form $y^* \circ A$ is continuous. Moreover, in such a case, we have seen that A^T is continuous. This fact is valid even if X and Y are just locally convex topological vector spaces (exercise).

Example* The *position operator* $Q : X \to X$, with $X := L_2(\mathbb{R})$, the Lebesgue space of square integrable functions on $\mathbb{R}$, is the map given by $Q(x)(r) := rx(r)$ for $x \in L_2(\mathbb{R})$. It plays an important role in quantum mechanics. Its domain is the set $D(Q) := \{x \in L_2(\mathbb{R}) : I_\mathbb{R} x \in L_2(\mathbb{R})\}$ where $I_\mathbb{R}(r) := r$ for $r \in \mathbb{R}$. It is unbounded since for $T_n := [n, n+1]$ one has $\|Q(1_{T_n})\|_2 \geq n$ while $\|1_{T_n}\|_2 = 1$. It is a *symmetric operator* since for $x, y \in D(Q)$ one has

$$\langle Qx \mid y \rangle = \int_\mathbb{R} rx(r)y(r)dr = \langle Qy \mid x \rangle.$$

Moreover, it is *self-adjoint*, i.e. one has $D(Q^*) \subset D(Q)$: if $y \in D(Q^*)$ there exists some $z \in L_2(\mathbb{R})$ such that $\langle Qx \mid y \rangle = \langle x \mid z \rangle$ for all $x \in D(Q)$, in particular for all $x \in C_c^\infty(\mathbb{R})$, which implies that $z(r) = ry(r)$ a.e. $r \in \mathbb{R}$ or $y \in D(Q)$. The *spectrum* $\sigma(Q)$ of Q is $\mathbb{R}$, i.e. for all $\lambda \in \mathbb{R}$ the operator $Q - \lambda I$ is not invertible: if T were a continuous inverse of $Q - \lambda I$, for all $x \in L_2(\mathbb{R})$ the relation $(Q - \lambda I)(Tx) = x$ would imply $(r - \lambda)(Tx)(r) = x(r)$ whereas for $x := 1_{]\lambda,\mu[}$ with $\mu > \lambda$ the function $r \mapsto (r - \lambda)^{-1}x(r)$ does not belong to $L_2(\mathbb{R})$. However, Q has no eigenvalue, as is easily seen. □

Proposition 3.33 *If $A : D(A) \to Y$ is a densely defined linear map, its transpose A^T is a closed map in the sense that its graph $G(A^\mathsf{T})$ is closed in $Y^* \times X^*$ (and even weakly* closed).*

For the domain $D(A^\mathsf{T})$ of A^T to be weakly dense in Y^* it is necessary and sufficient that A has a closed extension $\overline{A}$. If this is the case, the transpose $A^{\mathsf{T}\mathsf{T}} := (A^\mathsf{T})^\mathsf{T}$ of A^T is the smallest closed extension of A. In particular, if A is closed in the sense that $G(A)$ is closed, then $A^{\mathsf{T}\mathsf{T}} = A$.*

Proof Let $S : X^* \times Y^* \to Y^* \times X^*$ be the (symplectic) isomorphism defined by $S(x^*, y^*) = (y^*, -x^*)$. The definition of A^T shows that

$$G(A^\mathsf{T}) = S(G(A)^\perp). \tag{3.14}$$

In fact, $(y^*, x^*) \in G(A^\mathsf{T})$ if and only if for all $(x, y) \in G(A)$ one has $\langle y^*, y \rangle - \langle x^*, x \rangle = 0$ or $(y^*, x^*) = S(-x^*, y^*)$ with $(-x^*, y^*) \in G(A)^\perp$. Since S is an isomorphism, and since $G(A)^\perp$ is weakly* closed, $G(A^\mathsf{T})$ is weakly* closed.

Assume that $D(A^\mathsf{T})$ is weakly* dense in Y^*. The preceding shows that $A^{\mathsf{T}\mathsf{T}}$ considered as an operator from X into Y is closed and $G(A^{\mathsf{T}\mathsf{T}}) = T(G(A^\mathsf{T})^\perp)$ for $T := -S^\mathsf{T} : (y, x) \mapsto (x, -y)$. It is clearly an extension of A and, since for a linear subspace Z of $Y^* \times X^*$ one has $(S(Z))^\perp = T^{-1}(Z^\perp)$, one gets

$$G(A^{\mathsf{T}\mathsf{T}}) = T(G(A^\mathsf{T})^\perp) = T((S(G(A)^\perp)^\perp) = T(T^{-1}(G(A)^{\perp\perp})) = G(A)^{\perp\perp}.$$

Thus, by the bipolar theorem, $G(A^{\intercal\intercal})$ is the smallest closed subspace containing $G(A)$ and A has a closed extension $A^{\intercal\intercal}$.

Conversely, assuming that A has a closed extension $\overline{A}$, let us prove that $D(A^{\intercal})$ is dense. Since $D(\overline{A}^{\intercal}) \subset D(A^{\intercal})$, it suffices to prove that $D(\overline{A}^{\intercal})$ is dense in Y^* or that $D(B^{\intercal})$ is dense in Y^* when B is a closed densely defined operator. Thus, we assume that A is closed. Let $y \in D(A^{\intercal})^{\perp}$; then $(y, 0) \in G(A^{\intercal})^{\perp}$. Setting as above $T(v, u) := (u, -v)$ for $(v, u) \in Y \times X$, so that $T^{-1}(u, v) = (-v, u)$, for a subspace Z of $Y^* \times X^*$ we have $(S(Z))^{\perp} = T^{-1}(Z^{\perp})$, hence, by relation (3.14) and the bipolar theorem,

$$G(A^{\intercal})^{\perp} = (S(G(A)^{\perp}))^{\perp} = T^{-1}(G(A)^{\perp\perp}) = T^{-1}(G(A)).$$

Then $(y, 0) \in G(A^{\intercal})^{\perp}$ implies that $y \in D(A)^{\perp}$ and $y = 0$. Thus $D(A^{\intercal})$ is dense in Y^*. □

Other relationships between A and $A^{\intercal}$ are described in the next result.

Proposition 3.34 *For a closed, densely defined operator A between $D(A) \subset X$ and Y one has*

$$N(A^{\intercal}) = R(A)^{\perp}, \qquad N(A) = R(A^{\intercal})^{\perp}. \tag{3.15}$$

Proof Introducing the closed subspaces $G := G(A)$, $H := X \times \{0\}$ (the "horizontal subspace") of $E := X \times Y$, one has

$$X \times R(A) = G + H, \qquad N(A) \times \{0\} = G \cap H \tag{3.16}$$

and similarly, since $G(A^{\intercal}) = S(G(A)^{\perp}) = S(G^{\perp})$,

$$R(A^{\intercal}) \times Y^* = G^{\perp} + H^{\perp}, \qquad \{0\} \times N(A^{\intercal}) = G^{\perp} \cap H^{\perp}. \tag{3.17}$$

Thus, since $(G + H)^{\perp} = G^{\perp} \cap H^{\perp}$ by (3.5), we get

$$\{0\} \times N(A^{\intercal}) = G^{\perp} \cap H^{\perp} = (G + H)^{\perp} = \{0\} \times R(A)^{\perp}.$$

On the other hand, since $G \cap H = (G^{\perp} + H^{\perp})^{\perp}$ by (3.6), we get

$$N(A) \times \{0\} = G \cap H = (G^{\perp} + H^{\perp})^{\perp} = (R(A^{\intercal}) \times Y^*)^{\perp} = R(A^{\intercal})^{\perp} \times \{0\},$$

so that relation (3.15) holds. □

Remark From (3.15) and the relations $W^{\perp\perp} = \mathrm{cl}\, W$, $Z^{\perp\perp} = \mathrm{cl}^* Z$ for linear subspaces $W \subset X$, $Z \subset X^*$, we deduce $N(A^{\intercal})^{\perp} = \mathrm{cl}\, R(A)$, $N(A)^{\perp} = \mathrm{cl}^* R(A^{\intercal})$. □

The Closed Range Theorem can be generalized to densely defined closed maps.

Theorem 3.29 (Closed Range Theorem) *Let $A : D(A)(\subset X) \to Y$ be a densely defined closed operator. The following assertions are equivalent:*

$$R(A) = \mathrm{cl}\,(R(A)), \tag{3.18}$$

$$R(A^\intercal) = \mathrm{cl}\,R(A^\intercal), \tag{3.19}$$

$$R(A) = N(A^\intercal)^\perp, \tag{3.20}$$

$$R(A^\intercal) = N(A)^\perp. \tag{3.21}$$

Proof Setting $G := G(A)$, $H := X \times \{0\}$, we saw that

$$N(A) \times \{0\} = G \cap H, \qquad \{0\} \times N(A^\intercal) = G^\perp \cap H^\perp, \tag{3.22}$$

$$X \times R(A) = G + H, \qquad R(A^\intercal) \times Y^* = G^\perp + H^\perp. \tag{3.23}$$

Thus $R(A)$ is closed if and only if $G + H$ is closed, if and only if (by Theorem 3.25) $G^\perp + H^\perp$ is closed, if and only if $R(A^\intercal)$ is closed.

When $R(A) = N(A^\intercal)^\perp$, $R(A)$ is closed. Conversely, if $R(A)$ is closed, by relation (3.15) one has $R(A) = R(A)^{\perp\perp} = N(A^\intercal)^\perp$.

Similarly, when $R(A^\intercal) = N(A)^\perp$, $R(A^\intercal)$ is closed. Conversely, if $R(A^\intercal)$ is closed, then by (3.23) and Theorem 3.25, $G^\perp + H^\perp$ and $G+H$ are closed so that, by (3.23), $R(A)$ is closed. □

Alternate proof (for further training with the preceding results) (3.18)⇔(3.20) The implication (3.20)⇒(3.18) being obvious, let us prove the reverse implication. By definition of $A^\intercal$ we have $R(A) \subset N(A^\intercal)^\perp$ since for $x \in D(A)$, $y := Ax$, $y^* \in N(A^\intercal)$ we have $\langle y^*, y\rangle = \langle y^*, Ax\rangle = \langle A^\intercal y^*, x\rangle = 0$. To prove that $R(A) = N(A^\intercal)^\perp$ when $R(A)$ is closed we show that for $y \in Y\backslash R(A)$ we have $y \in Y\backslash N(A^\intercal)^\perp$. Since $R(A)$ is closed, the Hahn-Banach Theorem yields some $y^* \in Y^*$ such that $\langle y^*, z\rangle = 0$ for all $z \in R(A)$ and $\langle y^*, y\rangle = 1$. Thus $y^* \in D(A^\intercal)$ and $y^* \in N(A^\intercal)$. Since $\langle y^*, y\rangle = 1$, we have $y \notin N(A^\intercal)^\perp$.

(3.19)⇔(3.21) The implication (3.21)⇒(3.19) being obvious, let us prove the reverse implication. Again, by definition of $A^\intercal$, we have $R(A^\intercal) \subset N(A)^\perp$ since for $x^* := A^\intercal y^*$ with $y^* \in D(A^\intercal)$ and $x \in N(A)$ we have $\langle x^*, x\rangle = \langle A^\intercal y^*, x\rangle = \langle y^*, Ax\rangle = 0$. Let us prove the opposite inclusion when $R(A^\intercal)$ is closed. Given $x^* \in N(A)^\perp$, we observe that for every $y \in R(A)$ and any $x, x' \in A^{-1}(y)$ we have $\langle x^*, x\rangle = \langle x^*, x'\rangle$ since $x - x' \in N(A)$. Thus, we get a linear form f on $R(A)$ by setting $f(y) := \langle x^*, x\rangle$ for $x \in A^{-1}(y)$. Since $B : G(A) \to Y$ given by $B(x, Ax) = Ax$ is continuous when $G(A)$ is endowed with the norm induced by $X \times Y$ and since $R(B) = R(A)$ is closed, Corollary 3.26 provides some $c > 0$ such that for all $y \in R(A)$ there exists some $z \in G(A)$ such that $z \in B^{-1}(y)$ and $\|z\| \leq c\,\|y\|$. Taking $x \in D(A)$ such that $z := (x, Ax)$ we see that for any $y \in R(A)$ there exists some $x \in D(A)$ such that $\|x\| \leq c\,\|y\|$ and $y = A(x)$. Thus f is continuous. Taking a linear continuous extension $y^* \in Y^*$ of f we see that $y^* \in D(A^\intercal)$ and for all $x \in D(A)$ we have $\langle x^*, x\rangle = f(Ax) = \langle y^*, Ax\rangle$. Thus $x^* = A^\intercal y^* \in R(A^\intercal)$.

Let us reduce the proof of the equivalence (3.18)⇔(3.19) to the case of a continuous operator that is already known (Theorem 3.26). To do so, we provide the graph $G := G(A)$ of A with the norm induced by the norm of $X \times Y$, so that G is a Banach space. Its dual is isometric to the quotient space $(X^* \times Y^*)/G^\perp$ via the restriction map $S : X^* \times Y^* \to G^*$ and a subset H of G^* is closed if and only if $S^{-1}(H)$ is closed in $X^* \times Y^*$. Let P be the restriction to G of the second projection $(x, y) \mapsto y$. Since $A(x) = P(x, Ax)$ for all $x \in D(A)$ we have $R(P) = R(A)$. On the other hand, since $P^\top(z^*) = S(0, z^*)$ for all $z^* \in Y^*$, we have $(x^*, y^*) \in S^{-1}(R(P^\top))$ if and only if there exists a $z^* \in Y^*$ such that $S(x^*, y^*) = S(0, z^*)$, if and only if there exists $z^* \in Y^*$ such that $x^* = A^\top(z^* - y^*)$. Thus

$$S^{-1}(R(P^\top)) = R(A^\top) \times Y^*$$

and $R(P^\top)$ is closed if and only if $R(A^\top)$ is closed. Using the fact that $R(P^\top)$ is closed if and only if $R(P)$ is closed, we obtain the equivalence (3.18)⇔(3.19). □

The following consequence is often used to show the solvability of an equation $Ax = y$: one reduces the question to the search for a constant $c > 0$ such that $\|y^*\| \le c\,\|A^\top y^*\|$ for all $y^* \in D(A^\top)$. This method is called *the method of a priori estimates*.

Corollary 3.32 *Let $A : D(A)(\subset X) \to Y$ be a densely defined closed operator between Banach spaces. The following assertions are equivalent:*

(a) A is surjective: $R(A) = Y$;
(b) there exists some $c > 0$ such that $\|y^*\| \le c\,\|A^\top y^*\|$ for all $y^* \in D(A^\top)$;
(c) $R(A^\top)$ is closed and $N(A^\top) = \{0\}$.

Proof (a)⇒(b) By homogeneity it suffices to prove that the set

$$Z^* := \{y^* \in D(A^\top) : \|A^\top y^*\| \le 1\}$$

is bounded in Y^* or even (invoking the Uniform Boundedness Theorem), that for any $\overline{y} \in Y$, the set $Z^*(\overline{y}) := \{\langle y^*, \overline{y}\rangle : y^* \in Z^*\}$ is bounded in $\mathbb{R}$. Since $R(A) = Y$ we can pick some $\overline{x} \in D(A)$ such that $A\overline{x} = \overline{y}$. Then, for $y^* \in Z^*$ we have $|\langle y^*, \overline{y}\rangle| = |\langle A^\top y^*, \overline{x}\rangle| \le c\,\|A^\top y^*\|$ with $c := \|\overline{x}\|$.

(b)⇒(c) Let $\overline{x}^* \in \mathrm{cl}(R(A^\top))$. If (y_n^*) is a sequence in $D(A^\top)$ such that $(A^\top y_n^*) \to \overline{x}^*$ the estimate in (b) shows that (y_n^*) is a Cauchy sequence, hence has a limit $\overline{y}^*$ in Y^*. Since the graph of $A^\top$ is closed by Proposition 3.33, we get $\overline{x}^* = A^\top \overline{y}^* \in R(A^\top)$: $R(A^\top)$ is closed. The equality $N(A^\top) = \{0\}$ is a direct consequence in the relation $\|y^*\| \le c\,\|A^\top y^*\|$ for all $y^* \in D(A^\top)$.

(c)⇒(a) This follows from Theorem 3.29: $R(A) = N(A^\top)^\perp = Y$. □

In Exercise 1 a dual result is proposed; its use is more limited. For a densely defined closed operator A between Banach spaces we have the implications

$$A \text{ surjective} \Rightarrow A^\top \text{ injective},$$

$$A^\top \text{ surjective} \Rightarrow A \text{ injective}$$

and in finite dimensional spaces the reverse implications hold. In general, this is not the case in infinite dimensional spaces (see Exercises 2 and 3).

Exercises

1. Let $A : D(A)(\subset X) \to Y$ be a densely defined closed operator between Banach spaces. Show that the following assertions are equivalent:

   (a) A^{T} is surjective : $R(A^{\mathsf{T}}) = X^*$;
   (b) there exists a $c > 0$ such that $\|x^*\| \le c\,\|Ax^*\|$ for all $x^* \in D(A)$;
   (c) $R(A)$ is closed and $N(A^{\mathsf{T}}) = \{0\}$.

2. Let $X = Y = \ell_1$ and let $A \in L(X, Y)$ be given by $Ax := (a_n x_n)_n$ for $x := (x_n)_n$, where $(a_n) \to 0$ and $a_n > 0$ for all n. Verify that A is injective but that $R(A^{\mathsf{T}}) \neq X^*$.
3. With the same data as in the preceding exercise, verify that A^{T} is injective but that A is not surjective.
4. Let $A : D(A)(\subset X) \to Y$ be a densely defined closed operator between Banach spaces. Show that $R(A)$ is closed if and only if there exists some $c \in \mathbb{R}_+$ such that $d(x, N(A)) \le c\,\|Ax\|$ for all $x \in D(A)$. [Hint: first, consider the case $D(A) = X$ and use the quotient space $X/N(A)$. Then, reduce the general case to the preceding one by considering the operator $A_G : G(A) \to Y$ defined as the restriction to $G(A)$ of the second projection, $G(A)$ being endowed with the restriction of a product norm.]
5. Let $X := \ell_1$ and let $A : D(A)(\subset X) \to X$ be the densely defined closed operator defined by $D(A) := \{x := (x_n) \in \ell_1 : (nx_n) \in \ell_1\}$, $Ax = (nx_n)$ for $x := (x_n) \in D(A)$. Verify that A is densely defined and closed but that A^{T} is not densely defined in $X^* = \ell_\infty$.

3.5.3 *The Spectrum of a Linear Operator*

Given a Banach space X over $\mathbb{K} := \mathbb{R}$ or $\mathbb{C}$ and $A \in L(X) := L(X, X)$, it is of interest to consider the set $\rho(A)$ of $\lambda \in \mathbb{K}$ such that $A - \lambda I$ is an isomorphism; it is called the *resolvent set* of A. The complement of $\rho(A)$ is called the *spectrum* of A and is denoted by $\sigma(A)$. The spectrum of A obviously contains the set $\sigma_e(A)$ of eigenvalues of A, $\lambda \in \mathbb{K}$ being called an *eigenvalue* of A if there exists some $v \in X\backslash\{0\}$ such that $Av = \lambda v$. Then v is called an *eigenvector*. The set $\sigma_e(A)$ is also called the *point spectrum* and denoted by $\sigma_p(A)$. The *eigenspace* X_λ corresponding to the eigenvalue λ is the set of eigenvectors associated with λ, i.e. $N(A - \lambda I)$, the kernel of $A - \lambda I$.

If X is finite dimensional, $\sigma_e(A) = \sigma(A)$ since $A - \lambda I$ is an isomorphism whenever it is injective. However, in infinite dimensional spaces the inclusion $\sigma_e(A) \subset \sigma(A)$ may be strict, as one may have $N(A) := A^{-1}(0) = \{0\}$, i.e. $0 \notin \sigma_e(A)$ and $R(A) :=$

$A(X) \neq X$, hence $0 \in \sigma(A)$. This is the case for the *right shift* A on the space ℓ_∞ of bounded sequences given by $A(x_0, x_1, \ldots) = (0, x_0, x_1, \ldots)$.

As already shown by the finite dimensional case, the situation is much more interesting in complex vector spaces than in real spaces. In complex vector spaces it can be shown that $\sigma(A)$ is always nonempty (even if $\sigma_e(A)$ is empty, as in the above example), as in the finite dimensional setting. However, different techniques are involved and in this book we focus our attention on real vector spaces. In this setting we have the following elementary result.

Proposition 3.35 *The (real) spectrum of a linear continuous operator $A \in L(X)$ on a real Banach space X is a compact subset of the interval $[-c, c]$ for $c := \|A\|$. Moreover, if $\lambda_0 \in \rho(A)$, then for $|\lambda - \lambda_0| < 1/\left\|(\lambda_0 I - A)^{-1}\right\|$ one has $\lambda \in \rho(A)$ and*

$$(\lambda I - A)^{-1} = \sum_{n \geq 0} (\lambda - \lambda_0)^n (\lambda_0 I - A)^{-(n+1)}, \tag{3.24}$$

$$\left\|(\lambda I - A)^{-1}\right\| \leq \left\|(\lambda_0 I - A)^{-1}\right\| \exp(|\lambda - \lambda_0| . \left\|(\lambda_0 I - A)^{-1}\right\|). \tag{3.25}$$

Proof Since $\rho(A)$ is open by Proposition 3.17, $\sigma(A)$ is closed. We have to show that $\mathbb{R} \backslash [-c, c] \subset \rho(A)$. Now for $\lambda \in \mathbb{R} \backslash [-c, c]$, $I - \lambda^{-1} A$ is an isomorphism since $\left\|\lambda^{-1} A\right\| < 1$. Thus $A - \lambda I$ is an isomorphism and $\lambda \in \rho(A)$. Setting $R(\lambda) = (\lambda I - A)^{-1}$ and writing

$$\lambda I - A = \lambda_0 I - A + (\lambda - \lambda_0) I = [I - (\lambda_0 - \lambda) R(\lambda_0)](\lambda_0 I - A),$$

we see that $\lambda I - A$ is invertible when $|\lambda - \lambda_0| < 1/\|R(\lambda_0)\|$. Then the inverse is given by

$$R(\lambda) = R(\lambda_0)[I - (\lambda_0 - \lambda) R(\lambda_0)]^{-1} = \sum_{n=0}^{\infty} (\lambda - \lambda_0)^n R(\lambda_0)^{n+1}.$$

It follows that $\|R(\lambda)\| \leq \|R(\lambda_0)\| \exp(|\lambda - \lambda_0| \, \|R(\lambda_0)\|)$. □

Proposition 3.36 *If $(e_1, \ldots, e_n)$ is a finite family of eigenvectors corresponding to distinct eigenvalues of $A \in L(X)$, then $e_1, \ldots, e_n$ are linearly independent.*

Proof We prove the assertion by induction on n. For $n = 1$ the assertion is obvious. We assume the assertion is valid for $n-1$ and prove it for n. Let λ_i be the eigenvalue corresponding to e_i for $i \in \mathbb{N}_n$. Since $e_1, \ldots, e_{n-1}$ are linearly independent by our induction assumption, it suffices to prove that assuming that $e_n = c_1 e_1 + \ldots + c_{n-1} e_{n-1}$ for some $c_i \in \mathbb{R}$ leads to a contradiction. This relation implies that

$$\lambda_n (c_1 e_1 + \ldots + c_{n-1} e_{n-1}) = \lambda_n e_n = A e_n = c_1 \lambda_1 e_1 + \ldots + c_{n-1} \lambda_{n-1} e_{n-1},$$

hence that $c_i(\lambda_n - \lambda_i) = 0$ for $i \in \mathbb{N}_{n-1}$. It follows that $c_i = 0$ and $e_n = 0$, a contradiction. □

The notions of eigenvalue and eigenvector can be extended to any linear operator A whose domain $D(A)$ is smaller than the whole Banach space X. The *resolvent set* $\rho(A)$ of a closed (linear) operator A is defined as the set of $\lambda \in \mathbb{R}$ such that $\lambda I - A$ is a bijection of $D(A)$ onto X. Then, the *resolvent operator*

$$R_\lambda := R(\lambda) := (\lambda I - A)^{-1}$$

is a continuous linear operator since its graph is closed, as is the graph of $\lambda I - A$, as is easily seen. Thus, when $D(A) = X$, these definitions reduce to the former ones.

The following properties of the resolvent set and of the resolvent operator are useful, in particular for semigroups (see Chap. 10).

Proposition 3.37 *The resolvent set $\rho(A)$ of an unbounded closed linear operator A on a real Banach space X is open. Moreover, if $\lambda_0 \in \rho(A)$ and if $\lambda \in \mathbb{R}$ satisfies $|\lambda - \lambda_0| < 1/ \|R_{\lambda_0}\|$ we have $\lambda \in \rho(A)$ and*

$$R_\lambda = \sum_{n \geq 0} (\lambda - \lambda_0)^n (\lambda_0 I - A)^{-(n+1)}.$$

Proof Again, given $\lambda_0 \in \rho(A)$, for λ satisfying $|\lambda - \lambda_0| < 1/ \|R_\lambda\|$ we have

$$\lambda I - A = \lambda_0 I - A + (\lambda - \lambda_0)I = [I - (\lambda_0 - \lambda)R_{\lambda_0}](\lambda_0 I - A)$$

and we see that $\lambda I - A$ is invertible. Then the inverse is given as above by

$$R_\lambda = R_{\lambda_0}[I - (\lambda_0 - \lambda)R_{\lambda_0}]^{-1} = \sum_{n=0}^{\infty} (\lambda - \lambda_0)^n R_{\lambda_0}^{n+1}.$$

□

Proposition 3.38 *If R_λ is the resolvent of a closed linear operator A then one has $AR_\lambda = \lambda R_\lambda - I$ and*

(a) For $x \in D(A)$ and $\lambda \in \rho(A)$ one has $R_\lambda x \in D(A)$ and $R_\lambda Ax = AR_\lambda x$.
(b) For $\lambda, \mu \in \rho(A)$ one has $R_\lambda - R_\mu = (\mu - \lambda)R_\lambda R_\mu$ and $R_\mu R_\lambda = R_\lambda R_\mu$.
(c) The map $\lambda \mapsto R_\lambda$ is infinitely differentiable (in the usual sense that for all $n \in \mathbb{N}\backslash\{0\}$, the n-th derivative $R_\lambda^{(n)} := \lim_{\nu(\nu \neq 0) \to 0}(1/\nu)(R_{\lambda+\nu}^{(n-1)} - R_\lambda^{(n-1)})$ exists) and

$$R_\lambda^{(n)} = (-1)^n n! R_\lambda^{n+1}.$$

Proof For all $x \in X$, $\lambda \in \rho(A)$, setting $w := R_\lambda x$, one has $w \in D(A)$, $\lambda w - Aw = x$, hence $AR_\lambda x = \lambda R_\lambda x - x$.

(a) If $x \in D(A)$ and $\lambda \in \rho(A)$, for $w := R_\lambda x$ one has $Ax = A(\lambda w - Aw) = (\lambda I - A)(Aw)$, hence $R_\lambda Ax = Aw = AR_\lambda x$.
(b) Given $x \in X$ and $\lambda,\ \mu \in \rho(A)$, setting $w := R_\lambda x$, $z := R_\mu x$, we have $(\lambda I - A)(w-z) = x-(\mu I-A)(z)+(\mu-\lambda)z = (\mu-\lambda)z$, so that $w-z = (\mu-\lambda)R_\lambda z = (\mu-\lambda)R_\lambda R_\mu x$. The relation $R_\mu R_\lambda = R_\lambda R_\mu$ ensues.
(c) The map $\lambda \mapsto R_\lambda$ is differentiable: taking $\mu := \lambda + \nu$, and passing to the limit in the relation $\nu^{-1}(R_{\lambda+\nu} - R_\lambda) = -R_\lambda R_{\lambda+\nu}$, we get $R_\lambda^{(1)} = -R_\lambda^2$. An induction on n gives the relation $R_\lambda^{(n)} = (-1)^n n! R_\lambda^{n+1}$. It can also be obtained from the properties of power series.

□

Exercises

1. Prove that for any Banach space X and any $A \in L(X) := L(X,X)$ one has $\sigma(A^\intercal) = \sigma(A)$. Give examples showing that there is no general inclusion between $\sigma_e(A^\intercal)$ and $\sigma_e(A)$. [Hint: use the right shift and the left shift on a sequence space.]

2*. Let $T := [0,1]$, $X := L_p(T)$, with $p \in [1,\infty[$ and let $A \in L(X)$ be given by $(Ax)(t) := \int_0^t x(s)ds$. Determine $\sigma(A)$ and $\sigma_e(A)$. For $\lambda \in \rho(A)$, give an explicit expression of $(A - \lambda I)^{-1}$. Determine $A^\intercal$. [Hint: see Chap. 8 for the definition of $L_p(T)$.]

3. Let X be a Banach space and let $A \in L(X)$. Verify that for all $\lambda \in \rho(A)$ one has $(A-\lambda I)^{-1}A = A(A-\lambda I)^{-1}$ and $d(\lambda, \sigma(A)) \geq 1/\left\|(A-\lambda I)^{-1}\right\|$. [Hint: use the fact that all $\mu \in \mathbb{R}$ satisfying $|\lambda - \mu| . \left\|(A-\lambda I)^{-1}\right\| < 1$ are in $\rho(A)$]. Show that for all $\lambda \in \mathbb{R}$ satisfying $|\lambda| > \|A\|$ one has $\left\|I + \lambda(A-\lambda I)^{-1}\right\| \leq \|A\| / (|\lambda| - \|A\|)$. Prove that if $0 \in \rho(A)$ then one has $\sigma(A^{-1}) = 1/\sigma(A)$.

4. Let X be a Banach space and let $A \in L(X) := L(X,X)$. Show that for all $\lambda, \mu \in \rho(A)$ one has $(A-\lambda I)^{-1} - (A-\mu I)^{-1} = (\mu-\lambda)(A-\lambda I)^{-1}(A-\mu I)^{-1}$.

5. Let X be a Banach space and let $A \in L(X)$. Show that $(\|A^n\|^{1/n})$ converges to a limit denoted by $r_\sigma(A)$ and called the *spectral radius* of A. [Hint: show that $\limsup_n \|A^n\|^{1/n} \leq r := \inf_{n\geq 1} \|A^n\|^{1/n}$ by taking for all $s > r$ some $k \in \mathbb{N}$ such that $\left\|A^k\right\|^{1/k} < s$ and by noting that for $n = kq + p$ with $p, q \in \mathbb{N}, p < k$ one has $\|A^n\|^{1/n} \leq s^{kq/n} \|A\|^{p/n}$.] Verify that $r_\sigma(A) \leq \|A\|$ but for $A : \mathbb{R}^2 \to \mathbb{R}^2$ given by $A(x_1, x_2) = (0, x_1)$ one has $r_\sigma(A) = 0$, $\|A\| = 1$. Prove that $\lambda \in \rho(A)$ whenever λ satisfies $|\lambda| > r_\sigma(A)$ and that $(A - \lambda I)^{-1} = \Sigma_{n\geq 0}\lambda^{-n-1}A^{-n}$. Verify that for $A : \mathbb{R}^3 \to \mathbb{R}^3$ given by $A(x_1,x_2,x_3) = (-x_2, x_1, 0)$ one has $\sigma(A) = \{0\}$ but $r_\sigma(A) = 1$, as $A^3 = -A$. If X is a complex space one has $r_\sigma(A) = \max\{|\lambda| : \lambda \in \sigma(A)\}$.

6. Let X be a Banach space and let $A \in L(X)$ be such that $\{A^n : n \in \mathbb{N}\}$ is bounded. Show that $\mathrm{cl}(R(I-A)) = \{x \in X : (C_n x) \to 0\}$ for $C_n = (1/n)(A + A^2 + \ldots + A^n)$ (Cesàro's sum). Deduce from this characterization the relation $N(I-A) \cap \mathrm{cl}(R(I-A)) = \{0\}$.

Suppose that for some $x \in X$ and some infinite subset N of $\mathbb{N}$ $(C_n x)_{n \in N}$ weakly converges to some $\overline{x}$. Show that $A\overline{x} = \overline{x}$ and $(C_n x)_{n \geq 0} \to \overline{x}$ (*the mean ergodic theorem*). When X is reflexive, infer from the preceding that there exists some $M \in L(X)$ such that $(C_n x) \to Mx$ for all $x \in X$ and that $M = M^2 = A \circ M = M \circ A$, with $R(M) = N(I - A), N(M) = R(I - A) = \mathrm{cl}(R(I - A))$. [Hint: see [262].]

7*. Show that the Laplace operator Δ on $X := L_2(\mathbb{R}^d)$ with domain $H^2(\mathbb{R}^d)$ is self-adjoint and its spectrum is $\mathbb{R}_+$. [Hint: use the characterization of $H^2(\mathbb{R}^d)$ in terms of the Fourier transform; see [67, p. 266].]

3.5.4 Compact Operators

In this subsection X and Y are two Banach spaces. A continuous linear operator $A \in L(X, Y)$ is said to be *compact* if for any bounded subset B of X its image $A(B)$ is relatively compact, i.e. cl $(A(B))$ is compact. Such operators keep some familar properties of operators between finite dimensional normed spaces.

Example If $A \in L(X, Y)$ has finite rank, i.e. if the range $R(A)$ of A is finite dimensional, then A belongs to the set $K(X, Y)$ of compact (linear) operators. The converse is not true. It was proved by P. Enflo in 1972 that it is not even true that any element of $K(X, Y)$ can be approximated (for the norm topology of $L(X, Y)$) by a sequence of finite rank operators.

Exercise Prove that the set $K(X, Y)$ of compact operators from X into Y is closed in $L(X, Y)$. [Hint: if $A = \lim_n A_n$ with $A_n \in K(X, Y)$ for all n, prove that $A(B_X)$ is precompact.]

Exercise Prove that the set $K(X, Y)$ is a linear space and that for any Banach spaces W, Z, any $A \in L(W, X)$, $B \in K(X, Y)$, $C \in L(Y, Z)$ one has $B \circ A \in K(W, Y)$, $C \circ B \in K(X, Z)$.

Proposition 3.39 *For all $A \in L(X, Y)$ one has $A^{\intercal} \in K(Y^*, X^*)$ if and only if $A \in K(X, Y)$.*

Proof Given $A \in K(X, Y)$ let us show that $A^{\intercal}(B_{Y^*})$ is relatively compact in $(X^*, \|\cdot\|_*)$. Let $K := \mathrm{cl}(A(B_X))$ and let $F := A^{\intercal}(B_{Y^*}) \mid_K \subset C(K)$. Then K is compact, F is equicontinuous since for all $y^* \in B_{Y^*}, f := A^{\intercal}(y^*) \mid_K \in F$ and for y, $y' \in K$ one has $|f(y) - f(y')| = |y^*(Ay) - y^*(Ay')| \leq \|A\| . \|y - y'\|$. Moreover, for all $y \in K$ the set $F(y) = \{(y^* \circ A)(y) : y^* \in B_{Y^*}\}$ is bounded in $\mathbb{R}$. The Ascoli-Arzela theorem (Theorem 3.3) ensures that F is relatively compact in $C(K)$. Thus, any sequence (f_n) in F has a subsequence $(f_{k(n)})$ that converges to some $f \in C(K)$. Picking $y_n^* \in B_{Y^*}$ such that $f_n = A^{\intercal}(y_n^*) \mid_K$ we see that

$$\sup_{x \in B_X} \left| y_{k(m)}^*(Ax) - y_{k(n)}^*(Ax) \right| = \sup_{y \in K} \left| f_{k(n)}(y) - f_{k(m)}(y) \right| \to 0 \text{ as } m, n \to \infty$$

so that $(x^*_{k(n)}) := (A^\intercal(y^*_{k(n)}))$ is a Cauchy sequence in X^*, hence is convergent: $\mathrm{cl}(A^\intercal(B_{Y^*}))$ is compact in X^*.

Conversely, let $A \in L(X,Y)$ be such that $A^\intercal \in K(Y^*, X^*)$. Then, by the preceding, $A^{\intercal\intercal} \in K(X^{**}, Y^{**})$. Thus $\mathrm{cl}(A^{\intercal\intercal}(B_X)) \subset \mathrm{cl}(A^{\intercal\intercal}(B_{X^{**}}))$ is compact in Y^{**}. But since $A(B_X) = A^{\intercal\intercal}(B_X)$ and since Y is closed in Y^{**}, we get that $\mathrm{cl}(A(B_X))$ is compact in Y. □

The next result is of practical interest.

Theorem 3.30 (Fredholm Alternative) *For any $A \in K(X) := K(X,X)$ the spaces $R(I-A)$ and $R(I-A^\intercal)$ are closed and in fact*

$$R(I-A) = (N(I-A^\intercal))^\perp, \quad R(I-A^\intercal) = (N(I-A))^\perp.$$

Moreover, $N(I-A)$ and $N(I-A^\intercal)$ are finite dimensional with the same dimension. In particular, $N(I-A) = \{0\}$ if and only if $R(I-A) = X$.

This result entails a solvability criterion for the equation $x - Ax = y$: according to the two cases $\dim N(I-A^\intercal) = 0$ or $n := \dim N(I-A^\intercal) > 0$:

either for every $y \in X$ the equation $x - Ax = y$ has a unique solution

or the homogeneous equation $x - Ax = 0$ has n linearly independent solutions and the inhomogeneous equation $x - Ax = y$ is solvable if and only if y satisfies n orthogonality relations $\langle y, y_i^* \rangle = 0$, $i \in \mathbb{N}_n$, where $(y_i^*)_{i \in \mathbb{N}_n}$ is a base of $N(I-A^\intercal)$.

Proof Let $Z := N(I-A)$. Since $B_Z \subset A(B_X)$, B_Z is compact, hence Z is finite dimensional by Theorem 3.1. Since $A^\intercal$ is compact, $N(I-A^\intercal)$ is also finite dimensional.

Let us show that $R(I-A)$ is closed. By the Closed Range Theorem, this will imply that $R(I-A) = (N(I-A^\intercal))^\perp$. Given $y = \lim y_n$ with $y_n := x_n - Ax_n \in R(I-A)$ for all n, we have to show that $y \in R(I-A)$. Since $N(I-A)$ is finite dimensional, we can find $w_n \in N(I-A)$ such that

$$\|x_n - w_n\| = r_n := d(x_n, N(I-A)).$$

Let us first prove that the sequence (r_n) is bounded. Otherwise, taking a subsequence if necessary, we may assume $(r_n) \to \infty$. Setting $u_n := r_n^{-1}(x_n - w_n) \in B_X$ and taking another subsequence and relabelling it, we may assume $(A(u_n))$ converges to some $z \in \mathrm{cl}\,(A(B_X))$. Then, observing that $A(w_n) = w_n$, we see that

$$y_n = (x_n - w_n) - A(x_n - w_n) \tag{3.26}$$

and $(r_n^{-1} y_n) \to 0$. Thus $(u_n) \to z$ and $z = Az$. Therefore $z \in N(I-A)$ and $d(u_n, N(I-A)) \le \|u_n - z\| \to 0$, contradicting the fact that $d(u_n, N(I-A)) = r_n^{-1} d(r_n u_n, N(I-A)) = r_n^{-1}\,\|w_n - x_n\| = 1$.

Thus $(x_n - w_n)$ is bounded and since A is a compact operator, taking a subsequence if necessary, we may assume that $(A(x_n - w_n))$ has a limit v. Then, by (3.26), $(x_n - w_n) \to y + v$. Setting $x := y + v$ and passing to the limit in (3.26), we get $y = x - Ax \in R(I - A)$: this space is closed. Applying Theorem 3.26, we get that

$$R(I - A) = (N(I - A^{\mathsf{T}}))^{\perp}, \qquad R(I - A^{\mathsf{T}}) = (N(I - A))^{\perp}.$$

Let us prove the last assertion. We first show that assuming $N(I - A) = \{0\}$ and $R(I - A) \neq X$ leads to a contradiction. Setting $X_n := (I - A)^n(X)$ we see that X_n is closed and that the sequence (X_n) is strictly decreasing since taking $\overline{x} \in X \backslash R(I - A)$ we have $(I - A)^n(\overline{x}) \in X_n \backslash X_{n+1}$: if we could find $\overline{w} \in X$ such that $(I - A)^n(\overline{x}) = (I - A)^{n+1}(\overline{w})$, since $(I - A)^n$ is injective, we would have $\overline{x} = (I - A)(\overline{w})$ contradicting the fact that $\overline{x} \notin (I - A)(X)$. Since X_{n+1} is closed, Proposition 3.1 yields some $u_n \in X_n$ such that $\|u_n\| = 1$ and $d(u_n, X_{n+1}) \geq 1/2$. Given $k < n$ in $\mathbb{N}$ and observing that

$$Au_n - Au_k = v - u_k \text{ with } v := u_n - (I - A)(u_n) + (I - A)(u_k) \in X_{k+1}$$

we get $\|Au_n - Au_k\| \geq d(u_k, X_{k+1}) \geq 1/2$, contradicting the compactness of $A(B_X)$. Thus $R(I - A) = X$ when $N(I - A) = \{0\}$.

Conversely, suppose that $R(I - A) = X$. Then by relation (3.15) or Corollary 3.32 we have $N(I - A^{\mathsf{T}}) = (R(I - A))^{\perp} = \{0\}$. Since $A^{\mathsf{T}} \in K(X^*)$, the preceding step ensures that $R(I - A^{\mathsf{T}}) = X^*$. Using Corollary 3.32 once more, we get that $N(I - A) = (R(I - A^{\mathsf{T}}))^{\perp} = \{0\}$.

It remains to show that, for any $A \in K(X)$, $d := \dim N(I - A)$ is equal to $d' := \dim N(I - A^{\mathsf{T}})$. Suppose $d < d'$. Since $N(I - A)$ is finite dimensional there exists a continuous projector $P : X \to N(I - A)$ with $R(P) = N(I - A)$. Since $R(I - A) = (N(I - A^{\mathsf{T}}))^{\perp}$ has finite codimension d', it has a complement Y with dimension d'. Since we assume that $d < d'$, we can find some $C \in L(N(I - A), Y)$ that is injective but not surjective. Let $\overline{y} \in Y \backslash R(C)$ and let $B : X \to X$ be defined by $B := A + C \circ P$. Since $C \circ P$ has finite rank, we have $B \in K(X)$. Let us show that $N(I - B) = \{0\}$. Given $x \in N(I - B)$ we have

$$0 = x - Bx = (x - Ax) - (C \circ P)(x) \in R(I - A) \oplus Y,$$

so that $x - Ax = 0$ and $(C \circ P)(x) = 0$, hence $x \in N(I - A)$ and $Px = 0$. Thus $x = Px = 0$ and $N(I - B) = \{0\}$. By the preceding step we obtain that $R(I - B) = X$. However, if $\overline{x} \in (I - B)^{-1}(\overline{y})$ we have $\overline{y} = (\overline{x} - A\overline{x}) - C(P\overline{x}) \in R(I - A) \oplus Y$, hence $\overline{y} = -C(P\overline{x})$, contradicting $\overline{y} \in Y \backslash R(C)$. Thus $d' \leq d$.

Applying the same result to the compact operator A^{T}, we get

$$\dim N(I_{X^{**}} - A^{\mathsf{TT}}) \leq \dim N(I_{X^*} - A^{\mathsf{T}}) \leq \dim N(I_X - A).$$

But since A^{TT} is an extension of A, we have $N(I_X - A) \subset N(I_{X^{**}} - A^{\mathsf{TT}})$, hence equality. Thus $d' = d$. □

Corollary 3.33 *For every eigenvalue* $\lambda \neq 0$ *of a compact operator* A, λ *is an eigenvalue of* A^{T} *with the same multiplicity, i.e. the dimensions of the eigenspaces are the same.*

Proof It suffices to apply the second assertion of the theorem to $\lambda^{-1}A$. □

Now let us study the spectra of compact operators. A first result is as follows.

Proposition 3.40 *The set* $\sigma_e(A)\backslash\{0\}$ *of non-null eigenvalues of* $A \in K(X)$ *is formed of isolated points: if* $(\lambda_n) \to \lambda$ *with* $\lambda_n \in \sigma_e(A)\backslash\{0\}$ *for all* n *and* $\lambda_m \neq \lambda_n$ *for* $m \neq n$, *then one has* $\lambda = 0$.

Proof Let $(\lambda_n) \to \lambda$ with $\lambda_n \in \sigma_e(A)\backslash\{0\}$ for all n and $\lambda_m \neq \lambda_n$ for $m \neq n$. Let $e_n \in X\backslash\{0\}$ be such that $Ae_n = \lambda_n e_n$ and let E_n be the space spanned by $\{e_1, \ldots, e_n\}$. We know that E_n has dimension n (Proposition 3.36), so that $E_{n-1} \neq E_n$. For all $n > 1$ Riesz's lemma (Proposition 3.1) provides some $u_n \in E_n$ such that $\|u_n\| = 1$ and $d(u_n, E_{n-1}) > \frac{1}{2}$. Since $(A - \lambda_n I)(E_n) \subset E_{n-1}$, for $k < n$ we have

$$\begin{aligned}\left\|\lambda_k^{-1}Au_k - \lambda_n^{-1}Au_n\right\| &= \left\|\lambda_k^{-1}(Au_k - \lambda_k u_k) - \lambda_n^{-1}(Au_n - \lambda_n u_n) + u_k - u_n\right\| \\ &\geq d(u_n, E_{n-1}) \geq \frac{1}{2}.\end{aligned}$$

Since (Au_n) has a convergent subsequence and since $(\lambda_n) \to \lambda \neq 0$, we get a contradiction. □

Theorem 3.31 *If* $A \in K(X)$ *then* $\sigma_e(A)\backslash\{0\} = \sigma(A)\backslash\{0\}$. *If* X *is infinite dimensional one has* $0 \in \sigma(A)$ *and either* $\sigma(A)$ *is a finite set or there exists a sequence* (λ_n) *with limit* 0 *such that* $\sigma(A)\backslash\{0\} := \{\lambda_n\}$.

Proof If $\lambda \in \mathbb{R}\backslash\sigma_e(A)$ and $\lambda \neq 0$, we have $N(\lambda^{-1}A - I) = N(A - \lambda I) = \{0\}$. Then, by Theorem 3.30, we have $R(A - \lambda I) = X$. By the Banach isomorphism theorem, $A - \lambda I$ is an isomorphism: $\lambda \in \rho(A) = \mathbb{R}\backslash\sigma(A)$.

If $0 \notin \sigma(A)$, A is invertible and $I = A \circ A^{-1}$ is compact. Then B_X is compact, hence X is finite dimensional. Thus $0 \in \sigma(A)$ if X is infinite dimensional.

Suppose $\sigma(A)$ is an infinite set. For every $\varepsilon > 0$ the set $\{\lambda \in \sigma(A) : |\lambda| \geq \varepsilon\}$ must be finite since it is compact by Proposition 3.35 and formed of isolated points by the preceding proposition and the relation $\sigma_e(A)\backslash\{0\} = \sigma(A)\backslash\{0\}$. Thus $\sigma(A)\backslash\{0\}$ is countable and ordering $\sigma(A)\backslash\{0\} := \{\lambda_n\}$ in such a way that $(|\lambda_n|)$ is decreasing, we get a sequence with limit 0. □

Exercises

1. Given any sequence $(\lambda_n) \to 0$ in $\mathbb{R}$, show that there is a compact operator A in $X := \ell_\infty$ with $\sigma(A) = \{\lambda_n : n \in \mathbb{N}\}$. [Hint: take $A(x) = (\lambda_0 x_0, \lambda_1 x_1, \ldots)$ for

$x = (x_0, x_1, \ldots)$ and note that A is the limit in $L(X)$ of a sequence in finite rank operators.]

2. Prove that any compact operator $A \in K(X, Y)$ is *completely continuous* in the sense that for any weakly convergent sequence (x_n) in X the sequence $(A(x_n))$ is strongly convergent in Y (and its limit is Ax if x is the weak limit of (x_n)).
3. Let X and Y be Banach spaces, X being reflexive, and let $A \in L(X, Y)$. Show that $A(B_X)$ is closed and that $A(B_X)$ is compact when A is compact. Find a compact operator A such that $A(B_X)$ is not compact. [Hint: take $X = Y = C(T)$ for $T := [0, 1]$ and set $(Ax)(t) := \int_0^t x(s)ds$.]
4. **(J.-L. Lions)** Let $(X, \|\cdot\|_X)$, $(Y, \|\cdot\|_Y)$ and $(Z, \|\cdot\|_Z)$ be three Banach spaces and let $A \in K(X, Y)$, $B \in L(Y, Z)$ with B injective. Prove that for every $\varepsilon > 0$ there exists some $c_\varepsilon > 0$ such that for all $x \in X$ one has $\|Ax\|_Y \leq \varepsilon \|x\|_X + c_\varepsilon \|B(Ax)\|_Z$. [Hint: suppose that for some $\overline{\varepsilon} > 0$ and some sequence (x_n) of S_X one has $\|Ax_n\|_Y \geq \overline{\varepsilon} + n \|B(Ax_n)\|_Z$ and obtain a contradiction with the injectivity of B.]

   Apply the preceding to show that for every $\varepsilon > 0$ there exists some $c_\varepsilon > 0$ such that for all $x \in C^1(T)$, where T is a compact interval, one has $\|x\|_\infty \leq \varepsilon \|x'\|_\infty + c_\varepsilon \|x\|_1$.

   Can one interchange the assumptions on A and B?
5. Let $(X, \|\cdot\|_X)$, $(Y, \|\cdot\|_Y)$ and $(Z, \|\cdot\|_Z)$ be three Banach spaces and let $A \in K(X, Y)$, $C \in L(X, Z)$ with C injective. Suppose X is reflexive. Prove that for every $\varepsilon > 0$ there exists some $c_\varepsilon > 0$ such that for all $x \in X$ one has $\|Ax\|_Y \leq \varepsilon \|x\|_X + c_\varepsilon \|Cx\|_Z$. [Hint: use an argument similar to the one in the preceding exercise.]
6. Let S be a compact space endowed with a regular Borel measure μ. Given a continuous function $G : S \times S \to \mathbb{R}$, show that the operator $A : C(S) \to C(S)$ given by $(Ax)(s) := \int_S G(s, t)x(t)d\mu(t)$ is compact, $C(S)$ being equipped with the norm $\|\cdot\|_\infty$.
7. Prove the same conclusion when A is considered as an operator on $L_2(S)$.
8. In contrast with what occurs for operators of the form $I - A$ with A compact, show that a continuous linear operator on an infinite dimensional Banach space may be injective and not surjective or surjective and not injective. [Hint: take the right shift and the left shift on a sequence space such as ℓ_2, as defined in Exercise 11 below.]
9. Given two Banach spaces X and Y, let $A \in L(X, Y)$ be a continuous linear operator such that $R(A)$ has finite codimension, i.e. $Y = R(A) + Z$ for a finite dimensional subspace Z of Y. Show that $R(A)$ is closed. [Hint: apply the Open Mapping Theorem to the map $B : X \times Z \to Y$ given by $B(x, z) = Ax + z$.]
10. Given two Banach spaces X and Y, a continuous linear operator $A \in L(X, Y)$ is said to be a *Fredholm operator* if $N(A)$ is finite dimensional and $R(A)$ has finite codimension. The *index* of A in the set $F(X, Y)$ of Fredholm operators from X into Y is defined by

$$\operatorname{ind} A := \dim N(A) - \operatorname{codim} R(A).$$

Note that for any $K \in K(X,X)$ one has $A := I_X - K \in F(X,X)$ and that when X and Y are finite dimensional $F(X,Y) = L(X,Y)$ and that for any $A \in L(X,Y)$ one has $\operatorname{ind} A = \dim X - \dim Y$.

Show that the set $F(X,Y)$ is an open subset of $L(X,Y)$ and that the map $\operatorname{ind} : F(X,Y) \to \mathbb{N}$ is continuous, hence is locally constant.

Prove that $A \in L(X,Y)$ belongs to $F(X,Y)$ if and only if there exists some $B \in L(Y,X)$ such that $A \circ B - I_Y$ and $B \circ A - I_X$ are compact operators if and only if there exists some $B \in L(Y,X)$ such that $A \circ B - I_Y$ and $B \circ A - I_X$ are finite rank operators.

Prove that $A \in L(X,Y)$ belongs to $F(X,Y)$ if and only if $A^{\top}$ belongs to $F(Y^*,X^*)$ and then $\operatorname{ind} A^{\top} = \operatorname{ind} A$.

Show that for $A \in F(X,Y)$ and $B \in K(X,Y)$ one has $A + B \in F(X,Y)$ and $\operatorname{ind}(A+B) = \operatorname{ind} A$.

Given another Banach space Z and $B \in F(Y,Z)$, $A \in F(X,Y)$ show that $B \circ A \in F(X,Z)$ and that $\operatorname{ind}(B \circ A) = \operatorname{ind} A + \operatorname{ind} B$.

11. Let $X := \ell_2$ be the space of square summable sequences with its usual norm. Let $S_r : X \to X$ be the right shift defined by $S_r x := y$ with $y_0 = 0$, $y_n := x_{n-1}$ for $x := (x_n)$, $y := (y_n)$ and let $S_\ell : X \to X$ be the left shift defined by $S_\ell x = z$ with $z_n = x_{n+1}$ for $x := (x_n)$. Show that for $c \in \mathbb{R}\backslash\{-1,1\}$ the maps $S_r - cI_X$ and $S_\ell - cI_X$ are Fredholm operators and compute their indexes. Note that this is not the case for $c \in \{-1,1\}$.

3.6 Elementary Integration Theory

In this section we deal with some classes of one-variable maps that are regular enough and for which an integration theory can be given in a simple way.

3.6.1 *Regulated Functions and Their Integrals*

We start with functions that have simple discontinuities.

Definition 3.4 A function $f : T \to X$ from a compact interval $T := [a,b]$ of $\mathbb{R}$ into a real Banach space X is said to be *regulated* if, for all $t \in [a,b[$ (resp. $t \in]a,b]$), f has a limit on the right $f(t_+) := \lim_{r \to t, r>t} f(r)$ (resp. on the left $f(t_-) := \lim_{s \to t, s<t} f(s)$). The function f is said to be a *normalized regulated function* if it is regulated, if $f(b) = f(b_-)$ and if for all $t \in [a,b[$ one has $f(t) = f(t_+)$.

Continuous functions, monotone functions and stair functions are regulated functions. Here we say that $f : T \to X$ is an *order step* function, or in short a *stair function*, if there is a finite sequence $\sigma := (s_0, s_1, \ldots, s_k)$ with $s_0 = a < s_1 < \ldots < s_k = b$ called a *subdivision* of T such that f is constant on each open interval $]s_{i-1}, s_i[$ for $i \in \mathbb{N}_k$. We use this terminology in order to distinguish this notion from a more general one we shall use later. The stair function f is said to be a *normalized stair function* if f is constant on $[s_{i-1}, s_i[$ for $i = 1, \ldots, k-1$ and on $[s_{k-1}, b]$.

Proposition 3.41 *A function $f : T \to X$ is regulated (resp. normalized regulated) if and only if it is the uniform limit of a sequence (f_n) of stair functions (resp. normalized stair functions).*

Proof Let $f : T \to X$ be regulated. Given a sequence $(\varepsilon_n) \to 0_+$, for all $t \in T$ and all $n \in \mathbb{N}$ we pick an open interval $T_n(t) :=]a_n(t), b_n(t)[$ containing t such that $\|f(r) - f(s)\| \leq \varepsilon_n$ if $r, s \in T_n^-(t) := [a_n(t), t[\cap T$ or if $r, s \in T_n^+(t) :=]t, b_n(t)] \cap T$. Let $t_1, \ldots, t_m$ be such that $\{T_n(t_i) : i \in \mathbb{N}_m\}$ is a covering of T. Order the set $\{a_n(t_i), t_i, b_n(t_i), a, b : i \in \mathbb{N}_m\} \cap T$ into a strictly increasing sequence $s_{n,0} = a < s_{n,1} < \ldots < s_{n,k} = b$. Then, for $j \in \mathbb{N}_k$, the interval $]s_{n,j-1}, s_{n,j}[$ is contained either in some $T_n^-(t_i)$ or in some $T_n^+(t_i)$ with $i \in \mathbb{N}_m$, so that $\|f(r) - f(s)\| \leq \varepsilon_n$ whenever $r, s \in]s_{n,j-1}, s_{n,j}[$. Setting $f_n(r) := f((s_{n,j-1} + s_{n,j})/2)$ for $r \in]s_{n,j-1}, s_{n,j}[$ and $f_n(s_{n,j}) = f(s_{n,j})$, we get a stair function f_n satisfying $\|f_n - f\|_\infty \leq \varepsilon_n$. If f is normalized we may change $T_n^+(t)$ into $[t, b_n(t)] \cap T$ and set $f_n(r) := f(s_{n,j-1})$ if $r \in [s_{n,j-1}, s_{n,j}[$ or if $r = b$ and $j = k$.

Conversely, suppose f is the uniform limit of a sequence (f_n) of stair functions. Given $\varepsilon > 0$ we pick $k \in \mathbb{N}$ such that $\|f_n(t) - f(t)\| \leq \varepsilon/2$ for all $n \geq k$ and $t \in T$. Given $\bar{t} \in T$ we can find $\delta > 0$ such that f_n is constant on $[\bar{t} - \delta, \bar{t}[\cap T$ and $]\bar{t}, \bar{t} + \delta] \cap T$. Thus for s, t in one of these two intervals we have $\|f(s) - f(t)\| \leq \varepsilon$. Since X is complete, this proves that the one-sided limits of f at $\bar{t}$ exist. Moreover, if f_n is normalized, f is normalized.

Proposition 3.42 *The space $R(T, X)$ (resp. $R_n(T, X)$) of regulated functions (resp. normalized regulated functions) from T to X endowed with the norm $\|\cdot\|_\infty$ is a Banach space.*

Proof This follows from the fact that $R(T, X)$ is the closure of the set $S(T, X)$ of order step functions in $(B(T, X), \|\cdot\|_\infty)$, hence is complete. □

It is worth noting the following facts.

Proposition 3.43 *For any regulated function $f : T \to X$, the set $f(T)$ is relatively compact in X (i.e. $\mathrm{cl}(f(T))$ is compact). Moreover, the set of discontinuities of f is at most countable.*

Proof If $f \in R(T, X)$ and if (f_n) is a sequence in $S(T, X)$ that converges uniformly to f we see that $f(T)$ is precompact since it can be approximated by $f_n(T)$, which is finite. The second assertion stems from the fact that the set D of discontinuities of f is the union of the sets $D_n := \{t \in T : \|f(t_+) - f(t_-)\| \geq 1/n\}$ for $n \in \mathbb{N}\backslash\{0\}$. □

The *integral* of a stair function f can be defined unambiguously as follows: if $s_0 = a < s_1 < \ldots < s_k = b$ is such that $f(t) = c_i$ for all $t \in]s_{i-1}, s_i[$ for $i = 1, \ldots k$, then

$$\int_T f := \int_a^b f(t)dt := \sum_{i=1}^k (s_i - s_{i-1})c_i.$$

It is easy to show that this element of X does not depend on the subdivision of T. Moreover, for any stair function f from T to X, the triangle inequality ensures that

$$\left\| \int_T f \right\| \leq (b-a) \|f\|_\infty . \tag{3.27}$$

Since the space $S(T,X)$ of stair functions is dense in the space $R(T,X)$, the map $f \mapsto \int_T f$ can be extended by continuity from $S(T,X)$ to $R(T,X)$:

$$\int_T f = \lim_n \int_T f_n \quad \text{if } f = \lim_n f_n.$$

Again, this map is linear, continuous and with norm $b-a$ as (3.27) remains valid for $f \in R(T,X)$. Moreover, given $a \leq b \leq c$ in $\mathbb{R}$, for all $f \in R([a,c],X)$ one has the *Chasles' relation*

$$\int_a^c f = \int_a^b f + \int_b^c f \tag{3.28}$$

which easily follows from the case of stair functions by a passage to the limit.

The following composition property is crucial: using continuous linear forms e^* on X, it enables us to determine the integral of a regulated function $f \in R(T,X)$ with the help of the integrals of the real-valued functions $e^* \circ f$ which determine $\int_T f$ uniquely.

Proposition 3.44 *Given Banach spaces X and Y and $A \in L(X,Y)$, for every $f \in R(T,X)$ one has $A \circ f \in R(T,Y)$ and $\int_T A \circ f = A(\int_T f)$.*

Proof The first assertion is a direct consequence in the definition. It can also be checked by taking a sequence (f_n) in $S(T,X)$ which converges uniformly to f. Since the relation $\int_T A \circ f_n = A(\int_T f_n)$ is immediate, the second assertion follows from the definition of the integral of $A \circ f$ since $(A(\int_T f_n)) \to A(\int_T f)$, A being continuous and $(\int_T f_n)$ converging to $\int_T f$. □

This result can be completed.

Proposition 3.45 *Let E, F be Banach spaces, let U be an open subset of $E \times T$ and let $W := \{f \in R(T,E) : \mathrm{cl}(\{(f(t),t) : t \in T\} \subset U\}$. Then W is an open subset of $R(T,E)$. Moreover, if $g : U \to F$ is a continuous map, then the map $g^\diamond : W \to R(T,F)$ defined by $g^\diamond(f)(t) := g(f(t),t)$ is continuous.*

We encourage the reader to first consider the simpler case of the subset V of W formed by those $f \in C(T,E)$ such that $(f(t),t) \in U$ for all $t \in T$, or even suppose $U := U_0 \times T$ where U_0 is an open subset of E.

Proof For any $f_0 \in W$ the map $t \mapsto (f_0(t),t)$ is regulated on T with values in $E \times \mathbb{R}$. Thus the closure $K := \mathrm{cl}(\{(f_0(t),t) : t \in T\})$ of its image is compact. Thus

$$\delta := \mathrm{gap}(K, (E \times \mathbb{R}) \backslash U) := \inf\{\|k - z\| : k \in K,\ z \in (E \times \mathbb{R}) \backslash U\}$$

is positive. Thus, given $\delta' \in]0, \delta[$ and $f \in R(T, E)$ satisfying $\|f - f_0\|_\infty \leq \delta'$ one has

$$\{(f(t), t) : t \in T\} \subset B[K, \delta'] := \{z \in E \times \mathbb{R} : d(z, K) \leq \delta'\} \subset U$$

and since $B[K, \delta']$ is closed one gets $\mathrm{cl}(\{(f(t), t) : t \in T\} \subset U$, i.e. $f \in W$. Thus W is open.

The definition of a regulated function shows that for any $f \in W$ the map $g^\diamond(f) : t \mapsto g(f(t), t)$ is in $R(T, F)$. Moreover, for $f_0 \in W$ and K as above, since g is uniformly continuous around K in the sense that for any $\varepsilon > 0$ one can find some $\delta > 0$ such that $\|g(z) - g(k)\| \leq \varepsilon$ whenever $k \in K$ and $z \in B[k, \delta]$. Thus, given $f \in W$ satisfying $\|f - f_0\| \leq \delta$, one has $\|g^\diamond(f) - g^\diamond(f_0)\| \leq \varepsilon$. This proves that $g^\diamond$ is continuous on W. □

Corollary 3.34 *Given Banach spaces E, F, G and continuous maps $A : T \to L(E, F)$, $B : T \to L^2(E, F; G)$ the maps $A^\diamond : R(T, E) \to R(T, F)$ and $B^\diamond : R(T, E) \times R(T, F) \to R(T, G)$ given by*

$$A^\diamond(f)(t) := A(t).f(t), \qquad B^\diamond(f, g)(t) := B(t)(f(t), g(t))$$

are linear continuous and bilinear continuous respectively. Moreover, $\|A^\diamond\| = \|A\|_\infty$ and $\|B^\diamond\| = \|B\|_\infty$. In fact, the conclusions are valid whenever A (resp. B) is regulated.

Proof The linearity of $A^\diamond$ and the bilinearity of $B^\diamond$ are obvious. The fact that $A^\diamond$ and $B^\diamond$ are well-defined whenever A and B are regulated stems from the definition of a regulated map and the inequalities

$$\left\|A^\diamond(f)\right\|_\infty \leq \sup_{t \in T} \|A(t)\| \,.\, \sup_{t \in T} \|f(t)\|$$

$$\left\|B^\diamond(f, g)\right\|_\infty \leq \sup_{t \in T} \|B(t)\| \,.\, \sup_{t \in T} \|f(t)\| \,.\, \sup_{t \in T} \|g(t)\|$$

for $f \in R(T, E)$, $g \in R(T, F)$ are obvious and prove that $A^\diamond$ and $B^\diamond$ are continuous with $\|A^\diamond\| \leq \|A\|_\infty$, $\|B^\diamond\| \leq \|B\|_\infty$. The reverse inequalities (which are not as important) follow from appropriate choices of f and g and are left as exercises. □

3.6.2 *Functions of Bounded Variation and Integration

Let us turn to another class of functions that enables us to speak of the length of a curve. Here T still stands for a compact interval $[a, b]$ but we can replace the range space X or E with a metric space. Given $f : T \to X$ and a subdivision $\sigma := (s_0, s_1, \ldots, s_k)$ of T, i.e. an increasing finite sequence in T such that $s_0 = a$, $s_k = b$, we set

$$V_\sigma(f) := \sum_{i=1}^{k} d(f(s_{i-1}), f(s_i)).$$

Definition 3.5 A map $f : T \to X$ is said to be a *function of bounded variation* or a function of finite variation if the supremum $V(f) := \sup_\sigma V_\sigma(f)$ over the set $\Sigma(T)$ of subdivisions of T is finite. The number $V(f)$ is called the *total variation* of f. The set of functions of bounded variation from T to X is denoted by $BV(T, X)$.

If f is Lipschitzian with rate c on T, then for every $\sigma \in \Sigma(T)$ one has $V_\sigma(f) \le c(b-a)$, so that f is a function of bounded variation.

If $X = \mathbb{R}$ and if f is nondecreasing, then for every $\sigma \in \Sigma(T)$ one has $V_\sigma(f) \le f(b) - f(a)$, so that f belongs to $BV(T, X)$. Of course, the same inclusion holds if f is nonincreasing.

Let us endow $\Sigma(T)$ with the order induced by inclusion of the sets of values. Then $\Sigma(T)$ is a directed set and since the triangle inequality shows that $V_\sigma(f) \le V_\tau(f)$ if $\sigma \preceq \tau$, i.e. if τ is a refinement of σ, we see that $V(f)$ is the limit of the net $(V_\sigma(f))_{\sigma \in \Sigma(T)}$. When f is continuous, $V(f)$ can be interpreted as the limit of $V_\sigma(f)$ as $\mu(\sigma) \to 0$, where $\mu : \Sigma(T) \to \mathbb{R}_+$ is the mesh defined by $\mu(\sigma) := \sup\{s_i - s_{i-1} : i \in \mathbb{N}_k\}$ if $\sigma := (s_0, \ldots s_k)$ with $s_0 = a < s_1 < \ldots < s_k = b$.

Let S be another compact interval of $\mathbb{R}$ and let $f : S \to X$, $g : T \to X$ be equivalent in the sense that there exists an increasing bijection $h : S \to T$ such that $f = g \circ h$. Then one has $V(f) = V(g)$: if $\sigma := (s_0, \ldots, s_k) \in \Sigma(S)$, then for $\tau := h(\sigma) := (h(s_0), \ldots, h(s_k)) \in \Sigma(T)$ one has $V_\sigma(f) = V_\tau(g) \le V(g)$, hence $V(f) \le V(g)$ and similarly $V(g) \le V(f)$. This remark allows us to define the *length of an arc* of X, an *arc* being defined as the equivalence class $\tilde{f}$ of an element f of $BV(T, X)$: its length being defined by setting $\ell(\tilde{f}) := V(f)$.

Let us focus on the case when X is a normed space, in particular on the case $X = \mathbb{R}$.

Proposition 3.46 *If X is a normed vector space, the set $BV(T, X)$ of functions of bounded variation from the interval T to X is a vector space and $V(\cdot)$ is a seminorm on $BV(T, X)$. Moreover, if X is a Banach space, $BV(T, X)$ is a subset of $R(T, X)$.*

Proof Let $f, g \in BV(T, X)$ and let $h := f + g$. For every $\sigma := (s_0, \ldots, s_k) \in \Sigma(T)$ and $i \in \mathbb{N}_k$ we have

$$\|h(s_i) - h(s_{i-1})\| \le \|f(s_i) - f(s_{i-1})\| + \|g(s_i) - g(s_{i-1})\|,$$

hence $V_\sigma(h) \le V_\sigma(f) + V_\sigma(g)$, and $V(h) \le V(f) + V(g)$. The relation $V(rf) = |r| V(f)$ for $r \in \mathbb{R}$ and $f \in BV(T, X)$ is obvious.

Given $f \in BV(T, X)$ and $t \in T := [a, b]$, $t > a$, let (t_n) be an increasing sequence in T with limit t. Then $(f(t_n))$ must be a Cauchy sequence. Otherwise we can find $\varepsilon > 0$ and a subsequence $(t_{k(n)})_n$ of (t_n) such that $\left\|f(t_{k(2n)}) - f(t_{k(2n+1)})\right\| \ge \varepsilon$. Then for $m \in \mathbb{N}\backslash\{0\}$, taking $\sigma := (a, t_{k(0)}, \ldots, t_{k(2m+1)}, b)$ we get $V_\sigma(f) \ge \Sigma_{n=0}^m \left\|f(t_{k(2n)}) - f(t_{k(2n+1)})\right\| \ge m\varepsilon$, contradicting the fact that $V(f) < \infty$. Now, if (t_n) and (t'_n) are increasing sequences with limit t, we can order the terms of $\{t_n : n \in \mathbb{N}\} \cup \{t'_n : n \in \mathbb{N}\}$ in such a way that (t_n) and (t'_n) are subsequences of an increasing sequence (t''_n). Thus, when X is complete, $\lim_n f(t_n) = \lim_n f(t'_n)$. We deduce from this observation that the left limit $f(t_{-})$ exists. Similarly, we can show that the right limit $f(t_{+})$ exists for all $t \in [a, b[$: f is regulated. □

Theorem 3.32 *The space $BV(T, \mathbb{R})$ coincides with the set of differences of two nondecreasing functions.*

Proof Since $BV(T, \mathbb{R})$ is a vector space and contains the cone of nondecreasing functions, every difference of two nondecreasing functions belongs to $BV(T, \mathbb{R})$. Conversely, let $f \in BV(T, \mathbb{R})$. For $t \in T$, let f_t be the restriction of f to $T_t := [a, t]$ and let $g(t) := V(f_t)$. For $r < s$ in T one has $g(r) \leq g(s)$ and in fact $g(r) + |f(r) - f(s)| \leq g(s)$. Thus, for $h := f + g$ we get

$$h(r) - h(s) = g(r) - g(s) + f(r) - f(s) \leq - |f(r) - f(s)| + f(r) - f(s) \leq 0.$$

Thus h is nondecreasing and $f = h - g$. □

The integration of regulated functions presented in a previous subsection is a simplified approach to Riemann integration. In turn, the latter theory is a special case of the *Riemann-Stieltjes integration* theory as we intend to show. It is obtained by taking for integrator function g the identity map I_T given by $I_T(t) := t$. Let g be a given element of the space $BV_n(T, \mathbb{R})$ of normalized functions of bounded variation on T; we do not mention it in our notation as it is fixed. Let us denote by $\Sigma'(T)$ the set of pairs (σ, ρ) such that $\sigma := (s_0, \ldots, s_k) \in \Sigma(T)$, $\rho := (r_0, \ldots, r_{k-1})$ satisfy $s_{i-1} < r_{i-1} < s_i$ for $i \in \mathbb{N}_k$. We define a directed preorder on $\Sigma'(T)$ by setting $(\sigma, \rho) \preceq (\sigma', \rho')$ if $\sigma \preceq \sigma'$ (i.e. if σ' is obtained from σ by adding some points). For $f : T \to X$ let us say that f is integrable with respect to g if the net $(S_{\sigma,\rho}(f))_{(\sigma,\rho)\in\Sigma'(T)}$ defined as follows converges: for $(\sigma, \rho) \in \Sigma'(T)$ with $\sigma := (s_0, \ldots, s_k)$, $\rho := (r_0, \ldots, r_{k-1})$

$$S_{\sigma,\rho}(f) := \sum_{i=1}^{k} f(r_{i-1})(g(s_i) - g(s_{i-1})).$$

Then the limit is denoted by $\int f dg$ and is called the *Riemann-Stieltjes integral of* f with respect to g and, if $g = I_T$, the *Riemann integral* of f. This integrability requirement may seem stringent; however it is satisfied for continuous maps and even for regulated maps as we intend to show. Let f be an element of the set $S(T, X)$ of stair functions: for some $\overline{\sigma} = (\overline{s}_0, \ldots, \overline{s}_k) \in \Sigma(T)$, f is constant with value e_i on $]\overline{s}_{i-1}, \overline{s}_i[$. Then, for any $(\sigma, \rho) \in \Sigma'(T)$ such that $\overline{\sigma} \preceq \sigma$, gathering terms, we see that

$$S_{\sigma,\rho}(f) = \sum_{i=1}^{k} (g(\overline{s}_i) - g(\overline{s}_{i-1}))e_i,$$

so that $(S_{\sigma,\rho}(f))_{(\sigma,\rho)\in\Sigma'(T)}$ converges to this sum. Moreover, in such a case we have

$$\left\| \int fdg \right\| \leq \max_{1\leq i\leq k} \|e_i\| \sum_{i=1}^{k} |g(\overline{s}_i) - g(\overline{s}_{i-1})| \leq \|f\|_\infty V(g).$$

Now, given $f \in R_n(T,X)$ and a sequence (f_n) of $S_n(T,X)$ converging uniformly to f, the preceding inequality shows that for $n, p \in \mathbb{N}$ we have

$$\left\| \int f_n dg - \int f_p dg \right\| \leq \|f_n - f_p\|_\infty V(g),$$

so that $(\int f_n dg)_n$ is a Cauchy sequence in X. We denote by $\int f dg$ its limit. Since two Cauchy sequences appear as subsequences of the Cauchy sequence obtained by alternating terms, the limit is independent of the choice of (f_n). Moreover, the map $f \to \int f dg$ is linear and continuous from $R(T,X)$ into X and it satisfies

$$\left\| \int f dg \right\| \leq \|f\|_\infty V(g).$$

Remark If $f \in R_n(T,X)$ i.e. if $f \in R(T,X)$ is normalized, the preceding construction can be simplified: one takes a sequence (f_n) of normalized stair functions converging to f and replaces pairs $(\sigma, \rho) \in \Sigma'(T)$ with pairs $(\sigma, \rho(\sigma))$ where for $\sigma := (s_0, \ldots, s_k) \in \Sigma(S)$ $\rho(\sigma) := (s_0, \ldots, s_{k-1})$. However, the Riemann-Stieltjes construction addresses to a more general class of functions than $R_n(T,X)$. □

For $X := \mathbb{R}, f : T \to \mathbb{R}$ and g nondecreasing, it is usual to use the integrability criterion that the *Darboux's sums*

$$\overline{S}_\sigma(f) := \sum_{i=1}^{k} \overline{c}_{i-1}(g(s_i) - g(s_{i-1})), \qquad \underline{S}_\sigma(f) := \sum_{i=1}^{k} \underline{c}_{i-1}(g(s_i) - g(s_{i-1})),$$

where $\overline{c}_i := \sup\{f(t) : t \in]s_i, s_{i+1}[\}$ and $\underline{c}_i := \inf\{f(t) : t \in]s_i, s_{i+1}[\}$ are such that $(\overline{S}_\sigma(f) - \underline{S}_\sigma(f))_\sigma \to 0$. If f is Riemann-Stieltjes integrable this criterion is satisfied since for any subdivision σ the numbers $\overline{c_i}$ and $\underline{c}_i$ are arbitrarily close to actual values of f on the interval $]s_i, s_{i+1}[$. Conversely, if f is bounded the nets $(\overline{S}_\sigma(f))_\sigma$ and $\underline{S}_\sigma(f))_\sigma$ belong to a compact interval of $\mathbb{R}$ and if $\tau \in \Sigma(T)$ is a refinement of $\sigma \in \Sigma(T)$ in using refinements that consist in adding a single point and in making successive steps, one can show that

$$\underline{S}_\sigma(f) \leq \underline{S}_\tau(f) \leq \overline{S}_\tau(f) \leq \overline{S}_\sigma(f).$$

Theorem 1.6 shows that the increasing net $(\underline{S}_\sigma(f))_\sigma$ converges and the decreasing net $(\overline{S}_\sigma(f))_\sigma$ converges. Their limits are the same since $(\overline{S}_\sigma(f) - \underline{S}_\sigma(f))_\sigma \to 0$. Moreover, since for any $(\sigma, \rho) \in \Sigma'(T)$ one has

$$\underline{S}_\sigma(f) \leq S_{\sigma,\rho}(f) \leq \overline{S}_\sigma(f)$$

the net $(S_{\sigma,\rho}(f))_{(\sigma,\rho)\in\Sigma'(T)}$ converges.

We admit the following characterization (see [132, Thm 6.16] for instance).

Theorem 3.33 *A bounded function $f : T \to \mathbb{R}$ on a compact interval T is Riemann integrable if and only if there exists a subset N of T of Lebesgue measure 0 such that f is continuous at each point of $T \backslash N$.*

As above, one can show that given Banach spaces X and Y and $A \in L(X, Y)$, for every $f \in R(T, X)$ (or even for every Riemann integrable f) one has $\int (A \circ f) dg = A(\int_T f dg)$. Moreover, one can devise an analogue of the Chasles' relation (3.28).

Because the Riemann integral has poor properties in terms of convergence, we shall not continue its study. We just quote a comparison result with the Lebesgue integral to which we shall devote more attention, referring to [132, Thm 6.15] or [240, Thm 1.5 p. 57] for a proof.

Theorem 3.34 *If a bounded function $f : T \to \mathbb{R}$ on a compact interval T is Riemann integrable then it is Lebesgue integrable and its Riemann integral coincides with its Lebesgue integral.*

Exercises

1. Verify that the function f defined by $f(0) := 0, f(x) := x^2 \sin(1/x^2)$ for $x \in \mathbb{R} \setminus \{0\}$ is not of bounded variation on $T := [0, 1]$ although it has a derivative at each point of T.
2. Given $a < b < c$ in $\mathbb{R}$ and $v \in BV([a, c], X)$, show that $V_a^c(v) = V_a^b(v) + V_b^c(v)$ and that $s \mapsto V_a^s(v)$, the variation of v on the interval $[a, s]$, is a nondecreasing function.
3. Verify that the function $f \mapsto \|f(a)\| + V(f)$ is a norm on the space $BV(T, X)$, where $T := [a, b]$ and X is a normed space.
4. **(A generalization of the Stieltjes integral)** Given Banach spaces X, Y, Z, a continuous bilinear map $(x, y) \mapsto x * y$ from $X \times Y$ to Z, a function $v \in BV(T, Y)$ for $T := [a, b]$ and a (right-) normalized stair function f from T to X, given $(\sigma, \rho) \in \Sigma'(T)$ with $\sigma := (s_0, \ldots, s_k) \in \Sigma(T)$, $\rho := (r_0, \ldots, r_{k-1})$, set

$$S_{\sigma,\rho}(f) := \sum_{i=1}^{k} f(r_{i-1}) * (g(s_i) - g(s_{i-1})).$$

   (a) Show that the net $(S_{\sigma,\rho}(f))_{(\sigma,\rho)\in\Sigma'(T)}$ converges. [Hint: consider first the case when f is the step function $u_s e$ for $s \in]a, b[$, $e \in X$ and use linearity.] Its limit is denoted by $\int_T f * dg$.
   (b) Show that $\int_T f * dg$ does not depend on the decomposition of f. Verify that $\left\| \int_T f * dg \right\| \leq V(g) \|f\|_\infty$.
   (c) Deduce from this inequality that the map $f \mapsto \int_T f * dg$ can be extended to a continuous linear map from the space $R_n(T, X)$ of normalized regulated functions with values in X into Z satisfying the same inequality.

(**d**) Conversely, given a continuous linear form f^* on the space $R_n(T) := R_n(T, \mathbb{R})$, for $s \in T$, let $g(s) := f^*(u_s)$, where u_s is defined as above. Show that g is of bounded variation on T and that $V_a(g) \leq \|f^*\|$.

(**e**) Deduce from the preceding a correspondence between the (topological) dual of the space $R_n(T)$ and the space $BV(T)$. [See [189].]

5. **Integration by parts** Prove the following equality for $f, g \in BV(T, \mathbb{R})$, $T := [a, b]$:

$$\int_a^b f dg = f(b)g(b) - f(a)g(a) - \int_a^b g df.$$

6. Let f be the function on $[0, 1]$ given by $f(0) = 0, f(x) := \sin(1/x)$ for $x \in]0, 1]$. Prove that f is not regulated but is Riemann integrable.
7. Verify that the function f on $T := [0, 1]$ given by $f(x) = 0$ if $x \in T \setminus \mathbb{Q}$, $f(x) := 1/q$ for $x \in T \cap \mathbb{Q}$ and $x := p/q$ where $p, q \in \mathbb{N}$ have no common factor. Prove that f is regulated.
8. Show that the function f on $T := [0, 1]$ given by $f(x) = 0$ if $x \in T \setminus \mathbb{Q}, f(x) := 1$ for $x \in T \cap \mathbb{Q}$ is not Riemann integrable.
9. Deduce from the preceding exercise that a pointwise limit of a sequence of Riemann integrable function is not necessarily Riemann integrable. [Hint: taking a sequence (q_n) such that $\{q_n : n \in \mathbb{N}\} = T \cap \mathbb{Q}$ define f_n on T by $f_n(x) = 1$ if $x = q_k$ with $k \leq n, f_n(x) = 0$ otherwise and observe that f_n is a stair function and $(f_n) \to f$, where f is as in the preceding exercise.

3.6.3 *Application: The Dual of $C(T)$

Let us use the Stieltjes integral to identify the dual of $(C(T), \|\cdot\|_\infty)$ when T is a compact interval of $\mathbb{R}$.

Theorem 3.35 (Riesz) *For any element x^* of the dual of the space $X := C(T)$ of continuous functions on $T := [a, b]$ there exists some $g \in BV(T) := BV(T, \mathbb{R})$ such that $x^*(f) = \int f dg$ for all $f \in C(T)$.*

Proof We have seen that for all $g \in BV_n(T)$ the map $f \mapsto \int f dg$ is a continuous linear form on $C(T)$. Let us show that any element x^* in the dual of the space $C(T)$ is of this form. Using the Hahn-Banach Theorem, we pick some $x_R^* \in R(T)^*$ extending x^* and satisfying $\|x_R^*\| = \|x^*\|$. For $s \in T$, $s \neq b$ we denote by u_s the function $1_{[a,s[}$ defined by $1_{[a,s[}(t) := 1$ if $t \in [a, s[$, 0 otherwise and we set $u_b = 1_{[a,b]}$. Let us show that the function $g : T \to \mathbb{R}$ defined by

$$g(s) := x_R^*(u_s) \qquad s \in T$$

is of bounded variation. Let $\sigma := \{s_0 = a < s_1 \ldots < s_n = b\} \in \Sigma(T)$. Let $\varepsilon_i \in \{-1, 1\}$ be such that $\varepsilon_i(g(s_i) - g(s_{i-1})) = |g(s_i) - g(s_{i-1})|$. Then, by linearity of x_R^* we have

$$\sum_{i=1}^{n} |g(s_i) - g(s_{i-1})| = \sum_{i=1}^{n} \varepsilon_i(g(s_i) - g(s_{i-1})) = x_R^*(\sum_{i=1}^{n} \varepsilon_i(u_{s_i} - u_{s_{i-1}}))$$

hence, since $\|x_R^*\| = \|x^*\|$ and $\left\|\Sigma_{i=1}^n \varepsilon_i(u_{s_i} - u_{s_{i-1}})\right\|_\infty \leq 1$,

$$\sum_{i=1}^{n} |g(s_i) - g(s_{i-1})| \leq \|x^*\| .$$

Thus g is of bounded variation and $V(g) \leq \|x^*\|$. Now let us show that for all $f \in C(T)$ we have $x^*(f) = \int f dg$. Given $f \in C(T)$ and $\sigma := \{s_0 = a < s_1 < \ldots < s_n = b\} \in \Sigma(T)$ we set

$$f_\sigma := \sum_{i=1}^{n} (u_{s_i} - u_{s_{i-1}}) f(s_{i-1}),$$

so that f_σ is constant on $[s_{i-1}, s_i[$ with value $f(s_{i-1})$ and

$$\begin{aligned}\int f_\sigma dg &= \sum_{i=1}^{n} (g(s_i) - g(s_{i-1})) f(s_{i-1}) = \sum_{i=1}^{n} (x_R^*(u_{s_i}) - x_R^*(u_{s_{i-1}})) f(s_{i-1}) \\ &= x_R^*(f_\sigma).\end{aligned}$$

Since f is uniformly continuous, we have $(\|f - f_\sigma\|_\infty)_\sigma \to 0$ as the mesh of σ goes to 0. Since both x_R^* and $x \mapsto \int x dg$ are continuous on $R_n(T)$, we get $x^*(f) = \int f dg$. □

Additional Reading

[6, 9, 12, 14, 17, 22, 29, 33, 35, 52, 63, 67, 73, 75, 77, 78, 82, 97, 98, 100, 106, 108, 109, 111, 119–121, 160, 163, 172, 182, 187, 189, 196, 217, 220, 225, 227, 228, 239, 255, 262, 265].

Chapter 4
Hilbert Spaces

Riemann has shown us that proofs are better achieved through ideas than through long calculations.

David Hilbert, 1897.

Abstract Hilbert spaces form a major class of normed spaces. They offer geometric properties that are similar to those of Euclidean spaces. In particular one can identify them with their duals and, given a nonempty closed convex subset of such a space, to every point of the space corresponds a closest point in the set. When the set is a linear subspace, this correspondence defines an orthogonal projection. Hilbert spaces also serve as a models for important classes of function spaces. Since one can define Hilbert bases that generalize algebraic bases by using series, the study of Fourier series is set in such a framework.

Hilbert spaces form a special class of normed spaces of particular interest. They resemble Euclidean spaces as in them a notion of orthogonality can be defined. Moreover, the angle between two vectors can be given a meaning. Besides the interest in such generalizations of elementary geometry, they have nice properties in terms of duality, best approximation and form a convenient framework for the study of operators.

4.1 Hermitian Forms

In the sequel we suppose X is a vector space over the real or the complex numbers, using the familiar notation $\overline{\lambda}$ for the complex conjugate of $\lambda \in \mathbb{C}$ (so that $\overline{\lambda} = \lambda$ if λ is real).

Definition 4.1 A Hermitian form on X is a function $h : X \times X \to \mathbb{C}$ that is $\mathbb{C}$-linear in its second variable and such that for all $x, y \in X$ one has

$$h(y,x) = \overline{h(x,y)}.$$

It follows that for all $y \in X$ the function $h(\cdot, y)$ is *semi-linear* in the sense that for any $w, x \in X$, λ, $\mu \in \mathbb{C}$ one has $h(\lambda w + \mu x, y) = \overline{\lambda} h(w,y) + \overline{\mu} h(x,y)$. Moreover,

J.-P. Penot, *Analysis*, Universitext, DOI 10.1007/978-3-319-32411-1_4

one has $h(x, x) \in \mathbb{R}$ for all $x \in X$. If X is a real vector space, then a Hermitian form is just a bilinear symmetric form.

If for all $x \in X\backslash\{0\}$ one has $h(x, x) > 0$ (resp. $h(x, x) \geq 0$) one says that h is *positive definite* (resp. *positive*). Then h is also called a *scalar product*. We use *Dirac's notation* $\langle x \mid y\rangle$ for $h(x, y)$ (pronounced bra for $\langle x$ and ket for $y\rangle$) which is widely used in physics and in quantum mechanics, but a variety of notations can be encountered: among them are the original one $[x \mid y]$ by H. Grassmann (1862), its variant $(x \mid y)$ by N. Bourbaki and the simplified notation $\langle x, y\rangle$ that may introduce some confusion with the coupling between X and its dual. In general, the notation $x \cdot y$ is reserved for the scalar product of Euclidean spaces.

Example Let X be the space of continuous functions on $[0, 1]$ with values in $\mathbb{C}$. For $x, y \in X$ let $h(x, y) := \int_0^1 \overline{x(t)} y(t) dt$. This Hermitian form is the prototype of a useful class of Hermitian forms over infinite dimensional spaces.

Example Let I be an infinite set and let $X := \ell_2(I)$ (resp. $\ell_2(I, \mathbb{C})$) be the space of families $x := (x_i)_{i\in I}$ of real (resp. complex) numbers such that $(|x_i|^2)_{i\in I}$ is summable. Then, using the Minkowski inequality (4.1) in Euclidean spaces, one can show that for any $x := (x_i)_{i\in I}, y := (y_i)_{i\in I}$ the scalar product $\langle x \mid y\rangle := \sum_{i\in I} \overline{x_i} y_i$ is well defined. For $I := \mathbb{N}$ the space $\ell_2 := \ell_2(\mathbb{N})$ is a useful model. □

A vector $x \in X$ is said to be *orthogonal* to a family Y of vectors in X if for all $y \in Y$ one has $\langle x \mid y\rangle = 0$. Then one writes $x \perp Y$. Two subsets Y and Z of X are said to be orthogonal if $z \perp Y$ for all $z \in Z$ and then one writes $Y \perp Z$.

Theorem 4.1 (Bunyakovsky-Cauchy-Schwarz) *If h is a positive Hermitian form on X, for every $x, y \in X$ one has*

$$|h(x, y)|^2 \leq h(x, x)h(y, y).$$

Proof Let $a := h(x,x)$, $b := h(y, y)$, $c := h(x, y)$. Changing x into λx, with $\lambda \in \mathbb{C}$ satisfying $|\lambda| = 1$, $\overline{\lambda} c = |c|$, we may suppose $c \in \mathbb{R}_+$. The relation $0 \leq h(x-y, x-y)$, i.e. $2c \leq a + b$ yields $c \leq 1 = ab$ when $a = 1$, $b = 1$, hence, by homogeneity, $|c|^2 \leq ab$ when $a \neq 0$, $b \neq 0$. When $a = 0$, changing x into tx with $t \in \mathbb{R}_+$ we obtain the relation $2tc \leq b$ and $c = 0$ by taking the limit as $t \to +\infty$. The case $b = 0$ is similar. □

Corollary 4.1 (Minkowski) *If h is a positive Hermitian form on X, the function $p : x \mapsto h(x, x)^{1/2}$ is a semi-norm (and a norm if h is positive definite).*

Proof Clearly, for all $\lambda \in \mathbb{C}$, $x \in X$ one has $p(\lambda x) = |\lambda|\, p(x)$. Now, for all $x, y \in X$, the (Bunyakovsky-) Cauchy-Schwarz inequality entails

$$h(x, y) + h(y, x) \leq 2\,|h(x, y)| \leq 2p(x)p(y)$$

hence

$$p(x+y)^2 = h(x+y, x+y) = h(x, x) + h(x, y) + h(y, x) + h(y, y) \leq (p(x) + p(y))^2.$$

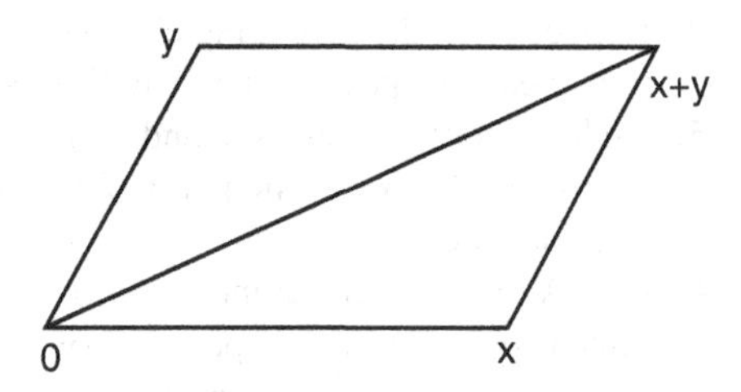

Fig. 4.1 The parallelogram law

Taking the square root of each side we get $p(x + y) \leq p(x) + p(y)$. □

Thus, when h is positive definite, setting $\|x\| := p(x)$, the Cauchy-Schwarz inequality can be written as

$$\forall x, y \in X \qquad |\langle x \mid y\rangle| \leq \|x\| \,.\, \|y\| \,, \tag{4.1}$$

a well-known inequality for the scalar product in Euclidean spaces. Expanding the square of the norm, the classical **parallelogram law** (Fig. 4.1) can also be obtained:

$$\forall x, y \in X \qquad \|x + y\|^2 + \|x - y\|^2 = 2\,\|x\|^2 + 2\,\|y\|^2 \,.$$

A similar expansion (made above) yields the famous Pythagoras' Theorem:

Theorem 4.2 (Pythagoras) *If x and y are orthogonal, then*

$$\|x + y\|^2 = \|x\|^2 + \|y\|^2 \,.$$

If X is endowed with a positive definite Hermitian form, one says that X is a *pre-Hilbertian space* or a *pre-Hilbert space*. If X is complete with respect to the associated norm, one says that X is a *Hilbert space*.

The following exercises show that in a pre-Hilbertian space the scalar product is determined by the associated semi-norm.

Exercises

1. Let h a Hermitian form on a real vector space. Show that

$$h(x, y) = \frac{1}{4}\left[h(x + y, x + y) - h(x - y, x - y)\right].$$

2. Let h a Hermitian form on a complex vector space. Show that $h(x, y) =$

$$\frac{1}{4}\left[h(x + y, x + y) - h(x - y, x - y) + ih(x + iy, x + iy) - ih(x - iy, x - iy)\right].$$

3. Let X be a complex linear space and let $b : X \times X \to \mathbb{C}$ be a $\mathbb{C}$-bilinear form on X. Show that there exist some $x \in X$, $x \neq 0$ such that $b(x, x) = 0$.

4. Deduce from the preceding exercises that the restriction of a Hermitian form h to a linear subspace Z of X is null if one has $h(z,z) = 0$ for all $z \in Z$.
5. Deduce from Exercises 1 and 2 that if the norm on a normed space $(X, \|\cdot\|)$ over $\mathbb{R}$ or $\mathbb{C}$ satisfies the parallelogram law, then it is the norm associated with a scalar product.
6. Let $(X, \|\cdot\|)$ be a normed space over $\mathbb{R}$ or $\mathbb{C}$. Suppose that for all two dimensional linear subspaces Y of X the induced norm is associated with a scalar product. Show the norm of X is associated with a scalar product.
7. Let us define a *sesquilinear form* on a normed space X over $\mathbb{C}$ as a map $f : X \times X \to \mathbb{C}$ such that, for all $x \in X, f(x,\cdot)$ is $\mathbb{C}$-linear and continuous and for all $y \in X, f(\cdot, y)$ is semi-linear and continuous. Let $b := \sup\{|f(x,y)| : x, \ y \in S_X\}$ and $c := \sup\{|f(x,x)| : x \in S_X\}$. Show that $c \leq b \leq 4c$. Compare with the case when f is Hermitian.
8. Show that when h is a positive definite Hermitian form, the Cauchy-Schwarz inequality is an equality for some vectors $x, y \in X$ if x and y are linearly dependent.
9. Prove that when h is a positive definite Hermitian form, Minkowski's inequality $\|x+y\| \leq \|x\| + \|y\|$ is an equality if and only if there exists $r \in \mathbb{R}_+$ such that $y = rx$ or $x = ry$.
10. Show that the complexified space of a real pre-Hilbertian space is a complex pre-Hilbertian space. [Hint: the complexified space of a real vector space X is $X \times X$ and, for $c := a + ib$ with $a, b \in \mathbb{R}$, $c(x,y) := (ax - by, bx + ay)$.]
11. Given two non-null vectors u, v in a pre-Hilbertian space prove that

$$\left\| \frac{u}{\|u\|^2} - \frac{v}{\|v\|^2} \right\| = \frac{\|u - v\|}{\|u\| \, . \, \|v\|}.$$

Give an interpretation in terms of surveying or astronomy: knowing the distance of the observer to two observed points and the angle under which they appear, one can compute their mutual distance.
12. Let X be a pre-Hilbertian space and let w, x, y, $z \in X$. Prove the *Ptolemae inequality*:

$$\|w - y\| \, . \, \|x - z\| \leq \|w - x\| \, . \, \|y - z\| + \|x - y\| \, . \, \|w - z\| \, .$$

[Hint: Reduce the question to the case $w = 0$ and use the preceding exercise.]
13. Let X, Y be two pre-Hilbertian spaces and let $A : X \to Y$ be isometric (i.e. such that $\|A(x) - A(x')\| = \|x - x'\|$ for all $x, x' \in X$) and such that $A(0) = 0$. Show that A is linear.
14. Let $f : \mathbb{R} \to \mathbb{C}$ be a continuous function such that $\int_{\mathbb{R}} |f(t)|^2 \, dt < +\infty$. Show that for all $r \in \mathbb{R}$ one has

$$\left| \int_{\mathbb{R}} f(t) f(t-r) dt \right| \leq \int_{\mathbb{R}} |f(t)|^2 \, dt.$$

4.2 Best Approximation

Given a closed convex subset C of a pre-Hilbertian space X and $w \in X \backslash C$, one may wonder whether there exists some $a \in C$ such that $\|a-w\| \leq \|x-w\|$ for all $x \in C$. Such a point a is called a *best approximation* of w in C or a *projection* of w on C. It is unique, as shown by the parallelogram law: given two such points a, a' one has

$$\left\|a-a'\right\|^2 = 2\|a-w\|^2 + 2\left\|a'-w\right\|^2 - 4\left\|\frac{1}{2}a+\frac{1}{2}a'-w\right\|^2 \leq 0,$$

since $(a+a')/2 \in C$ and $\|(a+a')/2-w\| \geq \inf_{x\in C}\|x-w\|$; hence $a=a'$.

It is easy to find the best approximation of w when C is a one-dimensional subspace $\mathbb{R}u$ or $\mathbb{C}u$ with $u \in X\backslash\{0\}$. It suffices to find $a \in C$ such that $w-a \perp C$: then, by Pythagoras' Theorem, for all $x \in C$ one has

$$\|w-a\|^2 \leq \|w-a\|^2 + \|a-x\|^2 = \|w-x\|^2 .$$

Such a point $a := \lambda u$ is obtained by taking λ such that $\langle w-\lambda u \mid u\rangle = 0$, i.e. $\lambda = \|u\|^{-2}\langle w \mid u\rangle$. The number λ is called the *Fourier coefficient* of w with respect to u.

A similar construction holds when C is the vector space spanned by a finite family $(u_1,\ldots,u_k)$ of orthogonal vectors. Taking

$$a := \lambda_1 u_1 + \ldots + \lambda_k u_k \qquad \text{with } \lambda_i := \|u_i\|^{-2}\langle w \mid u_i\rangle \quad \text{for } i=1,\ldots,k$$

we get that $w-a \perp u_i$ for $i=1,\ldots,k$ since $u_i \perp u_j$ for $j \neq i$, hence by linearity, $w-a \perp x$ for all $x \in C$ and again $\|w-a\|^2 \leq \|w-x\|^2$ by Pythagoras' Theorem.

A general existence result can be given.

Theorem 4.3 *Let C be a nonempty complete convex subset of a pre-Hilbertian space X and let $w \in X\backslash C$. Then there exists some $a \in C$ such that $\|a-w\| = \inf_{x\in C}\|x-w\|$. This point, called the projection of w on C is unique. It is denoted by $P_C(w)$ in the sequel and is characterized by the inequality*

$$\forall x \in C \qquad \mathrm{Re}\langle w-a \mid x-a\rangle \leq 0. \tag{4.2}$$

Proof Let (x_n) be a sequence in C such that $(\|w-x_n\|) \to d := \inf_{x\in C}\|x-w\|$. Setting $\varepsilon_n := \|x_n-w\|^2 - d^2$, the parallelogram law shows that (x_n) is a Cauchy sequence:

$$\begin{aligned}\frac{1}{2}\left\|x_n-x_p\right\|^2 &= \|x_n-w\|^2 + \left\|x_p-w\right\|^2 - 2\left\|\frac{1}{2}(x_n+x_p)-w\right\|^2 \\ &\leq d^2+\varepsilon_n+d^2+\varepsilon_p-2d^2 = \varepsilon_n+\varepsilon_p.\end{aligned}$$

Since C is complete with respect to the induced metric, (x_n) converges to some $a \in C$, and passing to the limit we get $\|w - a\| = d$.

Let us prove (4.2). For all $x \in C$ and all $t \in]0, 1]$ we have $x_t := (1-t)a + tx \in C$ hence

$$\|w - a\|^2 \leq \|w - x_t\|^2 = \|w - a\|^2 + t^2 \|x - a\|^2 - 2t \operatorname{Re}\langle w - a \mid x - a\rangle,$$
$$0 \leq t \|x - a\|^2 - 2 \operatorname{Re}\langle w - a \mid x - a\rangle$$

after simplification. Taking the limit as $t \to 0_+$, we get (4.2). Conversely, let us assume $a \in C$ satisfies (4.2). Then, for all $x \in C$, we have

$$\|w - x\|^2 = \|w - a\|^2 + \|a - x\|^2 - 2 \operatorname{Re}\langle w - a \mid x - a\rangle \geq \|w - a\|^2 .$$

Therefore a is a best approximation of w in C. □

Corollary 4.2 *Let C be a complete convex subset of a pre-Hilbertian space X. Then the map P_C is Lipschitzian with rate 1: for all $w, w' \in X$ one has $\|P_C(w) - P_C(w')\| \leq \|w - w'\|$.*

Proof Given $w, w' \in X$, let $a := P_C(w)$, $a' := P_C(w')$, $z := (w - w') - (a - a')$, so that

$$\|w - w'\|^2 = \|a - a'\|^2 + \|z\|^2 + 2 \operatorname{Re}\langle z \mid a - a'\rangle. \tag{4.3}$$

Relation (4.2) shows that

$$\operatorname{Re}\langle w - a \mid a' - a\rangle \leq 0, \qquad \operatorname{Re}\langle a' - w' \mid a' - a\rangle \leq 0,$$

hence, by addition, $\operatorname{Re}\langle z \mid a' - a\rangle = \operatorname{Re}\langle (w - a) + (a' - w') \mid a' - a\rangle \leq 0$. Plugging this estimate into (4.3) we get $\|w - w'\|^2 \geq \|a - a'\|^2$. □

Corollary 4.3 *Let Y be a complete linear subspace of a pre-Hilbertian space X. Then the map P_Y is linear and continuous. For all $x \in X$, $P_Y(x)$ is the unique point y of Y such that $x - y$ is orthogonal to Y. Thus X is the topological direct sum of Y and $Y^\perp$: $X = Y \oplus Y^\perp$ and $Y^\perp = \ker P_Y$, $Y^{\perp\perp} = Y$.*

Proof The characterization of $y := P_Y(x)$ deduced from (4.2) can be written as

$$\forall y' \in Y \qquad \operatorname{Re}\langle x - y \mid y' - y\rangle \leq 0.$$

For z arbitrary in Y, taking successively $y + z$, $y - z$ (and $y + iz$, $y - iz$ if X is a complex linear space) in place of y', we get $\langle x - y \mid z\rangle = 0$. The linearity of P_Y ensues. Writing $x = y + (x - y)$, we note that $X = Y + Y^\perp$. The sum is direct since for $u \in Y \cap Y^\perp$ we have $u = 0$. Since P_Y is continuous, this sum is a topological direct sum.

If $x \in \ker P_Y$ we have $x = x - P_Y(x) \in Y^{\perp}$. Conversely, if $x \in Y^{\perp}$ we have $x = P_Y(x) + (x - P_Y(x)) = 0 + x$ and the uniqueness of the decomposition ensures that $P_Y(x) = 0$. The inclusion $Y \subset Y^{\perp\perp}$ is obvious; if $x \in Y^{\perp\perp}$ we have in particular $\langle x - P_Y(x) \mid x \rangle = 0$ and since $\langle x - P_Y(x) \mid P_Y(x) \rangle = 0$, we get $\langle x - P_Y(x) \mid x - P_Y(x) \rangle = 0$, hence $x = P_Y(x) \in Y$. □

Corollary 4.4 *For any linear subspace Y of a Hilbert space X one has $Y^{\perp\perp} = \mathrm{cl}(Y)$.*

Proof Since $Y^{\perp\perp}$ is closed and contains Y, one has $\mathrm{cl}(Y) \subset Y^{\perp\perp}$. On the other hand, since $Y \subset \mathrm{cl}(Y)$, one has $Y^{\perp\perp} \subset (\mathrm{cl}(Y))^{\perp\perp} = \mathrm{cl}(Y)$. Thus $Y^{\perp\perp} = \mathrm{cl}(Y)$. □

Exercises

1. Let C be a closed convex cone of a Hilbert space. Show that the projection $x := P_C(w)$ of $w \in X$ is characterized by $w \in C$, $\mathrm{Re}\langle w - x \mid x \rangle = 0$, so that one has $\|w\|^2 = \|w - x\|^2 + \|x\|^2$.
2. Let C be a complete convex cone of a pre-Hilbertian space X. Show that the set $C^0 := \{x \in X : \forall y \in C\ \langle x \mid y \rangle \le 0\}$ is a closed convex cone and that $C^{00} := (C^0)^0 = C$.
3. Let C and D be two closed convex cones of a real Hilbert space X such that $D = C^0$, hence $C = D^0$ by the preceding exercise. Show that for all $(x, y, z) \in X^3$ the following two assertions are equivalent:

$$z = x + y, \quad x \in C, \quad y \in D, \quad x \perp y$$
$$x = P_C(z), \quad y = P_D(z).$$

4. Prove that the projection $x := P_Y(w)$ of a point w of a pre-Hilbertian space X on a complete linear subspace Y of X is the unique point $y \in Y$ such that $x - y$ is orthogonal to Y.
5. Let Y and Z be two linear subspaces of a pre-Hilbertian space X and for a, $b \in X$ let $A := a + Y$, $B := b + Z$. Show that the following two assertions are equivalent:

$$\mathrm{gap}(A, B) = \|a - b\| \quad \text{for } \mathrm{gap}(A, B) := \inf\{\|u - v\| : u \in A,\ v \in B\}$$
$$a - b \text{ is orthogonal to } Y \text{ and to } Z.$$

6. Let A and B be two nonempty complete convex subsets of a pre-Hilbertian space X, B being bounded. Show that there exists some $(a, b) \in A \times B$ such that $\|a - b\| = \mathrm{gap}(A, B) := \inf\{\|u - v\| : u \in A,\ v \in B\}$.
7. Let C be a convex subset of a closed affine subspace A of a Hilbert space X and let $w \in X$, $a := P_A(w)$. Show that if a has a projection x in C, then x is also the projection of w in C.
8. Show that the conclusion of the preceding exercise is no longer true if A is replaced with a general closed convex subset of X.

9. Let C be the closure of the union of an increasing family (C_n) of complete convex subsets of a pre-Hilbertian space X. Let $w \in X$ be such that w has a best approximation $x := P_C(w)$ in C. Show that $x = \lim_n P_{C_n}(w)$. [Hint: show that $(P_{C_n}(w))$ is a Cauchy sequence.]

10. Let (C_n) be a decreasing sequence of complete convex subsets of a pre-Hilbertian space X. For $x \in X$, let $d(x) := \lim_n d(x, C_n)$. Suppose that for some $\overline{x} \in X$ one has $d(\overline{x}) < +\infty$. Show that $d(x) < +\infty$ for all $x \in X$. Prove that the diameter of $C_n \cap B[x, d(x) + \varepsilon]$ tends to 0 as $(n, \varepsilon) \to (+\infty, 0)$. Deduce from this that the intersection C of the C_n's is nonempty and that $d(x) = d(x, C)$.

11. Let C be a bounded complete convex subset of a pre-Hilbertian space X and let $f : C \to \mathbb{R}$ be a lower semicontinuous convex function. Using Exercise 10 show that f attains its infimum on C. Prove the same conclusion when f is quasiconvex in the sense that for all $r \in \mathbb{R}$ the set $f^{-1}(]-\infty, r])$ is convex.

12. Let C be a complete convex subset of a real pre-Hilbertian space X. Using Theorem 4.3 show that C is the intersection of a family of closed affine half-spaces, i.e. of a family of subsets of the form $D_{f,r} := f^{-1}(]-\infty, r])$ with $f \in X^*$, $r \in \mathbb{R}$.

If C is a complete convex cone of X, show that C is the intersection of a family of closed half-spaces, i.e. subsets of the form $D_{f,0}$.

13. Let X be a Hilbert space, let a be the bilinear form associated with a continuous positive semidefinite symmetric linear map $A : X \to X$ and for a nonempty subset T of X let

$$V_0(T) := \{x \in X : a(x, x) \leq a(x - t, x - t) \ \forall t \in T\}$$

be the *Voronoi cell* of T with respect to the origin (taking $Y := X$ and $f := a$ in Exercise 8 of Sect. 2.3). Show that $V_0(T) = V_0(T \backslash \{0\}) = V_0(\mathrm{cl}(T))$ is a closed convex subset containing the origin and that $V_0(T)$ is polyhedral when T is finite. [Hint: show that $V_0(T) := \{x \in X : 2\langle At \mid x\rangle \leq \langle At \mid t\rangle \ \forall t \in T\}$.]

Suppose A is positive definite. Prove that $0 \in \mathrm{int} V_0(T)$ if and only if $0 \notin \mathrm{cl}(T)$. Prove that for any closed convex subset W containing the origin there exists some closed convex subset T of X such that $V_0(T) = W$. [Hint: use the bipolar theorem and write $T = \overline{\mathrm{co}}(\{0\} \cup \{\lambda_y A^{-1} y : y \in W^0 \backslash \{0\}\})$ for some $\lambda_y \in \mathbb{R}$.]

4.3 Orthogonal Families

The convenience of Cartesian coordinates in Euclidean spaces incites us to look for a similar device in Hilbert spaces. The use of orthogonal families will present such an analogy. However, some differences appear as finite families are not sufficient in infinite dimensional spaces.

A family $(b_i)_{i \in I}$ of elements of a pre-Hilbertian space is said to be *orthogonal* if for all $i, j \in I$ with $i \neq j$ one has $b_i \perp b_j$. It is said to be *orthonormal* if it is orthogonal and if for all $i \in I$ one has $\|b_i\| = 1$. Any orthogonal family of non-null vectors is linearly independent: if for some finite subset J of I and some family $(\lambda_j)_{j\in J}$ of numbers one has $\sum_{j\in J} \lambda_j b_j = 0$, then for all $k \in J$ one has $\lambda_k \|b_k\|^2 = \langle b_k \mid \sum_{j\in J} \lambda_j b_j\rangle = 0$ hence $\lambda_k = 0$.

One can pass from an orthogonal family $(b_i)_{i\in I}$ of non-null vectors to an orthonormal family $(e_i)_{i\in I}$ by setting $e_i := b_i/\|b_i\|$. Let us describe a process that allows us to pass from a linearly independent family to an orthogonal family.

Proposition 4.1 (Gram-Schmidt) *Let (b_n) be a finite or countable family of linearly independent vectors of a pre-Hilbertian space X and let X_n be the linear subspace spanned by $b_1,\ldots,b_n$. Setting $a_1 := b_1$, $a_{n+1} := b_{n+1} - P_{X_n}(b_{n+1})$ for $n \geq 1$, one gets an orthogonal family such that $a_1,\ldots,a_n$ generate X_n for all n.*

Let us note that if the family (b_n) is *total* in the sense that the union of the spaces X_n is dense in X, then (a_n) is total.

Proof Clearly a_1 generates X_1. Assume that $a_1,\ldots,a_n$ generate X_n. Then $a_{n+1} \in X_{n+1}$ and, on the other hand, any element of X_{n+1} can be written as a linear combination of $a_1,\ldots,a_n$ and $b_{n+1} = a_{n+1} + P_{X_n}(b_{n+1})$ hence as a linear combination of $a_1,\ldots,a_n$ and a_{n+1}. Thus $a_1,\ldots,a_{n+1}$ generate X_{n+1}. It remains to show that (a_n) is an orthogonal family. By Corollary 4.3, a_{n+1} is orthogonal to X_n, so that $a_{n+1} \perp a_i$ for $i = 1,\ldots,n$. □

In practice, one determines a_{n+1} by looking for coefficients $\lambda_1,\ldots,\lambda_n$ such that $a_{n+1} = b_{n+1} - \lambda_1 a_1 - \ldots - \lambda_n a_n$ satisfies $\langle a_j \mid a_{n+1}\rangle = 0$ for $j = 1,\ldots,n$ i.e. $\langle a_j \mid b_{n+1}\rangle - \lambda_j \|a_j\|^2 = 0$ so that $\lambda_j = \|a_j\|^{-2} \langle a_j \mid b_{n+1}\rangle$.

Proposition 4.2 (Bessel) *Let $(e_i)_{i\in I}$ be an orthonormal family of a pre-Hilbertian space X and for $x \in X$ let $x_i := \langle e_i \mid x\rangle$. Then the family $(|x_i|^2)_{i\in I}$ is summable and*

$$\sum_{i\in I} |x_i|^2 \leq \|x\|^2 . \tag{4.4}$$

Proof We have to show that for any finite subset J of I we have $\sum_{j\in J} |x_j|^2 \leq \|x\|^2$. Setting $x_J := \sum_{j\in J} x_j e_j$, this inequality stems from the relations

$$0 \leq \|x - x_J\|^2 = \|x\|^2 - \sum_{j\in J}(\langle x \mid x_j e_j\rangle + \langle x_j e_j \mid x\rangle) + \|x_J\|^2 = \|x\|^2 - \sum_{j\in J} |x_j|^2 , \tag{4.5}$$

since $\langle x \mid x_j e_j\rangle = x_j \overline{x_j} = |x_j|^2 = \langle x_j e_j \mid x\rangle$ and $\|x_J\|^2 = \sum_{j\in J} |x_j|^2$. □

Theorem 4.4 (Parseval) *For an orthonormal family $(e_i)_{i\in I}$ of a pre-Hilbertian space X the following assertions are equivalent:*

(a) *the family $(e_i)_{i\in I}$ is total in X;*
(b) *for all $x \in X$, setting $x_i := \langle e_i \mid x\rangle$, the family $(|x_i|^2)$ is summable with sum $\sum_{i\in I} |x_i|^2 = \|x\|^2$;*
(c) *for all $x \in X$ the family $(x_i e_i)_{i\in I}$ is summable with sum x.*

Proof (a)⇒(b) Let $x \in X$ and let $\varepsilon > 0$ be given. By assumption, there exists some finite subfamily $(e_j)_{j\in J}$ of $(e_i)_{i\in I}$ and some element y of the space X_J generated by $(e_j)_{j\in J}$ such that $\|x - y\| \leq \varepsilon$. Let z be the projection of x on X_J. One has $\|x - z\| \leq \|x - y\|$ and $x - \sum_{j\in J} x_j e_j$ is orthogonal to each e_j by the definition of x_j, so that $z = \sum_{j\in J} x_j e_j$. Therefore, by (4.5),

$$0 \leq \|x\|^2 - \sum_{j\in J} |x_j|^2 = \|x - z\|^2 \leq \varepsilon^2.$$

This shows that the family $\sum_{i\in I} |x_i|^2$ is summable, with sum $\|x\|^2$.

(b)⇒(c) We have to show that for any $\varepsilon > 0$ one can find a finite subset J_ε of I such that for all finite subsets J of I containing J_ε one has $\left\|\sum_{j\in J} x_j e_j - x\right\| \leq \varepsilon$. We take J_ε such that $\left|\sum_{j\in J} |x_j|^2 - \|x\|^2\right| \leq \varepsilon^2$ for any finite subset J of I containing J_ε. Since $\left\|\sum_{j\in J} x_j e_j - x\right\|^2 = \|x\|^2 - \sum_{j\in J} |x_j|^2$ by a computation of the preceding proof, this choice of J_ε is suitable.

(c)⇒(a) This implication is obvious since any $x \in X$ can be approximated by a finite linear combination of the family $(e_i)_{i\in I}$. □

An orthonormal family $(e_i)_{i\in I}$ satisfying the assertions of the preceding theorem is called an *orthonormal basis* or a *Hilbert basis*. One has to recall that such a family is not an algebraic basis if it is infinite.

Proposition 4.3 *An orthonormal family $(e_i)_{i\in I}$ of a Hilbert space X is a Hilbert basis if, and only if it is maximal in the set of orthonormal families with respect to set inclusion.*

Proof Let $(e_i)_{i\in I}$ be a Hilbert basis of a pre-Hilbertian space. If e is a vector orthogonal to all the e_i's, then e is orthogonal to the closed linear space generated by $(e_i)_{i\in I}$, i.e. e is orthogonal to X, hence is 0. This shows that there is no orthonormal family strictly containing $(e_i)_{i\in I}$: $(e_i)_{i\in I}$ is maximal.

Conversely, let $(e_i)_{i\in I}$ be a maximal orthonormal family of a Hilbert space X. If the family $(e_i)_{i\in I}$ is not total, one can find a non-null vector e orthogonal to the closed linear space generated by $(e_i)_{i\in I}$. Since we may suppose $\|e\| = 1$, adding the vector e to $(e_i)_{i\in I}$, we get a strictly larger family, a contradiction. Thus $(e_i)_{i\in I}$ is total. □

Corollary 4.5 *Any Hilbert space contains a Hilbert basis.*

Proof This follows from Zorn's Lemma, the union of an increasing set of orthonormal families being an orthonormal family. □

Corollary 4.6 *Any real (resp. complex) Hilbert space is isometric to some space $\ell_2(I)$ (resp. $\ell_2(I, \mathbb{C})$). Any real (resp. complex) separable Hilbert space is isometric to ℓ_2 (resp. $\ell_2(\mathbb{N}, \mathbb{C})$).*

Remark By the preceding corollary and the polarization identity, the assertions of Theorem 4.4 are equivalent to the following one:

(d) for all $x, y \in X$, setting $y_i := \langle e_i \mid y\rangle$, the family $(\overline{x}_i y_i)_{i\in I}$ is summable with sum $\langle x \mid y\rangle$.

Exercises

1. Let X and Y be two linear subspaces of a real pre-Hilbertian space Z. Suppose there exists some $c \in \mathbb{R}_+$ such that $|\langle x \mid y\rangle| = c\,\|x\|\,.\,\|y\|$ for all $x \in X$, $y \in Y$. Show that either X and Y are one-dimensional or $c = 0$ (i.e. $X \perp Y$).

2. Let $(x_i)_{i\in\mathbb{N}_k}$ be a finite sequence in a pre-Hilbertian space X. The *Gram determinant* of $(x_i)_{i\in\mathbb{N}_k}$ is the determinant $G(x_1,\ldots,x_k) := \det(\langle x_i \mid x_j\rangle)$.

(**a**) Show that $G(x_1,\ldots,x_k) \in \mathbb{R}_+$ and that $G(x_1,\ldots,x_k) = 0$ if and only if the family $(x_i)_{i\in\mathbb{N}_k}$ is linearly dependent. [Hint: use an orthonormal basis of the linear subspace generated by $(x_i)_{i\in\mathbb{N}_k}$.]

(**b**) Suppose the family $(x_i)_{i\in\mathbb{N}_k}$ is linearly independent and let Y be the linear subspace generated by $(x_i)_{i\in\mathbb{N}_k}$. Show that the distance $d(w, Y)$ of a point $w \in X$ to Y is equal to $\sqrt{G(w, x_1,\ldots,x_k)/G(x_1,\ldots,x_k)}$. [Hint: write the projection x of w on Y as a linear combination of the x_i's.]

3. Let X be the pre-Hilbertian space $C(T)$, where T is a compact interval of $\mathbb{R}$, endowed with the scalar product $\langle\cdot \mid \cdot\rangle$ given by $\langle x \mid y\rangle = \int_T x(t)y(t)dt$. Given a linearly independent family $(x_1,\ldots,x_n)$ of elements of X, let $d_n \in C(T)$ be such that $d_n(t)$ is the determinant of the matrix whose i-th line is $(\langle x_i \mid x_1\rangle,\ldots,\langle x_i \mid x_n\rangle)$ for $i = 1,\ldots,n-1$ and whose n-th line is $(x_1(t),\ldots,x_n(t))$. Prove that $d_1,\ldots,d_n$ form a linearly independent family spanning the space X_n generated by $(x_1,\ldots,x_n)$ and that $\langle d_n \mid d_n\rangle = G(x_1,\ldots,x_n)G(x_1,\ldots,x_{n-1})$ where $G(x_1,\ldots,x_k) := \det(\langle x_i \mid x_j\rangle)$.

4. Let X and Y be separable real Hilbert spaces with orthonormal basis $(e_n)_{n\in\mathbb{N}}$ and $(f_n)_{n\in\mathbb{N}}$ respectively. For $A \in L(X, Y)$ show that the following series have the same (possibly infinite) sum:

$$\sum_{m=0}^{\infty} \|A(e_m)\|^2,\ \sum_{m=0}^{\infty}\sum_{n=0}^{\infty}\langle A(e_m) \mid f_n\rangle^2,\ \sum_{n=0}^{\infty} \|A^*(f_n)\|^2.$$

When their sums are finite they are denoted by $\|A\|_{HS}^2$. Observe that $\|A\|_{HS}$ is independent of the choice of the basis $(e_m)_{m\in\mathbb{N}}$ and verify that $A \mapsto \|A\|_{HS}$ defines a norm on the space $HS(X, Y) := \{A \in L(X, Y) : \|A\|_{HS} < \infty\}$ called the space of *Hilbert-Schmidt operators*. Show that the norm $\|\cdot\|_{HS}$ is associated with the scalar product $\langle\cdot \mid \cdot\rangle_{HS}$ on $HS(X, Y)$ defined by

$$\langle A \mid B\rangle_{HS} := \sum_{m=1}^{\infty}\langle A(e_m) \mid B(e_m)\rangle.$$

Prove that $\|\cdot\| \le \|\cdot\|_{HS}$ and that $(HS(X, Y), \|\cdot\|_{HS})$ is complete.

5. Consider separable real Hilbert spaces X and Y with orthonormal basis $(e_n)_{n\in\mathbb{N}}$ and $(f_n)_{n\in\mathbb{N}}$ respectively. Given a bounded sequence $s := (s_n)_{n\in\mathbb{N}}$ of real numbers show that one defines a continuous linear map $A : X \to Y$ by setting $A(x) = \sum_{m=1}^{\infty} s_m (x \mid e_m) f_m$. Prove that A belongs to $HS(X, Y)$ as defined in the preceding exercise if and only if $s \in \ell_2$, i.e. $\|s\|_2^2 := \sum_{m=1}^{\infty} s_m^2 < \infty$ and that $\|A\|_{HS} = \|s\|_2$. Deduce from this relation that the norms $\|\cdot\|_{HS}$ and $\|\cdot\|$ on $HS(X, Y)$ are not equivalent.
6. Show that if W, X, Y, and Z are separable infinite dimensional Hilbert spaces and if $B \in L(W,X)$, $C \in L(Y,Z)$, $A \in HS(X, Y)$ as defined in Exercise 4, then $A \circ B \in HS(W, Y)$, $C \circ A \in HS(X, Z)$ and $\|A \circ B\|_{HS} \le \|A\|_{HS} \cdot \|B\|$, $\|C \circ A\|_{HS} \le \|A\|_{HS} \cdot \|C\|$.

4.4 The Dual of a Hilbert Space

A remarkable property of real Hilbert spaces is that they can be identified with their dual spaces.

Theorem 4.5 (Riesz) *If X is a Hilbert space, for every $w \in X$ the function $f_w : x \mapsto (w \mid x)$ is a continuous linear form with norm $\|w\|$. The map $w \mapsto f_w$ is an isometric semi-linear map from X onto the dual X^* of X.*

Proof The linearity of f_w is obvious and its continuity is a consequence of the Cauchy-Schwarz inequality: $|f_w(x)| \le \|w\| \cdot \|x\|$ for all $x \in X$, so that $\|f_w\| \le \|w\|$. Since $f_w(w) = \|w\|^2$, we get $\|f_w\| = \|w\|$. Clearly $w \mapsto f_w$ is semi-linear and injective. Let us show that this map is onto. Let $w^* \in X^*$. If $w^* = 0$ we have $w^* = f_w$ with $w = 0$. Assuming $w^* \neq 0$, let $Y := \ker w^*$. It is a closed linear subspace of X, hence a complete subspace. Corollary 4.3 ensures that $X = Y \oplus Y^{\perp}$. Since $Y \neq X$ as $w^* \neq 0$, we can pick some $b \in Y^{\perp}\backslash\{0\}$. Then we have $Y \subset \ker f_b$ and since both these subspaces are hyperplanes, we have $Y = \ker f_b$. It follows that there exists some scalar λ such that $w^* = \lambda f_b$. Then $w^* = f_{\overline{\lambda} b}$. □

It is very convenient to identify the dual of a real Hilbert space X with X itself. In particular, to the derivative of a differentiable function f at $x \in X$ corresponds a vector denoted by $\nabla f(x)$ and called the *gradient* of f at x. It is characterized by

$$\forall y \in X \qquad (\nabla f(x) \mid y) = \langle f'(x), y\rangle.$$

Corollary 4.7 *Any Hilbert space is reflexive.*

Proof The Riesz isometry $R : X \to X^*$ enables us to endow the dual X^* of X with a scalar product $(\cdot \mid \cdot)_*$ obtained by setting for $x^*, y^* \in X^*$

$$(x^* \mid y^*)_* := (R^{-1}(y^*) \mid R^{-1}(x^*)).$$

Thus, for all $x \in X$, taking $x^* = R(x)$ one gets $\langle x^* \mid y^* \rangle_* = \langle R^{-1}(y^*) \mid x \rangle$. The Riesz isometry $R_* : X^* \to X^{**}$ associated with the scalar product $\langle \cdot \mid \cdot \rangle_*$ on X^* is defined by $\langle R_*(x^*), y^* \rangle = \langle x^* \mid y^* \rangle_*$. Thus, for $y := R^{-1}(y^*)$ one gets

$$\langle R_*(x^*), y^* \rangle = \langle x^* \mid y^* \rangle_* = \langle y \mid x \rangle = \langle Ry, x \rangle = \langle y^*, x \rangle.$$

This string of equalities proves that $R_*(R(x))$ coincides with the image of x in the canonical injection $j : X \to X^{**}$. Since R and R_* are onto, $j = R_* \circ R$ is onto and X is reflexive. □

Corollary 4.8 *An element $\overline{x}$ of a Hilbert space X is the weak limit of a net $(x_i)_{i \in I}$ of X if and only if for all $y \in Y$ one has $\langle \overline{x} \mid y \rangle = \lim_{i \in I} \langle x_i \mid y \rangle$.*

Exercise Show that for a net $(x_i)_{i \in I}$ in a Hilbert space X and $\overline{x}$ one has $(x_i)_{i \in I} \to \overline{x}$ if and only if $\overline{x}$ is the weak limit of $(x_i)_{i \in I}$ and $(\|x_i\|)_{i \in I} \to \|\overline{x}\|$.

Given Hilbert spaces X and Y and an element A of the space $L(X, Y)$ of continuous linear maps from X into Y, the Riesz isomorphisms R_X and R_Y of X and Y respectively enable us to associate to the transpose $A^\intercal \in L(Y^*, X^*)$ of A an element A^* of $L(Y, X)$ called the *adjoint* of A. It is defined by $A^* := R_X^{-1} \circ A^\intercal \circ R_Y$ or through the relation

$$\forall x \in X,\ y \in Y \qquad \langle x \mid A^* y \rangle = \langle Ax \mid y \rangle.$$

We leave as an exercise the proof of the following proposition.

Proposition 4.4 *The map $A \mapsto A^*$ satifies the following properties:*

$$(\alpha A)^* = \overline{\alpha} A^*, \qquad (A + B)^* = A^* + B^*, \qquad (A^*)^* = A,$$

$$\|A^*\| = \|A\|, \qquad (A \circ B)^* = B^* \circ A^*, \qquad \|A^* A\| = \|A\|^2.$$

An operator $A \in L(X, X)$ such that $A^* = A$ is called *Hermitian* or *self-adjoint*. If X is a real Hilbert space, A is also said to be *symmetric* (since then the bilinear form $(x, y) \mapsto \langle Ax \mid y \rangle$ is symmetric). Then h_A given by $h_A(x, y) := \langle Ax \mid y \rangle$ is a Hermitian form on X.

Theorem 4.6 (Lax-Milgram) *Let X be a real Hilbert space and let $A \in L(X, X)$ be coercive in the sense that there exists some $c > 0$ such that $\langle Ax \mid x \rangle \geq c \|x\|^2$ for all $x \in X$. Then A is an isomorphism from X onto X.*

Conversely, if $A \in L(X, X)$ is an isomorphism and if A is symmetric and such that $\langle Ax \mid x \rangle \geq 0$ for all $x \in X$, then A is coercive.

Proof of the direct assertion For all $x \in X \backslash \{0\}$ we have $\|Ax\| = \sup_{u \in S_X} \langle Ax \mid u \rangle \geq \langle Ax \mid x / \|x\| \rangle \geq c \|x\|$. Thus A is an isomorphism from X onto $A(X)$. Thus $A(X)$ is complete, hence closed in X. On the other hand $A(X)$ is dense since $x = 0$ whenever $x \in A(X)^\perp$ as $c \|x\|^2 \leq \langle Ax \mid x \rangle = 0$. Thus $A(X) = X$. □

Proof in the case A is symmetric The norm $\|\cdot\|_A$ associated with the Hermitian form h_A introduced above satisfies

$$\forall x \in X \qquad \|x\|_A := (h_A(x,x))^{1/2} \geq c^{1/2} \|x\|$$

and also $\|x\|_A \leq \|A\|^{1/2} \|x\|$. Thus, $\|\cdot\|_A$ is equivalent to the norm associated with the scalar product $(\cdot \mid \cdot)$ of X and the dual of X with respect to $\|\cdot\|_A$ coincides with X^*. The Riesz isomorphism theorem ensures that for all $w^* \in X^*$ there exists some $w \in X$ such that $\langle w^*, x\rangle = h_A(w,x)$ for all $x \in X$. Given $y \in X$ and taking $w^* := (y \mid \cdot)$, we get $(y \mid x) = h_A(w,x) = (Aw \mid x)$ for all $x \in X$, hence $y = Aw$. Thus A is onto. The injectivity of A is immediate.

For the converse we introduce $a > 0$ such that $\|A^{-1}x\| \leq a\|x\|$ for all $x \in X$. The Cauchy-Schwarz inequality yields

$$|(Ay \mid x)|^2 = |(Ax \mid y)|^2 \leq (Ax \mid x)(Ay \mid y) \qquad \forall x, y \in X.$$

Taking $y := A^{-1}x$ we get $\|x\|^4 \leq (Ax \mid x)\|x\| \|y\| \leq a(Ax \mid x)\|x\|^2$, hence $\|x\|^2 \leq a(Ax \mid x)$ or $(Ax \mid x) \geq a^{-1}\|x\|^2$. □

Thus, when A is symmetric, the Lax-Milgram Theorem can be deduced from the Riesz representation theorem by introducing the new scalar product given by

$$(x \mid y)_A := (Ax \mid y).$$

Also, when A is symmetric, for all $b \in X$, setting $f(x) := \frac{1}{2}(Ax \mid x) - (b \mid x)$, we get a convex, continuous function f. Since it is coercive, it attains its infimum at some $\overline{x} \in X$ characterized by $A\overline{x} = b$ since $\nabla f(x) = Ax - b$; again this proves the surjectivity of A. This link with optimization is important.

Exercises

1. Using the map $w \mapsto f_w$ from a pre-Hilbertian space X into X^*, with f_w defined by $f_w(x) = (w \mid x)$ for $x \in X$, show that X is isometric to a dense linear subspace of X^*. Deduce from this fact that any pre-Hilbertian space is isometric to a dense linear subspace of a Hilbert space.
2. Let X be a Hilbert space and let $P : X \to X$ be a map satisfying $P \circ P = P$ and $(P(w) \mid x) = (w \mid P(x))$ for all $w, x \in X$. Show that P is linear and continuous. Setting $Y := \{y \in X : P(y) = y\}$ show that P coincides with the projection operator P_Y on Y.
3. Let $T := [0,1]$ and let $X := C(T, \mathbb{C})$ be endowed with the Hermitian form defined by $(x \mid y) = \int_0^1 \overline{x(t)}y(t)dt$. Given $a \in X$, let $A \in L(X,X)$ be given by $(Ax)(t) := a(t)x(t)$. Show that A has an adjoint $A^* \in L(X,X)$ although X is not complete. [Hint: Verify that $(A^*x)(t) := \overline{a(t)}x(t)$ for all $x \in X, t \in T$.] Compute $\|A\|$.

4. With the notation of the preceding exercise, let $K : T \times T \to \mathbb{C}$ be a continuous map and let $A \in L(X, X)$ be given by $(Ax)(t) := \int_0^1 K(s, t)x(s)ds$. Show that A has an adjoint and give an explicit expression for it.

5. Let X be a Hilbert space and let $A \in L(X, X)$. Prove that the following assertions are equivalent:

$$A^*A = I_X \text{ the identity map of } X$$
$$\langle Ax \mid Ay \rangle = \langle x \mid y \rangle \text{ for all } x, y \in X$$
$$\|Ax\| = \|x\| \text{ for all } x \in X.$$

An operator satisfying these three properties is called *unitary* if X is a complex space or *orthogonal* if X is a real space.

6. Let X be a pre-Hilbertian space and let $A \in L(X, X)$. One says that A is a *symmetry* if $P := (1/2)(A + I_X)$ is a projection operator. Using Exercises 2 and 5 show that A is a symmetry if and only if A is both Hermitian and unitary.

7. Let X be a Hilbert space and let $A \in L(X, X)$. Let $q : x \mapsto \frac{1}{2}b(x, x) := \frac{1}{2}\langle Ax \mid x \rangle$ be the quadratic form associated with A. Show that q is *coercive* (in the sense that $q(x) \to \infty$ as $\|x\| \to \infty$) if, and only if q is *supercoercive* (in the sense that $\liminf_{\|x\| \to \infty} q(x)/\|x\| > 0$) if, and only if q is *hypercoercive* (in the sense that $\lim_{\|x\| \to \infty} q(x)/\|x\| = \infty$), if, and only if there exists some $c > 0$ such that $q(x) \geq c\|x\|^2$ for all $x \in X$.

8. Let X be a Hilbert space, let W be a closed linear subspace of X and let $A \in L(X, X)$ be a coercive operator. Show that the operator $B \in L(W, W)$ given by $B := P_W \circ A \circ j_W$, where $j_W : W \to X$ is the canonical injection and $P_W : X \to W$ is the orthogonal projection onto W, is coercive and satisfies $b(w, w) = a(w, w)$ for all $w \in W$, where $b(w, w) := \langle Bw \mid w \rangle$, $a(x, x) := \langle Ax \mid x \rangle$ for $w \in W$, $x \in X$.

Given $\ell \in X^*$, prove that there exists some $c > 0$ such that the solutions $u \in X$, $v \in W$ of the equations $Au = \ell$, $Bv = \ell\,|_W$ satisfy $\|u - v\| \leq cd(u, W)$.

9. **(Galerkin method)** Let X be a Hilbert space, let (W_n) be a sequence of closed linear subspaces of X and let $A \in L(X, X)$ be a coercive operator. The map $P_{W_n} \circ A \circ j_{W_n}$ is denoted by $A_n \in L(W_n, W_n)$ and, given $\ell \in X^*$, the solution of the equation $A_n w = \ell\,|_{W_n}$ (resp. $Ax = \ell$) is denoted by u_n (resp. u). Let Y be a dense linear subspace of X such that $(d(y, W_n)) \to 0$ for all $y \in Y$. Show that $(\|u_n - u\|)_n \to 0$.

10. **(Stampacchia's Theorem)** Let C be a nonempty closed convex subset of a Hilbert space X and let a be a continuous bilinear form on X that is coercive in the sense that there exists some $c > 0$ such that $a(x, x) \geq c\|x\|^2$ for all $x \in X$. Prove that for all $f \in X^*$ there exists a unique $u \in C$ such that $a(u, x-u) \geq \langle f, x \rangle$ for all $x \in C$. [Hint: for $r := 2c/\|a\|^2$ consider the map $g : C \to C$ defined by $g(x) := P_C(rf - rAx + x)$, where $A \in L(X, X)$ is the operator associated with a and show that g is a contraction. Then the fixed point of g is the solution to the above *variational inequality*].

Many unilateral (i.e. one-sided) problems can be studied with such a model.

11. Show that the Lax-Milgram Theorem is a consequence of Stampacchia's Theorem. [Hint: take $C = X$]

12. Given separable infinite dimensional Hilbert spaces X and Y and elements A, B of the space $HS(X, Y)$ of Hilbert-Schmidt operators from X into Y, as defined in Exercise 4 of the preceding section, along with its scalar product, show that $\langle A^* \mid B^* \rangle_{HS} = \langle A \mid B \rangle_{HS}$, so that $\|A^*\|_{HS} = \|A\|_{HS}$.

4.5 Fourier Series

Let X be the space of continuous complex-valued functions that are periodic with period 1 (in short, 1-periodic). It can be identified with the space $C(\mathbb{T}, \mathbb{C})$ of continuous complex-valued functions on the *torus* $\mathbb{T} := \mathbb{R}/\mathbb{Z}$. For $x, y \in X$ let

$$\langle x \mid y \rangle := \int_0^1 \overline{x(t)} y(t) dt.$$

This positive Hermitian form is positive definite since the relation $\langle x \mid x \rangle = 0$ implies that x is null on $[0, 1]$ since x is continuous, hence is null on $\mathbb{R}$ since x is 1-periodic. The space X is not complete with respect to the associated norm, but its completion is the classical Lebesgue space $L_2(\mathbb{T}, \mathbb{C})$.

Proposition 4.5 *The family $(e_n)_{n \in \mathbb{Z}}$ given by $e_n(t) = e^{2\pi int}$ forms a Hilbert basis of $X := C(\mathbb{T}, \mathbb{C})$.*

For $x \in X$, $n \in \mathbb{Z}$, setting $x_n := c_n(x) := \langle e_n \mid x \rangle$, the series $\sum_{-\infty}^{+\infty} x_n e_n$ converges and its sum is x.

Proof The relations $\langle e_n \mid e_p \rangle = 0$ for $n, p \in \mathbb{Z}$, $n \neq p$ and $\langle e_n \mid e_n \rangle = 1$ are immediate.

It is a consequence in the Stone-Weierstrass Theorem that any $f \in X$ is the limit for the norm $\|\cdot\|_\infty$ of a sequence in linear combinations of the e_n's. Since one has $\langle x \mid x \rangle \leq \|x\|_\infty^2$ for all $x \in X$, the family (e_n) is total in X, hence forms a Hilbert basis of X.

The second assertion is a consequence in Theorem 4.4. □

In the space of continuous real-valued 1-periodic functions endowed with the restriction of the preceding scalar product, one can show that the family $(\sqrt{2} \sin 2\pi nt, \sqrt{2} \cos 2\pi nt)_{n \in \mathbb{N}}$ forms an orthonormal family and that a result similar to Proposition 4.5 holds for the Fourier series associated with a continuous 1-periodic real-valued function. Let us note that the convergence of the series holds for the norm $\|\cdot\|_2$ associated with the scalar product and not for the norm $\|\cdot\|_\infty$. One does not even have pointwise convergence in general.

The *Fourier series* of $x \in C(\mathbb{T}, \mathbb{C})$ is the series associated with the sequence $x_\Sigma := (x_n)_{n\in\mathbb{Z}}$ given by

$$x_n := \langle e_n \mid x \rangle = \int_0^1 e^{-2\pi i n t} x(t) dt.$$

We observe that the Fourier series of $f \in L_1(\mathbb{T}, \mathbb{C})$ is related to the Fourier transform $\widehat{1_S f}$ of the function $1_S f$, with $S := [0, 1]$, via the relation $f_\Sigma = \widehat{1_S f} \mid_\mathbb{Z}$. Here $1_S f$ denotes the function on $\mathbb{R}$ given by $1_S f(t) = f(t)$ for $t \in S$, $1_S f(t) = 0$ for $t \in \mathbb{R} \backslash S$ and the Fourier transform of $g \in R(\mathbb{R}, \mathbb{C}$ is given by

$$\hat{g}(y) := \int_{\mathbb{R}^d} e^{-2\pi i x y} g(x) dx.$$

If $f : \mathbb{R} \to \mathbb{C}$ is a periodic continuous function with period T, one defines the Fourier coefficients of f (with respect to the functions $t \mapsto (1/T)e^{2\pi i n t/T}$) as the Fourier coefficients of the 1-periodic function x given by $x(t) := f(Tt)$:

$$c_n(f, T) = \frac{1}{T} \int_0^T e^{-2\pi i n s/T} f(s) ds.$$

In particular, if f is 2π-periodic, then $c_n(f, 2\pi) = \frac{1}{2\pi} \int_0^{2\pi} e^{-ins} f(s) ds$. We denote by $S_n(f, T)$, or for short $S_n(f)$ when $T = 1$, the trigonometric polynomial defined by

$$S_n(f, T)(s) := \sum_{k=-n}^{n} c_k(f, T) e^{2\pi i k s/T} \qquad s \in \mathbb{R}.$$

When f is real, one has $c_{-n}(f, T) = \overline{c_n(f, T)}$ and it is of interest to gather terms and consider the trigonometric polynomial

$$S_n(f, T)(s) := c_0(f, T) + \sum_{k=1}^{n} a_k(f, T) \cos ks/T + \sum_{k=1}^{n} b_k(f, T) \sin ks/T \qquad s \in \mathbb{R},$$

with

$$a_k(f, T) := c_k(f, T) + c_{-k}(f, T) = (2/T) \int_0^T f(s) \cos nsds \in \mathbb{R},$$

$$b_k(f, T) := i(c_k(f, T) - c_{-k}(f, T)) = (2/T) \int_0^T f(s) \sin nsds \in \mathbb{R}.$$

Theorem 4.4 shows that for all $f \in C(\mathbb{T}, \mathbb{C})$ or $L_2(\mathbb{T}, \mathbb{C})$ one has $f_\Sigma \in \ell_2(\mathbb{Z}, \mathbb{C})$. More precisely, one has the following result.

Theorem 4.7 (Riemann-Lebesgue-Parseval) *For all* $f \in L_2(\mathbb{T}, \mathbb{C})$ *one has* $(\int_0^1 |S_n(f)(t) - f(t)|^2\, dt)_n \to 0$ *and* $(c_n(f)) \to 0$. *Moreover*

$$\sum_{n=-\infty}^{+\infty} |c_n(f)|^2 = \int_0^1 |f(t)|^2\, dt.$$

Conversely, one can associate to every $(c_n) \in \ell_2(\mathbb{Z}, \mathbb{C})$ an element f of $L_2(\mathbb{T}, \mathbb{C})$ such that $c_n(f) = c_n$ by setting $f(t) := \Sigma_{n\in\mathbb{Z}} c_n e^{2\pi int}$ for $t \in \mathbb{T}$. With a stronger assumption one gets a more regular function.

Theorem 4.8 (Fourier's Inversion Formula) *For every sequence* $c := (c_n)_{n\in\mathbb{Z}}$ *in* $\ell_1(\mathbb{Z}, \mathbb{C}) := \{(c_n)_{n\in\mathbb{Z}} : \Sigma_{n\in\mathbb{Z}} |c_n| < +\infty\}$ *the series* $\Sigma_{n\in\mathbb{Z}} c_n e^{2\pi int}$ *converges uniformly to some* $x \in C(\mathbb{T}, \mathbb{C})$.

Here, given a sequence $(z_n)_{n\in\mathbb{Z}} \in \mathbb{C}^{\mathbb{Z}}$, the series $\Sigma_{n\in\mathbb{Z}} z_n$ is said to converge if the series $\Sigma_{n\in\mathbb{N}} z_n$ and $\Sigma_{n\in\mathbb{N}} z_{-n}$ converge and its sum is the sum of the two sums. A similar convention holds for the uniform or pointwise convergence of a series of functions.

Proof For all $n \in \mathbb{Z}$ and all $t \in \mathbb{R}$ one has $\left|e^{2\pi int}\right| = 1$, so that $\Sigma_{n\in\mathbb{Z}} c_n e^{2\pi int}$ converges uniformly whenever $\Sigma_{n\in\mathbb{Z}} |c_n|$ converges. Since the trigonometric polynomial functions $\Sigma_{-k}^{k} c_n e^{2\pi int}$ are 1-periodic continuous functions, the sum of the series $\Sigma_{n\in\mathbb{Z}} c_n e^{2\pi int}$ is a 1-periodic continuous function. □

The growth property of the Fourier series of x reflects the regularity of x and the Fourier series of the derivative of x is obtained by a term-by-term differentiation of the Fourier series of x.

Proposition 4.6 *For any* x *in the space* $C^1(\mathbb{T}, \mathbb{C})$ *of* 1*-periodic continuously differentiable functions, the Fourier series of* x' *is the sequence* $(2\pi inx_n)$, *where* (x_n) *is the sequence of Fourier coefficients of* x *and* $\sup_{n\in\mathbb{Z}} |nx_n| < \infty$.

Conversely, if $(c_n) \in \mathbb{C}^{\mathbb{Z}}$ *is such that* $\Sigma_{n\in\mathbb{Z}} n |c_n| < +\infty$, *then the function* $t \mapsto \Sigma_{n\in\mathbb{Z}} c_n e^{2\pi int}$ *belongs to the space* $C^1(\mathbb{T}, \mathbb{C})$.

Proof An integration by parts shows that

$$x'_n := \int_0^1 e^{-2\pi int} x'(t)dt = 2\pi in \int_0^1 e^{-2\pi int} x(t)dt = 2\pi inx_n.$$

Moreover,

$$|nx_n| = \frac{1}{2\pi} \left|x'_n\right| := \frac{1}{2\pi} \left|\int_0^1 e^{-2\pi int} x'(t)dt\right| \leq \frac{1}{2\pi} \int_0^1 \left|x'(t)\right| dt.$$

The converse assertion is a consequence in Theorem 5.6 below since the series $2\pi i \Sigma_{n\in\mathbb{Z}} nc_n e^{2\pi int}$ converges uniformly. □

Example Let $f : \mathbb{R} \to \mathbb{R}$ be the 2π-periodic sawtooth function given by $f(s) = s$ for $s \in [-\pi, \pi[$. An integration by parts gives $c_n(f, 2\pi) = (-1)^{n+1}/in$ for $n \neq 0$ and $c_0(f, 2\pi) = 0$. Thus, the Fourier series of f is

$$\sum_{n \in \mathbb{Z}\setminus\{0\}} \frac{(-1)^{n+1}}{in} e^{ins} = 2\sum_{n=1}^{\infty}(-1)^{n+1}\frac{\sin ns}{n}.$$

It can be shown that this series converges to $f(s)$ for $s \in \mathbb{R}\setminus(2\mathbb{Z}+1)\pi$. However, for $s = (2k+1)\pi$ with $k \in \mathbb{Z}$, the sum of this series is 0, i.e. $(1/2)(f(s_-) + f(s_+))$, a general fact, as stated below. Then, Parseval's identity yields

$$\frac{1}{\pi}\int_{-\pi}^{\pi} s^2 ds = \sum_{n=1}^{\infty} |b_n(f, 2\pi)|^2 = 4\sum_{n=1}^{\infty}\frac{1}{n^2}.$$

We recover the result $\Sigma_{n\geq 1}(1/n^2) = \pi^2/6$ first established by Euler.

Example Considering the 2π-periodic function $f : \mathbb{R} \to \mathbb{R}$ given by $f(s) = s^2$ for $s \in [-\pi, \pi[$, we invite the reader to prove another identity due to Euler: $\Sigma_{n\geq 1}(1/n^4) = \pi^4/90$.

Example The 2π-periodic function $f : \mathbb{R} \to \mathbb{R}$ given by $f(s) = \pi/2 - s/2$ for $s \in [0, 2\pi[$ and $f(-\pi) = f(\pi) = 0$ is odd, so that its coefficients $a_n(f, 2\pi)$ are null, whereas an integration by parts shows that $b_n(f) := b_n(f, 2\pi) = 1/n$.

Example The *Dirichlet kernel* is the trigonometric polynomial D_n defined by

$$D_n(s) := \sum_{k=-n}^{k=n} e^{iks}.$$

It is of crucial importance in the study of Fourier series since for any 2π-periodic integrable function f one has

$$S_n(f, 2\pi)(s) := \frac{1}{2\pi}\sum_{k=-n}^{k=n} e^{iks}\int_0^{2\pi} f(t)e^{-ikt}dt = \frac{1}{2\pi}\int_0^{2\pi} f(t)D_n(s-t)dt.$$

Using the relation $e^{iks} + e^{-iks} = 2\cos ks$ we see that $D_n(s) = 1 + 2\Sigma_{k=1}^{n}\cos ks$. Moreover, considering the geometric progressions $\Sigma_{k=0}^{n}c^k$ and $\Sigma_{k=-n}^{-1}c^k$ with $c := e^{is}$ whose sums are $(1-c^{n+1})/(1-c)$ and $(c^{-n}-1)/(1-c)$ respectively for $s \in \mathbb{R}\setminus(\pi\mathbb{Z})$, and writing

$$(1-c^{n+1})/(1-c) + (c^{-n}-1)/(1-c) = (c^{-n-1/2} - c^{n+1/2})/(c^{-1/2} - c^{1/2}),$$

we get

$$D_n(s) = \frac{\sin((2n+1)s/2)}{\sin s/2}.$$

The question of convergence of Fourier series is a delicate and important subject which is beyond the scope of the book. In [239, pp. 83–87] one can find an example of a continuous periodic function whose Fourier series diverges. For the reader's information we quote some positive results. The last one is a recent and deep theorem.

Proposition 4.7 *For all $f \in \mathcal{L}_1(\mathbb{T}, \mathbb{C})$ its Fourier series converges to f in Cesàro's sense at every point of continuity t of f : $\tilde{S}_n(f)(t) := (1/n)[S_0(f)(t) + \ldots + S_{n-1}(f)(t)]$ converges to $f(t)$ as $n \to \infty$. If f is continuous then $(\tilde{S}_n(f))_n \to f$ uniformly.*

Proposition 4.8 *Let $f \in C(\mathbb{R}, \mathbb{C})$ be 1-periodic and stable at some $\bar{t} \in \mathbb{R}$ in the sense that there exists some $c > 0$ such that $|f(t) - f(\bar{t})| \le c\,|t - \bar{t}|$ for t near $\bar{t}$. Then the Fourier series of f converges to $f(\bar{t})$ at $\bar{t}$.*

Proposition 4.9 *Let f be 1-periodic and regulated. Then for all $t \in \mathbb{R}$ at which the Fourier series of f converges, it converges to $(1/2)(f(t_-) + f(t_+))$.*

Theorem 4.9 (Dirichlet-Jordan) *Let $f \in C(\mathbb{R}, \mathbb{C})$ be 1-periodic and with bounded variation on $[0, 1]$. Then, for all $t \in \mathbb{R}$ the Fourier series of f pointwise converges to $t \mapsto (1/2)(f(t_-) + f(t_+))$. In particular, the Fourier series of f pointwise converges to f on the set of continuity points of f.*

Theorem 4.10 (Carleson [65]) *For every $f \in \mathcal{L}_2(\mathbb{T}, \mathbb{C})$ its Fourier series converges to f almost everywhere.*

4.5.1 Application: The Dirichlet Problem for the Disk

The Dirichlet problem for the open unit disc $\Omega := B(0, 1)$ in the Euclidean plane is to solve the steady-state heat equation

$$\Delta u := \frac{\partial^2 u}{\partial x^2} + \frac{\partial^2 u}{\partial y^2} = 0 \quad \text{in } \Omega$$

$$u \mid \partial\Omega = f,$$

where $\partial\Omega$ is the boundary of Ω, i.e. the unit circle, and f is a given function on $\partial\Omega$. The geometry of the problem leads to use polar coordinates (r, θ), so that, writing

$u(r, \theta)$ instead of $u(rcos\theta, rsin\theta)$ by an abuse of notation,

$$\Delta u = \frac{\partial^2 u}{\partial r^2} + \frac{1}{r}\frac{\partial u}{\partial r} + \frac{1}{r^2}\frac{\partial^2 u}{\partial \theta^2}.$$

Writing $r^2 \Delta u = 0$ under the form

$$r^2 \frac{\partial^2 u}{\partial r^2} + r\frac{\partial u}{\partial r} = -\frac{\partial^2 u}{\partial \theta^2},$$

we look for solutions $(r, \theta) \mapsto u(r, \theta)$ with separable variables: $u(r, \theta) := v(r)w(\theta)$. We require that $v(1) = 1$, so that the boundary condition becomes $w = f$. Dividing both sides of the preceding equation by $v(r)w(\theta)$ we get

$$\frac{r^2 v''(r) + rv'(r)}{v(r)} = -\frac{w''(\theta)}{w(\theta)}.$$

Since the two sides depend on independent variables, they must be the same constant. We call it λ and we get two equations:

$$w''(\theta) + \lambda w(\theta) = 0 \tag{4.6}$$

$$r^2 v''(r) + rv'(r) = \lambda v(r). \tag{4.7}$$

Since w must be periodic, we look for solutions of the first equation of the form

$$w_n(\theta) = a_n \cos n\theta + b_n \sin n\theta$$

with $n^2 = \lambda$, $n \in \mathbb{N}$, $a_n, b_n \in \mathbb{R}$. In view of the linearity of the problem Fourier suggested to assume f is of the form

$$f(\theta) = \sum_{n \in \mathbb{N}} a_n \cos n\theta + \sum_{n \in \mathbb{N}} b_n \sin n\theta. \tag{4.8}$$

In fact, assuming f is 2π-periodic and of class C^2 Proposition 4.6 ensures that f can be expanded in a uniformly convergent Fourier series as in the right-hand side of (4.8), with $\Sigma_{n \geq 0} n^2(|a_n| + |b_n|) < +\infty$. Now we look for solutions v_n of (4.7) with $\lambda = n^2 > 0$ by setting $v(r) = r^n z(r)$ for $r \in]0, 1]$, with $z(1) = 1$. This leads to the equation $rz''(r) + (n + 1)z' = 0$. We discard the solutions of the form $r \mapsto cr^{-n}$ that are unbounded around 0 and we keep the solution $z = 1$, so that v_n is given by $v_n(r) := r^n$, $n \in \mathbb{N} \backslash \{0, 1\}$, in order that u be of class C^2 and we can take

$$u(r, \theta) = \sum_{n \in \mathbb{N}} v_n(r) w_n(\theta) = \sum_{n \in \mathbb{N}} r^n (a_n \cos n\theta + b_n \sin n\theta).$$

4.5.2 Application: Dido's Problem

Dido's problem is the most famous example of a so-called isoperimetric problem. It consists in determining a curve in $\mathbb{R}^2$ with a given length enclosing a figure of greatest area. It is connected with the legend of the foundation of Carthage, as told by Virgil, an instance of the guile of colonialists (or women, depending on your views).

Let $z := (x, y) \in C^1([0, 2\pi], \mathbb{R}^2)$ be a simple closed curve of $\mathbb{R}^2$. This means that $z(2\pi) = z(0)$ and that for any pair $s, t \in [0, 2\pi[$ one has $z(s) \neq z(t)$ whenever $s \neq t$. In fact, we identify two curves w and z if there exists an increasing function $h : [a, b] \to [0, 2\pi]$ of class C^1 such that $h(a) = 0$, $h(b) = 2\pi$ and $w(s) = z(h(s))$ for all $t \in [a, b]$. The length of z is given by

$$\ell(z) := \int_0^{2\pi} (x'(t)^2 + y'(t)^2)^{1/2} dt$$

and does not change if z is reparameterized into $z \circ h$ as above. Thus we assume that z is parameterized by arc length, i.e. that $x'(t)^2 + y'(t)^2 = 1$ for all $t \in [0, 1]$, so that $\ell(z) = 2\pi$. It can be shown (Jordan's theorem) that $\mathbb{R}^2 \backslash z([0, 2\pi])$ has two connected components, one bounded and one unbounded. Let Ω be the bounded one, called the region enclosed by z. Since $x(t)x'(t) + y(t)y'(t) = 0$, Green's formula asserts that the area a of Ω is given by

$$a := \frac{1}{2} \int_0^{2\pi} (x(t)y'(t) - y(t)x'(t))dt = \frac{1}{2i} \int_0^{2\pi} \overline{z(t)} z'(t)dt = \frac{1}{2i} \int_0^1 \overline{w(s)} w'(s)ds,$$

considering z as a complex-valued function and setting $w(s) := z(2\pi s)$ for $s \in [0, 1]$, taking into account the relations $x^2(0) = x^2(2\pi)$ and $y^2(0) = y^2(2\pi)$. Using Parseval's equality, since $c_n(w') = 2\pi i n c_n(w)$, we get

$$a = \frac{1}{2i} \sum_{n \in \mathbb{Z}} \overline{c_n(w)} c_n(w') = \pi \sum_{n \in \mathbb{Z}} n \left| c_n(w) \right|^2 .$$

$|z'| = 1$ and since $n^2 \geq n$ for all $n \in \mathbb{Z}$ we obtain

$$\begin{aligned} \ell^2(z) = 2\pi \int_0^{2\pi} \left| z'(t) \right|^2 dt &= \int_0^1 \left| w'(s) \right|^2 ds = \sum_{n \in \mathbb{Z}} \left| c_n(w') \right|^2 \\ &= 4\pi^2 \sum_{n \in \mathbb{Z}} n^2 \left| c_n(w) \right|^2 \geq 4\pi a, \end{aligned}$$

with strict inequality if $c_n(w) \neq 0$ for at least one $n \in \mathbb{Z} \backslash \{-1, 0, 1\}$. Equality holds for $w(s) = e^{2\pi i s}$ either by considering the Fourier coefficients of w or by noting that in such a case one has $\ell(z) = 2\pi$ and $a = \pi$. Thus, the circle is the solution of Dido's problem. The above proof is due to Hurwitz (1902).

The following exercises give a (very) short account of the recent theory of wavelets. It has numerous applications (for instance in the design of JPEG2000, the industrial standard for image compression replacing the older Fourier-based JPEG standard). See [31, 149, 158, 181, 212, 240, 257, 258].

Exercises

1*. Let $L_2(\mathbb{R})$ be the completion of the space of continuous functions f on $\mathbb{R}$ such that

$$\|f\|_2 := \left[\int_{\mathbb{R}} |f(r)|^2\, dr\right]^{1/2} < +\infty.$$

endowed with this norm $\|\cdot\|_2$.

(**a**) Show that for $m, n \in \mathbb{Z}$ the translation T_m and dilation D_n operators given as follows are unitary operators:

$$(T_m f)(r) := f(r-m), \qquad (D_n f)(r) := 2^{n/2} f(2^n r) \qquad r \in \mathbb{R}.$$

(**b**) Note that $D_n T_m = T_{2^{-n}m} D_n$.

(**c**) Show that for all $\psi \in L_2(\mathbb{R})$ one has $(D_n T_m(\hat{\psi}))(s) = g_{m,n}(s)\hat{\psi}(2^{-n}s)$ with $g_{m,n}(s) := 2^{-n/2} e^{-2i\pi m 2^{-n} s}$, $\hat{\psi} := \mathcal{F}(\psi)$ being the Fourier transform of ψ.

A (dyadic) *wavelet* is a function $\psi \in L_2(\mathbb{R})$ such that the family $W_\psi := \{D_n T_m \psi : m, n \in \mathbb{Z}\}$ forms an orthonormal basis of $L_2(\mathbb{R})$.

2*. (**a**) Let $S := [-1, -\frac{1}{2}[\cup[\frac{1}{2}, 1[$. Verify that the family $\{2^n S : n \in \mathbb{Z}\}$ is a partition of $\mathbb{R}\backslash\{0\}$.

(**b**) Show that the family $\{e_k 1_S : k \in \mathbb{Z}\}$ is an orthonormal basis of $L_2(S)$.

(**c**) Deduce from this that for all $n \in \mathbb{Z}$ the restrictions to the set $2^n S$ of the functions $g_{m,n}$ $(m \in \mathbb{Z})$ defined in the preceding exercise form an orthonormal basis of $L_2(2^n S)$.

(**d**) Conclude that $\psi^S := \mathcal{F}^{-1}(1_S)$ is a wavelet, called the *Shannon wavelet*.

3*. A general method to construct wavelets is the so-called *Multiresolution Analysis* (MRA) method. An MRA is a sequence $(V_n)_{n\in\mathbb{Z}}$ of closed linear subspaces of $L_2(\mathbb{R})$ whose union is dense in $L_2(\mathbb{R})$ and such that

(i) (V_n) is increasing: $V_n \subset V_{n+1}$ for all $n \in \mathbb{Z}$;
(ii) $\cap_n V_n = \{0\}$;
(iii) $V_{n+1} = D_1 V_n$ for all n: $f \in V_n$ if, and only if $f(2\cdot) \in V_{n+1}$;
(iv) there exists a $\varphi \in V_0$ such that $(T_k\varphi)_{k\in\mathbb{Z}}$ is an orthonormal basis of V_0.

(**a**) Given an MRA (V_n), let $W_n := V_n^{\perp} \cap V_{n+1}$. Show that the subspaces W_n are mutually orthogonal and that $\oplus_{n\in\mathbb{Z}} W_n = L_2(\mathbb{R})$.

(**b**) Prove that $D_n W_0 = W_n$ for all $n \in \mathbb{Z}$.

(c) Suppose that for some $\psi \in W_0$ the family $(T_k\psi)_{k\in\mathbb{Z}}$ forms an orthonormal basis of W_0. Deduce from the preceding that $(D_nT_k\psi)_{k\in\mathbb{Z}}$ forms an orthonormal basis of W_n and that $(D_nT_k\psi)_{k,n\in\mathbb{Z}}$ forms an orthonormal basis of $L_2(\mathbb{R})$: ψ is a wavelet.

(d) Let $\varphi := 1_{[0,1[}$ and let V_0 be the closed space spanned by the family $(T_k\varphi)_{k\in\mathbb{Z}}$. Show that for $V_n := D_nV_0$ the family $(V_n)_{n\in\mathbb{Z}}$ is an MRA.

(e) Let $\psi^H = 1_{[0,1/2[} - 1_{[1/2,1[}$. Show that $(D_nT_k\psi^H)_{k,n\in\mathbb{Z}}$ forms an orthonormal basis of $L_2(\mathbb{R})$: ψ^H is a called the *Haar wavelet*.

4*. Prove that a similar construction with $\varphi :=$ sinc given by $\mathrm{sinc}(r) := \frac{\sin r\pi}{r\pi}$ for $r \in \mathbb{R}\backslash\{0\}$ and $\sin c(0) := 0$ shows that the Shannon wavelet is associated with an MRA.

The function sinc is involved in the following result (*Whittaker-Shannon-Kotelnikov Sampling Theorem*): if $f \in L_2(\mathbb{R})$ is such that the support function of $\hat{f} := \mathcal{F}(f)$ is contained in $[-1/2, 1/2]$ then f is determined by its values on $\mathbb{Z}$:

$$\forall r \in \mathbb{R} \qquad f(r) = \sum_{k\in\mathbb{Z}} f(k) \sin c(r-k).$$

5. Let $(r_n)_n$ be a nonincreasing sequence in real numbers and let $(c_n)_n$ be a sequence in complex numbers such that there exists an $m \in \mathbb{R}_+$ satisfying $|c_0 + \ldots + c_n| \leq m$ for all $n \in \mathbb{N}$. Prove that for all $n \in \mathbb{N}$ one has

$$|r_0c_0 + \ldots + r_nc_n| \leq mr_0.$$

[Hint: set $s_n := c_0 + \ldots + c_n$ and write (*Abel transformation*)

$$r_0c_0 + \ldots + r_nc_n = s_0r_0 + (s_1 - s_0)r_1 + \ldots. + (s_n - s_{n-1})r_n.]$$

6. Using the preceding exercise, show that the series $\Sigma_{n\geq 1}(1/n)\sin nx$ converges uniformly on $[\theta, 2\pi - \theta]$ for all $\theta \in]0, \pi[$. [Hint: prove and use the relation

$$\sum_{1\leq k\leq p} \sin(n+k)x = \frac{1}{2\sin(x/2)}[\cos((2n+1)x/2) - \cos((2n+2p+1)x/2)]$$

for $x \in \mathbb{R}\backslash 2\pi\mathbb{Z}$.]

7. Denoting by $D_n : x \mapsto (\sin x/2)^{-1}\sin(2n+1)x/2$ for $x \in \mathbb{R}\backslash(2\pi\mathbb{Z})$ the Dirichlet kernel, **Fejér's kernel** is given by

$$F_n(x) := \frac{1}{n+1}(D_0(x) + D_1(x) + \ldots + D_n(x)).$$

Show that

$$F_{n-1}(x) = \frac{1}{n}\frac{\sin^2(nx/2)}{\sin^2(x/2)} \qquad x \in \mathbb{R}\backslash(2\pi\mathbb{Z}).$$

Given a 2π-periodic regulated function f one sets

$$R_n(f)(x) := \frac{1}{n+1}(S_0(f, 2\pi)(x) + \ldots + S_n(f, 2\pi)(x)).$$

Prove that

$$f(x) - R_n(f)(x) = \frac{1}{2\pi}\int_0^{2\pi} (f(x) - f(x-t))F_n(t)dt,$$

$$(R_n(f)(x))_n \to \frac{1}{2}(f(x_-) + f(x_+))$$

and that if f is continuous $(R_n(f))_n \to f$ uniformly. [Hint: cut the integration interval into the three pieces $[0, \alpha]$, $[\alpha, 2\pi - \alpha]$, $[2\pi - \alpha, 2\pi]$.]

4.6 Orthogonal Polynomials

Many mathematicians have proposed families of orthogonal polynomials. They are used for various concrete problems such as approximation, representation, and differential equations; see [20, 107] f.i.. They enjoy special properties. We just give a brief, general account. Given a closed interval T in $\overline{\mathbb{R}}$ (bounded or unbounded) and a continuous function w on T, positive on int$T$ considered as a weight, let X be the set of continuous functions $x(\cdot)$ on T such that

$$\int_T w(t)\,|x(t)|^2\,dt < +\infty.$$

The relations $2\,|ab| \leq |a|^2 + |b|^2$ and $|a+b|^2 \leq 2\,|a|^2 + 2\,|b|^2$ show that X is a linear space and that the function h given by

$$h(x,y) := \int_T w(t)\overline{x(t)}y(t)dt$$

is a Hermitian form on X. This Hermitian form is positive definite since $h(x,x) = 0$ implies that x is null on int$T$, hence is null on T. In the sequel we assume that for all

$n \in \mathbb{N}$ we have

$$\int_T w(t)\,|t^n|\,dt < +\infty,$$

so that X contains the restrictions to T of the polynomial functions. Since the coefficients of a polynomial function that is null on T are null, the monomials t^n are linearly independent and one can use the Gram-Schmidt process to construct from them an orthogonal family (p_n). Let us give some classical examples.

Example *Hermite polynomials* are obtained for $T := \mathbb{R}$ and $w(t) := e^{-t^2}$.

Example *Laguerre polynomials* are obtained for $T := \mathbb{R}_+$ and $w(t) := e^{-t}$.

Example *Jacobi polynomials* are obtained for $T := [-1, 1]$ and $w(t) := (1-t)^r(1+t)^s$ with $r, s \in]-1, +\infty[$. For $r = s = 0$ (so that $w(\cdot) = 1$) they are called *Legendre polynomials*. For $r = s = -1/2$ they are called *Chebyshev polynomials*.

Among the interesting properties of the scalar product associated with w and of the polynomials p_n we note the following, left as exercises (note that the degree of $p_1 p_n - p_{n+1}$ is at most n).

Proposition 4.10 *For all x, y, z in the space X introduced above, one has $\langle x \mid yz\rangle = \langle x\overline{y} \mid z\rangle = \langle x\overline{yz} \mid 1\rangle$.*
For all n one has $\langle p_1 p_n \mid p_{n+1}\rangle = \langle p_{n+1} \mid p_{n+1}\rangle$.

Exercises

1. Prove that the polynomials p_n satisfy an inductive relation of the form
$$p_n(t) = (t + b_n)p_{n-1}(t) - c_n p_{n-2}(t)$$
for $n \geq 2$, with $b_n \in \mathbb{R}$, $c_n > 0$.
2. Show that for all $n \in \mathbb{N}$ the polynomial p_n has n distinct real roots contained in int$T$.
3. Verify that for the Chebyshev polynomials p_n the functions $\sqrt{2/\pi}p_n$ form an orthonormal family.
4. Show that the first Legendre polynomials are given by $p_0(t) = 1$, $p_1(t) = t$, $p_2(t) = (1/2)(3t^2 - 1)$, $p_3(t) = (1/2)(5t^3 - 3t)$.
5. Prove that $\sum_n x^n p_n(t) = (1 - 2tx + x^2)^{-1/2}$ for the Legendre polynomials p_n.
6. Prove that the Legendre polynomial p_n is a solution of the differential equation
$$(t^2 - 1)p_n''(t) + 2tp_n'(t) - n(n+1)p_n = 0$$
and that one has $p_n(t) = (1/2^n n!)\frac{d^n}{dt^n}(t^2 - 1)^n$.

4.7 Elementary Spectral Theory for Self-Adjoint Operators

If X is a complex (resp. real) Hilbert space, an operator $A \in L_{\mathbb{C}}(X,X)$ (resp. $A \in L_{\mathbb{R}}(X,X)$) is called *Hermitian* (resp. *self-adjoint* or *symmetric*) if $A^* = A$. This relation is equivalent to the fact that h_A defined by $h_A(x,y) := \langle Ax \mid y \rangle$ for $x, y \in X$ is Hermitian (resp. symmetric). When X is a complex Hilbert space A is Hermitian if and only if $h_A(x,x) \in \mathbb{R}$ for all $x \in X$. In fact, if $A = A^*$, then for all $x \in X$ one has $\langle Ax \mid x \rangle = \langle x \mid A^*x \rangle = \langle x \mid Ax \rangle = \overline{\langle Ax \mid x \rangle}$. Conversely, if $h_A(x,x) \in \mathbb{R}$ for all $x \in X$ we have

$$\forall x \in X \qquad \langle A^*x \mid x \rangle = \langle x \mid Ax \rangle = \langle Ax \mid x \rangle,$$

hence $A^* = A$ in view of the second part of the next lemma, taking in it $A^* - A$ instead of A.

Lemma 4.1 *If a self-adjoint operator A on a real Hilbert space X satisfies $\langle Ax \mid x \rangle = 0$ for all $x \in X$ then $A = 0$. The same conclusion holds if X is a complex Hilbert space and if A is any $\mathbb{C}$-linear operator satisfying this property.*

Proof This stems from the so-called *polarization identity*: for all $x, y \in X$

$$\langle A(x+y) \mid x+y \rangle - \langle A(x-y) \mid x-y \rangle = 2\langle Ax \mid y \rangle + 2\langle Ay \mid x \rangle.$$

By assumption the left-hand side is 0. If A is self-adjoint the right-hand side is $4\langle Ax \mid y \rangle$, so that $Ax = 0$ for all $x \in X$.

If A is $\mathbb{C}$-linear, replacing x by ix we get

$$-i\langle Ax \mid y \rangle + i\langle Ay \mid x \rangle = 0$$

along with $\langle Ax \mid y \rangle + \langle Ay \mid x \rangle = 0$, hence $\langle Ax \mid y \rangle = 0$. □

Remark In the real case the conclusion does not hold if A is not self-adjoint, as shown by the rotation $(r,s) \mapsto (-s,r)$ in $\mathbb{R}^2$. □

On the space $L(X) := L_{\mathbb{R}}(X,X)$ one defines a preorder by setting $A \succeq B$ if $\langle Ax \mid x \rangle \geq \langle Bx \mid x \rangle$ for all $x \in X$ and one says that A is *positive* (or *positive semi-definite*) if $\langle Ax \mid x \rangle \geq 0$ for all $x \in X$. By the preceding lemma, this preorder induces an order on the space $L_s(X)$ of symmetric operators, or, when X is a complex space on the space $L_{\mathbb{C}}(X,X)$ of $\mathbb{C}$-linear operators. For $A \in L(X)$, setting

$$a := \inf_{x \in S_X} \langle Ax \mid x \rangle, \qquad b := \sup_{x \in S_X} \langle Ax \mid x \rangle \tag{4.9}$$

which are finite real numbers in the interval $[-\|A\|, \|A\|]$, one has

$$aI \preceq A \preceq bI,$$

where $I := I_X$ denotes the identity map of X. When A is symmetric one can refine Theorem 3.35 which asserts that the spectrum $\sigma(A)$ of A is contained in $[-\|A\|, \|A\|]$.

Proposition 4.11 *If A is a symmetric operator the numbers a and b defined in (4.9) are elements of the spectrum $\sigma(A)$ of A. Moreover, $\sigma(A) \subset [a, b]$ and $\|A\| = \max(-a, b)$.*

Proof We first observe that for $r > b$ one has $r \in \rho(A)$, i.e. $A - rI$ is invertible, in view of the Lax-Milgram Theorem and of the inequalities $r - b > 0$, $-\langle Ax \mid x\rangle \geq -b\|x\|^2$,

$$\forall x \in X \qquad \langle (rI - A)x \mid x\rangle \geq (r - b)\|x\|^2 .$$

Changing A into $-A$, we see that $]-\infty, a[\subset \rho(A)$. Gathering the two conclusions we get $\sigma(A) \subset [a, b]$.

Now let us show that $b \in \sigma(A)$. Let $\beta := \|bI - A\|^{1/2}$. Setting $h_A(x, y) := \langle (bI - A)x \mid y\rangle$ for $x, y \in X$, we define a positive symmetric bilinear form. The Cauchy-Schwarz inequality yields

$$|\langle (bI - A)x \mid y\rangle| \leq |\langle (bI - A)x \mid x\rangle|^{1/2} . |\langle (bI - A)y \mid y\rangle|^{1/2}$$

for all $x, y \in X$, so that, taking the supremum over $y \in S_X$ we see that

$$\|(bI - A)x\| \leq \beta\, |\langle (bI - A)x \mid x\rangle|^{1/2}$$

Taking a sequence (x_n) in S_X such that $(\langle Ax_n \mid x_n\rangle)_n \to b$, we get $(\|(bI - A)x_n\|)_n \to 0$: $bI - A$ is not invertible, i.e. $b \in \sigma(A)$. Similarly, $-a \in \sigma(-A) = -\sigma(A)$.

Let $c := \max(-a, b)$. Clearly, (4.9) shows that $b \leq \|A\|$ and $a \geq -\|A\|$. Thus $c \leq \|A\|$. Conversely, the polarization identity yields

$$\begin{aligned} 4\langle Ax \mid y\rangle &= \langle A(x + y) \mid x + y\rangle - \langle A(x - y) \mid x - y\rangle \\ &\leq b\|x + y\|^2 - a\|x - y\|^2 \\ &\leq c(\|x + y\|^2 + \|x - y\|^2) = 2c(\|x\|^2 + \|y\|^2). \end{aligned}$$

Assuming $x \neq 0$, $y \neq 0$ and replacing x with sx, y with ty and choosing $s := (\|y\| / \|x\|)^{1/2}$, $t := (\|x\| / \|y\|)^{1/2}$, we obtain

$$2\langle Ax \mid y\rangle \leq c(s^2\|x\|^2 + t^2\|y\|^2) = 2c\|x\| . \|y\| .$$

Taking the supremum for $x, y \in S_X$, we get $\|A\| \leq c$ and the announced equality. □

Remark If A is a Hermitian operator, one still has the relation $\|A\| = \max(|a|, |b|) = \sup\{|\langle Ax \mid x\rangle| : x \in S_X\}$; see [182, thm 10, section VII,2].

Corollary 4.9 *If a symmetric operator A is such that $\sigma(A) = \{0\}$, then $A = 0$.*

Starting with a symmetric operator A on X, for any polynomial p one can define a new operator $p(A)$ given by

$$p(A) := c_n A^n + \ldots + c_1 A + c_0 I_X \quad \text{if} \quad p(t) = c_n t^n + \ldots + c_1 t + c_0.$$

Then one obtains a ring homomorphism from the algebra $\mathbb{R}[t]$ of real polynomials into the subalgebra $\mathcal{A}$ of $L(X, X)$ generated by A. But one can achieve more in associating an operator $f(A)$ to any continuous function f on the interval $[a, b]$, a and b being given as above, so that one gets a linear ring homomorphism of the algebra $C([a, b])$ into the closure $\text{cl}\mathcal{A}$ of the algebra $\mathcal{A}$ in $L(X, X)$. We need a preliminary algebraic result.

Lemma 4.2 *Let p be a real polynomial that is nonnegative on $[a, b]$. Then there exist a finite family $(q_h)_{h \in H}$ of real polynomials and a partition $H = I \cup J \cup K$ such that*

$$p(t) = \sum_{i \in I} q_i(t)^2 + (t - a) \sum_{j \in J} q_j(t)^2 + (b - t) \sum_{k \in K} q_k(t)^2.$$

Proof We factor p into a product of real polynomials of degree one and two, the quadratic factors being irreducible polynomials of the form $(t-c)^2 + d^2$. The other factors are of the form $(t - r)$, where r belongs to the set R of real roots of p. If $r \in R$ belongs to the interior of $[a, b]$, its multiplicity is even since otherwise p would change sign around r. If $r \in R$ is not greater than a we write the linear factor $t - r$ as $(t - a) + (a - r)$ with $a - r \geq 0$. If $r \in R$ is not less than b we write the linear factor $r - t$ as $(r - b) + (b - t)$ and $r - b$ is a square. Since $p \geq 0$ on $[a, b]$, the coefficient in front of the product of such factors is positive (we assume $p \neq 0$, a trivial case). Multiplying out all these factors and noting that a product of a sum of squares is a sum of squares, we get an expression of the announced type, except that there still remains terms of the form $(t - r)(s - t)q(t)^2$, where $r \leq a$, $s \geq b$ and q is a real polynomial. However, the identity

$$(t - r)(s - t) = \frac{(s - t)(t - r)^2 + (t - r)(s - t)^2}{s - r}$$

enables us to reduce these terms to terms of the other types. □

From this lemma we can deduce that the map $p \mapsto p(A)$ preserves positivity.

Proposition 4.12 *If A is a symmetric operator, $a, b \in \mathbb{R}$ are such that $aI_X \preceq A \preceq bI_X$ and if p is a real polynomial that is positive on $[a, b]$, then one has $p(A) \succeq 0$. If p and q are real polynomials such that $p \geq q$ on $[a, b]$, then one has $p(A) \succeq q(A)$. Also $\|p(A)\| \leq \|p\|_\infty$ with $\|p\|_\infty := \sup_{t \in [a,b]} |p(t)|$.*

Proof Clearly, for any real polynomial q one has $q^2(A) = (q(A))^2 \succeq 0$. Moreover, if B and C are two symmetric operators with $B \succeq 0$ and if $BC = CB$, then one has $BC^2 \succeq 0$ since

$$\langle BC^2x \mid x\rangle = \langle CBCx \mid x\rangle = \langle BCx \mid Cx\rangle \geq 0.$$

The first assertion then follows from the preceding lemma. The second one ensues by considering $p - q$. Since for $c := \|p\|_\infty$ one has $-c \leq p \leq c$ on $[a, b]$, we get $-cI_X \preceq p(A) \preceq cI_X$, hence $\|p(A)\| \leq c$ by Proposition 4.7. □

Since the map $p \mapsto p(A)$ is linear and continuous from the set of restrictions to $[a, b]$ of polynomials to $L(X, X)$, the Stone-Weierstrass Theorem and the extension theorem ensure that this linear map can be extended to a linear map from $C([a, b])$ to $L(X, X)$ and that for all $f \in C([a, b])$ one has $\|f(A)\| \leq \|f\|_\infty$. Moreover, if $(p_n) \to f$ and $(q_n) \to g$ then $(p_nq_n) \to fg$ and so $(fg)(A) = f(A)g(A)$. Thus the extended map $\gamma : f \mapsto f(A)$ is again an algebra homomorphism from $C([a, b])$ to $\mathrm{cl}\mathcal{A}$.

Proposition 4.13 *If A is a positive symmetric operator, then there exist some S (called the square root of A and often denoted by $A^{1/2}$) in the closure of the algebra $\mathcal{A}$ generated by A such that $S^2 = A$. Moreover, one has $AS = SA$.*

If A and B are two commuting positive symmetric operators, then AB is again positive.

Proof The first assertion is obtained by using the function $t \mapsto t^{1/2}$ on $[0, \|A\|]$ and observing that for any polynomial p, A commutes with $p(A)$.

For the second assertion one introduces the square root S of A and uses the fact that $S^2B \succeq 0$, as observed above. □

The preceding analysis can be refined; this refinement can be bypassed in a first reading. Let K_A be the kernel of the map $\gamma : f \mapsto f(A)$:

$$K_A := \{f \in C([a, b]) : f(A) = 0\}.$$

It is an ideal of the ring $C([a, b])$: for all $h \in C([a, b])$, $k \in K_A$ one has $hk \in K_A$ since $(hk)(A) = h(A)k(A) = 0$. Let

$$Z_A := \{t \in [a, b] : \forall f \in K_A \ f(t) = 0\}$$

be the zero set of K_A called the *(Gelfand) spectrum* of A. Since Z_A is closed, any continuous real-valued function f on Z_A can be extended to a continuous function g on $[a, b]$, with $\|g\|_\infty = \|f\|_A := \sup\{|f(t)| : t \in Z_A\}$ by using Urysohn's theorem. If h is another extension of f, since $h - g$ is null on Z_A, the Ideal Theorem ensures that $h - g$ belongs to the ideal K_A, so that $h(A) = g(A)$. Thus, we may denote by $f(A)$ this unambiguously defined operator. The map $\alpha : C(Z_A) \to \mathrm{cl}\mathcal{A}$ thus obtained is easily seen to be an algebra homomorphism. If f is the restriction to Z_A of a function $\tilde{f} \in C([a, b])$, then, by construction, one has $f(A) = \tilde{f}(A)$. Thus the map

$\gamma : \tilde{f} \mapsto \tilde{f}(A)$ from $C([a, b])$ to $\mathrm{cl}\mathcal{A}$ can be factorized through the map $f \mapsto f(A)$ from $C(Z_A)$ to $\mathrm{cl}\mathcal{A}$ and the restriction map $\rho : \tilde{f} \mapsto \tilde{f} \mid_{Z_A}$ from $C([a, b])$ to $C(Z_A)$: $\gamma = \alpha \circ \rho$.

Theorem 4.11* (Spectral Theorem) *The map $\alpha : C(Z_A) \to \mathrm{cl}\mathcal{A}$ is an isomorphism of Banach algebras sending the cone of nonnegative functions on Z_A onto the cone of positive elements of $\mathrm{cl}\mathcal{A}$. Moreover, α is an isometry.*

Proof For $f \geq 0$ in $C(Z_A)$, we have $f(A) \succeq 0$ since f can be extended to some $\tilde{f} \geq 0$ in $C([a, b])$ and Proposition 4.12 can be used. Conversely, let us show that if for some $f \in C([a, b])$ we have $f(A) \succeq 0$, then necessarily $f \geq 0$ on Z_A. Suppose on the contrary that for some $\bar{t} \in Z_A$ we have $f(\bar{t}) < 0$. For some $\varepsilon > 0$ we have $f(t) < 0$ for all $t \in [\bar{t} - \varepsilon, \bar{t} + \varepsilon] \cap [a, b]$. Let $g \in C([a, b])$ be piecewise affine, $g \geq 0$, null off $[\bar{t} - \varepsilon, \bar{t} + \varepsilon]$ and such that $g(\bar{t}) = 1$. Then, since $-fg \geq 0$ we have $-f(A)g(A) \succeq 0$. But since $f(A) \succeq 0$, $g(A) \succeq 0$, Proposition 4.13 ensures that $f(A)g(A) \succeq 0$. Thus $f(A)g(A) = 0$ and $fg \in K_A$. This is impossible since $(fg)(\bar{t}) \neq 0$ and $K_A = \{h \in C([a, b]) : h(z) = 0 \forall z \in Z_A\}$ by the Ideal Theorem.

We have seen the inequality $\|f(A)\| \leq \|f\|_A$ for all $f \in C(Z_A)$. Let us prove the reverse inequality. Let $r := \|f(A)\|$, so that $rI_X - f(A) \succeq 0$ and $rI_X + f(A) \succeq 0$. The preceding shows that $r - f \geq 0$ and $r + f \geq 0$ on Z_A, hence $\|f\|_A \leq r$ and $\|f(A)\| = \|f\|_A$. Thus α is an isometry, hence is injective. Given $B \in \mathrm{cl}\mathcal{A}$ let (f_n) be a sequence in polynomial functions such that $(f_n(A)) \to B$. Since α is isometric, the restriction to Z_A of the sequence (f_n) is a Cauchy sequence, hence converges to some $f \in C(Z_A)$ in the norm $\|\cdot\|_A$. Then $B = f(A)$, so that α is a bijection. □

Let us give a characterization of the spectrum Z_A of A.

Proposition 4.14 *If A is a symmetric operator then its Gelfand spectrum Z_A coincides with its spectrum, i.e. the set $\sigma(A)$ of numbers $z \in \mathbb{C}$ (in fact $z \in \mathbb{R}$) such that $A - zI_X$ is not invertible.*

Proof For $z \in \mathbb{C} \backslash Z_A$ the function $g : t \mapsto (t - z)(t - \bar{z})$ is non-null on Z_A. Then $h := 1/g \in C(Z_A)$, so that $(A - zI_X)(A - \bar{z}I_X)h(A) = I_X$ and $A - zI_X$ is invertible. Thus, if $A - zI_X$ is non-invertible one has $z \in Z_A$.

Let us show that for any $z \in Z_A$ the operator $A - zI_X$ is non-invertible. Suppose on the contrary that $A - zI_X$ has an inverse B. For $n \in \mathbb{N}\backslash\{0\}$ let $g_n : \mathbb{R} \to \mathbb{R}$ be the continuous function defined by

$$g_n(t) := n \quad \text{for } t \in [z - \frac{1}{n}, z + \frac{1}{n}] \qquad g_n(t) := \frac{1}{|t - z|} \quad \text{for } t \in \mathbb{R}\backslash[z - \frac{1}{n}, z + \frac{1}{n}].$$

Since $|(t - z)g_n(t)| \leq 1$ for all t, we have $\|(A - zI_X)g_n(A)\| \leq 1$, hence

$$\|g_n(A)\| = \|B(A - zI_X)g_n(A)\| \leq \|B\|.$$

On the other hand, $\|g_n\|_A \geq |g_n(z)| = n$, so that $\|g_n(A)\| \geq n$, a contradiction. □

An *eigenvector* of an operator $A \in L(X, X)$ is a vector $v \in X \backslash \{0\}$ such that for some number c (called the associated *eigenvalue*) one has $Av = cv$. Any eigenvalue c of an operator A satisfies $|c| \leq \|A\|$ since $\|Av\| \leq \|A\| . \|v\|$. For a symmetric operator on a finite dimensional Hilbert space or, more generally, for a compact symmetric operator on a Hilbert space X, one can say more. Recall that A is said to be *compact* if the image $A(B_X)$ of the unit ball of X is contained in a compact set.

Proposition 4.15 *If A is a compact symmetric operator on a Hilbert space $X \neq \{0\}$, then either $a := \inf_{x \in S_X} \langle Ax \mid x \rangle$ or $b := \sup_{x \in S_X} \langle Ax \mid x \rangle$ is an eigenvalue of A.*

Proof Proposition 4.11 ensures the existence of a sequence (x_n) of S_X such that $(|\langle Ax_n \mid x_n \rangle|)_n \to c := \|A\|$. Since $r_n := \langle Ax_n \mid x_n \rangle \in \mathbb{R}$, we can find $\varepsilon_n \in \{-1, 1\}$ such that $r_n = \varepsilon_n |r_n|$. Taking a subsequence if necessary, we may suppose (ε_n) has a limit $\varepsilon \in \{-1, 1\}$. Then

$$\|Ax_n - \varepsilon_n cx_n\|^2 = \|Ax_n\|^2 - 2\varepsilon_n c\langle Ax_n \mid x_n \rangle + \varepsilon_n^2 c^2 \|x_n\|^2 \leq \|A\|^2 - 2c|r_n| + c^2.$$

The right-hand side converging to 0, we get that $(Ax_n - \varepsilon_n cx_n) \to 0$. Since A is compact, taking another subsequence if necessary, we may suppose (Ax_n) has a limit $y \in X$. Then $(\varepsilon_n cx_n)$ converges to y. If $c = 0$ one has $A = 0$ and the result is obvious. If $c \neq 0$, for $v := \varepsilon c^{-1} y$ one has $(x_n) \to v$, hence $v \in S_X$ and, passing to the limit, $Av - \varepsilon cv = 0$, so that v is an eigenvector of A. Since $\|A\| = \max(|a|, |b|)$ by Proposition 4.11, the corresponding eigenvalue εc is b if $\varepsilon = 1$ or a if $\varepsilon = -1$. □

Given an eigenvalue λ of a symmetric operator A, let E_λ be the subspace spanned by the corresponding eigenvectors. Clearly $E_\lambda = \{x \in X : Ax = \lambda x\}$. If λ and μ are two distinct eigenvalues of A one has $E_\lambda \perp E_\mu$ since for $x \in E_\lambda, y \in E_\mu$ one has

$$\overline{\lambda}\langle x \mid y \rangle = \langle Ax \mid y \rangle = \langle x \mid Ay \rangle = \overline{\mu}\langle x \mid y \rangle$$

hence $\langle x \mid y \rangle = 0$. If A is compact and if $\lambda \neq 0$, the dimension of the eigenspace E_λ is finite: otherwise, taking an infinite family (e_n) of orthonormal vectors of E_λ we could not find a convergent subsequence in $(A(e_n)) = (\lambda e_n)$, a contradiction with $\|e_n - e_{n+1}\|^2 = 2$. For a similar reason, for any $\varepsilon > 0$ there is only a finite number of eigenvalues λ satisfying $|\lambda| \geq \varepsilon$. Thus, if X is infinite dimensional, 0 is the unique limit of a sequence in distinct eigenvalues.

Given an operator A on a Hilbert space X, a linear subspace Y of X is said to be A-invariant if $A(Y) \subset Y$, i.e. $A(y) \in Y$ for all $y \in Y$. Then $\mathrm{cl}(Y)$ is A-invariant and if A is symmetric, $Y^\perp$ is invariant. Moreover, the restriction to Y is again symmetric.

Theorem 4.12 (Spectral Theorem for Compact Symmetric Operators) *Let A be a compact symmetric operator on a Hilbert space X. Then the set $S := \sigma_e$ of eigenvalues of A is a finite or countable subset of $\mathbb{R}$.*

If S is finite (in particular if X is finite dimensional), X is the direct sum of the mutually orthogonal eigenspaces E_λ for $\lambda \in S$.

If S is infinite, X is the Hilbertian sum of the eigenspaces E_λ for $\lambda \in S$.

The last assertion means that any $x \in X$ is the sum of a series $\sum_n x_n$ such that $x_n \in E_{\lambda_n}$ with $(\lambda_n) \to 0$ in S and $x_n \perp x_p$ for $n \neq p$. Taking a Hilbert basis of each subspace E_λ, including $E_0 = N(A)$, we get a Hilbert basis of X formed of eigenvectors.

Proof Let Y be the closure of the linear subspace generated by the eigenspaces E_λ for $\lambda \in S$ and let $Z := Y^\perp$. Then Z is invariant under A and the restriction A_Z of A to Z has no eigenvalue. Proposition 4.15 shows that $Z = \{0\}$. Taking orthonormal basis in all eigenspaces (including the kernel of A) and gathering them, we get the announced decomposition, even if we do not give more details about Hilbert sums of subspaces.

Exercises

1. Let X and Y be two pre-Hilbertian spaces. Given $a \in X$, $b \in Y$, one defines $A := a^* \otimes b \in L(X, Y)$ by $A(x) := \langle a \mid x \rangle b$ for $x \in X$. Describe the image and the kernel of A. Compute $\|A\|$. Determine the adjoint A^* of A. Find relations between the image (resp. kernel) of A and the kernel (resp. image) of A^*.

2. Let X and Y be two Hilbert spaces and let (e_n) (resp. (f_n)) be an orthonormal family of X (resp. Y). Given a bounded sequence (λ_n) of $\mathbb{R}$, let $A : X \to Y$ be defined by

$$A = \sum_n \lambda_n e_n^* \otimes f_n.$$

Show that A is a well defined continuous linear map and compute $\|A\|$ and A^*.

Assuming that $\lambda_n \neq 0$ for all $n \in \mathbb{N}$, show that A is injective if and only if (e_n) is a Hilbert basis of X. Show that $A(X)$ is dense in Y if and only if (f_n) is a Hilbert basis of Y. Find conditions ensuring that A is invertible and express A^{-1}.

In the case $X = Y$ find the eigenvalues of A.

3. Let X be a real Hilbert space and let $A \in L(X)$ be symmetric. Prove that $\sigma(A) \subset \mathbb{R}_+$ if and only if $\langle Ax \mid x \rangle \geq 0$ for all $x \in X$.

Prove the following equivalences:

$$\begin{aligned} \sigma(A) \subset [0, 1] &\Longleftrightarrow \|A\| \leq 1 \ \&\ \langle Ax \mid x \rangle \geq 0 \ \forall x \in X \\ &\Longleftrightarrow \langle Ax \mid x \rangle \geq \|Ax\|^2 \ \ \forall x \in X \\ &\Longleftrightarrow \|x\|^2 \geq \langle Ax \mid x \rangle \geq 0 \ \forall x \in X. \end{aligned}$$

Additional Reading

[14, 30, 33, 35, 67, 69, 75, 77, 100, 106, 157, 159, 189, 217, 225, 235, 239, 245, 257, 258, 262]

Chapter 5
The Power of Differential Calculus

> *...Leibniz quite rapidly developed formal analysis in the form in which we know it. That is, in a form specially suitable to teach analysis by people who do not understand it to people who will never understand it.*
>
> V.I. Arnol'd, Huygens and Barrow, Newton and Hooke, Birkhäuser Verlag, Basel, 1990.

Abstract This chapter is devoted to a rather complete exposition of classical differential calculus. However, we present some non-classical variants which are often easier to handle. Besides the idea of approximation carried by the notion of derivative, important existence theorems can be obtained: the inverse function theorem and its relative, the implicit function theorem. Several applications are studied: the Legendre transform, the method of characteristics and some partial differential equations. Applications to geometry and optimization are devised and the calculus of variations that served as an incentive to the development of the subject is evoked with numerous examples.

The geometrical counterpart to the derivative of a function, i.e. the concept of a tangent to a curve, has a long history since it was known to Euclid and Archimedes. Although we devote some attention to this geometrical concept in this chapter, it is the analytic version that is our main subject. The reason is that differential calculus is at the core of several sciences and techniques. Our world would not be the same without it: astronomy, electromagnetism, mechanics, optimization, thermodynamics among others use it as a fundamental tool.

The birth of differential calculus is usually attributed to I. Newton and G. Leibniz in the last part of the seventeenth century, with several other contributions. The pioneer work of P. de Fermat is seldom recognized, although he introduced the idea of approximation which is the backbone of differential calculus and which enabled him (and others) to treat many applications. During the eighteenth century, the topic reached its maturity and its achievements led to the principle of determinism in the beginning of the nineteenth century. But it is only with the appearance of functional analysis that it took its modern form.

Several notions of differentiability exist; they correspond to different needs or different situations. The most usual one is the notion of Fréchet differentiability that is displayed in Sect. 5.4. However, a weaker notion of directional differentiability due to Hadamard has some interest. We present it in Sect. 5.3 as a passage from the

J.-P. Penot, *Analysis*, Universitext, DOI 10.1007/978-3-319-32411-1_5

case of one-variable maps to the case of maps defined on open subsets of normed spaces. It is not as strong as Fréchet differentiability. Moreover, for some results, differentiability does not suffice and one needs some continuity property of the derivative or a stronger notion of approximation.

The main questions we treat are the invertibility of nonlinear maps, its applications to geometrical notions and its uses for optimization problems. We end the chapter with an introduction to the calculus of variations, which has been a strong incentive for the development of differential calculus since the end of the 17th century. Differentiability questions for convex functions will be considered in the next chapter. Differential equations will be studied in Chap. 10 devoted to evolution problems.

As an example showing the power of differential calculus, let us consider *Dido's problem* already evoked in the preceding chapter. It consists in finding a curve of given length enclosing the greatest area. Its origin can be found in the legend of the foundation of Carthage. According to the Eneid, the queen Dido made the modest request that a piece of land that could be delimited by the skin of a buoy be given to her and her men. It was accepted but she had the skin cut into a long, thin string and the piece of land became the city of Qart Hadasht (Carthage) (the first example of cunning colonialism). We take two opposite points considered as the beginning of the string and its middle and represented by $(0,0)$ and $(1,0)$ in $\mathbb{R}^2$ and we look for a function $x : [0,1] \to \mathbb{R}_+$ such that the area between its graph and the segment $[0,1]\times\{0\}$ is maximum for all curves of a given length ℓ. Thus, we have to maximize

$$\int_0^1 x(t)dt \qquad \text{subject to} \int_0^1 \sqrt{1+x'(t)^2}dt = \ell,\ x(0)=0, x(1)=0.$$

The theory of Lagrange multipliers and the Euler equation lead us to find a multiplier $\lambda \in \mathbb{R}$ and $x(\cdot)$ (an unknown in an infinite dimensional space!) such that

$$1 - \lambda \frac{d}{dt} \frac{x'(t)}{\sqrt{1+x'(t)^2}} = 0.$$

Integrating, we look for some constant c such that

$$t - \lambda \frac{x'(t)}{\sqrt{1+x'(t)^2}} = c.$$

Solving this equation in $x'(t)$ we get

$$x'(t) = \frac{t-c}{\sqrt{\lambda^2-(t-c)^2}}.$$

Another integration yields some constant c' such that

$$x(t) = -\sqrt{\lambda^2 - (t-c)^2} + c'.$$

Thus the graph of x lies on the circle with equation $(t-c)^2 + (y-c')^2 = \lambda^2$ in $\mathbb{R}^2$ with coordinates (t, y). The constraints $x(0) = 0$, $x(1) = 0$ yield $c = 1/2$, $\lambda^2 = c^2 + c'^2$ and the length constraint yields $2\lambda \arcsin(1/\lambda) = \ell$. Choosing the distance between the extremities of the string to be d, with $d := \pi^{-1}\ell$ rather than 1, one finds that the arc $w(\cdot) := dx(d\cdot)$ with length ℓ solving the problem is half a circle.

5.1 Differentiation of One-Variable Functions

The differentiation of one-variable vector-valued functions is not very different from the differentiation of one-variable real-valued functions. In both cases, the calculus relies on rules for limits. The aims are similar too. In both cases, the purpose consists in drawing some information about the behavior of the function from some knowledge concerning the derivative. In the vector-valued case, the direction of the derivative takes as great importance as its magnitude.

5.1.1 Derivatives of One-Variable Functions

In this section, T is an open interval of $\mathbb{R}$ and $f : T \to X$ is a map with values in a normed space X.

Definition 5.1 The map f is said to be *right differentiable* (resp. *left differentiable*) at $t \in T$ if the quotient $(f(t+s) - f(t))/s$ has a limit as $s \to 0_+$, i.e. $s \to 0$ with $s > 0$ (resp. as $s \to 0_-$, i.e. $s \to 0$ with $s < 0$). These limits, denoted by $f'_r(t)$ or $D_r f(t)$ and $f'_\ell(t)$ or $D_\ell f(t)$ respectively, are called the right and the left *derivatives* of f at t.

When these limits coincide, f is said to be *differentiable* at t and their common value $f'(t)$ is called the *derivative* of f at t.

Thus f is differentiable at t if and only if the quotient $(f(t+s)-f(t))/s$ has a limit as $s \to 0$, with $s \neq 0$, or, equivalently, if there exists some vector $v (= f'(t)) \in X$ and some functions $\alpha, r : T' := T - t \to X$ such that $\alpha(s) \to 0$ as $s \to 0$ and $r(s) = s\alpha(s)$ for which one has the expansion

$$f(t') = f(t) + (t'-t)v + r(t'-t), \tag{5.1}$$

as seen by setting $\alpha(0) := 0$, $\alpha(s) := (f(t+s) - f(t))/s - v$ for $s \in T' \backslash \{0\}$, $r(s) = f(t+s) - f(t) - sv$ for $s \in T'$. The function r is called a *remainder*.

The following rules are immediate consequences of the rules for limits.

Proposition 5.1 *If $f, g : T \to X$ are differentiable at $t \in T$ and $\lambda, \mu \in \mathbb{R}$, then $h := \lambda f + \mu g$ is differentiable at t and its derivative at t is $h'(t) = \lambda f'(t) + \mu g'(t)$.*

Proposition 5.2 *If $f : T \to X$ is differentiable at $t \in T$, if Y is another normed space and if $A : X \to Y$ is linear and continuous, then $g := A \circ f$ is differentiable at t and $g'(t) = A(f'(t))$.*

Similar rules hold for right derivatives and left derivatives. We will see later a more general composition rule (or chain rule). The following composition rule can be proved using quotients in the same way as for scalar functions. We prefer using expansions as in (5.1) because such expansions give the true flavor of differential calculus, i.e. approximations by continuous affine functions. Moreover, one does not need to take care of denominators taking the value 0.

Proposition 5.3 *If T, U are open intervals of $\mathbb{R}$, if $g : T \to U$ is differentiable at $\bar{t} \in T$ and if $h : U \to X$ is differentiable at $\bar{u} := g(\bar{t})$, then $f := h \circ g$ is differentiable at $\bar{t}$ and $f'(\bar{t}) = g'(\bar{t})h'(\bar{u})$.*

Proof Let $v := h'(\bar{u})$ and let $\alpha : T \to \mathbb{R}$, $\beta : U \to X$ be such that $\alpha(t) \to 0$ as $t \to \bar{t}$, $\beta(u) \to 0$ as $u \to \bar{u}$ with $g(t) - g(\bar{t}) = (t - \bar{t})g'(\bar{t}) + (t - \bar{t})\alpha(t)$, $h(u) - h(\bar{u}) = (u - \bar{u})v + (u - \bar{u})\beta(u)$. Then one has

$$f(t) - f(\bar{t}) = h(g(t)) - h(\bar{u}) = (g(t) - \bar{u})v + (g(t) - \bar{u})\beta(g(t))$$
$$= (t - \bar{t})g'(\bar{t})v + (t - \bar{t})\alpha(t)v + (t - \bar{t})(g'(\bar{t}) + \alpha(t))\beta(g(t)).$$

Since $g(t) \to \bar{u}$ as $t \to \bar{t}$, one sees that $\alpha(t)v + (g'(\bar{t}) + \alpha(t))\beta(g(t)) \to 0$ as $t \to \bar{t}$, so that f is differentiable at $\bar{t}$ and $f'(\bar{t}) = g'(\bar{t})v = g'(\bar{t})h'(\bar{u})$. □

Now let us devise a rule for the derivative of a product. It can be generalized to a finite number of factors.

Proposition 5.4 (Leibniz Rule) *Let X, Y, Z be normed spaces and let $b : X \times Y \to Z$ be a continuous bilinear map. If $f : T \to X$, $g : T \to Y$ are differentiable at t, then the function $h : r \mapsto b(f(r), g(r))$ is differentiable at t and*

$$h'(t) = b(f'(t), g(t)) + b(f(t), g'(t)).$$

Proof By assumption, there exist some $\alpha : (T - t) \to X$, $\beta : (T - t) \to Y$ satisfying $\alpha(s) \to 0$, $\beta(s) \to 0$ as $s \to 0$ such that

$$f(t + s) = f(t) + sf'(t) + s\alpha(s), \qquad g(t + s) = g(t) + sg'(t) + s\beta(s).$$

Plugging these expansions into b and setting

$$\gamma(s) := b(\alpha(s), g(t)) + b(f(t), \beta(s)) + sb(\alpha(s), \beta(s)),$$

so that $\gamma(s) \to 0$ as $s \to 0$, we get

$$h(t+s) - h(t) = sb(f'(t), g(t)) + sb(f(t), g'(t)) + s\gamma(s)$$

and $s^{-1}(h(t+s) - h(t)) \to b(f'(t), g(t)) + b(f(t), g'(t))$. □

For a real-valued function $f : [a, b] \to \mathbb{R}$, as substitutes for the derivative of f at $x \in]a, b[$, Ulisse Dini introduced the following four derivative numbers (called the *Dini derivatives* of f at x):

$$D_{-}f(t) := \liminf_{s \to 0_-} \frac{f(t+s) - f(t)}{s}, \qquad D^{-}f(t) := \limsup_{s \to 0_-} \frac{f(t+s) - f(t)}{s},$$

$$D_{+}f(t) := \liminf_{s \to 0_+} \frac{f(t+s) - f(t)}{s}, \qquad D^{+}f(t) := \limsup_{s \to 0_+} \frac{f(t+s) - f(t)}{s}.$$

They always exist in $\overline{\mathbb{R}}$. Of course, f is right differentiable at t if and only if $D_{+}f(t) = D^{+}f(t)$; a similar equivalence holds for the left derivative. They enable us to obtain estimates akin to the ones we will deal with in the next subsection. We also refer to Theorem 8.14 and its corollaries for some refinements.

Lemma 5.1 *Let $f : [a, b] \to \mathbb{R}$ be continuous and such that for some $c, d \in \mathbb{R}$ one of the Dini derivatives, denoted by Df, satisfies $c \leq Df(t) \leq d$ for all $t \in]a, b[$. Then one has $c(b-a) \leq f(b) - f(a) \leq d(b-a)$.*

Proof Changing f into $-f$ or f into $t \mapsto f(-t)$ we may suppose $Df = D^{+}f$. We first prove that $f(b) - f(a) \geq 0$ whenever f satisfies $Df(t) \geq 0$ for all $t \in]a, b[$. Suppose on the contrary that $f(b) - f(a) < 0$ and take $p \in \mathbb{P}$ such that $f(b) - f(a) < -p(b-a)$. For $a' \in]a, b[$ close to a we still have $f(b) - f(a') < -p(b-a')$. Setting $g(t) := f(t) - f(a') + p(t-a')$, we have $g(a') = 0$, $Dg(t) = Df(t) + p > 0$, so that there exists some $r \in]a', b[$ such that $g(r) > 0$. Since g is continuous and $g(b) < 0$ there exists some $s \in]r, b[$ such that $g(s) = 0$ and $g(t) < 0$ for all $t \in]s, b]$. Then we have $Dg(s) \leq 0$, contradicting $Dg(s) > 0$. Thus our assertion is proved.

Applying this assertion to the functions h and k given by $h(t) := f(t) - ct$, $k(t) := dt - f(t)$ we get the conclusion of the lemma. □

It follows from the lemma that the supremum (and the infimum) of the four Dini derivatives of a continuous function on any interval are the same. We leave to the reader the task of proving that when one of the four Dini derivatives of a continuous function f is continuous at some t, then f is differentiable at t.

The next result is well known.

Proposition 5.5 *Let T be an interval of $\mathbb{R}$, let $f : T \to \overline{\mathbb{R}}$, and let $\overline{x} \in \operatorname{int} T$ be such that f is differentiable at $\overline{x}$. If f attains its minimum over T at $\overline{x}$, then $f'(\overline{x}) = 0$.*

Proof After a passage to the limit, that is a consequence in the fact that for all $u \in \mathbb{P}$ one has $u^{-1}[f(\overline{x}+u) - f(\overline{x})] \geq 0$ and $(-u)^{-1}[f(\overline{x}-u) - f(\overline{x})] \leq 0$. □

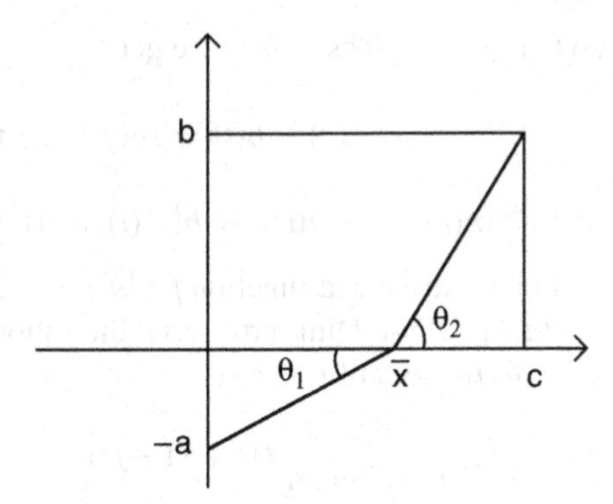

Fig. 5.1 The Descartes-Snell law of refraction

Application: The Descartes-Snell Law of Refraction Suppose two media are separated by the horizontal plane $z = 0$ and that the speed of light is s_1 in the first one and s_2 in the second one. Using the fact that the trajectory of light minimizes the travel time and is along a line in an isotropic medium, prove the relation

$$\frac{1}{s_1}\sin\theta_1 = \frac{1}{s_2}\sin\theta_2$$

in which θ_1 (resp. θ_2) is the angle of the first (resp. second) ray with the vertical line (Fig. 5.1). [Hint: assume the trajectory joins the point $(0, 0, -a)$ to the point $(c, 0, b)$, with $a, b, c \in \mathbb{P}$ and passes through the point $(\overline{x}, 0, 0)$. Using the fact that $\overline{x}$ minimizes the travel time

$$f(x) = \frac{1}{s_1}(x^2 + a^2)^{1/2} + \frac{1}{s_2}((c-x)^2 + b^2)^{1/2},$$

find an interpretation of the relation $f'(\overline{x}) = 0$.]

5.1.2 *The Mean Value Theorem*

The Mean Value Theorem is a precious tool for obtaining estimates. For this reason, it is a cornerstone of differential calculus. Let us note that the elementary version recalled in the following lemma is not valid when the function takes its values in a linear space of dimension greater than one.

Lemma 5.2 *Let $f : T \to \mathbb{R}$ be a continuous function on some interval $T := [a, b]$ of $\mathbb{R}$, with $a < b$. If f is differentiable on $]a, b[$ then there exists some $c \in]a, b[$ such that*

$$f(b) - f(a) = f'(c)(b - a).$$

Example Let $f : [0,1] \to \mathbb{R}^2$ be given by $f(t) := (t^2, t^3)$ for $t \in T := [0,1]$. Then one cannot find any $c \in \mathrm{int}T$ satisfying the preceding relation since the system $2c = 1, 3c^2 = 1$ has no solution. □

Instead, a statement in the form of an estimate is valid.

Theorem 5.1 *Let X be a normed space, let $T := [a,b]$ be a compact interval of $\mathbb{R}$ and let $f : T \to X$, $g : T \to \mathbb{R}$ be continuous on T and have right derivatives on $]a,b[$ such that $\|D_r f(t)\| \le D_r g(t)$ for every $t \in]a,b[$. Then*

$$\|f(b) - f(a)\| \le g(b) - g(a). \tag{5.2}$$

Proof It suffices to prove that for every given $\varepsilon > 0$, b belongs to the set

$$T_\varepsilon := \{t \in T : \|f(t) - f(a)\| \le g(t) - g(a) + \varepsilon(t-a)\}.$$

This set is nonempty since $a \in T_\varepsilon$ and it is closed, being defined by an inequality whose sides are continuous. Let $s := \sup T_\varepsilon \le b$. Then $s \in T_\varepsilon$.

We first suppose f and g have right derivatives on $[a,b[$ and we show that assuming $s < b$ leads to a contradiction. The existence of the right derivatives of f and g at s yields some $\delta \in]0, b-s[$ such that for $r \in]0,\delta]$ one has

$$\left\| \frac{f(s+r) - f(s)}{r} - D_r f(s) \right\| \le \frac{\varepsilon}{2}, \quad \left| \frac{g(s+r) - g(s)}{r} - D_r g(s) \right| \le \frac{\varepsilon}{2}.$$

It follows that for $r \in]0,\delta]$ one has

$$\|f(s+r) - f(s)\| \le r\,\|D_r f(s)\| + r\varepsilon/2, \quad g(s+r) - g(s) \ge r D_r g(s) - r\varepsilon/2.$$

Therefore, since $s \in T_\varepsilon$ and $\|D_r f(s)\| \le D_r g(s)$,

$$\begin{aligned} \|f(s+r) - f(a)\| &\le \|f(s+r) - f(s)\| + \|f(s) - f(a)\| \\ &\le r D_r g(s) + r\varepsilon/2 + g(s) - g(a) + \varepsilon(s-a) \\ &\le g(s+r) - g(s) + r\varepsilon + g(s) - g(a) + \varepsilon(s-a) \\ &\le g(s+r) - g(a) + \varepsilon(s+r-a). \end{aligned}$$

This string of inequalities shows that $s+r \in T_\varepsilon$, a contradiction with the definition of s. Thus $b \in T_\varepsilon$ and the result is established under the additional assumption that the right derivatives of f and g exist at a (note that we may have $s = a$ in the preceding).

When this additional assumption is not made, we take $a' \in]a,b[$ and we apply the preceding case to the interval $[a', b]$:

$$\left\|f(b) - f(a')\right\| \le g(b) - g(a').$$

Then, passing to the limit as $a' \to a_+$, we get the announced inequality. □

Remark Since we allow the possibility that the right derivatives do not exist at the extremities of the interval, we may assume that the derivatives do not exist (or do not satisfy the assumed inequality) at a finite number of points of T. To prove this, it suffices to subdivide the interval into subintervals and to gather the obtained inequalities by using the triangle inequality. In fact, one can exclude a countable set of points of T, but the proof is more involved; see [66], [100, p.153]. The result can also be deduced from Theorem 8.14, even by replacing $D_r g$ with one of its right Dini derivatives, observing that for $h : T \to \mathbb{R}$ given by $h(t) := \|f(t) - f(a)\|$ one has $D_r h(t) \leq \|D_r f(t)\|$ for all $t \in T$ since for $s > 0$ one has

$$\|f(t+s) - f(a)\| - \|f(t) - f(a)\| \leq \|f(t+s) - f(t)\|.$$

Thus $h - g$ is nonincreasing and one obtains the following refinement.

Theorem 5.2 *With the notation of Theorem 5.1, the estimate (5.2) holds when f and g are continuous on T and have right derivatives on $T \backslash D$, where D is countable, such that $\|D_r f(t)\| \leq D_r g(t)$ for every $t \in T \backslash D$.*

The most usual application is given in the following corollary in which we take $g(t) = mt$ for some $m \in \mathbb{R}_+$ and $t \in T$. The Lipschitz property is obtained in substituting an arbitrary pair t, t' (with $t \leq t'$) to a, b.

Corollary 5.1 *Let $f : T \to X$ be continuous on $T := [a, b]$, let $m \in \mathbb{R}_+$ and let D be a countable subset of T. Suppose that, for all $t \in]a, b[\backslash D$, f has a right derivative at t such that $\|D_r f(t)\| \leq m$. Then f is Lipschitzian with rate m on T and, in particular,*

$$\|f(b) - f(a)\| \leq m(b - a).$$

The case $m = 0$ yields the following noteworthy consequence.

Corollary 5.2 *Let $f : [a, b] \to X$ be continuous and such that f has a right derivative $D_r f$ on $]a, b[\backslash D$ that is null, D being countable. Then f is constant on $[a, b]$.*

The purpose of obtaining estimates often requires the introduction of auxiliary functions, as in the proof of the following useful corollary.

Corollary 5.3 *Let $f : T \to X$ be continuous on $T := [a, b]$, let $v \in X$, $r \in \mathbb{R}_+$ and let D be a countable subset of T. Suppose f has a right derivative on $]a, b[\backslash D$ such that $D_r f(t) \in v + rB_X$ for every $t \in]a, b[\backslash D$. Then*

$$f(b) \in f(a) + (b - a)v + (b - a)rB_X.$$

Proof Define $h : T \to X$ by $h(t) := f(t) - tv$. Then h is continuous and for $t \in]a, b[\backslash D$ one has $\|D_r h(t)\| = \|D_r f(t) - v\| \leq r$. Then Corollary 5.2 entails that

$$\|f(b) - f(a) - (b - a)v\| = \|h(b) - h(a)\| \leq (b - a)r,$$

an estimate equivalent to the inclusion of the statement. □

Remark The terminology of the theorem stems from the fact that the mean value $\overline{v} := (b-a)^{-1}(f(b)-f(a))$ is estimated by the approximate speed v, with an error r that is exactly the magnitude of the uncertainty of the estimate of the instantaneous speed $f'_r(t) := D_r f(t)$. Note that the shorter the lapse of time $(b-a)$, the more precise is the localization of $f(b)$ by $f(a)+(b-a)v$. Thus, if you lose your dog, be sure to have a rather precise idea of his speed and direction and do not lose time in looking for him.

Exercises

1. Prove Corollary 5.1 by deducing it from the classical Mean Value Theorem (Lemma 5.2) for real-valued functions, using the Hahn-Banach Theorem. [Hint: Take y^* with norm 1 such that $\langle y^*, y\rangle = \|y\|$ for $y := f(x) - f(w)$, set $g(t) := \langle y^*, f(x+t(w-x))\rangle$ and pick $\theta \in]0,1[$ such that $g(1)-g(0) = D_r g(\theta)$.]
2. Prove Theorem 5.2. [See [66, 100].]

5.2 Primitives and Integrals

The aim of this subsection is to present an inverse of the differentiation operator. In fact, as revealed by the Darboux property (Exercise 1), not all functions from some interval T of $\mathbb{R}$ to a real Banach space X are derivatives. Therefore, we will get a primitive g of a function f on T only if f is regular enough. Here we use the following terminology.

Definition 5.2 A function $g : T \to X$ is said to be a *primitive* of $f : T \to X$ if g is continuous and if there exists a countable subset D of T such that, for all $t \in T \backslash D$, g is differentiable at t and $g'(t) = f(t)$.

Uniqueness of the primitive taking an assigned value at some point of T is asserted by the next proposition.

Proposition 5.6 *If g_1 and g_2 are two primitives of an arbitrary function $f : T \to X$, then $g_1 - g_2$ is constant.*

Proof If g_1 and g_2 are two primitives of f there exist countable subsets D_1 and D_2 of T such that g_i is differentiable on $T\backslash D_i$ and $g'_i(t) = f(t)$ for all $t \in T\backslash D_i$ $(i = 1, 2)$. Then, for the countable set $D := D_1 \cup D_2$, the continuous function $g_1 - g_2$ is differentiable on $T\backslash D$ and its derivative is 0 there; thus $g_1 - g_2$ is constant. □

A partial inverse of the differentiation operator is given in the next result.

Theorem 5.3 *For $f : T \to X$ regulated, the map $g : t \mapsto \int_a^t f(s)ds$ is a primitive of f.*

Proof Given $t \in [a, b[$, $\varepsilon > 0$, let $\delta \in]0, b - t[$ be such that for $r \in]0, \delta]$ one has $\|f(t + r) - f(t_+)\| \leq \varepsilon$. Since for $c := f(t_+)$ one has $\int_t^{t+r} c = rc$, it follows from the Chasles' relation and (3.27) that

$$\left\| \int_a^{t+r} f - \int_a^t f - rc \right\| = \left\| \int_t^{t+r} (f - c) \right\| \leq r\varepsilon.$$

This relation shows that $g : t \mapsto \int_a^t f(s)ds$ has a right derivative at t whose value is c. Similarly, if $t \in]a, b]$, then g has $f(t_-)$ as a left derivative at t. Therefore, if f is continuous at $t \in]a, b[$, then g is differentiable at t and $g'(t) = f(t)$. Since the set D of discontinuities of f is countable, we get that g is differentiable on $T \backslash D$ with derivative f. Moreover, g is continuous on T in view of the Chasles' relation and (3.27). □

Corollary 5.4 *If $f : T \to X$ is continuous, then $g : t \mapsto \int_a^t f(s)ds$ is of class C^1 (i.e., differentiable with a continuous derivative) and its derivative is f.*

Let us give two rules that are useful for the computation of primitives.

Proposition 5.7 (Change of Variables) *Let $h : S = [\alpha, \beta] \to \mathbb{R}$ be the primitive of a regulated function h' such that $h(S) \subset T$ and let $f \in R(T, X)$. If either f is continuous or h is monotone, then $s \mapsto h'(s)f(h(s))$ is regulated and for all $r \in [\alpha, \beta]$ one has*

$$\int_\alpha^r h'(s)f(h(s))ds = \int_{h(\alpha)}^{h(r)} f(t)dt. \tag{5.3}$$

Proof When f is continuous, since h is continuous, $f \circ h$ is continuous and then $k : s \mapsto h'(s)f(h(s))$ is regulated; the same is true when h is either increasing or decreasing. Then, the left-hand side of equality (5.3) is the value at r of the primitive j of k satisfying $j(\alpha) = 0$. The right-hand side is $g(h(r))$, where g is the primitive of f satisfying $g(h(\alpha)) = 0$. Under each of our assumptions, for a countable subset D of S, the derivative of $g \circ h$ at $r \in S \backslash D$ exists and is $h'(r)g'(h(r)) = h'(r)f(h(r))$. The uniqueness of the primitive of k null at α gives the equality. □

Proposition 5.8 (Integration by Parts)) *Let X, Y and Z be Banach spaces, let $(x, y) \mapsto x * y$ be a continuous bilinear map from $X \times Y$ into Z and let $f : T \to X$, $g : T \to Y$ be primitives of regulated functions, with $T := [a, b]$. Then*

$$\int_a^b f(t) * g'(t)dt = f(b) * g(b) - f(a) * g(a) - \int_a^b f'(t) * g(t)dt.$$

Proof The functions $t \mapsto f(t)*g'(t)$ and $t \mapsto f'(t)*g(t)$ clearly have one-sided limits at all points of $T := [a, b]$. Moreover, their sum is the derivative of $h : t \mapsto f(t) * g(t)$ on $T \backslash D$ where D is the countable set of nondifferentiability of f or g. Thus the result amounts to the equality $\int_a^b h'(t)dt = h(b) - h(a)$ that stems from the uniqueness of the primitive of h' that takes the value 0 at a. □

Exercises

1. **(Darboux property)** Show that the derivative f of a differentiable function $g : T \to \mathbb{R}$ satisfies the intermediate value property: given $a, b \in T$ with $f(a) < f(b)$ and $r \in]f(a), f(b)[$, there exists some c between a and b such that $f(c) = r$.
2. Show that there exist a continuous function $f : \mathbb{R} \to \mathbb{R}$ and two continuous functions g_1, g_2 whose difference is not constant and which are such that g_1 and g_2 are differentiable on $\mathbb{R} \backslash N$, where N is a set of measure zero, with $g_1'(t) = g_2'(t) = f(t)$ for all $t \in \mathbb{R} \backslash N$. [Take $f = 0$, $g_1 = 0$ and for g_2 take an increasing function whose derivative is 0 a.e.]
3. Prove Theorem 5.2. [See [100, 8.5.1].]

5.3 Directional Differential Calculus

Now let us consider maps from an open subset W of a normed space X into another normed space Y. In order to reduce the study of differentiability to the one-variable case it is natural to take restrictions to line segments or to compose with regular curves in W.

Definition 5.3 Let X, Y be normed spaces, let W be an open subset of X, let $\overline{x} \in W$ and let $f : W \to Y$. We say that f has a *radial derivative* at $\overline{x}$ in the direction $u \in X$ if $(1/t)(f(\overline{x}+tu)-f(\overline{x}))$ has a limit when $t \to 0_+$. We denote by $d_r f(\overline{x}, u)$ this limit. If f has a radial derivative at $\overline{x}$ in every direction u, we say that f is *radially differentiable* at $\overline{x}$. If moreover the map $D_r f(\overline{x}) : u \mapsto d_r f(\overline{x}, u)$ is linear and continuous, we say that f is *Gateaux differentiable* at $\overline{x}$ and call $D_r f(\overline{x})$ the Gateaux derivative of f at $\overline{x}$.

One often says that f is directionally differentiable at $\overline{x}$ but we prefer to keep this terminology for a slightly more demanding notion we consider now. In fact, although the notion of radial differentiability is simple and useful, it has several drawbacks; the main one is the fact that this notion does not enjoy a chain rule. The variant that follows does enjoy such a rule and reflects a smoother behavior of f when the direction u is submitted to small changes.

Definition 5.4 Let X, Y be normed spaces, let W be an open subset of X, let $\overline{x} \in W$ and let $f : W \to Y$. We say that f has a *directional derivative* at $\overline{x}$ in the direction $u \in X$, or that f is differentiable at $\overline{x}$ in the direction u, if $(1/t)(f(\overline{x} + tv) - f(\overline{x}))$ has a limit when $(t, v) \to (0_+, u)$. We denote by $f'(\overline{x}, u)$ or $df(\overline{x}, u)$ this limit. If f has a directional derivative at $\overline{x}$ in every direction u, we say that f is *directionally differentiable* at $\overline{x}$. If moreover the map $f'(x) := Df(x) : u \mapsto f'(\overline{x}, u)$ is linear and continuous, we say that f is *Hadamard differentiable* at $\overline{x}$.

The concepts of directional derivative and radial derivative are different, as the next example shows. Thus, it is convenient to dispose of two notations.

Example-Exercise Let $f : \mathbb{R}^2 \to \mathbb{R}$ be given by $f(r,s) = (r^4 + s^2)^{-1} r^3 s$ for $(r,s) \in \mathbb{R}^2 \backslash \{(0,0)\}$, $f(0,0) = 0$. It is Gateaux differentiable at $(0,0)$ but not directionally differentiable at $(0,0)$. □

The (frequent) use of the same notation for the radial and directional derivatives is justified by the following observation showing the compatibility of the two notions.

Proposition 5.9 *If X and Y are normed spaces, if W is an open subset of X and if $f : W \to Y$ has a directional derivative at $\overline{x}$ in the direction u, then it has a radial derivative at $\overline{x}$ in the direction u and both derivatives coincide. In particular, if f is Hadamard differentiable at $\overline{x}$, then it is Gateaux differentiable at $\overline{x}$.*

Conversely, if f is Lipschitzian on a neighborhood V of $\overline{x}$, then f is directionally differentiable at $\overline{x}$ in any direction u in which f is radially differentiable.

Proof The first assertions stem from an application of the definition of a limit.

Let us prove the converse assertion. Let k be the Lipschitz rate of f on V and let $u \in X$ be such that f is radially differentiable at $\overline{x}$ in the direction u. Setting $r(t,v) := f(\overline{x} + tv) - f(\overline{x}) - tf'_r(\overline{x}, u)$ we have $t^{-1} r(t,u) \to 0$ as $t \to 0$ and since $\left\| t^{-1}(r(t,v) - r(t,u)) \right\| = \left\| t^{-1}(f(\overline{x} + tv) - f(\overline{x} + tu)) \right\| \leq k \|v - u\| \to 0$ as $(t,v) \to (0_+, u)$, we get $t^{-1} r(t,v) \to 0$ as $(t,v) \to (0_+, u)$. □

While radial differentiability of f at $\overline{x}$ in the direction u is equivalent to differentiability of the function $f_{\overline{x},u} : t \mapsto f(\overline{x} + tu)$ at 0, directional differentiability of f at $\overline{x}$ amounts to differentiability of the composition of f with curves issued from $\overline{x}$ with the initial direction u, as the next proposition shows.

Proposition 5.10 *The map $f : W \to Y$ is differentiable at $\overline{x}$ in the direction $u \in X \backslash \{0\}$ if and only if f is radially differentiable at $\overline{x}$ in the direction u and for any $\tau > 0$ and any map $c : [0, \tau] \to W$ that is right differentiable at 0 with $D_r c(0) = u$, $c(0) = \overline{x}$, the map $f \circ c$ is right differentiable at 0 and $D_r(f \circ c)(0) = d_r f(\overline{x}, u)$.*

The characterization is still valid if one makes the additional requirement that c be continuous on $[0, \tau]$.

Proof Suppose f is differentiable at $\overline{x}$ in the direction $u \in X$. Given $\tau > 0$ and $c : [0, \tau] \to W$ that is right differentiable at 0 with $c'_+(0) = u$ and $c(0) = \overline{x}$, let us set $v_t := (1/t)(c(t) - c(0))$, so that $v_t \to u$ as $t \to 0_+$. Then

$$\frac{f(c(t)) - f(c(0))}{t} = \frac{f(\overline{x} + tv_t) - f(\overline{x})}{t} \to df(\overline{x}, u) \text{ as } t \to 0_+.$$

Now let us prove the sufficient condition. Suppose f has a radial derivative at $\overline{x}$ in the direction u but is not differentiable at $\overline{x}$ in the direction $u \neq 0$. There exist $\varepsilon > 0$ and some sequence $(t_n, u_n) \to (0_+, u)$ such that $\overline{x} + t_n u_n \in W$ for all $n \in \mathbb{N}$ and

$$\left\| \frac{f(\overline{x} + t_n u_n) - f(\overline{x})}{t_n} - d_r f(\overline{x}, u) \right\| \geq \varepsilon. \tag{5.4}$$

We may assume that $t_{n+1} \leq (1/2)t_n$. Then, let us define $c : [0, t_0] \to X$ by $c(0) := \overline{x}$,

$$c(t) := \overline{x} + (t_n - t_{n+1})^{-1}[(t_n - t)t_{n+1}u_{n+1} + (t - t_{n+1})t_n u_n]$$

for $t \in [t_{n+1}, t_n[$. Then one sees that $(1/t)(c(t) - c(0)) \to u$, but since $c(t_n) = \overline{x} + t_n u_n$, in view of (5.4), $f \circ c$ is not differentiable at 0 with derivative $d_r f(\overline{x}, u)$. □

Corollary 5.5 *Let X, Y be normed spaces, let T, W be open subsets of $\mathbb{R}$ and X respectively, let $c : T \to X$ be differentiable at $\overline{t} \in T$ and $f : W \to Y$ be Hadamard differentiable at $\overline{x} := c(\overline{t}) \in W$ and such that $c(T) \subset W$. Then $f \circ c$ is differentiable at $\overline{t}$ and*

$$(f \circ c)'(\overline{t}) = Df(\overline{x})(c'(\overline{t})).$$

Thus, $Df(\overline{x})$ appears as the continuous linear map transforming velocities.

It is easy to show that any linear combination of maps having radial (resp. directional) derivatives at $\overline{x}$ in some direction u has a radial (resp. directional) derivative at $\overline{x}$ in direction u. In particular, any linear combination of two Gateaux (resp. Hadamard) differentiable maps is Gateaux (resp. Hadamard) differentiable. One also deduces from Proposition 5.2 that if f has a directional (resp. radial) derivative at $\overline{x}$ in the direction u and if $A : Y \to Z$ is a continuous linear map, then $A \circ f$ has a directional (resp. radial) derivative at $\overline{x}$ in the direction u and $(A \circ f)'(\overline{x}, u) = A(f'(\overline{x}, u))$.

The preceding example-exercise shows that the composition of two radially differentiable maps is not necessarily radially differentiable. However, one does have a chain rule for directionally differentiable maps. These facts show that Hadamard differentiability is a more interesting notion than Gateaux differentiability.

Theorem 5.4 *Let X, Y, Z be normed spaces, let U and V be open subsets of X and Y respectively and let $f : U \to Y$, $g : V \to Z$ be directionally differentiable maps at $\overline{x} \in W := f^{-1}(V)$ and $\overline{y} := f(\overline{x}) \in V$ respectively. Then $h := g \circ f$ is directionally differentiable at $\overline{x}$ and*

$$d\,(g \circ f)\,(\overline{x}, u) = dg(f(\overline{x}), df(\overline{x}, u)).$$

In particular, if f is Hadamard differentiable at $\overline{x}$ and if g is Hadamard differentiable at $\overline{y} := f(\overline{x})$, then $h := g \circ f$ is Hadamard differentiable at $\overline{x}$ and

$$D(g \circ f)(\overline{x}) = Dg(\overline{y}) \circ Df(\overline{x}).$$

Proof More generally, let us show that if f has a directional derivative at $\overline{x}$ in the direction $u \in X$ and if g has a directional derivative at $f(\overline{x})$ in the direction $v := df(\overline{x}, u)$, then $h := g \circ f$ has a directional derivative at $\overline{x}$ in the direction u. For (t, u') close enough to $(0, u)$ one has $\overline{x} + tu' \in W$. Let $q(t, u') := (1/t)(f(\overline{x} + tu') - f(\overline{x}))$.

Then $q(t, u') \to v := df(\overline{x}, u)$ as $(t, u') \to (0_+, u)$. Therefore

$$\frac{h(\overline{x} + tu') - h(\overline{x})}{t} = \frac{g(\overline{y} + tq(t, u')) - g(\overline{y})}{t} \to dg(\overline{y}, v) \text{ as } (t, u') \to (0_+, u).$$

The statement can also be proved by using Proposition 5.10. □

The notion of radial differentiability is sufficient to get a mean value theorem. Recall that the *line segment* $[a, b]$ (respectively $]a, b[$) with end points a, b in a normed space is the set $\{(1 - t)a + tb : t \in [0, 1]\}$ (respectively $\{(1 - t)a + tb : t \in]0, 1[\}$).

Proposition 5.11 *If $f : W \to Y$ is radially differentiable at each point of a segment $[w, x]$ contained in W, then*

$$\|f(x) - f(w)\| \leq \sup_{t \in]0,1[} \|d_r f(w + t(x - w), x - w)\|.$$

Proof Let $h : [0, 1] \to Y$ be given by $h(t) := f((1-t)w+tx)$; it is right differentiable on $]0, 1[$, with right derivative $D_r h(t) = d_r f((1 - t)w + tx, x - w)$ and continuous on $[0, 1]$. Then Corollary 5.1 yields the estimate. □

A variant can be derived when f is Gateaux differentiable at each point of $S :=]a, b[$, since then one has $\|d_r f(z, x - w)\| \leq \|D_r f(z)\| . \|x - w\|$ for all $z \in S$, w, $x \in X$.

Proposition 5.12 *Let X and Y be normed spaces, let W be an open subset of X containing the segment $[w, x]$ and let $f : W \to Y$ be continuous on $[w, x]$ and Gateaux differentiable at each point of $S :=]w, x[$, with $c := \sup_{z \in S} \|D_r f(z)\| < +\infty$. Then one has*

$$\|f(x) - f(w)\| \leq c \|x - w\| .$$

Corollary 5.6 *Let X and Y be normed spaces, let W be a convex open subset of X and let $f : W \to Y$ be Gateaux differentiable at each point of W and such that for some $c \in \mathbb{R}$ one has $\|D_r f(w)\| \leq c$ for every $w \in W$. Then, f is Lipschitzian with rate c: for all $x, x' \in W$ one has*

$$\left\|f(x) - f(x')\right\| \leq c \left\|x - x'\right\| .$$

In particular, if $D_r f(w) = 0$ for every $w \in W$, then f is constant on W. This result is also valid if W is connected instead of convex. An extension of the estimate of Proposition 5.12 is also valid in the case when W is connected, provided one replaces the usual distance with the geodesic distance d_W in W defined as in Exercise 5.

The following corollary gives an approximation of f in the case one disposes of an approximate value of the derivative of f around $\overline{x}$.

Corollary 5.7 *Let X and Y be normed spaces, let X_0 be a linear subspace of X, let W be a convex open subset of X and let $f : W \to Y$ be Gateaux differentiable at each point of W and such that for some $c \in \mathbb{R}$ and some $\ell \in L(X_0, Y)$ one has $\|D_r f(x)(u) - \ell(u)\| \leq c \|u\|$ for every $x \in W$, $u \in X_0$. Then, for any $x, x' \in W$ such that $x - x' \in X_0$, one has*

$$\|f(x) - f(x') - \ell(x - x')\| \leq c \|x - x'\|.$$

This result (obtained by changing f into $f - \ell$ in the preceding corollary) will serve to get Fréchet differentiability from Gateaux differentiability. For the moment, let us point out another passage from Gateaux differentiability to Hadamard differentiability.

Proposition 5.13 *Let W be an open subset of X. If $f : W \to Y$ is radially differentiable on a neighborhood V of $\overline{x}$ in W and if, for some $u \in X \backslash \{0\}$, its radial derivative $f_r' := d_r f : V \times X \to Y$ is continuous at $(\overline{x}, u)$, then f is directionally differentiable at $\overline{x}$ in the direction u.*

In particular, if f is Gateaux differentiable on V and if $f_r' : V \times X \to Y$ is continuous at each point of $\{\overline{x}\} \times X$, then f is Hadamard differentiable at $\overline{x}$.

Proof Without loss of generality, we may suppose u has norm 1. Given $\varepsilon > 0$, let $\delta \in]0, 1[$ be such that $|f_r'(x, v) - f_r'(\overline{x}, u)| \leq \varepsilon$ for all $(x, v) \in B(\overline{x}, 2\delta) \times B(u, \delta)$, with $B(\overline{x}, 2\delta) \subset V$. Setting $r(t, v) := f(\overline{x} + tv) - f(\overline{x}) - tf_r'(\overline{x}, u)$, we observe that for every $v \in B(u, \delta)$ the map $r_v := r(\cdot, v)$ is differentiable on $[0, \delta]$ and $\|r_v'(t)\| = \|f_r'(\overline{x} + tv, v) - f_r'(\overline{x}, u)\| \leq \varepsilon$. Since $r_v(0) = 0$, Corollary 5.1 yields $\|r(t, v)\| \leq \varepsilon t$. This shows that f has $f_r'(\overline{x}, u)$ as a directional derivative at $\overline{x}$ in the direction u. The last assertion is an immediate consequence. □

The importance of this continuity condition leads us to introduce a definition.

Definition 5.5 Given normed spaces X, Y and an open subset W of X, a map $f : W \to Y$ is said to be of *class* D^1 at $\overline{w} \in W$ (resp. on W) if it is Hadamard differentiable around $\overline{w}$ (resp. on W) and if $df : W \times X \to Y$ is continuous at $(\overline{w}, v)$ for all $v \in X$ (resp. on $W \times X$). We say that f is of class D^k with $k \in \mathbb{N}$, $k > 1$, if f is of class D^1 and if df is of class D^{k-1}.

We denote by $D^1(W, Y)$ the space of maps of class D^1 from W to Y and by $BD^1(W, Y)$ the space of maps $f \in D^1(W, Y)$ that are bounded and such that $f' : w \mapsto Df(w) := df(w, \cdot)$ is bounded from W to $L(X, Y)$. Let us note the following two properties.

Proposition 5.14 *For any $f \in D^1(W, Y)$ the map $f' : w \mapsto Df(w)$ is locally bounded.*

Proof Suppose, on the contrary, that there exist $w \in W$ and a sequence $(w_n) \to w$ such that $(r_n) := (\|Df(w_n)\|) \to +\infty$. For each $n \in \mathbb{N}$ one can pick some unit vector $u_n \in X$ such that $\|df(w_n, u_n)\| > r_n - 1$. Setting (for $n \in \mathbb{N}$ large) $x_n := r_n^{-1} u_n$, we see that $((w_n, x_n)) \to (w, 0)$ but $(\|df(w_n, x_n)\|) \to 1$, a contradiction. □

Corollary 5.8 *For $f : W \to Y$, where W is a convex open subset of X, the following assertions are equivalent:*

(a) f is of class D^1;

(b) there exists a continuous map $h : W \times X \times X \to Y$ such that for all $(w, x) \in W$ one has $h(w, x, \cdot) \in L(X, Y)$ and $f(w) - f(x) = h(w, x, w - x)$;

(c) f is Hadamard (or Gateaux) differentiable, f' is locally bounded and for all $u \in X$ the map $x \mapsto f'(x)u$ is continuous.

In particular, if $Y = \mathbb{R}$ and if $f \in D^1(W, \mathbb{R})$, the derivative of f is continuous when X^* is endowed with the topology of uniform convergence on compact sets.

Proof (a)⇒(b) Take $h(w, x, v) := \int_0^1 df((1-t)x + tw, w - x)dt$.

For the reverse implication, observe that for all $x \in W$, $u \in X$, setting $w := x + tu$ one has

$$\frac{1}{t}(f(x + tu) - f(x)) = h(x + tu, x, u) \underset{t(\neq 0)\to 0}{\to} h(x, x, u),$$

so that f is Gateaux differentiable and $df(x, u) = h(x, x, u)$. Thus df is continuous and f is of class D^1.

The implication (a)⇒(c) stems from the preceding proposition. For the reverse implication we assume that there exist an $m > 0$ and a neighborhood U of x in W such that $\|f'(w)\| \leq m$ for $w \in U$ and, given $u \in X$ and $\varepsilon > 0$ that there exists a neighborhood V_ε of x contained in U such that $\|f'(w)u - f'(x)u\| \leq \varepsilon/2$ for $w \in V_\varepsilon$. Then, for $v \in B(u, \varepsilon/2m)$ and $w \in V_\varepsilon$ we have

$$\left\|f'(w)v - f'(x)u\right\| \leq \left\|f'(w)(v - u)\right\| + \left\|f'(w)u - f'(x)u\right\| \leq \frac{m\varepsilon}{2m} + \frac{\varepsilon}{2} = \varepsilon$$

which shows that $(w, v) \mapsto f'(w, v)$ is continuous. □

Example Let E, F be Banach spaces, let T be a compact interval of $\mathbb{R}$, let U be an open subset of $E \times T$ and let $g : U \to F$ be continuous and such that for every $(t, e) \in U$ the Gateaux derivative $D_1 g(e, t) = Dg_t(e)$ of $g_t := g(\cdot, t)$ at e exists and the map $(e, t, v) \mapsto D_1 g(e, t).v$ is continuous on $U \times E$. Let $R(T, E)$ (resp. $R(T, F)$) be the space of regulated maps from T to E (resp. F). Denoting by W the set of $w \in R(T, E)$ such that $\mathrm{cl}(\{(w(t), t) : t \in T\}) \subset U$ and by $g^\diamond : W \to R(T, F)$ the map defined by $g^\diamond(f)(t) := g(w(t), t)$ for $w \in R(T, E)$, $t \in T$, we recall that Proposition 3.45 established that $g^\diamond$ is continuous. For a similar reason the map $(w, z) \mapsto D_1 g(w(\cdot), \cdot).z(\cdot)$ is continuous. Let

$$V := \{(e, e', t) : ((1 - s)e + se', t) \in U \ \forall s \in [0, 1]\}$$

and let $H : V \times E \to F$ be continuous and such that

$$g(e) - g(e') = G(e, e', t).(e - e') = H(e, e', e - e', t) \qquad \forall (e, e', t) \in V$$

where $G(e, e', t) = \int_0^1 D_1 g((1-s)e+se', t)ds$ and in particular $G(e, e, t) = D_1 g(e, t)$ for all $(e, t) \in U$. Plugging the values of $w, z \in W$ into g and H, for all $t \in T$, we get

$$g^{\diamond}(w)(t) - g^{\diamond}(z)(t) = H(w(t), z(t), w(t) - z(t), t) = H^{\diamond}(w, z, w - z)(t).$$

Since $H^{\diamond}$ is continuous and since $H^{\diamond}(w, z, \cdot)$ is linear and continuous by Proposition 3.45, we get that $g^{\diamond}$ is of class D^1 from W into $R(T, F)$. □

Proposition 5.15 *If X, Y, Z are normed spaces, if U and V are open subsets of X and Y respectively and if $f \in D^1(U, Y)$, $g \in D^1(V, Z)$, with $f(U) \subset V$ then $h := g \circ f \in D^1(U, Z)$.*

Proof This follows from the relation $dh(u, x) = dg(f(u), df(u, x))$ for all $(u, x) \in U \times X$. □

Under a differentiability assumption, convex functions, integral functionals and Nemytskii operators are important examples of maps of class D^1. See Proposition 6.16 and Sect. 8.5.2.

Exercises

1. Let X, Y be normed spaces and let W be an open subset of X. Prove that $f : W \to Y$ is Hadamard differentiable at $\overline{x}$ if and only if there exists a continuous linear map $\ell : X \to Y$ such that the map q_t given by $q_t(v) := (1/t)(f(\overline{x} + tv) - f(\overline{x}))$ converges to ℓ as $t \to 0_+$, uniformly on compact subsets of X. Deduce another proof of Proposition 5.12 below from this characterization.
2. Prove that if $f : W \to Y$ is radially differentiable at $\overline{x}$ in the direction u and if f is *directionally steady* at $\overline{x}$ in the direction u in the sense that $(1/t)(f(\overline{x} + tv) - f(\overline{x} + tu)) \to 0$ as $(t, v) \to (0_+, u)$, then f is directionally differentiable at $\overline{x}$ in the direction u. Give an example showing that this criterion is more general than the Lipschitz condition of Proposition 5.9.
3. Let $f : \mathbb{R}^2 \to \mathbb{R}$ be given by $f(r, s) := r^2 s(r^2 + s^2)^{-1}$ for $(r, s) \in \mathbb{R}^2 \backslash \{(0, 0)\}$, $f(0, 0) = 0$. Show that f has a radial derivative (which is, in fact, a bilateral derivative) but is not Gateaux differentiable at $(0, 0)$.
4. Let E be a Hilbert space and let $X := D^1(T, E)$, where $T := [0, 1]$. Endow X with the norm $\|x\| := \sup_{t \in T} \|x(t)\| + \sup_{t \in T} \|x'(t)\|$. Define the length of a curve $x : [0, 1] \to E$ by

$$\ell(x) := \int_0^1 \|x'(t)\| dt.$$

   (a) Show that ℓ is a continuous sublinear functional on X with Lipschitz rate 1.
   (b) Let W be the set of $x \in X$ such that $x'(t) \neq 0$ for all $t \in [0, 1]$. Show that W is an open subset of X and that ℓ is Gateaux differentiable on W.

(**c**) Show that ℓ is of class D^1 on W [Hint: use convergence results for integrals]. In order to prove that ℓ is of class C^1 one may use the results in the following questions.
(**d**) Let $E_0 := E\backslash\{0\}$ and let $D : E_0 \to E$ be given by $D(v) := \|v\|^{-1}v$. Given $u, v \in E_0$ show that $\|D(u) - D(v)\| \leq 2\|u\|^{-1}\|u - v\|$.
(**e**) Deduce from the preceding inequality that ℓ' is continuous.

5. Prove the assertions following Corollary 5.6, defining the geodesic distance $d_W(x, x')$ between two points x, x' of W as the infimum of the lengths of curves joining x to x'.
6. Prove that if $f : W \to Y$ has a directional derivative at some point $\overline{x}$ of the open subset W of X, then its derivative $Df(\overline{x}) : u \mapsto df(\overline{x}, u)$ is continuous if it is linear.
7. Prove Proposition 5.11 by deducing it from the classical Mean Value Theorem (Lemma 5.2) for real-valued functions, using the Hahn-Banach Theorem. [Hint: Take y^* with norm one such that $\langle y^*, y\rangle = \|y\|$ for $y := f(x) - f(w)$, set $g(t) := \langle y^*, f(x+t(w-x))\rangle$ and pick $\theta \in]0, 1[$ such that $g(1)-g(0) = D_r g(\theta)$.]
8. Show that the norm $x \mapsto \|x\| := \sup_{t\in T} |x(t)|$ on the Banach space $X := C(T)$ of continuous functions on $T := [0, 1]$ is Hadamard differentiable at $\overline{x} \in X$ if and only if the function $t \mapsto |\overline{x}(t)|$ attains its maximum on T at a single point.
9. (**a**) Let a, b be two points of a normed space X. Show that the function g given by $g(t) := \|a + tb\|$ has a right derivative and a left derivative at all points of $\mathbb{R}$.
(**b**) Let $f : T \to X$, where T is an interval of $\mathbb{R}$. Show that if f has a right derivative $D_r f(t)$ at some $t \in T$, then $g \circ f$ has a right derivative at t and $D_r(g \circ f)(t) \leq \|D_r f(t)\|$.
10. Use the preceding exercise to deduce a mean value theorem from Lemma 5.2.
11. Let $f : X \to Y$ be a map of class C^1 between two normed spaces such that $f(tx) = tf(x)$ for all $(t, x) \in \mathbb{R} \times X$. Show that f is linear, and in fact that $f(x) = f'(0)(x)$.
12. A map $f : X \to Y$ between two normed spaces is said to have a Schwarz derivative at $x \in X$ if for all $v \in X$ the quotient $(1/2t)(f(x + tv) - f(x - tv))$ has a limit $d_s f(x, v)$ as $t \to 0_+$. Show that if f has a radial derivative at x, then it has a Schwarz derivative at x and $d_s f(x, v) = (1/2)(d_r f(x, v) - d_r f(x, -v))$.

5.4 Classical Differential Calculus

The behaviors of nonlinear maps are difficult to control. Differential calculus can help. The main purpose of differential calculus consists in getting some information by using an affine approximation to a given nonlinear map around a given point.

5.4.1 The Main Concepts and Results of Differential Calculus

Of course, the meaning of the word "approximation" has to be made precise. For that purpose, we define remainders. Fréchet differentiability consists in an approximation by a continuous affine map, the error being a remainder.

Definition 5.6 Given normed spaces X and Y, we denote by $o(X, Y)$ the set of maps $r : X \to Y$ such that $r(0) = 0$ and $r(x)/\|x\| \to 0$ as $x \to 0$ in $X\backslash\{0\}$. The elements of $o(X, Y)$ will be called *remainders*.

Thus, $r : X \to Y$ is a *remainder* if and only if there exists some map $\alpha : X \to Y$ satisfying $\alpha(x) \to 0$ as $x \to 0$ and $r(\cdot) = \|\cdot\| \alpha(\cdot)$. Moreover, $r \in o(X, Y)$ if and only if there exists a modulus $\mu : \mathbb{R}_+ \to \overline{\mathbb{R}}_+ := \mathbb{R}_+ \cup \{\infty\}$ such that $\|r(x)\| \leq \mu(\|x\|)\,\|x\|$ for $x \in X$ (recall that $\mu : \mathbb{R}_+ \to \overline{\mathbb{R}}_+$ is a *modulus* when μ is nondecreasing, $\mu(0) = 0$ and μ is continuous at 0). Such a case occurs when there exist $c > 0$ and $p > 1$ such that $\|r(x)\| \leq c\,\|x\|^p$. Following Landau, remainders are often denoted by $o(\cdot)$ and different remainders are often represented by the same letters since they are considered as inessential for the assigned purposes.

If $r, s : X \to Y$ are two maps that coincide on some neighborhood V of 0 in X, then s belongs to $o(X, Y)$ if and only if r belongs to $o(X, Y)$. Thus if $q : V \to Y$ is defined on some neighborhood V of 0 in X, we consider that q is a remainder if some extension r of q to all of X is a remainder. The preceding observation shows that this property does not depend on the choice of the extension.

The following two results are direct consequences of the rules for limits.

Lemma 5.3 *For any normed spaces X, Y the set $o(X, Y)$ of remainders is a linear space.*

Lemma 5.4 *Given normed spaces $X, Y_1, \ldots, Y_k$, $Y := Y_1 \times \ldots \times Y_k$, a map $r : X \to Y$ is a remainder if and only if its components $r_1, \ldots, r_k$ are remainders.*

The class of remainders is stable under composition by continuous linear maps.

Lemma 5.5 *For any normed spaces W, X, Y, Z, for any $r \in o(X, Y)$ and any continuous linear maps $A : W \to X$, $B : Y \to Z$, one has $r \circ A \in o(W, Y)$ and $B \circ r \in o(X, Z)$ (hence $B \circ r \circ A \in o(W, Z)$).*

Proof Let $\alpha : X \to Y$ be such that $r(x) = \|x\|\,\alpha(x)$ and $\alpha(x) \to 0$ as $x \to 0$. Then, if $A : W \to X$ is *stable* at 0, i.e. is such that there exists some $c > 0$ for which $\|A(w)\| \leq c\,\|w\|$ for w in a neighborhood of 0 in W, in particular if A is linear and continuous, one has $\|r(A(w))\| = \|A(w)\|\,\|\alpha(A(w))\| \leq c\,\|w\|\,\|\alpha(A(w))\|$ and $\alpha(A(w)) \to 0$ as $w \to 0$, so that $r \circ A \in o(W, Y)$. Similarly, if $B : Y \to Z$ is stable at 0, then $B \circ r \in o(X, Z)$. The assertion about $B \circ r \circ A$ is a combination of the two other cases. □

We are ready to define differentiability in the Fréchet sense; this notion is so common that one often writes "differentiable" instead of "Fréchet differentiable".

Definition 5.7 Given normed spaces X, Y and an open subset W of X, a map $f : W \to Y$ is said to be *Fréchet differentiable* (or firmly differentiable, or just differentiable) at $\overline{x} \in W$ if there exist a continuous linear map $\ell : X \to Y$ and a remainder $r \in o(X, Y)$ such that for $w \in W$ one has

$$f(w) = f(\overline{x}) + \ell(w - \overline{x}) + r(w - \overline{x}). \tag{5.5}$$

It is often convenient to write the preceding relation in the form

$$f(\overline{x} + u) - f(\overline{x}) = \ell(u) + r(u)$$

for u close to 0. Here the continuous affine map $x \mapsto f(\overline{x}) + \ell(x - \overline{x})$ can be viewed as an approximation of f that essentially determines the behavior of f around $\overline{x}$. The continuous linear map ℓ is called the *derivative* of f at $\overline{x}$ and is denoted by $Df(\overline{x})$ or $f'(\overline{x})$. It is unique: given two approximations ℓ_1, ℓ_2 of $f(\overline{x} + \cdot) - f(\overline{x})$ around 0 and two remainders r_1, r_2 such that $f(\overline{x} + u) - f(\overline{x}) = \ell_1(u) + r_1(u) = \ell_2(u) + r_2(u)$ one has $\ell_1 = \ell_2$ since $\ell := \ell_1 - \ell_2$ is the remainder $r := r_2 - r_1$; in fact for every $u \in X$ and any $t > 0$ small enough one has

$$\ell(u) = \frac{1}{t} r(tu) = \frac{1}{t} \alpha(tu) \|tu\| = \alpha(tu) \|u\| \to 0 \text{ as } t \to 0,$$

so that $\ell(u) = 0$. Thus $L(X, Y) \cap o(X, Y) = \{0\}$. Uniqueness is also a consequence in Corollary 5.12 below and of the fact that the directional derivative is unique as it is obtained as a limit.

When $Y := \mathbb{R}$, the derivative $Df(\overline{x})$ of f at $\overline{x}$ belongs to the dual X^* of X. When X is a Hilbert space with scalar product $(\cdot \mid \cdot)$ it may be convenient to use the *Riesz isometry* $R : X \to X^*$ given by $\langle R(x), y \rangle = (x \mid y)$ to get an element $\nabla f(\overline{x})$ of X called the *gradient* of f at x by setting $\nabla f(\overline{x}) := R^{-1}(Df(\overline{x}))$. It allows us to visualize the derivative, but, in some respects, it is preferable to work with the derivative, in particular when dealing with composite maps and higher order derivatives.

Proposition 5.16 *If $f : W \to Y$ is differentiable at $\overline{x} \in W$, then it is continuous at $\overline{x}$.*

Proof This follows from the fact that any remainder is continuous at 0. □

Proposition 5.17 *If $f, g : W \to Y$ are differentiable at $\overline{x} \in W$, then for any $\lambda, \mu \in \mathbb{R}$ the map $h := \lambda f + \mu g$ is differentiable at $\overline{x}$ and $Dh(\overline{x}) = \lambda Df(\overline{x}) + \mu Dg(\overline{x})$.*

Proof If $r(x) := f(\overline{x} + x) - f(\overline{x}) - f'(\overline{x})(x)$, $s(x) := g(\overline{x} + x) - g(\overline{x}) - g'(\overline{x})x$, one has $h(\overline{x} + x) = h(\overline{x}) + \lambda f'(\overline{x})(x) + \mu g'(\overline{x})(x) + t(x)$, where $t := \lambda r + \mu s \in o(X, Y)$. Thus h is differentiable at $\overline{x}$ and $h'(\overline{x}) = \lambda f'(\overline{x}) + \mu g'(\overline{x})$. □

Examples

(a) A constant map is everywhere differentiable and its derivative is 0.
(b) A continuous linear map $\ell \in L(X, Y)$ is differentiable at any point $\overline{x}$ and its derivative at $\overline{x}$ is ℓ since $\ell(\overline{x} + x) = \ell(\overline{x}) + \ell(x)$.

(c) A continuous affine map $f := \ell + c$, where $\ell \in L(X, Y)$ and $c \in Y$, is differentiable at any $\overline{x} \in X$ and $Df(\overline{x}) = \ell$.

(d) If $f : X := X_1 \times X_2 \to Y$ is a continuous bilinear map then f is differentiable at any point $\overline{x} := (\overline{x}_1, \overline{x}_2) \in X$ and for $x = (x_1, x_2)$ one has $Df(\overline{x})(x) = f(x_1, \overline{x}_2) + f(\overline{x}_1, x_2)$ since $f(\overline{x} + x) - f(\overline{x}) = f(x_1, \overline{x}_2) + f(\overline{x}_1, x_2) + f(x_1, x_2)$. Here f is a remainder since $\|f(x)\| \leq \|f\| \|x_1\| \|x_2\| \leq \|f\| \|x\|^2$ whenever $\|x\| \geq \|x\|_\infty := \max(\|x_1\|, \|x_2\|)$.

(e) If $f : X \to Y$ is a continuous *quadratic map* in the sense that there exists a continuous bilinear map $b : X \times X \to Y$ such that $f(x) = b(x, x)$, then f is differentiable at any point $\overline{x} \in X$ and $Df(\overline{x})(x) = b(\overline{x}, x) + b(x, \overline{x})$ for $x \in X$. This follows from the chain rule below and the preceding example. Alternatively, one may observe that $r := f$ is a remainder since for every $x \in X$ one has $\|f(x)\| \leq \|b\| \|x\|^2$ and $f(\overline{x} + x) = f(\overline{x}) + b(\overline{x}, x) + b(x, \overline{x}) + f(x)$.

(f) If $p : X \to Y$ is a continuous homogeneous polynomial of degree k, i.e. if there exists a symmetric k-multilinear map $f : X^k \to Y$ such that $p(x) := f(x, \ldots, x)$ for all $x \in X$, then p is differentiable and $Dp(\overline{x})(x) = kf(\overline{x}, \ldots, \overline{x}, x)$ for all $x \in X$.

(g) If $f : T \to Y$ is defined on an open interval T of $\mathbb{R}$, f is differentiable at $\overline{x} \in T$ if and only if f has a derivative at $\overline{x}$ and $Df(\overline{x})$ is the linear map $r \mapsto rf'(\overline{x})$, hence $f'(\overline{x}) = Df(\overline{x})(1)$. The key point in this example is detailed in the following exercise. □

Exercise Show that for any normed space Y the space $L(\mathbb{R}, Y)$ is isomorphic (and even isometric) to Y via the evaluation map $\ell \mapsto \ell(1)$ whose inverse is the map $v \mapsto \ell_v$, where $\ell_v \in L(\mathbb{R}, Y)$ is defined by $\ell_v(r) := rv$ for $r \in \mathbb{R}$. □

The following characterization will be helpful.

Lemma 5.6 *Given an open subset W of X, a map $f : W \to Y$ is differentiable at $\overline{x} \in W$ if and only if there exists a map $F : W \to L(X, Y)$ that is continuous at $\overline{x}$ and such that $f(w) - f(\overline{x}) = F(w)(w - \overline{x})$ for all $w \in W$.*

Proof Assume that for a map $F : W \to L(X, Y)$ continuous at $\overline{x}$ we have $f(w) = f(\overline{x}) + F(w)(w - \overline{x})$ for all $w \in W$. Then $f(\overline{x} + x) - f(\overline{x}) = F(\overline{x})(x) + r(x)$, where $r(x) := (F(\overline{x} + x) - F(\overline{x}))(x)$ for x small. Since $\|r(x)\| \leq \|F(\overline{x} + x) - F(\overline{x})\| . \|x\|$, r is a remainder and f is differentiable at $\overline{x}$ with $Df(\overline{x}) = F(\overline{x})$.

Let us prove the converse. Using the Hahn-Banach Theorem, for $x \in X$ we pick $\ell_x \in X^*$ such that $\|\ell_x\| = 1$ and $\ell_x(x) = \|x\|$. Then, setting $A := Df(\overline{x})$, $\alpha(u) := r(u)/\|u\|$ for $u \in X \backslash \{0\}$, $\alpha(0) = 0$, where r is the remainder r appearing in (5.5), so that $r(u) = \alpha(u)\|u\| = \alpha(u)\ell_u(u)$ with $\alpha(u) \to 0$ as $u \to 0$, we get

$$f(\overline{x} + u) - f(\overline{x}) = (A + \alpha(u)\ell_u)(u) = F(w)(w - \overline{x}) \qquad \text{with } w := \overline{x} + u$$

for $F(w) := A + \alpha(w - \overline{x})\ell_{w - \overline{x}} \to A = F(\overline{x})$ as $w \to \overline{x}$. □

Corollary 5.9 *Given Banach spaces X, Y, an open subset W of X, a map $f : W \to Y$ is of class C^1 on W, i.e. differentiable on W with a continuous derivative $f' : W \to L(X, Y)$ if and only if there exists an open subset V of $W \times W$ containing the diagonal $\{(w, w) : w \in W\}$ and a continuous map $F : V \to L(X, Y)$ such that $f(w) - f(x) = F(w, x)(w - x)$ for all $(w, x) \in V$. In such a case, one has $f'(w) = F(w, w)$ for all $w \in W$.*

Proof The sufficient condition is a direct consequence of the proposition. For the necessary condition one can take $V := \{(w, x) \in W \times W : (1 - t)x + tw \in W \forall t \in [0, 1]\}$ and $F(w, x) := \int_0^1 f'((1 - t)x + tw)dt$. □

Example Let us keep the notation of the example following Cor. 5.8: E, F are Banach spaces, T is a compact interval of $\mathbb{R}$, U is an open subset of $E \times T$ and $g : U \to F$ is continuous and such that for every $(e, t) \in U$ the Gateaux derivative $D_1 g(e, t) = Dg_t(e)$ of $g_t := g(\cdot, t)$ at e exists. This time we assume that the map $D_1 g$ is continuous from U into $L(E, F)$. Again, let us denote by W the set of $w \in R(T, E)$ such that $\mathrm{cl}(\{(w(t), t) : t \in T\}) \subset U$ and by $g^\diamond : W \to R(T, F)$ the map defined by $g^\diamond(f)(t) := g(w(t), t)$ for $w \in R(T, E)$, $t \in T$, as in Proposition 3.45. Here $R(T, E)$ and $R(T, F)$ are the spaces of regulated functions from T to E and F, respectively. Then, $g^\diamond : W \to R(T, F)$ is of class C^1 and, with a slight abuse of notation, $Dg^\diamond(w) = (D_1 g)^\diamond(w) \in L(R(T, E), R(T, F))$ for all $w \in W$.

A similar (and simpler) result holds for $g^\diamond$ considered as a map from $W \cap C(T, E)$ into $C(T, F)$. □

Let us present the chain rule. It is a cornerstone of differential calculus.

Theorem 5.5 (Chain Rule) *Let X, Y, Z be normed spaces, let U, V be open subsets of X and Y respectively and let $f : U \to Y$, $g : V \to Z$ be differentiable at $\overline{x} \in U$ and $\overline{y} = f(\overline{x})$ respectively and such that $f(U) \subset V$. Then $h := g \circ f$ is differentiable at $\overline{x}$ and*

$$Dh(\overline{x}) = Dg(\overline{y}) \circ Df(\overline{x}). \tag{5.6}$$

Proof Let $\ell := Df(\overline{x})$, $m := Dg(\overline{y})$ and let $r \in o(X, Y)$, $s \in o(Y, Z)$ be defined by

$$r(x) := f(\overline{x} + x) - f(\overline{x}) - \ell(x), \quad s(y) := g(\overline{y} + y) - g(\overline{y}) - m(y).$$

Then, setting $y := \ell(x) + r(x)$ for $x \in U - \overline{x}$, so that $f(\overline{x} + x) = \overline{y} + y$, we get

$$h(\overline{x}+x)-h(\overline{x})-m(\ell(x)) = g(\overline{y}+y)-g(\overline{y})-m(y-r(x)) = s(y)+m(r(x)). \tag{5.7}$$

Lemma 5.5 ensures that $m \circ r \in o(X, Z)$. Now, given $c > \|\ell\|$, there exists some $\rho > 0$ such that for $x \in B(0, \rho)$ one has $\|r(x)\| \leq (c - \|\ell\|)\|x\|$ hence $\|\ell(x) + r(x)\| \leq c\|x\|$. Then the proof of Lemma 5.5 ensures that $s \circ (\ell + r) \in o(X, Z)$. Thus, the right-hand side $s \circ (\ell + r) + m \circ r$ of (5.7) is a remainder and we conclude that h is differentiable at $\overline{x}$ with derivative the continuous linear map $m \circ \ell$. □

Exercise Give a short proof of Theorem 5.5 using Lemma 5.4.

The following corollary is a consequence in the fact that the derivative of a continuous linear map ℓ at an arbitrary point is ℓ itself.

Corollary 5.10 *Let X, Y, Z be normed spaces, let U, V be open subsets of X and Y respectively and let $f : U \to Y$, $g : V \to Z$ be such that $f(U) \subset V$ and let $h := g \circ f$.*

(a) If f is differentiable at $\overline{x}$ and $V := Y$, $g \in L(Y, Z)$, then h is differentiable at $\overline{x}$ and $Dh(\overline{x}) = g \circ Df(\overline{x})$.
(b) If g is differentiable at $\overline{y} := f(\overline{x})$ and $U := X$, $f \in L(X, Y)$, then h is differentiable at $\overline{x}$ and $Dh(\overline{x}) = Dg(\overline{y}) \circ f$.

Corollary 5.11 *The differentiability of $f : W \to Y$ (with W open in X) at $\overline{x}$ does not depend on the choices of the norms on X and Y within their equivalences classes.*

In fact, changing the norm amounts to composing with the identity map, the source space and the range space being endowed with two different norms.

Corollary 5.12 *Let X, Y be normed spaces, let W be an open subset of X and let $f : W \to Y$. If f is Fréchet differentiable at $\overline{x} \in W$, then f is Hadamard differentiable at $\overline{x}$. If X is finite dimensional, the converse holds.*

Thus, the Mean Value theorems of Sect. 5.2 are in force for Fréchet differentiability. Also, the interpretation of the derivative as a rule for the transformation of velocities remains valid for the Fréchet derivative.

Proof The first assertion follows from the definitions or from Theorem 5.5 and Proposition 5.10.

Assuming the dimension of X is finite, let us prove that if f is directionally differentiable at $\overline{x}$, and if its directional derivative $f'(\overline{x}, \cdot)$ is continuous, then r given by

$$r(w) := f(\overline{x} + w) - f(\overline{x}) - f'(\overline{x}, w)$$

is a remainder. Adding the assumption that $f'(\overline{x}, \cdot)$ is linear, this will prove the converse assertion. Suppose, on the contrary, that there exist $\varepsilon > 0$ and a sequence $(w_n) \to 0$ such that, for all $n \in \mathbb{N}$, $\|r(w_n)\| > \varepsilon \|w_n\|$. Then $t_n := \|w_n\|$ is positive; setting $u_n := t_n^{-1} w_n$, we may suppose the sequence (u_n) converges to some vector u of the unit sphere of X. Then, given $\varepsilon' \in]0, \varepsilon[$, we can find $k \in \mathbb{N}$ such that for $n \geq k$ we have $\|f'(\overline{x}, u_n) - f'(\overline{x}, u)\| \leq \varepsilon - \varepsilon'$, so that

$$\left\| \frac{f(\overline{x} + t_n u_n) - f(\overline{x})}{t_n} - f'(\overline{x}, u) \right\| > \varepsilon \|u_n\| - \left\| f'(\overline{x}, u_n) - f'(\overline{x}, u) \right\| \geq \varepsilon',$$

contradicting the assumption that f is differentiable at $\overline{x}$ in the direction u. □

Another link between directional differentiability and firm differentiability is pointed out in the next statement. A direct proof using Corollary 5.7 is easy (Exercise 8). We present a proof in the case f' is continuous around $\overline{x}$.

Proposition 5.18 *If f is Gateaux differentiable on W and if $f' : W \to L(X, Y)$ is continuous at $\overline{x} \in W$, then f is Fréchet differentiable at $\overline{x}$.*

Proof Without loss of generality, replacing Y by its completion, we may suppose Y is complete; replacing W with a ball centered at $\overline{x}$, we may also suppose W is convex. Then, for $x \in W$ one has $f(x) - f(\overline{x}) = F(x)(x - \overline{x})$ with

$$F(x) := \int_0^1 Df(\overline{x} + t(x - \overline{x}))dt,$$

and F is continuous, so that the criterion of Lemma 5.6 applies. □

This result shows that it may be a sensible strategy to start with radial differentiability in order to prove that a map is of *class* C^1, i.e. that it is differentiable with a continuous derivative. For instance, if one is dealing with an integral functional

$$f(x) := \int_S F(s, x(s))ds,$$

where S is some measure space and x belongs to some space of measurable maps, it is advisable to use Lebesgue's Theorem to differentiate inside the integral (under appropriate assumptions) by taking the limit in the quotient

$$\frac{1}{t}[f(\overline{x} + tu) - f(\overline{x})] = \int_S \frac{1}{t}[F(s, \overline{x}(s) + tu(s)) - F(s, \overline{x}(s))]ds.$$

Continuity arguments may be invoked later, for instance by using Krasnoselski's criterion (see Subsection 8.5.2 and [12, 179, 254]).

However, there are cases in which such a map is Gateaux differentiable but not Fréchet differentiable. See [92, Example 15.2].

Let us note other consequences of Theorem 5.5.

Proposition 5.19 *Let X, $Y_1, \ldots, Y_n$ be normed spaces, let W be an open subset of X and let $f := (f_1, \ldots, f_n) : W \to Y := Y_1 \times \ldots \times Y_n$. Then f is differentiable at $\overline{x} \in W$ if and only if its components $f_i : W \to Y_i$ ($i = 1, \ldots, n$) are differentiable at $\overline{x}$ and for $v \in X$*

$$Df(\overline{x})(v) = (Df_1(\overline{x})(v), \ldots, Df_n(\overline{x})(v)).$$

Proof Let $p_i : Y \to Y_i$ denote the ith canonical projection. If f is differentiable at $\overline{x}$, then Corollary 5.10 ensures that $f_i := p_i \circ f$ is differentiable at $\overline{x}$ and $Df_i(\overline{x}) = p_i \circ Df(\overline{x})$. Conversely, suppose that $f_1, \ldots, f_n$ are differentiable at $\overline{x}$. Let $r_i \in o(X, Y_i)$ be given by $r_i(x) = f_i(\overline{x} + x) - f_i(\overline{x}) - Df_i(\overline{x})(x)$. Then, by Lemma 5.4, we have that $r := (r_1, \ldots, r_n) \in o(X, Y)$ and $r(x) = f(\overline{x}+x) - f(\overline{x}) - \ell(x)$ for $\ell \in L(X, Y)$ given by $\ell(x) := (Df_1(\overline{x})(x), \ldots, Df_n(\overline{x})(x))$. Thus f is differentiable at $\overline{x}$, with derivative ℓ. □

Now, let us consider the case when the source space X is a product $X_1 \times \ldots \times X_n$ and W is an open subset of X. One says that $f : W \to Y$ has a *partial derivative at* $\overline{x} \in W$ relative to X_i for some $i \in \mathbb{N}_n$ if the map $f_{i,\overline{x}} : x_i \mapsto f(\overline{x}_1, \ldots, \overline{x}_{i-1}, x_i, \overline{x}_{i+1}, \ldots, \overline{x}_n)$ is differentiable at $\overline{x}_i$. Then, one denotes by $D_i f(\overline{x})$ or $\frac{\partial f}{\partial x_i}(\overline{x})$ the derivative of the map $f_{i,\overline{x}}$ at $\overline{x}_i$. Let $j_i \in L(X_i, X)$ be the insertion given by $j_i(x_i) := (0, \ldots, 0, x_i, 0, \ldots, 0)$. Since the map $f_{i,\overline{x}}$ is just the composition of the affine map $x_i \mapsto j_i(x_i - \overline{x}_i) + \overline{x} = (\overline{x}_1, \ldots, \overline{x}_{i-1}, x_i, \overline{x}_{i+1}, \ldots, \overline{x}_n)$ with f, from Corollary 5.10 (b) and the fact that $v = j_1(v_1) + \ldots + j_n(v_n)$, while $D_i f(\overline{x}) = Df_{i,\overline{x}}(\overline{x}_i) = Df(\overline{x}) \circ j_i$ one gets the following proposition.

Proposition 5.20 *If $f : W \to Y$ is defined on an open subset W of a product space $X := X_1 \times \ldots \times X_k$ and if f is differentiable at $\overline{x}$, then for $i = 1, \ldots, k$, the map f has a partial derivative at $\overline{x}$ relative to X_i and*

$$\forall v := (v_1, \ldots, v_k) \qquad Df(\overline{x})(v) = D_1 f(\overline{x}) v_1 + \ldots + D_k f(\overline{x}) v_k.$$

When $X := \mathbb{R}^m$, $Y := \mathbb{R}^n$, the matrix $(D_i f_j(\overline{x}))$ of $Df(\overline{x})$ formed with the partial derivatives of the components $(f_j)_{1 \leq j \leq n}$ of f is called the *Jacobian matrix* of f at $\overline{x}$. It determines $Df(\overline{x})$.

Note that it may happen that f has partial derivatives at $\overline{x}$ with respect to all its variables but is not differentiable at $\overline{x}$.

Example Let $f : \mathbb{R}^2 \to \mathbb{R}$ be given by $f(r,s) := rs(r^2 + s^2)^{-1}$ for $(r,s) \neq (0,0)$ and $f(0,0) = 0$. Since $f(r,0) = 0 = f(0,s)$, f has partial derivatives with respect to its two variables at $(0,0)$. However, f is not continuous at $(0,0)$, hence is not differentiable at $(0,0)$. □

We will shortly present a partial converse of Proposition 5.20. For this purpose (among others) it will be useful to introduce a reinforced notion of differentiability that allows us to formulate several results with assumptions weaker than continuous differentiability.

Definition 5.8 Let X and Y be normed spaces, let W be an open subset of X and let $\overline{x} \in W$. A map $f : W \to Y$ is said to be *circa-differentiable* (or peri-differentiable, or strictly differentiable) at $\overline{x}$ if there exists some continuous linear map $\ell \in L(X,Y)$ such that for every $x, x' \in W$ one has

$$\frac{\|f(x) - f(x') - \ell(x - x')\|}{\|x - x'\|} \to 0 \text{ as } x, x' \to \overline{x} \text{ with } x' \neq x. \tag{5.8}$$

Taking $x' = \overline{x}$ in relation (5.8), one sees that if f is circa-differentiable at $\overline{x}$, then f is differentiable at $\overline{x}$ and $Df(\overline{x}) = \ell$. If X_0 is a linear subspace of X we say that f is *circa-differentiable* (or strictly differentiable) *at $\overline{x}$ with respect to X_0* if there exist some continuous linear map $\ell \in L(X_0, Y)$ such that (5.8) holds whenever $x, x' \in W$ satisfy $x - x' \in X_0$.

Let us relate the preceding notion to continuous differentiability.

Definition 5.9 The map $f : W \to Y$ will be said to be continuously differentiable at $\overline{x} \in W$, or of *class* C^1 at $\overline{x}$, and we write $f \in C^1_{\overline{x}}(W,Y)$, if f is differentiable on some neighborhood $V \subset W$ of $\overline{x}$ and if the derivative $f' : V \to L(X,Y)$ of f given by $f'(x) := Df(x)$ for $x \in V$ is continuous at $\overline{x}$. If f is of class C^1 at each point x of W, then f is said to be of class C^1 on W and one writes $f \in C^1(W, Y)$.

One says that f is of class C^k with $k \in \mathbb{N}$, $k > 1$, if f is of class C^1 and if f' is of class C^{k-1}. Then, one writes $f \in C^k(W, Y)$.

Proposition 5.21 *Let X and Y be normed spaces, let W be an open subset of X and let $\overline{x} \in W$. A map $f : W \to Y$ which is differentiable on a neighborhood $U \subset W$ of $\overline{x}$ is circa-differentiable at $\overline{x} \in W$ if and only if $f \in C^1_{\overline{x}}(W, Y)$.*

Proof Suppose $f \in C^1_{\overline{x}}(W, Y)$ and let $\ell := Df(\overline{x})$. Given $\varepsilon > 0$ one can find $\delta > 0$ such that $B(\overline{x}, \delta) \subset W$ and for $x \in B(\overline{x}, \delta)$ one has $\|Df(x) - \ell\| \leq \varepsilon$. Then, using Corollary 5.7, for $x, x' \in B(\overline{x}, \delta)$ one has

$$\left\| f(x') - f(x) - \ell(x' - x) \right\| \leq \varepsilon \left\| x' - x \right\|,$$

so that f is circa-differentiable at $\overline{x}$.

Conversely, suppose f is circa-differentiable at $\overline{x}$ and is differentiable on a neighborhood V of $\overline{x}$ contained in W. Given $u \in X$ and $\varepsilon > 0$, assuming that the preceding inequality holds whenever $x, x' \in B(\overline{x}, \delta) \subset V$, one gets for all $x \in B(\overline{x}, \delta)$,

$$\|Df(x)(u) - \ell(u)\| = \lim_{t \to 0_+} t^{-1} \|f(x + tu) - f(x) - \ell(tu)\| \leq \varepsilon \|u\|,$$

so that $\|Df(x) - \ell\| \leq \varepsilon$ and $f' : x \mapsto Df(x)$ is continuous at $\overline{x}$. □

We are now in a position to give a converse of Proposition 5.20.

Proposition 5.22 *If $f : W \to Y$ is defined on an open subset W of a product space $X := X_1 \times \ldots \times X_k$, if for $i \in \mathbb{N}_k$, f has a partial derivative at $\overline{x} \in W$ relative to X_i and if f is circa-differentiable at $\overline{x}$ with respect to $X_1, \ldots, X_{i-1}, X_{i+1}, \ldots, X_k$, then f is differentiable at $\overline{x}$. In particular, if f has partial derivatives on some neighborhood of $\overline{x}$, all of which but one being continuous at $\overline{x}$, then f is differentiable at $\overline{x}$.*

Proof It suffices to give the proof for $k = 2$; an induction yields the general case.

Thus, let f be circa-differentiable at $\overline{x}$ with respect to X_1 and have a partial derivative at $\overline{x}$ relatively to X_2. The first assumption means that there exists some $\ell_1 \in L(X_1, Y)$ such that for every $\varepsilon > 0$ one can find some $\delta > 0$ such that $B(\overline{x}, 2\delta) \subset W$ and for $x := (x_1, x_2) \in B(\overline{x}, \delta)$, $u_1 \in X_1$, $\|u_1\| \leq \delta$ one has

$$\|f(x_1 + u_1, x_2) - f(x_1, x_2) - \ell_1(u_1)\| \leq \varepsilon \|u_1\|. \tag{5.9}$$

Setting $\ell_2 := D_2 f(\overline{x})$ and taking a smaller $\delta > 0$ if necessary, we may suppose that

$$\|f(\overline{x}_1, \overline{x}_2 + u_2) - f(\overline{x}_1, \overline{x}_2) - \ell_2(u_2)\| \leq \varepsilon \|u_2\|$$

for any $u_2 \in X_2$ satisfying $\|u_2\| \leq \delta$. Then, taking $(x_1, x_2) := (\overline{x}_1, \overline{x}_2 + u_2)$ in (5.9) with $u := (u_1, u_2) \in B(0, \delta)$, we get

$$\begin{aligned}
&\|f(\overline{x}+u) - f(\overline{x}) - \ell_1(u_1) - \ell_2(u_2)\| \\
&\leq \|f(\overline{x}+u) - f(\overline{x}_1, \overline{x}_2 + u_2) - \ell_1(u_1)\| + \|f(\overline{x}_1, \overline{x}_2 + u_2) - f(\overline{x}) - \ell_2(u_2)\| \\
&\leq \varepsilon \|u_1\| + \varepsilon \|u_2\| = \varepsilon \|(u_1, u_2)\|
\end{aligned}$$

if one takes the norm on X given by $\|(u_1, u_2)\| := \|u_1\| + \|u_2\|$. □

Corollary 5.13 *A map $f : W \to Y$ defined on an open subset W of a product space $X := X_1 \times \ldots \times X_k$ is of class C^1 on W if and only if f has partial derivatives on W that are jointly continuous.*

Now, let us give a result dealing with the interchange of limits and differentiation.

Theorem 5.6 *Let (f_n) be a sequence of Fréchet (resp. Hadamard) differentiable functions from a bounded, convex, open subset W of a normed space X to a Banach space Y. Suppose*

(a) there exists some $\overline{x} \in W$ such that $(f_n(\overline{x}))$ converges in Y;
(b) the sequence (f_n') uniformly converges on W to some map $g : W \to L(X, Y)$.

Then (f_n) uniformly converges on W to some map f that is Fréchet (resp. Hadamard) differentiable on W. Moreover, $f' = g$.

Proof Let us prove the first assertion. Let $r > 0$ be such that W is contained in the ball $B(\overline{x}, r)$. Given n, p in $\mathbb{N}$, Corollary 5.6 yields, for every $x \in W$

$$\left\| f_p(x) - f_p(\overline{x}) - (f_n(x) - f_n(\overline{x})) \right\| \leq \left\| f_p' - f_n' \right\|_\infty \cdot \|x - \overline{x}\| \leq r \left\| f_p' - f_n' \right\|_\infty \tag{5.10}$$

$$\left\| f_p(x) - f_n(x) \right\| \leq \left\| f_p(\overline{x}) - f_n(\overline{x}) \right\| + r \left\| f_p' - f_n' \right\|_\infty . \tag{5.11}$$

Since $\left\| f_p' - f_n' \right\|_\infty \to 0$ as $n, p \to \infty$ and since $(f_p(\overline{x}) - f_n(\overline{x})) \to 0$ as $n, p \to \infty$, we see that $(f_n(x))$ is a Cauchy sequence, hence has a limit in the complete space Y; we denote it by $f(x)$. Passing to the limit on p in (5.11) we see that the limit is uniform on W.

Now, given $x \in W$, let us prove that f is differentiable at x with derivative $g(x)$. Given $\varepsilon > 0$, we can find $k \in \mathbb{N}$ such that for $p > n \geq k$ one has $\left\| f_p' - f_n' \right\|_\infty \leq \varepsilon/3$, hence $\left\| g - f_n' \right\|_\infty \leq \varepsilon/3$. Using again Corollary 5.6 with $x' := x + u \in W$, we get

$$\left\| (f_p(x+u) - f_p(x)) - (f_n(x+u) - f_n(x)) \right\| \leq (\varepsilon/3) \|u\|$$

and passing to the limit on p, we obtain

$$\|f(x+u) - f(x) - (f_n(x+u) - f_n(x))\| \leq (\varepsilon/3) \|u\| \tag{5.12}$$

In the Fréchet differentiable case, we can find $\delta > 0$ such that $B(x, \delta) \subset W$ and for all $u \in \delta B_X$

$$\|f_k(x+u) - f_k(x) - g(x)(u)\|$$
$$\leq \left\|f_k(x+u) - f_k(x) - f_k'(x)(u)\right\| + \left\|f_k'(x)(u) - g(x)(u)\right\| \leq \frac{\varepsilon}{3}\|u\| + \frac{\varepsilon}{3}\|u\|.$$

Combining this estimate with relation (5.12) in which we take $n = k$, we get

$$\forall u \in \delta B_X \qquad \|f(x+u) - f(x) - g(x)(u)\| \leq \varepsilon \|u\|,$$

so that f is Fréchet differentiable at x with derivative $g(x)$.

In the Hadamard differentiable case, given $\varepsilon > 0$ and a unit vector u, we take $\delta \in]0, 1[$ such that $B(x, 2\delta) \subset W$ and for $t \in]0, \delta[$, $v \in B(u, \delta)$

$$\|f_k(x+tv) - f_k(x) - g(x)(tu)\|$$
$$\leq \left\|f_k(x+tv) - f_k(x) - f_k'(x)(tu)\right\| + \left\|f_k'(x)(tu) - g(x)(tu)\right\| \leq \frac{\varepsilon}{3}t + \frac{\varepsilon}{3}t.$$

Gathering this estimate with relation (5.12), in which we take $n = k$, $u = tv$, we get

$$\forall (t, v) \in (0, \delta) \times B(u, \delta) \qquad \|f(x+tv) - f(x) - g(x)(tu)\| \leq \varepsilon t,$$

so that f is Hadamard differentiable at x and $f'(x) = g(x)$. □

Corollary 5.14 *Let X, Y be normed spaces, Y being complete, and let W be an open subset of X. The space $B^1(W, Y)$ (resp. $BC^1(W, Y)$) of bounded, Lipschitzian, differentiable (resp. of class C^1) maps from W to Y is complete with respect to the norm $\|\cdot\|_{1,\infty}$ given by*

$$\|f\|_{1,\infty} := \sup_{x \in W} \|f(x)\| + \sup_{x \in W} \left\|f'(x)\right\|.$$

Here we use the fact that if f is Lipschitzian and differentiable, its derivative is bounded.

Proof Let (f_n) be a Cauchy sequence in $\left(B^1(W, Y), \|\cdot\|_{1,\infty}\right)$. Then (f_n') is a Cauchy sequence in the space $B(W, L(X, Y))$ of bounded maps from W into $L(X, Y)$ with respect to the uniform norm; thus it converges and its limit is continuous if $f_n \in BC^1(W, Y)$. Similarly, (f_n) converges in $B(W, Y)$. The theorem ensures that the limit f of (f_n) is Fréchet differentiable and its derivative is the limit of (f_n'), hence is bounded. Thus f belongs to $B^1(W, Y)$ and $(f_n) \to f$ for $\|\cdot\|_{1,\infty}$. If (f_n) is contained in $BC^1(W, Y)$, then f' is continuous, hence $f \in BC^1(W, Y)$. □

A directional version follows similarly from Theorem 5.6.

Corollary 5.15 *Let X, Y be normed spaces, Y being complete and let W be an open subset of X. The space* $BH^1(W,Y)$ *of bounded, Lipschitzian, Hadamard differentiable maps from W to Y is complete with respect to the norm* $\|\cdot\|_{1,\infty}$. *The same is true for its subspace* $BD^1(W,Y)$ *formed by bounded, Lipschitzian maps of class* D^1.

The following theorem is a deep result we admit. It requires tools from measure theory. We refer to [118] for the proof.

Theorem 5.7 (Rademacher) *A locally Lipschitz function f on an open subset of* $\mathbb{R}^d$ *is differentiable almost everywhere.*

Exercises

1. **(a)** Show that $r : X \to Y$ is a remainder if and only if there exists a remainder ρ on $\mathbb{R}$ such that $\|r(x)\| \leq \rho(\|x\|)$ for all x close to 0.
   (b) Prove the other two characterizations of remainders that follow the definition.
2. Define a notion of directional remainder that could be used for the study of Hadamard differentiability.
3. Show that when $f : W \to Y$ is Fréchet differentiable at $\overline{x}$, then it is *stable* at $\overline{x}$ in the sense that there exists a $c > 0$ such that $\|f(\overline{x}+x) - f(\overline{x})\| \leq c\,\|x\|$ for $\|x\|$ small enough.
4. Give a direct proof that Fréchet differentiability implies Hadamard differentiability.
5. Show that if $f : X_1 \times X_2 \to Y$ is circa-differentiable at $\overline{x} := (\overline{x}_1, \overline{x}_2)$ with respect to X_1 and X_2, then it is circa-differentiable at $\overline{x}$.
6. In Theorem 5.6, when W is not bounded, assuming that (f'_n) converges to g uniformly on bounded subsets of W, obtain a similar interchange result in which the convergence of (f_n) to f is uniform on bounded subsets of the open convex set W.
7. In Theorem 5.6, assuming that W is a connected open subset of X and that the convergence of (f'_n) is locally uniform (in the sense that for every $x \in W$ there exists some ball with center x contained in W on which the convergence of (f'_n) is uniform), prove that (f_n) is locally uniformly convergent and that its limit f is differentiable with derivative g.
8. Give a proof of Proposition 5.18 assuming only continuity of f' at $\overline{x}$.
9. With the hypothesis of Proposition 5.18 show that the map f is circa-differentiable at $\overline{x}$. Is it of class C^1 at $\overline{x}$?
10. Express the chain rule for differentiable maps between $\mathbb{R}^m$, $\mathbb{R}^n$, $\mathbb{R}^p$ in terms of a matrix product for the Jacobians of f and g.
11. Using the Hahn-Banach Theorem, show that $f : W \to Y$ is circa-differentiable at $a \in W$ if and only if there exists a map $F : W \times W \to L(X,Y)$ continuous at (a,a) such that $f(u) - f(v) = F(u,v)(u-v)$. Then $f'(a) = F(a,a)$.

12. Show that if X is finite dimensional, $f : U \to Y$, with U open in X, is of class D^1 if and only if f is of class C^1. [Hint: for any element e of a basis of X the map $x \mapsto Df(x)(e)$ is continuous when f is of class D^1.]

13. Given normed spaces X, Y and a topology $\mathcal{T}$ (or a convergence) on the space of maps from B_X to Y, one can define the notion of a $\mathcal{T}$-semiderivative at $\overline{x}$ of a map $f : B(\overline{x}, r) \to Y$: it consists in requiring that the family of maps $(f_t)_{0<t<r}$ from B_X to Y given by $f_t(v) := t^{-1}(f(\overline{x} + tv) - f(\overline{x}))$ has a limit as $t \to 0_+$. If the limit is the restriction to B_X of a continuous linear map, one speaks of a $\mathcal{T}$-derivative. Interpret Gateaux, Hadamard and Fréchet derivatives with the help of the topologies of uniform convergence on the families of finite subsets, compact subsets and bounded subsets. Observe that such a process also applies to some other families of sets, such as the family of weakly compact subsets of B_X.

14. Show that the norm $x \mapsto \|x\| := \sup_{t\in T} |x(t)|$ on the Banach space $X := C(T)$ of continuous functions on $T := [0, 1]$ is not Fréchet differentiable at any point. Compare with Exercise 8 of the preceding section to conclude that there are Hadamard differentiable maps that are not Fréchet differentiable.

15. Let X and Y be normed spaces, let $\overline{x} \in X$, c, $r > 0$, $W := B(\overline{x}, r)$, $f : W \to Y$ be of class C^1 and such that $\|f'(x) - f'(\overline{x})\| \le c \|x - \overline{x}\|$ for all $x \in W$.

(**a**) Show that for all $x \in W$ one has $\|f(x) - f(\overline{x}) - f'(\overline{x})(x - \overline{x})\| \le (c/2) \|x - \overline{x}\|^2$.

(**b**) Suppose that f' is Lipschitzian with rate c on W. Show that for all $w, x \in W$ one has $\|f(x) - f(w) - f'(w)(x - w)\| \le (c/2) \|x - w\|^2$.

16. (**Valadier**, J. Convex Analysis, 21 (2), 449–452 (2014)) Let $(S, \mathcal{S}, \mu)$ be a finite measure space and let $p, q \in [1, \infty]$. Given $\overline{f} \in L_p(S, \mu)$ such that $Z := \{s \in S : \overline{f}(s) = 0\}$ has null measure, show that the map $P : L_p(S, \mu) \to L_q(S, \mu)$ defined by $P(f) := f^+ := \max(f, 0)$ is Fréchet differentiable at $\overline{f}$ if $p > q$ and is Hadamard differentiable at $\overline{f}$ if $p = q$, with $P'(\overline{f}).h = 1_T h$, where $T := \{t \in S : \overline{f}(t) > 0\}$ and $1_T(s) := 1$ for $s \in T$, $1_T(s) := 0$ for $s \in S \setminus T$. Give a counter-example showing that P is not necessarily Fréchet differentiable if $p = q$. [Hint: assume that there exists a sequence (S_n) such that $(\mu(S_n)) \to 0$ and $\mu(S_n) > 0$ for all $n \in \mathbb{N}$ and set $v_n := 1_{S_n}, \overline{f} := (1/2)1_S$ and show that $(P(\overline{f} + v_n) - P(\overline{f}) - v_n)/ \|v_n\|$ does not converge to 0 as $n \to \infty$.]

5.4.2 *Higher Order Derivatives*

Let X and Y be normed spaces and let $f : W \to Y$ be a differentiable map. Since the derivative $f' : W \to L(X, Y)$ of f takes its values in a normed space, requiring its differentiability at some $\overline{x} \in W$ has a meaning. Then we say that f is twice differentiable at $\overline{x}$ and we define the *second derivative* of f at $\overline{x}$ as the map

$$f''(\overline{x}) := Df'(\overline{x}) \in L(X, L(X, Y)).$$

Given $u, v \in X$ and observing that the evaluation map $e_v : L(X, Y) \to Y$ defined by $e_v(\ell) := \ell(v)$ is linear and continuous, the chain rule for derivatives yields

$$(f''(\overline{x})(u))(v) = e_v(Df'(\overline{x}).u) = D(e_v \circ f')(\overline{x})(u)$$
$$= \lim_{t\to 0_+} \frac{1}{t}(f'(x+tu)(v) - f'(x)(v)).$$

Such a relation allows us to compute $f''(\overline{x})$. The space $L(X, L(X, Y))$ is isometric to the space $L^2(X; Y) := L_2(X \times X; Y)$ of continuous bilinear maps from $X \times X$ to Y via the map $\theta : L(X, L(X, Y)) \to L^2(X; Y)$ given by $\theta(B) := b$ with $b(u, v) := B(u)(v)$; the equalities

$$\|b\|_{L^2(X;Y)} := \sup_{u,v\in S_X} |b(u, v)| = \sup_{u\in S_X}(\sup_{v\in S_X} B(u)(v)) = \sup_{u\in S_X} \|B(u)\|$$
$$= \|B\|_{L(X,L(X,Y))}$$

being consequences of the definitions. Thus, one can consider $f''(\overline{x})$ as a continuous bilinear map and write $f''(\overline{x})(u, v)$ instead of $(f''(\overline{x})(u))(v)$.

Definition 5.10 A map $f : W \to Y$, where W is an open subset of X is said to be n times differentiable at $\overline{x} \in X$ if f is $n-1$ times differentiable on an open neighborhood V of $\overline{x}$ contained in W and if the $(n-1)$-th derivative $f^{(n-1)}$ of f is differentiable at $\overline{x}$. Then one sets $f^{(n)}(\overline{x}) := Df^{(n-1)}(\overline{x}) \in L(X, L^{n-1}(X; Y))$, where $L^{n-1}(X; Y) := L_{n-1}(X \times \ldots \times X; Y)$.

Using the fact that the space $L(X, L^{n-1}(X; Y))$ is isometric to $L^n(X; Y)$ via the map θ given by $\theta(A)(x_1, \ldots, x_n) := A(x_1)(x_2, \ldots, x_n)$, $f^{(n)}(\overline{x})$ can be considered as a continuous n-linear map from X to Y. If f is n times differentiable at each point of W and if $f^{(n)}$ is continuous from W into $L^n(X; Y)$, then f is said to be of *class* C^n. It is of class C^∞ if it is of class C^n for all $n \in \mathbb{N}\backslash\{0\}$.

For $m \in \mathbb{N}\backslash\{0\}$, $n > m$, one can show that $f^{(n)}$ is the m-th derivative of $f^{(n-m)}$.

The next theorem exhibits an important symmetry property of $f''(\overline{x})$.

Theorem 5.8 (Schwarz) *Let $f : W \to Y$ be twice differentiable at $\overline{x} \in W$. Then, setting $V := \{(u, v) \in X \times X : \overline{x} + u \in W,\ \overline{x} + v \in W,\ \overline{x} + u + v \in W\}$,*

$$r(u, v) := f(\overline{x} + u + v) - f(\overline{x} + u) - f(\overline{x} + v) + f(\overline{x}) - f''(\overline{x})(u, v),$$

one has $r(u, v)/(\|u\| + \|v\|)^2 \to 0$ as $(u, v) \to (0, 0)$ in $V\backslash\{(0, 0)\}$. Therefore $f''(\overline{x})$ considered as a continuous bilinear map is symmetric.

By induction, for $n \geq 2$ one can prove that $f^{(n)}(\overline{x})$ is symmetric in its n variables.

Proof Given $\varepsilon > 0$, the differentiability of f' at $\overline{x}$ yields some $\delta > 0$ such that $\overline{x} + 2\delta B_X \subset W$ and

$$x \in 2\delta B_X \Longrightarrow \left\| f'(\overline{x} + x) - f'(\overline{x}) - Df'(\overline{x}).x \right\| \leq \frac{\varepsilon}{2} \|x\|. \tag{5.13}$$

Taking $u \in \delta B_X$ and setting

$$g_u(v) := r(u, v) \qquad\qquad v \in \delta B_X,$$

with r as in the statement, we compute $Dg_u(v)$ with the help of the rule $Db(u, \cdot) = b(u, \cdot)$ for the linear map $b(u, \cdot)$:

$$\begin{aligned} Dg_u(v) &= f'(\overline{x} + u + v) - f'(\overline{x} + v) - f''(\overline{x})(u, \cdot) \\ &= [f'(\overline{x} + u + v) - f'(\overline{x}) - f''(\overline{x})(u + v, \cdot)] \\ &\quad - [f'(\overline{x} + v) - f'(\overline{x}) - f''(\overline{x})(v, \cdot)], \end{aligned}$$

so that, by (5.13), for $v \in \delta B_X$, $\|Dg_u(v)\| \leq \varepsilon(\|u\| + \|v\|)$. Applying Corollary 5.6 we get $\|g_u(v)\| = \|g_u(v) - g_u(0)\| \leq \varepsilon(\|u\| + \|v\|)\,\|v\| \leq \varepsilon(\|u\| + \|v\|)^2$. This proves the first assertion. The second one follows from the symmetry of $s(u, v) := f(\overline{x} + u + v) - f(\overline{x} + u) - f(\overline{x} + v)$ in u and v and the relation $\|f''(\overline{x})(u, v) - f''(\overline{x})(v, u)\| = \|r(v, u) - r(u, v)\| \leq 2\varepsilon(\|u\| + \|v\|)^2$ for $u, v \in \delta B_X$ by a homogeneity argument. □

Corollary 5.16 *Given normed spaces X_1, X_2, Y, $\overline{x} := (\overline{x}_1, \overline{x}_2) \in W$, an open subset of $X := X_1 \times X_2$ and $f : W \to Y$ that is twice differentiable at $\overline{x}$, the partial derivatives $D_1 f : W \to L(X_1, Y)$ and $D_2 f : W \to L(X_2, Y)$ are differentiable and $D_2 D_1 f(\overline{x})(v_1)(v_2) = D_1 D_2 f(\overline{x})(v_2)(v_1)$ for all $v_1 \in X_1$, $v_2 \in X_2$.*

Proof Setting $u_1 := (v_1, 0)$, $u_2 := (0, v_2)$, one has $D_2 D_1 f(\overline{x})(v_1)(v_2) = D(f'(\cdot)u_1)(\overline{x}).u_2$ and $D_1 D_2 f(\overline{x})(v_2)(v_1) = D(f'(\cdot)u_2)(\overline{x}).u_1$. The symmetry of $f''(\overline{x})$ entails the result. □

When $X := \mathbb{R}^d$, for $f : W \to Y$, where W is an open subset of X, it is advisable to gather the partial derivatives. Namely, if f is k times differentiable at $\overline{x}$, if $\alpha := (\alpha_1, \ldots, \alpha_d) \in \mathbb{N}^d$ with $|\alpha| = k$ for $|\alpha| := \alpha_1 + \ldots + \alpha_d$ one writes

$$D^\alpha f(\overline{x}) := D^{\alpha_1} \ldots D^{\alpha_d} f(\overline{x}) := \frac{\partial^{\alpha_1}}{\partial x_1} \cdots \frac{\partial^{\alpha_d}}{\partial x_d} f(\overline{x})$$

for these partial derivatives.

Example If $f : X \to Y$ is linear and continuous, then f is of class C^∞ and $f^{(n)} = 0$ for $n \geq 2$ since $f'(x) = f$ for all $x \in X$, so that f' is constant.

Example Let $b : X_1 \times X_2 \to Y$ be a continuous bilinear map. Then b is of class C^∞ and $Db(x_1, x_2)(v_1, v_2) = b(v_1, x_2) + b(x_1, v_2)$, so that b' is linear from $X_1 \times X_2$ into $L(X_1 \times X_2, Y)$, $b^{(2)}(x_1, x_2)$ is independent of (x_1, x_2) and $b^{(n)} = 0$ for $n \geq 3$.

Proposition 5.23 *Let X, Y, Z be normed spaces, let U, V be open subsets of X and Y respectively and let $f : U \to Y$, $g : V \to Z$ be n times differentiable at $\overline{x} \in U$ and $\overline{y} := f(\overline{x}) \in V$ respectively, with $f(U) \subset V$. Then $h := g \circ f$ is n times differentiable at $\overline{x}$.*

If f and g are of class C^n, then $g \circ f$ is of class C^n.

Proof The result is known for $n = 1$ and one has $h'(x) = g'(f(x)) \circ f'(x)$ for $x \in U$. If $n = 2$, since h' is composed of the map $x \mapsto (f'(x), (g' \circ f)(x))$ with values in the product space $L(X, Y) \times L(Y, Z)$ and of the map $(A, B) \mapsto B \circ A$ from $L(X, Y) \times L(Y, Z)$ into $L(X, Z)$ which is bilinear and continuous, applying the Leibniz's rule and the chain rule one sees that h' is differentiable and for $u_1,\ u_2 \in X$ one has

$$h''(\overline{x})(u_1, u_2) = g''(f(\overline{x})).f'(\overline{x})u_1.f'(\overline{x})u_2 + g'(f(\overline{x})).f''(\overline{x})(u_1, u_2).$$

More generally, by induction we get that h' is $(n-1)$ times differentiable at $\overline{x}$ (or of class C^{n-1} if f and g are of class C^n). □

The preceding result and a change of notation give a means to compute the second derivative $f''(\overline{x})(u_1, u_2)$ at $\overline{x} \in U$ of a map $f : U \to Y$ that is twice differentiable at $\overline{x}$ by reducing the question to the computation of the second derivative of the map $k := f \circ j : \mathbb{R}^2 \to Y$ where $j : \mathbb{R}^2 \to X$ is the affine map defined by $j(r_1, r_2) := \overline{x} + r_1 u_1 + r_2 u_2$: setting $e_1 := (1, 0)$, $e_2 := (0, 1)$ one has

$$f''(\overline{x})(u_1, u_2) = k''(0, 0)(e_1, e_2).$$

Exercises

1. Let $f : \mathbb{R}^2 \to \mathbb{R}$ be given by $f(r, s) := rs(r^2 - s^2)/(r^2 + s^2)$ for $(r, s) \in \mathbb{R}^2 \backslash \{(0, 0)\}$ and $f(0, 0) = 0$. Show that the four partial derivatives $D_1 D_1 f$, $D_1 D_2 f$, $D_2 D_1 f$, $D_2 D_2 f$ exist at every $(r, s) \in \mathbb{R}^2$ but that $D_1 D_2 f(0, 0) \neq D_2 D_1 f(0, 0)$.
2. Let f_n, $g_n : \mathbb{R} \to \mathbb{R}$ be given by $g_n(r) := r/(1 + n\,|r|)$, $f_n(r) := \int_0^r g_n(s)ds$ for $r \in \mathbb{R}$, $n \in \mathbb{N}$. Let $X := c_0$ be the space of sequences $x := (x_n)_n$ such that $\lim_n x_n = 0$. Show that the map $f : (x_n) \mapsto (f_n(x_n))$ belongs to $C^1(X, X)$ and that for all $v \in V$ the map $f'(\cdot)(v)$ is differentiable at 0 but that f' is not differentiable at 0.
3. Let $f : X \to Y$ be a map of class C^2 between two normed spaces such that $f(tx) = t^2 f(x)$ for all $t \in \mathbb{R}$, $x \in X$. Show that f is quadratic and that in fact $f(x) = (1/2)f''(0)(x, x)$.
4. Let $\rho : \mathbb{R} \to \mathbb{R}$ be given by $\rho(r) = \exp(-(1 - r^2)^{-1})$ for $r \in]-1, 1[$, $\rho(r) = 0$ for $r \in \mathbb{R} \backslash]-1, 1[$. Verify that ρ is of class C^∞. Let $c := \int_{-\infty}^{\infty} \rho(r)dr$. For $f : \mathbb{R} \to \mathbb{R}$ continuous or regulated and $n \in \mathbb{N}$, let f_n be given by $f_n(t) = (n/c)\int_{-\infty}^{\infty} f(s)\rho(n(t - s))ds$. Show that f_n is of class C^∞. Prove that when f is continuous (f_n) converges to f uniformly. Find the limit of $(f_n(t))$ when f is regulated. [Hint: start with the case when f is a step function]
5. Let $f : \mathbb{R} \to \mathbb{R}$ be a twice differentiable function such that for some $a \in \mathbb{R}$ one has $f(r) \geq 0$, $f'(r) > 0$, and $f''(r) \leq 0$ for all $r \in [a, \infty[$. Show that $f(t) \geq f'(t)(t - a)/2$ for all $t \in [a, \infty[$ and $f(t) \geq f(t)t/4$ for $t \in [2a, \infty[$. [Hint: use the function g given by $g(t) := (t - a)f(t) - (1/2)(t - a)^2 f'(t)$ for $t \in [a, \infty[$.]

6. Given normed spaces Y_1, Y_2, Z, a bilinear product $b : Y_1 \times Y_2 \to Z$, and maps $f_1 : U \to Y_1, f_2 : U \to Y_2$ that are twice differentiable at $\overline{x} \in U$, show that $h := b \circ (f_1, f_2)$ is twice differentiable at $\overline{x}$ and give the expression of $h''(\overline{x})(u_1, u_2)$.
7. Given an open subset W of a normed vector space X, the *Lie bracket* of two vector fields f, g in $C^1(W, X)$ is the map $h := [f, g] \in C(W, X)$ given by

$$h(w) := Df(w)g(w) - Dg(w)f(w) \qquad w \in W.$$

Verify the following properties in which $f,\ g,\ h \in C^1(W, X)$:

(a) $[g, f] = -[f, g]$;
(b) $[rf + sg, h] = r[f, h] + s[g, h]$ for $r, s \in \mathbb{R}$;
(c) $[\varphi f, \psi g] = \varphi\psi[f, g] + \psi(D\varphi.g)f - \varphi(D\psi.f)g$ for $\varphi, \psi \in C^1(W, \mathbb{R})$;
(d) $[[f, g], h] + [[g, h], f] + [[h, f], g] = 0$ for $f,\ g,\ h \in C^2(W, X)$.

5.4.3 *Taylor's Formulas*

Higher order derivatives enable us to give precise approximations of a map. Such approximations that are so useful in applications can be given in different forms. We start with a result of a pointwise character that resembles the definition of Fréchet differentiability. Throughout this subsection X and Y are normed spaces, $\overline{x}$ is a point of an open subset W of X, and $f : W \to Y$ is a map.

Theorem 5.9 *If $f : W \to Y$ is n times differentiable at $\overline{x}$, with $n \in \mathbb{N}\backslash\{0\}$, then setting*

$$r_n(x) := f(\overline{x} + x) - f(\overline{x}) - \sum_{k=1}^{n} \frac{1}{k!} f^{(k)}(\overline{x})(x, \ldots, x) \tag{5.14}$$

one defines a remainder of order n, i.e. $\lim_{x(\neq 0) \to 0} r_n(x) / \|x\|^n = 0$, in symbols $r_n(x) = o(\|x\|^n)$.

Proof For $n = 1$, this is just the definition of differentiability. Let us prove the result by induction. Example (g) after the definition of differentiability ensures that for $k \geq 2$ the derivative of the polynomial function $x \mapsto (1/k!)f^{(k)}(\overline{x})(x^{(k)})$ (where $x^{(k)}$ stands for $(x, \ldots, x)$ with k entries) is the function $x \mapsto (1/(k-1)!)f^{(k)}(\overline{x})(x^{(k-1)}, \cdot) \in L(X, Y)$. Thus

$$Dr_n(x) = f'(\overline{x} + x) - f'(\overline{x}) - \sum_{k=2}^{n} \frac{1}{(k-1)!} f^{(k)}(\overline{x})(x^{(k-1)}, \cdot).$$

This is the remainder r'_{n-1} associated with f'. The induction ensures that for all $\varepsilon > 0$ one can find $\delta > 0$ such that $\left\|r'_{n-1}(x)\right\| \leq \varepsilon \|x\|^{n-1}$ for all $x \in \delta B_X$. Then, the Mean

Value Theorem applied to r_n in the ball $\|x\| B_X$ yields $\|r_n(x)\| = \|r_n(x) - r_n(0)\| \leq \varepsilon \|x\|^n$ for all $x \in \delta B_X$. □

One can obtain more precise estimates by making more stringent assumptions. We need some preliminaries.

Lemma 5.7 *Let E, F, G be normed spaces and let $b : E \times F \to G$ be a continuous bilinear map denoted by $(u, v) \mapsto u * v$. Given an open interval T of $\mathbb{R}$, $n + 1$ times differentiable functions $e : T \to E, f : T \to F$, the derivative of the function $g : T \to G$ defined by*

$$g(t) := \sum_{k=0}^{n} (-1)^{n-k} e^{(n-k)}(t) * f^{(k)}(t)$$

is $t \mapsto (-1)^n e^{(n+1)}(t) * f(t) + e(t) * f^{(n+1)}(t)$. In particular, the derivative of the function

$$g : t \mapsto \sum_{k=0}^{n} \frac{1}{k!}(1-t)^k f^{(k)}(t)$$

is $(1/n!)(1-t)^n f^{(n+1)}(t)$.

Proof Applying Leibniz's rule to $t \mapsto e^{(n-k)}(t) * f^{(k)}(t)$ and noting the cancellations we are left with the conclusion of the first assertion. For the second one we take $E := \mathbb{R}$, $e(t) := (1-t)^n/n!$, $b(r, x) := rx$ for $(r, x) \in \mathbb{R} \times F$. Then $(-1)^{n-k} e^{(n-k)}(t) = (1-t)^k/k!$ and $e^{(n+1)}(t) = 0$, hence the result. □

Theorem 5.10 (Taylor's Theorem with Lagrange's Remainder) *Let $f : W \to Y$ be $n + 1$ times differentiable on W and let $\overline{x} \in W$, $x \in X$ be such that the segment $[\overline{x}, \overline{x} + x]$ is contained in W. If $\left\| f^{(n+1)}(w) \right\|$ is bounded above by some $c \in \mathbb{R}_+$ for all $w \in W$, then the remainder r_n defined by relation (5.14) satisfies $\|r_n(x)\| \leq \frac{c}{(n+1)!} \|x\|^{n+1}$.*

Proof For $x \in X$ such that $[\overline{x}, \overline{x} + x] \subset W$ and $t \in [0, 1]$, let us set $f_x(t) := f(\overline{x} + tx)$. Then for $k \in \mathbb{N}_n$ we have $f_x^{(k)}(t) = f^{(k)}(\overline{x} + tx)(x^{(k)})$, hence $\left\| (1/n!)(1-t)^n f_x^{(n+1)}(t) \right\| \leq (c/n!)(1-t)^n \|x\|^{n+1}$. Applying the Mean Value Theorem to the functions $g_x : t \mapsto \sum_{k=0}^{n} (1/k!)(1-t)^k f_x^{(k)}(t)$ whose derivative is $(1/n!)(1-t)^n f^{(n+1)}(t)$ and $h_x : t \mapsto -(1-t)^{n+1}(c/(n+1)!) \|x\|^{n+1}$ whose derivative is $(c/n!)(1-t)^n \|x\|^{n+1}$, we get

$$\|r_n(x)\| = \|g_x(1) - g_x(0)\| \leq h_x(1) - h_x(0) = (c/(n+1)!) \|x\|^{n+1}$$

as required. □

Theorem 5.11 (Taylor's Theorem with Integral Remainder) *Let $f : W \to Y$ be $n + 1$ times continuously differentiable on W and let $\overline{x} \in W$, $x \in X$ be such that* *the segment* $[\overline{x}, \overline{x}+x]$ *is contained in* W. *Then the remainder* r_n *defined by (5.14) satisfies*

$$r_n(x) = \int_0^1 \frac{(1-t)^n}{n!} f^{(n+1)}(\overline{x}+tx)(x^{(n+1)})dt.$$

Proof With the notation of the preceding proof, that follows from the relations $r_n(x) = g_x(1) - g_x(0) = \int_0^1 g'_x(t)dt$, $g'_x(t) = (1/n!)(1-t)^n f_x^{(n+1)}(t) = (1/n!)(1-t)^n f^{(n+1)}(\overline{x}+tx)(x^{(n+1)})$. □

Let us quote the next result which is outside the scope of the present book.

Theorem 5.12 (Aleksandrov) *Let* $f : \mathbb{R}^d \to \mathbb{R}$ *be a convex function. Then for some set* N *of null measure in* $\mathbb{R}^d$ *and all* $\overline{x} \in \mathbb{R}^d \backslash N$, f *admits a second-order Taylor expansion at* $\overline{x}$.

Exercises

1. Given $k \in \mathbb{N}\backslash\{0\}$, Banach spaces X and Y and an open subset W of X, show that the space $BC^k(W, Y)$ of maps of class C^k from W to Y whose derivatives are bounded is a Banach space with respect to the norm $\|\cdot\|_{C^k}$ given by $\|f\|_{C^k} := \sup_{0 \le h \le k} \|D^h f\|_\infty$.
2. Let $g : T \to \mathbb{R}$, with $T :=]-a, a[$, be odd and five times differentiable. Show that for all $x \in T$ there exists some $y \in]0, a[$ such that
$$g(x) = \frac{x}{3}(g'(x) + 2g'(0)) - \frac{x^5}{180} g^{(5)}(y).$$
3. (*Simpson's formula*) Let $f : [a, b] \to \mathbb{R}$ be five time differentiable. Deduce from the preceding exercise that there exists some $c \in]a, b[$ such that
$$f(b) - f(a) = \frac{b-a}{6}[f'(a) + f'(b) + 4f'((a+b)/2)] - \frac{(b-a)^5}{2880} f^{(5)}(c).$$

5.4.4 Differentiable Partitions of Unity

We intend to show that the family of functions of class C^∞ on $\mathbb{R}^d$ is rich enough. We start with an analogue of Urysohn's lemma. Then we describe a useful tool known as a partition of unity.

Lemma 5.8 *Let* A, B *be two disjoint closed subsets of* $\mathbb{R}^d$ *such that*

$$\text{gap}(A, B) := \inf\{\|a - b\| : a \in A,\ b \in B\} > 0.$$

Then, there exists some $f \in C^{\infty}(\mathbb{R}^d)$ *such that* $f(a) = 1, f(b) = 0$ *for all* $a \in A$, $b \in B$ *and* $f(x) \in [0, 1]$ *for all* $x \in \mathbb{R}^d$.

Proof Let $\rho : \mathbb{R}^d \to \mathbb{R}$ be defined by

$$\rho(x) := ce^{1/(\|x\|^2-1)} \text{ for } x \in B(0,1), \qquad \rho(x) = 0 \text{ for } x \in \mathbb{R}^d \backslash B(0,1),$$

where $c > 0$ is chosen in such a way that $\int \rho = 1$. Given $r \in \mathbb{P}$, $r < (1/2)\text{gap}(A, B)$ and setting $C := B(A, r) := \{x \in \mathbb{R}^d : d(x, A) < r\}, f(x) = \int_C \rho((x - y)/r)dy$, we get the required function (we use here a derivation result for integrals depending on a parameter). □

Proposition 5.24 *Let* $(U_1, \ldots, U_k)$ *be a finite open covering of a compact subset* K *of* $\mathbb{R}^d$. *Then one can find nonnegative functions* $(p_i)_{i \in \mathbb{N}_k}$ *of class* C^{∞} *such that* $\Sigma_{i=1}^k p_i \leq 1$, $\Sigma_{i=1}^k p_i = 1$ *on* K *and* $\text{supp}\, p_i := \text{cl}(\{x : p_i(x) \neq 0\})$ *is a compact subset of* U_i *for* $i \in \mathbb{N}_k$.

Proof For each $x \in K$ we pick an open neighborhood W_x whose closure $\text{cl}(W_x)$ is a compact subset of some $U_{i(x)}$. Let $x_1, \ldots, x_n$ be a finite family of points of K such that $W_{x_1}, \ldots, W_{x_n}$ is a covering of K. For $i \in \mathbb{N}_k$ let V_i be the union of the sets W_{x_j} whose closures are contained in U_i. Then $A_i := \text{cl}(V_i)$ is a compact subset of U_i, so that $\text{gap}(A_i, \mathbb{R}^d \backslash U_i) > 0$. Let $q_i \in C^{\infty}(\mathbb{R}^d)$ be such that $q_i = 1$ on A_i, $q_i = 0$ on $\mathbb{R}^d \backslash U_i$ and $q_i(\mathbb{R}^d) \subset [0, 1]$; we may require that $\text{supp}\, q_i$ is compact. Setting $p_1 := q_1$,

$$p_2 := q_2(1 - q_1), \ldots\ldots, p_n := q_n(1 - q_1) \ldots (1 - q_{n-1})$$

by induction we see that

$$p_1 + \ldots + p_n = 1 - (1 - q_1) \ldots (1 - q_n).$$

This relation implies that $p_1 + \ldots + p_n = 1$ on K (and in fact on the union of the A_i's). □

A family $(S_i)_{i \in I}$ of subsets of $\mathbb{R}^d$ (or a topological space) is said to be *locally finite* if for all $x \in \mathbb{R}^d$ there exists a neighborhood V of x and a finite subset $I(x)$ of I such that $S_i \cap V = \varnothing$ for all $i \in I \backslash I(x)$. A family $(p_i)_{i \in I}$ of nonnegative functions of class C^{∞} on an open subset U of $\mathbb{R}^d$ is called a *partition of unity of* U if the family $(\text{supp}\, p_i)_{i \in I}$ is locally finite and if $\Sigma_{i \in I} p_i(x) = 1$ for all $x \in U$. A partition of unity $(p_i)_{i \in I}$ is said to be *subordinated* to a covering $(U_j)_{j \in J}$ of open subsets of U if for all $i \in I$ there exists some $j(i) \in J$ such that $\text{supp}\, p_i \subset U_{j(i)}$.

We admit the following result (see [182]). Some steps of its proof have been presented above.

Theorem 5.13 *Given an open subset* U *of* $\mathbb{R}^d$ *and an open covering* $(U_j)_{j \in J}$ *of* U *there exists a countable partition of unity* $(p_i)_{i \in I}$ *subordinated to the covering* $(U_j)_{j \in J}$.

5.5 Solving Equations and Inverting Maps

In this section, we show that simple methods linked with differentiability notions lead to efficient ways of solving nonlinear systems or vectorial equations

$$f(x) = y. \tag{5.15}$$

Here X and Y are Banach spaces, W is an open subset of X, $y \in Y$, and $f : W \to Y$ is a map. We start with a classical constructive algorithm.

5.5.1 *Newton's Method*

Newton's method is an iterative process that relies on a notion of approximation by a linear map that slightly differs from differentiability. We formulate it in the next definition.

Definition 5.11 The map $f : W \to Y$ has a *Newton approximation* at $\overline{x} \in W$ if there exist $r > 0$, $\alpha > 0$ and a map $A : B(\overline{x}, r) \to L(X, Y)$ such that $B(\overline{x}, r) \subset W$ and

$$\forall x \in B(\overline{x}, r) \qquad \|f(x) - f(\overline{x}) - A(x)(x - \overline{x})\| \leq \alpha \|x - \overline{x}\|. \tag{5.16}$$

A map $A : V \to L(X, Y)$ is a *slant derivative* of f at $\overline{x}$ if V is a neighborhood of $\overline{x}$ contained in W and if for any $\alpha > 0$ there exists some $r > 0$ such that $B(\overline{x}, r) \subset V$ and relation (5.16) holds.

Thus f is differentiable at $\overline{x}$ if and only if f has a slant derivative at $\overline{x}$ that is constant on some neighborhood of $\overline{x}$. But condition (5.16) is much less demanding, as the next lemma shows.

Lemma 5.9 *The following assertions about a map $f : W \to Y$ are equivalent:*

(a) f has a Newton approximation A that is bounded near $\overline{x}$;
(b) f is stable at $\overline{x}$, i.e. there exist $c > 0, r > 0$ such that

$$\forall x \in B(\overline{x}, r) \qquad \|f(x) - f(\overline{x})\| \leq c \|x - \overline{x}\|; \tag{5.17}$$

(c) f has a slant derivative A at $\overline{x}$ that is bounded on some neighborhood of $\overline{x}$.

Proof (a)⇒(b) If for some α, $\beta > 0$ and some $r > 0$ a map $A : B(\overline{x}, r) \to L(X, Y)$ is such that (5.16) holds with $\|A(x)\| \leq \beta$ for all $x \in B(\overline{x}, r)$, then, by the triangle inequality, relation (5.17) holds with $c := \alpha + \beta$.

(b)⇒(c) We use a corollary of the Hahn-Banach Theorem asserting the existence of some map $S : X \to X^*$ such that $S(x)(x) = \|x\|$ and $\|S(x)\| = 1$ for all $x \in X$. Suppose (5.17) holds. Then, setting $A(\overline{x}) = 0$ and, for $x \in W \backslash \{\overline{x}\}$, $u \in X$,

$$A(x)(u) = \langle S(x - \overline{x}), u \rangle \frac{f(x) - f(\overline{x})}{\|x - \overline{x}\|},$$

we easily check that $\|A(x)\| \leq c$ for all $x \in W$ and that $A(x)(x - \overline{x}) = f(x) - f(\overline{x})$ for all $x \in W$, so that (5.16) holds with $\alpha = 0$ and A is a slant derivative of f at $\overline{x}$.

(c)⇒(a) is clear since a slant derivative of f at $\overline{x}$ is a Newton approximation of f at $\overline{x}$. □

In the elementary method that follows, we first assume that (5.15) has a solution $\overline{x}$.

Proposition 5.25 *Let $\overline{x}$ be a solution to equation (5.15) with $y = 0$, let $\alpha, \beta, r > 0$ satisfying $\gamma := \alpha\beta < 1$ and let $A : B(\overline{x}, r) \to L(X, Y)$ be such that (5.16) holds, $A(x)$ being invertible with $\|A(x)^{-1}\| \leq \beta$ for all $x \in B(\overline{x}, r)$. Then, for any initial point $x_0 \in B(\overline{x}, r)$, the sequence (x_n) given by*

$$x_{n+1} := x_n - A(x_n)^{-1}(f(x_n)) \tag{5.18}$$

is well defined and converges linearly to $\overline{x}$ with rate γ in the sense that $\|x_{n+1} - \overline{x}\| \leq \gamma \|x_n - \overline{x}\|$ for all $n \in \mathbb{N}$.

Thus, setting $c := \|x_0 - \overline{x}\|$, one has $\|x_n - \overline{x}\| \leq c\gamma^n$. It follows that if A is a slant derivative of f at $\overline{x}$, (x_n) converges superlinearly to $\overline{x}$: for all $\varepsilon > 0$ there is some $k \in \mathbb{N}$ such that $\|x_{n+1} - \overline{x}\| \leq \varepsilon \|x_n - \overline{x}\|$ for all $n \geq k$.

Proof Using the fact that $f(\overline{x}) = 0$, so that

$$x_{n+1} - \overline{x} = A(x_n)^{-1} \left(f(\overline{x}) - f(x_n) + A(x_n)(x_n - \overline{x})\right),$$

we inductively obtain that

$$\|x_{n+1} - \overline{x}\| \leq \beta \|f(x_n) - f(\overline{x}) - A(x_n)(x_n - \overline{x})\| \leq \alpha\beta \|x_n - \overline{x}\|,$$

so that $x_{n+1} \in B(\overline{x}, r)$: the whole sequence (x_n) is well defined and converges to $\overline{x}$. □

Under reinforced assumptions, one can show the existence of a solution.

Theorem 5.14 (Kantorovich) *Let $x_0 \in W$, $\alpha, \beta > 0$, $r > 0$ with $\gamma := \alpha\beta < 1$, $B(x_0, r) \subset W$ and let $A : B(x_0, r) \to L(X, Y)$ be such that for all $x \in B(x_0, r)$ the map $A(x) : X \to Y$ has a right inverse $B(x) : Y \to X$ satisfying $\|B(x)(\cdot)\| \leq \beta \|\cdot\|$ and*

$$\forall w, x \in B(x_0, r) \qquad \|f(w) - f(x) - A(x)(w - x)\| \leq \alpha \|w - x\|. \tag{5.19}$$

If $\|f(x_0)\| < \beta^{-1}(1 - \gamma)r$ and if f is continuous, the sequence (x_n) given by the Newton iteration

$$x_{n+1} := x_n - B(x_n)(f(x_n)) \tag{5.20}$$

is well defined and converges to a solution $\overline{x}$ of equation (5.15) with $y = 0$. Moreover, one has $\|x_n - \overline{x}\| \leq r\gamma^n$ for all $n \in \mathbb{N}$ and $\|\overline{x} - x_0\| \leq \beta(1 - \gamma)^{-1} \|f(x_0)\| < r$.

Here $B(x)$ is a *right inverse* of $A(x)$ if $A(x) \circ B(x) = I_Y$; $B(x)$ is not supposed to be linear.

Proof Let us prove by induction that $x_n \in B(x_0, r)$, $\|x_{n+1} - x_n\| \le \beta\gamma^n \|f(x_0)\|$ and $\|f(x_n)\| \le \gamma^n \|f(x_0)\|$. For $n = 0$ these relations are obvious. Assuming they are valid for $n < k$, we get

$$\|x_k - x_0\| \le \sum_{n=0}^{k-1} \|x_{n+1} - x_n\| \le \beta \|f(x_0)\| \sum_{n=0}^{\infty} \gamma^n = \beta \|f(x_0)\| (1-\gamma)^{-1} < r,$$

or $x_k \in B(x_0, r)$ and, since $f(x_{k-1}) + A(x_{k-1})(x_k - x_{k-1}) = 0$, from (5.20), (5.19),

$$\begin{aligned}\|f(x_k)\| &= \|f(x_k) - f(x_{k-1}) - A(x_{k-1})(x_k - x_{k-1})\| \\ &\le \alpha \|x_k - x_{k-1}\| \le \gamma^k \|f(x_0)\|,\end{aligned}$$

$$\|x_{k+1} - x_k\| \le \beta \|f(x_k)\| \le \beta\gamma^k \|f(x_0)\|.$$

Since $\gamma < 1$, the sequence (x_n) is a Cauchy sequence, hence converges to some $\overline{x} \in X$ satisfying $\|\overline{x} - x_0\| \le \beta \|f(x_0)\| (1-\gamma)^{-1} < r$. Moreover, by continuity of f, we get $f(\overline{x}) = \lim_n f(x_n) = 0$. Finally,

$$\|x_n - \overline{x}\| \le \lim_{p\to+\infty} \|x_n - x_p\| \le \lim_{p\to+\infty} \sum_{k=n}^{p-1} \|x_{k+1} - x_k\| \le r\gamma^n.$$

□

From Kantorovich's Theorem we deduce a result that is the root of many important developments in nonlinear analysis.

Theorem 5.15 (Lyusternik-Graves Theorem) *Let X and Y be Banach spaces, let W be an open subset of X and let $g : W \to Y$ that is circa-differentiable at some $\overline{x} \in W$ with a surjective derivative $Dg(\overline{x})$. Then g is open at $\overline{x}$.*

More precisely, there exist some $\rho, \sigma, \kappa > 0$ such that g has a right inverse $h : B(g(\overline{x}), \sigma) \to W$ satisfying $\|h(y) - \overline{x}\| \le \kappa \|y - g(\overline{x})\|$ and

$$\forall (w, y) \in B(\overline{x}, \rho) \times B(g(\overline{x}), \sigma) \quad \exists x \in W : \; g(x) = y, \; \|x - w\| \le \kappa \|y - g(w)\|. \tag{5.21}$$

Proof Let $A : W \to L(X, Y)$ be the constant map with value $A := Dg(\overline{x})$ (we use a familiar abuse of notation). The Open Mapping Theorem yields some $\beta > 0$ and some right inverse $B : Y \to X$ of A such that $\|B(\cdot)\| \le \beta \|\cdot\|$. Let $\alpha, r > 0$ be such that $\gamma := \alpha\beta < 1$, $B(\overline{x}, 2r) \subset W$ and

$$\forall w, x \in B(\overline{x}, 2r) \qquad \|g(w) - g(x) - Dg(\overline{x})(w - x)\| \le \alpha \|w - x\|. \tag{5.22}$$

Let $\sigma, \tau > 0$ be such that $\sigma + \tau < \beta^{-1}(1-\gamma)r$, and let $\rho \in]0, r]$ be such that $g(w) \in B(g(\overline{x}), \tau)$ for all $w \in B(\overline{x}, \rho)$. Given $w \in B(\overline{x}, \rho)$, $y \in B(g(\overline{x}), \sigma)$, let us set $f(x) := g(x) - y$ for $x \in B(\overline{x}, \rho)$, so that

$$\|f(w)\| \leq \|g(w) - g(\overline{x})\| + \|g(\overline{x}) - y\| < \sigma + \tau < \beta^{-1}(1-\gamma)r$$

and, by (5.22), (5.19) holds in the ball $B(x_0, \rho)$, with $x_0 := w$. Using the estimate $\|x - x_0\| \leq \beta \|f(x_0)\| (1-\gamma)^{-1} < r$ obtained in the proof of Kantorovich's Theorem for a solution x of the equation $f(x) = 0$, we get some $x \in W$ such that $g(x) = y$, $\|x - w\| \leq \kappa \|g(w) - y\|$ with $\kappa := \beta(1-\gamma)^{-1}$. The right inverse h is obtained by taking $w := \overline{x}$ in (5.21). □

Exercises

1. Let X and Y be Banach spaces, let $\overline{x} \in X$, b, c, $r > 0$, $W := B(\overline{x}, r)$, $f : W \to Y$ be of class C^1 and such that f' is Lipschitzian with rate c on W and $\|f'(w)^{\mathsf{T}}(y^*)\| \geq b \|y^*\|$ for all $w \in W$, $y^* \in Y^*$. Let $b > cr$. Using Kantorovich's Theorem, prove that for all $y \in B(f(\overline{x}), (b - cr)r)$ there exists an $x \in W$ satisfying $f(x) = y$ and $\|x - \overline{x}\| \leq b^{-1} \|y - f(\overline{x})\|$. [Hint: use the Banach-Schauder Theorem to find a right inverse $B(w)$ of $A(w) := f'(w)$ for all $w \in W$ satisfying $\|B(w)(\cdot)\| \leq b^{-1} \|\cdot\|$ and use Exercise 15 of Sect. 2.2.4 to verify condition (5.19).]
2*. Using Exercise 15 of Sect. 2.2.4 establish a refined version of Kantorovich's Theorem and prove that the conclusion of the preceding result can be extended to any $y \in B(f(\overline{x}), br)$.
3*. **Convexity of images of small balls (Polyak)**. Let X be a Hilbert space, let Y be a normed space, let $a \in X$, c, ρ, $\sigma > 0$, $W := B(a, \rho)$, $f : W \to Y$ be differentiable and such that f' is Lipschitzian with rate c on W and $\|f'(a)^{\mathsf{T}}(y^*)\| \geq \sigma \|y^*\|$ for all $y^* \in Y^*$. Prove that for $r > 0$, $r < \min(\rho, \sigma/2c)$ the image $f(B)$ of $B := B(a, r)$ under the nonlinear map f is convex. [Hint: given $x_0, x_1 \in B$, $y_0 := f(x_0)$, $y_1 := f(x_1)$, $y := (1/2)(y_0 + y_1)$, $\overline{x} := (1/2)(x_0 + x_1)$, show that $\|f'(w)^{\mathsf{T}}(y^*)\| \geq b \|y^*\|$ for all $w \in W$, $y^* \in Y^*$ for $b := \sigma - cr$ and apply the preceding exercise.]
4*. Extend the (surprising!) result of the preceding exercise to the case when X is a Banach space with a uniformly convex norm.

5.5.2 *The Inverse Mapping Theorem*

The Inverse Mapping Theorem is a milestone of differential calculus. It shows the interest and the power of derivatives. It has numerous applications in differential geometry, differential topology and in the study of dynamical systems.

When $f : T \to \mathbb{R}$ is a continuous function on some open interval T of $\mathbb{R}$, one can use the order of $\mathbb{R}$ and the intermediate value theorem to obtain results about the invertibility of f. If, moreover, f is differentiable at some $r \in T$ and if $f'(r)$ is non-null, one can conclude that $f(T)$ contains some neighborhood of $f(r)$. When f is a map of several variables one would like to know whether such a conclusion is valid, and even more, whether f induces a bijection from some neighborhood of a given point $\overline{x}$ onto some neighborhood of $f(\overline{x})$. Of course, one cannot expect a global result without further assumptions since the derivative is a local notion.

Following R. Descartes' advice, we will reach our main results, concerning the possibility of inverting nonlinear maps, through several small steps; some of them are of independent interest.

First, given a bijection f between two metric spaces X, Y, we would like to know whether a map close enough to f is still a bijection. We have seen such a result for continuous linear maps between Banach spaces (Theorem 3.17).

Now let us turn to a nonlinear setting. Let us first observe that if $f : U \to V$ is a bijection between two open subsets of normed spaces X and Y respectively, it may occur that f is differentiable at some $a \in U$ whereas its inverse g is not differentiable at $b = g(a)$: take $U = V = \mathbb{R}, f$ given by $f(x) = x^3$ whose inverse $y \mapsto y^{1/3}$ is not differentiable at 0. However, if f is differentiable at some $a \in X$ and if its inverse g is differentiable at $b := g(a)$, then the derivative of g at b is the inverse $f'(a)^{-1}$ of the derivative $f'(a)$ of f at a. This fact simply follows from the chain rule: from $g \circ f = I_U$ and $f \circ g = I_V$ one deduces that $g'(b) \circ f'(a) = I_X$ and $f'(a) \circ g'(b) = I_Y$.

Our first step is not as obvious as the preceding observation since one of its assumptions is now a conclusion.

Lemma 5.10 *Let U and V be two open subsets of normed spaces X and Y respectively. Assume that $f : U \to V$ is a homeomorphism that is differentiable at $a \in U$ and such that $f'(a)$ is an isomorphism. Then the inverse g of f is differentiable at $b = f(a)$ and $g'(b) = f'(a)^{-1}$.*

Proof Using translations if necessary, we may suppose $a = 0, f(a) = 0$ without loss of generality. Changing f into $h^{-1} \circ f$, where $h := f'(a)$, we may also suppose $Y = X$ and $f'(a) = I_X$. Then, setting $s(y) := g(y) - y$, we have to show that $s(y)/\|y\| \to 0$ as $y \to 0, y \neq 0$. Let us set $r(x) := f(x) - x$. Given $\varepsilon \in]0, 1[$, we can find $\rho > 0$ such that $\|r(x)\| \leq (\varepsilon/2)\|x\|$ for $x \in \rho B_X$. Since g is continuous, we can find $\sigma > 0$ such that $\|g(y)\| \leq \rho$ for $y \in \sigma B_Y$. Then, for $y \in \sigma B_Y$ and $x := g(y)$, we have $y = f(x) = x + r(x)$, hence

$$\|y\| \geq \|x\| - \|r(x)\| \geq (1/2)\|x\|,$$

$$\|s(y)\| = \|g(y) - y\| = \|r(x)\| \leq (\varepsilon/2)\|x\| \leq \varepsilon\|y\|.$$

□

In order to get a stronger result in which the invertibility of f is part of the conclusion instead of being an assumption, we will use the reinforced differentiability property of Definition 5.8. Recall that a map $f : W \to Y$ from an open subset

W of a normed space X into another normed space Y is *circa-differentiable* (or strictly differentiable) at $a \in W$ if there exists a continuous linear map $\ell : X \to Y$ such that the map $r = f - \ell$ is Lipschitzian with arbitrary small Lipschitz rate on sufficiently small neighborhoods of a: for any $\varepsilon > 0$ there exists a $\rho > 0$ such that $B(a, \rho) \subset W$ and

$$\forall w, w' \in B(a, \rho) \qquad \left\| f(w) - f(w') - \ell(w - w') \right\| \leq \varepsilon \left\| w - w' \right\|.$$

The criterion for circa-differentiability given in Proposition 5.21 uses continuous differentiability or slightly less. Thus, the reader who is not interested in refinements may suppose throughout that f is of class C^1.

Our next step is a perturbation result. We formulate it in a general framework.

Lemma 5.11 *Let (U, d) be a metric space, let Y be a normed space, let $j, h : U \to Y$ be such that*

(a) j is injective and its inverse $j^{-1} : j(U) \to U$ is Lipschitzian with rate γ;
(b) h is Lipschitzian with rate λ.

Then, if $\gamma\lambda < 1$, the map $f := j + h$ is still injective and its inverse $f^{-1} : f(U) \to U$ is Lipschitzian with rate $\gamma(1 - \gamma\lambda)^{-1}$.

Note that the Lipschitz rate of the inverse of the perturbed map f is close to the Lipschitz rate γ of j^{-1} when λ is small. It may be convenient to reformulate this lemma by saying that a map $e : X \to Y$ between two metric spaces is *expansive with rate* $c > 0$ if for all $x, x' \in X$ one has

$$d(e(x), e(x')) \geq cd(x, x').$$

This property amounts to

$$d(e^{-1}(y), e^{-1}(y')) \leq c^{-1} d(y, y')$$

for any $y, y' \in e(X)$, i.e. e is injective and its inverse is Lipschitzian on the image $e(X)$ of e. Thus the lemma can be rephrased as follows:

Lemma 5.12 *Let X be a metric space and let Y be a normed spaces. Let $e : X \to Y$ be expansive with rate $c > 0$ and let $h : X \to Y$ be Lipschitzian with rate $\ell < c$. Then $g := e + h$ is expansive with rate $c - \ell$.*

Proof This ensues from the following relations which are valid for any $x, x' \in X$:

$$\left\| g(x) - g(x') \right\| \geq \left\| e(x) - e(x') \right\| - \left\| h(x) - h(x') \right\| \geq cd(x, x') - \ell d(x, x').$$

Note that for $c = \gamma^{-1}$, $\ell = \lambda$ one has $(c - \ell)^{-1} = \gamma(1 - \gamma\lambda)^{-1}$. □

Since we have defined differentiability only on open subsets, it will be important to ensure that $f(U)$ is open in order to apply Lemma 5.10. We reach this conclusion in two steps. The first one relies on the Banach-Picard contraction theorem.

Lemma 5.13 *Let W be an open subset of a Banach space Y and let $k : W \to Y$ be a Lipschitzian map with rate $c < 1$. Then the image of W under $f := I_W + k$ is open.*

Proof We will prove that for any $a \in W$ and for any closed ball $B[a, r]$ contained in W, the closed ball $B[f(a), (1-c)r]$ is contained in the set $f(W)$, and in fact in the set $f(B[a, r])$. Without loss of generality, we may suppose $a = 0$, $k(a) = 0$, using translations if necessary. Given $y \in (1-c)rB_Y$ we want to find $x \in rB_Y$ such that $y = f(x)$. This equation can be written as $y - k(x) = x$. We note that $x \mapsto y - k(x)$ is Lipschitzian with rate $c < 1$ and that it maps rB_Y into itself since

$$\|y - k(x)\| \leq \|y\| + \|k(x)\| \leq (1-c)r + cr = r.$$

Since rB_Y is a complete metric space, the contraction theorem yields some fixed point x of this map. Thus $y = f(x) \in f(W)$. □

Lemma 5.14 *Let (U, d) be a metric space, let Y be a Banach space, let $\gamma > 0, \lambda > 0$ with $\gamma\lambda < 1$, and let $j, h : U \to Y$ be such that $W := j(U)$ is open and*

(a) j is injective and its inverse $j^{-1} : W \to U$ is Lipschitzian with rate γ;
(b) h is Lipschitzian with rate λ.

Then, the map $f := j + h$ is injective, its inverse is Lipschitzian and $f(U)$ is open.

Proof Let $k := h \circ j^{-1}$, so that $f \circ j^{-1} = I_W + k$ and k is Lipschitzian with rate $\gamma\lambda < 1$. Then, Lemma 5.13 shows that $f(U) = f(j^{-1}(W)) = (I + k)(W)$ is open. □

We are ready to state the Inverse Mapping Theorem.

Theorem 5.16 (Inverse Mapping Theorem) *Let X and Y be Banach spaces, let W be an open subset of X and let $f : W \to Y$ be circa-differentiable at $a \in W$ and such that $f'(a)$ is an isomorphism from X onto Y. Then there exist neighborhoods U of a and V of $b := f(a)$ such that $U \subset W$ and such that f induces a homeomorphism from U onto V whose inverse is differentiable at b.*

Proof In the preceding lemma, let us take $j := f'(a)$, $h = f - j$. Since j is an isomorphism, its inverse is Lipschitzian with rate $\|j^{-1}\|$. Let U be a neighborhood of a such that h is Lipschitzian with rate $\lambda < 1/\|j^{-1}\|$. Then, by the preceding lemma, $V := f(U)$ is open and $f \mid U$ is a homeomorphism from U onto V and, by Lemma 5.10, its inverse is differentiable at b. □

Exercise Show that the inverse of f is in fact circa-differentiable at b.

Exercise (Square Root of an Operator) Let E be a Banach space and let $X := L(E, E)$. Considering the map $f : X \to X$ given by $f(u) := u^2 := u \circ u$, show that there exist a neighborhood V of I_E in X and a differentiable map $g : V \to X$ such that $g(v)^2 := g(v) \circ g(v) = v$ for all $v \in V$.

The following classical terminology is helpful.

Definition 5.12 A C^k-*diffeomorphism* between two open subsets of normed spaces is a homeomorphism that is of class C^k as is its inverse ($k \geq 1$).

The following example plays an important role in the sequel, so we make it a lemma.

Lemma 5.15 *Let X and Y be Banach spaces. Then the set* $\mathrm{Iso}(X, Y)$ *of isomorphisms from X onto Y is open in $L(X, Y)$ and the map $i : \mathrm{Iso}(X, Y) \to \mathrm{Iso}(Y, X)$ given by $i(u) = u^{-1}$ is a C^∞-diffeomorphism, i.e. a C^k-diffeomorphism for all $k \geq 1$.*

Proof The first assertion has been proved in Proposition 3.17. Let us prove the second assertion by first considering the case $X = Y$ and by showing that i is differentiable at the identity map I_X, with derivative $Di(I_X)$ given by $Di(I_X)(v) = -v$. Taking $\rho \in]0, 1[$, this follows from the expansion

$$\forall v \in L(X, X), \ \ \|v\| \leq \rho \qquad\qquad (I_X + v)^{-1} = I_X - v + s(v)$$

with $s(v) := v^2 \circ \sum_{k=0}^{\infty}(-1)^k v^k$: s defines a remainder since $\left\|(-1)^k v^k\right\| \leq \rho^k$ and $\|s(v)\| \leq (1 - \rho)^{-1} \|v\|^2$. Thus i is differentiable at I_X.

Now, in the general case, for $u \in \mathrm{Iso}(X, Y)$, and $w \in L(X, Y)$ satisfying $\|w\| < 1/\left\|u^{-1}\right\|$, $v := u^{-1} \circ w$, one has $u + w = u \circ (I_X + v) \in \mathrm{Iso}(X, Y)$,

$$\begin{aligned} i(u + w) &= \left[u \circ (I_X + u^{-1} \circ w)\right]^{-1} = (I_X + u^{-1} \circ w)^{-1} \circ u^{-1} \\ &= \left(I_X - u^{-1} \circ w + s(v)\right) \circ u^{-1} = i(u) - u^{-1} \circ w \circ u^{-1} + s(v) \circ u^{-1}, \end{aligned}$$

and since $s(\cdot) \circ u^{-1}$ is a remainder, one sees that i is differentiable at u, with

$$Di(u)(w) = -u^{-1} \circ w \circ u^{-1}. \tag{5.23}$$

Thus the derivative $i' : \mathrm{Iso}(X, Y) \to L(L(X, Y), L(Y, X))$ is obtained by composing i with the map $k : L(Y, X) \to L(L(X, Y), L(Y, X))$ given by $k(z)(w) := -z \circ w \circ z$ for $z \in L(Y, X)$, $w \in L(X, Y)$ that is continuous and quadratic, hence is of class C^1. It follows that i' is continuous and i is of class C^1. Then i' is of class C^1. By induction, we obtain that i is of class C^k for all $k \geq 1$. Since i is a bijection with inverse $i^{-1} : \mathrm{Iso}(Y, X) \to \mathrm{Iso}(X, Y)$ given by $i^{-1}(z) = z^{-1}$, we get that i is a C^∞-diffeomorphism. □

Note that formula (5.23) generalizes the usual case $i(t) = t^{-1}$ on $\mathbb{R}\backslash\{0\}$ for which $i'(u) = -u^{-2}$ and $Di(u)(w) = -u^{-2}w$.

Corollary 5.17 *Let X and Y be Banach spaces, let W be an open subset of X and let $f : W \to Y$ be of class C^k ($k \geq 1$) and such that $f'(a)$ is an isomorphism from X*

onto Y for some $a \in W$. Then there exist neighborhoods U of a and V of $b := f(a)$ such that $U \subset W$ and such that $f \mid U$ is a C^k-diffeomorphism between U and V.

Proof Let us first consider the case $k = 1$. The Inverse Mapping Theorem ensures that f induces a homeomorphism from a neighborhood U of a onto a neighborhood V of b. Since f' is continuous at a and since the set $\mathrm{Iso}(X, Y)$ of isomorphisms from X onto Y is open in $L(X, Y)$, taking a smaller U if necessary, we may assume that $f'(x)$ is an isomorphism for all $x \in U$. Then, Lemma 5.10 guarantees that $g := f^{-1}$ is differentiable at $f(x)$. Moreover, one has

$$g'(y) = (f'(g(y)))^{-1}.$$

Since the map $i : u \mapsto u^{-1}$ is of class C^1 on $\mathrm{Iso}(X, Y)$, $g' = i \circ f' \circ g$ is continuous. Thus g is of class C^1.

Now suppose by induction that g is of class C^k if f is of class C^k, and let us prove that when f is of class C^{k+1}, then g is of class C^{k+1}. This follows from the expression $g' = i \circ f' \circ g$ which shows that g' is of class C^k as a composite of maps of class C^k. □

Let us give a global version of the Inverse Mapping Theorem.

Corollary 5.18 *Let X and Y be Banach spaces, let W be an open subset of X and let $f : W \to Y$ be an injection of class C^k such that, for every $x \in W$, the linear map $f'(x)$ is an isomorphism from X onto Y. Then $f(W)$ is open and f is a C^k-diffeomorphism between W and $f(W)$.*

Proof The Inverse Mapping Theorem ensures that $f(W)$ is open in Y. Thus f is a continuous bijection from W onto $f(W)$ and its inverse is locally of class C^k, hence is of class C^k. □

Exercise Let $f : T \to \mathbb{R}$ be a continuous function on some open interval T of $\mathbb{R}$. Show that if f is differentiable at some $r \in T$ with $f'(r)$ non-null, then $f(T)$ contains some neighborhood of $f(r)$. Show by an example that it may happen that there is no neighborhood of r on which f is injective.

Example-Exercise (Polar Coordinates) Let $W :=]0, \infty[\times] - \pi, \pi[\subset \mathbb{R}^2$ and let $f : W \to \mathbb{R}^2$ be given by $f(r, \theta) = (r\cos\theta, r\sin\theta)$. Then f is a bijection from W onto $\mathbb{R}^2 \backslash D$, with $D :=] - \infty, 0] \times \{0\}$ and the Jacobian matrix of f at (r, θ) is

$$\begin{pmatrix} \cos\theta & -r\sin\theta \\ \sin\theta & r\cos\theta \end{pmatrix}.$$

Its determinant (called the *Jacobian* of f) is $r(\cos^2\theta + \sin^2\theta) = r > 0$, hence f is a diffeomorphism of class C^∞ from W onto $f(W)$. Using the relation $\tan(\theta/2) = 2\sin(\theta/2)\cos(\theta/2)/2\cos^2(\theta/2) = \sin\theta/(1 + \cos\theta)$, show that its inverse is given by

$$(x, y) \mapsto (\sqrt{x^2 + y^2}, 2\mathrm{Arc}\tan\frac{y}{x + \sqrt{x^2 + y^2}}).$$

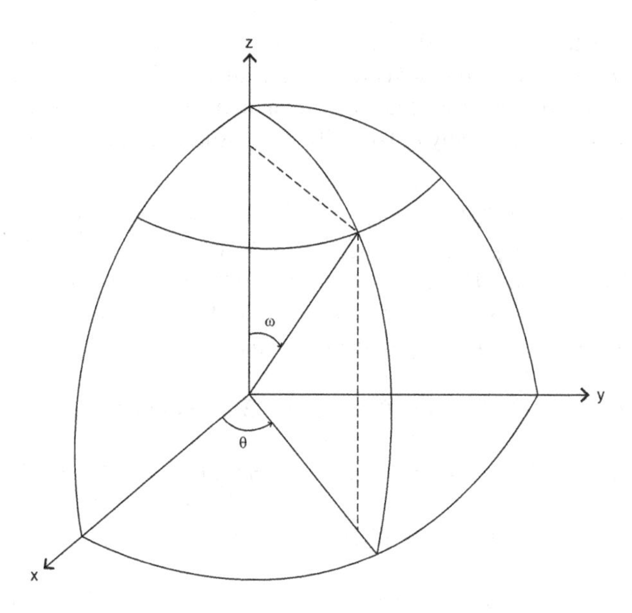

Fig. 5.2 Euler angles

Exercise (Spherical Coordinates) Let $W :=]0, \infty[\times] - \pi, \pi[\times] \frac{-\pi}{2}, \frac{\pi}{2}[$ and let $f : W \to \mathbb{R}^3$ be given by $f(r, \theta, \omega) = (r \cos\theta \sin\omega, r \sin\theta \sin\omega, r \cos\omega)$. Show that f is a diffeomorphism from W onto its image. The angles θ, ω are known as *Euler angles*. On the globe, they can serve to measure latitude and longitude (Fig. 5.2).

Exercise Is $f : \mathbb{R}^2 \to \mathbb{R}^2$ given by $f(x, y) := (x^2 - y^2, 2xy)$ a diffeomorphism? Give an interpretation by considering $z \mapsto z^2$, with $z := x + iy$, identifying $\mathbb{C}$ with $\mathbb{R}^2$.

Exercise* (Globalization) Let $f : X \to Y$ be a map of class C^1 between two Banach spaces. Suppose there exist $a, b \in \mathbb{R}_+$ such that for every $x \in X$ the linear map $f'(x)$ is invertible and satisfies $\left\| f'(x)^{-1} \right\| \leq a \|x\| + b$. Prove that f is a homeomorphism, in fact a diffeomorphism, from X onto Y. [See: [92, Thm 15.4].]

5.5.3 The Implicit Function Theorem

Functions are sometimes defined in an implicit, indirect way. For example, in economics, the famous Phillips curve is defined through the equation

$$1.39u(w + .9) = 9.64,$$

where u is the rate of unemployment and w is the annual rate of variation of nominal wages; in such a case one can express u in terms of w and vice versa. However, given Banach spaces X, Y, Z, an open subset W of $X \times Y$ and a map $f : W \to Z$, it is often impossible to determine an explicit map $h : X_0 \to Y$ from an open subset X_0 of X such that $(x, h(x)) \in W$ and $f(x, h(x)) = 0$ for all $x \in X_0$. When the existence of such a map is known, (but not necessarily in an explicit form) one says that it is an *implicit function* determined by f. The following result guarantees the existence and regularity of such a map.

Theorem 5.17 *Let X, Y, Z be Banach spaces, let W be an open subset of $X \times Y$ and let $f : W \to Z$ be a map of class C^1 at $(a, b) \in W$ such that $f(a, b) = 0$ and the second partial derivative $D_Y f(a, b)$ is an isomorphism from Y onto Z. Then, there exist open neighborhoods U of (a, b) and V of a in W and X respectively and a map $h : V \to Y$ of class C^1 at a such that $h(a) = b$ and*

$$((x, y) \in U,\ f(x, y) = 0) \Longleftrightarrow (x \in V,\ y = h(x)). \tag{5.24}$$

If f is of class C^k with $k \geq 1$ on W, then h is of class C^k on V. Moreover,

$$Dh(a) = -D_Y f(a, b)^{-1} \circ D_X f(a, b). \tag{5.25}$$

Proof Let $F : W \to X \times Z$ be the map given by $F(x, y) := (x, f(x, y))$. Then F is of class C^1 at (a, b) as are its components and

$$DF(a, b)(x, y) = (x, D_X f(a, b)x + D_Y f(a, b)y).$$

It is easy to see that $DF(a, b)$ is invertible and that its inverse is given by

$$(DF(a, b))^{-1}(x, z) = (x, -(D_Y f(a, b))^{-1} \circ D_X f(a, b)x + (D_Y f(a, b))^{-1} z).$$

Therefore, the Inverse Mapping Theorem yields open neighborhoods U of (a, b) in W and U' of $(a, 0)$ in $X \times Z$ such that F induces a homeomorphism from U onto U' of class C^1 at (a, b). Its inverse G is of class C^1 at $(a, 0)$, satisfies $G(a, 0) = (a, b)$, and has the form $(x, z) \mapsto (x, g(x, z))$. Let $V := \{x \in X : (x, 0) \in U'\}$ and let $h : V \to Y$ be given by $h(x) = g(x, 0)$. Then, the equivalence

$$((x, y) \in U,\ (x, z) = (x, f(x, y))) \Leftrightarrow \big((x, z) \in U',\ (x, y) = (x, g(x, z))\big)$$

entails, by definition of V and h,

$$((x, y) \in U,\ f(x, y) = 0) \Leftrightarrow (x \in V,\ y = h(x)).$$

When f is of class C^k on W, with $k \geq 1$, F is of class C^k, hence G and h are of class C^k on U' and V respectively. Moreover, the computation of the inverse $DF(a, b)^{-1}$ we have done shows that

$$Dh(a) = D_X g(a, 0) = -D_Y f(a, b)^{-1} \circ D_X f(a, b).$$

□

Example Let X be a Hilbert space and, for $Y := \mathbb{R}$, let $f : X \times Y \to \mathbb{R}$ be given by $f(x, y) = \|x\|^2 + y^2 - 1$. Then f is of class C^∞ and for $(a, b) := (0, 1)$ one has

$$Df(a, b)(u, v) = 2(a \mid u) + 2bv = 2v,$$

hence $D_Y f(a, b) = 2I_Y$ is invertible and $D_Y f(a, b)^{-1} = \frac{1}{2} I_Y$. Here we can take $U := B(a, 1) \times]0, +\infty[$, $V := B(a, 1)$ and the implicit function is given by $h(x) = (1 - \|x\|^2)^{1/2}$. As mentioned above, it is not always the case that U and h can be described explicitly as in this classical parameterization of the upper hemisphere.

□

When Z is finite dimensional, the regularity assumption on f can be relaxed in two ways.

Theorem 5.18* *Let X, Y, Z be Banach spaces, Y and Z being finite dimensional, let W be an open subset of $X \times Y$ and let $f : W \to Z$ be Fréchet differentiable at $(a, b) \in W$ such that $f(a, b) = 0$ and the partial derivative $D_Y f(a, b)$ is an isomorphism from Y onto Z. Then, there exist open neighborhoods U of (a, b) and V of a in W and X respectively and a map $h : V \to Y$ Fréchet differentiable at a such that $h(a) = b$ and*

$$\forall x \in V \qquad f(x, h(x)) = 0.$$

Differentiating this relation, we recover the value of $Dh(a)$:

$$Dh(a) = -D_Y f(a, b)^{-1} \circ D_X f(a, b).$$

The proof below is slightly simpler when $A := D_X f(a, b) = 0$; one can reduce it to that case by a linear change of variables.

Proof Using translations and composing f with $D_Y f(a, b)^{-1}$, we may suppose $(a, b) = (0, 0)$, $Z = Y$ and $D_Y f(a, b) = I_Y$. Let $r : W \to Y$ be a remainder such that

$$f(x, y) := Ax + y + r(x, y).$$

For $\varepsilon \in]0, 1/2]$ let $\delta := \delta(\varepsilon) > 0$ be such that $\delta B_{X \times Y} \subset W$, $\|r(x, y)\| \le \varepsilon(\|x\| + \|y\|)$ for all $(x, y) \in \delta B_{X \times Y}$. Let $\beta := \delta/2$, $\alpha := (2\|A\| + 1)^{-1}\beta$ and for $x \in \alpha B_X$ let $k_x : \beta B_Y \to Y$ be given by

$$k_x(y) := -Ax - r(x, y).$$

Then, k_x maps βB_Y into itself since for $y \in \beta B_Y$ we have $\|k_x(y)\| \le \|A\| \alpha + (1/2)(\alpha + \beta) \le \beta$. The Brouwer's Fixed Point Theorem ensures that k_x has a fixed point $y_x \in \beta B_Y$: $-Ax - r(x, y_x) = y_x$. Then, setting $h(x) := y_x$, we have $f(x, h(x)) = Ax + h(x) + r(x, h(x)) = 0$. It remains to show that h is differentiable at 0. Since

$$\|h(x)\| = \|k_x(h(x))\| \leq \|A\| \, \|x\| + \varepsilon \, \|x\| + \varepsilon \, \|h(x)\| \, ,$$

so that $\|h(x)\| \leq (1 - \varepsilon)^{-1}(\|A\| + \varepsilon) \, \|x\|$, we get

$$\|h(x) + Ax\| = \|r(x, h(x))\| \leq \varepsilon \, \|x\| + \varepsilon \, \|h(x)\| \leq \varepsilon(1 - \varepsilon)^{-1}(\|A\| + 1) \, \|x\| \, .$$

This shows that h is differentiable at 0 with derivative $-A$. □

A similar (and simpler) proof yields the first assertion of the next statement.

Theorem 5.19* *Let X and Y be normed spaces, Y being finite dimensional and let $f : X \to Y$ be continuous on a neighborhood of $a \in X$ and differentiable at a, with $f'(a)(X) = Y$. Then there exist a neighborhood V of $b := f(a)$ in Y and a right inverse $g : V \to X$ that is differentiable at a and such that $g(b) = a$.*

If C is a convex subset of X, if $a \in C$ and if $f'(a)(\mathrm{cl}(\mathbb{R}_+(C - a))) = Y$, one can even get that $g(V) \subset C$ if one does not require that the directional derivative of g at b is linear.

The second weakening of the assumptions concerns the kind of differentiability.

Theorem 5.20* *Let X, Y, Z be Banach spaces, Y and Z being finite dimensional, let W be an open subset of $X \times Y$ and let $f : W \to Z$ be a map of class D^1 at $(a, b) \in W$ such that $f(a, b) = 0$ and the partial derivative $D_Y f(a, b)$ is an isomorphism from Y onto Z. Then, there exist open neighborhoods U of (a, b) and V of a in W and X respectively and a map $h : V \to Y$ of class D^1 such that $h(a) = b$ and*

$$((x, y) \in U, \; f(x, y) = 0) \iff (x \in V, \; y = h(x)) \, .$$

Proof We may suppose W is a ball $B((a, b), \rho_0)$, $Y = Z$, $D_Y f(a, b) = I_Y$. With the notation of the preceding proof, using the compactness of the unit ball of Y, we may suppose the remainder r satisfies, for $\rho \in (0, \rho_0)$ and every $x \in \rho B_X$, $y, y' \in \rho B_Y$,

$$\begin{aligned} \left\| r(x, y) - r(x, y') \right\| &\leq \int_0^1 \left\| (D_Y f(x, (1 - t)y + ty') - I_Y)(y - y') \right\| dt \\ &\leq c(\rho) \left\| y - y' \right\| \end{aligned}$$

where $c(\rho) \to 0$ as $\rho \to 0_+$. Taking ρ_0 small enough, we see that the map k_x is a contraction with rate $c(\rho_0) \leq 1/2$. Picking $\alpha \in (0, \rho_0)$ so that, for $x \in \alpha B_X$, $\|k_x(0)\| = \|-f(x, 0)\| \leq \rho/2$, the Banach-Picard Contraction Theorem ensures that k_x has a unique fixed point y_x in the ball ρB_Y. Then, setting $h(x) := y_x$, we have $f(x, h(x)) = 0$ and y_x is the unique solution of the equation $f(x, y) = 0$ in the ball ρB_Y. Moreover, h is continuous as a uniform limit of continuous maps given by iterations. Restricting f to $X_1 \times Y$, where X_1 is an arbitrary finite dimensional subspace of X, we get that h is Gateaux differentiable. Since $\mathrm{Iso}(Y)$ is an open subset of $L(Y, Y)$ and since $(x, y) \mapsto D_Y f(x, y)$ is continuous with respect to the norm of $L(Y, Y)$ by the above argument, we obtain from the relation

$$Dh(x)v = -D_Y f(x, h(x))^{-1}(D_X f(x, h(x))v)$$

that $(x, v) \mapsto Dh(x)v$ is continuous. □

Exercises

1. Show that the Inverse Mapping Theorem can be deduced from the Implicit Mapping Theorem by considering the map $(x, y) \mapsto y - f(x)$.
2. Let $f : \mathbb{R}^4 \to \mathbb{R}^3$ be given by

$$f(w, x, y, z) = (w + x + y + z, w^2 + x^2 + y^2 + z - 2, w^3 + x^3 + y^3 + z).$$

Show that there is a neighborhood V of $a := 0$ in $\mathbb{R}$ and a map $h : V \to \mathbb{R}^3$ of class C^∞ such that $h(0) = (0, -1, 1)$ and $f(h(z), z) = 0$ for every $z \in V$. Compute the derivative of h at 0.
3. Let X be the space of square $n \times n$ matrices and let $f : X \times \mathbb{R} \to \mathbb{R}$ be given by $f(A, r) = \det(A - rI)$. Let $r \in \mathbb{R}$ be such that $f(A, r) = 0$ and $D_2 f(A, r) \neq 0$. Show that there is an open neighborhood U of A in X and a function $\lambda : U \to \mathbb{R}$ of class C^∞ such that, for each B in U, $\lambda(B)$ is a simple eigenvalue of B.
4. Given Banach spaces $W, X, Z, Y := Z^*$ and maps $f : W \times X \to \mathbb{R}$, $g : W \times X \to Z$ of class C^2, consider the parameterized mathematical programming problem

$$(\mathcal{P}_w) \quad \text{minimize } f(w, x) \text{ subject to } g(w, x) = 0$$

and let $p(w)$ be its value. Suppose that for some $\overline{w} \in W$ and a solution $\overline{x} \in X$ of $(\mathcal{P}_{\overline{w}})$ the derivative $B := D_X g(\overline{w}, \overline{x})$ is surjective and its kernel N has a topological supplement M. Let ℓ be the *Lagrangian* of $(\mathcal{P}_w)$:

$$\ell(w, x, y) := f(w, x) + \langle y, g(w, x) \rangle$$

and let $\overline{y}$ be a multiplier at $\overline{x}$, i.e. an element of Y such that $D_X \ell(\overline{w}, \overline{x}, \overline{y}) = 0$. Suppose $D_X^2 \ell(\overline{w}, \overline{x}, \overline{y}) \mid N$ induces an isomorphism from N onto $N^* \simeq M^\perp$. Let $A := D_X^2 \ell(\overline{w}, \overline{x}, \overline{y})$.

(a) Show that for any $(x^*, z) \in X^* \times Z$ the system

$$Au + B^\mathsf{T} v = x^*$$
$$Bu = z$$

has a unique solution $(u, v) \in X \times Y$ continuously depending on (x^*, z).

(b) Show that the Karush-Kuhn-Tucker system

$$\begin{aligned} D_X f(w,x) + y \circ D_X g(w,x) &= 0 \\ g(w,x) &= 0 \end{aligned}$$

determines $(x(w), y(w))$ as an implicit function of w in a neighborhood of $\overline{w}$ with $x(\overline{w}) = \overline{x}$, $y(\overline{w}) = \overline{y}$, the multiplier at $\overline{x}$.

(c) Suppose $x(w)$ is a solution to $(\mathcal{P}_w)$ for w close to $\overline{w}$. Show that p is of class C^1 near $\overline{w}$. Using the relations $p(w) = \ell(w, x(w), y(w))$, $D_X\ell(w, x(w), y(w)) = 0$, $D_Y\ell(w, x(w), y(w)) = 0$, show that $Dp(w) = D_W\ell(w, x(w), y(w))$.

(d) Deduce from the preceding that p is of class C^2 around $\overline{w}$ and give the expression of $D^2p(\overline{w}) := (p'(\cdot))'(\overline{w})$.

5.5.4 Geometric Applications

When looking at familiar objects such as forks, knifes, funnels, roofs, spires, one sees that some points are smooth, while some other points of the objects present ridges or peaks or cracks. Mathematicians have found concepts that enable one to deal with such cases (see [47], [78], [208], [221] for instance). In this subsection we essentially focus our attention on smooth objects.

The notions of (regular) curve, surface, hypersurface…can be embodied in a general framework in which some differential calculus can be done. The underlying idea is the possibility of straightening a piece of the set; for this purpose, some forms of the inverse map theorem will be appropriate.

We first define a notion of smoothness for a subset S of a normed space X around some point.

Definition 5.13 A subset S of a normed space X is said to be C^k*-smooth* around a point $a \in S$ if there exist normed spaces Y, Z, an open neighborhood U of a in X, an open neighborhood V of 0 in $Y \times Z$ and a C^k-diffeomorphism $\varphi : U \to V$ such that $\varphi(a) = 0$ and

$$\varphi(U \cap S) = (Y \times \{0\}) \cap V. \tag{5.26}$$

A subset S of a normed space X is said to be a *submanifold of class* C^k if it is C^k-smooth around each of its points.

Thus, φ straightens $U \cap S$ onto the piece $(Y \times \{0\}) \cap V$ of the linear space $Y \times \{0\}$ which can be identified with a neighborhood of 0 in Y. The map φ is called a *chart* and a collection $\{\varphi_i\}$ of charts whose domains form a covering of S is called an *atlas*. When Y is of dimension d, one says that S is of dimension d around a. When Z is of dimension c, one says that S is of codimension c around a.

The following example can be seen as a general model.

Example Let $X := Y \times Z$, where Y, Z are normed spaces, let W be an open subset of Y and let $f : W \to Z$ be a map of class C^k. Then, its graph $S := \{(w, f(w)) : w \in W\}$ is a C^k-submanifold of X: taking $U := V := W \times Z$, and setting $\varphi(w, z) := (w, z - f(w))$, we define a C^k-diffeomorphism from U onto V with inverse given by $\varphi^{-1}(w, z) = (w, z + f(w))$ for which (5.26) is satisfied. □

When in the preceding example $Z := \mathbb{R}$ and one takes the epigraph $E := \{(w, y) \in W \times \mathbb{R} : y \geq f(w)\}$ of f, one obtains a model for the notion of a submanifold with boundary. We just give a formal definition in which a subset Z_+ of a normed space Z is said to be a *half-space* of Z if there exists some $h \in Z^* \backslash \{0\}$ such that $Z_+ := h^{-1}(\mathbb{R}_+)$.

Definition 5.14 A subset S of a normed space X is said to be a C^k*-submanifold with boundary* if, for every point a of S, either S is C^k-smooth around a or there exist normed spaces Y, Z, a half-space Z_+ of Z, an open neighborhood U of a in X, an open neighborhood V of 0 in $Y \times Z$, and a C^k-diffeomorphism $\varphi : U \to V$ such that $\varphi(a) = 0$ and

$$\varphi(U \cap S) = (Y \times Z_+) \cap V.$$

Such a notion is useful when giving a precise meaning to the expression "S is a regular open subset of $\mathbb{R}^d$" (an improper expression, since usually one considers the closure of such a set).

There are two usual ways of obtaining submanifolds: either through equations or through parameterizations. For instance, the graph S of the preceding example can either be defined as the image under $(I_W, f) : w \mapsto (w, f(w))$ of the parameter space W or as the set of points $(y, z) \in Y \times Z$ satisfying $y \in W$ and the equation $z - f(y) = 0$. As a more concrete example, we observe that for given $a, b \in \mathbb{P}$, the ellipse

$$E := \{(x, y) \in \mathbb{R}^2 : \frac{x^2}{a^2} + \frac{y^2}{b^2} = 1\}$$

can be seen as the image of the parameterization $f : \mathbb{R} \to \mathbb{R}^2$ given by $f(t) := (a \cos t, b \sin t)$.

Exercise Give parameterizations for the ellipsoid $\{(x, y, z) \in \mathbb{R}^3 : \frac{x^2}{a^2} + \frac{y^2}{b^2} + \frac{z^2}{c^2} = 1\}$ and do the same for the other surfaces of $\mathbb{R}^3$ defined by quadratic forms.

Even if S is not smooth around $a \in S$, one can get an idea of its shape around a by using an approximation. The concept of tangent cone offers such an approximation; it can be seen as a geometric counterpart to the directional derivative.

Definition 5.15 The *tangent cone* (or *contingent cone*) to a subset S of a normed space X at some point a in the closure of S is the set $T(S, a)$ of vectors $v \in X$ such that there exist sequences $(v_n) \to v$, $(t_n) \to 0_+$ for which $a + t_n v_n \in S$ for all $n \in \mathbb{N}$.

Equivalently, one has $v \in T(S,a)$ if and only if there exist sequences (a_n) in S, $(t_n) \to 0_+$ such that $(v_n) := (t_n^{-1}(a_n - a)) \to v$: v is the limit of a sequence of secants to S issued from a.

Some rules for dealing with tangent cones are given in the next lemma, whose elementary proof is left as an exercise.

Lemma 5.16 *Let X be a normed space, let S, S' be subsets of X such that $S \subset S'$. Then for every $a \in S$ one has $T(S,a) \subset T(S',a)$.*

If U is an open subset of X, then for any $a \in S \cap U$ one has $T(S,a) = T(S \cap U, a)$.

If X' is another normed space, if $g : U \to X'$ is Hadamard differentiable at a, and if $S' \subset X'$ contains $g(S \cap U)$, then one has $Dg(a)(T(S,a)) \subset T(S', g(a))$.

If $\varphi : U \to V$ is a C^k-diffeomorphism between two open subsets of normed spaces X, X' and if S is a subset of X containing a, then, for $S' := \varphi(S \cap U)$ and $a' := \varphi(a)$, one has $T(S', a') = D\varphi(a)(T(S,a))$.

Exercise Deduce from the second assertion of the lemma that for $g : U \to X'$ Hadamard differentiable at a, $b := g(a)$, $S := g^{-1}(b)$ one has $T(S,a) \subset \ker Dg(a)$. Moreover, if for some $c > 0$, $\rho > 0$ one has $d(x, g^{-1}(b)) \leq cd(g(x), b)$ for all $x \in B(a, \rho)$, then one has $T(S,a) = \ker Dg(a)$.

Exercise Let $S := \{(x,y) \in \mathbb{R}^2 : x^3 = y^2\}$. Show that $T(S,(0,0)) = \mathbb{R}_+ \times \{0\}$.

When S is smooth around $a \in S$ in the sense of Definition 5.13 one can give an alternative characterization of $T(S,a)$ in terms of velocities.

Proposition 5.26 *If S is C^1-smooth around $a \in S$, then the tangent cone $T(S,a)$ to S at a coincides with the set $T^I(S,a)$ of $v \in X$ such that there exist $\tau > 0$ and $c : [0,\tau] \to X$ right differentiable at 0 with $c'_+(0) = v$ and satisfying $c(0) = a$, $c(t) \in S$ for all $t \in [0,\tau]$. Moreover, if $\varphi : U \to V$ is a C^1-diffeomorphism such that $\varphi(a) = 0$ and $\varphi(S \cap U) = (Y \times \{0\}) \cap V$, then one has $T(S,a) = (D\varphi(a))^{-1}(Y \times \{0\})$ and $T(S,a)$ is a closed linear subspace of X.*

Proof The result follows from Lemma 5.16 and the observation that if S is an open subset of some closed linear subspace L of X then $T(S,a) = L = T^I(S,a)$. □

Now let us turn to sets defined by equations. We need the following result.

Theorem 5.21 (Submersion Theorem) *Let X and Z be Banach spaces, let W be an open subset of X and let $g : W \to Z$ be a map of class C^k with $k \geq 1$, such that for some $a \in W$ the map $Dg(a)$ is surjective and its kernel N has a topological supplement M in X. Then, there exist an open neighborhood U of a in W, a diffeomorphism φ of class C^k from U onto a neighborhood V of $(0, g(a))$ in $N \times Z$ such that $\varphi(a) = (0, g(a))$,*

$$g \mid U = p \circ \varphi$$

where p is the canonical projection from $N \times Z$ onto Z. In particular, g is open around a in the sense that for every open subset U' of U, the image $g(U')$ is open.

This result shows that the nonlinear map g has been straightened into a simple continuous linear map, a projection, by using the diffeomorphism φ.

Proof Let $F : W \to N \times Z$ be given by $F(x) = (p_N(x) - p_N(a), g(x))$, where $p_N : X \to N$ is the projection on N associated with the isomorphism between X and $M \times N$. Then F is of class C^k and $DF(a)(x) = (p_N(x), Dg(a)(x))$. Clearly $DF(a)$ is injective: when $p_N(x) = 0$, $Dg(a)(x) = 0$, one has $x \in M \cap N$, hence $x = 0$. Let us show that $DF(a)$ is surjective: given $(y, z) \in N \times Z$, there exists $v \in X$ such that $Dg(a)(v) = z$ and since $y - p_N(v) \in N$, for $x := v + y - p_N(v)$, we have that $Dg(a)(x) = Dg(a)(v) = z$ and $p_N(x) = p_N(y) = y$. Thus, by the Banach isomorphism theorem, we have that $DF(a)$ is an isomorphism of X onto $N \times Z$. The Inverse Mapping Theorem ensures that the restriction φ of F to some open neighborhood U of a is a C^k-diffeomorphism onto some neighborhood V of $(0, g(a))$. □

Note that for $Z := \mathbb{R}$, the condition on g reduces to: g is of class C^k and $g'(a) \neq 0$. Note also that when $N := \{0\}$, we recover the inverse function theorem.

The application we have in view follows readily.

Corollary 5.19 *Let X and Z be Banach spaces, let W be an open subset of X and let $g : W \to Z$ be a map of class C^k with $k \geq 1$. Let*

$$S := \{x \in W : g(x) = 0\}.$$

Suppose that for some $a \in S$ the map $g'(a) := Dg(a)$ is surjective and its kernel N has a topological supplement in X. Then S is C^k-smooth around a. Moreover, $T(S, a) = \ker g'(a)$.

Proof Using the notation of the Submersion Theorem, setting $Y := N$, we see that Definition 5.13 is satisfied, noting that for $x \in U$ we have $x \in S \cap U$ if and only if $p(\varphi(x)) = g(x) = 0$, if and only if $\varphi(x) \in (Y \times \{0\}) \cap V$. Now, the preceding proposition asserts that $T(S, a) = (\varphi'(a))^{-1}(Y \times \{0\})$. But since $g \mid U = p \circ \varphi$, we have $g'(a) = p \circ \varphi'(a)$, $\ker g'(a) = (\varphi'(a))^{-1}(\ker p) = (\varphi'(a))^{-1}(Y \times \{0\})$. Hence $T(S, a) = \ker g'(a)$. □

The regularity condition on g can be relaxed thanks to the Lyusternik-Graves Theorem.

Proposition 5.27 (Lyusternik) *Let X and Y be Banach spaces, let W be an open subset of X and let $g : W \to Y$ be circa-differentiable at $a \in S := \{x \in W : g(x) = 0\}$, with $g'(a)(X) = Y$. Then $T(S, a) = \ker g'(a)$.*

Proof The inclusion $T(S, a) \subset \ker g'(a)$ follows from Lemma 5.16. Conversely, let $v \in \ker g'(a)$. Theorem 5.15 yields some κ, $\rho > 0$ such that for all $w \in B(a, \rho)$ there exists some $x \in W$ such that $g(x) = y := 0$, $\|x - w\| \leq \kappa \|g(w)\|$. Taking $w := a + tv$ with $t > 0$ so small that $w \in B(a, \rho)$, we get some $x_t \in S$ satisfying $\|x_t - (a + tv)\| \leq o(t) := \kappa \|g(\overline{x} + tv)\|$. Thus $v \in T(S, a)$ and even $v \in T^I(S, a)$. □

In the following example, we use the fact that when $Y = \mathbb{R}$, the surjectivity condition on $g'(a)$ reduces to $g'(a) \neq 0$ (or $\nabla g(a) \neq 0$ if X is a Hilbert space).

Example-Exercise For a Hilbert space X, let $g : X \to \mathbb{R}$ be given by $g(x) := \frac{1}{2}(\langle A(x) \mid x\rangle - 1)$, where A is a linear isomorphism from X onto X that is symmetric, i.e. such that $\langle A(x) \mid y\rangle = \langle A(y) \mid x\rangle$ for every $x, y \in X$. Let $S := g^{-1}(\{0\})$. For all $a \in S$ one has $\nabla g(a) = A(a) \neq 0$ since $\langle A(a) \mid a\rangle = 1$. Thus S is a C^∞-submanifold of X. Taking $X = \mathbb{R}^2$ and appropriate isomorphisms A, find the classical conic curves; then take $X = \mathbb{R}^3$ and find the classical conic surfaces, including the sphere, the ellipsoid, the paraboloid and the hyperboloid.

A variant of the submersion theorem can be given with differentiability instead of circa-differentiability when the spaces are finite dimensional. Its proof (which we skip) relies on the Brouwer's Fixed Point Theorem rather than on the contraction theorem.

Proposition 5.28 *Let X and Z be Banach spaces, Z being finite dimensional, let W be an open subset of X and let $g : W \to Z$ be Hadamard differentiable at $a \in W$, with $Dg(a)(X) = Z$. Then, there exist open neighborhoods U of a in W, V of $g(a)$ in Z and a map $h : V \to U$ that is differentiable at $g(a)$ and such that $h(g(a)) = a$, $g \circ h = I_V$. In particular, g is open at a.*

Now let us turn to representations via parameterizations. We need the following result.

Theorem 5.22 (Immersion Theorem) *Let P and X be Banach spaces, let O be an open subset of P and let $f : O \to X$ be a map of class C^k with $k \geq 1$, such that for some $\overline{p} \in O$ the map $Df(\overline{p})$ is injective and its image Y has a topological supplement Z in X. Then there exist open neighborhoods U of $a := f(\overline{p})$ in X, Q of $\overline{p}$ in O, W of 0 in Z and a C^k-diffeomorphism $\psi : V := Q \times W \to U$ such that $\psi(q, 0) = f(q)$ for all $q \in Q$.*

Again the conclusion can be written in the form of a commutative diagram, since $f \,|\, Q = \psi \circ j$, where $j : Q \to Q \times W$ is the canonical injection $y \mapsto (y, 0)$. Again the nonlinear map f has been straightened by ψ into a linear map $j = \psi^{-1} \circ (f \,|\, Q)$.

Proof Let $F : O \times Z \to X$ be given by $F(p, z) = f(p) + z$. Then F is of class C^k and $F'(\overline{p}, 0)(p, z) = f'(\overline{p})(p) + z$ for $(p, z) \in P \times Z$, so that $F'(\overline{p}, 0)$ is an isomorphism from $P \times Z$ onto $Y + Z = X$. The Inverse Mapping Theorem asserts that F induces a C^k-diffeomorphism ψ from some open neighborhood of $(\overline{p}, 0)$ onto some open neighborhood U of $f(\overline{p})$. Taking a smaller neighborhood of $(\overline{p}, 0)$ if necessary, we may suppose it has the form of a product $Q \times W$. Clearly, $\psi(q, 0) = f(q)$ for $q \in Q$. □

Example-Exercise Let $P := \mathbb{R}^2$, $O :=]-\pi, \pi[\times]-\pi/2, \pi/2[$, $X := \mathbb{R}^3$ and f be given by $f(\varphi, \theta) := (\cos\theta \cos\varphi, \cos\theta \sin\varphi, \sin\varphi)$. Identify the image of f.

Exercise Let us note that the image $f(O)$ of f is not necessarily a C^k-submanifold of X. Verify that a counterexample can be given by taking $P := \mathbb{R}$, $X := \mathbb{R}^2$, $f(p) = (p, 0)$ for $p \in]-\infty, 0[$, $f(p) := (p, 1 - (1 - p^2)^{1/2})$ for $p \in [0, 1[$, $f(p) := (2 - p, (p^2 - 3p + 2)^{1/2} + 1)$ for $p \in [1, 2[$, $f(p) := (0, e^{2-p})$ for $p \in [2, \infty[$.

A topological assumption ensures that the image $f(O)$ is a C^k-submanifold of X.

Corollary 5.20 (Embedding Theorem) *Let P and X be Banach spaces, let O be an open subset of P and let $f : O \to X$ be a map of class C^k with $k \geq 1$ such that for every $p \in O$ the map $f'(p)$ is injective and its image has a topological supplement in X. Then, if f is a homeomorphism from O onto $f(O)$, its image $S := f(O)$ is a C^k-submanifold of X.*

Moreover, for every $p \in O$ one has $T(S, f(p)) = f'(p)(P)$.

One says that f is an *embedding* of O into X and that S is parameterized by O.

Proof Given $a := f(p)$ in S, with $p \in O$, we take $Q_a \subset O$, $U_a \subset X$, $W_a \subset Z$ and a C^k-diffeomorphism $\psi_a : V_a := Q_a \times W_a \to U_a$ such that $\psi_a(q, 0) = f(q)$ for all $q \in Q_a$ as in the preceding theorem. Performing a translation in P, we may suppose $p = 0$. Using the assumption that f is a homeomorphism from O onto $S = f(O)$, we can find an open subset U'_a of X such that $f(Q_a) = S \cap U'_a$. Let $U := U_a \cap U'_a$, $V := \psi_a^{-1}(U)$, $\varphi := \psi_a^{-1} \mid U$, $Y := P$, so that $\varphi(a) = (0, 0)$. Let us check relation (5.26), i.e. $\varphi(S \cap U) = (Y \times \{0\}) \cap V$. For all $(y, 0) \in (Y \times \{0\}) \cap V$, we have $x := \varphi^{-1}(y, 0) = \psi_a(y, 0) = f(y) \in S$, hence $x \in S \cap U$; conversely, when $x \in S \cap U = f(Q_a)$ there is a unique $q \in Q_a$ such that $x = f(q)$, so that $x = \psi_a(q, 0) = \varphi^{-1}(q, 0)$ and $\varphi(x) = (q, 0) \in (Y \times \{0\}) \cap V$.

Then $T(S, a) = T(S \cap U, a) = (\varphi'(a))^{-1}(T((Y \times \{0\}) \cap V, 0))$, and, since $T((Y \times \{0\}) \cap V, 0) = \psi'_a(0)(P \times \{0\}) = Y \times \{0\}$, we get $T(S, a) = Y = f'(p)(P)$. □

Exercises

1. **(Conic section)** Let $S \subset \mathbb{R}^3$ be defined by the equations $x^2 + y^2 - 1 = 0$, $x - z = 0$. Show that S is a submanifold of $\mathbb{R}^3$ of class C^∞ (it has been known since Apollonius that S is an ellipse). Find an explicit diffeomorphism (in fact linear isomorphism) sending S onto an ellipse of the plane $\mathbb{R}^2 \times \{0\}$.
2. **(Viviani's window)** Let S be the subset of $\mathbb{R}^3$ defined by the system $x^2 + y^2 = x$, $x^2 + y^2 + z^2 - 1 = 0$. Show that S is a submanifold of $\mathbb{R}^3$ of class C^∞.
3. **(The torus)** Let $r > s > 0$, let $O :=]0, 2\pi[\times]0, 2\pi[$ and let $f : O \to \mathbb{R}^3$ be given by $f(\alpha, \beta) = ((r + s\cos\beta)\cos\alpha, (r + s\cos\beta)\sin\alpha, s\sin\beta)$. Show that f is an embedding onto the torus $\mathbb{T}$ deprived of its greatest circle and of the set $\mathbb{T} \cap (\mathbb{R}_+ \times \{0\} \times \mathbb{R})$, where

$$\mathbb{T} := \{(x, y, z) \in \mathbb{R}^3 : (\sqrt{x^2 + y^2} - r)^2 + z^2 = s^2\}.$$

4. Using the Submersion Theorem, show that $\mathbb{T}$ is a C^∞-submanifold of $\mathbb{R}^3$.
5. (a) (**Beltrami's tractricoid**) Let $f : \mathbb{R} \to \mathbb{R}^2$ be given by $f(t) := (1/\cosh t, t - \tanh t)$. Determine the points of $T := f(\mathbb{R})$ that are smooth.
   (b) (**Beltrami's pseudo-sphere**) Let $g(s,t) := (\cos s/\cosh t, \sin s/\cosh t, t - \tanh t)$. Determine the points of $S := g(\mathbb{R}^2)$ that are smooth. They form a surface of (negative) constant Gaussian curvature. This can serve as a model for hyperbolic geometry.
6. Study the **cross-cap surface** $\{((1 + \cos v)\cos u, (1 + \cos v)\sin u, \tanh(u - \pi)\sin v) : (u,v) \in [0, 2\pi] \times [0,1]\}$ and compare it with the **self-intersecting disc**, the image of $[0, 2\pi] \times [0,1]$ under the parameterization

$$(u,v) \mapsto (v\cos 2u, v\sin 2u, v\cos u).$$

7. Study **Whitney's umbrella** $\{(uv, u, v^2) : (u,v) \in \mathbb{R}^2\}$. Verify that it is determined by the equation $x^2 - y^2z = 0$. Such a surface is of interest in the theory of singularities. For this surface or the preceding one, make some drawings if you can or find some on the internet.
8. Let $O :=]0,1[\cup]1,\infty[\subset \mathbb{R}, f : O \to \mathbb{R}^2$ being given by $f(t) = (t + t^{-1}, 2t + t^{-2})$. Show that f is an embedding, but that its continuous extension to $]0, +\infty[$ given by $f(1) = (2,3)$ is of class C^k but is not an immersion.
9. Let X be a normed space and let $f : X \to \mathbb{R}$ be Lipschitzian around $x \in X$. Show that f is Hadamard differentiable at $x \in X$ if and only if the tangent cone to the graph G of f at $(x, f(x))$ is a hyperplane.
10. Show that the fact that the tangent cone at $(x, f(x))$ to the epigraph E of f is a half-space does not imply that f is Hadamard differentiable at x.

5.5.5 ***The Eikonal Equation**

The *eikonal equation* plays a role in wave propagation phenomena such as rings on the surface of water, or seismic techniques for petrol exploration. We present a study of this equation in the form of a detailed problem.

(A) Given a nonempty subset S of a Banach space X we denote by $d_S : X \to \mathbb{R}$ the *distance function* to S given by

$$d_S(x) := \inf_{y \in S} \|x - y\| \qquad x \in X.$$

1) Verify that d_S is Lipschitzian with rate 1 so that when d_S is differentiable at some $x \in X$ one has $\|d'_S(x)\| \leq 1$.
2) a) Suppose there exists a $\overline{y} \in S$ such that $\|x - \overline{y}\| \leq \|x - y\|$ for all $y \in S$. Setting $x_t := (1-t)x + t\overline{y}$ for $t \in [0,1]$, verify that $d_S(x_t) = (1-t)d_S(x)$.

b) Assuming moreover that d_S is differentiable at x, deduce from this the relation

$$d'_S(x)(\bar{y}-x) = -\|\bar{y}-x\|.$$

c) Conclude that in this case one has $\|d'_S(x)\| = 1$.

(B) Consider the example $S := \{(x_1, x_2) \in \mathbb{R}^2 : x_1^2 + x_2^2 = 1,\ x_1 > -1\} \subset X := \mathbb{R}^2$ with its Euclidean norm. Let $W :=]-\pi, \pi[$, $g : W \to \mathbb{R}^2$ be given by $g(w) = (\cos w, \sin w)$.

1) Show that $d_S(x) = |\|x\| - 1|$ for $x \in \mathbb{R}^2$ by first observing that for $x \in \mathbb{R}^2 \backslash D$ where $D :=]-\infty, 0] \times \{0\}$ there exists a $\bar{y} \in S$ such that $\|x - \bar{y}\| \le \|x - y\|$ for all $y \in S$.
2) Show that the map $f : (w, r) \mapsto ((r+1)\cos w, (r+1)\sin w)$ is a C^1-diffeomorphism from $U :=]-\pi, \pi[\times]-1, 1[$ onto a neighborhood V of S.
3) Set $f^{-1}(x_1, x_2) := (w(x_1, x_2), r(x_1, x_2))$, compute $r(x_1, x_2)$, and show that

$$(\frac{\partial r}{\partial x_1}(x_1, x_2))^2 + (\frac{\partial r}{\partial x_2}(x_1, x_2))^2 = 1.$$

(C) In the sequel S is a hypersurface of X parametrized by a C^1-embedding $g : W \to X$ where W is an open subset of a Banach space H, $g'(w)$ being injective for all $w \in W$. It is also assumed that there exists a C^1 map $n : W \to X \backslash \{0\}$ such that for all $w \in W$ the subspace $\mathbb{R}n(w) := \{rn(w) : r \in \mathbb{R}\}$ is a topological complement of $g'(w)(H)$.

1) Let $f : W \times \mathbb{R} \to X$ be given by $f(w, r) := g(w) + rn(w)$. Compute $f'(w, r)(u, s)$ for $(w, r) \in W \times \mathbb{R}$, $(u, s) \in H \times \mathbb{R}$.
2) Show that for all $w \in W$, $f'(w, 0)$ is an isomorphism from $H \times \mathbb{R}$ onto X.
3) Deduce from this that for all $w \in W$ there exists an open neighborhood U_w of $(w, 0)$ in $W \times \mathbb{R}$ and an open neighborhood V_w of $g(w)$ in X such that f induces a diffeomorphism from U_w onto V_w.
4) In the sequel it is assumed that there exists an open subset U_0 of $W \times \mathbb{R}$ containing $W \times \{0\}$ such that $f \mid U_0$ is injective. Show that there exists an open subset U of U_0 containing $W \times \{0\}$ and an open subset V of X containing S such that f induces a diffeomorphism from U onto V. The inverse of $f \mid U$ is denoted by $h : x \mapsto (w(x), r(x))$.

(D) From now on X is a Hilbert space whose norm is associated with the scalar product $\langle \cdot \mid \cdot \rangle$.

1) Assume that for all $w \in W$ one has $\|n(w)\| = 1$. Show that for all $(w, u) \in W \times H$ one has $\langle n'(w)(u) \mid n(w) \rangle = 0$.
2) Assuming moreover that for all $w \in W$ the vector $n(w)$ is orthogonal to $g'(w)(H)$, show that

$$\langle f'(w, r)(u, s) \mid n(w) \rangle = s \qquad \forall w \in W,\ (u, s) \in H \times \mathbb{R}.$$

3) Given $v \in X$, one takes $u := w'(x)(v)$, $s := r'(x)(v)$. Show that

$$\langle f'(w(x), r(x))(u, s) \mid n(w(x)) \rangle = s.$$

Conclude that

$$r'(x)(v) = \langle n(w(x)) \mid v \rangle,$$

and $\|r'(x)\|$ $(= \|n(w(x))\|) = 1$.

5.5.6 *Critical Points

For a function $f : W \to \mathbb{R}$ of class C^1 on an open subset W of a Banach space X a point $\overline{x} \in W$ such that $f'(\overline{x}) = 0$ is called a *critical point* of f. The set C_f of critical points of f plays an important role: for several phenomena it is more important than the set of minimizers of f or the set of maximizers of f. Of course, such sets are subsets of C_f; but even if $\overline{x} \in C_f$ is neither a local minimizer nor a local maximizer of f, one knows that f does not change much around $\overline{x}$ and that is already a noticeable property often called stationarity.

Under an additional nondegeneracy assumption, the behavior of f around a critical point can be described in a very simple manner. Let us note that it offers an analogy with the case $f'(\overline{x}) \neq 0$ for which one can find a neighborhood V of $\overline{x}$ and a C^1-diffeomorphism $h : V \to U$ onto a neighborhood U of 0 such that $h(\overline{x}) = 0$ and $f(x) = f(\overline{x}) + \ell(h(x))$ for $x \in V$ with $\ell := f'(\overline{x})$. Here, in contrast, we suppose $f'(\overline{x}) = 0$ and that $\overline{x}$ is a *nondegenerate critical point* in the sense that the second derivative $b := f''(\overline{x})$ of f at $\overline{x}$ is such that the map $x \mapsto b(x, \cdot)$ is a linear isomorphism from X onto X^*. Again, a change of variables yields a very simple form of the function.

Theorem 5.23 (Morse-Palais Lemma) *Let $f : W \to \mathbb{R}$ be a function of class C^n ($n \geq 3$) on an open subset W of a Hilbert space X with inner product $\langle \cdot \mid \cdot \rangle$. Let $\overline{x} \in W$ be a nondegenerate critical point of f and let $b := f''(\overline{x})$. Then there exist an open neighborhood $V \subset W$ of $\overline{x}$, an open neighborhood U of 0 in X and a C^{n-2}-diffeomorphism $h : V \to U$ such that $h(\overline{x}) = 0$, $h'(\overline{x}) = I_X$ and*

$$f(x) = f(\overline{x}) + b(h(x), h(x)) \qquad \forall x \in V.$$

Thus, by "the change of variables" h, f can be considered as a simple quadratic form.

Proof Without loss of generality we may assume $\overline{x} = 0$, $f(0) = 0$ and that W is a ball centered at 0. Then

$$f(x) = \int_0^1 f'(tx)x dt \qquad x \in W,$$

and, similarly, replacing f with f',

$$f'(tx) = \int_0^1 f''(stx)txds \qquad x \in W, \ t \in [0, 1],$$

so that $f(x) = \langle G(x)(x) \mid x \rangle$ with

$$G(x) = \int_0^1 \int_0^1 f''(stx)tdsdt.$$

Here G is a map of class C^{n-2} from W into the space $L_s^2(X, X^*)$ of symmetric linear maps from X into X^* identified with the space of continuous symmetric bilinear forms on X. By definition, $B := G(0) = f''(0)$ is an isomorphism of X onto X^*. Thus, for x close to 0, $G(x)$ is an isomorphism of X onto X^*. Let $H : W \to L_s(X, X)$ be given by

$$H(x) := B^{-1} \circ G(x),$$

so that, for x close to 0, $H(x)$ is an isomorphism of X onto X close to I_X and since B and $G(x)$ are symmetric

$$H(x)^* = G(x) \circ B^{-1},$$
$$H(x)^* \circ B = B \circ H(x).$$

Let $R(x)$ be the square root of $H(x)$, so that $R(x) \circ R(x) = H(x)$. Since $R(x)$ is the sum of a convergent series in $H(x)$ and $R(x)^*$ is the sum of a convergent series in $H(x)^*$, we have

$$R(x)^* \circ B = B \circ R(x),$$

hence

$$R(x)^* \circ B \circ R(x) = B \circ R(x) \circ R(x) = B \circ H(x) = G(x).$$

Setting $h(x) := R(x)(x)$ we get

$$\begin{aligned} \langle B(h(x)) \mid h(x) \rangle &= \langle B(R(x)(x)) \mid R(x)(x) \rangle = \langle (R(x)^* \circ B \circ R(x))(x) \mid x \rangle \\ &= \langle G(x)(x) \mid x \rangle = f(x). \end{aligned}$$

Finally, h is of class C^{n-2} and $h'(0) = R(0)$, the square root of $H(0) = I_X$, so that h defines a diffeomorphism as required. □

Exercise Using the spectral decomposition of B, show that one can find two closed subspaces Y and Z of X endowed with compatible norms such that $X = Y \oplus Z$,

$Z = Y^{\perp}$ and for $x := y + z$ with $(y, z) \in Y \times Z$

$$f(h^{-1}(x)) = f(\overline{x}) + \|y\|^2 - \|z\|^2 .$$

Exercise For $r \in \mathbb{R}$, classify the critical points of $f_r : \mathbb{R}^2 \to \mathbb{R}$ defined by $f_r(x, y) := x^3 - y^3 + 3rxy$ as local maximizers, minimizers or saddle points.

The search of critical points is facilitated by the following condition.

Definition 5.16 A differentiable function $f : W \to \mathbb{R}$ is said to satisfy the *Palais-Smale condition* (PS_c) for some value $c \in \mathbb{R}$ if any sequence (x_n) of W such that $(f(x_n)) \to c$ and $(f'(x_n)) \to 0$ has a convergent subsequence. It satisfies condition (PS) if for all $c \in \mathbb{R}$ it satisfies condition (PS_c).

Such a condition is a kind of compactness condition that involves f itself rather than the space X: of course, a coercive differentiable function on a finite dimensional space satisfies condition (PS). But condition (PS) can also be satisfied for interesting functions on infinite dimensional spaces. Such a fact has been used for the study of partial differential equations.

An important consequence of condition (PS_c) is the following *deformation property* around a level c which is not a critical value of f, i.e. an element of $f(C_f)$, where C_f is the set of critical points of f. It is obtained by taking the flow of an appropriate vector field. We admit it. Here we say that f is of *class* $C^{1,1}$ if f is of class C^1 and if its derivative is Lipschitzian on bounded subsets. Also, for $r \in \mathbb{R}$, we set $W_r := \{w \in W : f(w) \leq r\}$.

Theorem 5.24 *Suppose that $f : W \to \mathbb{R}$ is of class $C^{1,1}$ and satisfies condition (PS_c) for some $c \in \mathbb{R}$ such that $C_f \cap f^{-1}(c) = \varnothing$. Then for all $\varepsilon > 0$ sufficiently small there exists some $\delta \in]0, \varepsilon[$ and a homotopy $h : W \times [0, 1] \to W$, i.e. a continuous map satisfying $h(\cdot, 0) = I_W(\cdot)$, the identity map and the following properties:*

(a) $h(w, 1) = w$ for all $w \in W \backslash f^{-1}([c - \varepsilon, c + \varepsilon])$;
(b) $h(w, 1) \in W_{c-\delta} := f^{-1}(] - \infty, c - \delta])$ for all $w \in W_{c+\delta}$;
(c) $f(h(w, t)) \leq f(w)$ for all $(w, t) \in W \times [0, 1]$.

This deformation property can be used to obtain the following rather intuitive result known as the *Mountain Pass Theorem*. Hikers will not be surprised by it (Fig. 5.3).

Theorem 5.25 (Ambrosetti-Rabinowitz) *Let W be an open subset of a Banach space X and let $f : W \to \mathbb{R}$ be of class $C^{1,1}$ and such that for some $w_0 \in W$, $r \in]0, d(w_0, X \backslash W)[$, $w_1 \in W \backslash B[w_0, r]$, $m > \max(f(w_0), f(w_1))$ one has $f(w) \geq m$ for all $w \in S(w_0, r) := \{w \in X : \|w - w_0\| = r\}$. If f satisfies condition (PS_c) for*

$$c := \inf_{g \in G} \max_{t \in T} f(g(t))$$

with $T := [0, 1]$, $G := \{g \in C(T, W) : g(0) = w_0,\ g(1) = w_1\}$, then c is a critical value of f, i.e. $c = f(\overline{w})$ for some critical point $\overline{w}$ of f.

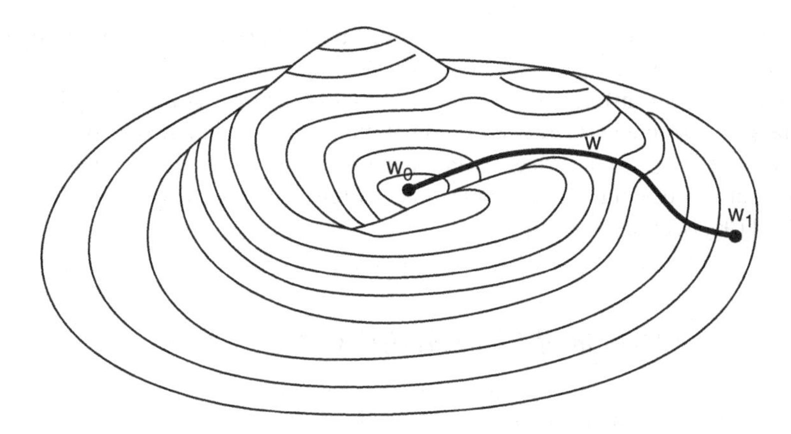

Fig. 5.3 The Mountain Pass Theorem

Proof The Custom Lemma 2.5 (or just the intermediate value theorem) ensures that for all $g \in G$ there exists some $t_g \in T$ such that $g(t_g) \in S(w_0, r)$. Thus, for all $g \in G$ one has $\max_{t \in T} f(g(t)) \geq m$, hence $c \geq m$.

Assuming that $C_f \cap f^{-1}(c) = \varnothing$, i.e. that c is not a critical value of f, we will obtain a contradiction with the deformation property in which we take $\varepsilon > 0$ satisfying $\varepsilon < m - \max(f(w_0), f(w_1))$. Let $\delta \in]0, \varepsilon[$ and let $h : W \times [0, 1] \to W$ be a homotopy satisfying conditions (a), (b), (c) of Theorem 5.24. Let $g_\delta \in G$ be such that $\max_{t \in T} f(g_\delta(t)) \leq c + \delta$ and let $g := h(g_\delta(\cdot), 1)$. Since h is continuous and $h(w_0, 1) = w_0$, $h(w_1, 1) = w_1$, we have $g \in G$, whence $\max_{t \in T} f(g(t)) \geq c$. However, since $g_\delta(T) \subset W_{c+\delta}$ by our choice of g_δ and since $h(w, 1) \in W_{c-\delta}$ for all $w \in W_{c+\delta}$ by condition (b) of Theorem 5.24, we get $\max_{t \in T} f(g(t)) \leq c - \delta$, a contradiction with the definition of c. □

The following example shows that one cannot drop assumption (PS_c).

Example (Brézis-Nirenberg) Let $f : \mathbb{R}^2 \to \mathbb{R}$ be given by $f(x, y) = x^2 + (1-x)^3 y^2$, let $w_0 := (0, 0)$, $w_1 := (2, 2)$, $r := 1/2$. Then one can show that $f(w_0) = 0$, $f(w_1) = 0$, $\inf f(r\mathbb{S}_1) > 0$ but 0 is the unique critical point of f. Note that one can find a sequence (w_n) such that $(f'(w_n)) \to 0$, $(f(w_n)) \to c$ but $(\|w_n\|) \to \infty$.

In Sect. 9.4 an application of the Mountain Pass Theorem will be given to semilinear boundary-value problems.

Exercises

1. Let S be a subset of a proper affine subspace A of $\mathbb{R}^d$ contained in a ball with radius r and for $\varepsilon > 0$ let $S_\varepsilon := S + \varepsilon B$, where B is the closed unit ball for the norm $\|\cdot\|_\infty$. Prove the following estimate for the outer Lebesgue measure

λ_d of S_ε: $\lambda_d(S_\varepsilon) \leq 2^d(r+\varepsilon)^{d-1}\varepsilon$. [Hint: use a translation and an orthogonal transformation.]

2. **(Sard's Theorem)** Let W be an open subset of $\mathbb{R}^d$ and let $f : W \to \mathbb{R}^d$ be of class C^1. Denote by C the set of critical points of f, i.e. the set of points $z \in W$ such that $Df(z)$ is not surjective. Prove that the set $f(C)$ of critical values of f has measure zero. [Hint: use the preceding exercise and the Mean Value Theorem. See [190, Section 13.5.1].]

5.5.7 *The Method of Characteristics

Let us consider the nonlinear partial differential equation

$$F(w, Du(w), u(w)) = 0, \qquad w \in W_0, \tag{5.27}$$

where W is a reflexive Banach space, W_0 is an open subset of W whose boundary ∂W_0 is a submanifold of class C^2 and $F : (w,p,z) \mapsto F(w,p,z)$ is a function of class C^2 on $W_0 \times W^* \times \mathbb{R}$. We look for a solution u of class C^2 satisfying the boundary condition

$$u \mid \partial W_0 = g, \tag{5.28}$$

where $g : \partial W_0 \to \mathbb{R}$ is a given function of class C^2. We leave apart the question of compatibility conditions for the data (F, g). The method of characteristics consists in associating to (5.27) a system of ordinary differential equations (in which W^{**} is identified with W) called the *system of characteristics*:

$$w'(s) = D_pF(w(s), p(s), z(s)) \tag{5.29}$$

$$p'(s) = -D_wF(w(s), p(s), z(s)) - D_zF(w(s), p(s), z(s))p(z) \tag{5.30}$$

$$z'(s) = \langle D_pF(w(s), p(s), z(s)), p(s)\rangle. \tag{5.31}$$

Suppose a smooth solution u of (5.27) is known. Let us relate it to a solution $s \mapsto (w(s), p(s), z(s))$ of the system (5.29)–(5.31). Let

$$q(s) := Du(y(s)), \qquad r(s) := u(y(s))$$

where $y(\cdot)$ is the solution of the differential equation

$$y'(s) := D_pF(y(s), Du(y(s)), u(y(s))), \qquad y(0) = w_0.$$

Then

$$r'(s) = Du(y(s)).y'(s) = \langle q(s), D_pF(y(s), q(s), r(s))\rangle$$

For all $e \in W$, identifying W^{**} and W we have

$$q'(s).e = D^2u(y(s)).y'(s).e = \langle D_pF(y(s), q(s), r(s)), D^2u(y(s)).e\rangle.$$

Now, taking the derivative of the function $F(\cdot, Du(\cdot), u(\cdot))$ and writing u, Du instead of $u(w)$, $Du(w)$, we have

$$D_wF(w, Du, u)e + D_pF(w, Du, u)D^2u(w).e + D_zF(w, Du, u)Du(w)e = 0.$$

Thus, replacing (w, Du, u) by $(y(s), q(s), r(s))$ and noting that e is arbitrary in W, we get

$$q'(s) = -D_wF(w(s), q(s), r(s)) - D_zF(w(s), q(s), r(s))q(s).$$

It follows that $s \mapsto (y(s), q(s), r(s))$ is a solution of the characteristic system. Taking the same initial data $(w_0, p_0, g(w_0))$, by uniqueness of the solution of the characteristic system, we get $y(s) = w(s)$, $p(s) = q(s)$ and $z(s) = r(s) := u(w(s))$. This means that knowing the solution of the characteristic system, we get the value of u at $w(s)$. If around some point $\overline{w} \in W_0$ we can represent any point w of a neighborhood of $\overline{w}$ as the value $w(s)$ for the solution of (5.29)–(5.31) issued from some initial data, then we get u around $\overline{w}$. In the following classical example, the search for the initial data is particularly simple.

Example: quasilinear equations Let $W := \mathbb{R}^n$, $W_0 := \mathbb{R}^{n-1} \times \mathbb{P}$, F being given by $F(w, p, z) := p.b(w, z) - c(w, z)$, where $b : W_0 \times \mathbb{R} \to W$, $c : W_0 \times \mathbb{R} \to \mathbb{R}$. Then, taking into account the relation $D_pF(w(s), p(s), z(s)).p(s) = p(s).b(w(s), z(s)) = c(w(s), z(s))$, equations (5.29), (5.31) of the characteristic system read as a system in (w, z):

$$w'(s) = b(w(s), z(s))$$
$$z'(s) = c(w(s), z(s)).$$

In the case $b := (b_1, \ldots, b_n)$ is constant with $b_n \neq 0$ and $c(w, z) := z^{k+1}/k$, with $k > 0$, the solution of this system with initial data $((v, 0), g(v)) \in \mathbb{R}^n \times \mathbb{P}$ is given by

$$w_i(s) = b_is + v_i \ (i = 1, \ldots, n-1), \quad w_n(s) = b_ns, \qquad z(s) = \frac{g(v)}{(1 - g(v)^k s)^{1/k}}.$$

It is defined for s in the interval $S := [0, g(v)^{-k}[$. Given $x := (x_1, \ldots, x_n) \in W_0$ near $\overline{x} \in W_0$, the initial data v is found by solving the equations $b_is + v_i = x_i$ $(i \in \mathbb{N}_{n-1})$, $x_n = b_ns$: $v_i = x_i - a_ix_n$ with $a_i := b_i/b_n$. The preceding shows that u is given by

$$u(x) = \frac{g(x_1 - a_1x_n, \cdots, x_{n-1} - a_{n-1}x_n)}{(1 - g(x_1 - a_1x_n, \cdots, x_{n-1} - a_{n-1}x_n)^k x_n/b_n)^{1/k}}$$

for x in the set $\{(x_1, \cdots, x_n) : x_ng(x_1 - a_1x_n, \cdots, x_{n-1} - a_{n-1}x_n)^k < b_n\}$. □

A special case of equation (5.27) is of great importance. It corresponds to the case $w := (x, t) \in W_0 := U \times]0, \tau[$ for some $\tau \in]0, +\infty]$ and some open subset U of a hyperplane X of W and $F((x, t), (y, v), z) := v + H(x, t, y, z)$, so that equation (5.27) and the boundary condition (5.28) take the form

$$D_t u(x, t) + H(x, t, D_x u(x, t), u(x, t)) = 0 \qquad (x, t) \in W_0 \times]0, \tau[\tag{5.32}$$

$$u(x, 0) = g(x) \qquad x \in W_0. \tag{5.33}$$

Such a system is called a *Hamilton-Jacobi equation*.

Let us note that, as in the example of quasilinear equations, the general case can be reduced to this form under a mild condition. First, since W_0 is the interior of a smooth manifold with boundary, taking a chart, we may assume for a local study that $W_0 = U \times]0, \tau[$ for some $\tau > 0$ and some open subset U of a hyperplane X of W. Now, using the implicit function theorem around $\overline{w} \in \partial W_0$, F can be reduced to the form $F((x, t), (y, v), z) := v + H(x, t, y, z)$ provided $D_v F(\overline{w}, \overline{p}, \overline{z}) \neq 0$. Such a condition can be expressed intrinsically (i.e., without using the chart) by finding a vector $\overline{v}$ transverse to ∂W_0 at $\overline{w}$ such that $D_w F(\overline{w}, \overline{y}, \overline{z}).\overline{v} \neq 0$.

The characteristic system associated with (5.32) can be reduced to

$$x'(s) = D_y H(x(s), s, y(s), z(s)) \tag{5.34}$$

$$y'(s) = -D_x H(x(s), s, y(s), z(s)) - D_z H(x(s), s, y(s), z(s)) y(s) \tag{5.35}$$

$$z'(s) = D_y H(x(s), s, y(s), z(s)).y(s) - H(x(s), s, y(s), z(s)) \tag{5.36}$$

by dropping the equation $t'(s) = 1$ and noting that an equation for $D_t u(x(s), t(s))$ is not needed since this derivative is known to be $-H(x(s), s, y(s), z(s))$. In order to take into account the dependence on the initial condition $(v, Dg(v), g(v))$, the one-jet of g at $v \in U \subset X$, let us denote by $s \mapsto (\hat{x}(s, v), \hat{y}(s, v), \hat{z}(s, v))$ the solution to the system (5.34)–(5.36). Since the right-hand side of this system is of class C^1, the theory of differential equations ensures that the solution is a mapping of class C^1 in (s, v). In view of the initial data, we have

$$\forall v \in U,\ v' \in X \qquad D_v \hat{x}(0, v) v' = v'.$$

It follows that for all $\overline{v} \in U$ there exist a neighborhood V of $\overline{v}$ in U and some $\sigma \in]0, \tau[$ such that, for $s \in]0, \sigma[$, the map $\hat{x}_s : v \mapsto \hat{x}(s, v)$ is a diffeomorphism from V onto $V_s := \hat{x}(s, V)$. From the preceding analysis, we get that for $x \in V_s$ one has $u(x, s) = \hat{z}(s, v)$ with $v := (\hat{x}_s)^{-1}(x)$. Thus we get a local solution to the system (5.32)–(5.33). In general, one cannot get a global solution with such a method: it may happen that for two values v_1, v_2 of v the characteristic curves issued from v_1 and v_2 take the same value for some $t > 0$.

Exercises

1. Write down the characteristic system for the *conservation law*

$$D_t u(x,t) + D_x u(x,t).b(u(x,t)) = 0, \qquad u(v,0) = g(v),$$

where $b : \mathbb{R} \to X$, $g : X \to \mathbb{R}$ are of class C^1. Verify that its solution satisfies $\hat{x}(s,v) = v + sb(g(v))$, $z(s,v) = g(v)$. Compute $D_v\hat{x}(s,v)$ and show that for all $\overline{v} \in X$, this element of $L(X,X)$ is invertible for (s,v) close enough to $(0,\overline{v})$. Deduce a local solution of the conservation law equation from this property.

2. (**Haar's Uniqueness Theorem**) Suppose $X = \mathbb{R}$ and $H : X \times \mathbb{R} \times X^* \times \mathbb{R} \to \mathbb{R}$ satisfies the Lipschitz condition with constants k, ℓ

$$\left|H(x,t,y,z) - H(x,t,y',z')\right| \le k\left|y - y'\right| + \ell\left|z - z'\right|,$$

for $(x,t,y,y',z,z') \in T \times \mathbb{R}^4$, where, for some constants a, b, c, T is the triangle $T := \{(x,t) \in X \times [0,a] : x \in [b + \ell t, c - \ell t]\}$. Show that if u_1, u_2 are two solutions of class C^1 in T of the system (5.32)–(5.33), then $u_1 = u_2$.

3. Suppose $X = \mathbb{R}$, $g = I_X$ and $H : X \times \mathbb{R} \times X^* \times \mathbb{R} \to \mathbb{R}_\infty$ is given by $H(x,t,y,z) := |t-1|^{-1/2} y$ for $t \in [0,1[$, $+\infty$ otherwise. Using the method of characteristics, show that a solution to the system (5.32)–(5.33) is given by $u(x,t) = x - 2 + 2\sqrt{1-t}$ for $(x,t) \in X \times]0,1[$.

4. Suppose $X = \mathbb{R}$ and g and H are given by $g(x) := x^2/2$ and $H(x,t,y,z) := -y^2/2$. Using the method of characteristics, show that a solution to the system (5.32)–(5.33) is given by $u(x,t) = x^2/2(1-t)$ for $(x,t) \in X \times (0,1)$.

5. Suppose $X = \mathbb{R}$, g and H are given by $g(x) := x$, $H(x,t,y,z) := e^{-3t}yz(a'(t)e^{2t} + b'(t)z^2) - z$, where a and b are nonnegative functions of class C^1 satisfying $a(0) = 1$, $b(0) = 0$, $a + b > 0$. Show that the characteristics associated with the system (5.32)–(5.33) satisfy $\hat{x}(t,v) = a(t)v + b(t)v^3$, $\hat{z}(t) = e^t v$, so that $v \mapsto \hat{x}(t,v)$ is a bijection. Assuming that there exists some $\tau > 0$ such that $a(t) = 0$ for $t \ge \tau$, show that $u(x,t) = e^t b(t)^{-1/3} x^{1/3}$ for $(x,t) \in X \times [\tau, \infty[$ so that u is not differentiable at $(0,t)$.

6. Suppose $X = \mathbb{R}$ and g and H are given by $g(x) := x^2/2$, $H(x,t,y,z) := a'(t)e^{-t}y^2/2 + b'(t)e^{-3t}y^4 - z$, where a and b are as in the preceding exercise. Show that the characteristics associated with the system (5.32)–(5.33) satisfy $\hat{x}(t,v) = a(t)v + 4b(t)v^3$, $\hat{z}(t) = e^t(a(t)v^2/2 + 3b(t)v^4)$, so that for $t \ge \tau$, $v \mapsto \hat{x}(t,v)$ is a bijection on a neighborhood of 0, in spite of the fact that $D_v\hat{x}(t,0) = 0$ and $u(x,t) = 3.4^{-4/3}b(t)x^{4/3}$, so that u is of class C^1 but not C^2 around $(0,t)$.

7*. Let W be an open subset of a Banach space X, let $F : W \to X$ be of class C^1, and let $u : \mathbb{R} \times W \to W$ be its flow, so that $D_1 u = F \circ u$. Show that $D_2 u.F = F \circ u$. Given $g \in C^1(W,\mathbb{R})$, let $f := g \circ u$. Show that f satisfies $D_1 f(t,x) - D_2 f(t,x)F(x) = 0$ and $f(0,x) = g(x)$ for all $x \in W$, $t \in \mathbb{R}$.

For $X := \mathbb{R}^d$, $x := (x_1, \ldots, x_d)$, solve the equation $\frac{\partial f}{\partial t}(t,x) = \Sigma_{i=1}^d x_i \frac{\partial f}{\partial x_i}(t,x)$.

5.6 Applications to Optimization

For unconstrained optimization, differential calculus offers easily obtained criteria. Constrained minimization (or maximization) requires more attention and we have to deal with some preliminaries. We consider it in the form of the problem

$$(\mathcal{P}) \qquad \text{minimize } f(x) \text{ under the constraint } x \in F,$$

where F is a nonempty subset of a normed space X called the *feasible set* or the *admissible set* and where $f : X \to \mathbb{R}$ is the objective function.

5.6.1 Unconstrained Minimization

In the case when the feasible set is the whole space or an open subset W of a normed space X, differential calculus can readily be applied to the minimization problem (for the maximization problem one uses $-f$ instead of f).

Proposition 5.29 *Let $\overline{x} \in W$ be a local minimizer of a function $f : W \to \mathbb{R}$, i.e. such that for some neighborhood V of $\overline{x}$ one has $f(\overline{x}) \leq f(v)$ for all $v \in V$. Then, if f is differentiable at $\overline{x}$ one has $f'(\overline{x}) = 0$.*

If f is twice differentiable at $\overline{x}$ then one has $f''(\overline{x})(x,x) \geq 0$ for all $x \in X$.

Proof For the first assertion it suffices to assume f is Gateaux differentiable at $\overline{x}$ since for all $v \in X$ one has $(1/t)(f(\overline{x}+tv)-f(\overline{x})) \geq 0$ for $t > 0$ small enough, hence, passing to the limit as $t \to 0_+$ $d_r f(\overline{x}, v) \geq 0$. Since similarly we have $d_r f(\overline{x}, -v) \geq 0$ we conclude that $D_r f(x) = 0$.

When f is twice differentiable at $\overline{x}$, Theorem 5.9 asserts that

$$r(x) = f(\overline{x}+x) - f(\overline{x}) - f'(\overline{x})(x) - (1/2)f''(\overline{x})(x,x)$$

is a remainder of order two. Thus, for all $v \in X$, $\lim_{t \to 0_+} t^{-2} r(tx) = 0$ and

$$f''(\overline{x})(v,v) = \lim_{t \to 0_+} \frac{2}{t^2}(f(\overline{x}+tv) - f(\overline{x})) \geq 0$$

since $f'(\overline{x}) = 0$. □

A corresponding sufficient condition requires a reinforced assumption.

Proposition 5.30 *Let $f : W \to \mathbb{R}$ be twice differentiable at $\overline{x} \in W$ and such that $f'(\overline{x}) = 0$ and, for some $c > 0$, $f''(\overline{x})(x,x) \geq c \|x\|^2$ for all $x \in X$. Then $\overline{x}$ is a local strict minimizer of f.*

Note that the assumption on $f''(\overline{x})$ is satisfied if $f''(\overline{x})$ is a nondegenerate positive bilinear form.

Proof Given $\varepsilon \in]0, c/2[$ one can find some $\delta > 0$ such that the remainder r given by $r(x) := f(\overline{x}+x) - f(\overline{x}) - f'(\overline{x})x - (1/2)f''(\overline{x})(x,x)$ satisfies $|r(x)| \leq \varepsilon \|x\|^2$ when $x \in \delta B_X$. Then, our assumptions entail that $f(v) \geq f(\overline{x}) + (c/2 - \varepsilon)\|v - \overline{x}\|^2$ for $v \in B[\overline{x}, \delta]$, hence $f(v) > f(\overline{x})$ for $v \in B[\overline{x}, \delta]\backslash\{\overline{x}\}$. □

Other criteria could be given using higher order derivatives.

Hereafter we will formulate optimality conditions for the problem with constraints. These conditions will involve the concept of a normal cone.

Exercises

1. Give examples of functions and of critical points that are not local minimizers.
2. Give an example of a function f on $\mathbb{R}^2$ such that $Df(0) = 0$, $D^2f(0).v.v \geq 0$ but 0 is not a local minimizer of f.
3. Given $(x, y) \in \mathbb{R}^2$, find $t \in \mathbb{R}$ such that $(\cos t - x)^2 + (\sin t - y)^2 \leq (\cos r - x)^2 + (\sin r - y)^2$ for all $r \in \mathbb{R}$. Give a geometric interpretation.
4. Let X be a Euclidean space or a Hilbert space, let $a_1, a_2 \in X$, and let $u_1, u_2 \in X\backslash\{0\}$. Solve the minimization problem of $\|x_1 - x_2\|$ for $x_1 \in a_1 + \mathbb{R}u_1$, $x_2 \in a_2 + \mathbb{R}u_2$. Consider in particular the case $X = \mathbb{R}^3$.
5. For $d \geq 2$, let f be the polynomial function given by
$$f(x) = (x_d + 1)^3(x_1^2 + \ldots + x_{d-1}^2) + x_d^2.$$
Show that 0 is the unique critical point of f and that 0 is a local strict minimizer of f, but not a global minimizer.
6. Let X be a Euclidean space or a Hilbert space and let $f : X \to \mathbb{R}$ be a function of class C^2. Suppose that for some $c > 0$ and some $\overline{x} \in X$ one has $f(x) \geq f(\overline{x}) + c\|x - \overline{x}\|^2$. Show that $Df(\overline{x}) = 0$ and $D^2f(\overline{x}).v.v \geq 2c\|v\|^2$ for all $v \in X$. Prove that there exist $a > 0$ and a neighborhood U of $\overline{x}$ such that $\|\nabla f(x)\| \geq a\|x - \overline{x}\|$ for all $x \in U$. Deduce from this the existence of some $b > 0$ and some neighborhood V of $\overline{x}$ such that $f(x) \leq \inf f(X) + b\|\nabla f(x)\|^2$ for all $x \in V$.
7. **Gradient algorithm**. Let X be a Euclidean space or a Hilbert space and let $f : X \to \mathbb{R}$ be a function of class C^1 whose gradient is Lipschitzian with rate c. Suppose that for some $b > 0$ one has $f(x) \leq m + b\|\nabla f(x)\|^2$ for all $x \in X$, with $m := \inf f(X)$. Given $x_0 \in X$ and a sequence (t_n) in $]0, 1/c[$ one defines inductively a sequence (x_n) by setting $x_{n+1} = x_n - t_n\nabla f(x_n)$. Show that $f(x_{n+1}) - f(x_n) \leq -\frac{1}{2}t_n\|\nabla f(x_n)\|^2$ for all $n \in \mathbb{N}$ and that if for some $q \in]0, 1[$ one has $t_n \geq 2(1 - q)$ for all $n \in \mathbb{N}$, then one has $f(x_n) - m \leq q^n(f(x_0) - m)$ and $(f(x_n)) \to m$.

   Prove that $\|x_{n+1} - x_n\|^2 \leq 2t_n(f(x_n) - f(x_{n+1})) \leq 2t_n(f(x_n) - m)$ for all $n \in \mathbb{N}$. Setting $a := (\frac{2}{c}(f(x_0) - m))^{1/2}$ deduce from the preceding inequalities that $\|x_{n+1} - x_n\| \leq aq^{n/2}$ for all n. Conclude that (x_n) converges to some minimizer $\overline{x}$ of f.

8. **Gradient algorithm with optimal step.** Let $f : \mathbb{R}^d \to \mathbb{R}$ be the quadratic function given by $f(x) := \frac{1}{2}\langle Ax \mid x\rangle + \langle b \mid x\rangle + c$, where A is a positive definite matrix, $b \in \mathbb{R}^d$ and $c \in \mathbb{R}$. Show that f has a unique minimizer $\overline{x}$. Given $x_0 \in \mathbb{R}^d$, consider the sequence defined inductively by $x_{n+1} = x_n + t_n d_n$, where $d_n := -\nabla f(x_n)$ and $t_n := \|d_n\| / \langle Ax_n \mid x_n\rangle$ minimizes the function $t \mapsto f(x_n + td_n)$ on $\mathbb{R}$. Verify that $d_{n+1} = d_n - t_n A d_n$ and $\langle d_{n+1} \mid d_n\rangle = 0$ for all $n \in \mathbb{N}$.

   Let $m := \inf f(\mathbb{R}^d)$ and let $\gamma := \lambda_1/\lambda_d$ be the *conditioning* of A, where $\lambda_1 \geq \lambda_2 \geq \ldots \geq \lambda_d$ is the sequence in eigenvalues of A. Show that

$$f(x_{n+1}) - m = (f(x_n) - m)\left[1 - \frac{\|d_n\|^2}{\langle Ad_n \mid d_n\rangle\langle A^{-1}d_n \mid d_n\rangle}\right]$$

$$f(x_n) - m \leq (f(x_0) - m)\frac{(\gamma-1)^{2n}}{(\gamma+1)^{2n}}$$

$$\|x_n - \overline{x}\| \leq 2^{1/2}\lambda_d^{-1/2}(f(x_0) - m)^{1/2}\frac{(\gamma-1)^n}{(\gamma+1)^n}.$$

   [Hint: use Kantorovich's inequality: $\langle Ax \mid x\rangle\langle A^{-1}x \mid x\rangle \leq \frac{1}{4}(\gamma^{1/2} + \gamma^{-1/2})^2 \|x\|^4$ for all $x \in \mathbb{R}^d$.]

9. Let $p : \mathbb{R} \to \mathbb{R}$ be a polynomial function with positive values. If n is its degree, show that the polynomial function $q := p + p' + \ldots + p^{(n)}$ takes its values in $\mathbb{R}_+$. [Hint: observe that the term of highest degree in q is the term of highest degree in p, so that q attains its minimum at some $\overline{r} \in \mathbb{R}$ and that $q'(r) = q(r) - p(r)$, so that $q(r) \geq q(\overline{r}) = p(\overline{r})$.]

5.6.2 *Normal Cones, Tangent Cones, and Constraints*

In fact we will use some variants of the concept of normal cone that fit different differentiability assumptions on the function f. When the feasible set is a convex set these variants coincide (Exercise 6) and the concept is very simple (Fig. 5.4).

Definition 5.17 The *normal cone* $N(C, \overline{x})$ to a convex subset C of X at $\overline{x} \in C$ is the set of $\overline{x}^* \in X^*$ which attain their maximum on C at $\overline{x}$:

$$N(C, \overline{x}) := \{\overline{x}^* \in X^* : \forall x \in C \quad \langle \overline{x}^*, x - \overline{x}\rangle \leq 0\}.$$

Thus, when C is a linear subspace, $N(C, \overline{x}) = C^{\perp}$, where $C^{\perp}$ is the orthogonal of C (or annihilator of C) in X^*

$$C^{\perp} := \{\overline{x}^* \in X^* : \forall x \in C \quad \langle \overline{x}^*, x\rangle = 0\}.$$

When C is a cone, one has $N(C, 0) = C^0$, where C^0 is the polar cone of C.

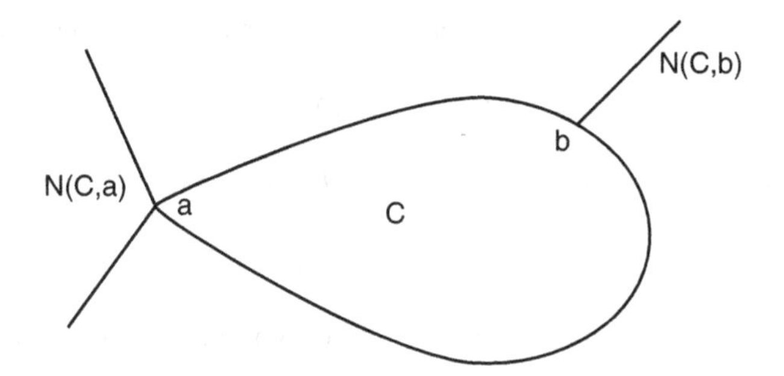

Fig. 5.4 The normal cone to a convex subset

In the nonconvex case the preceding definition has to be modified by introducing a remainder in the inequality in order to allow a certain curvature or inaccuracy.

Definition 5.18 The *firm or Fréchet normal cone* $N_F(S, \overline{x})$ to a subset S of X at $\overline{x} \in S$ is the set of $\overline{x}^* \in X^*$ for which there exists a remainder $r(\cdot)$ such that $\overline{x}^*(\cdot) - r(\cdot - \overline{x})$ attains its maximum on S at $\overline{x}$:

$$\overline{x}^* \in N_F(S, \overline{x}) \Longleftrightarrow \exists r \in o(X, \mathbb{R}) \quad \forall x \in S \quad \langle \overline{x}^*, x - \overline{x} \rangle \leq r(x - \overline{x}).$$

In other terms, $\overline{x}^* \in X^*$ is a firm normal to S at $\overline{x}$ iff for every $\varepsilon > 0$ there exists a $\delta > 0$ such that for all $x \in S \cap B(\overline{x}, \delta)$ one has $\langle \overline{x}^*, x - \overline{x} \rangle \leq \varepsilon \|x - \overline{x}\|$.

Equivalently

$$\overline{x}^* \in N_F(S, \overline{x}) \quad \Longleftrightarrow \quad \limsup_{x (x \in S) \to \overline{x},\ x \neq \overline{x}} \frac{1}{\|x - \overline{x}\|} \langle \overline{x}^*, x - \overline{x} \rangle \leq 0.$$

We will give some properties and calculus rules in the next subsection. For the moment it is important to convince oneself that this notion corresponds to the intuitive idea of an "exterior normal" to a set, for instance by making drawings in simple cases. We present a necessary condition using this concept without any delay. In it we say that f attains a *local maximum* (resp. *local minimum*) on F at $\overline{x}$ if $f(x) \leq f(\overline{x})$ (resp. $f(x) \geq f(\overline{x})$) for all x in some neighborhood of $\overline{x}$ in F. It is convenient to say that $\overline{x}$ is a *local maximizer* (resp. *local minimizer*) of f on F.

Theorem 5.26 (Fermat's Rule) *Suppose f attains a local maximum on F at $\overline{x}$ and is Fréchet differentiable at $\overline{x}$. Then*

$$f'(\overline{x}) \in N_F(F, \overline{x}).$$

If f attains a local minimum on F at $\overline{x}$ and is Fréchet differentiable at $\overline{x}$ then

$$0 \in f'(\overline{x}) + N_F(F, \overline{x}).$$

Proof Suppose f attains a local maximum on F at $\overline{x}$ and is differentiable at $\overline{x}$. Set

$$f(x) = f(\overline{x}) + \langle \overline{x}^*, x - \overline{x} \rangle + r(x - \overline{x})$$

with r a remainder, $\overline{x}^* := f'(\overline{x})$, so that for $x \in F$ close enough to $\overline{x}$ one has

$$\langle \overline{x}^*, x - \overline{x} \rangle + r(x - \overline{x}) = f(x) - f(\overline{x}) \leq 0.$$

Hence $\overline{x}^* \in N_F(F, \overline{x})$. Changing f into $-f$, one obtains the second assertion. □

The second formula shows how the familiar rule $f'(\overline{x}) = 0$ of unconstrained minimization has to be changed by introducing an additional term involving the normal cone. Without such an additional term the condition would be utterly invalid.

Example The identity map $f = I_{\mathbb{R}}$ on $\mathbb{R}$ attains its minimum on $F := [0, 1]$ at 0 but $f'(0) = 1$.

Example Suppose F is the unit sphere of the Euclidean space $\mathbb{R}^3$ representing the surface of the earth and suppose f is a smooth function representing the temperature. If f attains a local minimum on F at $\overline{x}$, in general $\nabla f(\overline{x})$ is not 0; however $\nabla f(\overline{x})$ is on the downward vertical at $\overline{x}$ and, if one can increase one's altitude at that point, one usually experiences a decrease of temperature. □

When the objective function f is not Fréchet differentiable but just Hadamard differentiable, an analogue of Fermat's Rule can still be given by introducing a variant of the notion of a firm normal cone. It goes as follows; although this variant appears to be more technical than the concept of a Fréchet normal cone, it is a natural and important notion. It can be formulated with the help of the notion of a directional remainder: $r : X \to Y$ is a *directional remainder* if for all $u \in X \backslash \{0\}$ one has $r(tv)/t \to 0$ as $t \to 0_+$, $v \to u$; we write $r \in o_D(X, Y)$.

Definition 5.19 The *normal cone* (or *directional normal cone*) to the subset F at $\overline{x} \in \text{cl}(F)$ is the set $N(F, \overline{x}) := N_D(F, \overline{x})$ of $\overline{x}^* \in X^*$ for which there exists a directional remainder $r(\cdot)$ such that $\overline{x}^*(\cdot) - r(\cdot - \overline{x})$ attains its maximum on F at $\overline{x}$:

$$\overline{x}^* \in N(F, \overline{x}) \Longleftrightarrow \exists r \in o_D(X, \mathbb{R}) \quad \forall x \in F \quad \langle \overline{x}^*, x - \overline{x} \rangle \leq r(x - \overline{x}).$$

In other terms, $\overline{x}^* \in X^*$ is a normal to F at $\overline{x}$ if and only if for all $u \in X \backslash \{0\}$ and $\varepsilon > 0$ there exists a $\delta > 0$ such that $\langle \overline{x}^*, v \rangle \leq \varepsilon$ for any $(t, v) \in]0, \delta] \times B(u, \delta)$ satisfying $\overline{x} + tv \in F$:

$$\overline{x}^* \in N(F, \overline{x}) \Longleftrightarrow \forall u \in X \qquad \limsup_{(t,v) \to (0_+, u),\ \overline{x} + tv \in F} \frac{1}{t} \langle \overline{x}^*, (\overline{x} + tv) - \overline{x} \rangle \leq 0.$$

Let us note that the case $u = 0$ can be discarded in the preceding reformulation because the condition is automatically satisfied in this case with $\delta = \varepsilon \min(1, \|\overline{x}^*\|^{-1})$. This cone often coincides with the Fréchet normal cone and it always contains it, as the preceding reformulations show.

Lemma 5.17 *For any subset F and any $\overline{x} \in \mathrm{cl}(F)$ one has $N_F(F,\overline{x}) \subset N(F,\overline{x})$.*

The duality property we prove now compensates the complexity of the definition of the (directional) normal cone compared to the definition of the firm normal cone.

Proposition 5.31 *The normal cone to F at $\overline{x}$ is the polar cone to the tangent cone to F at $\overline{x}$:*

$$\left(\overline{x}^* \in N(F,\overline{x})\right) \Leftrightarrow \left(\forall u \in T(F,\overline{x}) \quad \langle \overline{x}^*, u\rangle \leq 0\right).$$

Proof Given $\overline{x}^* \in N(F,\overline{x})$ and $u \in T(F,\overline{x})\backslash\{0\}$, for any $\varepsilon > 0$, taking $\delta \in]0,\varepsilon[$ such that $\langle \overline{x}^*, v\rangle \leq \varepsilon$ for any $(t,v) \in]0,\delta] \times B(u,\delta)$ satisfying $\overline{x} + tv \in F$ and observing that such a pair (t,v) exists since $u \in T(F,\overline{x})$, we get $\langle \overline{x}^*, u\rangle \leq \langle \overline{x}^*, v\rangle + \|\overline{x}^*\| \, \|u - v\| \leq \varepsilon + \varepsilon \, \|\overline{x}^*\|$. As ε is arbitrarily small, we get $\langle \overline{x}^*, u\rangle \leq 0$.

Conversely, given $\overline{x}^*$ in the polar cone of $T(F,\overline{x})$, given $u \in T(F,\overline{x})$ and given $\varepsilon > 0$, taking $\delta > 0$ such that $\delta \, \|\overline{x}^*\| \leq \varepsilon$, the inequality $\langle \overline{x}^*, v\rangle \leq \varepsilon$ holds whenever $t \in]0,\delta[$ and $v \in t^{-1}(F - \overline{x}) \cap B(u,\delta)$ since

$$\langle \overline{x}^*, v\rangle \leq \langle \overline{x}^*, u\rangle + \langle \overline{x}^*, v - u\rangle \leq \|\overline{x}^*\| \, \|u - v\| \leq \delta \, \|\overline{x}^*\| \leq \varepsilon.$$

If $u \in X\backslash T(F,\overline{x})$ we can find $\delta > 0$ such that no such pair (t,v) exists. Thus, we have $\langle \overline{x}^*, v\rangle \leq \varepsilon$ for any $(t,v) \in]0,\delta] \times B(u,\delta)$ satisfying $\overline{x} + tv \in F$: $\overline{x}^* \in N(F,\overline{x})$. □

Theorem 5.27 (Fermat's Rule) *Suppose f attains a local maximum on F at $\overline{x} \in F$ and is Hadamard differentiable at $\overline{x}$. Then, for all $v \in T(F,\overline{x})$ one has $f'(\overline{x})v \leq 0$:*

$$f'(\overline{x}) \in N(F,\overline{x}).$$

If f attains a local minimum on F at $\overline{x}$ then, for all $v \in T(F,\overline{x})$ one has $f'(\overline{x})v \geq 0$:

$$0 \in f'(\overline{x}) + N(F,\overline{x}).$$

Proof Let V be an open neighborhood of $\overline{x}$ in X such that $f(x) \leq f(\overline{x})$ for all $x \in F \cap V$. Given $v \in T(F,\overline{x})$, let $(v_n) \to v$, $(t_n) \to 0_+$ be sequences such that $\overline{x} + t_n v_n \in F$ for all $n \in \mathbb{N}$. For n large enough, we have $x_n := \overline{x} + t_n v_n \in F \cap V$, hence $f(\overline{x} + t_n v_n) - f(\overline{x}) \leq 0$. Dividing by t_n and passing to the limit, the (Hadamard) differentiability of f at $\overline{x}$ yields $f'(\overline{x})(v) \leq 0$. □

It is possible to give a third version of Fermat's Rule that does not assume that f is differentiable; it is set in the space X instead of its dual X^*. In it, we use the *directional (lower) derivative* (or *contingent derivative*) of f given by

$$f^D(\overline{x}, u) := \liminf_{(t,v)\to(0_+,u)} \frac{1}{t}(f(\overline{x} + tv) - f(\overline{x}))$$

and the tangent cone to F at $\overline{x}$ as introduced in Definition 5.15.

In view of their fundamental character, we will return to these notions of tangent and normal cones. For the moment, the definition itself suffices to give the primal version of the Fermat rule we announced. Note that this version entails the preceding theorem since $f^D(\overline{x}, \cdot) = f'(\overline{x})$ when f is Hadamard differentiable at $\overline{x}$.

Theorem 5.28 *Suppose f attains a local maximum on F at $\overline{x}$. Then*

$$f^D(\overline{x}, u) \leq 0 \textit{ for all } u \in T(F, \overline{x}).$$

Proof Let $u \in T(F, \overline{x})$. There exist $(t_n) \to 0_+$, $(u_n) \to u$ such that $\overline{x} + t_n u_n \in F$ for all $n \in \mathbb{N}$. For n large enough we have $f(\overline{x} + t_n u_n) \leq f(\overline{x})$, so that

$$f^D(\overline{x}, u) \leq \liminf_n \frac{1}{t_n} (f(\overline{x} + t_n u_n) - f(\overline{x})) \leq 0.$$

□

For minimization problems, a variant of the tangent cone is required since the rule $f^D(\overline{x}, u) \geq 0$ for $u \in T(F, \overline{x})$ is not valid in general.

Example Let $F := \{0\} \cup \{2^{-2n} : n \in \mathbb{N}\} \subset \mathbb{R}$ and let $f : \mathbb{R} \to \mathbb{R}$ be even and given by $f(x) = 0$ for every $x \in F$, $f(2^{-2k+1}) = -2^{-2k+1}$, f being affine on each interval $[2^{-j}, 2^{-j+1}]$. Show that $f^D(\overline{x}, 1) = -1$ for $\overline{x} := 0$, although $f(\overline{x}) = \min f(F)$.

Definition 5.20 The *incident cone* (or *adjacent cone*) to F at $\overline{x} \in \mathrm{cl}(F)$ is the set

$$\begin{aligned} T^I(F, \overline{x}) &:= \{u \in X : \forall (t_n) \to 0_+, \exists (u_n) \to u, \quad \overline{x} + t_n u_n \in F \quad \forall n\} \\ &= \left\{u \in X : \forall (t_n) \to 0_+, \exists (x_n) \to \overline{x},\ (\frac{x_n - \overline{x}}{t_n}) \to u,\ x_n \in F\ \forall n\right\}. \end{aligned}$$

It is easy to show that

$$u \in T^I(F, \overline{x}) \Leftrightarrow \lim_{t \to 0_+} \frac{1}{t} d(\overline{x} + tu, F) = 0.$$

Let us also introduce the *incident derivative* of a function f at $\overline{x}$ by

$$f^I(\overline{x}, u) := \inf\{r \in \mathbb{R} : (u, r) \in T^I(E_f, \overline{x}_f)\},$$

where E_f is the epigraph of f and $\overline{x}_f := (\overline{x}, f(\overline{x}))$.

Proposition 5.32 *Suppose f is directionally stable at $\overline{x}$ in the sense that for all $u \in X \backslash \{0\}$ one has $(1/t)(f(\overline{x} + tv) - f(\overline{x} + tu)) \to 0$ as $(t, v) \to (0, u)$. If f attains a local minimum on F at $\overline{x}$ then*

$$f^I(\overline{x}, u) \geq 0 \textit{ for all } u \in T(F, \overline{x}),$$

$$f^D(\overline{x}, u) \geq 0 \textit{ for all } u \in T^I(F, \overline{x}).$$

Proof Suppose on the contrary that there exists some $u \in T(F,\overline{x})$ such that $f^I(\overline{x},u) < 0$. Then, there exists some $r < 0$ such that $(u,r) \in T^I(E_f,\overline{x}_f)$; thus, if $(t_n) \to 0_+$ and $(u_n) \to u$ are such that $\overline{x} + t_n u_n \in F$ for all $n \in \mathbb{N}$, one can find a sequence $((v_n,r_n)) \to (u,r)$ such that $\overline{x}_f + t_n(v_n,r_n) \in E_f$ for all $n \in \mathbb{N}$. Then $f(\overline{x}) + t_n r_n \geq f(\overline{x} + t_n v_n)$ for all $n \in \mathbb{N}$ and

$$0 > r \geq \limsup_n \frac{1}{t_n}(f(\overline{x} + t_n v_n) - f(\overline{x})) = \limsup_n \frac{1}{t_n}(f(\overline{x} + t_n u_n) - f(\overline{x}) \geq 0,$$

a contradiction. The proof of the second assertion is similar. □

Exercises

1. Given an element $\overline{x}$ of the closure of a subset F of a normed space X, show that the tangent cone and the incident cone can be expressed as follows:

$$T(F,\overline{x}) = \{v \in X : \liminf_{t\to 0_+} \frac{1}{t} d(\overline{x} + tv, F) = 0\},$$

$$T^I(F,\overline{x}) = \{v \in X : \lim_{t\to 0_+} \frac{1}{t} d(\overline{x} + tv, F) = 0\}.$$

2. Deduce from Exercise 1 that $v \in T(F,\overline{x})$ if and only if $v \in \limsup_{t\to 0_+} \frac{1}{t}(F - \overline{x})$ and that $v \in T^I(F,\overline{x})$ if and only if $v \in \liminf_{t\to 0_+} \frac{1}{t}(F - \overline{x})$, the limits of a family of sets being defined as in Exercise 8 of Sect. 2.3.1.
3. Find a subset F of $\mathbb{R}$ such that $1 \in T(F,0)$ but $T^I(F,0) = \{0\}$.
4. Show that if X is a finite dimensional normed space, then, for any subset F of X and any $\overline{x} \in \mathrm{cl}(F)$, one has $N(F,\overline{x}) = N_F(F,\overline{x})$.
5. Show that for any subset F of a normed space and any $\overline{x} \in \mathrm{cl}(F)$, the cones $N(F,\overline{x})$ and $N_F(F,\overline{x})$ are convex and closed.
6. Show that for any convex subset C of a normed space X and any $\overline{x} \in \mathrm{cl}(C)$ the cones $N(C,\overline{x})$ and $N_F(C,\overline{x})$ coincide with the normal cone in the sense of convex analysis described in Definition 5.17.
7. Let $f : \mathbb{R} \to \mathbb{R}$ be differentiable at $a \in \mathbb{R}$ and such that a is a minimizer of f on some interval $[a,b]$ with $b > a$. Verify that $f'(a) \geq 0$.
8. Show that the incident cone $T^I(F,\overline{x})$ can be called the velocity cone of F at $\overline{x}$ since $v \in T^I(F,\overline{x})$ if and only if there exists some $c : [0,1] \to F$ such that $c(0) = \overline{x}$, c is right differentiable at 0 and $c'_+(0) = v$.
9. Give an example of a function f on a normed space X that attains its infimum on a subset F of X at some $a \in F$ and of some $u \in T(F,a)$ such that $f^D(a,u) < 0$. [Hint: use Exercise 3.]

5.6.3 Calculus of Tangent and Normal Cones

We devote this subsection to some calculus rules for normal cones. These rules will enable us to compute the normal cones to sets defined by equalities and inequalities, an important topic for the application to concrete optimization problems.

In order to show that the two notions of normal cone we introduced correspond to the classical notion in the smooth case, let us make some easy but useful observations.

Proposition 5.33 *The notions of normal cone and of Fréchet normal cone are local notions: if F and G are two subsets such that $F \cap V = G \cap V$ for some neighborhood V of $\overline{x}$, then $N(F,\overline{x}) = N(G,\overline{x})$ and $N_F(F,\overline{x}) = N_F(G,\overline{x})$.*

Proposition 5.34 *Given normed spaces X, Y and $\overline{x} \in F \subset X$, $\overline{y} \in G \subset Y$, one has*

$$N(F \times G, (\overline{x},\overline{y})) = N(F,\overline{x}) \times N(G,\overline{y}),$$
$$N_F(F \times G, (\overline{x},\overline{y})) = N_F(F,\overline{x}) \times N_F(G,\overline{y}).$$

Proposition 5.35 *The normal cone and the Fréchet normal cone are antitone: for $F \subset G$ and any $\overline{x} \in \mathrm{cl}F$ one has $N(G,\overline{x}) \subset N(F,\overline{x})$ and $N_F(G,\overline{x}) \subset N_F(F,\overline{x})$. Moreover, if F is a finite union, $F = \bigcup_{i \in I} F_i$, then*

$$N(F,\overline{x}) = \bigcap_{i \in I} N(F_i,\overline{x}), \qquad N_F(F,\overline{x}) = \bigcap_{i \in I} N_F(F_i,\overline{x}).$$

This fact helps in the computation of normal cones, as the next example shows.

Example Let $F := \{(r,s) \in \mathbb{R}^2 : rs = 0\}$, so that $F = F_1 \cup F_2$ with $F_1 := \mathbb{R} \times \{0\}$, $F_2 := \{0\} \times \mathbb{R}$. Then, as F_i is a linear subspace, one has $N(F_i, 0) = F_i^{\perp}$, hence $N(F,0) = F_1^{\perp} \cap F_2^{\perp} = \{0\}$.

However, the computations of normal cones to intersections are not obvious. One may just have the inclusions

$$N(F \cap G,\overline{x}) \supset N(F,\overline{x}) \cup N(G,\overline{x}), \qquad N_F(F \cap G,\overline{x}) \supset N_F(F,\overline{x}) \cup N_F(G,\overline{x}).$$

Example Let $X := \mathbb{R}^2$ with its usual Euclidean norm and let $F := B_X + e$, $G := B_X - e$, where $e = (0,1)$. Then $N(F \cap G, 0) = \mathbb{R}^2$ whereas $N(F,0) \cup N(G,0) = \{0\} \times \mathbb{R}$.

Now let us show that the notions of normals and firm normals are invariant under differentiable transformations (diffeomorphisms).

Proposition 5.36 *Let $g : U \to V$ be a map between two open subsets of the normed spaces X and Y, respectively, and let $B \subset U$, $C \subset V$ be such that $g(B) \subset C$. Then,* *if g is Fréchet differentiable (resp. Hadamard differentiable) at $\overline{x} \in B$, for $\overline{y} := g(\overline{x})$, one has*

$$N_F(C, \overline{y}) \subset (g'(\overline{x})^{\intercal})^{-1}(N_F(B, \overline{x})) \tag{5.37}$$

$$(\textit{resp.} \quad N(C, \overline{y}) \subset (g'(\overline{x})^{\intercal})^{-1}(N(B, \overline{x})) \quad).$$

Relation (5.37) is an equality when $C = g(B)$ and there exist $\rho > 0$, $c > 0$ such that

$$\forall y \in C \cap B(\overline{y}, \rho) \qquad d(\overline{x}, g^{-1}(y) \cap B) \leq c d(y, \overline{y}). \tag{5.38}$$

Proof Let $\overline{y}^*$ be an element of $N_F(C, \overline{y})$: for some remainder $r(\cdot)$ and for all $y \in C$ we have $\langle \overline{y}^*, y - \overline{y} \rangle \leq r(\|y - \overline{y}\|)$. The differentiability of g at $\overline{x}$ can be written in the following form for some remainder s

$$g(x) - g(\overline{x}) = A(x - \overline{x}) + s(\|x - \overline{x}\|), \tag{5.39}$$

where $A := g'(\overline{x})$. Taking $x \in B$, since $y := g(x) \in C$, we get

$$\begin{aligned} \langle A^{\intercal}(\overline{y}^*), x - \overline{x} \rangle &= \langle \overline{y}^*, g(x) - g(\overline{x}) - s(\|x - \overline{x}\|) \rangle \\ &\leq r(\|g(x) - g(\overline{x})\|) - \langle \overline{y}^*, s(\|x - \overline{x}\|) \rangle := t(\|x - \overline{x}\|) \end{aligned}$$

where t is a remainder since $\|g(x) - g(\overline{x})\| \leq (\|A\| + 1)\|x - \overline{x}\|$ for x close enough to $\overline{x}$. The proof for the normal cone is similar. It can also be deduced from the inclusion $g'(\overline{x})(T(B, \overline{x})) \subset T(C, \overline{y})$.

Now suppose $C = g(B)$ and relation (5.38) holds for some $\rho > 0$, $c > 0$. Then, for all $y \in C \cap B(\overline{y}, \rho)$, there exists some $x_y \in g^{-1}(y) \cap B$ satisfying $\|x_y - \overline{x}\| \leq 2c \|y - \overline{y}\|$. Let $\overline{y}^* \in Y^*$ be such that $\overline{x}^* := g'(\overline{x})^{\intercal}(\overline{y}^*) \in N_F(B, \overline{x})$. Then, there exists a remainder $r(\cdot)$ such that

$$\forall x \in B \qquad \langle \overline{y}^*, g'(\overline{x})(x - \overline{x}) \rangle = \langle \overline{x}^*, x - \overline{x} \rangle \leq r(x - \overline{x}).$$

Taking into account (5.39), we get, for all $y \in C \cap B(\overline{y}, \rho)$

$$\langle \overline{y}^*, y - \overline{y} \rangle = \langle \overline{y}^*, g(x_y) - g(\overline{x}) \rangle \leq r(\|x_y - \overline{x}\|) + \|\overline{y}^*\| s(\|x_y - \overline{x}\|)$$

and, since $\|x_y - \overline{x}\| \leq 2c \|y - \overline{y}\|$, we conclude that $\overline{y}^* \in N_F(C, \overline{y})$. □

Corollary 5.21 *Let $g : U \to V$ be a bijection between two open subsets of the normed spaces X and Y respectively such that g and $h := g^{-1}$ are Hadamard differentiable (resp. Fréchet differentiable) at $\overline{x}$ and $\overline{y} := g(\overline{x})$ respectively and let $B \subset U$, $C = g(B)$. Then*

$$N(B, \overline{x}) = g'(\overline{x})^{\intercal}(N(C, \overline{y}))$$

$$(\textit{resp.} \quad N_F(B, \overline{x}) = g'(\overline{x})^{\intercal}(N_F(C, \overline{y})) \quad).$$

Proof Since $h'(\overline{y})^{\mathsf{T}}$ is the inverse of $g'(\overline{x})^{\mathsf{T}}$, one has the inclusions of Proposition 5.36 and their analogues in which $h, \overline{y}, C$ take the roles of $g, \overline{x}, B$, respectively. □

For an inverse image, it is possible to ensure equality in the inclusions of Proposition 5.36. However a technical assumption called a qualification condition should be added, otherwise the result may be invalid, as the following example shows.

Example Let $X = Y = \mathbb{R}$, $g(x) = x^2$, $C = \{0\}, B = g^{-1}(C)$. Then $N(B, 0) = \mathbb{R} \neq g'(0)^{\mathsf{T}}(N(C, 0)) = \{0\}$.

The factorization of Lemma 3.20 will be helpful for handling inverse images.

Proposition 5.37 (Lyusternik) *Let X, Y be Banach spaces, let U be an open subset of X and let $g : U \to Y$ be circa-differentiable at $\overline{x} \in U$ with $g'(\overline{x})(X) = Y$. Then, for $S := g^{-1}(\overline{y})$ with $\overline{y} := g(\overline{x})$ one has $N(S, \overline{x}) = N_F(S, \overline{x}) = g'(\overline{x})^{\mathsf{T}}(Y^*)$.*

Proof Proposition 5.36 ensures that $g'(\overline{x})^{\mathsf{T}}(Y^*) \subset N_F(S, \overline{x}) \subset N(S, \overline{x})$. Now, given $x^* \in N(S, \overline{x})$, for all $v \in T(S, \overline{x}) = \ker g'(\overline{x}) = T^l(S, \overline{x})$ we have $\langle v, x^* \rangle = 0$, so that Lemma 3.20 yields some $y^* \in Y^*$ such that $x^* = y^* \circ g'(\overline{x}) = g'(\overline{x})^{\mathsf{T}}(y^*)$. □

A more general case is treated in the next theorem.

Theorem 5.29 *Let X, Y be Banach spaces, let U be an open subset of X and let $g : U \to Y$ be a map that is circa-differentiable at $\overline{x} \in U$ with $A := g'(\overline{x})$ surjective. Then, if C is a subset of Y and if $\overline{x} \in B := g^{-1}(C)$, $\overline{y} := g(\overline{x}) \in C$, one has*

$$N(B, \overline{x}) = g'(\overline{x})^{\mathsf{T}}(N(C, \overline{y})),$$
$$N_F(B, \overline{x}) = g'(\overline{x})^{\mathsf{T}}(N_F(C, \overline{y})).$$

Proof We prove the Fréchet case only, leaving the directional case to the reader. The Lyusternik-Graves Theorem (Theorem 5.15) asserts the existence of $\sigma > 0$, $c > 0$ such that for all $y \in B(\overline{y}, \sigma)$ there exists a $x_y \in g^{-1}(y)$ satisfying $\|x_y - \overline{x}\| \leq c \|y - \overline{y}\|$. When $y \in C \cap B(\overline{y}, \sigma)$ we have $x_y \in g^{-1}(C) = B$, hence $d(\overline{x}, g^{-1}(y) \cap B) \leq d(\overline{x}, x_y) \leq cd(\overline{y}, y)$. Moreover, setting $V := B(\overline{y}, \sigma)$, $U := g^{-1}(V)$, $B' := B \cap U$, $C' := C \cap V$, we have $g(B') = C'$ and $N_F(B, \overline{x}) = N_F(B', \overline{x})$ and $N_F(C, \overline{y}) = N_F(C', \overline{y})$. Thus, we can replace B with B' and C with C'. Then Proposition 5.36 ensures that $N_F(B, \overline{x}) = g'(\overline{x})^{\mathsf{T}}(N_F(C, \overline{y}))$. □

Exercise With the notation and the assumption of the theorem show that $T(B, \overline{x}) := (g'(x))^{-1}(T(C, \overline{y}))$.

Exercise Apply the theorem and the preceding exercise to the case $Y := \mathbb{R}^d$, $C := \mathbb{R}^d_+$.

5.6.4 Multiplier Rules

As observed above, the usual necessary condition $f'(a) = 0$ in order that a function $f : X \to \mathbb{R}$ attains its minimum at a when it is differentiable there, has to be modified when some restrictions are imposed. In the present section we consider the frequent case of constraints defined by equalities and we present a practical rule. The case of inequalities is dealt with in the exercises. The famous Lagrange multiplier rule is a direct consequence of Fermat's Rule and of Proposition 5.37.

Theorem 5.30 (Lagrange Multiplier Rule) *Let X, Y be Banach spaces, let W be an open subset of X, let $f : W \to \mathbb{R}$ be differentiable at a and let $g : W \to Y$ be circa-differentiable at a with $g'(a)(X) = Y$. Let $b := g(a)$. Suppose that f attains on $S := g^{-1}(b)$ a local minimum at a. Then there exists some $y^* \in Y^*$ (called the Lagrange multiplier) such that*

$$f'(a) = y^* \circ g'(a).$$

Example Let us find the shape of a box having a given volume $v > 0$ and a minimum area. Denoting by x, y, z the sizes of the sides of the box, we are led to minimize

$$f(x, y, z) := 2(xy + yz + zx) \quad \text{subject to } g(x, y, z) := xyz - v = 0, \ x, y, z > 0.$$

First, we secure the existence of a solution by showing that f is coercive on $S := g^{-1}(0)$. In fact, if $w_n := (x_n, y_n, z_n) \in S$ and $(\|w_n\|) \to +\infty$, one of the components of w_n, say x_n, converges to $+\infty$; then, since $y_n + z_n \geq 2\sqrt{y_n z_n} = 2\sqrt{v/x_n}$, we get

$$f(w_n) \geq 2x_n(y_n + z_n) \geq 4\sqrt{vx_n} \to +\infty.$$

Now let (x, y, z) be a minimizer of f on S. Since the derivative of g is non-null at (x, y, z), the Lagrange multiplier rule yields some $\lambda \in \mathbb{R}$ such that

$$2\,(y + z) = \lambda yz$$
$$2\,(z + x) = \lambda zx$$
$$2\,(x + y) = \lambda xy.$$

Then, multiplying each side of the first equation by x, and doing similar operations with the other two equations, we get

$$\lambda v = \lambda xyz = 2x(y + z) = 2y(z + x) = 2z(x + y),$$

hence, by summation, $3\lambda v = 4(xy + yz + zx) > 0$. Subtracting sides by sides the equations expressing the Lagrange multiplier rule, we get

$$2\,(y - x) = \lambda z(y - x), \quad 2(z - y) = \lambda x(z - y), \quad 2(x - z) = \lambda y(x - z).$$

Since λ, x, y, z are positive, considering the various cases, we get $x = y = z$. Since the unique solution of the necessary condition is $w := (v^{1/3}, v^{1/3}, v^{1/3})$, we conclude that w is the solution of the problem and the optimal box is a cube. We also note that the least area is $a(v) := f(w) = 6v^{2/3}$ and that $\lambda = 4v^{-1/3}$ is exactly the derivative of the function $v \mapsto a(v)$, a general fact we will explain later on which shows that the artificial multiplier λ has in fact an important interpretation as a measure of the change of the optimal value when the parameter v varies.

Example-Exercise Let X be some Euclidean space and let $A \in L(X, X)$ be symmetric. Let f and g be given by $f(x) = (Ax \mid x)$, $g(x) = \|x\|^2 - 1$. Take $v \in S_X$ such that f attains its minimum on the unit sphere S_X at v. Then show that there exists some $\lambda \in \mathbb{R}$ such that $Av = \lambda v$. Deduce from this result that any symmetric square matrix is diagonalizable.

Exercises

1. (Simplified **Karush-Kuhn-Tucker Theorem**) Let X, Y be Banach spaces, let $g : X \to Z$ be circa-differentiable at $\overline{x}$ with $g'(\overline{x})(X) = Z$ and let $C \subset Z$ be a closed convex cone of Z. Suppose $\overline{x} \in F := g^{-1}(C)$ is a minimizer on F of a function $f : X \to \mathbb{R}$ that is differentiable at $\overline{x}$. Use Theorem 5.29 and the Fermat rule in order to obtain the existence of some $\overline{y}^* \in C^0$ such that $\langle \overline{y}^*, g(\overline{x}) \rangle = 0$, $f'(\overline{x}) + \overline{y}^* \circ g'(\overline{x}) = 0$.
2. (**a**) Compute the tangent cone at $(0, 0)$ to the set

$$F := \left\{(r, s) \in \mathbb{R}^2 : s \geq \mid r \mid (1 + r^2)^{-1}\right\}.$$

   (**b**) Use the Fermat rule to give a necessary condition in order that $(0, 0)$ be a local minimizer of a function f on F, assuming that f is differentiable at $(0, 0)$.
   (**c**) Rewrite F as $F = \left\{(r, s) \in \mathbb{R}^2 : g_1(r, s) \leq 0, g_2(r, s) \leq 0\right\}$ with g_1, g_2 given by $g_1(r, s) = r(1 + r^2)^{-1} - s$, $g_2(r, s) = -r(1 + r^2)^{-1} - s$ and apply the Karush-Kuhn-Tucker Theorem to get the condition obtained in (**b**).
3. (**a**) Compute the tangent cone to the set $F = F' \cup F''$ at $a \in F$ where

$$F' := \left\{(r, s) \in \mathbb{R}^2 : r^4 + s^4 - 2rs = 0\right\}$$

$$F'' := \left\{(r, s) \in \mathbb{R}^2 : r^4 + s^4 + 2rs = 0\right\}$$

first for some point $a \neq (0,0)$, then for $a = (0,0)$. [Hint: first study the symmetry properties of F and set $s = tr$].

(**b**) Write a necessary condition in order that a differentiable function $f : \mathbb{R}^2 \to \mathbb{R}$ attains on F a local minimizer at $(0,0)$. Assuming that f is twice differentiable at $(0,0)$, write a second order necessary condition.

4. Give the dimensions of a cylindrical can that has a given volume v and the least area $a(v)$. Give an interpretation of the multiplier in terms of the derivative of $a(\cdot)$.
5. Give the dimensions of a cylindrical can that has a given area a and the greatest volume $v(a)$. Give an interpretation of the multiplier in terms of the derivative of $v(\cdot)$.
6. Give the dimensions of a box without lid that has a given volume v and the least area $a(v)$. Give an interpretation of the multiplier in terms of the derivative of $a(\cdot)$.
7. Give the dimensions of a box without lid that has a given area a and the greatest volume $v(a)$. Give an interpretation of the multiplier in terms of the derivative of $v(\cdot)$.

5.7 Introduction to the Calculus of Variations

The importance of the calculus of variations stems from its role in the history of the development of analysis and from its ability to present general principles that govern a number of physical phenomena. Among these are the Fermat Principle ruling the route of light and the Euler-Maupertuis Principle of Least Action governing mechanics. Historically, the calculus of variations appeared at the end of the 17th century with the *brachistochrone problem*, solved in 1696 by Johann Bernoulli (Fig. 5.5).

This problem consists in determining a curve joining two given points of space along which a frictionless bead slides under the action of gravity in a minimal time. The novelty of such a problem lies in the fact that the unknown is a geometrical object, a curve or a function, not a real number or a finite sequence in real numbers.

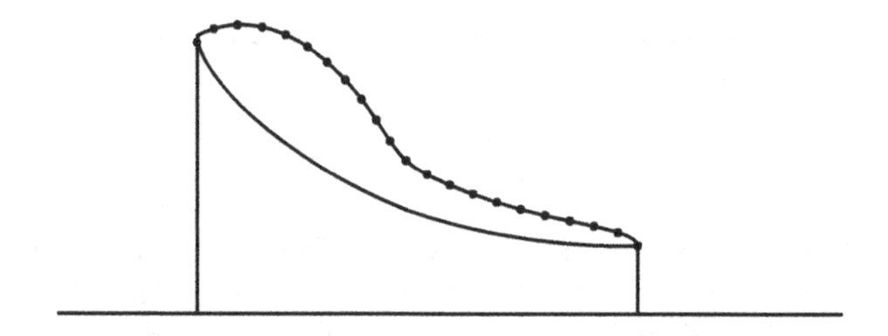

Fig. 5.5 The brachistochrone problem

Thus, such a topic puts to the fore the use of functional spaces, even for one-dimensional problems. We do not limit our attention to this case because many partial differential equations are derived from problems in the calculus of variations.

5.7.1 The One-Variable Case

In fact, the choice of an appropriate space of functions in which a solution is to be found is part of the problem. Several choices are possible. The simplest one is the space of functions of class C^1, but it rules out piecewise C^1 functions. The most general one involves absolutely continuous maps and Lebesgue null sets and is a bit technical; for many problems piecewise C^1 curves would suffice but the space of piecewise C^1 functions is not complete. We adopt an intermediate choice close to the choice of the class C^1.

Let E be a Banach space and let $T := [a, b]$ be a compact interval of $\mathbb{R}$. We will use the space $X := R^1(T, E)$ of functions $x : T \to E$ that are primitives of (normalized) regulated functions from T to E; this means that there exists a function $x' : T \to E$ that is right continuous on $[0, 1[$ and has a left limit $x'(t_-)$ for all $t \in]0, 1]$ with $x'(b) = x'(b_-)$ such that

$$x(t) = x(0) + \int_a^t x'(s)ds \qquad\qquad t \in T.$$

Then x' is determined by x since for each $t \in [a, b[$, $x'(t)$ is the right derivative of x at t and $x'(b)$ is the left derivative of x at b. We endow X with the norm

$$\|x\| = \sup_{t\in T} \|x(t)\| + \sup_{t\in T} \left\|x'(t)\right\| .$$

It is equivalent to the norm $x \mapsto \|x(a)\| + \sup_{t\in T} \|x'(t)\|$, as is easily seen. Then X is a Banach space (use Theorem 5.6). Without loss of generality, we may suppose $T := [0, 1]$; passing from $x \in R^1(T, E)$ to $w \in R^1([0, 1], E)$ by setting $w(s) := x(a + s(b - a))$ enables us to reduce a problem set on $R^1(T, E)$ to a problem set in $R^1([0, 1], E)$.

Given $(e_0, e_1) \in E \times E$, an open subset U of $E \times E \times T$ and a continuous function $L : U \to \mathbb{R}$, the problem consists in minimizing the function j given by

$$j(x) = \int_0^1 L(x(t), x'(t), t)dt$$

over the set $W_U(e_0, e_1)$ of elements x of X such that $x(0) = e_0$, $x(1) = e_1$ and $(x(t), x'(t), t) \in U$ for each $t \in T$. We note that since L is continuous the function $t \mapsto L(x(t), x'(t), t)$ is regulated, so that the integral is well defined.

Let us note that our choice of the space $R^1(T,E)$ for the solution is larger than the space $C^1(T,E)$, so that it may happen that a solution exists in $R^1(T,E)$, but not in $C^1(T,E)$.

Example Let $L : \mathbb{R} \times \mathbb{R} \to \mathbb{R}$ be given by $L(e,v) := e^2(v^2-4)^2$ and let $e_0 := 0$, $e_1 := 1$. Then $\overline{x} \in R^1(T) := R^1(T,\mathbb{R})$ given by $\overline{x}(t) = (2t-1)^+ := \max(2t-1,0)$ is a solution of the problem since $j(\overline{x}) = 0$ and $j(x) \geq 0$ for all $x \in R^1(T)$. However, the problem has no solution in $C^1(T)$. To see this, we note that if $x \in C^1(T)$ satisfies the conditions $x(0) = 0$, $x(1) = 1$, x' must take the value 1 for at least one $\overline{t} \in T$ since otherwise we can find $\alpha > 0$ such that either $x'(t) \geq 1+\alpha$ or $x'(t) \leq 1-\alpha$ for all $t \in T$ and in each case the boundary conditions are not satisfied. Then, on some neighborhood of $\overline{t}$ one has $x'(t) \in]0,2[$ and x vanishes at most once on this neighborhood, so that $j(x) > 0$. □

Even by using a larger space than $C^1(T,E)$ and taking a very regular Lagrangian L, the minimization problem of j over $W_U(e_0,e_1)$ may have no solution, as shown by the next example. The existence of a minimizer can be ensured by the use of compactness and lower semicontinuity arguments. Such a method is usually called the *direct method* of the calculus of variations (see the last example of Sect. 2.2.3 and f.i. [61, 78, 86]).

Example Let $L : U := \mathbb{R} \times \mathbb{R} \to \mathbb{R}$ be given by $L(e,v) = e^2 + (v^2-1)^2$ and let $e_0 = 0 = e_1$. Let us show that $\inf\{j(w) : w \in W(e_0,e_1)\} = 0$ but that there is no $w \in W_U(e_0,e_1)$ such that $j(w) = 0$. The latter fact is obvious since $j(w) = 0$ implies that $w(t) = 0$ for all $t \in [0,1]$ (since w is continuous) and $|w'(t)| = 1$ for almost every $t \in [0,1]$ and these two requirements are incompatible. To demonstrate the first fact we note that $j(w) \geq 0$ for all $w \in W_U(e_0,e_1)$ and we construct a sequence (w_n) in $W_U(e_0,e_1)$ such that $(j(w_n)) \to 0$. We determine w_n by requiring that $w_n(k/n) = 0$ for all $k \in \mathbb{N}_n \cup \{0\}$ and $w_n'(t) = 1$ for $t \in](2k-2)/2n, (2k-1)/2n[$ and $w_n'(t) = -1$ for $t \in](2k-1)/2n, 2k/2n[$ for $k \in \mathbb{N}_n$. Then we have $w_n'(t)^2 = 1$ for almost all t and $w_n(t)^2 \leq 1/4n^2$ for all $t \in [0,1]$, so that $j(w_n) \leq 1/4n^2$. □

In fact, in the sequel we replace the set $W_U(e_0,e_1)$ with $W \cap H(e_0,e_1)$, where $H(e_0,e_1)$ is the affine subspace $H(e_0,e_1) := \{x \in X : x(0) = e_0,\ x(1) = e_1\}$ and

$$W := \{x \in X : \mathrm{cl}((J^1x)(T)) \subset U\}$$

with $(J^1x)(T) := \{(x(t),x'(t),t) : t \in T\}$. Such a choice brings some preliminary results of interest.

Lemma 5.18 *Given U, L, W and j as above, the set W is open in X and j is continuous on W.*

Proof This can be deduced from Proposition 3.45 by using the embedding $x \mapsto (x,x')$ from $X := R^1(T,E)$ into $R(T,E^2)$, or more directly as follows. By Proposition 3.43, for all $x \in X$, the set $\mathrm{cl}((J^1x)(T))$ is a compact subset of $E \times E \times T$. Thus, if $x \in W$, there exists some $r > 0$ such that $B((J^1x)(T),r) \subset U$. Then for all $w \in X$ satisfying $\|w-x\| < r$ one has $w \in W$. Thus W is open in X.

Moreover, L being continuous, it is uniformly continuous around $\mathrm{cl}((J^1x)(T))$ in the sense that for every $\varepsilon > 0$ one can find a $\delta > 0$ such that for all $(e, v, t) \in \mathrm{cl}((J^1x)(T))$ and all $(e', v', t') \in B((e, v, t), \delta)$ one has $|L(e', v', t') - L(e, v, t)| \leq \varepsilon$. Therefore, for all $w \in X$ satisfying $\|w - x\| \leq \delta$, one has

$$\left|L(w(t), w'(t), t) - L(x(t), x'(t), t)\right| \leq \varepsilon,$$

hence $|j(w) - j(x)| \leq \varepsilon$. □

Proposition 5.38 *Suppose L is continuous on U and has partial derivatives with respect to its first and second variables that are continuous on U. Then j is of class C^1 on W and for $\overline{x} \in W$, $x \in X$ one has*

$$j'(\overline{x})x = \int_0^1 [D_1L(\overline{x}(t), \overline{x}'(t), t)x(t) + D_2L(\overline{x}(t), \overline{x}'(t), t)x'(t)]dt.$$

Proof Using an interchange between integration and differentiation one can prove that j is Gateaux differentiable with a derivative given by the above formula. Then the continuity of the derivative shows that in fact j is Fréchet differentiable. Alternatively, using the embedding $x \mapsto (x, x')$ from $R^1(T, E)$ into $R(T, E^2)$ considered above and the linear map $f \mapsto \int_0^1 f(t)dt$ from $R(T, E)$ into E, differentiability can be deduced from the chain rule and Corollary 5.8. In order to be more explicit, let us set $L_t(e, v) = L(e, v, t)$ for $(e, v, t) \in U$,

$$Y := \{(e_1, e_2, v_1, v_2, t) : \forall s \in [0, 1],\ ((1 - s)e_1 + se_2, (1 - s)v_1 + sv_2, t) \in U\},$$

$$Z := \{(w_1, w_2) \in W^2 : \forall t \in T,\ (w_1(t), w_2(t), w_1'(t), w_2'(t), t) \in Y\},$$

and, for $(e_1, e_2, v_1, v_2, t) \in Y$

$$K(e_1, e_2, v_1, v_2, t) := \int_0^1 DL_t((1 - s)e_1 + se_2, (1 - s)v_1 + sv_2)ds,$$

so that

$$L(e_1, v_1, t) - L(e_2, v_2, t) = K(e_1, e_2, v_1, v_2, t)(e_1 - e_2, v_1 - v_2). \tag{5.40}$$

The compactness of $T := [0, 1]$ easily yields that Y is open in $E^2 \times E^2 \times T$. Then a proof similar to that of Lemma 5.18 shows that Z is open in $X \times X$ and that the map

$$h : (w_1, w_2, x) \mapsto \int_0^1 K(w_1(t), w_2(t), w_1'(t), w_2'(t), t)(x(t), x'(t))dt$$

from $Z \times X$ into $\mathbb{R}$ is continuous. Since this map is linear in x and such that

$$j(w_1) - j(w_2) = h(w_1, w_2, w_1 - w_2)$$

we get from Corollary 5.8 that j is of class D^1 and that for $w \in W$ the derivative of j at w is $h(w, w, \cdot)$. Such a conclusion is enough for the necessary condition we have in view.

With an additional effort we can show that j is of class C^1. For $(w_1, w_2) \in Z$, $t \in T$ setting

$$k(w_1, w_2)(t) := K(w_1(t), w_2(t), w_1'(t), w_2'(t), t),$$

we define a continuous map k from Z into $R(T, E^* \times E^*)$. Now the map $\varphi : R(T, E^* \times E^*) \to (R^1(T, E))^*$ defined by

$$\varphi(k_1, k_2)(x) := \int_0^1 [k_1(t)x(t) + k_2(t)x'(t)]dt$$

is linear and continuous. Composing k with φ we obtain a continuous map from Z into $(R^1(T, E))^*$. For $(w_1, w_2) \in Z$, by substituting $(w_i(t), w_i'(t))$ to (e_i, v_i) for $i = 1$, 2 in (5.40) and integrating over T, we see that

$$j(w_1) - j(w_2) = (\varphi \circ k)(w_1, w_2)(w_1 - w_2)$$

and we deduce from Lemma 5.6 that j is differentiable at $w \in W$ and that $j'(w) = (\varphi \circ k)(w, w)$. Thus j is of class C^1; see also Cor. 5.9. □

Let us look for necessary minimality conditions.

Proposition 5.39 *Suppose L satisfies the assumptions of the preceding proposition and $\overline{x}$ is a local minimizer of j on $W(e_0, e_1)$. Then $\overline{x}$ is a critical point of j on $W(e_0, e_1)$ in the following sense:*

$$j'(\overline{x})v = 0 \qquad \forall v \in X_0 := W(0, 0) := \{x \in X : x(0) = 0 = x(1)\}.$$

Proof Let N be a neighborhood of $\overline{x}$ in X such that $j(w) \geq j(\overline{x})$ for every $w \in N \cap W(e_0, e_1)$. Given $v \in X_0$, for $r \in \mathbb{R}$ with $|r|$ small enough, we have $w := \overline{x} + rv \in W$ by Lemma 5.18 and $w(0) = e_0$, $w(1) = e_1$. Thus $w \in N \cap W(e_0, e_1)$, hence $j(\overline{x} + rv) \geq j(\overline{x})$ for $|r|$ small enough. It follows that $j'(\overline{x})x = 0$. □

Such a condition can be given a more explicit and more efficient form.

Theorem 5.31 (Euler-Lagrange Condition) *Suppose L satisfies the assumptions of Proposition 5.38 and $\overline{x} \in W$ is a critical point of j on $W(e_0, e_1)$. Then the function $D_2L(\overline{x}(\cdot), \overline{x}'(\cdot), \cdot)$ is a primitive of $D_1L(\overline{x}(\cdot), \overline{x}'(\cdot), \cdot)$: for every $t \in [0, 1[$ the right derivative of $D_2L(\overline{x}(\cdot), \overline{x}'(\cdot), \cdot)$ exists and is such that*

$$\frac{d}{dt}\left(D_2L(\overline{x}(t), \overline{x}'(t), t)\right) = D_1L(\overline{x}(t), \overline{x}'(t), t). \tag{5.41}$$

Here, by an abuse of notation, we denote the right derivative as the derivative. In fact, for a countable subset D of T this relation holds for all $t \in T \backslash D$ with the usual bilateral derivative.

The solutions of equation 5.41 are called *extremals*.

We break the proof into three steps of independent interest. Taking $A(t) := D_1L(\overline{x}(t), \overline{x}'(t), t)$, $B(t) := D_2L(\overline{x}(t), \overline{x}'(t), (t))$, $B(t) := D_1L(t, \overline{x}(t), \overline{x}'(t))$ in the last one, we shall obtain the result. The first step is as follows.

Lemma 5.19 *Let f be an element of the space $R_n(T, \mathbb{R})$ of normalized regulated functions on T such that $f(t) \geq 0$ for all $t \in T$ and $\int_0^1 f(t)dt = 0$. Then $f = 0$.*

Proof Suppose, on the contrary, that there exists some $r \in T$ such that $f(r) > 0$. When $r < 1$, by the right continuity of f at r, we can find some $\alpha, \delta > 0$ such that $r + \delta < 1$ and $f(s) \geq \alpha$ for $s \in [r, r + \delta]$. Then, we get $\int_0^1 f(t)dt \geq \int_r^{r+\delta} f(t)dt \geq \alpha\delta > 0$, a contradiction. If $r = 1$, a similar argument using the left continuity of f at 1 also leads to a contradiction. □

Lemma 5.20 *Let $F \in R_n(T, E^*)$ be such that for all $x \in X_0 := \{x \in X : x(0) = 0 = x(1)\}$ one has $\int_0^1 F(t).x'(t)dt = 0$. Then $F(\cdot)$ is constant.*

More precisely, for $e^* := \int_0^1 F(t)dt$ one has $F(t) = e^*$ for all $t \in T$.

Proof Changing F into $G := F - e^*$, it suffices to show that $G(\cdot) = 0$ when $\int_0^1 G(t)dt = 0$ and $\int_0^1 G(t).x'(t)dt = 0$ for every $x \in X_0$. Given $e \in E$, let us introduce f, $g : T \to \mathbb{R}$, and $v, x : T \to E$ given by $g(t) = G(t)(e) := \langle G(t), e \rangle$, $f(t) = (g(t))^2$, $v(t) := G(t)(e)e := \langle G(t), e \rangle e$, $x(t) = \int_0^t v(s)ds$. We see that $x(0) = 0$, $x'_r(t) = v(t)$ for $t \in [0, 1[$, $x(1) = \int_0^1 v(t)dt = \langle \int_0^1 G(t)dt, e \rangle e = 0$ since the mean of G is 0. Thus $x \in X_0$. Our assumption yields

$$\int_0^1 f(t)dt = \int_0^1 G(t)\,(e)\,\langle G(t), e \rangle dt$$
$$= \int_0^1 G(t)\,(\langle G(t), e \rangle e)\,dt = \int_0^1 G(t).x'(t)dt = 0.$$

Lemma 5.19 entails $f(\cdot) = 0$. Since e is arbitrary in E, we get $G(\cdot) = 0$. □

Lemma 5.21 (Dubois-Reymond) *Let $A, B \in R_n(T, E^*)$ be such that*

$$\forall x \in X_0 \qquad \int_0^1 \left[A(t)x(t) + B(t)x'(t)\right] dt = 0.$$

Then B is a primitive of A: for every $t \in T$ one has $B(t) = B(0) + \int_0^t A(s)ds$.

Proof Let us set $C(t) := B(0) + \int_0^t A(s)ds$. Then, for each $x \in X_0$ the function $t \mapsto C(t)x(t)$ has a right derivative $t \mapsto A(t)x(t) + C(t)x'(t)$ and by assumption

$$0 = \int_0^1 \left[A(t)x(t) + B(t)x'(t)\right] dt = \int_0^1 [\frac{d}{dt}(C(t)x(t)) + (B(t) - C(t))x'(t)]dt$$

$$= C(1)x(1) - C(0)x(0) + \int_0^1 (B(t) - C(t))x'(t)dt = \int_0^1 (B(t) - C(t))x'(t)dt.$$

Lemma 5.20 ensures that $B - C$ is constant. Since $B(0) - C(0) = 0$, $B = C$. □

Let us consider some cases for which equation (5.41) can be simplified. In such cases, one can obtain *first integrals*, i.e. functions of $x(\cdot)$ that are constant along an extremal.

Corollary 5.22 *Suppose the Lagrangian L is independent of e* : $L(e, v, t) = \widehat{L}_t(v)$. *Then, for every extremal* $\overline{x}(\cdot)$, *the map* $t \mapsto D\widehat{L}_t(\overline{x}'(t))$ *is constant. More generally, if for some* $\overline{e} \in E$ *one has* $D_1L(e, v, t).\overline{e} = 0$ *for all* $(e, v, t) \in E \times E \times T$, *then the function* $t \mapsto D_2L(\overline{x}(t), \overline{x}'(t), t).\overline{e}$ *is constant.*

Proof Since $D_1L(\overline{x}(t), \overline{x}'(t), t).\overline{e} = 0$, the Euler-Lagrange relation (5.41) means that $\frac{d}{dt}D_2L(\overline{x}(t), \overline{x}'(t), t).\overline{e} = 0$, hence $t \mapsto D_2L(\overline{x}(t), \overline{x}'(t), t).\overline{e}$ is constant. If $D_1L = 0$, this conclusion holds for every $\overline{e} \in E$ and $t \mapsto D\widehat{L}_t(\overline{x}'(t))$ is constant. □

Example Let $E := \mathbb{R}$ and let L be given by $L(e, v, t) := v^3/3$. Then every extremal $\overline{x}$ must satisfy $\overline{x}'(t)^2 = c^2$ for some $c \in \mathbb{R}$. If we choose to minimize j on $C^1(T, \mathbb{R})$ with the boundary conditions $x(0) = x_0$, $x(1) = x_1$ we find a unique solution $\overline{x}(t) = x_0 + (x_1 - x_0)t$ for $t \in T$. If we minimize j on $X := R^1(T, \mathbb{R})$, we find infinitely many extremals that are piecewise C^1, hence in X.

Example Let $E := \mathbb{R}$ and let L be given by $L(e, v, t) := (1/t)\sqrt{v^2 + 1}$. Then every extremal x satisfies for some $c \in \mathbb{R}$ the equation $D_2L(x(t), x'(t), t) = c$, i.e.

$$\frac{x'(t)}{t\sqrt{x'(t)^2 + 1}} = c,$$

so that $x'(t)^2(1 - c^2t^2) = c^2t^2$ or, assuming x' takes nonnegative values,

$$x'(t) = \frac{ct}{\sqrt{1 - c^2t^2}}.$$

Integrating, we obtain $x(t) = -c^{-1}\sqrt{1 - c^2t^2} + b$ for some $b \in \mathbb{R}$, so that

$$(x(t) - b)^2 + t^2 = c^{-2}$$

and the graph of x lies on a circle. The constants b and c can be determined by the initial and final values e_0 and e_1.

Exercise In a number of classical problems the Lagrangian has the form $L(e, v, t) := g(t, e)\sqrt{\|v\|^2 + 1}$. Show that in such a case, with $E := \mathbb{R}$, the Euler equation takes the form

$$D_1L(\overline{x}(t), \overline{x}'(t), t) - D_3L(\overline{x}(t), \overline{x}'(t), t)\overline{x}'(t) - \frac{\overline{x}''(t)}{\overline{x}'(t)^2 + 1}L(\overline{x}(t), \overline{x}'(t), t) = 0.$$

Under some regularity conditions, higher-order necessary conditions can be obtained.

Corollary 5.23 (Erdmann) *Suppose the Lagrangian L is autonomous, i.e. independent of t : $L(e, v, t) = L(e, v)$. Then, for every twice differentiable extremal $\overline{x}(\cdot)$, the function $h : t \mapsto D_2L(\overline{x}(t), \overline{x}'(t)).\overline{x}'(t) - L(\overline{x}(t), \overline{x}'(t))$ is constant.*

Proof It suffices to check that the two terms $D_2L(\overline{x}(t), \overline{x}'(t)).\overline{x}''(t)$ cancel in the computation of h' so that:
$h'(t) = \frac{d}{dt}[D_2L(\overline{x}(t), \overline{x}'(t))].\overline{x}'(t) - D_1L(\overline{x}(t), \overline{x}'(t)).\overline{x}'(t) = 0.$ □

Proposition 5.40 (Legendre) *Suppose L is of class C^2. If $\overline{x}$ is a minimizer of j on some ball centered at $\overline{x}$ for the norm of $R^1(T, E)$, then for all $t \in T$ the operator $D_2^2L(\overline{x}(t), \overline{x}'(t), t)$ is positive semi-definite.*

Proof As in Proposition 5.38, given $y \in R^1(T, E)$ one can show that the second derivative $j''(\overline{x}).y.y$ of j at $\overline{x}$ in the direction y is given by

$$\begin{aligned} &j''(\overline{x}).y.y \\ &= \int_0^1 [D_1^2L(z(t)).y(t).y(t) + 2D_{1,2}^2L(z(t)).y(t)y'(t) + D_2^2L(z(t)).y'(t).y'(t)]dt \end{aligned}$$

where $z(t) := (\overline{x}(t), \overline{x}'(t), t)$. For $a, b \in T$ with $a < b$, $e \in E$, and $n \geq 2$ let $h_n : [0, 1] \to \mathbb{R}$ be the sawtooth function in $R^1(T, \mathbb{R})$ satisfying $h_n(t) = 0$ for $t \in [0, a] \cup [b, 1]$, $h_n'(t) = (-1)^k$ for $t \in [a + k(b-a)/n, a + (k+1)(b-a)/n]$ and let $y_n(t) := h_n(t)e$, so that $\|y_n'\|_\infty = \|e\|$ and $\|y_n\| = \|e\|(b-a)/n$. Taking $y := y_n$ in the relation $j''(\overline{x}).y.y \geq 0$ and passing to the limit we get

$$\int_a^b D_2^2L(z(t)).e.edt \geq 0.$$

Using a proof by contradiction as in Lemma 5.19, we conclude that for all $t \in T$ we have $D_2^2L(z(t)).e.e \geq 0$. □

5.7.2 Some Examples

Geodesics These are the curves of minimal length joining two points. The simplest case is that of a Euclidean space $E := \mathbb{R}^d$. Then the length of a curve $x : [0, 1] \to E$, $x := (x_1, \ldots, x_d)$ of class C^1 (or in $R^1(T, E)$) is given by

$$j(x) = \int_0^1 \|x'(t)\| \, dt = \int_0^1 \sqrt{x_1'(t)^2 + \ldots + x_d'(t)^2 dt}.$$

Thus the Lagrangian L is given by $L(e, v, t) := \widehat{L}(v) := \|v\|$ and Corollary 5.22 applies. Since $D\widehat{L}(v) = v/\|v\|$ for $v \in \mathbb{R}^d \backslash \{0\}$, along an extremal $\overline{x}$ whose derivative does not vanish, the direction $\overline{x}'(t)/\|\overline{x}'(t)\|$ of the derivative is a constant vector u. This can be seen by using coordinates since for $i \in \mathbb{N}_d$ the function $t \mapsto D_i\widehat{L}(\overline{x}(t)) = \overline{x}_i'(t)/\sqrt{\overline{x}_1'(t)^2 + \ldots + \overline{x}_d'(t)^2}$ is constant. Parameterizing the extremal $\overline{x}$ joining two points e_0, e_1 of E by the arc length $s(t) := \int_0^t \|\overline{x}'(r)\| \, dr$, we find that the curve is a line segment: $\overline{x}(t) = e_0 + s(t)u$ with $u = (e_1 - e_0)/s(1)$, so that $\overline{x}$ runs along the segment $[e_0, e_1]$.

The Trajectories of Light in Isotropic Media Suppose that according to the *Fermat Principle* the trajectory in an isotropic medium minimizes the travel time. Saying that the medium is isotropic means that the speed of light does not depend on the direction. However, it may depend on the point: in deserts, the air is warmer near the ground and the trajectory of light is curved, giving rise to possible mirages. Let us take the Lagrangian in the form

$$L(e, v) := c(e) \|v\|,$$

where c is a smooth function representing the physical properties of the medium. If c does not depend on its ith variable, one obtains that along an extremal $\overline{x}$ the function

$$t \mapsto c(\overline{x}(t)) \frac{\overline{x}'(t).e_i}{\|\overline{x}'(t)\|} = c(\overline{x}(t)) \frac{\overline{x}_i'(t)}{\sqrt{\overline{x}_1'(t)^2 + \ldots + \overline{x}_d'(t)^2}}$$

is constant. From this one can recover the *Descartes-Snell law of refraction* (Fig. 5.1). If the plane $x_3 = 0$ separates two media whose refractive indexes are constants n_1 and n_2 respectively, the angles θ_1 and θ_2 of $\overline{x}'(t_-)$ and $\overline{x}'(t_+)$ with the plane $x_3 = 0$ at the point where the trajectory crosses it satisfy the relation

$$\frac{1}{n_1} \cos \theta_1 = \frac{1}{n_2} \cos \theta_2.$$

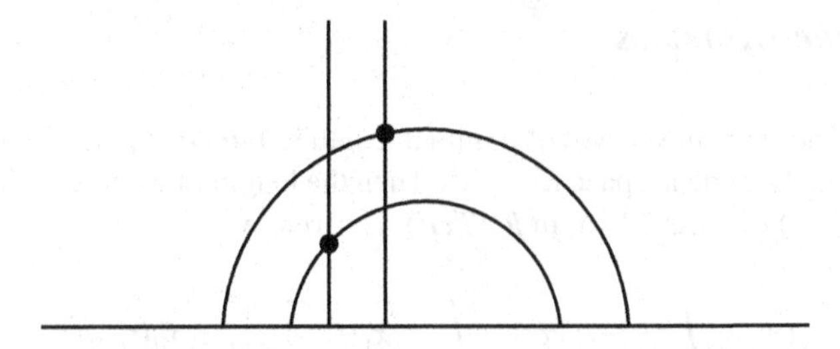

Fig. 5.6 Lobatchevski's geometry in Poincaré's half-space

Lobatchevski's Geometry In the half-space $P := \mathbb{R} \times \mathbb{P}$ endowed with the Riemannian metric given by $g_e(v) := (1/e_2) \|v\|$, the length of a curve $x(\cdot) : [0,1] \to P$ is given by

$$\ell(x) := \int_0^1 \frac{\sqrt{x_1'(t)^2 + x_2'(t)^2}}{x_2(t)} dt$$

The Lagrangian L is as in the preceding example with $c(e) := 1/e_2$, hence is independent of e_1. Thus, for any extremal $\overline{x}$ we obtain from Corollary 5.22 that there exists some $c \in \mathbb{R}$ such that

$$\frac{\overline{x}_1'(t)}{\overline{x}_2(t)\sqrt{\overline{x}_1'(t)^2 + \overline{x}_2'(t)^2}} = c.$$

The solutions of such equations are either the lines $x_1(t) = x_0$ or the arcs of the circle $x_1(t) = r\cos t + a$, $x_2(t) = r\sin t$ centered on $\mathbb{R}\times\{0\}$. Two points of P can be joined by a geodesic, but such lines do not satisfy all the axioms of Euclid (Fig. 5.6).

The Brachistochrone Problem (J. Bernoulli, 1696) This consists in finding a curve joining two points of $\mathbb{R}^3$ so that a particle moving along this curve starting from the highest of these two points reaches the other one in the shortest time, friction and resistance of the medium being neglected. We may assume the highest point is $(0,0,h)$ and the lowest point is $(1,0,0)$. We admit the curve remains in the plane $\mathbb{R}\times\{0\} \times \mathbb{R}$ and can be parameterized as $(x(r), 0, z(r))$, $r \in [0,1]$, with $x(r) = r$, i.e. that it is the graph of a function $z(\cdot)$. According to Galileo's law, the velocity $\frac{ds}{dt} = \frac{ds}{dr}\frac{dr}{dt}$ of the particle is independent of the shape of the curve and is given by $\sqrt{2gz(r)}$ where g is the acceleration due to gravity. The travel time is given by

$$T(z) := \frac{1}{\sqrt{2g}} \int_0^1 \frac{\sqrt{1 + z'(r)^2}}{\sqrt{z(r)}} dt, \qquad z \in R^1([0,1], \mathbb{R}).$$

Since the Lagrangian is autonomous, i.e. does not depend on r (which replaces t here in order to avoid confusion with time), Corollary 5.23 yields some $c \in \mathbb{R}$ such that

$$\frac{\sqrt{1+z'(r)^2}}{\sqrt{z(r)}} - \frac{z'(r)^2}{\sqrt{z(r)(1+z'(r)^2)}} = c$$

or equivalently $1/\sqrt{z(r)(1+z'(r)^2)} = c$, or $z(r)(1+z'(r)^2) = c^{-2}$, with the boundary conditions $z(0) = h$, $z(1) = 0$. Introducing the slope $s(\cdot) : r \mapsto s(r)$ of the tangent $(1, z'(r))$ to the curve by $s(r) := \tan^{-1} z'(r)$, so that $z'(r) = \tan s(r)$, we get

$$z(r) = \frac{c}{1+\tan^2 s(r)} = c\cos^2 s(r).$$

Differentiating each side of this relation, we obtain

$$z'(r) = -2cs'(r)\cos s(r)\sin s(r),$$

hence, since $z'(r) = \tan s(r)$,

$$1 = -2cs'(r)\cos^2 s(r) = -cs'(r)(1+\cos 2s(r)).$$

Thus, by integration, $x = r = r_0 - c(s + (1/2)\sin 2s)$, $z = (1/2)c(1+\cos 2s)$. The curve has the shape of a *cycloid*, a curve traced out by a point on the edge of a wheel, as the wheel is rolling along a straight line.

The Minimal Surface Problem Given a function $x : [0,1] \to \mathbb{R}_+$ of class C^2, the area of the surface obtained by rotating its graph about the z-axis (considering $t := z$) is given by

$$A(x) := 2\pi \int_0^1 x(t)\sqrt{x'(t)^2+1}dt.$$

Setting $L(e, v) := e\sqrt{v^2+1}$, we can find extremals of $A(\cdot)/2\pi$ by applying Corollary 5.23. The latter asserts the existence of a constant k such that

$$\frac{x(t)x'(t)^2}{\sqrt{x'(t)^2+1}} - x(t)\sqrt{x'(t)^2+1} = k$$

or (if $x(t) > 0$) $k\sqrt{x'(t)^2+1} = x(t)x'(t)^2 - x(t)(x'(t)^2+1) = -x(t)$, hence

$$x'(t)^2 = k^{-2}x(t)^2 - 1.$$

This differential equation written by separation of variables as

$$\int \frac{kdx}{\sqrt{x^2-k^2}} = \int dt$$

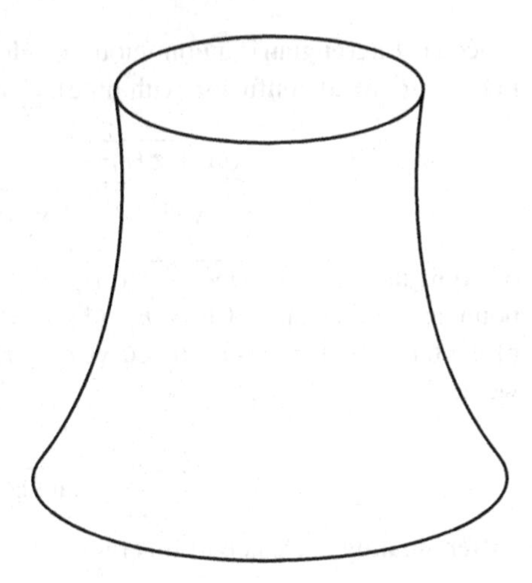

Fig. 5.7 A minimal surface of revolution

can be solved by finding a primitive in the left-hand side: $k\cosh^{-1}(x/k)$. Thus $x(t) = k\cosh(t/k + c)$, where the constants c and k have to be determined by the values $x(0)$ and $x(1)$. The graph of such a function is called a *catenary*. Its shape can be seen in the cooling towers of power plants or in soap bubbles between two rings (Fig. 5.7).

Classical Mechanics Let us consider a solid with mass m whose position is determined by parameters $(q_1, \ldots, q_n) \in E := \mathbb{R}^n$. It is subject to a force $F(q_1, \ldots, q_n)$ deriving from a potential $U(q_1, \ldots, q_n)$ in the sense that $F(q_1, \ldots, q_n) = \nabla U(q_1, \ldots, q_n)$. Its kinetic energy is given by $T(v_1, \ldots, v_n) = (1/2)m(v_1^2 + \ldots + v_n^2)$. Let L be the Lagrangian given by

$$L(q, v) := L(q_1, \ldots, q_n, v_1, \ldots, v_n) = T(v_1, \ldots, v_n) + U(q_1, \ldots, q_n).$$

Using the isomorphism between E and E^*, the Euler-Lagrange equations turn to be

$$mq''(t) = F(q(t)),$$

the *Newton equation,* in which $q''(t) := (q_1''(t), \ldots, q_n''(t))$ is the acceleration.

5.7.3 *The Legendre Transform*

The Legendre transform is a classical method used in the calculus of variations and in the study of differential equations. We give a short account of it as an illustration of inversion techniques. We present a slight refinement of it, replacing a continuous differentiability assumption or a local Lipschitz assumption with a stability assumption. We say that a map $g : U \to V$ between two metric spaces is *stable* or is *Stepanovian* if for any $\overline{u} \in U$ there exist some $r > 0$, $c \in \mathbb{R}_+$ such that for every $u \in B(\overline{u}, r)$ one has

$$d(g(u), g(\overline{u})) \leq cd(u, \overline{u}).$$

Definition 5.21 A function $f : U \to \mathbb{R}$ on an open subset U of a Banach space X is a (classical) *Legendre function* if it is differentiable, if its derivative $f' : U \to Y := X^*$ is a stable bijection onto an open subset V of Y whose inverse h is stable too.

Then one defines the *Legendre transform* of f as the function $f^L : V \to \mathbb{R}$ given by

$$f^L(y) := \langle h(y), y \rangle - f(h(y)) \qquad y \in V.$$

If h is of class C^1, f^L is clearly of class C^1. However, we do not make this assumption.

Proposition 5.41 *If f is a Legendre function on U, then its Legendre transform f^L is of class C^1 on $V := f'(U)$. It is of class C^k ($k \geq 1$) if f is of class C^k. Moreover, f^L is a Legendre function, $\left(f^L\right)^L = f$ and for all $(u, v) \in U \times V$ one has*

$$v = Df(u) \Leftrightarrow u = Df^L(v).$$

Furthermore, for $k \geq 2$ one has $D^2 f^L(v) = (D^2 f(u))^{-1}$ for $v \in V$, $u = Df^L(v)$.

Here $D^2 f(u)$ is considered as an element of $L(X, X^*)$ and $D^2 f^L(v)$ as an element of $L(X^*, X)$.

Proof Given $u \in U$, $v := Df(u) \in V$, let $y \in V - v$, let $x := x(y) := h(v + y) - h(v) \in U - u$ and let $r(x) = f(u + x) - f(u) - Df(u)x$. Then, since $h(v) = u$, $h(v + y) = u + x$, the definition of f^L yields

$$\begin{aligned} f^L(v + y) - f^L(v) - \langle u, y \rangle &= \langle u + x, v + y \rangle - f(u + x) - \langle u, v \rangle + f(u) - \langle u, y \rangle \\ &= \langle x, v + y \rangle - Df(u)(x) - r(x) = \langle x(y), y \rangle - r(x(y)). \end{aligned}$$

Since there exists a $c \in \mathbb{R}_+$ such that $\|x(y)\| \leq c\,\|y\|$ for $\|y\|$ small enough, the last right-hand side is a remainder as a function of y. Thus f^L is differentiable at v and $Df^L(v) = u = h(v)$. Therefore $(f^L)' = h$ is a bijection with inverse f' and f^L is a Legendre function. Now

$$\left(f^L\right)^L(u) = \langle Df^L(v), v \rangle - f^L(v) = \langle u, v \rangle - (\langle u, v \rangle - f(u)) = f(u).$$

Suppose now that f is of class C^2. Let $g := f'$, $u \in U$, $v := g(u) \in V$, $x \in X$, $A := g'(u)$. For $t > 0$ small enough to ensure $u + tx \in U$, we set $y_t := t^{-1}(g(u + tx) - g(u))$, so that $v + ty_t = g(u + tx)$, $h(v + ty_t) = u + tx$, $h(v) = u$ and $\|tx\| = \|h(v + ty_t) - h(v)\| \leq tc\,\|y_t\|$ for t small enough. Since $(y_t) \to y := A(x)$ as $t \to 0_+$, we get $\|x\| \leq c\,\|A(x)\|$. Thus A is injective and its image is a complete subspace of Y, as is easily seen. Let us show that this image is dense, which will prove that A is an isomorphism. Given $y \in Y$, let us set $x_t := t^{-1}(h(v+ty) - h(v))$, so that $v + ty = g(h(v) + tx_t)$ and $ty = g(u + tx_t) - v = g(u + tx_t) - g(u) = tA(x_t) + tz_t$, where $(z_t) \to 0$ as $t \to 0_+$ since g is differentiable at u and (x_t) is bounded by the assumption that h is stable at v. Thus $d(y, A(X)) \leq \lim_t \|z_t\| = 0$ and $y \in \mathrm{cl}(A(X)) = A(X)$ which is closed in Y. Thus A is an isomorphism and the inverse mapping theorem shows that the inverse h of g is differentiable at v with derivative $A^{-1} := g'(u)^{-1}$. Thus h is of class C^1 and since $(f^L)' = h$, we get that f^L is of class C^2. Moreover,

$$D^2 f^L(v) = Dh(v) = (Dg(u))^{-1} = (D^2 f(u))^{-1} = (D^2 f(h(v)))^{-1}.$$

When f is of class C^k, $(f^L)' = h = g^{-1}$ is of class C^{k-1}. Thus, f^L is of class C^k. □

It will be shown that the Legendre transform can be used to reduce the nonlinear second order partial differential equation of *minimal surfaces* in $\mathbb{R}^3$

$$\operatorname{div}\left(\frac{\nabla f(x)}{(1 + \|\nabla f(x)\|^2)^{1/2}}\right) = 0 \qquad x \in \Omega \subset \mathbb{R}^2 \tag{5.42}$$

to a linear equation.

Exercise Let X be a Banach space, let $A : X \to X^*$ be a linear isomorphism, let $b \in X^*$, $c \in \mathbb{R}$ and let f be given by $f(x) := (1/2)\langle Ax, x\rangle + \langle b, x\rangle + c$ for $x \in X$. Show that f is a Legendre function and compute f^L.

Exercise Given $a, b \in \mathbb{R}$, let $f : \mathbb{R} \to \mathbb{R}$ be given by $f(x) := (1/4)x^4 + ax + b$. Show that f is a Legendre function and find its Legendre transform.

5.7.4 The Hamiltonian Formalism

When L is of class C^2, the Euler-Lagrange equation (5.41) is an implicit ordinary differential equation of order two. Let us show how it can be reduced to an explicit first order differential system under the assumption that for $(e,t) \in E \times T$ the function $L_{e,t} : v \mapsto L(e, v, t)$ is a Legendre function on $U_{e,t} := \{v \in E : (e, v, t) \in U\}$. We set $V_{e,t} := DL_{e,t}(U_{e,t})$ and we denote by

$$V := \bigcup_{(e,t) \in E \times T} \{e\} \times V_{e,t} \times \{t\}$$

the image of U under the diffeomorphism $(e, v, t) \mapsto (e, DL_{e,t}(v), t)$. We introduce the *Hamiltonian* $H : V \to \mathbb{R}$ by

$$H(e, p, t) = \langle p, v\rangle - L(e, v, t) \qquad \text{for } p := D_2 L(e, v, t), \tag{5.43}$$

so that $H_{e,t} := H(e, \cdot, t)$ is the Legendre transform of $L_{e,t}$. The passage from the Euler-Lagrange equation to the Hamiltonian system is described in the next result.

Theorem 5.32 (Hamilton) *Suppose L is continuous on U, has continuous partial derivatives D_1L and D_2L and that for all $(e,t) \in T \times E$, the map $D_2L(e,\cdot,t)$ is a diffeomorphism of class C^1 from $U_{e,t}$ onto its image $V_{e,t}$. Let $\overline{x}$ be an extremal and let $\overline{y}(t) := D_2L(\overline{x}(t), \overline{x}'(t), t)$. Then the pair $(\overline{x}, \overline{y})$ satisfies the Hamilton differential system*

$$\overline{x}'(t) = D_2H(\overline{x}(t), \overline{y}(t), t) \tag{5.44}$$

$$\overline{y}'(t) = -D_1H(\overline{x}(t), \overline{y}(t), t). \tag{5.45}$$

Proof Since $L_{e,t}$ is a Legendre function, the relation $p = D_2L(e, v, t)$ is equivalent to the relation $v = D_2H(e, p, t)$:

$$v = D_2H(e, p, t) \Longleftrightarrow p = D_2L(e, v, t). \tag{5.46}$$

Plugging $e := \overline{x}(t)$, $v := \overline{x}'(t)$, $p := \overline{y}(t) := D_2L(\overline{x}(t), \overline{x}'(t), t)$ in (5.46), we get (5.44).

Since for all $(e,t) \in T \times E$, the map $D_2L(e,\cdot,t)$ is a diffeomorphism of class C^1 from $U_{e,t}$ onto $V_{e,t}$, the map $w \mapsto D_2^2L(e, v, t).w$ is invertible and its inverse is the derivative at $p := D_2L(e, v, t)$ of $D_2L(e, \cdot, t)^{-1} = D_2H(e, \cdot, t)$. Thus, the map $v_t : (e, p) \mapsto v_t(e, p)$ defined by the implicit equation

$$p - D_2L(e, v_t(e, p), t) = 0$$

is of class C^1. The expression of H in (5.43) yields

$$H_t(e, p) := H(e, p, t) = \langle p, v_t(e, p)\rangle - L_t(e, v_t(e, p)),$$

where $L_t(e, v_t(e, p)) = L(e, v_t(e, p), t)$. Differentiating both sides with respect to e, abbreviating $v_t(e, p)$ as v_t, one has

$$\begin{aligned} D_1H_t(e, p)e' &= \langle p, D_1v_t(e, p).e'\rangle - D_1L_t(e, v_t)e' - D_2L_t(e, v_t)(D_1v_t(e, p)e') \\ &= -D_1L(e, v(e, p, t), t)e', \end{aligned}$$

for all $e' \in E$, or

$$D_1H(e, p, t) = -D_1L(e, v(e, p, t), t). \tag{5.47}$$

Plugging $e = \overline{x}(t)$, $v = \overline{x}'(t)$, $p := \overline{y}(t)$ into this relation and taking into account the Euler-Lagrange equation (5.41) and relation (5.47), we get

$$\overline{y}'(t) := \frac{d}{dt}\left(D_2L(\overline{x}(t), \overline{x}'(t), t)\right) = D_1L(\overline{x}(t), \overline{x}'(t), t) = -D_1H(\overline{x}(t), \overline{y}(t), t).$$

□

Corollary 5.24 (Legendre) *If L is of class C^2 and if for an extremal $\overline{x}$ and all $t \in T$ the second derivative $D_2^2L(\overline{x}(t), \overline{x}'(t), t)$ is an invertible element of $L(E, E^*)$, then $\overline{x}$ is of class C^2.*

Proof When L is of class C^2 the definition of H in (5.43) shows that the Hamiltonian H is of class C^2. Moreover, our assumption ensures that for all $t \in T$ there is an open neighborhood U of $(\overline{x}(t), \overline{x}'(t), t)$ such that $v \mapsto D_2L(\overline{x}(t), v, t)$ is a C^1 diffeomorphism from $U_{\overline{x}(t),t}$ onto its image $V_{\overline{x}(t),t}$, where $U_{e,t} := \{v \in E : (e, v, t) \in U\}$. We may suppose U contains the compact set $\{(\overline{x}(t), \overline{x}'(t), t) : t \in T\}$. Thus, we can apply Theorem 5.32. By the regularity and uniqueness results for ordinary differential equations we get that $(\overline{x}(\cdot), \overline{y}(\cdot))$ is of class C^1 and in fact of class C^2 since the right-hand sides of (5.44), (5.45) are then of class C^1. □

Exercises

1. Let $(e, v) \mapsto L(e, v)$ be a nonnegative autonomous Lagrangian on some open subset U of $E \times E$. Show that if $\overline{x}(\cdot)$ is an extremal of $\sqrt{L}$ such that the Lagrangian $L(\overline{x}(\cdot), \overline{x}'(\cdot))$ is constant, then $\overline{x}$ is an extremal of L.

   Suppose that for all $(e, v) \in U$ and all $r > 0$ one has $(e, rv) \in U$ and $L(e, rv) = r^2L(e, v)$. Show that any extremal of L is also an extremal of $\sqrt{L}$. [Hint: use Corollary 5.23 and the homogeneity assumption in order to get that L is constant along an extremal.]
2. Let E be a Euclidean space and let L be the Lagrangian given by $L(e, v) := \|v\|^2$. Show that if $\overline{x}$ minimizes $j : x \mapsto \int_0^1 \|x'(t)\|^2 \, dt$ over the set $W(e_0, e_1) := \{x \in R^1(T, E) : x(0) = e_0,\ x(1) = e_1,\ x'(T) \subset E \backslash \{0\}\}$ with $T := [0, 1]$, then $t \mapsto \overline{x}'(t)$ is constant on T and $\overline{x}$ is also an extremal of the length functional $\ell : x \mapsto \int_0^1 \|x'(t)\| \, dt$ over the set $W(e_0, e_1)$. Use the preceding exercise to show that conversely, if $\overline{x}$ is an extremal of the length functional ℓ and if for some change of parameter θ the function $s \mapsto \left\|\overline{x}'(\theta(s))\right\|$ is constant, then $\overline{x} \circ \theta$ is an extremal of j.

5.7.5 *The Several Variables Case*

The Euler-Lagrange necessary condition for the minimization of integral functionals with unknown functions of several variables is the source of numerous partial differential equations. However, we just give a short treatment, considering that the one-variable case gives the main ideas.

Now we suppose that Ω is a bounded open subset of $\mathbb{R}^d$ with smooth boundary, that E is still a Banach space (usually $E := \mathbb{R}$), and that

$$L : E^d \times E \times \Omega \to \mathbb{R},$$

the so-called *Lagrangian* gives rise to the functional

$$j(w) = \int_{\Omega} L(Dw(x), w(x), x)dx \qquad w \in C^2(\Omega, E).$$

Note that here we exchange the place of w and $Dw(x)$ for the convenience of notation. We look for conditions implied by the assumption that a function $u : \Omega \to E$ of class C^2 is a local minimizer of j on a space of functions from Ω to E that contains the affine subspace $C_c^\infty(\Omega, E) + u$, where $C_c^\infty(\Omega, E)$ is the space of functions of class C^∞ with compact support from Ω into E. Taking an arbitrary $v \in C_c^\infty(\Omega, E)$ and assuming that L is of class C^2 and that the conditions for differentiating the integral $j(u + tv)$ with respect to $t \in \mathbb{R}$ for $t = 0$ are satisfied, we obtain the relation $j'(u)(v) = 0$, with

$$j'(u)(v) = \int_{\Omega} [\sum_{i=1}^{d} D_i L(Du(x), u(x), x) D_i v(x) + D_{d+1} L(Du(x), u(x), x) v(x)] dx.$$

Since v has compact support, with our smoothness assumption the Green's formula of Theorem 9.9 yields

$$\int_{\Omega} [-\sum_{i=1}^{d} D_i(D_i L(Du(x), u(x), x)) v(x) + D_{d+1} L(Du(x), u(x), x) v(x)] dx = 0.$$

Since v is arbitrary in $C_c^\infty(\Omega, E)$ we obtain the following second-order partial differential equation in divergence form

$$-\sum_{i=1}^{d} D_i(D_i L(Du(x), u(x), x)) + D_{d+1} L(Du(x), u(x), x) = 0 \qquad x \in \Omega$$

known as the *Euler-Lagrange equation*.

In the next examples we take $E := \mathbb{R}$. In the case $E := \mathbb{R}^n$ we would obtain systems.

Example Assuming that $L(p, e, x) = \frac{1}{2} \|p\|^2 := \frac{1}{2} p_1^2 + \ldots + \frac{1}{2} p_d^2$ for $(p, e, x) \in \mathbb{R}^d \times \mathbb{R} \times \Omega$ we obtain

$$\Delta u(x) = 0 \quad x \in \Omega$$

where Δ is the *Laplacian*; this is *Dirichlet's Principle*.

Example Assuming that for a matrix $A(x) := (a_{i,j}(x))$ and a continuous function $f : \Omega \to \mathbb{R}$ the Lagrangian is the quadratic form $L(p, e, x) = \frac{1}{2} \Sigma_{i,j=1}^d a_{i,j}(x) p_i p_j - f(x) e$ for $(p, e, x) \in \mathbb{R}^d \times \mathbb{R} \times \Omega$ we obtain

$$-\Sigma_{i,j=1}^d D_i(a_{i,j}(x) D_j u(x)) = f(x) \quad x \in \Omega$$

a linear (or rather affine) second-order equation.

It can be shown that the second-order necessary condition

$$j''(u)vv \geq 0 \qquad \forall v \in C_c^\infty(\Omega)$$

leads to the *Legendre condition*

$$\sum_{i,j=1}^{d} \frac{\partial^2 L}{\partial p_i \partial p_j}(Du(x), u(x), x) q_i q_j \geq 0 \qquad x \in \Omega,\ q := (q_i) \in \mathbb{R}^d.$$

Such a condition is related to the assumption that for all $(e, x) \in \mathbb{R} \times \Omega$ the function $L(\cdot, e, x)$ is convex. Such an assumption can be used to get existence results. For such results it is advisable to use the Sobolev space $W_q^1(\Omega)$ (with $q \in]1, \infty[$) considered in Chap. 9 rather than $C^1(\Omega)$: not only it is a larger space but it is also reflexive. The first step in this method (called the *direct method*) is the following semicontinuity result.

Proposition 5.42 *Assume that for some $q \in]1, \infty[$ and some $a \in \mathbb{P}$, $b \in \mathbb{R}_+$ one has $L(p, e, x) \geq a \|p\|^q - b$ for all $(p, e, x) \in \mathbb{R}^d \times \mathbb{R} \times \Omega$ and that $L(\cdot, e, x)$ is convex. Then the functional $w \mapsto j(w)$ is sequentially weakly lower semicontinuous on $W_q^1(\Omega)$.*

Note that the assumption on L ensures that for all $w \in W_q^1(\Omega)$

$$j(w) \geq a \|Dw\|_{L_q(\Omega)}^q - b\lambda_d(\Omega)$$

where $\lambda_d(\Omega)$ is the Lebesgue measure of Ω. Thus $j(w) \to \infty$ as $\|Dw\|_{L_q(\Omega)} \to \infty$. Fixing the value g of w on the boundary $\partial\Omega$ of Ω, one can deduce from the theory of traces that j is coercive on the affine space $W_q^1(\Omega)_g$ of those $w \in W_q^1(\Omega)$ whose traces on $\partial\Omega$ are g. Existence ensues (see [117, Section 8.2]).

Theorem 5.33 *Assuming that the space $W_q^1(\Omega)_g$ is not empty and that the assumptions of the preceding proposition are satisfied, the functional j attains its minimum on $W_q^1(\Omega)_g$.*

Additional Reading

[4, 5, 7, 12, 16, 27, 28, 40, 60, 61, 66, 78, 86, 100, 102, 110, 124, 134–138, 142, 146, 148, 170, 182, 199, 200, 224, 229, 237, 243, 263, 270]

Chapter 6
A Touch of Convex Analysis

A thing of beauty is a joy for ever:
Its loveliness increases; it will never
Pass into nothingness.

John Keats, *Endymion*

Abstract This chapter can be conceived as a substantial course on convex analysis. But it appears here in view of its relationships with other subjects such as optimization and differential calculus. Convex functions have remarkable continuity and differentiability properties. They offer a substitute to the derivative, the subdifferential, whose calculus rules are delineated. Moreover, convexity allows rich duality properties that are displayed along two classical lines: the Lagrangian one and the perturbational one.

The beauty of convex sets (essentially polytopes) has attracted mathematicians since early antiquity. But it is only from the second half of the twentieth century that the class of convex functions, as introduced in Sect. 3.3.1, has been recognized as an important class that enjoys striking, specific, and useful properties. The field of convex analysis is still very active as testified by recent publications [44, 81, 265]. A homogenization procedure enables one to reduce the class of convex functions to the subclass of sublinear functions. This subclass is next to the family of linear functions in terms of simplicity: the epigraph of a sublinear function is a convex cone, a notion almost as simple and useful as the notion of a linear subspace. These two facts explain the rigidity of the class, and its importance: convexity is a simple notion with much power and complexity…

Our interest in this topic arises from the fact that it is the domain of an important transform, the *Young-Fenchel transform* or *Legendre-Fenchel transform*. It enjoys pleasant properties, such as change of infimal convolution into addition. It leads to useful duality properties that are helpful for optimization problems. It relies on separation properties we have already visited.

Convex analysis presents strong analogies with differential calculus. Both subjects enjoy nice calculus rules and lead to necessary conditions for optimization problems. However, in the convex case, these conditions are also sufficient, a remarkable feature. In particular, both approaches are useful for the calculus of variations. Besides striking continuity and differentiability properties, the class of

J.-P. Penot, *Analysis*, Universitext, DOI 10.1007/978-3-319-32411-1_6

convex functions exhibits a substitute for the derivative called the *subdifferential*. It serves as a prototype for nonsmooth analysis. The main differences with classical analysis are the one-sided character of the subdifferential and the fact that a set of linear forms is substituted for the derivative. Still, nice calculus rules can be devised. Some of them, for instance for the subdifferential of the maximum of two functions, go beyond usual calculus rules. Besides classical rules of convex analysis, we sketch some fuzzy rules for the calculus of subdifferentials in the exercises of Sect. 6.3.1 (see also [208, Section 3.5]).

Subdifferentials are closely linked with duality, so we provide a short account of this important topic. We also gather some elements of the geometry of normed spaces that may be useful. Even if we do not insist on that point, it appears that duality plays some role in the interplay between convexity and differentiability of norms or powers of norms.

Convex analysis illustrates a typical feature of nonsmooth analysis that shows a spectacular difference with classical analysis: the study of functions of this class is intimately tied to the study of a class of sets. The many passages from functions to sets and vice versa represent a fruitful and attractive approach that exemplifies the unity and the flexibility of mathematics.

The usefulness of convex analysis in various fields (combinatorics, game theory, geometry, mathematical economics, mechanics, optimization...) is undeniable. Here we illustrate it with just two examples.

Example (Linear Programming) Given $A \in L(\mathbb{R}^m, \mathbb{R}^n)$, $b \in \mathbb{R}^n$, $c \in \mathbb{R}^m$, let us consider the problem

$$(\mathcal{P}) \quad \text{minimize } \langle c|x\rangle \quad \text{subject to } x \in \mathbb{R}^m, \quad x \geq 0, \quad Ax = b.$$

If m is large and n is small, it may be of interest to consider the dual problem

$$(\mathcal{D}) \quad \text{maximize } \langle b|y\rangle \quad \text{subject to } y \in \mathbb{R}^n, \quad A^*y \leq c$$

for which the number n of unknowns is reduced. Here A^* is the adjoint of A. In the case $n = 2$, one can even even solve $(\mathcal{D})$ graphically: the constraint set $C := \{y \in \mathbb{R}^n : A^*y \leq c\}$ is a polyhedron of $\mathbb{R}^2$ and it suffices to move the line $L_r := \{y \in \mathbb{R}^2 : \langle b|y\rangle = r\}$ as much as possible so that it still meets C. In the case $m = 12$ and C is the convex hull of the twelve vertices of a regular planar cristal or of a clock, one obtains in such a way either a vertex or a segment joining two consecutive vertices. If for instance $b = (1, 1)$ is the direction of the light, the segment joining the vertices I and II of the clock in Fig. 6.1 is the set of solutions to $(\mathcal{D})$. Such a graphical solution can be found by young children.

It remains to apply the known relations between the solutions of $(\mathcal{D})$ and the solutions of $(\mathcal{P})$ provided by the theory of linear programming, a special case of convex programming. In particular, the values of the two problems are equal unless one of the feasible sets is empty, and if a solution $\overline{y}$ of $(\mathcal{D})$ is such that $(A^*\overline{y})_i < c_i$ then any solution $\overline{x}$ of $(\mathcal{P})$ must satisfy $\overline{x}_i = 0$. This is often enough to find a

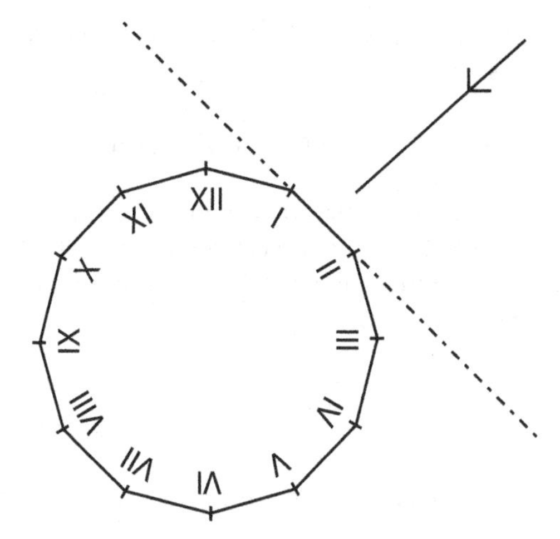

Fig. 6.1 The dual of a linear programming problem

solution of the given problem $(\mathcal{P})$. Of course, in the case of industrial or trading problems the values of m and n may be very large (say 10^5 or 10^6) and one has to use algorithms. Such algorithms exist (for instance the *simplex algorithm* and *interior point algorithms*) and are very efficient.

Example (The Fermat-Torricelli Problem) Given a finite family $a_1,\ldots,a_n$ of distinct points in a normed vector space $(X, \|\cdot\|)$ this problem consists in finding a point $\overline{x} \in X$ that minimizes the function f given by

$$f(x) := \|x - a_1\| + \ldots + \|x - a_n\| \qquad\qquad x \in X.$$

It appears when one wants to find the best location for a warehouse that serves several shops or factories. It can be generalized to the case when some weights w_i affect the terms $\|x - a_i\|$ or when such terms are replaced by $h_i(\|x - a_i\|)$, where $h_i : \mathbb{R}_+ \to \mathbb{R}_+$. It was solved by Torricelli in the case $X = \mathbb{R}^2$ and $n = 3$. The methods of the present chapter can efficiently give the solution to this problem. One first remarks that f is a convex continuous function and that the solution $\overline{x}$ is characterized by the relation

$$0 \in \partial f(\overline{x}) = \partial \|\cdot\| (\overline{x} - a_1) + \partial \|\cdot\| (\overline{x} - a_2) + \partial \|\cdot\| (\overline{x} - a_3),$$

where ∂ is the subdifferential introduced in this chapter. Since $\partial \|\cdot\| (0) = B_{\mathbb{R}^2}$ and $\partial \|\cdot\| (z) = z/ \|z\|$ for $z \in \mathbb{R}^2\backslash\{0\}$, $\overline{x} = a_1$ satisfies this relation if and only if one has

$$0 \in B_{\mathbb{R}^2} + u_2(a_1) + u_3(a_1)$$

where $u_i(x) := (x - a_i)/\|x - a_i\|$ for $x \in \mathbb{R}^2\backslash\{a_i\}$. Since $u_i(x)$ is a unit vector, this relation can be written as

$$\|u_2(a_1) + u_3(a_1)\|^2 \leq 1 \quad \text{or} \quad \langle u_2(a_1) \mid u_3(a_1)\rangle \leq -1/2.$$

This means that the cosine of the angle between the vectors $a_1 - a_2$ and $a_1 - a_3$ is not greater than $-1/2$ or that the angle between these two vectors is at least 120°.

When all the angles of the triangle $a_1a_2a_3$ with vertices a_1, a_2, a_3 are less than 120° the solution $\overline{x}$ is characterized by

$$u_1 + u_2 + u_3 = 0 \qquad \text{with } u_i := u_i(\overline{x}) := \frac{\overline{x} - a_i}{\|\overline{x} - a_i\|}.$$

Since $\|u_i\|^2 = 1$ for $i \in \mathbb{N}_3$, taking scalar products, this relation implies that

$$\begin{aligned}
\langle u_2 \mid u_1\rangle + \langle u_3 \mid u_1\rangle &= -1 \\
\langle u_1 \mid u_2\rangle + \langle u_3 \mid u_2\rangle &= -1 \\
\langle u_1 \mid u_3\rangle + \langle u_2 \mid u_3\rangle &= -1.
\end{aligned}$$

Solving this system by taking the sums of the sides of two different lines yields $\langle u_i \mid u_j\rangle = -1/2$ for $i, j \in \mathbb{N}_3$ with $i \neq j$. Conversely, when these relations hold one gets

$$\|u_1 + u_2 + u_3\|^2 = \|u_1\|^2 + \|u_2\|^2 + \|u_3\|^3 + \sum_{i \neq j = 1}^{3} \langle u_i \mid u_j\rangle = 0$$

since $\|u_i\|^2 = 1$ for $i \in \mathbb{N}_3$. Thus $\overline{x}$ is characterized by these relations and can be constructed by drawing equilateral triangles on each side of the triangle $a_1a_2a_3$ and taking the point $\overline{x}$ common to the 3 segments joining the vertex a_i to the vertex a'_i of the new equilateral triangle opposite to the segment a_ja_k with $j \neq i,\ k \neq i$ as in the next figure (Fig. 6.2).

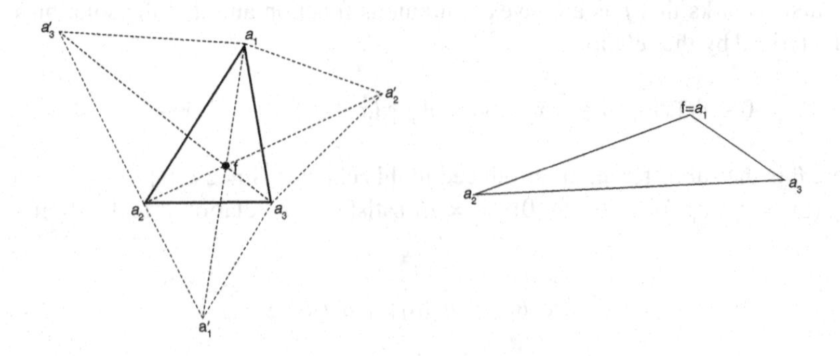

Fig. 6.2 The Fermat-Torricelli problem

6.1 Continuity Properties of Convex Functions

A nice semicontinuity property of convex functions is given in the following statement.

Theorem 6.1 *If $f : X \to \mathbb{R}_\infty$ is a convex function that is lower semicontinuous on a normed space X for the topology associated with the norm, then f is lower semicontinuous on X endowed with the weak topology.*

Proof This is an immediate consequence of Mazur's Theorem: for every real number r the sublevel set $[f \leq r] := \{x \in X : f(x) \leq r\}$ of f is closed and convex, hence weakly closed. □

The preceding proof shows that the same property holds for quasiconvex functions, i.e. functions whose sublevel sets are convex. The following consequence follows from Lemma 2.6, since bounded, closed, convex subsets of a reflexive Banach space are weakly compact.

Corollary 6.1 *A coercive lower semicontinuous convex function f on a reflexive Banach space X attains its infimum.*

Since the epigraph of a lower semicontinuous convex function is the intersection of a family of closed half-spaces, one may guess that such a function is the supremum of a family of continuous affine forms. However, some care is in order. Hereafter we say that a convex function is *closed* if it is lower semicontinuous and either it is identically equal to $-\infty$ (in that case we denote it by $-\infty^X$) or it takes its values in $\mathbb{R}_\infty := \mathbb{R} \cup \{+\infty\}$. Recall that $f \in \overline{\mathbb{R}}^X$ is *proper* if f does not take the value $-\infty$ and if it is not the constant function ∞^X. Then its epigraph is a proper subset of $X \times \mathbb{R}$ (i.e. is nonempty and different from the whole space).

We observed that a lower semicontinuous convex function assuming the value $-\infty$ cannot take a finite value (see Sect. 3.3). Thus a closed convex function $f \in \overline{\mathbb{R}}^X$ is either proper, or $-\infty^X$ or ∞^X. Note that, given a nonempty proper closed convex subset C of X, the *valley function* v_C given by $v_C(x) = -\infty$ for $x \in C$, $v_C(x) = +\infty$ for $x \in X \backslash C$ is an example of a lower semicontinuous convex function that is not closed and not proper. Note that the lower semicontinuous hull $\overline{f}$ of a proper function may be not proper: consider the function $f : \mathbb{R} \to \mathbb{R}$ given by $f(0) := 0$ and $f(x) = -1/|x|$ for $x \in \mathbb{R} \backslash \{0\}$.

If f is the supremum of a nonempty family of continuous affine functions, then f is either ∞^X or a closed proper convex function. In both cases, and in the case of $f = -\infty^X$ (which corresponds to the empty family), it is a closed convex function. A remarkable converse holds.

Theorem 6.2 *Any closed convex function is the supremum of a family of continuous affine functions (the ones it majorizes). If f is proper, this family is nonempty.*

Clearly, if $f = \infty^X$, one can take the family of all continuous affine functions on X, while if $f = -\infty^X$ one takes the empty family. The following lemma is the first step of the proof of this result for the case $f \neq -\infty^X$.

Lemma 6.1 *For any lower semicontinuous convex function $f : X \to \mathbb{R}_\infty$ there exists a continuous affine function g such that $g \leq f$. Moreover, if $w \in \mathrm{dom} f$ and $r < f(w)$ we may require that $g(w) > r$.*

Proof The case $f = \infty^X$ is obvious. Let us suppose $f \neq \infty^X$, so that the epigraph E_f of f is nonempty. Let $w \in \mathrm{dom} f$ and $r < f(w)$. The Hahn-Banach Theorem allows us to separate the compact set $\{(w, r)\}$ from the closed convex set E_f : there exist $(h, c) \in X^* \times \mathbb{R} = (X \times \mathbb{R})^*$ and $b \in \mathbb{R}$ such that

$$\forall (x, s) \in E_f \qquad \langle h, x\rangle + cs > b > \langle h, w\rangle + cr. \tag{6.1}$$

Taking $x = w$, $s > f(w) > r$, we see that $c > 0$. Dividing each side of these inequalities by c, we get

$$s > -c^{-1}h(x) + c^{-1}b \qquad \forall x \in \mathrm{dom} f, \ \ \forall s \geq f(x).$$

It follows that $f \geq g$ for g given by $g(x) := -c^{-1}h(x) + c^{-1}b$. Moreover, the second inequality in relation (6.1) can be written as $g(w) > r$. □

Now let us prove Theorem 6.2. Again, the cases $f = \infty^X, f = -\infty^X$ being obvious, we may suppose f is proper. Let $w \in X$ and $r < f(w)$. If $w \in \mathrm{dom} f$, the preceding lemma provides us with a continuous affine function $g \leq f$ with $g(w) > r$. Now, let us consider the case $w \in X \backslash \mathrm{dom} f$. Separating $\{(w, r)\}$ from E_f, we get some $(h, c) \in (X \times \mathbb{R})^*$ and $b \in \mathbb{R}$ such that relation (6.1) holds. Taking $x \in \mathrm{dom} f$ and s large, we see that $c \geq 0$. If $c > 0$, we can conclude as in the preceding proof. If $c = 0$, observing that $b - h(w) > 0$, taking a continuous affine function k such that $k \leq f$ (such a function exists, by the lemma) and setting

$$g := k + n(b - h),$$

with $n > (b - h(w))^{-1}(r - k(w))$, we see that $g(w) > r$ and $g \leq f$ as $k \leq f$ and $b - h(x) \leq 0$ for $x \in \mathrm{dom} f$ by relation (6.1) with $c = 0$. □

Since lower semicontinuity is stable by the operation of taking suprema, one can deduce Theorem 6.1 from Theorem 6.2.

For convex functions, one disposes of remarkably simple continuity criteria.

Proposition 6.1 *Let $f : X \to \mathbb{R}_\infty$ be a convex function on a normed space (or topological vector space) X. If f is finite at some $\overline{x} \in X$, the following assertions are equivalent:*

(a) f is bounded above on some neighborhood V of $\overline{x}$;
(b) f is upper semicontinuous at $\overline{x}$;
(c) f is continuous at $\overline{x}$.

Proof The implications (c)⇒(b)⇒(a) are obvious. Let us show (a)⇒(b) and (b)⇒(c). We may suppose that $\overline{x} = 0$, $f(\overline{x}) = 0$ by performing a translation and adding a constant. Given $\varepsilon > 0$, let $m \geq \sup f(V)$, $m \geq \varepsilon$. Let $U := \varepsilon m^{-1} V$. Then, for $u \in U$, setting $v := \varepsilon^{-1} m u \in V$, by convexity we have

$$f(u) \leq \varepsilon m^{-1} f(v) + (1 - \varepsilon m^{-1}) f(0) \leq \varepsilon$$

and, as U is a neighborhood of $0, f$ is upper semicontinuous at 0.

In order to deduce (c) from (b), i.e. that f is continuous at 0, we note that for $w \in W := U \cap (-U)$ we have $0 = f(0) \leq \frac{1}{2} f(w) + \frac{1}{2} f(-w) \leq \frac{1}{2} f(w) + \frac{1}{2}\varepsilon$ hence $f(w) \geq -\varepsilon$. □

Remark If V is the ball $B[\overline{x}, r]$ and $\sup f(V) \leq m$, for $c := r^{-1}(m - f(\overline{x}))$ one has

$$\forall x \in B[0, r] \qquad f(\overline{x} + x) - f(\overline{x}) \leq c \left\| x \right\|$$

since for $x \in B[0, r]$, setting $t := r^{-1} \left\| x \right\|$, taking u such that $\left\| u \right\| = r$, $x = tu$, one gets

$$\begin{aligned} f(\overline{x} + x) - f(\overline{x}) &= f((1 - t)\overline{x} + t(\overline{x} + u)) - f(\overline{x}) \\ &\leq t\,(f(\overline{x} + u) - f(\overline{x})) \leq r^{-1}(m - f(\overline{x})) \left\| x \right\|, \end{aligned}$$

a property called *quietness* at $\overline{x}$. In fact, for all $x \in B[0, r]$, since $f(\overline{x} + x) - f(\overline{x}) \geq -(f(\overline{x} - x) - f(\overline{x})) \geq -c \left\| x \right\|$, we have $|f(\overline{x} + x) - f(\overline{x})| \leq c \left\| x \right\|$, and we say that f is *stable at* $\overline{x}$. Later on this property will be reinforced into a local Lipschitz property. □

The following results illustrate the uses of the preceding criteria.

Proposition 6.2 *Suppose $f : X \to \mathbb{R}_\infty$ is a convex function on a finite dimensional space X. Then f is continuous on the interior of its domain* $D_f := \operatorname{dom} f := f^{-1}(\mathbb{R})$.

Proof Given $\overline{x} \in \operatorname{int} D_f$, let $x_1, \ldots, x_n \in D_f$ be such that $\overline{x}$ belongs to the interior of the convex hull C of $\{x_1, \ldots, x_n\}$ (for instance, one can take for C a ball with center $\overline{x}$ for some polyhedral norm, X being identified with some $\mathbb{R}^d$). Then f is bounded above on C by $m := \max(f(x_1), \ldots, f(x_n))$, hence is continuous at $\overline{x}$. □

Proposition 6.3 *Let $f : X \to \mathbb{R}_\infty$ be a lower semicontinuous convex function on a Banach space X. Then f is continuous on the core of its domain D_f (which coincides with the interior of D_f).*

Proof Given $\overline{x} \in \operatorname{core} D_f$, let $m > f(\overline{x})$ and let $C := \{x \in X : f(x) \leq m\}$. Again we may suppose $\overline{x} = 0$. Then C is a closed convex subset of X that is absorbing: for all $x \in X$ we can find $r > 0$ such that $rx \in D_f$ and for $s > 0$ small enough we have $f(rsx) \leq (1 - s)f(0) + sf(rx) < m$ so that $rsx \in C$. Thus C is a neighborhood of 0 by Lemma 3.31 and f is continuous at 0 by Proposition 6.1 □

Convex functions enjoy a "miraculous" propagation property.

Proposition 6.4 *Let $f : X \to \mathbb{R}_\infty$ be a convex function on a normed space X. If f is continuous at some $\overline{x} \in D_f :=$ dom$f$ then f is continuous on the interior of D_f.*

Proof Given $x_0 \in \mathrm{int} D_f$, let us prove that f is continuous at x_0. Using a translation, we may suppose $x_0 = 0$. Then, as D_f is a neighborhood of 0, there exists some $r > 0$ such that $\overline{y} := -r\overline{x} \in D_f$. Let V be a neighborhood of 0 such that f is bounded above by some m on $\overline{x} + V$. Then, by convexity, f is bounded above on

$$r(1+r)^{-1}(\overline{x}+V) + (1+r)^{-1}\overline{y} = r(1+r)^{-1}V \in \mathcal{N}(0)$$

by $r(1+r)^{-1}m + (1+r)^{-1}f(\overline{y})$. Then, by Proposition 6.1, f is continuous at x_0. □

Local boundedness of a convex function entails a regularity property stronger than continuity: a local Lipschitz property. In fact, the result is not just a local one: the following statement and its corollary give a precise content to this assertion: the corollary shows that a Lipschitz property is available on balls that may be big provided the function is bounded above on a larger ball. One even gets a quantitative estimate of the Lipschitz rate.

Proposition 6.5 *Let f be a convex function on a convex subset C of a normed space X and let $\alpha, \beta \in \mathbb{R}$, $\rho > 0$. Suppose f is bounded below by β on a subset B of C and is bounded above by α on a subset A of C such that $B + \rho U_X \subset A$, where U_X is the open unit ball of X. Then f is Lipschitzian on B with rate $\rho^{-1}(\alpha - \beta)$.*

Proof Given $x, y \in B$ and $\delta > \|x - y\|$, let $z := y + \rho\delta^{-1}(y - x) \in A$, since $B + \rho U_X \subset A$. Then $y = x + t(z - x)$ where $t := \delta(\delta + \rho)^{-1} \in [0, 1]$, hence

$$f(y) - f(x) \le t(f(z) - f(x)) \le t(\alpha - \beta) \le \delta\rho^{-1}(\alpha - \beta).$$

Interchanging the roles of x and y and taking the infimum on δ in $]\|x - y\|, \infty[$, we get $|f(y) - f(x)| \le \rho^{-1}(\alpha - \beta)\|x - y\|$. □

The preceding statement is versatile enough to apply in a variety of geometric cases. The simplest one is the case of balls.

Corollary 6.2 *Suppose the convex function f on the normed space X is bounded above by α on some ball $B(\overline{x}, r)$. Then, for any $s \in]0, r[$ the function f is Lipschitzian on the ball $B(\overline{x}, s)$ with rate $2(r - s)^{-1}(\alpha - f(\overline{x}))$.*

Proof Taking $A := B(\overline{x}, r), B := B(\overline{x}, s)$, $\rho := r - s$, $\beta := 2f(\overline{x}) - \alpha$ it suffices to observe that for all $x \in B$ one has $f(x) \ge \beta$ by convexity. □

Corollary 6.3 *Any convex function which is continuous on an open convex subset U of a normed space is locally Lipschitzian on U.*

Convex functions enjoy nice properties for what concerns optimization. A simple example is as follows.

Proposition 6.6 *Any local minimizer of a convex function $f : X \to \mathbb{R}_\infty := \mathbb{R} \cup \{+\infty\}$ on a normed space (or topological vector space) X is a global minimizer.*

Proof Let $\overline{x} \in X$ and let V be a neighborhood of $\overline{x}$ such that $f(\overline{x}) \leq f(v)$ for all $v \in V$. Given $x \in X$, one can find $t \in]0, 1[$ such that $v := \overline{x} + t(x - \overline{x}) \in V$. Then, by convexity, we have $tf(x) + (1 - t)f(\overline{x}) \geq f(v) \geq f(\overline{x})$, hence $f(x) \geq f(\overline{x})$. □

Exercises

1. Let X be a separable Hilbert space with Hilbertian basis $\{e_n : n \in \mathbb{N}\}$ and let the function $f : X \to \mathbb{R}$ be given by

$$f(x) := \sum_{n=0}^{\infty} | x_n |^{n+2} \qquad \text{for } x = \sum_{n=0}^{\infty} x_n e_n.$$

   (a) Show that f is well defined on X, bounded above by 1 on the unit ball and everywhere bounded below by 0.
   (b) Show that the Lipschitz rate of f around e_k is at least $k + 2$.
   (c) Deduce from the preceding that f is not Lipschitzian on the ball rB_X with $r > 1$. Observe that f is not bounded above on such a ball.

2. Using the data and the notation of Corollary 6.2 and noting that f is bounded above on $B(\overline{x}, s)$ by $(1 - r^{-1}s)f(\overline{x}) + r^{-1}s\alpha$, hence is bounded below by $\beta := (1 + r^{-1}s)f(\overline{x}) - r^{-1}s\alpha$ on this ball, show that the Lipschitz rate of f on $B(\overline{x}, s)$ is at most $(1 + r^{-1}s)(r - s)^{-1}(\alpha - f(\overline{x}))$.
3. Prove a similar estimate of the Lipschitz rate of f when one supposes that f is bounded above by some α on the sphere with center $\overline{x}$ and radius r.
4. **(a)** Let $f : X \to \mathbb{R}$ be a uniformly continuous function on a normed space. Show that for any $\delta > 0$ there exists a $k > 0$ such that $d(f(x), f(y)) \leq kd(x, y)$ for all $x, y \in X$ satisfying $d(x, y) \geq \delta$. [Hint: use a subdivision of the segment $[x, y]$ by points u_i such that $d(u_i, u_{i+1}) \leq \alpha$, where $\alpha > 0$ is such that $d(f(u), f(v)) \leq 1$ whenever $u, v \in X$ satisfy $d(u, v) \leq \alpha$.]
   (b) Prove that any uniformly continuous convex function f on X is Lipschitzian. [Hint: use (a) and Proposition 6.5.]
5. (**The log barrier**) Prove that $f : \mathbb{R}^{n^2} \to \mathbb{R}_\infty$ given by $f(u) = -\log(\det u)$ if u is a symmetric positive definite matrix, $+\infty$ otherwise, is a convex function.
6. Deduce from Proposition 6.3 that for any closed convex subset of a Banach space one has $\text{int}C = \text{core}\, C$. [Hint: use the indicator function ι_C of C.]
7. Prove that on the dual X^* of a non-reflexive Banach space X one can find a convex function f that is continuous in the topology associated with the dual norm, but that is not lower semicontinuous in the weak* topology. [Hint: take $f \in X^{**} \backslash X$.]

6.2 Differentiability Properties of Convex Functions

Convexity of a function entails particular differentiability properties. The case of a one-variable function, which is our starting point, will provide our first evidence. However, it is a substitute for the derivative that will be the main point of this section. Later on, we will see that this new object, called the subdifferential, enjoys useful calculus rules.

6.2.1 Derivatives of Convex Functions

We first observe that if $f : T \to \mathbb{R}$ is a finite convex function on some interval T of $\mathbb{R}$, then for $r < s < t$ in T the following inequalities hold:

$$\frac{f(s) - f(r)}{s - r} \leq \frac{f(t) - f(r)}{t - r} \leq \frac{f(t) - f(s)}{t - s}. \tag{6.2}$$

They express that the slope of a secant to the graph of f is a nondecreasing function of the abscissas of its extremities and stem from the convexity inequality

$$f(s) = f\left(\frac{t-s}{t-r}r + \frac{s-r}{t-r}t\right) \leq \frac{t-s}{t-r}f(r) + \frac{s-r}{t-r}f(t)$$

(since the coefficients of $f(r)$ and $f(t)$ are in $[0, 1]$ and have sum 1), which yields

$$f(s) - f(r) \leq \frac{s-r}{t-r}(f(t) - f(r)), \quad f(t) - f(s) \geq \frac{t-s}{t-r}(f(t) - f(r)).$$

Lemma 6.2 *If $f : T \to \mathbb{R}$ is a finite convex function on some interval T of $\mathbb{R}$, then, for any $s \in T \backslash \{\sup T\}$ the right derivative $f'_r(s) := D_r f(s)$ of f at s exists in $\mathbb{R} \cup \{-\infty\}$ and is given by*

$$D_r f(s) := \lim_{t \to s+} \frac{f(t) - f(s)}{t - s} = \inf_{t>s} \frac{f(t) - f(s)}{t - s}.$$

If moreover, s is in the interior of T then $D_r f(s)$ is finite, the left derivative $D_\ell f(s)$ exists, is finite and $D_\ell f(s) \leq D_r f(s)$. Furthermore, the functions $s \mapsto D_r f(s)$ and $s \mapsto D_\ell f(s)$ are nondecreasing.

Proof The first assertion is a direct consequence in the existence of a limit for the nondecreasing function $t \mapsto (t - s)^{-1}(f(t) - f(s))$ on $]s, \sup T[$. Changing f into g given by $g(u) := f(-u)$, we get the assertions about the left derivative. The second assertion stems from the fact that when $s \in \text{int}T$, the limit is finite since, by (6.2), for $r < u < s$ the quotient $(s - u)^{-1}(f(s) - f(u))$ is bounded below by

$(s-r)^{-1}(f(s)-f(r))$. Thus

$$\frac{f(s)-f(r)}{s-r} \leq D_\ell f(s) = \sup_{u<s} \frac{f(u)-f(s)}{u-s} \leq \inf_{t>s} \frac{f(t)-f(s)}{t-s} = D_r f(s).$$

Similarly, changing (s,t) into (r,s), we have $D_r f(r) \leq (s-r)^{-1}(f(s)-f(r)) \leq D_r f(s)$. □

It may happen that the left derivative $D_\ell f$ of a convex function f does not coincide with the right derivative (consider $t \mapsto |t|$). Relation (6.2) shows that for $r < t$ one has $D_r f(r) \leq D_\ell f(t)$. Thus, if $D_\ell f(t) < D_r f(t)$ one gets $\lim_{r \to t_-} D_r f(r) \leq D_\ell f(t) < D_r f(t)$ and $D_r f(\cdot)$ has a jump at t. Since $D_r f(\cdot)$ is nondecreasing, such points of discontinuity of $D_r f(\cdot)$ are at most countable. Since f is nondifferentiable at t if and only if $D_\ell f(t) < D_r f(t)$, we get the next result.

Proposition 6.7 *Let $f : T \to \mathbb{R}$ be a convex function on an open interval of $\mathbb{R}$. Then the set of points at which f is not differentiable is at most countable.*

The following characterizations of convexity are classical and useful.

Proposition 6.8 *Let $f : T \to \mathbb{R}$ be a differentiable function on an open interval of $\mathbb{R}$. Then f is convex if and only if its derivative is nondecreasing.*

If f is twice differentiable, f is convex if and only if for all $r \in T$ one has $f''(r) \geq 0$.

Proof The necessary condition is a consequence in Lemma 6.2. Let us prove the sufficient condition. Let f be differentiable with a nondecreasing derivative. Given $r, t \in T$ and $s \in]r,t[$, we have $s = ar + bt$ with $a = (t-r)^{-1}(t-s) \geq 0$, $b = (t-r)^{-1}(s-r) \geq 0$, $a+b = 1$. The Mean Value Theorem yields some $p \in]r,s[$ and some $q \in]s,t[$ such that $(s-r)^{-1}(f(s)-f(r)) = f'(p)$, $(t-s)^{-1}(f(t)-f(s)) = f'(q)$. Since $f'(p) \leq f'(q)$, rearranging terms, we get

$$(t-r)f(s) \leq (t-s)f(r) + (s-r)f(t)$$

or equivalently $f(s) \leq af(r) + bf(t)$. Thus f is convex. The last assertion is given by elementary calculus. □

Now suppose $f : X \to \mathbb{R}_\infty$ is defined on a vector space X.

Proposition 6.9 *If $f : X \to \mathbb{R}_\infty := \mathbb{R} \cup \{+\infty\}$ is a convex function on a vector space X, then for all $x \in \operatorname{dom} f$ and for all $v \in X$ the radial derivative*

$$d_r f(x,v) := \lim_{t \to 0_+} \frac{f(x+tv)-f(x)}{t}$$

exists and is equal to $\inf_{t>0} t^{-1}[f(x+tv)-f(x)]$. It is finite if $x \in \operatorname{core}(\operatorname{dom} f)$. If X is a Banach space, if $x \in \operatorname{int}(\operatorname{dom} f)$, and if f is lower semicontinuous, then the *directional derivative*

$$df(x,v) := \lim_{(t,w)\to(0_+,v)} \frac{f(x+tw)-f(x)}{t}$$

exists and coincides with the radial derivative. In particular, if $u : [0,a] \to X$ *with* $a > 0$ *is right differentiable at* 0 *with* $u(0) = x$*, then* $f \circ u$ *is right differentiable at* 0 *and* $(f \circ u)'_r(0) = d_r f(x, u'_r(0))$.

Proof Let g be given by $g(t) = f(x+tv)$. Then g is convex and its right derivative at 0 is $d_r f(x,v)$. It exists in $[-\infty, +\infty[$ if $]x + \mathbb{P}v[\cap \operatorname{dom} f$ is nonempty and it is $+\infty$ otherwise. Even in the latter case, this right derivative is $\inf_{t>0} t^{-1}(g(t) - g(0)) = \inf_{t>0} t^{-1}(f(x+tv) - f(x))$. When x belongs to core(dom$f$), for any $v \in X$, 0 is in the interior of dom g and we can conclude with Lemma 6.2.

When X is a Banach space, f is lower semicontinuous, and $x \in \operatorname{int}(\operatorname{dom} f)$ the function f is Lipschitzian on a neighborhood of x, so that there exists some $c \in \mathbb{R}_+$ such that $|(1/t)(f(x+tw) - f(x+tv))| \leq c\,\|w - v\| \to 0$ as $(t,w) \to (0_+, v)$. Thus $df(x,v) = d_r f(x,v)$. The last assertion ensues. □

Proposition 6.10 *If* $f : X \to \mathbb{R}_\infty$ *is a convex function on a vector space* X*, then for all* $x \in \operatorname{dom} f$*, the radial derivative* $d_r f(x,\cdot)$ *is a sublinear function.*

Proof Clearly $d_r f(x,\cdot)$ is positively homogeneous. Let us prove it is subadditive: for any $v, w \in X$ we have $f(x + \frac{1}{2}t(v+w)) \leq \frac{1}{2}f(x+tv) + \frac{1}{2}f(x+tw)$, hence

$$\begin{aligned}
d_r f(x, v+w) &= \lim_{t\to 0_+} \frac{2}{t}[f(x + \frac{t}{2}(v+w)) - f(x)] \\
&\leq \lim_{t\to 0_+} \frac{1}{t}(f(x+tv) - f(x)) + \lim_{t\to 0_+} \frac{1}{t}(f(x+tw) - f(x)) \\
&= d_r f(x,v) + d_r f(x,w).
\end{aligned}$$

□

The preceding statement can also be justified by checking that

$$d_r f(x,v) = \inf\{s : (v,s) \in T^r(E_f, x_f)\},$$

where E_f is the epigraph of f, $x_f := (x, f(x))$ and $T^r(E_f, x_f)$ is the radial tangent cone to E_f at x_f, where the *radial tangent cone* to a convex set C at $z \in C$ is the set

$$T^r(C,z) := \mathbb{R}_+(C - z).$$

When X is a normed space, $T^r(C,z)$ is not closed in general, as simple examples show. Therefore, it is advisable to replace it with the *tangent cone* $T(C,z)$ to C at z. In the case when C is convex, $T(C,z)$ is just the closure of $T^r(C,z)$. In the case when C is the epigraph of a convex function f finite at x and $z := x_f := (x, f(x))$, it can be shown that $T(C, z)$ is the epigraph of the (lower) *directional derivative* of f at x defined by

$$f'(x, v) := df(x, v) := \liminf_{(t,u)\to(0_+,v)} \frac{f(x+tu)-f(x)}{t}.$$

Since $f'(x, \cdot) = df(x, \cdot)$ is lower semicontinuous, it has better duality properties than $d_r f(x, \cdot)$. Moreover, it is as closely connected to the notion of a subdifferential of f at x as $d_r f(x, \cdot)$. We consider this notion in the next subsection.

6.2.2 *Subdifferentials of Convex Functions*

Since a general convex function f may have kinks, it may happen that there is not just one affine function minorizing f and taking the value $f(x)$ at a given point $x \in \operatorname{dom} f$. For example, this occurs with $f := |\cdot| : \mathbb{R} \to \mathbb{R}$ and $x := 0$: every linear form $x^* \in [-1, 1]$ satisfies $x^* \leq f$, $x^*(0) = f(0)$. It is worth considering the set of such continuous linear forms.

Definition 6.1 (Fenchel, Moreau) If $f : X \to \mathbb{R}_\infty$ is a function on a normed space X and $x \in X$, then the *subdifferential* of f at x is the empty set if $x \in X \setminus \operatorname{dom} f$ and if $x \in \operatorname{dom} f$, it is the set $\partial f(x)$ of $x^* \in X^*$ such that

$$\forall w \in X \qquad f(w) \geq f(x) + \langle x^*, w - x \rangle. \tag{6.3}$$

This is a global notion which is very restrictive for an arbitrary function. For a convex function it turns into a crucial tool that is a useful substitute for the derivative, as we will shortly see. A strong advantage of the subdifferential is that it yields a characterization of minimizers.

Proposition 6.11 *A function f on a normed space X attains its minimum at $x \in \operatorname{dom} f$ if and only if $0 \in \partial f(x)$.*

The result is an immediate consequence in the definition. Calculus rules will make it efficient. In particular, they enable us to give optimality conditions for problems with constraints.

A first consequence in the next result is that the subdifferential of a convex function f is not just a global notion, but also a local notion. Let us recall that f is said to be *Gateaux differentiable* at x with derivative $Df(x) := \ell \in X^*$ if f is finite at x and if for all $v \in X$

$$\frac{f(x+tv)-f(x)}{t} \to \ell(v) \text{ as } t \to 0,\ t \neq 0.$$

Theorem 6.3 *If f is a convex function on a normed space X and $x \in \mathrm{dom} f$ then*

$$x^* \in \partial f(x) \Longleftrightarrow \forall v \in X \ \ \langle x^*, v\rangle \le df(x,v)$$
$$\Longleftrightarrow \forall v \in X \ \ \langle x^*, v\rangle \le d_r f(x,v).$$

If $x \in \mathrm{core}(\mathrm{dom} f)$ and f is Gateaux differentiable at x, then $\partial f(x) = \{Df(x)\}$.

Proof Given $x^* \in \partial f(x)$, for any $t \in \mathbb{P}$, $u \in X$ we have

$$\langle x^*, tu\rangle \le f(x+tu) - f(x).$$

Dividing by t and taking the lower limit as $(t,u) \to (0_+, v)$, we get $\langle x^*, v\rangle \le df(x,v) \le d_r f(x,v)$. Now, if f is convex and if x^* satisfies the inequality $\langle x^*, v\rangle \le d_r f(x,v)$ for all $v \in X$, then, for $v \in X$, $t \in]0,1[$, by the monotonicity observed in relation (6.2), we have

$$\langle x^*, v\rangle \le d_r f(x,v) \le \frac{1}{t}(f(x+tv) - f(x)) \le f(x+v) - f(x).$$

Setting $v = w - x$, we obtain relation (6.3). The last assertion is obvious: if $x^* \le \ell := Df(x)$ then one has $x^* = \ell$. □

A geometric interpretation of the subdifferential of a function can be given in terms of the normal cone to its epigraph. Recall that the *normal cone* to a convex subset C of a normed space X at some $z \in C$ is defined as the set $N(C,z)$ of $z^* \in X^*$ such that $\langle z^*, w - z\rangle \le 0$ for every $w \in C$; thus it is the polar cone to the radial tangent cone $T^r(C,z)$ and also, by density, it is the polar cone to $T(C,z)$.

Proposition 6.12 *For a convex function f on a normed space X and $x \in \mathrm{dom} f$, one has the following equivalence in which E_f is the epigraph of f and $x_f := (x, f(x))$:*

$$x^* \in \partial f(x) \Longleftrightarrow (x^*, -1) \in N(E_f, x_f).$$

The proof is immediate from the definition of $\partial f(x)$:

$$x^* \in \partial f(x) \Leftrightarrow \forall (w,r) \in E_f \ \ \langle x^*, w - x\rangle \le r - f(x) \Leftrightarrow (x^*, -1) \in N(E_f, x_f).$$

On the other hand, the normal cone to a convex set can be described in terms of subdifferentials.

Proposition 6.13 *For a convex subset C of a normed space X, the normal cone to C at $x \in C$ is the subdifferential of the indicator function ι_C to C at x. It is also the cone $\mathbb{R}_+ \partial d_C(x)$ generated by the subdifferential of the distance function to C at x.*

Proof By definition, $x^* \in N(C,x)$ if and only if $\langle x^*, w - x\rangle \le 0$ for all $w \in C$. Since $\iota_C(w) = 0$ for $w \in C$ and $\iota_C(w) = \infty$ for $w \in X \backslash C$, this property is equivalent to $x^* \in \partial \iota_C(x)$.

The inclusion $\mathbb{R}_+\partial d_C(x) \subset N(C,x)$ is obvious: when $r \in \mathbb{R}_+$, $x^* \in \partial d_C(x)$, one has

$$\forall w \in C \qquad \langle rx^*, w - x\rangle \leq rd_C(w) - rd_C(x) = 0.$$

Conversely, when $x^* \in N(C,x)$, the function $-x^*$ attains its infimum on C at x, and is Lipschitzian with rate $c = \|x^*\|$, so that, by the Penalization Lemma, $-x^* + cd_C$ attains its infimum on X at x; then $0 \in \partial(-x^* + cd_C)(x)$, which is equivalent to $x^* \in c\partial d_C(x)$. □

The last argument shows the interest of disposing of calculus rules. Such rules will be considered in the next section.

A simple consequence of the *subdifferentiability* of a convex function f at a point x (i.e., of the nonemptiness of $\partial f(x)$) is the lower semicontinuity of f at x. The converse is not true, as shows the example of $f : \mathbb{R} \to \mathbb{R}_\infty$ given by $f(r) := -\sqrt{1-r^2}$ for $r \in [-1,1]$, $+\infty$ otherwise and $x := 1$. Still continuity entails subdifferentiability. This is a remarkable criterion!

Theorem 6.4 (Moreau) *If a convex function f on a normed space X is finite and continuous at x, then $\partial f(x)$ is nonempty and weak* compact. Moreover, for all $u \in X$*

$$df(x,u) = \max\{\langle x^*, u\rangle : x^* \in \partial f(x)\}.$$

Proof For every $r > f(x)$ there exists a neighborhood V of x such that $V \times [r, \infty[$ is contained in the epigraph E_f of f. Thus, the interior of E_f is convex and nonempty. It does not contain $x_f := (x, f(x))$ since for $s < f(x)$ close to $f(x)$ one has $(x,s) \notin E_f$. The geometric Hahn-Banach theorem yields some $(u^*, c) \in (X \times \mathbb{R})^*$ such that

$$\langle u^*, w\rangle + cr > \langle u^*, x\rangle + cf(x) \qquad \forall (w,r) \in \text{int}E_f.$$

This implies (by taking $w = x$, $r = f(x) + 1$) that $c > 0$ and, by Lemma 3.12, that

$$\langle u^*, w - x\rangle + c(r - f(x)) \geq 0 \qquad \forall (w,r) \in E_f.$$

In turn, this relation, which can be written as

$$f(w) - f(x) \geq \langle -c^{-1}u^*, w - x\rangle \qquad \forall w \in X,$$

shows that $x^* := -c^{-1}u^* \in \partial f(x)$. Thus $\partial f(x)$ is nonempty.

Since $\partial f(x)$ is the intersection of the weak* closed half-spaces

$$D_w := \{x^* \in X^* : \langle x^*, w - x\rangle \leq f(w) - f(x)\}, \qquad w \in \text{dom}f,$$

it is always weak* closed. When f is continuous at x, taking $\rho > 0$ such that $\sup f(B(x,\rho)) \le f(x) + 1$, for all $x^* \in \partial f(x)$ we have

$$\|x^*\| = \rho^{-1} \sup\{\langle x^*, w - x\rangle : w \in B(x,\rho)\} \le \rho^{-1}.$$

The second assertion will be proved with the alternative proof that follows. □

Alternative proof By the remark following Proposition 6.1 we can find $c \in \mathbb{R}_+$ and $r > 0$ such that $|f(x+v) - f(x)| \le c\|v\|$ for $v \in B(0,r)$. It follows that $|df(x,w)| \le c\|w\|$ for $w \in X$. Given $u \in X$ the Hahn-Banach Theorem yields some linear form x^* such that $x^* \le df(x,\cdot) \le c\|\cdot\|$ and $\langle x^*, u\rangle = df(x,u)$. Thus, x^* is continuous and $x^* \in \partial f(x)$. □

Remark Without the continuity assumption, $\partial f(x)$ may be unbounded. This is the case for the indicator function of $\mathbb{R}_+$ on $X = \mathbb{R}$, for which $\partial f(0) = -\mathbb{R}_+$.

Examples

(a) For $f := \|\cdot\|$ one has $\partial\|\cdot\|(0) = B_{X^*}$ and $\partial\|\cdot\|(x) = \{x^* \in X^* : \|x^*\| = 1, \langle x^*, x\rangle = \|x\|\}$ for $x \in X\backslash\{0\}$.
(b) Let X be a normed space and let $j(\cdot) := \frac{1}{2}\|\cdot\|^2$. Then $\partial j(x) = J(x)$, the *duality (multi)map* defined by

$$J(x) := \{x^* \in X^* : \|x^*\| = \|x\|, \langle x^*, x\rangle = \|x\|^2\},$$

and $J(x)$ is nonempty, as shown by applying Corollary 3.14 or Theorem 6.4. □

Corollary 6.4 *Let $f : X \to \mathbb{R}_\infty$ be a convex function finite and continuous at $x \in X$. Then f is Gateaux and Hadamard differentiable at x if and only if $\partial f(x)$ is a singleton $\{x^*\}$. Moreover, $Df(x) = x^*$.*

Proof Suppose $\partial f(x) = \{x^*\}$. The preceding theorem ensures that $df(x,\cdot) = x^*$. Thus f is Gateaux differentiable. Since f is Lipschitzian around x, it is Hadamard differentiable. The converse is an obvious consequence in Theorem 6.3. □

Corollary 6.5 *Let f be a convex function on a normed space X. Suppose the restriction of f to the affine subspace A generated by* dom$f$ *is continuous at $x \in$ dom$f$. Then $\partial f(x)$ is nonempty.*

Proof Without loss of generality, we may suppose $x = 0$, so that A is the vector subspace generated by dom$f$. The preceding theorem ensures that the restriction $f \mid A$ of f to A is subdifferentiable at 0. Then, any continuous linear extension of any element of $\partial(f \mid A)(0)$ belongs to $\partial f(0)$, and such extensions exist by the Hahn-Banach Theorem. □

Recall that for a subset D of a normed vector space X, ri$D$ is the set of points that belong to the interior of D in the affine subspace Y generated by D.

Corollary 6.6 *Let f be a convex function on a finite dimensional normed space X and let $x \in \operatorname{ri}\operatorname{dom}f$ (i.e., be such that $\mathbb{R}_+(\operatorname{dom}f - x)$ is a linear subspace). Then $\partial f(x)$ is nonempty.*

Proof Taking $D = \operatorname{dom}f$, by Prop. 6.2 we have that the restriction g of f to Y is continuous at x. The preceding corollary applies. □

In the general case of a closed proper convex function f on a Banach space, a density result for the set $\operatorname{dom}\partial f$ of subdifferentiability points of f can be given.

Theorem 6.5 (Brøndsted-Rockafellar) *For a closed proper convex function f on a Banach space X, the set of points $x \in X$ such that $\partial f(x)$ is nonempty is dense in* $\operatorname{dom}f$. *More precisely, for any $\overline{x} \in \operatorname{dom}f$ there exists a sequence $(x_n) \to \overline{x}$ such that $(f(x_n)) \to f(\overline{x})$ and $\partial f(x_n) \neq \varnothing$ for all $n \in \mathbb{N}$.*

Proof Given $\overline{x} \in \operatorname{dom}f$ and a sequence $(\varepsilon_n) \to 0_+$, Lemma 6.1 provides some $x_n^* \in X^*$, $r_n \in \mathbb{R}$ such that $g_n := \langle x_n^*, \cdot \rangle - r_n \leq f$ and $g_n(\overline{x}) > f(\overline{x}) - \varepsilon_n$. Then we have $\inf(f - g_n) \geq 0$ and $f(\overline{x}) - g_n(\overline{x}) < \varepsilon_n$. The Ekeland's variational principle yields some $x_n \in B(\overline{x}, \varepsilon_n)$ such that

$$f(x) - g_n(x) + \|x - \overline{x}_n\| \geq f(x_n) - g_n(x_n).$$

Thus, $0 \in \partial(f - g_n + \|\cdot - x_n\|)(x_n)$. The sum rule we shall see shortly (Theorem 6.8) ensures that there exists some $u_n^* \in \partial \|\cdot - x_n\|(x_n) = \partial \|\cdot\|(0) = B_{X^*}$ such that $x_n^* - u_n^* \in \partial f(x_n)$.

Exercises

1. Establish the inequality $xy \leq p^{-1}x^p + q^{-1}y^q$ for any $x, y \in \mathbb{R}_+, p, q > 1$ satisfying $p^{-1} + q^{-1} = 1$ by minimizing the function $x \mapsto p^{-1}x^p - xy$ for a fixed $y > 0$. Deduce from this inequality the *Hölder's inequality*:

$$\forall a := (a_i),\ b := (b_i) \in \mathbb{R}^n \qquad \sum_{i=1}^{n} |a_i b_i| \leq \left(\sum_{i=1}^{n} |a_i|^p\right)^{1/p} \left(\sum_{i=1}^{n} |b_i|^q\right)^{1/q}.$$

   [Hint: set $s_i := a_i / \|a\|_p$, $t_i := b_i / \|b\|_q$, with $\|a\|_p := (\Sigma_{1 \leq i \leq n} |a_i|^p)^{1/p}$, $\|b\|_q := (\Sigma_{1 \leq i \leq n} |b_i|^q)^{1/q}$ and note that $\Sigma_{1 \leq i \leq n} |s_i|^p = 1$, $\Sigma_{1 \leq i \leq n} |t_i|^q = 1$, $\Sigma_{1 \leq i \leq n} |s_i t_i| \leq \Sigma_{1 \leq i \leq n}(p^{-1} \|s\|_p + q^{-1} \|t\|_q)$.]
2. **(a)** Let A be a positive definite matrix, let λ_1 (resp. λ_n) be its smallest (resp. largest) eigenvalue and let $\lambda := \sqrt{\lambda_1 . \lambda_n}$. Verify that the function $f : t \mapsto t/\lambda + \lambda/t$ is convex on $[\lambda_1, \lambda_n]$ and satisfies $f(\lambda_1) = \sqrt{\lambda_1/\lambda_n} + \sqrt{\lambda_n/\lambda_1} = f(\lambda_n)$, hence $f(t) \leq \sqrt{\lambda_1/\lambda_n} + \sqrt{\lambda_n/\lambda_1}$ for all $t \in [\lambda_1, \lambda_n]$.

(b) Show that μ is an eigenvalue of $\lambda^{-1}A+\lambda A^{-1}$ if and only if μ is an eigenvalue of A. [Hint: reduce A to a diagonal form.]
(c) Prove that $2\sqrt{\langle Ax,x\rangle.\langle A^{-1}x,x\rangle} \leq \lambda^{-1}\langle Ax,x\rangle + \lambda\langle A^{-1}x,x\rangle$ for all $x \in \mathbb{R}^n$. [Hint: use the inequality $2\sqrt{ab} \leq a+b$ for $a,b > 0$.]
(d) Deduce from this *Kantorovich's inequality*:

$$\forall x \in \mathbb{R}^n \quad \langle Ax,x\rangle.\langle A^{-1}x,x\rangle \leq (1/4)(\sqrt{\lambda_1/\lambda_n} + \sqrt{\lambda_n/\lambda_1})\|x\|^4.$$

3. (**Calmness subdifferentiability criterion**). A function $f : X \to \mathbb{R}_\infty$ finite at $\overline{x} \in X$ is said to be *calm at* $\overline{x}$ if $-f$ is quiet at $\overline{x}$, i.e. if there exist $c \in \mathbb{R}_+$ and a neighborhood V of $\overline{x}$ such that $f(x) - f(\overline{x}) \geq -c\,\|x - \overline{x}\|$ for all $x \in V$. The *calmness rate* of f at $\overline{x}$ is the infimum $\gamma_f(\overline{x})$ of the constants $c > 0$ for which the preceding inequality is satisfied on some neighborhood of $\overline{x}$.

Show that a convex function $f : X \to \mathbb{R}_\infty$ finite at some $\overline{x} \in X$ is subdifferentiable at $\overline{x}$ if and only if it is calm at $\overline{x}$. Verify that the calmness rate of f at $\overline{x}$ is equal to the remoteness $\rho(\partial f(\overline{x}))$ of $\partial f(\overline{x})$, where the *remoteness* of a nonempty subset S of X or X^* is the number $\rho(S) := \inf\{\|s\| : s \in S\}$.

4. (**Subdifferential determination of convex functions**). Given two continuous proper convex functions f, g on an open convex subset W of a Banach space X satisfying $\partial f \subset \partial g$, prove that there exists some $c \in \mathbb{R}$ such that $f(\cdot) = g(\cdot) + c$ on W. [Hint: reduce the question to the case $X = \mathbb{R}$: given $w, x \in W$ show that $f(w) - f(x) = g(w) - g(x)$ by taking compositions of f and g with the affine map $h : t \mapsto h(t) := w + t(x - w)$ from $\mathbb{R}$ to X and note that the functions $f \circ h$ and $g \circ h$ have nondecreasing derivatives satisfying $(f \circ h)' \leq (g \circ h)'$.]

5. Prove that the subdifferential $M := \partial f$ of a proper convex function f is *cyclically monotone* in the sense that for all $n \in \mathbb{N}\backslash\{0\}$ and any family $\{(x_i, x_i^*) : i \in \{0\} \cup \mathbb{N}_n\}$ in (the graph) of M one has

$$\langle x_0 - x_1, x_0^*\rangle + \cdots + \langle x_{n-1} - x_n, x_{n-1}^*\rangle + \langle x_n - x_0, x_n^*\rangle \geq 0.$$

6. Deduce from Theorem 9.23 that ∂f is *maximally cyclically monotone* in the sense that any multimap $T : X \rightrightarrows X^*$ whose graph is cyclically monotone and contains the graph of ∂f coincides with ∂f.

7. Prove that if a multimap $T : X \rightrightarrows X^*$ is cyclically monotone there exists a proper convex function f on X such that $T \subset \partial f$. Deduce from this and the two preceding exercises that a multimap $T : X \rightrightarrows X^*$ is the subdifferential ∂f of a proper convex function if and only if it is maximally cyclically monotone. [See [218].]

6.2.3 Differentiability of Convex Functions

Convex functions also enjoy particular differentiability properties. A first instance is the next result displaying an easy differentiability test using the functions

$$r_x(w) := f(x+w) + f(x-w) - 2f(x) \tag{6.4}$$

$$r_{x,u}(t) := f(x+tu) + f(x-tu) - 2f(x) = r_x(tu) \tag{6.5}$$

This criterion enables us to prove differentiability without knowing the derivative.

Proposition 6.14 *Let $f : X \to \mathbb{R}_\infty$ be a convex function finite and continuous (or more generally subdifferentiable) at some point $x \in X$. Then f is Fréchet (resp. Hadamard) differentiable at x if and only if r_x is a remainder (resp. if for all $u \in S_X$ the one-variable function $r_{x,u}$ is a remainder).*

Proof Necessity is obtained by addition directly from the definitions. Let us prove sufficiency in the Fréchet case. Let $x^* \in \partial f(x)$. Then the definition of $\partial f(x)$ and (6.4) yield

$$0 \le f(x+w) - f(x) - \langle x^*, w\rangle = f(x) - f(x-w) + \langle x^*, -w\rangle + r_x(w) \le r_x(w).$$

This shows that f is Fréchet differentiable at x with derivative x^*. The Gateaux case follows by a reduction to one-dimensional subspaces. Since f is continuous at x, it is Lipschitzian around x, so that Gateaux differentiability coincides with Hadamard differentiability. □

Other instances arise with continuity properties of derivatives or closure properties of subdifferentials. Hereafter, for a net $(x_i^*)_{i\in I}$ weak* converging to some x^* in X^* we write $(x_i^*)_{i\in I} \overset{*}{\to} x^*$.

Proposition 6.15 *Let f be a convex function on a normed space X, let $x \in \operatorname{dom} f$, $(x_i)_{i\in I} \to x$ and let $x_i^* \in \partial f(x_i)$ be such that $(f(x_i))_{i\in I} \to f(x)$, $(x_i^*)_{i\in I} \overset{*}{\to} x^*$, and $(\langle x_i^*, x_i - x\rangle)_{i\in I} \to 0$. Then $x^* \in \partial f(x)$.*

Note that the assumption $(\langle x_i^*, x_i - x\rangle)_{i\in I} \to 0$ is satisfied when $(x_i^*)_{i\in I}$ is bounded.

Proof It suffices to observe that for all $w \in X$ one has

$$\langle x^*, w-x\rangle = \lim_i \langle x_i^*, w-x\rangle = \lim_i \langle x_i^*, w-x_i\rangle \le \lim_i (f(w) - f(x_i)) = f(w) - f(x).$$

□

Taking for f the indicator function of a convex set C, we get the following consequence which can be given an easy direct proof.

Corollary 6.7 *Let C be a convex subset of a normed space X, let $(x_i)_{i\in I}$ be a net in C with limit $x \in C$ and let $x_i^* \in N(C, x_i)$ be such that $(x_i^*)_{i\in I}$ weak* converges to some x^* and $(\langle x_i^*, x_i - x\rangle)_{i\in I} \to 0$. Then $x^* \in N(C, x)$.*

Proposition 6.16 *If $f : W \to \mathbb{R}$ is continuous and convex on an open convex subset W of a normed space X, then df is upper semicontinuous on $W \times X$. If, moreover, f is Gateaux differentiable at $x \in W$ then f is Hadamard differentiable at x and for all $v \in X$ df is continuous at (x, v).*

If, moreover, f is Gateaux differentiable around x, then f is of class D^1 around x.

Proof For any $r > df(x, v)$ one can find a positive number s such that $r > s^{-1}[f(x+sv) - f(x)]$. Thus, for (x', v') close enough to (x, v) one has $r > s^{-1}[f(x' + sv') - f(x')] \geq df(x', v')$, so that

$$df(x, v) \geq \limsup_{(x',v')\to(x,v)} df(x', v').$$

If f is Gateaux differentiable at x, since $df(x', v') \geq -df(x', -v')$, the linearity of $df(x, \cdot)$ implies that

$$\liminf_{(x',v')\to(x,v)} df(x', v') \geq -\limsup_{(x',v')\to(x,v)} df(x', -v') \geq -df(x, -v) = df(x, v),$$

These inequalities prove our continuity assertion. Hadamard differentiability ensues (and can be deduced from the local Lipschitz property of f). □

In the next statement the continuity of the derivative of f is reinforced and for a subset A of X^* and $r \in \mathbb{P}$, we use the notation $B(A, r) := \{x^* : d(x^*, A) < r\}$.

Proposition 6.17 *Let $f : W \to \mathbb{R}$ be a convex function on some open convex subset W of a normed space X. If f is Fréchet differentiable at some $x \in W$ and Gateaux differentiable on W, then its derivative is continuous at x.*

More generally, if f is Fréchet differentiable at some $x \in W$, then its subdifferential ∂f is continuous at x in the following sense: for all $\varepsilon > 0$, there exists an $\eta > 0$ such that $\partial f(w) \cap B(f'(x), \varepsilon) \neq \varnothing$ and $\partial f(w) \subset B(f'(x), \varepsilon)$ for all $w \in B(x, \eta)$.

Proof It suffices to prove the second assertion. The differentiability of f at x entails the continuity of f on W, hence that $\partial f(w) \neq \varnothing$ for all $w \in W$. Let $x^* := Df(x)$. Given $\varepsilon \in]0, d(x, X\backslash W)[$, $\alpha \in]0, \varepsilon[$, let $\delta > 0$ be such that

$$\forall u \in B(0, \delta) \qquad f(x + u) - f(x) - \langle x^*, u\rangle \leq \alpha \|u\|. \tag{6.6}$$

Let $c := \alpha\varepsilon^{-1} \in]0, 1[$. For all $w \in B(x, (1 - c)\delta)$, $w^* \in \partial f(w)$, $v \in X$ one has

$$f(w) - f(w + v) + \langle w^*, v\rangle \leq 0.$$

Setting $u := w - x + v$ in (6.6) with $v \in B(0, c\delta)$, one has $u \in B(0, \delta)$, $x + u = w + v$ and, adding the respective sides of the preceding inequalities, one gets

$$f(w) - f(x) - \langle x^*, u \rangle + \langle w^*, v \rangle \leq \alpha \|u\| .$$

Using the relation $\langle x^*, u - v \rangle = \langle x^*, w - x \rangle \leq f(w) - f(x)$, this inequality yields

$$\langle w^* - x^*, v \rangle \leq \alpha \|u\| \leq \alpha \delta.$$

Taking the supremum over $v \in B(0, c\delta)$, one gets $\|w^* - x^*\| \leq c^{-1}\alpha = \varepsilon$. □

Corollary 6.8 *A Fréchet differentiable convex function on an open convex subset of a normed space is of class C^1.*

Let us mention some density properties of the set of points of differentiability of a convex function; see [95, 119, 120, 187, 209].

Theorem 6.6 (Asplund, Lindenstrauss) *Let $f : W \to \mathbb{R}$ be a continuous convex function on some open convex subset W of a Banach space X whose dual is separable, then the set F of points of differentiability of f is dense in W.*

Theorem 6.7 (Mazur) *If X is a separable Banach space, the set H of Hadamard differentiability points of a continuous convex function $f : W \to \mathbb{R}$ is dense in W.*

Let us recall that for what concerns subdifferentiability, no restriction on the space is required in view of the Brøndsted-Rockafellar Theorem.

Exercises

1. **(a)** Let $f : \mathbb{R} \to \mathbb{R}$ be given by $f(x) = |x|$. Show that $\partial f(0) = [-1, 1]$.
   (b) Verify that the subdifferential at 0 of a sublinear function f on a normed space X is given by $\partial f(0) = \{x^* \in X^* : x^* \leq f\}$. Prove that $\partial f(x) = \{x^* \in X^* : x^* \leq f, \langle x^*, x \rangle = f(x)\}$ for $x \in X$.
2. For a convex function f on $\mathbb{R}$ finite at x show that $\partial f(x) = [D_\ell f(x), D_r f(x)]$.
3. Prove that the closure of the radial tangent cone at $x \in C$ to a convex subset of a normed space coincides with the tangent cone to C as defined in Chap. 5.
4. Prove that the normal cone $N(C, x)$ to a convex subset C of a normed space at $x \in C$ coincides with the normal cone to C as defined in Chap. 5.
   Compute $N(C, x)$ for $X := \mathbb{R}^m$, $C := \mathbb{R}^m_+$, $x \in C$.
5. **(Ubiquitous convex sets)** Exhibit a proper convex subset C of a Banach space X such that $T(C, \overline{x}) = X$ for some $\overline{x} \in C$. Show that X must be infinite dimensional. [Hint: take for X a separable Hilbert space with Hilbert basis (e_n) and set $C := \{x = \Sigma_n x^n e_n : |x^n| \leq 2^{-n}\ \forall n\}, \overline{x} = 0$.]
6. Let $f : \mathbb{R}^2 \to \mathbb{R}_\infty$ be given by $f(x_1, x_2) := \max(|x_1|, 1 - \sqrt{x_2})$ for $(x_1, x_2) \in \mathbb{R} \times \mathbb{R}_+$, $+\infty$ otherwise. Prove that f is convex but that dom ∂f is not convex.

7. Let $f : X \to \mathbb{R}_\infty$ be a proper convex function on a normed space X and let $x, y \in \operatorname{dom} f$. Show that $df(x, y - x) + df(y, x - y) \leq 0$. Deduce from this relation that when f is Gateaux differentiable at x and y one has $\langle f'(x) - f'(y), x - y \rangle \geq 0$.
8. Let X be a Hilbert space, let C be a nonempty closed convex subset of X and let $f : X \to \mathbb{R}$ be given by $f(x) := (1/2)[\|x\|^2 - \|x - P(x)\|^2]$, where P is the *metric projection* of X onto C: $P(x) := \{u\}$, where $u \in C$, $\|x - u\| = d(x, C)$. Show that f is convex and that f is everywhere Fréchet differentiable, with gradient given by $\nabla f(x) = P(x)$ for all $x \in X$. [Hint: note that $f(x) = \sup\{\langle x, y \rangle - (1/2)\|y\|^2 : y \in C\}$, i.e. f is the conjugate of $(1/2)\|\cdot\|^2 + \iota_C(\cdot)$; use the estimates $\|x + u - P(x + u)\|^2 \leq \|x + u - P(x)\|^2$ and $\|x - P(x)\|^2 \leq \|x - P(x + u)\|^2$ to prove f is differentiable at x.]
9. Let $f : W \to \mathbb{R}$ be convex on some open convex subset of $\mathbb{R}^d$ and such that the partial derivatives $D_i f(\overline{x})$ $(i \in \mathbb{N}_d)$ of f at some $\overline{x} \in W$ exist. Show that f is differentiable at $\overline{x}$.
10*. Prove that a convex function $f : W \to \mathbb{R}$ on an open subset W of $\mathbb{R}^d$ is differentiable almost everywhere on W.
11. Show that for a convex function $f : X \to \mathbb{R}_\infty$ on a normed space X, the multimap $\partial f : X \rightrightarrows X^*$ is *monotone*, i.e. satisfies $\langle w^* - x^*, w - x \rangle \geq 0$ for all $w, x \in X$, $w^* \in \partial f(w)$, $x^* \in \partial f(x)$.

6.2.4 Elementary Calculus Rules for Subdifferentials

The fact that the calculus of subdifferentials satisfies structured rules is part of the value of convex analysis. We first consider simple rules. Then we turn to more general rules.

Convex functions enjoy several subdifferential calculus rules that are akin to the classical rules of differential calculus. Nonetheless, there are some differences: in general a technical assumption is needed to get the interesting inclusion. Moreover, one does not have $\partial(-f)(x) = -\partial f(x)$ in general. On the other hand, some rules of convex analysis have no analogues in the differentiable case. An example of such a new rule is the following obvious observation.

Lemma 6.3 *Suppose $f \leq g$ and $f(\overline{x}) = g(\overline{x})$ for some $\overline{x} \in X$. Then $\partial f(\overline{x}) \subset \partial g(\overline{x})$.*

This observation easily yields the following (rather inessential) rule for infima.

Lemma 6.4 *Let $(f_i)_{i \in I}$ be a family of functions and let $\overline{x} \in \bigcap_{i \in I} \operatorname{dom} f_i$. If $f := \inf_{i \in I} f_i$ and if $f_i(\overline{x}) = f(\overline{x})$ for all $i \in I$, then $\partial f(\overline{x}) = \bigcap_{i \in I} \partial f_i(\overline{x})$.*

Proof The inclusion $\partial f(\overline{x}) \subset \bigcap_{i \in I} \partial f_i(\overline{x})$ stems from Lemma 6.3. For the opposite inclusion, we note that for all $\overline{x}^* \in \bigcap_{i \in I} \partial f_i(\overline{x})$, for all $i \in I$ and all $x \in X$ one has $f_i(\overline{x}) + \langle \overline{x}^*, x - \overline{x} \rangle \leq f_i(x)$, hence $f(\overline{x}) + \langle \overline{x}^*, x - \overline{x} \rangle \leq f(x)$ as $f(\overline{x}) = f_i(\overline{x})$. □

A general rule can be given for value functions of parameterized problems.

Proposition 6.18 *Let $f : W \times X \to \mathbb{R}_\infty$, where W and X are normed spaces. Let p be the performance function given by $p(w) := \inf\{f(w,x) : x \in X\}$ and let $S : W \rightrightarrows X$ be the solution multimap given by $S(w) := \{x \in X : f(w,x) = p(w)\}$. Suppose that for some $\overline{w} \in X$ one has $S(\overline{w}) \neq \varnothing$. Then one has the equivalence*

$$\overline{w}^* \in \partial p(\overline{w}) \Longleftrightarrow \forall \overline{x} \in S(\overline{w}) \quad (\overline{w}^*, 0) \in \partial f(\overline{w}, \overline{x})$$

$$\Longleftrightarrow \exists \overline{x} \in S(\overline{w}) \quad (\overline{w}^*, 0) \in \partial f(\overline{w}, \overline{x}).$$

Proof For all $\overline{x} \in S(\overline{w})$, $(w,x) \in W \times X$, one has $p(\overline{w}) = f(\overline{w}, \overline{x})$, $p(w) \leq f(w,x)$, whence

$$\overline{w}^* \in \partial p(\overline{w}) \Leftrightarrow \forall w \in W \qquad p(w) \geq p(\overline{w}) + \langle \overline{w}^*, w - \overline{w} \rangle$$
$$\Rightarrow \forall (w,x) \in W \times X \;\; f(w,x) \geq f(\overline{w}, \overline{x}) + \langle (\overline{w}^*, 0), (w - \overline{w}, x - \overline{x}) \rangle,$$

or $(\overline{w}^*, 0) \in \partial f(\overline{w}, \overline{x})$.

Conversely, if this last relation holds for some $\overline{x} \in S(\overline{w})$ and some $\overline{w}^* \in W^*$, then, taking the infimum over $x \in X$ in the last inequality, one gets

$$\forall w \in W \qquad p(w) \geq p(\overline{w}) + \langle \overline{w}^*, w - \overline{w} \rangle,$$

i.e. $\overline{w}^* \in \partial p(\overline{w})$. □

Corollary 6.9 *Given functions g, $h : X \to \mathbb{R}_\infty$, $\overline{w} \in \operatorname{dom}(g \Box h)$, $\overline{x} \in X$ such that $(g \Box h)(\overline{w}) = g(\overline{w} - \overline{x}) + h(\overline{x})$, one has $\partial (g \Box h)(\overline{w}) = \partial g(\overline{w} - \overline{x}) \cap \partial h(\overline{x})$.*

Proof Setting $f(w,x) := g(w - x) + h(x)$, $p := g \Box h$, it suffices to see that for any $\overline{w}^* \in \partial g(\overline{w} - \overline{x}) \cap \partial h(\overline{x})$ one has $(\overline{w}^*, 0) \in \partial f(\overline{w}, \overline{x})$ and that conversely for $(\overline{w}^*, 0) \in \partial f(\overline{w}, \overline{x})$ one has $\overline{w}^* \in \partial g(\overline{w} - \overline{x})$ and by symmetry $\overline{w}^* \in \partial h(\overline{x})$. □

The case of the supremum of a finite family of convex functions is more likely to occur than the case of the infimum. For the case of an infinite family we refer to [208, Section 3.3.1].

Proposition 6.19 *Let $(f_i)_{i \in I}$ be a finite family of convex functions on a normed space X and let $f := \sup_{i \in I} f_i$. Let $\overline{x} \in \bigcap_{i \in I} \operatorname{dom} f_i$ and let $I(\overline{x}) := \{i \in I : f_i(\overline{x}) = f(\overline{x})\}$. Suppose that for all $i \in I$ the function f_i is continuous at $\overline{x}$. Then one has*

$$df(\overline{x}, \cdot) = \max_{i \in I(\overline{x})} df_i(\overline{x}, \cdot), \tag{6.7}$$

$$\partial f(\overline{x}) = \operatorname{co}\Big(\bigcup_{i \in I(\overline{x})} \partial f_i(\overline{x})\Big). \tag{6.8}$$

Proof Let $u \in X$. Since f_i and f are continuous at $\overline{x}$, by Proposition 6.9 $df(\overline{x}, \cdot)$ coincides with the radial derivative. For $i \in I(\overline{x})$, since $f_i \leq f$ and $f_i(\overline{x}) = f(\overline{x})$, we have $df_i(\overline{x}, u) \leq df(\overline{x}, u)$. Thus $s := \max_{i \in I(\overline{x})} df_i(\overline{x}, u) \leq df(\overline{x}, u)$ and equality holds when $s = \infty$. Let us suppose that $s < \infty$ and let us show that for every $r > s$ we have $r \geq f'(\overline{x}, u)$; this will prove that $s = f'(\overline{x}, u)$. For $i \in I(\overline{x})$, let $t_i > 0$ be such that

$$(1/t)\,(f_i(\overline{x} + tu) - f_i(\overline{x})) < r \qquad \text{for } t \in]0, t_i[.$$

Since for $j \in I \backslash I(\overline{x})$ the function f_j is continuous at $\overline{x}$, given $\varepsilon > 0$ such that $f_i(\overline{x}) + \varepsilon < f(\overline{x})$ for all $i \in I \backslash I(\overline{x})$, we can find $t_j > 0$ such that

$$f_j(\overline{x} + tu) < f(\overline{x}) - \varepsilon \qquad \text{for } t \in]0, t_j[.$$

Then, for $t \in]0, t_0[$, with $t_0 := \min(|r|^{-1}\varepsilon, \min_{j \in I \backslash I(\overline{x})} t_j)$ we have $-\varepsilon \leq tr$, hence

$$f(\overline{x} + tu) = \max_{i \in I} f_i(\overline{x} + tu) \leq \max(\max_{i \in I(\overline{x})}(f_i(\overline{x}) + tr), f(\overline{x}) - \varepsilon) = f(\overline{x}) + tr.$$

Thus $df(\overline{x}, u) \leq r$ and $df(\overline{x}, u) = \max_{i \in I(\overline{x})} df_i(\overline{x}, u)$.

For $i \in I(\overline{x})$, the inclusion $\partial f_i(\overline{x}) \subset \partial f(\overline{x})$ follows from Lemma 6.3 or from the inequality $df_i(\overline{x}, \cdot) \leq df(\overline{x}, \cdot)$. Denoting by C the right-hand side of (6.8), and observing that $\partial f(\overline{x})$ is convex, the inclusion $C \subset \partial f(\overline{x})$ ensues. Let us show that assuming there exists some $\overline{w}^* \in \partial f(\overline{x}) \backslash C$ leads to a contradiction. Since C is weak* closed (in fact weak* compact), the Hahn-Banach theorem yields some $c \in \mathbb{R}$ and $u \in X$ (the dual of X^* endowed with the weak* topology in view of Proposition 3.18) such that

$$\langle \overline{w}^*, u \rangle > c \geq \langle x^*, u \rangle \qquad \forall x^* \in C.$$

Since $df(\overline{x}, u) \geq \langle \overline{w}^*, u \rangle$ we get

$$df(\overline{x}, u) > c \geq \sup_{x^* \in C} \langle x^*, u \rangle = \sup_{i \in I(\overline{x})} \sup_{x^* \in \partial f_i(\overline{x})} \langle x^*, u \rangle = \sup_{i \in I(\overline{x})} df_i(\overline{x}, u),$$

contradicting the equality we established. □

Now we state classical and convenient sum and composition rules. They are special cases of a mixed rule given later in Corollary 6.14. We incite the reader to devise direct proofs using the Hahn-Banach Theorem. Such proofs are available in most books dealing with convex analysis. Moreover, each result can be deduced from the other one.

Theorem 6.8 *Let f and g be convex functions on a normed space X. If f and g are finite at $\overline{x}$ and if g is continuous at some point of* $\operatorname{dom} f \cap \operatorname{dom} g$ *then*

$$\partial(f + g)(\overline{x}) = \partial f(\overline{x}) + \partial g(\overline{x}).$$

Theorem 6.9 (Chain Rule) *Let X and Y be normed spaces, let $A : X \to Y$ be a linear continuous map and let $g : Y \to \mathbb{R}_\infty$ be finite at $\overline{y} := A(\overline{x})$ and continuous at some point of $A(X)$. Then, for $f := g \circ A$ one has*

$$\partial f(\overline{x}) = A^\intercal(\partial g(\overline{y})) := \partial g(\overline{y}) \circ A.$$

In Banach spaces, one can get rid of the continuity assumptions in the preceding two rules, replacing them by a so-called "qualification condition", as will be shown in a forthcoming subsection devoted to duality results. The statement we present gathers a chain rule and a sum rule. It will be proved in Theorem 6.19.

Theorem 6.10 (Attouch-Brézis) *Let X, Y be Banach spaces, let $A \in L(X, Y)$, and let $f : X \to \mathbb{R}_\infty$, $g : Y \to \mathbb{R}_\infty$ be closed proper convex functions. If the cone*

$$Z := \mathbb{R}_+ (A(\mathrm{dom} f) - \mathrm{dom}\, g)$$

is closed and symmetric (i.e., $\mathrm{cl}\, Z = Z = -Z$), in particular if $Z = Y$, then, for all $x^* \in X^*$ and $x \in \mathrm{dom} f \cap A^{-1}(\mathrm{dom}\, g)$ one has

$$\partial(f + g \circ A)(x) = \partial f(x) + A^\intercal(\partial g(Ax)).$$

Exercises

1. Let f, g be two convex functions on a normed space X that are finite at some $\overline{x} \in X$. Suppose g is Fréchet differentiable at $\overline{x}$ and show that $\partial(f + g)(\overline{x}) = \partial f(\overline{x}) + g'(\overline{x})$.
2. Let f be a convex function on a normed space X that is finite at some $\overline{x} \in X$. Suppose there exists some $\ell \in X^*$ such that r defined by $r(x) := \max(f(\overline{x} + x) - f(\overline{x}) - \ell(x), 0)$ is a remainder. Show that f is Fréchet differentiable at $\overline{x}$.
3. Prove that a convex function f on a normed space X finite at some $\overline{x} \in X$ is subdifferentiable at $\overline{x}$ if and only if it is calm at $\overline{x}$ in the sense that there exist $c > 0$ and a neighborhood V of $\overline{x}$ such that $f(w) \geq f(\overline{x}) - c \left\| w - \overline{x} \right\|$ for all $w \in V$ if and only if it is globally calm at $\overline{x}$ in the sense that the preceding inequality is valid for $V = X$. Show that in such a case one has $\partial f(x) \cap cB_{X^*} \neq \varnothing$ but that one may have $\partial f(x) \not\subset cB_{X^*}$.
4. Prove that a differentiable function $f : W \to \mathbb{R}$ defined on an open convex subset of a normed space X is convex if and only if $f' : W \to X^*$ is *monotone*, i.e. satisfies $\langle f'(w) - f'(x), w - x \rangle \geq 0$ for all $w, x \in W$.
5. Assuming that X is complete (resp. X and Y are complete) show that Theorem 6.8 (resp. 6.9) is a consequence in the Attouch-Brezis theorem.
6. Let X and Y be reflexive Banach spaces, let $A : X \to Y$ be linear and continuous and let $C := A^{-1}(D)$, where D is a closed convex subset of Y. Let $\overline{x} \in C$, $\overline{y} := A(\overline{x})$. Show that $\overline{x}^* \in N(C, \overline{x})$ if and only if there exist sequences $(x_n) \to \overline{x}$,

$(y_n) \to \overline{y} := A\overline{x}$ in D, (y_n^*) in Y^* such that $y_n^* \in N(D, y_n)$ for all n and

$$(\|A^\top y_n^* - \overline{x}^*\|)_n \to 0, \quad \left(\|y_n^*\| \cdot \|y_n - Ax_n\|\right)_n \to 0.$$

[Hint: introduce the penalized decoupling function $p_n : X \times Y \to \mathbb{R}_\infty$ given by

$$p_n(x, y) := \iota_D(y) - \langle \overline{x}^*, x \rangle + n \|Ax - y\|^2 + \|x\|^2$$

and take a minimizer (x_n, y_n) of p_n on $B_{X \times Y}$.]

7. Let X and Y be reflexive Banach spaces, let $A \in L(X, Y)$ and let $f := g \circ A$, where $g : Y \to \mathbb{R}_\infty$ is lower semicontinuous and convex. Let $\overline{x} \in \operatorname{dom} f$, $\overline{x}^* \in X^*$. Show that $\overline{x}^* \in \partial f(\overline{x})$ if and only if there exist sequences $(x_n) \to \overline{x}$ in X, $(y_n) \to \overline{y} := A\overline{x}$ in Y, (y_n^*) in Y^* such that $y_n^* \in \partial g(y_n)$ for all n, $(g(y_n)) \to g(\overline{y})$, and

$$(\|A^\top y_n^* - \overline{x}^*\|)_n \to 0, \quad \left(\|y_n^*\| \cdot \|y_n - Ax_n\|\right)_n \to 0.$$

8. Let h, k be lower semicontinuous proper convex functions on a reflexive Banach space X and let $f := h + k$ be finite at $\overline{x} \in X$. Show that $\overline{x}^* \in X^*$ belongs to $\partial f(\overline{x})$ if and only if there exist sequences (w_n), $(z_n) \to \overline{x}$ in X, (w_n^*), (z_n^*) in X^* such that $w_n^* \in \partial h(w_n)$, $z_n^* \in \partial k(z_n)$ for all n, $(h(w_n)) \to h(\overline{x})$, $(k(z_n) \to k(\overline{x})$ and

$$(w_n^* + z_n^*)_n \overset{*}{\to} \overline{x}^*, \qquad ((\|w_n - \overline{x}\| + \|z_n - \overline{x}\|).(\|w_n^*\| + \|z_n^*\|))_n \to 0.$$

6.2.5 *Application to Optimality Conditions*

Let us apply the above calculus rules to the constrained optimization problem

$$(\mathcal{C}) \quad \text{minimize } f(x) \qquad \text{subject to } x \in C,$$

where $f : X \to \mathbb{R}_\infty$ is convex and C is a convex subset of X. We assume that f takes at least one finite value on C, so that $\inf(\mathcal{C})$ is not $+\infty$. Then $(\mathcal{C})$ is equivalent to the minimization of $f_C := f + \iota_C$ on X.

Optimality conditions for problem $(\mathcal{C})$ involve the notion of a normal cone to C at some $\overline{x} \in C$; in the convex case we are dealing with presently, its simple definition has been given before Proposition 6.12. We recall it for the reader's convenience: the *normal cone* to C at $\overline{x} \in C$ is the set $N(C, \overline{x})$ of continuous linear forms on X which attain their maximum on C at $\overline{x}$:

$$N(C, \overline{x}) := \partial \iota_C(\overline{x}) := \{x^* \in X^* : \forall x \in C \ \langle \overline{x}^*, x - \overline{x} \rangle \leq 0\}.$$

Example If $\overline{x}$ is in the interior of C, one has $N(C, \overline{x}) = \{0\}$ since a continuous linear form that has a local maximum is null. □

Example Let $g \in X^*\backslash\{0\}$, $c \in \mathbb{R}$ and let $D := \{x \in X : g(x) \leq c\}$. Then, if $\overline{x}$ is such that $g(\overline{x}) < c$ one has $\overline{x} \in \mathrm{int}D$, hence $N(D, \overline{x}) = \{0\}$, while for all $\overline{x}$ such that $g(\overline{x}) = c$, one has $N(D, \overline{x}) = \mathbb{R}_+ g$. In fact, for any $r \in \mathbb{R}_+$ and all $x \in D$ one has $rg(x - \overline{x}) \leq 0$, hence $rg \in N(D, \overline{x})$. Conversely, let $h \in N(D, \overline{x})$. Then, for all $u \in \mathrm{Ker}\, g$, one has $\overline{x} + u \in D$, hence $h(u) \leq 0$. Changing u into $-u$, we see that $\mathrm{Ker}\, g \subset \mathrm{Ker}\, h$, so that there exists an $r \in \mathbb{R}$ such that $h = rg$: picking $u \in X$ satisfying $g(u) = 1$ (this is possible since $g \neq 0$), we have $r = h(u)$, and since $\overline{x} - u \in D$ we get that $-r = h(-u) = h((\overline{x} - u) - \overline{x}) \leq 0$, hence $r \in \mathbb{R}_+$. □

Theorem 6.11 *A sufficient condition for $\overline{x} \in C$ to be a solution to $(\mathcal{C})$ is*

$$0 \in \partial f(\overline{x}) + N(C, \overline{x}).$$

Under one of the following assumptions, this condition is necessary:

(a) f is finite and continuous at some point of C;
(b) f is finite at some point of the interior of C;
(c) f is lower semicontinuous, $\mathbb{R}_+(\mathrm{dom} f - C) = -\mathrm{cl}(\mathbb{R}_+(\mathrm{dom} f - C))$, C is closed and X is complete.

Proof Suppose $\overline{x} \in C$ is such that $0 \in \partial f(\overline{x}) + N(C, \overline{x})$. Let $\overline{x}^* \in \partial f(\overline{x})$ be such that $-\overline{x}^* \in N(C, \overline{x})$. Then, $f(\overline{x})$ is finite and, for all $x \in C$, one has $f(x) - f(\overline{x}) \geq \langle \overline{x}^*, x - \overline{x} \rangle \geq 0$: $\overline{x}$ is a solution to $(\mathcal{C})$.

The necessary condition stems from the relations $0 \in \partial(f + \iota_C)(\overline{x}) = \partial f(\overline{x}) + \iota_C(\overline{x})$ valid under each of the assumptions of (a)–(c). □

Using a fuzzy sum rule one can give a necessary and sufficient optimality condition that does not require additional assumptions (see [205, 251]).

In order to apply the conditions of Theorem 6.11 to the important case in which C is defined by inequalities, let us give a means to compute the normal cone to C in such a case. We start with the case of a single inequality, generalizing the second example of this subsection.

Lemma 6.5 *Let $g : X \to \mathbb{R}_\infty$ be a convex function and let $C := \{x \in X : g(x) \leq 0\}$, $\overline{x} \in g^{-1}(0)$. Suppose $C' := \{x \in X : g(x) < 0\}$ is nonempty and g is continuous at $\overline{x}$ and at some point $x' \in C'$. Then one has $N(C, x') = \{0\}$ and $N(C, \overline{x}) = \mathbb{R}_+ \partial g(\overline{x})$.*

Proof For all $x' \in C'$ the set C is a neighborhood of x', so that $N(C, x') = \{0\}$. The inclusion $N(C, \overline{x}) \supset \mathbb{R}_+ \partial g(\overline{x})$ is obvious: given $r \in \mathbb{R}_+$ and $\overline{x}^* \in \partial g(\overline{x})$, for all $x \in C$ one has $\langle r\overline{x}^*, x - \overline{x} \rangle \leq r(g(x) - g(\overline{x})) \leq 0$, hence $r\overline{x}^* \in N(C, \overline{x})$.

Conversely, let $\overline{x}^* \in N(C, \overline{x})\backslash\{0\}$. The interior of C is nonempty since it contains x'. Since $\langle \overline{x}^*, x \rangle \leq \langle \overline{x}^*, \overline{x} \rangle$ for all $x \in C$, we have $\langle \overline{x}^*, x \rangle < \langle \overline{x}^*, \overline{x} \rangle$ for all $x \in \mathrm{int}(C)$ (otherwise $\overline{x}^*$ would have a local maximum, hence would be 0). In particular, $g(x) < 0$ implies $\langle \overline{x}^*, x \rangle < \langle \overline{x}^*, \overline{x} \rangle$ since for y in the segment $]x, \overline{x}[$ we have $g(y) < 0$ and g is bounded above near y, hence $y \in \mathrm{int}(C)$. Thus, $g(x) \geq 0$ for all x such that $\langle \overline{x}^*, x \rangle \geq \langle \overline{x}^*, \overline{x} \rangle$. Therefore $\overline{x}$ is a minimizer of g on $D := \{x \in X : \langle \overline{x}^*, x \rangle \geq \langle \overline{x}^*, \overline{x} \rangle\}$. Since g is continuous at $\overline{x} \in D$, we have $0 \in \partial g(\overline{x}) + N(D, \overline{x})$ by assertion (a) of the preceding theorem. But the second example of the present section, with

$c := \langle -\overline{x}^*, \overline{x} \rangle$ and $-\overline{x}^*$ instead of g, ensures that $N(D, \overline{x}) = -\mathbb{R}_+ \overline{x}^*$. Since $0 \notin \partial g(\overline{x})$ because $\overline{x}$ is not a minimizer of g, we get some $s > 0$ such that $s\overline{x}^* \in \partial g(\overline{x})$, hence $\overline{x}^* \in s^{-1} \partial g(\overline{x})$. □

The case of a finite number of inequalities is a consequence in Lemma 6.5 and of the following rule for the calculus of normal cones.

Lemma 6.6 *Let $C_1, \ldots, C_k$ be convex subsets of X and let $\overline{x} \in C := C_1 \cap \ldots \cap C_k$. Then*

$$N(C, \overline{x}) = N(C_1, \overline{x}) + \ldots + N(C_k, \overline{x})$$

whenever one of the following assumptions is satisfied:

(a) there exist $j \in \mathbb{N}_k$ and some $z \in C_j$ that belongs to $\operatorname{int} C_i$ for all $i \neq j$;
(b) X is complete, $C_1, \ldots, C_k$ are closed and for $D := \{(x, \ldots, x) : x \in X\}$, $P := C_1 \times \ldots \times C_k$, the cone $\mathbb{R}_+(P - D)$ is a closed linear subspace of X^k.

Proof Assumption (a) ensures that $\partial(\iota_{C_1} + \ldots + \iota_{C_k})(\overline{x}) = \partial \iota_{C_1}(\overline{x}) + \ldots + \partial \iota_{C_k}(\overline{x})$ since for $i \neq j$ the function ι_{C_i} is finite and continuous at $z \in \operatorname{dom} \iota_{C_j}$. The Attouch-Brézis theorem gives the conclusion under assumption (b) since, if A is the diagonal map $x \mapsto (x, \ldots, x)$ from X into X^k, one has $C = A^{-1}(P)$, hence $\iota_C = \iota_P \circ A$ and

$$\partial \iota_C(\overline{x}) = A^\mathsf{T}(\partial \iota_P(\overline{x})) = A^\mathsf{T}(\partial \iota_{C_1}(\overline{x}) \times \ldots \times \partial \iota_{C_k}(\overline{x})) = \partial \iota_{C_1}(\overline{x}) + \ldots + \partial \iota_{C_k}(\overline{x}),$$

as easily checked. □

The next example shows the necessity of requiring some additional assumptions traditionally called "qualification conditions".

Example For $i = 1, 2$, let $C_i := B[c_i, 1]$ with $c_i := (0, (-1)^i)$ in $X := \mathbb{R}^2$ endowed with the Euclidean norm. Then $C := C_1 \cap C_2 = \{\overline{x}\}$, with $\overline{x} := (0, 0)$, hence $N(C, \overline{x}) = X^*$, but $N(C_i, \overline{x}) = \{0\} \times (-1)^{i+1} \mathbb{R}_+$ and $N(C_1, \overline{x}) + N(C_2, \overline{x}) = \{0\} \times \mathbb{R}$. □

Lemma 6.7 *Let $g_i : X \to \mathbb{R}_\infty$ be convex, let $C_i := \{x \in X : g_i(x) \leq 0\}$ for $i \in I := \mathbb{N}_k$, let $\overline{x} \in C := C_1 \cap \ldots \cap C_k$ and let $I(\overline{x}) := \{i \in I : g_i(\overline{x}) = 0\}$. Suppose that for all $i \in I$ g_i is continuous at $\overline{x}$. Assume Slater's condition: there exists some $x_0 \in C_i' := \{x \in X : g_i(x) < 0\}$ for all $i \in I(\overline{x})$. Then, for $\overline{x}^* \in X^*$, one has $\overline{x}^* \in N(C, \overline{x})$ if and only if there exist $y_1, \ldots, y_k$ in $\mathbb{R}_+$ such that*

$$\overline{x}^* \in y_1 \partial g_1(\overline{x}) + \ldots + y_k \partial g_k(\overline{x}), \quad y_1 g_1(\overline{x}) = 0, \ldots, y_k g_k(\overline{x}) = 0. \tag{6.9}$$

Proof The sufficient condition is immediate: if $\overline{x}^* = y_1 \overline{x}_1^* + \ldots + y_k \overline{x}_k^*$ with $\overline{x}_i^* \in \partial g_i(\overline{x})$ and $y_i \in \mathbb{R}_+$ with $y_i g_i(\overline{x}) = 0$, for all $x \in C$ we get $\langle \overline{x}^*, x - \overline{x} \rangle \leq 0$ as the sum of the terms $y_i \langle \overline{x}_i^*, x - \overline{x} \rangle \leq 0$ since $x \in C_i$ and $\overline{x}_i^* \in \partial g_i(\overline{x})$.

Let us suppose now that $\overline{x}^* \in N(C, \overline{x})$. For $i \in I \backslash I(\overline{x})$, since g_i is continuous at $\overline{x}$ and $g_i(\overline{x}) < 0$, one has $\overline{x} \in \mathrm{int}(C_i)$, hence $N(C, \overline{x}) = N(C', \overline{x})$ where C' is the intersection of the family (C_i) for $i \in I(\overline{x})$. Given $x_0 \in C_i'$ for all $i \in I(\overline{x})$, since g_i is continuous at $\overline{x}$, hence for $t \in \,]0, 1[$, $z := (1-t)x_0 + t\overline{x} \in \mathrm{int}C_i$ since g_i is bounded above around z and $g_i(z) < 0$, hence $z \in \mathrm{int}(C')$. Thus Lemma 6.6 yields some $\overline{w}_i^* \in N(C_i, \overline{x})$ such that $\overline{x}^* = \overline{w}_1^* + \ldots + \overline{w}_k^*$ (with $\overline{w}_i^* = 0$ for $i \in I \backslash I(\overline{x})$). For $i \in I(\overline{x})$, Lemma 6.5 provides some $y_i \in \mathbb{R}_+$ and some $\overline{x}_i^* \in \partial g_i(\overline{x})$ satisfying $\overline{w}_i^* = y_i \overline{x}_i^*$. Since, for $i \in I \backslash I(\overline{x})$, g_i is continuous at $\overline{x}$, we can write $\overline{w}_i^* = y_i \overline{x}_i^*$ with $y_i = 0, \overline{x}_i^*$ arbitrary in $\partial g_i(\overline{x})$ which is nonempty by Theorem 6.4. Thus relation (6.9) holds. □

This characterization and Theorem 6.11 give immediately a necessary and sufficient optimality condition for the mathematical programming problem

$$(\mathcal{M}) \quad \text{minimize} \, f(x) \quad \text{subject to} \, x \in C := \{x \in X : g_1(x) \leq 0, \ldots, g_k(x) \leq 0\},$$

where f is convex and $g_1, \ldots, g_k$ are as above.

Theorem 6.12 (Karush-Kuhn-Tucker Theorem) *Let $f : X \to \mathbb{R}_\infty$, $g_1, \ldots, g_k$ be as in the preceding lemma and let $\overline{x} \in C$. Suppose f is convex, continuous at some point of C, and* Slater's condition *holds: there exists some x_0 such that $g_i(x_0) < 0$ for $i \in I(\overline{x})$. Then $\overline{x}$ is a solution to $(\mathcal{M})$ if and only if there exist $\overline{y}_1, \ldots, \overline{y}_k$ in $\mathbb{R}_+$ such that*

$$0 \in \partial f(\overline{x}) + \overline{y}_1 \partial g_1(\overline{x}) + \ldots + \overline{y}_k \partial g_k(\overline{x}), \qquad \overline{y}_1 g_1(\overline{x}) = 0, \ldots, \overline{y}_k g_k(\overline{x}) = 0.$$

Introducing the *Lagrangian function* ℓ by

$$\ell(x, y) := \ell_y(x) := f(x) + y_1 g_1(x) + \ldots + y_k g_k(x) \qquad x \in X, \; y \in \mathbb{R}^k$$

and the set $K(\overline{x})$ of *Karush-Kuhn-Tucker multipliers* at $\overline{x}$,

$$K(\overline{x}) := \{y := (y_1, \ldots, y_k) \in \mathbb{R}_+^k, \; 0 \in \partial \ell_y(\overline{x}), \; y.g(\overline{x}) = 0\},$$

the above condition can be written as $\overline{y} \in K(\overline{x})$. Here we use the fact that $\overline{y}_i g_i(\overline{x}) \leq 0$ for all i, so that $\overline{y}_1 g_1(\overline{x}) + \ldots + \overline{y}_k g_k(\overline{x}) = 0$ is equivalent to $\overline{y}_i g_i(\overline{x}) = 0$ for all i; we also use the continuity assumption on the g_i's at $\overline{x}$. Thus, in order to take the constraints into account, the condition $0 \in \partial f(\overline{x})$ of the unconstrained problem has been replaced by a similar condition with $\ell_{\overline{y}}$ in place of f. Despite this justification, the multipliers $\overline{y}_i$ seem to be artificial ingredients. However they cannot be neglected, as shown by Exercise 1 below, even if in solving practical problems one is led to get rid of them as soon as possible. In fact, the "marginal" interpretation we provide below shows that their knowledge is not without interest, as they provide useful information about the behavior of the value of perturbed problems. In order to shed some light on such an interpretation, let us introduce for $w := (w_1, \ldots, w_k) \in \mathbb{R}^k$ the perturbed problem

$$(\mathcal{M}_w) \quad \text{minimize}\, f(x) \quad \text{subject to}\; x \in C_w := \{x \in X : g_i(x) + w_i \leq 0, \;\; i \in \mathbb{N}_k\}$$

and set $G := \{(x, w) \in X \times \mathbb{R}^k : g_1(x) + w_1 \leq 0, \ldots, g_k(x) + w_k \leq 0\}$,

$$p(w) := \inf\{f(x) : x \in C_w\}.$$

Since $p(w) = \inf_{x \in X} P(w, x)$, with $P(w, x) := f(x) + \iota_G(x, w)$, p is convex, G and P being convex.

Let us also introduce the set M of *Lagrange multipliers*:

$$M := \{y \in \mathbb{R}^k_+ : \inf_{x \in C} f(x) = \inf_{x \in X} \ell_y(x)\}.$$

Theorem 6.13 *Suppose $p(0)$ is finite. Then, the set M of Lagrange multipliers is contained in $\partial p(0)$. If the functions g_i are finite, then $M = \partial p(0)$ and for all $\overline{x}$ in the set S of solutions to $(\mathcal{M})$ one has $K(\overline{x}) = M$.*

It follows that the set $K(\overline{x})$ is independent of the choice of $\overline{x}$ in S.

Proof Let $y \in M$. Given $w \in \mathbb{R}^k$, for all $x \in C_w$ and $i = 1, \ldots, k$, we have $y_i g_i(x) \leq -y_i w_i$ since $y_i \in \mathbb{R}_+$ and $g_i(x) \leq -w_i$. Thus, by definition of M,

$$p(0) = \inf_{x \in X} \ell_y(x) \leq \inf_{x \in X} f(x) - \langle y, w \rangle \leq \inf_{x \in C_w} f(x) - \langle y, w \rangle = p(w) - \langle y, w \rangle,$$

so that $y \in \partial p(0)$.

Conversely, assume the functions g_i are finite and let $y \in \partial p(0)$. We first observe that $y \in \mathbb{R}^k_+$ since for $w \in \mathbb{R}^k_+$ we have $C \subset C_{-w}$, hence $p(-w) \leq p(0)$, so that, taking for w the elements of the canonical basis of $\mathbb{R}^k$, the inequalities $\langle y, -w \rangle \leq p(-w) - p(0) \leq 0$ imply that the components of y are nonnegative. Now, given $x \in X$, taking $w_i := -g_i(x)$ for $i = 1, \ldots, k$, one has $x \in C_w$, hence $f(x) \geq p(w)$ and

$$f(x) + \langle y, g(x) \rangle \geq p(w) + \langle y, g(x) \rangle \geq p(0) + \langle y, w \rangle + \langle y, g(x) \rangle = p(0),$$

so that $\inf_{x \in X} \ell_y(x) \geq p(0)$. Since for $x \in C$ one has $\langle y, g(x) \rangle \leq 0$, hence $\inf_{x \in X} \ell_y(x) \leq \inf_{x \in C} \ell_y(x) \leq \inf_{x \in C} f(x) = p(0)$, we get $\inf_{x \in X} \ell_y(x) = p(0)$, hence $y \in M$.

Finally, let $\overline{x} \in S$ and let $\overline{y} \in K(\overline{x})$. Then, as $0 \in \partial \ell_{\overline{y}}(\overline{x})$ or $\ell_{\overline{y}}(\overline{x}) = \inf_{x \in X} \ell_{\overline{y}}(x)$ and $\overline{y}.g(\overline{x}) = 0$, we have $\ell_{\overline{y}}(\overline{x}) = f(\overline{x}) = \inf_{x \in C} f(x)$ and we get $\overline{y} \in M$. Conversely, let $\overline{y} \in M$. Then $f(\overline{x}) = p(0) = \inf_{x \in X} \ell_{\overline{y}}(x) \leq \ell_{\overline{y}}(\overline{x})$, so that $\overline{y}.g(\overline{x}) \geq 0$. Since for all $i = 1, \ldots, k$ we have $\overline{y}_i \geq 0$ and $g_i(\overline{x}) \leq 0$, the reverse inequality holds, hence $\overline{y}.g(\overline{x}) = 0$. Moreover, the relations $\inf_{x \in X} \ell_{\overline{y}}(x) = p(0) = f(\overline{x}) = \ell_{\overline{y}}(\overline{x})$ imply that $0 \in \partial \ell_{\overline{y}}(\overline{x})$. Therefore $\overline{y} \in K(\overline{x})$. □

Exercises

1. **(a)** Compute the normal cone to $\mathbb{R}_+$.
   (b) Given a convex function $f : \mathbb{R} \to \mathbb{R}$ give a necessary and sufficient condition in order that it attains its minimum on $C := \{x \in \mathbb{R} : -x \leq 0\}$ at 0. Taking $f(x) = x$, note that the condition $f'(0) = 0$ is not satisfied.
   (c) Compute the normal cone to $\mathbb{R}^n_+$ at some $\overline{x} \in \mathbb{R}^n_+$.
2. Show that the sufficient condition of the Karush-Kuhn-Tucker Theorem holds without the Slater condition and continuity assumptions.
3. State and prove a necessary and sufficient optimality condition for a program including equality constraints given by continuous affine functions.
4. **(a)** Use the Lagrangian $\ell : X \times \mathbb{R}^k \to \mathbb{R} \cup \{-\infty, +\infty\}$ given by

$$\ell(x, y) = f(x) + y_1 g_1(x) + \ldots + y_k g_k(x) \quad \text{for } (x, y) \in X \times \mathbb{R}^k_+,$$

$$\ell(x, y) = -\infty \qquad \text{if } (x, y) \in X \times (\mathbb{R}^k \backslash \mathbb{R}^k_+),$$

   to formulate optimality conditions for the problem $(\mathcal{M})$.
   (b) Introduce a Lagrangian $\ell : X \times Y \to \overline{\mathbb{R}}$ adapted to the problem of minimizing a convex function f under the constraint $x \in C := \{x \in X : g(x) \in -Z_+\}$, where Z_+ is a closed convex cone in a Banach space Z and $g : X \to Z$ is a map whose epigraph $E := \{(x, z) \in X \times Z : z \in Z_+ + g(x)\}$ is closed and convex.
5. Let $X := \mathbb{R}$ and let $f : X \to \overline{\mathbb{R}}$, $g : X \to \mathbb{R}$ be given by $f(x) := -x^\alpha$ for $x \in \mathbb{R}_+$, with $\alpha \in]0, 1[$, $f(x) := +\infty$ for $x < 0$, $g(x) = x$ for $x \in \mathbb{R}$. Show that there is no Karush-Kuhn-Tucker multiplier at the solution of $(\mathcal{M})$ with such data.
6. (*Minimum volume ellipsoid problem*) Let $(e_1, \ldots, e_n)$ be the canonical basis of $\mathbb{R}^n$ and let S^n_{++} be the set of positive definite matrices of size (n, n).

   (a) Show that the identity matrix I is the unique optimal solution of the problem

$$\text{Minimize } -\log \det u, \quad u \in S^n_{++}, \quad \|u(e_i)\|^2 - 1 \leq 0, \ i = 1, \ldots, n.$$

   [Hint: Use Theorem 6.12 and a compactness argument; see [44, pp. 32, 48].]
   (b) Deduce from (a) the following special form of *Hadamard's inequality*: for $u \in S^n_{++}$ and $u_i := u(e_i)$, one has $\det(u_1, \ldots, u_n) \leq \|u_1\| \ldots \|u_n\|$.

7*. Characterize the tangent cone to the positive cone $L_p(S)_+$ of $L_p(S)$ for $p \in [1, \infty[$, S being a finite measure space.

6.3 The Legendre-Fenchel Transform and Its Applications

There are several instances in mathematics in which a duality can be used to transform a given problem into an associated one called the dual problem. The dual problem may appear to be more tractable and may yield useful information about the original problem and even help to solve it entirely. For an abstract approach, see [143]. For optimization problems, the Legendre-Fenchel conjugacy is certainly the most useful duality. We present its main properties and we show how it can be used to get duality results for minimization problems. In the last subsection we show that duality enables us to give calculus rules for subdifferentials.

6.3.1 The Legendre-Fenchel Transform

Given a normed space X in duality with its topological dual X^* through the usual pairing $\langle \cdot, \cdot \rangle$ and a function $f : X \to \overline{\mathbb{R}}$, the knowledge of the performance function

$$f_*(x^*) := \inf_{x \in X} (f(x) - \langle x^*, x \rangle) \tag{6.10}$$

associated with the natural perturbation of f by continuous linear forms is likely to give precious information about f, at least when f is closed proper convex.

Notice that a pair $(x^*, r) \in X^* \times \mathbb{R}$ is in the hypograph $H_{f_*} := \{(x^*, r) : f_*(x^*) \geq r\}$ of f_* if and only if one has $f \geq x^* + r$. Thus, when f is closed and proper convex, f_* is related to the characterization of the epigraph of f as the intersection of the upper half-spaces determined by the continuous affine forms that are less than f, as in Theorem 6.2.

Since f_* is concave (it is called the *concave conjugate* of f) and upper semicontinuous, one usually prefers to deal with the convex conjugate or *Legendre-Fenchel*

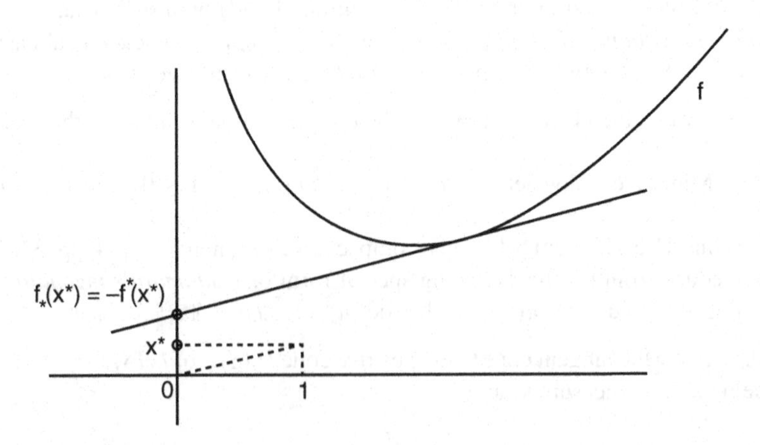

Fig. 6.3 The Young-Fenchel transform in one-dimension

conjugate (or simply Fenchel conjugate) f^* of f given by $f^* = -f_*$:

$$f^*(x^*) := \sup_{x \in X}(\langle x^*, x\rangle - f(x)). \tag{6.11}$$

We note that whenever the domain dom$f$ of f is nonempty, f^* takes its values in $\mathbb{R}_\infty := \mathbb{R} \cup \{+\infty\}$; in contrast, if $f = \infty^X$ then $f^* = -\infty^{X^*}$, the function whose sole value is $-\infty$. We also observe that f^* is convex and lower semicontinuous with respect to the weak* topology on X^* as a supremum of continuous affine functions. Notice that we could replace X^* with another space Y in duality with X. Then, for $g : Y \to \overline{\mathbb{R}}$ we use a similar notation for the conjugate $g^* : X \to \overline{\mathbb{R}}$ of g defined by

$$g^*(x) := \sup_{y \in Y}(\langle x, y\rangle - g(y)) \qquad x \in X.$$

The computation of conjugates is eased by the calculus rules we give below. The following examples illustrate the interest of this transformation.

Examples

(a) Let f be the indicator function ι_C of a subset C of X. Then f^* is the *support function* h_C or σ_C of C given by $h_C(x^*) := \sigma_C(x^*) := \sup_{x \in C}\langle x^*, x\rangle$.
(b) Let $h_S : X^* \to \mathbb{R}_\infty$ be the support function of $S \subset X$, S nonempty. Then $h_S^* = \iota_C$, where $C := \mathrm{clco}(S)$ is the closed convex hull of S.
(c) If f is linear and continuous, then f^* is the indicator function of $\{f\}$.
(d) For $f = \frac{1}{p}\|\cdot\|^p$ with $p \in]1, \infty[$, $q := (1 - \frac{1}{p})^{-1}$, one has $f^* = \frac{1}{q}\|\cdot\|^q$.
(e) If $f = \|\cdot\|$, then $f^* = \iota_{B^*}$, the indicator function of the closed unit ball B^* of X^*.
(f) More generally, if f is positively homogeneous and $f(0) = 0$, then f^* is the indicator function of $\partial f(0)$.

Other examples are given in the exercises. Examples (e) and (f) point out a connection between subdifferentials and conjugates; we will consider this question with more generality later on. Examples (a) and (b) illustrate the close relationships between functions and sets. Let us point out the potential generality of Example (a), which shows that the computation of conjugate functions can be reduced to the calculus of support functions: for any function $f : X \to \overline{\mathbb{R}}$ with epigraph $E := E_f$, the value at x^* of f^* satisfies the relation

$$f^*(x^*) = h_E(x^*, -1), \tag{6.12}$$

as an immediate interpretation of the definition shows (see also Exercise 1 below). The next example in the simple one-dimensional case points out a link with the Legendre transform (Fig. 6.3).

Example-Exercise Let $g : \mathbb{R}_+ \to \mathbb{R}$ be an increasing continuous function such that $g(0) = 0$ and $g(r) \to \infty$ as $r \to \infty$. Let h be the inverse of g and let f be given by $f(x) := \int_0^x g(r)dr$. Show that f^* is given by $f^*(y) = \int_0^y h(s)ds$.

This transform enjoys nice properties. We leave their easy proofs as exercises. Here the infimal convolution $f\Box g$ of two functions f, g is defined by $(f\Box g)(x) := \inf_w(f(x-w) + g(w))$.

Proposition 6.20 *The Fenchel transform satisfies the following properties:*
It is antitone: for any functions f, g with $f \leq g$ one has $f^* \geq g^*$.
For any function f and $c \in \mathbb{R}$, $(f+c)^* = f^* - c$.
For any function f and $c > 0$, $(cf)^*(x^*) = cf^*(c^{-1}x^*)$ for all $x^* \in X^*$.
For any function f and $c > 0$, if $g := f(c\,\cdot)$, then $g^* = f^*(c^{-1}\,\cdot)$.
For any function f and $\ell \in X^*$, $(f+\ell)^* = f^*(\cdot - \ell)$.
For any function f and $\overline{x} \in X$, $(f(\cdot + \overline{x}))^* = f^* - \langle \cdot, \overline{x}\rangle$.
For all functions f, g one has $(f\Box g)^* = f^* + g^*$.
For $p(\cdot) := \inf_{x\in X} f(\cdot, x)$, where $f : W \times X \to \overline{\mathbb{R}}$, one has $p^*(\cdot) = f^*(\cdot, 0)$.
If h is positively homogeneous one has $h^* = \iota_S$, where

$$S := \{x^* \in X^* : x^* \leq h\}.$$

Let us examine whether f^* enables one to recover f. With this aim in mind we introduce the *biconjugate* of f as the function $f^{**} := (f^*)^*$. As above we use the same symbol for the conjugate g^* of a function g on X^* :

$$g^*(x) := \sup_{x^*\in X^*} (\langle x^*, x\rangle - g(x^*)).$$

In doing so we commit some abuse of notation since in fact we consider the restriction of g^* to $X \subset X^{**}$. However, the notation is compatible with the choice of the pairing between X and X^*. In fact, our study could be cast in the framework of topological vector spaces X, Y in separated duality; taking for Y the dual of X endowed with the weak* topology, one would get X as the dual of Y.

Theorem 6.14 *For any function $f : X \to \overline{\mathbb{R}}$ one has $f^{**} \leq f$. If f is closed proper convex one has $f^{**} = f$. This relation also holds if $f = \infty^X$ or $f = -\infty^X$, the constant functions with values $+\infty$ and $-\infty$, respectively.*

Proof Given $x \in X$, for any function $f : X \to \overline{\mathbb{R}}$ and any $x^* \in X^*$ we have $-f^*(x^*) \leq f(x) - \langle x^*, x\rangle$ hence $f^{**}(x) = \sup\{\langle x^*, x\rangle - f^*(x^*) : x^* \in X^*\} \leq f(x)$.

Let us suppose f is closed proper convex. For any $w \in X$ and $r < f(w)$ we can find $x^* \in X^*$ and $c \in \mathbb{R}$ such that $r < \langle x^*, w\rangle - c$, $\langle x^*, x\rangle - c \leq f(x)$ for all $x \in X$. Then we have $f^*(x^*) \leq c$, hence $f^{**}(w) \geq \langle x^*, w\rangle - c > r$ for all $w \in X$. Therefore $f^{**} \geq f$, hence $f^{**} = f$. The cases of the constant functions $-\infty^X$, ∞^X with values $-\infty$ and $+\infty$, respectively, are immediate. □

Corollary 6.10 *For any function $f : X \to \overline{\mathbb{R}}$ bounded below by a continuous affine function and with nonempty domain, the greatest closed proper convex function on X bounded above by f is $f^{**} \mid X$.*

If f is not bounded below by a continuous affine function, then $f^{**} = -\infty^X$.

Proof Let us note that the epigraph of f^* is the set of $(w^*, r) \in X^* \times \mathbb{R}$ such that $f(x) \geq \langle w^*, x\rangle - r$ for all $x \in X$. The last assertion is an immediate consequence of this observation. If g is a closed proper convex function satisfying $g \leq f$, we have $g^* \geq f^*$ since the Fenchel transform is antitone; then $g = g^{**} \leq f^{**}$. Thus, when $f \neq \infty^X$ and f is bounded below by a continuous affine function, f^{**} is proper and clearly lower semicontinuous and convex, hence closed proper convex and f^{**} is the greatest such function bounded above by f. □

Corollary 6.11 *For any function $f : X \to \overline{\mathbb{R}}$ one has $f^{***} = f^*$.*

Proof The result is obvious if $f^* = \infty^{X^*}$ or if $f^* = -\infty^{X^*}$; otherwise f^* is closed proper convex. □

A crucial relationship between the Fenchel conjugate and the Fenchel-Moreau subdifferential is given by the Young-Fenchel relation that follows.

Theorem 6.15 (Young-Fenchel) *For any function $f : X \to \overline{\mathbb{R}}$ and for any $x \in X$, $x^* \in X^*$ one has $f(x) + f^*(x^*) \geq \langle x^*, x\rangle$.*

When $f(x) \in \mathbb{R}$ equality holds if and only if $x^* \in \partial f(x)$.

Moreover, $x^* \in \partial f(x)$ implies $x \in \partial f^*(x^*)$.

Proof The first assertion is a direct consequence in the definition. When $f(x) \in \mathbb{R}$ the equality $f(x) + f^*(x^*) = \langle x^*, x\rangle$ is equivalent to each of the following assertions

$$f(x) + f^*(x^*) \leq \langle x^*, x\rangle$$
$$f(x) - f(w) + \langle x^*, w\rangle \leq \langle x^*, x\rangle \qquad \forall w \in X$$
$$x^* \in \partial f(x).$$

Moreover, they imply the inequality $f^{**}(x) + f^*(x^*) \leq \langle x^*, x\rangle$ equivalent to $x \in \partial f^*(x^*)$. □

Theorem 6.16 *For any function $f : X \to \overline{\mathbb{R}}$ and $x \in X$ one has $f^{**}(x) = f(x)$ whenever $\partial f(x) \neq \varnothing$.*

Moreover, when $f^{**}(x) = f(x) \in \mathbb{R}$, one has $\partial f(x) = \partial f^{**}(x)$ and $x^* \in \partial f(x)$ if and only if $x \in \partial f^*(x^*)$.

Proof Given $x^* \in \partial f(x)$, let $g : w \mapsto \langle x^*, w - x\rangle + f(x)$. Then g is a continuous affine function satisfying $g \leq f$, so that $g \leq f^{**}$ and $g(x) = f(x) \geq f^{**}(x)$, so that $f^{**}(x) = f(x)$ and $x^* \in \partial f^{**}(x)$. Moreover, when $f^{**}(x) = f(x) \in \mathbb{R}$, the reverse inclusion $\partial f^{**}(x) \subset \partial f(x)$ follows from the relations $f^{**} \leq f, f^{**}(x) = f(x)$. The last assertion is a consequence of the last assertion of Theorem 6.15 and of the preceding. □

Corollary 6.12 *When $f = f^{**}$ the multimap ∂f^* is the inverse of the multimap ∂f:*

$$x^* \in \partial f(x) \Leftrightarrow x \in \partial f^*(x^*).$$

The following special case is of great importance for dual problems.

Corollary 6.13 *When $f^{**}(0) = f(0) \in \mathbb{R}$ the set of minimizers of f^* is $\partial f(0)$.*

For any function g with finite infimum, the set $\partial g^*(0)$ is the set of minimizers of g^{**}.

When $f^{**} = f$ and $f^*(0)$ is finite, the set $\partial f^*(0)$ is the set of minimizers of f.

Proof The first assertion follows from the equivalences $x^* \in \partial f(0) \Leftrightarrow 0 \in \partial f^*(x^*) \Leftrightarrow x^*$ is a minimizer of f^*. The second one ensues, as $g^*(0) = -\inf g(X)$ and $g^{***} = g^*$. Taking $g := f$, one gets the last assertion. □

Exercises

1. Show that for any function $f : X \to \overline{\mathbb{R}}$ with nonempty domain, the support function of the epigraph $E := E_f$ of f satisfies $h_E(x^*, -1) = f^*(x^*)$ and

$$h_E(x^*, r) = -rf^*(-r^{-1}x^*) \quad \text{for } r < 0,$$
$$h_E(x^*, 0) = h_{\text{dom} f}(x^*),$$
$$h_E(x^*, r) = +\infty \qquad \text{for } r > 0.$$

2. Given a function $f : X \to \mathbb{R}_\infty$ with nonempty domain, verify that

$$\text{epi} f^* \times \{-1\} = (S(Q))^0 \cap (X^* \times \mathbb{R} \times \{-1\}),$$

where $Q := \mathbb{R}_+(\text{epi} f \times \{-1\})$ and S is the map $(x, r, s) \mapsto (x, s, r)$, a linear isometry.
3. Show that for any function $f : X \to \overline{\mathbb{R}}$, the greatest lower semicontinuous convex function bounded above by f is either f^{**} or the "*valley function*" υ_C associated with the closed convex hull C of $\text{dom} f$, given by $\upsilon_C(x) = -\infty$ if $x \in C$, $\upsilon_C(x) = +\infty$ if $x \notin C$.
4. If X is a normed space and $f = g \circ \|\cdot\|$, where $g : \mathbb{R}_+ \to \mathbb{R}_\infty$ is extended by $+\infty$ on $\mathbb{R}_-$, show that $f^* = g^* \circ \|\cdot\|_*$, where $\|\cdot\|_*$ is the dual norm of $\|\cdot\|$.
5. For $X = \mathbb{R}$ and $f(x) = \exp x$, verify that $f^*(y) = y \log y - y$ for $y > 0, f^*(0) = 0$, $f(y) = +\infty$ for $y < 0$.
6. Let $f : \mathbb{R} \to \mathbb{R}_\infty$ be given by $f(x) := -\log x$ for $x \in \mathbb{P}, f(x) := +\infty$ for $x \in \mathbb{R}_-$. Verify that $f^*(x^*) = -\log |x^*| - 1$ for $x^* \in -\mathbb{P}, f(x) := +\infty$ for $x \in \mathbb{R}_+$.
7. Let $f : X \to \mathbb{R}_\infty$ and let g be the convex hull of f. Show that $g^* = f^*$.
8. Let $f : X \to \mathbb{R}_\infty$ and let h be the lower semicontinuous hull of f. Show that $h^* = f^*$.
9. (*Legendre transform*) Let $f : X \to \mathbb{R}_\infty$ be a lower semicontinuous proper convex function that is differentiable on its open domain W and such that its derivative $f' : W \to X^*$ realizes a bijection between W and $W^* := f'(W)$, with inverse

h. Let $f^L : W^* \to \mathbb{R}$ be the Legendre transform of f: $f^L(w^*) := \langle w^*, h(w^*)\rangle - f(h(w^*))$. Show that f^L coincides with the restriction to W^* of the conjugate f^* of f.

6.3.2 A Brief Account of Convex Duality Theories

There are two main theories of convex duality. The perturbational approach is probably the most natural one, so we shall start with it. An alternative is the Lagrangian theory, which can also be cast in a general framework. We relate the two approaches and we apply them to classical problems.

Given a nonempty set X and an objective function $f : X \to \mathbb{R}_\infty$ which may incorporate constraints, we consider the problem

$$(\mathcal{P}) \quad \text{minimize} f(x) \qquad \text{subject to } x \in X.$$

A perturbation of $(\mathcal{P})$ is the data of a normed space W named the space of parameters coupled with a normed space Y and of a function $P : X \times W \to \overline{\mathbb{R}}$ called a *perturbation function* satisfying

$$\forall x \in X, \qquad P(x, 0) = f(x).$$

If $P(x, 0) \leq f(x)$ holds for all $x \in X$ one says that P is a *sub-perturbation* of $(\mathcal{P})$. The *performance function* $p : W \to \overline{\mathbb{R}}$ (or *value function*) is given by

$$\forall w \in W, \qquad p(w) := \inf_{x \in X} P(x, w).$$

The relation $p^{**}(0) := \sup\{-p^*(y) : y \in Y\}$ is an incentive to introduce the *dual problem*:

$$(\mathcal{D}_P) \quad \text{maximize } d_P(y) := -p^*(y) \qquad \text{subject to } y \in Y,$$

so that $\sup(\mathcal{D}_P) = p^{**}(0)$ and $\inf(\mathcal{P}) = p(0)$. The so-called *weak duality property*

$$\sup(\mathcal{D}_P) \leq \inf(\mathcal{P})$$

consists in the obvious inequality $p^{**}(0) \leq p(0)$.

Strong duality is said to hold whenever $\sup(\mathcal{D}_P) = \inf(\mathcal{P})$ and $(\mathcal{D}_P)$ has a solution. Let us note that the "opposite" of $(\mathcal{D}_P)$, that is the *adjoint problem*

$$(\mathcal{P}^*) \quad \text{minimize } -d_P(y) = p^*(y) \qquad y \in Y$$

is a convex problem even if f is nonconvex.

To obtain duality results, one may require that X be a normed vector space and P be convex. Then

$$p^*(y) = \sup_{w \in W}(\langle y, w\rangle - \inf_{x \in X} P(w,x))$$
$$= \sup_{(w,x)\in W\times X}(\langle (y,0),(w,x)\rangle - P(w,x)) = P^*(y,0)$$

and $p^{**}(0) = \sup\{-P^*(y,0) : y \in Y\}$. But duality results can be obtained under the weaker assumption that the performance function p associated with P is convex, X being an arbitrary set.

Let us give an interpretation of the set S^* of solutions to the dual problem $(\mathcal{D}_P)$. Here, although p is not necessarily convex, we set

$$\partial p(0) := \{y \in Y : \forall w \in W \quad p(w) \geq p(0) + \langle w, y\rangle\}.$$

Proposition 6.21 *Suppose* $\sup(\mathcal{D}_P) = \inf(\mathcal{P}) \in \mathbb{R}$. *Then one has* $\partial p(0) = S^*$ *the set of solutions to* $(\mathcal{D}_P)$.

Moreover, if $p(0)$ *is finite the relation* $\sup(\mathcal{D}_P) = \inf(\mathcal{P})$ *holds whenever* $\partial p(0)$ *is nonempty.*

Proof Suppose $\sup(\mathcal{D}_P) = \inf(\mathcal{P})$, i.e. $p^{**}(0) = p(0)$. Given $\overline{y} \in S^*$, we have $-p^*(\overline{y}) = \sup_{y\in Y} -p^*(y) = p^{**}(0) = p(0)$, so that $\inf_{w\in W}(p(w) - \langle w, \overline{y}\rangle) = p(0)$ or $p(w) \geq p(0) + \langle w, \overline{y}\rangle$ for all $w \in W$ and $\overline{y} \in \partial p(0)$.

Given $\overline{y} \in \partial p(0)$, we have $-\sup_{w\in W}(\langle w, \overline{y}\rangle - p(w)) \geq p(0)$, hence $p^{**}(0) \geq -p^*(\overline{y}) \geq p(0)$. Since the inequality $p^{**}(0) \leq p(0)$ is always satisfied, we obtain $p^{**}(0) = -p^*(\overline{y}) = p(0)$ and $\overline{y} \in S^*$ since $p^{**}(0) = \sup_{y\in Y} -p^*(y)$. □

Example-Exercise (Linear Programming) Given $A \in L(\mathbb{R}^m, \mathbb{R}^n)$, $b \in \mathbb{R}^n$, $c \in \mathbb{R}^m$, consider the problem

$$(\mathcal{P}) \quad \text{minimize } \langle c|x\rangle \quad \text{subject to } x \in \mathbb{R}^m, \quad x \geq 0, \quad Ax = b.$$

Define a perturbation whose associated dual problem is as in the first example of this chapter:

$$(\mathcal{D}) \quad \text{maximize } \langle b|y\rangle \quad \text{subject to } y \in \mathbb{R}^n, \quad A^*y \leq c.$$

Compare it to the Lagrangian dual problem described hereafter. □

The *Lagrangian scheme* consists in replacing $(\mathcal{P})$ with a family $(\mathcal{P}_y)_{y\in Y}$ of simpler problems

$$(\mathcal{P}_y) \quad \text{minimize } L_y(x) \qquad \text{subject to } x \in X.$$

where $L : X \times Y \to \overline{\mathbb{R}}$ is called a *Lagrangian* if $\sup_{y \in Y} L(x, y) = f(x)$ for all $x \in X$ and a *sub-Lagrangian* if $\sup_{y \in Y} L(x, y) \leq f(x)$ for all $x \in X$. Here $L_y := L(\cdot, y)$ for all $y \in Y$. In both cases, setting

$$d_L(y) := \inf_{x \in X} L(x, y), \tag{6.13}$$

the value $\sup(\mathcal{D}_L)$ of the dual problem

$$(\mathcal{D}_L) \quad \text{maximize } d_L(y) \qquad \text{subject to } y \in Y$$

satisfies the estimate $\sup(\mathcal{D}_L) \leq \inf(\mathcal{P})$ (*weak duality*). One says that there is no *duality gap* when $\sup(\mathcal{D}_L) = \inf(\mathcal{P})$ and one says that *strong duality* holds when there is no duality gap and $(\mathcal{D}_L)$ has a solution. An element $\overline{y}$ of Y is called a *multiplier* if $\inf_{x \in X} L(x, \overline{y}) = \inf_{x \in X} f(x)$, i.e. if $d_L(\overline{y}) = \inf_{x \in X} f(x)$. Since for all $y \in Y$ one has $d_L(y) \leq \inf_{x \in X} f(x)$, we see that $\overline{y} \in Y$ is a multiplier if and only if $\overline{y}$ is a solution to the dual problem $(\mathcal{D}_L)$ and there is no duality gap. Multipliers can be used to detect solutions of $(\mathcal{P})$, as shown by the next statement.

Proposition 6.22 *Let L be a Lagrangian or a sub-Lagrangian of $(\mathcal{P})$. If $\overline{y}$ is an element of the set M of multipliers, then the set S of solutions of $(\mathcal{P})$ is contained in the set $S_{\overline{y}}$ of solutions of $(\mathcal{P}_{\overline{y}})$ and* $\inf(\mathcal{P}) = \inf(\mathcal{P}_{\overline{y}})$.

Conversely, if for some $\overline{y} \in Y$ and some $\overline{x} \in S_{\overline{y}}$ one has $L(\overline{x}, \overline{y}) = f(\overline{x})$, then $\overline{x} \in S$ and $\overline{y} \in M$.

Under each of the preceding conditions, $(\overline{x}, \overline{y})$ is a saddle point of L:

$$L(\overline{x}, y) \leq L(\overline{x}, \overline{y}) \leq L(x, \overline{y}) \qquad \forall (x, y) \in X \times Y. \tag{6.14}$$

Proof Let $\overline{x} \in S$ and $\overline{y} \in M$. Since $L(\overline{x}, \overline{y}) \leq f(\overline{x}) = \inf(\mathcal{P}) = \inf_{x \in X} L(x, \overline{y})$ by definition of a multiplier, we see that $\overline{x}$ minimizes $L(\cdot, \overline{y})$ and $L(\overline{x}, \overline{y}) = f(\overline{x}) = \inf(\mathcal{P}_{\overline{y}})$.

Conversely, given $\overline{y} \in Y$ such that $f(\overline{x}) = L(\overline{x}, \overline{y})$ for some $\overline{x} \in S_{\overline{y}}$, i.e. such that $L(\overline{x}, \overline{y}) \leq L(x, \overline{y})$ for all $x \in X$, then, since $L(x, \overline{y}) \leq f(x)$ for all $x \in X$, we get that $\overline{x}$ minimizes f and $f(\overline{x}) = \inf_{x \in X} L(x, \overline{y})$: $\overline{x} \in S$ and $\overline{y} \in M$.

Since in such a case the inequalities $L(\overline{x}, \overline{y}) \leq L(x, \overline{y})$ for all $x \in X$ can be completed with the relations $L(\overline{x}, y) \leq f(\overline{x}) = L(\overline{x}, \overline{y})$ by definition of a sub-Lagrangian, $(\overline{x}, \overline{y})$ is a saddle point of L. □

Now let us show that any perturbation or sub-perturbation P of $(\mathcal{P})$ yields a sub-Lagrangian of $(\mathcal{P})$ by setting

$$L(x, y) := \inf_{w \in W} (P(x, w) - \langle w, y \rangle) \qquad \forall (x, y) \in X \times Y, \tag{6.15}$$

or $L_x := -P_x^*$ for all $x \in X$, where $L_x := L(x, \cdot)$ and $P_x := P(x, \cdot)$.

Proposition 6.23

(a) *The function L deduced from a sub-perturbation P via (6.15) is a sub-Lagrangian of $(\mathcal{P})$. If P is a perturbation and if $P_x^{**}(0) = P_x(0)$ for all $x \in X$, in particular, if the function $P_x := P(x, \cdot)$ is closed proper convex, then L is a Lagrangian of $(\mathcal{P})$.*
(b) *Moreover, the objective $d_L : Y \to \overline{\mathbb{R}}$ of the sub-Lagrangian dual problem given by (6.13) coincides with the objective $d_P := -p^* : Y \to \overline{\mathbb{R}}$ of the perturbational dual problem. Thus, the values and the optimal solutions of the two dual problems are the same.*
(c) *The set M of multipliers of the sub-Lagrangian L coincides with $\partial p(0)$.*

Proof

(a) For all $x \in X$, by definition of L, we have

$$\sup_{y \in Y} L(x, y) = \sup_{y \in Y} -P_x^*(y) = P_x^{**}(0) \le P_x(0) \le f(x)$$

and these relations are equalities whenever $P_x^{**}(0) = P_x(0) = f(x)$ for all $x \in X$.
(b) For all $y \in Y$ the definitions of d_L, L and d_P yield

$$d_L(y) := \inf_{x \in X} L(x, y) := \inf_{x \in X} \inf_{w \in W} (P(x, w) - \langle w, y \rangle)$$
$$= \inf_{w \in W} (p(w) - \langle w, y \rangle) = -p^*(y) := d_P(y).$$

(c) If $\overline{y} \in M$ we have $\sup(\mathcal{D}_P) = \sup(\mathcal{D}_L) = \inf(\mathcal{P})$ and $\overline{y} \in S^* = \partial p(0)$ in view of Proposition 6.21. Conversely, if $\overline{y} \in \partial p(0)$ Proposition 6.21 shows that $\sup(\mathcal{D}_P) = \inf(\mathcal{P})$ and $\overline{y} \in S^*$, so that $\overline{y} \in M$ since $\sup(\mathcal{D}_L) = \sup(\mathcal{D}_P)$. □

Now let us consider the passage from a Lagrangian to a perturbation. Given a sub-Lagrangian L let us set

$$P(x, w) := \sup_{y \in Y}(\langle y, w \rangle + L(x, y)) = (-L_x)^*(w) \qquad (x, w) \in X \times W. \qquad (6.16)$$

Proposition 6.24 *Given a Lagrangian (resp. a sub-Lagrangian) L, the function P defined by (6.16) is a perturbation (resp. a sub-perturbation) of $(\mathcal{P})$.*

Moreover, the function $K : X \times Y \to \overline{\mathbb{R}}$ given by $K_x := -(-L_x)^{**}$ for all $x \in X$ is a sub-Lagrangian of $(\mathcal{P})$ and K is the sub-Lagrangian associated with P. In particular, the dual objective function of K coincides with the dual objective function of P. Thus, when $(-L_x)^{**} = -L_x$ for all $x \in X$ the dual problem $(\mathcal{D}_P)$ associated with P coincides with the Lagrangian dual problem $(\mathcal{D}_L)$.

Proof Since for all $x \in X$ one has $P(x, 0) := \sup_{y \in Y} L(x, y) \le f(x)$ the first assertion is an immediate consequence in the definitions. Since for all $(x, y) \in X \times Y$, one has $-L_x(\cdot) \ge -f(x)$, taking biconjugates one gets $-K(x, y) := (-L_x)^{**}(y) \ge -f(x)$, so that K is a sub-Lagrangian of $(\mathcal{P})$ and $K \geq L$. Since for all $x \in X$ one has $P_x = (-L_x)^*$ and $K_x = -(-L_x)^{**}$, we have $K_x = -(P_x)^*$, so that K is the sub-Lagrangian associated with P. Then, by Proposition 6.23 (b), the dual objective function d_K associated with K coincides with d_P.

When $(-L_x)^{**} = -L_x$ for all $x \in X$ one has $K = L$, so that the dual problem associated with $(\mathcal{P})$, which coincides with the dual problem associated with K, also coincides with the dual problem associated with L. □

Thus, the Lagrangians L such that $-L_x = -(-L_x)^{**}$ for all $x \in X$ are in one-to-one correspondence with the perturbations P of $(\mathcal{P})$ such that $P_x = (P_x)^{**}$.

Mathematical programming problems can be cast in such frameworks. Given a nonempty set X, a normed vector space W, a multimap $G : X \rightrightarrows W$ with inverse $F : W \rightrightarrows X$ and a function $f : X \to \mathbb{R}_\infty$, let us consider the problem

$$(\mathcal{M}) \quad \text{minimize} f(x) \qquad \text{subject to} \; x \in F(0) := G^{-1}(0).$$

It is natural to take as a perturbation of $(\mathcal{M})$ the function $P : X \times W \to \mathbb{R}_\infty$ given by

$$\forall (x,w) \in X \times W, \qquad P(x,w) := f(x) + \iota_{F(w)}(x) = f(x) + \iota_G(x,w),$$

where ι_G is the indicator function of (the graph of) G. In such a case, the (sub-) Lagrangian L associated with P is given by

$$L(x,y) = -(P_x)^*(y) = \inf_{w \in W} (f(x) + \iota_{G(x)}(w) - \langle w, y \rangle)) = f(x) - \sigma_{G(x)}(y),$$

where for a subset C of W, σ_C is the support function of C. If the values of G are closed convex, then $P_x = P_x^{**}$ for all $x \in X$ and L is a Lagrangian. When G is given by $G(x) := C - g(x)$, where C is a subset of W and $g : X \to W$ is a map, one gets $P(x,w) = f(x) + \iota_C(g(x) + w)$ and

$$L(x,y) = f(x) + \langle g(x), y \rangle - \sigma_C(y),$$

a familiar form when C is a convex cone of W, since in such a case one has $\sigma_C = \iota_{C^0}$ where C^0 is the polar cone of C. The classical mathematical programming problem

$$(\mathcal{M}_{f,g}) \; \text{minimize} f(x) \; \text{subject to} \; g_i(x) = 0, \; g_j(x) \leq 0, \; i \in \mathbb{N}_k, \; j \in \mathbb{N}_m + k$$

corresponds to the case $W := \mathbb{R}^{k+m}$ and $C := -\{0_{\mathbb{R}^k}\} \times \mathbb{R}^m_+$; then $C^0 := \mathbb{R}^k \times \mathbb{R}^m_+$ and the Lagrangian is given by

$$L(x,y) = f(x) + \sum_{i=1}^{k+m} y_i g_i(x) \qquad (x,y) := (x, y_1, \ldots y_{k+m}) \in X \times \mathbb{R}^k \times \mathbb{R}^m_+,$$

$L(x,y) = -\infty$ if $y \in \mathbb{R}^{k+m} \backslash (\mathbb{R}^k \times \mathbb{R}^m_+)$.

Example Let us consider the *linear programming* problem

$$(\mathcal{P}) \text{ minimize } \langle c, x\rangle \text{ under the constraints } x \in \mathbb{R}^n_+, Ax \leq b$$

where $A \in L(\mathbb{R}^n, \mathbb{R}^m)$, $b \in \mathbb{R}^m$ and $c \in \mathbb{R}^n$ identified with the dual of $\mathbb{R}^n$. Setting $f := c$, $C := -\mathbb{R}^m_+$, $X := \mathbb{R}^n_+$, $g(x) := Ax - b$, since $C^0 = \mathbb{R}^n_+$ and since for $y \in \mathbb{R}^m_+$ one has

$$d(y) := \inf_{x \in \mathbb{R}^n_+} \langle c, x\rangle + \langle y, Ax - b\rangle = -\langle y, b\rangle \quad \text{if } c + A^{\mathsf{T}} y \in \mathbb{R}^m_+, \; -\infty \text{ otherwise}$$

the dual problem can be written as

$$(\mathcal{D}) \text{ maximize } \langle -b, y\rangle \text{ under the constraints } y \in \mathbb{R}^m_+, \; -A^{\mathsf{T}} y \leq c.$$

Thus the dual problem has a form similar to the form of the primal problem. If m is smaller than n, the dual problem may be easier to solve than the primal problem. Then, using Proposition 6.22, one may use a solution to $(\mathcal{D})$ to solve the primal problem. Moreover, the interpretation of the set of multipliers in terms of $\partial p(0)$ yields precious sensitivity results. In the exercises other examples are provided.

A case of special interest is the minimization problem

$$(\mathcal{P}_{f,g,h}) \qquad \text{minimize} f(x) + h(g(x)), \quad x \in D, \tag{6.17}$$

where $f : X \to \mathbb{R}_\infty$, $g : D \to W$ and $h : W \to \mathbb{R}_\infty$, X, W being Banach spaces, D being a subset of X. When h is the indicator function ι_C of a subset C of W, $(\mathcal{P}_{f,g,h})$ amounts to the minimization of f over $D \cap g^{-1}(C)$. A natural perturbation of problem $(\mathcal{P}_{f,g,h})$ is given by $P(x, w) := f(x) + h(g(x) + w)$.

The objective function $-p^*$ of $(\mathcal{D})$ can easily be expressed in terms of the data:

$$\begin{aligned}
-p^*(y) &= \inf_{w \in W} (p(w) - \langle y, w\rangle) = \inf_{w \in W} \inf_{x \in D} (f(x) + h(g(x) + w) - \langle y, w\rangle) \\
&= \inf_{x \in D} \inf_{w \in W} (f(x) + \langle y, g(x)\rangle + h(g(x) + w) - \langle y, g(x) + w\rangle) \\
&= \inf_{x \in D} [f(x) + \langle y, g(x)\rangle + \inf_{z \in W} (h(z) - \langle y, z\rangle)] \\
&= \inf_{x \in D} [f(x) + \langle y, g(x)\rangle] - h^*(y).
\end{aligned}$$

When $D := X$ and $h = \iota_C$, with C a convex cone in W, as in problem $(\mathcal{M})$ above with $G(x) := C - g(x)$, h^* is the indicator function of the polar cone C^0 and the function ℓ given by

$$\ell(x, y) := f(x) + \langle y, g(x)\rangle - \iota_{C^0}(y)$$

called the *Lagrangian* of $(\mathcal{M})$ corresponds to the Lagrangian L introduced in Proposition 6.23.

Exercises

1. Let C be a closed convex subset of a normed space X, and let σ_C be the support function of C given by $\sigma_C(x^*) := \sup\{\langle x^*, x\rangle : x \in C\}$ for $x^* \in X^*$. Prove that $d_C = (\sigma_C + \iota_{B_{X^*}})^*$.

2. If C is a subset of a normed space X, the *signed distance* to C is the function $d_C^{\pm}$ given by $d_C^{\pm}(x) := d_C(x)$ if $x \in X\backslash C$, $d_C^{\pm}(x) := -d_{X\backslash C}(x)$ for $x \in C$.

(a) Show that $d_C^{\pm}$ is convex when C is convex.

(b) Let C be a closed convex subset of X, let ι_S be the indicator function of the unit sphere in X^*, and let σ_C be the support function of C. Prove that $d_C^{\pm} = (\sigma_C + \iota_S)^*$.

(c) Suppose C is a nonempty open convex subset of X and let $w \in W\backslash C$. Let $s : x \mapsto 2w - x$. Verify that $C \cap s(C) = \varnothing$ and use a separation theorem. Prove the relation

$$\inf\{\|w - x\| : x \in X\backslash C\} = -\sup\{\langle x^*, w\rangle - \sigma_C(x^*) : x^* \in X^*\backslash B(0,1)\}.$$

(d) Show that if the infimum is attained at some $\overline{x} \in X\backslash C$, then there exists some $\overline{x}^* \in X^*$ such that $\overline{x}^* \in S(\overline{x} - w) := \{x^* \in S_{X^*} : \langle x^*, \overline{x} - w\rangle = \|\overline{x} - w\|\}$ (Fig. 6.4).

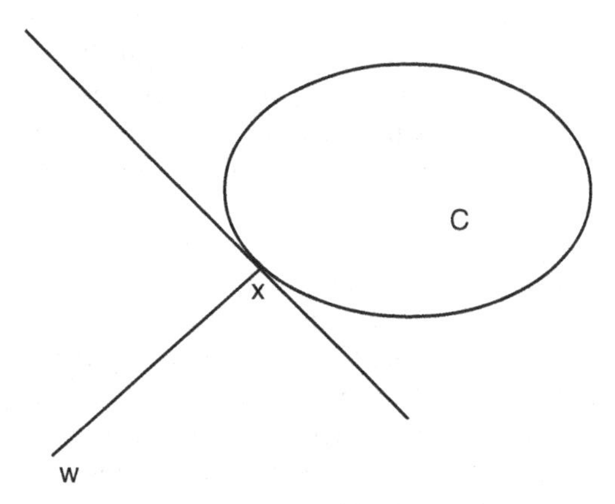

Fig. 6.4 Duality between distance and support functions

3. Assuming W is a normed vector space, show that the weak duality inequality $\inf(\mathcal{P}) + \inf(\mathcal{P}^*) \geq 0$ stems from the Fenchel inequality $P(0,x) + P^*(w^*,0) \geq \langle 0,x\rangle + \langle w^*,0\rangle = 0$.
4. Verify that there is no Lagrange multiplier for the problem of minimizing the function $f : D := \mathbb{R}_+ \to \mathbb{R}$ given by $f(x) = \sqrt{x}$ under the constraint $g(x) \leq 0$ for $g(x) = x$ although it has 0 as optimal solution.
5. Show that the dual problem of the *quadratic programming* problem

$$(\mathcal{P}) \text{ minimize } \frac{1}{2}\langle Qx, x\rangle + \langle c, x\rangle \text{ under the constraints } x \in \mathbb{R}^n,\ Ax \leq b$$

when Q is positive definite can be written as

$$(\mathcal{D}) \text{ maximize } -\frac{1}{2}\langle AQ^{-1}A^{\mathsf{T}}y, y\rangle - \langle b + AQ^{-1}c, y\rangle - \frac{1}{2}\langle Q^{-1}c, c\rangle.$$

Show how the solution to the dual problem can be used to solve the primal problem.
6. (**Bourass-Giner**) Let p be the performance function associated to the perturbation P of problem $(\mathcal{P}_{f,g,h})$. A criterion for the convexity of p can be given by considering the set

$$E_{f,g} := \{(w,r) \in W \times \mathbb{R} : \exists x \in g^{-1}(w) \cap D,\ f(x) < r\}.$$

The pair (f,g) is said to be *convex-like* if $E_{f,g}$ is convex. Verify that $E_{f,g}$ is the strict epigraph of the performance function q given by

$$q(w) := \inf\{f(x) : x \in D, g(x) = w\},$$

so that (f,g) is convex-like if and only if q is convex. Show that if (f,g) is convex-like and if h is convex, then p is convex. [Hint: p is the infimal convolution $h \Box \tilde{q}$ of h and $\tilde{q}$, where $\tilde{q}(w) := q(-w)$ for $w \in W$.]
7. (*Geometric programming*) Let $G(X)$ be the class of functions on X that are finite sums of functions of the form $x \mapsto c\log(\exp\langle a_1^*, x\rangle + \ldots + \exp\langle a_m^*, x\rangle)$ for some $a_i^* \in X^*$ $(i \in \mathbb{N}_m)$, $c > 0$. Given $g_0, g_1, \ldots, g_k$ in $G(X)$, write down a Lagrangian dual problem for the problem of minimizing $g_0(x)$ under the constraints $g_i(x) \leq 0$ $(i \in \mathbb{N}_k)$ and give a duality result.
8. (*Allocation problem*) Assume that a given quantity x_0 of some commodity has to be allocated among n distinct activities or agents. Suppose the return of activity i for an allocation x_i is $g_i(x_i)$, where g_i is an increasing concave function (in view of the law of diminishing marginal returns). The allocation problem $(\mathcal{A})$ consists in finding some $x := (x_1, \ldots, x_n) \in \mathbb{R}^n$ that maximizes the total return $g(x) := g(x_1) + \ldots + g(x_n)$ subject to $x_1 + \ldots + x_n = x_0$, $x_i \in \mathbb{R}_+$. Convert this problem into a convex minimization problem and use duality to show that it can be reduced to a one-variable maximization problem $(\mathcal{D})$ once the conjugates of the functions $f_i := -g_i$ have been computed and then choosing x_i in such a way

that $yx_i - g_i(x_i) = \inf_s\{ys - g_i(s)\}$, where $-y$ is the solution to $(\mathcal{D})$. [See [189, p. 202].]

6.3.3 Duality and Subdifferentiability Results

Let us first give criteria for duality results. Then we shall apply them to subdifferentiability rules. We start with a perturbation P of problem $(\mathcal{P})$ and its associated performance function p.

Proposition 6.25 *Suppose p is convex and $\inf(\mathcal{P})$ is finite. Let V be the vector space generated by $\operatorname{dom} p$. Suppose there exist some $r > 0$, $m \in \mathbb{R}$ and some map $w \mapsto x(w)$ from $rB_W \cap V$ to X such that $P(w, x(w)) \leq m$ for all $w \in rB_W \cap V$. Then $p \mid V$ is continuous, p is subdifferentiable at 0 and strong duality holds.*

In particular, if for some $r > 0$, $m \in \mathbb{R}$, and $\overline{x} \in X$ one has $P(w, \overline{x}) \leq m$ for all $w \in rB_W \cap V$, then strong duality holds.

Proof Under our assumption, p is bounded above by m on $rB_W \cap V$ since for $w \in W$ one has $p(w) \leq P(w, x(w))$. Thus $p \mid V$ is continuous at 0 and, by Corollary 6.5, p is subdifferentiable at 0. □

From the preceding proposition one can get the subdifferentiability rules under continuity assumptions we have seen previously (exercise). We rather deduce new subdifferentiability rules under closedness assumptions and algebraic assumptions that are quite convenient.

Theorem 6.17 *Let W, X be Banach spaces and let p be the performance function associated with a perturbation $P : W \times X \to \mathbb{R}_\infty$ that is convex, lower semicontinuous and such that*

$$Z := \bigcup_{x \in X} \mathbb{R}_+ \operatorname{dom} P(\cdot, x) = -Z = \operatorname{cl}(Z). \tag{6.18}$$

Then, if $p(0) \in \mathbb{R}$, p is subdifferentiable at 0 and strong duality holds.

Note that assumption (6.18) means that Z is a closed vector subspace of W. It is obviously satisfied if $Z = W$, i.e. if $\operatorname{dom} p$ is absorbant or $0 \in \operatorname{core}(\operatorname{dom} p)$.

Proof By Corollary 6.5, we may suppose $Z = W$. The set

$$F := \{(x, r, w) \in X \times \mathbb{R} \times W : P(w, x) \leq r\}$$

is closed and convex as the image of the epigraph of P under the isomorphism $(x, w, r) \mapsto (x, r, w)$. Relation (6.18) means that the projection $C := p_W(F)$ of F is absorbing, i.e. $0 \in \operatorname{core} C$. The Robinson-Ursescu Theorem ensures that F, considered as a multimap from $X \times \mathbb{R}$ to W, is open at any $(\overline{x}, \overline{r}) \in X \times \mathbb{R}$ such that $(\overline{x}, \overline{r}, 0) \in F$; in particular, there exists some $c > 0$ such that

$$cB_W \subset F((\overline{x}, \overline{r}) + B_{X \times \mathbb{R}})).$$

Thus, for all $w \in cB_W$ there exists some $(x_w, r_w) \in B[(\overline{x}, \overline{r}), 1]$ such that $(x_w, r_w, w) \in F$, i.e. $P(w, x_w) \leq r_w \leq m := |\overline{r}| + 1$. Thus p is bounded above by m on cB_W, hence p is continuous at 0 and subdifferentiable at 0. The preceding proposition entails that strong duality holds. □

Let us present some consequences for subdifferential calculus and the calculus of conjugates. In the next theorem we gather a sum rule and a composition rule. It generalizes Theorems 6.8 and 6.9 (Exercise 1). Again, for a function $h : Z \to \mathbb{R}$ and $r \in \mathbb{R}$, we set $\{h \leq s\} := h^{-1}(]-\infty, s])$. Here we have to make a change of notation in order to capture this classical result.

Theorem 6.18 (Fenchel-Rockafellar) *Let X, Y be normed spaces, let $A : X \to Y$ be a continuous linear map and let $f : X \to \mathbb{R}_\infty$, $g : Y \to \mathbb{R}_\infty$ be convex functions such that there exist $r > 0$, $s \in \mathbb{R}_+$ for which*

$$rB_Y \subset A(\{f \leq s\} \cap sB_X) - \{g \leq s\}. \tag{6.19}$$

Then, with the usual convention that an infimum is denoted by min *whenever it is attained when finite, for all $x^* \in X^*$ one has*

$$(f + g \circ A)^*(x^*) = \min_{y^* \in Y^*} \left(f^*(x^* - A^\mathsf{T} y^*) + g^*(y^*)\right). \tag{6.20}$$

Moreover, for any $x \in \operatorname{dom} f \cap A^{-1}(\operatorname{dom} g)$ one has

$$\partial(f + g \circ A)(x) = \partial f(x) + A^\mathsf{T}(\partial g(Ax)). \tag{6.21}$$

Proof Let $W := Y$, let $x^* \in X^*$ and let $P : W \times X \to \mathbb{R}_\infty$ be given by

$$P(w, x) := f(x) - \langle x^*, x\rangle + g(Ax + w).$$

For all $w \in rB_W$ (6.19) yields $x_w \in \{f \leq s\} \cap sB_X$ and $y_w \in \{g \leq s\}$ such that $w = y_w - Ax_w$. Then, the performance function p given by $p(w) := \inf_{x \in X} P(w, x)$ satisfies

$$p(w) \leq P(w, x_w) = f(x_w) - \langle x^*, x_w\rangle + g(y_w) \leq 2s + s\,\|x^*\|,$$

and strong duality holds. Now, for $y^* \in Y^*$, setting $y := Ax + w$, one has

$$\begin{aligned} P^*(y^*, 0) &= \sup_{(w,x) \in W \times X} \left(\langle y^*, w\rangle + \langle x^*, x\rangle - f(x) - g(Ax + w)\right) \\ &= \sup_{x \in X} \left(\langle x^*, x\rangle - \langle y^*, Ax\rangle - f(x)\right) + \sup_{y \in W} \left(\langle y^*, y\rangle - g(y)\right) \\ &= f^*(x^* - A^\mathsf{T} y^*) + g^*(y^*), \end{aligned}$$

so that (6.20) follows from the relation

$$(f + g \circ A)^*(x^*) = -\inf_{x \in X} P(0, x) = -\inf(\mathcal{P}) = \min(\mathcal{P}^*)$$
$$= \min_{y^* \in Y^*} P^*(y^*, 0) = \min_{y^* \in Y^*} \left(f^*(x^* - A^\intercal y^*) + g^*(y^*)\right).$$

Now if $x^* \in \partial j(x)$, with $j := f + g \circ A$, one has $j(x) + j^*(x^*) - \langle x^*, x\rangle = 0$ and there exists some $y^* \in Y^*$ such that $j^*(x^*) = f^*(x^* - A^\intercal y^*) + g^*(y^*)$, hence

$$0 = \left(f(x) + f^*(x^* - A^\intercal y^*) - \langle x^* - A^\intercal y^*, x\rangle\right) + \left(g(Ax) + g^*(y^*) - \langle A^\intercal y^*, x\rangle\right).$$

Since both terms between brackets are nonnegative, they are null. Thus $x^* - A^\intercal y^* \in \partial f(x)$, $A^\intercal y^* \in \partial g(Ax)$, and the non-trivial inclusion of equality (6.21) holds. □

Corollary 6.14 *Let X, Y be normed spaces, let $A : X \to Y$ be a continuous linear map and let $f : X \to \mathbb{R}_\infty$, $g : Y \to \mathbb{R}_\infty$ be convex functions such that for some $\overline{x} \in \operatorname{dom} f$ the function g is finite and continuous at $A\overline{x}$. Then, for all $x^* \in X^*$ and $x \in \operatorname{dom} f \cap A^{-1}(\operatorname{dom} g)$ one has*

$$(f + g \circ A)^*(x^*) = \min_{y^* \in Y^*} \left(f^*(x^* - A^\intercal y^*) + g^*(y^*)\right), \tag{6.22}$$

$$\partial(f + g \circ A)(x) = \partial f(x) + A^\intercal(\partial g(Ax)). \tag{6.23}$$

Proof Let $s > \max(\|\overline{x}\|, f(\overline{x}), g(A\overline{x}))$. Since g is continuous at $A\overline{x}$, one can find $r > 0$ such that $g(y) < s$ for all $y \in B[A\overline{x}, r]$. Then $rB_Y \subset A(\overline{x}) - \{g \leq s\}$ and condition (6.19) is satisfied. The preceding theorem applies. □

In Banach spaces, under semicontinuity assumptions a similar conclusion can be obtained under a transversality condition that is often just of an algebraic nature.

Theorem 6.19 (Attouch-Brézis) *Let X, Y be Banach spaces, let $A \in L(X, Y)$, and let $f : X \to \mathbb{R}_\infty$, $g : Y \to \mathbb{R}_\infty$ be closed proper convex functions. If the cone*

$$Z := \mathbb{R}_+ \left(A(\operatorname{dom} f) - \operatorname{dom} g\right)$$

is closed and symmetric (i.e., $Z = -Z = \operatorname{cl} Z$) then, for all $x^* \in X^*$ and $x \in \operatorname{dom} f \cap A^{-1}(\operatorname{dom} g)$ relations (6.22) and (6.23) hold.

Note that the assumption on Z means that Z is a closed linear subspace. It is obviously satisfied when the simple algebraic condition that follows is fulfilled

$$Y = \mathbb{R}_+ \left(A(\operatorname{dom} f) - \operatorname{dom} g\right). \tag{6.24}$$

Proof Taking $W := Y$, we define the perturbation function P as in the preceding proof. Then, for $x \in X$, we have $w \in \operatorname{dom} P(\cdot, x)$ if and only if $x \in \operatorname{dom} f$ and $w \in \operatorname{dom} g - Ax$, so that the cone generated by the union over x of $\operatorname{dom} P(\cdot, x)$ is

$\mathbb{R}_+ (\mathrm{dom}\, g - A(\mathrm{dom} f))$, the closed linear subspace Z. Then Corollary 6.17 ensures that strong duality holds and the proof can be finished in the same way as the preceding one. □

As an application of the Fenchel-Rockafellar theorem let us give a dual expression of the distance function to a closed convex subset.

Proposition 6.26 *The distance function d_C to a nonempty closed convex subset C of a normed space X and its support function $h_C := \sigma_C$ are linked by the following relations for all $\overline{x} \in X$, B^* denoting the closed unit ball B_{X^*} of X^**

$$d_C(\overline{x}) = \max_{x^* \in B^*} (\langle \overline{x}, x^* \rangle - h_C(x^*)), \quad \frac{1}{2} d_C^2(\overline{x}) = \max_{x^* \in B^*} \inf_{x \in C} (\langle \overline{x} - x, x^* \rangle - \frac{1}{2} \|x^*\|^2).$$

If C is a cone with polar cone C^0 then

$$d_C(\overline{x}) = \max_{x^* \in B^* \cap C^0} \langle \overline{x}, x^* \rangle, \qquad \frac{1}{2} d_C^2(\overline{x}) = \max_{x^* \in C^0} (\langle \overline{x}, x^* \rangle - \frac{1}{2} \|x^*\|^2).$$

Proof Considering the closed convex functions $f_1, f_2 : X \to \mathbb{R}$ given by $f_1(x) := \|x - \overline{x}\|$, $f_2(x) := (1/2) \|x - \overline{x}\|^2$, the Fenchel-Rockafellar theorem yields

$$\begin{aligned} d_C(\overline{x}) &= \inf_{x \in X} (f_1(x) + \iota_C(x)) = \max_{x^* \in X^*} (-f_1^*(-x^*) - \iota_C^*(x^*)) \\ &= \max_{x^* \in B^*} (\langle \overline{x}, x^* \rangle - h_C(x^*)), \\ \frac{1}{2} d_C^2(\overline{x}) &= \inf_{x \in X} (f_2(x) + \iota_C(x)) = \max_{x^* \in X^*} (-f_2^*(-x^*) - \iota_C^*(x^*)) \\ &= \max_{x^* \in B^*} (\langle \overline{x}, x^* \rangle - \frac{1}{2} \|x^*\|^2 - h_C(x^*)) \\ &= \max_{x^* \in B^*} \inf_{x \in C} (\langle \overline{x} - x, x^* \rangle - \frac{1}{2} \|x^*\|^2). \end{aligned}$$

When C is a cone, one has $h_C(x^*) = \iota_{C^0}(x^*)$ and the expressions can be simplified as stated. □

Proposition 6.27 *For a nonempty closed convex subset C of X the conjugates and the subdifferentials of the functions d_C and $\frac{1}{2} d_C^2$ are such that*

$$(d_C)^* = \iota_{B_{X^*}} + h_C, \quad (\frac{1}{2} d_C^2)^* = \frac{1}{2} \|\cdot\|_*^2 + h_C,$$

$$\partial d_C(x) = \partial \|\cdot\| (x - w) \cap N(C, w), \quad \partial(\frac{1}{2} d_C^2)(x) = J(x - w) \cap N(C, w)$$

for $w \in P_C(x) := \{z \in C : \|x - z\| = d_C(x)\}$ assumed to be nonempty and $J(x - w) = \frac{1}{2} \partial \|\cdot\|^2 (x - w)$.

In particular, for $x \in C$ one has

$$\partial d_C(x) = B_{X^*} \cap N(C,x), \quad \partial(\frac{1}{2}d_C^2)(x) = \{0\}.$$

Proof Since $d_C = \|\cdot\| \Box \iota_C$, $\frac{1}{2}d_C^2 = (\frac{1}{2}\|\cdot\|^2)\Box\iota_C$, we have $d_C^* = \iota_{B_{X^*}} + h_C$, $(\frac{1}{2}d_C^2)^* = \frac{1}{2}\|\cdot\|_*^2 + h_C$ by Proposition 6.20 and the relation $(\frac{1}{2}\|\cdot\|^2)^* = \frac{1}{2}\|\cdot\|_*^2$. For the calculus of ∂d_C and $\partial(\frac{1}{2}d_C^2)$ one applies Corollary 6.9. □

Let us give a generalization of a result about infimal convolution we shall use later on. Given Banach spaces X, Y and $h,\ k : X \times Y \to \mathbb{R}_\infty$ we introduce the partial convolutions $h\Box_1 k$ and $h\Box_2 k$ by

$$(h\Box_1 k)(x,y) := \inf\{h(u,y) + k(v,y) : u + v = x\} \qquad (x,y) \in X \times Y,$$

$$(h\Box_2 k)(x,y) := \inf\{h(x,w) + k(x,z) : w + z = y\} \quad (x,y) \in X \times Y$$

and, denoting by $p_X : X \times Y \to X$ the canonical projection, the condition

$$X_0 := \mathbb{R}_+(p_X(\operatorname{dom} h) - p_X(\operatorname{dom} k)) = -X_0 = \operatorname{cl} X_0. \tag{6.25}$$

meaning that X_0 is a closed linear subspace.

Proposition 6.28* *Let X, Y be Banach spaces, let $h,\ k : X \times Y \to \mathbb{R}_\infty$ be proper, convex, lower semicontinuous functions satisfying condition (6.25) and $(h\Box_2 k)(x,y) > -\infty$ for all $(x,y) \in X \times Y$. Then for $(x^*,y^*) \in X^* \times Y^*$*

$$(h\Box_2 k)^*(x^*,y^*) = (h^*\Box_1 k^*)(x^*,y^*)$$

and if this value is finite there exists some u^*, $v^* \in X^*$ such that $u^* + v^* = x^*$ and $(h\Box_2 k)^*(x^*,y^*) = h^*(u^*,y^*) + k^*(v^*,y^*)$.

Proof Since $h\Box_2 k$ is the performance function associated with a convex function of (x,w,z), it is convex. Moreover, since Lemma 3.19 and (6.25) imply that $p_X(\operatorname{dom} h) \cap p_X(\operatorname{dom} k) \neq \varnothing$, we can find some $(x,y) \in X \times Y$ such that $(h\Box_2 k)(x,y) < \infty$. Furthermore, for all $(x^*,y^*) \in X^* \times Y^*$ and all u^*, $v^* \in X^*$ satisfying $u^* + v^* = x^*$ we have

$$\begin{aligned}
&(h\Box_2 k)^*(x^*,y^*)\\
&\le \sup_{x\in X}\sup_{w,z\in Y}(\langle x,u^*\rangle + \langle x,v^*\rangle + \langle w,y^*\rangle + \langle z,y^*\rangle - h(x,w) - k(x,z))\\
&\le h^*(u^*,y^*) + k^*(v^*,y^*),
\end{aligned}$$

hence $(h\Box_2 k)^*(x^*,y^*) \le (h^*\Box_1 k^*)(x^*,y^*)$. It remains to prove that when $(h\Box_2 k)^*(x^*,y^*) < \infty$ there exists some u^*, $v^* \in X^*$ such that $u^* + v^* = x^*$

and

$$h^*(u^*, y^*) + k^*(v^*, y^*) \leq (h\Box_2 k)^*(x^*, y^*). \tag{6.26}$$

For such a purpose, we introduce the functions $f,\ g : X \times Y \times Y \to \mathbb{R}_\infty$ given by

$$f(u, w, z) := h(u, w) - \langle u, x^* \rangle - \langle w, y^* \rangle,$$
$$g(u, w, z) := k(u, z) - \langle z, y^* \rangle + (h\Box_2 k)^*(x^*, y^*).$$

We note that $\operatorname{dom} f = \operatorname{dom} h \times Y$ and $\operatorname{dom} g = \{(u, w, z) : (u, z) \in \operatorname{dom} k,\ w \in Y\}$, so that for any $(x, y_1, y_2) \in X_0 \times Y \times Y$ we can find $r \in \mathbb{P},\ u, v \in X,\ w, z \in Y$ such that $(u, w) \in \operatorname{dom} h$, $(v, z) \in \operatorname{dom} k$ and

$$(x, y_1, y_2) = r(u, w, r^{-1}y_2 + z) - r(v, w - r^{-1}y_1, z).$$

Thus $X_0 \times Y \times Y \subset \mathbb{P}(\operatorname{dom} f - \operatorname{dom} g)$. The reverse inequality being obvious, we see that $\mathbb{P}(\operatorname{dom} f - \operatorname{dom} g)$ is a closed linear subspace. Moreover, for all $(u, w, z) \in X \times Y \times Y$ we have

$$\begin{aligned}
&(f + g)(u, w, z)\\
&= h(u, w) + k(u, z) - \langle u, u^* \rangle - \langle w, y^* \rangle - \langle z, y^* \rangle + (h\Box_2 k)^*(u^*, y^*)\\
&\geq (h\Box_2 k)(u, w + z) + (h\Box_2 k)^*(u^*, y^*) - \langle u, u^* \rangle - \langle w + z, y^* \rangle \geq 0.
\end{aligned}$$

The sandwich theorem (or the Attouch-Brezis theorem) ensures that there exists some $(u^*, w^*, z^*) \in X^* \times Y^* \times Y^*$ such that

$$f^*(u^*, w^*, z^*) + g^*(-u^*, -w^*, -z^*) \leq 0.$$

Then both terms of this sum are finite, so that $z^* = 0$, $w^* = 0$ and

$$f^*(u^*, w^*, z^*) = h^*(u^* + x^*, w^* + y^*),$$
$$g^*(-u^*, -w^*, -z^*) = k^*(-u^*, y^* - w^*) - (h\Box_2 k)^*(x^*, y^*).$$

Thus, the preceding inequality amounts to relation (6.26). □

Exercises

1. Show that Corollary 6.14 generalizes Theorems 6.8 and 6.9.
2. Let P be a closed convex cone of a Banach space X, let Q be its polar cone and let B be the closed unit ball. Prove that the distance function to Q and the support function to $P \cap B$ are equal.

3. Show that if C is a nonempty closed convex subset of X containing the origin, the conjugate μ_C^* of the gauge function μ_C of C is given by $\mu_C^* = \iota_{C^0}$ where $C^0 := \{x^* \in X^* : \langle x^*, x\rangle \leq 1\}$ is the polar set of C and μ_C is given by $\mu_C(x) := \inf\{r \in \mathbb{P} : x \in rC\}$.

4. Let $A : X \to W$ be a continuous linear operator. Suppose W is ordered by a closed convex cone W_+ and $Y := W^*$ is ordered by the cone $Y_+ = -W_+^0$. Let $b \in W$ and let $f : X \to \mathbb{R}_\infty$.

(**a**) Find the dual problem of the mathematical programming problem

$$(\mathcal{P}) \qquad \text{minimize} f(x) \qquad x \in X,\ Ax \leq b$$

by using the perturbation function P given by $P(x, w) := f(x) + \iota_{F(w)}(x)$ where $F(w) := \{x \in X : Ax \leq b - w\}$.

(**b**) Show that the function $L : X \times Y \to \overline{\mathbb{R}}$ given by $L(x, y) := -(P_x)^*(y)$ is a Lagrangian of $(\mathcal{P})$ in the sense that $\sup\{L(x, y) : y \in Y\} = f(x) + \iota_{F(0)}(x)$.

(**c**) (*Quadratic programming*) Give an explicit form of the dual problem when $X = \mathbb{R}^n$, $W := \mathbb{R}^m$, $W_+ = \mathbb{R}^m_+$ and f is a quadratic form: $f(x) = (1/2)\langle Qx, x\rangle + \langle q, x\rangle$, with Q positive definite. Generalize to the case when Q is positive semidefinite. [See [189].]

5. (*General Fenchel equality*) Given a family $f_1, \ldots, f_k$ of convex lower semicontinuous functions that are finite and continuous at some point of X, prove that

$$\inf_{x \in X}(f_1(x) + \ldots + f_k(x)) = \inf\{f_1^*(x_1^*) + \ldots + f_k^*(x_k^*) : x_1^* + \ldots + x_k^* = 0\}.$$

6.3.4 The Interplay Between a Function and Its Conjugate

Much information can be drawn from the study of the conjugate of a convex function. In the present subsection we consider growth properties versus boundedness properties. Then we deal with reinforced convexity properties versus smoothness properties. In the next subsection we give applications to optimization problems and in a following subsection we display some applications to the geometry of normed spaces.

We first study the correspondence between some simple properties of one-variable functions. For that purpose, we introduce some classes of functions on $\mathbb{R}_+$ that will play a role in the sequel. Each such function is extended by ∞ on $]-\infty, 0[$ (or by setting $\alpha(r) := \alpha(-r)$ for $r < 0$) so that, for $\alpha \in \mathcal{A}$, where

$$\mathcal{A} := \{\alpha : \mathbb{R}_+ \to \overline{\mathbb{R}}_+ : \alpha(0) = 0\}$$

one has

$$\alpha^*(s) = \sup\{rs - \alpha(r) : r \in \mathbb{R}_+\} \qquad s \in \mathbb{R}_+$$

and $\alpha^* \in \mathcal{A}$. We denote by $\mathcal{A}_0$ the set of *gages*, i.e. the set of $\alpha \in \mathcal{A}$ that are nondecreasing and *firm*, i.e. such that $(r_n) \to 0$ whenever $(\alpha(r_n)) \to 0$, and we denote by $\mathcal{A}_c$ the set of $\alpha \in \mathcal{A}_0$ that are convex and lower semicontinuous. Note that $\alpha \in \mathcal{A}$ is in $\mathcal{A}_0$ if and only if α is nondecreasing and $\alpha^{-1}(0) = \{0\}$. A function $\alpha \in \mathcal{A}$ is said to be *starshaped* if $\alpha(r)/r \leq \alpha(s)/s$ for $r \leq s$ in $\mathbb{P}$. Any $\alpha \in \mathcal{A}_c$ is starshaped and increasing.

Lemma 6.8 *For any starshaped $\alpha \in \mathcal{A}_0$, in particular for any $\alpha \in \mathcal{A}_c$, α^* is a remainder on $\mathbb{R}_+$: $\alpha^*(s)/s \to 0$ as $s \to 0_+$.*

For any remainder $\rho : \mathbb{R}_+ \to \overline{\mathbb{R}}_+$ one has $\rho^* \in \mathcal{A}_c$ (hence $\rho^* \in \mathcal{A}_0$ and ρ^* is convex, hence starshaped).

For any starshaped $\alpha \in \mathcal{A}_0$, one has $\alpha^{**} \in \mathcal{A}_c$.

This lemma shows that one can often replace an element α of $\mathcal{A}_0$ by an element $\beta = \alpha^{**} \in \mathcal{A}_c$.

Proof Let $\alpha \in \mathcal{A}_0$ be starshaped and let $\varepsilon > 0$, so that $\delta := \alpha(\varepsilon)/\varepsilon$ is positive. We claim that for all $s \in]0, \delta]$ we have $\alpha^*(s)/s \leq \varepsilon$. Indeed, since for $r \geq \varepsilon$ we have $\alpha(r) \geq r\alpha(\varepsilon)/\varepsilon$, we get

$$\begin{aligned}\alpha^*(s) &\leq \max\Big(\sup_{r\in[0,\varepsilon]} (rs - \alpha(r)),\ \sup_{r\in]\varepsilon,\infty[} (rs - r\alpha(\varepsilon)/\varepsilon)\Big)\\ &\leq \max(\varepsilon s, \sup\{rs - r\delta : r \in]\varepsilon, \infty[\}) = \varepsilon s.\end{aligned}$$

Now, let $\rho : \mathbb{R}_+ \to \overline{\mathbb{R}}_+$ be a remainder. Since $\rho \in \mathcal{A}$, we have $\rho^* \in \mathcal{A}$. Since ρ^* is lower semicontinuous and convex with $\rho^*(0) = 0$, it suffices to verify that for all $r > 0$ we have $\rho^*(r) > 0$. Let $\varepsilon \in]0, r[$. Since ρ is a remainder, we can find $\delta > 0$ such that $\rho(s)/s \leq \varepsilon$ for all $s \in [0, \delta]$. Setting $\alpha(s) = \varepsilon s$ for $s \in [0, \delta]$ and $\alpha(s) = \infty$ for $s \in]\delta, \infty[$ we have $\rho \leq \alpha$, hence $\rho^*(r) \geq \alpha^*(r) = \sup\{rs - \varepsilon s : s \in [0, \delta]\} = (r - \varepsilon)\delta > 0$. The last assertion is a consequence in the two previous ones. □

Now let us turn to functions defined on normed vector spaces. In order to obtain symmetry in the properties below, we assume that we have two normed spaces X, Y in *metric duality*, i.e. that there exists a continuous bilinear coupling $c := \langle \cdot, \cdot \rangle : X \times Y \to \mathbb{R}$ such that $\|y\| = \sup\{\langle x, y\rangle : x \in B_X\}$ for all $y \in Y$ and $\|x\| = \sup\{\langle x, y\rangle : y \in B_Y\}$ for all $x \in X$. That is the case when Y is the dual of X or when X is the dual of Y. In order to fix the terminology, we say that $f : X \to \overline{\mathbb{R}}$ is *coercive* if $\lim_{\|x\|\to\infty} f(x) = \infty$; it is called *supercoercive* if its coercivity rate

$$c_f := \liminf_{\|x\|\to+\infty} f(x)/\|x\|$$

is positive and *hypercoercive* if $c_f = \infty$.

The distinctness of these notions allows us to give different perturbation results.

Lemma 6.9 *For f, $g : X \to \mathbb{R}_\infty$ one has $c_{f+g} \geq c_f + c_g$. In particular, if f is hypercoercive and if $c_g > -\infty$ i.e. if there exist some $c \in \mathbb{R}$ such that $g(x) \geq -c\|x\|$ for $\|x\|$ large enough, then $f + g$ is hypercoercive.*

If f is supercoercive and if $c_g > -c_f$ then $f + g$ is supercoercive.

Proof The first assertion is a consequence in the sum rule for a lower limit. In fact, given $a < c_f$, $b < c_g$ for $x \in X$ with $\|x\|$ large enough one has $f(x) > a\|x\|$, $g(x) > b\|x\|$, hence $(f+g)(x) > (a+b)\|x\|$. The other two assertions are direct consequences. □

In the next lemma, for $g : Y \to \overline{\mathbb{R}}$, we set

$$r_g := \sup\{s \in \mathbb{R}_+ : \sup g(sB_Y) < \infty\}.$$

This lemma completes the fact that f is bounded below on X if and only if $f^*(0) < \infty$. We recall that the assumption that f is bounded below on bounded subsets is automatically fulfilled if f is closed proper convex since then f is bounded below by a continuous affine function.

Lemma 6.10 *Let $f : X \to \mathbb{R}_\infty$ be proper and let $r, s \in \mathbb{R}_+$, $a, b \in \mathbb{R}$.*

(a) If f is such that $f \geq a$ on rB_X and $f(\cdot) \geq s\|\cdot\| - b$ on $X \backslash rB_X$ then, for $y \in sB_Y$ one has $f^*(y) \leq r\|y\| - \min(a, rs - b)$. In particular, if $f(\cdot) \geq s\|\cdot\| - b$ on X then $f^*(\cdot) \leq b$ on sB_Y.
(b) If f is supercoercive and bounded below on bounded sets, then for all $s \in]0, c_f[$ there exist some $c \in \mathbb{R}$, $r \in \mathbb{R}_+$ such that $f^*(\cdot) \leq r\|\cdot\| - c$ on sB_Y so that f^* is bounded above on sB_Y. Moreover $r_{f^*} \geq c_f$.
(c) If f is hypercoercive and bounded below on bounded sets, then f^* is bounded above on bounded sets.
(d) If f is such that $f(\cdot) \leq s\|\cdot\| - b$ on rB_X, then one has $f^* \geq b$ on sB_Y and $f^*(\cdot) \geq r\|\cdot\| + b - rs$ on $Y \backslash sB_Y$. If f is such that $f^*(\cdot) \leq s\|\cdot\| - b$ on rB_Y, then one has $f \geq b$ on sB_X and $f(\cdot) \geq r\|\cdot\| + b - rs$ on $X \backslash sB_X$. In particular, if f^* is bounded above by $c \in \mathbb{R}$ on rB_Y then $f(\cdot) \geq r\|\cdot\| - c$ on X and $c_f \geq r_{f^*}$.
(e) If f^* is bounded above on bounded sets then f is hypercoercive and bounded below on bounded sets.

Proof

(a) For $y \in sB_Y$, $x \in X$, separating the cases $x \in rB_X$ and $x \in X \backslash rB_X$, we have

$$\begin{aligned} f^*(y) &\leq \max\Big(\sup_{x \in rB_X} (\langle y, x\rangle - a), \sup_{x \in X \backslash rB_X} (\langle y, x\rangle - s\|x\| + b)\Big) \\ &\leq \max\big(r\|y\| - a, \sup_{t \geq r} t(\|y\| - s) + b\big) = r\|y\| + \max(-a, b - rs). \end{aligned}$$

Taking $r = 0$, $a = -b$, we get $f^*(\cdot) \leq b$ on sB_Y.
(b) For all $s \in]0, c_f[$ one can find $r > 0$ such that $f(x)/\|x\| \geq s$ for all $x \in X \backslash rB_X$. Setting $b = 0$ and $a := \inf f(rB_X)$ in (a), one gets $f^*(\cdot) \leq r\|\cdot\| - \min(a, sr) \leq \max(sr - a, 0)$ on sB_Y.
(c) This is an immediate consequence of (b) with $c_f = \infty$.

(d) If f is such that $f(\cdot) \leq s\|\cdot\| - b$ on rB_X, then for $y \in Y$ one has $f^*(y) \geq \sup_{x \in rB_X}(\langle x, y\rangle - s\|x\| + b) = \sup_{t \leq r}(t\|y\| - st + b)$, hence $f^*(y) \geq b$ for $y \in sB_Y$ and $f^*(y) \geq r\|y\| + b - rs$ for $y \in Y \backslash sB_Y$. Interchanging the roles of X and Y and applying the obtained implication to f^* we get $f \geq f^{**} \geq b$ on sB_X and $f(\cdot) \geq f^{**}(\cdot) \geq r\|\cdot\| + b - rs$ on $X \backslash sB_X$. Taking $s := 0, b := -c$, we obtain that $f(\cdot) \geq r\|\cdot\| - c$ on X when f is such that $f^* \leq c$ on rB_Y.

(e) This is an immediate consequence of (d) since r and s can be arbitrary in $\mathbb{R}_+$.

□

For a closed proper convex function, the relationships between growth properties of f and boundedness properties of f^* are more striking.

Proposition 6.29 *For $f : X \to \mathbb{R}_\infty$ closed, proper and convex the following assertions are equivalent:*

(a) f is coercive;
(b) the sublevel sets of f are bounded;
(c) there exists $b \in \mathbb{R}$, $c \in \mathbb{P}$ such that $f \geq c\|\cdot\| + b$;
(d) f is supercoercive: $c_f := \liminf_{\|x\| \to +\infty} f(x)/\|x\| > 0$;
(e) f^* is bounded above on a neighborhood of 0.

If Y is complete the preceding assertions are equivalent to the next one:

(f) $0 \in \operatorname{int}(\operatorname{dom} f^*)$.

Proof (a)⇔(b) is easy and (c)⇒(a) is obvious.

(d)⇒(c) Since f is bounded below by a continuous affine function, it is bounded below on balls. Given $c \in]0, c_f[$, we can find $r > 0$ such that $f(\cdot) \geq c\|\cdot\|$ on $X \backslash rB_X$ and $a \in \mathbb{R}$ such that $f(\cdot) \geq a$ on rB_X. Taking $b := \min(a - cr, 0)$, we get $f(\cdot) \geq c\|\cdot\| + b$ on rB_X and $X \backslash rB_X$ hence on X.

(a)⇒(d) Suppose $c_f \leq 0$. Given a sequence $(\varepsilon_n) \to 0_+$ in $]0, 1[$, one can find $x_n \in X$ such that $\|x_n\| \geq n/\varepsilon_n$ and $f(x_n) \leq \varepsilon_n\|x_n\|$. Let $t_n := 1/(\varepsilon_n\|x_n\|) \leq 1/n$. Then, given $w \in \operatorname{dom} f$, for $u_n := (1 - t_n)w + t_n x_n$, one has

$$f(u_n) \leq (1 - t_n)f(w) + t_n f(x_n) \leq |f(w)| + 1$$

but (u_n) is unbounded since $\|u_n\| \geq t_n\|x_n\| - (1 - t_n)\|w\| \geq 1/\varepsilon_n - \|w\|$, a contradiction with (a).

(d)⇔(e) has been proved in the preceding lemma and (e)⇒(f) is obvious.

(f)⇒(e) in the case when Y is complete is a consequence in Proposition 6.3 since f^* is convex and lower semicontinuous. □

Now let us point out relationships between the rotundity properties of f and the smoothness of f^*. This can be done in a quantitative manner. The rotundity properties we introduce are strengthenings of the notion of *strict convexity*: f is said to be *strictly convex* if for all $t \in]0, 1[$ and distinct $x_0, x_1 \in X$, one has

$$(1 - t)f(x_0) + tf(x_1) > f((1 - t)x_0 + tx_1).$$

Given $\rho \in \mathcal{A}$ and $x_0 \in X$, a function $f : X \to \mathbb{R}_\infty$ is said to be *ρ-convex* at x_0 if for all $x_1 \in X$, $t \in]0,1[$ one has

$$(1-t)f(x_0) + tf(x_1) \geq f((1-t)x_0 + tx_1) + t(1-t)\rho(\|x_0 - x_1\|). \tag{6.27}$$

It is *ρ-convex* if it is *ρ-convex* at x_0 for all $x_0 \in X$. The greatest function ρ satisfying (6.27) is called the *index of (uniform) convexity* of f and is denoted by ρ_f. Note that ρ_f is given by

$$\rho_f(r) := \sup_{\substack{x_0,\, x_1 \in \mathrm{dom} f,\ \|x_1 - x_0\| = r,\\ t \in]0,1[}} \frac{(1-t)f(x_0) + tf(x_1) - f((1-t)x_0 + tx_1)}{t(1-t)}.$$

If $\rho_f \in \mathcal{A}_0$ one says that f is *uniformly convex*. When $\rho_f = c|\cdot|^2$ for some $c \in \mathbb{P}$ one says that f is *strongly convex*. The function f is said to be *ρ-convex on a subset B* of X if $f_B := f + \iota_B$ is ρ-convex. It is *uniformly convex on a subset B* of X if $f_B := f + \iota_B$ is ρ-convex for some $\rho \in \mathcal{A}_0$.

Given $\sigma \in \mathcal{A}$, a function $f : X \to \mathbb{R}_\infty$ is said to be *σ-smooth* at $x_0 \in X$, if for all $x_1 \in X$, $t \in [0,1]$ one has

$$f((1-t)x_0 + tx_1) + t(1-t)\sigma(\|x_0 - x_1\|) \geq (1-t)f(x_0) + tf(x_1). \tag{6.28}$$

The function $f : X \to \mathbb{R}_\infty$ is said to be *σ-smooth* if for all $x_0 \in X$ it is *σ-smooth* at x_0. The least function σ satisfying such a property is called the *index of (uniform) smoothness* of f and is denoted by σ_f. Note that setting $x_t := (1-t)x_0 + tx_1$, the function σ_f is given by

$$\sigma_f(r) := \inf_{\substack{x_0, x_1 \in X,\ \|x_0 - x_1\| \geq r\\ t \in]0,1[}} \left\{\frac{(1-t)f(x_0) + tf(x_1) - f(x_t)}{t(1-t)} : x_t \in \mathrm{dom} f\right\}.$$

If $\sigma_f \in o(\mathbb{R}_+)$, the set of remainders on $\mathbb{R}_+$, one says that f is *uniformly smooth.*

Theorem 6.20 *If for some $\sigma \in \mathcal{A}$ a function $f : X \to \mathbb{R}_\infty$ is σ-smooth, then f^* is σ^*-convex.*

If f is uniformly smooth, then f^* is uniformly convex and $\rho_{f^*} \geq (\sigma_f)^*$.

If for some $\rho \in \mathcal{A}$ the function f is ρ-convex, then f^* is ρ^*-smooth.

If f is uniformly convex, then f^* is uniformly smooth and $\sigma_{f^*} \leq (\rho_f)^*$.

Proof Suppose f is σ-smooth. Given $y_0, y_1 \in Y$, $t \in]0,1[$ and $w, x \in X$, let $x_t := x + tw$, $y_t := (1-t)y_0 + ty_1$. Then, using (6.28) we have

$$\begin{aligned}
&(1-t)f^*(y_0) + tf^*(y_1)\\
&\geq (1-t)\langle x, y_0\rangle + t\langle x_1, y_1\rangle - (1-t)f(x) - tf(x_1)\\
&\geq \langle x_t, y_t\rangle + t(1-t)\langle w, y_1 - y_0\rangle - f(x_t) - t(1-t)\sigma(\|w\|).
\end{aligned}$$

Setting $w := \|w\| \, u$ with $u \in S_X$, taking the supremum on $x \in X$, $u \in S_X$, then on $\|w\| \in \mathbb{R}_+$, we get

$$(1-t)f^*(y_0) + tf^*(y_1) \geq f^*(y_t) + t(1-t)\sigma^*(\|y_1 - y_0\|),$$

so that f^* is σ^*-convex. If f is uniformly smooth, $\sigma := \sigma_f$ is a remainder on $\mathbb{R}_+$, so that, by Lemma 6.8, σ^* is in $\mathcal{A}_c$ and f^* is uniformly convex.

Now suppose f is ρ-convex. Given $y_0, y_1 \in Y, t \in [0, 1]$, for any $r_0 < f^*(y_0)$, $r_1 < f^*(y_1)$ we can pick $x_0, x_1 \in X$ such that

$$r_0 < \langle x_0, y_0 \rangle - f(x_0), \qquad r_1 < \langle x_1, y_1 \rangle - f(x_1).$$

Multiplying both sides of the first (resp. second) inequality by $(1-t)$ (resp. t) and adding to both sides of the Young-Fenchel inequality

$$0 \leq f^*(y_t) + f(x_t) - \langle x_t, y_t \rangle$$

with $x_t := (1-t)x_0 + tx_1$, $y_t := (1-t)y_0 + ty_1$, we get that $(1-t)r_0 + tr_1$ is bounded above by

$$\begin{aligned}
&f^*(y_t) + f(x_t) + t(1-t)\langle x_0 - x_1, y_0 - y_1 \rangle - (1-t)f(x_0) - tf(x_1) \\
&\leq f^*(y_t) + t(1-t) \, \|x_0 - x_1\| \, . \, \|y_0 - y_1\| - t(1-t)\rho(\|x_0 - x_1\|) \\
&\leq f^*(y_t) + t(1-t)\rho^*(\|y_0 - y_1\|).
\end{aligned}$$

Since r_0 and r_1 are arbitrarily close to $f^*(y_0)$ and $f^*(y_1)$ respectively, we get

$$(1-t)f^*(y_0) + tf^*(y_1) \leq f^*(y_t) + t(1-t)\rho^*(\|y_0 - y_1\|),$$

so that f^* is ρ^*-smooth. If f is uniformly convex, $(\rho_f)^*$ being a remainder by Lemma 6.8, f^* is uniformly smooth.

The estimate for σ_{f^*} stems from its minimality property; similarly, the estimate for ρ_{f^*} stems from its maximality property. □

Since X and Y play symmetric roles, we get the following corollary.

Corollary 6.15 *A function $f : X \to \mathbb{R}_\infty$ is uniformly smooth (resp. uniformly convex) if and only if f^* is uniformly convex (resp. uniformly smooth).*

In [17, 46, 256, 264, 265] the reader will find more information about rotundity properties and smoothness properties of convex functions. We shall return to such properties in the next subsection after considering some applications to minimization.

Exercises

1. Let X be a Hilbert space with scalar product $\langle \cdot \mid \cdot \rangle$ and let $A : X \to X$ be a symmetric, linear, continuous map such that the quadratic form q associated with A is positive on $X\backslash\{0\}$. Let $b \in X$ and let f be given by $f(x) = q(x) - \langle b \mid x \rangle$.

   (a) Show that A and the square root $A^{1/2}$ of A are injective and that their images satisfy $R(A) \subset R(A^{1/2})$.
   (b) Using Theorem 6.9 and the relation $q = g \circ A^{1/2}$ for $g := \frac{1}{2} \|\cdot\|^2$ show that $q^*(x^*) = \frac{1}{2} \left\|(A^{1/2})^{-1}(x^*)\right\|^2$ for $x^* \in R(A^{1/2})$ and $q^*(x^*) = +\infty$ otherwise.
   (c) Show that if $b \in R(A)$, then f attains its minimum at $A^{-1}(b)$.
   (d) Show that if $b \in R(A^{1/2})\backslash R(A)$, then f is bounded below but does not attain its infimum.
   (e) Show that if $b \notin R(A^{1/2})$, then $\inf_{x \in X} f(x) = -\infty$.
   (f) Deduce from the preceding questions that when $R(A)$ is closed, then $R(A^{1/2}) = R(A)$. [Hint: when $R(A) \neq X$ take $b \in X\backslash R(A)$ and pick some $u \in R(A)^{\perp}$ such that $\langle b \mid u \rangle > 0$; then verify that $\inf_{r>0} f(ru) = -\infty$.]

2. **(a)** Show that if $f : X \to \mathbb{R}_\infty$ is such that $f \geq b$ and $\mathrm{dom} f \subset rB_X$ then one has $f^*(\cdot) \leq r \|\cdot\| - b$.
   (b) Show that if f is such that $f \geq b$ and $f(\cdot) \geq c \|\cdot\|$ on $X\backslash rB_X$ then one has $cB_{X^*} \subset \mathrm{dom} f^*$.

3. Give an example of a coercive function that is not supercoercive.

4. Give an example of a supercoercive function that is not hypercoercive.

5. If $f(x) = \frac{1}{p} \|x\|^p$ with $p \in]1, \infty[$ show that $f^*(x^*) = \frac{1}{q} \|x^*\|_*^q$ with $q := (1 - \frac{1}{p})^{-1}$, where $\|\cdot\|_*$ is the dual norm. Observe that for $p = 2$ one has $q = 2$.

6. Let X be a Hilbert space identified with its dual. Show that $f^* = f$ if and only if $f := \frac{1}{2} \|\cdot\|^2$.

7. Let $h : X \to \mathbb{R}_\infty$ be a positively homogeneous function such that $h(0) = 0$, let $b \in \mathbb{R}$, and let $f : X \to \mathbb{R}_\infty$ be such that $f \geq h - b$. Verify that $f^* \leq b$ on $S := \partial h(0)$. Show that conversely if for some $g : X^* \to \mathbb{R}_\infty$ and some $S \subset X^*$ one has $g \leq b$ on S, then $g^* \leq h_S - b$ where $h_S(x) := \sup\{\langle x, x^* \rangle : x^* \in S\}$. Conclude that for $b \in \mathbb{R}$ and some weak* closed convex subset S of X^* one has $f \geq h_S - b$ if and only if $f^* \leq b$ on S.

8. Given a Banach space X and a convex function f defined on it, show that the Fenchel conjugate f^* of f is Gateaux differentiable at some $x^* \in X^*$ if and only if any sequence (x_n) such that $(f(x_n) - x^*(x_n)) \to \inf(f - x^*)$ converges.

9*. (Figiel [121]) If f is convex, show that the function $r \mapsto \rho_f(r)/r$ is starshaped, i.e. ρ_f is such that $\rho_f(0) = 0$ and $r \mapsto \rho_f(r)/r^2$ is nondecreasing [see [121] or [265, Prop. 3.5.1]].

10. For f convex and $\overline{x} \in \mathrm{dom} f$, one defines the index of uniform convexity of f at $\overline{x}$ as

$$\rho_{\overline{x}}(r) := \inf\{\frac{(1-t)f(\overline{x}) + tf(x) - f((1-t)\overline{x} + tx)}{t(1-t)} : t \in]0,1[,\ x \in rS_X + \overline{x}\},$$

so that f is uniformly convex at $\overline{x}$ if and only if $\rho_{\overline{x}} \in \mathcal{A}_0$. One also defines $\theta_{\overline{x}}$ by

$$\theta_{\overline{x}}(r) := \inf\{f(x) - f(\overline{x}) - f'(\overline{x}, x - \overline{x}) : x \in (\mathrm{dom} f) \cap (rS_X + \overline{x})\}.$$

Show that $\theta_{\overline{x}} \geq \rho_{\overline{x}}$, so that $\theta_{\overline{x}} \in \mathcal{A}_0$ whenever f is uniformly convex at $\overline{x}$.

11. With the notation of the preceding exercise, prove that when $\theta_{\overline{x}} \in \mathcal{A}_0$ and f is Fréchet differentiable at $\overline{x}$, then f is uniformly convex at $\overline{x}$. [See [265, Prop. 3.4.5].]

6.3.5 Conditioning and Well-Posedness

Given a nonempty closed subset S of a normed space X and $\alpha \in \mathcal{A}$, we say that $f : X \to \mathbb{R}_\infty$ whose set of minimizers $\arg\min f$ contains S is *α-conditioned for S* if

$$\forall x \in X \qquad f(x) \geq \inf f(X) + \alpha(d_S(x)), \tag{6.29}$$

where $d_S(x) := \inf\{d(w,x) : w \in S\}$, and that f is *α-conditioned* if this relation holds and S is the set of minimizers of f. This is the case whenever (6.29) holds with α firm, or such that $\alpha^{-1}(0) = \{0\}$. We define the *conditioning index* γ_f of f with respect to S by

$$\gamma_f(r) := \inf\{f(x) - \inf f(X) : x \in X,\ d_S(x) \geq r\} \qquad r \in \mathbb{R}_+.$$

Clearly, $\gamma_f \in \mathcal{A}$ and γ_f is nondecreasing. Moreover, γ_f is the greatest nondecreasing element α of $\mathcal{A}$ such that f is α-conditioned. When $\gamma_f \in \mathcal{A}_0$ we say that f is *well-conditioned.*

Let us first note the following observation.

Proposition 6.30 *The conditioning index γ_f of a convex function f with a nonempty set S of minimizers is starshaped.*

Proof Without loss of generality we suppose $\inf f(X) = 0$. Given $r > 0$ and $c > 1$ we have to prove that $\gamma_f(cr)/c \geq \gamma_f(r)$. Let $x \in X$ be such that $s := d_S(x) \geq cr$. Given $(q_n) \to 1$ with $q_n \in]0,1[$ for all $n \in \mathbb{N}$, we can find $a_n \in S$ such that $q_n \|x - a_n\| \leq s$. Let $s_n := \|x - a_n\|$, let $t_n := 1 - q_n(1 - 1/c) \in [0,1[$, and let

$w_n = (1 - t_n)a_n + t_n x$. Then $\|w_n - x\| = (1 - t_n)\|x - a_n\| = (1 - t_n)s_n$, so that

$$\begin{aligned} d_S(w_n) &\geq d_S(x) - \|w_n - x\| = s - (1 - t_n)s_n \\ &= s - s_n q_n(1 - 1/c) \geq s - s(1 - 1/c) = s/c \geq r. \end{aligned}$$

Since $f(a_n) = 0$, by convexity of f we have

$$\gamma_f(r) \leq f(w_n) \leq t_n f(x).$$

Taking the limit as $n \to \infty$ we get $\gamma_f(r) \leq (1/c)f(x)$. Since x is arbitrary in $\{x : d_S(x) \geq cr\}$, taking the infimum we obtain $\gamma_f(r) \leq (1/c)\gamma_f(cr)$. □

As usual, in the next statement, a sequence (x_n) of X such that $(f(x_n)) \to m := \inf f(X)$ is said to be *minimizing*.

Lemma 6.11 ([206]) *The following assertions on $f : X \to \mathbb{R}_\infty$ and a nonempty closed subset S of* $\arg\min f$ *are equivalent and imply that $S :=$* $\arg\min f$*:*

(a) any minimizing sequence (x_n) satisfies $(d_S(x_n)) \to 0$;
(b) γ_f belongs to $\mathcal{A}_0$, the set of nondecreasing firm elements of $\mathcal{A}$;
(c) there exists some $\alpha \in \mathcal{A}_0$ such that f is α-conditioned for S.

Proof (a)⇒(b) Suppose $\gamma_f(r) = 0$ for some $r > 0$. Then there exists a sequence (x_n) such that $d_S(x_n) \geq r$ and $(f(x_n)) \to m := \inf f(X)$, a contradiction with (a).

The implication (b)⇒(c) is obvious since f is γ_f-conditioned.

(c)⇒(a) Suppose $f(x) \geq m + \alpha(d_S(x))$ for some $\alpha \in \mathcal{A}_0$. Let (x_n) be a minimizing sequence in f and let $\varepsilon > 0$ be given. For $r \geq \varepsilon$ we have $\alpha(r) \geq \alpha(\varepsilon) > 0$. Thus, taking $k \in \mathbb{N}$ such that $f(x_n) - m < \alpha(\varepsilon)$ for all $n \geq k$, we have $d_S(x_n) < \varepsilon$ for all $n \geq k$.

We already observed that $S = \arg\min f$ when f is α-conditioned for S with $\alpha \in \mathcal{A}_0$. □

In the convex case one disposes of a larger list of characterizations.

Proposition 6.31 *For $f : X \to \mathbb{R}_\infty$ convex and $S \subset X$ closed convex, the three assertions of the preceding lemma are equivalent to the following ones:*

(d) there exists some convex lower semicontinuous remainder ρ such that for all $x^* \in X^*$ one has $f^*(x^*) \leq f^*(0) + \iota_S^*(x^*) + \rho(\|x^*\|)$;
(e) there exists some starshaped $\alpha \in \mathcal{A}_0$ such that for all $z \in S$, $(x, x^*) \in \partial f$ one has $\alpha(d_S(x)) \leq \langle x^*, x - z\rangle$;
(f) there exists some $\beta \in \mathcal{A}_0$ such that for all $(x, x^*) \in \partial f$ one has $\beta(d_S(x)) \leq \|x^*\|$;
(g) $(d_S(x_n)) \to 0$ for any sequence (x_n) in X satisfying $(d_{\partial f(x_n)}(0)) \to 0$.

Proof (c)⇒(d) Given $\alpha \in \mathcal{A}_0$ such that $f \geq \inf f(X) + \alpha \circ d_S$, using the relation

$$(\alpha \circ d_S)^* = \iota_S^* + \alpha^* \circ \|\cdot\|$$

obtained by observing that for all $x \in S$ there exist sequences $(x_n) \to x$, w_n in S satisfying $\|x_n - w_n\| \leq d_S(x)$, so that

$$(\alpha \circ d_S)^*(x^*) = \sup_{(x,w)\in X\times S} (\langle x^*, x\rangle - \alpha(\|x-w\|)) = \sup_{(w,z)\in S\times X} (\langle x^*, w+z\rangle - \alpha(\|z\|)),$$

passing to the conjugates and taking $\rho := \alpha^*$, we get the inequality in (d).

(d)⇒(c) Given ρ as in (d), using the relation

$$(\iota_S^*(\cdot) + \rho \circ \|\cdot\|)^* = \rho^* \circ d_S$$

obtained by writing

$$\sup_{x^*\in X^*} \inf_{w\in S}(\langle x^*, x-w\rangle - \rho(\|x^*\|)) = \sup_{r\in\mathbb{R}_+} \inf_{w\in S}(r\|x-w\| - \rho(r)),$$

we get the inequality in (c) since $f \geq f^{**}$ and since $\alpha := \rho^*$ belongs to $\mathcal{A}_c$ by Lemma 6.8.

(c)⇒(e) Given $\alpha \in \mathcal{A}_0$ such that f is α-conditioned:

$$\forall x \in X \qquad f(x) \geq \inf f(X) + \alpha(d_S(x)),$$

adding to each side of this inequality the respective sides of the relation $f(z) - f(x) \geq \langle x^*, z-x\rangle$, we get the inequality in (e).

(e)⇒(f) Since $\alpha \in \mathcal{A}_0$ is starshaped, β defined by $\beta(r) := \alpha(r)/r$ for $r \in \mathbb{P}$, $\beta(0) := 0$ is in $\mathcal{A}_0$ and assertion (e) implies that $d_S(x)\beta(d_S(x)) \leq \inf_{z\in S}\|x^*\| . \|x-z\|$ for all $(x,x^*) \in \partial f$. Assertion (f) follows after simplification, the case $x \in S$ being obvious.

(f)⇒(g) being obvious, let us show that (g)⇒(f). For this purpose, let us set

$$\beta(r) := \inf\{\|x^*\| : \exists x \in X,\ d_S(x) \geq r,\ x^* \in \partial f(x)\}.$$

Clearly, β is nondecreasing and for any $(x,x^*) \in \partial f$ we have $\beta(d_S(x)) \leq \|x^*\|$. Let us show that $\beta \in \mathcal{A}_0$, i.e. that β is firm. Suppose on the contrary that $\beta(\overline{r}) = 0$ for some $\overline{r} > 0$. Then, there exists a sequence $((x_n, x_n^*))$ in ∂f such that $(x_n^*) \to 0$ and $d_S(x_n) \geq \overline{r}$ for all $n \in \mathbb{N}$. This is a contradiction with (g). □

The following consequence reminds us of the contents of the preceding subsection since a reinforced convexity property of f at some $\overline{x}$ corresponds to a smoothness property of f^* at $\overline{x}^*$.

Corollary 6.16 *For* $f : X \to \mathbb{R}_\infty$ convex, $\overline{x} \in \text{dom} f$, $\overline{x}^* \in X^*$, *the following assertions are equivalent:*

(a) $\exists \alpha \in \mathcal{A}_c : \forall x \in X \qquad f(x) \geq f(\overline{x}) + \langle x - \overline{x}, \overline{x}^*\rangle + \alpha(\|x - \overline{x}\|)$

(b) $\exists \rho \in o(\mathbb{R}) : \forall x^* \in X^* \quad f^*(x^*) \leq f^*(\overline{x}^*) + \langle \overline{x}, x^* - \overline{x}^*\rangle + \rho(\|x^* - \overline{x}^*\|)$.

Moreover, one can take $\rho := \alpha^*$ *and conversely* $\alpha := \rho^*$ *and each of these assertions implies that* $\overline{x}^* \in \partial f(\overline{x})$.

Proof Using Lemmas 6.8, 6.11, this equivalence can be deduced from the equivalence (c)$\Leftrightarrow$(d) of the preceding proposition by taking $S := \{\overline{x}\}$ and changing f into $g := f - \langle \cdot, \overline{x}^* \rangle$, observing that $S := \{\overline{x}\}$ is the set of minimizers of g and that $f^{**} \geq f$ with $f^{**}(\overline{x}) = f(\overline{x})$. One can also give a direct proof. This equivalence can be completed with other assertions (see the exercises). □

For f convex and $\overline{x} \in \mathrm{dom} f$, condition (a) is satisfied whenever f is *uniformly convex at* $\overline{x}$ in the following sense: there exists an $\alpha \in \mathcal{A}_c$ such that for all $x \in X$, $t \in [0, 1]$ one has

$$(1-t)f(\overline{x}) + tf(x) \geq f((1-t)\overline{x} + tx) + t(1-t)\alpha(\|x - \overline{x}\|).$$

In fact, this condition implies that for all $x \in X$, $t \in]0, 1]$ one has

$$f(x) - f(\overline{x}) \geq (1/t)[f(\overline{x} + t(x - \overline{x})) - f(\overline{x})] + (1-t)\alpha(\|x - \overline{x}\|),$$

whence, for all $\overline{x}^* \in \partial f(\overline{x})$, $x \in X$

$$f(x) - f(\overline{x}) \geq df(\overline{x}, x - \overline{x}) + \alpha(\|x - \overline{x}\|) \geq \langle \overline{x}^*, x - \overline{x} \rangle + \alpha(\|x - \overline{x}\|).$$

Theorem 6.21 *Let* f *be a closed proper convex function finite and continuous at* $\overline{x} \in X$. *If* f^* *is strictly convex (resp. uniformly convex), then* f *is Hadamard (resp. Fréchet) differentiable at* $\overline{x}$.

Proof For $\overline{x}^* \in \partial f(\overline{x})$ one has $\overline{x} \in \partial f^*(\overline{x}^*)$ by Theorem 6.16, hence $0 \in \partial(f^* - \overline{x})(\overline{x}^*)$ and $\overline{x}^*$ is a minimizer of $f^* - \overline{x}$. When f^* is strictly convex, $f^* - \overline{x}$ is strictly convex too and it has at most one minimizer. Thus $\partial f(\overline{x})$ is a singleton and f is Hadamard differentiable at $\overline{x}$ in view of Corollary 6.4.

When $g := f^*$ is uniformly convex, taking $\overline{x}^* \in \partial f(\overline{x})$, so that $\overline{x} \in \partial g(\overline{x}^*)$, using the preceding observation and the implication (a)$\Rightarrow$(b) of the preceding corollary we obtain that g^* is Fréchet differentiable at $\overline{x}$, the reverse inequality in (b) being a consequence of subdifferentiability. Since f is the restriction to X of g^*, we get that f is Fréchet differentiable at $\overline{x}$. □

Exercises

1. Let $c \in \mathbb{P}$ and let $\alpha \in \mathcal{A}_c$ be given by $\alpha(r) := cr$ for $r \in \mathbb{R}_+$. Verify that $\alpha^* = \iota_{[0,c]}$. If for $f : X \to \mathbb{R}$ one has $\gamma_f = \alpha$ one says that f is linearly conditioned and one says that c is the (linear) *rate of conditioning* of f.
2. **(Conditioning of a matrix)** Let $Q \in L(X, X)$ be a symmetric, positive linear operator on a (finite dimensional) Euclidean space X and let $f := \sqrt{q}$, where

$q(x) := \langle Qx \mid x \rangle$ for $x \in X$. Show that the rate of conditioning of f is $\sqrt{\lambda}$ where λ is the smallest eigenvalue of Q. Note that when $\|Q\| = 1$, λ is the conditioning rate of Q defined as the ratio between its smallest eigenvalue and its largest eigenvalue.

3. Show that the assertions of Proposition 6.31 are equivalent to the following one:

   (h) there exists a $\gamma \in \mathcal{A}$ continuous at 0 such that $d_S(x) \leq \gamma(\|x^*\|)$ for all $(x, x^*) \in \partial f$.

4. Deduce from Proposition 6.31 that each of the two assertions of Corollary 6.16 is equivalent to any of the following ones:

   (c) there exists a $\gamma \in \mathcal{A}_c$ such that $\langle x - \overline{x}, x^* - \overline{x}^* \rangle \geq \gamma(\|x - \overline{x}\|)$ for all $(x, x^*) \in \partial f$;
   (d) there exists a $\gamma \in \mathcal{A}_0$ such that $\|x^* - \overline{x}^*\| \geq \gamma(\|x - \overline{x}\|)$ for all $(x, x^*) \in \partial f$;
   (e) there exists a $\beta \in \mathcal{A}$ continuous at 0 such that $\|x - \overline{x}\| \leq \beta(\|x^* - \overline{x}^*\|)$ for all $(x, x^*) \in \partial f$;
   (f) for any sequence (x_n) of X $(d_{\partial f(x_n)}(\overline{x}^*)) \to 0$ implies that $(x_n) \to \overline{x}$;
   (g) $\overline{x}^* \in \mathrm{int}(\mathrm{dom} f)$ and f^* is Fréchet differentiable at $\overline{x}^*$.

6.4 *Applications to the Geometry of Normed Spaces

Several interesting properties of normed spaces depend on the geometry of their unit balls. We already noted the fact that uniform convexity implies reflexivity. In this section we give various complements, some of which use functions associated with the norm through weight functions. A function $h : \mathbb{R}_+ \to \mathbb{R}_+$ will be called a *weight function* if it is continuous, increasing, and such that $h(0) = 0$, $h(r) \to \infty$ as $r \to \infty$. To any such function h we associate the function $j_h : X \to \mathbb{R}_+$ by

$$j_h(x) := k(\|x\|), \quad \text{where } k(s) := \int_0^s h(r)dr \qquad s \in \mathbb{R}_+. \tag{6.30}$$

The choice $h(r) := r^{p-1}$ for $r \in \mathbb{R}_+$ is convenient for the study of L_p spaces, with $p \in [1, \infty[$; then $j_h(x) = (1/p) \|x\|^p$. In particular, for $p := 2$ one has $j_h(\cdot) = (1/2) \|\cdot\|^2$.

Lemma 6.12 *Let h be a weight function and let k and j_h be defined by (6.30). Then k is strictly convex, differentiable with $k' = h$ on $\mathbb{R}_+$ and $k^*(t) = \int_0^t h^{-1}(s)ds$. Moreover, $k(s) + k^*(t) = st$ if and only if $t = h(s)$.*

The function j_h is convex continuous on X and $(j_h)^* = k^* \circ \|\cdot\|_*$ where $\|\cdot\|_*$ is the dual norm of $\|\cdot\|$ (or the norm of Y if X and Y are in metric duality). Moreover,

$$y \in \partial j_h(x) \Leftrightarrow y \in h(\|x\|)\partial \|\cdot\| (x) \Leftrightarrow \langle x, y \rangle = \|x\| \cdot \|y\|_*, \ \|y\|_* = h(\|x\|).$$

Thus, since h^{-1} is a weight function, one sees that $(j_h)^* = k^* \circ \|\cdot\|_*$, being the function associated with h^{-1} and the dual norm, has the same structure as j_h.

Proof The fundamental theorem of calculus (Corollary 5.4) ensures that k is differentiable with $k' = h$. Since h is positive and increasing on $\mathbb{P}$, k is increasing and strictly convex. The relation $k(s) + k^*(t) = st$ means that $t \in \partial k(s)$ or $t = k'(s) = h(s)$. It is equivalent to $s \in \partial k^*(t)$. Since for a given t there exists only one s ($s = h^{-1}(t)$) such that this relation holds, we get that k^* is differentiable and $(k^*)'(t) = s = h^{-1}(t)$. Since h^{-1} is continuous, we have $k^*(t) = \int_0^t h^{-1}(s)ds$.

Since k is convex, increasing, and continuous, it follows that j_h is convex and continuous. Let us compute $(j_h)^*$. Assuming $(Y, \|\cdot\|_*)$ is in metric duality with $(X, \|\cdot\|)$, for $y \in Y$ with X one has

$$\begin{aligned}(j_h)^*(y) &:= \sup\{\langle x, y\rangle - k(\|x\|) : x \in X\} \\ &= \sup\{\langle x, y\rangle - k(r) : r \in \mathbb{R}_+,\ x \in rS_X\} \\ &= \sup\{r\,\|y\|_* - k(r)\} = k^*(\|y\|_*).\end{aligned}$$

Finally

$$y \in \partial j_h(x) \Leftrightarrow j_h(x) + k^*(\|y\|_*) = \langle x, y\rangle \Leftrightarrow k(\|x\|) + k^*(\|y\|_*) \leq \langle x, y\rangle$$

since the reverse inequality is always valid in view of the relation $\langle x, y\rangle \leq \|x\| . \|y\|_*$. This means that $\langle x, y\rangle = \|x\| . \|y\|_*$ and $\|y\|_* \in \partial k(\|x\|) = \{h(\|x\|)\}$. □

The following notions are weakened versions of uniform convexity, as is easily seen.

Definition 6.2 A norm $\|\cdot\|$ on a vector space $X \neq \{0\}$ is said to be *rotund* (or *strictly convex*) if for every $w \neq x$ in its unit sphere S_X and $t \in]0, 1[$ one has $(1-t)w + tx \notin S_X$.

A norm $\|\cdot\|$ on a vector space X is said to be *locally uniformly rotund* (LUR), or *locally uniformly convex*, if for all $x, x_n \in X$ satisfying $(\|x_n\|) \to \|x\|$, $(\|x + x_n\|) \to 2\,\|x\|$ one has $(x_n) \to x$.

Thus $(X, \|\cdot\|)$ is rotund if any $u \in S_X$ is an *extremal point* of the unit ball B_X in the sense that u cannot be the midpoint of a segment of B_X not reduced to a singleton. Taking $x_n = w$, we see that a locally uniformly rotund norm is rotund.

Let us display characterizations of these properties.

Lemma 6.13 *For a normed space $(X, \|\cdot\|)$ the following assertions are equivalent:*

(a) $\|\cdot\|$ is rotund;
(b) if $x, w \in S_X$ satisfy $\|x + w\| = 2$, then $x = w$;
(c) if $x, w \in X$ satisfy $\|x + w\|^2 = 2\,\|x\|^2 + 2\,\|w\|^2$, then $x = w$;

(d) if $x, w \in X \backslash \{0\}$ satisfy $\|x + w\| = \|x\| + \|w\|$, then $x = \lambda w$ for some $\lambda \in \mathbb{R}_+$;
(e) for any weight function h, the function j_h is strictly convex.

Proof (a)⇔(b): (a)⇒(b) is immediate. For the reverse implication suppose that for some $t \in]0, 1[$, $w, x \in S_X$ we have $\|z_t\| = 1$ for $z_t := (1-t)w + tx$. Let $s := \min(t, 1-t) > 0$, so that $t + s \in [0, 1]$, $t - s \in [0, 1]$ and $z_{t+s} + z_{t-s} = 2z_t$. Since $\|z_{t+s}\| \leq 1$, $\|z_{t-s}\| \leq 1$ our assumption implies that $z_{t+s} = z_{t-s}$, hence $2sw = 2sx$ and $w = x$.

(b)⇔(c): (c)⇒(b) is immediate. (b)⇒(c). For $x, w \in X$, since

$$2\|x\|^2 + 2\|w\|^2 - \|x + w\|^2 \geq 2\|x\|^2 + 2\|w\|^2 - (\|x\| + \|w\|)^2 = (\|x\| - \|w\|)^2,$$

the relation $2\|x\|^2 + 2\|w\|^2 - \|x + w\|^2 = 0$ implies $\|x\| = \|w\|$. Setting $x := ru$, $w := rv$ with $r := \|x\| = \|w\|$, $u, v \in S_X$, for $r > 0$ we get $\|u + v\| = 2$, so that $u = v$ and $x = w$, whereas for $r = 0$ we have $x = w = 0$.

(d)⇒(b) is immediate. Let us prove (b)⇒(d). Suppose $\|x + w\| = \|x\| + \|w\|$ for $x, w \in X \backslash \{0\}$ and $r := \|x\| \leq s := \|w\|$. Then

$$\begin{aligned} 2 &\geq \left\|r^{-1}x + s^{-1}w\right\| \geq r^{-1}\|x + w\| - \left\|r^{-1}w - s^{-1}w\right\| \\ &= r^{-1}(\|x\| + \|w\|) - (r^{-1} - s^{-1})\|w\| = r^{-1}\|x\| + s^{-1}\|w\| = 2. \end{aligned}$$

Thus $\left\|r^{-1}x + s^{-1}w\right\| = 2$ and $r^{-1}x = s^{-1}w$.

(a)⇒(e) For $w, x \in X$, $t \in]0, 1[$ the relation $(1-t)j_h(w) + tj_h(x) = j_h(\|(1-t)w + tx\|)$ implies that $\|w\| = \|x\| = \|(1-t)w + tx\|$ since k is strictly convex and increasing and since $\|(1-t)w + tx\| \leq (1-t)\|w\| + t\|x\|$. Then, either $\|w\| = \|x\| = 0$ and $w = x = 0$, or $0 < \|w\| = \|x\| = \|(1-t)w + tx\|$ and $w = x$ by strict convexity of the norm.

(e)⇒(b) Let $x, w \in S_X$ be such that $\|x + w\| = 2$. Then we have $j_h((1/2)x + (1/2)w) = k((1/2)\|x + w\|) = k(1) = (1/2)j_h(x) + (1/2)j_h(w)$. Since j_h is strictly convex we must have $x = w$. □

Let us turn to characterizations of locally uniformly convex normed spaces. Again, we recall that a function f is said to be *uniformly convex* at $w \in X$ if there exists some $\alpha \in \mathcal{A}_0$ such that f is α-convex at w, i.e. if

$$\forall x \in X \qquad (1-t)f(w) + tf(x) \geq f((1-t)w + tx) + t(1-t)\alpha(\|w - x\|)$$

and f is said to be *locally uniformly convex* if for all $w \in X$ it is uniformly convex at w. A normed space $(X, \|\cdot\|)$ is said to be locally uniformly convex or locally uniformly rotund (LUR) if the square of its norm is locally uniformly rotund.

Lemma 6.14 *For a normed space $(X, \|\cdot\|)$ the following assertions are equivalent:*

(a) $(x_n) \to x$ whenever $(\|x_n\|) \to \|x\|$ and $(\|x + x_n\|) \to 2\|x\|$;
(b) if $x, x_n \in S_X$ for $n \in \mathbb{N}$ satisfy $(\|x + x_n\|) \to 2$, then $(x_n) \to x$;

(c) if $x, x_n \in X$ satisfy $(2\|x\|^2 + 2\|x_n\|^2 - \|x + x_n\|^2) \to 0$, then $(x_n) \to x$;
(d) for any weight function h the function j_h is locally uniformly convex.

Proof (a)⇒(b) is obvious. The converse is obtained by considering (in the non-trivial case $x \neq 0$) $u := x/\|x\|$, $u_n := x_n/\|x_n\|$ (for n large enough).

(c)⇒(a) is immediate. (a)⇒(c) For $x, x_n \in X$, since

$$2\|x\|^2 + 2\|x_n\|^2 - \|x + x_n\|^2 \geq 2\|x\|^2 + 2\|x_n\|^2 - (\|x\| + \|x_n\|)^2 = (\|x\| - \|x_n\|)^2,$$

the relation $\lim_n(2\|x\|^2 + 2\|x_n\|^2 - \|x + x_n\|^2) = 0$ implies $(\|x_n\|) \to \|x\|$. Since we may suppose $x \neq 0$, we can write $x_n = t_n w_n$, with $\|w_n\| = \|x\|$ and $(t_n) \to 1$, observing that $(2\|x\|^2 + 2\|w_n\|^2 - \|x + w_n\|^2) \to 0$, we obtain that $(w_n) \to x$ and $(x_n) \to x$.

The equivalence (a)⇔(d) can be obtained by fixing w in the proof of Proposition 6.33 below. □

The LUR property has interesting consequences, as the next proposition shows.

Proposition 6.32 *If $\|\cdot\|$ is a LUR norm, then X has the (sequential) Kadec-Klee Property: a sequence (x_n) of X converges to $x \in X$ whenever it weakly converges to x and $(\|x_n\|) \to \|x\|$.*

Proof Let $x \in X$ and let $(x_n)_{n \in \mathbb{N}}$ be a weakly convergent sequence whose limit x is such that $(\|x_n\|) \to \|x\|$. Then, $\limsup_n \|x + x_n\| \leq \limsup_n(\|x\| + \|x_n\|) = 2\|x\|$. On the other hand, since the norm is weakly lower semicontinuous, we have $\liminf_n \|x + x_n\| \geq \|2x\|$. Thus $(\|x + x_n\|) \to 2\|x\|$ and, since the norm is LUR, we get $(x_n) \to x$. □

Let us turn to uniform convexity. The next result enables us to obtain several characterizations by using Corollary 6.16 and the exercises following it.

Proposition 6.33 *For any weight function h, j_h is uniformly convex on bounded subsets if and only if $(X, \|\cdot\|)$ is uniformly convex.*

Proof Suppose that for some $r > 0$ and some weight function h the function j_h is uniformly convex on rB_X. Let $\rho \in \mathcal{A}_0$ be such that for $w, x \in rB_X$ and $t \in [0, 1]$ one has

$$(1 - t)j_h(w) + tj_h(x) \geq j_h((1 - t)w + tx) + t(1 - t)\rho(\|x - w\|).$$

Given $\varepsilon > 0$ and $u, v \in S_X$ satisfying $\|u - v\| \geq \varepsilon$, taking $w := ru$, $x := rv$, $t := \frac{1}{2}$ we get $k(r) - \frac{1}{4}\rho(r\|u - v\|) \geq k(\frac{1}{2}r\|u + v\|)$. Taking $r' \in \mathbb{R}_+$ such that $k(r') = k(r) - \frac{1}{4}\rho(r\varepsilon)$ and $\delta := 1 - r'/r$ we see that $\left\|\frac{1}{2}(u + v)\right\| \leq 1 - \delta$ because k is increasing. Thus $(X, \|\cdot\|)$ is uniformly convex.

Conversely, suppose $(X, \|\cdot\|)$ is uniformly convex. Let h be a weight function and let $r \in \mathbb{P}$. If j_h is not uniformly convex on rB_X, there exists some $\varepsilon > 0$ such that for any $\delta > 0$ one can find $w, x \in rB_X$ such that

$$\|w - x\| = \varepsilon \text{ and } j_h(\frac{1}{2}w + \frac{1}{2}x) > \frac{1}{2}j_h(w) + \frac{1}{2}j_h(x) - \delta.$$

Taking a sequence $(\delta_n) \to 0_+$, one can find sequences (w_n), (x_n) in rB_X such that $\|w_n - x_n\| = \varepsilon$ and $j_h(\frac{1}{2}w_n + \frac{1}{2}x_n) > \frac{1}{2}j_h(w_n) + \frac{1}{2}j_h(x_n) - \delta_n$ for all $n \in \mathbb{N}$. Let s_n, $t_n \in [0, r]$, u_n, $v_n \in S_X$ be such that $w_n = s_n u_n$, $x_n = t_n v_n$. The preceding inequality implies that

$$k(\frac{1}{2}s_n + \frac{1}{2}t_n) \geq k(\frac{1}{2}\|s_n u_n + t_n v_n\|) > \frac{1}{2}k(s_n) + \frac{1}{2}k(t_n) - \delta_n \tag{6.31}$$

for all $n \in \mathbb{N}$. Taking subsequences if necessary, one may assume that (s_n) and (t_n) converge to s and t respectively in $[0, r]$ and that $(\frac{1}{2}\|u_n + v_n\|)$ converges to some $q \in [0, 1]$. Passing to the limits in the preceding inequalities one gets $k(\frac{1}{2}s + \frac{1}{2}t) \geq \frac{1}{2}k(s) + \frac{1}{2}k(t)$. By strict convexity of k one obtains $s = t$. Since

$$\varepsilon = \|s_n u_n - t_n v_n\| \leq s_n \|u_n - v_n\| + |s_n - t_n| \leq 2s_n + |s_n - t_n|$$

and $(s_n - t_n) \to 0$, one must have $s > 0$. Since $(\|s_n u_n + t_n v_n\| - \|s u_n + t v_n\|) \to 0$ and since k is continuous, the second inequality in (6.31) yields $k(sq) \geq k(s)$; if $q < 1$ this inequality contradicts the assumption that k is increasing. Thus $(\frac{1}{2}\|u_n + v_n\|) \to 1$, contradicting the assumption that $(X, \|\cdot\|)$ is uniformly convex. □

Now let us turn to differentiability properties of the norm.

Definition 6.3 The space $(X, \|\cdot\|)$ is said to be *smooth* if for all $x \in X\backslash\{0\}$ there is only one $x^* \in X^*$ such that $\|x^*\|_* = 1$ and $\langle x^*, x\rangle = \|x\|$.

This condition means that the *normalized duality mapping* $S : X \to \mathcal{P}(X^*)$ given by

$$S(x) := \{x^* \in X^* : \langle x^*, x\rangle = \|x\|, \ \|x^*\| = 1\}$$

and the *duality map* $J := \|\cdot\| S(\cdot)$ are single-valued, or equivalently, by Corollary 6.4, that the norm and $(1/2)\|\cdot\|^2$ are Gateaux (or Hadamard) differentiable on $X\backslash\{0\}$.

In order to give some versatility to the following famous differentiability test for a norm, we adopt the framework of normed spaces X, Y in metric duality for a continuous bilinear coupling $c := \langle\cdot,\cdot\rangle : X \times Y \to \mathbb{R}$. We say that a sequence (y_n) in Y *c-weakly converges* (or simply, weakly converges) to $y \in Y$ if for every $x \in X$ we have $(\langle x, y_n\rangle) \to \langle x, y\rangle$. This notion coincides with weak* convergence when $Y := X^*$ and with weak convergence when $X := Y^*$.

Proposition 6.34 (Šmulian Test) *Let X and Y be normed spaces in metric duality and let $\overline{x} \in S_X$. The following assertions (a) and (b) are equivalent and are implied by (c). If Y is the dual of X, then (a), (b) and (c) are equivalent:*

(a) the norm of X is Fréchet (resp. Hadamard) differentiable at $\overline{x}$;
(b) for any sequences (y_n), (z_n) in S_Y such that $(\langle \overline{x}, y_n \rangle) \to 1$, $(\langle \overline{x}, z_n \rangle) \to 1$, one has $(\|y_n - z_n\|) \to 0$ (resp. $(y_n - z_n)$ c-weakly converges to 0);
(c) a sequence (y_n) of S_Y is convergent (resp. c-weakly convergent) whenever $(\langle \overline{x}, y_n \rangle) \to 1$.

Proof (a)$\Rightarrow$(b) Suppose the norm $\|\cdot\|$ of X is Hadamard differentiable at $\overline{x} \in S_X$. By Lemma 6.14, for any given $\varepsilon > 0$ and any $u \in S_X$ there exists some $\delta > 0$ such that $\|\overline{x} + tu\| + \|\overline{x} - tu\| \leq 2 + \varepsilon t$ when $t \in [-\delta, \delta]$. Let (y_n) and (z_n) be sequences in S_Y such that $(\langle \overline{x}, y_n \rangle) \to 1$ and $(\langle \overline{x}, z_n \rangle) \to 1$. Then, for $t := \delta$, one can find $k \in \mathbb{N}$ such that for all $n \geq k$ one has

$$\begin{aligned} t\langle u, y_n - z_n \rangle &= \langle \overline{x} + tu, y_n \rangle + \langle \overline{x} - tu, z_n \rangle - \langle \overline{x}, y_n \rangle - \langle \overline{x}, z_n \rangle \\ &\leq \|\overline{x} + tu\| + \|\overline{x} - tu\| - 2 + 2\delta\varepsilon \leq 3\delta\varepsilon. \end{aligned}$$

Thus $\langle u, y_n - z_n \rangle \leq 3\varepsilon$ for $n \geq k$. Changing u into $-u$, we see that $(\langle u, y_n - z_n \rangle) \to 0$. The Fréchet case is similar, using uniformity in $u \in S_X$.

(b)$\Rightarrow$(a) Suppose the norm $\|\cdot\|$ of X is not Hadamard differentiable at $\overline{x} \in S_X$. By Lemma 6.14 there exist some $u \in S_X$, some $\varepsilon > 0$ and some sequence $(t_n) \to 0_+$ such that $\|\overline{x} + t_n u\| + \|\overline{x} - t_n u\| - 2 \geq 3t_n\varepsilon$ for all n. Let us pick y_n, z_n in S_Y such that

$$\langle \overline{x} + t_n u, y_n \rangle \geq \|\overline{x} + t_n u\| - t_n\varepsilon, \qquad \langle \overline{x} - t_n u, z_n \rangle \geq \|\overline{x} - t_n u\| - t_n\varepsilon. \tag{6.32}$$

Then $\langle \overline{x}, y_n \rangle \geq \|\overline{x} + t_n u\| - t_n\varepsilon - t_n \|u\| . \|y_n\|$ and $\langle \overline{x}, y_n \rangle \leq 1$, so that $(\langle \overline{x}, y_n \rangle) \to 1$ and similarly, $(\langle \overline{x}, z_n \rangle) \to 1$. Since $\|\overline{x}\| = 1$, $\|y_n\| = 1$, $\|z_n\| = 1$, we get

$$\begin{aligned} t_n\langle u, y_n - z_n \rangle &= \langle \overline{x} + t_n u, y_n \rangle + \langle \overline{x} - t_n u, z_n \rangle - \langle \overline{x}, y_n \rangle - \langle \overline{x}, z_n \rangle \\ &\geq \|\overline{x} + t_n u\| + \|\overline{x} - t_n u\| - 2t_n\varepsilon - \|\overline{x}\| (\|y_n\| + \|z_n\|) \geq t_n\varepsilon, \end{aligned}$$

hence $\langle u, y_n - z_n \rangle \geq \varepsilon$, contradicting the assumption that $(y_n - z_n)$ c-weakly converges to 0.

When the norm $\|\cdot\|$ of X is not Fréchet differentiable at $\overline{x} \in S_X$ one can find $\varepsilon > 0$ and sequences $(t_n) \to 0_+$, (u_n) in S_X such that $\|\overline{x} + t_n u_n\| + \|\overline{x} - t_n u_n\| - 2 \geq 3t_n\varepsilon$ for all $n \in \mathbb{N}$. Then, taking (y_n), (z_n) $\in S_Y$ as in relation (6.32) with u replaced with u_n, the preceding computation reads $\langle u_n, y_n - z_n \rangle \geq \varepsilon$, hence $\|y_n - z_n\| \geq \varepsilon$, a contradiction with the assumption that $(y_n - z_n) \to 0$.

(c)$\Rightarrow$(b) Let (y_n), (z_n) be sequences of S_Y such that $(\langle x, y_n \rangle) \to 1$, $(\langle x, z_n \rangle) \to 1$. Let $w_n = y_p$ when $n := 2p$, $w_n = z_p$ when $n := 2p + 1$. Then $(\langle x, w_n \rangle) \to 1$, so that, by (c), (w_n) converges (resp. c-weakly converges). Thus $(y_n - z_n) \to 0$ (resp. c-weakly converges to 0).

(a)⇒(c) when $Y = X^*$. Let $\overline{y} := \|\cdot\|' (\overline{x})$. One has $\|\overline{y}\| \leq 1$ since the norm is Lipschitzian with rate 1 and, by homogeneity, $\langle \overline{y}, \overline{x} \rangle = \lim_{t \to 0}(1/t)(\|\overline{x} + t\overline{x}\| - \|\overline{x}\|) = 1$, so that $\overline{y} \in S_Y$ and we can take $z_n := \overline{y}$ in assertion (b). Thus (c) holds. □

Let us turn to duality results.

Proposition 6.35 *Let $\|\cdot\|$ be a norm on X and let $\|\cdot\|_*$ be its dual norm.*

(a) If $\|\cdot\|_*$ is a rotund norm, then $\|\cdot\|$ is Hadamard differentiable on $X \backslash \{0\}$.
(b) If $\|\cdot\|_*$ is Hadamard differentiable on $X^* \backslash \{0\}$, then $\|\cdot\|$ is a rotund norm.

In particular, a compatible norm on a reflexive Banach space X is Hadamard differentiable on $X \backslash \{0\}$ if and only if its dual norm is rotund.

Proof

(a) By Corollary 6.4, it suffices to show that for every $x \in X \backslash \{0\}$,

$$S(x) := \partial \|\cdot\| (x) = \{x^* \in X^* : \|x^*\|_* = 1, \langle x^*, x \rangle = \|x\|\}$$

is a singleton. Let $x^*, y^* \in S(x)$. Then

$$2 \|x\| = \langle x^*, x \rangle + \langle y^*, x \rangle \leq \|x^* + y^*\|_* . \|x\| \leq 2 \|x\| ,$$

hence $\|x^* + y^*\|_* = 2$, and by assertion (b) of Lemma 6.13, we have $x^* = y^*$.

(b) If $\|\cdot\|$ is not strictly convex, one can find $x, y \in S_X$ such that $x \neq y$ and $z := (1-t)x + ty \in S_X$ with $t := 1/2$. Taking $f \in S_{X^*}$ such that $f(z) = 1$, we see that $1 = f(z) = (1-t)f(x) + tf(y) \leq 1$, so that this inequality is an equality and $f(x) = f(y) = 1$. Viewing x and y as elements of X^{**}, we have $x, y \in \partial \|\cdot\|_* (f)$, so that $\|\cdot\|_*$ is not differentiable at f. □

Proposition 6.36 *Let $(X, \|\cdot\|)$ be a normed space. If the dual norm $\|\cdot\|_*$ is LUR, then $\|\cdot\|$ is Fréchet differentiable on $X \backslash \{0\}$.*

Proof We use Šmulian Test (c). Let $x \in S_X$. Using a corollary of the Hahn-Banach theorem, we pick $f \in S_{X^*}$ such that $f(x) = 1$. Let (f_n) be a sequence in S_{X^*} such that $(f_n(x)) \to 1$. Since

$$2 \geq \|f + f_n\|_* \geq (f + f_n)(x) \to 2,$$

we have $\lim_n(2 \|f\|_*^2 + 2 \|f_n\|_*^2 - \|f + f_n\|_*^2) = 0$, hence, by the LUR property, $(f_n) \to f$. Then, by Proposition 6.34, $\|\cdot\|$ is Fréchet differentiable at x, hence on $]0, +\infty[x$. □

So, it will be useful to detect when a norm on the dual of X is a dual norm.

Lemma 6.15 *An equivalent norm $\|\cdot\|$ on the dual X^* of a Banach space X is the dual norm of an equivalent norm $\|\cdot\|_X$ on X if and only if it is weak* lower semicontinuous.*

Proof If $\|\cdot\|$ is the dual norm of an equivalent norm $\|\cdot\|_X$, then $\|\cdot\| = \sup\{\langle x, \cdot\rangle : x \in X,\ \|x\|_X = 1\}$ is weak* lower semicontinuous as a supremum of weak* continuous linear forms.

Conversely, if $\|\cdot\|$ is weak* lower semicontinuous, its unit ball B^* is convex and weak* closed, hence coincides with its bipolar. Then one can see that $\|\cdot\|$ is the dual norm of the Minkowski gauge of the polar set of B^*, a compatible norm on X. □

In order to deal with quantitative properties, it is useful to introduce the function $\rho_X : \mathbb{R}_+ \to \mathbb{R}_+$ associated with a norm $\|\cdot\|$ on X given by

$$\rho_X(t) := \sup\{(1/2)(\|x + tu\| + \|x - tu\|) - \|x\| : x, u \in S_X\} \qquad t \in \mathbb{R}_+.$$

It is called the *modulus of smoothness* of $(X, \|\cdot\|)$; it is a modulus since $\rho_X(t) \leq t$ for $t \in \mathbb{R}_+$. Moreover, since $\|x + tu\| + \|x - tu\| \geq \|2tu\|$, one has $\rho_X(t) \geq t - 1$ for all $t \in \mathbb{R}_+$.

Definition 6.4 The space $(X, \|\cdot\|)$ is *uniformly smooth* if the function ρ_X is a remainder (i.e., $\rho_X(t)/t \to 0$ as $t \to 0_+$).

Proposition 6.37 *The function ρ_X is starshaped, i.e. $t \mapsto \rho_X(t)/t$ is nondecreasing.*

It can be shown (see [265, Prop. 3.5.1]) that even $t \mapsto \rho_X(t)/t^2$ is nondecreasing.

Proof For $t > s > 0$, $x, u \in S_X$, by a property of convex functions, one has

$$\frac{1}{s}(\|x + su\| - \|x\|) \leq \frac{1}{t}(\|x + tu\| - \|x\|)$$

and a similar inequality with u changed into $-u$, hence the result. □

In order to give a quantitative measure of rotundity, it is useful to introduce the function $\gamma_X : \mathbb{R} \to \mathbb{R}_\infty$ given by $\gamma_X(s) := +\infty$ for $s \in \mathbb{R}\backslash[0, 1]$ and

$$\gamma_X(s) := \inf\{1 - \|(x + y)/2\| : x, y \in S_X, \|(x - y)/2\| \geq s\} \qquad s \in [0, 1].$$

It can be shown that γ_X can be given other expressions among which are

$$\begin{aligned}\gamma_X(s) &= \inf\{1 - \|(x + y)/2\| : x, y \in S_X, \|(x - y)/2\| = s\} \qquad s \in [0, 1]\\ &= \inf\{1 - \|(x + y)/2\| : x, y \in B_X, \|(x - y)/2\| \geq s\} \qquad s \in [0, 1].\end{aligned}$$

Let us relate this function to the definition of uniform convexity of the norm we gave in Sect. 3.4.3:

$$\forall \varepsilon > 0\ \exists \delta > 0 : x, y \in B_X, \|(x - y)/2\| \geq \varepsilon \Rightarrow \|(x + y)/2\| < 1 - \delta. \qquad (6.33)$$

Proposition 6.38 *A norm* $\|\cdot\|$ *on a vector space* $X \neq \{0\}$ *is* uniformly rotund *(or* uniformly convex*) if and only if* γ_X *is firm, i.e. such that* $\gamma_X(s) > 0$ *for all* $s > 0$.

Proof If γ_X is firm, for $\varepsilon > 0$ we can take $\delta := \gamma_X(\varepsilon)$ in (6.33): given $x, y \in B_X$, such that $1 - \|(x+y)/2\| < \gamma_X(\varepsilon)$ we must have $\|(x-y)/2\| < \varepsilon$.

Conversely, if the norm $\|\cdot\|$ is uniformly rotund, given $s \in]0,1]$ we can find $\delta > 0$ such that for all $x, y \in B_X$ satisfying $\|(x-y)/2\| \geq s$ we have $1 - \|(x+y)/2\| > \delta$, hence $\gamma_X(s) \geq \delta$: γ_X is firm. □

Note that since γ_X is nondecreasing, $(X, \|\cdot\|)$ is uniformly rotund if and only if γ_X is *forcing*, i.e. $(s_n) \to 0$ whenever $(\gamma_X(s_n)) \to 0$.

We shall see (Theorem 8.27) that the usual norm on $L_p(S)$ for $p > 1$ and S a measure space is uniformly rotund.

Exercise Show that for $p \in]1, \infty[$ the space ℓ_p of real sequences $x := (x_n)$ such that $\|x\|_p := (\Sigma_{n\geq 0} |x_n|^p)^{1/p}$ is finite is uniformly rotund for $\|\cdot\|_p$.

The function $\delta_X : [0,2] \to \mathbb{R}$ given by $\delta_X(t) := \gamma_X(t/2)$ for $t \in [0,2]$ is classically called the *modulus of rotundity* of $(X, \|\cdot\|)$ but it seems to us that γ_X is preferable to δ_X in view of the following remarkable property. We phrase it in the framework of normed spaces in metric duality which enables us to take for Y either the dual space of X or a predual of X. In fact, we use the crucial properties $\|x\| = \sup\{\langle x, y\rangle : y \in S_Y\}$, $\|y\| = \sup\{\langle x, y\rangle : x \in S_X\}$.

Proposition 6.39 (Lindenstrauss) *If X and Y are normed vector spaces in metric duality, then* $\rho_Y = \gamma_X^*$ *and* $\rho_X = \gamma_Y^*$, *the Fenchel conjugate* γ^* *of a function* γ *on* $\mathbb{R}_+$ *being the conjugate of the extension of* γ *by* $+\infty$ *on* $\mathbb{R}\backslash\mathbb{R}_+$, *i.e.* γ^* *is given by* $\gamma^*(t) := \sup\{st - \gamma(s) : s \in \mathbb{R}_+\}$.

Proof By metric duality, for $t > 0$ one has

$$\begin{aligned}
\rho_Y(t) &:= (1/2)\sup\{\|y + tv\| + \|y - tv\| - 2 : y, v \in S_Y\} \\
&= (1/2)\sup\{\langle y + tv, x\rangle + \langle y - tv, w\rangle - 2 : w, x \in S_X,\ y, v \in S_Y\} \\
&= (1/2)\sup\{\langle y, x + w\rangle + \langle tv, x - w\rangle - 2 : w, x \in S_X,\ y, v \in S_Y\} \\
&= (1/2)\sup\{\|x + w\| + t\|x - w\| - 2 : w, x \in S_X\} \\
&= \sup\{\|(x + w)/2\| + ts - 1 : w, x \in S_X,\ s \in \mathbb{R}_+,\ s \leq \|(x - w)/2\|\} \\
&= \sup\{st - \inf\{1 - \left\|\frac{x + w}{2}\right\| : w, x \in S_X,\ \left\|\frac{x - w}{2}\right\| \geq s\} : s \in \mathbb{R}_+\}.
\end{aligned}$$

This last expression is nothing but $\sup\{st - \gamma_X(s) : s \in \mathbb{R}_+\} = \gamma_X^*(t)$. Finally, the roles of X and Y are symmetric. □

Corollary 6.17 (Šmulian) *A normed space is uniformly rotund if and only if its dual space is uniformly smooth.*

A normed space is uniformly smooth if and only if its dual space is uniformly rotund.

Proof It suffices to show that if X and Y are two normed spaces in metric duality, then X is uniformly rotund if and only if Y is uniformly smooth. This follows from the last proposition and Lemma 6.8. □

One can deduce from the last corollary an analogue to Proposition 6.33.

Proposition 6.40 *The space $(X, \|\cdot\|)$ is uniformly smooth if and only if for any weight function h, j_h is uniformly smooth.*

Proof Since the unit sphere of X is dense in the unit sphere of the completion $\widehat{X}$ of X we may suppose X is complete. Given an arbitrary weight function h, the function j_h is uniformly smooth if and only if $(j_h)^*$ is uniformly convex (Theorem 6.20). By Lemma 6.12, $(j_h)^* = k^* \circ \|\cdot\|_*$ and k^* is a weight function. Then, by Proposition 6.33, $(j_h)^*$ is a uniformly convex function if and only if $(X^*, \|\cdot\|_*)$ is uniformly convex. Combining these equivalences we get the announced assertion. □

The restriction of the duality map J to the unit sphere of a uniformly smooth Banach space is uniformly continuous. More precisely, one can give a modulus of local uniform continuity of $S(\cdot) := J(\cdot)/\|\cdot\|$ on $X\backslash\{0\}$, remembering that ρ_X is a remainder on $\mathbb{R}_+$ or that γ_X is firm.

Proposition 6.41 *The duality map J of a uniformly smooth Banach space X is uniformly continuous on bounded subsets of X.*

Proof By Corollary 6.17, X^* is uniformly convex. Thus, given $\varepsilon > 0$, $x^*, y^* \in B_X$ such that $1 - \|(x^* + y^*)/2\| < \gamma_{X^*}(\varepsilon/2)$ we have $\|x^* - y^*\| < \varepsilon$. Given $x, y \in S_X$ such that $\|x - y\| < \delta := 2\gamma_{X^*}(\varepsilon/2)$ we have

$$\begin{aligned}\|J(x) + J(y)\| &\geq \langle y, J(x) + J(y)\rangle = \langle x, J(x)\rangle + \langle y, J(y)\rangle - \langle x - y, J(x)\rangle \\ &\geq 2 - \|x - y\| > 2 - \delta = 2(1 - \gamma_{X^*}(\varepsilon/2)),\end{aligned}$$

hence $\|J(x) - J(y)\| < \varepsilon$. This shows that J is uniformly continuous on S_X.

Let us prove that J is uniformly continuous on B_X. Given $\varepsilon > 0$, for $x, y \in (\varepsilon/2)B_X$ we have $\|J(x) - J(y)\| \leq \|J(x)\| + \|J(y)\| \leq \varepsilon$ since $\|J(w)\| = \|w\|$ for all $w \in X$. Assuming $x \in B_X\backslash(\varepsilon/2)B_X$, $y \in B_X$, with $\|x - y\| < \delta\varepsilon/4 \leq \varepsilon/2$ (since $\delta := 2\gamma_{X^*}(\varepsilon/2) \leq 2$), we have $y \neq 0$, so that, setting $u := x/\|x\|$, $v := y/\|y\|$ and using the inequalities $|\|y\| - \|x\|| \leq \|x - y\|$,

$$\|u - v\| \leq \frac{1}{\|x\|}\|x - y\| + \frac{1}{\|x\|}|\|y\| - \|x\|| \leq \frac{4}{\varepsilon}\|x - y\| < \delta,$$

we see that

$$\|J(x) - J(y)\| \leq \|x\|\,\|S(u) - S(v)\| + \|\|x\|\,S(v) - \|y\|\,S(v)\| \leq 2\varepsilon$$

since $J(x) = \|x\| S(u)$ and $\|S(v)\| = 1$. This shows that J is uniformly continuous on B_X. By homogeneity, the same holds on any bounded set. □

Exercise Verify that for $\gamma_J(r) := (r/4)\gamma_X(r/4)$ one has $\|J(x) - J(y)\| \leq \varepsilon$ whenever $x, y \in B_X$ are such that $\|x - y\| \leq \gamma_J(\varepsilon)$.

Remark The unit duality map $S(\cdot)$ of a uniformly smooth Banach space $(X, \|\cdot\|)$ satisfies for all $x, y \in X\backslash\{0\}$ the following relation in which $u_x := x/\|x\|$, $u_y := y/\|y\|$:

$$\|S(x) - S(y)\| \leq \frac{2\rho_X(2\|u_x - u_y\|)}{\|u_x - u_y\|}. \tag{6.34}$$

In particular, for $J(\cdot) := \|\cdot\| S(\cdot)$, for $x, y \in S_X$ one has

$$\|J(x) - J(y)\| \leq \frac{2\rho_X(2\|x - y\|)}{\|x - y\|}.$$

Proof Since $S(x) = J(u_x)$ and $S(y) = J(u_y)$, to prove (6.34) we may assume that $x, y \in S_X$. By a general property of convex functions, since $S(\cdot) = \|\cdot\|'$ and $\langle S(u), v\rangle \leq \|u + v\| - \|u\|$ for all $u \in X\backslash\{0\}$, $v \in X$, we have

$$\langle S(x), y - x\rangle \leq \|y\| - \|x\| = 0.$$

Let $r := \|x - y\|$ and let $z \in rB_X$. By the preceding inequalities we have

$$\begin{aligned}
\langle S(y), z\rangle - \langle S(x), z\rangle &\leq \|y + z\| - \|x\| - \langle S(x), z\rangle \\
&\leq \|y + z\| - 1 + \langle S(x), x - y - z\rangle \\
&\leq \|y + z\| - 1 + \|2x - y - z\| - \|y\| \\
&\leq \|x + (y - x + z)\| - \|x - (y - x + z)\| - 2 \\
&\leq 2\rho_X(\|y - x + z\|) \leq 2\rho_X(2r).
\end{aligned}$$

Taking the supremum over $z \in rB_X$ we get $r\|S(y) - S(x)\| \leq 2\rho_X(2r)$ and relation (6.34). □

Example For a measure space S and $p \in]1, \infty[$, the duality map of $L_p(S)$ with its usual norm is given by $J(x)(s) := \|x\|_p^{2-p} |x(s)|^{p-2} x(s)$ for $x \in L_p(S)$, $s \in S$. For $p = 1$ one has $J(x) = \{y \in L_\infty(S) : y(s) \in \|x\|_1 \operatorname{sign} x(s)\}$.

Example For $X = W^1_{p,0}(\Omega)$, where Ω is a bounded open subset of $\mathbb{R}^d$, $J(x) = -\|x\|_{1,p}^{2-p} \Sigma_{i=1}^d D_i(|D_i x|^{p-2} D_i x)$.

Finally, let us note that in any normed space X one can define a kind of substitute to an inner product by setting

$$\langle x \mid y\rangle_+ := \sup_{x^* \in J(x)} \langle x^*, y\rangle = \lim_{t \to 0_+} \frac{1}{2t}(\|x+ty\|^2 - \|x\|^2) \qquad x, y \in X,$$

as $J(x) = \partial j(x)$, with $j(\cdot) := \frac{1}{2}\|\cdot\|^2$, using Theorem 6.4. It is called the *semi-inner product* of X. This definition is related to the notion of a semi-scalar product introduced in Proposition 3.15 by the inequality

$$[x, y] \le \langle x \mid y\rangle_+ \qquad \forall x, y \in X$$

since $[x, \cdot]$ is an element of $J(x) = \partial j(x)$ for all $x \in X$. When X is a smooth space one has $\langle x \mid y\rangle_+ = \langle J(x), y\rangle$ for all $x,\ y \in X$ and $\langle x \mid \cdot\rangle_+$ is linear and continuous, not just sublinear and continuous. In the general case the following properties hold. They are left as exercises stemming from Proposition 6.9.

Lemma 6.16 *For any Banach space $(X, \|\cdot\|)$ the following properties hold.*

(a) $|\langle x \mid y\rangle_+| \le \|x\| \, . \, \|y\|$ *and* $(x, y) \mapsto \langle x \mid y\rangle_+$ *is upper semicontinuous.*
(b) $\langle x \mid y + z\rangle_+ \le \langle x \mid y\rangle_+ + \langle x \mid z\rangle_+$.
(c) $\langle x \mid y + rx\rangle_+ = \langle x \mid y\rangle_+ + r\|x\|^2$ *for all* $r \in \mathbb{R}$.
(d) *If* $u : T \to X$ *is right differentiable at some* t *in the interior of an interval* T, *then* $f(\cdot) := (1/2)\|u(\cdot)\|^2$ *is right differentiable at* t *and its right derivative is* $f'_r(t) = \langle u(t) \mid u'_r(t)\rangle_+$.
(e) *If* X *is uniformly convex, then* $\langle \cdot \mid \cdot\rangle_+$ *is uniformly continuous on bounded subsets of* $X \times X$.

The following renorming theorem is of interest. We refer to specialized monographs (f.i. [119, Thm 8.20] as a recent reference) for the proof of its second assertion.

Theorem 6.22

(a) Every separable Banach space X has an equivalent norm that is Hadamard differentiable on $X \setminus \{0\}$.
(b) Every Banach space X whose dual is separable has an equivalent norm that is Fréchet differentiable on $X \setminus \{0\}$.

Proof

(a) Let $(e_n)_{n \in \mathbb{N}}$ be a countable dense subset of B_X. Define a norm on X^* by

$$\|f\| = \left[\|f\|_0^2 + \sum_{n=0}^{\infty} 2^{-n} f^2(e_n)\right]^{1/2} \qquad f \in X^*$$

where $\|\cdot\|_0$ is the original norm of X^*. The norm $\|\cdot\|$ is easily seen to be weak* lower semicontinuous, so that it is the dual norm of some norm $\|\cdot\|_*$ on X. In view of Lemma 6.35, it suffices to show that $\|\cdot\|$ is strictly convex. Let $f, g \in X^*$ be such that $\|f+g\|^2 = 2\|f\|^2 + 2\|g\|^2$. Since $2\|f\|_0^2 + 2\|g\|_0^2 \geq \|f+g\|_0^2$ and $2f^2(e_n) + 2g^2(e_n) \geq (f+g)^2(e_n)$ for all n, we get that these last inequalities are equalities, so that

$$(f-g)^2(e_n) = 2f^2(e_n) + 2g^2(e_n) - (f+g)^2(e_n) = 0$$

for all n. Thus $f(e_n) = g(e_n)$ for all n and, by density, $f = g$. □

The proofs of following theorems are beyond the scope of this book. However the results may be useful.

Theorem 6.23 (Asplund) *Every reflexive Banach space can be provided with an equivalent norm for which both X and X^* endowed with the dual norm are strictly rotund.*

Theorem 6.24 (Kadec, Troyanski) *Every reflexive Banach space can be provided with an equivalent norm for which both X and X^* are locally uniformly rotund, and uniformly smooth.*

The notion of an obtuse angle between two vectors can be extended to normed spaces in a number of equivalent ways, as shown in the next lemma.

Lemma 6.17 (Kato) *For given elements x, y of a Banach space X the following assertions are equivalent:*

(a) $\|x\| \leq \|x - ty\|$ *for all* $t \in \mathbb{R}_+$;
(b) *there exists an* $x^* \in J(x)$ *such that* $\langle x^*, y\rangle \leq 0$;
(c) *there exists a semi-scalar product* $[\cdot,\cdot]$ *on X such that* $[x, y] \leq 0$;
(d) $\langle x \mid -y\rangle_+ \geq 0$, *where* $\langle u \mid v\rangle_+ := \lim_{t\to 0_+}(1/2t)(\|u+tv\|^2 - \|u\|^2)$.

Proof (a)⇒(b) Without loss of generality we may suppose $\|x\| = 1$. For $t > 0$ small enough let us take $u_t^* \in S(x-ty) := \|x-ty\|^{-1} J(x-ty)$, so that $\|u_t^*\| = 1$ and

$$\|x\| \leq \|x-ty\| = \langle u_t^*, x-ty\rangle = \langle u_t^*, x\rangle - t\langle u_t^*, y\rangle \leq \|x\| - t\langle u_t^*, y\rangle.$$

By the Banach-Alaoglu theorem there exists a weak* limit point $u^* \in B_{X^*}$ of the net $(u_t^*)_{t>0}$ and the preceding relations imply $\langle u^*, y\rangle \leq 0$ and $\langle u^*, x\rangle = \|x\|$ (replacing y with $-x$), so that $u^* \in S(x)$ and $x^* := \|x\|\, u^* \in J(x)$ satisfies $\langle x^*, y\rangle \leq 0$.

(b)⇒(a) Given $x^* \in J(x)$ such that $\langle x^*, y\rangle \leq 0$, $t \in \mathbb{R}_+$, we have

$$\|x^*\| \,.\, \|x\| = \langle x^*, x\rangle \leq \langle x^*, x-ty\rangle \leq \|x^*\| \,.\, \|x-ty\|$$

hence $\|x\| \leq \|x-ty\|$ if $x \neq 0$. If $x = 0$ the inequality is obvious.

(b)⇒(c) This follows from the fact that we can choose a selection j of J such that $j(x) = x^*$; then, setting $[u, v] = \langle j(u), v \rangle$ for $(x, v) \in X \times X$, we have $[x, y] = \langle x^*, y \rangle$.

(c)⇒(b) Given a semi-scalar product $[\cdot, \cdot]$ on X and $x \in X$, we have $x^* := [x, \cdot] \in J(x)$.

(c)⇒(d) Since for all $(u, v) \in X \times X$ we have $\langle u \mid v \rangle_+ \geq [u, v]$, taking $u = x$, $v := -y$ we get $\langle x \mid -y \rangle_+ \geq [x, -y] = -[x, y] \geq 0$.

(d)⇒(a) Since $f : t \mapsto (1/2) \|x - ty\|^2$ is convex, for $t \in \mathbb{R}_+$ we have $\|x - ty\|^2 - \|x\|^2 \geq 2tdf(x, -y) = 2t\langle x, -y \rangle_+ \geq 0$. □

Exercises

1. Show that the spaces ℓ_1, c_0 and ℓ_∞ are not strictly convex. [Hint: for $e_1 := (1, 0, \ldots)$ and $e_2 := (0, 1, 0, \ldots)$ one has $\|(1/2)(e_1 + e_2)\|_1 = 1$ and for $u := e_1 + e_2$, $v := e_1 - e_2$ one has $u, v \in S_X$ for $X := c_0$ and $(1/2)(u + v) \in S_X$.]

2. Show that the space $C([0, 1])$ is not strictly convex for the norm $\|\cdot\|_\infty$.

3. Show that for a measure space S, the spaces $L_1(S)$ and $L_\infty(S)$ are not strictly convex.

4. Show that a Hilbert space H is uniformly convex and $\gamma_H(s) = 1 - \sqrt{1 - s^2}$. It can be shown that for any normed space X one has $\gamma_X \leq \gamma_H$, a result due to Nörlander (1960).

5. Using the Šmulian Test, show that if the norm of a normed space is Fréchet differentiable on $X \setminus \{0\}$, then it is of class C^1 there.

6. Show that a normed space X is strictly convex if and only if each point x of its unit sphere S_X is an *exposed point* of the unit ball B_X, i.e. for each $x \in S_X$ there exists an $f \in X^*$ such that $f(x) > f(u)$ for all $u \in B_X \setminus \{x\}$.

7. Show that a normed space $(X, \|\cdot\|)$ is uniformly rotund if for any sequences (x_n), (y_n) in B_X such that $(\|x_n + y_n\|) \to 2$ one has $(\|x_n - y_n\|) \to 0$.

8*. Let S be a locally compact topological space and let $X := C_0(S)$ be the space of bounded continuous functions on S converging to 0 at infinity: $x \in C_0(S)$ if and only if $x(\cdot)$ is bounded, continuous on S and if, for any $\varepsilon > 0$, one can find a compact subset K of S such that $\sup |x(S \setminus K)| \leq \varepsilon$.

(**a**) Show that the supremum norm $\|\cdot\|_\infty$ is Hadamard differentiable at $x \in X$ if and only if $M_x := \{s \in S : |x(s)| = \|x\|_\infty\}$ is a singleton.

(**b**) Show that $\|\cdot\|_\infty$ is Fréchet differentiable at $x \in X$ if and only if M_x is a singleton $\{\bar{s}\}$ such that $\bar{s}$ is an isolated point of S.

9*. Let $X := \ell_1(I)$ be the space of absolutely summable families $x := (x_i)_{i \in I}$ endowed with the norm $\|x\|_1 := \sum_{i \in I} |x_i|$.

(**a**) Show that $\|\cdot\|_1$ is nowhere Hadamard differentiable if I is uncountable.

(**b**) If $I := \mathbb{N}$, show that $\|\cdot\|_1$ is Hadamard differentiable at x if and only if $x_i \neq 0$ for all $i \in I$.

(**c**) If $I := \mathbb{N}$, show that $\|\cdot\|_1$ is nowhere Fréchet differentiable.

10. Show that the space $X := L_p(S, \mu)$ $(p > 1)$ is uniformly rotund and uniformly smooth (Hanner, 1956) with
$\gamma_X(s) = (p-1)s^2/2 + o(s^2)$ for $p \in]1, 2]$, $\gamma_X(s) = s^p/p + o(s^p)$ for $p \geq 2$.
$\rho_X(t) = t^p/p + o(t^p)$ for $p \in]1, 2]$, $\rho_X(t) = (p-1)t^2/2 + o(t^2)$ for $p \geq 2$.

11. Let S be a subset of a normed space X, let $x \in X$ and let $\overline{x} \in S$ be such that $\|x - \overline{x}\| \leq \|x - w\|$ for all $w \in S$. Let $T(S, \overline{x})$ be the set of $v \in X$ such that there exist sequences $(v_n) \to v$, $(t_n) \to 0_+$ satisfying $\overline{x} + t_n v_n \in S$ for all $n \in \mathbb{N}$. Show that $\langle x - \overline{x} \mid -v \rangle_+ \geq 0$ for all $v \in T(S, \overline{x})$. In this sense one can say that $x - \overline{x}$ is *normal* to S at $\overline{x}$.

12*. Show that for $p \in]1, \infty[$ there exists some $\alpha_p \in \mathbb{P}$ such that for all $r, s \in \mathbb{R}$ one has

$$\left| |r|^{p-2} r - |s|^{p-2} s \right| \leq \alpha_p (|r| + |s|)^p \, |r - s| \, .$$

Verify that the function $g : t \mapsto (1+t)^{2-p}(1-t)^{-1}(1-t^{p-1})$ on $[0, 1[$ has limit $2^{2-p}(p-1)$ as $t \to 1_-$ and is bounded on $[0, 1[$. Deduce from this the fact that there exist constants c_p, c'_p such that for all $a \in \mathbb{P}$, $b \in]0, a]$ one has

$$c'_p (a+b)^{p-2}(a-b)^2 \leq (a^{p-1} - b^{p-1})(a-b) \leq c_p (a+b)^{p-2}(a-b)^2.$$

Conclude that for $p \in [2, \infty[$, the function $k_p : \mathbb{R}^d \times \mathbb{R}^d \to \mathbb{R}$ given by $k_p(x, y) := (\|x\| + \|y\|)^{p-2} \|x - y\|^2$ for $x, y \in \mathbb{R}^d$ endowed with the Euclidean norm satisfies the inequalities

$$2^{1-p} k_p(x, y) \leq \langle \|x\|^{p-2} x - \|y\|^{p-2} y \mid x - y \rangle \leq c_p k_p(x, y).$$

[See [74, Prop. 17.3].] Find some consequences for the space $L_p(S, \mu)$, where $(S, \mathcal{S}, \mu)$ is a finite measure space.

6.5* Regularization of Convex Functions

It may be useful to approximate a function by a sequence in regular functions. For a locally integrable function on $\mathbb{R}^d$ a standard means is to use integral convolution with a smooth bump function. In the case of a convex function on a normed space, classical processes are the Baire regularization and the Moreau regularization. Although some of the following results can be extended to the case of functions defined on the dual of a Banach space, for the sake of simplicity we only consider functions defined on a reflexive space. In fact, our first result can be given in the framework of metric spaces.

Proposition 6.42 (Baire, Hausdorff, McShane, Pasch) *Let $f : X \to \mathbb{R}_\infty$ be a proper, lower semicontinuous function on a metric space X. Suppose there exist b,*

$c \in \mathbb{R}$, $\overline{x} \in X$ such that $f(\cdot) \geq b - cd(\cdot, \overline{x})$. Then, for all $n \in \mathbb{N}$ the function $f_n : X \to \mathbb{R}$ given by $f_n(x) := \inf_{u \in X}(f(u) + nd(u,x))$ for $x \in X$ is Lipschitzian and the sequence (f_n) pointwise converges to f on X.

Since a lower semicontinuous proper convex function on a normed space is bounded below by a continuous affine function, this approximation result holds for such a function. Also, note that when X is a normed space, one has $f_n := f \square n \|\cdot\|$.

Proof For $n \in \mathbb{N}$, $u \in X$, let $g_{n,u} : X \to \mathbb{R}_\infty$ be given by $g_{n,u}(x) = f(u) + nd(u,x)$ for $x \in X$. Since $f_n(x) := \inf_{u \in X} g_{n,u}(x)$, for all $x \in X$, taking $u = x$ in the infimum, we see that $f_n(x) \leq f(x)$. Moreover, taking some $\overline{u} \in \operatorname{dom} f$, we see that $f_n(x) \leq f(\overline{u}) + nd(\overline{u}, x) < \infty$ for all $x \in X$. Since for $n \geq c$

$$g_{n,u}(x) \geq b - cd(u, \overline{x}) + nd(u,x) \geq b + (n-c)d(u, \overline{x}) - nd(x, \overline{x}) \tag{6.35}$$

we have $f_n(x) := \inf_u g_{n,u}(x) \geq b - nd(x, \overline{x}) > -\infty$ and since the Lipschitz rate of $g_{n,u}$ is n, the function $f_n := \inf_{u \in X} g_{n,u}$ is Lipschitzian with rate at most n. Given $x \in X$ and $s < f(x)$, since f is lower semicontinuous at x, we can find $r > 0$ such that $f(u) > s$ for all $u \in B(x,r)$. Then, for $u \in B(x,r)$ we have $g_{n,u}(x) \geq f(u) > s$ whereas for $u \in X \backslash B(x,r)$ we have $g_{n,u}(x) \geq b - cd(x, \overline{x}) + nr \geq s$ provided n is large enough. Thus, for such an n we get $f_n(x) \geq s$. This shows that $\lim_n f_n(x) = f(x)$. □

Exercise Let $f : X \to \mathbb{R}_\infty$ be the indicator function ι_C of a nonempty subset C of X. Identify f_n and transcribe the conclusion of the proposition.

Exercise Let $f : X \to \mathbb{R}_\infty$ be a closed convex function on a uniformly convex Banach space and let $C := \operatorname{cl}(\operatorname{dom} f))$. Show that for every $x \in X$ and $n \in \mathbb{N} \backslash \{0\}$ there exists a unique point $w_n \in X$ such that $f(w_n) + n \|x - w_n\| = \inf\{f(w) + n\|x - w\| : w \in X\}$ and that $(w_n) \to P_C(x)$, the unique point $w \in C$ such that $\|w - x\| = \inf_{u \in C} \|u - x\|$. □

Exercise (Subdifferential Determination: The Approach of Zlateva [269]) Let $f : X \to \mathbb{R}_\infty$ be a closed convex function on a Banach space X. For $n \in \mathbb{N} \backslash \{0\}$ and $\varepsilon \in \mathbb{P}$, setting $f_n := f \square n \|\cdot\|$, let

$$M_{n,\varepsilon}(x) := \{w \in X : f(w) + n\|x - w\| \leq \inf_{u \in X}(f(u) + n\|x - u\| + \varepsilon)\},$$

$$\partial_\varepsilon f(x) := \{x^* \in X^* : f - x^* \geq f(x) - \langle x^*, x \rangle - \varepsilon\}.$$

(a) Show that for all $x \in \operatorname{dom} f$, $n \in \mathbb{N} \backslash \{0\}$, $\varepsilon \in \mathbb{P}$, $w \in M_{n,\varepsilon}(x)$ one has

$$\partial f_n(x) \subset \partial_\varepsilon f(w) \cap \partial_\varepsilon n \|\cdot\| (x - w).$$

(b) Assume that $f(0) = 0$ and $0 \in \partial f(0)$. Show that for all $r \in \mathbb{P}$, $x \in B[0,r]$, $\varepsilon \in]0,1]$, $n \geq 1/r$ one has $M_{n,\varepsilon}(x) \subset B[0, 3r]$.

(c) Assume that f is as in (b) and let $g : X \to \mathbb{R}_\infty$ be closed proper convex and such that $g(0) = 0$ and $\partial f \subset \partial g$, i.e. $\partial f(x) \subset \partial g(x)$ for all $x \in X$. Show that for all $r \in \mathbb{P}$, $x \in B[0, r]$, $n \geq 1/r$ one has $\partial f_n(x) \subset \partial g_n(x)$ with $g_n := g \Box n \|\cdot\|$.

(d) Deduce from Exercise 4 of Sect. 2.2 that for all $r > 0$ and $n \geq 1/r$ there exist some $c_{r,n} \in \mathbb{R}$ such that $f_n = g_n + c_{r,n}$ and $f = g + c_r$ on $B(0, r)$. Conclude that there exists some $c \in \mathbb{R}$ such that $f = g + c$ on X. □

Let us turn to the most useful approximation result for convex functions. We use a parameter $r \in \mathbb{P}$ and an increasing function $h : \mathbb{R}_+ \to \mathbb{R}_+$ satisfying $h(0) = 0$ and the condition

$$\forall c \in \mathbb{R} \qquad \liminf_{t \to \infty} \frac{k(t-c)}{k(t)} > 0 \tag{6.36}$$

for $k(t) := \int_0^t h(s)ds$. Such a condition is fulfilled if, for some $p > 1$, $k(t) := (1/p)t^p$ or if $k(t) := e^t$ for $t \in \mathbb{R}_+$.

Given a function $f : X \to \mathbb{R}_\infty$, we set $j_h := k \circ \|\cdot\|$, $f_r := f \Box r^{-1} j_h$:

$$f_r(x) := \inf_{u \in X} f_{r,x}(u) \quad \text{with} \quad f_{r,x}(u) := f(u) + \frac{1}{r} j_h(x - u) \qquad x \in X.$$

In the usual case of the *Moreau regularization*, one takes $h(t) = t$ for $r \in \mathbb{R}_+$, so that $f_r(x) = \inf_{u \in X}(f(u) + \frac{1}{2r}\|x - u\|^2)$, but the choice $h(t) := t^{p-1}$ can be convenient, for instance when dealing with L_p spaces, with $p \in]1, \infty[$. Also, the use of a weight h enables us to take into account the growth properties of f. In the sequel we denote by $r_{f,h}$ or simply by r_f the extended real given by

$$r_f := \sup\{r \in \mathbb{P} : \exists m \in \mathbb{R} \; f \geq m - \frac{1}{r} j_h\}.$$

Exercise Show that, when f is bounded below on bounded sets, $-1/r_{f,h}$ coincides with the h-coercivity rate

$$c_{f,h} := \liminf_{\|x\| \to \infty} \frac{f(x)}{j_h(x)}.$$

Proposition 6.43 *Let $f : X \to \mathbb{R}_\infty$ be a proper function on a Banach space X and let $h : \mathbb{R} \to \mathbb{R}$ be a weight function satisfying condition (6.36). Assume that $\inf(f - cj_h) > -\infty$ for some $c \in \mathbb{R}$. For $r \in]0, r_f[$ let $f_r := f \Box r^{-1} j_h$:*

$$f_r(x) := \inf_{u \in X}(f(u) + \frac{1}{r} j_h(x - u)) \qquad \text{with } j_h(v) := \int_0^{\|v\|} h(s)ds.$$

Then, denoting by $\overline{f}$ the lower semicontinuous hull of f, one has

$$\lim_{r \to 0+} f_r(x) = \sup_{r>0} f_r(x) = \overline{f}(x) := \liminf_{w \to x} f(w) \qquad x \in X.$$

For any bounded subset B of X there exists some $r_B \in]0, r_f[$ such that for all $r \in]0, r_B]$ the function f_r is finite and Lipschitzian on B.

Moreover, for $x \in X$, denoting by $P_r(x)$ the set of minimizers of $f_{r,x} : u \mapsto f(u) + r^{-1}j_h(x-u)$, given $x \in \operatorname{dom}\overline{f}$ one has

$$e(P_r(x), x) := \sup\{\|u - x\| : u \in P_r(x)\} \to 0 \text{ as } r \to 0_+.$$

The map P_r is called the *proximal map* of f. When f is convex P_r is characterized by the relation $0 \in \partial f_{r,x}(P_r(x))$ or, in view of Theorem 6.8, j_h being convex continuous,

$$\frac{1}{r}J_h(x - P_r(x)) \in \partial f(P_r(x)) \tag{6.37}$$

where $J_h := \partial j_h$ is characterized by $\langle v, J_h(v)\rangle = h(\|v\|)$, $\|J_h(v)\| = h(\|v\|)$ (see Lemma 6.12). Moreover, since the function $(x, u) \mapsto f_{r,x}(u)$ is convex, f_r is convex.

Proof We first observe that for all $r \in \mathbb{P}$ the function f_r is bounded above on bounded subsets: fixing $\overline{x} \in f^{-1}(\mathbb{R})$, $c \in \mathbb{R}_+$, for all $x \in cB_X$ we have $f_r(x) \leq m_{r,c} := f(\overline{x}) + r^{-1}k(\|\overline{x}\| + c) < \infty$.

Let us show that for all $x \in X$, $r > 0$ we have $f_r(x) \leq \overline{f}(x)$: given a sequence (x_n) converging to x such that $\lim_n f(x_n) = \overline{f}(x)$, since j_h is continuous at 0, we have

$$f_r(x) \leq \liminf_n f_{r,x}(x_n) = \liminf_n (f(x_n) + r^{-1}j_h(x - x_n)) = \overline{f}(x).$$

Let us prove that for any bounded subset B there exists some $r_B \in]0, r_f[$ such that for all $r \in]0, r_B[$ the function f_r is bounded below on B. Let $c > 0$ be such that $B \subset cB_X$ and let $\overline{r} \in]0, r_f[$. By definition of r_f there exists an $m \in \mathbb{R}$ such that $f + (1/\overline{r})j_h \geq m$. Also, there exist some $a \in]0, 1]$ and $\overline{s} \in \mathbb{P}$ such that for $s \geq \overline{s}$ one has $k(s - c) \geq ak(s)$. Thus, setting $r_B := a\overline{r}$, for $r \in]0, r_B[$, $x \in cB_X$, $u \in X$, and $s := \|u\| \geq \overline{s}$, we have $j_h(u-x) \geq k(\|u\| - \|x\|) \geq k(s-c)$ since k is nondecreasing, $j_h(u) = k(s)$, and

$$\begin{aligned} f(u) + \frac{1}{r}j_h(u - x) &\geq f(u) + \frac{1}{\overline{r}}j_h(u) - \frac{1}{\overline{r}}k(s) + \frac{1}{r}k(s - c) \\ &\geq m + (\frac{a}{r} - \frac{1}{\overline{r}})k(s) \geq m + (\frac{a}{r} - \frac{1}{\overline{r}})k(\overline{s}). \end{aligned} \tag{6.38}$$

For $s := \|u\| \leq \overline{s}$, by definition of r_f one has

$$f(u) + (1/r)j_h(x - u) \geq m - (1/\overline{r})j_h(u) \geq m - (1/\overline{r})k(\overline{s}).$$

Thus $f_{r,x}$ is bounded below on X by $m - (1/\overline{r})k(\overline{s})$, uniformly on $x \in cB_X$: $\inf_{x \in cB_X} f_r(x) > -\infty$.

Since $k(s) \geq h(\overline{s})(s - \overline{s}) + k(\overline{s})$, with $h(\overline{s}) > 0$, estimate (6.38) shows that the function $f_{r,x} : u \mapsto f(u) + (1/r)j_h(x - u)$ is supercoercive, uniformly for $x \in cB_X$. Since for $x \in cB_X$ we have $f_r(x) \leq m_{r,c}$, there exists some $\rho > 0$ such that

$$f_r(x) = \inf\{f_{r,x}(u) : u \in \rho B_X\} \qquad \forall x \in cB_X. \tag{6.39}$$

Since j_h is Lipschitzian on balls, we deduce from this relation that f_r is Lipschitzian on cB_X, hence on B.

Given $x \in X$, let us show that $(f_r(x)) \to \overline{f}(x)$ as $r \to 0_+$ and that $e(P_r(x), x) \to 0$ as $r \to 0_+$ if $x \in \mathrm{dom}\overline{f}$. Let $c \geq \|x\|$, $B := cB_X$, $r \in]0, r_B[$ and let $\varepsilon > 0$, $\lambda \in \mathbb{R}$ be given. Let $u \in X$ be such that $t := \|u - x\| \geq \varepsilon$. If $t > \overline{s} + c$ we have $s := \|u\| \geq \overline{s}$ hence, for $r > 0$ small enough, namely $r < r_\lambda := ak(\overline{s})(\lambda - m + k(\overline{s})/\overline{r})^{-1}$, by (6.38) we get

$$f(u) + (1/r)j_h(u - x) \geq m + (a/r - 1/\overline{r})k(\overline{s}) > \lambda.$$

If $t := \|u - x\| \leq \overline{s} + c$ we have $\|u\| \leq \overline{s} + 2c$ and the assumption $f \geq m - (1/\overline{r})j_h$ yields

$$f(u) + (1/r)j_h(u - x) \geq m - (1/\overline{r})k(\overline{s} + 2c) + (1/r)k(\varepsilon) > \lambda$$

for r small enough, namely $r < r_{\lambda,\varepsilon}$ for some $r_{\lambda,\varepsilon} \in \mathbb{P}$. Taking $\lambda < \overline{f}(x)$ and choosing $\varepsilon > 0$ such that $f(w) \geq \lambda$ for all $w \in B(x, \varepsilon)$, we get $f_r(x) := \inf_{u \in X} f_{r,x}(u) \geq \lambda$ for r as above. Thus $\liminf_{r \to 0_+} f_r(x) \geq \overline{f}(x)$ and since $f_r(x) \leq \overline{f}(x)$ we obtain $(f_r(x)) \to \overline{f}(x)$. Now, for $x \in \mathrm{dom}\overline{f}$, taking $\lambda \geq \overline{f}(x)$, we see that for $r < \min(r_\lambda, r_{\lambda,\varepsilon})$, for all $u \in P_r(x)$ we must have $u \in B(x, \varepsilon)$. □

Additional assumptions enhance the interest of this regularization.

Theorem 6.25 *Suppose $(X, \|\cdot\|)$ is reflexive and that f is a weakly lower semicontinuous function such that* $\inf(f - cj_h) > -\infty$ *for some $c \in \mathbb{R}$. Then, for any bounded subset B of X there exists some $r_B > 0$ such that for $r \in]0, r_B[$ the restriction to B of the proximal multimap P_r is closed and with nonempty values. If, moreover, $(X, \|\cdot\|)$ is strictly convex and f is convex, then P_r is a continuous map.*

Furthermore, if $(X^*, \|\cdot\|_*)$ satisfies the sequential dual Kadec-Klee Property, then, for r large enough, f_r is of class C^1 with derivative given by $Df_r(x) = (1/r)J(x - P_r(x))$.

Proof When f is weakly lower semicontinuous, $f_{r,x}$ is weakly lower semicontinuous too and coercive, hence $f_{r,x}$ attains its infimum when X is reflexive. Moreover, when $(X, \|\cdot\|)$ is strictly convex and f is convex, $f_{r,x}$ is strictly convex too and its set of minimizers $P_r(x)$ is a singleton.

Given a bounded subset B of X, $x \in B$, a convergent sequence $(x_n) \to x$ in B, $r \in]0, r_B[$, the proof of relation (6.39) shows that any sequence (u_n) satisfying $u_n \in P_r(x_n)$ for all n is bounded. From any subsequence in (u_n) we extract a subsequence $(u_{k(n)})$ that has a weak limit u. Passing to the limit in the relation $f_r(x_n) = f(u_n) +$

$rj_h(x_n - u_n)$, we get

$$f_r(x) \geq \liminf_n f(u_{k(n)}) + \liminf_n (1/r) j_h(x_{k(n)} - u_{k(n)})$$
$$\geq f(u) + (1/r) j_h(u - x).$$

This shows that $u \in P_r(x)$. When $P_r(x)$ is a singleton $\{z\}$, we get that $(u_n) \to u := z$.

Given $x \in \operatorname{dom} f$, we set $Q_r(x) := x - P_r(x)$. Proposition 6.18 ensures that for all $x' \in X$

$$f_r(x') - f_r(x) - \frac{1}{r}\langle J_h(Q_r(x)), x' - x\rangle \geq 0.$$

In order to prove that f_r is differentiable at x with derivative $(1/r)J(Q_r(x))$ let us denote by $o_r(x')$ the left-hand side of this inequality and let us show that $o_r(\cdot)$ is a remainder at x. Using relation (6.37) and thus the inequalities

$$f(P_r(x)) - f(P_r(x')) \geq \frac{1}{r}\langle J_h(Q_r(x')), P_r(x) - P_r(x')\rangle$$
$$\frac{1}{r} j_h(Q_r(x)) - \frac{1}{r} j_h(Q_r(x')) \geq \frac{1}{r}\langle J_h(Q_r(x')), Q_r(x) - Q_r(x')\rangle$$

and using the relation $f_r(x) = f(P_r(x)) + \frac{1}{r} j_h(Q_r(x))$ and the similar one with x changed into x', by adding side by side we obtain

$$f_r(x') - f_r(x) \leq \frac{1}{r}\langle J_h(Q_r(x')), (P_r(x') - P_r(x)) + (Q_r(x') - Q_r(x))\rangle$$
$$o_r(x') \leq \frac{1}{r}\langle J_h(Q_r(x')) - J_h(Q_r(x)), x' - x\rangle.$$

Since P_r and J_h are continuous (as easy changes in the proof of Proposition 3.29 show), we get that $|o_r(x')| \leq \varepsilon(x') \|x' - x\|$ where $\varepsilon(x') \to 0$ as $x' \to x$. That proves that f_r is differentiable at x with derivative $(1/r)J(Q_r(x))$. □

Several algorithms use the properties of proximal maps. The Yosida regularization process for monotone operators described in Sect. 9.4.3 is related to this Moreau type regularization via subdifferentials of convex functions.

Additional Reading

[12–15, 17, 21, 22, 29, 43–48, 81, 95, 112, 119–121, 162, 165, 166, 198, 205, 208, 209, 218, 220, 221, 232, 234, 251, 256, 264, 265, 269]

[illegible]

[illegible] is described in [illegible] is reflected in this [illegible] differentials of convex function.

Additional Readings

[illegible]

Chapter 7
Integration

In the case of continuous functions, the notion of integral coincides with the notion of primitive. Riemann has defined the integral of some discontinuous functions, but not all derivative functions are integrable in the Riemann sense. Thus, the problem of searching for primitive functions through integration is not solved, and one may wish for a definition of an integral including Riemann's which allows one to solve the problem of primitive functions.

Henri Lebesgue, Sur une généralisation de l'intégrale définie, Comptes-rendus de l'Académie des Sciences de Paris 132, pp. 128–132, April 29th 1901.

Abstract Using the notions of measure theory introduced in Chap. 1, an integration process is introduced for functions with values in normed vector spaces. Such an extension does not require much supplementary effort but can be bypassed in a first reading. Convergence results and calculus rules form the bulk of the chapter.

In this chapter we deal with a crucial tool of analysis, namely integration. Its birth is contemporary with the surge of differential calculus at the end of the seventeenth century. Its first appearances concerned the calculation of areas or volumes. Probability questions gave it a further impetus. But it is during the twentieth century that firm grounds were given to the topic by the use of measure theory.

Throughout this chapter $(S, \mathcal{S}, \mu)$ is a measure space. We treat the case of vector-valued functions in order to obtain in a single stroke the case of complex-valued functions and the case of real-valued functions. In a first reading the reader may assume that E is $\mathbb{R}$ or $\mathbb{C}$. However the construction in the case when E is a general Banach space is similar to that for the scalar case: one starts with a class of simple functions for which the definition of the integral is undeniable. Then one passes to a more general class by a kind of completion process. We suggest that in a first step the reader replaces the notation for the norm of the Banach space $(E, \|\cdot\|_E)$ by the notation $|\cdot|_E$ or even $|\cdot|$ in order to be easily convinced that the construction of the integral in the vectorial case is not different from the construction in the scalar case.

J.-P. Penot, *Analysis*, Universitext, DOI 10.1007/978-3-319-32411-1_7

7.1 Step Functions and μ-Measurable Functions

Given a Banach space $(E, \|\cdot\|_E)$, a map $f : (S, \mathcal{S}, \mu) \to E$ is called a *step map* or a simple map if it is measurable and if it takes its values in a finite subset $\{e_1, \ldots, e_n\}$ of E. It is a *μ-step map* if, moreover, it is 0 outside a set of finite measure. We denote by $St(\mu, E)$ the set of μ-step maps from S to E.

The functions we shall integrate are not necessarily measurable, but they are *μ-measurable* in the sense of the next definition.

Definition 7.1 A function $f : S \to E$ is said to be μ-measurable if there exists a sequence (f_n) of $St(\mu, E)$ that converges a.e. to f. We denote by $\mathcal{L}_0(\mu, E)$ the set of μ-measurable functions from S to E.

It is easy to show that the set $\mathcal{L}_0(\mu, E)$ is a vector space and that $\mathcal{L}_0(\mu, \mathbb{R})$ has pleasant stability properties. Let us give a characterization. We recall that $T \in \mathcal{S}$ is said to be of *σ-finite measure* if there exists a sequence (T_n) of $\mathcal{S}$ with union T such that all T_n's have finite measures. We also recall that a subset of a metric space is said to be *separable* if it contains a dense countable subset.

Proposition 7.1 *A function $f : S \to E$ is μ-measurable if and only if it satisfies the following conditions:*

(a) there exists a measurable map $g : S \to E$ such that $f = g$ a.e.;
(b) there exists a $T \in \mathcal{S}$ with σ-finite measure such that $f = 0$ on $S \backslash T$;
(c) there exists a null set N of S such that $f(S \backslash N)$ is separable.

Proof Let $f \in \mathcal{L}_0(\mu, E)$ so that there exist a null set N of S and a sequence (f_n) of $St(\mu, E)$ that converges to f on $S \backslash N$. Replacing N with a larger set if necessary, we may suppose $N \in \mathcal{S}$ and $\mu(N) = 0$. Setting $g_n := 1_{S\backslash N} f_n$ we see that $g_n \in St(\mu, E)$ and $(g_n) \to g := 1_{S\backslash N} f$. Then g is measurable and $f = g$ a.e., so that (a) is satisfied. Since $g_n \in St(\mu, E)$ there exists an element S_n of $\mathcal{S}$ with finite measure such that g_n vanishes on $S \backslash S_n$. Then g vanishes on $S \backslash \cup_n S_n$, hence f vanishes on $S \backslash T$ for $T := \cup_n S_n \cup N$ and $\mu(S_n \cup N)$ is finite. Thus (b) holds. Since g_n is a μ-step function, the set $F_n := g_n(S)$ is finite. Then the union $F := \cup_n F_n$ is countable and its closure contains $f(S \backslash N)$, so that (c) holds.

Let us prove the converse. First, we assume that $\mu(S)$ is finite. Let $f : S \to E$ satisfying conditions (a), (b), (c) and let $D := \{e_k : k \in \mathbb{N}\}$ be a countable dense subset of $f(S)$. Given a decreasing sequence (r_k) of positive numbers with limit 0, for all $n \in \mathbb{N}$ the union $\cup_k B(e_k, r_n)$ of the open balls with radius r_n and centers in D contains $f(S)$, so that $S = \cup_k f^{-1}(B(e_k, r_n))$ and $(\mu(\cup_{k=0}^m f^{-1}(B(e_k, r_n))) \to \mu(S)$ as $m \to +\infty$. Thus, we can find some $m(n) \in \mathbb{N}$ such that $\mu(S \backslash S_n) \leq 2^{-n}$ for

$$S_n := \cup_{k=0}^{m(n)} f^{-1}(B(e_k, r_n)).$$

Let

$$Z_n := \bigcup_{p>n} (S \backslash S_p) \qquad Z := \bigcap_{n \geq 0} Z_n,$$

so that $\mu(Z_n) \leq 2^{-n+1}$ and $\mu(Z) = 0$. Let us define a μ-step map $f_n : S \to E$ by setting $f_n(s) = 0$ for $s \in S \backslash S_n$ and $f_n(s) = e_0$ if $f(s) \in B(e_0, r_n)$, $f_n(s) = e_k$ if $f(s) \in B(e_k, r_n) \backslash (\cup_{i=0}^{k-1} B(e_i, r_n))$ for $k = 1, \ldots, m(n)$. Then, for $s \in S_n$ we have $\|f_n(s) - f(s)\|_E < r_n$. Given $s \in S \backslash Z$ we can find $n_s \in \mathbb{N}$ such that $s \in S \backslash Z_{n_s}$ hence $s \in S_n$ for $n \geq n_s$: this shows that $(f_n(s)) \to f(s)$ and that f is μ-measurable.

Now let us reduce the general case to the case when S has finite measure. By assumption (b) there exists a sequence (T_k) in $\mathcal{S}$ such that $f = 0$ on $S \backslash T$ with $T := \cup_k T_k$ and $\mu(T_k) < \infty$ for all $k \in \mathbb{N}$. We may suppose $T_k \subset T_{k+1}$ for all k. Using the special case we established, for each $k \in \mathbb{N}$ we can find a sequence $(f_{k,n})_n$ of μ-step functions on T_k converging to $f \mid T_k$ on $T_k \backslash Y_k$ with Y_k a null set of T_k. Let us define inductively a sequence (f_n) of μ-step functions on S by setting $f_n(s) = 0$ for $s \in S \backslash T_n$,

$$f_n(s) := f_{0,n}(s) \text{ for } s \in T_0, \ldots, f_n(s) = f_{k,n}(s) \text{ for } s \in T_k \backslash T_{k-1} \quad (k \leq n).$$

Then $(f_n) \to f$ on $(S \backslash \cup_k Y_k)$. □

Corollary 7.1 *Let (f_n) be a sequence in μ-measurable maps, converging almost everywhere to a map $f : S \to E$. Then f is μ-measurable.*

Proof This follows from Lemma 1.3 and from the characterization we have just proved using the following facts: a countable union of null subsets is a null subset, a countable union of σ-finite subsets is σ-finite, and the union of a countable family (A_n) of subsets of E having countable dense subsets D_n has a countable dense subset $D := \cup_n D_n$ since $\mathrm{cl}(D) \supseteq \cup_n \mathrm{cl}(D_n) = \cup_n A_n$. □

Corollary 7.2* *If $(S, \mathcal{S}, \mu)$ is complete and σ-finite and if E is separable, then $f : S \to E$ is μ-measurable if and only if f is measurable.*

Proof The sufficient condition is immediate. The necessary condition stems from the fact that if $g : S \to E$ is measurable and if $f = g$ a.e. then f is measurable when μ is complete. □

Besides the preceding stability result, one can show that the family of μ-measurable functions with values in $\overline{\mathbb{R}}$ is a complete lattice. We omit the proof.

Proposition 7.2* *Any family $F := \{f_i : i \in I\}$ of μ-measurable functions from $(S, \mathcal{S}, \mu)$ to $\overline{\mathbb{R}}$ has a supremum f called the essential supremum of F. This means that there exists some μ-measurable function $f : S \to \overline{\mathbb{R}}$ such that $f \geq f_i$ a.e. for all $i \in I$ and $f \leq g$ a.e. for any μ-measurable function $g : S \to \overline{\mathbb{R}}$ satisfying $g \geq f_i$ a.e. for all $i \in I$.*

Of course, a similar result holds for the essential infimum.

7.2 Integrable Functions and Their Integrals

One defines the *integral* of a μ-step map f with distinct values $\{e_1, \ldots, e_m\}$ by

$$\int_S f d\mu := \int_S f(s) d\mu(s) := \sum_{k=1}^{m} \mu(S_k) e_k \qquad \text{for } S_k := f^{-1}(e_k),$$

using a classical notation. Thus $\int_S f d\mu$ is a weighted sum of the values of f and if f is constant with value e on a subset T of S with finite measure and is null on $S \backslash T$, one has $\int_S f d\mu = \mu(T)e$. More generally, the additivity of μ shows that if $(S_i)_{i \in I}$ is a finite partition of S (i.e. $S = \cup_{i \in I} S_i$ and $S_i \cap S_j = \varnothing$ for $i \neq j$ in I) by elements of $\mathcal{S}$ and if f is constant on S_i, with value $e_i \in E$ satisfying $e_i = 0$ if $\mu(S_i) = +\infty$, using the convention $(+\infty) \times 0 = 0$, one has

$$\int_S f d\mu = \sum_{i \in I} \mu(S_i) e_i.$$

The map $f \mapsto \int_S f d\mu$ is easily seen to be linear from the space $St(\mu, E)$ of μ-step maps from $(S, \mathcal{S}, \mu)$ to E (if $f := \Sigma_{i \in I} 1_{A_i} a_i$, $g := \Sigma_{j \in J} 1_{B_j} b_j$ with I, J finite, a_i, $b_j \in E$, and A_i, $B_j \in \mathcal{S}$, take a finite family $(C_k)_{k \in K}$ of disjoint elements of $\mathcal{S}$ such that all A_i's and all B_j's are unions of subfamilies of $(C_k)_{k \in K}$). If $E = \mathbb{R}$ and if $f \geq 0$ one has $\int_S f d\mu \geq 0$.

If $T \in \mathcal{S}$ and if f is a μ-step function on $(S, \mathcal{S}, \mu)$, the function $1_T f$ is a μ-step function on $(S, \mathcal{S}, \mu)$ and $f\,|_T$ is a μ_T-step function on $(T, \mathcal{S}_T, \mu_T)$, where $\mathcal{S}_T$ is the σ-algebra induced by $\mathcal{S}$ on T, μ_T is the induced measure by μ on $\mathcal{S}_T$, and we have

$$\int_T f d\mu := \int_T (f\,|_T) d\mu_T = \int_S 1_T f d\mu.$$

If (T, T') is a measurable partition of S (i.e. T, $T' \in \mathcal{S}$, $S = T \cup T'$, and $T \cap T' = \varnothing$) one has

$$\int_S f d\mu = \int_T f d\mu + \int_{T'} f d\mu.$$

Observing that for all $f \in St(\mu, E)$ the function $\|f(\cdot)\|_E$ is a μ-step function, one easily sees that the function $\|\cdot\|_1 : St(\mu, E) \to \mathbb{R}$ given by

$$\|f\|_1 := \int_S \|f(s)\|_E d\mu(s)$$

is a seminorm. Then the map $f \mapsto \int_S f d\mu$ is continuous from $St(\mu, E)$ to E since for all $f := \Sigma_{i=1}^m 1_{S_i} e_i \in St(\mu, E)$ the triangle inequality $\left\| \Sigma_{i=1}^m \mu(S_i) e_i \right\|_E \leq \Sigma_{i=1}^m \mu(S_i) \left\| e_i \right\|_E$ yields

$$\left\| \int_S f d\mu \right\|_E \leq \int_S \|f(s)\|_E \, d\mu(s) = \|f\|_1 . \tag{7.1}$$

It follows from Corollary 3.2 that the map $f \mapsto \int_S f d\mu$ can be extended into a linear continuous map to the completion of $(St(\mu, E), \|\cdot\|_1)$. However this completed space is abstract. It is the purpose of the following analysis to represent it by a concrete space $L_1(\mu, E)$ of (equivalence classes of) functions from S to E called *μ-integrable functions*.

Definition 7.2 A function $f : S \to E$ (resp. $f : S \to \mathbb{R}$) is said to be *μ-integrable* (or in short *integrable*) if there exists a Cauchy sequence (f_n) in $St(\mu, E)$ (resp. $St(\mu, \mathbb{R})$) for the seminorm $\|\cdot\|_1$ which converges a.e. to f.

The space of μ-integrable functions from S to E is denoted by $\mathcal{L}_1(S, \mathcal{S}, \mu, E)$ or $\mathcal{L}_1(\mu, E)$ or $\mathcal{L}_1(S, E)$ if there is no ambiguity. For $E := \mathbb{R}$ one writes $\mathcal{L}_1(\mu)$ or $\mathcal{L}_1(S)$.

Remark If $f \in \mathcal{L}_1(\mu, E)$ and if $g : S \to E$ coincides with f a.e. then $g \in \mathcal{L}_1(\mu, E)$ since if (f_n) is a Cauchy sequence in $St(\mu, E)$ converging a.e. to f then (f_n) converges to g a.e. □

To prove that the map $f \mapsto \int f d\mu$ can be extended from $St(\mu, E)$ to $\mathcal{L}_1(\mu, E)$ we need some preliminary results. First, we observe that $\mathcal{L}_1(\mu, E)$ is a linear space: if (f_n) and (g_n) are Cauchy sequences in $St(\mu, E)$ and if $(f_n) \to f$, $(g_n) \to g$ a.e., and $c \in \mathbb{R}$ (or $c \in \mathbb{C}$ is E is a complex space), then $(f_n + cg_n) \to f + cg$ a.e. and, by the triangle inequality, $(f_n + cg_n)$ is a Cauchy sequence. Next, we give a refined convergence property.

Proposition 7.3 *Any Cauchy sequence in $St(\mu, E)$ has a subsequence which converges almost everywhere. More precisely, if (f_n) is an Abel sequence in $St(\mu, E)$, then (f_n) converges almost uniformly to some $f \in \mathcal{L}_1(S, E)$ in the sense that $(f_n) \to f$ a.e. and for every $\varepsilon > 0$ there exists $T \in \mathcal{S}$ such that $\mu(T) < \varepsilon$ and (f_n) converges uniformly to f on $S \backslash T$.*

Proof Since any Cauchy sequence has an Abel subsequence, it suffices to prove the second assertion. Suppose (f_n) is an Abel sequence in $St(\mu, E)$: let $c > 0$ and $q \in [0, 1[$ be such that $\|f_{n+1} - f_n\|_1 \leq cq^n$ for all $n \in \mathbb{N}$. Let $r \in]q, 1[$ and let

$$R_n := \{s \in S : \|f_{n+1}(s) - f_n(s)\|_E \geq r^n\}.$$

Since

$$r^n \mu(R_n) \leq \int_{R_n} \|f_{n+1}(s) - f_n(s)\|_E \, d\mu(s) \leq \int_S \|f_{n+1} - f_n\|_E \, d\mu \leq cq^n,$$

one has $\mu(R_n) \leq c(q/r)^n$. For $m \in \mathbb{N}$ let $T_m := \cup_{n \geq m} R_n$, so that $T_m \in \mathcal{S}$. For any given $\varepsilon > 0$, for m large enough one has $\mu(T_m) \leq cr(r-q)^{-1}(q/r)^m \leq \varepsilon$. Let $T := T_m$. For $s \in S \backslash T$ and $n \geq m$ one has $\|f_{n+1}(s) - f_n(s)\|_E < r^n$, so that (f_n) converges uniformly on $S \backslash T$. Then (f_n) converges on $S \backslash N$, where $N \in \mathcal{S}$ is the intersection of the family (T_m), so that N has measure 0. By definition, the limit f of (f_n) (extended by 0 on N) is in $\mathcal{L}_1(S, E)$. □

Lemma 7.1 *Let (f_n) and (g_n) be Cauchy sequences of $St(\mu, E)$, converging a.e. to the same map f. Then $(\int_S f_n d\mu)$ and $(\int_S g_n d\mu)$ converge and their limits are equal. Moreover, $(\|f_n - g_n\|_1) \to 0$.*

Proof The sequence $(\int_S f_n d\mu)$ is a Cauchy sequence in E since for $n, p \in \mathbb{N}$

$$\left\| \int_S f_n d\mu - \int_S f_p d\mu \right\|_E \leq \int_S \|f_n - f_p\|_E \, d\mu \underset{n,p \to +\infty}{\to} 0.$$

Thus $(\int_S f_n d\mu)$ converges in E. Similarly $(\int_S g_n d\mu)$ converges. Let us show the limits are the same. By the triangle inequality, the sequence $(h_n) := (f_n - g_n)$ is a Cauchy sequence in $(St(\mu, E), \|\cdot\|_1)$ and it converges almost everywhere to 0. Let us prove that $(\|h_n\|_1) \to 0$. Since

$$\left\| \int_S f_n d\mu - \int_S g_n d\mu \right\|_E \leq \int_S \|h_n\|_E \, d\mu = \|h_n\|_1$$

this will prove that $\lim_n \int_S f_n d\mu = \lim_n \int_S g_n d\mu$. We may suppose (h_n) is an Abel sequence rather than a Cauchy sequence since $(\|h_n\|_1) \to 0$ whenever $(\|h_{k(n)}\|_1) \to 0$ for some subsequence $(h_{k(n)})_n$ of $(h_n)_n$.

Given $\varepsilon > 0$ there exists an $m \in \mathbb{N}$ such that $\|h_n - h_p\|_1 < \varepsilon$ for $n, p \geq m$. Let $A \in \mathcal{S}$ be a set of finite measure outside of which h_m vanishes. Then, for $n \geq m$ we have

$$\int_{S \backslash A} \|h_n\|_E \, d\mu = \int_{S \backslash A} \|h_n - h_m\|_E \, d\mu \leq \int_S \|h_n - h_m\|_E \, d\mu \leq \varepsilon. \tag{7.2}$$

Let $c > \|h_m\|_\infty := \sup_{s \in S} \|h_m(s)\|$. The preceding proposition yields a subset $T \in \mathcal{S}$ such that $\mu(T) < \varepsilon/c$ and such that (h_n) converges to 0 uniformly on $A \backslash T$. Let $m' \geq m$ be such that for $n \geq m'$ we have

$$\int_{A \backslash T} \|h_n\|_E \, d\mu \leq \varepsilon \tag{7.3}$$

(we use the fact that $\mu(A \backslash T) < +\infty$). Then, for $n \geq m'$ we have

$$\int_T \|h_n\|_E \, d\mu \leq \int_T (\|h_n - h_m\|_E + \|h_m\|_E) d\mu \tag{7.4}$$

$$\leq \|h_n - h_m\|_1 + \mu(T) \|h_m\|_\infty < 2\varepsilon. \tag{7.5}$$

Gathering relations (7.2), (7.3), (7.5), for $n \geq m'$ we get

$$\|h_n\|_1 = \int_{S\backslash A} \|h_n\|_E \, d\mu + \int_{A\backslash T} \|h_n\|_E \, d\mu + \int_T \|h_n\|_E \, d\mu < 4\varepsilon$$

so that $(\|h_n\|_1) \to 0$. □

The last lemma implies that we can set without ambiguity

$$\int_S f(s) d\mu(s) := \lim_n \int_S f_n(s) d\mu(s)$$

since the right-hand side does not depend on the choice of the Cauchy sequence (f_n) in $St(\mu, E)$ converging to f a.e.; moreover this definition is compatible with the definition of $\int_S f d\mu$ for $f \in St(\mu, E)$ since one can take $f_n = f$ for all n in this case. It implies more.

Proposition 7.4 *Let $f \in \mathcal{L}_1(\mu, E)$ and let (f_n) be a Cauchy sequence of μ-step maps of S into E, converging a.e. to f. Then $\|f(\cdot)\|_E$ is integrable and*

$$\int_S \|f(s)\|_E \, d\mu(s) = \lim_n \int_S \|f_n(s)\|_E \, d\mu(s) = \lim_n \|f_n\|_1 \,, \tag{7.6}$$

$$\left\| \int_S f(s) d\mu(s) \right\|_E \leq \int_S \|f(s)\|_E \, d\mu(s). \tag{7.7}$$

Proof The sequence $(\|f_n(\cdot)\|_E)$ clearly converges to $\|f(\cdot)\|_E$ a.e. and it is a Cauchy sequence in $St(\mu, \mathbb{R})$ since

$$\left| \|f_n(s)\|_E - \|f_p(s)\|_E \right| \leq \|f_n(s) - f_p(s)\|_E \qquad s \in S,$$

$$\left| \int_S \|f_n\|_E \, d\mu - \int_S \|f_p\|_E \, d\mu \right| \leq \int_S \left| \|f_n\|_E - \|f_p\|_E \right| d\mu \leq \int_S \|f_n - f_p\|_E \, d\mu.$$

Thus $\|f(\cdot)\|_E$ is integrable and the definition of $\int_S \|f(\cdot)\|_E \, d\mu$ yields the first announced relation. The second one follows by a passage to the limit in the relation $\left\| \int_S f_n d\mu \right\|_E \leq \int_S \|f_n\|_E \, d\mu$. □

Proposition 7.5 *The function $f \mapsto \|f\|_1 := \int_S \|f(s)\|_E \, d\mu(s)$ is a semi-norm on $\mathcal{L}_1(\mu, E)$. Moreover, $St(\mu, E)$ is dense in $\mathcal{L}_1(\mu, E)$ equipped with $\|\cdot\|_1$: if $f \in \mathcal{L}_1(\mu, E)$ and if (f_n) is a Cauchy sequence in $St(\mu, E)$ converging a.e. to f, then one has $(\|f - f_n\|_1) \to 0$.*

Proof The relations $\|cf\|_1 = |c| \, \|f\|_1$, $\|f + g\|_1 \leq \|f\|_1 + \|g\|_1$ for $c \in \mathbb{R}$, f, $g \in \mathcal{L}_1(\mu, E)$ are obtained from the similar relations in $St(\mu, E)$ by a passage to the limit using relation (7.6). Let $f \in \mathcal{L}_1(\mu, E)$ and let (f_n) be a Cauchy sequence in $St(\mu, E)$ converging to f a.e. Given $\varepsilon > 0$ we can find $k \in \mathbb{N}$ such that

$\|f_m - f_n\|_1 \leq \varepsilon$ for $n \geq m \geq k$. Then, by relation (7.6), for $m \geq k$ we have $\|f - f_m\|_1 = \lim_{n\to\infty} \|f_n - f_m\|_1 \leq \varepsilon$ so that $(\|f - f_m\|_1)_m \to 0$. □

By (7.6) the semi-norm $\|\cdot\|_1$ on $\mathcal{L}_1(\mu, E)$ extends the semi-norm $\|\cdot\|_1$ on $St(\mu, E)$. Since $St(\mu, E)$ is dense in $\mathcal{L}_1(\mu, E)$, the next result shows that $\mathcal{L}_1(\mu, E)$ can be considered as the completion of $St(\mu, E)$.

Theorem 7.1 (Fisher-Riesz) *The space $(\mathcal{L}_1(\mu, E), \|\cdot\|_1)$ is complete, as is its quotient space $(L_1(\mu, E), \|\cdot\|_1)$ by the subspace*

$$\mathcal{N}(\mu, E) := \{f \in \mathcal{L}_1(\mu, E) : \int_S \|f\| \, d\mu = 0\}.$$

Proof Let (f_n) be an Abel sequence in $(\mathcal{L}_1(\mu, E), \|\cdot\|_1)$: for some $c > 0$, $r \in]0, 1[$ we have $\|f_n - f_{n+1}\|_1 \leq cr^n$. Since $St(\mu, E)$ is dense in $\mathcal{L}_1(\mu, E)$, for all $n \in \mathbb{N}$ we can pick some $g_n \in St(\mu, E)$ such that $\|f_n - g_n\|_1 \leq r^n$. Then, since

$$\|g_n - g_{n+1}\|_1 \leq \|g_n - f_n\|_1 + \|f_n - f_{n+1}\|_1 + \|f_{n+1} - g_{n+1}\|_1 \leq (c+2)r^n$$

the sequence (g_n) is an Abel sequence in $(St(\mu, E), \|\cdot\|_1)$. Proposition 7.3 ensures that (g_n) converges a.e. to some function f in $\mathcal{L}_1(\mu, E)$. Then

$$\|f - f_n\|_1 \leq \|f - g_n\|_1 + \|g_n - f_n\|_1 \leq \|f - g_n\|_1 + r^n,$$

and since $(\|f - g_n\|_1) \to 0$ by Proposition 7.5, we get that (f_n) converges to f in $\mathcal{L}_1(\mu, E)$. The last assertion is a general fact about complete spaces. □

It is useful to characterize the subspace $\mathcal{N}(\mu, E)$ and the equivalence relation it induces. We need an extension of Proposition 7.3.

Proposition 7.6 *Let (f_n) be an Abel sequence in $(\mathcal{L}_1(\mu, E), \|\cdot\|_1)$ with limit f. Then (f_n) converges a.e. to f, and given $\varepsilon > 0$, there exists a subset T of S of measure less than ε such that the convergence is uniform on $S \backslash T$.*

Proof Let $c > 0$, $r \in]0, 1[$ be such that $\|f - f_n\|_1 \leq cr^{2n}$ for all $n \in \mathbb{N}$. Changing f and all the f_n's on sets of measure 0, we may assume that they are all measurable. Let S_n be the set of $s \in S$ such that $\|f(s) - f_n(s)\|_E \geq cr^n$. Then

$$cr^n \mu(S_n) \leq \int_{S_n} \|f - f_n\|_E \, d\mu \leq \int_S \|f - f_n\|_E \, d\mu \leq cr^{2n},$$

so that $\mu(S_n) \leq r^n$. Setting $T_k := \cup_{n \geq k} S_n$ and $N := \cap_k T_k$ we have $\mu(T_k) < r^k/(1-r)$ and $\mu(N) = 0$. For $s \in S \backslash T_k$ and $n \geq k$ we have $\|f(s) - f_n(s)\|_E \leq cr^n$ so that (f_n) converges to f uniformly on $S \backslash T_k$ and (f_n) converges pointwise to f on $S \backslash N$. □

We are ready to give a characterization of $\mathcal{N}(\mu, E)$.

Corollary 7.3 *For a function $f : S \to E$, the following assertions are equivalent:*

(a) $f \in \mathcal{N}(\mu, E)$, i.e. $f \in \mathcal{L}_1(S, E)$ and $\|f\|_1 = 0$;
(b) $f = 0$ almost everywhere.

Proof If $f = 0$ a.e., the sequence (f_n) of $St(\mu, E)$ given by $f_n = 0$ for all n is a Cauchy sequence and converges to f a.e. so that $f \in \mathcal{L}_1(S, E)$ and $\|f\|_1 = \lim_n \|f_n\|_1 = 0$ by Proposition 7.5.

Conversely, if $f \in \mathcal{N}(\mu, E)$, the sequence (f_n) with $f_n = 0$ for all n is an Abel sequence in $St(\mu, E)$ and $(\|f - f_n\|_1) \to 0$ since $\|f - f_n\|_1 = \|f\|_1 = 0$ for all n. Proposition 7.6 ensures that $(f - f_n) \to 0$ a.e., so that $f = 0$ a.e. □

Corollary 7.4 *The space $L_1(S, E)$ is the quotient space of $\mathcal{L}_1(S, E)$ by the equivalence relation: $f, g \in \mathcal{L}_1(S, E)$ are equivalent if and only if $f = g$ almost everywhere.*

It is usual and convenient to identify a function f with its equivalence class with respect to the relation of equality a.e. (although that should not be done).

If $T \in \mathcal{S}$ and if $f \in \mathcal{L}_1(\mu, E)$, the function $1_T f$ belongs to $\mathcal{L}_1(\mu, E)$ as is easily seen. Setting

$$\int_T f d\mu = \int_S 1_T f d\mu,$$

and using the linearity of the integral and the relation $1_{A \cup B} f = 1_A f + 1_B f$ when $A, B \in \mathcal{S}$, $A \cap B = \varnothing$, we can see that the property

$$A, B \in \mathcal{S}, A \cap B = \varnothing \Longrightarrow \int_{A \cup B} f d\mu = \int_A f d\mu + \int_B f d\mu$$

is still valid for $f \in \mathcal{L}_1(\mu, E)$. Let us give some calculus rules.

Proposition 7.7 *If $A : E \to F$ is a continuous linear map with values in another Banach space F, then for any $f \in \mathcal{L}_1(S, E)$, the function $A \circ f$ belongs to $\mathcal{L}_1(S, F)$ and*

$$\int_S A(f(s)) d\mu(s) = A\left(\int_S f(s) d\mu(s)\right),$$

$$\left\| \int_S A(f(s)) d\mu(s) \right\|_F \leq \|A\| \cdot \|f\|_1 .$$

Proof When $f \in St(\mu, E)$ the result is an easy consequence in the definition and of the triangle inequality. The general case is obtained by a passage to the limit. □

Corollary 7.5 *Given Banach spaces $E_1, \ldots, E_k$, a function $f : S \to E := E_1 \times \ldots \times E_k$ belongs to $\mathcal{L}_1(\mu, E)$ if and only if its components $f_1, \ldots, f_k$ belong to $\mathcal{L}_1(\mu, E_1), \ldots, \mathcal{L}_1(\mu, E_k)$, respectively.*

Proof Since $f_i = p_i \circ f$, where $p_i : E \to E_i$ is the canonical projection, one has $f_i \in \mathcal{L}_1(\mu, E_i)$ whenever $f \in \mathcal{L}_1(\mu, E)$. The converse is a consequence in the definition, observing that if $(f_{i,n})_n$ is a Cauchy sequence in $St(\mu, E_i)$ converging a.e. to f_i for $i = 1, \ldots, k$ then $f_n := (f_{1,n}, \ldots, f_{k,n})$ is a μ-step function, (f_n) is a Cauchy sequence and $(f_n) \to f$ a.e. □

Thus, the integration of functions with values in $\mathbb{R}^d$ or $\mathbb{C}$ can be reduced to the integration of real-valued functions.

The next consequence shows that the vectorial integral is determined by scalar integrals.

Corollary 7.6 *If $f, g \in \mathcal{L}_1(S, E)$ and if for all $e^* \in E^*$ one has $\int_S e^* \circ f d\mu = \int_S e^* \circ g d\mu$ then one has $\int_S f d\mu = \int_S g d\mu$.*

Proof Since for all $h \in \mathcal{L}_1(S, E)$ and all $e^* \in E^*$ one has $e^*(\int_S h d\mu) = \int_S e^* \circ h d\mu$, the result follows from a consequence in the separation theorem ensuring that for $e_1, e_2 \in E$ one has $e_1 = e_2$ whenever $e^*(e_1) = e^*(e_2)$ for all $e^* \in E^*$. □

Let us end this subsection with an extension of the construction of the integral to the case of functions with values in $\overline{\mathbb{R}}$. Such an extension is useful, in particular for convergence questions.

Definition 7.3 A function $f : S \to \overline{\mathbb{R}}$ is said to be *integrable* if there exists some $g \in \mathcal{L}_1(S, \mathbb{R})$ such that $f = g$ almost everywhere.

More generally, given a Banach space E, a null subset N of S and a function $f : S \backslash N \to E$ one says that f is *integrable* if there exists some $g \in \mathcal{L}_1(S, E)$ such that $f = g$ a.e. on $S \backslash N$.

In both cases one sets

$$\int_S f d\mu = \int_S g d\mu.$$

Clearly this value does not depend on the choice of g. One can verify that the properties of the present subsection can be extended to the present case. However, one has to be careful with calculus rules and take into account the convention $(+\infty) + (-\infty) := +\infty$.

If $f : S \to \overline{\mathbb{R}}$ is an *almost measurable* function (in short, a.m. function) in the sense that it coincides a.e. with a real-valued measurable function, we define $\int_S f d\mu$ to be $\int_S g d\mu$ if there is an integrable function $g : S \to \mathbb{R}$ such that $f = g$ a.e. and $\int_S f d\mu = +\infty$ if there is no such function. This definition of the (upper) integral of f does not depend on the choice of g in $\mathcal{L}_1(S, \mathbb{R})$ among those satisfying $g = f$ a.e. Moreover, one sees that for a.m. functions f, f' with $f \leq f'$ one has $\int_S f d\mu \leq \int_S f' d\mu$.

Alternatively, one sees that the integral of an a.m. function $f : S \to \overline{\mathbb{R}}$ is

$$\int_S f d\mu := \int_S f(s)\mu(ds) = \inf\{\int_S h(s) d\mu(s) : h \in \mathcal{L}_1(S, \mathbb{R}),\ h \geq f \text{ a.e.}\},$$

with our standing convention $\inf \varnothing := +\infty$. In fact, if $g \in \mathcal{L}_1(S, \mathbb{R})$ is such that $f = g$ a.e. we can take $h = g$ in the right-hand side, so that the infimum is less than or equal to $\int_S f d\mu$; on the other hand, if $h \in \mathcal{L}_1(S, \mathbb{R})$ is such that $h \geq f$ a.e. and if $g : S \to \mathbb{R}$ is integrable with $f = g$ a.e., changing g, h to some measurable functions g', h' such that $g' = g$, $h' = h$ a.e., the preceding corollary ensures that $\int_S h d\mu \geq \int_S g d\mu$, so that the infimum is greater than or equal to $\int_S f d\mu$. Moreover, with the convention $(+\infty) - (+\infty) = +\infty$, one can easily verify that

$$\int_S f(s)\mu(ds) = \int_S \max(f(s), 0) d\mu(s) - \int_S \max(-f(s), 0) d\mu(s).$$

Thus, the only case in which $\int_S f d\mu$ differs from $-\int_S (-f) d\mu$ is when both $\int_S f d\mu$ and $\int_S (-f) d\mu$ are $+\infty$.

The next lemmas will be useful for the study of convergence properties. They are also of independent interest.

Lemma 7.2 *For $f \in \mathcal{L}_1(\mu, E)$, $c > 0$ let $S_c := \{s \in S : \|f(s)\|_E \geq c\}$. Then there exists a $T \in \mathcal{S}$ with $\mu(T) < +\infty$ and a null set N such that $S_c = T \backslash N$. Moreover, if f is measurable then S_c is measurable with a finite measure.*

Taking a sequence $(c_n) \to 0_+$ we see that for every $f \in \mathcal{L}_1(\mu, E)$ there exist a sequence (T_n) in $\mathcal{S}$ and a null set N such that $\mu(T_n) < +\infty$ for all $n \in \mathbb{N}$ and $f = 0$ on $S \backslash (T \cup N)$ where $T = \cup_n T_n$. When $(S, \mathcal{S}, \mu)$ is complete, the preceding assertions can be simplified.

Proof Let us first suppose $f \in \mathcal{L}_1(\mu, E)$ is measurable. Then $\|f(\cdot)\|_E$ is measurable, so that S_c is measurable. Let (f_n) be an Abel sequence in $St(\mu, E)$ converging a.e. to f. Proposition 7.3 ensures that for every $\varepsilon > 0$ there exists some $Z \in \mathcal{S}$ with $\mu(Z) < \varepsilon$ such that (f_n) converges to f uniformly on $S \backslash Z$. Thus, for n large enough and all $s \in S_c \backslash Z$, we have $\|f_n(s)\|_E \geq c/2$ so that

$$\int_S \|f_n\|_E d\mu \geq \int_{S_c \backslash Z} \|f_n\|_E d\mu \geq (c/2)\mu(S_c \backslash Z)$$

and $S_c \backslash Z$ and S_c have finite measures.

Now let f be an arbitrary element of $\mathcal{L}_1(\mu, E)$. Since f is μ-measurable there exist a measurable map $g : S \to E$ and a null set $M \in \mathcal{S}$ such that $f = g$ on $S \backslash M$. Let $T_c := \{s \in S : \|g(s)\|_E \geq c\}$, so that $T_c \in \mathcal{S}$ and $\mu(T_c)$ is finite. We have $S_c \backslash M = T_c \backslash M$, so that, setting $T := T_c \backslash M \in \mathcal{S}$, we get $\mu(T) < \infty$, $S_c = T \cup (S_c \cap M)$, and $N := S_c \cap M$ is a null set. □

Lemma 7.3 *Let $f : S \to \overline{\mathbb{R}}_+$ and let (f_n) be an increasing sequence of μ-step functions converging a.e. to f. Then f is integrable if and only if the sequence $(\int_S f_n d\mu)$ is bounded above.*

In such a case, one has $\int_S f d\mu = \sup_n \int_S f_n d\mu = \lim_n \int_S f_n d\mu$.

Proof Let (f_n) be an increasing sequence of μ-step functions converging a.e. to f. Suppose first that $m := \sup_n(\int_S f_n d\mu) < +\infty$. Then, for $p > n$ in $\mathbb{N}$ one has $\int_S |f_p - f_n|\, d\mu = \int_S f_p d\mu - \int_S f_n d\mu$, so that (f_n) is a Cauchy sequence and f is integrable. Then, by definition, $\int_S f d\mu = \lim_n \int_S f_n d\mu$. Now suppose $m = +\infty$ and let us show that assuming that f is integrable leads to a contradiction. In fact, if (g_n) is a Cauchy sequence of μ-step functions converging a.e. to f, fixing $k \in \mathbb{N}$ and setting $h_n := \max(f_k, g_n)$, for $p \geq n$ one has $|h_p - h_n| \leq |g_p - g_n|$, so that (h_n) is a Cauchy sequence, hence is convergent in $\mathcal{L}_1(S, \mathbb{R})$, and $\left(\int_S h_n d\mu\right)$ converges in $\mathbb{R}$. Since $\int_S h_n d\mu \geq \int_S f_k d\mu$ and $\sup_k \int_S f_k d\mu = +\infty$, we get a contradiction. □

Remark If $f : S \to \overline{\mathbb{R}}_+ := [0, +\infty]$ is measurable, one can always find an increasing sequence (f_n) of step functions which converges a.e. to f: setting for $n \in \mathbb{N}$, $k \in \mathbb{N}$ with $0 \leq k \leq 4^n$

$$S_{k,n} := f^{-1}([k2^{-n}, (k+1)2^{-n}[) \text{ if } k < 4^n, \quad S_{k,n} := f^{-1}([2^n, +\infty]) \text{ if } k = 4^n,$$

$$f_n(s) = k2^{-n} \text{ if } s \in S_{k,n},$$

one gets a partition $(S_{k,n})_k$ of S by measurable sets and a step function f_n. The sequence (f_n) is increasing. The step function f_n is a μ-step function if $\mu(S_{k,n}) < +\infty$ for $k \in [1, 4^n]$. □

Corollary 7.7 *If $f : S \to \overline{\mathbb{R}}_+$ is measurable, if $g : S \to \overline{\mathbb{R}}_+$ is integrable and if $f \leq g$ a.e., then f is integrable and $\int_S f d\mu \leq \int_S g d\mu$.*

Proof Let us first observe that if $k : S \to \overline{\mathbb{R}}_+$ is integrable, then $\int_S k d\mu \geq 0$ since for any Cauchy sequence $(k_n) \to k$ a.e. the sequence (k_n^+) with $k_n^+ := \max(k_n, 0)$ is still Cauchy, $(k_n^+) \to k$ a.e., and $k_n^+ \in St(\mu, \mathbb{R})$ if $k_n \in St(\mu, \mathbb{R})$.

Now, let h be a step function such that $0 \leq h \leq f$ and let $T := \{s : h(s) > 0\}$, $\alpha := \min\{h(t) : t \in T\}$. Since $h \leq g$ and g is integrable, $T \subset \{s \in S : g(s) \geq \alpha\}$ and T has finite measure. Hence h is a μ-step function. Thus there exists an increasing sequence (f_n) of μ-step functions that converges a.e. to f. Let $k_n := g - f_n$. Then k_n is integrable by the preceding remark, and $\int_S k_n d\mu \geq 0$ by the first observation of the proof. Thus, the sequence $(\int_S f_n d\mu) = (\int_S g d\mu - \int_S k_n d\mu)$ is bounded above by $\int_S g d\mu$. The preceding lemma asserts that f is integrable and $\int_S f d\mu = \sup_n \int_S f_n d\mu \leq \int_S g d\mu$. □

Exercises

1. Using a common refinement of two finite partitions, show that the integral of a μ-step map f does not depend on the choice of the partition used to write f.
2. Verify that the set $St(\mu, E)$ of μ-step maps from S into a normed space E is a linear space and that the integral is linear on $St(\mu, E)$.

3. If $(S, \mathcal{S}, \mu)$ is a σ-finite measure space, show that there exists an integrable function $u : S \to \mathbb{R}$ with positive values such that $\int_S u(s)d\mu(s) = 1$. [Hint: Given a sequence (S_n) in $\mathcal{S}$ such that $S = \cup_n S_n$ and $c_n := \mu(S_n) \in]0, +\infty[$ for all n, set $u = \Sigma_n (2^{-n}/c_n) 1_{S_n}$.]

4. Let $S := \mathbb{R}$ equipped with the Lebesgue measure λ. Give an example of a sequence (f_n) of integrable functions uniformly converging to a function f that is not integrable.

Show that if all the functions f_n are null on the complement of a measurable subset T of S of finite measure, then f is integrable.

Give an example of a sequence of integrable functions that are null on the complement of a measurable subset T of S of finite measure and that pointwise converges to a function f that is not integrable.

5. Let $(S, \mathcal{S}, \mu)$ be a measure space and let f, g be two integrable functions on S such that

$$\mu(\{f > r\} \Delta \{g > r\}) = 0 \qquad \text{a.e. } r \in \mathbb{R}.$$

Show that $f = g$ a.e.

6. Let μ_c be the *counting measure* of $(\mathbb{N}, \mathcal{P}(\mathbb{N}))$ defined by $\mu_c(A) = \text{card}(A)$ if $A \in \mathcal{P}(\mathbb{N})$ is finite and $\mu_c(A) := +\infty$ if A is infinite. Show that for $f : \mathbb{N} \to \mathbb{R}_+$ one has $\int f d\mu_c = \Sigma_n f(n)$.

7. (*Jensen's Inequality*) Let $(S, \mathcal{S}, \mu)$ be a measure space with finite measure, let T be an open interval of $\mathbb{R}$ and let $g : T \to \mathbb{R}$ be a convex function. Prove that for all integrable function $g : S \to \mathbb{R}$ with values in T the following inequality holds:

$$g(\frac{1}{\mu(S)} \int_S f(s)d\mu(s)) \leq \frac{1}{\mu(S)} \int_S g(f(s))d\mu(s)).$$

Interpret it in terms of averages and reduce it to the case μ is a probability. [Hint: assuming $\mu(S) = 1$, let $a := \int_S f d\mu$ and let $c := g'_\ell(a)$. Verify that $g(t) - g(a) \geq c(t - a)$ for all $t \in T$; take $t := f(s)$ and integrate.]

8. Let $g : \mathbb{R} \to \mathbb{R}$ be continuous and such that for any bounded measurable function f on a bounded interval S of $\mathbb{R}$ endowed with the measure μ induced by the Lebesgue measure the inequality of the preceding exercise holds. Prove that g is convex.

9. (*Markov's Inequality*). Let $(S, \mathcal{S}, \mu)$ be a measure space and let $f : S \to \mathbb{R}_+$ be measurable. Show that for all $c > 0$ one has $\mu(\{f \geq c\}) \leq (1/c) \int f d\mu$.

Deduce from this inequality that if f is integrable then $\mu(\{f = +\infty\}) = 0$. [Hint: use the continuity property of a measure and note that $\{f = +\infty\} = \cap_n \{f \geq n\}$.]

7.3 Approximation of Integrable Functions

It is natural to wonder whether $\mathcal{L}_1(\mu, E) := \mathcal{L}_1(S, \mathcal{S}, \mu, E)$ remains the same if we replace $\mathcal{S}$ with a smaller family $\mathcal{R}$ of subsets of S and μ with the restriction λ of μ to $\mathcal{R}$. If $S = \mathbb{R}^d$ for instance, instead of taking for $\mathcal{S}$ the Borel algebra, one may wish to use the ring $\mathcal{A}$ generated by the semi-ring $\mathcal{R}$ of semi-closed rectangles, i.e. by Lemma 1.4, the ring $\mathcal{A}$ formed by the unions of finite families of disjoint elements of the semi-ring $\mathcal{R}$. Let us denote by $St(\lambda, E)$ (resp. $St(\lambda, \mathbb{R})$) the set of step maps from S to E (resp. $\mathbb{R}$) with respect to the restriction λ of μ to $\mathcal{R}$. We say that μ is *σ-finite* with respect to λ (or $\mathcal{R}$) if every $T \in \mathcal{S}$ with finite measure is contained in the union of a countable family of elements of $\mathcal{R}$ with finite measures. An answer to the above question is as follows.

Theorem 7.2 *Let $\mathcal{R}$ be a semi-ring generating the σ-algebra $\mathcal{S}$. If μ is σ-finite with respect to its restriction λ to $\mathcal{R}$, then $St(\lambda, E)$ is dense in $\mathcal{L}_1(\mu, E)$ and $\mathcal{L}_1(\lambda, E) = \mathcal{L}_1(\mu, E)$.*

The proof relies on two lemmas.

Lemma 7.4 *Let R be an element of $\mathcal{R}$ with finite measure and let*

$$\mathcal{S}_R := \{T \in \mathcal{S} : T \subset R, \ 1_T \in \mathrm{cl}(St(\lambda, \mathbb{R}))\}.$$

Then $\mathcal{S}_R$ considered as a family of subsets of R is a σ-algebra.

Proof Since $1_R \in St(\lambda, \mathbb{R})$ as $\mu(R) < +\infty$ we have $R \in \mathcal{S}_R$. Since $1_{R\backslash T} = 1_R - 1_T$ for $T \subset R$ and since $\mathcal{L}_R := \mathrm{cl}(St(\lambda, \mathbb{R}))$ is a linear subspace of $\mathcal{L}_1(\mu, \mathbb{R})$, we have $R\backslash T \in \mathcal{S}_R$ whenever $T \in \mathcal{S}_R$. By Proposition 1.5 it suffices to prove that $\cup_n T_n \in \mathcal{S}_R$ whenever $T_n \in \mathcal{S}_R$ for all $n \in \mathbb{N}$. We first show that $T \cup T' \in \mathcal{S}_R$ if $T \in \mathcal{S}_R$ and $T' \in \mathcal{S}_R$. Given $\varepsilon > 0$, we pick $g, g' \in St(\lambda, \mathbb{R})$ such that $\|1_T - g\|_1 < \varepsilon/2$, $\|1_{T'} - g'\|_1 < \varepsilon/2$. We use the relation

$$\left|\sup(r, s) - \sup(r', s')\right| \leq \left|r - r'\right| + \left|s - s'\right|$$

for all $r, r', s, s' \in \mathbb{R}_+$ to get that $\|\sup(1_T, 1_{T'}) - \sup(g, g')\|_1 < \varepsilon$. Since $\sup(1_T, 1_{T'}) = 1_{T \cup T'}$ and $\sup(g, g') \in St(\lambda, \mathbb{R})$ we see that $T \cup T' \in \mathcal{S}_R$. Thus $T \cap T' = R\backslash(R\backslash T) \cup (R\backslash T')) \in \mathcal{S}_R$ and $T'\backslash T = T' \cap (R\backslash T)$. Therefore, when considering $T = \cup_n T_n$ we may suppose the sets $T_n \in \mathcal{S}_R$ are disjoint. Then, given $\varepsilon > 0$, we pick $g_n \in St(\lambda, \mathbb{R})$ such that

$$\|1_{T_n} - g_n\|_1 < \varepsilon/2^n,$$

so that, for n large enough we have $|\mu(T) - \mu(T_1 \cup \ldots \cup T_n| \leq \varepsilon/2$ and

$$\left\|1_T - \sum_{k=1}^{n} g_k\right\|_1 \leq \|1_T - 1_{T_1 \cup \ldots \cup T_n}\|_1 + \left\|1_{T_1 \cup \ldots \cup T_n} - \sum_{k=1}^{n} g_k\right\|_1 \leq \varepsilon.$$

Thus $T \in \mathcal{S}_R$ and $\mathcal{S}_R$ is a σ-algebra. □

Let us denote by $\mathcal{S}'$ the family of subsets T of S such that $T \cap R \in \mathcal{S}_R$ for all $R \in \mathcal{R}$. Lemma 1.2 shows that $\mathcal{S}'$ is a σ-algebra. Since it obviously contains $\mathcal{R}$, we have $\mathcal{S}' = \mathcal{S}$. Let us prove more.

Lemma 7.5 *If μ is σ-finite with respect to λ, then for all $T \in \mathcal{S}$ with finite measure one has $1_T \in \mathrm{cl}(St(\lambda, \mathbb{R}))$.*

Proof Let $T \in \mathcal{S}$ with finite measure. By assumption one can find a sequence (R_n) in $\mathcal{S}$ such that $T \subset \cup_n R_n$ and $\mu(R_n) < +\infty$ for all $n \in \mathbb{N}$. Since for all n we have $T \cap R_n \in \mathcal{S}_R$ by the preceding remark, given $\varepsilon > 0$ we can find $g_n \in St(\lambda, \mathbb{R})$ such that

$$\|1_{T\cap R_n} - g_n\|_1 \le \varepsilon/2^n.$$

Since $T = \cup_n (T \cap R_n)$ we can find some n such that $\mu(T \backslash (R_1 \cup \ldots \cup R_n)) < \varepsilon/2$. Thus

$$\left\| 1_T - \sum_{k=1}^{n} g_k \right\|_1 \le \left\| 1_T - \sum_{k=1}^{n} 1_{T\cap R_k} \right\|_1 + \left\| \sum_{k=1}^{n} 1_{T\cap R_k} - \sum_{k=1}^{n} g_k \right\|_1 \le \varepsilon$$

as $\|1_T - (1_{T\cap R_1} + \ldots + 1_{T\cap R_n})\|_1 = \mu(T \backslash (R_1 \cup \ldots \cup R_n)) < \varepsilon/2$. Therefore $1_T \in \mathrm{cl}(St(\lambda, \mathbb{R}))$. □

Now let us show that $St(\mu, E) \subset \mathrm{cl}(St(\lambda, E))$. Since $St(\mu, E)$ is dense in $\mathcal{L}_1(\mu, E)$, this will prove that $\mathrm{cl}(St(\lambda, E)) = \mathcal{L}_1(\mu, E)$ and that $\mathcal{L}_1(\lambda, E) = \mathcal{L}_1(\mu, E)$. Let $f \in St(\mu, E)$:

$$f = 1_{S_1} e_1 + \ldots + 1_{S_m} e_m$$

with $e_i \in E \backslash \{0\}$ and $S_i \in \mathcal{S}$ with $\mu(E_i) < +\infty$ for $i \in \mathbb{N}_m$. Given $\varepsilon > 0$ for all $i \in \mathbb{N}_m$ we can find some $g_i \in St(\lambda, \mathbb{R})$ such that

$$\|1_{S_i} - g_i\|_1 \le \varepsilon/(m\,\|e_i\|).$$

Then we get $\|f - (g_1 e_1 + \ldots + g_m e_m)\|_1 \le \varepsilon$. Thus $f \in \mathrm{cl}(St(\lambda, E))$. □

Let us apply Theorem 7.2 to the approximation of integrable functions on $\mathbb{R}^d$ by smooth functions. Given $r < s$ in $\mathbb{R}$ we denote by b the bell-shaped function on $\mathbb{R}$ defined by

$$b(t) := e^{\frac{-1}{(t-r)(s-t)}} \quad \text{if } t \in]r, s[\qquad b(t) = 0 \quad \text{otherwise}.$$

It can be checked that b is of class C^∞ and $a := \int b > 0$. The function $c : \mathbb{R} \to \mathbb{R}$ given by

$$c(x) := \frac{1}{a} \int_{-\infty}^{x} b(t)dt$$

starts with the value 0 untill r and then climbs to the constant value 1 reached for $x \geq s$. Thus, for $q \in [r, s]$, it can be considered as an approximation of $1_{[q,+\infty[}$, the shifted Heaviside function. Changing r and s we can make the climb as steep as necessary. Combining c with a shifted function of $x \mapsto c(-x)$, we get an approximation g of the characteristic function of $[r, s]$ which differs from $1_{[r,s]}$ on small intervals around the extremities of $[r, s]$ and takes its values in $[0, 1]$. Therefore, given $\varepsilon > 0$, we can choose g in such a way that $\left\| 1_{[r,s]} - g \right\|_1 \leq \varepsilon$ and g is of class C^∞ with a compact support. If R is the rectangle $[r_1, s_1] \times \cdots \times [r_d, s_d]$ in $\mathbb{R}^d$, using approximations g_i of $1_{[r_i,s_i]}$ for $i \in \mathbb{N}_d$ and setting $g(x) := g_1(x_1) \cdots g_d(x_d)$ for $x := (x_1, \cdots, x_d)$, we get an approximation of 1_R. If D belongs to the ring $\mathcal{R}$ generated by the class $\mathcal{C}$ of semi-closed rectangles, we easily get an approximation of 1_D since D is the union of a finite family of disjoint semi-closed rectangles. Combining this construction with Theorem 7.2, we get the following useful result.

Theorem 7.3 *The space $C_c^\infty(\mathbb{R}^d)$ of functions of class C^∞ with compact support is dense in the space $(L_1(\mathbb{R}^d), \|\cdot\|_1)$.*

Exercises

1. Let $(X, \mathcal{S}, \mu)$ be a measure space, let E be a Banach space, and let $\mathcal{R}$ be a semi-ring generating the σ-algebra $\mathcal{S}$. Assume that μ is σ-finite with respect to the restriction λ of μ to $\mathcal{R}$. Let $f \in \mathcal{L}_1(\mu, E)$ be such that $\int_R f d\mu = 0$ for all $R \in \mathcal{R}$. Show that $f = 0$ a.e. [Hint: use Theorem 7.2 and the Dominated Convergence Theorem of the next section.]
2. Let λ_d be the Lebesgue measure on $\mathbb{R}^d$ and let E be a Banach space. Prove that if $f \in \mathcal{L}_1(\lambda_d, E)$ is such that $\int fg d\lambda_d = 0$ for all $g \in C_c^\infty(\mathbb{R}^d)$ then $f = 0$ a.e. [Hint: use the preceding exercise, the Dominated Convergence Theorem, and observe that the characteristic function of a semi-closed rectangle R can be approximated by functions of $C_c^\infty(\mathbb{R}^d)$ taking their values in $[0, 1]$ and pointwise converging to 1_R.]
3. Prove that when $\mu(S) > 0$, for $u \in L_1(S, \mathbb{R})$ with $u(s) > 0$ for all $s \in S$, one has $\int_S u(s)d\mu(s) > 0$. [Hint: consider the family of sets $S_n := \{s \in S : u(s) \geq 2^{-n}\}$ for $n \in \mathbb{N}$ and observe that $\int_S u(s)d\mu(s) \geq 2^{-n}\mu(S_n)$ and for at least one $n \in \mathbb{N}$ one has $\mu(S_n) > 0$.]
4. (*Lusin's Theorem*) Let μ be a regular Borel measure on a metric space (X, d) and let $f \in L_1(\mu)$. Prove that for every $\varepsilon > 0$ there exists some $g \in C(X) \cap L_1(\mu)$ such

that $\|f - g\|_1 \leq \varepsilon$, $\|g\|_\infty \leq \|f\|_\infty$ and $\mu(\{f \neq g\}) \leq \varepsilon$. [Hint: use Urysohn's Lemma.]

7.4 Convergence Results

For the study of integrals of real-valued functions on a measure space $(S, \mathcal{S}, \mu)$ one can make use of order properties. This is also the case for functions with values in $\overline{\mathbb{R}} := \mathbb{R} \cup \{-\infty, +\infty\}$. In view of the compactness of $\overline{\mathbb{R}}$ it is convenient to consider such functions, in particular for convergence questions. Then one can use the fact that any increasing sequence (f_n) of functions with values in $\overline{\mathbb{R}}$ has a limit, with $\lim_n f_n(x) = +\infty$ if $(f_n(x))_n$ is unbounded. We devote this section to convergence results for such functions. However, in order to avoid complications with null sets, we start with the simple case of measurable functions for a notion that is more often used in probability theory than in analysis.

Definition 7.4 Let $(S, \mathcal{S}, \mu)$ be a measure space and let $(E, \|\cdot\|)$ be a Banach space. A sequence (f_n) of measurable maps from S to E is said *to converge to* $f : S \to E$ *in measure* if for all $c > 0$ one has $\lim_n \mu(\{\|f_n - f\| > c\}) = 0$, where $\{\|f_n - f\| > c\}$ stands for $\{x \in S : \|f_n(x) - f(x)\| > c\}$.

Proposition 7.8 *If* $f \in \mathcal{L}_1(S, E)$ *and if* (f_n) *is a sequence in* $\mathcal{L}_1(S, E)$ *such that* $(\|f_n - f\|_1) \to 0$, *then* (f_n) *converges to* f *in measure.*

Proof This stems from the inequality $\mu(\{\|f_n - f\| > c\}) \leq (1/c)\|f_n - f\|_1$ for all n. □

Proposition 7.9 *If* $(f_n) \to f$ *a.e. and if* $\mu(S) < +\infty$, *then* $(f_n) \to f$ *in measure.*

Proof Given $c > 0$, let $A_n := \{\|f_n - f\| > c\}$ and let $B_k := \cup_{n \geq k} A_n$. The sequence (B_k) is decreasing and its intersection is contained in the set N of $s \in S$ such that $(f_n(s))$ does not converge to $f(s)$. Thus $\mu(\cap_k B_k) = 0$, hence $\lim_k \mu(B_k) = 0$ (Lemma 1.5). Since $A_k \subset B_k$ we get $\lim_k \mu(A_k) = 0$: (f_n) converges to f in measure. □

Proposition 7.10 *If* $(f_n) \to f$ *in measure then a subsequence of* (f_n) *converges to* f *a.e.*

Proof Using the definition of convergence in measure, we construct inductively an increasing sequence $(n(k))_{k \geq 1}$ of $\mathbb{N}$ such that

$$n \geq n(k) \Longrightarrow \mu(\{\|f_n - f\| > 1/k\}) \leq 2^{-k}.$$

Setting $A_k := \{\|f_{n(k)} - f\| > 1/k\}$ and $A := \cap_{j \geq 1} \cup_{k \geq j} A_k$ we see that for all $s \in S \backslash A$ there exists $j \geq 1$ such that $s \notin A_k$ for all $k \geq j$, i.e. $\|f_{n(k)}(s) - f(s)\| \leq 1/k$ for $k \geq j$.

This means that $(f_{n(k)}(s))_k \to f(s)$ for all $s \in S\backslash A$. Since for all $j \geq 1$

$$\mu(\bigcup_{k\geq j} A_k) \leq \sum_{k\geq j} \mu(A_k) \leq \sum_{k\geq j} 2^{-k} = 2^{-j+1}$$

we get that $\mu(A) = \lim_j \mu(\cup_{k\geq j} A_k) = 0$: $(f_{n(k)}) \to f$ almost everywhere. □

Let us gather some more classical convergence results. They are most useful. They can be extended to the case when the functions (f_n) are almost measurable with values in $\overline{\mathbb{R}}$.

Theorem 7.4 (Monotone Convergence Theorem) *Let (f_n) be a sequence in $\mathcal{L}_1(S, \mathbb{R})$ such that $f_n \leq f_{n+1}$ a.e. and let $f(s) := \lim_n f_n(s)$. Then f is integrable if and only if $(\int_S f_n d\mu)$ is bounded in $\mathbb{R}$. In such a case $(\|f_n - f\|_1) \to 0$ as $n \to \infty$ and $(\int_S f_n d\mu) \to \int_S f d\mu$ as $n \to \infty$.*

Proof Suppose f is integrable (in the sense of Definition 7.3). Then for all $n \in \mathbb{N}$ one has $\int_S f_0 d\mu \leq \int_S f_n d\mu \leq \int_S f d\mu$, so that $(\int_S f_n d\mu)$ is bounded. Conversely, if the nondecreasing sequence $(\int_S f_n d\mu)$ is bounded, it is convergent, hence Cauchy. Thus, given $\varepsilon > 0$, for n, p in $\mathbb{N}$ large enough with $p \geq n$, one has

$$\|f_p - f_n\|_1 = \int_S (f_p - f_n) d\mu = \int_S f_p d\mu - \int_S f_n d\mu \leq \varepsilon.$$

Since $\mathcal{L}_1(S, \mathbb{R})$ is complete, (f_n) converges to some $g \in \mathcal{L}_1(S, \mathbb{R})$ for the semi-norm $\|\cdot\|_1$. Proposition 7.6 ensures that (f_n) has a subsequence converging a.e. to g. The monotonicity of (f_n) implies that the whole sequence (f_n) converges a.e. to g. Thus $g = f$ a.e. and f is integrable. The convergence $(\int_S f_n d\mu) \to \int_S f d\mu$ follows from $\left|\int_S f_n d\mu - \int_S f d\mu\right| \leq \int_S |f_n - f| \, d\mu$ and $(\|f_n - f\|_1) \to 0$ as $n \to \infty$. □

Corollary 7.8 (Fatou's Lemma) *Let (f_n) be a sequence in $\mathcal{L}_1(S, \mathbb{R})$ such that $f_n(s) \geq 0$ a.e. for all $n \in \mathbb{N}$. If $\liminf_n \|f_n\|_1 < +\infty$, then $f := \liminf_n f_n$ is integrable and*

$$\int_S f(s) d\mu(s) \leq \liminf_n \int_S f_n(s) d\mu(s).$$

Consequently, the semi-norm $\|\cdot\|_1$ is lower semicontinuous on the positive cone of $\mathcal{L}_1(S, \mathbb{R})$ with respect to a.e. convergence.

Proof For $m, p \in \mathbb{N}$ with $m \leq p$, let $g_{m,p} := \inf_{m \leq n \leq p} f_n$, let $g_m := \inf_{p \geq m} g_{m,p} = \inf_{n \geq m} f_n$, so that $f = \lim_m g_m$. Since $(g_{m,p})_{p \geq m}$ is decreasing and $(\int_S g_{m,p} d\mu)_{p \geq m}$ is bounded below by 0, the Monotone Convergence Theorem ensures that g_m is integrable. Moreover,

$$\int_S g_m d\mu \leq \inf_{n \geq m} \int_S f_n d\mu = \inf_{n \geq m} \|f_n\|_1 \leq c := \liminf_n \|f_n\|_1 .$$

Again the Monotone Convergence Theorem ensures that $f = \lim_m g_m$ is integrable and $\int_S f d\mu \leq c$. □

The next result is a cornerstone of Lebesgue integration theory.

Theorem 7.5 (Lebesgue's Dominated Convergence Theorem) *Let (f_n) be a sequence in $\mathcal{L}_1(S, E)$ such that there exists an $h \in \mathcal{L}_1(S, \mathbb{R})$ satisfying $\|f_n(s)\|_E \leq h(s)$ a.e. $s \in S$ for all $n \in \mathbb{N}$. If (f_n) converges a.e. to some map f, then f is in $\mathcal{L}_1(S, E)$ and (f_n) converges to f in $\mathcal{L}_1(S, E)$. Moreover, $(\int_S f_n d\mu) \to \int_S f d\mu$.*

Proof For $k, p \in \mathbb{N}$ with $k \leq p$, let $g_{k,p} := \sup_{k \leq m,n \leq p} \|f_m - f_n\|_E$, $g_k := \sup_{m,n \geq k} \|f_m - f_n\|_E$, so that $0 \leq g_{k,p} \leq g_k \leq 2h$. Since $(g_{k,p})_p$ is nondecreasing and (g_k) is nonincreasing, the Monotone Convergence Theorem ensures that $g_k = \sup_{p \geq k} g_{k,p}$ and $g := \inf_k g_k$ are integrable and $(\|g_k - g\|_1) \to 0$ as $k \to \infty$. However, since (f_n) converges a.e. to some map f, we have $g = 0$ a.e. Thus $(\|g_k\|_1) \to 0$ as $k \to \infty$ and (f_n) is a Cauchy sequence in $\mathcal{L}_1(S, E)$. By Proposition 7.6 (f_n) has a subsequence which converges a.e. and in $\mathcal{L}_1(S, E)$ to some element f' of $\mathcal{L}_1(S, E)$. Then $f' = f$ a.e. and $f \in \mathcal{L}_1(S, E)$. □

Corollary 7.9 *Let $f : S \to E$ be μ-measurable. If there exists some $g \in \mathcal{L}_1(S, \mathbb{R})$ such that $\|f(\cdot)\|_E \leq g(\cdot)$ almost everywhere then $f \in \mathcal{L}_1(S, E)$.*

In particular, $f \in \mathcal{L}_1(S, E)$ if and only if $\|f(\cdot)\|_E$ is in $\mathcal{L}_1(S, \mathbb{R})$.

Proof Changing f and g on a set of measure 0, we may assume g is measurable. Let N be a measurable set of measure 0 and let (f_n) be a sequence in $St(\mu, E)$ such that $(f_n) \to f$ on $S \backslash N$. Given $a > 1$, for $n \in \mathbb{N}$ let

$$S_n := \{x \in S : ag(x) - \|f_n(x)\| \geq 0\}$$

and let $f'_n := 1_{S_n} f_n$. Then S_n is measurable and one has $f'_n \in \mathcal{L}_1(S, E)$. Let us observe that for all $x \in S \backslash N$ we have $(f'_n(x)) \to f(x)$: if $g(x) = 0$ we have $f(x) = 0$ and either $x \in S_n$ so that $f'_n(x) = f_n(x) = 0$ or $x \notin S_n$ and $f'_n(x) = 0$ whereas if $g(x) > 0$, for n large enough we have $x \in S_n$ and $f'_n(x) = f_n(x)$. Since $\|f'_n(x)\| \leq ag(x)$ for all $x \in S$ the Dominated Convergence Theorem entails that $f \in \mathcal{L}_1(S, E)$.

The second assertion is a consequence in the first one and of Proposition 7.4. □

Corollary 7.10 *Let (f_n) be a sequence in $\mathcal{L}_1(S, E)$ which converges a.e. to some map f. If there exists some $c \in \mathbb{R}$ such that $\|f_n\|_1 \leq c$ for all $n \in \mathbb{N}$, then $f \in \mathcal{L}_1(S, E)$ and $\|f\|_1 \leq c$.*

Proof Since $(g_n) := (\|f_n(\cdot)\|_E) \to g := \|f(\cdot)\|_E$ a.e. and $\|g_n\|_1 = \|f_n\|_1 \leq c$ for all n, Fatou's lemma ensures that $g \in \mathcal{L}_1(S, \mathbb{R})$ and $\|g\|_1 \leq c$. Since all f_n are μ-measurable, f is μ-measurable by Corollary 7.1. Then, the preceding corollary implies that $f \in \mathcal{L}_1(S, E)$. Moreover, $\|f\|_1 = \|g\|_1 \leq c$. □

Corollary 7.11 (Beppo Levi) *Let $\Sigma_{n=1}^{\infty} f_n$ be an infinite series whose terms are $\overline{\mathbb{R}}_+$-valued μ-measurable functions on S. Then $f := \Sigma_{n=1}^{\infty} f_n$ is μ-measurable and*

$$\int \sum_{n=1}^{\infty} f_n d\mu = \sum_{n=1}^{\infty} \int f_n d\mu. \tag{7.8}$$

If E is a Banach space and if the terms of a series $\Sigma_{n=1}^{\infty} f_n$ are E-valued integrable functions such that $\Sigma_{n=1}^{\infty} \|f_n\|_1$ converges, and if the series $\Sigma_{n=1}^{\infty} f_n$ converges a.e., then its sum f is integrable and the preceding relation holds.

Proof The first assertion stems from the Monotone Convergence Theorem applied to the partial sums of the series.

If the terms of the series $\Sigma_{n=1}^{\infty} f_n$ are E-valued integrable functions such that $c := \Sigma_{n=1}^{\infty} \|f_n\|_1 < +\infty$, the partial sums g_n satisfy $\|g_n\|_1 \leq \Sigma_{k=1}^{n} \|f_k\|_1 \leq c$ and the result follows from the preceding corollary and from Theorem 7.5. □

Corollary 7.12 *Let F, G, H be Banach spaces and let $b : F \times G \to H$ be a continuous bilinear map. Let $f \in \mathcal{L}_1(S, F)$ and let $g : S \to G$ be a bounded μ-measurable map. Then $h(\cdot) := b(f(\cdot), g(\cdot))$ is in $\mathcal{L}_1(S, H)$ and*

$$\|h\|_1 \leq \|b\| \cdot \|g\|_\infty \|f\|_1 \, .$$

In particular, one has $gf \in \mathcal{L}_1(S, E)$ whenever $f \in \mathcal{L}_1(S, E)$ and $g : S \to \mathbb{R}$ is bounded and μ-measurable.

Proof Let $c := \|g\|_\infty := \sup_{x \in S} \|g(x)\|_G$, let (f_n) be a Cauchy sequence of μ-step maps converging a.e. to f and let (g_n) be a sequence of step functions converging a.e. to g. As in the proof of Corollary 7.9, given $a > 1$ we may suppose $\|g_n(x)\|_G \leq a \|g(x)\|_G$ for all $x \in S$. Then $h_n(\cdot) := b(f_n(\cdot), g_n(\cdot))$ is a μ-step map, $(h_n) \to h$ a.e., and $\|h_n\|_1 \leq ac \|b\| \|f_n\|_1$ for all n. Corollary 7.10 enables us to conclude that $h \in \mathcal{L}_1(S, H)$ and $\|h\|_1 \leq ac \|b\| \|f\|_1$. Since a is arbitrary close to 1, we get the announced estimate. □

The next theorem has a probabilistic interpretation in terms of the averages

$$m_f(A) := \frac{1}{\mu(A)} \int_A f d\mu.$$

Theorem 7.6 *Let $f \in \mathcal{L}_1(S, E)$ and let F be a closed subset of E. If for all $A \in \mathcal{S}$ with positive finite measure one has $m_f(A) \in F$, then $f(x) \in F$ for almost all $x \in S$.*

Proof Changing f on a set of measure 0 and replacing E with a separable subspace E' and F with $F \cap E'$, we may suppose E is separable and that f is null outside of a set S' which is a countable union of measurable subsets of finite measures. Thus we are reduced to the case when S has finite measure. Given $e \in E \backslash F$ we pick $r > 0$

such that the closed ball $B := B[e, r]$ with center e and radius r does not meet F. Let $A := f^{-1}(B)$. If $\mu(A)$ is positive we have

$$\|m_f(A) - e\|_E = \left\| \frac{1}{\mu(A)} (\int_A f d\mu - \int_A e d\mu) \right\|_E \leq \frac{1}{\mu(A)} \int_A \|f - e\|_E d\mu \leq r,$$

a contradiction with $m_f(A) \in F \subset E \backslash B$. Thus $\mu(A) = 0$. Taking a countable dense subset D of $E \backslash F$ and closed balls with centers in D and rational radius contained in $E \backslash F$ we get that the set $S \backslash f^{-1}(F)$ has measure 0. □

Important consequences are obtained by taking $F := B[0, b]$, $F := \{0\}$, $F := [a, b]$ in $\mathbb{R}$, $F := \mathbb{R}_+$ respectively.

Corollary 7.13 *For $f \in \mathcal{L}_1(S, E)$, $b \in \mathbb{R}_+$ one has $\|f\|_E \leq b$ a.e. if and only if*

$$\forall A \in \mathcal{S} \qquad \left\| \int_A f d\mu \right\|_E \leq b\mu(A).$$

Corollary 7.14 *If $f \in \mathcal{L}_1(S, E)$ and if for all measurable subsets A of S with finite positive measure one has $m_f(A) = 0$, then $f = 0$ almost everywhere.*

Corollary 7.15 *If $f \in \mathcal{L}_1(S, \mathbb{R})$ and if for all measurable subsets A of S with finite positive measure one has $a \leq m_f(A) \leq b$, then $f(x) \in [a, b]$ almost everywhere.*

If for all measurable subsets A of S with finite positive measure one has $0 \leq \int_A f d\mu$, then $f(x) \geq 0$ almost everywhere.

A kind of converse is also of interest.

Corollary 7.16 *If $f \in \mathcal{L}_1(S, \mathbb{R})$, if $f \geq 0$ a.e., and if $\int_S f d\mu = 0$ then $f = 0$ almost everywhere.*

Proof This follows from Corollary 7.14 since for all $A \in \mathcal{S}$ one has $1_A f \leq f$ hence

$$0 \leq \int_A f d\mu = \int_S 1_A f d\mu \leq \int_S f d\mu = 0.$$

□

The next result is in the same vein as the preceding ones. It will be used later on.

Theorem 7.7 *Let $(S, \mathcal{S}, \mu)$ be a measure space and let $h : \mathcal{S} \to \mathbb{R}_+$ be a μ-measurable function. A measure $\nu : \mathcal{S} \to \mathbb{R}_+$ coincides with $A \mapsto \int_A h d\mu$ if and only if for all $A \in \mathcal{S}$ one has (with the convention $+\infty.0 = 0$)*

$$\mu(A) \inf h(A) \leq \nu(A) \leq \mu(A) \sup h(A). \qquad (7.9)$$

Proof Condition (7.9) is clearly necessary. Let us prove it is sufficient. We may suppose h is measurable and in a first step we assume h is positive everywhere. We fix $c \in]0, 1[$ for a moment and for $A \in \mathcal{S}$ and $n \in \mathbb{Z}$ we set

$$A_n := \{x \in A : c^{n+1} \leq h(x) < c^n\}.$$

These sets form a measurable partition of A. Relation (7.9) yields

$$c^{n+1}\mu(A_n) \leq \nu(A_n) \leq c^n\mu(A_n)$$

for all $n \in \mathbb{N}$. On the other hand, integrating h on A_n we have

$$c^{n+1}\mu(A_n) \leq \int_{A_n} hd\mu \leq c^n\mu(A_n).$$

Summing these inequalities over $n \in \mathbb{Z}$ we get

$$c\int_A hd\mu \leq \sum_n c^{n+1}\mu(A_n) \leq \sum_n \nu(A_n) \leq \sum_n c^n\mu(A_n) \leq \frac{1}{c}\int_A hd\mu.$$

Using the σ-additivity of ν, these relations imply $c\int_A hd\mu \leq \nu(A) \leq \frac{1}{c}\int_A hd\mu$. Since c is arbitrarily close to 1 we get $\nu(A) = \int_A hd\mu$.

In the general case we apply this partial result to the induced measure on $S_0 := \{x \in S : h(x) > 0\}$ and we obtain that $\nu(A) = \int_A hd\mu$ for all $A \in \mathcal{S}$ contained in S_0. Now (7.9) ensures that $\nu(A) = 0 = \int_A hd\mu$ for all $A \in \mathcal{S}$ contained in $S \backslash S_0$. Using the additivity of ν and of $A \mapsto \int_A hd\mu$ we get the result in the general case. □

Exercises

1. Show that under the assumption of Corollary 7.10 it is not always true that (f_n) converges to f in $\mathcal{L}_1(S, E)$. [Hint: take $f_n := 2^n 1_{I_n}$ with $I_n := [2^{-n}, 2^{-n+1}]$ and $f = 0$.]
2. If $(S, \mathcal{S}, \mu)$ is a σ-finite measure space, show that there exists an integrable function $u : S \to \mathbb{R}$ with positive values such that $\int_S u(s)d\mu(s) = 1$. [Hint: Given a sequence (S_n) of disjoint subsets in $\mathcal{S}$ such that $S = \cup_n S_n$ and $c_n := \mu(S_n) \in]0, +\infty[$ for all n, set $u = \Sigma_n(2^{-n}/c_n)1_{S_n}$.]
3. Let $(S, \mathcal{S}) := ([0, 1], \mathcal{B}([0, 1]))$ and let μ be the restriction of the Lebesgue measure to $\mathcal{B}([0, 1])$. For $n \geq 1$ let f_n be given by $f_n(r) := \min(r^{-1/2}e^{-nr}, n)$. Show that $(f_n) \to 0$ a.e. and that $|f_n| \leq h$ with $h(r) := r^{-1/2}$, so that $(\int_S f_n d\mu) \to 0$ by Theorem 7.5 but that (f_n) does not converge uniformly to 0.
4. The following example shows that Theorem 7.5 is not a universal tool. Let $(S, \mathcal{S}, \mu) := (\mathbb{R}, \mathcal{B}(\mathbb{R}), \lambda)$ and let $g : S \to \mathbb{R}_+$ be a continuous function null on $\mathbb{R}\backslash[0, 1]$ with a non-null integral and for $n \in \mathbb{N}\backslash\{0\}$ let f_n be given by

$f_n(r) := g(r+n)/n$. Show that $(f_n) \to 0$ uniformly, that $(\int_{\mathbb{R}} f_n d\lambda) \to 0$ but that $\int_{\mathbb{R}} h d\lambda = +\infty$ for any measurable function h satisfying $f_n \leq h$ for all $n \in \mathbb{N}$.

5. Let $(S, \mathcal{S}, \mu)$ be a measure space, and let (f_n) be a sequence in $\mathcal{L}_1(\mu, E)$ that pointwise converges to a function f and such that $(\|f_n\|_1)$ is bounded. Show that $f \in \mathcal{L}_1(\mu, E)$. Give an example showing that the sequence $(\int f_n d\mu)$ may not converge and another one showing that this sequence converges but that its limit is not $\int f d\mu$.
6. Show the convergence of the sequences whose general terms are:

$$\int_0^n (1 - \frac{x}{n})^n e^{-x/2} dx, \quad \int_0^n (1 + \frac{x}{n})^n e^{-2x} dx, \quad \int_0^n (1 - \frac{x}{n})^n \cos^2 nx dx.$$

7. Let (r_n) be a sequence in $[0, 1]$ such that $\{r_n : n \in \mathbb{N}\}$ is dense in $[0, 1]$ and let $S_n := \{r_k : k \in \mathbb{N}_n\}$. Show that the function $f_n := 1_{S_n}$ is Riemann-integrable with null integral and that the pointwise limit f of the increasing sequence (f_n) is Riemann-integrable with integral equal to 1.
8. Show that convergence in measure satisfies the conditions of a space with limits (Definition 2.1). Is this the case for convergence almost everywhere?
9. Verify that given a measure space $(S, \mathcal{S}, \mu)$, for any $A \in \mathcal{S}$ with finite measure the function p_A on the space $\mathcal{L}_0(\mu, E)$ of μ-measurable maps from S into E given by

$$p_A(f) := \inf\{c \in \mathbb{R}_+ : \mu(\{|1_A f| > c\}) \leq c\}$$

is a seminorm. The convergence associated with this family of seminorms is called *local convergence in measure*. Show that when μ is finite, this convergence coincides with convergence in measure.
10. With the notation of the preceding exercise, show that for all $f \in \mathcal{L}_0(\mu, E)$ the intersection of the family of semi-balls $\{g \in \mathcal{L}_0(\mu, E) : p_A(g - f) < r\}$ for $r > 0, A \in \mathcal{S}$ with finite measure, is $\{f\}$.
11. With the notation of Exercise 9, let p be the generalized semi-norm on $\mathcal{L}_0(\mu, E)$ given by $p(f) := p_S(f)$ if there exists $c \in \mathbb{R}_+$ such that $\mu(\{|f| > c\}) \leq c$, $p(f) = +\infty$ otherwise. Show that the convergence associated to p coincides with convergence in measure.
12. Show that if $(S, \mathcal{S}, \mu)$ is a probability space, i.e. a measure space with $\mu(S) = 1$, a sequence $(f_n) \to f$ in measure if and only if one has $(\int \frac{|f_n - f|}{1+|f_n - f|} d\mu) \to 0$.

7.5 Integrals Depending on a Parameter

Since many functions are defined by integrals, it is useful to learn how one can ensure continuity or differentiability of integrals depending on some parameters.

Proposition 7.11 *Let $(S, \mathcal{S}, \mu)$ be a measure space, let (T, d) be a metric space, let $\bar{t} \in T$, let E be a Banach space and let $f : S \times T \to E$ be a map such that*

(a) for all $t \in T$ the partial map $f_t := f(\cdot, t)$ belongs to $\mathcal{L}_1(S, \mu, E)$,
(b) for a.e. $s \in S$ the partial map $t \mapsto f(s, t)$ is continuous at $\bar{t} \in T$,
(c) there exists a $h \in \mathcal{L}_1(S, \mu, \mathbb{R})$ such that $\|f(s, t)\| \leq h(s)$ for all $(s, t) \in S \times T$.

Then the map $g : T \to E$ given by $g(t) := \int_S f_t d\mu$ is continuous at $\bar{t}$.

Proof Given a sequence $(t_n) \to \bar{t}$ in T we have to show that $(g(t_n)) \to g(\bar{t})$. We have

$$\|g(t_n) - g(\bar{t})\| = \left\| \int_S (f_{t_n} - f_{\bar{t}}) d\mu \right\| \leq \int_S \|f_{t_n} - f_{\bar{t}}\| \, d\mu,$$

$(f_{t_n} - f_{\bar{t}}) \to 0$ a.e. and $\|f_{t_n} - f_{\bar{t}}\| \leq 2h$. The Dominated Convergence Theorem ensures that $(\int_S \|f_{t_n} - f_{\bar{t}}\| \, d\mu) \to 0$. Thus $(g(t_n)) \to g(\bar{t})$. □

For simplicity, the differentiability result we give now is presented for one-variable functions. It leads to a differentiability result in the sense of Hadamard. A Fréchet differentiability result can be devised as in the following exercise.

Proposition 7.12 *Let $(S, \mathcal{S}, \mu)$ be a measure space, let T be an open interval of $\mathbb{R}$, let $\bar{t} \in T$, let E be a Banach space and let $f : S \times T \to E$ be a map such that*

(a) for all $t \in T$ the partial map $f_t := f(\cdot, t)$ belongs to $\mathcal{L}_1(S, \mu, E)$,
(b) for all $s \in S$ the partial map $t \mapsto f(s, t)$ is differentiable at $\bar{t} \in T$,
(c) there exists some $h \in \mathcal{L}_1(S, \mu, \mathbb{R})$ such that $\|f(s, t) - f(s, \bar{t})\| \leq h(s) |t - \bar{t}|$ for all $(s, t) \in S \times T$.

Then the map $g : T \to E$ given by $g(t) := \int_S f_t d\mu$ is differentiable at $\bar{t}$ and

$$g'(\bar{t}) = \int_S \frac{\partial}{\partial t} f(s, \bar{t}) d\mu(s).$$

Proof Let us denote by v the right-hand side of this relation. Given a sequence $(t_n) \to \bar{t}$ in $T \backslash \{\bar{t}\}$ we have to show that $((1/r_n)(g(t_n) - g(\bar{t}))_n \to v$ for $r_n := t_n - \bar{t}$. Now

$$(q_n(s)) := (\frac{1}{r_n}(f(s, \bar{t} + r_n) - f(s, \bar{t}))) \to_{n \to +\infty} q(s) := \frac{\partial}{\partial t} f(s, \bar{t})$$

and $\|q_n(s)\| \leq h(s)$ for all $n \in \mathbb{N}$ and $s \in S$. Invoking again Theorem 7.5, we get $(\int_S q_n d\mu)_n \to \int_S q d\mu$. □

Exercise Suppose W is an open convex subset of a normed space X, $\overline{w} \in W$, the map $f : S \times W \to E$ is such that for all $w \in W$ one has $f_w := f(\cdot, w) \in \mathcal{L}_1(S, \mu, E)$, and such that for all $s \in S$ the partial map $w \mapsto f(s, w)$ is Fréchet differentiable on W (or around $\overline{w}$) with $\frac{\partial}{\partial w} f(\cdot, \overline{w}) \in \mathcal{L}_1(S, \mu, L(X, E))$. Assume there exists some $h \in \mathcal{L}_1(S, \mu, \mathbb{R})$ such that $\|f(s, w) - f(s, \overline{w})\|_E \leq h(s) \|w - \overline{w}\|_X$ for all $(s, w) \in S \times W$. Then the map $g : W \to E$ given by $g(w) := \int_S f_w d\mu$ is Fréchet differentiable at $\overline{w}$

and

$$Dg(\overline{w}) = \int_S \frac{\partial}{\partial w} f(s, \overline{w}) d\mu(s).$$

[Hint: using Theorem 7.5, show that for every sequence $(x_n) \to 0$ in $X \backslash \{0\}$ one has $\lim_n \frac{1}{\|x_n\|}(g(\overline{w} + x_n) - g(\overline{w}) - \int_S \frac{\partial}{\partial w} f(s, \overline{w}).x_n d\mu(s)) = 0$.]

7.6 Integration on a Product

In elementary calculus, integrals of continuous functions of several variables are often computed by iterating one-variable integrals. It is the purpose of the present section to consider a similar device in a general framework. The subject has a long history. Cavalieri (1598–1647) observed that if two bodies in space have the same height and are such that their horizontal sections have the same area, then they have the same volume. But such a result was known to Chinese mathematicians about one millennium earlier. Moreover, Archimedes (circa 287–212 BC) was proud to have established that the volume of a ball is 2/3 the volume of the circumscribed cylinder with the same height. Cicero claimed that he discovered an engraving on Archimedes' grave with a verse and a drawing of a ball and a circumscribed cylinder. Use such a drawing to deduce that the volume of the ball is equal to the volume of the cylinder deprived from two cones with apex at the center of the ball and bases the bottom and the top of the cylinder (Fig. 7.1).

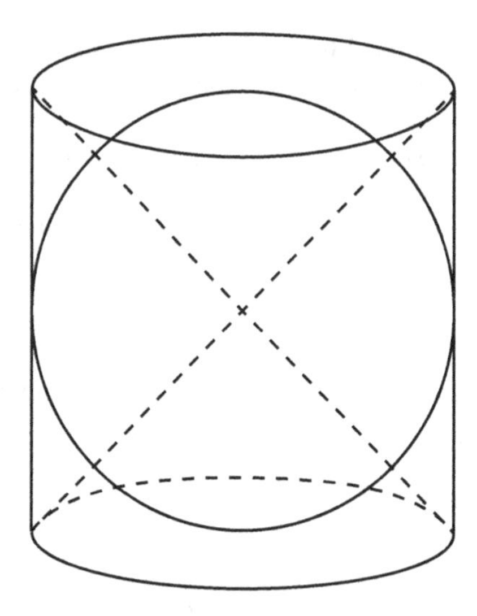

Fig. 7.1 The ball and the circumscribed cylinder

We first consider product measures with a fresh (and simpler) approach. Given sets X, Y and rings $\mathcal{A}$, $\mathcal{B}$ on X and Y respectively, we denote by $\mathcal{C}$ the collection of sets of the form $C := A \times B$ with $A \in \mathcal{A}$, $B \in \mathcal{B}$ and by $\mathcal{A} \boxtimes \mathcal{B}$ the collection of unions of finite families of disjoint elements of $\mathcal{C}$. The relations

$$\begin{aligned}(A \times B) \cap (A' \times B') &= (A \cap A') \times (B \cap B'),\\ (A \times B)\backslash(A' \times B') &= [(A\backslash A') \times B] \cup [(A \cap A') \times (B\backslash B')],\\ (A \times B)^c &= (A^c \times Y) \cup (X \times B^c)\end{aligned}$$

show that $\mathcal{C}$ is a semi-ring and Lemma 1.4 entails that $\mathcal{A} \boxtimes \mathcal{B}$ is the ring generated by $\mathcal{C}$. Moreover, $\mathcal{A} \boxtimes \mathcal{B}$ is an algebra if $\mathcal{A}$ and $\mathcal{B}$ are algebras. The following proposition shows that $\mathcal{A} \boxtimes \mathcal{B}$ is a σ-algebra whenever $\mathcal{A}$ and $\mathcal{B}$ are σ-algebras. Given a class $\mathcal{G}$ of subsets of a set Z, we denote by $\mathcal{G}_\sigma$ the σ-algebra generated by $\mathcal{G}$. Let us state again Proposition 1.16.

Proposition 7.13 *Given rings $\mathcal{A}$, $\mathcal{B}$ on the sets X and Y respectively, the σ-algebra $\mathcal{A} \otimes \mathcal{B} := (\mathcal{A} \boxtimes \mathcal{B})_\sigma$ generated by the ring $\mathcal{A} \boxtimes \mathcal{B}$ coincides with the σ-algebra $\mathcal{A}_\sigma \otimes \mathcal{B}_\sigma$ generated by the ring $\mathcal{A}_\sigma \boxtimes \mathcal{B}_\sigma$.*

Let us also recall a fundamental measurability result (Lemma 1.8). If $f : X \times Y \to Z$ is a map with values in a measurable space $(Z, \mathcal{S})$ and if $P \in \mathcal{P}(X \times Y)$, for $x \in X$ we denote by $f_x : Y \to Z$ the partial map of f and by $P_x \in \mathcal{P}(Y)$ the slice of P defined by

$$f_x(y) := f(x, y), \; y \in Y \qquad P_x := \{y \in Y : (x, y) \in P\}.$$

Lemma 7.6 *Let X, Y be sets and let $\mathcal{A}$, $\mathcal{B}$ be rings (resp. σ-algebras) on X and Y, respectively. Then, for all $P \in \mathcal{A} \boxtimes \mathcal{B}$ (resp. $\mathcal{A} \otimes \mathcal{B}$) and for all $x \in X$ one has $P_x \in \mathcal{B}$.*

For any measurable map $f : (X \times Y, \mathcal{A} \otimes \mathcal{B}) \to (Z, \mathcal{S})$ and any $x \in X$, f_x is measurable. If $(W, \mathcal{W})$ is a measurable space, a map $g : W \to X \times Y$ is measurable with respect to $\mathcal{W}$ and $\mathcal{A} \otimes \mathcal{B}$ if and only if $p_X \circ g$ and $p_Y \circ g$ are measurable.

For the rest of this section, $(X, \mathcal{M}, \mu)$ and $(Y, \mathcal{N}, \nu)$ are σ-finite measure spaces. We want to construct a natural measure on $(X \times Y, \mathcal{M} \times \mathcal{N})$. We denote by $\mathcal{A}$ (resp. $\mathcal{B}$) the ring formed by those $A \in \mathcal{M}$ (resp. $B \in \mathcal{N}$) such that $\mu(A) < +\infty$ (resp. $\nu(B) < +\infty$). We observe that for $A \in \mathcal{A}$, $B \in \mathcal{B}$, $e \in E$, $x \in X$ and $f := 1_{A \times B} e$ we have $f_x = 1_A(x) 1_B e$,

$$\int_Y f_x(y) d\nu(y) = 1_A(x) \nu(B) e,$$

$$\int_X \left(\int_Y f_x(y) d\nu(y)\right) d\mu(x) = \mu(A) \nu(B) e.$$

By linearity, we conclude that if f is a step function with respect to $\mathcal{A} \boxtimes \mathcal{B}$ then f_x is a step function with respect to $\mathcal{B}$ and $g : x \mapsto \int_Y f_x(y) d\nu(y)$ is a step function with respect to $\mathcal{A}$. Moreover, the map

$$f \mapsto \int_X (\int_Y f d\nu) d\mu := \int_X (\int_Y f_x(y) d\nu(y)) d\mu(x)$$

is linear on the space of step functions with respect to $\mathcal{A} \boxtimes \mathcal{B}$ since each of the integrals is linear. These observations lead to a simple proof of the existence result.

Theorem 7.8 *Given σ-finite measure spaces* $(X, \mathcal{M}, \mu)$ *and* $(Y, \mathcal{N}, \nu)$ *there exists a unique measure* $\mu \otimes \nu$ *on the σ-algebra* $\mathcal{M} \otimes \mathcal{N}$ *generated by the semi-ring* $\mathcal{C}$ *of subsets of the form* $A \times B$ *with* $A \in \mathcal{A}$, $B \in \mathcal{B}$ *i.e.,* $A \in \mathcal{M}$, $B \in \mathcal{N}$ *of finite measures such that for all* $A \in \mathcal{A}$, $B \in \mathcal{B}$ *one has*

$$(\mu \otimes \nu)(A \times B) = \mu(A)\nu(B).$$

Proof Let $\mu \boxtimes \nu : \mathcal{C} \to \mathbb{R}_+$ be the function given by

$$(\mu \boxtimes \nu)(A \times B) := \mu(A)\nu(B) \qquad A \in \mathcal{A},\ B \in \mathcal{B}.$$

Let us check that $\mu \boxtimes \nu$ is additive on $\mathcal{C}$. Let $A \in \mathcal{A}$, $B \in \mathcal{B}$, $A_i \in \mathcal{A}$, $B_i \in \mathcal{B}$ for $i = 1, \dots, n$ be such that $A_j \times B_j$ and $A_k \times B_k$ are disjoint for $j \neq k$ and

$$A \times B = \bigcup_{i=1}^{n} (A_i \times B_i),$$

Then for $f := 1_{A \times B}, f_i = 1_{A_i} 1_{B_i}$ we have $f = \Sigma_{i=1}^n f_i$ hence

$$(\mu \boxtimes \nu)(A \times B) = \int_X (\int_Y f d\nu) d\mu = \sum_{i=1}^{n} \int_X (\int_Y f_i d\nu) d\mu = \sum_{i=1}^{n} (\mu \boxtimes \nu)(A_i \times B_i).$$

Then, by Lemma 1.6, $\mu \boxtimes \nu$ has a unique extension to an additive function (still denoted by $\mu \boxtimes \nu$) on the ring $\mathcal{A} \boxtimes \mathcal{B}$ generated by $\mathcal{C}$. Let us show that in fact $\mu \boxtimes \nu$ is countably additive. We make use of Lemma 1.5. Let (P_n) be an increasing sequence in $\mathcal{A} \boxtimes \mathcal{B}$ whose union P is in $\mathcal{A} \boxtimes \mathcal{B}$. Let f_n be the characteristic function of P_n. Then $(f_n)_n$ is increasing and converges to the characteristic function f of P. Moreover, for all $x \in X$ the sequence $(f_{n,x})_n$ of partial functions is increasing and converges to f_x. Furthermore, $f_{n,x}$ (resp. f_x) is in $St(\nu, E)$. The Monotone Convergence Theorem ensures that

$$(g_n(x))_n := (\int_Y f_{n,x}(y) d\nu(y))_n \underset{n\to+\infty}{\to} g(x) := \int_Y f_x(y) d\nu(y).$$

Since g_n and g are μ-step functions as we observed above and since (g_n) is increasing, another application of the Monotone Convergence Theorem yields that

$$(\int_X g_n(x) d\mu(x))_n \underset{n\to+\infty}{\to} \int_X g(x) d\mu(x).$$

Since $\int_X g_n(x) d\mu(x) = (\mu \boxtimes \nu)(P_n)$ and $\int_X g(x) d\mu(x) = (\mu \boxtimes \nu)(P)$ by the observations above, we get $(\mu \boxtimes \nu)(P) = \lim_n (\mu \boxtimes \nu)(P_n)$. Then Hahn's Theorem ensures that $\mu \boxtimes \nu$ can be extended to a measure on the σ-algebra $\mathcal{M} \otimes \mathcal{N}$ generated by $\mathcal{A} \boxtimes \mathcal{B}$ or $\mathcal{C}$. □

The values of $\mu \otimes \nu$ can be computed. We need a preliminary result.

Lemma 7.7 *Given σ-finite measure spaces $(X, \mathcal{M}, \mu)$ and $(Y, \mathcal{N}, \nu)$, for all $P \in \mathcal{M} \otimes \mathcal{N}$ the function $x \mapsto \nu(P_x)$ is measurable.*

Proof First suppose ν is finite. Let $\mathcal{D}$ be the class of $P \in \mathcal{M} \otimes \mathcal{N}$ such that for all $x \in X$ the function $f_P : x \mapsto \nu(P_x)$ is measurable (we know that $P_x \in \mathcal{N}$, so that $\nu(P_x)$ is well defined in $\overline{\mathbb{R}}_+$). For all $A \in \mathcal{M}$ and all $B \in \mathcal{N}$ one has $A \times B \in \mathcal{D}$ since $\nu((A \times B)_x) = 1_A(x)\nu(B)$. In particular $X \times Y \in \mathcal{D}$. Let $P, Q \in \mathcal{D}$ with $Q \subset P$. Since for all $x \in X$ we have $\nu((P \setminus Q)_x) = \nu(P_x \setminus Q_x) = \nu(P_x) - \nu(Q_x)$ we see that $P \setminus Q \in \mathcal{D}$. If P is the union of an increasing sequence (P^n) of $\mathcal{D}$ we have $P_x = \cup_n P^n_x$ hence $\nu(P_x) = \lim_n \nu(P^n_x)$, so that $x \mapsto \nu(P_x)$ is measurable and $P \in \mathcal{D}$. Thus $\mathcal{D}$ is a complemented increasing class containing the class $\mathcal{C}$ of rectangles $A \times B$ with $A \in \mathcal{M}$, $B \in \mathcal{N}$. Since

$$(A \times B) \cap (A' \times B') = (A \cap A') \times (B \cap B'),$$

$\mathcal{C}$ is closed under finite intersections and Proposition 1.9 ensures that $\mathcal{M} \otimes \mathcal{N} \subset \mathcal{D}$: for all $P \in \mathcal{M} \otimes \mathcal{N}$ the function $x \mapsto \nu(P_x)$ is measurable.

Now let us suppose μ and ν are σ-finite. Let (Y_n) be a sequence of disjoint measurable subsets of Y with finite measures and with union Y. For each $n \in \mathbb{N}$ define a finite measure ν_n on $\mathcal{N}$ by setting $\nu_n(B) := \nu(B \cap Y_n)$. According to the first part of the proof, for all $P \in \mathcal{M} \otimes \mathcal{N}$ the function $x \mapsto \nu_n(P_x)$ is measurable. Since $\nu(P_x) = \Sigma_n \nu_n(P_x)$ the function $x \mapsto \nu(P_x)$ is measurable too. □

Proposition 7.14 *Given σ-finite measure spaces $(X, \mathcal{M}, \mu)$ and $(Y, \mathcal{N}, \nu)$, for all $P \in \mathcal{M} \otimes \mathcal{N}$ one has*

$$(\mu \otimes \nu)(P) = \int_X \nu(P_x) d\mu(x).$$

Of course, the roles of (X, μ) and (Y, ν) can be interchanged.

Proof We already know that for all $P \in \mathcal{M} \otimes \mathcal{N}$ the function $x \mapsto \nu(P_x)$ is measurable so that the function $\pi : \mathcal{M} \otimes \mathcal{N} \to \overline{\mathbb{R}}_+$ given by

$$\pi(P) = \int_X \nu(P_x) d\mu(x)$$

is well defined. By additivity of ν and linearity of integration, π is additive. If P is the union of an increasing sequence (P^n) of $\mathcal{D}$ we have seen that $\nu(P_x) = \lim_n \nu(P^n_x)$ and the Monotone Convergence Theorem ensures that

$$\pi(P) = \int_X \nu(P_x) d\mu(x) = \lim_n \int_X \nu(P^n_x) d\mu(x) = \lim_n \pi(P^n).$$

Thus, by Lemma 1.5, π is σ-additive. Since $\pi(P) = \mu(A)\nu(B) = (\mu \otimes \nu)(P)$ for $P = A \times B \in \mathcal{C}$, π and $\mu \otimes \nu$ coincide on the algebra $\mathcal{A} \boxtimes \mathcal{B}$ generated by $\mathcal{C}$, hence on the σ-algebra $\mathcal{M} \otimes \mathcal{N}$ generated by $\mathcal{A} \boxtimes \mathcal{B}$ in view of Theorem 1.11. □

Lemma 7.8 *Let Z be a set of $(\mu \otimes \nu)$-measure 0 in $X \times Y$. Then, for almost all $x \in X$ one has $\nu(Z_x) = 0$.*

Proof We may suppose Z belongs to $\mathcal{M} \otimes \mathcal{N}$. Then for all x the set Z_x is measurable and $f : x \mapsto \nu(Z_x)$ is measurable. Since $\int_X f d\mu = (\mu \otimes \nu)(Z) = 0$ and $f \geq 0$ we have $f = 0$ a.e. by Corollary 7.16. □

Alternative proof Since by Theorem 1.12 $\mu \otimes \nu$ is the restriction to $\mathcal{M} \otimes \mathcal{N}$ of the outer measure deduced from $\mu \boxtimes \nu$, the construction of Proposition 1.9 yields for every $\varepsilon > 0$ and $k \in \mathbb{N}\backslash\{0\}$ a sequence (C_n) of elements of $\mathcal{C}$ whose union contains Z and satisfies

$$\sum_{n=1}^{\infty} (\mu \boxtimes \nu)(C_n) \leq \frac{\varepsilon}{2^k k}.$$

For all $x \in X$ the slice Z_x is contained in the union of the slices $C_{n,x} := (C_n)_x$. Let

$$S_k := \{x \in X : \nu(Z_x) \geq \frac{1}{k}\}, \quad T_k := \{x \in X : f(x) := \sum_{n=1}^{\infty} \nu(C_{n,x}) \geq \frac{1}{k}\}.$$

Then f is measurable and by the Beppo Levi's Theorem one has

$$\frac{1}{k}\mu(T_k) \leq \int_X f(x)d\mu(x) = \sum_{n=1}^{\infty}\int_X \nu(C_{n,x})d\mu(x) = \sum_{n=1}^{\infty}(\mu \boxtimes \nu)(C_n) \leq \frac{\varepsilon}{2^k k}$$

hence $\mu(T_k) \leq 2^{-k}\varepsilon$. Since $\nu(Z_x) \leq f(x)$ we have $S_k \subset T_k$ hence $\mu(S_k) \leq 2^{-k}\varepsilon$. Thus, since $S := \{x \in X : \nu(Z_x) > 0\}$ is $\cup_{k\geq 1} S_k$ we get $\mu(S) \leq \varepsilon$ hence $\mu(S) = 0$, ε being arbitrarily small. □

Now, let us turn to functions. We start with nonnegative functions.

Theorem 7.9 (Tonelli) *Given σ-finite measure spaces $(X, \mathcal{M}, \mu)$, $(Y, \mathcal{N}, \nu)$ and a $(\mu \otimes \nu)$-measurable function $f : X \times Y \to \overline{\mathbb{R}}_+$, for almost all $x \in X$ the function f_x is ν-measurable and*

$$\int_{X\times Y} fd(\mu \otimes \nu) = \int_X (\int_Y f_x d\nu)d\mu(x).$$

Proof Let us first consider the case when f is the characteristic function of some $P \in \mathcal{M} \otimes \mathcal{N}$. Then, for all $x \in X$, f_x is the characteristic function of P_x, $\int_Y f_x d\nu = \nu(P_x)$ and Proposition 7.14 yields the result. By linearity we get the result for every nonnegative step function.

Now if f is $(\mu \otimes \nu)$-measurable, using the assumption that $X \times Y$ is σ-finite and the remark following Lemma 7.3 we can find an increasing sequence of μ-step functions (f_n) converging a.e. to f. Then, using the Monotone Convergence Theorem, we get the result by a passage to the limit.

Finally, let $Z \in \mathcal{M} \otimes \mathcal{N}$ be a null set and let $g : X \times Y \to \mathbb{R}_+$ be $(\mu \otimes \nu)$-measurable and such that $f = g$ off Z. The preceding lemma shows that there exists a null set N of $(X, \mathcal{M}, \mu)$ such that for $x \in X \backslash N$ we have $f_x = g_x$ off the ν-null set Z_x, so that f_x is ν-measurable and $\int_Y f_x d\nu = \int_Y g_x d\nu$ for all $X \backslash N$. Thus

$$\int_X (\int_Y f_x d\nu)d\mu(x) = \int_X (\int_Y g_x d\nu)d\mu(x) = \int_{X\times Y} gd(\mu \otimes \nu) = \int_{X\times Y} fd(\mu \otimes \nu).$$

Corollary 7.17 (Tonelli) *Given σ-finite measure spaces $(X, \mathcal{M}, \mu)$, $(Y, \mathcal{N}, \nu)$ and a $(\mu \otimes \nu)$-measurable function $f : X \times Y \to \overline{\mathbb{R}}_+$, if for almost all $x \in X$ the ν-measurable function $|f_x|$ is such that $\int_Y |f_x(y)| d\nu(y) < +\infty$, and if $\int_X(\int_Y |f_x| d\nu)d\mu(x) < +\infty$ then f is integrable on $(X \times Y, \mu \otimes \nu)$ and*

$$\int_{X\times Y} fd(\mu \otimes \nu) = \int_X (\int_Y f_x d\nu)d\mu(x).$$

Theorem 7.10 (Fubini) *Given σ-finite measure spaces $(X, \mathcal{M}, \mu)$, $(Y, \mathcal{N}, \nu)$, a Banach space E and $f \in \mathcal{L}_1(X \times Y, \mu \otimes \nu, E)$ then for almost all $x \in X$ the function*

f_x is in $\mathcal{L}_1(Y, \nu, E)$ and the map $x \mapsto \int_Y f_x d\nu$ is integrable with

$$\int_{X\times Y} fd(\mu \otimes \nu) = \int_X (\int_Y f_x d\nu) d\mu(x). \tag{7.10}$$

Proof Since $f \in \mathcal{L}_1(X\times Y, \mu\otimes\nu, E)$ we have $\|f\|_E \in \mathcal{L}_1(X\times Y, \mu\otimes\nu, \mathbb{R})$. Tonelli's Theorem yields a null set N in (X, μ) such that for $x \in X\backslash N$ the function $\|f_x\|_E$ is ν-measurable and

$$\int_{X\times Y} \|f\|_E \, d(\mu \otimes \nu) = \int_X (\int_Y \|f_x\|_E \, d\nu) d\mu(x).$$

Thus, by Corollary 7.9, for all $x \in X\backslash N$ we have $f_x \in \mathcal{L}_1(Y, \nu, E)$. Let (f^n) be a sequence in $St(\mu \otimes \nu, E)$ converging to f on $(X \times Y)\backslash Z$, where Z is a null set of $(X \times Y, \mu \otimes \nu)$. Setting $S_n := \{w \in X \times Y : \|f^n(w)\|_E \leq 2 \|f(w)\|_E\}$ and replacing f^n with $f_n' := 1_{S_n} f^n$ if necessary, we may suppose $\|f^n\|_E \leq 2 \|f\|_E$ for all n. We may suppose the μ-null set N of X is such that $\nu(Z_x) = 0$ for all $x \in X\backslash N$, enlarging N if necessary. Then, for all $x \in X\backslash N$ and all $y \in Y\backslash Z_x$ we have $(f_x^n(y)) \to f_x(y)$ and $\|f_x^n\|_E \leq 2 \|f_x\|_E$. Theorem 7.5 ensures that

$$\forall x \in X\backslash N \qquad (g^n(x))_n := (\int_Y f_x^n d\nu)_n \to g(x) := \int_Y f_x d\nu.$$

Moreover,

$$\forall x \in X\backslash N \qquad \|g^n(x)\|_E \leq \int_Y \|f_x^n\|_E \, d\nu \leq 2 \int_Y \|f_x\|_E \, d\nu.$$

Since the function $x \mapsto \int_Y \|f_x\|_E \, d\nu$ is integrable, another application of Theorem 7.5 yields

$$(\int_X g^n(x) d\mu(x))_n \to \int_X g(x) d\mu(x).$$

Since $\int_{X\times Y} f^n d(\mu \otimes \nu) = \int_X g^n d\mu$ and $(\int_{X\times Y} f^n d(\mu \otimes \nu))_n \to \int_{X\times Y} fd(\mu \otimes \nu)$, we get relation (7.10). □

The roles of X and Y being symmetric, relation (7.10) ensures that the two iterated integrals coincide. However, this coincidence may not occur if f is not supposed to be integrable for $\mu \otimes \nu$ (Exercise 3).

Fubini's Theorem allows an interpretation of the integral of a nonnegative integrable function $f : X \to \mathbb{R}_+$ as the "area" of the subset of $X \times \mathbb{R}_+$ under the graph of f.

Corollary 7.18 *Let $(X, \mathcal{M}, \mu)$ be a σ-finite measure space, let $f \in \mathcal{L}_1(X, \mu)$ be nonnegative, and let $H := \{(x, r) \in X \times \mathbb{R}_+ : r \leq f(x)\}$. Then H is measurable if* *and only if f is measurable. If this occurs, providing* $(\mathbb{R}, \mathcal{B}(\mathbb{R}))$ *with the Lebesgue measure* λ, *one has*

$$(\mu \otimes \lambda)(H) = \int_X f(x) d\mu(x). \tag{7.11}$$

Proof If f is measurable, one can show that $g : X \times \mathbb{R} \to \mathbb{R}$ given by $g(x, y) := f(x) - y$ is measurable. Since $H = g^{-1}(\mathbb{R}_+) \cap (X \times \mathbb{R}_+)$, H is measurable. Conversely, suppose H is measurable. Then, by Proposition 7.14 and the lemma preceding it, for all $x \in X$, $r \in \mathbb{R}$ the slices $H_x := [0, f(x)]$ and $H^r := \{x \in X : (x, r) \in H\} = f^{-1}([r, \infty[)$ are measurable, so that f is measurable and since $\lambda(H_x) = f(x)$, relation (7.11) stems from the relation $(\mu \otimes \lambda)(H) = \int_X \lambda(H_x) d\mu(x)$ of Proposition 7.14. □

Tonelli's Theorem and the Stieltjes measure can be used to give a practical means to compute integrals.

Proposition 7.15 (Integration by Parts) *Let* $S := [a, b]$ *with* $a, b \in \mathbb{R}$, $a < b$ *and let* $g : [a, b] \to \mathbb{R}$ *be nondecreasing and left-continuous. Setting* $g(r) := g(a)$ *for* $r < a$, $g(r) = g(b)$ *for* $r > b$, *let* μ_g *be the Stieltjes measure associated with* g. *Given* $f \in \mathcal{L}_1(S, \lambda, \mathbb{R})$ *let* $F : S \to \mathbb{R}$ *be defined by* $F(r) := \int_a^r f(s) d\lambda(s)$. *Then*

$$\int_a^b f(s) g(s) d\lambda(s) = F(b) g(b) - \int_a^b F(r) d\mu_g(r).$$

Note that since g is bounded and measurable, $fg \in \mathcal{L}_1(S, \lambda, \mathbb{R})$ and since F is continuous it is μ_g-integrable.

Proof Let us endow $S^2 := S \times S$ with its Borel σ-algebra and the measure $\mu_g \otimes \lambda$ and let us define $h : S \times S \to \mathbb{R}$ by

$$h(r, s) := f(s) \text{ for } s \leq r, \qquad h(r, s) := 0 \text{ for } s > r.$$

Tonelli's Theorem ensures that h is in $\mathcal{L}_1(S^2, \mu_g \otimes \lambda, \mathbb{R})$ since

$$\int_{S \times S} |h| \, d\mu_g \otimes \lambda = \int_a^b \left(\int_a^r |f(s)| \, d\lambda(s) \right) d\mu_g(r) \leq \|f\|_1 \left(g(b) - g(a) \right).$$

Then, Fubini's Theorem applied to h yields

$$\int_a^b F(r) d\mu_g(r) = \int_a^b \left(\int_a^r f(s) d\lambda(s) \right) d\mu_g(r) = \int_a^b \left(\int_s^b f(s) d\mu_g(r) \right) d\lambda(s)$$
$$= \int_a^b f(s)(g(b) - g(s)) d\lambda(s) = F(b) g(b) - \int_a^b f(s) g(s) d\lambda(s).$$

□

Another application of Fubini's Theorem is noteworthy.

Proposition 7.16 *In $\mathbb{R}^d$ endowed with the Lebesgue measure λ_d every hyperplane H has measure* 0.

Proof We use an induction on d. For $d = 1$, the result is obvious since a hyperplane is just a point. We assume the result holds for $d - 1$. Since λ_d is invariant by translation, we may suppose the hyperplane H contains 0. If $(e_1, \ldots, e_d)$ is the canonical basis, there exists a $k \in \mathbb{N}_d$ such that $e_k \notin H$. Since λ_d is invariant under the isomorphism u given by $u(e_d) = e_k$, $u(e_k) = e_d$, $u(e_i) = e_i$ for $i \in \mathbb{N}_d \backslash \{k, d\}$, as a product of intervals is changed into a product of intervals with the same measure, we may suppose $k = d$. For $r \in \mathbb{R}$ the slice $H_r := H \cap (\mathbb{R}^{d-1} \times \{r\})$ is a hyperplane of $\mathbb{R}^{d-1}$, hence is a null set. Integrating over r, we get $\lambda_d(H) = 0$ by Proposition 7.14. □

Another application of Fubini's Theorem concerns the *convolution* operation.

Proposition 7.17 *Let $f, g \in \mathcal{L}_1(\mathbb{R}^d, \lambda_d, \mathbb{R})$, where λ_d is the Lebesgue measure on $\mathcal{B}(\mathbb{R}^d)$. Then there exists a measurable subset S of $\mathbb{R}^d$ whose complement is a null set such that for all $x \in S$ the function $y \mapsto f(x-y)g(y)$ belongs to $\mathcal{L}_1(\mathbb{R}^d, \lambda_d, \mathbb{R})$ and the function $h := f * g : \mathbb{R} \to \mathbb{R}$ given by $h(x) = \int_{\mathbb{R}^d} f(x-y)g(y)d\lambda_d(y)$ for $x \in S$, $h(x) = 0$ for $x \in \mathbb{R}^d \backslash S$ is integrable. Moreover, $\|f * g\|_1 \leq \|f\|_1 \|g\|_1$.*

Proof We begin by showing that the function $k : (x, y) \mapsto f(x-y)g(y)$ is measurable when f and g are measurable. In fact, k is the composition of the continuous map $(x, y) \mapsto (x-y, y)$ with $f \times g$ and $p : (r, s) \mapsto rs$. Then, Tonelli's Theorem and the translation invariance of the Lebesgue measure yield

$$\int_{\mathbb{R}^d \times \mathbb{R}^d} |k| \, d(\lambda_d \otimes \lambda_d) = \int_{\mathbb{R}^d} (\int_{\mathbb{R}^d} |f(x-y)g(y)| \, d\lambda_d(x)) d\lambda_d(y)$$

$$= \int_{\mathbb{R}^d} (\int_{\mathbb{R}^d} |f(x)| \, d\lambda_d(x)) \, |g(y)| \, d\lambda_d(y) = \|f\|_1 \int_{\mathbb{R}^d} |g| \, d\lambda_d$$

or $\||k|\|_1 = \|f\|_1 \|g\|_1$. Then, by Proposition 7.9, $k \in \mathcal{L}_1(\mathbb{R}^d \times \mathbb{R}^d, \lambda_d \otimes \lambda_d, \mathbb{R})$ and Fubini's Theorem implies that for almost every $x \in \mathbb{R}^d$ the function $k_x : y \mapsto f(x-y)g(y)$ belongs to $\mathcal{L}_1(\mathbb{R}^d, \lambda_d, \mathbb{R})$. Since $\left|\int_{\mathbb{R}^d} k_x(y) d\lambda_d(y)\right| \leq \int_{\mathbb{R}^d} |k_x(y)| \, d\lambda_d(y)$ we deduce from the preceding equalities that

$$\int_{\mathbb{R}^d} |h(x)| \, d\lambda_d(x) \leq \int_{\mathbb{R}^d} \int_{\mathbb{R}^d} |f(x-y)g(y)| \, d\lambda_d(x) d\lambda_d(y) = \|f\|_1 \|g\|_1 .$$

□

Let us note that the preceding result is valid when $(\mathbb{R}^d, \lambda_d)$ is replaced by a topological group G endowed with a translation invariant measure. Let us also observe that the class a.e. of $f * g$ depends only on the classes a.e. of f and g so that the convolution can be considered as an operation from $L_1(\mathbb{R}^d) \times L_1(\mathbb{R}^d)$ into $L_1(\mathbb{R}^d)$.

Exercises

1. (**Commutativity of products**) Given σ-finite measure spaces $(X, \mathcal{M}, \mu)$ and $(Y, \mathcal{N}, \nu)$ and an element P of $\mathcal{M} \otimes \mathcal{N}$, show that $P^{\mathsf{T}} := \{(y,x) : (x,y) \in P\}$ belongs to $\mathcal{N} \otimes \mathcal{M}$ and that $(\nu \otimes \mu)(P^{\mathsf{T}}) = (\mu \otimes \nu)(P)$.
2. (**Associativity of products**) Given σ-finite measure spaces $(X, \mathcal{M}, \mu)$, $(Y, \mathcal{N}, \nu)$ and $(Z, \mathcal{P}, \varpi)$ show that $(\nu \otimes \mu) \otimes \varpi = \nu \otimes (\mu \otimes \varpi)$.
3. Let μ be the Lebesgue measure on $X := (\mathbb{R}, \mathcal{B}(\mathbb{R}))$, let ν be the counting measure on $Y := (\mathbb{R}, \mathcal{B}(\mathbb{R}))$, and let f be the characteristic function of the line $L := \{(x,x) : x \in \mathbb{R}\}$. Show that
$$\int_{X \times Y} fd(\mu \otimes \nu) \neq \int_X (\int_Y f_x d\nu) d\mu(x).$$
4. Let $f : [0,1]^2 \to \mathbb{R}$ be given by $f(0,0) = 0, f(x,y) := (x^2 - y^2)(x^2 + y^2)^{-2}$. Prove that $\int_0^1 (\int_0^1 f(x,y)dy)dx = \pi/4$. Deduce from this result that f is not integrable on $[0,1]^2$.
5. Let X and Y be two topological spaces and let $\mathcal{B}(X)$ and $\mathcal{B}(Y)$ be the associated Borelian σ-algebras. Verify that $\mathcal{B}(X) \otimes \mathcal{B}(Y) \subset \mathcal{B}(X \times Y)$. Prove that equality holds when X and Y have countable bases of open sets.
6. Let $(S, \mathcal{S}, \mu)$ be a measure space, μ being σ-finite and let $f \in L_1(S)$. For $t \in \mathbb{R}_+$ one sets $E_t := \{x \in S : |f(x)| \geq t\}$, $m(t) := \mu(E_t)$, so that $m(\cdot)$ is nonincreasing. Show that for every $p \in \mathbb{P} :=]0, +\infty[$ one has $\int |f|^p d\mu = \int_0^{+\infty} pt^{p-1} m(t)dt$. [Hint: use Fubini's Theorem for $(x,t) \mapsto pt^{p-1}m(t)$ on $E := \{(x,t) \in S \times \mathbb{R}_+ : t \leq |f(x)|\}$.]

   Deduce from the preceding relation that for all $t > 0$ and all $p \geq 1$ one has *Tchebychev's inequality* $t^p m(t) \leq \|f\|_p^p$.

   Verify that for $S :=]0,1]$ endowed with λ_1, for f given by $f(x) := 1/x \log x$ one has $\sup_{t \geq 0} tm(t) < +\infty$, but $f \notin L_1(S)$.

7.7 Change of Variables

Let us start with a general overview of image measures. This notion is important in probability theory.

Proposition 7.18 *Let $h : (X, \mathcal{A}) \to (Y, \mathcal{B})$ be a measurable map between two measurable spaces and let μ be a measure on $(X, \mathcal{A})$. Then the map $\nu : \mathcal{B} \to \overline{\mathbb{R}}_+$*

given by $\nu(B) := \mu(h^{-1}(B))$ *is a measure on* $(Y, \mathcal{B})$ *called the image measure of* μ *by* h*; it is denoted by* $h(\mu)$ *or* $h\sharp\mu$.

Proof Clearly, $\nu(\varnothing) := \mu(h^{-1}(\varnothing)) = \mu(\varnothing) = 0$. If (B_n) is a sequence of disjoint elements of $\mathcal{B}$ with union B, one has $h^{-1}(B) = \cup_n h^{-1}(B_n)$ and for $m \neq n$ $h^{-1}(B_m) \cap h^{-1}(B_n) = h^{-1}(B_m \cap B_n) = \varnothing$. Thus $\nu(B) = \Sigma_n \mu(h^{-1}(B_n)) = \Sigma_n \nu(B_n)$. Thus ν is a measure on $(Y, \mathcal{B})$ and even a measure on $\mathcal{A}_h := \{B \subset Y : h^{-1}(B) \in \mathcal{A}\}$. □

Theorem 7.11 *Let* h *and* μ *be as in the preceding proposition and let* $f : (Y, \mathcal{B}) \to \mathbb{R}$ *be* $h(\mu)$*-integrable. Then* $f \circ h$ *is* μ*-integrable and*

$$\int_Y f dh(\mu) = \int_X f \circ h d\mu. \tag{7.12}$$

Proof Both assertions are obvious when $f := 1_B$ with $B \in \mathcal{B}$ and $h(\mu)(B) < \infty$ since $1_{h^{-1}(B)} = 1_B \circ h$. By linearity these assertions are extended to the case when f is a μ-step function on $(Y, \mathcal{B})$. If f is the limit a.e. of a Cauchy sequence of $h(\mu)$-step functions of Y, then $f \circ h$ is the limit a.e. of the sequence $(f_n \circ h)$ which is a Cauchy sequence of μ-step functions on X. Relation (7.12) is obtained by a passage to the limit. □

Remark If $(Y, \mathcal{B}) = (X, \mathcal{A})$ and if μ is invariant under h in the sense that $h(\mu) = \mu$, then for every $f \in \mathcal{L}_1(X, \mu)$ one has $\int f d\mu = \int f \circ h d\mu$.

Fubini's Theorem can be used to show that the Lebesgue measure λ_d on $\mathbb{R}^d$ is invariant under linear isometries. We need a preliminary algebraic result.

Lemma 7.9 *For* $d \geq 2$*, any linear map* $u : \mathbb{R}^d \to \mathbb{R}^d$ *is obtained as the composition of a finite family* $\{u_1, \ldots, u_k\}$ *of linear maps of the following types described in terms of the canonical basis* $e_1, \ldots, e_d$*:*

(a) for some permutation σ *of* $\mathbb{N}_d := \{1, \ldots, d\}$ *one has* $u(e_i) = e_{\sigma(i)}$*;*
(b) for some $r \in \mathbb{R}$ *one has* $u(e_1) = re_1$*,* $u(e_i) = e_i$ *for* $i \in \mathbb{N}_d \backslash \{1\}$*;*
(c) $u(e_1) = e_1 + e_2$*,* $u(e_i) = e_i$ *for* $i \in \mathbb{N}_d \backslash \{1\}$*.*

Proof If $u = 0$ one can write $u = p_d \circ p_1$ where p_1 (resp. p_d) is the first (resp. the last) projection. If $u \neq 0$ one of the coefficients of the matrix of u is non-null. Using permutations if necessary we may suppose it is $u_{1,1}$. Composing u with maps of the type (a), (b) and (c) in order to get $u_{i,1} = 0$ for $i \in \mathbb{N}_d \backslash \{1\}$ we can suppose there exist $c := u_{1,1}$, $v \in L(\mathbb{R}^{d-1}, \mathbb{R}^{d-1})$, $w \in L(\mathbb{R}^{d-1}, \mathbb{R})$ such that for $(r, y) \in \mathbb{R} \times \mathbb{R}^{d-1}$ we have

$$u(r, y) = (cr + w(y), v(y)).$$

Let $u' \in L(\mathbb{R}^d, \mathbb{R}^d)$, $v' \in L(\mathbb{R}^d, \mathbb{R}^d)$ be defined by

$$u'(s, z) := (s, v(z)), \qquad v'(r, y) := (cr + w(y), y),$$

so that $u = u' \circ v'$. Clearly v' is a composition of maps of types (b) and (c). Then, using an induction assumption on d applied to v, we get the result. □

Theorem 7.12 *For any linear map $u : \mathbb{R}^d \to \mathbb{R}^d$ and any (Lebesgue) measurable subset S of $\mathbb{R}^d$ the set $u(S)$ is Lebesgue measurable and*

$$\lambda_d(u(S)) = |\det(u)|\, \lambda_d(S).$$

In particular, the Lebesgue measure λ_d is invariant under the orthogonal group.

When u is an isomorphism, the preceding relation means that the image measure of λ_d by u^{-1} is $|\det(u)|\, \lambda_d$.

Proof The result follows from Proposition 7.16 when $\det(u) = 0$ since then $u(S)$ is contained in a hyperplane. Thus, we may suppose u is an isomorphism. Since u^{-1} is continuous, it is measurable, so that $u(S) = (u^{-1})^{-1}(S)$ is measurable whenever S is measurable. Since for two linear isomorphisms v, w from $\mathbb{R}^d$ into itself we have $\det(v \circ w) = \det(v)\det(w)$, it suffices to prove the result for each map described in the preceding lemma. We already observed that λ is invariant under u when u permutes two coordinates, hence when u is any permutation of the coordinates. For isomorphisms of the type (b) and (c) of the lemma, it suffices to prove that the measures λ_d and $B \mapsto (1/\det(u))\lambda_d(u(B))$ coincide on the class $\mathcal{C}$ of products of intervals. For type (b) this is obvious and the case of type (c) can be reduced to the case $d = 2$. Now if $u : \mathbb{R}^2 \to \mathbb{R}^2$ is the map $(x, y) \mapsto (x + y, y)$ and if A, B are intervals of $\mathbb{R}$ we observe that the slice $(u(A \times B))_y := \{r : (r, y) \in u(A \times B)\}$ is just $A + y$. Tonelli's Theorem and the invariance of the Lebesgue measure under translations yield the conclusion. □

Corollary 7.19 *Let E be a Banach space, let $f \in \mathcal{L}_1(\mathbb{R}^d, \lambda_d, E)$ and let $u : \mathbb{R}^d \to \mathbb{R}^d$ be a linear map. Then $f \circ u \in \mathcal{L}_1(\mathbb{R}^d, \lambda_d, E)$ and*

$$\int_{\mathbb{R}^d} f(x) d\lambda_d(x) = \int_{\mathbb{R}^d} f(u(w))\, |\det(u)|\, d\lambda_d(w). \tag{7.13}$$

Proof For $f := 1_T e$ with $e \in E$, T measurable, relation (7.13) follows from the preceding theorem. By additivity, relation (7.13) holds for $f \in St(\lambda_d, E)$. Taking a Cauchy sequence (f_n) in $St(\lambda_d, E)$ such that $(f_n) \to f$ a.e. we see that $(f_n \circ u\, |\det(u)|)$ is a Cauchy sequence in $St(\lambda_d, E)$ converging a.e. to $f \circ u\, |\det(u)|$. Then relation (7.13) is obtained by a passage to the limit. □

Now let us pass to nonlinear changes of variables. We take on $\mathbb{R}^d$ the norm given by $\|x\| := \|x\|_\infty := \max(|x_1|, \ldots, |x_d|)$ for $x := (x_1, \ldots, x_d) \in \mathbb{R}^d$.

Theorem 7.13 *Let $h : W \to X$ be a C^1-diffeomorphism between two open subsets of $\mathbb{R}^d$ and let $f \in \mathcal{L}_1(X, \lambda_X, \mathbb{R})$ where λ_X is the induced Lebesgue measure on $(X, \mathcal{B}(X))$. Then $w \mapsto f(h(w))\, |\det(Dh(w)|$ is integrable on W with respect to the* *measure* λ_W *on* $(W, \mathcal{B}(W))$ *induced by the Lebesgue measure and*

$$\int_X f(x) d\lambda_X(x) = \int_W f(h(w)) \, |\det(Dh(w)| \, d\lambda_W(w). \tag{7.14}$$

Proof Using Proposition 7.5 it suffices to prove the result in the case when f is the characteristic function of some $B \in \mathcal{B}(X)$. In such a case, setting $A := h^{-1}(B)$ and introducing the *Jacobian* J_h of h given by $J_h(w) = \det(Dh(w))$, relation (7.14) takes the form

$$\lambda_X(h(A)) = \int_A |J_h| \, d\lambda_W \qquad \forall A \in \mathcal{B}(W).$$

Let us set $\nu(A) := \lambda_X(h(A)) = h^{-1}(\lambda_X)(A)$ for $A \in \mathcal{B}(W)$, so that $\nu = h^{-1}(\lambda_X)$ is a measure on $\mathcal{B}(W)$. In view of Theorem 7.7, the preceding relation is a consequence of the estimates

$$|\inf J_h(A)| \, \lambda_W(A) \leq \nu(A) \leq |\sup J_h(A)| \, \lambda_W(A) \qquad \forall A \in \mathcal{B}(W).$$

Taking into account the relations $h(\nu) = \lambda_X$, $J_{h^{-1}}(x) = 1/J_h(w)$ for $x := h(w)$, it suffices to prove the right-hand inequality and to apply it to the map h^{-1}.

We first prove that for any closed cube C (i.e. a ball for the norm $\|\cdot\|_\infty$) contained in W we have $\nu(C) \leq |\sup J_h(C)| \, \lambda_W(C)$. Suppose, on the contrary, that we have $\nu(C) > |\sup J_h(C)| \, \lambda_W(C)$ for some closed cube C contained in W. Let $c > |\sup J_h(C)|$ be such that $\nu(C) > c\lambda_W(C)$. Taking 2^d cubes the edges of which have lengths that are half the length of the edges of C, we get a λ_W-partition of C i.e., a covering of C by measurable subsets whose mutual intersections are null sets. By additivity we get that one of the new cubes we call C_1 is such that $\nu(C_1) > c\lambda_W(C_1)$. Repeating this division, we inductively get a closed cube C_n contained in C such that $\mathrm{diam}(C_n) = 2^{-n}\mathrm{diam}(C)$ and

$$c\lambda_W(C_n) < \nu(C_n). \tag{7.15}$$

The intersection $\cap_n C_n$ is a singleton $\{\overline{w}\}$. The derivative $u := Dh(\overline{w})$ of h at $\overline{w}$ is a linear isomorphism satisfying $|\det u| < c$ since $\overline{w} \in C$. Let $k : W \to \mathbb{R}^d$ be given by

$$k(w) = \overline{w} + u^{-1}(h(w) - h(\overline{w})) \qquad w \in W.$$

Since k is of class C^1, $k(\overline{w}) = \overline{w}$ and $Dk(w) = u^{-1} \circ Dh(w)$ is close to $I_{\mathbb{R}^d}$ for w close to $\overline{w}$, for all $\varepsilon > 0$ we can find $\delta > 0$ such that for all $w \in B(\overline{w}, \delta)$, by the Mean Value Theorem applied to $k - I_{\mathbb{R}^d}$ we have

$$\|k(w) - w\| \leq \varepsilon \, \|w - \overline{w}\| \, .$$

Since $\overline{w} \in C_n$ for all n and $(\text{diam}(C_n)) \to 0$, for n large enough we have $C_n \subset B(\overline{w}, \delta)$ and $k(C_n) \subset C_n + \varepsilon B[0, \text{diam}(C_n)/2]$. This last set is a cube whose diameter is $(1+\varepsilon)\text{diam}(C_n)$. Thus

$$\lambda(k(C_n)) \leq (1+\varepsilon)^d \lambda_W(C_n).$$

Using the invariance by translation of the Lebesgue measure λ on $\mathbb{R}^d$ and Theorem 7.12 we get

$$\lambda(k(C_n)) = \lambda(u^{-1}(h(C_n))) = \left|\det u^{-1}\right| \lambda(h(C_n)) = \frac{\nu(C_n)}{|\det u|}.$$

Then, relation (7.15) yields

$$c\lambda_W(C_n) < \nu(C_n) = |\det u| \, \lambda(k(C_n)) \leq (1+\varepsilon)^d \, |\det u| \, \lambda_W(C_n).$$

Thus $c < (1+\varepsilon)^d |\det u|$ for all $\varepsilon > 0$, hence $c \leq |\det u|$, contradicting $c > |\sup J_h(C)|$. Therefore, for all closed ball C of W we have $\nu(C) \leq |\sup J_h(C)| \, \lambda_W(C)$.

Since any open subset O of W is the union of a countable disjoint family of closed balls contained in W, we get the estimate $\nu(O) \leq |\sup J_h(O)| \, \lambda_W(O)$ for every open subset O of W. Now, by the regularity of the Lebesgue measure, for all measurable subsets A of W we get

$$\begin{aligned}\nu(A) &\leq \inf\{\nu(O) : O \in \mathcal{O}_W, \ A \subset O\} \\ &\leq \inf\{|\sup J_h(O)| \, \lambda_W(O) : O \in \mathcal{O}_W, \ A \subset O\} = \lambda(A) \, |\sup J_h(A)|\end{aligned}$$

since $w \mapsto |J_h(w)|$ is continuous. This proves the required inequality and the theorem. □

Example (Polar Coordinates) Let $W :=]-\pi, \pi[\times \mathbb{P}$, $X := \mathbb{R}^2 \backslash (\mathbb{R}_- \times \{0\})$, with $\mathbb{P} :=]0, \infty[$ and let $h : W \to X$ be given by $h(\theta, r) := (r\cos\theta, r\sin\theta)$. Since $\lambda_2(\mathbb{R}^2 \backslash X) = 0$ and since $|\det(Dh(\theta, r)| = r(\cos^2\theta + \sin^2\theta) = r$, for any $f \in L_1(\mathbb{R}^2)$ one has

$$\int_{\mathbb{R}^2} f(x,y)dxdy = \int_{-\pi}^{\pi} \int_0^{+\infty} f(h(\theta, r))rdrd\theta.$$

□

As an application, let us show that $\int_{\mathbb{R}} e^{-x^2} dx = \sqrt{\pi}$. This follows from the use of the Fubini-Tonelli Theorem and of polar coordinates:

$$\begin{aligned}(\int_{\mathbb{R}} e^{-x^2} dx)^2 &= \int_{\mathbb{R}^2} e^{-(x^2+y^2)} dxdy \\ &= \int_{-\pi}^{\pi} \int_0^{+\infty} e^{-r^2} rdrd\theta = 2\pi[-\frac{1}{2}e^{-r^2}]_0^{+\infty} = \pi.\end{aligned}$$

By a similar calculation, for any $\delta > 0$, using the inclusion $(\mathbb{R}_+\backslash[0,\delta])^2 \subset \mathbb{R}_+^2\backslash\delta B_2$, where B_2 is the Euclidean unit ball of $\mathbb{R}^2$, we get

$$(\int_\delta^\infty e^{-x^2}dx)^2 = \int_{(\mathbb{R}_+\backslash[0,\delta])^2} e^{-(x^2+y^2)}dxdy \leq \int_\delta^{+\infty}\int_0^{\pi/2} e^{-r^2}rd\theta dr = \frac{\pi}{4}e^{-\delta^2}.$$

Then, for $t > 0$ one has $\int_\delta^\infty t^{-1/2}e^{-u^2/t}du = \int_{\delta/\sqrt{t}}^\infty e^{-x^2}dx \leq \frac{\sqrt{\pi}}{2}e^{-\delta^2/2t}$. Moreover, for $d \in \mathbb{N}\backslash\{0,1\}$, using the fact that for $x \in \mathbb{R}^d\backslash\delta B_d$ there exists $k \in \mathbb{N}_d$ such that $|x_k| > \delta/d^{1/2}$, we get

$$\int_{\mathbb{R}^d\backslash\delta B_d} t^{-d/2}e^{-\|x\|^2/t}dx \leq (\int_{\mathbb{R}} t^{-\frac{1}{2}}e^{-u^2/t}du)^{d-1}\int_{\delta/d^{1/2}}^\infty t^{-\frac{1}{2}}e^{-u^2/t}du$$
$$\leq \frac{1}{2}\pi^{\frac{d-1}{2}}(\pi e^{-\delta^2/td})^{1/2} = \frac{1}{2}\pi^{\frac{d}{2}}e^{-\delta^2/2td}.$$

Hence $\int_{\mathbb{R}^d\backslash\delta B_d} t^{-d/2}e^{-\|x\|^2/t}dx \to 0$ as $t \to 0_+$.

Example (Spherical Coordinates) Let $W :=]-\pi,\pi[\times]-\pi/2,\pi/2[$, $X := \mathbb{R}^3\backslash(\mathbb{R}_- \times \{0\} \times \mathbb{R})$, and let $h : W \to X$ be given by $h(\theta,r,\varphi) := (r\cos\theta\cos\varphi, r\sin\theta\cos\varphi, r\sin\varphi)$. It can be proved that h is a diffeomorphism of class C^∞ of W onto X and $|\det(Dh(\theta,r,\varphi)| = r^2\cos\varphi$. Then, for any $f \in L_1(\mathbb{R}^3)$ one has

$$\int_{\mathbb{R}^3} f(x,y,z)dxdydz = \int_{-\pi}^{\pi}\int_0^{+\infty}\int_{-\pi/2}^{\pi/2} f(h(\theta,r,\varphi))r^2\cos\varphi d\varphi drd\theta.$$

Exercises

1. Let E be the ellipsoid $E = \{x \in \mathbb{R}^d : (x_1/a_1)^2 + \ldots + (x_d/a_d)^2 \leq 1\}$. Compute $\lambda_d(E)$. [Hint: use either an induction and Fubini's Theorem or a linear isomorphism transforming E into the unit ball B_d of $\mathbb{R}^d$.]
2. (*Cylindrical coordinates*). Let $X := (\mathbb{R}^2\backslash(\mathbb{R}_-\times\{0\}))\times\mathbb{R}$, $W :=]-\pi,\pi[\times\mathbb{P}\times\mathbb{R}$, and let $h : W \to X$ be given by $h(\theta,r,z) := (r\cos\theta, r\sin\theta, z)$. Show that for any $f \in L_1(\mathbb{R}^3)$ one has

$$\int_{\mathbb{R}^3} f(x,y,z)dxdydz = \int_{-\pi}^{\pi}\int_0^{+\infty}\int_{-\infty}^{+\infty} f(h(\theta,r,z))dzrdrd\theta.$$

3. Given $r > s > 0$, compute the volume of the solid *torus*

$$T := \{(x,y,z) \in \mathbb{R}^3 : ((x^2+y^2)^{1/2} - r)^2 + z^2 \leq s^2\}.$$

Compare $\lambda_3(T)$ with the volume of the cylinder $C := B(0, r-s) \times [0, 2\pi s]$.

4. Show that the measure b_d of the unit ball $\mathbb{B}_d := B_{\mathbb{R}^d}$ of $\mathbb{R}^d$ is given by $b_d = (1/(d/2)!)\pi^{d/2}$ if d is even and $b_d := (k!/d!)2^d\pi^k$ if d is odd, $d = 2k+1$. [Hint: for $t > -1$, using polar coordinates, compute $I(t) := \int_{\mathbb{R}^2}(1 - x^2 - y^2)^t dxdy$ and show that for $d \geq 1$ one has $b_{d+2} = I(d/2)b_d$.]
5. (*First Guldin's Theorem*). Let S be a Borel subset of $\mathbb{R}^2$ and let $\overline{s} := (\overline{x}, \overline{y})$ be its center of inertia given by $\overline{s} := \lambda_2(S)^{-1}(\int_S x d\lambda_2(x,y), \int_S y d\lambda_2(x,y))$. Let $V := \{(x,y,z) \in \mathbb{R}^3 : \exists(u,v) \in S : x = u, y^2 + z^2 = v^2\}$. Prove that $\lambda_3(V) = 2\pi\,|\overline{y}|\,\lambda_2(S)$. [Hint: use cylindrical coordinates with axis $\mathbb{R} \times \{(0,0)\}$.]
6. Show that if $h : W \to X$ is a diffeomorphism of class C^1 between two open subsets of $\mathbb{R}$ and if $C := [a,b]$ is a compact interval of W one has

$$\lambda_X(h(C)) = \int_a^b \left|h'(w)\right| d\lambda(w).$$

[Hint: use the fact that h' does not vanish on C and that h can be assumed to be increasing or decreasing.]
7. (*Euler's beta function*) For $s, t \in \mathbb{P}$, let $B(s,t) := \int_0^1 x^{s-1}(1-x)^{t-1}dx$. Check that $B(s,t) = \int_0^\infty w^{s-1}(1+w)^{-s-t}dw$ by making the substitution $x = w(1+w)^{-1}$.
8. (*Euler's gamma function*) For $t \in \mathbb{P}$, let $\Gamma(t) := \int_0^\infty x^{t-1}e^{-x}dx$. Verify that $\Gamma(t)$ is well defined and that Γ satisfies the relation $\Gamma(t+1) = t\Gamma(t)$ for $t > 0$, so that $\Gamma(n) = (n-1)!$ for $n \in \mathbb{N}$.
9. Given an open subset W of $\mathbb{R}^d$, with $d \geq 2$, and a C^1-diffeomorphism $h : W \to h(W) \subset \mathbb{R}^d$, show that for all $\overline{x} \in W$ there exists a neighborhood V of $\overline{x}$ such that $h \mid V$ is obtained as the composition of permutations of coordinates and of C^1-diffeomorphisms of the form $(x_1, \ldots, x_d) \mapsto (j(x_1, \ldots, x_d), x_2, \ldots, x_d)$ and $(v_1, \ldots, v_d) \mapsto (v_1, k_2(v_1, \ldots, v_d), \ldots, k_d(v_1, \ldots, v_d))$, where j and k_i are of class C^1.

   Using the preceding exercise, show that one can avoid the use of Theorem 7.7 in a proof of Theorem 7.13.

7.8 Measures on Spheres

We intend to define a natural measure on the unit sphere $\mathbb{S}^{d-1}$ of $\mathbb{R}^d$ endowed with its Euclidean structure. In fact $\mathbb{S}^{d-1}$ is a compact Riemannian manifold i.e., a compact manifold on which the tangent spaces are given a smoothly varying scalar product and on any such manifold one can associate a canonical measure. In the present case one can use the homeomorphism $h : \mathbb{P} \times \mathbb{S}^{d-1} \to \mathbb{R}^d \backslash \{0\}$ given by $h(r,s) := rs$, with $\mathbb{P} :=]0, +\infty[$. Given a subset B of $\mathbb{S}^{d-1}$, we set $C(B) := h(T \times B)$ with $T :=]0,1[$. It is a Borel subset of $\mathbb{R}^d$ if and only if B is a Borel subset of $\mathbb{S}^{d-1}$ since a subset G of $\mathbb{S}^{d-1}$ is open if and only if the set $C(G)$ is open in $\mathbb{R}^d \backslash \{0\}$ or in $\mathbb{R}^d$. Let us define a Borel measure on $\mathbb{S}^{d-1}$ by setting for $B \in \mathcal{B}(\mathbb{S}^{d-1})$

$$\sigma_{d-1}(B) := d\lambda_d(C(B))$$

where λ_d is the Lebesgue measure on $\mathbb{R}^d$. Since λ_d is invariant under the orthogonal group O_d and since $C(u(B)) = u(C(B))$ for $u \in O_d$, $B \in \mathcal{B}(\mathbb{S}^{d-1})$, σ_{d-1} is invariant under O_d. Note that taking $B = \mathbb{S}^{d-1}$ we get that $\sigma_{d-1}(\mathbb{S}^{d-1}) = db_d$, where $b_d := \lambda_d(B_{\mathbb{R}^d})$ is the measure of the unit ball of $\mathbb{R}^d$.

Lemma 7.10 *Let π_{d-1} be the measure on $(\mathbb{P}, \mathcal{B}(\mathbb{P}))$ with density $r \mapsto r^d$ on $\mathbb{P}$, i.e. the measure induced by the Stieltjes measure on $(\mathbb{R}, \mathcal{B}(\mathbb{R}))$ associated with the function $r \mapsto (r^+)^d$ on $\mathbb{R}$, with $r^+ := \max(r, 0)$, so that $\pi_{d-1}([a, b[) := b^d - a^d$ for $a < b$ in $\mathbb{P}$. Then, for any Borel subset A of $\mathbb{P} \times \mathbb{S}^{d-1}$, one has $\lambda_d(h(A)) = (\pi_{d-1} \otimes \sigma_{d-1})(A)$.*

Proof We first observe that $\mathcal{B}(\mathbb{P} \times \mathbb{S}^{d-1})$ is generated by the class $\mathcal{C}$ of products of the form $A := [a, b[\times B$ with $a, b \in \mathbb{P}$, $a \leq b$, $B \in \mathcal{B}(\mathbb{S}^{d-1})$. For such a product A we have $h(A) = bC(B) \backslash aC(B)$, hence

$$\begin{aligned}
\lambda_d(h(A)) &= \lambda_d(bC(B)) - \lambda_d(aC(B)), \\
&= b^d \sigma_{d-1}(B) - a^d \sigma_{d-1}(B) \\
&= \pi_{d-1}([a, b[) \times \sigma_{d-1}(B) = (\pi_{d-1} \otimes \sigma_{d-1})(A).
\end{aligned}$$

Since the family $\mathcal{A}$ of sets $A \in \mathcal{B}(\mathbb{P} \times \mathbb{S}^{d-1})$ satisfying the relation $\lambda_d(h(A)) = (\pi_{d-1} \otimes \sigma_{d-1})(A)$ is a σ-algebra, $\mathcal{A}$ coincides with $\mathcal{B}(\mathbb{P} \times \mathbb{S}^{d-1})$. □

Proposition 7.19 *For any integrable function f on $\mathbb{R}^d$ one has*

$$\int_{\mathbb{R}^d} f d\lambda_d = \int_{\mathbb{P} \times \mathbb{S}^{d-1}} f(rs) r^{d-1} d\lambda_1(r) d\sigma_{d-1}(s).$$

Proof This follows from Theorem 7.7 and the fact that the measure $A \mapsto \lambda_d(h(A))$ on $\mathcal{B}(\mathbb{P} \times \mathbb{S}^{d-1})$ has the density $(r, s) \mapsto r^{d-1}$ with respect to the measure $\lambda_1 \otimes \sigma_{d-1}$. □

Exercises

1. Given a function $g : \mathbb{R}_+ \to \mathbb{R}_+$, let $f := g \circ \|\cdot\| : \mathbb{R}^d \to \mathbb{R}$. Verify that if g is measurable (resp. integrable with respect to the measure $r^{d-1}dr$) so is f for λ_d and

$$\int_{\mathbb{R}^d} f d\lambda_d = db_d \int_0^{+\infty} g(r) r^{d-1} dr \quad \text{with } b_d := \lambda_d(B_{\mathbb{R}^d}).$$

2. Given $d \in \mathbb{N} \backslash \{0\}$ and $t \in \mathbb{R}$, let $f := \|\cdot\|^t$. Show that for $t > -d$ the function f is integrable on the unit ball B_d of $\mathbb{R}^d$ but f is not integrable on $\mathbb{R}^d \backslash B_d$. Show that

for $t < -d$ the function f is integrable on $\mathbb{R}^d \backslash B_d$ but f is not integrable on B_d. Verify that for $t = d$ the function f is not integrable on B_d or on $\mathbb{R}^d \backslash B_d$.

3. Let $g : \mathbb{R} \to \mathbb{R}_+$ be given by $g(r) := \exp(-r^2)$ and let $f := g \circ \|\cdot\| : \mathbb{R}^2 \to \mathbb{R}$. Show that the functions f and g are integrable and that $\|f\|_1 = \|g\|_1^2$. Compute $\|f\|_1$ using polar coordinates. Deduce from this the relation $\|g\|_1 = \sqrt{\pi}$.
4. Let $f_t(x, y) := (1 - x^2 - y^2)^t 1_{B_2}(x, y)$ with $t \in]-1, +\infty[$. Using polar coordinates compute $\int f_t d\lambda_2$. Prove that for $d \geq 1$ one has $b_{d+2} = b_d \left\|f_{d/2}\right\|_1$. Deduce from this the values of b_{2k} and b_{2k+1}.

Additional Reading

[34, 35, 41, 54, 63, 80, 100, 106, 115, 126, 132, 159, 177, 182, 193, 224, 247, 262]

Chapter 8
Differentiation and Integration

Abstract The aim of this chapter is twofold. In the first part vectorial measures are introduced and the question of the representation of a measure in terms of another one is tackled. In the second part, Lebesgue L_p spaces are studied and their main properties established. The main properties of the Fourier transform and the Radon transform are displayed in view of their important applications.

In this chapter we consider some advanced subjects of measure theory and integration such as vectorial measures and the derivatives of a measure with respect to another one. We also introduce and study important spaces of functions known as Lebesgue spaces. They serve as models for several questions in functional analysis. We devote attention to some useful transforms, the most important one being the Fourier transform, another being the Radon transform used in medical tomography.

8.1 Vectorial Measures

In the sequel $(S, \mathcal{S})$ is a fixed measurable space. What we called "measure" will be called "positive measure" whenever a risk of confusion may appear. The reason is that we intend to deal with measures with values in $\overline{\mathbb{R}}$, $\mathbb{C}$ or even a Banach space E. A map

$$\nu : \mathcal{S} \to E$$

is called a *vectorial measure* or an E-valued measure if $\nu(\varnothing) = 0$ and if it is *countably additive* in the sense that for any $A \in \mathcal{S}$ and any *countable (measurable) partition* $\{A_n : n \in \mathbb{N}\}$ of A, i.e. any sequence in disjoint measurable sets whose union is A, the family $(\nu(A_n))$ is summable and one has

$$\nu(A) = \sum_{n=1}^{\infty} \nu(A_n).$$

A map $\nu : \mathcal{S} \to \overline{\mathbb{R}}$ is called a *signed measure* if it takes its values in $\mathbb{R}_\infty :=\,]-\infty, +\infty]$ or in $[-\infty, +\infty[$, if $\nu(\varnothing) = 0$ and if it is *countably additive* in the

J.-P. Penot, *Analysis*, Universitext, DOI 10.1007/978-3-319-32411-1_8

sense that the preceding relation holds for any countable partition of A and any $A \in \mathcal{S}$. Note that the sum is unambiguously defined since we exclude the case when ν takes both values $-\infty$ and $+\infty$. For simplicity, in the sequel we always assume a signed measure takes its values in $\mathbb{R}_\infty$.

The *total variation* $|\nu|$ of a vectorial measure ν is the function $|\nu| : \mathcal{S} \to \overline{\mathbb{R}}_+$ given by

$$|\nu|(A) := \sup \sum_{n=0}^{\infty} \|\nu(A_n)\|_E \qquad A \in \mathcal{S},$$

the supremum being taken over all countable partitions $\{A_n : n \in \mathbb{N}\}$ of A. If ν is a signed measure we replace $\|\nu(A_n)\|_E$ with $|\nu(A_n)|$, with the convention that $|+\infty| = +\infty$. If ν is a positive measure the σ-additivity of ν yields $|\nu| = \nu$. If ν is a vectorial measure and $A \in \mathcal{S}$, taking the partition $\{A_n : n \in \mathbb{N}\}$ of A given by $A_0 := A$, $A_n := \varnothing$ for $n \geq 1$ we see that

$$|\nu|(A) \geq \|\nu(A)\|_E . \tag{8.1}$$

Similarly, if ν is a signed measure, for all $A \in \mathcal{S}$ we have $|\nu|(A) \geq |\nu(A)|$.

It is easy to see that the space $\mathcal{M}(S, \mathcal{S}, E)$ (also denoted by $\mathcal{M}(S, E)$ if there is no risk of confusion) of E-valued measures on $(S, \mathcal{S})$ and its subset $\mathcal{M}_b(S, E) := \{\nu \in \mathcal{M}(S, E) : |\nu|(S) < +\infty\}$ are linear spaces and that the function $\nu \mapsto \|\nu\| := |\nu|(S)$ is a norm on $\mathcal{M}_b(S, E)$ (see Exercise 3). In Exercise 1 you are invited to prove that if E is finite dimensional then any E-valued measure ν is *bounded* in the sense that $\|\nu\| := |\nu|(S)$ is finite, i.e. $\nu \in \mathcal{M}_b(S, E)$.

Proposition 8.1 *Let $\nu : \mathcal{S} \to E$ be an E-valued measure (resp. $\nu : \mathcal{S} \to \overline{\mathbb{R}}$ be a signed measure). Then $|\nu|$ is a positive measure. Moreover, if μ is a positive measure satisfying $\|\nu(A)\|_E \leq \mu(A)$ for all $A \in \mathcal{S}$, then one has $|\nu| \leq \mu$.*

Proof Clearly $|\nu|(\varnothing) = 0$. Let us show $|\nu|$ is countably additive. Let $\{A_k\}$ be a countable partition of $A \in \mathcal{S}$ by sets in $\mathcal{S}$. For any sequence (r_k) of nonnegative numbers satisfying $r_k = 0$ if $|\nu|(A_k) = 0$ and $r_k < |\nu|(A_k)$ whenever $|\nu|(A_k) > 0$, and for any $k \in \mathbb{N}$ we pick a countable partition $\{A_{k,n} : n \in \mathbb{N}\}$ of A_k such that $r_k \leq \Sigma_n \|\nu(A_{k,n})\|_E$ (taking $A_{k,0} = A_k$, $A_{k,n} = \varnothing$ for $n \geq 1$ when $|\nu|(A_k) = 0$). Then

$$\sum_k r_k \leq \sum_k \sum_n \|\nu(A_{k,n})\|_E \leq |\nu|(A)$$

since $\{A_{k,n} : (k, n) \in \mathbb{N}^2\}$ can be viewed as a countable partition of A. Taking the supremum over the sequences (r_k) chosen as above we get $\Sigma_k |\nu|(A_k) \leq |\nu|(A)$.

To get the reverse inequality we consider an arbitrary countable partition $\{B_n\}$ of A. Then $\{B_n \cap A_k : k \in \mathbb{N}\}$ is a countable partition of B_n, so that $\nu(B_n) = \Sigma_k \nu(B_n \cap A_k)$, $\|\nu(B_n)\|_E \leq \Sigma_k \|\nu(B_n \cap A_k)\|_E$ and

$$\sum_n \|\nu(B_n)\|_E \leq \sum_n \sum_k \|\nu(B_n \cap A_k)\|_E$$

$$\leq \sum_k \sum_n \|\nu(A_k \cap B_n)\|_E \leq \sum_k |\nu| (A_k)$$

since $\{A_k \cap B_n : n \in \mathbb{N}\}$ is a countable partition of A_k. Taking the supremum over the set of countable partitions $\{B_n\}$ of A we get $|\nu| (A) \leq \Sigma_k |\nu| (A_k)$ and equality holds.

If μ is a positive measure satisfying $\|\nu(A)\|_E \leq \mu(A)$ for all $A \in \mathcal{S}$, then for any countable partition $\{A_n\}$ of $A \in \mathcal{S}$ one has $\Sigma_n \|\nu(A_n)\|_E \leq \Sigma_n \mu(A_n) = \mu(A)$, hence, taking the supremum over all partitions of A, $|\nu| (A) \leq \mu(A)$. □

It follows from this result that if $A, B \in \mathcal{S}$ are such that $A \subset B$ then one has $|\nu| (A) \leq |\nu| (B)$. In particular, if $|\nu| (B)$ is finite, then for all $A \in \mathcal{S}$ included in B, $|\nu| (A)$ is finite.

Let us give an important example of vectorial measure.

Example Given a Banach space E, a positive measure μ on $(S, \mathcal{S})$, and $h \in \mathcal{L}_1(\mu, E)$, one defines an E-valued measure on $\mathcal{S}$ by setting for $A \in \mathcal{S}$

$$\mu_h(A) := \int_A h d\mu. \tag{8.2}$$

Clearly $\mu_h(\varnothing) = 0$ and the countable additivity of μ_h follows from Corollary 7.11. Moreover, μ_h is absolutely continuous with respect to μ in the sense of the next definition.

Definition 8.1 A vectorial measure ν or a signed measure ν is said to be *absolutely continuous* with respect to a vectorial measure or a signed measure μ if for all $A \in \mathcal{S}$ one has $\nu(A) = 0$ whenever $|\mu| (A) = 0$. Then one writes $\nu \ll \mu$ and one also says that ν is μ-continuous.

Let us observe that one has $\nu \ll \mu$ if and only if, for all $A \in \mathcal{S}$, $|\mu| (A) = 0$ implies $|\nu| (A) = 0$. In fact, if $\nu \ll \mu$, when $|\mu| (A) = 0$, for any measurable partition (A_n) of A one has $\nu(A_n) = 0$ for all n, hence $|\nu| (A) = 0$. The converse stems from the relation $\|\nu(A)\|_E \leq |\nu| (A)$ for all $A \in \mathcal{S}$. Another characterization explains the terminology.

Proposition 8.2 *If μ is a finite positive measure, a vectorial measure ν or a signed measure ν is absolutely continuous with respect to μ if and only if for every $\varepsilon > 0$ there exists a $\delta > 0$ such that for all $A \in \mathcal{S}$ satisfying $|\mu| (A) < \delta$ one has $|\nu| (A) < \varepsilon$.*

Proof Since $\|\nu(A)\|_E \leq |\nu|(A)$ for all $A \in \mathcal{S}$, the condition implies that $\nu \ll \mu$. Conversely, assume that $\nu \ll \mu$. If the condition fails, one can find $\varepsilon > 0$ such that for any positive integer n there exists some $A_n \in \mathcal{S}$ with $|\mu|(A_n) < 2^{-n}$ but $|\nu|(A_n) \geq \varepsilon$. Let $B_n := \cup_{p \geq n} A_p$ and let $B := \cap_n B_n$. Then, $|\mu|$ being a positive measure, we have

$$|\mu|(B) \leq |\mu|(B_n) \leq \sum_{p \geq n} |\mu|(A_p) \leq 2^{-n+1},$$

hence $|\mu|(B) = 0$. But since $A_n \subset B_n$ we have $|\nu|(B_n) \geq |\nu|(A_n) \geq \varepsilon$, hence $|\nu|(B) = \lim_n |\nu|(B_n) \geq \varepsilon$. By definition of $|\nu|$ one can find some $C \in \mathcal{S}$ such that $C \subset B$ and $\nu(C) \neq 0$, contradicting the assumption that ν is absolutely continuous with respect to μ and the fact that $0 \leq |\mu|(C) \leq |\mu|(B) = 0$. □

Let us study more closely the preceding example in order to get some familiarity with the notions we have introduced.

Proposition 8.3 *Given a Banach space E, a positive σ-finite measure μ on $(S, \mathcal{S})$, and $h \in \mathcal{L}_1(\mu, E)$, let μ_h be the E-valued measure defined by (8.2). Then one has $|\mu_h|$. More generally, for all $A \in \mathcal{S}$ one has*

$$|\mu_h|(A) = \int_A \|h\|_E \, d\mu. \tag{8.3}$$

Moreover, if $g \in \mathcal{L}_1(\mu, E)$ is such that $\mu_g = \mu_h$, then $g = h$ almost everywhere.

Proof We have already observed that relation (8.2) implies that $\mu_h \ll \mu$: if $A \in \mathcal{S}$ is such that $|\mu|(A) = 0$, then we have $\mu(A) = 0$ as μ is a positive measure, hence $\mu_h(A) := \int_A h d\mu = 0$.

Let us prove relation (8.3). Given $A \in \mathcal{S}$ and a countable partition (A_n) of A by members of $\mathcal{S}$, since $\|\mu_h(A_n)\|_E = \left\| \int 1_{A_n} h d\mu \right\|_E \leq \int 1_{A_n} \|h\|_E \, d\mu$, Corollary 7.11 yields

$$\sum_n \|\mu_h(A_n)\|_E \leq \sum_n \int 1_{A_n} \|h\|_E \, d\mu = \int \sum_n 1_{A_n} \|h\|_E \, d\mu = \int_A \|h\|_E \, d\mu.$$

Taking the supremum over all countable partitions of A we get $|\mu_h|(A) \leq \int_A \|h\|_E \, d\mu$.

The opposite inequality $|\mu_h|(A) \geq \int_A \|h\|_E \, d\mu$ holds if h is a μ-step function, $h := 1_{B_1} e_1 + \ldots + 1_{B_k} e_k$ with $e_i \in E$, $B_i \in \mathcal{S}$, $B_i \cap B_j = \varnothing$ for $i \neq j$ since any (measurable) countable partition $(A_n)_{n \geq 0}$ of A can be refined in a partition $(A_n \cap B_i)_{(i,n) \in \mathbb{N}_k \times \mathbb{N}}$ of A such that h is constant on each $A_n \cap B_i$ so that

$$|\mu_h|(A) \geq \sum_n \sum_{i=1}^{k} \left\| \int_{A_n \cap B_i} h d\mu \right\|_E = \sum_n \sum_{i=1}^{k} \int_{A_n \cap B_i} \|h\|_E \, d\mu = \int_A \|h\|_E \, d\mu.$$

In order to prove the inequality $|\mu_h|(A) \geq \int_A \|h\|_E \, d\mu$ in the case $h \in \mathcal{L}_1(\mu, E)$, let us observe that we can reduce the task to the case when $\mu(A)$ is finite: if we had $\int_A \|h\|_E \, d\mu \geq |\mu_h|(A) + \varepsilon$ for some $\varepsilon > 0$, using the assumption that μ is σ-finite, we would pick $B \in \mathcal{S}$ with $B \subset A$, $\mu(B) < +\infty$ and $\int_B \|h\|_E \, d\mu > \int_A \|h\|_E \, d\mu - \varepsilon$, so that, since $|\mu_h|(A) \geq |\mu_h|(B)$, the inequality $|\mu_h|(B) \geq \int_B \|h\|_E \, d\mu$ would be impossible. Assuming now that $\alpha := \mu(A)$ is finite, given $\varepsilon > 0$, it suffices to show that $|\mu_h|(A) \geq \int_A \|h\|_E \, d\mu - \varepsilon$. Let $k(\cdot) := \|h(\cdot)\|_E$. Since $\mu_k \ll \mu$ we can find $\delta > 0$ such that for all $Z \in \mathcal{S}$ satisfying $\mu(Z) < \delta$ we have $|\mu_k|(Z) < \varepsilon/3$. By Proposition 7.3 we can find a μ-step function g and $Z \in \mathcal{S}$ such that $\mu(Z) < \delta$ and

$$\|g - h\|_1 < \frac{\varepsilon}{3}, \qquad \sup_{s \in A \setminus Z} \|g(s) - h(s)\|_E < \frac{\varepsilon}{3\alpha}.$$

Then, since we can easily show that

$$|\mu_h|(A \setminus Z) \geq |\mu_g|(A \setminus Z) - \sup_{s \in A \setminus Z} \|g(s) - h(s)\|_E \, \mu(A \setminus Z) \geq |\mu_g|(A \setminus Z) - \frac{\varepsilon}{3},$$

and since $|\mu_h|(A) \geq |\mu_h|(A \setminus Z)$, $|\mu_g|(A \setminus Z) \geq \int_{A \setminus Z} \|g\|_E \, d\mu$, $\int_Z \|h\|_E = \int_Z k = \mu_k(Z) \leq |\mu_k|(Z) < \varepsilon/3$, we get

$$|\mu_h|(A) \geq \int_{A \setminus Z} \|g\|_E \, d\mu - \frac{\varepsilon}{3} \geq \int_{A \setminus Z} \|h\|_E \, d\mu - \frac{2\varepsilon}{3} \geq \int_A \|h\|_E \, d\mu - \varepsilon.$$

Finally, let $g, h \in \mathcal{L}_1(\mu, E)$ be such that $\mu_g = \mu_h$. Then $f := g - h \in \mathcal{L}_1(\mu, E)$ and $\int_A f d\mu = 0$ for all $A \in \mathcal{S}$. Then Corollary 7.14 ensures that $f = 0$ a.e. □

A vectorial measure can be used for integration, following the scheme we expounded in Sect. 7.2.

Theorem 8.1 *Let E, F, G be Banach spaces, let $b : E \times F \to G$ be a continuous bilinear map and let $\nu : \mathcal{S} \to F$ be a vectorial measure on the measurable space $(S, \mathcal{S})$ with total variation $|\nu|$. Then, there is a unique continuous linear map $\hat{b} : L_1(|\nu|, E) \to G$ denoted by $f \mapsto \int_S f d\nu$ such that $\int_S 1_A e d\nu = b(e, \nu(A))$ for all $A \in \mathcal{S}$, $e \in E$. Moreover*

$$\left\| \int_S f d\nu \right\|_G \leq \|b\| \int_S \|f\|_E \, d|\nu| \qquad \textit{for all } f \in L_1(|\nu|, E).$$

Proof Changing the norm on E, F, or G and assuming $b \neq 0$, we could suppose $\|b\| = 1$ but that would not simplify the proof much. For $f \in St(|\nu|, E)$, $f = \Sigma_{i=1}^k 1_{A_i} e_i$ with $A_i \in \mathcal{S}$, $e_i \in E$ for $i \in \mathbb{N}_k$, we set

$$\int_S f d\nu := \sum_{i=1}^k b(e_i, \nu(A_i)).$$

We observe that since $\|\nu(A)\|_F \leq |\nu|(A)$ for all $A \in \mathcal{S}$, we have

$$\left\| \int_S f d\nu \right\|_G \leq \|b\| \sum_{i=1}^{k} \|e_i\|_E \|\nu(A_i)\|_F$$
$$\leq \|b\| \sum_{i=1}^{k} \|e_i\|_E |\nu|(A_i) = \|b\| \int_S \|f\|_E d|\nu|.$$

Given $f \in \mathcal{L}_1(|\nu|, E)$, we pick a Cauchy sequence (f_n) in $St(|\nu|, E)$ for the semi-norm $\|\cdot\|_1$ associated with $|\nu|$ that satisfies $(f_n) \to f$ a.e., so that for $n, p \in \mathbb{N}$ we deduce from the preceding inequalities that

$$\left\| \int_S f_n d\nu - \int_S f_p d\nu \right\|_G = \left\| \int_S (f_n - f_p) d\nu \right\|_G \leq \|b\| \left\| f_n - f_p \right\|_1.$$

Thus $(\int_S f_n d\nu)$ is a Cauchy sequence in G, hence has a limit we denote by $\int_S f d\nu$. It is easy to see that this limit does not depend on the choice of the Cauchy sequence (f_n). Moreover, the map $f \mapsto \int_S f d\nu$ is linear and continuous from $\mathcal{L}_1(|\nu|, E)$ into G since

$$\left\| \int_S f d\nu \right\|_G \leq \|b\| \int_S \|f\|_E d|\nu|.$$

Uniqueness on $St(|\nu|, E)$ follows from linearity; the last inequality ensures uniqueness of the extension to $L_1(|\nu|, E)$. □

The preceding construction can be used for the case $E = \mathbb{R}$ (or $\mathbb{C}$ if G is a complex space) with $F = G$ and $b(r, y) := ry$ for $(r, y) \in \mathbb{R} \times F$, for the case $F := L(E, G)$ and $b(x, u) := u(x)$ for $(x, u) \in E \times L(E, G)$, and of course for the usual case $F := \mathbb{R}$, $G = E$. When $E = \mathbb{R}$, $F = G$ is a Hilbert space, using the scalar product, we shall see later that all vectorial measures $\nu : \mathcal{S} \to F$ of bounded variation are quite simple: for some $h \in L_1(|\nu|, F)$ one has

$$\nu(A) = \int_A h d|\nu| \qquad \text{for all } A \in \mathcal{S}. \tag{8.4}$$

Then, for $f \in L_1(\mu, \mathbb{R})$ one has

$$\int_S f d\nu = \int_S f h d|\nu|$$

since by linearity both sides coincide on $St(\mu, \mathbb{R})$ and both sides are linear and continuous with respect to the norm $\|\cdot\|_1$.

Exercises

1. Prove that if E is a finite dimensional Banach space and if $\nu : \mathcal{S} \to E$ is an E-valued measure, then $|\nu|$ is a finite measure.
2. Prove that if E is an infinite dimensional Banach space then the conclusion of the preceding exercise may fail. [Hint: Let $S := \mathbb{R}_+$ equipped with its Borel σ-algebra $\mathcal{B}$ and the Lebesgue measure λ and let E be a separable Hilbert space with Hilbert basis (e_n); set $\nu(A) = \Sigma_n \lambda(A \cap [n, n+1[)e_n$ for $A \in \mathcal{B}$.]
3. Show that the space of bounded E-valued measures is a vector space and that $\nu \mapsto \|\nu\|$ is a norm on this space.
4. Prove that the space of bounded E-valued measures is complete with respect to the norm $\|\cdot\|$.

8.2 Decomposition and Differentiation of Measures

In this section we present some results concerning the structure of signed measures or vectorial measures. We also gather some related results of interest.

8.2.1 *Decompositions of Measures*

Let us turn to decompositions of signed measures. Given a signed measure ν on $\mathcal{S}$ we denote by $\mathcal{S}_+$ (resp. $\mathcal{S}_-$) the family of sets $A \in \mathcal{S}$ such that for all $B \in \mathcal{S}$ contained in A one has $\nu(B) \geq 0$ (resp. $\nu(B) \leq 0$). The richness of $\mathcal{S}_+$ and $\mathcal{S}_-$ is guaranteed by the next lemma.

Lemma 8.1 *Let ν be a signed measure on $\mathcal{S}$ and let $A \in \mathcal{S}$ be such that $\nu(A) < 0$. Then there exists some $B \in \mathcal{S}_-$ such that $B \subset A$, $\nu(B) \leq \nu(A)$.*

Proof We define B as $A \backslash (\cup_n C_n)$ where the sequence (C_n) is defined as follows. Setting

$$r_0 := \sup\{\nu(C) : C \in \mathcal{S},\ C \subset A\},$$

we pick $C_0 \in \mathcal{S}$ such that $C_0 \subset A$ and $\nu(C_0) \geq \min(r_0/2, 1)$ (note that $r_0 \in \overline{\mathbb{R}}_+$ as $\nu(\varnothing) = 0$). We proceed by induction, assuming $C_0, \ldots, C_{n-1}$ are chosen and taking $C_n \in \mathcal{S}$ such that $C_n \subset A \backslash (\cup_{0 \leq k \leq n-1} C_k)$, $\nu(C_n) \geq \min(r_n/2, 1)$, where

$$r_n := \sup\{\nu(C) : C \in \mathcal{S},\ C \subset A \backslash (\cup_{0 \leq k \leq n-1} C_k)\}.$$

Let us verify that $B := A \backslash (\cup_n C_n)$ has the required properties. Since the sets C_n are disjoint and satisfy $\nu(C_n) \geq 0$, we have $\nu(\cup_n C_n) \geq 0$, hence

$$\nu(A) = \nu(B) + \nu(\cup_n C_n) \geq \nu(B).$$

Since $\nu(A)$ is finite, $\nu(\cup_n C_n)$ is finite too and since $\nu(\cup_n C_n) = \Sigma_n \nu(C_n)$ we have $(\nu(C_n)) \to 0$. Thus $(r_n) \to 0$. Now, for all $C \in \mathcal{S}$ contained in B and all $n \in \mathbb{N}$ we have $\nu(C) \leq r_n$, hence $\nu(C) \leq 0$ and $B \in \mathcal{S}_-$. □

Theorem 8.2 (Hahn) *Let $(S, \mathcal{S})$ be a measurable space and let $\nu : \mathcal{S} \to \mathbb{R}_\infty$ be a signed measure. Then there is a partition (S_+, S_-) of S with $S_+ \in \mathcal{S}_+$, $S_- \in \mathcal{S}_-$. Thus, for every $A \in \mathcal{S}$ one has $\nu(A \cap S_+) \geq 0$, $\nu(A \cap S_-) \leq 0$, and $A = (A \cap S_+) \cup (A \cap S_-)$, $\nu(A) = \nu(A \cap S_+) + \nu(A \cap S_-)$.*

Proof One has $r := \inf\{\nu(A) : A \in \mathcal{S}_-\} < +\infty$ since $\mathcal{S}_-$ is nonempty ($\varnothing \in \mathcal{S}_-$). Take a sequence (A_n) in $\mathcal{S}_-$ such that $r = \lim_n \nu(A_n)$ and let $S_- := \cup_n A_n$. Since each $C \in \mathcal{S}$ contained in S_- is the union of a sequence of measurable disjoint subsets, each of which is contained in some A_n, we have $S_- \in \mathcal{S}_-$. Thus $r \leq \nu(S_-) \leq \nu(A_n)$ for all n. Consequently $\nu(S_-) = r \in \mathbb{R}$ (since ν does not take the value $-\infty$).

Let $S_+ := S \backslash S_-$ and let us show that $S_+ \in \mathcal{S}_+$. By the preceding lemma, if S_+ contained some $A \in \mathcal{S}$ satisfying $\nu(A) < 0$, then A would contain a subset $B \in \mathcal{S}_-$ with $\nu(B) \leq \nu(A) < 0$ and we would get $\nu(S_- \cup B) = \nu(S_-) + \nu(B) < r$, contradicting the definition of r and the fact that $S_- \cup B \in \mathcal{S}_-$. □

Remark The decomposition we obtained is not unique, but it is essentially unique in the sense that if (S'_+, S'_-) is another such decomposition, then $S_+ \cap S'_-$ and $S_- \cap S'_+$ are both in $\mathcal{S}_-$ and $\mathcal{S}_+$, hence are null sets.

Corollary 8.1 (Jordan) *Every signed measure ν is the difference $\nu^+ - \nu^-$ of two positive measures, at least one of which is finite, and $|\nu| = \nu^+ + \nu^-$.*

Every $\mathbb{C}$-measure ν can be written as $\nu = \nu_1 - \nu_2 + i\nu_3 - i\nu_4$ where ν_1, ν_2, ν_3 and ν_4 are finite positive measures.

Proof Given a signed measure ν with values in $\mathbb{R}_\infty$, let (S_+, S_-) be a Hahn decomposition of ν. Setting

$$\nu^+(A) := \nu(A \cap S_+), \qquad \nu^-(A) := -\nu(A \cap S_-) \qquad A \in \mathcal{S},$$

we get two positive measures ν^+ and ν^- on $\mathcal{S}$ and ν^- takes its values in $\mathbb{R}_+$. Thus $\nu(A) = \nu(A \cap S_+) + \nu(A \cap S_-) = \nu^+(A) - \nu^-(A)$ for all $A \in \mathcal{S}$. Let $\mu = \nu^+ + \nu^-$. Then, for all $A \in \mathcal{S}$ we have $|\nu(A)| \leq \mu(A)$, hence $|\nu|(A) \leq \mu(A)$ by Proposition 8.1. On the other hand, the definition of $|\nu|$ yields $\mu(A) := \nu(A \cap S_+) + \nu^-(A \cap S_-) \leq |\nu|(A)$, so that $|\nu| = \mu$.

If ν is a $\mathbb{C}$-measure, its real part and its imaginary part are finite signed measures. Each of them can be decomposed as a difference of two finite positive measures. □

Remark From the preceding remark one can see that the pair (ν^+, ν^-) does not depend on the Hahn decomposition (S_+, S_-) of S. Moreover, one has $\|\nu\| := |\nu|(S) = \nu(S_+) - \nu(S_-)$.

Corollary 8.2 *If μ and ν are signed measures on $\mathcal{S}$ the following assertions are equivalent:*

(a) $\nu \ll \mu$;
(b) $\nu^+ \ll \mu$ *and* $\nu^- \ll \mu$;
(c) $\nu^+ \ll |\mu|$ *and* $\nu^- \ll |\mu|$;
(d) $|\nu| \ll |\mu|$.

Proof Assume (a) and let (S_+, S_-) be a Hahn decomposition of S with respect to ν. Then, if for some $A \in \mathcal{S}$ we have $|\mu|(A) = 0$, we obtain

$$0 \le |\mu|(A \cap S_+) \le |\mu|(A) = 0,$$

$$0 \le |\mu|(A \cap S_-) \le |\mu|(A) = 0,$$

hence $\nu^+(A) := \nu(A \cap S_+) = 0$ and $\nu^-(A) := \nu(A \cap S_-) = 0$, so that (b) holds.

The implications (b)⇒(c) and (d)⇒(a) are consequences of the relations $|\mu(A)| \le |\mu|(A)$ and $|\nu(A)| \le |\nu|(A)$ for all $A \in \mathcal{S}$. The implication (c)⇒(d) stems from the relation $|\nu|(A) = \nu^+(A) + \nu^-(A)$. □

We also present a decomposition of a signed or complex measure with respect to a given measure μ. Hereafter $(S, \mathcal{S})$ is a fixed measurable space.

A positive measure μ on $(S, \mathcal{S})$ is said to be *concentrated* on $T \in \mathcal{S}$ if $\mu(S \setminus T) = 0$. A signed or vectorial measure ν on $(S, \mathcal{S})$ is said to be concentrated on $T \in \mathcal{S}$ if its variation $|\nu|$ is concentrated on T or equivalently if for all $A \in \mathcal{S}$ contained in $S \setminus T$ one has $\nu(A) = 0$. Two signed or vectorial measures μ, ν on $\mathcal{S}$ are said to be *singular* (or ν is said to be singular with respect to μ) if for some $T \in \mathcal{S}$ the measure μ is concentrated on T and ν is concentrated on $S \setminus T$. In such a case, we write $\mu \perp \nu$.

Example If μ is a signed measure, its positive part and its negative part are singular.

Theorem 8.3 (Lebesgue Decomposition Theorem) *Let $(S, \mathcal{S}, \mu)$ be a measure space and let ν be a signed (resp. positive, resp. vectorial) measure whose total variation $|\nu|$ is σ-finite. Then there are unique signed (resp. positive, resp. vectorial) measures ν_a and ν_s on $\mathcal{S}$ such that $\nu = \nu_a + \nu_s$, $\nu_a \ll \mu$, $\nu_s \perp \mu$, $\nu_s \perp \nu_a$.*

Proof We start with the case when ν is a finite positive measure. Then we set

$$\mathcal{N}_\mu := \{A \in \mathcal{S} : \mu(A) = 0\}.$$

We pick a sequence (N_n) of $\mathcal{N}_\mu$ such that

$$\lim_n \nu(N_n) = \sup\{\nu(A) : A \in \mathcal{N}_\mu\}$$

and we set $N := \cup_n N_n$, so that $\mu(N) = 0$. Then we define measures ν_a and ν_s by setting $\nu_a(A) := \nu(A \setminus N)$, $\nu_s(A) := \nu(A \cap N)$. Clearly $\nu = \nu_a + \nu_s$ and $\nu_s \perp \mu$, $\nu_s \perp \nu_a$. Moreover, $\nu(N) \ge \sup \nu(N_n)$, hence

$$\nu(N) = \sup\{\nu(A) : A \in \mathcal{N}_\mu\}.$$

It follows that any $A \in \mathcal{S}$ satisfying $\mu(A) = 0$ also satisfies $\nu_a(A) = 0$ since otherwise, setting $B := A \backslash N$ one would have $\nu(B) > 0$, $\mu(B) = 0$, hence $\nu(N \cup B) > \nu(N)$ and $N \cup B \in \mathcal{N}_\mu$, an impossibility. Thus $\nu_a \ll \mu$.

In the case when ν is a vectorial measure, we can apply the preceding to the finite measure $|\nu|$, obtaining a μ-null set N such that the Lebesgue decomposition of $|\nu|$ is given by $|\nu| = |\nu|_a + |\nu|_s$ with $|\nu|_a(A) = |\nu|(A \backslash N)$, $|\nu|_s = |\nu|(A \cap N)$. Then one can verify that the vectorial measures ν_a and ν_s defined by $\nu_a(A) := \nu(A \backslash N)$ and $\nu_s(A) := \nu(A \cap N)$ form a Lebesgue decomposition of ν.

Now let us consider the case when ν is a σ-finite positive measure. Let (S_n) be a partition of S into elements of $\mathcal{S}$ with finite measures and let $\mathcal{S}_n := \{A \in \mathcal{S} : A \subset S_n\}$. The preceding construction for $\nu \mid \mathcal{S}_n$ yields μ-null subsets $N_n \in \mathcal{S}_n$ and corresponding measures. Let $N := \cup_n N_n$, a μ-null subset of S. Setting $\nu_a(A) := \nu(A \cap N)$, $\nu_s(A) := \nu(A \backslash N)$ for $A \in \mathcal{S}$, we get measures ν_a and ν_s forming a Lebesgue decomposition of ν.

Finally, we turn to uniqueness of the Lebesgue decomposition. Let $\nu = \nu_a + \nu_s$ and $\nu = \nu'_a + \nu'_s$ be two such decompositions. Then $\nu_a - \nu'_a \ll \mu$ and $\nu'_s - \nu_s \perp \mu$, so that there exists some $N \in \mathcal{N}_\mu$ such that $(\nu'_s - \nu_s)(A) = (\nu'_s - \nu_s)(A \cap N)$ for all $A \in \mathcal{S}$. But since $\nu_a - \nu'_a \ll \mu$ we have $\nu'_s(A \cap N) - \nu_s(A \cap N) = (\nu_a - \nu'_a)(A \cap N) = 0$. Thus $\nu'_s = \nu_s$ and $\nu'_a = \nu_a$. In the case when ν is a σ-finite positive measure we use a partition (S_n) of S as above and we apply uniqueness of the Lebesgue decomposition for the restriction of ν to each measure space $(S_n, \mathcal{S}_{S_n}, \mu \mid \mathcal{S}_{S_n})$. □

Theorem 8.4 (Bartle-Graves-Schwartz) *The range $\nu(\mathcal{S})$ of a vectorial measure $\nu : \mathcal{S} \to E$ is relatively weakly compact in E.*

Proof Let R be the weak closure of $\nu(\mathcal{S})$. For every $e^* \in E^*$, $e^* \circ \nu$ is a signed measure on $\mathcal{S}$. Let (S_+, S_-) be a Hahn decomposition of S for $e^* \circ \nu$: (S_+, S_-) is a partition of S and $e^* \circ \nu$ (resp. $-e^* \circ \nu$) is a positive measure on S_+ (resp. S_-). Since

$$\sup e^*(R) = \sup\{e^*(\nu(A)) : A \in \mathcal{S}\} = (e^* \circ \nu)(S_+),$$

James' theorem ensures that R is weakly compact. □

Let us end this section with a famous result which has several applications. We just state it (see [13, 186, Theorem 8.7.4] for the proof). Here we say that $A \in \mathcal{S}$ is an *atom* of a measure space $(S, \mathcal{S}, \mu)$ if $\mu(A) > 0$ and if for every $B \in \mathcal{S}$ contained in A one has either $\mu(B) = 0$ or $\mu(B) = \mu(A)$. The measure space $(S, \mathcal{S}, \mu)$ is said to be *non-atomic* if $\mathcal{S}$ contains no atom. The Lebesgue measure is non-atomic, but Dirac measures are atomic.

Theorem 8.5 (Lyapunov) *Let F be a finite dimensional Banach space and let $\nu := \mu_h : \mathcal{S} \to F$ be the vector measure associated with an integrable map $h \in L_1(S, F)$ by*

$$\nu(A) := \int_A h d\mu.$$

If $(S, \mathcal{S}, \mu)$ is non-atomic, then $\nu(\mathcal{S}) := \{\nu(A) : A \in \mathcal{S}\}$ is convex.

When F is an infinite dimensional Banach space the conclusion is that the closure of $\nu(\mathcal{S})$ is convex. See [13, Thm 8.7.4].

8.2.2 Differentiation of Measures

We devote this subsection to a crucial result giving a representation of absolutely continuous measures with respect to a given measure μ. Its most classical version is as follows.

Theorem 8.6 (Radon-Nikodým) *Let $(S, \mathcal{S}, \mu)$ be a σ-finite measure space and let ν be a positive measure on $(S, \mathcal{S})$ that is absolutely continuous with respect to μ. Then, there exists a measurable function $h : S \to \mathbb{R}_+$ such that $\nu(A) = \int_A h d\mu$ for all $A \in \mathcal{S}$.*

If ν is σ-finite (resp. finite) the function h is unique up to a set of μ-measure zero (resp. and is integrable).

Proof First, let us consider the case when μ is finite. Let F be the set of all μ-measurable functions $f : S \to \overline{\mathbb{R}}_+$ such that $\int_A f d\mu \leq \nu(A)$ for all $A \in \mathcal{S}$. It is nonempty since $0 \in F$. We first note that for $f, g \in F$ one has $f \vee g := \max(f, g) \in F$ since for all $A \in \mathcal{S}$, setting $B := \{s \in A : f(s) > g(s)\}$, $C := A \backslash B$ we have

$$\int_A (f \vee g) d\mu = \int_B f d\mu + \int_C g d\mu \leq \nu(B) + \nu(C) = \nu(A).$$

Let (f_n) be a sequence in F such that $\lim_n \int f_n d\mu = \sup\{\int f d\mu : f \in F\}$. Replacing f_n with $g_n := f_1 \vee \ldots \vee f_n$ we may assume that (f_n) is increasing. Let $h = \lim_n f_n$. The Monotone Convergence Theorem implies that $\int_A h d\mu = \lim_n \int_A f_n d\mu \leq \nu(A)$ for all $A \in \mathcal{S}$, hence that $h \in F$. Moreover, $\int_S h d\mu = \lim_n \int_S f_n d\mu = \sup\{\int_S f d\mu : f \in F\}$.

Now let us show that $\nu(A) = \int_A h d\mu$ for all $A \in \mathcal{S}$. By definition of F, setting $\nu'(A) := \nu(A) - \mu_h(A)$ with $\mu_h(A) := \int_A h d\mu$, we get a positive measure. We want to show that $\nu' = 0$. Assume on the contrary that $\nu'(S) > 0$. Since $\mu(S) < +\infty$ there is some $c > 0$ such that

$$\nu'(S) > c\mu(S).$$

Let (S_+, S_-) be a Hahn decomposition of S for the signed measure $\nu' - c\mu$. For all $A \in \mathcal{S}$ we have $\nu'(A \cap S_+) \geq c\mu(A \cap S_+)$, $\nu'(A) \geq \nu'(A \cap S_+)$, and $\nu(A) := \nu'(A) + \mu_h(A)$ hence

$$\nu(A) \geq \nu'(A \cap S_+) + \int_A h d\mu \geq c\mu(A \cap S_+) + \int_A h d\mu = \int_A (h + c 1_{S_+}) d\mu,$$

so that $h' := h + c1_{S_+} \in F$. Moreover, since ν' is absolutely continuous with respect to μ we have $\mu(S_+) > 0$ (otherwise we would have $\nu'(S_+) = 0$ and $\nu'(S) - c\mu(S) = (\nu' - c\mu)(S_-) \leq 0$, a contradiction with $\nu'(S) > c\mu(S)$). Then $\int h' d\mu = \int h d\mu + c\mu(S_+) > \int h d\mu = \sup\{\int f d\mu : f \in F\}$, a contradiction.

Since $\int h d\mu$ is finite, changing h on a set of μ-measure 0 we may assume h takes only finite values.

Now let us consider the case where μ is σ-finite. Then S is the union of a sequence in disjoint subsets (S_n) in $\mathcal{S}$ with finite measures under μ. The first part of the proof provides measurable functions $h_n : S_n \to \mathbb{R}_+$ such that $\nu(A) = \int_A h_n d\mu$ for all $A \in \mathcal{S}$ contained in S_n. The function $h : S \to \mathbb{R}_+$ that agrees with h_n on S_n for all n is easily seen to be measurable and (by σ-additivity) such that $\mu_h = \nu$.

The uniqueness assertion is contained in Proposition 8.3 in the case when ν is finite; the general case ensues as a countable union of null sets is a null set. □

Corollary 8.3 *Let $(S, \mathcal{S})$ be a measurable space and let μ be a σ-finite positive measure on $(S, \mathcal{S})$. Let $\nu : \mathcal{S} \to \mathbb{C}$ be a complex measure that is absolutely continuous with respect to μ. Then, there exists an integrable function $h : S \to \mathbb{C}$ such that $\nu = \mu_h$, i.e. $\nu(A) = \int_A h d\mu$ for all $A \in \mathcal{S}$. The function h is unique up to a set of μ-measure zero.*

Proof Let ν_r and ν_i be the signed measures such that $\nu = \nu_r + i\nu_i$. Clearly $\nu_r \ll \mu$ and $\nu_i \ll \mu$. Then by Corollary 8.2 we have $\nu_r^+ \ll \mu$, $\nu_r^- \ll \mu$, $\nu_i^+ \ll \mu$, and $\nu_i^- \ll \mu$. Applying Theorem 8.6 to each of these positive measures we get h as required. □

The function h (or rather its class) appearing in the preceding corollary is called the *Radon-Nikodým derivative* of ν with respect to μ. It is usually denoted by $\frac{D\nu}{D\mu}$ or $\frac{d\nu}{d\mu}$

It is not true that for any Banach space E, any finite measure space $(S, \mathcal{S}, \mu)$, and any measure $\nu : \mathcal{S} \to E$ that is μ-continuous and of bounded variation there exists some $h \in L_1(\mu, E)$ such that $\nu = \mu_h$. See Exercises 7 and 8. Thus, a definition is in order.

Definition 8.2 A Banach space E is said to have the *Radon-Nikodým Property* (RNP) with respect to a measure space $(S, \mathcal{S}, \mu)$ if for any vectorial measure $\nu : \mathcal{S} \to E$ that is absolutely continuous with respect to μ and is of bounded variation there exists some $h \in L_1(\mu, E)$ such that $\nu = \mu_h$, i.e. such that $\nu(A) = \int_A h d\mu$ for all $A \in \mathcal{S}$.

The space E has the *Radon-Nikodým Property* (RNP) if for any finite measure space $(S, \mathcal{S}, \mu)$ it has the *Radon-Nikodým Property* (RNP) with respect to $(S, \mathcal{S}, \mu)$.

A number of characterizations of this property can be given. Here we just mention that any reflexive Banach space and any separable dual space has the Radon-Nikodým Property. More generally, if X is the dual space of a space X_*, X has the RNP if and only if X_* is an Asplund space, i.e. a Banach space whose separable subspaces have separable dual spaces (see for example [209, Thm 5.7]). Spaces with the RNP have remarkable permanence properties as the following quotation from [99, p. 81] shows.

Theorem 8.7 *If a Banach space has the Radon-Nikodým Property, so does each of its closed linear subspaces.*

A Banach space has the Radon-Nikodým Property if each of its closed separable linear subspace has this property.

Here we contend to show that any Hilbert space has the Radon-Nikodým Property. Before doing that, let us compare the Radon-Nikodým Property with another important property.

Definition 8.3 Given a Banach space E and a measure space $(S, \mathcal{S}, \mu)$, a continuous linear operator $T : L_1(\mu) \to E$ is said to be (Riesz) *representable* if there exists some $h \in L_\infty(\mu, E)$ such that

$$T(f) = \int_S fhd\mu \qquad \text{for all } f \in L_1(\mu) := L_1(\mu, \mathbb{R}).$$

Proposition 8.4* *Given a finite measure space $(S, \mathcal{S}, \mu)$, a Banach space E has the Radon-Nikodým Property with respect to $(S, \mathcal{S}, \mu)$ if and only if every continuous linear operator $T : L_1(\mu) \to E$ is representable.*

Proof Suppose the Banach space E has the RNP with respect to $(S, \mathcal{S}, \mu)$ and let $T : L_1(\mu) \to E$ be a continuous linear map. Let $\nu : \mathcal{S} \to E$ be given by $\nu(A) := T(1_A)$ for $A \in \mathcal{S}$. Then, for all $A \in \mathcal{S}$ we have $\|\nu(A)\| \leq \|T\| \, \mu(A)$, so that ν is μ-continuous and is countably additive, not just additive, as shown by Corollary 7.11 and the continuity of T. The RNP yields some $h \in L_1(\mu, E)$ such that $\nu(A) = \int_A hd\mu$ for all $A \in \mathcal{S}$. Moreover, the total variation $|\nu|$ of ν satisfies $|\nu|(A) \leq \|T\| \, \mu(A)$ for all $A \in \mathcal{S}$ by Proposition 8.1 and $|\nu|(A) = |\mu_h|(A) = \int_A \|h\|_E \, d\mu$ for all $A \in \mathcal{S}$. Then, by Corollary 7.13, we get $h \in L_\infty(\mu, E)$ and $\|h\| \leq \|T\|$ almost everywhere. The relation $T(f) = \int fhd\mu$ obviously holds for $f \in St(\mu, E)$ and can be extended to $L_1(\mu)$ by continuity: T is representable.

Conversely, suppose every continuous linear operator $T : L_1(\mu) \to E$ is representable. Let $\nu : \mathcal{S} \to E$ be a μ-continuous vectorial measure of bounded variation. By its very definition, the measure $|\nu|$ is also μ-continuous. The Radon-Nikodým Theorem yields some $h \in L_1(\mu)$ such that $|\nu| = \mu_h$, i.e. $|\nu|(A) = \int_A hd\mu$ for all $A \in \mathcal{S}$. For $n \in \mathbb{N}\backslash\{0\}$ let $S_n := h^{-1}(]n-1, n])$ and let $T_n : St(\mu, \mathbb{R}) \to E$ be the linear map defined by

$$T_n(f) := \sum_{i=1}^{k} r_i \nu(S_n \cap A_i) = \int_{S_n} fd\nu \qquad \text{for } f := \sum_{i=1}^{k} r_i 1_{A_i} \in St(\mu, \mathbb{R}).$$

For f as above, $T_n(f)$ satisfies

$$\|T_n(f)\| \le \sum_{i=1}^{k} |r_i| \, \|\nu(S_n \cap A_i)\| \le \sum_{i=1}^{k} |r_i| \, |\nu| \, (S_n \cap A_i)$$

$$\le \sum_{i=1}^{k} |r_i| \, n\mu(S_n \cap A_i) \le n \, \|f\|_1 .$$

Thus T_n can be extended to $L_1(\mu)$ into a continuous linear operator still denoted by T_n. Since T_n is representable there exists a $g_n \in L_\infty(\mu, E)$ such that $T_n(f) = \int_S f g_n d\mu$ for all $f \in L_1(\mu)$. In particular, if $A \in \mathcal{S}$ we have

$$\int_A g_n d\mu = T_n(1_A) = \nu(A \cap S_n).$$

Let $g : S \to E$ be given by $g(s) := g_n(s)$ if $s \in S_n$. Setting $B_n := \cup_{k=1}^{n} S_k$, for $A \in \mathcal{S}$ we have

$$\nu(A) = \lim_n \nu(A \cap B_n) = \lim_n \int_{A \cap B_n} g d\mu.$$

Since ν is of bounded variation, we have

$$\int_{B_n} \|g\| \, d\mu = \sum_{k=1}^{n} \int_{S_k} \|g_k\| \, d\mu = \sum_{k=1}^{n} |\nu| \, (S_k) \le |\nu| \, (B_n) \le |\nu| \, (S).$$

For all $A \in \mathcal{S}$, replacing B_n by $A \cap B_n$ in these relations and passing to the limit over n, we see that $\int_A \|g\| \, d\mu \le |\nu| \, (A)$ and Corollary 7.13 ensures that $\|g\| \in \mathcal{L}_\infty(\mu)$ and $g \in \mathcal{L}_\infty$. Moreover $\nu(A) = \lim_n \int_A 1_{B_n} g d\mu = \int_A g d\mu$. Thus E has the Radon-Nikodým Property. □

For spaces with the Radon-Nikodým Property, the Lebesgue decomposition theorem yields a representation of vectorial measures of bounded variation.

Theorem 8.8 *Let E be a Banach space with the Radon-Nikodým Property and let $(S, \mathcal{S}, \mu)$ be a finite measure space. For any vectorial measure $\nu : \mathcal{S} \to E$ whose total variation is finite there exist $h \in L_1(\mu, E)$ and a measure $\nu_s : \mathcal{B} \to E$ that is singular with respect to μ such that $\nu = \mu_h + \nu_s$.*

Proof Let $\nu = \nu_a + \nu_s$ be the Lebesgue decomposition of ν, with $\nu_a \ll \mu$ and $\nu_s \perp \mu$. Since E has the RNP and the total variation of ν_a is the total variation of ν, for $h := \frac{D\nu_a}{D\mu} \in L_1(\mu, E)$ one has $\nu_a = \mu_h : B \mapsto \int_B h d\mu$. □

Exercises

1. Let ν be a signed measure on $(S, \mathcal{S})$. Set $\nu^+ := \frac{1}{2}(|\nu| + \nu)$ and $\nu^- := \frac{1}{2}(|\nu| - \nu)$. Prove that there is a partition (S_+, S_-) of S into two measurable disjoint subsets such that $\nu^+(B) = 0$ for all $B \subset S_-$, $B \in \mathcal{S}$ and

$\nu^-(C) = 0$ for all $C \subset S_+$, $C \in \mathcal{S}$ and that

$$\nu^+(A) = \sup\{\nu(C) : C \in \mathcal{S},\ C \subset A \cap S_+\}$$
$$\nu^-(A) = \inf\{\nu(B) : B \in \mathcal{S},\ A \cap B \subset S_-\}.$$

2. Let $(S, \mathcal{S}, \mu)$ be a measure space and let $\mathcal{A} := \{A \in \mathcal{S} : \mu(A) < \infty\}$. For A, $B \in \mathcal{A}$ set $d(A, B) := \mu(A \backslash B) + \mu(B \backslash A)$. Show that d is a semimetric on $\mathcal{A}$. Verify that a measure $\nu : \mathcal{A} \to \mathbb{R}_+$ is absolutely continuous with respect to μ if and only if ν is continuous on $(\mathcal{A}, d)$.

3. Let $(S, \mathcal{S}, \mu)$ be a measure space, and let $\nu = \nu_a + \nu_s$ be the Lebesgue decomposition of a vectorial measure ν on $(S, \mathcal{S})$. Show that $\|\nu\| = \|\nu_a\| + \|\nu_s\|$.

4. Let $(S, \mathcal{S}, \mu)$ be a measure space. Show that the set $\{\nu \in \mathcal{M}(S, \mathcal{S}) : \nu \perp \mu\}$ is a closed linear subspace of the set $\mathcal{M}(S, \mathcal{S})$ of finite signed measures on $(S, \mathcal{S})$.

5. Let μ and ν be two positive measures on $(S, \mathcal{S})$ such that for all $\varepsilon > 0$ there exists some $A \in \mathcal{S}$ satisfying $\mu(A) < \varepsilon$ and $\nu(S \backslash A) < \varepsilon$. Show that $\nu \perp \mu$. [Hint: consider a sequence (A_n) of $\mathcal{S}$ such that $\mu(A_n) < 2^{-n}$, $\nu(S \backslash A_n) < 2^{-n}$ and set $T := \bigcap_m \bigcup_{n \geq m} A_n$.]

6. Let ν be a finite signed or complex measure on a measurable space $(S, \mathcal{S})$. Show that the Radon-Nikodým derivative $\frac{d\nu}{d|\nu|}$ of ν with respect to $|\nu|$ satisfies $\left|\frac{d\nu}{d|\nu|}\right| = 1$ $|\nu|$-almost everywhere on S.

Let ν', ν'' be finite signed or complex measures on a measure space $(S, \mu, \mathcal{S})$ with μ σ-finite. Verify that if $\nu' \ll \mu$ and $\nu'' \ll \mu$ one has $\nu := \nu' + \nu'' \ll \mu$ and $\frac{d\nu}{d\mu} = \frac{d\nu'}{d\mu} + \frac{d\nu''}{d\mu}$.

7. Let λ and μ be σ-finite measures on a measurable space $(S, \mathcal{S})$ such that $\mu \ll \lambda$ and let ν be a σ-finite signed measure such that $\nu \ll \mu$. Show that $\nu \ll \lambda$ and that $\frac{d\nu}{d\lambda} = \frac{d\nu}{d\mu} \cdot \frac{d\mu}{d\lambda}$ (λ-almost everywhere).

8. Let $E := \ell_1$ be the space of sequences $x := (x_n)$ of real numbers such that $\|x\|_1 := \Sigma_n |x_n| < +\infty$. Prove that E has the RNP. [Hint: note that if e_n is the element of E whose components are all 0 except for the nth which is 1, the following property holds: if (r_n) is a sequence of real numbers such that $\sup_n \|\Sigma_{k \leq n} r_k e_k\|_1 < \infty$ then $\Sigma_{k \leq n} r_k e_k$ converges in ℓ_1. Then, given a measure space $(S, \mathcal{S}, \mu)$ and a measure $\nu : \mathcal{S} \to E$ of bounded variation satisfying $\nu \ll \mu$ apply the Radon-Nikodým Theorem to the measures ν_n given by $\nu(A) := \Sigma_n \nu_n(A) e_n$ to get a Radon-Nikodým derivative of ν.]

9*. Prove that the space c_0 of sequences $x := (x_n)$ with limit 0 endowed with the supremum norm does not have the RNP.

10*. Let $S := [0, 1]$ endowed with the restriction μ of the Lebesgue measure and let $E := C(S)$ endowed with the supremum norm. Define $\nu : \mathcal{S} \to E$ by $\nu(A)(t) := \mu(A \cap [0, t])$ for $t \in S$. Verify that $\nu \ll \mu$ and that ν is of bounded variation. Prove that ν has no Radon-Nikodým derivative. [Hint: for this counterexample due to Lewis, see [99, p. 73].]

8.3 Differentiation of Measures on $\mathbb{R}^d$

In this section we want to give a concrete notion of the derivative of a measure μ on the Borel algebra $\mathcal{B} := \mathcal{B}_d$ of $\mathbb{R}^d$ with respect to the Lebesgue measure $\lambda := \lambda_d$ on $\mathcal{B}$. We shall use the notion of a Vitali covering of $\mathbb{R}^d$. Recall that a family $\mathcal{F}$ of subsets of a set S is said to be *disjoint* if distinct members of $\mathcal{F}$ are disjoint. If, moreover, $\mathcal{F}$ is a covering of S in the sense that the union of the members of $\mathcal{F}$ is S, then $\mathcal{F}$ is called a *partition* of S.

Definition 8.4 A *Vitali covering* of a subset A of $\mathbb{R}^d$ is a family $\mathcal{V}$ of measurable bounded subsets of $\mathbb{R}^d$ such that there exists some $c > 0$ for which one has $(\lambda(V))^{1/d} \geq c\,\mathrm{diam}(V) > 0$ for all $V \in \mathcal{V}$ and such that for all $x \in A$ and all $r > 0$ one can find some $V \in \mathcal{V}$ containing x with diameter $\mathrm{diam}(V) < r$.

The condition $\lambda(V) \geq c\,\mathrm{diam}(V) > 0$ for all $V \in \mathcal{V}$ is satisfied if $\mathcal{V}$ is a family of balls or cubes (which are balls for the norm $\|\cdot\|_\infty$). This condition discards sets that are too thin. Sometimes Vitali coverings are defined as families of balls, but we prefer to dispose of a more versatile definition. We observe that if $\mathcal{V}$ is a Vitali covering of A, then $\{\mathrm{cl}(V) : V \in \mathcal{V}\}$ is also a Vitali covering of A because for any subset V of $\mathbb{R}^d$ one has $\mathrm{diam}(\mathrm{cl}(V)) = \mathrm{diam}(V)$. We also note that if $\mathcal{V}$ is a Vitali covering of A with associated constant c, for any $c' \in]0, c[$ one can find an open Vitali covering $\mathcal{V}'$ of A with constant at least c' by replacing each member V of $\mathcal{V}$ with $V' := V + B(0, \varepsilon)$ where $\varepsilon > 0$ is such that $c\,\mathrm{diam}(V) \geq c'(\mathrm{diam}(V) + 2\varepsilon)$.

The following result is crucial but its proof is not easy. We advise the reader to consider first the case $d = 1$, even if the general case is not much different.

Theorem 8.9.*(Vitali) *Let A be an arbitrary nonempty subset of $\mathbb{R}^d$ and let $\mathcal{V}$ be a Vitali covering of A. Then there exist a finite or countable subfamily $\mathcal{W} := \{V_n : n \in \mathbb{N}\}$ of $\mathcal{V}$ and a null set N of $\mathbb{R}^d$ such that the sets $\mathrm{cl}(V_n)$ are disjoint and $A \subset (\cup_n \mathrm{cl}(V_n)) \cup N$.*

Proof Let us first suppose A is bounded; let $b > \sup\{\|x\| : x \in A\}$. Taking into account the preceding remarks we may assume the members of $\mathcal{V}$ are closed. Taking a subfamily of $\mathcal{V}$ if necessary, we may assume that all the members of $\mathcal{V}$ meet A and are contained in $B(0, b)$. If for a finite subfamily $\mathcal{U} := \{V_k : k \in \mathbb{N}_n\}$ of $\mathcal{V}$ the members of $\mathcal{U}$ are disjoint and $A \subset A_{\mathcal{U}} := \cup_k V_k$ we can take $\mathcal{W} := \mathcal{U}$. Thus we suppose henceforth that there is no such family.

Then, for any finite disjoint subfamily $\mathcal{U}$ of $\mathcal{V}$ there is some point $x \in A \backslash A_{\mathcal{U}}$. Since $A_{\mathcal{U}}$ is closed, we can find $\delta > 0$ such that $B(x, \delta) \cap A_{\mathcal{U}} = \varnothing$ and $B(x, \delta) \subset B(0, b)$. Taking some $V \in \mathcal{V}$ containing x with diameter less than δ we get a member of $\mathcal{V}$ disjoint from the members of $\mathcal{U}$, hence a larger disjoint subfamily of $\mathcal{V}$. In order to make this construction more precise and more efficient, we pick $V \in \mathcal{V}$ such that $\mathrm{diam} V \geq \gamma/2$, where

$$\gamma := \sup\{\mathrm{diam} V : V \in \mathcal{V},\ V \cap A_{\mathcal{U}} = \varnothing\},$$

noting that $0 < \gamma \leq 2b$ since each element of $\mathcal{V}$ is contained in $B(0, b)$. We set $\gamma_0 := \sup\{\operatorname{diam} V : V \in \mathcal{V}\}$, and we start with $\mathcal{U}_0 := \{V_0\}$, where $V_0 \in \mathcal{V}$ is such that $\operatorname{diam} V_0 \geq \gamma_0/2$. Assuming inductively that a disjoint subfamily $\mathcal{U}_{n-1} := \{V_0, \ldots, V_{n-1}\}$ of $\mathcal{V}$ has been chosen, we set

$$\gamma_n := \sup\{\operatorname{diam} V : V \in \mathcal{V},\ V \cap A_{\mathcal{U}_{n-1}} = \varnothing\},$$

and we pick some V_n in $\{V \in \mathcal{V} : V \cap A_{\mathcal{U}_{n-1}} = \varnothing\}$ such that $\operatorname{diam} V_n > \gamma_n/2$. Taking $\mathcal{U}_n := \mathcal{U}_{n-1} \cup \{V_n\}$ completes our induction step.

Since the family $\mathcal{W} := \{V_n : n \in \mathbb{N}\}$ is disjoint and its members are contained in $B(0, b)$, we have $\Sigma_n \lambda(V_n) \leq \lambda(B(0, b)) < \infty$, hence $(\lambda(V_n)) \to 0$. It follows that $(\operatorname{diam} V_n) \to 0$ and $(\gamma_n) \to 0$. For each $n \in \mathbb{N}$ we pick a closed ball B_n with center in V_n and radius $2\gamma_n$. Since $\operatorname{diam} V_n \leq \gamma_n$ we have $V_n + B[0, \gamma_n] \subset B_n$. Since $\lambda(V_n) \geq c^d (\operatorname{diam} V_n)^d \geq 2^{-d} c^d (\gamma_n)^d$ we even have $\Sigma_n (\gamma_n)^d < +\infty$ and $\Sigma_n \lambda(B_n) < +\infty$ since $\lambda(B_n) = b_d (2\gamma_n)^d$ where $b_d := \lambda(B[0, 1])$.

We claim that for all $m \in \mathbb{N}$ we have

$$A \backslash \bigcup_{j \leq m} V_j \subset \bigcup_{k > m} B_k. \tag{8.5}$$

In fact, given $x \in A \backslash \cup_{j \leq m} V_j$, we can find some $V \in \mathcal{V}$ such that $x \in V$ and $V \cap \cup_{j \leq m} V_j = \varnothing$. Since $(\gamma_n) \to 0$ and $\operatorname{diam} V > 0$, there are integers n such that $V \cap (\cup_{j \leq n} V_j) \neq \varnothing$. Let k be the smallest such integer, so that $k > m$ and $V \cap V_k \neq \varnothing$. Let $y \in V \cap V_k$, so that $d(x, y) \leq \operatorname{diam} V \leq \gamma_k$ by definition of γ_k and $x \in B[y, \gamma_k] \subset V_k + B[0, \gamma_k] \subset B_k$. Thus relation (8.5) holds.

Denoting by λ^* the outer measure associated with λ, we deduce from (8.5) that for all $m \in \mathbb{N}$ we have

$$\lambda^*(A \backslash \bigcup_{j \geq 0} V_j) \leq \lambda^*(A \backslash \bigcup_{j \leq m} V_j) \leq \sum_{k > m} \lambda(B_k).$$

Since the series $\Sigma_k \lambda(B_k)$ is convergent, we have $(\Sigma_{k > m} \lambda(B_k)) \to 0$ as $m \to \infty$, hence $\lambda^*(N) = 0$ for $N := A \backslash (\cup_{j \geq 0} V_j)$.

When A is unbounded, taking a countable disjoint family $\{G_i : i \in \mathbb{N}\}$ of bounded open subsets such that $\lambda(\mathbb{R}^d \backslash \cup_i G_i) = 0$, for all $i \in \mathbb{N}$ we get a countable disjoint subfamily $\{V_{i,n} : n \in \mathbb{N}\}$ of members of $\mathcal{V}$ contained in G_i covering $A \cap G_i$ up to a null set. Merging these families into a single family we obtain the required countable family covering A up to a null set. □

Definition 8.5 Given a Vitali covering $\mathcal{V}$ of $\mathbb{R}^d$ and a signed measure μ on $\mathcal{B}$, the *upper derivative* and the *lower derivative* of μ at $x \in \mathbb{R}^d$ are defined respectively by

$$\overline{D}\mu(x) = \inf_{r>0} \sup\{\frac{\mu(V)}{\lambda(V)} : V \in \mathcal{V}, x \in \text{int}(V), \text{ diam}(V) < r\},$$

$$\underline{D}\mu(x) = \sup_{r>0} \inf\{\frac{\mu(V)}{\lambda(V)} : V \in \mathcal{V}, x \in \text{int}(V), \text{ diam}(V) < r\}.$$

The measure μ is said to be *differentiable at* x when $\overline{D}\mu(x) = \underline{D}\mu(x) \in \mathbb{R}_+$. Then this value is denoted by $D\mu(x)$ and is called the *derivative* of μ at x.

Note that in this definition $\inf_{r>0}$ and $\sup_{r>0}$ can be replaced with $\lim_{r\to 0_+}$ and when μ is differentiable at x, for $\mathcal{V}(x) := \{V \in \mathcal{V} : x \in \text{int}(V)\}$, one has

$$D\mu(x) = \lim_{\text{diam}(V)\to 0_+} \{\frac{\mu(V)}{\lambda(V)} : V \in \mathcal{V}(x)\}.$$

When μ is a vectorial measure we adopt this definition for the derivative of μ at x.

Lemma 8.2 *For any signed measure μ on $\mathcal{B}$ and any open Vitali covering $\mathcal{V}$ of $\mathbb{R}^d$ the functions $\overline{D}\mu$ and $\underline{D}\mu$ are measurable.*

Proof Since $\underline{D}\mu(x) = -\overline{D}(-\mu)(x)$ for all $x \in \mathbb{R}^d$, it suffices to prove the measurability of $\overline{D}\mu$. Taking a sequence (r_n) of positive rational numbers with limit 0, we have $\overline{D}\mu = \lim_n \overline{D}_{r_n}\mu$ with

$$\overline{D}_r\mu(x) := \sup\{\frac{\mu(V)}{\lambda(V)} : V \in \mathcal{V},\ x \in \text{int}(V), \text{ diam}(V) < r\}.$$

For all $t \in \mathbb{R}$ the set $\{x \in \mathbb{R}^d : \overline{D}_r\mu(x) > t\}$ is open. Thus $\overline{D}_r\mu$ is a lower semicontinuous function, hence is measurable. Then $\overline{D}\mu$ is measurable as the limit of a sequence in measurable functions. □

The definitions of lower and upper derivatives enable us to obtain estimates.

Lemma 8.3 *Let μ be a positive measure on $\mathcal{B}$ finite on compact sets and let $B \in \mathcal{B}$.*

(a) If for some $c > 0$ one has $\overline{D}\mu(x) \geq c$ for all $x \in B$ then one has $\mu(B) \geq c\lambda(B)$.
(b) If μ is absolutely continuous with respect to the Lebesgue measure λ of $\mathbb{R}^d$ and if for some $c > 0$ one has $\underline{D}\mu(x) \leq c$ for all $x \in B$, then one has $\mu(B) \leq c\lambda(B)$.

Proof

(a) Since μ is regular by Theorem 1.14, it suffices to show that for all open subsets U of $\mathbb{R}^d$ containing B and all $b \in]0, c[$ one has $\mu(U) \geq b\lambda(B)$.
 Given a Vitali covering $\mathcal{V}$ of $\mathbb{R}^d$ let $\mathcal{V}_b$ be the family of those $V \in \mathcal{V}$ contained in U and such that $\mu(V) \geq b\lambda(V)$. Since $\overline{D}\mu(x) \geq c$ for all $x \in B$, the

definition of $\overline{D}\mu(x)$ ensures that $\mathcal{V}_b$ is a Vitali covering of B. Theorem 8.9 provides a countable disjoint subfamily $\mathcal{W} := \{V_n : n \in \mathbb{N}\}$ of $\mathcal{V}_b$ such that $\lambda(B \setminus \cup_n V_n) = 0$. Since the sets V_n are disjoint, contained in U, and such that $\mu(V_n) \geq b\lambda(V_n)$, we get

$$\mu(U) \geq \sum_n \mu(V_n) \geq \sum_n b\lambda(V_n) = b\lambda(\bigcup_n V_n) = b\lambda(B).$$

(b) Given $a > c$ and a Vitali covering $\mathcal{V}$ of $\mathbb{R}^d$ let $\mathcal{V}_a$ be the family of those $V \in \mathcal{V}$ such that $\mu(V) \leq a\lambda(V)$. Since $\underline{D}\mu(x) < a$ for all $x \in B$, the definition of $\underline{D}\mu(x)$ ensures that $\mathcal{V}_a$ is a Vitali covering of B. Theorem 8.9 provides a countable disjoint subfamily $\mathcal{W} := \{V_n : n \in \mathbb{N}\}$ of $\mathcal{V}_a$ such that $\lambda(B \setminus \cup_n V_n) = 0$. Since the sets V_n are disjoint and such that $\mu(V_n) \leq a\lambda(V_n)$ and since $\lambda(B \setminus \cup_n V_n) = 0$, hence $\mu(B \setminus \cup_n V_n) = 0$ as $\mu \ll \lambda$, we get

$$a\lambda(B) \geq a\sum_n \lambda(V_n) \geq \sum_n \mu(V_n) = \mu(\bigcup_n V_n) \geq \mu(B).$$

Since $a > c$ is arbitrarily close to c, we get $\mu(B) \leq c\lambda(B)$. □

Let us turn to differentiability results. It is natural to compare the definition of derivative we just introduced with the notion of Radon-Nikodým derivative. We start with some special cases.

Lemma 8.4 *Let μ be a finite Borel measure on $\mathbb{R}^d$ that is singular with respect to the Lebesgue measure λ. Then μ is differentiable, with derivative 0 almost everywhere.*

Proof Let $N \in \mathcal{B}$ be such that $\lambda(N) = 0$ and $\mu(N^c) = 0$ for $N^c := \mathbb{R}^d \setminus N$. Let $B := \{x \in \mathbb{R}^d : \overline{D}\mu(x) > 0\}$ and for $n \in \mathbb{N}\setminus\{0\}$ let $B_n := \{x \in N^c : \overline{D}\mu(x) \geq 1/n\} \in \mathcal{B}$. Lemma 8.3 ensures that

$$\lambda(B_n) \leq n\mu(B_n) \leq n\mu(N^c) = 0.$$

Since $B \subset N \cup (\cup_n B_n)$ we have $\lambda(B) = 0$. Since $0 \leq \underline{D}\mu \leq \overline{D}\mu$, we get that $\underline{D}\mu = \overline{D}\mu = 0$ on $\mathbb{R}^d \setminus B$. □

Theorem 8.10 *Given $h \in L_1(\mathbb{R}^d)$, let μ_h be the associated measure with density h given by $\mu_h(A) := \int_A h d\lambda$. Then μ_h is differentiable almost everywhere and $D\mu_h = h$ almost everywhere.*

Proof Without loss of generality we suppose h is measurable. For $t \in \mathbb{R}$ we denote by μ_t the measure with density $(h-t)^+$ with respect to λ :

$$\mu_t(B) = \int_B (h(x) - t)^+ d\lambda(x) \qquad \text{for } B \in \mathcal{B}.$$

Then, for $S_t := \{h < t\}$, μ_t is a positive measure satisfying $\mu_t(B) = 0$ for all $B \in \mathcal{B}$ contained in S_t. Taking a sequence $(\alpha_n) \to 0_+$ and setting $B_n := \{x \in S_t : \overline{D}\mu_t(x) > \alpha_n\} \in \mathcal{B}$, we deduce from Lemma 8.3 that $\lambda(B_n) = 0$. Thus $\overline{D}\mu_t = 0$ a.e. on S_t. Moreover, since $h \le (h-t)^+ + t$, for any $B \in \mathcal{B}$ we have

$$\mu_h(B) := \int_B h d\lambda \le \mu_t(B) + t\lambda(B).$$

In particular, for $x \in \mathbb{R}^d$ and $V \in \mathcal{V}(x)$ we have

$$\frac{\mu_h(V)}{\lambda(V)} \le \frac{\mu_t(V)}{\lambda(V)} + t,$$

so that $\overline{D}\mu_h(x) \le \overline{D}\mu_t(x) + t$. Since $\overline{D}\mu_t = 0$ a.e. on S_t, for all $t \in \mathbb{Q}$, the set $N_t := \{x \in \mathbb{R}^d : h(x) < t < \overline{D}\mu_h(x)\}$ is a λ-null set. Since $\mathbb{Q}$ is countable, the set $\{h < \overline{D}\mu_h\}$ is a λ-null set. Since $\mu_{-h} = -\mu_h$ and $\underline{D}\mu_h = -\overline{D}(-\mu_h)$, we see that $\{h > \underline{D}\mu_h\}$ is a λ-null set and $\overline{D}\mu_h \le h \le \underline{D}\mu_h$ a.e. Since $\underline{D}\mu_h \le \overline{D}\mu_h$, as is easily seen, we obtain that $\underline{D}\mu_h = h = \overline{D}\mu_h$ a.e. and μ_h is differentiable a.e. with derivative h. □

The preceding result can be extended to vectorial measures.

Theorem 8.11 *Let $h \in L_1(\mathbb{R}^d, \lambda, E)$, where E is a Banach space. Then the measure μ_h with density h is almost everywhere differentiable on $\mathbb{R}^d$ and its derivative is h: $D\mu_h = h$.*

Proof Modifying h on a null set if necessary, we may suppose h is measurable and that $h(\mathbb{R}^d)$ is separable. Let $\mathcal{V}$ be a Vitali covering of $\mathbb{R}^d$. For $w \in \mathbb{R}^d$ and $r > 0$ let $\mathcal{V}_r(w) := \{V \in \mathcal{V} : w \in \mathrm{int}(V),\ \mathrm{diam} V < r\}$. Let

$$W := \{w \in \mathbb{R}^d : \forall \varepsilon > 0\ \exists r > 0\ \forall V \in \mathcal{V}_r(w) \int_V \|h - h(w)\|\, d\lambda \le \varepsilon\lambda(V)\}.$$

For $w \in W$, since

$$\left\| \frac{\mu_h(V)}{\lambda(V)} - h(w) \right\| = \left\| \frac{1}{\lambda(V)} \int_V h d\lambda - h(w) \right\| \le \frac{1}{\lambda(V)} \int_V \|h - h(w)\|\, d\lambda$$

the measure μ_h is differentiable at w with derivative $h(w)$. It remains to show that $\lambda(\mathbb{R}^d \backslash W) = 0$. Let $\{e_n : n \in \mathbb{N}\}$ be a dense countable subset of $h(\mathbb{R}^d)$. For $n \in \mathbb{N}$ let $W_n := \cap_{\varepsilon > 0} W_{n,\varepsilon}$, where

$$W_{n,\varepsilon} := \{w \in \mathbb{R}^d : \exists r > 0\ \forall V \in \mathcal{V}_r(w) \left| \int_V \frac{\|h - e_n\|}{\lambda(V)} d\lambda - \|h(w) - e_n\| \right| \le \frac{\varepsilon}{3}\}.$$

Applying Theorem 8.10 to the function $\|h - e_n\|$ we get that $\lambda(\mathbb{R}^d \setminus W_n) = 0$. Let us show that $\cap_n W_n \subset W$; this will prove that $\lambda(\mathbb{R}^d \setminus W) = 0$. For $w \in \cap_n W_n$, given $\varepsilon > 0$ let $k \in \mathbb{N}$ be such that $\|h(w) - e_k\| < \varepsilon/3$. Then there exists an $r > 0$ such that for all $V \in \mathcal{V}_r(w)$ we have

$$\int_V \|h - h(w)\| \, d\lambda \leq \int_V \|h - e_k\| \, d\lambda + \lambda(V) \, \|e_k - h(w)\|$$

$$\leq \left| \int_V \|h - e_k\| \, d\lambda - \lambda(V) \, \|h(w) - e_k\| \right| + 2\lambda(V) \, \|e_k - h(w)\| \leq \varepsilon \lambda(V).$$

Thus $w \in W$. □

Another differentiability result for a vectorial measure follows.

Proposition 8.5 *Let $\mu : \mathcal{B} \to E$ be a vectorial measure with values in a Banach space E. Suppose the total variation $|\mu|$ is finite on every compact subset of $\mathbb{R}^d$. If μ is singular with respect to the Lebesgue measure λ then μ is almost everywhere differentiable and its derivative is* 0.

Proof Let us first show that if $B \in \mathcal{B}$ is such that $|\mu|(B) = 0$ then μ is differentiable a.e. on B and its derivative is 0. For $n \in \mathbb{N}$ let

$$B_n := \{x \in B : \overline{D} \, |\mu| \, (x) \geq 2^{-n}\}.$$

Lemma 8.3 shows that $\lambda(B_n) = 0$. It follows that for $A := \{x \in B : \overline{D} \, |\mu| \, (x) > 0\}$ one has $\lambda(A) = 0$. Thus μ is differentiable a.e. on B and its derivative is 0.

If $\mu \perp \lambda$ there exists a $B \in \mathcal{B}$ such that $|\mu|(B) = 0$ and $\lambda(\mathbb{R}^d \setminus B) = 0$. The preceding shows that μ is differentiable a.e. on B, hence a.e. on $\mathbb{R}^d$ with derivative 0. □

Theorem 8.12 *Let $\mu : \mathcal{B} \to \mathbb{R}$ be a finite measure on the Borel σ-algebra of $\mathbb{R}^d$. Then there exists a Lebesgue null set N such that μ is differentiable on $\mathbb{R}^d \setminus N$ and the function h given by $h(x) = D\mu(x)$ for $x \in \mathbb{R}^d \setminus N$, $h(x) = 0$ for $x \in N$ is a Radon-Nikodým derivative of the absolutely continuous part μ_a of μ.*

Proof Let us first suppose $\mu \ll \lambda$ and let h be a Radon-Nikodým derivative of μ with respect to λ. For rational numbers r, s satisfying $r < s$ let

$$A(r,s) := \{x \in \mathbb{R}^d : \underline{D}\mu(x) \leq r < s \leq h(x)\}.$$

Lemma 8.3 ensures that

$$s\lambda(A(r,s)) \leq \int_{A(r,s)} h d\lambda = \mu(A(r,s)) \leq r\lambda(A(r,s)).$$

The first inequality shows that $\lambda(A(r,s)) < +\infty$. Then, since $r < s$, we get that $\lambda(A(r,s)) = 0$. Since $A := \{x \in \mathbb{R}^d : \underline{D}\mu(x) < h(x)\}$ is the countable union of the sets $A(r,s)$ with $r, s \in \mathbb{Q}$, $r < s$, we get that $\lambda(A) = 0$. Similarly, for $B := \{x \in \mathbb{R}^d : h(x) < \overline{D}\mu(x)\}$ we have $\lambda(B) = 0$. Since $\underline{D}\mu(x) \le \overline{D}\mu(x)$, we conclude that $\underline{D}\mu(x) = h(x) = \overline{D}\mu(x)$ almost everywhere.

Now suppose μ is an arbitrary finite Borel measure and let $\mu = \mu_a + \mu_s$ be its Lebesgue decomposition. Let h be a Radon-Nikodým derivative of μ_a with respect to λ. Since $D\mu_a = h$ by what precedes and $D\mu_s = 0$ by Proposition 8.5, by additivity we have

$$D\mu = D\mu_a + D\mu_s = h$$

almost everywhere. □

Given a Lebesgue measurable subset M of $\mathbb{R}^d$, let us consider the Borel measure μ defined by $\mu(B) := \lambda(B \cap M)$. We say that a point x of $\mathbb{R}^d$ is a *point of density* of M if μ is differentiable at x and $D\mu(x) = 1$. We say that x is a *point of dispersion* of M if x is a point of density of $\mathbb{R}^d \backslash M$, or, equivalently, if μ is differentiable at x and $D\mu(x) = 0$.

Corollary 8.4 *Let M be a Lebesgue measurable subset of $\mathbb{R}^d$. Then λ-almost every point of M is a point of density of M and λ-almost every point of $M^c := \mathbb{R}^d \backslash M$ is a point of dispersion of M.*

For spaces with the Radon-Nikodým property, a representation of vectorial measures of finite variation on the Borel σ-algebra $\mathcal{B}$ of $\mathbb{R}^d$ can be given.

Theorem 8.13 *Let E be a Banach space with the Radon-Nikodým Property. Any vectorial measure $\nu : \mathcal{B} \to E$ on the Borel σ-algebra $\mathcal{B}$ of $\mathbb{R}^d$ whose total variation is finite is almost everywhere differentiable and there exists a measure $\nu_s : \mathcal{B} \to E$ that is singular with respect to the Lebesgue measure λ such that, denoting by h the derivative of ν with respect to λ, one has $\nu = \mu_h + \nu_s$.*

Proof Let $\nu = \nu_a + \nu_s$ be the Lebesgue decomposition of ν, with $\nu_a \ll \lambda$ and $\nu_s \perp \lambda$. Since E has the RNP and the total variation of ν_a is finite, ν_a is differentiable a.e. and for $h := \frac{D\nu_a}{D\lambda} \in L_1(\mu, E)$ one has $\nu_a = \mu_h : B \mapsto \int_B h d\lambda$. By Proposition 8.5 ν_s, whose total variation is finite too, is differentiable a.e. with a null derivative. Since $\frac{D\nu}{D\lambda} = \frac{D\nu_a}{D\lambda} + \frac{D\nu_s}{D\lambda}$, we get that ν is differentiable a.e. and $\frac{D\nu}{D\lambda} = \frac{D\nu_a}{D\lambda} = h$. □

Exercises

1. Given a Vitali covering $\mathcal{V}$ of $\mathbb{R}^d$ show that there exists some $b > 0$ such that for any finite subfamily $\{V_i : i \in I\}$ of $\mathcal{V}$ one can find a subfamily $\{V_j : j \in J\}$ of $\{V_i : i \in I\}$ satisfying $V_j \cap V_{j'} = \varnothing$ for $j \neq j'$ in J and $\lambda(\cup_{j \in J} V_j) \ge b\lambda(\cup_{i \in I} V_i)$.
2. Let $(S, \mathcal{S}, \mu)$ be a finite measure space without atoms and let $E := L_1(\mu)$. Show that the identity map $I : L_1(\mu) \to E$ is not representable with respect to $(S, \mathcal{S}, \mu)$. [Hint: see [99, p. 61].]

8.4 Derivatives of One-Variable Functions

In this section we use the preceding notions of derivative of a measure to deal with the usual concept of derivative for a one-variable function. Let us first recall the definitions of the Dini derivatives of a function $f : [a, b] \to \mathbb{R}$ at $x \in]a, b[$:

$$D_{-}f(x) := \liminf_{u \to 0-} \frac{f(x+u) - f(x)}{u}, \quad D_{+}f(x) := \liminf_{u \to 0+} \frac{f(x+u) - f(x)}{u},$$

$$D^{-}f(x) := \limsup_{u \to 0-} \frac{f(x+u) - f(x)}{u}, \quad D^{+}f(x) := \limsup_{u \to 0+} \frac{f(x+u) - f(x)}{u}.$$

The left (resp. right) derivative of f at x exists if and only if $D_{-}f(x) = D^{-}f(x)$ (resp. $D_{+}f(x) = D^{+}f(x)$) and f is differentiable at x if and only if these four quantities are finite and coincide.

In the next lemma we use the fact that the set S of strict local minimizers of a real one-variable function g is countable. Here we say that x is a *strict local minimizer* of g if there exists some $\varepsilon > 0$ such that $g(w) > g(x)$ for all $w \in [x - \varepsilon, x + \varepsilon] \backslash \{x\}$. To prove the assertion, for $n \in \mathbb{N} \backslash \{0\}$ let

$$S_n := \{x : g(w) > g(x)\ \forall w \in [x - 1/n, x + 1/n] \backslash \{x\}\}$$

so that for $x, y \in S_n$ we have $|x - y| > 1/n$ if $x \neq y$, and S_n is countable. Since $S = \cup_{n \geq 1} S_n$ our assertion ensues.

Lemma 8.5 *For an arbitrary one-variable function f the following sets are at most countable:*

$$E := \{x : D^{+}f(x) < D_{-}f(x)\}, \qquad F := \{x : D^{-}f(x) < D_{+}f(x)\}.$$

Moreover, the set of points at which the right and left derivatives of f exist, but are not equal, is at most countable.

Proof For each $r \in \mathbb{Q}$ let $f_r(x) := f(x) - rx$ and let

$$F_r := \{x : D^{-}f(x) < r < D_{+}f(x)\},$$

so that $F = \cup_{r \in \mathbb{Q}} F_r$. Then, each $x \in F_r$ is a strict local minimizer of f_r since $D^{-}f_r(x) < 0 < D_{+}f_r(x)$. Thus F_r is countable and F is countable too. The proof for E is similar.

The set G of points at which the right and left derivatives of f exist, but are not equal is contained in $E \cup F$, hence is at most countable. □

Dini's derivatives can be used to prove that a function is nondecreasing.

Theorem 8.14 *Let T be an interval of $\mathbb{R}$, let $f : T \to \mathbb{R}$ be continuous and let D be a countable subset of T. Then f is nondecreasing on T if and only if for all $x \in T \backslash D$ one has $D^+ f(x) \geq 0$.*

The condition $D^+ f(x) \geq 0$ (and even $D_+ f(x) \geq 0$) for all $x \in T \backslash \{\sup T\}$ is obviously necessary. We derive the sufficiency of the condition from the following lemma.

Lemma 8.6 (Zygmund) *Let $f : T \to \mathbb{R}$ be a continuous function and let D be a subset of T such that* $\operatorname{int} f(D)$ *is empty. If for all $x \in T \backslash D$ one has $D^+ f(x) > 0$ then f is nondecreasing on T.*

Assuming D is countable, for all $\varepsilon > 0$ the function f_ε given by $f_\varepsilon(x) = f(x) + \varepsilon x$ satisfies the assumptions of the lemma since $D^+ f_\varepsilon(x) = D^+ f(x) + \varepsilon > 0$ for all $x \in T \backslash D$ and since $f_\varepsilon(D)$ is countable. Thus, for all $u, v \in T$ with $u < v$ we have $f_\varepsilon(u) \leq f_\varepsilon(v)$, hence $f(u) \leq f(v)$, ε being arbitrarily small.

Proof of the lemma Given $u < v$ in T, let us prove that $f(u) \leq f(v)$, or equivalently that for any $c < f(u)$ we have $c < f(v)$. Since $f(D)$ does not contain an interval, we may assume that $c \notin f(D)$, replacing c by some $c' \in]c, f(u)[$ if necessary.

Let $S := \{t \in [u, v] : f(t) \geq c\}$, so that $u \in S$, and $s := \sup S$. Since S is closed it suffices to show that $s = v$ or that the inequality $s < v$ leads to a contradiction. By continuity of f, we have $f(s) \geq c$ and we cannot have $f(s) > c$ since otherwise s would be in an open interval contained in S. Thus $f(s) = c$ and since $c \notin f(D)$ we have $D^+ f(s) > 0$. That implies that there exists a sequence $(s_n) \to s$ in $]s, v[$ such that $f(s_n) > f(s) = c$ for all $n \in \mathbb{N}$, contradicting the definition of s as $\sup S$. Thus $s = v$ and $f(v) \geq c$ and we conclude that $f(v) \geq f(u)$. □

Corollary 8.5 (Dini) *Let $f : T \to \mathbb{R}$ be a continuous function on an interval T of $\mathbb{R}$ such that for some $c \in \mathbb{R}$ and some countable subset D of T one has $D^{\pm} f(t) \geq c$ for all $t \in T \backslash D$, where $D^{\pm} f$ is one of the four Dini derivatives of f. Then, for any pair (s, t) of distinct points of T one has*

$$\frac{f(t) - f(s)}{t - s} \geq c$$

Applying the result to the function $-f$, one obtains an upper bound for the quotient from an upper bound of $D^{\pm} f$.

Proof We may suppose $s < t$ and $D^{\pm} f = D^+ f$ (otherwise we use the function g given by $g(r) = -f(-r)$ for $r \in -T$). Then the result stems from the theorem applied to the function $f_c : r \mapsto f(r) - cr$. □

Exercise Assume that one of the four Dini derivatives of f is finite and continuous at some $r \in \operatorname{int} T$. Show that f is differentiable at r. □

We need to make clear some continuity properties of nondecreasing functions.

Lemma 8.7 *Let $f : \mathbb{R} \to \mathbb{R}$ be a nondecreasing function. Then, for all $x \in \mathbb{R}$ the one-sided limits $f(x_+) := \lim_{y \to x,\ y > x} f(y)$ and $f(x_-) := \lim_{y \to x,\ y < x} f(y)$ exist and the set C of points at which f fails to be continuous is at most countable.*

Moreover, the function $g : \mathbb{R} \to \mathbb{R}$ defined by $g(x) := f(x_-)$ is left-continuous and agrees with f at each point at which f is left-continuous.

Furthermore, for all $x \in \mathbb{R}$ we have $g(x_+) = f(x_+)$ and if g is right-continuous at $r \in \mathbb{R}$, we have $g(r) = f(r)$.

A similar result holds if f is defined on some interval T of $\mathbb{R}$ as one can extend f to all of $\mathbb{R}$ into a nondecreasing function that is constant on each side of T.

Proof The existence of the limit $\lim_{y \to x,\ y > x} f(y)$ is an easy consequence of the fact that f is nondecreasing, so that this limit $f(x_+)$ is $\inf f(]x, \infty[) \geq f(x)$. A similar argument holds for $f(x_-)$. Thus f is continuous at x if and only if $f(x_-) = f(x_+)$. For each $x \in C$ we pick a rational number $q_x \in]f(x_-), f(x_+)[$. Since for $x < y$ we have $q_x < q_y$, the countability of C stems from the countability of $\mathbb{Q}$.

The function g is such that $g(x) = \sup f(]-\infty, x[)$, so that g is nondecreasing and left-continuous since for any sequence (x_n) in $]-\infty, x[$ with limit x we have $]-\infty, x[= \cup_n]-\infty, x_n[$. If f is left-continuous at x we have $g(x) := f(x_-) = f(x)$.

Since $g \leq f$, for all $x \in \mathbb{R}$ we have $g(x_+) \leq f(x_+)$ and in fact $g(x_+) = f(x_+)$ since for all $c > g(x_+) = \inf g(]x, \infty[)$, we can find some $y \in]x, \infty[$ such that $c > g(y)$, so that for all $z \in]x, y[$ we have $c > f(z)$, hence $c > f(x_+)$. If g is right-continuous at r, we have $g(r) = g(r_+) = f(r_+) \geq f(r)$, hence $g(r) = f(r)$. □

Let us pass to differentiability properties. We need a comparison of the derivative of the Stieltjes measure associated with a nondecreasing function f and the derivative of f when it exists.

Lemma 8.8 *Let μ be a finite signed measure on $\mathcal{B}(\mathbb{R})$ and let $f : \mathbb{R} \to \mathbb{R}$ be given by $f(x) := \mu(]-\infty, x[)$ for $x \in \mathbb{R}$. If μ is differentiable at some $t \in \mathbb{R}$, then f is differentiable at t and $f'(t) = D\mu(t)$.*

Proof By definition, for $s \in \mathbb{R}$, $r > 0$ we have

$$\frac{f(s+r) - f(s)}{r} = \frac{\mu([s, s+r[)}{\lambda([r, r+s[)},$$

$$\frac{f(s-r) - f(s)}{-r} = \frac{-\mu([s-r, s[)}{-\lambda([s-r, s[)}.$$

Taking the Vitali covering $\mathcal{V} := \{[s, s+r[: s \in \mathbb{R},\ r > 0\}$, setting $T_{r,s} := [s, s+r[$, $T'_{r,s} := [s-r, s[$, when t is a point of differentiability of μ we have

$$D\mu(t) = \lim_{\substack{r \to 0_+ \\ s < t < s+r}} \frac{\mu(T_{r,s})}{\lambda(T_{r,s})} = \lim_{\substack{r \to 0_+ \\ s-r < t < s}} \frac{\mu(T'_{r,s})}{\lambda(T'_{r,s})}.$$

Since $\mu(T_{r,s}) = f(s+r) - f(s)$, $\mu(T'_{r,s}) = f(s) - f(s-r)$ and $\lambda(T_{r,s}) = r = \lambda(T'_{r,s})$, the existence of these limits means that for any $\varepsilon > 0$ one can find some $\delta > 0$ such that

$$|f(s+r) - f(s) - rD\mu(t)| \leq \varepsilon r \quad \text{for } s < t < s + r \text{ and } r \in]0, \delta[,$$

$$|f(s-r) - f(s) + rD\mu(t)| \leq \varepsilon r \quad \text{for } s - r < t < s \text{ and } r \in]0, \delta[.$$

Since f is continuous, this implies that

$$|f(t+r) - f(t) - rD\mu(t)| \leq \varepsilon r \qquad \forall r \in]0, \delta[,$$

$$|f(t-r) - f(t) + rD\mu(t)| \leq \varepsilon r \qquad \forall r \in]0, \delta[,$$

so that f is differentiable at t with derivative $D\mu(t)$. □

We are ready to prove one of the basic results of differentiation theory.

Theorem 8.15 (Lebesgue) *A function $f : T \to \mathbb{R}$ that is nondecreasing on an interval T of $\mathbb{R}$ is differentiable a.e. for the Lebesgue measure.*

Proof We first assume that $T = \mathbb{R}$, that f is bounded, nondecreasing, left-continuous and $\inf f(\mathbb{R}) = 0$. Let μ be the finite measure on $\mathcal{B}(\mathbb{R})$ characterized by $\mu([a, b[) = f(b) - f(a)$ (Proposition 1.13). Then, for all $x \in \mathbb{R}$ we have $f(x) = \mu(] - \infty, x[)$. Combining Theorem 8.12 with the preceding lemma, we see that f is differentiable a.e.

Now let us remove the left-continuity assumption. We introduce the function g given by $g(x) := f(x_-)$. Since it satisfies the assumptions of the case we just considered, it is differentiable on a subset D of $\mathbb{R}$ such that $\lambda(\mathbb{R} \setminus D) = 0$. By Lemma 8.7, for all $x \in \mathbb{R}$ we have $g(x_+) = f(x_+)$. If $r \in D$, or if g is just continuous at r, we have $g(r) = f(r)$ and for all $s \in \mathbb{R}$ with $s > r$

$$g(s) - g(r) \leq f(s) - f(r) \leq g(s_+) - g(r).$$

Setting $t := s-r$ and observing that $t^{-1}(g(s_+)-g(r)) \leq t^{-1}(g(r+t+t^2)-g(r))$ tends to $g'(r)$ as $t \to 0$ with $t \neq 0$, we see that $(s-r)^{-1}(f(s) - f(r)) \to g'(r)$ as $s \to r$, $s > r$, Thus f is right differentiable a.e. The function h given by $h(x) := -f(-x)$ also being right differentiable a.e., f is differentiable a.e. with $f' = g'$.

Finally, let f be an arbitrary nondecreasing function on an arbitrary interval T. We can write the interior of T as the union of a countable family of open intervals $]a_n, b_n[$. Let f_n the function given by $f_n(x) = f(x) - f(a_n)$ for $x \in]a_n, b_n[$, $f_n(x) = 0$ for $x \in] - \infty, a_n]$ and $f_n(x) = f(b_n) - f(a_n)$ for $x \in [b_n, \infty[$. By the preceding case, the set N_n of non differentiability points of f_n has measure 0. It follows that the set $N := \mathbb{R} \setminus D$ of non differentiability points of f has measure 0. □

From this theorem we can deduce the following consequence.

Theorem 8.16 *A function of bounded variation on an interval T of $\mathbb{R}$ is differentiable a.e. Moreover, its derivative f' is Lebesgue measurable and for $a, x \in T$,* *denoting by $V_a(x)$ the variation of f on the interval with extremities a and x one has*

$$\int_a^x |f'(t)|\, d\lambda(t) \leq V_a(x_+). \tag{8.6}$$

In particular, if f is nondecreasing, for $a, x \in T$ one has

$$\int_a^x |f'(t)|\, d\lambda(t) \leq f(x_+) - f(a).$$

Proof The first assertion follows from the fact that a function of bounded variation is the difference of two nondecreasing functions. Its derivative f' is Lebesgue measurable since it is the limit a.e. of the sequence (q_n) of functions defined by

$$q_n(t) := n[f(t + n^{-1}) - f(t)].$$

Let us assume f is nondecreasing and prove the last assertion. In such a case the functions q_n are nonnegative and Fatou's lemma implies that

$$\begin{aligned}
\int_a^x f'(t)d\lambda(t) &\leq \liminf_n \int_a^x q_n(t)d\lambda(t) \\
&= \liminf_n n[\int_{a+1/n}^{x+1/n} f(t)d\lambda(t) - \int_a^x f(t)d\lambda(t)] \\
&= \liminf_n n[\int_x^{x+1/n} f(t)d\lambda(t) - \int_a^{a+1/n} f(t)d\lambda(t)] \\
&\leq \liminf_n n[\frac{1}{n} f(x + 1/n) - \frac{1}{n} f(a)] = f(x_+) - f(a)
\end{aligned}$$

since the last inequality is a consequence of the fact that f is nondecreasing.

Now let us assume f is an arbitrary function of bounded variation. Recall that $g := (1/2)(V_a + f)$ and $h := (1/2)(V_a - f)$ are nondecreasing, hence have nonnegative derivatives, so that $f = g - h$ and

$$|f'| = |g' - h'| \leq g' + h' = V_a' \qquad \text{a.e.}$$

Since V_a is nondecreasing, relation (8.6) ensues. □

In the following definition we introduce the important class $AC(T)$ of *absolutely continuous functions* on an interval T of $\mathbb{R}$ for which the preceding result can be made more precise.

Definition 8.6 A function $f : T \to \mathbb{R}$ on an interval T of $\mathbb{R}$ is said to be *absolutely continuous* if for each $\varepsilon \in \mathbb{P}$ there exists some $\delta \in \mathbb{P}$ such that for any finite family

$(T_i)_{i\in\mathbb{N}_m}$ of disjoint open intervals $T_i :=]a_i, b_i[$ of T satisfying $\Sigma_{i=1}^m (b_i - a_i) < \delta$ one has $\Sigma_{i=1}^m |f(b_i) - f(a_i)| < \varepsilon$.

Taking a family consisting of a single interval, we see that an absolutely continuous function is uniformly continuous. The converse is not true, as the example of the Cantor-Lebesgue function shows (see Exercise 1).

A Lipschitzian function is clearly absolutely continuous. Another example is described in the following lemma.

Lemma 8.9 *Let $h : T \to \mathbb{R}$ be an integrable function on an interval T, let $a \in T$, and let $f : T \to \mathbb{R}$ be given by $f(x) = \int_a^x h(t)dt$. Then f is absolutely continuous.*

Moreover, f is differentiable a.e. on T and $f' = h$ a.e.

Proof Let μ be the measure induced by the Lebesgue measure λ on T and let $\nu := \mu_h$ be the measure with density h with respect to μ. Since for any μ-null set N of T we have $\int_N |h|\, d\mu = 0$, μ_h is absolutely continuous with respect to μ. Then Proposition 8.2 ensures that given $\varepsilon > 0$ one can find some $\delta > 0$ such that for any measurable subset A of T satisfying $\mu(A) < \delta$ one has $|\mu_h(A)| < \varepsilon$. In particular, taking for A the union of a finite family $(T_i)_{i\in\mathbb{N}_m}$ of disjoint open intervals $T_i :=]a_i, b_i[$ of T satisfying $\Sigma_i(b_i - a_i) < \delta$ we have $\Sigma_i |f(b_i) - f(a_i)| \leq |\mu_h|(A) < \varepsilon$.

For the second assertion we may assume T is a compact interval $[a, b]$ and extend f' to a function h on $\mathbb{R}$ by 0 on $\mathbb{R}\backslash T$. Then the assertion follows from Lemma 8.8 applied to the measure μ_h with density h with respect to the Lebesgue measure, so that for $x \in [a, b]$ one has $\mu_h(] - \infty, x[) = \mu_h([a, x[) = \int_a^x h(t)d\lambda(t) = f(x)$. □

Proposition 8.6 *If T is the compact interval $[a, b]$ for some $a < b$ in $\mathbb{R}$, any absolutely continuous function f on T is of bounded variation: $AC(T) \subset BV(T)$.*

The Cantor-Lebesgue function shows that the reverse inclusion does not hold (see Exercise 1).

Proof Let $f \in AC(T)$ and let $\delta \in \mathbb{P}$ correspond to $\varepsilon = 1$ in the preceding definition. Introduce a subdivision $\tau := \{t_0 = a < t_1 < \cdots < t_n = b\}$ such that $t_i - t_{i-1} < \delta$ for $i \in \mathbb{N}_n$ and $n < (b - a)/\delta + 1$. Given a subdivision $\sigma := \{s_0 = a < s_1 < \cdots < s_m = b\}$ since the sum

$$S_\sigma := \sum_{i=1}^{m} |f(s_i) - f(s_{i-1})|$$

does not decrease when some additional points are introduced, we may assume that σ contains all the points of τ. Gathering the terms of this sum corresponding to intervals contained in some interval $[t_{j-1}, t_j]$, we see that $S_\sigma \leq n$. Thus f is in $VB(T)$. □

The following condition characterizes absolutely continuous functions, but we only prove it is a necessary condition.

Proposition 8.7 (Lusin) *Let f be an absolutely continuous function on $T := [a, b]$. Then f satisfies the following condition:*

(N) $S \subset T, \lambda(S) = 0 \Longrightarrow \lambda(f(S)) = 0.$

Proof Let $f \in AC(T)$. Given $\varepsilon \in \mathbb{P}$, let $\delta \in \mathbb{P}$ correspond to ε as in the definition of absolute continuity. Given $S \subset T$ such that $\lambda(S) = 0$, we consider an open subset G containing S with $\lambda(G) < \delta$. Now G is the union of a countable family (T_n) of open intervals. Since f is continuous, $f(\text{cl}(T_n))$ is an interval whose endpoints are points $f(a_n), f(b_n)$ with $a_n, b_n \in \text{cl}(T_n)$, $\{f(a_n), f(b_n)\} = \{\min f(\text{cl}(T_n)), \max f(\text{cl}(T_n))\}$, $a_n \leq b_n$, so that $\lambda(f(T_n)) \leq |f(b_n) - f(a_n)|$. Since

$$\sum_n (b_n - a_n) \leq \sum_n \lambda(T_n) = \lambda(G) < \delta,$$

an extension of Definition 8.6 to countable families shows that

$$\lambda(f(S)) \leq \lambda(f(G)) \leq \sum_n \lambda(f(T_n)) = \sum_n |f(b_n) - f(a_n)| \leq \varepsilon.$$

Since ε is arbitrarily small, we have $\lambda(f(S)) = 0$. □

Proposition 8.8 *Let $f : T \to \mathbb{R}$ be a continuous function on $T := [a, b]$ satisfying condition (N). Let S be the set of $s \in]a, b[$ such that f is differentiable at s with derivative 0. Then $\lambda(f(S)) = 0$.*

Proof Given $\varepsilon \in \mathbb{P}$, for each $s \in S$ we can find some $\delta_s \in \mathbb{P}$ such that $T_s := [s - \delta_s, s + \delta_s] \subset T$ and

$$|f(s + r) - f(s)| < \varepsilon |r| \qquad \forall r \in [-\delta_s, \delta_s].$$

For arbitrary points a, b in T_s we have

$$|f(b) - f(a)| \leq |f(b) - f(s)| + |f(a) - f(s)| \leq 2\varepsilon\delta_s = \varepsilon\lambda(T_s). \tag{8.7}$$

The family $\mathcal{V} := \{T_s : s \in S\}$ is a Vitali covering of S, so that by the Vitali's theorem there exists a finite or countable subfamily $\mathcal{W} := \{T_s : s \in C\}$ and a null set N such that the sets T_s with $s \in C$ are disjoint and

$$S \subset (\bigcup_{s \in C} T_s) \cup N.$$

By condition (N) we have $\lambda(f(N)) = 0$ and by relation (8.7), for all $s \in C$ we have $\lambda(f(T_s)) \leq \varepsilon\lambda(T_s)$. Then, the preceding inclusion yields

$$\lambda(f(S)) \leq \lambda((\bigcup_{s \in C} f(T_s)) \cup f(N)) = \sum_{s \in C} \lambda(f(T_s)) \leq \sum_{s \in C} \varepsilon\lambda(T_s) \leq \varepsilon\lambda(T)$$

since the intervals T_s with $s \in C$ are disjoint and contained in T. Since ε is arbitrarily small we see that $\lambda(f(S)) = 0$. □

Corollary 8.6 *Let f be an absolutely continuous function on an interval T. If the derivative f' of f is nonnegative a.e. then f is nondecreasing.*

Proof Given $\varepsilon > 0$ let f_ε be given by $f_\varepsilon(t) := f(t) + \varepsilon t$. Let D be the set of $t \in T$ such that either $f_\varepsilon'(t)$ does not exists or is negative. By assumption $\lambda(D) = 0$. Then $\lambda(f_\varepsilon(D)) = 0$, so that the interior of $f_\varepsilon(D)$ is empty. Then Zygmund's lemma applies and f_ε is nondecreasing. It follows that f is nondecreasing. □

Corollary 8.7 *Let $f : T \to \mathbb{R}$ be an absolutely continuous function on $T := [a, b]$ whose derivative f' is 0 a.e. Then f is constant.*

Such an assertion is not valid for an arbitrary function, even if it is nondecreasing (see Exercise 1).

Proof Let S the set of points of $]a, b[$ at which f is differentiable, so that, by the inclusion $AC(T) \subset BV(T)$ and Theorem 8.16, $T = S \cup N$, where N is of measure 0. Since f satisfies condition (N) the preceding proposition shows that $\lambda(f(T)) \leq \lambda(f(S)) + \lambda(f(N)) = 0$. Thus, the interval $f(T)$ is a singleton and f is constant. □

The following theorem is often called the *Fundamental Theorem of Calculus.* It shows the power of Lebesgue integration theory.

Theorem 8.17 *A function $f : T \to \mathbb{R}$ on an interval T of $\mathbb{R}$ is absolutely continuous if and only if it is differentiable a.e., if its derivative f' is locally integrable and for all $a, x \in T$*

$$f(x) = f(a) + \int_a^x f'(t)dt.$$

Proof By Lemma 8.9 the condition is sufficient.

Let us prove it is necessary. Since the restriction of f to any compact interval is of bounded variation, f is differentiable a.e. and its derivative f' is Lebesgue measurable. Relation (8.6) shows that f' is Lebesgue integrable on any compact interval, so that it is meaningful to set

$$g(x) := f(a) + \int_a^x f'(t)dt.$$

Lemma 8.9 ensures that g is absolutely continuous and a.e. differentiable with derivative f'. Thus the absolutely continuous function $f - g$ is a.e. differentiable with derivative 0. By the preceding corollary it is constant with value $f(a) - g(a) = 0$: $f = g$. □

Let us end this section by quoting the following result for which we refer to [60, 226].

Theorem 8.18 (Lusin) *Let $g : [a, b] \to \overline{\mathbb{R}}$ be a measurable function that is finite almost everywhere. Then there exists a continuous function $f : [a, b] \to \mathbb{R}$ that is differentiable a.e. and such that $f' = g$ almost everywhere.*

Exercises

1. (*The Cantor-Lebesgue function*) Recall that the Cantor set C is the image of the set $\{0, 1\}^{\mathbb{N}}$ of sequences $(k_n)_n$ with $k_n = 0$ or 1 under the map $(k_n) \mapsto \Sigma_n 2k_n/3^{n+1}$. Let $f : C \to \mathbb{R}$ be given by $f(x) = \Sigma_n k_n/2^{n+1}$ for $x = \Sigma_n 2k_n/3^{n+1}$. Show that f is well defined (in spite of the nonuniqueness of the representation of x) and is continuous, with $f(0) = 0, f(1) = 1$. Show that $f(C) = [0, 1]$. Check that if $]a, b[$ is an open interval in $[0, 1]\backslash C$, then $f(a) = f(b)$, so that f can be extended into a continuous function on $[0, 1]$ by giving it a constant value on each such interval $]a, b[$, so that $f'(x) = 0$ on $[0, 1]\backslash C$ with $\lambda(C) = 0$, in spite of the fact that f is increasing on C.
2. Point out two results of the present subsection showing that the function of the preceding exercise is not absolutely continuous.
3. Let $f : [0, 1] \to \mathbb{R}$ be given by $f(0) = 0, f(x) = x^2 \sin x^{-2}$ for $x \in]0, 1]$. Show that f is differentiable everywhere, but is not absolutely continuous.
4. Let T be a compact interval of $\mathbb{R}$ and let $f : T \to \mathbb{R}$ be a continuous function that is such that f is differentiable at all except countably many of the points of T, with an integrable derivative f'. Prove that f is absolutely continuous.
5. Let μ be a finite signed measure on $(\mathbb{R}, \mathcal{B}(\mathbb{R}))$, and let $f_\mu : \mathbb{R} \to \mathbb{R}$ be given by $f_\mu(x) = \mu(] - \infty, x[)$. Show that f_μ is absolutely continuous if and only if μ is absolutely continuous with respect to Lebesgue measure. [See Lemma 8.9 and [80, Prop.4.4.5].]
6. Let f and g be absolutely continuous functions on the compact interval $[a, b]$. Prove the following version of *integration by parts*:

$$f(b)g(b) - f(a)g(a) = \int_a^b f(t)g'(t)dt + \int_a^b f'(t)g(t)dt.$$

8.5 Lebesgue $L_p(S, E)$ Spaces

We devote the present section and the next chapter to two classes of normed spaces that play a crucial role in analysis. They are closely related. In this section, unless otherwise specified, p is an element of the interval $[1, +\infty]$ and $q \in [1, +\infty]$ is the so-called *conjugate exponent* given by $q = (1 - 1/p)^{-1}$ if $p \in]1, +\infty[$, $q = +\infty$ if $p = 1$, and $q = 1$ if $p = +\infty$, so that the relation $1/p + 1/q = 1$ holds by convention.

8.5.1 Basic Facts About Lebesgue Spaces

We start with some classical inequalities.

Lemma 8.10 *For $p \in]1, +\infty[$ let $q := (1 - \frac{1}{p})^{-1}$. Then for $r, s \in \mathbb{R}_+$ one has*

$$rs \leq \frac{1}{p}r^p + \frac{1}{q}s^q, \tag{8.8}$$

$$(\frac{1}{2}r + \frac{1}{2}s)^p \leq \frac{1}{2}r^p + \frac{1}{2}s^p. \tag{8.9}$$

Proof Since relation (8.8) is satisfied if $r = 0$ or $s = 0$, we may assume r and s are positive. Setting $u = r^p$ and $v = s^q$ we are reduced to showing that

$$u^{1/p}v^{1/q} \leq \frac{1}{p}u + \frac{1}{q}v. \tag{8.10}$$

Let us consider the function $g :]0, +\infty[\to \mathbb{R}$ given by $g(t) = t/p + 1/q - t^{1/p}$. Its derivative g', given by $g'(t) = (1/p)(1 - t^{-1/q})$ is negative on $]0, 1[$ and positive on $]1, +\infty[$, so that g attains its minimum at $t = 1$. Thus $g(t) \geq 0$ for all $t > 0$. Setting $t := u/v$ and then multiplying by v, we get inequality (8.10), the so-called *Young's inequality*.

Relation (8.9) is a consequence in the convexity of the function $t \mapsto t^p$ (its derivative $t \mapsto pt^{p-1}$ is nondecreasing). Relation (8.9) is an improvement of the obvious estimate $(\frac{1}{2}r + \frac{1}{2}s)^p \leq (r \vee s)^p \leq r^p + s^p$. □

Definition 8.7 Given a measure space $(S, \mathcal{S}, \mu)$, $p \in]1, +\infty[$, and a Banach space E, let $\mathcal{L}_p(S, \mathcal{S}, \mu, E)$, or in short $\mathcal{L}_p(S, E)$ or $\mathcal{L}_p(\mu, E)$, be the set of μ-measurable maps $f : S \to E$ such that $\|f(\cdot)\|_E^p$ is integrable. For $E := \mathbb{R}$ the notation $\mathcal{L}_p(S)$ replaces $\mathcal{L}_p(S, \mathbb{R})$.

This set is a vector space since $\|cf\|_E^p = |c|^p \|f\|_E^p$ for $c \in \mathbb{R}$ (or $c \in \mathbb{C}$) and $f \in \mathcal{L}_p(S, E)$ and since $\|f + g\|_E^p \leq 2^{p-1} \|f\|_E^p + 2^{p-1} \|g\|_E^p$ for $f, g \in \mathcal{L}_p(S, E)$ in view of relation (8.9). In the sequel, for $f \in \mathcal{L}_p(S, \mathcal{S}, \mu, E)$, we set

$$\|f\|_p := \left(\int_S \|f\|_E^p \, d\mu \right)^{1/p}.$$

Given a μ-measurable map $f : S \to E$, let $\|f\|_\infty$ be the infimum of the set of $c \in \mathbb{R}_+$ such that $\|f(s)\|_E \le c$ a.e. (with $\|f\|_\infty = \infty$ if there is no such c). When finite, this infimum is attained since for any sequence $(c_n) \to c := \|f\|_\infty$ with $c_n > c$ for all n one can find a μ-null set N_n such that $\|f(s)\|_E \le c_n$ for all $s \in S \backslash N_n$, so that $\|f(s)\|_E \le c$ for all $s \in S \backslash N$ with $N := \cup_n N_n$. We denote by $\mathcal{L}_\infty(S, \mathcal{S}, \mu, E)$, or in short $\mathcal{L}_\infty(S, E)$, the set of μ-measurable maps $f : S \to E$ such that $\|f\|_\infty < +\infty$. The set $\mathcal{L}_\infty(S, E)$ is a vector space and $f \mapsto \|f\|_\infty$ is easily seen to be a semi-norm on $\mathcal{L}_\infty(S, E)$.

The next proposition prepares the proof that $f \mapsto \|f\|_p$ is a semi-norm on $\mathcal{L}_p(S, E)$.

Proposition 8.9 (Hölder's Inequality) *Let $p, q \in [1, +\infty]$ satisfying $1/p + 1/q = 1$ in the extended sense described above and let $(S, \mathcal{S}, \mu)$ be a measure space. Given $f \in \mathcal{L}_p(S, \mathcal{S}, \mu)$, $g \in \mathcal{L}_q(S, \mathcal{S}, \mu)$ one has $fg \in \mathcal{L}_1(S, \mathcal{S}, \mu)$ and*

$$\int |fg| \, d\mu \le \|f\|_p \, \|g\|_q \, . \tag{8.11}$$

If F, G, H are Banach spaces, if $b : F \times G \to H$ is a continuous bilinear map and if $f \in \mathcal{L}_p(S, F)$, $g \in \mathcal{L}_q(S, G)$ one has $b \circ (f, g) \in \mathcal{L}_1(S, H)$ and

$$\|b \circ (f, g)\|_1 \le \|b\| \, \|f\|_p \, \|g\|_q \, . \tag{8.12}$$

Proof The first assertion is a special case of the second one with $F = G = H = \mathbb{R}$, $b(r, s) = rs$. The map $h := b \circ (f, g)$ is μ-measurable when f and g are μ-measurable. In view of Corollary 7.12 we may suppose $p, q \in]1, +\infty[$. Since $f = 0$ a.e. and $h = 0$ a.e. when $\|f\|_p = 0$, we may suppose $\alpha := \|f\|_p \ne 0$ and $\beta := \|g\|_q \ne 0$. Then, for $t \in S$, the relation $\|h(t)\|_H \le \|b\| \, \|f(t)\|_F \, \|g(t)\|_G$ and inequality (8.8) with $r := (1/\alpha) \, \|f(t)\|_F$, $s := (1/\beta) \, \|g(t)\|_G$ yield

$$\frac{1}{\alpha} \frac{1}{\beta} \, \|h(t)\|_H \le \|b\| \, (\frac{1}{p\alpha^p} \, \|f(t)\|_F^p + \frac{1}{q\beta^q} \, \|g(t)\|_G^q).$$

Integrating over S, we get $(1/\alpha\beta) \int_S \|h(t)\|_H \, d\mu(t) \le \|b\|$ and relation (8.12). □

Corollary 8.8 *For $k \in \mathbb{N} \setminus \{0, 1\}$ and $p, p_1, \cdots, p_k \in [1, +\infty]$ satisfying $1/p = 1/p_1 + \cdots + 1/p_k$ and $f_i \in \mathcal{L}_{p_i}(S)$ for $i \in \mathbb{N}_k$, one has $f := f_1 \cdots f_k \in \mathcal{L}_p(S)$ and*

$$\|f\|_p \le \|f_1\|_{p_1} \cdots \|f_k\|_{p_k} \, .$$

Proof Setting $q_i := p_i/p$ and $g_i := |f_i|^p$ we reduce the result to the case $p = 1$. Then an induction starting with the case $k = 2$ and the Hölder's inequality gives the result. □

Example (Interpolation) For $m, p, q \in [1, +\infty]$ and $t \in [0, 1]$ with $m \leq p \leq q$, $1/p = t/m + (1-t)/q$ and $f \in \mathcal{L}_m(S) \cap \mathcal{L}_q(S)$ one has $f \in \mathcal{L}_p(S)$ and $\|f\|_p \leq \|f\|_m^t \|f\|_q^{1-t}$.

In the sequel we write $\|\cdot\|$ instead of $\|\cdot\|_E$ when no confusion may arise.

Corollary 8.9 *If μ is a finite measure and if $f \in \mathcal{L}_p(S, E)$ for some $p \in]1, +\infty]$, then for all $p' \in [1, p]$ one has $f \in \mathcal{L}_{p'}(S, E)$ and*

$$\|f\|_{p'} \leq \mu(S)^{1/p' - 1/p} \|f\|_p .$$

Proof Setting $r := p/p'$, $s = (1-1/r)^{-1}$, $f' := \|f\|^{p'}$, $g := 1_S$, relation (8.11) with (p, q, f) changed into (r, s, f') yields $\int_S \|f\|^{p'} d\mu \leq \|f\|_p^{p'} \|1_S\|_s \leq \infty$. □

Proposition 8.10 *The function $\|\cdot\|_p : f \mapsto (\int \|f\|^p d\mu)^{1/p}$ is a semi-norm on $\mathcal{L}_p(S, E)$. In particular, for $f, g \in \mathcal{L}_p(S, E)$ we have the Minkowski inequality*

$$\|f + g\|_p \leq \|f\|_p + \|g\|_p . \tag{8.13}$$

Proof The relation $\|cf\|_p = |c| \|f\|_p$ for $c \in \mathbb{C}$ (or $\mathbb{R}$) and $f \in \mathcal{L}_p(S, E)$ is immediate. Let $f, g \in \mathcal{L}_p(S, E)$. Writing

$$\|f + g\|^p \leq \|f\| \|f + g\|^{p-1} + \|g\| \|f + g\|^{p-1} ,$$

integrating over S and using the Hölder's inequality on each term we get

$$\int_S \|f + g\|^p d\mu \leq \|f\|_p (\int_S \|f + g\|^{q(p-1)} d\mu)^{\frac{1}{q}} + \|g\|_p (\int_S \|f + g\|^{q(p-1)} d\mu)^{\frac{1}{q}}.$$

When $\|f + g\|_p$ is non-null, dividing both sides by $(\int_S \|f + g\|^p d\mu)^{p/q}$ and using the relations $p - p/q = 1$, $q(p-1) = p$, we get $(\int_S \|f + g\|^p d\mu)^{1/p} \leq \|f\|_p + \|g\|_p$. When $\|f + g\|_p = 0$, the inequality $\|f + g\|_p \leq \|f\|_p + \|g\|_p$ is obvious. □

If $f \in \mathcal{L}_p(S, E)$ is such that $\|f\|_p = 0$ we have $\|f\|^p = 0$ a.e. by Corollary 7.16, hence $f = 0$ a.e. Such a fact incites us to consider the space $L_p(S, E)$ of equivalence classes of maps in $\mathcal{L}_p(S, E)$ with respect to the relation of equality almost everywhere. Then we dispose of properties similar to those in $\mathcal{L}_1(S, E)$. Moreover, the semi-norm $\|\cdot\|_p$ induces a norm on $L_p(S, E)$. We first prove an analogue of Egoroff's Theorem.

Proposition 8.11 *Let (f_n) be an Abel sequence in $(\mathcal{L}_p(\mu, E), \|\cdot\|_p)$ with $p \in [1, +\infty[$. Then (f_n) converges a.e. to some μ-measurable function f, and given $\varepsilon > 0$, there exists a subset T of S of measure less than ε such that the convergence is uniform on $S \backslash T$. Moreover, f is in $\mathcal{L}_p(\mu, E)$ and $(f_n) \to f$ in $\mathcal{L}_p(\mu, E)$.*

Proof Let $c > 0$ and $r \in]0, 1[$ be such that $\|f_{n+1} - f_n\|_p \leq cr^{2n}$ for all $n \in \mathbb{N}$. Changing all the f_n's on a set of measure 0, we may assume that they are all measurable. Let S_n be the set of $s \in S$ such that $\|f_{n+1}(s) - f_n(s)\|_E \geq cr^n$. Then

$$c^p r^{np} \mu(S_n) \leq \int_{S_n} \|f_{n+1} - f_n\|_E^p \, d\mu \leq \int_S \|f_{n+1} - f_n\|_E^p \, d\mu \leq c^p r^{2np},$$

so that $\mu(S_n) \leq r^{np}$. Setting $T_k := \cup_{n \geq k} S_n$ and $N := \cap_k T_k$ we have $\mu(T_k) < r^{kp}/(1 - r^p)$ and $\mu(N) = 0$. For $s \in S \backslash T_k$ and $n \geq k$ we have $\|f_{n+1}(s) - f_n(s)\|_E \leq cr^n$ so that (f_n) converges uniformly on $S \backslash T_k$ and (f_n) converges pointwise to some function f on $S \backslash N$. Extending f by 0 on N, we get a μ-measurable function on S. We still denote it by f.

Moreover, for $n \geq k$ in $\mathbb{N}$ we have

$$\left(\int_S \|f_n - f_k\|^p \, d\mu \right)^{1/p} = \|f_n - f_k\|_p \leq \sum_{j=k}^{\infty} \left\| f_{j+1} - f_j \right\|_p \leq c \frac{r^{2k}}{1 - r^2}.$$

Applying Fatou's Lemma to the sequence $(\|f_n - f_k\|^p)_{n \geq k}$ which converges almost everywhere to $\|f - f_k\|^p$, we get that

$$\int_S \|f - f_k\|^p \, d\mu \leq \text{liminf}_n \int_S \|f_n - f_k\|^p \, d\mu \leq c^p \frac{r^{2kp}}{(1 - r^2)^p}.$$

Since $f_k \in \mathcal{L}_p(S, E)$ and $\|f - f_k\|_p < +\infty$ we obtain that $f \in \mathcal{L}_p(S, E)$ and $\|f - f_k\|_p \leq cr^{2k}/(1 - r^2)$, so that $(f_k) \to f$ in $\mathcal{L}_p(S, E)$. □

Theorem 8.19 *For $p \in [1, +\infty]$, any measure space $(S, \mathcal{S}, \mu)$, and any Banach space E, the spaces $(\mathcal{L}_p(S, E), \|\cdot\|_p)$ and $(L_p(S, E), \|\cdot\|_p)$ are complete.*

Proof Since completeness can be established by using Abel sequences, the case $p \in [1, +\infty[$ is established by Proposition 8.11. Thus, we turn to the case $p = +\infty$. Let (f_n) be a Cauchy sequence in $(\mathcal{L}_\infty(S, E), \|\cdot\|_\infty)$. From the comments following the definition of $\|\cdot\|_\infty$ we know that for all $m, n \in \mathbb{N}$ there exists a null set $N_{m,n}$ such that $\|f_m - f_n\|_\infty = \sup\{\|f_m(s) - f_n(s)\| : s \in S \backslash N_{m,n}\}$. Let $N := \cup_{m,n} N_{m,n}$; it is a null set and on $S \backslash N$ the sequence (f_n) is a Cauchy sequence for the norm of uniform convergence. Since E is complete, this sequence converges to some function f. We extend f by 0 on N. Then f is μ-measurable and $f \in \mathcal{L}_\infty(S, E)$. Moreover, since $(\sup\{\|f_n(s) - f(s)\| : s \in S \backslash N\}) \to 0$, we get $(\|f_n - f\|_\infty)_n \to 0$. □

Theorem 8.20 (Monotone Convergence Theorem in L_p) *For $p \in [1, +\infty[$, let (f_n) be an increasing sequence in $\mathcal{L}_p(S, \mathbb{R})$ such that there exists some $c \in \mathbb{R}_+$ satisfying $\|f_n\|_p \leq c$ for all $n \in \mathbb{N}$. Then there exists some $f \in \mathcal{L}_p(S, \mathbb{R})$ such that $(f_n) \to f$ a.e. and $(\|f_n - f\|_p) \to 0$.*

Proof Let $g_n := f_n - f_0$, so that (g_n) is increasing and g_n takes its values in $\mathbb{R}_+$. Moreover, the Minkowski inequality yields $\|g_n\|_p \leq \|f_n\|_p + \|f_0\|_p \leq 2c$. By the Monotone Convergence Theorem, the sequence $(g_n^p)_n$ converges a.e. to some element $h \in \mathcal{L}_1(S, \mathbb{R})$ and $(\left\|g_n^p - h\right\|_1)_n \to 0$. Since $h \geq 0$ a.e. we can set $g := h^{1/p}$

and $f := f_0 + g$. Then $(g_n) \to g$ a.e., so that g is μ-measurable and even $g \in \mathcal{L}_p(S, \mathbb{R})$ since $g^p \in \mathcal{L}_1(S, \mathbb{R})$. For $c \in \mathbb{R}_+$ the function $r \mapsto r^p - (r-c)^p$ being nondecreasing on $[c, +\infty[$, we have $r^p \geq (r-c)^p + c^p$ for all $r \geq c$ hence

$$(g - g_n)^p \leq g^p - g_n^p.$$

Integrating over S, we get $(\|g - g_n\|_p)^p \leq \|g^p\|_1 - \left\|g_n^p\right\|_1$. Since $(\left\|g_n^p\right\|_1)_n \to \|h\|_1 = \|g^p\|_1$ we get $(\|g - g_n\|_p)_n \to 0$, hence $(\|f - f_n\|_p)_n \to 0$. □

Theorem 8.21 (Dominated Convergence Theorem in L_p) *For $p \in [1, +\infty[$, let (f_n) be a sequence in $\mathcal{L}_p(S, E)$ such that there exists some $h \in \mathcal{L}_p(S, \mathbb{R})$ satisfying $\|f_n\| \leq h$ for all $n \in \mathbb{N}$. If $(f_n) \to f$ a.e. for some function f, then $f \in \mathcal{L}_p(S, E)$ and $(\|f_n - f\|_p) \to 0$.*

Proof Since $(f_n) \to f$ a.e. and all f_n's are μ-measurable functions, f is μ-measurable and satisfies $\|f\| \leq h$ a.e., so that $\|f\|_p \leq \|h\|_p < \infty$ and $f \in \mathcal{L}_p(S, E)$. Now $\|f_n - f\|^p \in \mathcal{L}_1(S, \mathbb{R})$ and since $\|f_n - f\|^p \leq 2^p h^p$, the Dominated Convergence Theorem ensures that $(\|f_n - f\|^p)_n \to 0$ in $\mathcal{L}_1(S, \mathbb{R})$, so that $(\|f_n - f\|_p)_n \to 0$. □

Let us give a kind of converse.

Theorem 8.22 *For $p \in [1, +\infty[$, let $(f_n) \to f$ be a convergent sequence in $(\mathcal{L}_p(S, E), \|\cdot\|_p)$. Then there exist $h \in \mathcal{L}_p(S, \mathbb{R})$ and a subsequence $(f_{k(n)})$ of (f_n) such that $(f_{k(n)}) \to f$ a.e. and $\left\|f_{k(n)}(x)\right\| \leq h(x)$ a.e. for all $n \in \mathbb{N}$.*

Proof In fact, the conclusion is valid for any Abel subsequence in (f_n). Thus we may suppose that the sequence (f_n) satisfies

$$\|f_n - f_{n+1}\|_p \leq 2^{-n} \qquad \forall n \in \mathbb{N}.$$

Setting

$$g_n(x) := \sum_{k=1}^{n} \|f_{k+1}(x) - f_k(x)\|$$

we see that $\|g_n\|_p \leq 1$. The Monotone Convergence Theorem implies that there exists some $g \in \mathcal{L}_p(S, \mathbb{R})$ such that $(g_n) \to g$ a.e. and $(\|g_n - g\|_p) \to 0$. For $n > m \geq 2$ we have

$$\|f_n(x) - f_m(x)\| \leq \sum_{j=m}^{n-1} \left\|f_{j+1}(x) - f_j(x)\right\| \leq g(x) - g_{m-1}(x),$$

so that $(f_n(x))$ is a.e. a Cauchy sequence in $\mathbb{R}$, hence has a limit. Denoting this limit by $f_\infty(x)$ and passing to the limit on n in the preceding inequalities we get

$$\|f_\infty(x) - f_m(x)\| \leq g(x) - g_{m-1}(x) \leq g(x).$$

Thus $f_\infty \in \mathcal{L}_p(S, \mathbb{R})$. Using again the Dominated Convergence Theorem we get $(\|f_\infty - f_m\|_p) \to 0$ as $m \to \infty$, so that $f_\infty = f$ a.e. and $\|f_m(x)\| \le \|f(x)\| + g(x)$ with $h := \|f\| + g \in \mathcal{L}_p(S, \mathbb{R})$. □

Exercises

1. Let $S := [0, 1]$ be endowed with the Lebesgue measure. For $m \in \mathbb{N}\backslash\{0\}$ and $k \in \{0, \dots, 2^m - 1\}$ let $T_{m,k} := [2^{-m}k, 2^{-m}(k + 1)[$ and let $f_{2^m+k} = 1_{T_{m,k}}$. Show that $\|f_n\|_p = 2^{-m/p}$ for $n = 2^m+k$ with $k \in \{0, \dots, 2^m-1\}$, hence that $(f_n) \to 0$ in $L_p(S)$. Given $x \in [0, 1[$ and $m \in \mathbb{N}\backslash\{0\}$, let $k_m(x) \in \{0, \dots, 2^m - 1\}$ be such that $k_m(x) \le 2^m x < k_m(x)+1$, so that $f_{n(x)}(x) = 1$ for $n(x) := 2^m+k_m(x)$. Show that for all $x \in S$ there exists a subsequence in $(f_n(x))_n$ that does not converge to 0.
2. For $p \ge 1$ give an example of a measure space $(S, \mathcal{S}, \mu)$ and of a sequence (f_n) of $L_p(S)$ such that $(f_n) \to 0$ a.e. that does not converge to 0 in $L_p(S)$.
3. Show that the Monotone Convergence Theorem does not hold in $L_\infty(S, \mathbb{R})$.
4. Show that the Dominated Convergence Theorem does not hold in $L_\infty(S, \mathbb{R})$.
5. Given a finite measure space $(S, \mathcal{S}, \mu)$ and p, q, r such that $1 \le p \le q \le r$ and $t \in [0, 1]$ such that $1/q = t/p + (1 - t)/r$ show that $L_p(S) \cap L_r(S) \subset L_q(S)$ and $\|f\|_q \le \|f\|_p^t \|f\|_r^{1-t}$ for all $f \in L_p(S) \cap L_r(S)$. [Hint: set $s := p/tq$, $s' := r/(1 - t)q$ so that $1/s + 1/s' = 1$ and apply Hölder's inequality to $|f|^q = |f|^{tq} |f|^{(1-t)q}$.]
6. Given $p \in [1, \infty[$, a finite measure space $(S, \mathcal{S}, \mu)$, and $f \in L_p(S)$ show that

$$\|f\|_\infty = \lim_{p\to\infty} \frac{1}{\mu(S)} \|f\|_p .$$

   [Hint: set $h(p) := \mu(S)^{-1/p} \|f\|_p$ so that h is increasing and bounded above by $\|f\|_\infty$; given $r < \|f\|_\infty$ let $S_r := \{s \in S : |f(s)| \ge r\}$ and observe that $\mu(S_r) < +\infty$, $h(p) \ge \mu(S)^{-1/p}(\int_{S_r} |f|^p)^{1/p} \ge \mu(S)^{-1/p}\mu(S_r)^{1/p} r$.]

   For $r \in \mathbb{R}_+$ let $f \in L_p(S)$ be such that $\|f\|_p \le r$ for all $p \in [1, \infty[$. Show that $f \in L_\infty(S)$.

   Exhibit an example of a function $f \in L_p(S)\backslash L_\infty(S)$ for all $p \in [1, \infty[$. [Hint: take $f := \ln$ on $S := [0, 1]$.]
7. For $d \in \mathbb{N}$, $d \ge 2$, let $f_1, \cdots, f_d \in \mathcal{L}_{d-1}(\mathbb{R}^{d-1})$. Then, setting $x_i' := (x_1, \cdots, x_{i-1}, x_{i+1}, \cdots, x_d) \in \mathbb{R}^{d-1}$ for $x \in \mathbb{R}^d$ and $i \in \mathbb{N}_d$, for f given by $f(x) := f_1(x_1') \cdots f_d(x_d')$ show that $f \in \mathcal{L}_1(\mathbb{R}^d)$ and

$$\|f\|_1 \le \|f_1\|_{d-1} \cdots \|f_d\|_{d-1} .$$

8. (*Young's inequality*) Given p, q, $r \in [1, +\infty]$ satisfying $1/p + 1/q = 1 + 1/r$ and $f \in \mathcal{L}_p(\mathbb{R})$, $g \in \mathcal{L}_q(\mathbb{R})$ show that $f*g \in \mathcal{L}_r(\mathbb{R})$ and $\|f * g\|_r \le \|f\|_p \cdot \|g\|_q$.

9. Show that the convolution $f * g$ of $f \in \mathcal{L}_p(\mathbb{R})$ with $g \in \mathcal{L}_q(\mathbb{R})$ may not exist if the condition $1/p + 1/q \geq 1$ is not satisfied. [Hint: one may have $\int_{\mathbb{R}} f(x-w)g(w)dw = +\infty$ for all $x \in \mathbb{R}$.]

10. Given a σ-finite measure space $(X, \mathcal{S}, \mu)$ and a measurable function $f : X \to \overline{\mathbb{R}}$ one defines $m_f : \mathbb{R}_+ \to \overline{\mathbb{R}}$ by $m_f(r) := \mu(\{x \in X : |f(x)| > r\})$. Verify that m_f is nonincreasing. Show that for all $p \geq 1$ one has $\int |f|^p \, d\mu = \int_0^{+\infty} pr^{p-1} m_f(r)dr$. [Hint: apply Fubini's Theorem to the function $(x, r) \mapsto pr^{p-1} m_f(r)$ on the set $S := \{(x, r) \in X \times \mathbb{R} : r \in [0, f(x)]\}$]. Deduce from this *Tchebychev's inequality* $r^p m_f(r) \leq \|f\|_p$.

11. With the notation of the preceding exercise let $M_p(\mu)$ be the set of measurable functions f such that $r \mapsto r^p m_f(r)$ is bounded on $\mathbb{R}_+$. Verify that for $X :=]0, 1]$ endowed with the restriction μ of the Lebesgue measure the function $f : X \to \mathbb{R}$ given by $f(x) := 1/x \ln x$ belongs to $M_1(\mu) \backslash L_1(\mu)$.

12. With the notation of the two preceding exercises, for $f \in L_1(\lambda)$, where λ is the Lebesgue measure on $\mathbb{R}$, $x \in \mathbb{R}$ and if T is an open interval of $\mathbb{R}$ containing x one sets $T_f(x) := (1/\lambda(T)) \int_T f d\lambda$. Let $h_f : \mathbb{R} \to \overline{\mathbb{R}}$ be the *Hardy-Littlewood function* given by $h_f(x) = \sup\{T_f(x) : T \in \mathcal{T}(x)\}$, where $\mathcal{T}(x)$ is the set of open intervals containing x. Show that h_f is measurable. [Hint: check that for all $r > 0$ the set $H_r := \{h_f > r\}$ is open as for all $x \in H_r$ there exists some $T \in \mathcal{T}(x)$ such that $r\lambda(T) \leq \int_T f d\lambda$.]

Given a Borel measure μ on $\mathbb{R}$ finite on bounded intervals, show that for every finite family $(T_i)_{i \in I}$ of bounded open intervals of $\mathbb{R}$ one can find a subfamily $(T_j)_{j \in J}$ of disjoint intervals such that $\mu(\cup_{i \in I} T_i) \leq 2\Sigma_{j \in J} \mu(T_j)$.

Show that for all $r \in \mathbb{R}_+$ one has $r\lambda(H_r) \leq 2 \|f\|_1$. [Hint: given a compact subset K of H_r take a finite covering $(T_i)_{i \in I}$ of K by open intervals T_i satisfying $r\lambda(T_i) \leq \int_{T_i} f d\lambda$ and apply the preceding question.] Conclude that h_f is finite a.e. and belongs to $M_1(\lambda)$.

13. Prove that for all $p \in [1, \infty]$ the space $L_p(S)$ is a lattice and that the map $(f, g) \mapsto f \vee g := \max(f, g)$ is continuous. [Hint: if $h_n := f_n \vee g_n$, note that $h_n = (1/2)(|f_n - g_n| + f_n + g_n)$.]

14. Given $p \in [1, \infty[$, a sequence $(f_n) \to f$ in $L_p(S)$, a bounded sequence (g_n) in $L_\infty(S)$ such that $(g_n) \to g$ a.e., prove that $(f_n g_n) \to fg$ in $L_p(S)$. [Hint: note that $f_n g_n - fg = (f_n - f)g_n + f(g_n - g)$ and that $f(g_n - g) \to 0$ in $L_p(S)$ by dominated convergence.]

15. Given $p \in [1, \infty[, f \in L_p(\mathbb{R}^d)$ show that, setting $f_u(x) := f(x - u)$ for $u, x \in \mathbb{R}^d$, one has $f_u \to f$ as $u \to 0$ in $\mathbb{R}^d$.

16. (*Riesz convexity theorem*) Let $(S, \mathcal{S}, \mu)$ be a measure space, let X be a Banach space and let A be a linear map from X into the space $L_0(S)$ of classes a.e. of μ-measurable functions on S. Let R be an interval of $[0, 1]$ such that for all $r \in R$ the map A is continuous from X into $L_{1/r}(S)$. Prove that $r \mapsto \ln(\|A\|_{1/r})$ is convex. Compare this result with Exercise 5. [See [106, p.524].]

8.5.2 Nemytskii Maps

Given $p, q \in]1, \infty[$ with $p \geq q$, separable Banach spaces E, F, a measure space $(S, \mathcal{S}, \mu)$, and a map $g : S \times E \to F$, let us consider conditions ensuring that for $u \in \mathcal{L}_p(S, E)$ the map $v := g \diamond u := g(\cdot, u(\cdot))$ belongs to $\mathcal{L}_q(S, F)$. We assume that g is a *Caratheodory map*, i.e. for all $e \in E$ the map $g(\cdot, e)$ is measurable and for a.e. $s \in S$ the map $g_s := g(s, \cdot)$ is continuous from E to F.

Lemma 8.11 *If $g : S \times E \to F$ is a Caratheodory map, then, for all μ-measurable maps $u : S \to E$, the map $g \diamond u := g(\cdot, u(\cdot))$ is μ-measurable.*

Proof Let $u : S \to E$ be a μ-measurable map. By definition there is a Cauchy sequence (u_n) of μ-step functions that converges to u a.e., so that $(g \diamond u_n)$ converges to $g \diamond u$ a.e. In view of Corollary 7.1, it suffices to prove that $g \diamond v$ is μ-measurable for each μ-step function v. Writing

$$v = \sum_{i=1}^{k} 1_{A_i} e_i$$

with $e_i \in E$, $A_i \in \mathcal{S}$ with $\mu(A_i) < \infty$, and $A_i \cap A_j = \varnothing$ for $i \neq j$, we see that

$$(g \diamond v)(s) = \sum_{i=1}^{k} 1_{A_i}(s) g(s, e_i) \qquad s \in S,$$

so that $g \diamond v$ is μ-measurable. □

We need the growth condition:

(G) there exist $a \in \mathcal{L}_q(S, \mathbb{R})$, $b \in \mathcal{L}_\infty(S, \mathbb{R})$ and a null set N such that

$$\forall (s, e) \in (S \setminus N) \times E \qquad \|g(s, e)\| \leq a(s) + b(s) \|e\|^{p/q}.$$

Theorem 8.23 (Krasnoselskii) *Let $p, q \in [1, \infty[$ with $p \geq q$. For a Caratheodory map $g : S \times E \to F$ the following assertions are equivalent:*

(a) g satisfies the growth condition (G);
(b) for all $u \in \mathcal{L}_p(S, E)$ the map $v := g \diamond u := g(\cdot, u(\cdot))$ belongs to $\mathcal{L}_q(S, F)$ and the Nemytskii map $G : \mathcal{L}_p(S, E) \to \mathcal{L}_q(S, F)$ given by $G(u) = g(\cdot, u(\cdot))$ is bounding, i.e. maps bounded subsets into bounded subsets;
(c) the Nemytskii map $G : \mathcal{L}_p(S, E) \to \mathcal{L}_q(S, F)$ given by $G(u) = g(\cdot, u(\cdot))$ is well defined and continuous.

Proof We just prove the most useful implications. We use the inequality $|r + s|^q \leq 2^{q-1} |r|^q + 2^{q-1} |s|^q$ for all $r, s \in \mathbb{R}$ we have derived from the convexity of $t \mapsto |t|^q$.

(a)⇒(b) Clearly, for $u \in \mathcal{L}_p(S,E)$ the map $v := g(\cdot, u(\cdot))$ is μ-measurable and satisfies for $s \in S$

$$\|g(s,u(s))\|^q \leq \left|a(s) + b(s)\,\|u(s)\|^{p/q}\right|^q \leq 2^{q-1}\,|a(s)|^q + 2^{q-1}\,|b(s)|^q\,\|u(s)\|^p .$$

Integrating, we get

$$\int_S \|v(s)\|^q\,d\mu(s) \leq 2^{q-1}\,\|a\|_q^q + 2^{q-1}\,\|b\|_\infty^q\,\|u\|_p^p < \infty$$

so that $v \in \mathcal{L}_q(S,F)$. Moreover, this inequality shows that G maps bounded subsets into bounded subsets.

(a)⇒(c) To prove that G is continuous, we use Proposition 2.19. Given $u \in \mathcal{L}_p(S,E)$ and a sequence $(u_n) \to u$ in $\mathcal{L}_p(S,E)$, we pick an Abel subsequence which we do not relabel. Then by Theorem 8.22 $(u_n) \to u$ a.e. and there exists some $h \in \mathcal{L}_p(S,\mathbb{R})$ such that $\|u_n(s)\| \leq h(s)$ a.e. Using assumption (G) in which we may assume that a and b are nonnegative, we get

$$\begin{aligned}\|g(s,u_n(s)) - g(s,u(s))\|^q &\leq 2^{q-1}\,\|g(s,u_n(s))\|^q + 2^{q-1}\,\|g(s,u(s))\|^q \\ &\leq 2^{q-1}(2^q a(s)^q + 2^{q-1} b(s)^q(\|u_n(s)\|^p + \|u(s)\|^p)) \\ &\leq 2^{2q-1} a(s)^q + 2^{2q-1} b(s)^q h(s)^p .\end{aligned}$$

Observing that $(g(s,u_n(s))) \to g(s,u(s))$ a.e. and applying the Dominated Convergence Theorem, we obtain that $(g \diamond u_n) \to g \diamond u$ in $\mathcal{L}_q(S,F)$. □

Corollary 8.10 *Let $g : S \times E \to F$ be a Caratheodory map satisfying the growth condition (G) for some $p \geq 1$ and $q = 1$. Then the map I_g given by*

$$I_g(u) := \int_S g(s,u(s))d\mu(s) \qquad u \in L_p(S,E)$$

is well defined and continuous from $L_p(S,E)$ into F.

Proof The map I_g is just the composition of $G : u \mapsto g \diamond u$ with integration $v \mapsto \int_S v d\mu$. □

Remark For $p \in [1,\infty[$ and $q = \infty$, and g satisfying for some $a \in \mathcal{L}_\infty(S)$ the condition $\|g(s,e)\| \leq a(s)$ for all $(s,e) \in S \times E$, clearly one has $G(u) := g \diamond u \in \mathcal{L}_\infty(S,F)$ for all $u \in \mathcal{L}_p(S,E)$. However, unless G is constant, G is not continuous from $\mathcal{L}_p(S,E)$ into $\mathcal{L}_\infty(S,F)$ in general. A counterexample can be given whenever for any measurable subset T of S with $\mu(T) > 0$ one can find a sequence (T_n) of measurable subsets of T such that $\mu(\cap_n T_n) = 0$ and $\mu(T_n) > 0$ for all n, as is the case when $(S,\mathcal{S},\mu)$ is non-atomic. In fact, for $u,\ v \in \mathcal{L}_p(S,E)$ such that $G(u) \neq G(v)$, taking $\varepsilon > 0$ satisfying $\varepsilon < \|G(u) - G(v)\|_\infty$, setting $T := \{s \in S : \|G(u)(s) - G(v)(s)\| \geq \varepsilon\}$, taking an associated sequence (T_n), and setting $u_n := 1_{S\backslash T_n}u + 1_{T_n}v$, we see that $(u_n) \to u$ in $\mathcal{L}_p(S, E)$ but since $T_n \subset \{s : \|G(u_n)(s) - G(u)(s)\| \geq \varepsilon\}$, we do not have $(G(u_n)) \to G(u)$ in $\mathcal{L}_\infty(S, F)$. □

Exercise Let $g : S \times E \to \mathbb{R}$ be a measurable map. Suppose g takes nonnegative values and is lower semicontinuous with respect to its second variable. Using Fatou's Lemma, show that the map I_g given by

$$I_g(u) := \int_S g(s, u(s))d\mu(s) \qquad u \in L_p(S, E)$$

is well defined and lower semicontinuous from $L_p(S, E)$ into $\overline{\mathbb{R}}_+$.

Exercise Prove the same conclusion when g is lower semicontinuous with respect to its second variable and satisfies the growth condition

(G_-) there exist $a \in \mathcal{L}_1(S, \mathbb{R})$, $b \in \mathcal{L}_\infty(S, \mathbb{R})$ and a null set N such that

$$\forall (s, e) \in (S \setminus N) \times E \qquad g(s, e) \geq a(s) - b(s) \|e\|^p .$$

See Sect. 2.2.3. □

Let us turn to differentiability properties. The case of a map of class D^1 is easier than the case of a map of class C^1. We just treat the case $p = q$, the case $p > q$ being obtained by slight changes from the proof of Theorem 8.25.

Theorem 8.24 *Let $p \in [1, \infty[$. Let $g : S \times E \to F$ be a Caratheodory map that is of class D^1 with respect to its second variable and such that $D_2 g : S \times E \to L(E, F)$ is a Caratheodory map satisfying for some $b \in \mathcal{L}_\infty(S)$ the conditions $a := g(\cdot, 0) \in \mathcal{L}_p(S, F)$ and for a.e. s*

$$\|D_2 g(s, \cdot)\| \leq b(s).$$

Then the Nemytskii map $G : u \mapsto g \diamond u$ is of class D^1 on $\mathcal{L}_p(S, E)$, hence is Hadamard differentiable, and

$$DG(u)(v) = D_2 g(\cdot, u(\cdot))v(\cdot) \qquad \forall u,\ v \in \mathcal{L}_p(S, E).$$

Proof It follows from the growth assumption on $D_2 g$ and from the Mean Value Theorem that g satisfies condition (G). Since g is of class D^1 with respect to its second variable, the map $h : S \times E^3 \to F$ given by

$$h(s, e, e', e'') := \int_0^1 D_2 g(s, (1 - t)e + te')e'' dt$$

is a Caratheodory map such that $\|h(s, e, e', e'')\| \leq b(s) \|e''\|$ and for a.e. s,

$$g(s, e) - g(s, e') = h(s, e, e', e - e') \qquad \forall e, e' \in E. \tag{8.14}$$

Taking $q = 1$ in the preceding theorem, we get that the Nemytskii operator H associated with h maps $\mathcal{L}_p(S, E^3)$ continuously into $\mathcal{L}_1(S, F)$. Since for $w, x \in \mathcal{L}_p(S, E)$ one has

$$G(w) - G(x) = H(w, x, w - x)$$

with H continuous and $H(w, x, \cdot)$ linear, we get from the characterization of Corollary 5.8 that G is of class D^1 and that $DG(u)(v) = H(u, u, v) = D_2g(\cdot, u(\cdot))v(\cdot)$ for all $u, v \in \mathcal{L}_p(S, E)$. □

Remark For $p = q \in [1, \infty[$, unless for some null subset N of S for all $s \in S \backslash N$ the map $D_2g(s, \cdot)$ is constant, the Nemytskii map G is not of class C^1 since its Hadamard derivative $DG(u)$ at $u \in \mathcal{L}_p(S, E)$ is the map $D_2g \diamond u \in \mathcal{L}_\infty(S, L(E, F))$ considered as a linear subspace of $L(\mathcal{L}_p(S, E), \mathcal{L}_p(S, F))$ and $u \mapsto D_2g \diamond u$ is not continuous in view of the preceding remark. Note that the fact that $\mathcal{L}_\infty(S, L(E, F))$ is isometrically embedded in $L(\mathcal{L}_p(S, E), \mathcal{L}_p(S, F))$ requires a measurable selection theorem, so we admit it.

Theorem 8.25 *Let $p, q \in [1, \infty[$ with $p > q$ and let $r := pq/(p - q)$. Let $g : S \times E \to F$ be a Caratheodory map that is Fréchet differentiable with respect to its second variable and such that $D_2g : S \times E \to L(E, F)$ is a Caratheodory map satisfying for some $a_1 \in \mathcal{L}_r(S)$ and $b_1 \in \mathcal{L}_\infty(S)$ the conditions $g(\cdot, 0) \in L_q(S, F)$ and*

$$\|D_2g(s, e)\| \leq a_1(s) + b_1(s) \|e\|^{p/r} \qquad \text{a.e. } s, \ \forall e \in E.$$

Then the Nemytskii map $G : u \mapsto g \diamond u$ is Fréchet differentiable from $L_p(S, E)$ to $L_p(S, F)$ and

$$DG(u)(v) = D_2g(\cdot, u(\cdot))v(\cdot) \qquad \forall u, \ v \in L_p(S, E).$$

Proof It follows from the growth assumption on D_2g, from the relation $p/r + 1 = p/q$, and from the Mean Value Theorem that g satisfies condition (G) with $a = g(\cdot, 0)$ and $b = (q/p)b_1$. Since g is of class C^1 with respect to its second variable, the map $h : S \times E^2 \to L(E, F)$ given by $h(s, e, e') := \int_0^1 D_2g(s, (1 - t)e + te')dt$ is a Caratheodory map such that for a.e. s,

$$g(s, e) - g(s, e') = h(s, e, e')(e - e') \qquad \forall e, e' \in E. \tag{8.15}$$

Then h induces a Nemytskii operator $H : \mathcal{L}_p(S, E^2) \to \mathcal{L}_r(S, L(E, F))$ given by $H(u, v) := h(\cdot, u(\cdot), v(\cdot))$. Now, by a result similar to Corollary 8.8 the bilinear evaluation map $L(E, F) \times E \to F$ given by $(\ell, e) \mapsto \ell(e)$ induces a continuous bilinear map $\mathcal{L}_r(S, L(E, F)) \times \mathcal{L}_p(S, E) \to \mathcal{L}_q(S, F)$ given by $(z, w) \mapsto z.w$ with $(z.w)(s) := z(s).w(s)$. Thus, an element z of $\mathcal{L}_r(S, L(E, F))$ can be considered as an element of $L(\mathcal{L}_p(S,E), \mathcal{L}_q(S,F))$. Taking $z := h(\cdot, u(\cdot), v(\cdot)) := H(u,v)$, we deduce from (8.15) that

$$G(u) - G(v) = H(u,v).(u-v)$$

and considering $H(u,v)$ as an element of $L(\mathcal{L}_p(S,E), \mathcal{L}_q(S,F))$ and observing that $(u,v) \mapsto H(u,v)$ is continuous, we obtain that G is of class C^1 in view of Lemma 5.6. □

Exercise Let S be a compact interval of $\mathbb{R}$ and let X be the space of Lipschitzian functions on S endowed with the norm given by $\|x\| := \sup_{s\in S}|x(s)| + \sup_{(s,t),\, s\neq t} \frac{|x(s)-x(t)|}{|s-t|}$. For $g : S \times \mathbb{R} \to \mathbb{R}$ such that $G : x \mapsto g(\cdot, x(\cdot))$ sends X into X and is Lipschitzian, prove that there are $a, b \in X$ such that $g(s,t) = a(s)t + b(s)$.

8.6 Duality and Reflexivity of Lebesgue Spaces

In this section, the measure space $(S, \mathcal{S}, \mu)$ being fixed, we simplify the notation $\mathcal{L}_p(S, \mathcal{S}, \mu, \mathbb{R})$ into $\mathcal{L}_p(S)$ and we pass from $\mathcal{L}_p(S)$ to $L_p(S)$ without making the necessary comments about equivalence classes. Let us first establish a geometric property of Lebesgue spaces.

Lemma 8.12 (Clarkson) *For $p \in [2, +\infty[$ and for all $f, g \in L_p(S)$ one has*

$$\left\| \frac{1}{2}(f+g) \right\|_p^p + \left\| \frac{1}{2}(f-g) \right\|_p^p \leq \frac{1}{2}\|f\|_p^p + \frac{1}{2}\|g\|_p^p. \tag{8.16}$$

For $p \in]1,2]$, $q := (1-1/p)^{-1}$, and for all $f, g \in L_p(S)$ one has

$$\left\| \frac{1}{2}(f+g) \right\|_p^q + \left\| \frac{1}{2}(f-g) \right\|_p^q \leq \left(\frac{1}{2}\|f\|_p^p + \frac{1}{2}\|g\|_p^p \right)^{q/p}. \tag{8.17}$$

Proof For $p \in [2, +\infty[$ and $a \in \mathbb{R}_+$, $b \in \mathbb{P}$, setting $t := a/b$ and using the fact that $h(t) := (t^2+1)^{p/2} - t^p \geq h(0) = 1$ for $t \in \mathbb{R}_+$ since $h' \geq 0$, we have

$$a^p + b^p \leq (a^2+b^2)^{p/2}.$$

This relation is still valid if $b = 0$. On the other hand, given $r, s \in \mathbb{R}_+$, the convexity of $t \mapsto |t|^{p/2}$ yields the relation

$$(\left|\frac{1}{2}(r+s)\right|^2 + \left|\frac{1}{2}(r-s)\right|^2)^{p/2} = (\frac{1}{2}r^2 + \frac{1}{2}s^2)^{p/2} \leq \frac{1}{2}r^p + \frac{1}{2}s^p.$$

Taking $a := \frac{1}{2}(r+s)$, $b := \frac{1}{2}(r-s)$, relation (8.16) follows from the properties of the integral.

Relation (8.17) is more delicate. We refer the reader to [97, 159]. □

Theorem 8.26 *Given a coupling function $c : E \times F \to \mathbb{R}$ between two separable Banach spaces and $p \in [1, +\infty[$, $q := (1-1/p)^{-1}$, the mapping $\overline{c} : (f, g) \mapsto \int_S c(f(s), g(s))d\mu(s)$ is a coupling between $L_p(S, E)$ and $L_q(S, F)$.*

If c is a metric coupling between two Banach spaces, then $\overline{c}$ is a metric coupling.

Proof Since $c : E \times F \to \mathbb{R}$ is bilinear and continuous, for $f \in \mathcal{L}_p(S, E)$ and $g \in \mathcal{L}_q(S, F)$ the function $h := c \circ (f, g) : s \mapsto c(f(s), g(s)) = \langle f(s), g(s)\rangle$ is μ-measurable by Proposition 7.1. Moreover, Hölder's inequality (8.11) yields

$$|\overline{c}(f,g)| \leq \int_S |h|\, d\mu \leq \|c\| \int_S \|f\| \cdot \|g\|\, d\mu \leq \|c\| \, \|f\|_p \cdot \|g\|_q \,. \tag{8.18}$$

Let $g \in \mathcal{L}_q(S, F)$ be such that $\overline{c}(f, g) = 0$ for all $f \in \mathcal{L}_p(S, E)$. Then, for all $A \in \mathcal{S}$ with finite positive measure and all $e \in E$ we have $\int_A c(e, g(s))d\mu(s) = 0$. Then, by Corollary 7.14, for all $e \in E$ we get $c(e, g(s)) = 0$ a.e. Taking a countable dense subset of E and using the fact that c is a coupling we get $g = 0$ a.e. A similar implication holds for f. Thus $\overline{c}$ is a coupling function between $X := L_p(S, E)$ and $Y := L_q(S, F)$.

Now let us suppose c is a metric coupling. Given a μ-step function $f = \Sigma_{i \in I} 1_{A_i} a_i$ as above with $\|f\|_p \neq 0$, for each $i \in I$ we pick a sequence $(b_{i,n})_{n \geq 0}$ in the unit sphere S_Y of Y such that $(c(a_i, b_{i,n}))_n \to \|a_i\|$ and we set $g_n := \Sigma_{i \in I} \|a_i\|^{p-1} b_{i,n} 1_{A_i}$. Then, since $q(p-1) = p$, we have

$$\|g_n\|_q^q = \sum_{i \in I} \|a_i\|^{q(p-1)} \|b_{i,n}\|^q \mu(A_i) = \sum_{i \in I} \|a_i\|^p \mu(A_i) = \|f\|_p^p \,,$$

$$\overline{c}(f, g_n) = \int_S \sum_{i \in I} \|a_i\|^{p-1} c(a_i, b_{i,n}) 1_{A_i} d\mu \to \sum_{i \in I} \|a_i\|^p \mu(A_i) = \|f\|_p^p \,.$$

It follows that $\|g_n\|_q = \|f\|_p^{p/q}$ and

$$\sup_n \frac{\overline{c}(f, g_n)}{\|g_n\|_q} \geq \frac{\|f\|_p^p}{\|f\|_p^{p/q}} = \|f\|_p \,.$$

Then $\|\overline{c}(f, \cdot)\|_{L_q(S,F)^*} \geq \|f\|_p$ and since the reverse inequality follows from (8.18), we get $\|\overline{c}(f, \cdot)\|_{L_q(S,F)^*} = \|f\|_p$. Since both sides of this equality are continuous functions of f and since the space $St(\mu, E)$ is dense in $L_p(S, E)$ with respect to the norm $\|\cdot\|_p$, this equality holds for all $f \in L_p(S, E)$. Similarly, one can show that $\|\overline{c}(\cdot, g)\|_{L_p(S,E)^*} = \|g\|_q$ for all $g \in L_q(S, F)$. Thus $\overline{c}$ is a metric coupling. □

Exercise Give a simplified proof of the last assertion of the theorem in the case $E = F = \mathbb{R}$. [Hint: given $f \in L_p(S)$ take $g := |f|^{p-2} f$ and note that $\int_S fg d\mu = \|f\|_p^p$ and $\|g\|_q = \|f\|_p^{p-1}$.]

Theorem 8.27 (Clarkson) *For $p \in]1, +\infty[$ the space $L_p(S)$ is uniformly convex, hence reflexive.*

Proof Let us first consider the case $p \in [2, +\infty[$. By Clarkson's inequality (8.16) for $f, g \in L_p(S)$ satisfying $\|f\|_p \leq 1$, $\|g\|_p \leq 1$, $\|f - g\|_p > \varepsilon$ we have $\|(f+g)/2\|_p^p < 1 - (\varepsilon/2)^p$. This shows that $\gamma_p : t \mapsto 1 - (1 - (t/2)^p)^{1/p}$ is a gage of convexity of $(L_p(S), \|\cdot\|_p)$.

When $p \in]1, 2]$ we use the second Clarkson's inequality (8.17) showing that for $f, g \in L_p(S)$ satisfying $\|f\|_p \leq 1$, $\|g\|_p \leq 1$, $\|f - g\|_p > \varepsilon$ we have $\|(f+g)/2\|_p^q < 1 - (\varepsilon/2)^q$. Thus, γ_p given by $\gamma_p(t) := 1 - (1 - (t/2)^q)^{1/q}$ is a gage of convexity of $(L_p(S), \|\cdot\|_p)$. The reflexivity of $L_p(S)$ then follows from Theorem 3.19. □

Remark If E is a Banach space of type p in the sense that there exists some $c > 0$ such that

$$\forall u, v \in E \qquad \|u + v\|^p + \|u - v\|^p \leq 2\|u\|^p + c\|v\|^p$$

then $L_p(S, E)$ is of type p: one has

$$\forall f, g \in L_p(S, E) \qquad \|f + g\|_p^p + \|f - g\|_p^p \leq 2\|f\|_p^p + c\|g\|_p^p.$$

It can be shown that any reflexive Banach space can be endowed with an equivalent norm that is of type p.

Remark If S is an open subset of $\mathbb{R}^d$ and μ is the restriction of the Lebesgue measure λ, then the space $L_1(S)$ is not reflexive (see Exercise 1).

It is important to identify the dual of $L_p(S)$.

Theorem 8.28 (Riesz) *For $p \in]1, +\infty[$, $q := (1 - 1/p)^{-1}$, $Y := L_q(S)$ can be identified with the dual of $X := L_p(S)$ via the map $\overline{c}_Y$ given by $\overline{c}_Y(g)(f) = \int_S fg d\mu$ for $f \in L_p(S)$, $g \in L_q(S)$.*

When $(S, \mathcal{S}, \mu)$ is σ-finite, $Y := L_\infty(S)$ can be identified with the dual of $L_1(S)$ via the map $\overline{c}_Y$.

Proof For $p \in]1, +\infty[$ this is a consequence in Proposition 3.23 and Theorem 8.26.

Let us consider the case $p = 1$. We start with the additional assumption that $\mu(S) < +\infty$. We already know that the map $\overline{c}_Y$ is an isometry from $L_\infty(S)$ onto its image in the dual of $L_1(S)$. Hölder's inequality ensures that for $p \in]1, +\infty[$ the space $L_p(S)$ is contained in $L_1(S)$ and the canonical injection $j_p : L_p(S) \to L_1(S)$ is continuous since for $q := (1 - 1/p)^{-1}$ one has $\|f\|_1 \leq \|1_S\|_q \|f\|_p = (\mu(S))^{1/q} \|f\|_p$ for all $f \in L_p(S)$. Given $\ell \in (L_1(S))^*$, by the preceding case we can find $g_q \in L_q(S)$ such that $(\ell \circ j_p)(f) = \int_S f g_q d\mu$ for all $f \in L_p(S)$; moreover g_q is unique and $\|g_q\|_q = \|\ell \circ j_p\| \leq c^{1/q} \|\ell\|$ with $c := \mu(S)$. For $r > p$ and $s := (1 - 1/r)^{-1}$, denoting by

$j_{q,s} : L_q(S) \to L_s(S)$ the canonical injection of Corollary 8.9, by uniqueness we get that $j_{q,s}(g_q) = g_s$. Taking a representant g of the common class of g_q and g_s, let us show that $g \in \mathcal{L}_\infty(S)$ and $\|g\|_\infty \leq \|\ell\|$. Given $b > \|\ell\|$, let $T := \{t \in S : |g(t)| \geq b\}$. If $a := \mu(T)$ is positive, we have $\|g\|_q \geq a^{1/q}b$; since for q large enough we have $a^{1/q}b > c^{1/q}\|\ell\|$, which is impossible since $\|g_q\|_q \leq c^{1/q}\|\ell\|$. Thus $\mu(T) = 0$, $g \in \mathcal{L}_\infty(S)$ and $\ell(f) = \int_S fgd\mu$ for all $p > 1$ and all $f \in L_p(S)$. Now, given $f \in \mathcal{L}_1(S)$ and $k \in \mathbb{N}$, setting $f_k(s) = f(s)$ whenever $|f(s)| \leq k$ and $f_k(s) = 0$ otherwise, we see that $f_k \in \mathcal{L}_p(S)$ and that $|f_k g| \leq |fg|$ for all $k \in \mathbb{N}$ with $fg \in \mathcal{L}_1(S)$ and that $(f_k g) \to fg$ a.e., so that, by the Dominated Convergence Theorem we have $(f_k g) \to fg$ in $\mathcal{L}_1(S)$. Thus

$$\ell(f) = \lim_k \ell(f_k) = \lim_k \int_S f_k g d\mu = \int_S fgd\mu$$

and $\ell = \overline{c}_Y(g)$ since f is arbitrary in $\mathcal{L}_1(S)$.

Now let us consider the case when $(S, \mathcal{S}, \mu)$ is σ-finite. Let $(S_n)_n$ be a partition of S into sets of finite measure. Given $\ell \in (L_1(S))^*$, for any $n \in \mathbb{N}$ we can find some $g_n \in \mathcal{L}_\infty(S_n)$ such that for all $h \in \mathcal{L}_1(S_n)$ we have $\int_{S_n} hg_n = \ell(j_n(h))$, where $j_n(h) \in \mathcal{L}_1(S)$ is the extension of h by 0 on $S \backslash S_n$ and moreover $\|g_n\|_\infty \leq \|\ell\|$ since $\|j_n(h)\|_1 = \|h\|_1$. Let $g \in \mathcal{L}_\infty(S)$ be such that $g_n = g \mid S_n$. Then, for all $f \in \mathcal{L}_1(S)$, observing that for $f_n := f \mid S_n \in \mathcal{L}_1(S_n)$ we have $j_n(f_n) = 1_{S_n}f$, $f = \Sigma_n 1_{S_n}f$, since ℓ is linear and continuous, we get

$$\ell(f) = \sum_n \ell(1_{S_n}f_n) = \sum_n \int_{S_n} f_n g_n = \sum_n \int_S 1_{S_n} fg = \int_S fg$$

by Corollary 7.11, using the fact that $\Sigma_n 1_{S_n} fg = fg$, $\Sigma_n |1_{S_n} fg| = |fg|$, and the fact that an absolutely convergent series is convergent and $\Sigma_{n \leq k} |1_{S_n} fg| \leq |fg|$. □

The following general equivalence result is outside the scope of this book. We quote it from [99, p. 98] in order to give perspective. We shall just prove an easier result.

Theorem 8.29 *Let $(S, \mathcal{S}, \mu)$ be a finite measure space, $p \in [1, \infty[$, $q \in]1, \infty]$ with $1/p + 1/q = 1$, and let E be a Banach space. Then $L_p(\mu, E)^* = L_q(\mu, E^*)$ if and only if E^* has the Radon-Nikodým property.*

Note that the surjectivity of the map $\overline{c}_Y : L_\infty(S) \to L_1(S)^*$ when $(S, \mathcal{S}, \mu)$ is finite can be rephrased by saying that any continuous linear form $\ell : L_1(S) \to \mathbb{R}$ is representable. Theorem 8.30 below gives a generalization yielding an alternative proof (due to J. Von Neumann) to this statement in view of the equivalence of Proposition 8.4. We start with an analogue of the surjectivity result for $\overline{c}_Y$ valid for Hilbert spaces.

Proposition 8.12 *If E is a Hilbert space, for every σ-finite measure space $(S, \mathcal{S}, \mu)$, $L_\infty(\mu, E)$ can be identified with the dual space of $L_1(\mu, E)$.*

Proof Using the arguments of the last proof, it suffices to assume that μ is finite and to show that every continuous linear map $\ell : L_1(\mu, E) \to \mathbb{R}$ is of the form $f \mapsto \int \langle f \mid g \rangle d\mu$ for some $g \in L_\infty(\mu, E)$. As already observed, by Corollary 8.9 (or the Cauchy-Schwarz inequality) we have a continuous linear injection j of $L_2(\mu, E)$ into $L_1(\mu, E)$ since

$$\int |f|\, d\mu \leq \|f\|_2 \|1_S\|_2 = \mu(S)^{1/2} \|f\|_2 \qquad \text{for all } f \in L_2(\mu, S).$$

Thus $\ell \circ j$ is a continuous linear form on the Hilbert space $L_2(\mu, E)$. The Riesz representation theorem in Hilbert spaces provides some $g \in L_2(\mu, E)$ such that $\|g\|_2 = \|\ell \circ j\|$ and

$$\ell(j(f)) = \int_S \langle f \mid g \rangle d\mu \qquad \text{for all } f \in L_2(\mu, E). \tag{8.19}$$

In particular, for all $A \in \mathcal{S}$, $e \in E$ one has $1_A e \in L_2(\mu, E)$ and $j(1_A e) \in L_1(\mu, E)$, hence

$$\left| \int_A \langle e \mid g \rangle d\mu \right| = \left| \int_S \langle 1_A e \mid g \rangle d\mu \right| = |\ell(j(1_A e))| \leq \|\ell\| \|e\| \mu(A).$$

For all $e \in E$ Corollary 7.13 ensures that $|\langle e \mid g \rangle| \leq \|\ell\| \|e\|$ a.e. Using the fact that for a representant g_0 of g the set $g_0(S)$ is contained in a separable subspace of E, we get that $\|g\| \leq \|\ell\|$ a.e. so that $g \in L_\infty(\mu, E)$. Since $L_2(\mu, E)$ (and even $St(\mu, E)$) is dense in $L_1(\mu, E)$, relation (8.19) can be extended into $\ell(f) = \int_S \langle f \mid g \rangle d\mu$ for all $f \in L_1(\mu, E)$. □

Theorem 8.30 *Hilbert spaces satisfy the Radon-Nikodým Property: if E is a Hilbert space, if $(S, \mathcal{S}, \mu)$ is a finite measure space, then for any μ-continuous vectorial measure $\nu : \mathcal{S} \to E$ of bounded variation, there exists an $h \in L_1(\mu, E)$ such that $\nu = \mu_h$, i.e. $\nu(A) = \int_A h d\mu$ for all $A \in \mathcal{S}$.*

Proof Let $\nu : \mathcal{S} \to E$ be a μ-continuous vectorial measure on the finite measure space $(S, \mathcal{S}, \mu)$ with total variation $|\nu|$. Since $|\nu|$ is μ-continuous, it has a Radon-Nikodým derivative $k \in L_\infty(\mu)$ with respect to μ : for all $\varphi \in L_1(|\nu|)$ one has

$$\int \varphi d|\nu| = \int \varphi k d\mu. \tag{8.20}$$

Now, by Theorem 8.1, one can associate to the scalar product $b = \langle \cdot \mid \cdot \rangle : E \times E \to \mathbb{R}$ a unique continuous linear map $\hat{b} : L_1(|\nu|, E) \to \mathbb{R}$ denoted by $f \mapsto \int_S f d\nu$ such that $\hat{b}(1_A e) = \langle e \mid \nu(A) \rangle$ for all $A \in \mathcal{S}$, $e \in E$ and $\left|\hat{b}(f)\right| \leq \|f\|_1$ for all $f \in L_1(|\nu|, E)$. The preceding proposition yields a unique $g \in L_\infty(|\nu|, E)$ such that $\hat{b}(f) = \int \langle f \mid g \rangle d|\nu|$ for all $f \in L_1(|\nu|, E)$. In particular, for all $A \in \mathcal{S}$, $e \in E$ we have

$$\langle e \mid \nu(A)\rangle = \hat{b}(1_A e) = \int_S \langle 1_A e \mid g\rangle d\,|\nu| = \langle e \mid \int_A g d\,|\nu|\rangle.$$

Setting $h := kg \in L_\infty(\mu, E)$ and taking $\varphi := \langle f \mid g\rangle$ in relation (8.20) we get $\langle e \mid \nu(A)\rangle = \langle e \mid \int_A h d\mu\rangle$ for all $A \in \mathcal{S}$, $e \in E$, hence $\nu(A) = \int_A h d\mu$ for all $A \in \mathcal{S}$. □

Now let us turn to approximation properties of Lebesgue spaces. Given $p \in [1, +\infty[$ and an open subset Ω of $\mathbb{R}^d$ (with $d \in \mathbb{N}\backslash\{0\}$) equipped with the measure λ induced by the Lebesgue measure λ_d, we want to show that the elements of $L_p(\Omega)$ can be approximated by some simple functions. For this purpose, let us introduce the space $\mathcal{L}_{1,loc}(\Omega)$ of *locally integrable* functions on Ω, that is, the space of λ-measurable functions such that for every compact subset K of Ω one has $1_K f \in \mathcal{L}_1(\Omega, \lambda)$. It is easy to see that $\mathcal{L}_{1,loc}(\Omega)$ is a linear space. It contains $\mathcal{L}_p(\Omega)$ for all $p \in [1, +\infty]$ since for $f \in \mathcal{L}_p(\Omega)$ and a compact subset K of Ω one has $\int |1_K f|\, d\lambda \leq \lambda(K)^{1/q} \|f\|_p$ with $q := (1 - 1/p)^{-1}$. We start with a preliminary result of interest.

Lemma 8.13 *A function $f \in \mathcal{L}_{1,loc}(\Omega)$ is null a.e. if and only if it is such that $\int fg = 0$ for all g in the space $C_c(\Omega)$ of continuous functions with compact support in Ω.*

Proof If $f \in \mathcal{L}_{1,loc}(\Omega)$ is null a.e., obviously for any $g \in C_c(\Omega)$ we have $fg = 0$ a.e. and $\int fg = 0$. Conversely, suppose $f \in \mathcal{L}_{1,loc}(\Omega)$ is such that $\int fg = 0$ for all $g \in C_c(\Omega)$. In a first step we assume that $f \in \mathcal{L}_1(\Omega)$ and that $\lambda(\Omega) < +\infty$. By Theorem 7.3, given $\varepsilon > 0$ we can find $h \in C_c(\Omega)$ such that $\|f - h\|_1 < \varepsilon$. Let $K := K_+ \cup K_-$ with

$$K_+ := \{x \in \Omega : h(x) \geq \varepsilon\}, \qquad K_- := \{x \in \Omega : h(x) \leq -\varepsilon\}.$$

Taking a compact subset L of Ω whose interior contains K, the Tietze-Urysohn Theorem (Theorem 2.8) yields some continuous function g on Ω such that

$$g \mid (\Omega\backslash L) = 0, \quad g \mid K_+ = 1, \quad g \mid K_- = -1, \quad \sup_{x\in\Omega} |g(x)| \leq 1.$$

Since $|hg|\,(x) \leq |h(x)| \leq \varepsilon$ for all $x \in \Omega\backslash K$, we have

$$\int_{\Omega\backslash K} |hg| \leq \int_{\Omega\backslash K} |h| \leq \varepsilon\lambda(\Omega\backslash K),$$

$$\left|\int_\Omega hg\right| = \left|\int_\Omega (h - f)g\right| \leq \|g\|_\infty \|h - f\|_1 \leq \varepsilon,$$

$$\int_K |h| = \int_K hg = \int_\Omega hg - \int_{\Omega\backslash K} hg \leq \varepsilon + \left|\int_{\Omega\backslash K} hg\right| \leq \varepsilon + \varepsilon\lambda(\Omega\backslash K),$$

so that

$$\int_{\Omega} |h| = \int_{\Omega \setminus K} |h| + \int_{K} |h| \leq \varepsilon + 2\varepsilon\lambda(\Omega \setminus K) \leq \varepsilon(2\lambda(\Omega) + 1).$$

Thus $\|f\|_1 \leq \|f - h\|_1 + \|h\|_1 \leq 2\varepsilon(\lambda(\Omega) + 1)$. Since ε is arbitrarily small, we get $\|f\|_1 = 0$ and $f = 0$ a.e.

In the general case we take a sequence (Ω_n) of open subsets of Ω whose closures are compact and whose union is Ω (for instance $\Omega_n := \{x \in \Omega : \|x\| < n,\ d(x, \mathbb{R}^d \setminus \Omega) > 1/n\}$). Then $f_n := f \mid \Omega_n \in \mathcal{L}_1(\Omega_n)$ and $\int_{\Omega_n} f_n g = 0$ for all $g \in C_c(\Omega_n)$, so that $f_n = 0$ a.e. It follows that $f = 0$ a.e. □

Theorem 8.31 *For $p \in [1, +\infty[$ and for an open subset Ω of $\mathbb{R}^d$, the space $C_c(\Omega)$ of continuous functions with compact support in Ω is dense in $L_p(\Omega)$.*

Proof In view of Theorem 7.3, it suffices to consider the case $p \in]1, +\infty[$. Since for $q := (1 - 1/p)^{-1}$ the dual of $L_p(\Omega)$ is $L_q(\Omega)$, it also suffices to prove that for all $h \in \mathcal{L}_q(\Omega)$ satisfying $\int hg = 0$ for all $g \in C_c(\Omega)$ we have $h = 0$ a.e. Since $h \in \mathcal{L}_{1,loc}(\Omega)$, this follows from the preceding lemma. □

Let us give a useful application. Given $f \in \mathcal{L}_p(\mathbb{R}^d)$ with $p \in [1, +\infty]$ and $w \in \mathbb{R}^d$ we denote by $T_w f$ the function defined by $T_w f := f \circ t_w$ where $t_w : \mathbb{R}^d \to \mathbb{R}^d$ is the translation $x \mapsto x - w$. Since t_w is an isometry, $f \circ t_w$ is μ-measurable when f is measurable. Moreover, if $f = g$ a.e. then $\{x : T_w f(x) \neq T_w g(x)\} = \{y : f(y) \neq g(y)\} + w$, so that $T_w f = T_w g$ a.e. and we can regard T_w as operating on equivalence classes of functions with respect to equality a.e. We observe that T_w is a linear isometry from $L_p(\mathbb{R}^d)$ onto $L_p(\mathbb{R}^d)$. Moreover, we have a convergence result.

Lemma 8.14 *For $p \in [1, +\infty[$ and $f \in L_p(\mathbb{R}^d)$ one has $\|T_w f - f\|_p \to 0$ as $w \to 0$.*

Proof Let us first suppose $f \in C_c(\mathbb{R}^d)$, so that f is uniformly continuous: given $\varepsilon > 0$ there exists some $\delta \in]0, 1[$ such that $|f(x - w) - f(x)| \leq \varepsilon$ when $\|w\| \leq \delta$. Denoting by K the support of f, for $w \in \delta B_{\mathbb{R}^d}$ we have

$$\|T_w f - f\|_p^p = \int_{K \cup (K+w)} |f(x - w) - f(x)|^p \, d\lambda_d(x) \leq 2\lambda_d(K)\varepsilon^p,$$

so that $\|T_w f - f\|_p \to 0$ as $w \to 0$.

Now let us suppose $f \in \mathcal{L}_p(\mathbb{R}^d)$. Given $\varepsilon > 0$, since $C_c(\mathbb{R}^d)$ is dense in $L_p(\mathbb{R}^d)$, there exists a $g \in C_c(\mathbb{R}^d)$ such that $\|g - f\|_p < \varepsilon/3$. Then we choose $\delta > 0$ such that $\|T_w g - g\|_p < \varepsilon/3$ whenever $w \in \delta B_{\mathbb{R}^d}$, so that we have

$$\|T_w f - f\|_p \leq \|T_w f - T_w g\|_p + \|T_w g - g\|_p + \|g - f\|_p < \varepsilon$$

since $\|T_w f - T_w g\|_p = \|f - g\|_p$ by the change of variables theorem. □

Proposition 8.13 *For $p \in [1, +\infty[$ and for an open subset Ω of $\mathbb{R}^d$, the space $L_p(\Omega)$ is separable.*

Proof Let (K_n) be an increasing sequence of compact subsets of Ω whose interiors cover Ω (take f.i. $K_n := \{x \in \Omega \cap nB_{\mathbb{R}^d} : d(x, \mathbb{R}^d \backslash \Omega) \geq 1/n\}$). Let L_n be the set of functions on Ω which are null on $\Omega \backslash K_n$ and whose restrictions to K_n are restrictions of polynomial functions over $\mathbb{Q}$. Then $L := \cup_n L_n$ is countable. Since $C_c(\Omega)$ is dense in $L_p(\Omega)$, it suffices to prove that for any $g \in C_c(\Omega)$ and any $\varepsilon > 0$ we can find some $h \in L$ such that $\|h - g\|_p \leq \varepsilon$. There is some $m \in \mathbb{N}$ such that the support K of g is contained in the interior of K_m. Using Weierstrass' Theorem we can find some $h \in L_m$ such that $\|(h - g) \mid K_m\|_\infty \leq \varepsilon / \lambda_d(K_m)^{1/p}$. Since $g \mid \Omega \backslash K_m = 0$ and $h \mid \Omega \backslash K_m = 0$ we have $\|h - g\|_p \leq \varepsilon$. □

Exercises

1. Prove that if Ω is an open subset of $\mathbb{R}^d$, if $\mathcal{S}$ is the Borel σ-algebra of Ω and if λ is the restriction of the Lebesgue measure, then $L_1(\Omega)$ is not reflexive. [Hint: Given $a \in \Omega$ and a sequence $(r_n) \to 0_+$ with $r_n < d(a, \mathbb{R}^d \backslash \Omega)$, let $f_n := c_n 1_{B(a,r_n)}$ with $c_n := 1/\lambda(B(a, r_n))$. Assuming $L_1(\Omega)$ is reflexive one can find $f \in L_1(\Omega)$ and a subsequence $(f_{k(n)})_n$ of (f_n) such that $(\int_\Omega f_{k(n)} g) \to \int_\Omega fg$ for all $g \in L_\infty(\Omega)$. Taking for g an element of the space $C_c(\Omega_a)$ of continuous functions with compact support in $\Omega_a := \Omega \backslash \{a\}$, by density this entails that $f \mid \Omega_a = 0$, hence that $f = 0$ a.e., a contradicting $\int_\Omega f 1_\Omega d\lambda = 1$.]

2. Prove that if Ω is an open subset of $\mathbb{R}^d$, if $\mathcal{S}$ is the Borel σ-algebra of Ω and if λ is the restriction of the Lebesgue measure, then $L_\infty(\Omega)$ is not separable. [Hint: For all $a \in \Omega$, given $r_a > 0$, $r_a < d(a, \mathbb{R}^d \backslash \Omega)$, let $f_a := 1_{B(a,r_a)}$ and let G_a be the open ball of $(L_\infty(\Omega), \|\cdot\|_\infty)$ with center f_a and radius $1/2$. Then for $a \neq b$ in Ω one has $G_a \cap G_b = \varnothing$ and since the family $(G_a)_{a \in \Omega}$ is uncountable, $L_\infty(\Omega)$ cannot have a countable base of open sets.]

3. Let $(S, \mathcal{S}, \mu)$ and $(T, \mathcal{T}, \nu)$ be two measure spaces with σ-finite measures. Let $K \in L_2(S \times T)$, $S \times T$ being endowed with the measure $\mu \otimes \nu$. Given $f \in L_2(S)$, show that for almost every $t \in T$ the function $s \mapsto K(s, t)f(s)$ is in $L_2(T)$. [Hint: use Fubini's Theorem to show that for almost every $t \in T$ one has $K(\cdot, t) \in L_2(S)$ and apply the Cauchy-Schwarz inequality to f and $K(\cdot, t)$.]

Prove that g given by $g(t) := \int_S K(s, t)f(s)d\mu(s)$ is in $L_2(T)$, that $A : f \mapsto g$ is linear from $L_2(S)$ to $L_2(T)$ and that $\|g\|_2 \leq \|K\|_2 \|f\|_2$. Under appropriate assumptions on S and T, Hilbert-Schmidt operators between $L_2(S)$ and $L_2(T)$ can be represented by operators as above.

8.7 Compactness in Lebesgue Spaces

The following notion can be used for convergence criteria; it is also useful for compactness theorems and existence results. In this section, $(E, \|\cdot\|)$ is a Banach space; for simplicity we also denote by $|\cdot|$ the norm of E, observing that the whole study can be reduced to the case $E = \mathbb{R}$.

Definition 8.8 Let $p \in [1, +\infty[$. A subset F of $L_p(S, E)$ is said to be *p-equi-integrable* if it satisfies the following two conditions:

(a) for every $\varepsilon > 0$ there exists a $\delta > 0$ such that for every $T \in \mathcal{S}$ satisfying $\mu(T) < \delta$ one has $\sup_{f\in F}(\int_T |f|^p \, d\mu)^{1/p} \leq \varepsilon$;
(b) for every $\varepsilon > 0$ there exists a $B \in \mathcal{S}$ such that $\mu(B) < +\infty$ and $\sup_{f\in F}(\int_{S\setminus B} |f|^p \, d\mu)^{1/p} \leq \varepsilon$.

When $\mu(S) < +\infty$, condition (b) (which is omitted by some authors) is trivially satisfied. For $p = 1$ one simply says that F is *equi-integrable*. Thus F is p-equi-integrable if and only if the family $\{\|f\|^p : f \in F\}$ is equi-integrable.

Let us give some examples.

Example 1 If $F := \{f\}$ with $f \in L_p(S, \mathbb{R})$, then F is equi-integrable. Let us verify this. Since $|f|^p$ is integrable, by Proposition 8.2, condition (a) is satisfied. Moreover, since by Lemma 7.2 $|f|^p$ is null off a set of σ-finite measure $S_f := \cup_n S_n$, where (S_n) is an increasing sequence in $\mathcal{S}$ with $\mu(S_n) < \infty$ for all n, taking $B := S_n$ with n large enough, we get assertion (b).

Example 2 Let $F \subset L_p(S, E)$ be such that there exists a $h \in L_p(S, \mathbb{R})$ satisfying $|f| \leq h$ for all $f \in F$. Then F is p-equi-integrable. This easily follows from the preceding example.

Our next example is important, so we state it as a proposition.

Proposition 8.14 *Let (f_n) be a convergent sequence in $L_p(S, E)$. Then $F := \{f_n : n \in \mathbb{N}\}$ is p-equi-integrable.*

Proof Given $\varepsilon > 0$ let $k \in \mathbb{N}$ be such that $\|f_n - f\|_p < \varepsilon/2$ for $n > k$, where $f := \lim_n f_n$. Then, for $n > k$ and for any $T \in \mathcal{S}$ the Minkowski inequality yields

$$(\int_T |f_n|^p \, d\mu)^{1/p} \leq (\int_T |f_n - f|^p \, d\mu)^{1/p} + (\int_T |f|^p \, d\mu)^{1/p} \leq \frac{\varepsilon}{2} + (\int_T |f|^p \, d\mu)^{1/p}.$$

Using Example 1 we can find $\delta > 0$ such that for any $T \in \mathcal{S}$ satisfying $\mu(T) < \delta$ we have $(\int_T |f|^p \, d\mu)^{1/p} < \varepsilon/2$, hence $\sup_{n>k}(\int_T |f_n|^p \, d\mu)^{1/p} \leq \varepsilon$. Taking $h := \max_{n\leq k}\{|f_n|\}$ and using Example 2, we get $\sup_{n\leq k}(\int_T |f_n|^p \, d\mu)^{1/p} \leq \varepsilon$ if $T \in \mathcal{S}$ satisfies $\mu(T) < \delta$ (with a possibly smaller δ). Thus condition (a) is satisfied. Condition (b) can be established similarly. □

We have a converse, provided $(f_n) \to f$ a.e. or $(f_n) \to f$ in measure or converges locally in measure.

Proposition 8.15 (Vitali) *For a sequence (f_n) of $L_p(S, E)$ and $f \in L_p(S, E)$ one has $(f_n) \to f$ in $L_p(S, E)$ if and only if $(f_n) \to f$ in measure and the set $F := \{f_n : n \in \mathbb{N}\}$ is p-equi-integrable.*

Proof If $(\|f_n - f\|_p)_n \to 0$, Proposition 7.8 implies that $(f_n) \to f$ in measure and the preceding proposition shows that the set F is p-equi-integrable.

Conversely, suppose the set F is p-equi-integrable and $(f_n) \to f$ in measure. Then $G := F \cup \{f\}$ is p-equi-integrable in view of Example 1. Given $\varepsilon > 0$, by condition (a) of the preceding definition there exists a $\delta > 0$ such that for all $T \in \mathcal{S}$ satisfying $\mu(T) < \delta$ one has $\int_T |g|^p \, d\mu \leq \varepsilon^p/2^p$ for all $g \in G$, hence $(\int_T |f_n - f|^p \, d\mu)^{1/p} \leq \varepsilon$ for all $n \in \mathbb{N}$ by the Minkowski inequality. Also, by condition (b) there exists a $B \in \mathcal{S}$ such that $\mu(B) < +\infty$ and $(\int_{S \setminus B} |f_n - f|^p \, d\mu)^{1/p} \leq \varepsilon$ for all $n \in \mathbb{N}$. Since $(f_n) \to f$ in measure, given $c := \varepsilon/\mu(B)^{1/p}$, setting

$$T_n := \{s \in B : |f_n(s) - f(s)| > c\}$$

we have $\mu(T_n) < \delta$ for n large enough. Then

$$\begin{aligned}\int_S |f_n - f|^p \, d\mu &= \int_{S \setminus B} |f_n - f|^p \, d\mu + \int_{B \setminus T_n} |f_n - f|^p \, d\mu + \int_{T_n} |f_n - f|^p \, d\mu \\ &\leq \varepsilon^p + \mu(B \setminus T_n)c^p + \varepsilon^p \leq 3\varepsilon^p\end{aligned}$$

since $\mu(B \setminus T_n)c^p \leq \mu(B)c^p \leq \varepsilon^p$. This shows that $(\|f_n - f\|_p) \to 0$. □

The following characterization sheds more light on the notion of p-equi-integrability.

Proposition 8.16 (De la Vallée Poussin) *Let $p \in [1, +\infty[$ and let F be a subset of $L_p(S, E)$. Among the following assertions one has (a)⇔(b)⇒(c). If F is bounded, these assertions are all equivalent:*

(a) for some increasing function $\gamma : \mathbb{R}_+ \to \overline{\mathbb{R}}_+$ such that $\lim_{r \to \infty} \frac{\gamma(r)}{r} = \infty$ one has $m := \sup\{\int_S \gamma(|f|^p) d\mu : f \in F\} < \infty$;
(b) F is uniformly p-integrable: $\lim_{r \to \infty} \sup\{\int_{\{|f| > r\}} |f|^p \, d\mu : f \in F\} = 0$;
(c) assertion (a) of Definition 8.8 holds: for some function $\delta : \mathbb{P} \to \mathbb{P}$ one has $\sup_{f \in F} (\int_T |f|^p \, d\mu)^{1/p} \leq \varepsilon$ whenever $T \in \mathcal{S}$ satisfies $\mu(T) < \delta(\varepsilon)$.

Proof Without loss of generality we may suppose $p = 1$ and $E = \mathbb{R}$.

(a)⇒(b) Since $\lim_{r \to \infty} \gamma(r)/r = \infty$, for every $\varepsilon > 0$ there exists a $r_\varepsilon > 0$ such that

$$r \geq r_\varepsilon \Longrightarrow \gamma(r) \geq (m/\varepsilon)r.$$

Then, for $r \geq r_\varepsilon$ and all $f \in F$ we have

$$\int_{\{|f| > r\}} |f| \, d\mu \leq \int_{\{|f| > r\}} \frac{\varepsilon}{m} \gamma(|f|) d\mu \leq \frac{\varepsilon}{m} \int_S \gamma(|f|) d\mu \leq \varepsilon.$$

(b)⇒(a) We choose an increasing sequence $(k(n))_n$ of $\mathbb{N}$ such that, for all $n \in \mathbb{N}$,

$$a_n := \sup_{f \in F} \int_{\{|f| > k(n)\}} |f| \, d\mu \leq 2^{-n}.$$

For $i \in \mathbb{N}$ let $N_i := \{n \in \mathbb{N} : i - 1 \leq k(n) < i\}$ and let b_i be the number of elements of N_i, $c_i := b_0 + \ldots + b_i$. Note that $\lim_i c_i = +\infty$. Define $\gamma : \mathbb{R}_+ \to \overline{\mathbb{R}}_+$ by

$$\gamma(r) := rc_i \quad \text{for } r \in [i, i+1[.$$

Then, denoting by $[r]$ the integer part of r, so that $[r] \leq r < [r] + 1$ and $[r] \in \mathbb{N}$, we have

$$\gamma(r)/r \geq c_{[r]} \to +\infty \text{ as } r \to +\infty.$$

Using associativity for series of nonnegative numbers and the fact that for $n \in N_j$ we have $k(n) < j$, for all $f \in F$ we get

$$\begin{aligned}
\int_S \gamma(|f|)d\mu &= \sum_{i=0}^{\infty} \int_{\{i \leq |f| < i+1\}} \gamma(|f|)d\mu = \sum_{i=0}^{\infty} c_i \int_{\{i \leq |f| < i+1\}} |f|\, d\mu \\
&= \sum_{i=0}^{\infty} \sum_{j=0}^{i} b_j \int_{\{i \leq |f| < i+1\}} |f|\, d\mu = \sum_{j=0}^{\infty} \sum_{i=j}^{\infty} b_j \int_{\{i \leq |f| < i+1\}} |f|\, d\mu \\
&= \sum_{j=0}^{\infty} b_j \int_{\{j \leq |f|\}} |f|\, d\mu = \sum_{j=0}^{\infty} \sum_{n \in N_j} \int_{\{j \leq |f|\}} |f|\, d\mu \\
&\leq \sum_{j=0}^{\infty} \sum_{n \in N_j} \int_{\{k(n) < |f|\}} |f|\, d\mu \leq \sum_{j=0}^{\infty} \sum_{n \in N_j} a_n = \sum_{n=0}^{\infty} a_n < +\infty.
\end{aligned}$$

(b)$\Rightarrow$(c) Given $\varepsilon > 0$ we fix $r_\varepsilon > 0$ such that

$$r \geq r_\varepsilon \Longrightarrow \sup\{\int_{\{|f| > r\}} |f|\, d\mu : f \in F\} < \varepsilon/2.$$

Then, for all $T \in \mathcal{S}$ satisfying $\mu(T) \leq \varepsilon/2r_\varepsilon$ and all $f \in F$ we have

$$\int_T |f|\, d\mu = \int_{T \cap \{|f| > r_\varepsilon\}} |f|\, d\mu + \int_{T \cap \{|f| \leq r_\varepsilon\}} |f|\, d\mu \leq \varepsilon/2 + r_\varepsilon \mu(T) \leq \varepsilon.$$

(c)$\Rightarrow$(b) when F is bounded. Let $m := \sup\{\|f\|_1 : f \in F\}$. Given $\varepsilon > 0$, let $\delta(\varepsilon) > 0$ be such that for all $T \in \mathcal{S}$ satisfying $\mu(T) < \delta(\varepsilon)$ we have $\sup\{\int_T |f|\, d\mu : f \in F\} \leq \varepsilon$. Let $r_\varepsilon := m/\delta(\varepsilon)$. Then for $r \geq r_\varepsilon$ and $f \in F$ we have

$$\mu(\{|f| > r\}) \leq \|f\|_1 / r \leq m/r \leq \delta(\varepsilon)$$

hence

$$\sup\{\int_{\{|f|>r\}} |f|\, d\mu : f \in F\} \leq \varepsilon.$$

□

The next result explains the importance of equi-integrability for existence questions. We just quote it; see [106, 112].

Theorem 8.32 (Dunford-Pettis Criterion) *If $p \in [1, +\infty[$ and if E is a reflexive Banach space, the weak closure of a p-equi-integrable subset F of $L_p(S, E)$ is weakly compact.*

For $p > 1$ every bounded subset of $L_p(S, E)$ is weakly relatively compact.

In order to give a compactness criterion for the strong topology in $L_p(S)$, where S is a measurable subset of $\mathbb{R}^d$, let us introduce a notation for the shift $f_u := T_u f$ of a function $f \in L_p(S)$ by $u \in \mathbb{R}^d$ given by $(T_u f)(x) := f(x - u)$ for $x \in S_u := S + u$.

Theorem 8.33 (Fréchet-Kolmogorov) *Let $p \in [1, +\infty[$, let S be a bounded measurable subset of $\mathbb{R}^d$ and let $\Omega \subset \mathbb{R}^d$ be a measurable subset of $\mathbb{R}^d$ such that $S + rB_{\mathbb{R}^d} \subset \Omega$ for some $r > 0$. Let G be a bounded subset of $L_p(\Omega)$ satisfying the condition:*

$$\forall \varepsilon > 0\ \exists \delta \in]0, r] : g \in G,\ u \in \delta B_{\mathbb{R}^d} \Longrightarrow \|(T_u g - g) \mid S\|_p < \varepsilon. \tag{8.21}$$

Then the set F of restrictions to S of the functions g in G is relatively compact in $L_p(S)$.

Proof We observe that for $\delta \in]0, r]$ and $u \in \delta B_{\mathbb{R}^d}$, $x \in S$ we have $x - u \in \Omega$ so that, setting $g_u := T_u g$, $g_u \mid S$ is well defined. Shrinking Ω if necessary, we may assume Ω is bounded. Moreover, extending the functions in G or $L_p(\Omega)$ by 0 on $\mathbb{R}^d \backslash \Omega$, we may consider G as a bounded subset of $L_p(\mathbb{R}^d) \cap L_1(\mathbb{R}^d)$. Let $R : L_p(\Omega) \to L_p(S)$ be the restriction map: $g \mapsto g \mid S$. For all $f \in F := R(G)$ we pick some $g_f \in G$ such that $R(g_f) = f$.

Let (j_n) be a mollifier: $j_n \in C_c^{\infty}(\mathbb{R}^d)$ has its support in $r_n B_{\mathbb{R}^d}$, where $(r_n) \to 0$ in $]0, r]$ and $\int j_n = 1$ for all n. For $f \in F$ and $n \in \mathbb{N}$, writing $\left|T_u g_f - f\right| j_n = \left|T_u g_f - f\right| j_n^{1/p} j_n^{1/q}$, using Hölder's inequality, and noting that $\int j_n = 1$, for $x \in S$ we get

$$\begin{aligned} \left|(g_f * j_n)(x) - f(x)\right| &\leq \int_{\mathbb{R}^d} \left|g_f(x - u) j_n(u) - g_f(x) j_n(u)\right| du \\ &\leq \left(\int_{\mathbb{R}^d} \left|g_f(x - u) - g_f(x)\right|^p |j_n(u)|\, du\right)^{1/p}. \end{aligned}$$

Thus, given $\varepsilon > 0$, taking $\delta > 0$ as in (8.21) and $n \in \mathbb{N}$ such that $r_n \leq \delta$, denoting by B_d the closed unit ball of $\mathbb{R}^d$, and using Fubini's Theorem, we get

$$\left|(g_f * j_n)(x) - f(x)\right|^p \leq \int_{r_n B_d} \left|g_f(x-u) - g_f(x)\right|^p |j_n(u)|\, du,$$

$$\int_S \left|(g_f * j_n)(x) - f(x)\right|^p dx \leq \int_{r_n B_d} |j_n(u)| \int_S \left|g_f(x-u) - g_f(x)\right|^p dxdu \leq \varepsilon^p,$$

hence $\left\|g_f * j_n - f\right\|_{L_p(S)} \leq \varepsilon$.

Let us show that the family $H_n := \{(g_f * j_n) \mid \mathrm{cl}(S) : f \in F\}$ satisfies the assumptions of Ascoli's theorem: for all $x, x' \in K := \mathrm{cl}(S), f \in F$ we have

$$\sup_{f\in F} \left|(g_f * j_n)(x)\right| \leq \sup_{f \in F} \|j_n\|_\infty \left\|g_f\right\|_1 < +\infty,$$

$$\left|(g_f * j_n)(x) - (g_f * j_n)(x')\right| \leq c_n \left\|g_f\right\|_1 \left\|x - x'\right\|,$$

where c_n is the Lipschitz constant of j_n. Thus, H_n is relatively compact in $C(K)$, hence in $L_p(S)$. Given $\varepsilon > 0$, we can cover H_n by a finite family of balls with radius ε. Since $\left\|g_f * j_n - f\right\|_{L_p(S)} \leq \varepsilon$ for all $f \in F$, we see that F is covered by a finite number of balls with radius 2ε: F is precompact. Since $L_p(S)$ is complete, F is relatively compact. □

Corollary 8.11 *Let $p \in [1, +\infty[$, let G be a bounded subset of $L_p(\mathbb{R}^d)$ satisfying the condition:*

$$\forall \varepsilon > 0\ \exists \delta > 0 : g \in G,\ u \in \delta B_{\mathbb{R}^d} \Longrightarrow \|T_u g - g\|_p < \varepsilon. \tag{8.22}$$

Suppose that for every $\varepsilon > 0$ there exists a bounded measurable subset S_ε of $\mathbb{R}^d$ such that $\left|\int_{\mathbb{R}^d \setminus S_\varepsilon} |g|^p\right| < \varepsilon^p$ for all $g \in G$. Then G is relatively compact in $L_p(\mathbb{R}^d)$.

Proof Given $\varepsilon > 0$, in the preceding theorem let us take $\Omega := \mathbb{R}^d$, $S := S_\varepsilon$. Since for all $g \in G$ we have $\|(T_u g - g) \mid S\|_p \leq \|T_u g - g\|_p$, condition (8.22) entails condition (8.21) and the set F_ε of restrictions to S_ε of the elements of G is precompact. Let $g_1, \ldots, g_n \in G$ be such that F_ε is covered by the balls $B(f_i, \varepsilon)$ with radius ε and centers $f_i := g_i \mid S_\varepsilon$. Given $g \in G$ we can find $i \in \mathbb{N}_n$ such that $f := g \mid S_\varepsilon \in B(f_i, \varepsilon)$. Then, since $\left|\int_{\mathbb{R}^d \setminus S_\varepsilon} |g|^p\right| < \varepsilon^p$ and $\left|\int_{\mathbb{R}^d \setminus S_\varepsilon} |g_i|^p\right| < \varepsilon^p$ we have

$$\|g - g_i\|_p^p = \int_{\mathbb{R}^d \setminus S_\varepsilon} |g - g_i|^p + \int_{S_\varepsilon} |f - f_i|^p \leq 2^{p-1}\varepsilon^p + 2^{p-1}\varepsilon^p + \varepsilon^p,$$

so that G is covered by a finite number of balls of radius $(2^p + 1)^{1/p}\varepsilon$, hence is precompact or relatively compact. □

Exercises

1. Show that condition (b) of Definition 8.8 is not a consequence in condition (a). [Hint: take $S := \mathbb{R}$ with the Lebesgue measure, $f_n := (1/n)1_{[n,2n]}$.]
2. Show that the two conditions of Definition 8.8 imply the following assertion: (c) for any decreasing sequence (S_n) of $\mathcal{S}$ satisfying $\mu(\cap S_n) = 0$ and for any $\varepsilon > 0$ there exists a $k \in \mathbb{N}$ such that $\sup_{f \in F}(\int_{S_k} |f|^p \, d\mu)^{1/p} < \varepsilon$.

   Conversely, when μ is σ-finite or when F is countable, show that assertion (c) implies F is equi-integrable.
3. Let (g_n), (h_n) be convergent sequences of $L_p(S, \mathbb{R})$. Show that the following set is p-equi-integrable: $F := \{f \in L_p(S, \mathbb{R}) : \exists n \in \mathbb{N}, g_n \leq f \leq h_n\}$.
4. Let F be a p-equi-integrable subset of $L_p(S, E)$ and let (f_n), (g_n) be two sequences of F. Show that $(\|f_n - g_n\|_p) \to 0$ if and only if $(f_n - g_n) \to 0$ in measure.

5*. (*Di Perna-Lions*) Let $(S, \mathcal{S}, \mu)$ be a measure space with finite measure, let $(f_n) \to f$ weakly in $L_1(S, \mu)$, and let (g_n) be a bounded sequence in $L_\infty(S, \mu)$ such that $(g_n) \to g$ a.e. Prove that $(f_n g_n) \to fg$ weakly in $L_1(S, \mu)$. [Hint: Use Egorov's theorem.]

6*. (*Chacon's Biting lemma*) Let $(S, \mathcal{S}, \mu)$ be a measure space with finite measure and let (f_n) be a bounded sequence in $L_1(S, \mu)$. Prove that there exist a subsequence $(f_{k(n)})$ of (f_n) and $f \in L_1(S, \mu)$ such that for every $\varepsilon > 0$ there exists some $T \in \mathcal{S}$ such that $\mu(S \backslash T) < \varepsilon$ and $(f_{k(n)})_n \to f$ weakly in $L_1(T, \mu_T)$. [See: [55, 124, p. 184].]

7. Given a measure space $(S, \mathcal{S}, \mu)$ with finite measure, prove that a subset F of $L_1(S)$ is uniformly integrable if and only if F is bounded and equi-integrable.
8. Given a measure space $(S, \mathcal{S}, \mu)$ with finite measure, prove that a sequence (f_n) of $L_1(S)$ converges to some $f \in L_1(S)$ if and only if it converges to f in measure and $\{f_n : n \in \mathbb{N}\}$ is uniformly integrable.

8.8 Convolution and Regularization

Given a (Lebesgue) measurable function $f : \mathbb{R}^d \to \overline{\mathbb{R}}$ and $w \in \mathbb{R}^d$ we denote by f_w the function obtained by composing the translated function $T_w f : x \mapsto f(x - w)$ with the symmetry $S : f \mapsto \tilde{f}$ defined by $\tilde{f}(x) := f(-x)$ for $x \in \mathbb{R}^d$:

$$f_w(x) := f(w - x) \qquad\qquad x \in \mathbb{R}^d.$$

Clearly, f_w is measurable and $f_w \in \mathcal{L}_p(\mathbb{R}^d)$ if $f \in \mathcal{L}_p(\mathbb{R}^d)$ with $p \in [1, +\infty]$. If f and g are equal a.e., then f_w and g_w are equal a.e., so that we regard f_w as an element of $L_p(\mathbb{R}^d)$ if $f \in L_p(\mathbb{R}^d)$. Here $\mathbb{R}^d$ is equipped with the Lebesgue measure λ_d, but we write $\int f(x)dx$ rather than $\int_{\mathbb{R}^d} f(x) d\lambda_d(x)$.

Given two nonnegative measurable functions f, g on $\mathbb{R}^d$ we define their (integral) *convolution* $h := f * g$ by

$$(f * g)(w) := \int_{\mathbb{R}^d} f(w-x)g(x)dx. \tag{8.23}$$

Such a definition is justified (but one may have $(f * g)(w) = +\infty$) because the function $x \mapsto f(w-x)g(x)$ is measurable. In fact, it is easy to see that $(w, x) \mapsto f(w-x)g(x)$ is measurable. Then the Fubini-Tonelli Theorem shows that the function $w \mapsto \int f(w-x)g(x)dx$ is measurable and

$$\begin{aligned}\int\int f(w-x)g(x)dxdw &= \int\int f(w-x)g(x)dwdx \\ &= \int g(x)(\int f(w-x)dw)dx = \int g(x)dx \int f(w)dw\end{aligned}$$

or

$$\int (f * g)(w)d\lambda_d(w) = (\int fd\lambda_d)(\int gd\lambda_d). \tag{8.24}$$

Given f, $g \in \mathcal{L}_1(\mathbb{R}^d)$ and $w \in \mathbb{R}^d$ the function $f_w g$ is integrable if and only if $\int_{\mathbb{R}^d} |f(w-x)g(x)|\,dx < +\infty$, if and only if one has $|f| * |g|\,(w) < +\infty$. Then one defines $(f * g)(w)$ by relation (8.23). One easily sees that $(f * g)(w)$ is independent of the choice of f and g in their equivalence classes with respect to a.e. equality. Moreover, one has $(f * g)(w) = (g * f)(w)$ but the operation $*$ is not associative (Exercise 1).

In order to present an existence criterion for $f * g$ on the whole of $\mathbb{R}^d$, let us make precise the notion of the *support* (or essential support) $\mathrm{supp} f$ of a measurable function f defined up to a null set N. It is the complement of the greatest open subset O of $\mathbb{R}^d$ such that $f = 0$ a.e. on O. Such an open set exists: choosing a countable base $\mathcal{G}$ of open subsets of $\mathbb{R}^d$, O is the union of the family $\mathcal{G}_f$ whose members are the members G of $\mathcal{G}$ such that $f = 0$ a.e. on G. This set does not depend on the choice of the base $\mathcal{G}$. It is also independent of the choice of a representant of f in its equivalence class with respect to a.e. equality. If f is continuous, this notion of support coincides with the usual one: the support of f is then the closure of the set of points at which f is non-null.

Lemma 8.15 *Let f and g be two measurable functions such that $f * g$ is defined almost everywhere. Then* $\mathrm{supp}\,(f * g) \subset \mathrm{cl}(\mathrm{supp} f + \mathrm{supp}\, g)$.

Proof Let $F := \mathrm{supp} f$, $G := \mathrm{supp}\, g$ and let $U := \mathbb{R}^d \backslash \mathrm{cl}(F + G)$. Let N be the (null) set of points $w \in \mathbb{R}^d$ such that $f_w g : x \mapsto f(w-x)g(x)$ is not integrable. If $w \in U \backslash N$ and $x \in \mathbb{R}^d$ either we have $x \notin G$ (and then $g(x) = 0$) or $x \in G$ and then $w - x \notin F$ since $(w-x) + x \notin (F + G)$; in both cases we have $f(w-x)g(x) = 0$. Since

$f_w g = 0$ for $w \in U \backslash N$, we have $(f * g)(w) = \int f_w g = 0$ for $w \in U \backslash N$, hence $\operatorname{supp}(f * g) \subset \mathbb{R}^d \backslash U = \operatorname{cl}(F + G)$. □

Proposition 8.17 *Let $f \in \mathcal{L}_{1,loc}(\mathbb{R}^d)$ and let $g \in \mathcal{L}_\infty(\mathbb{R}^d)$ with compact support. Then $(f * g)(w)$ is defined for all $w \in \mathbb{R}^d$. The same conclusion holds when $f \in \mathcal{L}_1(\mathbb{R}^d)$ and $g \in \mathcal{L}_\infty(\mathbb{R}^d)$ and then $f * g \in \mathcal{L}_\infty(\mathbb{R}^d)$ with $\|f * g\|_\infty \le \|f\|_1 \|g\|_\infty$.*

Proof Let K be the support of g and let $w \in \mathbb{R}^d$. Then $w - K$ is compact and

$$(|f| * |g|)(w) = \int_K |f(w-x)|\,|g(x)|\,dx$$

$$\le \|g\|_\infty \int_K |f(w-x)|\,dx = \|g\|_\infty \int_{w-K} |f(u)|\,du < +\infty,$$

so that the function $f_w g$ is integrable. For the second assertion, in the preceding inequalities one replaces K with $\mathbb{R}^d \backslash N$, where N is the null set $N := \{x \in \mathbb{R}^d : |g(x)| > \|g\|_\infty\}$ and one gets the estimate $\|f * g\|_\infty \le \|f\|_1 \|g\|_\infty$. □

Another result asserting that the convolution is well defined is as follows.

Theorem 8.34 *Let $p \in [1, +\infty]$, $f \in L_1(\mathbb{R}^d)$, and $g \in L_p(\mathbb{R}^d)$. Then, for almost every $w \in \mathbb{R}^d$ the function $f_w g$ is integrable and for $h := f * g$ defined by relation (8.23), i.e. $h(w) := \int f_w g$, one has $h \in L_p(\mathbb{R}^d)$ and*

$$\|f * g\|_p \le \|f\|_1 \|g\|_p . \tag{8.25}$$

Proof We take functions rather than classes since $f * g$ depends on the classes of f and g. The case $p = +\infty$ is treated by Proposition 8.17. Let us first consider the case $p = 1$. Setting $k(w,x) := f_w(x)g(x) := f(w-x)g(x)$, for $x \in \mathbb{R}^d$ we have

$$\int |k(w,x)|\,dw = |g(x)| \int |f(w-x)|\,dw = |g(x)|\,\|f\|_1 < +\infty,$$

hence $k \in \mathcal{L}_1(\mathbb{R}^d \times \mathbb{R}^d)$ by Corollary 7.17. Then Fubini's Theorem yields

$$\int dw \int |k(w,x)|\,dx = \int dx \int |k(w,x)|\,dw \le \|f\|_1 \|g\|_1 ,$$

so that for almost every $w \in \mathbb{R}^d$ one has $f_w g \in \mathcal{L}_1(\mathbb{R}^d)$ and inequality (8.25) holds for $p = 1$.

Now let $p \in]1, \infty[$, $q := (1 - 1/p)^{-1}$, and let $f \in \mathcal{L}_1(\mathbb{R}^d)$, $g \in \mathcal{L}_p(\mathbb{R}^d)$. By the preceding, $|f_w|\,|g|^p \in \mathcal{L}_1(\mathbb{R}^d)$ for almost every $w \in \mathbb{R}^d$. Thus $|f_w|^{1/p}\,|g| \in \mathcal{L}_p(\mathbb{R}^d)$, $|f_w|^{1/q} \in \mathcal{L}_q(\mathbb{R}^d)$, and since $|f_w g| = |f_w|^{1/q}\,(|f_w|^{1/p}\,|g|)$, Hölder's inequality yields

$$\int |f_w(x)|\,|g(x)|\,dx \le \left\| |f_w|^{1/q} \right\|_q \left\| |f_w|^{1/p}\,|g| \right\|_p = \|f_w\|_1^{1/q}\,\|f_w\,|g|^p\|_1^{1/p} < +\infty.$$

Using the relations $\|f_w\,|g|^p\|_1 = \int |f_w| \cdot |g|^p = (|f| * |g|^p)(w)$ and $\|f_w\|_1 = \|f\|_1$ we get

$$|h(w)|^p \leq \left(\int |f_w(x)|\,|g(x)|\,dx\right)^p \leq \|f_w\|_1^{p/q}\,\|f_w\,|g|^p\|_1 = \|f\|_1^{p/q}.(|f| * |g|^p)(w).$$

The relation $\int (|f| * |g|^p)(w)dw \leq \|f\|_1\,\||g|^p\|_1$ obtained in the case $p = 1$ yields

$$\|f * g\|_p^p = \int |h(w)|^p\,dw \leq \|f\|_1^{p/q} \int (|f| * |g|^p)(w)dw \leq \|f\|_1^{p/q}\,\|f\|_1\,\|g\|_p^p.$$

Since $p/q + 1 = p$, inequality (8.25) ensues. □

Now let us start to describe some regularizing effects of convolution.

Theorem 8.35 *Let $p, q \in [1, +\infty]$ with $1/p + 1/q = 1$, $f \in \mathcal{L}_p(\mathbb{R}^d)$, and $g \in \mathcal{L}_q(\mathbb{R}^d)$. Then, for all $w \in \mathbb{R}^d$ the function $f_w g$ is integrable and the function $h := f * g$ defined by relation (8.23), i.e. $h(w) := \int f_w g$, is uniformly continuous and bounded and one has*

$$\|f * g\|_\infty \leq \|f\|_p\,\|g\|_q. \tag{8.26}$$

If, moreover, $p, q \in]1, +\infty[$ one has $(f * g)(w) \to 0$ as $\|w\| \to +\infty$.

Proof For all $w \in \mathbb{R}^d$ Hölder's inequality and a change of variables yield

$$(|f| * |g|)(w) = \int |f_w|\,|g|\,d\lambda_d \leq \|f_w\|_p\,\|g\|_q = \|f\|_p\,\|g\|_q.$$

Since p or q is finite, we may suppose p is finite. Then, for all u, w, x we have $f_{w-u}(x) = f(w - u - x) = (T_u f)(w - x) = (T_u f)_w(x)$, so that, by the preceding inequality,

$$|(f * g)(w - u) - (f * g)(w)| \leq \int |((T_u f)_w - f_w)g|\,d\lambda_d \leq \|T_u f - f\|_p\,\|g\|_q.$$

Since $\|T_u f - f\|_p \to 0$ as $u \to 0$, this proves that $f * g$ is uniformly continuous.

Now let us prove the second assertion, assuming that $p, q \in]1, +\infty[$. Let us first suppose f belongs to the space $C_c(\mathbb{R}^d)$ of continuous functions with compact support. Let K be the support of f, so that $f = 0$ on $\mathbb{R}^d \backslash K$ and by Hölder's inequality, for $c := \lambda_d(w - K) = \lambda_d(K)$, we have

$$|f * g|\,(w) = \int_{w-K} |f(w - x)|\,|g(x)|\,dx \leq c^{1/p} \left(\int_{\mathbb{R}^d} |f(w - x)|^q\,|g(x)|^q\,dx\right)^{1/q}.$$

Now, noting that $|f_w|^q |g|^q \leq \|f\|_\infty^q |g|^q \in \mathcal{L}_1(\mathbb{R}^d)$ and that for any sequence (w_n) satisfying $(\|w_n\|) \to +\infty$ we have $(|f_{w_n}(x)|^q |g(x)|^q)_n \to 0$, the Dominated Convergence Theorem yields $(|f * g| (w_n))_n \to 0$.

Finally, let $f \in \mathcal{L}_p(\mathbb{R}^d)$. Since $C_c(\mathbb{R}^d)$ is dense in $\mathcal{L}_p(\mathbb{R}^d)$, we pick a sequence (f_n) in $C_c(\mathbb{R}^d)$ with limit f in $\mathcal{L}_p(\mathbb{R}^d)$. Since

$$|(f_n * g)(w) - (f * g)(w)| = |(f - f_n) * g| (w) \leq \|f - f_n\|_p \|g\|_q \to 0,$$

given $\varepsilon > 0$ we pick $m \in \mathbb{N}$ such that $\|f - f_n\|_p \|g\|_q \leq \varepsilon/2$ for $n \geq m$. Then we choose $r > 0$ such that $|f_m * g| (w) < \varepsilon/2$ for $\|w\| > r$ and for such a $w \in \mathbb{R}^d$ we get

$$|(f * g)(w)| \leq |(f * g)(w) - (f_m * g)(w)| + |(f_m * g)(w)| \leq \varepsilon.$$

□

Let us turn to differentiability properties. In the sequel, $\alpha := (\alpha_1, \ldots, \alpha_d) \in \mathbb{N}^d$ being a multi-index, we denote by $D^\alpha f$ the partial derivative

$$D^\alpha f(x_1, \ldots, x_d) := D_1^{\alpha_1} \ldots D_d^{\alpha_d} f(x_1, \ldots, x_d) = (\frac{\partial}{\partial x_1})^{\alpha_1} \ldots . (\frac{\partial}{\partial x_d})^{\alpha_d} f(x_1, \ldots, x_d),$$

with the convention $D_i^{\alpha_i} g = g$ if $\alpha_i = 0$. We set $|\alpha| := \alpha_1 + \ldots + \alpha_d$. We denote by $C^k(\Omega)$ the space of functions of class C^k (i.e. that have continuous partial derivatives of order at most k) on the open subset Ω of $\mathbb{R}^d$ and we set

$$C_c^k(\Omega) := C^k(\Omega) \cap C_c(\Omega), \ \ C^\infty(\Omega) := \bigcap_{k \geq 0} C^k(\Omega), \ \ C_c^\infty(\Omega) := C^\infty(\Omega) \cap C_c(\Omega).$$

Proposition 8.18 *For $f \in C_c^k(\mathbb{R}^d)$, $g \in \mathcal{L}_{1,loc}(\mathbb{R}^d)$ one has $f * g \in C^k(\mathbb{R}^d)$ and*

$$\forall \alpha \in \mathbb{N}^d, \ |\alpha| \leq k \qquad D^\alpha (f * g) = (D^\alpha f) * g.$$

In particular, $f * g \in C^\infty(\mathbb{R}^d)$ when $f \in C_c^\infty(\mathbb{R}^d)$ and $g \in \mathcal{L}_{1,loc}(\mathbb{R}^d)$.

Proof As an induction shows, it suffices to prove the case $k = 1$. Let $i \in \mathbb{N}_d$ and $\alpha_i = 1$, $\alpha_j = 0$ for $j \neq i$ and let $e_i \in \mathbb{R}^d$ be such that all its components are null except the i-th which is 1. Given $f \in C_c^k(\mathbb{R}^d)$, $g \in \mathcal{L}_{1,loc}(\mathbb{R}^d)$, and a fixed $w \in \mathbb{R}^d$, we denote by K the support of f and for $x \in \mathbb{R}^d$ and $t \in T := [-1, 1]$ we write

$$f(w + te_i - x) - f(w - x) = th_i(w - x, t)$$

with $h_i(w - x, t) := \int_0^1 D_i f(w + ste_i - x)ds$, a continuous function of t, x (w being fixed) that is 0 if $x \notin K_i := w + Te_i - K$, a compact subset. Let

$$m_i := \sup\{|h_i(w - x, t)| : x \in K_i, t \in T\}.$$

Since $g \in \mathcal{L}_{1,loc}(\mathbb{R}^d)$ the function $x \mapsto m_i g(x) 1_{K_i}(x)$ is integrable and we have $|h_i(w-x,t)g(x)| \leq m_i |g(x)| 1_{K_i}(x)$ for almost every x. Applying Theorem 7.12, we get that $t \mapsto (f * g)(w + te_i)$ is differentiable at $t = 0$ and its derivative is $\int_{\mathbb{R}^d} D_i f(w-x)g(x)dx = (D_i f * g)(w)$. □

We take advantage of the preceding result with the aim of regularization. We say that a sequence (g_n) of $C_c^\infty(\mathbb{R}^d)$ is a *mollifier* if for all n the support $\operatorname{supp} g_n$ of g_n is contained in $B(0, r_n)$ with $(r_n) \to 0_+$, if $g_n \geq 0$ and if $\int g_n = 1$. A common means to get a mollifier consists in taking a nonnegative function $g \in C_c^\infty(\mathbb{R}^d)$ with support in $B_{\mathbb{R}^d}$ and, for a given sequence $(r_n) \to 0_+$, in setting

$$g_n(x) = c_n g(x/r_n)$$

where $c_n := 1/(r_n^d \int g)$. For g one can take the function given by

$$g(x) := \exp(1/(\|x\|^2 - 1)) \text{ for } x \in B(0,1), \qquad g(x) = 0 \text{ for } x \in \mathbb{R}^d \backslash B(0,1).$$

Lemma 8.16 *Given $f \in C(\mathbb{R}^d)$ and a mollifier (g_n) one has $(f * g_n) \to f$ uniformly on every compact subset of $\mathbb{R}^d$.*

Proof Let $f \in C(\mathbb{R}^d)$, let (g_n) be a mollifier, and let K be a compact subset of $\mathbb{R}^d$. Since f is uniformly continuous around K in the sense that for every $\varepsilon > 0$ there exists a $\delta > 0$ such that for all $w \in K$ and all $x \in B(0,\delta)$ one has $|f(w-x) - f(w)| \leq \varepsilon$, hence

$$\begin{aligned}(f * g_n)(w) - f(w) &= \int (f(w-x) - f(w))g_n(x)dx \\ &= \int_{B(0,r_n)} (f(w-x) - f(w))g_n(x)dx.\end{aligned}$$

For $w \in K$ and $r_n < \delta$ we get $|(f * g_n)(w) - f(w)| \leq \varepsilon \int g_n = \varepsilon$. □

Theorem 8.36 *For $p \in [1, +\infty[$, $f \in L_p(\mathbb{R}^d)$, one has $(f * g_n)_n \to f$ in $L_p(\mathbb{R}^d)$.*

Proof Given $\varepsilon > 0$, Theorem 8.31 yields some $h \in C_c(\mathbb{R}^d)$ such that $\|f - h\|_p < \varepsilon$. Let $(r_n) \to 0_+$ be such that $\operatorname{supp} g_n \subset r_n B_{\mathbb{R}^d}$, so that, for $r := \max(r_n)$, by Lemma 8.15,

$$\operatorname{supp}(h * g_n) \subset \operatorname{supp} h + r_n B_{\mathbb{R}^d} \subset K := \operatorname{supp} h + r B_{\mathbb{R}^d}.$$

Since $(h * g_n) \to h$ uniformly on K by the preceding lemma, we have $(\varepsilon_n) := (\|h * g_n - h\|_p) \to 0$. Then, since $\|g_n\|_1 = 1$, relation (8.25) entails the conclusion:

$$\begin{aligned}\|f * g_n - f\|_p &\leq \|(f-h) * g_n\|_p + \|h * g_n - h\|_p + \|h - f\|_p \\ &\leq \|f - h\|_p \|g_n\|_1 + \varepsilon_n + \varepsilon \leq 2\varepsilon + \varepsilon_n\end{aligned}$$

and $\|f * g_n - f\|_p \leq 3\varepsilon$ for n large enough. □

Corollary 8.12 *For $p \in [1, +\infty[$ and an open subset Ω of $\mathbb{R}^d$, the space $C_c^\infty(\Omega)$ is dense in $L_p(\Omega)$.*

Proof Given $\varepsilon > 0$ and $f \in L_p(\Omega)$, using Theorem 8.31 we pick $g \in C_c(\Omega)$ such that $\|f - g\|_p < \varepsilon$. Extending g by 0 on $\mathbb{R}^d \backslash \Omega$ we get an element $h \in L_p(\mathbb{R}^d)$. Taking a mollifier (g_n) and $(r_n) \to 0_+$ such that $\operatorname{supp} g_n \subset r_n B_{\mathbb{R}^d}$ we see that for n large enough we have $\operatorname{supp} h * g_n \subset \operatorname{supp} h + r_n B_{\mathbb{R}^d} \subset \Omega$. The restriction f_n of $h * g_n$ to Ω belongs to $C_c^\infty(\Omega)$ by Proposition 8.18 and $\|f_n - g\|_{L_p(\Omega)} = \|h * g_n - h\|_p \leq \varepsilon$ for n large enough. Thus $\|f - f_n\|_p \leq 2\varepsilon$ for n large enough. □

Let us note a convergence result for the convolution with the Gaussian functions $g_t(\cdot) := t^{-d/2} e^{-\pi \|\cdot\|^2/t}$ that will be used in the next section. It bears some analogy with the preceding theorem, albeit g_t does not have a compact support. From the example at the end of Sect. 7.7 we know that for all $\delta > 0$ the following two properties are satisfied:

$$\int_{\mathbb{R}^d \backslash \delta B_d} g_t(x) dx \to 0 \text{ as } t \to 0_+ \tag{8.27}$$

$$\int_{\mathbb{R}^d} g_t(x) dx = 1. \tag{8.28}$$

These facts enable us to prove the following result.

Lemma 8.17 *For all $f \in L_1(\mathbb{R}^d)$ one has $\|f * g_t - f\|_1 \to 0$ as $t \to 0_+$.*

Proof By (8.28), we have

$$(f * g_t)(w) - f(w) = \int_{\mathbb{R}^d} f(w - x) g_t(x) dx - f(w) = \int_{\mathbb{R}^d} (f(w - x) - f(w)) g_t(x) dx.$$

Using Fubini's Theorem we get the estimate

$$\begin{aligned}
\|f * g_t - f\|_1 &= \int_{\mathbb{R}^d} |(f * g_t)(w) - f(w)| \, dw \\
&\leq \int_{\mathbb{R}^d} \Big(\int_{\mathbb{R}^d} |f(w - x) - f(w)| \, dw \Big) g_t(x) dx \\
&= \int_{\mathbb{R}^d} \|T_x f - f\|_1 \, g_t(x) dx.
\end{aligned}$$

Given $\varepsilon > 0$, Lemma 8.14 yields some $\delta > 0$ such that $\|T_x f - f\|_1 \leq \varepsilon$ for every $x \in \delta B_d$. On the other hand, by (8.27) we can find $\tau > 0$ such that for $t \in]0, \tau]$ we have

$$\int_{\mathbb{R}^d \backslash \delta B_d} g_t(x) dx \leq \varepsilon.$$

Using this inequality, noting that $\|T_x f - f\|_1 \le \|T_x f\|_1 + \|f\|_1 \le 2\|f\|_1$, and taking (8.27) into account, we get

$$\begin{aligned}\|f * g_t - f\|_1 &\le \int_{\delta B_d} \|T_x f - f\|_1 g_t(x)dx + \int_{\mathbb{R}^d \setminus \delta B_d} \|T_x f - f\|_1 g_t(x)dx \\ &\le \int_{\delta B_d} \varepsilon g_t(x)dx + \int_{\mathbb{R}^d \setminus \delta B_d} 2\|f\|_1 g_t(x)dx \le \varepsilon(1 + 2\|f\|_1).\end{aligned}$$

Since $\varepsilon > 0$ is arbitrarily small, we have $\|f * g_t - f\|_1 \to 0$ as $t \to 0_+$. □

Exercises

1. Consider the following example showing that the convolution operation is not associative. For $d = 1$ take the functions $f := 1_{\mathbb{R}_+}$, $g := 1_{[-1,0]} - 1_{[0,1]}$, $h := 1$. Verify that $f * g$, $g * h$ are everywhere defined and that $(f * g) * h = 1$ whereas $f * (g * h) = 0$.
2. Prove that if f and g are nonnegative measurable functions, then $f * g$ is lower semicontinuous.
3. Show that if $f, g \in C_c(\mathbb{R}^d)$, then $f * g \in C_c(\mathbb{R}^d)$, the space of continuous functions with compact supports.
4. Given p, q, r in $\mathbb{R}_+$ with $p, q \in [1, \infty]$ and $1/p + 1/q - 1/r = 1$ and $f \in L_p(\mathbb{R}^d)$, $g \in L_q(\mathbb{R}^d)$ show that $f * g \in L_r(\mathbb{R}^d)$ and that $\|f * g\|_r \le \|f\|_p \|g\|_q$. [Hint: see [185, p. 98].]
5. Show that $L_1(\mathbb{R}^d)$ is a Banach algebra with respect to the convolution operation but without a unit element.
6. (**Müntz**) Given an increasing sequence $(s_n)_{n \ge 0}$ of positive numbers, let L be the linear subspace of $X := C([0,1], \mathbb{R})$ generated by the functions $x_n : t \mapsto t^{s_n}$. Show that L is dense in X for the norm induced by $L_2([0,1], \mathbb{R})$.
7. Let $p \in [1, \infty[$, let $h \in L_1(\mathbb{R}^d)$, and let G be a bounded subset of $L_p(\mathbb{R}^d)$. Show that $F := G * h := \{g * h : g \in G\}$ is such that for any measurable subset S of $\mathbb{R}^d$ with finite measure, the set $F_S := \{f \mid S : f \in F\}$ is relatively compact in $L_p(S)$. [Hint: use Exercise 15 of Sect. 8.5.1 and the Fréchet-Kolmogorov theorem.]

8.9 Some Useful Transforms

We devote this section to two important transformations using integration processes: the Fourier transform and the Radon transform. In a later section we deal with the Laplace Transform.

8.9.1 The Fourier Transform

The Fourier transform we introduce in this section is a widely used tool. It can be defined on various spaces. We limit our study to the most elementary properties of this transform. We denote the scalar product of two vectors $x := (x_1, \ldots, x_d)$, $y := (y_1, \ldots, y_d)$ of $\mathbb{R}^d$ by $x.y := \langle x \mid y \rangle = \Sigma_{k=1}^{k=d} x_k y_k$. The functions we consider take their values in $\mathbb{C}$; for simplicity, in the present section we just write $L_p(\mathbb{R}^d)$ instead of $L_p(\mathbb{R}^d, \mathbb{C})$. In some sources, the definitions of the Fourier transform $\mathcal{F}$ differ by some scaling factors. That does not change the essence of the transform. Our choice is dictated by the properties that $\mathcal{F}$ can be extended to an isometry of $L_2(\mathbb{R}^d)$ changing convolution into product.

Definition 8.9 The Fourier transform of a function $f \in \mathcal{L}_1(\mathbb{R}^d)$ is the function $\mathcal{F}f := \hat{f}$ given by

$$\hat{f}(y) := \int_{\mathbb{R}^d} e^{-2\pi i \langle x|y \rangle} f(x) dx \qquad y \in \mathbb{R}^d.$$

Since $\left|e^{-2\pi i \langle x|y \rangle} f(x)\right| = |f(x)|$ for all $x, y \in \mathbb{R}^d$, the right-hand side is the integral of an integrable function. If $f = g$ a.e. then $\hat{f}(y) = \hat{g}(y)$ for all $y \in \mathbb{R}^d$, so that $\hat{f}(y)$ just depends on the class of f in $L_1(\mathbb{R}^d)$. The Dominated Convergence Theorem and an elementary change of variables justify the following property in which $(T_w f)(x) := f(x - w)$ as above.

Proposition 8.19 *For all $f \in L_1(\mathbb{R}^d)$ the function $\hat{f}$ is continuous and bounded and the map $f \mapsto \hat{f}$ is linear and continuous from $L_1(\mathbb{R}^d)$ into the space $C_b(\mathbb{R}^d)$ of bounded continuous functions endowed with the norm $\|\cdot\|_\infty$. Moreover, for all $f \in L_1(\mathbb{R}^d)$, $w, y \in \mathbb{R}^d$, $t > 0$ one has* $\left\|\hat{f}\right\|_\infty \leq \|f\|_1$,

$$\widehat{T_w f}(y) = e^{-2\pi i \langle w|y \rangle} \hat{f}(y), \quad \widehat{f(t\cdot)}(y) = t^{-d} \hat{f}(y/t), \quad \widehat{f(\cdot/t)}(y) = t^d \hat{f}(ty).$$

The next properties explain the success of the Fourier transform.

Proposition 8.20 *For all $f, g \in L_1(\mathbb{R}^d)$ and all $y \in \mathbb{R}^d$ one has $\widehat{f * g}(y) = \hat{f}(y)\hat{g}(y)$.*

Proof This follows from Fubini's Theorem:

$$\begin{aligned} \widehat{f * g}(y) &= \int_{\mathbb{R}^d} e^{-2\pi i \langle x|y \rangle} \Big(\int_{\mathbb{R}^d} f(x - u) g(u) du \Big) dx \\ &= \int_{\mathbb{R}^d} e^{-2\pi i \langle u|y \rangle} g(u) \Big(\int_{\mathbb{R}^d} e^{-2\pi i \langle x-u|y \rangle} f(x - u) dx \Big) du = \hat{f}(y)\hat{g}(y) \end{aligned}$$

since $(u, x) \mapsto \left|e^{-2\pi i \langle x|y \rangle} f(x - u) g(u)\right|$ is integrable on $\mathbb{R}^d \times \mathbb{R}^d$. □

In the sequel, given $k \in \mathbb{N}_d$ and $\alpha := (\alpha_1, \dots, \alpha_d) \in \mathbb{N}^d$, we denote by m_k and m_α the functions $x \mapsto -2\pi i x_k$ and $x \mapsto (-2\pi i)^{|\alpha|} x_1^{\alpha_1} \dots x_d^{\alpha_d}$ respectively. Our notation stems from the fact that we want to see the effect of the Fourier transform after multiplication by one of these functions. For $\alpha := (\alpha_1, \dots, \alpha_d)$, $\beta := (\beta_1, \dots, \beta_d) \in \mathbb{N}^d$ we write $\beta \leq \alpha$ if $\beta_k \leq \alpha_k$ for $k \in \mathbb{N}_d$.

Proposition 8.21 *If $f \in \mathcal{L}_1(\mathbb{R}^d)$ and $m_k f \in \mathcal{L}_1(\mathbb{R}^d)$ for some $k \in \mathbb{N}_d$, then $\widehat{m_k f} = D_k \hat{f}$.*

If $m_\beta f \in \mathcal{L}_1(\mathbb{R}^d)$ for every $\beta \in \mathbb{N}^d$ such that $\beta \leq \alpha$, then one has $\widehat{m_\alpha f} = D^\alpha \hat{f}$.

Proof The first assertion is obtained by applying the differentiability criterion for a parameterized integral (Proposition 7.12). The second one is obtained by iterating the first relation. □

Corollary 8.13 *If f is measurable and such that for all $n \in \mathbb{N}_m$ (resp. $n \in \mathbb{N}$) one has $\int_{\mathbb{R}^d} (1 + \|x\|^n) |f(x)| \, dx < +\infty$, then $\hat{f} \in C_b^m(\mathbb{R}^d)$ (resp. $C_b^\infty(\mathbb{R}^d)$), the space of functions of class C^m (resp. C^∞) with bounded partial derivatives of order not greater than m.*

Example Let us show that the Fourier transform of the Gaussian density $g_t(x) := t^{-d/2} e^{-\pi \|x\|^2/t}$ is given by $\widehat{g_t}(y) = h_t(y)$ with $h_t(y) := e^{-\pi t \|y\|^2}$ and the Fourier transform of h_t is given by $\widehat{h_t}(z) = g_t(z) := t^{-d/2} e^{-\pi \|z\|^2/t}$.

In view of Proposition 8.19, it suffices to prove the case $t = 1$. Since $g_1(x) = e^{-\pi x_1^2} \cdots e^{-\pi x_d^2}$ we may suppose that $d = 1$. Setting $k(y) := \int_{\mathbb{R}} e^{-\pi (x+iy)^2} dx$, we have

$$\widehat{g_1}(y) = \int_{\mathbb{R}} e^{-2\pi i x y} e^{-\pi x^2} dx = g_1(y) k(y).$$

Using the criterion for differentiating a parameterized integral we get

$$\begin{aligned} k'(y) &= -2\pi i \int_{\mathbb{R}} (x + iy) e^{-\pi (x+iy)^2} dx \\ &= i \int_{\mathbb{R}} \frac{d}{dx} e^{-\pi (x+iy)^2} dx = [i e^{-\pi (x+iy)^2}]_{-\infty}^{+\infty} = 0. \end{aligned}$$

Thus, $k(y) = k(0) = \int_{\mathbb{R}} e^{-\pi x^2} dx = 1$ as computed in an example at the end of Sect. 7.7 and $\widehat{g_1} = g_1 = h_1$. □

Proposition 8.22 *For all $f, g \in L_1(\mathbb{R}^d)$ one has $\int_{\mathbb{R}^d} f \hat{g} = \int_{\mathbb{R}^d} \hat{f} g$.*

Proof Since for $f, g \in L_1(\mathbb{R}^d)$ the functions $\hat{f}$ and $\hat{g}$ are continuous and bounded, the functions $f\hat{g}$ and $\hat{f}g$ are integrable. Now the function h on $\mathbb{R}^d \times \mathbb{R}^d$ given by $h(x, y) := e^{-2\pi i \langle x | y \rangle} f(x) g(y)$ is measurable and

$$\int_{\mathbb{R}^d} \int_{\mathbb{R}^d} |h(x, y)| \, dx dy = \int_{\mathbb{R}^d} |f(x)| \, dx \int_{\mathbb{R}^d} |g(y)| \, dy < +\infty.$$

Applying Fubini's Theorem in writing $\int_{\mathbb{R}^d}\int_{\mathbb{R}^d} h(x,y)dxdy$ as an iterated integral in two different orders, we get the relation $\int_{\mathbb{R}^d} f\hat{g} = \int_{\mathbb{R}^d} \hat{f}g$. □

The question of the inversion of the Fourier transform is crucial. A first answer follows.

Theorem 8.37 *If $f \in \mathcal{L}_1(\mathbb{R}^d)$ is such that $\hat{f} \in \mathcal{L}_1(\mathbb{R}^d)$ then $f(x) = \hat{\hat{f}}(-x)$ a.e.*

Proof When $f \in \mathcal{L}_1(\mathbb{R}^d)$ is such that $\hat{f} \in \mathcal{L}_1(\mathbb{R}^d)$, for all $x \in \mathbb{R}^d$ one has

$$\hat{\hat{f}}(-x) := \int_{\mathbb{R}^d} e^{2\pi i \langle x|y\rangle} \hat{f}(y)dy.$$

Since for $k_{t,x}(y) := e^{2\pi i\langle x|y\rangle} e^{-\pi t\|y\|^2}$ we have $\left|\hat{f}(y)k_{t,x}(y)\right| \leq \left|\hat{f}(y)\right|$ and since $k_{t,x}(y)\hat{f}(y) \to e^{2\pi i\langle x|y\rangle}\hat{f}(y)$ as $t \to 0_+$, the Dominated Convergence Theorem yields

$$\hat{\hat{f}}(-x) = \lim_{t\to 0+} \int_{\mathbb{R}^d} k_{t,x}(y)\hat{f}(y)dy.$$

Applying the preceding proposition to f and the function $k_{t,x}$ we get

$$\int_{\mathbb{R}^d} k_{t,x}(y)\hat{f}(y)dy = \int_{\mathbb{R}^d} f(w)\widehat{k_{t,x}}(w)dw.$$

Since the preceding example shows that the Fourier transform of the function $y \mapsto h_t(y) := e^{-\pi t\|y\|^2}$ is given by $\widehat{h_t}(z) = g_t(z) := t^{-d/2}e^{-\pi\|z\|^2/t}$, we have

$$\widehat{k_{t,x}}(w) = \int_{\mathbb{R}^d} e^{-2\pi i\langle w-x|y\rangle} e^{-\pi t\|y\|^2} dy = \widehat{h_t}(w-x) = g_t(x-w),$$

$$\hat{\hat{f}}(-x) = \lim_{t\to 0+} \int_{\mathbb{R}^d} f(w)\widehat{k_{t,x}}(w)dw = \lim_{t\to 0+} \int_{\mathbb{R}^d} f(w)g_t(x-w)dw.$$

Since $(f * g_t) \to f$ in $L_1(\mathbb{R}^d)$ by Lemma 8.17, any sequence $(t_n) \to 0$ has a subsequence $(t_{k(n)})$ such that $(f * g_{t_{k(n)}}) \to f$ a.e. by Proposition 7.6. Thus $\hat{\hat{f}}(-x) = f(x)$ a.e. □

Corollary 8.14 *Let $f \in L_1(\mathbb{R}^d)$ be such that $\hat{f} = 0$. Then $f = 0$.*

Corollary 8.15 *If $f \in C^k(\mathbb{R}^d)$ is such that $D^\beta f \in L_1(\mathbb{R}^d)$ and $\left|x^\beta\right|\hat{f} \in L_1(\mathbb{R}^d)$ for all $\beta \in \mathbb{N}^d$ satisfying $|\beta| \leq k$, then one has $\widehat{D^\alpha f}(y) = (2\pi i)^k y^\alpha \hat{f}(y)$ for all $\alpha \in \mathbb{N}^d$ satisfying $|\alpha| = k$ and all $y \in \mathbb{R}^d$.*

Proof Setting $g := \hat{f}$ in the relation $\widehat{m_\alpha f} = D^\alpha \hat{f}$ and applying the inverse Fourier transform, we get $\widehat{D^\alpha g}(x) = m_\alpha(-x)f(-x) = (2\pi i)^k x^\alpha \hat{g}(x)$, a relation equivalent to the one in the statement. □

Theorem 8.38 (Plancherel) *The map $f \mapsto \hat{f}$ from $L_2(\mathbb{R}^d) \cap L_1(\mathbb{R}^d)$ into $L_2(\mathbb{R}^d)$ has a unique extension as a linear isometry from $L_2(\mathbb{R}^d)$ into $L_2(\mathbb{R}^d)$ still denoted by $\mathcal{F} : f \mapsto \hat{f}$.*

Proof Let us first show that for $f \in L_2(\mathbb{R}^d) \cap L_1(\mathbb{R}^d)$ we have $\hat{f} \in L_2(\mathbb{R}^d)$ and $\left\| \hat{f} \right\|_2 = \|f\|_2$. By Proposition 8.19, $\hat{f}$ is bounded, so that, for $t \in \mathbb{P}$, $h_t(y) := e^{-\pi t \|y\|^2}$, the function $\left| \hat{f} \right|^2 h_t$ is integrable. Since $f \in L_1(\mathbb{R}^d)$ the function $(w, x, y) \mapsto \bar{f}(x)f(y)h_t(w)$ is in $\mathcal{L}_1(\mathbb{R}^{3d})$. Since $\widehat{h_t}(z) = g_t(z) = t^{-d/2}e^{-\pi\|z\|^2/t}$ as we have seen, applying Fubini's Theorem, we get

$$\begin{aligned}\int_{\mathbb{R}^d} \left| \hat{f}(y) \right|^2 h_t(y)dy &= \int_{\mathbb{R}^d} (\int_{\mathbb{R}^d} \bar{f}(w)e^{2\pi i \langle w|y\rangle}dw)(\int_{\mathbb{R}^d} f(x)e^{-2\pi i\langle x|y\rangle}dx)h_t(y)dy \\ &= \int_{\mathbb{R}^{3d}} \bar{f}(w)f(x)e^{2\pi i\langle w-x|y\rangle}h_t(y)dwdxdy \\ &= \int_{\mathbb{R}^{2d}} \bar{f}(w)f(x)g_t(x-w)dwdx.\end{aligned}$$

Lemma 8.17 ensures that $f * g_t \to f$ in $L_1(\mathbb{R}^d)$ as $t \to 0_+$. Thus there exists a sequence $(t_n) \to 0_+$ such that $((f*g_{t_n})(w) \to f(w)$ for almost every w. We also have $(\bar{f}(w)(f * g_{t_n})(w) \to \bar{f}(w)f(w)$ for almost every w and since $\left|\bar{f}(w)(f * g_{t_n})(w)\right| = \left|\int_{\mathbb{R}^d} \bar{f}(w)f(x)g_{t_n}(w-x)dx\right| \le \|f\|_1 \left|\bar{f}(w)\right|$ and since $\|f\|_1 \left|\bar{f}(\cdot)\right|$ is integrable, the Dominated Convergence Theorem yields

$$\int_{\mathbb{R}^d} \left| \hat{f}(y) \right|^2 h_{t_n}(y)dy = \int_{\mathbb{R}^{2d}} \overline{f(w)}f(x)g_{t_n}(x-w)dxdw \to \int_{\mathbb{R}^d} \overline{f(w)}f(w)dw.$$

Since $h_t \to 1$ pointwise as $t \to 0_+$, by the Monotone Convergence Theorem we get

$$\left\| \hat{f} \right\|_2^2 = \int_{\mathbb{R}^d} \left| \hat{f}(y) \right|^2 dy = \lim_n \int_{\mathbb{R}^d} \left| \hat{f}(y) \right|^2 h_{t_n}(y)dy = \int_{\mathbb{R}^d} \overline{f(w)}f(w)dw = \|f\|_2^2.$$

Thus $\hat{f} \in L_2(\mathbb{R}^d)$. Since $L_2(\mathbb{R}^d) \cap L_1(\mathbb{R}^d)$ (and even $C_c(\mathbb{R}^d)$) is dense in $L_2(\mathbb{R}^d)$, Theorem 3.2 entails that $f \mapsto \hat{f}$ can be extended to $L_2(\mathbb{R}^d)$ in such a way that $\left\| \hat{f} \right\|_2 = \|f\|_2$. □

Remark Using the polarization identity for the scalar product $\langle f \mid g\rangle := \int f\bar{g}$, i.e. the relation

$$4\langle f \mid g\rangle = \|f+g\|^2 - \|f-g\|^2 + i\,\|f+ig\|^2 - i\,\|f-ig\|^2,$$

one deduces from the relation $\left\|\hat{f}\right\|_2 = \|f\|_2$ the *Parseval identity*:

$$\forall f, g \in L_2(\mathbb{R}^d) \qquad \langle f \mid g\rangle = \langle \hat{f} \mid \hat{g}\rangle.$$

The preceding theorem can be completed.

Theorem 8.39 *The Fourier transform $\mathcal{F} : L_2(\mathbb{R}^d) \to L_2(\mathbb{R}^d)$ defined by $\mathcal{F}(f) := \hat{f}$ is an isometry onto $L_2(\mathbb{R}^d)$ and its inverse is $S \circ \mathcal{F} : g \mapsto \check{g}$, where S is the symmetry defined by $S(f)(x) := f(-x)$ for $f \in L_2(\mathbb{R}^d)$, $x \in \mathbb{R}^d$.*

Proof Let us use again the Gaussian functions $g_t := t^{-d/2}e^{-\pi\|\cdot\|^2/t}$ whose Fourier transform $h_t := \widehat{g_t}$ is given by $h_t(y) = e^{-\pi t\|y\|^2}$. We claim that for all $t > 0$, all $y \in \mathbb{R}^d$, and all $f \in L_2(\mathbb{R}^d)$ we have

$$\int_{\mathbb{R}^d} f(x)g_t(y-x)dx = \int_{\mathbb{R}^d} \hat{f}(w)e^{2\pi i\langle w|y\rangle}h_t(w)dw. \tag{8.29}$$

Since $e^{2\pi i\langle\cdot|y\rangle}h_t(\cdot) \in L_1(\mathbb{R}^d)$ and its Fourier transform is $T_y\widehat{h_t} := g_t(\cdot - y)$, Proposition 8.22 asserts that this relation holds when $f \in L_1(\mathbb{R}^d)$. Taking $f \in L_2(\mathbb{R}^d)$ and a sequence (f_n) in $L_1(\mathbb{R}^d) \cap L_2(\mathbb{R}^d)$ with L_2-limit f and using the fact that $(\widehat{f_n}) \to \hat{f}$ in $L_2(\mathbb{R}^d)$ since $\mathcal{F}$ is continuous, we see that relation (8.29) still holds, both sides of it being continuous functions of f in $L_2(\mathbb{R}^d)$ since $T_y g_t := g_t(\cdot - y)$ and $e^{2\pi i\langle\cdot|y\rangle}h_t(\cdot)$ are in $L_2(\mathbb{R}^d)$.

As $t \to 0_+$ the left-hand side $(f * g_t)(y)$ of relation (8.29) considered as a function of y converges to f in $L_1(\mathbb{R}^d)$ by Lemma 8.17. Thus, we can find a sequence $(t_n) \to 0_+$ such that $((f * g_{t_n})(y)) \to f(y)$ for almost every $y \in \mathbb{R}^d$. By the Dominated Convergence Theorem, the right-hand side converges to $\int_{\mathbb{R}^d} \hat{f}(w)e^{2\pi i\langle w|y\rangle}dw = \mathcal{F}(\hat{f})(-y) = (S \circ \mathcal{F})(\hat{f})(y)$. Thus $f = (S \circ \mathcal{F})(\hat{f})$ a.e. □

As an application of the Fourier transform, let us point out its use for the partial differential equation

$$-\Delta u + u = f \tag{8.30}$$

where $f \in L_2(\mathbb{R}^d)$ is given. Taking the Fourier transform of both sides of this equation, we get

$$(1 + 4\pi^2\|y\|^2)\hat{u}(y) = \hat{f}(y) \quad y \in \mathbb{R}^d.$$

Thus

$$u = \left(\frac{\hat{f}}{1 + 4\pi^2\|\cdot\|^2}\right)^{\vee} = f * b,$$

where b, the Fourier transform of $1/(1 + 4\pi^2 \|\cdot\|^2)$, is called the *Bessel potential*. For the computation of b, see [117, p. 187] for instance.

Exercises

1. Prove the properties asserted in Proposition 8.19 using the given hints.
2. (**Riemann-Lebesgue Lemma**) Show that for all $f \in L_1(\mathbb{R}^d)$ one has $\hat{f}(y) \to 0$ as $\|y\| \to +\infty$.
3. Show that for $p \in]2, +\infty[$ there are no $q \in [1, +\infty[$ and $c \in \mathbb{R}_+$ such that $\left\|\hat{f}\right\|_q \leq c \|f\|_p$ for all $f \in L_p(\mathbb{R}^d) \cap L_1(\mathbb{R}^d)$. [Hint: use the function $h_z := e^{-\pi z \|\cdot\|^2}$ with $z := a + ib$, $a > 0$.]
4. Show that if for $p \in]1, 2[$ there are some $q \in [1, +\infty[$ and $c \in \mathbb{R}_+$ such that $\left\|\hat{f}\right\| \leq c \|f\|_p$ for all $f \in L_p(\mathbb{R}^d) \cap L_1(\mathbb{R}^d)$ then one must have $q = (1 - 1/p)^{-1}$. [Hint: use a scaling argument, replacing f with tf. The fact that such a constant c exists for $q := (1 - 1/p)^{-1}$ is true, but not easy to prove.]
5. Show that for $f \in L_1(\mathbb{R}^d)$ satisfying $D_k f \in L_1(\mathbb{R}^d)$ one has $\widehat{D_k f}(x) = (m_k f)(-x)$. [Hint: use an integration by parts.] Generalize this result to any partial derivative D^α.
6. Let $\mathcal{S}(\mathbb{R}^d)$ be the space of functions of class C^∞ such that for all $\alpha \in \mathbb{N}^d$ and all $n \in \mathbb{N}$ the function $x \mapsto \|x\|^n D^\alpha f(x)$ is bounded. Show that the Fourier transform maps $\mathcal{S}(\mathbb{R}^d)$ into $\mathcal{S}(\mathbb{R}^d)$.

8.9.2 *Introduction to the Radon Transform*

The Radon transform is used in a number of fields, in particular in tomography, as is the Ray or X-ray transform (see the exercises for the latter), [89, 90, 122, 157, 201], (Fig. 8.1).

Given $u \in \mathbb{S}^{d-1}$, the unit sphere in $\mathbb{R}^d$, we denote by H_u the hyperplane

$$H_u := u^\perp := \{x \in \mathbb{R}^d : \langle u \mid x \rangle = 0\}$$

of $\mathbb{R}^d$. Taking an orthonormal basis $b := (b_1, \ldots, b_{d-1})$ of H_u we get an isometry $h_b : \mathbb{R}^{d-1} \to H_u$ given by $h_b(x_1, \ldots, x_{d-1}) := x_1 b_1 + \ldots + x_{d-1} b_{d-1}$. The image measure μ_u on H_u by h_b does not depend on the basis b: if $a := (a_1, \ldots, a_{d-1})$ is another orthonormal basis, then for every measurable subset A of H_u we have $\mu_u(A) := \lambda_{d-1}(h_b^{-1}(A)) = \lambda_{d-1}(T(h_a^{-1}(A))) = \lambda_{d-1}(h_a^{-1}(A))$, where T is the isometry $T := h_b^{-1} \circ h_a$ of $\mathbb{R}^{d-1}$ as λ_{d-1} is invariant under linear isometries.

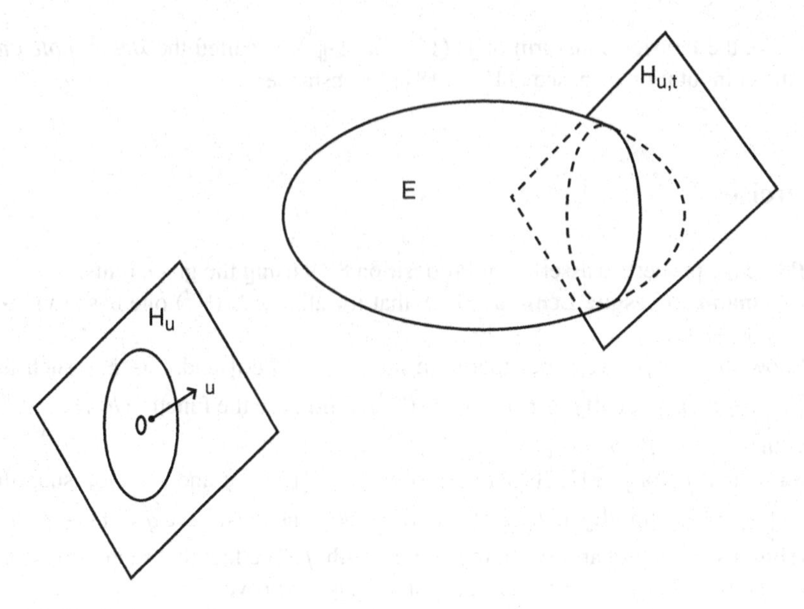

Fig. 8.1 The Radon transform

For $t \in \mathbb{R}$ we denote by $H_{u,t}$ the affine hyperplane of $\mathbb{R}^d$ given by

$$H_{u,t} := H_u + tu = \{x \in \mathbb{R}^d : \langle u \mid x \rangle = t\}.$$

Since for any orthonormal basis $b := (b_1, \ldots, b_{d-1})$ of $H_u := H_{u,0}$ the map $h_{b,tu} : \mathbb{R}^{d-1} \to H_{u,t}$ given by $h_{b,tu}(x) := h_b(x) + tu$ is an isometry, the set $H_{u,t}$ can be equipped with the measure $\mu_{u,t}$, the image of λ_{d-1} by $h_{b,tu}$.

Definition 8.10 If f is a measurable function on $\mathbb{R}^d$, its Radon transform is the function $\tilde{f} := Rf : \mathbb{S}^{d-1} \times \mathbb{R} \to \overline{\mathbb{R}}$ given by

$$\tilde{f}(u,t) := \int_{H_{u,t}} (f \mid_{H_{u,t}}) d\mu_{u,t} = \int_{\mathbb{R}^{d-1}} f \circ h_{b,tu} d\lambda_{d-1}$$

If f is in the space $C_c(\mathbb{R}^d)$ of continuous functions with compact support, then $\tilde{f}(u,t)$ is defined for all (u,t) and is a continuous function of (u,t). If f is just in $L_1(\mathbb{R}^d)$, then $\tilde{f}$ is not defined for all (u,t).

Proposition 8.23 *For $f \in C_c(\mathbb{R}^d)$ and for all $(u,t) \in \mathbb{S}^{d-1} \times \mathbb{R}$ one has $\tilde{f}(-u,-t) = \tilde{f}(u,t)$ and if $f(x) = 0$ for $\|x\| > r$, then $\tilde{f}(u,t) = 0$ for $|t| > r$.*

For $f \in C_c^\infty(\mathbb{R}^d)$, a multi-index α, and $(u,t) \in \mathbb{S}^{d-1} \times \mathbb{R}$, one has

$$\widetilde{(D^\alpha f)}(u,t) = u^\alpha \frac{\partial^{|\alpha|}}{\partial t^{|\alpha|}} \tilde{f}(u,t).$$

In particular, for $f \in C_c^\infty(\mathbb{R}^d)$ one has $\widetilde{\Delta f} = \frac{\partial^2}{\partial t^2}\tilde{f}$.

Proof The first assertion is obvious since $H_{-u,-t} = H_{u,t}$ for all $(u,t) \in \mathbb{S}^{d-1} \times \mathbb{R}$ and since $H_{u,t} \subset \mathbb{R}^d \backslash B[0,r]$ if $t > r$.

Given $(u,t) \in \mathbb{S}^{d-1} \times \mathbb{R}$, taking an orthonormal basis $(b_1, \ldots, b_{d-1})$ of $H_u := H_{u,0}$, since $(b_1, \ldots, b_{d-1}, u)$ is an orthonormal basis of $\mathbb{R}^d$ we have

$$D_i f = e_i.\nabla f = e_i.(\sum_{j=1}^{d-1} (\nabla f.b_j)b_j + (\nabla f.u)u).$$

Since the function $(\nabla f.b_j) \circ h_{b,tu} = D_j(f \circ h_{b,tu})$ has compact support and $\int D_j(f \circ h_{b,tu})dx_j = 0$ we get

$$\begin{aligned}
\widetilde{D_i f}(u,t) &= \sum_{j=1}^{d-1} e_i.b_j \int_{\mathbb{R}^{d-1}} (\nabla f.b_j) \circ h_{b,tu} d\lambda_{d-1} + (e_i.u) \int_{\mathbb{R}^{d-1}} (\nabla f.u) \circ h_{b,tu} d\lambda_{d-1} \\
&= u_i \int_{\mathbb{R}^{d-1}} (\nabla f.u)(h_b(x) + tu) d\lambda_{d-1}(x) \\
&= u_i \frac{d}{dt} \int_{\mathbb{R}^{d-1}} f(h_b(x) + tu) d\lambda_{d-1}(x)
\end{aligned}$$

by differentiating an integral depending on a parameter. Iterating this formula, we get the second assertion. □

Let us point out the interplay between the Fourier transform and the Radon transform. We just consider the case $f \in C_c(\mathbb{R}^d)$. For fixed $u \in \mathbb{S}^{d-1}$ we denote by $\widehat{Rf_u}$ the Fourier transform of the function $(Rf)_u := \tilde{f}_u : t \mapsto \tilde{f}(u,t)$:

$$\widehat{Rf_u}(s) := \int_{\mathbb{R}} e^{-2\pi i s t} \tilde{f}(u,t) dt.$$

Lemma 8.18 *If $f \in C_c(\mathbb{R}^d)$ the following relation holds between the partial Fourier Transform of the Radon transform Rf of f and the Fourier Transform $\hat{f}$ of f :*

$$\forall u \in \mathbb{S}^{d-1}, \ \forall s \in \mathbb{R} \qquad \widehat{Rf_u}(s) = \hat{f}(su). \tag{8.31}$$

Proof For $u \in \mathbb{S}^{d-1}$ and a basis $b := (b_1, \ldots, b_{d-1})$ of H_u, let us use the isomorphism $h_{b,u} : \mathbb{R}^{d-1} \times \mathbb{R} \to \mathbb{R}^{d-1}$ given by

$$h_{b,u}(x,t) := h_{b,u}(x_1, \ldots, x_{d-1}, t) := x_1 b_1 + \ldots + x_{d-1} b_{d-1} + tu$$

for $x := (x_1, \ldots, x_{d-1}) \in \mathbb{R}^{d-1}$, $t \in \mathbb{R}^d$. Applying Fubini's Theorem and this orthogonal change of coordinates, since $t = \langle h_{b,u}(x,t) \mid u \rangle$, setting $w := h_{b,u}(x,t)$, we get

$$\begin{aligned}\widehat{Rf_u}(s) &= \int_{\mathbb{R}} e^{-2\pi ist} \int_{\mathbb{R}^{d-1}} f(h_{b,u}(x,t))dxdt \\ &= \int_{\mathbb{R}^d} e^{-2\pi is\langle h_{b,u}(x,t)|u\rangle} f(h_{b,u}(x,t)) \left|\det h_{b,u}\right| dxdt \\ &= \int_{\mathbb{R}^d} e^{-2\pi i\langle w|su\rangle} f(w)dw = \hat{f}(su)\end{aligned}$$

since $\left|\det h_{b,u}\right| = 1$. □

Lemma 8.19 *If $f \in C_c(\mathbb{R}^d)$ and if σ_{d-1} denotes the measure on $\mathbb{S}^{d-1}$ defined in Sect. 7.8 one has*

$$\int_{\mathbb{S}^{d-1}} \int_{\mathbb{R}} \left|\widehat{Rf_u}(s)\right|^2 |s|^{d-1} dsd\sigma_{d-1}(u) = 2 \int_{\mathbb{R}^d} |f(x)|^2 dx. \tag{8.32}$$

Proof Using Plancherel's formula $\|f\|_2 = \left\|\hat{f}\right\|_2$ and polar coordinates $y = su$ with $(s,u) \in \mathbb{P} \times \mathbb{S}^{d-1}$ to compute $\left\|\hat{f}\right\|_2$ with the help of Proposition 7.19 yields

$$\int_{\mathbb{R}^d} |f(x)|^2 dx = \int_{\mathbb{S}^{d-1}} \int_0^{+\infty} \left|\hat{f}(su)\right|^2 s^{d-1} dsd\sigma_{d-1}(u).$$

Now, using the change of variables $r := -s$, $v := -u$, we have

$$\int_{\mathbb{S}^{d-1}} \int_0^{+\infty} \left|\hat{f}(su)\right|^2 s^{d-1} dsd\sigma_{d-1}(u) = \int_{\mathbb{S}^{d-1}} \int_{-\infty}^0 \left|\hat{f}(rv)\right| |r|^{d-1} drd\sigma_{d-1}(v).$$

Adding the two sides and invoking relation (8.31), we get relation (8.32). □

We can deduce from Lemma 8.18 an inversion formula for the Radon transform. Thus f can be recovered from its Radon transform.

Theorem 8.40 *If $f \in C_c(\mathbb{R}^d)$ then for $x \in \mathbb{R}^d$ one has*

$$f(x) = \frac{1}{2} \int_{\mathbb{R}} \int_{\mathbb{S}^{d-1}} \widehat{Rf_u}(s) |s|^{d-1} e^{2\pi isu.x} d\sigma_{d-1}(u)ds.$$

Proof Let us denote by $g(x)$ the right-hand side of this relation. By relation (8.31), using again the change of variables $r := -s$, $v := -u$ and then polar coordinates, by Theorem 7.19, we have

$$\begin{aligned} g(x) &= \frac{1}{2}\int_{\mathbb{R}}\int_{\mathbb{S}^{d-1}} \hat{f}(su)\,|s|^{d-1}\,e^{2\pi i su.x} d\sigma_{d-1}(u)ds \\ &= \int_0^{\infty}\int_{\mathbb{S}^{d-1}} \hat{f}(su)s^{d-1}e^{2\pi i su.x} d\sigma_{d-1}(u)ds \\ &= \int_{\mathbb{R}^d} \hat{f}(y)e^{2\pi i x.y}dy = f(x) \end{aligned}$$

by the Fourier inversion theorem. □

Corollary 8.16 *For* $d = 2k + 1$ *and* $f \in C_c(\mathbb{R}^d)$, *for* $x \in \mathbb{R}^d$ *one has*

$$f(x) = \frac{(-1)^k}{2(2\pi)^{2k}} \int_{\mathbb{S}^{d-1}} (\widehat{Rf_u})^{(2k)}(u.x)d\sigma_{d-1}(u).$$

Proof Using Corollary 8.15 we have

$$(\widehat{Rf_u})^{(2k)}(s) = (-1)^k(2\pi)^{2k}\int_{\mathbb{R}} t^{2k}e^{2\pi i st}\widehat{Rf_u}(t)dt.$$

Given $x \in \mathbb{R}^d$, setting $s = u.x$ for $u \in \mathbb{S}^{d-1}$ and integrating over $\mathbb{S}^{d-1}$ we get

$$\begin{aligned} &\int_{\mathbb{S}^{d-1}} (\widehat{Rf_u})^{(2k)}(u.x)d\sigma_{d-1}(u) \\ &= (-1)^k(2\pi)^{2k}\int_{\mathbb{R}}\int_{\mathbb{S}^{d-1}} |t|^{d-1}\,e^{2\pi i tu.x}\widehat{Rf_u}(t)d\sigma_{d-1}(u)dt = (-1)^k(2\pi)^{2k}2f(x) \end{aligned}$$

in view of Theorem 8.40. □

Application. It follows from this corollary that if d is odd and $(\widehat{Rf_u})(s) = 0$ for $|s| \le r$, then $f \mid B(0,r) = 0$.

We observe that if f is the characteristic function 1_E of a measurable subset E of $\mathbb{R}^d$ then $\tilde{f}(u,t) = \mu_{u,t}(E_{u,t})$, where $E_{u,t}$ is the slice $E_{u,t} := E \cap H_{u,t}$. Thus $\tilde{f}(u,t)$ gives precious information about the size of E or rather $E_{u,t}$ (in particular when E is a tumor).

It follows from Lemma 7.7 that for all $u \in \mathbb{S}^{d-1}$ the function $t \mapsto \mu_{u,t}(E_{u,t})$ is measurable if E is measurable. In general, not much more can be said about this function. However, for $d \ge 3$ we have the following remarkable regularity result.

Theorem 8.41 *For $d \geq 3$ and for any measurable subset E of $\mathbb{R}^d$ with finite measure there exists a subset N of $\mathbb{S}^{d-1}$ with null measure such that for all $u \in \mathbb{S}^{d-1} \backslash N$ and for all $t \in \mathbb{R}$ the set $E_{u,t}$ is $\mu_{u,t}$-measurable and $t \mapsto \mu_{u,t}(E_{u,t})$ satisfies a Hölder condition for any $\alpha \in]0, 1/2[$ (hence is continuous): for some $c := c(u, \alpha)$ one has*

$$\forall s,\ t \in \mathbb{R} \qquad |\mu_{u,s}(E_{u,s}) - \mu_{u,t}(E_{u,t})| \leq c\,|s-t|^{\alpha}\,.$$

It has been shown by Besicovitch that such a result is not valid for $d = 2$. We shall deduce it from a similar result pertaining to functions.

Theorem 8.42 *For $d \geq 3$ and for any $f \in L_1(\mathbb{R}^d) \cap L_2(\mathbb{R}^d)$ there exists a subset N of $\mathbb{S}^{d-1}$ with null measure such that for all $u \in \mathbb{S}^{d-1} \backslash N$ the function $f\,|_{H_{u,t}}$ is $\mu_{u,t}$-integrable. Moreover, there exists some $c > 0$ such that for all $f \in C_c(\mathbb{R}^d)$ one has*

$$\int_{\mathbb{S}^{d-1}} \sup_t Rf(u,t) d\sigma_{d-1}(u) \leq c\,\|f\|_1 + c\,\|f\|_2\,.$$

Furthermore, for any $u \in \mathbb{S}^{d-1} \backslash N$, $\alpha \in]0, 1/2[$, $\tilde{f}_u := Rf_u : t \mapsto Rf(u,t)$ is continuous and satisfies a Hölder condition: for some $c := c(u, \alpha)$ one has

$$\forall t,\ t' \in \mathbb{R} \qquad \left|\tilde{f}_u(t) - \tilde{f}_u(t')\right| \leq c\left|t - t'\right|^{\alpha}.$$

To prove this result we need a criterion for the Hölderian behavior of $\tilde{f}_u := Rf_u$ or, more generally for a function $h : \mathbb{R} \to \mathbb{R}$.

Lemma 8.20 *Let $h : \mathbb{R} \to \mathbb{R}$ be such that $\hat{h}$ is defined, belongs to $\mathcal{L}_1(\mathbb{R})$ and satisfies $h(t) = \hat{\hat{h}}(-t)$ a.e. Then for $d > 2$ there exists some $c = c(d) > 0$ such that whenever*

$$\int_{-1}^{1} \left|\hat{h}(s)\right| ds \leq a, \qquad \left(\int_{\mathbb{R}} \left|\hat{h}(s)\right|^2 |s|^{d-1}\, ds\right)^{1/2} \leq b$$

for some $a, b > 0$ one has

$$\sup_{t \in \mathbb{R}} |h(t)| \leq a + bc. \tag{8.33}$$

Moreover, for all $\alpha \in]0, d/2 - 1[$, $\alpha \leq 1$, there exists some $c_\alpha > 0$ such that

$$t,\ t' \in \mathbb{R} \implies \left|h(t) - h(t')\right| \leq 4\pi(a + bc_\alpha)\left|t - t'\right|^{\alpha}. \tag{8.34}$$

Proof Since $h(t) = \widehat{\hat{h}}(-t)$, setting $S := \mathbb{R}\backslash[-1, 1]$ we have

$$h(t) = \int_{\mathbb{R}} \hat{h}(s)e^{2\pi ist}ds = \int_{-1}^{1} \hat{h}(s)e^{2\pi ist}ds + \int_{S} \hat{h}(s)e^{2\pi ist}ds.$$

The modulus of the first term is bounded by a. We estimate the second one by applying the Cauchy-Schwarz inequality:

$$\int_S \left|\hat{h}(s)\right| ds \leq \left(\int_S \left|\hat{h}(s)\right|^2 |s|^{d-1} ds\right)^{1/2} \left(\int_S |s|^{1-d} ds\right)^{1/2} \leq bc$$

for $c := \left(\int_S |s|^{1-d} ds\right)^{1/2} < +\infty$ since $d - 1 > 1$. Thus relation (8.33) holds.

To get relation (8.34) we use the estimate $\left|e^{ir} - 1\right| \leq |r| \leq |r|^{\alpha}$ for all $r \in [-1, 1]$ obtained by applying the Mean Value Theorem, identifying $\mathbb{C}$ with $\mathbb{R}^2$ and the estimate $\left|e^{ir} - 1\right| \leq 2 \leq 2\,|r|^{\alpha}$ for all $r \in \mathbb{R}\backslash[-1, 1]$. Then, for s, t, $t' \in \mathbb{R}$ we have

$$\left|e^{2\pi ist} - e^{2\pi ist'}\right| \leq 4\pi \left|s\right|^{\alpha}.\left|t - t'\right|^{\alpha},$$

$$\left|\int_{-1}^{1} \hat{h}(s)(e^{2\pi ist} - e^{2\pi ist'})ds\right| \leq 4\pi \left|t - t'\right|^{\alpha} \int_{-1}^{+1} \left|\hat{h}(s)\right| |s|^{\alpha} ds \leq 4\pi a \left|t - t'\right|^{\alpha}.$$

We estimate the second term $\left|\int_S \hat{h}(s)(e^{2\pi ist} - e^{2\pi ist'})ds\right|$ in the decomposition of $|h(t) - h(t')|$ by applying again the Cauchy-Schwarz inequality

$$4\pi \left|t - t'\right|^{\alpha} \left(\int_S \left|\hat{h}(s)\right|^2 |s|^{d-1} ds\right)^{1/2} \left(\int_S |s|^{1-d+2\alpha} ds\right)^{1/2} \leq 4\pi b c_{\alpha} \left|t - t'\right|^{\alpha}$$

for $c_{\alpha} := \left(\int_S |s|^{1-d+2\alpha} ds\right)^{1/2} < +\infty$ since $1 - d + 2\alpha < -1$. Gathering the two estimates, we get relation (8.34). □

Proof of Theorem 8.42 We replace h with $h_u := Rf_u$ in the preceding lemma, remembering that $\widehat{Rf_u}(s) = \hat{f}(su)$ by (8.31), and we set

$$a_u := \left|\int_{-1}^{1} \widehat{h_u}(s)ds\right| \leq 2 \sup_s \left|\widehat{h_u}(s)\right| = 2 \sup_s \left|\widehat{Rf_u}(s)\right| = 2 \sup_s \left|\hat{f}(su)\right| \leq 2 \left\|f\right\|_1,$$

$$b_u := (\int_{\mathbb{R}} \left|\widehat{h_u}(s)\right|^2 |s|^{d-1} ds)^{1/2}.$$

Relation (8.32) ensures that

$$\int_{\mathbb{S}^{d-1}} b_u^2 d\sigma_{d-1}(u) := \int_{\mathbb{S}^{d-1}} \int_{\mathbb{R}} \left|\widehat{h_u}(s)\right|^2 |s|^{d-1} ds d\sigma_{d-1}(u) = 2 \left\|f\right\|_2^2.$$

Thus, there exists a set N of null measure in $\mathbb{S}^{d-1}$ such that $b_u < +\infty$ for all $u \in \mathbb{S}^{d-1}\backslash N$. Therefore, for $u \in \mathbb{S}^{d-1}\backslash N$ we get

$$\sup_{t\in\mathbb{R}} |h_u(t)| \leq a_u + b_u c,$$

hence, by the Cauchy-Schwarz inequality,

$$\begin{aligned} \int_{\mathbb{S}^{d-1}} \sup_{t\in\mathbb{R}} |h_u(t)|\, d\sigma_{d-1}(u) &\leq \int_{\mathbb{S}^{d-1}} a_u d\sigma_{d-1}(u) + c\int_{\mathbb{S}^{d-1}} b_u d\sigma_{d-1}(u) \\ &\leq 2\sigma_{d-1}(\mathbb{S}^{d-1})\, \|f\|_1 + 2^{1/2} c(\sigma_{d-1}(\mathbb{S}^{d-1}))^{1/2}\, \|f\|_2 \,. \end{aligned}$$

The last assertion of Theorem 8.42 is a consequence of relation (8.34). □

Exercises

1. The Ray or X-ray transform of a measurable function f on $\mathbb{R}^d$ is defined on the tangent bundle $T\mathbb{S}^{d-1}$ of the unit sphere $\mathbb{S}^{d-1}$ of $\mathbb{R}^d$. This set is the set of pairs (u, v) with $u \in \mathbb{S}^{d-1}$ and $v \in T_u\mathbb{S}^{d-1}$, the tangent space to $\mathbb{S}^{d-1}$ at u, i.e. the orthogonal subspace $H_u := u^\perp$ to u. For a measurable function f on $\mathbb{R}^d$ it is given by

$$(Pf)(u, v) := (P_u f)(v) := \int_{\mathbb{R}} f(v + tu)dt.$$

Thus, Pf takes into account the behavior of f on the ray issued from v and passing through $u + v$.

Given a function f on $T\mathbb{S}^{d-1}$ and $u \in \mathbb{S}^{d-1}$, denote by f_u the restriction of f to the tangent space to $\mathbb{S}^{d-1}$ at u: $f_u(v) := f(u, v)$. If g is another measurable function on $T\mathbb{S}^{d-1}$ define the partial convolution of f and g by

$$(f *_u g)(u, v) = \int_{H_u} f_u(v - w)g_u(w)d\mu_u(w) \qquad (u, v) \in T\mathbb{S}^{d-1},$$

$H_u = T_u\mathbb{S}^{d-1}$ being endowed with the image measure μ_u of λ_{d-1} described above. For $f, g \in C_c^\infty(\mathbb{R}^d)$, $u \in \mathbb{S}^{d-1}$ show that $Pf *_u Pg = P(f *_u g)$.

2. With the notation of the preceding exercise, define the Fourier transform of f_u by:

$$\widehat{f_u}(w) := \int_{H_u} e^{-2\pi i(v|w)} f_u(v)d\mu_u(v) \qquad (u, v) \in T\mathbb{S}^{d-1}.$$

Prove that for $f \in C_c^\infty(\mathbb{R}^d)$, $(u, w) \in T\mathbb{S}^{d-1}$ one has $\widehat{P_u f}(w) = \hat{f}(w)$.

3. Given $f, g \in C_c^\infty(\mathbb{R}^d)$, let $h := f * g$. Compute Rh_u in terms of Rf_u and Rg_u.

Additional Reading

[2, 18, 20, 31, 35, 38, 42, 54, 55, 64, 80, 88–90, 99, 106, 118, 124, 127, 132, 140, 146, 150, 158, 168, 185, 190, 193, 212, 226, 239, 240, 256, 258]

Chapter 9
Partial Differential Equations

The most practical solution is a good theory.

Albert Einstein

Abstract A large part of this chapter is devoted to Sobolev spaces, which are convenient spaces for handling partial differential equations. The weakened notion of derivative they convey is related to the question of transposition. Such a notion gives a natural approach to the concept of a weak solution to a partial differential equation. The question of regularity for such a solution is given a concise treatment. On the other hand, some nonlinear problems are considered. In particular, monotone operators are viewed through recent advances using representations by convex functions.

We devote this chapter to an introduction to the study of partial differential equations. Such equations or systems are numerous and serve the modeling of various physical or biological phenomena: impressive lists of such equations are given in the books [88, 117, 248] among many others. They are constantly completed with the studies of new phenomena or processes. As an example, the mathematical study of hydraulic fracture (fracking) emerges from the knowledge of equations for porous media and thin film equations. Also, more and more examples stem from the progress of mathematical biology.

Such equations involve partial derivatives of order one or higher of an unknown function u on an open subset Ω of $\mathbb{R}^d$ or a finite family of unknown functions. Thus, we use again the notation

$$D_i u = \frac{\partial u}{\partial x_i}, \qquad D^\alpha u = \frac{\partial^{|\alpha|} u}{\partial x_1^{\alpha_1} \cdots \partial x_d^{\alpha_d}}$$

for $i \in \mathbb{N}_d$ or $\alpha := (\alpha_1, \cdots, \alpha_d) \in \mathbb{N}^d$, with $|\alpha| := \alpha_1 + \cdots + \alpha_d$.

The most famous operator obtained by combining such partial derivatives is the *Laplacian*

$$\Delta := \frac{\partial^2}{\partial x_1^2} + \cdots + \frac{\partial^2}{\partial x_d^2}.$$

J.-P. Penot, *Analysis*, Universitext, DOI 10.1007/978-3-319-32411-1_9

It plays an important role in Riemannian geometry and in physics and it can be considered as the prototype of a large class of linear partial differential equations, the class of elliptic equations. Among the deep questions centered around the Laplacian is the following surprising one: can one hear the shape of a drum ([141, 171])? Such a question is motivated by the fact that one hears some harmonics that are linked with the eigenvalues of the Laplacian. Thus one may wonder whether two bounded domains Ω and Ω' of $\mathbb{R}^2$ are isometric when the eigenvalues of Δ on Ω and Ω' are the same. The answer is positive if Ω is a disc but negative if Ω has corners or if Ω and Ω' are smooth domains of the sphere $\mathbb{S}^{d-1}$ of $\mathbb{R}^d$; it is still open for smooth domains. A general problem consists in finding as much information as possible on Ω from the knowledge of the spectrum of Δ (see [32, 36, 69, 141, 203, 235] among hundreds of studies).

Solving such equations is often difficult and it is rare that explicit solutions can be found. Thus, since one must seek approximate solutions, this topic is closely connected to numerical analysis.

Because nonlinear equations require particular tools, we essentially restrict our approach to linear equations. Even for this class of equations, the proofs of the fundamental results are not simple. Thus, we just present the main lines of some of these results.

The choice of the class of functions in which one would like to find a solution is part of the problem. The classical classes of continuously differentiable functions on Ω are not the most appropriate classes. Some results can be given in spaces of functions satisfying a Hölder property. But the most important class of functions on Ω for the study of partial differential equations is the class of Sobolev functions. They are (a.e. equality equivalence classes of) functions in $L_p(\Omega)$ that have partial derivatives in $L_p(\Omega)$ in a weak sense we make precise in Sect. 9.1. Such classes of spaces have good compactness and completeness properties and can be embedded in some classical spaces such as $C^k(\Omega)$ or $L_q(\Omega)$. Moreover, the reflexivity of the Sobolev space $W_p^m(\Omega)$ is a great advantage.

In the sequel Ω denotes an open subset of $\mathbb{R}^d$ and $\mathcal{K}(\Omega)$ stands for the family of compact subsets of Ω. In some cases we require boundedness or smoothness of Ω.

9.1 Definition and Basic Properties of Sobolev Spaces

This section deals with the definition of a class of functions spaces that is well suited for the study of partial differential equations and for the calculus of variations. In the case $\Omega = \mathbb{R}^d$, one can introduce this class by using the Fourier transform. However, we are interested in the case of an arbitrary open subset Ω of $\mathbb{R}^d$.

9.1.1 Test Functions and Weak Derivatives

Let us recall that for $k \in \mathbb{N} \cup \{\infty\}$ the space of functions of class C^k with compact support in an open subset Ω of $\mathbb{R}^d$ is denoted by $C_c^k(\Omega)$, the *support* $\operatorname{supp} \varphi$ of a continuous function φ being the closure of the set $\varphi^{-1}(\mathbb{R} \backslash \{0\})$. Such functions are called *test functions*. The notation $\mathcal{D}(\Omega)$ is also classical for $C_c^\infty(\Omega)$. This vector space can be endowed with a topology induced by a family of seminorms (Exercise 1), but it is easier to use the associated convergence defined by:

$(\varphi_i)_{i \in I} \to \varphi$ if and only if there exist $K \in \mathcal{K}(\Omega)$, $\bar{i} \in I$ and $k \in \mathbb{N}$ such that $\operatorname{supp} \varphi_i \subset K$ for all $i \geq \bar{i}$ and $(p_{K,k}(\varphi_i - \varphi))_{i \in I} \to 0$, where

$$p_{K,k}(\psi) := \sup_{|\alpha| \leq k} \sup_{x \in K} |D^\alpha \psi(x)| .$$

One can verify the axioms of convergence (Definition 2.2). It is also easy to verify that for every multi-index $\alpha := (\alpha_1, \cdots, \alpha_d) \in \mathbb{N}^d$ the map $D^\alpha : \varphi \mapsto D^\alpha \varphi$ is continuous from $C_c^\infty(\Omega)$ into $C_c^\infty(\Omega)$ endowed with the convergence defined above. One must be aware that the convergence on $C_c^\infty(\Omega)$ is *not* the convergence associated with the seminorms $p_{K,k}$, even if for all $K \in \mathcal{K}(\Omega)$ the induced convergence on the subspace $\mathcal{D}(K)$ formed by the functions $\varphi \in C_c^\infty(\Omega)$ with support in K coincides with the convergence associated with the seminorms $p_{K,k}$. The construction of a family of seminorms on $C_c^\infty(\Omega)$ inducing the above convergence is rather sophisticated; it is proposed as an exercise (Exercise 1 at the end of this section).

A *distribution* on Ω is a continuous linear form on $C_c^\infty(\Omega)$ for the convergence just defined. The space of distributions on Ω will be denoted by $C_c^\infty(\Omega)^*$; it is often also denoted by $\mathcal{D}'(\Omega)$.

Proposition 9.1 *A linear form T on $C_c^\infty(\Omega)$ is a distribution if and only if for all $K \in \mathcal{K}(\Omega)$ there exist $c > 0$ and $k \in \mathbb{N}$ such that $|T(\varphi)| \leq c \Sigma_{|\alpha| \leq k} \sup_K |D^\alpha \varphi|$ for all $\varphi \in C_c^\infty(\Omega)$ satisfying* $\operatorname{supp} \varphi \subset K$.

If the same k can be used for all $K \in \mathcal{K}(\Omega)$, T is said to be a distribution of *order k*.

Proof The condition is obviously sufficient. Let us show it is necessary. If it is not satisfied, there exist some $K \in \mathcal{K}(\Omega)$ and a sequence (φ_n) in the space $C_c^\infty(K)$ of functions in $C_c^\infty(\Omega)$ satisfying $\operatorname{supp} \varphi \subset K$ such that $p_{K,n}(\varphi_n) \leq 1$ and $T(\varphi_n) \geq n$ for all $n \in \mathbb{N}$. Then $(\frac{1}{n}\varphi_n) \to 0$ in $C_c^\infty(\Omega)$ but $(T(\frac{1}{n}\varphi_n))$ does not converge to 0. □

Example Let μ be a Radon measure, i.e. a continuous linear form on $C_c^0(\Omega)$ equipped with a convergence similar to the one on $C_c^\infty(\Omega)$, but with the seminorms $p_{K,0}$ instead of the seminorms $p_{K,k}$. Then $\mu|_{C_c^\infty(\Omega)}$ is a distribution of order 0.

Example Let $f \in L_{1,loc}(\Omega)$ and let T_f be given by $T_f(\varphi) := \int_\Omega f(x)\varphi(x)dx$ for $\varphi \in C_c^\infty(\Omega)$. Then T_f is a Radon measure, hence a distribution. By Corollary 8.12

and Theorem 8.26, for any $f \in L_q(\Omega)$ with $q \in [1, \infty[$ we have $f = 0$ whenever $T_f(\varphi) = 0$ for all $\varphi \in C_c^\infty(\Omega)$. Thus one can identify f and T_f.

Example The *Dirac measure* δ_a associated with $a \in \Omega$ is the distribution $\varphi \mapsto \varphi(a)$.

Transposition enables us to define the derivative of a distribution.

Definition 9.1 Given a multi-index α and a distribution T, the α-derivative $D^\alpha T$ is the distribution defined by

$$(D^\alpha T)(\varphi) := (-1)^{|\alpha|} T(D^\alpha \varphi) \qquad \varphi \in C_c^\infty(\Omega).$$

Since $\varphi \mapsto D^\alpha \varphi$ is continuous, the linear form $\varphi \mapsto T(D^\alpha \varphi)$ is continuous on $C_c^\infty(\Omega)$, so that $D^\alpha T$ is indeed a distribution. The sign in front of $T(D^\alpha \varphi)$ is justified by the following coherence result.

Proposition 9.2 *If $f \in C^k(\Omega)$ with $k \in \mathbb{N} \setminus \{0\}$ and if α is a multi-index satisfying $|\alpha| \leq k$, then $D^\alpha T_f = T_g$ with $g := D^\alpha f$.*

Proof By the next exercise, it suffices to prove the result for $k = 1$ and $\alpha := (0, \cdots, 0, 1, 0, \cdots, 0)$. This follows from the integration by part formula:

$$\forall \varphi \in C_c^\infty(\Omega) \qquad \int_\Omega D_i f(x) \varphi(x) dx = -\int_\Omega f(x) D_i \varphi(x) dx.$$

Exercise For any $T \in C_c^\infty(\Omega)^*$ and any multi-index α, β, verify that $D^\beta(D^\alpha T) = D^{\alpha+\beta} T$.

Definition 9.2 Given a multi-index α and $u \in L_{1,loc}(\Omega)$, an element w of $L_{1,loc}(\Omega)$ is said to be the *weak α-partial derivative* of u if for every $\varphi \in C_c^\infty(\Omega)$ one has

$$\int_\Omega w(x) \varphi(x) dx = (-1)^{|\alpha|} \int_\Omega u(x) D^\alpha \varphi(x) dx. \tag{9.1}$$

The preceding definition avoids distributions. Nonetheless, one recognizes in it that the distribution T_w associated with w is just $D^\alpha T_u$ in the sense of distributions.

Example For any open interval $\Omega :=]a, b[$ of $\mathbb{R}$ containing 0, the (a.e. equality equivalence class of the) *Heaviside function* $w : \mathbb{R} \to \mathbb{R}$ given by $w(x) = -1$ for $x < 0$, $w(x) = 1$ for $x > 0$, $w(0)$ being arbitrary, is the weak derivative Du of $u(\cdot) = |\cdot|$. Indeed, for every $\varphi \in C_c^\infty(\Omega)$, since $a\varphi(a) = 0 = 0\varphi(0)$ and $b\varphi(b) = 0$ when v is extended by 0 on $\mathbb{R} \setminus \Omega$, one has

$$\begin{aligned} \int_a^b u(x) D\varphi(x) dx &= -\int_a^0 x D\varphi(x) dx + \int_0^b x D\varphi(x) dx \\ &= \int_a^0 \varphi(x) dx - \int_0^b \varphi(x) dx = -\int_a^b w(x) \varphi(x) dx. \end{aligned}$$

Example Not all elements $u \in L_{1,loc}(\Omega)$ have a weak derivative in $L_{1,loc}(\Omega)$. Taking $\Omega := \mathbb{R}$, for u the function given by $u(x) = 0$ for $x \in \mathbb{R}_- \cup [1, \infty[$, $u(x) = 1$ for $x \in [0, 1[$, we easily see that if $w \in L_{1,loc}(\Omega)$ is a weak derivative of u, then we must have $w = 0$ on $\mathbb{R}_- \cup [1, \infty[,]0, 1[$, hence $\int_\Omega w(x)\varphi(x)dx = 0$ for all $\varphi \in C_c^\infty(\Omega)$. However, for $\varphi \in C_c^\infty(\Omega)$ such that $D\varphi = 1$ on $[0, 1]$ we have $\int_\Omega u(x)D\varphi(x)dx > 0$.

The assertions of the following lemma are left as exercises.

Lemma 9.1 *If $u \in C^{|\alpha|}(\Omega)$, then $D^\alpha u$ is the weak α-partial derivative of u.*

For $u \in L_{1,loc}(\Omega)$ there is at most one weak α-partial derivative of u.

Let $u \in L_{1,loc}(\Omega)$ be such that the weak derivatives $D_i u$ exist and belong to $L_{1,loc}(\Omega)$ for $i \in \mathbb{N}_d$. Then for $f \in C^1(\Omega)$ the weak derivatives $D_i(fu)$ exist in $L_{1,loc}(\Omega)$ and $D_i(fu) = uD_i f + fD_i u$.

These assertions justify the notation $D^\alpha u$ for the weak α-partial derivative of u.

For $p \in [1, \infty[$, $u, v \in L_{p,loc}(\Omega)$, and a multi-index α, one says that $v = D^\alpha u$ in the *strong L_p sense*, if for any $K \in \mathcal{K}(\Omega)$ there exists a sequence (u_n) in $C^{|\alpha|}(\Omega)$ such that

$$(\int_K |u_n(x) - u(x)|^p dx) \to 0, \qquad (\int_K |D^\alpha u_n(x) - v(x)|^p dx) \to 0. \tag{9.2}$$

Proposition 9.3 *For $p \in [1, \infty[$, $u, v \in L_{p,loc}(\Omega)$, and a multi-index α, one has $v = D^\alpha u$ in the strong L_p sense if and only if $v = D^\alpha u$ in the weak sense.*

Proof Suppose $v = D^\alpha u$ in the strong L_p sense. Given $\varphi \in C_c^\infty(\Omega)$, let $K := \operatorname{supp}\varphi$ and let (u_n) be a sequence in $C^{|\alpha|}(\Omega)$ satisfying (9.2). Since $w \mapsto \int_K w\varphi$ is continuous with respect to the norm of $L_p(\Omega)$, we have $\int_\Omega v\varphi = \int_K v\varphi = \lim_n \int_K D^\alpha u_n \varphi$ and similarly $\int_\Omega uD^\alpha\varphi = \int_K uD^\alpha\varphi = \lim_n \int_K u_n D^\alpha\varphi$. Since $\int_K u_n D^\alpha\varphi = (-1)^{|\alpha|}\int_K D^\alpha u_n\varphi$, we get (9.1).

The converse is obtained in a more precise form in the next theorem. □

Theorem 9.1 (Friedrich) *Let $p \in [1, \infty[$ and let $u, v_\alpha \in L_p(\Omega)$ for $\alpha \in \mathbb{N}^d$ satisfying (9.1) with $w := v_\alpha$. Then, there exists a sequence (u_n) in $C_c^\infty(\mathbb{R}^d)$ such that for all $K \in \mathcal{K}(\Omega)$ one has*

$$(u_n \mid_\Omega)_n \to u \text{ in } L_p(\Omega),$$

$$(D^\alpha u_n \mid_K)_n \to v_\alpha \mid_K \text{ in } L_p(K).$$

Proof We extend every $w \in L_p(\Omega)$ by 0 on $\mathbb{R}^d \setminus \Omega$. Let $\rho : \mathbb{R}^d \to \mathbb{R}$ be defined by

$$\rho(x) = ce^{1/(\|x\|^2-1)} \text{ for } x \in B_d := B_{\mathbb{R}^d}(0, 1), \ \rho(x) = 0 \text{ for } x \in \mathbb{R}^d \setminus B_d,$$

the constant c being adjusted so that $\int_{\mathbb{R}^d} \rho = 1$. Given a sequence $(r_n) \to 0_+$, let $\rho_n(\cdot) := r_n^{-d}\rho(\cdot/r_n)$. Define the *regularization operators* $R_n : L_p(\Omega) \to C(\mathbb{R}^d)$ by

$$(R_n w)(x) := \int_\Omega \rho_n(x-y)w(y)dy = \int_{B_d} \rho(z)w(x-r_n z)dz \quad w \in L_p(\Omega),\ x \in \mathbb{R}^d.$$

Given $K \in \mathcal{K}(\Omega)$, let $n_K \in \mathbb{N}$ be such that for $n \geq n_K$ one has $r_n < r := \text{gap}(K, \mathbb{R}^d \backslash \Omega) := \inf\{\|x-y\| : x \in K,\ y \in \mathbb{R}^d \backslash \Omega\}$, hence $x - r_n z \in \Omega$ for $x \in K, z \in B_d$. By Hölder's inequality and the relation $\int_{B_d} \rho = 1$, for $n \geq n_K, x \in K$ one has

$$\begin{aligned}|(R_n w)(x)|^p &\leq \left(\int_{B_d} \rho(z)dz\right)^{p-1} \int_{B_d} \rho(z)\,|w(x-r_n z)|^p\,dz \\ &= \int_{B_d} \rho(z)\,|w(x-r_n z)|^p\,dz,\end{aligned}$$

hence, using the Fubini-Tonelli Theorem,

$$\int_K |(R_n w)(x)|^p\,dx \leq \int_{B_d} (\int_K |w(x-r_n z)|^p\,dx)\rho(z)dz,$$

so that

$$\forall n \geq n_K,\ w \in L_p(\Omega) \qquad \|R_n w\|_{L_p(K)} \leq \|w\|_{L_p(\Omega)}\,. \tag{9.3}$$

Given $\varepsilon > 0$, using the density of $C(\Omega)$ in $L_p(\Omega)$ (Corollary 8.12), we pick $v \in C(\Omega)$ such that $\|u - v\|_{L_p(\Omega)} < \varepsilon$. The estimate (9.3) with $w := v - u$ yields

$$\|R_n v - R_n u\|_{L_p(K)} \leq \|v - u\|_{L_p(\Omega)} < \varepsilon.$$

Observing that for $x \in K$ we have $v(x) = \int_{B_d} \rho(z)v(x)dz$ and

$$|(R_n v)(x) - v(x)| \leq \int_{B_d} \rho(z)\,|v(x-r_n z) - v(x)|\,dz$$

so that $(R_n v - v)_n \to 0$ uniformly on K and $\|R_n v - v\|_{L_p(K)} \leq \varepsilon$ for n large enough. Thus, for n large enough,

$$\|R_n u - u\|_{L_p(K)} \leq \|R_n u - R_n v\|_{L_p(K)} + \|R_n v - v\|_{L_p(K)} + \|v - u\|_{L_p(K)} \leq 3\varepsilon.$$

Now, for $x \in K$ the function $\rho_{n,x} : y \mapsto \rho_n(x-y)$ belongs to $C_c^\infty(\Omega)$ and $(-1)^{|\alpha|} r_n^d D^\alpha \rho_{n,x}(y) = r_n^{-|\alpha|} D^\alpha \rho((x-y)/r_n)$ so that the definition of the weak derivative yields after differentiating under the integral symbol

$$D^\alpha(R_n u)(x) = \frac{r_n^{-|\alpha|}}{r_n^d} \int_\Omega D^\alpha \rho(\frac{x-y}{r_n})u(y)dy = \int_\Omega \rho_{n,x}(y)v_\alpha(y)dy = (R_n v_\alpha)(x).$$

Replacing u with v_α in the previous estimate we get $(\|R_n v_\alpha - v_\alpha\|_{L_p(K)})_n \to 0$ and

$$(\|D^\alpha R_n u - v_\alpha\|_{L_p(K)})_n \to 0.$$

□

9.1.2 Definition and First Properties of Sobolev Spaces

We are ready to describe the important class of Sobolev spaces.

Definition 9.3 Given $m \in \mathbb{N}\backslash\{0\}$ and $p \in [1, \infty]$, the Sobolev space $W_p^m(\Omega)$ is the set of $w \in L_p(\Omega)$ such that for every multi-index α satisfying $|\alpha| \leq m$ the weak derivative $D^\alpha w$ of w is an element of $L_p(\Omega)$. The norm of $W_p^m(\Omega)$ is given by

$$\|w\|_{m,p} := \|w\|_{W_p^m(\Omega)} := \left[\|w\|_p^p + \sum_{|\alpha| \leq m} \|D^\alpha w\|_p^p \right]^{1/p}.$$

For $p = 2$, $W_p^m(\Omega)$ is often denoted by $H^m(\Omega)$.

Theorem 9.2 *For all $m \in \mathbb{N}\backslash\{0\}$ and $p \in [1, \infty]$ the Sobolev space $W_p^m(\Omega)$ is a Banach space and $H^m(\Omega)$ is a Hilbert space.*

Moreover, for $p \in [1, \infty[$ the space $W_p^m(\Omega)$ is separable and for $p \in]1, \infty[$, it is reflexive.

Proof In view of the properties of the $L_p(\Omega)$ space, it suffices to show that the map $J_{m,p} : w \mapsto (w, D^\alpha w)_{|\alpha| \leq m}$ is an isometry of $W_p^m(\Omega)$ onto a closed subspace of $L_p(\Omega)^{m(d)}$, where $m(d) := \operatorname{card} \mathbb{N}(m, d)$, with

$$\mathbb{N}(m, d) := \{\alpha \in (\{0\} \cup \mathbb{N}_m)^d : 0 \leq |\alpha| \leq m\}.$$

The fact that $J_{m,p}$ is isometric is obvious. Let (w_n) be a sequence in $W_p^m(\Omega)$ such that $(J_{m,p}(w_n))_n$ converges to some $(w, u_\alpha) \in L_p(\Omega)^{m(d)}$. Then the sequences (w_n) and $(D^\alpha w_n)$ are Cauchy sequences in $L_p(\Omega)$ and

$$\forall \varphi \in C_c^\infty(\Omega),\ \forall \alpha \in \mathbb{N}(m, d) \qquad \int_\Omega w_n D^\alpha \varphi = (-1)^{|\alpha|} \int_\Omega D^\alpha w_n \varphi,$$

so that, passing to the limit in this relation, we get

$$\forall \varphi \in C_c^\infty(\Omega),\ \forall \alpha \in \mathbb{N}(m, d) \qquad \int_\Omega w D^\alpha \varphi = (-1)^{|\alpha|} \int_\Omega u_\alpha \varphi.$$

This shows that u_α is the weak α-derivative of w, hence that $w \in W_p^m(\Omega)$. Finally, the norm of $H^m(\Omega)$ is clearly associated with the scalar product induced by the one in $L_2(\Omega)^{m(d)}$. □

Friedrich's Theorem can be refined. Hereafter, for $k \in \mathbb{N} \cup \{\infty\}$, we adopt the usual (but somewhat queer) notation $C^k(\overline{\Omega})$ for the space of restrictions to Ω of functions in $C^k(\mathbb{R}^d)$. Note that for $k = 0$ the space $C^k(\overline{\Omega})$ is just the space $C(\overline{\Omega})$ of continuous functions on $\overline{\Omega}$ and that when Ω is smooth, $C^k(\overline{\Omega})$ can be given an intrinsic definition.

Theorem 9.3 (Meyers-Serrin) *For $m \in \mathbb{N}\backslash\{0\}$ and $p \in [1, \infty[$, the space $C^\infty(\overline{\Omega}) \cap W_p^m(\Omega)$ is dense in $W_p^m(\Omega)$.*

Proof Let $(\Omega_n)_n$ be a sequence of open subsets of Ω covering Ω such that for all n the set $K_n := \mathrm{cl}(\Omega_n)$ is compact and contained in Ω_{n+1}. For instance, one can take $\Omega_n := \{x \in \Omega : d(x, \mathbb{R}^d\backslash\Omega) > 2^{-n}\} \cap B(0,n)$. Let $K_n' := K_n\backslash\Omega_{n-1}$ and $V_n := \Omega_{n+1}\backslash K_{n-1}$, so that $K_n' \subset V_n$. Since K_n' is compact and V_n is open, there exists a function q_n of class C^∞ whose support is contained in V_n satisfying $q_n \mid_{K_n} = 1$. Since $(V_n)_n$ is a covering of Ω and since $V_{n+1} \cap V_{n-1} = \varnothing$, the families $(V_n)_n$ and $(\mathrm{supp}(q_n))_n$ are locally finite. Thus $q := \Sigma_n q_n$ is of class C^∞ and $q \geq 1$ on Ω. Let $p_n := q_n/q$, so that $(p_n)_n$ is a partition of unity subordinated to $(V_n)_n$.

Given $u \in W_p^m(\Omega)$ and $\varepsilon > 0$ we have $\mathrm{supp}(p_n u) \subset V_n$ so that there exists some $r_n > 0$ such that $\mathrm{supp}(R_{r_n}(p_n u)) \subset V_n$ and $\|R_{r_n}(p_n u) - p_n u\|_p < \varepsilon/2^n$, $\|R_{r_n}(D_i(p_n u)) - D_i(p_n u)\|_p < \varepsilon/2^n$ for $i \in \mathbb{N}_d$, $n \in \mathbb{N}\backslash\{0\}$. Let us set

$$u_\varepsilon := \sum_{n\geq 1} R_{r_n}(p_n u).$$

Since the family $(\mathrm{supp}(R_{r_n}(p_n u)))_n$ is locally finite, $u_\varepsilon \in C^\infty(\mathbb{R}^d)$ and since $u = \Sigma_{n\geq 1} p_n u$ we have

$$\|u_\varepsilon \mid_\Omega - u\|_p \leq \sum_{n\geq 1} \|R_{r_n}(p_n u) - p_n u\|_p \leq \varepsilon,$$

$$\|D_i u_\varepsilon \mid_\Omega - D_i u\|_p \leq \sum_{n\geq 1} \|R_{r_n}(D_i(p_n u)) - D_i(p_n u)\|_p \leq \varepsilon$$

since $D_i(R_{r_n}(p_n u)) = R_{r_n}(D_i(p_n u))$ by Lemma 9.1 and the end of the proof of Theorem 9.1. Thus $u_\varepsilon \in W_p^m(\Omega)$ and $(u_\varepsilon) \to u$ in $W_p^m(\Omega)$ as $\varepsilon \to 0_+$. □

The following characterizations will be useful for the study of regularity properties of solutions to elliptic partial differential equations.

Proposition 9.4 *For $p \in]1, \infty]$, $q := (1 - 1/p)^{-1}$ (with $q = 1$ when $p = \infty$) and $u \in L_p(\Omega)$ the following assertions are equivalent:*

(a) $u \in W_p^1(\Omega)$;
(b) there exists some constant $c \in \mathbb{R}_+$ such that for all $i \in \mathbb{N}_d$

$$\left| \int_\Omega u D_i \varphi \right| \leq c \left\| \varphi \right\|_{L_q(\Omega)} \qquad \forall \varphi \in C_c^\infty(\Omega);$$

(c) there exists some constant $c \in \mathbb{R}_+$ such that for all $K \in \mathcal{K}(\Omega)$ and all $w \in \delta B_{\mathbb{R}^d}$ satisfying $K + \delta B_{\mathbb{R}^d} \subset \Omega$, one has for $t_w(x) := x - w$

$$\left\| u \circ t_w - u \right\|_{L_p(K)} \leq c \left\| w \right\|.$$

Moreover, one can take $c = \|u\|_{W_p^1(\Omega)}$ in (b) and (c) when (a) holds.

Proof (a)⇒(b) Since $\int_\Omega u D_i \varphi = -\int_\Omega \varphi D_i u$ this follows from Hölder's inequality with $c = \max_{i \in \mathbb{N}_d} \|D_i u\|_p$. This implication is also a consequence of the other ones below.

(a)⇒(c) We first observe that by Theorem 9.1 and a passage to the limit it suffices to prove (c) with $c := \|u\|_{W_p^1(\Omega)}$ when $u \in C_c^\infty(\Omega)$. Then, given $K \in \mathcal{K}(\Omega)$ and $w \in \mathbb{R}^d$ satisfying $t_w(K) \subset \Omega$, for $x \in K$ we have

$$u(x - w) - u(x) = -\int_0^1 \nabla u(x - tw).w dt,$$

hence, for $p \in]1, \infty[$

$$|u(x - w) - u(x)|^p \leq \|w\|^p \int_0^1 \|\nabla u(x - tw)\|^p \, dt,$$

$$\int_K |u(x - w) - u(x)|^p \, dx \leq \|w\|^p \int_K \int_0^1 \|\nabla u(x - tw)\|^p \, dt dx$$
$$\leq \|w\|^p \int_\Omega \|\nabla u(x)\|^p \, dx.$$

Passing to the limit as $p \to \infty$ in the inequality $\|u \circ t_w - u\|_{L_p(K)} \leq \|\nabla u\|_{L_p(\Omega)} \|w\|$ we just obtained, we extend this inequality to the case $p = \infty$.

(c)⇒(b) Given $i \in \mathbb{N}_d$, for $t > 0$ small enough, a change of variables shows that

$$\int_\Omega u(x) \frac{1}{t} (\varphi(x + te_i) - \varphi(x)) dx = -\int_\Omega \frac{1}{t} (u(x) - u(x - te_i)) \varphi(x) dx$$

for all $\varphi \in C_c^\infty(\Omega)$. By Hölder's inequality and (c), the absolute value of the right-hand side is bounded above by $c \|\varphi\|_{L_q(\Omega)}$. Since the left-hand side converges to $\int_\Omega u D_i \varphi$ as $t \to 0_+$, we get $\left|\int_\Omega u D_i \varphi\right| \leq c \|\varphi\|_{L_q(\Omega)}$.

(b)$\Rightarrow$(a) Assertion (b) shows that the linear map $\varphi \mapsto \int_\Omega u D_i \varphi$ is continuous on $C_c^\infty(\Omega)$ endowed with the L_q norm. Since $C_c^\infty(\Omega)$ is dense in $L_q(\Omega)$, this map can be extended to $L_q(\Omega)$ into a continuous linear map ℓ. Then, the Riesz representation theorem yields some $v_i \in L_p(\Omega)$ such that $\ell(\varphi) = \int_\Omega v_i \varphi$ for all $\varphi \in L_q(\Omega)$, in particular for all $\varphi \in C_c^\infty(\Omega)$. This shows that $-v_i$ is the i-partial weak derivative of u, so that $u \in W_p^1(\Omega)$. □

Remark The preceding proof shows that if $w \in \mathbb{R}^d$ is such that $t_w(\Omega) \subset \Omega$, then

$$\|u \circ t_w - u\|_{L_p(\Omega)} \leq \|w\| . \|\nabla u\|_{L_p(\Omega)} \qquad \forall u \in W_p^1(\Omega).$$

The next result presents another characterization in the case $p := \infty$. We admit it (see [118] for instance).

Proposition 9.5 (Rademacher) *For a bounded open subset Ω of class C^1 of $\mathbb{R}^d$, the space $W_\infty^1(\Omega)$ is the space of Lipschitzian functions on Ω. Moreover, every element $u \in W_\infty^1(\Omega)$ is differentiable a.e. and its partial derivatives are its weak partial derivatives.*

A characterization of $H^k(\mathbb{R}^d)$ using the Fourier transform can be given.

Proposition 9.6 *Given $k \in \mathbb{N}\backslash\{0\}$, $u \in L_2(\mathbb{R}^d)$ one has $u \in H^k(\mathbb{R}^d)$ if and only if*

$$(1 + \|\cdot\|^k)\hat{u}(\cdot) \in L_2(\mathbb{R}^d),$$

where $\hat{u}$ is the Fourier transform of u. Moreover, there exists some $c > 0$ such that

$$\frac{1}{c} \|u\|_{H^k(\mathbb{R}^d)} \leq \left\|(1 + \|\cdot\|^k)\hat{u}(\cdot)\right\|_{L_2(\mathbb{R}^d)} \leq c \|u\|_{H^k(\mathbb{R}^d)}.$$

Proof Let $u \in H^k(\mathbb{R}^d)$. For any multi-index α satisfying $|\alpha| \leq k$, approximating u with a sequence in $C_c^k(\mathbb{R}^d)$ we get

$$y \mapsto \widehat{D^\alpha u}(y) = (2\pi i y)^\alpha \hat{u}(y)$$

is in $L_2(\mathbb{R}^d)$ by Plancherel's theorem, hence, taking $\alpha := (0\ldots,0,k,0,\ldots,0)$ with k at the j-th place, $k \in \mathbb{N}_d$, we obtain

$$\int_{\mathbb{R}^d} (1 + \|y\|^k)^2 |\hat{u}(y)|^2 dy \leq c^2 \int_{\mathbb{R}^d} \sum_{j=1}^d \left|D_j^k u(x)\right|^2 dx \leq c^2 \|u\|_{H^k(\mathbb{R}^d)}^2$$

for some $c > 0$ and $(1 + \|\cdot\|^k)\hat{u}(\cdot) \in L_2(\mathbb{R}^d)$.

Conversely, let $u \in L_2(\mathbb{R}^d)$ be such that $(1 + \|\cdot\|^k)\hat{u}(\cdot) \in L_2(\mathbb{R}^d)$. Denoting by $v := \check{u}$ the inverse image of u by the Fourier transform, so that $u = \hat{v}$, setting $m_\alpha(y) := (-2\pi i)^{|\alpha|} y^\alpha$, $u_\alpha := \widehat{m_\alpha v}$, the relation $\widehat{m_\alpha v} = D^\alpha \hat{v}$ obtained in Proposition 8.21 yields $u_\alpha = D^\alpha u$. Then, for all $\varphi \in C_c^\infty(\mathbb{R}^d)$, by Proposition 8.22 we have

$$\int_{\mathbb{R}^d} D^\alpha \varphi u = \int_{\mathbb{R}^d} D^\alpha \varphi \hat{v} = \int_{\mathbb{R}^d} \widehat{D^\alpha \varphi} v = \int_{\mathbb{R}^d} \overline{m_\alpha} \hat{\varphi} v$$

$$= \int_{\mathbb{R}^d} \hat{\varphi} (-1)^{|\alpha|} m_\alpha v = (-1)^{|\alpha|} \int_{\mathbb{R}^d} \varphi \widehat{m_\alpha v}.$$

This means that $u_\alpha := \widehat{m_\alpha v}$ is the weak α-partial derivative of u. Since

$$\left\|\widehat{m_\alpha v}\right\|_2^2 = \|m_\alpha v\|_2^2 = (2\pi)^{2|\alpha|} \int_{\mathbb{R}^d} |y|^{2|\alpha|} |v(y)|^2 \, dy \leq (2\pi)^{2|\alpha|} \left\|(1 + \|\cdot\|^k)\hat{u}(\cdot)\right\|_2$$

we have $u_\alpha := \widehat{m_\alpha v} \in L_2(\mathbb{R}^d)$ and $u \in H^k(\mathbb{R}^d)$. □

The preceding characterization shows that the definition that follows for fractional Sobolev spaces is compatible with the one we gave for entire values of s.

Definition 9.4 Given $s \in \mathbb{P} :=]0, \infty[$ the space $H^s(\mathbb{R}^d)$ is the set of $u \in L_2(\mathbb{R}^d)$ such that $(1 + \|\cdot\|^s)\hat{u} \in L_2(\mathbb{R}^d)$.

Properties of such spaces, which can be used for trace on boundaries results, can be found in more advanced books.

9.1.3 Calculus Rules in Sobolev Spaces

Up to now we have used some product rules when one of the factors is in $C^1(\Omega)$. More general rules should be given.

Proposition 9.7 *Given $p \in [1, \infty]$ and $u, v \in W_p^1(\Omega) \cap L_\infty(\Omega)$, one has $uv \in W_p^1(\Omega) \cap L_\infty(\Omega)$ and $D_i(uv) = uD_i v + vD_i u$ for $i \in \mathbb{N}_d$.*

Proof Let us first suppose $p \in [1, \infty[$. The Meyers-Serrin Theorem yields sequences $(u_n) \to u$, $(v_n) \to v$ in $W_p^1(\Omega)$ such that $u_n, v_n \in C^\infty(\overline{\Omega}) \cap W_p^1(\Omega)$ for all $n \in \mathbb{N}$. In fact, for some sequence $(r_n) \to 0_+$, one can take $u_n = R_{r_n} u$, $v_n = R_{r_n} v$ and then $D_i u_n = R_{r_n} D_i u$, $D_i v_n = R_{r_n} D_i v$ for all $n \in \mathbb{N}$ and $i \in \mathbb{N}_d$. The family of all these functions is bounded in $L_\infty(\Omega)$ since $\|R_r w\|_\infty \leq \|w\|_\infty$ for all $r > 0$ and $w \in L_\infty(\Omega)$. Thus, for all $\varphi \in C_c^\infty(\Omega)$, using the estimates

$$\|u_n v_n D_i \varphi - uv D_i \varphi\|_1 \leq \|u_n - u\|_p \|v_n\|_\infty \|D_i \varphi\|_q + \|u\|_\infty \|v_n - v\|_p \|D_i \varphi\|_q$$

with $q := p/(p-1)$ and similar ones, we can pass to the limit in the relations

$$\forall \varphi \in C_c^\infty(\Omega),\ i \in \mathbb{N}_d \qquad \int_\Omega u_n v_n D_i\varphi = -\int_\Omega (u_n D_i v_n + v_n D_i u_n)\varphi$$

and get similar relations with u, v replacing u_n, v_n proving the assertion for $p \in [1, \infty[$, taking $q = \infty$ when $p = 1$.

Now let us suppose $p = \infty$. Then, given $\varphi \in C_c^\infty(\Omega)$, we pick some bounded open subset Ω' of Ω satisfying $\operatorname{supp}(\varphi) \subset \Omega'$ so that for all $p \geq 1$ we have u, $v \in W_p^1(\Omega') \cap L_\infty(\Omega')$, the measure of Ω' being finite, and

$$\forall i \in \mathbb{N}_d \qquad \int_{\Omega'} uvD_i\varphi = -\int_{\Omega'} (uD_i v + vD_i u)\varphi$$

by what precedes. In this equality we can replace Ω' with Ω and we get the result.

□

Let us give a composition result.

Proposition 9.8 *Let $g : \mathbb{R} \to \mathbb{R}$ be a Lipschitz function and let $p \in [1, \infty]$. Then, for all $u \in W_p^1(\Omega)$ such that $g \circ u \in L_p(\Omega)$ (this occurs when $p = \infty$ or when $g(0) = 0$) one has $f := g \circ u \in W_p^1(\Omega)$ and $D_i f = (g' \circ u)D_i u$ for $i \in \mathbb{N}_d$.*

Proof Such a result relies on Rademacher's Theorem (Proposition 9.5). We prove it in the case when g is of class C^1 with a bounded derivative and $g(0) = 0$. Let $c := \|g'\|_\infty$. Then for all $r \in \mathbb{R}$ we have $|g(r)| \leq c\,|r|$ by the Mean Value Theorem, hence $|g(u(x))| \leq c\,|u(x)|$ for $x \in \Omega$ and $g \circ u \in L_p(\Omega)$. Also $(g' \circ u)D_i u \in L_p(\Omega)$ since $|g'(u(x))D_i u(x)| \leq c\,|D_i u(x)|$. When $p \in [1, \infty[$ we pick a sequence (u_n) in $C^\infty(\Omega) \cap W_p^m(\Omega)$ such that $(\|u_n - u\|_{m,p}) \to 0$. The inequality $|g(u_n(x)) - g(u(x))| \leq c\,|u_n(x) - u(x)|$ ensures that $(\|g \circ u_n - g \circ u\|_p)_n \to 0$ whereas the estimate

$$|g' \circ u_n D_i u_n - g' \circ u D_i u| \leq |g' \circ u_n| . |D_i u_n - D_i u| + |g' \circ u_n - g' \circ u| . |D_i u|$$

implies that $(\|g' \circ u_n D_i u_n - g' \circ u D_i u\|_p)_n \to 0$ by dominated convergence. Thus, for all $\varphi \in C_c^\infty(\Omega)$, using Hölder's inequality, one can pass to the limit in the relation

$$\int_\Omega g \circ u_n D_i\varphi = -\int_\Omega g' \circ u_n D_i u_n \varphi$$

and obtain that $D_i f = (g' \circ u)D_i u$ in the weak sense.

For $p = \infty$, given $\varphi \in C_c^\infty(\Omega)$ one takes a bounded open subset $\Omega' \subset \Omega$ such that $\operatorname{supp}(\varphi) \subset \Omega'$ so that, for all $p \geq 1$, one has $u \in W_p^1(\Omega')$, $(g' \circ u)D_i u \in L_p(\Omega')$ and

$$\int_{\Omega'} g \circ u D_i\varphi = -\int_{\Omega'} g' \circ u D_i u \varphi$$

by the preceding case. In this relation one can replace Ω' with Ω and get $D_i f = (g' \circ u) D_i u$, φ being arbitrary. □

Corollary 9.1 *Given $p \in [1, \infty]$, for all $u \in W_p^1(\Omega)$ one has $u^+ := \max(u, 0) \in W_p^1(\Omega)$ and $D_i u^+ = D_i u$ a.e. in $\Omega^+ := \{x : u(x) > 0\}$, $D_i u^+ = 0$ a.e. in $\Omega^- := \Omega \backslash \Omega^+$. Similar assertions are valid for $u^- := (-u)^+$ and $|u|$.*

Proof Apply the proposition with $g(r) := r^+$. One can also use the C^1 version, taking $g_\varepsilon(r) := (r^2 + \varepsilon^2)^{1/2} - \varepsilon$ for $r \in \mathbb{R}_+$, $g_\varepsilon(r) = 0$ for $r \in \mathbb{R}_-$ and passing to the limit as $\varepsilon \to 0_+$. For $|u|$ one uses the fact that $|u| = u^+ + u^-$. □

We deduce from the preceding corollary that $W_p^1(\Omega)$ has a lattice structure.

Corollary 9.2 *Given $p \in [1, \infty]$, for all u, $v \in W_p^1(\Omega)$ one has $u \vee v := \max(u, v) \in W_p^1(\Omega)$, $u \wedge v := \min(u, v) \in W_p^1(\Omega)$.*

If Ω is bounded, then for all $c \in \mathbb{R}$ one has $u \wedge c \in W_p^1(\Omega)$, $u \vee c \in W_p^1(\Omega)$.

Proof The first assertion stems from the relation $u \vee v = (1/2)(u + v + |u - v|)$. The second one follows from $u \wedge v = -(-u) \vee (-v)$. If Ω is bounded, then any constant c belongs to $W_p^1(\Omega)$. □

Now let us give a change of variable result.

Theorem 9.4 *Let $h : \Omega' \to \Omega$ be a bijection between two open subsets of $\mathbb{R}^d$, h and h^{-1} being Lipschitzian. Then for all $u \in W_p^1(\Omega)$ one has $v := u \circ h \in W_p^1(\Omega')$ and $D_i v = \Sigma_{j=1}^d (D_j u \circ h) D_i h^j$ for $h := (h^1, \ldots, h^d)$.*

Proof We give the proof in the case h and h^{-1} are Lipschitzian and of class C^1. When $p \in [1, \infty[$ one picks a sequence $(u_n)_n$ in $C^\infty(\Omega) \cap W_p^1(\Omega)$ such that $(\|u_n - u\|_{1,p})_n \to 0$. Then, by the change of variable theorem (Theorem 7.13) one has $(u_n \circ h) \to u \circ h$ and $((D_j u_n \circ h) D_i h^j)_n \to (D_j u \circ h) D_i h^j$ in $L_p(\Omega')$. For all $\varphi \in C_c^\infty(\Omega')$ we have

$$\int_{\Omega'} (u_n \circ h) D_i \varphi = -\int_{\Omega'} \sum_{j=1}^d (D_j u_n \circ h) D_i h^j \varphi. \tag{9.4}$$

Passing to the limit, one gets a similar relation with u_n replaced by u. It shows that the weak derivative $D_i v$ of $v := u \circ h$ is $\Sigma_{j=1}^d (D_j u \circ h) D_i h^j$. Thus $v \in W_p^1(\Omega')$.

When $p = \infty$, given $\varphi \in C_c^\infty(\Omega)$ one takes a bounded open subset $\Omega'' \subset \Omega'$ such that $\text{supp}(\varphi) \subset \Omega''$ and using the fact that $u_n \mid_{h(\Omega'')} \in C^\infty(h(\Omega'')) \cap W_p^1(h(\Omega''))$ one has a relation similar to (9.4) with Ω' replaced by Ω'' and the result follows as above. □

9.1.4 Extension

The study of the Sobolev spaces $W_p^m(\Omega)$ is easier in the case $\Omega = \mathbb{R}^d$ than in the case of a general open subset Ω. For example, we have seen that for $\Omega = \mathbb{R}^d$ and $p = 2$ we can use the Fourier transform. Thus, it may be useful to derive properties of $W_p^m(\Omega)$ by extending its elements to functions in $W_p^m(\mathbb{R}^d)$. This is possible provided Ω is smooth enough.

We start with a simple situation that captures the essence of the process, especially in the case $m = 1$. We recommend a rewriting of the proof for this special case. Then the extended function $E(u)$ of $u \in W_p^1(\mathbb{R}^{d-1} \times \mathbb{P})$, where $\mathbb{P} :=]0, \infty[$, is obtained by reflection: $(E(u))(s,t) = -u(s,-t)$ for $t < 0$.

Proposition 9.9 *Given $m \in \mathbb{N}\backslash\{0\}$, $p \in [1, \infty]$ and r, $b > 0$, let $U := B_{\mathbb{R}^{d-1}}(0, r)\times] - b, b[$, $U^+ := U \cap (\mathbb{R}^{d-1} \times \mathbb{P})$. Then there is a continuous linear map $E := E_{m,p} : W_p^m(U^+) \to W_p^m(U)$ such that for all $u \in W_p^m(U^+)$ one has $E(u) \mid_{U^+} = u$. In other terms, $E(u)$ is an extension of u.*

Proof Since $C^\infty(\overline{U^+}) \cap W_p^m(U^+)$ is dense in $W_p^m(U^+)$, it suffices to define $E(u)$ for $u \in C^\infty(\overline{U^+}) \cap W_p^m(U^+)$. Let $(c_j)_{j\in\mathbb{N}_m}$ be the solution of the linear system

$$c_1(-1)^k + c_2(-2)^k + \ldots + c_m(-m)^k = m^k \qquad k \in \mathbb{N}_m$$

whose determinant is non-zero, as it is a Vandermonde determinant. Then, for $u \in C^\infty(\overline{U^+}) \cap W_p^m(U^+)$, we set $E(u)(s,t) = u(s,t)$ for $(s,t) \in U^+$,

$$E(u)(s,t) = \sum_{j=1}^{m} c_j \tilde{u}(s, -jt/m) \text{ for } (s,t) \in U\backslash U^+,$$

where $\tilde{u} \in C^\infty(\mathbb{R}^d)$ is an extension of u. Then, for $k \in \mathbb{N}_m$ we have

$$D_d^k E(u)(s,0) = \frac{1}{m^k} \sum_{j=1}^{m} c_j(-j)^k D_d^k \tilde{u}(s,0) = D_d^k \tilde{u}(s,0) = \lim_{t\to 0_+} D_d^k u(s,t)$$

and since for $\alpha = (\beta, k) \in \mathbb{N}^d$ with $|\alpha| = |\beta| + k \leq m$ we have $D^\alpha E(u) = D_d^k E(D^\beta u)$, we see that $E(u) \in C^m(U)$ and that for some $c > 0$ we have $\|E(u)\|_{m,p} \leq c\,\|u\|_{m,p}$ for all $u \in C^\infty(\overline{U^+}) \cap W_p^m(U^+)$, hence for all $u \in W_p^m(U^+)$. This assertion can be checked directly or by using Theorem 9.4 with the map $(s,t) \mapsto (s, -jt/m)$. □

Theorem 9.5 *Let Ω be an open subset of class C^m of $\mathbb{R}^d$, i.e. the interior of a submanifold with boundary of dimension d and of class C^m of $\mathbb{R}^d$. Assume the boundary Γ of Ω is bounded. Then there exists a continuous linear map $E : W_p^m(\Omega) \to W_p^m(\mathbb{R}^d)$ such that $E(u) \mid_\Omega = u$ for all $u \in W_p^m(\Omega)$.*

Proof We first suppose that for some $r, b > 0$ and some function $g : V \to \mathbb{R}$ of class C^m whose derivatives of order not greater than m are bounded, with $V := B_{\mathbb{R}^{d-1}}(0, r)$, the set Ω is given by

$$\Omega := \{s, t) \in V \times \mathbb{R} : g(s) < t < g(s) + b\}. \tag{9.5}$$

Then, the map $h : (s, t) \mapsto (s, t + g(s))$ is a bijection from $U' := V \times T$, with $T :=]-b, b[$, onto an open subset U that is of class C^m with bounded derivatives of order not greater than m, as is its inverse $(s, t') \mapsto (s, t' - g(s))$ and such that $\Omega = U \cap h(V \times T^+)$ with $T^+ :=]0, b[$. Setting $E(u) := \tilde{v} \circ h^{-1}$ for $v := u \circ h$, $\tilde{v} := E_{U'}(v)$, $E_{U'}$ being the extension operator defined in the preceding proposition, we get an extension $\tilde{u} \in W_p^m(U)$ of u.

In the general case we take an open covering $(U_i)_{i \in \mathbb{N}_k}$ of Γ by open subsets that are the images by some orthogonal maps ℓ_i of some sets

$$U_i' := \{(s, g_i(s) + t) : s \in V_i := B(a_i, r_i),\ t \in]-b_i, b_i[\}$$

with $g_i : V_i \to \mathbb{R}$ of class C^m with bounded derivatives of order not greater than m such that $\Omega \cap U_i = \ell_i(U_i'^+)$ with

$$U_i'^+ := \{(s, g_i(s) + t) : s \in V_i,\ t \in]0, b_i[\}$$

as in (9.5). We take a C^∞ partition of unity $(p_i)_{i \in \mathbb{N}_k}$ subordinated to this covering. We complete it with $p_0 := 1 - \Sigma_{i=1}^k p_i$ and we set

$$E(u) = p_0 u + \sum_{i=1}^{k} p_i E_i(u \mid_{\Omega \cap U_i})$$

where E_i is the extension operator $W_p^m(\Omega \cap U_i) \to W_p^m(U_i)$ defined in the first step. Here $p_0 u$ stands for $p_0 u$ on Ω and 0 on $\mathbb{R}^d \backslash \Omega$, so that $p_0 u \in L_p(\mathbb{R}^d)$ and $D_i(p_0 u) = u D_i p_0 + p_0 D_i u$ in the weak sense, and $p_0 u \in W_p^m(\mathbb{R}^d)$ with $\|p_0 u\|_{m,p} \le c \|u\|_{m,p}$ for some constant $c > 0$. Since $\|u \mid_{\Omega \cap U_i}\|_{m,p} \le \|u\|_{m,p}$ and since for some constant c_i we have $\|E_i(w_i)\|_{m,p} \le c_i \|w_i\|_{m,p}$ for all $w \in W_p^m(\Omega \cap U_i)$, the rule for products completes the proof since $p_i \in C_c^\infty(\mathbb{R}^d)$ and obviously $E(u) \mid_\Omega = u$. □

9.1.5 *Traces*

Again, let Γ be the boundary $\partial\Omega$ of Ω. In order to define the trace on Γ of $u \in W_p^m(\Omega)$ one has to assume some regularity of Ω. For the sake of simplicity, we assume that Ω is of class C^1. For a Lipschitz open set one has to use Hausdorff measures and we wish to leave aside this refinement. Some care is required since Γ

has Lebesgue measure 0 whereas u is defined up to a set of measure 0! Again, we start with a simple case for which

$$\Omega := \{(s,t) \in V \times \mathbb{R} : g(s) < t < g(s) + b\}, \tag{9.6}$$

$$U := \{(s,t) \in V \times \mathbb{R} : g(s) - b < t < g(s) + b\}, \tag{9.7}$$

where $V := B_{\mathbb{R}^{d-1}}(0, r)$, $g : V \to \mathbb{R}$ is of class C^1. We endow $\Gamma_g := \{(s, g(s)) : s \in V\}$ with the measure σ_g transported by the diffeomorphism $s \mapsto (s, g(s))$ from the measure with density $(1 + \|\nabla g\|^2)^{1/2}$ with respect to the measure induced by the Lebesgue measure λ_{d-1} on V.

Lemma 9.2 *For $p \in [1, \infty[$ there exists a $c > 0$ such that for all $u \in C^1_c(U) \cap W^1_p(\Omega)$ one has*

$$\left\| u \mid_{\Gamma_g} \right\|_{L_p(\Gamma_g)} \leq c \left\| D_d u \right\|^{1/p}_{L_p(\Omega)} \left\| u \right\|^{1-1/p}_{L_p(\Omega)}, \tag{9.8}$$

$$\left\| u \mid_{\Gamma_g} \right\|_{L_p(\Gamma_g)} \leq c \left\| u \right\|_{W^1_p(\Omega)}. \tag{9.9}$$

Proof Given $\varphi \in C^1_c(]-r, r[)$ we note that

$$|\varphi^p(0)| = \left| -\int_0^r (\varphi^p)'(t)dt \right| \leq p \int_0^r |\varphi(t)|^{p-1} . \left|\varphi'(t)\right| dt.$$

For $u \in C^1_c(U) \cap W^1_p(\Omega)$, applying this relation to $\varphi : t \mapsto u(s, g(s) + t)$, integrating and using Hölder's inequality, since $c := p \sup_{s \in V}(1 + \|\nabla g(s)\|^2)^{1/2} < \infty$, g being Lipschitzian, we get

$$\int_V |u(s, g(s))|^p \, (1 + \|\nabla g(s)\|^2)^{1/2} ds \leq c \int_V \int_0^r (|u|^{p-1} . |D_d u|)(s, g(s) + t) dt ds$$

$$(\int_{\Gamma_g} |u|^p \, d\sigma_g)^{1/p} \leq c \left\| D_d u \right\|^{1/p}_{L_p(\Omega)} \left\| u \right\|^{1-1/p}_{L_p(\Omega)}.$$

The inequality $\left\| u \mid_{\Gamma_g} \right\|_{L_p(\Gamma_g)} \leq c \left\| u \right\|_{W^1_p(\Omega)}$ follows from the relations

$$\left\| D_d u \right\|_{L_p(\Omega)} \leq \left\| u \right\|_{m,p}, \qquad \left\| u \right\|_{L_p(\Omega)} \leq \left\| u \right\|_{m,p}.$$

Given $u \in C_c(U) \cap W^1_p(\Omega)$, we can use a mollifier to get a sequence $(u_n) := (R_n u) \to u$ pointwise and in $W^1_p(\Omega)$. Then $(u_n \mid_{\Gamma_g})_n$ is a Cauchy sequence in $L_p(\Gamma)$ pointwise converging to $u \mid_{\Gamma_g}$. Taking limits in the inequality $\left\| u_n \mid_{\Gamma_g} \right\|_{L_p(\Gamma_g)} \leq c \left\| u_n \right\|_{W^1_p(\Omega)}$ we obtain (9.9). □

Now let us suppose Ω is an open subset of class C^1 with a bounded boundary $\Gamma := \partial\Omega$. As in the preceding subsection, we take an open covering $(U_i)_{i \in \mathbb{N}_k}$ of Γ

by open subsets that are the images under some orthogonal maps ℓ_i of some sets

$$U_i' := \{(s, g_i(s) + t) : s \in V_i := B(a_i, r_i),\ t \in]-b_i, b_i[\}$$

with $g_i : V_i \to \mathbb{R}$ Lipschitz such that $\Omega \cap U_i = \ell_i(U_i'^{+})$ with

$$U_i'^{+} := \{(s, g_i(s) + t) : s \in V_i,\ t \in]0, b_i[\}.$$

We admit that there is a Borel measure σ on Γ inducing on all $U_i \cap \Gamma$ the measure $\ell_i(\sigma_{g_i})$ transported by ℓ_i of the measure σ_{g_i} on the graph of g_i, the measure σ_{g_i} being itself transported by the diffeomorphism $s \mapsto (s, g_i(s))$ from the measure with density $(1 + \|\nabla g_i\|^2)^{1/2}$ with respect to $\lambda_{d-1} \mid_{V_i}$. Let $(p_i)_{i \in \mathbb{N}_k}$ be a partition of unity of class C^∞ subordinated to the covering $(U_i)_{i \in \mathbb{N}_k}$. A measurable function $f : \Gamma \to \mathbb{R}$ is integrable for σ if for all $i \in \mathbb{N}_k$ the function $p_i f$ is integrable with respect to $\ell_i(\sigma_{g_i})$ and then

$$\begin{aligned}\int_\Gamma f d\sigma &= \sum_{i=1}^{k} \int_{U_i \cap \Gamma} p_i f d\ell_i(\sigma_{g_i}) \\ &= \sum_{i=1}^{k} \int_{V_i} ((p_i f) \circ \ell_i)(s, g_i(s))(1 + \|\nabla g_i(s)\|^2)^{1/2} d\lambda_{d-1}(s).\end{aligned}$$

It can be shown (see [202] for example) that this definition does not depend on the choice of the covering $(U_i)_{i \in \mathbb{N}_k}$ or on $(p_i)_{i \in \mathbb{N}_k}$. For $p \in [1, \infty[$, the space $L_p(\Gamma)$ is the space of measurable (equivalence classes of) functions f such that $|f|^p$ is integrable and $\|f\|_p := (\int_\Gamma |f|^p \, d\sigma)^{1/p}$.

The (unit) *normal vector* $n(x)$ to Γ at $x \in U_i \cap \Gamma$ is the vector $\ell_i(n_i(\ell_i^{-1}(x)))$, with

$$n_i(s, g_i(s)) := \frac{(\nabla g_i(s), -1)}{(\|\nabla g_i(s)\|^2 + 1)^{1/2}}.$$

The operator T of the next statement is called the *trace operator*.

Theorem 9.6 *Let Ω be an open subset of class C^1 of $\mathbb{R}^d$ whose boundary Γ is bounded and let $p \in [1, \infty[$. Then there exists a unique continuous linear map $T := T_\Gamma : W_p^1(\Omega) \to L_p(\Gamma)$ such that $T(u) = u \mid_\Gamma$ for all $u \in C^1(\overline{\Omega}) \cap W_p^1(\Omega)$.*

Proof Since $C^1(\overline{\Omega}) \cap W_p^1(\Omega)$ is dense in $W_p^1(\Omega)$ by the Meyers-Serrin Theorem, it suffices to show that there exists some $c > 0$ such that for all $u \in C^1(\overline{\Omega}) \cap W_p^1(\Omega)$ one has $\|u \mid_\Gamma\|_{L_p(\Gamma)} \le c \, \|u\|_{W_p^1(\Omega)}$ or even

$$\|u \mid_\Gamma\|_{L_p(\Gamma)} \le c \, \|\nabla u\|_{L_p(\Omega)}^{1/p} \, \|u\|_{L_p(\Omega)}^{1-1/p}.$$

We pick a covering $(U_i)_{i\in\mathbb{N}_k}$ of Γ by open subsets as above and we take a C^∞ partition of unity $(p_i)_{i\in\mathbb{N}_k}$ subordinated to this covering. We complete it with $p_0 := 1 - \Sigma_{i=1}^k p_i$. For $i \in \mathbb{N}_k$, the preceding lemma yields some $c_i > 0$ such that

$$\|p_i u \mid_\Gamma \|_{L_p(\Gamma)} \le c_i \, \|\nabla(p_i u)\|_{L_p(\Omega)}^{1/p} \, \|p_i u\|_{L_p(\Omega)}^{1-1/p}$$

since $p_i u \in C_c^1(U_i) \cap W_p^1(\Omega)$. Now, $\|p_i u\|_{L_p(\Omega)} \le \|u\|_{L_p(\Omega)}$ and for some constant c_i' (depending on p_i)

$$\|\nabla(p_i u)\|_{L_p(\Omega)} \le \|u \nabla p_i\|_{L_p(\Omega)} + \|p_i \nabla u\|_{L_p(\Omega)} \le c_i' \, \|u\|_{W_p^1(\Omega)} \, .$$

Since $(p_0 u) \mid_\Gamma = 0$ and $u \mid_\Gamma = \Sigma_{i=1}^k (p_i u) \mid_\Gamma$, by definition of the measure σ on Γ, we get

$$\|u \mid_\Gamma \|_{L_p(\Gamma)} \le \sum_{i=1}^k \|p_i u \mid_\Gamma \|_{L_p(\Gamma)} \le c \, \|\nabla u\|_{L_p(\Omega)}^{1/p} \, \|u\|_{L_p(\Omega)}^{1-1/p}$$

for $c := c_1 c_1'^{1/p} + \ldots + c_k c_k'^{1/p}$. □

The operator T is not surjective from $W_p^1(\Omega)$ onto $L_p(\Gamma)$. Characterizing its image would require the definition of Sobolev spaces $W_p^s(\Gamma)$ with $s \in \mathbb{R}_+ \backslash \mathbb{N}$ since the image of T is $W_p^{1-1/p}(\Gamma)$. In the model case $\Omega := \mathbb{R}^{d-1} \times \mathbb{P}$, $\Gamma := \mathbb{R}^{d-1} \times \{0\}$, the space $W_p^{1-1/p}(\Gamma)$ can be defined as the set of $u \in L_p(\mathbb{R}^{d-1})$ such that

$$\|u\|_{1-1/p,p} := (\|u\|_p^p + \int_{\mathbb{R}^{d-1}\times\mathbb{R}^{d-1}} \frac{|u(x) - u(y)|^p}{\|x - y\|^{p+d-2}} dxdy)^{1/p} < \infty.$$

It can be proved that $W_p^{1-1/p}(\Gamma)$ endowed with this norm is a Banach space and a Hilbert space $H^{1/2}(\Gamma)$ for $p = 2$ (see [94, Thm 3.5 p. 115]; the relation $T(W_p^1(\Omega)) = W_p^{1-1/p}(\Gamma)$ is proved in [94, Thm 3.9 p. 117]).

Definition 9.5 Given an open subset Ω of $\mathbb{R}^d$, the closure of $C_c^m(\Omega)$ in $W_p^m(\Omega)$ (resp. $H^m(\Omega)$) is denoted by $W_{p,0}^m(\Omega)$ (resp. $H_0^m(\Omega)$).

Since one can approach in the norm $\|\cdot\|_{m,p}$ any function of $C_c^m(\Omega)$ by a sequence of functions in $C_c^\infty(\Omega)$, the space $W_{p,0}^m(\Omega)$ is also the closure of $C_c^\infty(\Omega)$ in $W_p^m(\Omega)$.

Proposition 9.10 *For all $p \in [1, \infty[$ one has $W_{p,0}^1(\mathbb{R}^d) = W_p^1(\mathbb{R}^d)$.*

Proof For $n \in \mathbb{N}\backslash\{0\}$, let $\psi_n \in C_c^\infty(\mathbb{R}^d)$ be given by $\psi_n(x) = 1$ for $x \in B(0, n-1)$, $\psi_n(x) = c_n e^{1/(\|x\|^2 - n^2)}$ for $x \in B(0,n)\backslash B(0, n-1)$, $\psi_n(x) = 0$ for $x \in \mathbb{R}^d \backslash B(0,n)$, with c_n adjusted so that $c_n e^{1/(1-2n)} = 1$. For $w \in C^\infty(\mathbb{R}^d) \cap W_p^1(\mathbb{R}^d)$, by dominated convergence, we have $(\|\psi_n w - w\|_p) \to 0$ as $n \to \infty$ and

$$\|D_i(\psi_n w) - D_i w\|_p \le \|\psi_n D_i w - D_i w\|_p + \|w D_i \psi_n\|_p \to 0 \quad \text{as } n \to \infty$$

since both $((\psi_n - 1)D_i w)$ and $(wD_i\psi_n) \to 0$ a.e. and are dominated by a function in $L_p(\mathbb{R}^d)$.

Since $C^\infty(\mathbb{R}^d) \cap W_p^1(\mathbb{R}^d)$ is dense in $W_p^1(\mathbb{R}^d)$, we get that $C_c^\infty(\mathbb{R}^d)$ is dense in $W_p^1(\mathbb{R}^d)$. □

The following characterization of $W_{p,0}^1(\Omega)$ is valid if Ω is Lipschitz, but we prove it when Ω is of class C^1.

Theorem 9.7 *If Ω is an open subset of class C^1 of $\mathbb{R}^d$, then $W_{p,0}^1(\Omega) = \ker T$.*

Proof Since $C_c^\infty(\Omega)$ is contained in the kernel $\ker T$ of the trace map T, and since $\ker T$ is closed in $W_p^1(\Omega)$, we have $W_{p,0}^1(\Omega) \subset \ker T$.

Conversely, let us prove that any $u \in \ker T$ belongs to $W_{p,0}^1(\Omega)$ when Ω is of class C^1. Taking a partition of unity, and using a diffeomorphism of class C^1, we reduce the problem to the case $\Omega := V \times \mathbb{P}$, where V is the open unit ball of $\mathbb{R}^{d-1}$ and $\mathbb{P} :=]0, \infty[$. Since the space $C^\infty(\overline{\Omega}) \cap W_p^1(\Omega)$ is dense in $W_p^1(\Omega)$ by the Meyers-Serrin Theorem (Theorem 9.3), we take a sequence $(u_n)_n$ in $C^\infty(\overline{\Omega}) \cap W_p^1(\Omega)$ that converges to u in $W_p^1(\Omega)$. For $k \in \mathbb{N}\backslash\{0\}$, $(x', r) \in \Omega_k := V\times]0, 1/k[$, we have $u_n(x', r) = u_n(x', 0) + v_n(x', r)$ with $v_n(x', r) := \int_0^r D_d x_n(x', t)dt = \int_0^{1/k} 1_{[0,r]}(x', t)D_d x_n(x', t)dt$, and, by Hölder's inequality with $q := p/(p-1)$,

$$\left|v_n(x', r)\right| \le r^{1/q}(\int_0^{\frac{1}{k}} \left|D_d u_n(x', t)\right|^p dt)^{1/p},$$

$$\int_{\Omega_k} \left|v_n(x', r)\right|^p dx'dr \le \int_0^{\frac{1}{k}} r^{\frac{p}{q}} dr \int_V \int_0^{\frac{1}{k}} \left|D_d u_n(x', t)\right|^p dt = \frac{1}{pk^p} \left\|D_d u\right\|_{L_p(\Omega_k)}^p,$$

hence

$$\|u_n\|_{L_p(\Omega_k)} \le (\int_0^{\frac{1}{k}} \|u_n(\cdot, 0)\|_{L_p(V)}^p dt)^{1/p} + \frac{1}{p^{1/p}k} \|D_d u_n\|_{L_p(\Omega_k)}.$$

Passing to the limit as $n \to \infty$, since $(T(u_n))_n \to 0$ in $L_p(V)$ by continuity of T, we get

$$\|u\|_{L_p(\Omega_k)} \le \frac{1}{p^{1/p}k} \|D_d u\|_{L_p(\Omega_k)}. \tag{9.10}$$

Now, let $h \in C^\infty(\mathbb{R})$ be such that $h(\mathbb{R}) \subset [0, 1]$, $h\ |_{[0,1/2]}= 1$, $h\ |_{[1,\infty[}= 0$ and let us set

$$w_n(x', t) := u(x', t)(1 - h_n(t)),$$

with $h_n(t) := h(nt)$, so that

$$D_i w_n = D_i u(1 - h_n),\ i \in \mathbb{N}_{d-1} \qquad D_d w_n = D_d u(1 - h_n) - u h_n'.$$

Thus

$$\int_\Omega |D_i w_n - D_i u|^p \le \int_\Omega h_n\, |D_i u|^p\,, \qquad i \in \mathbb{N}_{d-1}$$

$$\int_\Omega |D_d w_n - D_d u|^p \le \int_\Omega (h_n\, |D_d u| + |u h_n'|)^p.$$

Since the support of h_n is contained in $\Omega \times [0, 1/n]$, the first integrals converge to 0 by the Dominated Convergence Theorem. Since $w_n = u$ on $[1/n, \infty[$, the last integral can be estimated with the help of (9.10) with $k = n$, $c := \|h'\|_\infty$,

$$\|D_d w_n - D_d u\|_{L_p(\Omega)} = \|D_d w_n - D_d u\|_{L_p(\Omega_n)}$$

$$\le \|h_n D_d u\|_{L_p(\Omega_n)} + \|h_n' u\|_{L_p(\Omega_n)}$$

$$\le \|D_d u\|_{L_p(\Omega_n)} + cn\, \|u\|_{L_p(\Omega_n)} \le \|D_d u\|_{L_p(\Omega_n)} + \frac{c}{p^{1/p}}\, \|D_d u\|_{L_p(\Omega_n)}$$

and each of these terms tends to 0. Since we also have $(w_n) \to u$ in $L_p(\Omega)$, we see that $(w_n) \to u$ in $W_p^1(\Omega)$. Now, since $w_n\ |_{V \times]0,1/2n[} = 0$, using a mollifier we see that $w_n \in W_{p,0}^1(\Omega)$. Thus $u \in W_{p,0}^1(\Omega)$ since this subspace of $W_p^1(\Omega)$ is closed. □

In the sequel we say that a subset Ω of $\mathbb{R}^d$ has a *bounded width* if there exist some $b > 0$ and some unit vector $e \in \mathbb{R}^d$ such that $|e.x| \le b$ for all $x \in \Omega$. Then the width of Ω is $2b$. Of course, any bounded subset has a bounded width. For such open subsets the following inequality is often useful; it was used by Poincaré in his study of tides.

Theorem 9.8 (Poincaré) *Let Ω be an open subset of $\mathbb{R}^d$ with a bounded width and let $p \in [1, \infty[$. Then there exists some $c > 0$ such that*

$$\|w\|_{L_p(\Omega)} \le c\, \|\nabla w\|_{L_p(\Omega)} \qquad \forall w \in W_{p,0}^1(\Omega).$$

Thus $w \mapsto \|\nabla w\|_{L_p(\Omega)}$ is a norm on $W_{p,0}^1(\Omega)$ equivalent to the usual norm.

Proof By density, it suffices to prove this inequality for $w \in C_c^\infty(\Omega)$. Using an orthogonal transformation applying e_1 onto the vector e above, we may suppose $|x_1| \le b$ for all $x := (x_1, x') \in \Omega$ with $x' \in \mathbb{R}^{d-1}$. Given $w \in C_c^\infty(\Omega)$ we have

$$w(x_1, x') = \int_{-b}^{x_1} D_1 w(t, x')dt,$$

hence, by Hölder's inequality with $q := (1 - 1/p)^{-1}$,

$$\left|w(x_1, x')\right|^p \le (2b)^{p/q} \int_{-b}^{b} \left|D_1 w(t, x')\right|^p dt,$$

$$\int_{\mathbb{R}^{d-1}} \left|w(x_1, x')\right|^p dx' \le (2b)^{p/q} \int_{\mathbb{R}^{d-1}} \int_{-b}^{b} \left|D_1 w(t, x')\right|^p dtdx'.$$

Since $|D_1 w| \le \|\nabla w\|$, integrating on x_1 from $-b$ to b, we get the announced inequality

$$\int_{\Omega} |w|^p = \int_{-b}^{b} \int_{\mathbb{R}^{d-1}} \left|w(x_1, x')\right|^p dx' dx_1 \le (2b)^{1+p/q} \int_{\mathbb{R}^d} |D_1 w|^p$$
$$\le (2b)^p \int_{\Omega} \|\nabla w\|^p .$$

Thus we can take $c = 2b$, where $2b$ is the width of Ω. □

We admit the following formulas (which can be generalized to the case when $\partial\Omega$ is just Lipschitzian).

Theorem 9.9 (Green's Formulas) *Let Ω be a bounded open subset of class C^1 of $\mathbb{R}^d$. Then*

$$\int_{\Omega} v \nabla u = -\int_{\Omega} u \nabla v + \int_{\Gamma} uvn d\sigma \qquad \forall u,\ v \in H^1(\Omega),$$
$$-\int_{\Omega} v \Delta u = \int_{\Omega} \nabla u \nabla v - \int_{\Gamma} \frac{\partial u}{\partial n} v d\sigma \qquad \forall u,\ v \in H^2(\Omega).$$

Taking the scalar product of each side by the vector e_i from the canonical basis of $\mathbb{R}^d$, the first relation can be written componentwise as

$$\int_{\Omega} u D_i v + \int_{\Omega} v D_i u = \int_{\Gamma} uvn_i d\sigma \qquad \forall u,\ v \in H^1(\Omega)$$

where $n_i := n.e_i$ is the i-th component of the normal vector n. Replacing u by $D_i u$ and summing over $i \in \mathbb{N}$ we get the second relation of the statement. Also, taking $v = 1$, this relation yields the *Gauss-Green's formula*

$$\int_{\Omega} D_i u = \int_{\Gamma} un_i d\sigma \qquad \forall u \in H^1(\Omega).$$

In turn, replacing u by uv we recover the preceding relation. Moreover, replacing u by $D_i u$ in the Gauss-Green formula and summing over $i \in \mathbb{N}$ we obtain

$$\int_{\Omega} \Delta u = \int_{\Gamma} \frac{\partial u}{\partial n} d\sigma.$$

Given $p,\ q \in [1, d[$, with $1/p + 1/q = 1 + 1/d$ and $u \in W_p^1(\Omega)$, $v \in W_q^1(\Omega)$, Green's formula can be extended to:

$$\int_\Omega u D_i v + \int_\Omega v D_i u = \int_\Gamma T(u)T(v) n_i d\sigma \qquad i \in \mathbb{N}_d.$$

Exercises

1*. A family of seminorms on $C_c^\infty(\Omega)$ inducing the convergence we defined can be described as follows. Take a sequence (K_n) in $\mathcal{K}(\Omega)$ such that $K_n \subset \mathrm{int}(K_{n+1})$, set $\Omega_n := \Omega \backslash K_n$, take a sequence $(k_n) \to \infty$, and set, for $\varphi \in C_c^\infty(\Omega)$

$$p_{(K_n),(k_n)}(\varphi) := \sup_n \sup\{|D^\alpha \varphi(x)| : |\alpha| \le k_n,\ x \in \Omega_n\}.$$

Note that such a family is uncountable. Show that the associated convergence on $C_c^\infty(\Omega)$ is the one described in this section. [Hint: show that for every $K \in \mathcal{K}(\Omega)$ the family of seminorms $p_{(K_n),(k_n)}$ induces on $\mathcal{D}(K) := \{\varphi \in C_c^\infty(\Omega) : \mathrm{supp}(\varphi) \subset K\}$ the same topology as the one defined by the seminorms $p_{K,k}$].

2. Show that for $f, g \in L_{1,loc}(\Omega)$ satisfying $T_f = T_g$ one has $f = g$.

3. Show that the weak derivative of the Heaviside function h defined by $h(r) = 0$ for $r \in \mathbb{R}_-$, $h(r) = 1$ for $r \in \mathbb{P}$ is the Dirac measure δ_0.

4. Given $f \in C^\infty(\Omega)$ and a distribution T, verify that setting $(fT)(\varphi) := T(f\varphi)$ one defines a distribution fT. Show that $f(T_1 + T_2) = fT_1 + fT_2$ and that $(f_1 + f_2)T = f_1T + f_2T, f_1(f_2T) = (f_1f_2)T$ for $T_1, T_2 \in C_c^\infty(\Omega)^*, f_1, f_2 \in C^\infty(\Omega)$.

5. For $u \in L_p(\mathbb{R}^d)$ show that $u \in W_p^1(\Omega)$ if and only if u has a representative v that is absolutely continuous in each of its variables and whose partial derivatives are in $L_p(\mathbb{R}^d)$. [See [268, Thm 2.1.4].]

6. Prove that if Ω is an open subset of class C^1 of $\mathbb{R}^d$ with a bounded boundary, for every $\beta \in \mathbb{N}^d$ satisfying $0 < |\beta| < m$ and every $\varepsilon > 0$ there exists some $c > 0$ such that

$$\forall u \in W_p^m(\Omega) \qquad \left\|D^\beta u\right\|_p \le \varepsilon \sum_{|\alpha|=m} \|D^\alpha u\|_p + c\,\|u\|_p.$$

Deduce from this that the following norm is equivalent to the usual norm on $W_p^m(\Omega)$

$$u \mapsto \|u\|_p + \sum_{|\alpha|=m} \|D^\alpha u\|_p.$$

[See [1, 3].]

7*. Prove the following characterization of $W^m_{p,0}(\Omega)$ when Ω is of class C^1 and $p \in]1,\infty[$, $q := (1-1/p)^{-1}$. The following assertions are equivalent:

(a) $u \in W^1_{p,0}(\Omega)$;
(b) $u \in L_p(\Omega)$ and there exists some constant $c \in \mathbb{R}_+$ such that

$$\left| \int_\Omega u D_i \varphi \right| \leq c \, \|\varphi\|_{L_q(\Omega)} \qquad \forall \varphi \in C_c^\infty(\mathbb{R}^d), i \in \mathbb{N}_d;$$

(c) the function $\bar{u}$ given by $\bar{u} \mid_\Omega = u$, $\bar{u} \mid_{\mathbb{R}^d \setminus \Omega} = 0$ belongs to $W^1_p(\mathbb{R}^d)$ and, in this case, $D_i \bar{u}$ is given by a similar extension of $D_i u$. [See [52, Prop. 9.18].]

9.2 Embedding Results

The purpose of this section is to prove that, for an open subset Ω of $\mathbb{R}^d$, $m \in \mathbb{N}$, $p \in [1,\infty]$, the functions in $W^m_p(\Omega)$ belong to more usual spaces such as $L_q(\Omega)$ or $C^k(\Omega)$. In particular, we start by showing that for an appropriate q, the norm $\|f\|_q$ of a function $f \in C^1_c(\mathbb{R}^d)$ is dominated by $\|\nabla f\|_p$ where $\nabla f := (D_1 f, \ldots, D_d f)$ denotes the gradient of f. For $d = 1$, $p = 1$, $q := +\infty$, this follows from the relations $f(x) = \int_{-\infty}^x f'(r)dr = -\int_x^{+\infty} f'(r)dr$, which imply

$$\|f\|_\infty = \sup_{x \in \mathbb{R}} |f(x)| \leq \frac{1}{2} \int_{-\infty}^{+\infty} |f'(r)| \, dr. \tag{9.11}$$

In the case $d > 1$, the following general estimate is of tremendous importance.

Theorem 9.10 (Gagliardo-Nirenberg-Sobolev) *For $f \in C^1_c(\mathbb{R}^d)$, $p \in [1,d[$, $q := q_d(p) := (1/p - 1/d)^{-1}$, $c := p(d-1)/(d-p)$ one has the following inequalities:*

$$\|f\|_q \leq \frac{c}{2} \|D_1 f\|_p^{1/d} \cdots \|D_d f\|_p^{1/d} \leq \frac{c}{2} \|\nabla f\|_p \,. \tag{9.12}$$

Thus for any Lipschitzian open subset Ω of $\mathbb{R}^d$ there exists a continuous linear injection of $W^1_p(\Omega)$ into $L_q(\Omega)$ extending the identity map on $C^1_c(\mathbb{R}^d)$.

The value $(1/p - 1/d)^{-1} = (d-p)/dp$ of the Sobolev's exponent $q_d(p)$ can be recovered by using a scaling argument. Taking $u_\lambda := u(\lambda \cdot)$ with $\lambda > 0$ instead of u in the relation $\|f\|_q \leq \frac{c}{2} \|\nabla f\|_p$ we obtain

$$\|f\|_q \leq \frac{c}{2} \lambda^{1+d/q-d/p} \|\nabla f\|_p$$

which implies $1 + d/q - d/p = 0$, i.e. $q = (1/p - 1/d)^{-1}$.

We start the proof with the case $p = 1$, so that $c = 1$, $q := d/(d-1)$.

Lemma 9.3 *Let $d \in \mathbb{N}\backslash\{0,1\}$, let $q := d/(d-1)$, and let $f \in C_c^1(\mathbb{R}^d)$. Then*

$$\Big(\int_{\mathbb{R}^d} |f(x)|^q\,dx\Big)^{1/q} \leq \frac{1}{2}\Big(\int_{\mathbb{R}^d} |D_1 f(x)|\,dx\Big)^{1/d} \cdots \Big(\int_{\mathbb{R}^d} |D_d f(x)|\,dx\Big)^{1/d}. \tag{9.13}$$

Proof We know the result for $d = 1$ and we prove it by induction on d, admitting the result for a function of $d-1 \geq 1$ variables. For $k \in \mathbb{N}_{d-1}$, $x := (s,t) \in \mathbb{R}^{d-1} \times \mathbb{R}$ we set

$$I_k(t) := \int_{\mathbb{R}^{d-1}} |D_k f(s,t)|\,ds \qquad J_d(s) := \int_{\mathbb{R}} |D_d f(s,t)|\,dt.$$

Besides the exponent $q := d/(d-1)$, we need the exponent $r := (d-1)/(d-2)$ corresponding to the dimension $d-1$, so that the induction assumption can be written as

$$\Big(\int_{\mathbb{R}^{d-1}} |f(s,t)|^r\,ds\Big)^{1/r} \leq \frac{1}{2} I_1(t)^{\frac{1}{d-1}} \cdots I_{d-1}(t)^{\frac{1}{d-1}} \tag{9.14}$$

since $D_k f(\cdot,t) \in C_c^1(\mathbb{R}^{d-1})$ for $k \in \mathbb{N}_{d-1}$. By (9.11), for all $(s,t) \in \mathbb{R}^{d-1} \times \mathbb{R}$ we have $|f(s,t)| \leq (1/2)J_d(s)$, hence $|f(s,t)|^q \leq 2^{-(q-1)}|f(s,t)|\,J_d(s)^{1/(d-1)}$ since $q-1 = 1/(d-1)$. Applying Hölder's inequality with the exponent r and using the relation $1 - 1/r = 1/(d-1)$, we get

$$\int_{\mathbb{R}^{d-1}} |f(s,t)|^q\,ds \leq 2^{1-q}\Big(\int_{\mathbb{R}^{d-1}} |f(s,t)|^r\,ds\Big)^{\frac{1}{r}}\Big(\int_{\mathbb{R}^{d-1}} J_d(s)ds\Big)^{\frac{1}{d-1}}.$$

Taking into account the induction assumption (9.14) and integrating with respect to t, we obtain

$$\int_{\mathbb{R}^{d-1}\times\mathbb{R}} |f(s,t)|^q\,dsdt \leq 2^{-q}\Big(\int_{\mathbb{R}} I_1(t)^{\frac{1}{d-1}} \cdots I_{d-1}(t)^{\frac{1}{d-1}}\,dt\Big)\Big(\int_{\mathbb{R}^{d-1}} J_d(s)ds\Big)^{\frac{1}{d-1}}.$$

Using Hölder's inequality for $d-1$ functions (Corollary 8.8 with $p_i := d-1$), we estimate the first integral and obtain a relation equivalent to (9.13):

$$\int_{\mathbb{R}^d} |f(x)|^q\,dx \leq 2^{-q}\Big(\int_{\mathbb{R}} I_1(t)dt\Big)^{\frac{1}{d-1}} \cdots \Big(\int_{\mathbb{R}} I_{d-1}(t)dt\Big)^{\frac{1}{d-1}}\Big(\int_{\mathbb{R}^{d-1}} J_d(s)ds\Big)^{\frac{1}{d-1}}.$$

□

Proof of the theorem It remains to consider the case $p \in]1, d[$, for which $c := p(d-1)/(d-p) > 1$. We apply the lemma to the function $g := f^c$ rather than f, obtaining, since $D_k g = cf^{c-1}D_k f$ for $k \in \mathbb{N}_d$

$$\begin{aligned}(\int_{\mathbb{R}^d} |g|^{\frac{d}{d-1}})^{\frac{d-1}{d}} &\leq \frac{1}{2}(\int_{\mathbb{R}^d} |D_1 g|)^{\frac{1}{d}} \ldots (\int_{\mathbb{R}^d} |D_d g|)^{\frac{1}{d}} \\ &\leq \frac{c}{2}(\int_{\mathbb{R}^d} |f|^{c-1}.|D_1 f|)^{\frac{1}{d}} \ldots (\int_{\mathbb{R}^d} |f|^{c-1}.|D_d f|)^{\frac{1}{d}}.\end{aligned}$$

We use Hölder's inequality with exponent p and the relations $(d-1)/d = c/q$ and $(c-1)p/(p-1) = q$ to estimate each integral in the last product and we obtain

$$(\int_{\mathbb{R}^d} |f|^q)^{\frac{c}{q}} \leq \frac{c}{2}(\int_{\mathbb{R}^d} |f|^q)^{\frac{p-1}{p}}.(\int_{\mathbb{R}^d} |D_1 f|^p)^{\frac{1}{dp}} \ldots (\int_{\mathbb{R}^d} |D_d f|^p)^{\frac{1}{dp}}.$$

Since $c/q-(p-1)/p = (d-p)/dp = 1/q$, simplifying by $(\int_{\mathbb{R}^d} |f(x)|^q\, dx)^{(p-1)/p}$ we get the first inequality of (9.12). The second one follows since $|D_k f(x)| \leq \|\nabla f(x)\|$ for $k \in \mathbb{N}_d$.

The last assertion is obtained by using the extension theorem 9.5. □

The next embedding theorem shows the interest of using $W_p^m(\Omega)$ spaces instead of just $H^m(\Omega)$ spaces.

Theorem 9.11 (Morrey) *Let Ω be a Lipschitzian open subset of $\mathbb{R}^d$ whose boundary Γ is bounded and let $p > d$, $h := 1 - d/p$. Then there exists some $c > 0$ such that every $w \in W_p^1(\Omega)$ is the class for a.e. equality of a continuous function still denoted by w satisfying the Hölder condition*

$$\forall x, x' \in \Omega \qquad \left|w(x) - w(x')\right| \leq c \,\|\nabla w\|_{L_p(\mathbb{R}^d)} \left\|x - x'\right\|^h.$$

Proof In this proof we denote by B the closed ball with center 0 and radius $r/2 > 0$ for the norm $\|\cdot\|_\infty$. We first suppose $w \in C_c^1(\mathbb{R}^d) \cap W_p^1(\mathbb{R}^d)$. For $x, y \in B$, using the relation

$$w(x) - w(y) = \int_0^1 \nabla w(x + t(y-x)).(x-y)dt,$$

the inequality $|\nabla w(z).(x-y)| \leq r\Sigma_{i=1}^d |D_i w(z)|$, and integrating over B, we get

$$\begin{aligned}\left|r^d w(x) - \int_B w(y)dy\right| &\leq r\int_B\int_0^1 \sum_{i=1}^d |D_i w(x + t(y-x))|\, dtdy \\ &\leq r\sum_{i=1}^d \int_0^1 \int_B |D_i w(x + t(y-x))|\, dydt.\end{aligned}$$

Using the change of variables $z = ty$ and Hölder's inequality, we obtain

$$\begin{aligned}\left| r^d w(x) - \int_B w \right| &\le r \sum_{i=1}^d \int_0^1 \int_{tB} |D_i w((1-t)x + z)|\, t^{-d} dz dt \\ &\le r \sum_{i=1}^d \int_0^1 \left(\int_{tB} |D_i w((1-t)x + z)|^p \, dz \right)^{\frac{1}{p}} (tr)^{\frac{d(p-1)}{p}} t^{-d} dt \\ &\le \frac{r^{1+d-d/p}}{1-d/p} \sum_{i=1}^d \left(\int_B |D_i w|^p \right)^{1/p} .\end{aligned}$$

The triangle inequality ensures that for $x, x' \in B$ we have

$$|w(x) - w(x')| \le \left| w(x) - \frac{1}{r^d} \int_B w \right| + \left| w(x') - \frac{1}{r^d} \int_B w \right| \le 2 \frac{r^{1-d/p}}{1-d/p} \|\nabla w\|_p .$$

Now, given $x, x' \in \mathbb{R}^d$, applying what precedes to the translated function $w_a := w(\cdot - a)$ with $a := (x - x')/2$ and setting $r = \|x - x'\|$, we get the announced inequality with $c := 2/(1 - d/p)$. Since the space $C_c^1(\mathbb{R}^d)$ is dense in $W_p^1(\mathbb{R}^d)$, taking a sequence in $C_c^1(\mathbb{R}^d)$ that converges in $W_p^1(\mathbb{R}^d)$ and a.e., this inequality is valid for $w \in W_p^1(\mathbb{R}^d)$ on the complement of a null set N, hence on $\mathbb{R}^d$ since $\mathbb{R}^d \backslash N$ is dense in $\mathbb{R}^d$ and w is uniformly continuous on $\mathbb{R}^d \backslash N$.

When Ω is a Lipschitzian open subset of $\mathbb{R}^d$ with a bounded boundary, we use Theorem 9.5 to get the result, with another constant depending on Ω. □

The preceding embedding results can be completed with some compactness properties.

Theorem 9.12 (Rellich-Kondrachov) *Suppose Ω is bounded and of class C^1. Then the following injections are compact:*

$$W_p^1(\Omega) \to L_r(\Omega) \quad \text{for } p \in [1, d[,\ r \in [1, q[\text{ with } q = (\frac{1}{p} - \frac{1}{d})^{-1} = \frac{pd}{d-p},$$

$$W_p^1(\Omega) \to L_r(\Omega) \quad \text{for } p = d,\ r \in [p, \infty[,$$

$$W_p^1(\Omega) \to C(\overline{\Omega}) \quad \text{for } p > d.$$

The proof relies on extension and regularization methods and on the Arzela-Ascoli theorem (see [52, 117]).

9.3 Elliptic Problems

In this section we consider linear partial differential equations that do not involve time. Since our study relies on the Lax-Milgram Theorem, we only use the Hilbertian Sobolev spaces $H^m(\Omega) := W^m_p(\Omega)$ with $p = 2$, assuming Ω is an open subset of $\mathbb{R}^d$ whose boundary is of class C^1. Because $p = 2$ is fixed, we simply denote by $\|\cdot\|_m$ the norm $\|\cdot\|_{m,p}$ when no confusion may arise. For $m = 0$, for the sake of clarity, we often write $\|\cdot\|_{L_2(\Omega)}$ or $\|\cdot\|_{L_2}$ instead of $\|\cdot\|_0$. Also, we concentrate on elliptic problems of the form

$$(Lu)(x) := \sum_{|\alpha|\le 2m} c_\alpha(x) D^\alpha u(x) = f(x) \tag{9.15}$$

where f and the coefficients c_α are given functions on Ω, with $m \in \mathbb{N}\backslash\{0\}$. To such equations are usually adjoined boundary conditions. For $m = 1$ they are usually of one of the forms

$$u \mid_{\partial\Omega} = g \qquad \text{(Dirichlet condition)}$$

$$\text{or} \quad \frac{\partial u}{\partial n} \mid_{\partial\Omega} = g \qquad \text{(Neumann condition),}$$

where g is a given function on $\Gamma := \partial\Omega$ and $\frac{\partial u}{\partial n} := n.\nabla u$ is the normal derivative of u. For $m > 1$ the traces of higher order derivatives may be involved.

In the sequel we assume that the operator L can be written in *divergence form*:

$$(Lu)(x) = \sum_{|\alpha|,\ |\beta|=0}^{m} (-1)^{|\alpha|} D^\alpha (a_{\alpha\beta}(x) D^\beta u(x)).$$

For $m = 1$ the multi-indices are just indices and L can be written as

$$Lu = -\mathrm{div}(A(\cdot)\nabla u(\cdot)) + a(\cdot).\nabla u(\cdot) + a_0(\cdot)u(\cdot), \tag{9.16}$$

where $A := (a_{ij}(\cdot)) : \Omega \to L(\mathbb{R}^d, \mathbb{R}^d)$, $a : \Omega \to \mathbb{R}^d$, and $a_0 : \Omega \to \mathbb{R}$ are in $\mathcal{L}_\infty(\Omega, L(\mathbb{R}^d, \mathbb{R}^d))$, $\mathcal{L}_\infty(\Omega, \mathbb{R}^d)$, and $\mathcal{L}_\infty(\Omega)$ respectively. Also, for differentiable maps $u : \Omega \to \mathbb{R}$, $v := (v^i) : \Omega \to \mathbb{R}^d$ or $u \in H^1(\Omega)$, $v \in H^1(\Omega)^d$, we write

$$\nabla u := (D_i u)_{i\in\mathbb{N}_d}, \qquad \mathrm{div}(v(\cdot)) := \sum_{i=1}^{d} D_i v^i(\cdot).$$

Then L is a continuous linear map from $H^1(\Omega)$ into the dual $H^{-1}(\Omega)$ of $H^1_0(\Omega)$. The operator L is associated with the restriction to $H^1(\Omega) \times H^1_0(\Omega)$ of the bilinear form $b : H^1(\Omega) \times H^1(\Omega) \to \mathbb{R}$ defined for $u, v \in H^1(\Omega)$ by

$$b(u, v) := \sum_{i,j=1}^{d} \int_\Omega a_{ij} D_i u D_j v + \sum_{i=1}^{d} \int_\Omega a_i D_i u v + \int_\Omega a_0 u v, \tag{9.17}$$

so that for $(u, v) \in H^1(\Omega) \times H_0^1(\Omega)$ one has

$$b(u, v) = \langle Lu, v \rangle.$$

9.3.1 Ellipticity

The operator L is said to be *uniformly elliptic*, or just elliptic (over Ω) if the coefficients $(a_{\alpha\beta})_{|\alpha|,|\beta|\le m}$ are measurable and essentially bounded and if there exists a constant $c_E > 0$ such that

$$\sum_{|\alpha|=|\beta|=m} a_{\alpha\beta}(x) v_\alpha v_\beta \ge c_E \|v\|^{2m} \qquad \forall v := (v_\alpha) \in \mathbb{R}^{d(m)}, x \in \Omega \tag{9.18}$$

where $d(m) := \operatorname{card}\{\alpha : |\alpha| = m\}$; moreover, in the case $m > 1$ we require that the coefficients $(a_{\alpha\beta})_{|\alpha|,|\beta|=m}$ are uniformly continuous with some modulus of uniform continuity c_Ω. For the sake of simplicity, we only treat the case $m = 1$.

We are essentially interested in the model problem in which $L = -\Delta + a_0 I$, where Δ is the *Laplacian*

$$\Delta u := \sum_{i=1}^{m} \frac{\partial^2 u}{\partial x_i^2}$$

for which the preceding conditions are clearly satisfied. Note that here and in the rest of this section we commit an abuse of notation, writing I instead of the injection

$$I_1 : H^1(\Omega) \to H^{-1}(\Omega)$$

defined by

$$\langle I_1(u), v \rangle := \langle u | v \rangle_{L_2(\Omega)} := \langle j(u) | j_0(v) \rangle_{L_2(\Omega)}$$

for $u \in H^1(\Omega), v \in H_0^1(\Omega)$, j (resp. j_0) being the canonical injection of $H^1(\Omega)$ (resp. $H_0^1(\Omega)$) into $L_2(\Omega)$. Thus,

$$I_1 = j_0^{\mathsf{T}} \circ R \circ j = R_1 \circ j_0^* \circ j,$$

where R (resp. R_1) is the Riesz isometry from $L_2(\Omega)$ (resp. $H_0^1(\Omega)$) onto its dual space:

$$\begin{array}{ccccc} & & L_2(\Omega)^* & \stackrel{j_0^{\mathsf{T}}}{\to} & H^{-1}(\Omega) \\ & & \uparrow R & & \uparrow R_1 \\ H^1(\Omega) & \stackrel{j}{\to} & L_2(\Omega) & \stackrel{j_0^*}{\to} & H_0^1(\Omega) \end{array}$$

Note that since $j_0(H_0^1(\Omega))$ is dense in $L_2(\Omega)$, the maps j_0^{T} and j_0^* are injective as is I_1, so that identifying the spaces with their images is not a great abuse. Such an abuse is justified when for $u \in H^1(\Omega)$ one views Δu and u as distributions.

9.3.2 Energy Estimates and Existence Results

As already mentioned, existence results for the equation $Lu = f$ along with some boundary condition rely on the Lax-Milgram Theorem, so that we are led to consider the bilinear form b associated with L. The proofs we give are for the case $m = 1$, even if we state a more general result (see for instance [3, 216] for the first assertion).

Theorem 9.13 (Gårding) *If L is uniformly elliptic of order $2m$, there exist constants c_m, $c_0 > 0$ such that*

$$\sum_{|\alpha|,\,|\beta|\le m} \int_\Omega a_{\alpha\beta} D^\alpha u D^\beta u \ge c_m \|u\|_m^2 - c_0 \|u\|_0^2 \qquad \forall u \in H_0^m(\Omega). \tag{9.19}$$

If $m = 1$, if the operator L of (9.16) is uniformly elliptic and if for some $\beta > 0$ one has $a_0 \ge \beta$ a.e., then there exists some $c > 0$ such that

$$\sum_{i,j=1}^d \int_\Omega a_{ij} D_i u D_j u + \int_\Omega a_0 uu \ge c \|u\|_1^2 \qquad \forall u \in H_0^1(\Omega). \tag{9.20}$$

If $m = 1$ and if the operator L of (9.16) is uniformly elliptic, then there exist $c_0 > 0$, $c > 0$ such that for all $u \in H_0^1(\Omega)$

$$\int_\Omega \sum_{i,j=1}^d a_{ij} D_i u D_j u + \int_\Omega \sum_{i=1}^d a_i u D_i u + \int_\Omega a_0 uu \ge c \|u\|_1^2 - c_0 \|u\|_0^2 . \tag{9.21}$$

Proof For $m = 1$, $u \in H_0^1(\Omega)$, $x \in \Omega$, taking $\alpha := (0, \dots, 1, 0, \dots 0)$, $v_\alpha := v_i := D_i u(x)$ in the ellipticity condition (9.18) we obtain, with $c_1 := c_E$, $c_0 := c_E + \|a_0\|_\infty$

$$\sum_{i,j=1} \int_\Omega a_{ij} D_i u D_j u + \int_\Omega a_0 uu \ge c_E \int_\Omega (D_i u)^2 + \int_\Omega a_0 uu \ge c_1 \|u\|_1^2 - c_0 \|u\|_0^2 .$$

When for some $\beta > 0$ we have $a_0 \ge \beta$ a.e., setting $c := \min(c_E, \beta)$, we have the estimate

$$\sum_{i,j=1} \int_\Omega a_{ij} D_i u D_j u + \int_\Omega a_0 uu \ge c_E \int_\Omega (D_i u)^2 + \beta \int_\Omega uu \ge c \|u\|_1^2 .$$

To get the final assertion we take $\varepsilon > 0$ such that $\varepsilon d \|a_i\|_\infty < 2c_E$, we use the estimate

$$\left|\int_\Omega a_i u D_i u\right| \leq \|a_i\|_\infty \|u\|_0 \|D_i u\|_0 \leq \frac{1}{2} \|a_i\|_\infty (\varepsilon^{-1} \|u\|_0^2 + \varepsilon \|D_i u\|_0^2),$$

and we take $c_0 > \|a_0\|_\infty + (1/2\varepsilon)\Sigma_{i=1}^d \|a_i\|_\infty$, $c := c_E - \frac{1}{2}\varepsilon d \|a_i\|_\infty$. □

The preceding inequalities incite us to transform the given problem into a more tractable one. We first consider the case of a Dirichlet condition with Ω bounded and of class C^1. We assume that for all $i, j \in \mathbb{N}_d$ we have $a_{ij} \in C_b^1(\Omega)$ and $a_i \in \mathcal{L}_\infty(\Omega)$, and that f and g are given.

We say that $u \in C^2(\Omega) \cap C(\overline{\Omega})$ is a *classical solution* of the system $Lu = f$, $u \mid_{\partial\Omega} = g$ if these equalities are pointwise relations. A *weak solution* of this system is an element $u \in H^1(\Omega)$ such that the trace $T_\Gamma(u)$ of u on $\Gamma := \partial\Omega$ satisfies $T_\Gamma(u) = g$ and, for all $v \in H_0^1(\Omega)$, $b(u, v) = \int_\Omega fv$, where b is the bilinear form introduced in (9.17). Let us note that the relation $b(u, v) = \int_\Omega fv$ or $b(u, v) = \langle f \mid j_0 v\rangle_{L_2(\Omega)} = \langle Rf, j_0 v\rangle$ can be written

$$\langle Lu, v\rangle = \langle j_0^\mathsf{T} Rf, v\rangle$$

for the coupling between $H_0^1(\Omega)$ and its dual. When it is satisfied for all $v \in H_0^1(\Omega)$, it means that $Lu = j_0^\mathsf{T} Rf$ or $Lu = f$ when we identify f with $j_0^\mathsf{T} Rf \in H^{-1}(\Omega)$.

Proposition 9.11 *Let $f \in L_2(\Omega)$, $g \in C^1(\Gamma)$ and let L be a uniformly elliptic differential operator of order 2 in divergence form, Ω being bounded and of class C^1.*

A classical solution $u \in C_b^2(\Omega) \cap C^1(\overline{\Omega})$ of the system $Lu = f$, $u \mid_\Gamma = g$ is a weak solution of this system.

Conversely, if $u \in H^1(\Omega)$ is a weak solution belonging to $C_b^2(\Omega) \cap C^1(\overline{\Omega})$, then u is a classical solution of this system.

Proof If $u \in C_b^2(\Omega) \cap C^1(\overline{\Omega})$ is a classical solution, then $u \in H^1(\Omega)$ (and even $W_\infty^1(\Omega)$) and $T_\Gamma(u)$ is just the restriction of u to Γ. Since $a_{ij} D_j u \in C^1(\Omega)$ we can use Green's formula:

$$\int_\Omega D_i(a_{ij} D_j u)v + \int_\Omega a_{ij} D_j u D_i v = 0 \qquad i, j \in \mathbb{N}_d$$

for all $v \in C_c^1(\Omega)$, hence also for all $v \in H_0^1(\Omega)$ by density since the left-hand side is a continuous linear form on $C_c^1(\Omega)$ with respect to the topology induced by $H^1(\Omega)$ as a function of v; here we use the relations $D_i a_{ij} D_j u \in C_b(\Omega) \subset \mathcal{L}_2(\Omega)$ and $a_{ij} D_i D_j u \in \mathcal{L}_2(\Omega)$. Summing on i, j and using the relation $Lu = f$ we get $b(u, v) = \int_\Omega fv$ for all $v \in H_0^1(\Omega)$.

Conversely, let $u \in H^1(\Omega)$ belonging to $C_b^2(\Omega) \cap C^1(\overline{\Omega})$ and satisfying $T_\Gamma(u) = g$ and $b(u, v) = \int_\Omega fv$ for all $v \in H_0^1(\Omega)$. Then we have $u \mid_\Gamma = g$. Using Green's formula again we get

$$\int_\Omega \sum_{i,j=1}^{d} D_i(a_{ij}D_ju)v + \int_\Omega \sum_{i=1}^{d}(a_iD_iu + a_0u)v - \int_\Omega fv = 0 \qquad \forall v \in C_c^1(\Omega)$$

or $\langle Lu - f \mid v\rangle_{L_2} = 0$ for all $v \in C_c^1(\Omega)$, where $\langle\cdot \mid \cdot\rangle_{L_2}$ is the scalar product in $L_2(\Omega)$. Since $C_c^1(\Omega)$ is dense in $L_2(\Omega)$, we get $Lu = f$ in $L_2(\Omega)$ and Lu is a continuous representative of f. □

Now we observe that given $f_0 \in L_2(\Omega)$ and $g \in T_\Gamma(H^1(\Omega))$, the *inhomogeneous weak problem*

find $u \in H^1(\Omega)$ such that $T_\Gamma(u) = g$ and $b(u, v) = \langle f_0 \mid v\rangle_{L_2} \forall v \in H_0^1(\Omega)$

can be reduced to the *homogeneous weak problem*

find $w \in H_0^1(\Omega)$ such that $b(w, v) = \langle f \mid v\rangle_{L_2}$ for all $v \in H_0^1(\Omega)$ (9.22)

where $f := f_0 - Lh$, with $h \in H^1(\Omega)$ such that $T_\Gamma(h) = g$. Setting $w := u - h$, this stems from the linearity of $b(\cdot, v)$ and from the fact that for all $v \in H_0^1(\Omega)$ one has $b(h, v) = \langle Lh \mid v\rangle_{L_2}$, as shown above. Thus, we concentrate on the Dirichlet problem (9.22) in a case incorporating the *Laplace equation*

$$-\Delta u + u = f \qquad u \in H_0^1(\Omega).$$

Theorem 9.14 (Dirichlet, Riemann, Hilbert) *For all $f \in L_2(\Omega)$ the problem $Lu = f$, $u \in H_0^1(\Omega)$ in which $a = 0$ and $a_0 \geq \beta > 0$ has a unique weak solution.*

If, moreover, $a_{ij} = a_{ji}$ for all $(i,j) \in \mathbb{N}_d^2$ then the solution u minimizes on $H_0^1(\Omega)$ the functional

$$v \mapsto \frac{1}{2}b(v, v) - \langle f \mid v\rangle_{L_2}.$$

Proof The bilinear form b is obviously continuous on $H_0^1(\Omega) \times H_0^1(\Omega)$, and even on $H^1(\Omega) \times H^1(\Omega)$. Moreover, when $a = 0$ and $a_0 \geq \beta > 0$, Gårding's inequality shows that b is coercive on $H_0^1(\Omega)$. The Lax-Milgram Theorem ensures that there is a unique $u \in H_0^1(\Omega)$ such that $b(u, v) = \langle f \mid v\rangle_{L_2}$ for all $v \in H_0^1(\Omega)$. Thus u is a solution of (9.22) with $h := 0$, hence is a weak solution of $Lu = f$, $u \in H_0^1(\Omega)$. The characterization of u in the symmetric case is part of the Lax-Milgram Theorem. □

Corollary 9.3 *For all $f \in L_2(\Omega)$ the problem $Lu + \lambda u = f$, $u \in H_0^1(\Omega)$ in which $a = 0$ and $\lambda > \|a_0\|_\infty$ has a unique weak solution.*

Proof This follows from the fact that when $\lambda > \|a_0\|_\infty$ there exists some $\beta > 0$ such that $a_0 + \lambda \geq \beta$. □

Let us return to the inhomogeneous case.

Corollary 9.4 *Given $f \in L_2(\Omega)$, $g := T_\Gamma(h)$ with $h \in H^1(\Omega)$, the problem $Lu = f$, $u \in H^1(\Omega)$, $T_\Gamma(u) = g$ in which $a = 0$ and $a_0 \geq \beta > 0$ has a unique weak solution. If, moreover, $a_{ij} = a_{ji}$ for all $(i,j) \in \mathbb{N}_d^2$ then the solution u minimizes on the affine space $H_0^1(\Omega) + h$ the functional*

$$F : v \mapsto \frac{1}{2}b(v,v) - \langle f \mid v\rangle_{L_2}.$$

Proof The first assertion follows from the theorem using the translation $v \mapsto v + h$. The second one stems from the fact that $u \in H_0^1(\Omega) + h$, $F(u) \leq F(v)$ for all $v \in H_0^1(\Omega) + h$ is equivalent to $w := u - h \in H_0^1(\Omega)$, $F(w) \leq F(z)$ for all $z \in H_0^1(\Omega)$, as one can see by using the relation $b(w,\cdot) = \langle f \mid \cdot\rangle_{L_2}$. □

Now let us consider the general (homogeneous) case in which L is given by (9.16) and the bilinear form b is given by (9.17) with $a \neq 0$. The operator $L : u \mapsto \Sigma_{i,j=1}^d D_i(a_{ij}D_j u) + \Sigma_{i=1}^d a_i D_i u + a_0 u$ can be seen as an unbounded operator with domain $H_0^2(\Omega) \subset L_2(\Omega)$. We rather view it as a continuous linear map $L : H_0^1(\Omega) \to H_0^1(\Omega)^*$ associated with the bilinear form b^T.

Theorem 9.15 *For $f \in L_2(\Omega)$ and L given by (9.16), the set of $u \in H_0^1(\Omega)$ such that $Lu = f$ is either a singleton for all $f \in L_2(\Omega)$ or else a finite dimensional affine subspace of $H_0^1(\Omega)$ provided f satisfies a finite number of linear relations; otherwise it is empty.*

Proof We fix $\gamma \in [c_0, \infty[$, where c_0 is as in the last assertion of Theorem 9.13 and we set

$$L_\gamma := L + \gamma I, \qquad b_\gamma(u,v) := b(u,v) + \gamma\langle u \mid v\rangle_{L_2},$$

writing I instead of the injection $I_1 : H^1(\Omega) \to H^{-1}(\Omega)$ or its restriction to $H_0^1(\Omega)$ and u, v instead of $j_0(u), j_0(v) \in L_2(\Omega)$. Since b_γ is coercive on $H_0^1(\Omega)$, by the Lax-Milgram Theorem, for all $f_0 \in L_2(\Omega)$ there exists a unique $u_\gamma \in H_0^1(\Omega)$ such that

$$b_\gamma(u_\gamma, v) = \langle f_0 \mid v\rangle_{L_2} \qquad \forall v \in H_0^1(\Omega).$$

Equivalently, $u_\gamma = L_\gamma^{-1}(j_0^\mathsf{T}(Rf_0))$ but we simply write $u_\gamma = L_\gamma^{-1}f_0$, identifying f_0 with its image under $j_0^\mathsf{T} \circ R$ in $H^{-1}(\Omega)$. We note that, given $f \in L_2(\Omega)$, $u \in H_0^1(\Omega)$ is a weak solution of $Lu = f$, $T_\Gamma(u) = 0$ if and only if

$$b_\gamma(u,v) = \langle f + \gamma u \mid v\rangle_{L_2} \qquad \forall v \in H_0^1(\Omega)$$

or $u = L_\gamma^{-1}(f + \gamma u)$, or

$$u - \gamma L_\gamma^{-1}u = L_\gamma^{-1}f.$$

Now the operator $L_\gamma^{-1} : L_2(\Omega) \to H_0^1(\Omega)$ (standing for $L_\gamma^{-1} \circ j_0^{\mathsf{T}} \circ R = L_\gamma^{-1} \mid_{L_2(\Omega)}$) is continuous since by the coercivity property of Theorem 9.13 there exists some $c > 0$ such that

$$c \left\| u_\gamma \right\|_1^2 \leq b_\gamma(u_\gamma, u_\gamma) = \langle f \mid u_\gamma \rangle_{L_2} \leq \|f\|_{L_2} \left\| u_\gamma \right\|_1,$$

hence $\left\| u_\gamma \right\|_1 \leq c^{-1} \|f\|_{L_2}$.

Composing L_γ^{-1} (or rather $L_\gamma^{-1} \circ j_0^{\mathsf{T}} \circ R = L_\gamma^{-1} \mid_{L_2(\Omega)}$) with the canonical injection $j_0 : H_0^1(\Omega) \to L_2(\Omega)$, by the Rellich-Kondrachov Theorem we get a compact operator $A_\gamma := \gamma j_0 \circ L_\gamma^{-1}$ in $L_2(\Omega)$. Thus u is a weak solution if and only if for $f_\gamma := j_0(L_\gamma^{-1} f)$ we have

$$u - A_\gamma u = f_\gamma,$$

u being considered as the element $j_0(u)$ of $L_2(\Omega)$ and we can apply the Fredholm alternative: $N(I - A_\gamma) = R(I - A_\gamma^{\mathsf{T}})^\perp$ is finite dimensional and if $N(I - A_\gamma) = \{0\}$, then $I - A_\gamma$ is an isomorphism. Moreover, $R(I - A_\gamma)$ is a finite codimensional subspace, and the Fredholm alternative asserts that $f_\gamma \in R(I - A_\gamma)$ if and only if

$$\langle f_\gamma \mid v \rangle_{L_2} = 0 \qquad \forall v \in N(I - A_\gamma^{\mathsf{T}}).$$

□

Corollary 9.5 *Let L be given by (9.16). Then there exists a finite or countable subset C of $\mathbb{R}$ such that for all $f \in L_2(\Omega)$ the equation $Lu - \lambda u = f$ has a unique weak solution in $H_0^1(\Omega)$ if and only if $\lambda \in \mathbb{R} \backslash C$. If C is infinite it is of the form $C = \{\lambda_n : n \in \mathbb{N}\}$ where (λ_n) is an increasing sequence with limit $+\infty$.*

Proof We take $\gamma \in [c_0, \infty[$ and set $L_\gamma := L + \gamma I$ as in the preceding proof. The Fredholm alternative asserts that the equation $Lu - \lambda u = f$ has a unique weak solution in $H_0^1(\Omega)$ if and only if $N(L - \lambda I) = \{0\}$. This relation is equivalent to $N(L_\gamma - (\gamma + \lambda)I) = \{0\}$. Setting $K_\gamma := L_\gamma^{-1}$ considered as a compact linear operator from $L_2(\Omega)$ into itself, the relation $L_\gamma(u) = (\gamma + \lambda)u$ is equivalent to $u = (\gamma + \lambda)K_\gamma(u)$. Thus $u \in N(L - \lambda I) \backslash \{0\}$ if and only if $\gamma + \lambda \neq 0$ and u is an eigenvector of K_γ associated with the eigenvalue $(\gamma + \lambda)^{-1}$. The spectral analysis of the compact operator K_γ expounded in Theorem 3.31 yields the conclusion. □

We can derive from the preceding corollary a characterization of the spectrum of the principal part $u \mapsto L_0(u) := \Sigma_{i,j=1}^d - D_i(a_{ij} D_j u)$ of the operator L in the case $a_{ij} = a_{ji}$ and Ω is bounded. Then, by ellipticity and by Poincaré's inequality, the operator L_0 is symmetric and positive. The eigenvalues of $K_0 := L_0^{-1}$ form a decreasing sequence $(\mu_n) \to 0_+$. Thus, the eigenvalues of L_0 form the increasing sequence (λ_n) with $\lambda_n := 1/\mu_n$. Taking a basis of the eigenspace associated with λ_n, we obtain the following result.

Proposition 9.12 *When Ω is bounded, $a_{ij} = a_{ji}$, $a_i = 0$, $a_0 = 0$, the eigenvalues of the elliptic operator $L = L_0$ form an increasing sequence of positive numbers with limit $+\infty$ and there exists an orthonormal sequence in $L_2(\Omega)$ formed of eigenvectors of L.*

Exercise Interpret the adjoint of A_γ with the help of the formal adjoint operator:

$$L^*u = -\sum_{i,j=1}^{d} D_i(a_{ji}D_j u) + a_0 u \qquad u \in H_0^1(\Omega)$$

that is associated with the bilinear form b^T given by

$$b^\mathsf{T}(u, v) := b(v, u) \qquad \forall u,\ v \in H_0^1(\Omega).$$

[Hint: use Green's formula

$$\langle Lu \mid v\rangle_{L_2} = b(u, v) = b^\mathsf{T}(v, u) = \langle L^*v \mid u\rangle_{L_2} \qquad \forall u,\ v \in H_0^1(\Omega).]$$

9.3.3 *Regularity of Solutions*

The question of regularity of the weak solutions of the equation $Lu = f$ or even of the solutions to the Laplace equation

$$-\Delta u + u = f \tag{9.23}$$

is delicate. We just state typical results and give an idea of the proofs.

Theorem 9.16 (Agmon-Douglis-Nirenberg) *If Ω is an open subset of class C^2 of $\mathbb{R}^d$ and $p \geq 2$, there exists some constant $c > 0$ such that if $f \in L_p(\Omega) \cap L_2(\Omega)$ then the unique weak solution $u \in H_0^1(\Omega)$ of equation (9.23) belongs to $W_p^2(\Omega)$ and one has*

$$\|u\|_{2,p} \leq c\,\|f\|_p \qquad \forall f \in L_p(\Omega).$$

In particular, if Ω is bounded and if $f \in C(\overline{\Omega})$, then $u \in C^1(\overline{\Omega})$.

Proof Let us give a proof in the case $p = 2$, $\Omega = \mathbb{R}^d$ and sketch the case $\Omega := \mathbb{R}^{d-1} \times \mathbb{P}$ and the case of an arbitrary open subset of class C^2. The existence and uniqueness of the weak solution u stems from Theorem 9.14 or Corollary 9.3.

For $v \in L_2(\Omega)$, $w \in \mathbb{R}^d \setminus \{0\}$ let us set

$$q_w(v) := \frac{1}{\|w\|}(T_w(v) - v),$$

where $T_w(v)(x) := (v \circ t_w)(x) := v(x-w)$. By Proposition 9.4 it suffices to prove that the norm $\|q_w(u)\|_p$ in $L_p(\Omega)$ of the difference quotient $q_w(u)$ is bounded above by $c\,\|f\|_p$ for some $c > 0$. Since $\Omega = \mathbb{R}^d$, for all $v \in L_2(\Omega)$ we have

$$\int_\Omega uT_w(v) = \int_\Omega vT_{-w}(u), \tag{9.24}$$

$$\int_\Omega uq_w(v) = \int_\Omega vq_{-w}(u). \tag{9.25}$$

Taking $v := q_{-w}(u)$ in relation (9.25) and then observing that $q_w(D_iu) = D_iq_w(u)$ for $i \in \mathbb{N}_d$, we get

$$\int_\Omega uq_w(q_{-w}(u)) = \int_\Omega (q_{-w}(u)).(q_{-w}(u)),$$

$$\int_\Omega \nabla u.\nabla q_w(q_{-w}(u)) = \int_\Omega \nabla q_{-w}(u).\nabla q_{-w}(u)$$

hence, using the relation

$$\int_\Omega \nabla u \nabla \varphi + \int_\Omega u\varphi = \int_\Omega f\varphi \qquad \forall \varphi \in H_0^1(\Omega) \tag{9.26}$$

expressing that $u \in H_0^1(\Omega)$ is the weak solution of equation (9.23), taking $\varphi := q_w(q_{-w}(u))$, we obtain

$$\int_\Omega \|\nabla q_{-w}(u)\|^2 + \int_\Omega |q_{-w}(u)|^2 = \int_\Omega fq_w(q_{-w}(u)).$$

Thus

$$\|q_{-w}(u)\|_{1,2}^2 \le \|f\|_2\,\|q_w(q_{-w}(u))\|_2\,.$$

By Proposition 9.4, for all $v \in H^1(\Omega)$ we have $\|q_w(v)\|_2 \le \|v\|_{1,2}$. Taking $v := q_{-w}(u)$ and simplifying by $\|q_{-w}(u)\|_{1,2}$, we obtain

$$\|q_{-w}(u)\|_{1,2} \le \|f\|_2\,. \tag{9.27}$$

Since w is arbitrary in $\mathbb{R}^d$, applying again Proposition 9.4, and observing that $\|q_{-w}(D_iu)\|_2 = \|D_iq_{-w}(u)\|_2 \le \|f\|_2$, we get $D_iu \in H^1(\Omega)$ for all $i \in \mathbb{N}_d$, hence $u \in H^2(\Omega)$ and $\|u\|_{H^2(\Omega)} \le c\,\|f\|_2$ for some $c > 0$.

In the case when $\Omega := \mathbb{R}^{d-1} \times \mathbb{P}$, for w horizontal, i.e. $w \in \mathbb{R}^{d-1} \times \{0\}$, by the remark following Proposition 9.4, relation (9.27) is still valid. Since for $i \in \mathbb{N}_d$ one has $q_{-w}(D_iu) = D_iq_{-w}(u)$ one obtains

$$\|q_{-w}(D_iu)\|_{L_2(\Omega)} \le \|f\|_2\,.$$

Given $\varphi \in C_c^\infty(\Omega)$, taking $v := D_i\varphi$ in relation (9.25) we see that

$$\left|\int_\Omega uq_{-w}(D_i\varphi)\right| = \left|\int_\Omega \varphi q_{-w}(D_iu)\right| \le \|\varphi\|_2 \, \|f\|_2 \, .$$

Replacing w with $-te_k$, $k \in \mathbb{N}_{d-1}$, $t > 0$ and passing to the limit as $t \to 0_+$, we get

$$\left|\int_\Omega uD_kD_i\varphi\right| \le \|\varphi\|_2 \, \|f\|_2 \qquad \forall \varphi \in C_c^\infty(\Omega),\ i \in \mathbb{N}_d. \tag{9.28}$$

Let us prove that

$$\left|\int_\Omega uD_d^2\varphi\right| \le \|\varphi\|_2 \, \|f\|_2 \qquad \forall \varphi \in C_c^\infty(\Omega).$$

Returning to equation (9.26) and using relation (9.28) we get

$$\left|\int_\Omega uD_d^2\varphi\right| \le \sum_{k=1}^{d-1}\left|\int_\Omega uD_k^2\varphi\right| + \left|\int_\Omega (f-u)\varphi\right| \le c\,\|\varphi\|_2 \, \|f\|_2 \qquad \forall \varphi \in C_c^\infty(\Omega)$$

for some $c > 0$. Combining these inequalities with relation (9.28) we obtain

$$\left|\int_\Omega uD_jD_i\varphi\right| \le \|\varphi\|_2 \, \|f\|_2 \qquad \forall \varphi \in C_c^\infty(\Omega),\ i,\ j \in \mathbb{N}_d.$$

Riesz's Theorem yields some $u_{i,j} \in L_2(\Omega)$ such that $\int_\Omega uD_jD_i\varphi = \int_\Omega u_{i,j}\varphi$ for all $\varphi \in C_c^\infty(\Omega)$. It follows that $u \in H^2(\Omega)$.

The case of a general open subset is treated by using a partition of unity. We refer to specialized books about partial differential equations for the proof. □

This result can be generalized to the case of a weak solution of equation $Lu = f$ for L a uniformly elliptic differential operator of order two. We start with a glance at interior regularity. We denote by $H^m_{loc}(\Omega)$ the space of $u \in L_{2,loc}(\Omega)$ whose weak derivatives up to order m are in $L_{2,loc}(\Omega)$. For $m = 0$ we set $H^m(\Omega) := L_2(\Omega)$.

Theorem 9.17 *Assume that for some $m \in \mathbb{N}$ the coefficients of the uniformly elliptic operator L are in $C^{m+1}(\Omega)$ and $f \in H^m(\Omega)$. If $u \in H^1(\Omega)$ is a weak solution of the equation $Lu = f$, then u belongs to $H^{m+2}_{loc}(\Omega)$ and for any open set Ω' whose closure is compact and contained in Ω, for some $c > 0$ depending only on m, Ω, Ω', and the coefficients of L one has the estimate*

$$\|u\|_{H^{m+2}(\Omega')} \le c\,\|f\|_{H^m(\Omega)} + c\,\|u\|_{L_2(\Omega)} \, .$$

Note that since $u \in H^2_{loc}(\Omega)$, we can use Green's formula to integrate by parts and transform the relation $b(u, v) = \langle f \mid v\rangle$ for all $v \in C_c^\infty(\Omega)$ into

$$\langle Lu \mid v\rangle = \langle f \mid v\rangle \qquad \forall v \in C_c^\infty(\Omega). \tag{9.29}$$

This relation implies that $Lu = f$ a.e. in Ω. Thus, under additional assumptions as in the next corollary, u is a classical solution, the relation $Lu = f$ holding everywhere.

Corollary 9.6 *Assume that the coefficients of the uniformly elliptic operator L are in $C^\infty(\Omega)$ and $f \in C^\infty(\Omega)$. If $u \in H^1(\Omega)$ is a weak solution of the equation $Lu = f$, then u belongs to $C^\infty(\Omega)$.*

Under refined assumptions, one can get regularity up to the boundary; see [3, 52, 117, 128, 139, 202] for instance. We denote by $\overline{\Omega}$ the closure of Ω.

Theorem 9.18 *Let Ω be of class C^{m+2} for some $m \in \mathbb{N}$. Suppose the coefficients of L belong to $C^{m+1}(\overline{\Omega})$, $f \in H^m(\Omega)$, and $u \in H^1_0(\Omega)$ is a weak solution of the equation $Lu = f$. Then $u \in H^{m+2}(\Omega)$ and for some $c > 0$ depending only on m, Ω, and the coefficients of L, one has the estimate*

$$\|u\|_{H^{m+2}(\Omega)} \leq c\,\|f\|_{H^m(\Omega)} + c\,\|u\|_{L_2(\Omega)}\,.$$

If, moreover, the weak solution is unique, as in Theorem 9.14, one has the estimate

$$\|u\|_{H^{m+2}(\Omega)} \leq c\,\|f\|_{H^m(\Omega)}\,.$$

Also one can get regularity results in the class $C^{2,s}(\overline{\Omega})$ of functions that are twice differentiable and whose second-order derivatives are *s-Hölderian* with $s \in]0,1[$, i.e. satisfy inequalities of the form $|v(x) - v(x')| \leq c\,\|x - x'\|^s$ for $x, x' \in \overline{\Omega}$. See [136], [139].

Theorem 9.19 (Schauder) *If Ω is of class $C^{2,s}$, with $s \in]0,1[$ and if $f \in C^{0,s}(\overline{\Omega})$, then a weak solution u of the equation $Lu = f$ is a classical solution in $C^{2,s}(\overline{\Omega})$ and there exists some $c > 0$ such that*

$$\|u\|_{C^{2,s}(\overline{\Omega})} \leq c\,\|f\|_{C^{0,s}(\overline{\Omega})}\,.$$

9.3.4 Maximum Principles

We have seen that for an open subset Ω of $\mathbb{R}^d$, when a function $u : \Omega \to \mathbb{R}$ attains a local maximum at some point $\overline{x} \in \Omega$ and is twice differentiable there, one has

$$Du(\overline{x}) = 0, \qquad D^2u(\overline{x}) \leq 0. \tag{9.30}$$

The results we present in this subsection make use of such a fact. They have important consequences; see [52, 117, 214, 248]. Again, we consider an elliptic differential operator of order two:

$$(Lu)(x) := -\sum_{i,j=1}^{d} a_{i,j}(x)D^2_{i,j}u(x) + \sum_{i=1}^{d} a_i(x)D_iu(x) + a_0(x) \qquad \forall x \in \Omega,$$

As for Poincaré's theorem, we say that the set Ω has a bounded width in a direction $y \in \mathbb{R}^d \backslash \{0\}$ if for some $b \in \mathbb{R}_+$ and all $x \in \Omega$ one has $|x.y| \leq b \|y\|$.

Theorem 9.20 (Weak Maximum Principle)

(a) *Let $u \in C^2(\Omega) \cap C(\overline{\Omega})$ be such that $(Lu)(x) < 0$ for all $x \in \Omega$. Then, if $a_0(\cdot) = 0$, one has*

$$u(x) < \sup u(\partial\Omega) \qquad \forall x \in \Omega. \tag{9.31}$$

(b) *If Ω has a bounded width in some direction y, if $a_0(\cdot) = 0$ on Ω, and if $u \in C^2(\Omega) \cap C(\overline{\Omega})$ is such that $(Lu)(x) \leq 0$ for all $x \in \Omega$ then one has*

$$\sup u(\Omega) \leq \sup u(\partial\Omega). \tag{9.32}$$

(c) *If Ω has a bounded width in some direction $y \in \mathbb{R}^d$, if $a_0(\cdot) \geq 0$ on Ω, and if $u \in C^2(\Omega) \cap C(\overline{\Omega})$ is such that $(Lu)(x) \leq 0$ for all $x \in \Omega$ then one has*

$$\sup u(\Omega) \leq \sup u^+(\partial\Omega). \tag{9.33}$$

Proof

(a) We assume that there is some $\overline{x} \in \Omega$ such that $u(\overline{x}) \geq \sup u(\partial\Omega)$ and we show that a contradiction occurs. Changing u into $u - \sup u(\partial\Omega)$ we may suppose $u(\overline{x}) \geq 0$ and $\sup u(\partial\Omega) = 0$. Then, taking $\varphi \in C_c^\infty(\Omega)$ with $\varphi = 1$ around $\overline{x}$ and changing u into φu and $\overline{x}$ into another point, we may suppose $u(\overline{x}) = \sup u(\overline{\Omega})$. Since $D_{i,j}^2 u(x) = D_{j,i}^2 u(x)$, there is no loss of generality in assuming $(a_{i,j})$ is symmetric, replacing $a_{i,j}$ with $\frac{1}{2}(a_{i,j} + a_{j,i})$ if necessary. Then relation (9.30) holds. Since the matrix $A(\overline{x})$ is symmetric, there exist an orthogonal matrix Q and a diagonal matrix B such that $A(\overline{x}) = QBQ^\intercal$. Let $(e_1, \ldots, e_d)$ be the canonical basis of $\mathbb{R}^d$ and let $H := (D_{i,j}^2 u(\overline{x}))$. Since for two matrices M, N one has $\mathrm{tr}(MN) = \mathrm{tr}(NM)$, the relation $Lu(\overline{x}) < 0$ reads

$$0 < \mathrm{tr}(A(\overline{x})H) = \mathrm{tr}(QBQ^\intercal H) = \mathrm{tr}(BQ^\intercal HQ).$$

Since B is diagonal and its elements are positive by ellipticity, for some $b_i > 0$ we obtain $\Sigma_{1 \leq i \leq d} b_i \langle HQe_i \mid Qe_i \rangle > 0$. This contradicts the fact that $D^2 u(\overline{x}) \leq 0$.

(b) Suppose $u \in C^2(\Omega) \cap C(\overline{\Omega})$ is such that $(Lu)(x) \leq 0$ for all $x \in \Omega$. Given $r > 0$ such that $rc_E > \|a\|_\infty$, where c_E is the uniform ellipticity constant of L, $a := (a_1, \ldots, a_d)$, and $y := (y^1, \ldots, y^d)$, for $\varepsilon > 0$ we set

$$u_\varepsilon(x) := u(x) + \varepsilon e^{rx.y} \qquad x \in \Omega.$$

Then, assuming without loss of generality that $\|y\| = 1$, for all $x \in \Omega$ we have $A(x)y.y \geq c_E$ hence $-\Sigma_{1 \leq i,j \leq d} r^2 a_{i,j}(x) y^i y^j + \sum_{1 \leq i \leq d} r a_i(x).y^i < 0$ and

$$(Lu_\varepsilon)(x) = (Lu)(x) + \varepsilon y^{rx.y}(-r^2 \sum_{1 \leq i,j \leq d} a_{i,j}(x) y^i y^j + r \sum_{1 \leq i \leq d} a_i(x).y^i) < 0.$$

Part (a) ensures that $\sup u_\varepsilon(\Omega) < \sup u_\varepsilon(\partial\Omega)$. Since the function $x \mapsto e^{rx.y}$ is bounded on $\overline{\Omega}$, passing to the limit as $\varepsilon \to 0_+$, we get $\sup u(\Omega) \leq \sup u(\partial\Omega)$.

(c) See [74]. □

One can give a more striking maximum principle (see [117]).

Theorem 9.21 (Hopf) *Let Ω be a connected, bounded, open subset of $\mathbb{R}^d$. Suppose $a_0 \geq 0$ on Ω and $u \in C^2(\Omega) \cap C(\overline{\Omega})$ satisfies $Lu \leq 0$ on Ω. Then, if u attains its maximum over $\overline{\Omega}$ at some $\overline{x} \in \Omega$ and if $u(\overline{x}) \geq 0$, then u is constant on Ω.*

Of course, one has the same conclusion if $Lu \geq 0$ on Ω and if u attains its minimum on $\overline{\Omega}$ at some $\overline{x} \in \Omega$, with $u(\overline{x}) \leq 0$.

Exercises

1. Write down explicitly the passages from the general form of a partial differential operator to its divergence form and vice versa in the case when the operator is of order 2 and the coefficients are smooth enough.
2. For the **homogeneous Neumann problem** of finding a function $u : \overline{\Omega} \to \mathbb{R}$ satisfying

$$-\Delta u + u = f \quad \text{in } \Omega, \qquad \frac{\partial u}{\partial n} = 0 \quad \text{on } \partial\Omega,$$

where Ω is a bounded open subset of $\mathbb{R}^d$ of class C^1, one says that u is a classical solution if $u \in C^2(\overline{\Omega})$ and the preceding relations are satisfied, where $\frac{\partial u}{\partial n}(x) := \nabla u(x).n(x)$, $n(x)$ being the outward normal to Ω at $x \in \partial\Omega$. A weak solution is an element u of $H^1(\Omega)$ satisfying

$$\int_\Omega \nabla u.\nabla v + \int_\Omega uv = \int_\Omega fv \qquad \forall v \in H^1(\Omega).$$

Show that a classical solution is a weak solution.

Prove that for any given $f \in L_2(\Omega)$ there exists a unique weak solution u of the Neumann problem and that u is given as the solution to the minimization problem of the function $v \mapsto \frac{1}{2}\int_\Omega(|\nabla v|^2 + v^2) - \int_\Omega fv$ on $H^1(\Omega)$.
3. (**Poisson's integral formula** for the upper half-space) Given a sufficiently smooth function $g : \mathbb{R} \to \mathbb{R}$, show that the function $u : \mathbb{R} \times \mathbb{P} \to \mathbb{R}$ given by

$$u(x, y) := \frac{y}{\pi} \int_{-\infty}^{\infty} \frac{g(r)}{(x-r)^2 + y^2} dr$$

satisfies Laplace's equation and can be continuously extended to $\mathbb{R} \times \mathbb{R}_+$ so that it satisfies the Dirichlet condition $u(x, 0) = g(x)$.

4. (**Poisson's integral formula** for the unit ball B_d of $\mathbb{R}^d$) Given a continuous function g on the unit sphere $\mathbb{S}^{d-1}$ show that u given by

$$u(x) := \int_{\mathbb{S}^{d-1}} \frac{1 - \|x\|^2}{\lambda_d(B_d)\, \|x - y\|^d} d\sigma(y)$$

satisfies the equation $\Delta u = 0$ in B_d and $\lim_{x \to \bar{x}} u(x) = g(\bar{x})$ for $\bar{x} \in \mathbb{S}^{d-1}$. [See [151, Section 4.1.3].]
5. (**Hopf's Lemma**) Let $a_0 = 0$ and let $u \in C^2(\Omega) \cap C^1(\Omega)$ be such that $Lu \leq 0$ on Ω. Suppose that for some $\bar{x} \in \partial\Omega$ there exists an open ball $B \subset \Omega$ such that $\bar{x} \in \partial B$ and $u(\bar{x}) > u(x)$ for all $x \in \Omega$. Prove that $\frac{\partial u}{\partial n}(\bar{x}) > 0$, where n is the outer normal to Ω at $\bar{x}$. [See [117, p. 330].]

9.4 Nonlinear Problems

Nonlinear problems are much more intricate than linear problems. For such a reason, some results are obtained via linearization; however, in general, they are just local results. In some cases it is possible to reduce nonlinear problems to linear problems by using adapted transformations. Besides such reductions, in view of the abundance of techniques for dealing with nonlinear problems (see [7, 8, 58, 62, 68, 77, 92, 103, 188, 266]), we restrict our attention to two important methods: order methods and dissipativity methods.

9.4.1 Transforming Equations

In some cases the *Legendre transform* is a powerful means to pass from a nonlinear equation to a linear equation. This is so for the *minimal surface equation*

$$\operatorname{div}\Big(\frac{\nabla u}{(1 + \|\nabla u\|^2)^{1/2}}\Big) = 0.$$

In dimension 2, setting $p := p(r,s) := \frac{\partial u}{\partial r}(r,s)$, $q := q(r,s) := \frac{\partial u}{\partial s}(r,s)$ this equation can be rewritten as

$$(1 + q^2)\frac{\partial^2 u}{\partial r^2}(r,s) - 2pq\frac{\partial^2 u}{\partial r \partial s}(r,s) + (1 + p^2)\frac{\partial^2 u}{\partial s^2}(r,s) = 0. \tag{9.34}$$

Assume that on some open subset U of $\mathbb{R}^2$ the map $(r,s) \mapsto \nabla u(r,s) := (p(r,s), q(r,s))$ is a C^1 diffeomorphism from U onto the open subset $V = \nabla u(U)$ of $\mathbb{R}^2$. We denote by $(p,q) \mapsto x(p,q) := (r(p,q), s(p,q))$ its inverse. The Legendre transform v of u is given by

$$v(p,q) := pr(p,q) + qs(p,q) - u(x(p,q)) \qquad (p,q) \in V.$$

We know from Theorem 5.41 that $Dv(p,q) = (r(p,q), s(p,q))$ for all $(p,q) \in V$ and $D^2u(r,s) = (D^2v(p,q))^{-1}$ for all $(r,s) \in U$. Thus, setting $D(r,s) := \frac{\partial r}{\partial p}(p,q)\frac{\partial s}{\partial q}(p,q) - (\frac{\partial r}{\partial q}(p,q))^2$, one has

$$\frac{\partial^2 u}{\partial r^2}(r,s) = \frac{1}{D(r,s)}\frac{\partial^2 v}{\partial q^2}(p,q) \qquad \frac{\partial^2 u}{\partial s^2}(r,s) = \frac{1}{D(r,s)}\frac{\partial^2 v}{\partial p^2}(p,q),$$
$$\frac{\partial^2 u}{\partial r \partial s}(r,s) = -\frac{1}{D(r,s)}\frac{\partial^2 v}{\partial p \partial q}(p,q).$$

Substituting this expression for the second-order partial derivatives of u in equation (9.34) we get the linear equation

$$(1+p^2)\frac{\partial^2 v}{\partial p^2}(p,q) + 2pq\frac{\partial^2 v}{\partial p \partial q}(p,q) + (1+q^2)\frac{\partial^2 v}{\partial q^2}(p,q) = 0.$$

If one obtains a solution v to this equation, one gets u as the Legendre transform of v as seen in Theorem 5.41.

The *hodograph transform* consists in reversing the roles of unknown functions and independent variables in order to convert certain quasilinear systems of partial differential equations into linear systems. As an example, let us consider the case of the equations of *steady, irrotational fluid flow* in two dimensions:

$$(\sigma^2(w) - u^2)\frac{\partial u}{\partial r} - uv(\frac{\partial u}{\partial s} + \frac{\partial v}{\partial r}) + (\sigma^2(w) - v^2)\frac{\partial v}{\partial s} = 0 \tag{9.35}$$

$$\frac{\partial u}{\partial s} - \frac{\partial v}{\partial r} = 0. \tag{9.36}$$

Here we have omitted the variables (r,s) and the unknown function is the velocity field $w := (u,v)$ whereas the sound speed $\sigma(w)$ is given.

Let us assume the map $x := (r,s) \mapsto w := (u(r,s), v(r,s))$ defines a diffeomorphism from an open subset U of $\mathbb{R}^2$ onto an open subset V of $\mathbb{R}^2$. By the inverse function theorem this occurs (locally) when the Jacobian J satisfies

$$J := J(r,s) := \det Dw(r,s) := \frac{\partial u}{\partial r}\frac{\partial v}{\partial s} - \frac{\partial u}{\partial s}\frac{\partial v}{\partial r} \neq 0.$$

Then $Dx(u,v) = (Dw(r,s))^{-1}$ for $(r,s) := (r(u,v), s(u,v))$, i.e.

$$\frac{\partial r}{\partial u} = \frac{1}{J}\frac{\partial v}{\partial s}, \quad \frac{\partial r}{\partial v} = -\frac{1}{J}\frac{\partial u}{\partial r}, \quad \frac{\partial s}{\partial u} = -\frac{1}{J}\frac{\partial v}{\partial r}, \quad \frac{\partial s}{\partial v} = \frac{1}{J}\frac{\partial u}{\partial s}.$$

We intend to look for the equations satisfied by the inverse map $(u, v) \mapsto (r(u, v), s(u, v))$. Inserting the preceding expressions of the elements of the matrix $Dx(u, v)$ into equations (9.35)–(9.36), we get the linear system

$$(\sigma^2(w) - u^2)\frac{\partial s}{\partial v} + uv(\frac{\partial r}{\partial v} + \frac{\partial s}{\partial u}) + (\sigma^2(w) - v^2)\frac{\partial r}{\partial u} = 0$$

$$\frac{\partial r}{\partial v} - \frac{\partial s}{\partial u} = 0.$$

The last equation suggests that we look for a function $(u, v) \mapsto z(u, v)$ such that

$$r = \frac{\partial z}{\partial u}(u, v), \qquad s = \frac{\partial z}{\partial v}(u, v).$$

Then, the first equation above becomes the equation

$$(\sigma^2(w) - u^2)\frac{\partial^2 z}{\partial v^2} + 2uv\frac{\partial^2 z}{\partial u \partial v} + (\sigma^2(w) - v^2)\frac{\partial^2 z}{\partial u^2} = 0,$$

a linear second-order partial differential equation.

In some cases, it may be useful to associate to the unknown solution u of a nonlinear equation a related function w that is a solution to a simpler equation, for instance a linear equation. Such a process can be applied to evolution problems as well as to stationary problems. Taking a smooth function $h : \mathbb{R} \to \mathbb{R}$, let us set $w := h \circ u$. The case $h(r) := e^{cr}$ for $r \in \mathbb{R}$ for some constant $c \in \mathbb{R} \setminus \{0\}$ is called the *Hopf-Cole transformation*. Let us consider its effect on the nonlinear parabolic equation

$$\frac{\partial u}{\partial t}(x, t) + a\Delta u(x, t) + b \|\nabla u(x, t)\|^2 = 0 \qquad (x, t) \in \mathbb{R}^d \times \mathbb{R}_+ \tag{9.37}$$

$$u(x, 0) = g(x) \quad x \in \mathbb{R}^d, \tag{9.38}$$

where g is a given function, $a, b \in \mathbb{R}$, the Laplacian Δ and the gradient ∇ bearing on the space variable x. Using the relations

$$\frac{\partial w}{\partial t}(x, t) = h'(u(x, t))\frac{\partial u}{\partial t}(x, t), \tag{9.39}$$

$$\Delta w(x, t) = h'(u(x, t))\Delta u(x, t) + h''(u(x, t)) \|\nabla u(x, t)\|^2 \tag{9.40}$$

we see by multiplication of both sides of (9.37) by $h'(u(x, t))$ that

$$\frac{\partial w}{\partial t}(x, t) + a\Delta w(x, t) - [ah''(u(x, t)) - bh'(u(x, t))] \|\nabla u(x, t)\|^2 = 0.$$

Thus, if we choose h in such a way that $ah''(r) - bh'(r) = 0$ for all $r \in \mathbb{R}$, the equation satisfied by w is simply the *heat equation*:

$$\frac{\partial w}{\partial t}(x,t) + a\Delta w(x,t) = 0, \tag{9.41}$$

$$w(x,0) = h(g(x)) \tag{9.42}$$

with conductivity a and initial condition $h \circ g$.

Assuming $a \neq 0$, $b \neq 0$, setting $c := b/a$ and taking $h(r) = e^{cr}$ the equation $ah''(r) - bh'(r) = 0$ is satisfied. Then, the unique bounded solution of the equation satisfied by w is given by

$$w(x,t) = \frac{1}{(4\pi at)^{d/2}} \int_{\mathbb{R}^d} e^{\|x-y\|^2/4at} e^{bg(y)/a} dy$$

and we have $u(x,t) = (a/b)\log w(x,t)$, with w as above, an explicit solution.

A further transformation allows us to solve the viscous *Burger's equation*

$$\frac{\partial v}{\partial t}(x,t) - a\frac{\partial^2 v}{\partial x^2}(x,t) + v(x,t)\frac{\partial v}{\partial x}(x,t) = 0 \qquad (x,t) \in \mathbb{R} \times \mathbb{P}$$

$$v(x,0) = k(x) \qquad x \in \mathbb{R}$$

that serves as a model in fluid dynamics. Here k is a given smooth function. This equation can be reduced to (9.37) with $d = 1$ and $b := 1/2$ by setting

$$u(x,t) = \int_{-\infty}^{x} v(s,t)ds, \qquad g(x) = \int_{-\infty}^{x} k(s)ds,$$

so that $u(\cdot,0) = g$. In fact, if u is a solution to the system

$$\frac{\partial u}{\partial t}(x,t) + a\Delta u(x,t) + \frac{1}{2}(\nabla u(x,t))^2 = 0 \qquad (x,t) \in \mathbb{R}^d \times \mathbb{R}_+,$$

$u(\cdot,0) = g$, then $v(x,t) := \frac{\partial u}{\partial x}(x,t)$ is a solution to Burger's equation. Thus, Burger's equation can be solved explicitly, a rather exceptional situation for partial differential equations.

Exercises

1. Let $g : \mathbb{P}\times]-\pi,\pi[\to W := \mathbb{R}^2\backslash D$, with $D := \mathbb{R}_- \times \{0\}$, be the homeomorphism given by $g(r,\theta) := (r\cos\theta, r\sin\theta)$. Let $h : W \to \mathbb{R}$ be a *harmonic function*, i.e. a function of class C^∞ satisfying $D_1^2 h + D_2^2 h = 0$. Verify that $f := h \circ g$ satisfies

$D_{11}^2 f + r^{-2} D_{22}^2 f + r^{-1} D_1 f = 0$. Characterize those harmonic functions h such that $f = h \circ g$ is independent of θ.

2. (*The vibrating string equation*) Find those functions $f : \mathbb{R}^2 \to \mathbb{R}$ of class C^2 satisfying the equation $D_{11}^2 f - D_{22}^2 f = 0$.
3. Let $v : \mathbb{R}^2 \to \mathbb{R}_+$ be a function of class C^3 satisfying the *heat equation* $D_2 v(x,t) = D_{11}^2 v(x,t)$. Given $c \in \mathbb{R}$, set $u = 2cD_1 v/(1 - cv)$. Show that u is of class C^2 on the open set $W := \{(x,t) \in \mathbb{R} \times \mathbb{P} : v(x,t) \neq 1/c\}$ and satisfies on W *Burgers' equation* $D_2 u = D_{11}^2 u - uD_1 u$. Study the reverse passage. Deduce from this particular solutions of Burgers' equation of the form $2cD_1 v/(1 - cv)$ where v is a solution of the heat equation.

9.4.2 Using Potential Functions

Using a potential function may also enable one to reduce a system of nonlinear equations to a single linear equation. We illustrate this method with *Euler equation* for inviscid, incompressible fluid flows. We denote by $p : \mathbb{R}^3 \times \mathbb{R} \to \mathbb{R}$ the *pressure of the flow*, by $u : \mathbb{R}^3 \times \mathbb{R} \to \mathbb{R}^3$ the *velocity vector field*, the two unknown functions, by $f : \mathbb{R}^3 \times \mathbb{R} \to \mathbb{R}^3$ the external force and by $g : \mathbb{R}^3 \to \mathbb{R}^3$ the initial velocity that are given. The Euler system is as follows:

$$\frac{\partial u}{\partial t}(x,t) + Du(x,t).u(x,t) = -\nabla p(x,t) + f(x,t) \quad (x,t) \in \mathbb{R}^3 \times \mathbb{R} \tag{9.43}$$

$$\operatorname{div} u(x,t) = 0 \quad (x,t) \in \mathbb{R}^3 \times \mathbb{R} \tag{9.44}$$

$$u(x,0) = g(x) \quad x \in \mathbb{R}^3. \tag{9.45}$$

Here the gradient ∇, the divergence div, and the derivative D are taken with respect to the spatial variable $x = (x_1, x_2, x_3)$, so that (9.43) means that for $i \in \mathbb{N}_3$

$$\frac{\partial u^i}{\partial t}(x,t) + \sum_{j=1}^{3} D_j u^i(x,t).u^j(x,t) = -D_i p(x,t) + f^i(x,t) \qquad (x,t) \in \mathbb{R}^3 \times \mathbb{R}.$$

It is natural to assume that $\operatorname{div} g(x) = 0$ in order to ensure compatibility of (9.44) and (9.45). We consider the case when the external force f is derived from a *potential* $h : \mathbb{R}^3 \times \mathbb{R} \to \mathbb{R}$: $f(x,t) = \nabla h(x,t)$. We look for a solution (u,p) of this system for which the velocity u is also derived from a potential $v : \mathbb{R}^3 \times \mathbb{R} \to \mathbb{R}$:

$$u(x,t) := \nabla v(x,t).$$

Then equation (9.44) requires v to be a harmonic function:

$$\Delta v = \operatorname{div} u = 0.$$

Besides this condition, let us see how equation (9.43) is transformed. Assuming v is twice differentiable, we note that

$$D_j u^i . u^j = D_j D_i v . D_j v = D_i D_j v . D_j v = \frac{1}{2} D_i (D_j v)^2,$$

so that (9.43) turns out to be

$$\nabla(\frac{\partial v}{\partial t}(x,t) + \frac{1}{2}\left\|\nabla v(x,t)\right\|^2) = -\nabla(p(x,t) - h(x,t)).$$

Thus, assuming that we have found a function v that is harmonic in x and satisfies $\nabla v(\cdot, 0) = g(\cdot)$, we can take

$$p(x,t) = h(x,t) - \frac{\partial v}{\partial t}(x,t) - \frac{1}{2}\left\|\nabla v(x,t)\right\|^2.$$

This is *Bernoulli's law.*

Now let us illustrate the utility of the Mountain Pass Theorem by sketching an investigation of the semilinear boundary-value problem consisting in finding a weak solution $u \in H_0^1(\Omega)$ (in the sense of Chap. 9) to the equation

$$-\Delta u = \varphi \circ u \tag{9.46}$$

where Ω is a bounded open subset of $\mathbb{R}^d$ $(d \geq 3)$ and $\varphi : \mathbb{R} \to \mathbb{R}$ is a smooth function satisfying for some $p \in]1, \frac{d+2}{d-2}[$ and $a, b \in \mathbb{R}_+, \alpha, \ \beta \in \mathbb{P}, \ \gamma \in]0, 1/2[$ the growth conditions

$$|\varphi(t)| \leq a + a\,|t|^p, \quad |\varphi'(t)| \leq b + b\,|t|^{p-1} \qquad \forall t \in \mathbb{R} \tag{9.47}$$

$$0 \leq \psi(t) := \int_0^t \varphi(s)ds \leq \gamma t \varphi(t) \qquad \forall t \in \mathbb{R} \tag{9.48}$$

$$\alpha\,|t|^{p+1} \geq |\psi(t)| \geq \beta\,|t|^{p+1} \qquad \forall t \in \mathbb{R}. \tag{9.49}$$

Note that the function φ given by $\varphi(t) := t\,|t|^{p-1}$ satisfies the preceding conditions and that the last one implies that $\varphi(0) = 0$, so that $u = 0$ is a solution to equation (9.46). Let us show that there exists a weak solution $u \neq 0$ in the space $X := H_0^1(\Omega)$ endowed with the scalar product given by $\langle u \mid v \rangle := \int_\Omega DuDv$ and the associated norm. Let us note that the Sobolev injection theorem (Theorem 9.10) ensures that $H_0^1(\Omega)$ is embedded in $L_q(\Omega)$ with $q := 2d/(d-2)$. Thus, using Sect. 8.5.2, we see that condition (9.47) and the inequality

$$\frac{p}{q} = p\frac{d-2}{2d} < \frac{d+2}{d-2}\frac{d-2}{2d} = \frac{d+2}{2d}$$

ensure that for $u \in H_0^1(\Omega)$ or even $L_{2d/d-2}(\Omega)$ one has $\varphi \circ u \in L_s(\Omega)$ for $s := \frac{2d}{d+2}$. Now, since $1/q + 1/s = 1$, i.e. $(d-2)/2d + (d+2)/2d = 1$, one has $L_s(\Omega) = (L_q(\Omega))^*$, so that $\varphi \circ u$ can be considered as a continuous linear form on $H_0^1(\Omega)$, i.e. an element of $X^* := H^{-1}(\Omega)$. On the other hand, by condition (9.49), we have $\psi \circ u \in L_1(\Omega)$ since $p + 1 < \frac{2d}{d-2}$. Thus the function f given by

$$f(w) := \int_\Omega (\frac{1}{2} |Dw|^2 - \psi \circ w) \qquad w \in H_0^1(\Omega)$$

is well defined on the space $X := H_0^1(\Omega)$. We shall show that the Mountain Pass Theorem can be applied to f and yields some critical point $u \neq 0$ of f. We leave to the reader the task of proving (with the help of the Sobolev inequalities as in [117, pp. 483–484]) that f is of class $C^{1,1}$ with derivative given by

$$f'(u)v = \int_\Omega (DuDv - (\varphi \circ u)v) \qquad u, v \in H_0^1(\Omega).$$

Thus, if u is a critical point of f one has $\int_\Omega (DuDv - (\varphi \circ u)v) = 0$ for all $v \in H_0^1(\Omega)$, i.e. u is a weak solution to (9.46). We denote by $G : H^{-1}(\Omega) := X^* \to X$ the isometry defined by $G(w^*) = w$ for $w^* \in H^{-1}(\Omega)$, where w is the unique solution of the equation

$$-\Delta w = w^* \qquad w \in H_0^1(\Omega).$$

Thus, the relation $f'(u) = 0$ can be written as $-\Delta u = \varphi \circ u$. Here $-\Delta = D^\intercal D$, where $D : H_0^1(\Omega) \to L_2(\Omega)^d$ is the map $w \mapsto \nabla w$ and $D^\intercal : (L_2(\Omega)^d)^* \to H^{-1}(\Omega)$ is the transpose map of D when $(L_2(\Omega)^d)^*$ is identified with $L_2(\Omega)^d$.

We set $W := X$, $w_0 = 0$. For $w \in X$ with $\|w\| = r$, for some $\alpha', \alpha'' \geq \alpha$ we have

$$\left| \int_\Omega \psi \circ w \right| \leq \alpha \int_\Omega |w|^{p+1} \leq \alpha' (\int_\Omega |w|^q)^{\frac{p+1}{q}} \leq \alpha'' \|w\|^{p+1} = \alpha'' r^{p+1}$$

hence $f(w) \geq \frac{1}{2} r^2 - \alpha'' r^{p+1} \geq \frac{1}{4} r^2$ for $r > 0$ small enough, since $p + 1 > 2$. Now let us fix $w \in X$ with $\|w\| = r$ and look for some $t \geq 1$ such that $w_1 := tw$ satisfies $f(w_1) < m := \frac{1}{4} r^2$. Such a t can be found since

$$f(tw) = \int_\Omega (\frac{1}{2} t^2 |Dw|^2 - \psi \circ (tw)) \leq t^2 \int_\Omega \frac{1}{2} |Dw|^2 - t^{p+1} \beta \int_\Omega |w|^{p+1} < 0$$

for $t \geq 1$ large enough.

Finally, let us verify condition (PS$_c$). Let (x_n) be a sequence in X such that $(f(x_n)) \to c$ and $(f'(x_n)) \to 0$. Thus, setting $\varepsilon_n := \|f'(x_n)\|$, for all $v \in X$ we have

$$\left| \int_\Omega (Dx_n Dv - (\varphi \circ x_n)v) \right| \leq \varepsilon_n \|v\|.$$

Taking $v = x_n$ we get

$$\int_{\Omega} (\varphi \circ x_n)x_n \leq \varepsilon_n \|x_n\| + \|x_n\|^2 .$$

But since $(c_n) := (f(x_n)) \to c$ and since $f(x_n) = \frac{1}{2}\|x_n\|^2 - \int_{\Omega} \psi \circ x_n$ we have

$$\|x_n\|^2 = 2c_n + 2\int_{\Omega} \psi \circ x_n \leq 2c_n + 2\gamma \int_{\Omega} (\varphi \circ x_n)x_n \leq 2c_n + 2\gamma\varepsilon_n \|x_n\| + 2\gamma \|x_n\|^2 .$$

Since $2\gamma < 1$ and $(\varepsilon_n) \to 0$ we get that $(\|x_n\|)$ is bounded. Since X is reflexive, taking a subsequence of (x_n) if necessary, we may suppose (x_n) has a weak limit x. By compactness of the Sobolev embedding we get $(x_n) \to x$ in $L_q(\Omega)$. Since $(p+1)/q \leq 1$, the continuity of the Nemytskii operator $v \mapsto \varphi \circ v$ from $L_q(\Omega)$ into $L_s(\Omega)$ implies that $(\varphi \circ x_n) \to \varphi \circ x$ in $L_s(\Omega) = (L_q(\Omega))^*$, hence in $H^{-1}(\Omega)$. Let $z_n := D^{\mathsf{T}} D x_n - \varphi \circ x_n = f'(x_n)$. Since $(z_n) \to 0$ in $H^{-1}(\Omega)$ and $(\varphi \circ x_n) \to \varphi \circ x$ in $H^{-1}(\Omega)$, we have $(D^{\mathsf{T}} D x_n) \to \varphi \circ x$ in $H^{-1}(\Omega)$. Since $x_n = G(D^{\mathsf{T}} D x_n)$ and since G is continuous, we get that $(x_n) \to G(\varphi \circ x)$. Thus condition (PS$_c$) is satisfied. □

9.4.3 Order Methods

For nonlinear problems, uniqueness is usually lost. For instance, if $\Omega =]0, \pi[^2$ the equation

$$-\Delta v + 2\,|v| = 0, \qquad v \in H_0^1(\Omega)$$

admits an infinity of solutions given by $v_n(x, y) = -n \sin x \sin y$ for all $n \in \mathbb{N}$. Another elementary example is the equation $x^3 + px + q = 0$ in $\mathbb{R}$ which may have 1, 2, or 3 solutions. In this subsection we give a uniqueness result and a useful notion of subsolution and supersolution. For an existence result for the so-called *logistic equation* see [74, Thm 11.3].

Given a bounded open subset Ω of $\mathbb{R}^d$, $f \in H^{-1}(\Omega) := H_0^1(\Omega)^*$ and $h : \mathbb{R} \to \mathbb{R}$ (globally) Lipschitzian with rate ℓ, let us consider the equation

$$-\Delta v + h \circ v = f \qquad v \in H_0^1(\Omega). \tag{9.50}$$

Let us note that since $|h \circ v| \leq \ell\,|v| + |h(0)|$, for $v \in H_0^1(\Omega)$ or even $v \in L_2(\Omega)$ we have $h \circ v \in L_2(\Omega)$ since Ω is bounded.

Given $v_1, v_2 \in H^1(\Omega)$ we write $v_1 \leq v_2$ on $\partial\Omega$ if $(v_1 - v_2)^+ \in H_0^1(\Omega)$, with $t^+ := \max(t, 0)$. We endow $H_0^1(\Omega)$ with the order induced by $L_2(\Omega)$ and we provide $H^{-1}(\Omega) := H_0^1(\Omega)^*$ with the dual order, i.e. $v_1^* \leq v_2^*$ if and only if $\langle v_1^*, v\rangle \leq \langle v_2^*, v\rangle$ for all $v \in H_0^1(\Omega),\ v \geq 0$.

Definition 9.6 One says that u is a *supersolution* to equation (9.50) if $u \in H^1(\Omega)$, $u \geq 0$ on $\partial\Omega$ and $-\Delta u + h \circ u \geq f$ in the weak sense that

$$\int_\Omega \nabla u \nabla v + \langle h \circ u, v \rangle \geq \langle f, v \rangle \qquad \forall v \in H_0^1(\Omega),\ v \geq 0.$$

One says that w is a *subsolution* to equation (9.50) if $w \in H^1(\Omega)$, $w \leq 0$ on $\partial\Omega$ and $-\Delta w + h \circ w \leq f$ in the weak sense that

$$\int_\Omega \nabla w \nabla v + \langle h \circ w, v \rangle \leq \langle f, v \rangle \qquad \forall v \in H_0^1(\Omega),\ v \geq 0.$$

Let us first show the existence of a greatest solution and a smallest solution lying between a subsolution and a supersolution.

Theorem 9.22 *Let u be a supersolution and let w be a subsolution of (9.50) such that $u \geq w$. Then there exist solutions $\hat{u}$ and $\hat{w}$ of (9.50) such that $u \geq \hat{u} \geq \hat{w} \geq w$ and even such that for any solution v of (9.50) satisfying $u \geq v \geq w$ one has*

$$u \geq \hat{u} \geq v \geq \hat{w} \geq w.$$

Proof We pick $c > \ell$ and we consider the map $S : L_2(\Omega) \to L_2(\Omega)$ which assigns to every $y \in L_2(\Omega)$ the weak solution $z \in H_0^1(\Omega) \subset L_2(\Omega)$ of the equation

$$-\Delta z + cz = cy - h \circ y + f. \tag{9.51}$$

A solution to equation (9.50) is a fixed point of S. We note that S is well defined since $cy - h \circ y + f \in H^{-1}(\Omega)$ for all $y \in L_2(\Omega)$. Moreover, S is *homotone*, i.e. order preserving: given $y \leq y'$ in $L_2(\Omega)$ we have $S(y) \leq S(y')$ by the weak maximum principle since for $z := S(y)$, $z' := S(y')$ we have

$$-\Delta z + cz \leq -\Delta z' + cz'$$

as

$$(cy' - h(y')) - (cy - h \circ y) \geq c(y' - y) - \ell(y' - y) \geq 0.$$

Furthermore, S is continuous from $L_2(\Omega)$ to itself: given y, y' in $L_2(\Omega)$, for $z := S(y)$, $z' := S(y')$ we have

$$-\Delta(z - z') + c(z - z') = (cy - cy') - (h \circ y - h \circ y'))$$

and applying the weak form of equation (9.51) with $z - z'$ as test function we get

$$\begin{aligned} c\left\|z - z'\right\|_2^2 &\leq \int_\Omega (c(y - y') - (h \circ y - h(y')))(z - z') \\ &\leq (c\left\|y - y'\right\|_2 + \left\|h \circ y - h \circ y')\right\|_2)\left\|z - z'\right\|_2 \end{aligned}$$

and it follows that $\|z - z'\|_2 \leq (1 + \ell/c)\ \|y - y'\|_2$: S is even Lipschitzian.

Starting with $y_0 := w$ we inductively define a sequence (y_n) by $y_{n+1} := S(y_n)$. Since w is a subsolution, we have

$$-\Delta y_1 + cy_1 = cy_0 - h \circ y_0 + f \geq -\Delta y_0 + cy_0$$

and the maximum principle entails that $y_1 \geq y_0$. Applying $S^{(n)} := S \circ \ldots \circ S$ to each side of this inequality, we get $y_{n+1} \geq y_n$ for all n. Similarly, using the fact that u is a supersolution, we get $y_n \leq u$ for all n and even $y_n \leq z_n := S^{(n)}(u)$.

Passing to the limit using the Dominated Convergence Theorem, we get elements $\hat{u}, \hat{w}$ of $L_2(\Omega)$ such that $(y_n) \to \hat{w}$ and $(z_n) \to \hat{u}$ in $L_2(\Omega)$. Since S is continuous, we see that $\hat{u}, \hat{w}$ are fixed points of S, hence are solutions. Moreover, if v is a solution to (9.50) satisfying $u \geq v \geq w$, applying $S^{(n)}$ we get

$$z_n \geq S^{(n)}(v) = v \geq y_n$$

and passing to the limit we obtain $\hat{u} \geq v \geq \hat{w}$. □

A uniqueness result can be obtained under an additional assumption.

Corollary 9.7 *Let $f \in L_2(\Omega)$ and let $h : \mathbb{R} \to \mathbb{R}$ be nondecreasing, Lipschitzian with rate ℓ and such that $h(0) = 0$. Then equation (9.50) has at most one solution.*

Proof Let u and w be the weak solutions in $H_0^1(\Omega)$ of

$$-\Delta u = f^+, \qquad -\Delta w = -f^-$$

respectively, with $f^+ = \max(f, 0)$, $f^- = \max(-f, 0)$. By the weak maximum principle (Theorem 9.20) we have $u \geq 0 \geq w$, hence, since h is nondecreasing,

$$-\Delta u + h \circ u \geq -\Delta u = f^+ \geq f \geq -f^- = -\Delta w \geq -\Delta w + h \circ w.$$

If v_1 and v_2 are two solutions to equation (9.50), by subtraction we have

$$-\Delta(v_1 - v_2) = -(h \circ v_1 - h \circ v_2)$$

in Ω, i.e.

$$\int_\Omega \nabla(v_1 - v_2)\nabla v = -\int_\Omega (h \circ v_1 - h \circ v_2)v \qquad \forall v \in H_0^1(\Omega).$$

Taking $v := v_1 - v_2$, since h is nondecreasing we get $\int_\Omega |\nabla(v_1 - v_2)|^2 = 0$ and $v_1 = v_2$ by Poincaré's inequality. □

9.4.4 Monotone Multimaps

We devote this subsection to a class of multivalued maps of great importance (see [43, 51, 207–209, 219, 230, 233, 234] f.i.). This class is used in the modern theory of partial differential equations. We encourage the reader to return to Sect. 1.3 on multivalued analysis when necessary. In the sequel we often identify a multimap M with its graph $G(M)$.

Definition 9.7 A multimap (or multivalued operator) $M : D(M) \rightrightarrows X^*$ from its domain $D(M) := \operatorname{dom} M$, a subset of a Banach space X, into the dual X^* of X is said to be *monotone* if for all $w, x \in D(M)$ and $w^* \in M(w)$, $x^* \in M(x)$ one has

$$\langle w^* - x^*, w - x \rangle \geq 0.$$

A multimap $M : D(M) \rightrightarrows X^*$ is said to be *dissipative* if $-M$ is monotone.

Definition 9.8 A multimap $M : D(M) \rightrightarrows X^*$ is said to be *maximally monotone* if for any monotone multimap $N : D(N) \rightrightarrows X^*$ whose graph $G(N)$ satisfies $G(M) \subset G(N)$ one has $G(M) = G(N)$.

A similar definition holds for maximally dissipative multimaps: maximality means maximality in terms of inclusion of graphs (Fig. 9.1).

In Hilbert spaces there is a tight relationship between monotone multimaps and nonexpansive maps. Here a multimap $F : W \rightrightarrows Z$ between two metric spaces is said to be *nonexpansive* if

$$d_Z(z_1, z_2) \leq d_W(w_1, w_2) \quad \text{for all } w_1, w_2 \in D(F),\ z_1 \in F(w_1),\ z_2 \in F(w_2).$$

Such a condition implies that F is single-valued (take $w_1 = w_2$).

Proposition 9.13 *Given a Hilbert space X and $c > 0$, let $S : X^2 \to X^2$ be defined by*

$$S(x, y) := (c(x + y), c(x - y)), \quad S^{-1}(w, z) = (\frac{1}{2c}(w + z), \frac{1}{2c}(w - z)).$$

Let $M : X \rightrightarrows X$, $P : X \rightrightarrows X$ be such that $G(P) = S(G(M))$. Then P is a nonexpansive map if and only if M is a monotone multimap.

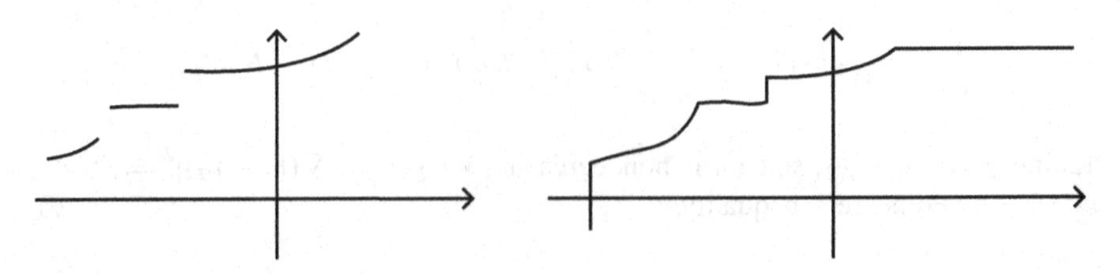

Fig. 9.1 A monotone multimap and a maximally monotone multimap

Note that for $c := 1/\sqrt{2}$ the bijection S is an isometry.

Proof Given $(x_i, y_i) \in G(M)$ for $i = 1, 2$, let $(w_i, z_i) := S(x_i, y_i)$. Then

$$\begin{aligned}\|w_1 - w_2\|^2 - \|z_1 - z_2\|^2 &= \langle (w_1 - w_2) + (z_1 - z_2) \mid (w_1 - w_2) - (z_1 - z_2)\rangle \\ &= \langle (w_1 + z_1) - (w_2 + z_2) \mid (w_1 - z_1) - (w_2 - z_2)\rangle \\ &= 4c^2 \langle x_1 - x_2 \mid y_1 - y_2 \rangle\end{aligned}$$

and thus

$$\|z_1 - z_2\| \leq \|w_1 - w_2\| \iff \langle x_1 - x_2 \mid y_1 - y_2 \rangle \geq 0.$$

In particular, $z_1 = z_2$ whenever $w_1 = w_2$: P is a single-valued map on its domain $D(P) := \{x + y : (x, y) \in G(M)\}$. This equivalence means that P is nonexpansive if and only if M is monotone. □

Example Let $f : \mathbb{R} \to \mathbb{R}$ be nondecreasing. Then f is monotone. Note that this is not the case if f is nonincreasing! Thus the terminology "monotone" is not really satisfactory; but it is well established.

Example Let $M := \partial f$ be the subdifferential of a convex function. Then the very definition of ∂f shows (by an addition sides by sides) that M is monotone. The maximal monotonicity is not as obvious, however. In the case $f := (1/2)\|\cdot\|^2$, the norm being smooth on $X \backslash \{0\}$, the result is a consequence of a general fact pointed out in Proposition 9.14. In the general case, in view of the importance of this example, we give two different proofs. The first one is in the next theorem; the second one is in Corollary 9.8.

Theorem 9.23 (Rockafellar) *Let $f : X \to \mathbb{R}_\infty$ be a closed proper convex function. Then $M := \partial f$ is maximally monotone.*

Proof (M. Ivanov and N. Zlateva [167]) Let $(w, w^*) \in X \times X^*$ be monotonically related to M in the sense that

$$\forall (x, x^*) \in M \qquad \langle x - w, x^* - w^* \rangle \geq 0. \tag{9.52}$$

We have to prove that $(w, w^*) \in \partial f$. Changing f into $g : x \mapsto f(w + x) - \langle x, w^* \rangle + m$ we may suppose $(w, w^*) = (0, 0)$ and $f(0) > 0$. Then relation (9.52) reads

$$\forall (x, x^*) \in \partial f \qquad \langle x, x^* \rangle \geq 0. \tag{9.53}$$

Since f is the supremum of the family of continuous affine functions bounded above by f, there exists some $z^* \in X^*$ such that $f \geq z^*$. Let $c := \|z^*\|$ and let (ε_n) be a sequence in $]0, c/2[$ with limit 0. Let $g_n : X \to \mathbb{R}$ be given by

$$g_n(x) := 2\varepsilon_n \|x\| \text{ for } x \in B_X, \qquad g_n(x) := 2\varepsilon_n + c(\|x\| - 1) \text{ for } x \in X \backslash B_X,$$

so that $f_n := f + g_n$ is bounded below. It can be checked that g_n is convex continuous ($g_n = k_n \circ \|\cdot\|$ where $k_n : \mathbb{R} \to \mathbb{R}$ has an increasing right derivative). Let $w_n \in X$ be such that $f_n(w_n) \leq \inf_{x \in X} f_n(x) + \varepsilon_n$ for all $n \in \mathbb{N}$. The Brøndsted-Rockafellar theorem (Theorem 6.5) with $\delta_n := 1$, $w_n^* := 0$ for all n yields sequences (x_n) in X, (x_n^*) in X^* such that $x_n \in B[w_n, \delta_n]$ and $x_n^* \in \partial f_n(x_n) \cap B[0, \varepsilon_n]$ for all $n \in \mathbb{N}$. By the sum rule we have $x_n^* \in \partial f(x_n) + \partial g_n(x_n)$, and by definition of g_n, for all $y^* \in \partial g_n(x_n)$ we have $\langle x_n, y^* \rangle \geq 2\varepsilon_n \|x_n\|$. Thus, taking (9.53) into account, we get $\langle x_n, x_n^* \rangle \geq 2\varepsilon_n \|x_n\|$. This relation implying $\|x_n^*\| \geq 2\varepsilon_n$ if $x_n \neq 0$, the inclusion $x_n^* \in B[0, \varepsilon_n]$ leads to the conclusion that we must have $x_n = 0$. Since $\partial f_n(0) = \partial f(0) + 2\varepsilon_n B_{X^*}$ and since $\partial f(0)$ is closed, we obtain $0 \in \partial f(0)$, the expected conclusion. □

Another criterion for maximal monotonicity can be given under a mild continuity assumption. We say that a map $F : X \to X^*$ is *radially (weak*) continuous* (often called *hemicontinuous*) if for all $x_0, x_1 \in X$, $t \in \mathbb{R}$, $x_t := (1-t)x_0 + tx_1$, one has $F(x_t) \to F(x_0)$ weakly* when $t \to 0$. If J is the duality map of a smooth Banach space X, this condition is satisfied since $J(x_t)$ is bounded for t in any compact interval and any weak* limit point x^* of $J(x_t)$ as $t \to 0$ satisfies $\|x^*\|_* \leq \liminf_{t \to 0} \|J(x_t)\|_* = \liminf_{t \to 0} \|x_t\| = \|x_0\|$ and $\langle x^*, x_0 \rangle \geq \liminf_{t \to 0} \langle J(x_t), x_t \rangle = \lim_{t \to 0} \|x_t\|^2 = \|x_0\|^2$, hence $x^* = J(x_0)$.

Proposition 9.14 *Let $F : X \to X^*$ be a radially weak* continuous (single-valued) monotone map. Then F is maximally monotone.*

In particular, if X^* is strictly convex, then the duality map $J : X \to X^*$ is maximally monotone.

In fact, by Theorem 9.23, the duality multimap $J : X \rightrightarrows X^*$ is always maximally monotone since $J = \partial j$ with $j(\cdot) := (1/2) \|\cdot\|^2$.

Proof (Standard Trick for Monotone Operators) Let $(\overline{x}, \overline{x}^*) \in X \times X^*$ be such that $\langle \overline{x} - u, \overline{x}^* - u^* \rangle \geq 0$ for all $(u, u^*) \in F$. Fixing $x \in X$ and taking $u := u_t := \overline{x} + t(x - \overline{x})$ with $t \in]0, 1]$, after simplification by t we get

$$\langle \overline{x} - x, \overline{x}^* - F(u_t) \rangle \geq 0.$$

Passing to the limit as $t \to 0_+$, by radial weak* continuity of F, we get $\langle \overline{x} - x, \overline{x}^* - F(\overline{x}) \rangle \geq 0$. Since x is arbitrary in X, we obtain $\overline{x}^* = F(\overline{x})$. □

Let us display some remarkable properties of monotone and maximally monotone multimaps. We first observe that if $M : X \rightrightarrows X^*$ is monotone, then M^{-1} is monotone when considering it takes its values in X^{**}. Moreover, if M is maximally monotone and if X is reflexive, then M^{-1} is maximally monotone. The next result we present is reminiscent of the fact that the subdifferential of a convex function f on a Banach space is locally bounded on the interior of the domain of ∂f.

Theorem 9.24 *Any monotone multimap $M : X \rightrightarrows X^*$ is locally bounded on* $\operatorname{int} D(M)$.

In particular, any linear monotone map $A : X \to X^*$ is continuous.

The proof we present (due to Brezis-Crandall-Pazy) uses a surprising lemma.

Lemma 9.4 (Fitzpatrick) *Let $(x_n) \to 0$ in X, (y_n^*) in X^* with $(\|y_n^*\|) \to \infty$. Then, for all $r > 0$ there exist $w \in rB_X$ and subsequences $(x_{k(n)})$, $(y_{k(n)}^*)$ of (x_n) and (y_n^*) respectively such that*

$$\lim_{n\to\infty} \langle w - x_{k(n)}, y_{k(n)}^* \rangle = +\infty.$$

Proof Suppose on the contrary that there exists an $r > 0$ such that for all $u \in rB_X$ one can find some $c_u \in \mathbb{R}_+$ such that

$$\forall n \in \mathbb{N} \qquad \langle u - x_n, y_n^* \rangle \leq c_u.$$

For $k \in \mathbb{N}$, let $C_k := \{u \in rB_X : \forall n \in \mathbb{N}\ \langle u - x_n, y_n^* \rangle \leq k\}$. Our assumption ensures that $rB_X = \cup_{k\in\mathbb{N}} C_k$. Since C_k is closed for all k, Baire's Theorem ensures that for some $j \in \mathbb{N}$ the interior of C_j in rB_X is nonempty. Let $u_0 \in rB_X$ and $s_0 > 0$ be such that $B[u_0, s_0] \cap rB_X \subset C_j$. We can find $s \in]0, s_0[$ and $q \in]0, 1[$ such that $B[qu_0, s] \subset B[u_0, s_0] \cap rB_X$. Then, $u_1 := qu_0 \in rB_X$ and for all $v \in sB_X$ we have

$$\forall n \in \mathbb{N} \qquad \langle u_1 + v - x_n, y_n^* \rangle \leq j, \qquad \langle -u_1 - x_n, y_n^* \rangle \leq c := c_{-u_1}.$$

Combining these inequalities, for all $v \in sB_X$ we get

$$\langle v - 2x_n, y_n^* \rangle \leq j + c.$$

For n large enough we have $\|2x_n\| \leq s/2$ and we get

$$(s/2)\|y_n^*\| \leq (-\|2x_n\| + s)\|y_n^*\| \leq \sup_{v\in sB_X} \langle v - 2x_n, y_n^* \rangle \leq j + c,$$

a contradiction with $(\|y_n^*\|) \to \infty$. □

Proof of Theorem 9.24 Replacing M with $M_w = \{(y - w, y^*) : (y, y^*) \in M\}$, we may suppose $0 \in \operatorname{int} D(M)$. Let $r_0 > 0$ be such that $r_0 B_X \subset D(M)$. Let us show that there exists an $r \in]0, r_0]$ such that $M(rB_X)$ is bounded. Assume on the contrary that there exist sequences $(x_n) \to 0$, (y_n^*) in Y such that $(\|y_n^*\|) \to \infty$ and $y_n^* \in M(x_n)$ for all $n \in \mathbb{N}$. The lemma provides $w \in r_0 B_X$ and subsequences $(x_{k(n)})$, $(y_{k(n)}^*)$ of (x_n) and (y_n^*) respectively such that

$$\lim_{n\to\infty} \langle w - x_{k(n)}, y_{k(n)}^* \rangle = +\infty.$$

Then $w \in D(M)$ and the monotonicity of M implies that for $z \in M(w)$ one has

$$\liminf_{n\to\infty} \langle w - x_{k(n)}, z \rangle \geq \lim_{n\to\infty} \langle w - x_{k(n)}, y_{k(n)}^* \rangle = +\infty,$$

a contradiction. □

It is easy to see that (the graph of) a maximally monotone multimap is closed. In general it is not closed in the weak×weak* topology. However, some related properties are presented in the next proposition.

Proposition 9.15 *Let $M : X \rightrightarrows X^*$ be maximally monotone. Then*

(a) For all $x \in D(M)$ the set $M(x)$ is weak closed and convex.*
(b) For any sequence $(x_n) \to x$ weakly in X and any sequence $(x_n^*) \to x^*$ weakly in X^* with $(x_n, x_n^*) \in M$ for all n one has $(x, x^*) \in M$ and $\langle x, x^* \rangle = \lim_n \langle x_n, x_n^* \rangle$ whenever one of the next equivalent conditions is satisfied:*

$$\limsup_n \langle x_n, x_n^* \rangle \leq \langle x, x^* \rangle, \tag{9.54}$$

$$\limsup_n \langle x_n - x, x_n^* - x^* \rangle \leq 0. \tag{9.55}$$

(c) For any sequence $(x_n) \to x$ weakly in X and any sequence $(x_n^*) \to x^*$ strongly in X^* with $(x_n, x_n^*) \in M$ for all n, one has $(x, x^*) \in M$.
(d) For any sequence $(x_n) \to x$ in X and any sequence $(x_n^*) \to x^*$ weakly in X^* such that $x_n^* \in M(x_n)$ for all n, one has $(x, x^*) \in M$.*
(e) For any sequence $(x_n) \to x$ in X, any bounded sequence (x_n^*) in X^* such that $x_n^* \in M(x_n)$ for all n, and any weak limit point x^* of (x_n^*), one has $(x, x^*) \in M$. In particular, if M is single-valued at $x \in \operatorname{int} D(M)$, then M is demi-continuous at x, i.e. continuous at x from $D(M)$ endowed with its strong topology into X^* endowed with its weak* topology.*

Proof

(a) Given $x \in D(M)$, by maximal monotonicity of M we have

$$M(x) = \{x^* \in X^* : \forall (u, u^*) \in M \ \langle x - u, x^* \rangle \geq \langle x - u, u^* \rangle\}.$$

These inequalities clearly define a weak* closed convex subset of X^*.

(b) Given sequences $(x_n) \to x$ weakly in X, $(x_n^*) \to x^*$ weakly* in X^* with $(x_n, x_n^*) \in M$ for all n, we first observe that (9.54) and (9.55) are equivalent since

$$\langle x_n - x, x_n^* - x^* \rangle = \langle x_n, x_n^* \rangle - \langle x, x_n^* - x^* \rangle - \langle x_n, x^* \rangle.$$

Now, passing to the limit in the relation

$$0 \leq \langle x_n, x_n^* \rangle - \langle x_n, u^* \rangle - \langle u, x_n^* \rangle + \langle u, u^* \rangle \qquad \forall (u, u^*) \in M$$

and assuming that (9.54) is satisfied, we get

$$0 \leq \langle x, x^* \rangle - \langle x, u^* \rangle - \langle u, x^* \rangle + \langle u, u^* \rangle \qquad \forall (u, u^*) \in M$$

or $0 \leq \langle x - u, x^* - u^* \rangle$ for all $(u, u^*) \in M$. By maximal monotonicity we get $(x, x^*) \in M$.

(c) Since (x_n) is bounded when $(x_n) \to x$ weakly, if $(x_n^*) \to x^*$ strongly we have $\limsup_n |\langle x_n - x, x_n^* - x^* \rangle| \leq \limsup_n \|x_n - x\| . \|x_n^* - x^*\| = 0$ and (9.55) holds.

(d) When $(x_n) \to x$ and $(x_n^*) \to x^*$ weakly*, the sequence (x_n^*) being bounded, we have $(\langle x_n - x, x_n^* \rangle) \to 0$, so that $\limsup_n \langle x_n, x_n^* \rangle \leq \limsup_n \langle x, x_n^* \rangle = \langle x, x^* \rangle$.

(e) Given sequences $(x_n) \to x$ in X, (x_n^*) bounded in X^* such that $x_n^* \in M(x_n)$ for all n, and the limit x^* of a subnet $(x_{n(i)}^*)_{i \in I}$ of (x_n^*), we have $(\langle x_{n(i)} - x, x_{n(i)}^* \rangle)_{i \in I} \to 0$, hence, for all $(u, u^*) \in M$

$$0 \leq \liminf_{i \in I} \langle x_{n(i)} - u, x_{n(i)}^* - u^* \rangle = \liminf_{i \in I} \langle x - u, x_{n(i)}^* - u^* \rangle = \langle x - u, x^* - u^* \rangle.$$

By maximality we get $(x, x^*) \in M$. If $x \in \operatorname{int} D(M)$, for any sequences (x_n) in $D(M)$ with limit x, (x_n^*) in X^* such that $x_n^* \in M(x_n)$ for all n, the sequence (x_n^*) is bounded by Theorem 9.24 and the preceding argument shows that any weak* limit point x^* of (x_n^*) is in $M(x)$. The last assertion ensues. □

Single-valued monotone maps enjoy remarkable continuity properties.

Proposition 9.16

(a) Let $F : X \to X^*$ be a radially continuous monotone single-valued map on a finite dimensional space X. Then F is continuous.

(b) For any Banach space X, any radially weak continuous monotone map $F : X \to X^*$ is demicontinuous, i.e. continuous from X endowed with its strong topology into X^* endowed with its weak* topology.*

Proof

(a) Let us first prove that any monotone map $F : X \to X^*$ is *bounding*, i.e. bounded on bounded subsets (we prefer this term to the more usual term "bounded" which is confusing since the image of F is not necessarily bounded). If F is not bounding there exists a bounded sequence (x_n) in X such that $(\|F(x_n)\|) \to \infty$. We may suppose (x_n) has a limit $\overline{x}$ and that $(u_n^*) := (F(x_n)/\|F(x_n)\|)$ has a limit $u^* \in S_{X^*}$. Passing to the limit in the relation

$$\frac{1}{\|F(x_n)\|} \langle x_n - x, F(x_n) - F(x) \rangle \geq 0 \qquad \forall x \in X, \ \forall n \in \mathbb{N},$$

we obtain

$$\langle \overline{x} - x, u^* \rangle \geq 0 \qquad \forall x \in X.$$

Thus $u^* = 0$, contradicting $\|u^*\| = 1$. Thus F is bounding.

Now let us suppose a sequence (x_n) converges to $x \in X$ and let y^* be a limit point of the bounded sequence $(F(x_n))$. The monotonicity of F yields

$$\langle x - w, y^* - F(w) \rangle \geq 0 \qquad \forall w \in X.$$

Taking u arbitrary in X, $t \in]0, 1]$ and $w := x - t(x - u)$, we get

$$\langle x - u, y^* - F(x) \rangle \geq 0$$

by the radial continuity of F. Since u is arbitrary in X, this means that $y^* = F(x)$. Thus $(F(x_n)) \to F(x)$.

(b) The map F is maximally monotone by Proposition 9.14 and locally bounded by Theorem 9.24. Thus, the assertion is a consequence in Proposition 9.15(e).

□

Exercises

1. Let $F : C \to X^*$ be a single-valued monotone map with domain a convex subset of a Banach space X. Suppose that F is (radially) differentiable at some point x of the interior of C. Show that for all $v \in X$ one has $\langle F'(x).v, v \rangle \geq 0$.
2. Let $L : X \to X$ be a linear monotone operator on a Hilbert space. Show that L is maximally monotone if and only if its graph is closed and L^* is monotone. [See: [152, Thm 10 p. 48].]
3. (**Debrunner-Flor**) Let $M : X \rightrightarrows X$ be a monotone operator with graph $G(M)$ in a Hilbert space and let C be a closed convex subset of X containing the domain $D(M)$ of M. Show that for all $y \in X$ there exists some $x \in C$ such that

$$\langle v + x \mid u - x \rangle \geq \langle y \mid u - x \rangle \qquad \forall (u, v) \in G(M).$$

[See [29, Thm 21.7].]

9.4.5 Representation of Monotone Multimaps

It is the purpose of this subsection to show that convex analysis can be used to study monotonicity. Thus, following [207], given a monotone operator M, we look for a convex function $f : X \times X \to \overline{\mathbb{R}}$ representing M in such a way that some properties of f can be transferred to M and some operations on monotone operators correspond to operations on functions.

In the sequel c and b denote the coupling functions $c : X \times X^* \to \mathbb{R}$, $b : (X \times X^*) \times (X \times X^*) \to \mathbb{R}$ given by

$$c(x, x^*) := \langle x^*, x \rangle \qquad (x, x^*) \in X \times X^*,$$

$$b((w, w^*), (x, x^*)) = \langle w^*, x \rangle + \langle x^*, w \rangle \qquad (w, w^*),\ (x, x^*) \in X \times X^*.$$

Note that b is a symmetric bilinear function that realizes a metric coupling of $X \times X^*$ with itself (exercise). Such a fact expresses the particular structure of $X \times X^*$.

If f is a function on $X \times X^*$, we denote by f^b its conjugate with respect to the coupling function b :

$$f^b(x,x^*) := \sup\{b((w,w^*),(x,x^*)) - f(w,w^*) : (w,w^*) \in X \times X^*\}$$

for $(x,x^*) \in X \times X^*$ and we set $f^\intercal(x^*,x) := f(x,x^*)$. Considering $X \times X^*$ as a subset of $X^{**} \times X^* = (X^* \times X)^*$, we note that f^b is the restriction to $X \times X^*$ of $(f^\intercal)^*$, where for $g : X^* \times X \to \mathbb{R}$, $g^* : X^{**} \times X^* \to \overline{\mathbb{R}}$ is the usual conjugate of g for the usual coupling function between $X^* \times X$ and $X^{**} \times X^*$. Equivalently, $f^b = h^\intercal$, where h is the restriction to $X^* \times X$ of the usual conjugate f^* of f and where $h^\intercal(x,x^*) := h(x^*,x)$.

In the sequel we find it convenient to use f^b rather than f^* in particular because it is defined on the same space as f) and to identify a multimap with its graph.

Since a multimap $M : X \rightrightarrows X^*$ is characterized by (or even identified with) its graph $G(M)$ or $M \subset X \times X^*$, it is faithfully represented by the indicator function ι_M of its graph M. However, ι_M has no differentiability or convexity property. Thus it is natural to replace ι_M or closely related functions such as $c_M := \iota_M + c$ with their convex envelopes or their closed convex envelopes. The following observation enlightens our route. It can be formulated as follows: if $M : X \rightrightarrows X^*$ is monotone, then c is convex on (the graph of) M, a striking fact. Thus, it is natural to use convexity to study monotone multimaps.

Proposition 9.17 *If $M : X \rightrightarrows X^*$ is a monotone multimap, then the restriction to M of the convex envelope $g_M := \operatorname{co} c_M$ of $c_M := c + \iota_M$ coincides with the restriction to M of c.*

Proof Given $(x_i,x_i^*) \in M$ for i in a finite set I and $(t_i)_{i\in I} \in \mathbb{R}_+^I$ satisfying $\Sigma_{i\in I}t_i = 1$ and $(x,x^*) := (\Sigma_{i\in I}t_ix_i, \Sigma_{i\in I}t_ix_i^*) \in M$, by monotonicity we have

$$\begin{aligned}\Sigma_{i\in I}t_i\langle x_i,x_i^*\rangle - \langle x,x^*\rangle &= \Sigma_{i\in I}t_i\langle x_i,x_i^* - x^*\rangle \\ &\geq \Sigma_{i\in I}t_i\langle x,x_i^* - x^*\rangle = \langle x, \Sigma_{i\in I}t_ix_i^* - x^*\rangle = 0.\end{aligned}$$

By construction of $\operatorname{co} c_M$, this shows that for any $(x,x^*) \in M$ we have $(\operatorname{co} c_M)(x,x^*) \geq \langle x,x^*\rangle$. Since the reverse inequality $\operatorname{co} c_M \leq c_M$ holds, we get $\operatorname{co} c_M \mid_M = c \mid_M$. □

Let us describe an easy way of obtaining monotone multimaps that is a kind of converse of the preceding observation.

Lemma 9.5 *Let $\mathcal{G}$ be the set of proper convex functions $g : X \times X^* \to \mathbb{R}_\infty$ such that $g \geq c$. For $g \in \mathcal{G}$ let*

$$M_g := G(M_g) := \{(x,x^*) \in X \times X^* : g(x,x^*) = c(x,x^*)\}.$$

Then M_g is monotone, as is any multimap M such that $M \subset M_g$.

Proof Clearly, for $g \in \mathcal{G}$ one has

$$M_g := \{(x,x^*) \in X \times X^* : g(x,x^*) \leq c(x,x^*)\}.$$

Thus, given $(x,x^*), (y,y^*) \in M_g$, by convexity of g and this relation we have

$$\begin{aligned}\langle \frac{1}{2}(x+y), \frac{1}{2}(x^*+y^*)\rangle &\leq g(\frac{1}{2}(x+y), \frac{1}{2}(x^*+y^*)) = g(\frac{1}{2}(x,x^*) + \frac{1}{2}(y,y^*)) \\ &\leq \frac{1}{2}g(x,x^*) + \frac{1}{2}g(y,y^*) = \frac{1}{2}\langle x,x^*\rangle + \frac{1}{2}\langle y,y^*\rangle.\end{aligned}$$

Whence $\langle x-y, x^*-y^*\rangle \geq 0$ and M_g is monotone; so is any multimap M such that $M \subset M_g$. □

The next proposition completes the preceding lemma.

Proposition 9.18 *For a nonempty subset M of $X \times X^*$ the following assertions are equivalent:*

(a) M is the graph of a monotone multimap;
(b) $c_M := c + \iota_M \geq c_M^b$;
(c) $g_M := \mathrm{co}\, c_M \geq c$;
(d) the function $g_M := \mathrm{co}\, c_M$ belongs to $\mathcal{G}$ and satisfies $g_M + \iota_M = c_M$;
(e) there exists some $g \in \mathcal{G}$ such that $g + \iota_M = c_M$, i.e. $M \subset M_g$.

Proof (a)⇒(b) Given $(x,x^*) \in M$, for all $(w,w^*) \in M$ we have

$$\langle x,x^*\rangle \geq \langle w,x^*\rangle + \langle x,w^*\rangle - \langle w,w^*\rangle = b((w,w^*),(x,x^*)) - c(w,w^*)$$

hence $\langle x,x^*\rangle \geq (c+\iota_M)^b(x,x^*) := c_M^b(x,x^*)$. Thus $c_M \geq c_M^b$.

(b)⇒(c) Let $g_M := \mathrm{co} c_M$, the convex envelope of c_M. Then $g_M^* = c_M^*$ and $g_M^b = c_M^b$. Moreover, since c_M^b is convex and $c_M \geq c_M^b$, by definition of a convex envelope, we have $g_M \geq c_M^b = (c_M^*)^\intercal$. Thus, for all $(x,x^*) \in X \times X^*$ we have

$$\begin{aligned}2g_M(x,x^*) &\geq g_M(x,x^*) + (c_M^*)^\intercal(x,x^*) = g_M(x,x^*) + g_M^*(x^*,x) \\ &\geq \langle (x,x^*),(x^*,x)\rangle = 2\langle x,x^*\rangle.\end{aligned}$$

Therefore $g_M \geq c$.

(c)⇒(d) We observe that (c) means that $g_M := \mathrm{co}\, c_M \in \mathcal{G}$. Thus $g_M + \iota_M \geq c + \iota_M = c_M$; on the other hand, $g_M + \iota_M \leq c_M + \iota_M = c_M$, so that $g_M + \iota_M = c_M$.

(d)⇒(e) Taking $g = g_M$, it suffices to observe that $M \subset M_g$ since $g + \iota_M = c_M$, for $(x,x^*) \in M$ we have $g(x,x^*) = c(x,x^*)$, hence $M \subset M_g$.

(e)⇒(a) is a consequence in the preceding lemma. □

The set $\mathcal{G}$ of proper convex functions on $X \times X^*$ bounded below by c can serve to define maximally monotone multimaps, as the next theorem shows. We need some preliminary results in which we set

$$q(x,x^*) := \frac{1}{2}(\|x\|^2 + \|x^*\|^2) \qquad (x,x^*) \in X \times X^*.$$

We observe that for all $(x,x^*) \in X \times X^*$ we have $q(x,x^*) \geq c(x,x^*)$, $q(x,x^*) \geq c(x,-x^*)$. Moreover, if $q(x,x^*) = c(x,x^*)$ we have

$$0 = q(x,x^*) - c(x,x^*) \geq \frac{1}{2}(\|x\|^2 - 2\|x\| . \|x^*\| + \|x^*\|^2) \geq 0$$

hence $\|x\| = \|x^*\|$ and $\langle x,x^*\rangle = \|x\|^2 = \|x^*\|^2$. Therefore

$$M_q = J,$$

the duality multimap, the inclusion $J \subset M_q$ being a consequence in the definition of J. This fact explains the importance of the relation $q(x,x^*) = c(x,x^*)$.

Lemma 9.6 *Let $g \in \mathcal{G}$ and let $M \subset M_g$ be such that for all $(x,x^*) \in X \times X^*$ there exists some $(w,w^*) \in M$ such that*

$$g(w,w^*) - \langle w,w^*\rangle + q(w-x,w^*-x^*) + c(w-x,w^*-x^*) \leq 0. \tag{9.56}$$

Then M is maximally monotone (hence $M = M_g$).

Note that condition (9.56) implies that $(g-c)\Box(q+c) \leq 0$.

Proof Since $M \subset M_g$, M is monotone. Let (x,x^*) be *monotonically related* to M in the sense that

$$\langle w-x,w^*-x^*\rangle \geq 0 \qquad \forall (w,w^*) \in M.$$

Picking $(w,w^*) \in M$ as in the statement and using the assumption $g \geq c$ we get $q(w-x,w^*-x^*) \leq 0$, hence $w = x$, $w^* = x^*$. Thus $(x,x^*) \in M$ and M is maximally monotone. □

Theorem 9.25 *Let $g \in \mathcal{G}$ be such that $g^{bb} = g$ and $g^b \in \mathcal{G}$. Then $M_g := \{(x,x^*) \in X \times X^* : g(x,x^*) = c(x,x^*)\}$ is (the graph of) a maximally monotone multimap.*

Proof Given $(x,x^*) \in X \times X^*$, setting

$$h(u,u^*) := q(u-x,u^*-x^*) + c(u-x,u^*-x^*) - \langle u,u^*\rangle \qquad (u,u^*) \in X \times X^*$$

and performing an easy computation, we see that

$$h^b(u,u^*) = h(-u,-u^*) \qquad \forall (u,u^*) \in X \times X^*.$$

Since $g^b \geq c$, for all $(u,u^*) \in X \times X^*$ we have

$$g^b(u,u^*) + h(u,u^*) \geq q(u-x, u^*-x^*) + c(u-x, u^*-x^*) \geq 0.$$

The sandwich theorem and the continuity of h (or Corollary 6.14) yield some $(u,u^*) \in X \times X^*$ such that $(g^b)^*(u^*,u) + h^*(-u^*,-u) \leq 0$. This amounts to $g^{bb}(u,u^*) + h(u,u^*) \leq 0$ and means that (9.56) is satisfied since $g^{bb} = g$. The preceding lemma ensures that M_g is maximally monotone. □

Corollary 9.8 *Let $f : X \to \mathbb{R}_\infty$ be a closed, proper convex function. Then $M := \partial f$ is a maximally monotone multimap.*

Proof Let $g : X \times X^* \to \mathbb{R}_\infty$ be given by $g(x,x^*) := f(x) + f^*(x^*)$. Then, for all $(x,x^*) \in X \times X^*$ we have $g(x,x^*) \geq c(x,x^*)$ and

$$\begin{aligned} g^b(x,x^*) &= \sup_{(w,w^*)\in X\times X^*} (\langle x,w^*\rangle + \langle w,x^*\rangle - f(w) - f^*(w^*)) \\ &= f^{**}(x) + f^*(x^*) = g(x,x^*). \end{aligned}$$

Thus $g^{bb} = g$ and $g^b \in \mathcal{G}$. Since $M_g = G(\partial f)$, Theorem 9.25 ensures that ∂f is maximally monotone. □

Now let us consider the reverse passage, from monotone subsets of $X \times X^*$ to functions. Since in Banach spaces closed proper convex functions have a rather nice calculus, it is sensible to pass to the families

$$\begin{aligned} \mathcal{H} &:= \{h \in \mathcal{G} : h = h^{bb}\} = \{h : h = h^{bb}, h \geq c\}, \\ \mathcal{H}_M &:= \{h \in \mathcal{H} : h + \iota_M = c_M := c + \iota_M\}. \end{aligned}$$

The set $\mathcal{H}_M$ is called the set of *representative functions* of M. Note that $h \in \mathcal{H}_M$ if and only if $h = h^{bb}$, $h \geq c$ and $M \subset M_h$.

In Proposition 9.18, to a nonempty subset M of $X \times X^*$ we have associated the function $g_M := \operatorname{co}(c + \iota_M)$. If M is monotone g_M belongs to $\mathcal{G}$, hence satisfies $g_M \geq c$, so that its lower semicontinuous hull

$$p_M := g_M^{bb} := (\operatorname{co}(c + \iota_M))^{bb} = (c + \iota_M)^{bb} := c_M^{bb}$$

also satisfies $p_M \geq c$ and belongs to $\mathcal{H}$. It seems to be closely related to c_M: its construction is given by simple operations and in some cases it is possible to give an explicit expression for it. We also consider its conjugate f_M :

$$f_M(x,x^*) := p_M^b(x,x^*) \qquad p_M(x,x^*) = f_M^b(x,x^*)$$

$$f_M(x,x^*) := c_M^b(x,x^*) := \sup\{\langle x,w^*\rangle + \langle w,x^*\rangle - \langle w,w^*\rangle : (w,w^*) \in M\}.$$

The function f_M is called the *Fitzpatrick function* of M. We call the function p_M the *predominant function* of M in view of the following proposition.

Proposition 9.19 *For any subset M of $X \times X^*$ the function p_M is the greatest closed proper convex function on $X \times X^*$ bounded above by c_M. In fact, p_M is the lower semicontinuous hull of $g_M := \mathrm{co}\, c_M$.*

For any monotone multimap M one has $p_M \geq c$, hence $p_M \in \mathcal{H}_M$,

$$p_M \geq f_M \qquad c_M \geq f_M + \iota_M. \tag{9.57}$$

The function p_M is also the greatest element of $\mathcal{H}_M$.

Proof The first assertion is a general fact about the biconjugate of a function applied to c_M.

Let M be monotone. The implication (a)⇒(c) of Proposition 9.18 ensures that $g_M \geq c$ hence $p_M = g_M^{bb} \geq c$ since c is lower semicontinuous and in fact continuous. The implication (a)⇒(b) of Proposition 9.18 yields $c_M \geq c_M^b = f_M$ hence $c_M = c_M + \iota_M \geq f_M + \iota_M$. Since f_M is closed proper convex, from the inequality $c_M \geq f_M$ we deduce the relations $p_M := c_M^{bb} \geq f_M^{bb} = f_M$.

Finally, for $h \in \mathcal{H}_M$ we have $h \leq h + \iota_M = c_M$, hence $h = h^{bb} \leq c_M^{bb} = p_M$. □

Note that for a monotone multimap M the function f_M is not necessarily a representative function of M. However, f_M is of interest since it satisfies the properties $f_M^{bb} = f_M$, $f_M = c_M$ on M if M is monotone since

$$f_M(x,x^*) = \langle x, x^* \rangle - \inf_{(w,w^*) \in M} \langle w - x, w^* - x^* \rangle.$$

Moreover, when M is monotone, one has $f_M(x,x^*) \leq \langle x, x^* \rangle$ if and only if $M \cup \{(x,x^*)\}$ is monotone.

The function p_M is often close to g_M, hence is often easier to compute than f_M.

Examples

(a) Let $M := \{(0,0)\}$. Then M is monotone, $p_M = g_M = c_M = \iota_M$ and $f_M = \iota_M$. More generally, when M is (the graph of) a continuous linear map satisfying $M = -M^{\mathsf{T}} \mid_X$, where M^{T} is the transpose of M, one has $p_M = \iota_M = f_M$ since for all $(x,x^*) \in M$ one has $c(x,x^*) = \langle x, M(x) \rangle = \langle -M(x), x \rangle$ hence $c(x,x^*) = 0$ or $c_M = \iota_M$.

(b) More generally, let M be a linear subspace of $X \times X^*$ such that $\langle x, x^* \rangle \geq 0$ for all $(x,x^*) \in M$. In such a case M is the graph of a monotone multimap since for all (w,w^*), $(x,x^*) \in M$ one has $(w - x, w^* - x^*) \in M$. Proposition 9.17 asserts that c_M is convex, so that $g_M = c_M$ and $p_M = c_{\mathrm{cl}(M)}$ since $\mathrm{cl}(M)$ also is a linear subspace of $X \times X^*$ and $c \mid_{\mathrm{cl}(M)}$ is convex and continuous. In particular, when M is the graph of a monotone, continuous, linear map, one has $p_M = c_M$, a remarkable simple fact, whereas, setting $q_M(w) = (1/2)\langle w, Mw \rangle$,

$$f_M(x,x^*) = \sup_{w \in X} (\langle w, x^* + M^{\mathsf{T}} x \rangle - \langle w, Mw \rangle) = (2q_M)^*(x^* + M^{\mathsf{T}} x).$$

(c) A special case of the preceding example concerns the identity map I on a Hilbert space X identified to X^* by the Riesz' isomorphism. Then $p_I(x, x^*) = g_I(x, x^*) = c_I(x, x^*)$ and $f_I(x, x^*) = (1/4)\,\|x + x^*\|^2$.
(d) Let $M := J$, the duality map J of X and let q be the quadratic form on $X \times X^*$ given by $q(x, x^*) := (1/2)\,\|x\|^2 + (1/2)\,\|x^*\|_*^2$ as above. The definitions of p_M and $f_M := c_M^b$ yield $p_M \geq q$ and $f_M \leq q$.
(e) Let $M := \partial s$, where $s : X \to \mathbb{R}$ is a continuous sublinear function on X. Let $S := \partial s(0)$, so that $s(x) = \sup\{\langle x, w^*\rangle : w^* \in S\}$, $\partial s(w) = \{w^* \in S : c(w, w^*) = s(w)\}$ and $s^*(w^*) = \iota_S(w^*)$. Thus one has $c_M(x, x^*) = s(x) + \iota_S(x^*) = s(x) + s^*(x^*)$, hence $p_M(x, x^*) = s(x) + s(x^*) = f_M(x, x^*)$. □

Exercise Let M be a monotone multimap with nonempty graph and let $(x, x^*) \in X \times X^*$, $r \in \mathbb{P}$. Show that $p_{rM}(x, x^*) = rp_M(x, x^*/r)$ and $f_{rM}(x, x^*) = rf_M(x, x^*/r)$.

Exercise Let X be reflexive and let $M : X \rightrightarrows X^*$. Show that for all $(x, x^*) \in X \times X^*$ one has $p_{M^{-1}}(x^*, x) = p_M^{\mathsf{T}}(x, x^*)$, $f_{M^{-1}}(x^*, x) = f_M^{\mathsf{T}}(x, x^*)$.

Let us show that a maximally monotone multimap M is represented by the functions f_M and p_M.

Theorem 9.26 *For a nonempty subset M of $X \times X^*$ and $h : X \times X^* \to \mathbb{R}_\infty$, the following assertions are equivalent:*

(a) M is maximally monotone;
(b) $c_M \geq p_M \geq f_M \geq c$ and $M = M_{f_M} = M_{p_M}$;
(c) if h satisfies $h = h^{bb}$ and $p_M \geq h \geq f_M$, then $h \in \mathcal{H}_M$ and $M_h = M$;
(d) for $h := f_M$ one has $h \in \mathcal{H}_M$ and $M_h = M$;
(e) the function $h := f_M$ satisfies $h \geq c$ and $M_h = M$.

Proof (a)⇒(b) When M is monotone we already know that $c_M \geq c_M^{bb} =: p_M \geq f_M$ and $\inf\{\langle x - w, x^* - w^*\rangle : (w, w^*) \in M\} = 0$ for all $(x, x^*) \in M$. Since

$$f_M(x, x^*) := c_M^b(x, x^*) := \sup\{\langle x, w^*\rangle + \langle w, x^*\rangle - \langle w, w^*\rangle : (w, w^*) \in M\}$$
$$= \langle x, x^*\rangle - \inf\{\langle x - w, x^* - w^*\rangle : (w, w^*) \in M\}$$

we get $f_M(x, x^*) = \langle x, x^*\rangle$ for all $(x, x^*) \in M$. When M is maximally monotone, for $(x, x^*) \in X \times X^* \backslash M$ we have $\inf\{\langle x - w, x^* - w^*\rangle : (w, w^*) \in M\} < 0$ hence $f_M(x, x^*) > \langle x, x^*\rangle$. Thus we have $M = M_{f_M}$ hence $M_{p_M} \subset M_{f_M} = M$; conversely, for $(x, x^*) \in M$ we have $c(x, x^*) = c_M(x, x^*) \geq p_M(x, x^*) \geq c(x, x^*)$, hence $(x, x^*) \in M_{p_M}$.

(b)⇒(c) and (c)⇒(d) are obvious.

(d)⇒(e) This follows from the inclusion $\mathcal{H}_M \subset \mathcal{G}$.

(e)⇒(a) For $g := h := f_M$ one has $g^{bb} = g$ and $g^b = p_M \in \mathcal{H}_M \subset \mathcal{G}$ and $M = M_g$, so that Theorem 9.25 ensures that M is maximally monotone. □

Corollary 9.9 *If M is maximally monotone, then for $h : X \times X^* \to \mathbb{R}_\infty$ satisfying $h = h^{bb}$, $p_M \geq h \geq f_M$, one has $p_M \geq h^b \geq f_M$ and, denoting by π_X the canonical projection from $X \times X^*$ onto X, one has*

$$\operatorname{co} D(M) \subset \pi_X(\operatorname{dom} h) \cap \pi_X(\operatorname{dom} h^b).$$

Proof Since $p_M = f_M^b$ and $f_M = p_M^b$, the relations $p_M \geq h^b \geq f_M$ follow from the fact that $h \mapsto h^b$ is antitone. Thus, assertion (c) of the preceding theorem ensures that h, $h^b \in \mathcal{H}_M$ and h and h^b play symmetric roles. Since $\operatorname{dom} h$ is convex, it suffices to prove that $D(M) \subset \pi_X(\operatorname{dom} h)$. Given $x \in D(M)$, we pick $x^* \in M(x)$; then $(x, x^*) \in M = M_h$, so that $h(x, x^*) = \langle x, x^* \rangle \in \mathbb{R}$ and $x \in \pi_X(\operatorname{dom} h)$. □

9.4.6 Surjectivity of Maximally Monotone Multimaps

Let us give important characterizations of maximal monotonicity using the duality multimap J of X or rather its graph $G(J)$.

Theorem 9.27 (Simons) *Let $M : X \rightrightarrows X^*$ be a monotone multimap such that $G(M) + G(-J) = X \times X^*$. Then M is maximally monotone.*

If X is reflexive, the converse holds.

Proof Let $(x, x^*) \in X \times X^*$ be monotonically related to M in the sense that

$$\langle x - w, x^* - w^* \rangle \geq 0 \qquad \forall (w, w^*) \in M.$$

Assuming M is monotone and $G(M) + G(-J) = X \times X^*$, we can find $(u, u^*) \in M$ and $(v, v^*) \in G(J)$ such that $(u, u^*) + (v, -v^*) = (x, x^*)$. Then we have

$$0 \leq \langle x - u, x^* - u^* \rangle = \langle v, -v^* \rangle = - \|v\|^2 = - \|v^*\|^2$$

hence $v = 0$, $v^* = 0$ and $(x, x^*) = (u, u^*) \in M$. Therefore M is maximally monotone.

Conversely, let M be maximally monotone and let $(z, z^*) \in X \times X^*$, X being reflexive. Since N given by $G(N) := G(M) - (z, z^*)$ is maximally monotone, replacing M with N, it suffices to show that $(0, 0) \in G(N) + G(-J)$.

Observing that q given by $q(x, x^*) := (1/2) \|x\|^2 + (1/2) \|x^*\|_*^2$ for $(x, x^*) \in X \times X^*$ is in $\mathcal{H}$ and $(1/2) \|x\|^2 + (1/2) \|x^*\|_*^2 \geq \|x\| . \|x^*\|_* \geq -\langle x, x^* \rangle$, by maximal monotonicity of M and the implication (a)⇒(b) of Theorem 9.26, we get that

$$f_N(x, x^*) + q(x, x^*) \geq c(x, x^*) - \langle x, x^* \rangle = 0.$$

Since q is convex, finite and continuous, the sandwich theorem yields some $(w^*, w) \in X^* \times X = (X \times X^*)^*$, $r \in \mathbb{R}$ such that

$$f_N(x, x^*) \geq \langle x, w^* \rangle + \langle w, x^* \rangle + r \geq -q(x, x^*) \qquad \forall (x, x^*) \in X \times X^*,$$

or $-f_N^b(w,w^*) \geq r \geq q^b(-w,-w^*)$ (this also follows from the Fenchel-Rockafellar theorem or the Attouch-Brézis theorem). Thus $f_N^b(w,w^*)+q^b(-w,-w^*) \leq 0$. Since $\langle w,w^*\rangle \leq p_N(w,w^*) = f_N^b(w,w^*)$ and $\langle -w,w^*\rangle \leq q^b(-w,-w^*) = q(-w,w^*)$, we get

$$(p_N(w,w^*) - \langle w,w^*\rangle) + (q(-w,w^*) - \langle -w,w^*\rangle) \leq 0,$$

whence $(w,w^*) \in M_{p_N} = G(N)$ and $(-w,w^*) \in M_q = G(J)$. Therefore $(0,0) \in G(N) + G(-J)$ and $(z,z^*) \in G(M) + G(-J)$. □

Let us give a more striking form to the preceding result in the reflexive case.

Theorem 9.28 (Minty, Rockafellar) *Let X be a reflexive Banach space. Then a monotone multimap $M : X \rightrightarrows X^*$ is maximally monotone if and only if one has $R(M+J) = X^*$.*

Proof We first note that the relation $G(M) + G(-J) = X \times X^*$ implies the equality $R(M+J) = X^*$ since for all $x^* \in X^*$ there exist some $(u,u^*) \in G(M)$ and $(v,v^*) \in G(-J)$ such that $(u,u^*)+(v,v^*) = (0,x^*)$, hence $v = -u$ and $v^* \in -J(v) = J(u)$ with $x^* = u^* + v^* \in (M+J)(u)$. Thus, if X is reflexive and if M is maximally monotone, the preceding theorem entails that $R(M+J) = X^*$.

Now let us suppose $R(M+J) = X^*$. In order to prove that M is maximally monotone, we assume that the reflexive space X is endowed with a compatible strictly convex norm, as the Kadec-Troyanski Renorming Theorem (Theorem 6.24) allows us to do. Then, the duality map J is single-valued and strictly monotone in the sense that for $u,\ v \in X$ with $u \neq v$ one has

$$\langle u - v, J(u) - J(v)\rangle > 0.$$

Let (x,x^*) be monotonically related to M :

$$\langle w - x, w^* - x^*\rangle \geq 0 \qquad \forall (w,w^*) \in M.$$

Since $R(M+J) = X^*$ there exists some $(w,w^*) \in M$ such that

$$x^* + J(x) = w^* + J(w). \tag{9.58}$$

Then

$$\langle w - x, J(w) - J(x)\rangle = -\langle w - x, w^* - x^*\rangle \leq 0.$$

By strict monotonicity of J we get $w = x$, hence $w^* = x^*$ by (9.58) and $(x,x^*) \in M$. Thus M is maximally monotone. □

Remark Let us give a direct proof of the relation $R(J+M) = X$, assuming that M is maximally monotone and X is reflexive. By Theorem 9.26 there exists $h \in \mathcal{H}$

such that $M = M_h$. Since $h \geq c$, Cauchy inequality $\langle x, x^* \rangle \geq -\frac{1}{2}(\|x\|^2 + \|x^*\|^2)$ implies that

$$h(x, x^*) \geq -\frac{1}{2}(\|x\|^2 + \|x^*\|^2) \qquad \forall (x, x^*) \in X \times X^*.$$

The sandwich theorem yields some $(w, w^*) \in X \times X^*$ and $r \in \mathbb{R}$ such that

$$h(x, x^*) \geq \langle x, w^* \rangle + \langle w, x^* \rangle + r \geq -\frac{1}{2}(\|x\|^2 + \|x^*\|^2) \qquad \forall (x, x^*) \in X \times X^*. \tag{9.59}$$

Choosing $(x, x^*) \in (-J^{-1}w^*, -Jw)$ (recall that J is onto), we get

$$r \geq \frac{1}{2}(\|w\|^2 + \|w^*\|^2).$$

Then, for any $(x, x^*) \in M$, relation (9.59) implies that

$$\langle x, x^* \rangle \geq \langle x, w^* \rangle + \langle w, x^* \rangle + \frac{1}{2}(\|w\|^2 + \|w^*\|^2).$$

Adding $\langle w, w^* - x^* \rangle - \langle x, w^* \rangle$ to both sides of this inequality, we get

$$\langle x - w, x^* - w^* \rangle \geq \langle w, w^* \rangle + \frac{1}{2}(\|w\|^2 + \|w^*\|^2) \geq 0 \qquad \forall (x, x^*) \in M.$$

By maximality of M we conclude that $(w, w^*) \in M$. Now, taking $(x, x^*) = (w, w^*)$ in the last inequalities, we get

$$0 \geq \frac{1}{2}(\|w\|^2 + \|w^*\|^2) + \langle w, w^* \rangle \geq 0,$$

hence $0 \geq \|w\|^2 + \|w^*\|^2 - 2\|w\| . \|-w^*\| = (\|w\| - \|-w^*\|)^2$ or $\|-w^*\| = \|w\|$ and $\langle w, -w^* \rangle \geq \|w\|^2$, so that $-w^* \in J(w)$. Thus $0 \in (M + J)(w)$. Given $z^* \in X^*$ and applying what precedes to the maximally monotone multimap with graph $M + (0, -z^*)$, we get $z^* \in (M + J)(X)$, hence $R(M + J) = X^*$. □

The following consequence clarifies the structure of the graph of a maximally monotone multimap.

Corollary 9.10 (Minty, Rockafellar) *The graph $G(M)$ of a maximally monotone multimap on a Hilbert space X is a Lipschitz submanifold of $X \times X$. More precisely, setting $P := (I + M)^{-1}$, the map*

$$w \mapsto (\frac{1}{2}(w + P(w)), \frac{1}{2}(w - P(w)))$$

is a bijective parameterization of $G(M)$ by X that is Lipschitzian as is its inverse.

Proof Since M and M^{-1} are maximally monotone, we know from Theorem 9.28 and Proposition 9.13 that P given by

$$G(P) := \{(x+y, x-y) : (x,y) \in M\}$$

is defined on the whole of X and is nonexpansive. For any $w \in X$ there exists some $(x,y) \in M$ such that $w = x + y$. Setting $z := x - y = P(w)$ we have $(\frac{1}{2}(w+z), \frac{1}{2}(w-z)) = (x,y) \in G(M)$. Conversely, for any $(x,y) \in G(M)$, setting $w := x+y, z := x-y$ we have $(x,y) = (\frac{1}{2}(w+P(w)), \frac{1}{2}(w-P(w)))$. Moreover, $w \mapsto (\frac{1}{2}(w+P(w)), \frac{1}{2}(w-P(w)))$ is nonexpansive and its inverse $(x,y) \mapsto (x+y, x-y)$ is also Lipschitzian. □

Remark The result can also be deduced from Proposition 9.13 since $G(M)$ is the image of $G(P)$ under the inverse S^{-1} of the isomorphism of Proposition 9.13 and since $G(P)$ is a Lipschitzian submanifold of $X \times X$ naturally parameterized by X as is the graph of any nonexpansive map. □

The following theorem is the prototype of several results in the same vein. Here a map $F : X \to X^*$ is said to be *scalarly coercive* if $\langle F(x), x\rangle / \|x\| \to \infty$ as $\|x\| \to \infty$. This condition is satisfied when F is a continuous linear map $A : X \to X^*$ that is positive definite in the sense that there exists some $c > 0$ such that $\langle Ax, x\rangle \geq c\|x\|^2$ for all $x \in X$. Note that a scalarly coercive map is *coercive* in the sense that $\|F(x)\| \to \infty$ as $\|x\| \to \infty$ since $\|F(x)\| . \|x\| \geq \langle F(x), x\rangle$.

Theorem 9.29 (Browder, Minty) *Let X be a reflexive Banach space, let $F : X \to X^*$ be a monotone, radially continuous map, and let b, $r \in \mathbb{R}_+$, $y^* \in bB_{X^*}$. If $\langle F(x), x\rangle / \|x\| \geq b$ for $x \in rS_X$, then the equation $F(x) = y^*$ has a solution $x \in rB_X$.*

In particular, a monotone, radially continuous, scalarly coercive map F is surjective.

In fact, any monotone, radially continuous, and coercive map F is surjective.

For the sake of simplicity we give the proof in the case when X is separable and reflexive, using the **Galerkin method**.

Proof When X is finite dimensional, the first assertion is just a rephrasing of Corollary 2.11. Let $(X_n)_{n\geq 0}$ be an increasing sequence of finite dimensional linear subspaces whose union W is dense in X. Let $j_n : X_n \to X$ be the canonical injection and let $p_n := j_n^{\mathsf{T}} : X^* \to X_n^*$ be its transpose. We identify the elements of X_n with their images by j_n when it is convenient. Setting $F_n := p_n \circ F \circ j_n$, we see that F_n is monotone and radially continuous, hence is continuous and for $x \in rS_{X_n}$ we have $\langle F_n(x), x\rangle = \langle F(j_n(x)), j_n(x)\rangle \geq b\|j_n(x)\| = b\|x\|$. By Corollary 2.11, given $y^* \in rB_{X^*}$, there exists some $x_n \in X_n$ such that $F_n(x_n) = p_n(y^*)$ and $\|x_n\| \leq r$. Since X^2 is reflexive and since F is bounding, we can find a subsequence $((x_{n(k)}, F(x_{n(k)})))$ of $((x_n, F(x_n)))$ that weakly converges to some $(x, x^*) \in X \times X^*$. Let us show that $F(x) = y^*$.

For $w \in W$, let $m \in \mathbb{N}$ be such that $w \in X_m$, so that, for $n \geq m$, we have

$$0 \leq \langle F(x_n) - F(w), j_n(x_n) - j_n(w) \rangle = \langle F_n(x_n) - F_n(w), x_n - w \rangle$$
$$= \langle p_n(y^*) - F_n(w), x_n - w \rangle = \langle y^* - F(w), j_n(x_n) - w \rangle.$$

Taking the limit over the subsequence, we get

$$0 \leq \langle y^* - F(w), x - w \rangle.$$

Since W is dense in X, for any $z \in X$ we can find a sequence (w_n) in W with limit z; moreover, since F is maximally monotone by Proposition 9.14 and demicontinuous by Proposition 9.15, we may suppose that $(F(w_n))$ weak* converges to $F(z)$. Passing to the limit in the last inequality with w changed into w_n we obtain

$$0 \leq \langle y^* - F(z), x - z \rangle.$$

By maximal monotonicity of F we conclude that $y^* = F(x)$.

Now, let us suppose F is just coercive (and monotone, radially continuous). Using Theorem 6.24 and endowing X with a norm whose dual norm is rotund, we see that for all $\varepsilon \in]0, 1]$ the map $F_\varepsilon := F + \varepsilon J$ is scalarly coercive since for all $x \in X$

$$\langle F_\varepsilon(x), x \rangle \geq \langle F(x) - F(0), x \rangle + \langle F(0), x \rangle + \varepsilon \|x\|^2 \geq - \|F(0)\| . \|x\| + \varepsilon \|x\|^2 .$$

Since J is demicontinuous and monotone, F_ε is radially continuous and monotone. By the preceding we can find $x_\varepsilon \in X$ such that $F(x_\varepsilon) + \varepsilon J(x_\varepsilon) = 0$. Then, the relation $\langle F(x_\varepsilon) - F(0), x_\varepsilon \rangle \geq 0$ implies that

$$\varepsilon \|x_\varepsilon\|^2 = \varepsilon \langle J(x_\varepsilon), x_\varepsilon \rangle = \langle -F(x_\varepsilon), x_\varepsilon \rangle \leq \|F(0)\| . \|x_\varepsilon\|$$

so that $\|F(x_\varepsilon)\| = \|\varepsilon J(x_\varepsilon)\| = \|\varepsilon x_\varepsilon\|$ is bounded. Since F is coercive, we get that (x_ε) is bounded for $\varepsilon \in]0, 1]$ and $(F(x_\varepsilon)) \to 0$ as $\varepsilon \to 0_+$. Then, for any weak limit point x of (x_ε), by the standard trick for monotone operators we get $F(x) = 0$. Changing F into $F - y^*$ we see that F is onto. □

Remark The assumption that X is reflexive cannot be dropped as the case $F = J$ shows. □

Exercise If X is a finite dimensional space, show that a surjective monotone map $F : X \to X^*$ is coercive. [Hint: assuming the contrary, find sequences $(r_n) \to \infty$, $(u_n) \to u$ in S_X such that $(F(r_n u_n))$ converges to some $v \in X^*$, take $x \in X$ such that $u + v = F(x)$ and get a contradiction to the relation $\langle F(r_n u_n) - F(x), r_n u_n - x \rangle \geq 0$.] □

Exercise Let X be a uniformly convex and uniformly smooth Banach space and let $F : X \to X^*$ be a continuous monotone map such that F^{-1} is locally bounded in the sense that each $x^* \in X^*$ has a neighborhood V such that $F^{-1}(V)$ is bounded. Prove that $F(X) = X^*$. [See [57, Thm 7].] □

Let us apply the preceding general solvability result to the case of partial differential equations in divergence form. For simplicity, we restrict our attention to the case of operators of order two over an open subset Ω of $\mathbb{R}^d$:

$$F(u)(x) := \sum_{i=1}^{d} -D_i F_i(x, Du(x)) \qquad x \in \Omega,\ u \in W_p^1(\Omega)$$

where for $i \in \mathbb{N}_d$ the function $F_i : \Omega \times \mathbb{R}^d \to \mathbb{R}$ is measurable in its first variable, continuous in its second variable, and satisfies the following conditions for some $p > 1,\ b, c > 0$, some $g, h \in L_q(\Omega)$ with $q := p/(p-1)$:

$$|F_i(x, y)| \le b\, \|y\|^{p-1} + g(x) \quad x \in \Omega,\ y \in \mathbb{R}^d \tag{9.60}$$

$$\sum_{i=1}^{d} F_i(x, y) y_i \ge c\, \|y\|^p - h(x), \quad x \in \Omega,\ y \in \mathbb{R}^d \tag{9.61}$$

$$\sum_{i=1}^{d} (F_i(x, y) - F_i(x, z))(y_i - z_i) \ge 0 \quad x \in \Omega,\ y,\ z \in \mathbb{R}^d, \tag{9.62}$$

with $y := (y_i)$, $z := (z_i)$. Note that (9.62) replaces the ellipticity condition of linear elliptic equations.

We introduce the generalized *Dirichlet form* $a(\cdot,\cdot)$ on $W_p^1(\Omega)$ given by

$$a(u, v) = \sum_{i=1}^{d} \int_{\Omega} F_i(x, Du(x)) D_i v(x) dx = \sum_{i=1}^{d} \langle F_i(\cdot, Du(\cdot)), D_i v(\cdot)\rangle,$$

for $u, v \in W_p^1(\Omega)$. Our assumptions ensure that it is well defined since $F_i(\cdot, Du(\cdot)) \in L_q(\Omega)$. Moreover, it satisfies the inequality

$$|a(u, v)| \le k(\|u\|_{1,p})\, \|v\|_{1,p} \tag{9.63}$$

for some function k depending on b and g. In particular, for each $u \in W_p^1(\Omega)$ the map $F(u) := a(u, \cdot)$ is a continuous linear form on $W_p^1(\Omega)$ and on any closed linear subspace X containing $W_{p,0}^1(\Omega)$. Note that the choice of X incorporates boundary conditions (Dirichlet conditions for $X = W_{p,0}^1(\Omega)$).

Let us show that the map $F : X \to X^*$ defined above satisfies the assumptions of Theorem 9.29. Inequality (9.60) ensures that $G : w \mapsto (F_i(\cdot, w(\cdot)))$ is *bounding*, i.e. maps bounded subsets of $L_p(\Omega)$ into bounded subsets of $L_q(\Omega)^d$. By a general property of Nemytskii operators, this implies that G is continuous from $L_p(\Omega)$ into

$L_q(\Omega)^d$, hence that F is continuous from X into X^*. Assumption (9.61) entails scalar coercivity of F:

$$a(u,u) \geq c \int_\Omega \|Du(x)\|^p \, dx - \|h\|_q \|Du\|_p$$

$$\frac{\langle F(u), u\rangle}{\|u\|_{1,p}} = \frac{a(u,u)}{\|u\|_{1,p}} \geq c \|u\|_{1,p}^{p-1} - \|h\|_q \to \infty \quad \text{as } \|u\|_{1,p} \to \infty.$$

Finally, the monotonicity of F can be derived from (9.62): for $u, v \in X$

$$\langle F(u) - F(v), u - v\rangle = a(u, u - v) - a(v, u - v)$$

$$\geq \sum_{i=1}^d \int_\Omega (F_i(\cdot, D_i u(\cdot)) - F_i(\cdot, D_i v(\cdot)))(D_i u(\cdot) - D_i v(\cdot)) \geq 0.$$

Thus Theorem 9.29 can be applied. We can conclude as in the following statement.

Corollary 9.11 *Under assumptions (9.60)–(9.61)–(9.62), for any f in the dual of $W^1_{p,0}(\Omega)$ the equation $F(u) = f$ has a solution.*

In particular, this result applies to the nonlinear model equation

$$-\sum_{i=1}^d D_i(|D_i u|^{p-2} D_i u) + u = f.$$

Exercises

1. Let H be a Hilbert space, let $X := L_2(\mathbb{R}_+, H)$, and let A be the densely defined operator with domain $D(A) := H^1(\mathbb{R}_+, H)$ given by $A(x) := x'$. Show that A is maximally monotone. [Hint: to prove maximality show that for any $u \in X$ there exists some $x \in D(A)$ such that $x' + x = u$.]
2*. (**Debrunner-Flor**) Let X be a Banach space, let $M : X \rightrightarrows X^*$ be a monotone operator and let C be a weak* compact, convex subset of X containing the range $R(M)$ of M. Let $f : C \to X$ be a weak* to norm continuous map. Show that there exists some $x^* \in C$ such that

$$\langle u + f(x^*), v - x^*\rangle \geq 0 \qquad \forall (u, v) \in M.$$

[See [209].]
3. Prove that the result of the preceding exercise implies Brouwer's Fixed Point Theorem. [Hint: given a compact convex subset K of a Euclidean space E and a continuous map $g : K \to K$, take for C a closed ball of E containing K in its

interior, $M := I_C$ and $f := g \circ p_K$, where p_K is the projection map onto K and show that if $x^* \in C$ is such that $\langle u + f(x^*), u - x^* \rangle \geq 0$ for all $u \in C$, then one has $x^* = f(x^*)$ and $x^* \in K$, $g(x^*) = x^*$.]

4*. (**Hammerstein equation**) Let $f : \mathbb{R}_+ \times \mathbb{R} \to \mathbb{R}$ be continuous, nondecreasing in its second variable and measurable in its first variable, with $f(\cdot, 0) \in X := L_2(\mathbb{R}_+)$, let $g \in X$ and let $k \in L_1(\mathbb{R})$. Prove that the integral equation

$$u(s) + \int_0^\infty k(r-s)f(r, u(r))dr = g(s) \qquad s \in \mathbb{R}_+$$

has a solution $u \in X$ when the following assumptions are satisfied:

(a) there exist $a \in X$ and $b \in \mathbb{R}$ such that $|f(r,t)| \leq a(r) + b\,|t|$ for all $(r,t) \in \mathbb{R}_+ \times \mathbb{R}$;
(b) there exist $c \in \mathbb{P}$ and $a' \in X$ such that $f(r,t) \geq ct^2 - a'(r)\,|t|$ for all $(r,t) \in \mathbb{R}_+ \times \mathbb{R}$;
(c) setting $K(v)(s) := \int_0^\infty k(r-s)v(r)dr$ for $v \in X$, $s \in \mathbb{R}_+$ one has $\int_0^\infty K(v)(s)v(s)ds \geq 0$.

[See [92, Example 11.2] [57, 58, 266].]

5*. (**Kenderov**) Let $M : X \rightrightarrows X$ be a maximally monotone multimap on a reflexive Banach space such that int($D(M)$) is nonempty. Show that there exists a dense $\mathcal{G}_\delta$ subset, i.e. a countable intersection of open subsets, G of int($D(M)$) such that M is single-valued on G and every selection of M is continuous at each point of G.

9.4.7 Sums of Maximally Monotone Multimaps

Now we intend to show that representative functions can be used to deal with sums of maximally monotone operators. They are also useful for compositions with linear maps, but we skip such a (related) subject. We recall that for $h, k : X \times X^* \to \mathbb{R}_\infty$ the function $(h \Box_2 k) : X \times X^* \to \mathbb{R}_\infty$ is defined by

$$(h \Box_2 k)(x, x^*) := \inf\{h(x, u^*) + k(x, v^*) : u^* + v^* = x^*\}.$$

Theorem 9.30 *Let X be a reflexive Banach space, let $M, N : X \rightrightarrows X^*$ be maximally monotone operators satisfying the condition*

$$\mathbb{R}_+(\operatorname{co} D(M) - \operatorname{co} D(N)) = X, \tag{9.64}$$

and let $h \in \mathcal{H}_M$, $k \in \mathcal{H}_N$ be such that $f_M \leq h \leq p_M$, $f_N \leq k \leq p_N$. Then $h \Box_2 k = (h^b \Box_2 k^b)^b$ is in $\mathcal{H}_{M+N}$ and $x^* \in (M+N)(x)$ if and only if $(h \Box_2 k)(x, x^*) = \langle x, x^* \rangle$. Moreover, $M + N$ is maximally monotone.

Proof Since $h \in \mathcal{H}_M$, $k \in \mathcal{H}_N$ we have $h\Box_2 k \geq c\Box_2 c = c$. Moreover, for $x \in X$, $x^* \in (M+N)(x)$, say $x^* = u^* + v^*$ with $u^* \in M(x)$, $v^* \in N(x)$, we have $(h\Box_2 k)(x,x^*) \leq h(x,u^*) + k(x,v^*) = c(x,u^*) + c(x,v^*) = c(x,x^*)$, hence $(h\Box_2 k)(x,x^*) = c(x,x^*)$. In Proposition 6.28 we take $Y := X^*$. Since $\operatorname{co} D(M) \subset \pi_X(\operatorname{dom} h^b)$ and $\operatorname{co} D(N) \subset \pi_X(\operatorname{dom} k^b)$ as seen by Corollary 9.9, condition (9.64) entails condition (6.25). Proposition 6.28 ensures that

$$(h^b\Box_2 k^b)^b = h^{bb}\Box_2 k^{bb} = h\Box_2 k, \tag{9.65}$$

so that $h\Box_2 k = (h\Box_2 k)^{bb}$ and $h\Box_2 k \in \mathcal{H}_{M+N}$. If $(h\Box_2 k)(x,x^*) = c(x,x^*)$, by Proposition 6.28 there exist u^*, v^* such that $x^* = u^* + v^*$ and $(h\Box_2 k)(x,x^*) = h(x,u^*) + k(x,v^*)$. Then

$$0 = (h\Box_2 k)(x,x^*) - \langle x,x^*\rangle = \{h(x,u^*) - \langle x,u^*\rangle\} + \{k(x,v^*) - \langle x,v^*\rangle\}$$

and since each of the bracketed terms is nonnegative, both of them are zero and so, by the implication (a)⇒(b) of Theorem 9.26, $u^* \in M(x)$, $v^* \in N(x)$ and $x^* \in (M+N)(x)$.

We also have $p_M = f_M^b \geq h^b \geq p_M^b = f_M$ and similarly $p_N \geq k^b \geq f_N$. Then $h^b\Box_2 k^b \geq c\Box_2 c = c$ and since $\operatorname{co} D(M) \subset \pi_X(\operatorname{dom} h^b)$, $\operatorname{co} D(N) \subset \pi_X(\operatorname{dom} k^b)$, as above we have $h^b\Box_2 k^b = (h^b\Box_2 k^b)^{bb}$, hence $h^b\Box_2 k^b \in \mathcal{H}$. Since $(h^b\Box_2 k^b)^b = h\Box_2 k \in \mathcal{H}$ it follows from Theorem 9.25 that $M + N = \{(x,x^*) : (h^b\Box_2 k^b)^b(x,x^*) = c(x,x^*)\}$ is maximally monotone. □

Corollary 9.12 (Rockafellar) *Let X be a reflexive Banach space, let $M, N : X \rightrightarrows X^*$ be maximally monotone multimaps. If the following condition is satisfied, in particular if $D(M) \cap \operatorname{int}(D(N)) \neq \varnothing$, then $M + N$ is maximally monotone:*

$$\operatorname{co}(D(M)) \cap \operatorname{int}\operatorname{co}(D(N)) \neq \varnothing.$$

Proof Let $C := \operatorname{co}(D(M))$, $D := \operatorname{co}(D(N))$ and let $a \in C \cap \operatorname{int}(D)$. Let $r > 0$ be such that $a + rB_X \subset D$. Then we have $rB_X \subset C - D$, hence $\mathbb{R}_+(C - D) = X$ and the theorem applies. □

Let us give an extension of Theorem 9.29 to multivalued maps. For this purpose, we say that a multimap $M : X \rightrightarrows X^*$ is *coercive* if there exists a function $\gamma : \mathbb{R}_+ \to \mathbb{R}$ satisfying $\lim_{r\to\infty} \gamma(r) = \infty$ such that

$$\|x^*\| \geq \gamma(\|x\|) \text{ for all } (x,x^*) \in M \text{ with } \|x\| \text{ large.} \tag{9.66}$$

This condition is obviously a generalization of coercivity to the multivalued case.

Theorem 9.31 *Let X be a reflexive space. If $M : X \rightrightarrows X^*$ is maximally monotone and coercive then M is surjective.*

Moreover, if $F : X \to X^*$ is a single-valued, monotone, radially weak continuous map and if $M + F$ is coercive, then, $M + F$ is surjective: for all $\overline{x}^* \in X^*$*

there exists some $\overline{x} \in D(M)$ *such that*

$$\overline{x}^* \in M(\overline{x}) + F(\overline{x}).$$

Proof We endow X with a strictly convex norm whose dual norm is strictly convex, so that the duality map J is single-valued. Without loss of generality we assume that $(0,0) \in M$. Given $\overline{x}^* \in X^*$ we have to find some $\overline{x} \in D(M)$ such that $\overline{x}^* \in M(\overline{x}) + F(\overline{x})$. By Theorem 9.28, since $\varepsilon^{-1}M$ is maximally monotone, for any $\varepsilon > 0$ there exists some $x_\varepsilon \in D(M)$ such that

$$\overline{x}^* \in M(x_\varepsilon) + \varepsilon J(x_\varepsilon).$$

Then, since M is monotone and $(0,0) \in M$, we have $\langle \overline{x}^* - \varepsilon J(x_\varepsilon), x_\varepsilon \rangle \geq 0$, hence

$$\varepsilon \left\| x_\varepsilon \right\|^2 \leq \left\| \overline{x}^* \right\| . \left\| x_\varepsilon \right\|$$

and $\varepsilon \left\| x_\varepsilon \right\| \leq \left\| \overline{x}^* \right\|$. Then, the net $(\overline{x}^* - \varepsilon J(x_\varepsilon))_{\varepsilon>0}$ is bounded, and, since $\overline{x}^* - \varepsilon J(x_\varepsilon) \in M(x_\varepsilon)$, by coercivity, we infer that (x_ε) is bounded as $\varepsilon \to 0$. We can find a sequence $(\varepsilon_n) \to 0$ and a weak limit $\overline{x}$ of the sequence (x_{ε_n}). Then, by Proposition 9.15 (c), $\overline{x}^* = \lim_n (\overline{x}^* - \varepsilon_n J(x_{\varepsilon_n}))$ belongs to $M(\overline{x})$.

The second assertion follows from the fact that F is maximally monotone by Proposition 9.14 and that $M + F$ is maximally monotone too by Corollary 9.12 since the domain of F is X. □

Exercises

1. Show that in Theorem 9.28 the duality map J can be replaced with $J_p := \frac{1}{p}\partial \left\| \cdot \right\|^p$ with $p \in]1,\infty[$. Such a modified duality map is adapted to spaces such as L_p spaces.
2. For a maximally monotone multimap $M : X \rightrightarrows X^*$, show that for all $h \in \mathcal{H}_M$ one has $f_M \leq h \leq p_M$.
3. Suppose that for some monotone multimap M one has $f_M \in \mathcal{H}_M$. Then prove that for some $(x, x^*) \in X \times X^*$ one has $f_M(x, x^*) = \langle x, x^* \rangle$ if and only if $p_M(x, x^*) = \langle x, x^* \rangle$. [Hint: use (9.57).]
4. Let $h \in \mathcal{H}$ be such that $h^b \in \mathcal{H}$. Show that $M_h := \{(x, x^*) : h(x, x^*) = c(x, x^*)\}$ is maximally monotone.
5. (**Kirszbraun-Valentine**) Let D be a nonempty subset of a Hilbert space X. Show that for any nonexpansive map $F : D \to X$ there exists a nonexpansive map $\widehat{F} : X \to X$ whose restriction to D is F. [Hint: associate to F a monotone multimap M as in Proposition 9.13 and take a maximally monotone multimap $\widehat{M}$ extending M and the associated nonexpansive map $\widehat{F}$.]
6. Let $X := L_2(\mathbb{R})$, $D(F) := \{x \in W_2^1(\mathbb{R}) : \lim_{|t|\to\infty} x(t) = 0\}$ and let $F : x \mapsto x'$ for $x \in D(F)$, $G := -F$. Verify that F and G are maximally monotone but $F + G$

is not maximally monotone. [Hint: given $y \in X$ one has $x + F(x) = y$ for x given by $x(t) = \int_{-\infty}^{t} e^{s-t} y(s)ds$, so that F is maximally monotone, whereas $F + G$ has a strict extension by 0 on X.]

9.4.8 Variational Inequalities

Several phenomena or processes have a one-sided character, either because time is involved or because obstacles are present. Models of such phenomena cannot be designed in the form of equations. Inequalities are more appropriate: see the motivation concerning industrial mathematics and the historical remarks provided in [88, 144, 145, 175, 188], [231, Section 9.1]. In this subsection we study such inequalities in which a subset C of a Banach space X and a map $F : C \to X^*$ are involved.

Hereafter we say that $F : C \to X^*$ is *C-(scalarly) coercive* if there exists some $\overline{v} \in C$ such that

$$\frac{\langle F(w) - F(\overline{v}), w - \overline{v}\rangle}{\|w - \overline{v}\|} \to +\infty \text{ as } \|w\| \to +\infty, \; w \in C. \tag{9.67}$$

This condition is independent of the choice of $\overline{v} \in C$ since it is equivalent to

$$(1/\|w\|)\langle F(w), w\rangle \to +\infty \text{ as } \|w\| \to +\infty, \; w \in C.$$

Thus, when $C = X$ it coincides with scalar coercivity. As already observed, this condition implies that F is coercive on C, i.e. that $\|F(x)\| \to \infty$ as $\|x\| \to \infty$ with $x \in C$. It is satisfied when F is the restriction to C of a continuous linear map $A : X \to X^*$ that is positive definite in the sense that there exists some $c > 0$ such that $\langle Ax, x\rangle \geq c \|x\|^2$ for all $x \in X$. This condition plays a role in one of the steps of the proof of the main result concerning such inequations. It is as follows.

Theorem 9.32 *Let C be a nonempty, closed, convex subset of a reflexive Banach space X, let $f \in X^*$, and let $F : C \to X^*$ be a monotone map that is radially weak* continuous. If either C is bounded or F is C-coercive, there exists some $u \in C$ such that*

$$\langle F(u), w - u\rangle \geq \langle f, w - u\rangle \qquad \forall w \in C. \tag{9.68}$$

Such an inequation is called a *variational inequality*. It can be interpreted as the search of $u \in C$ satisfying $f - F(u) \in N(C, u)$, the normal cone to C at u. When $C = X$ relation (9.68) is reduced to the equation $F(u) = f$.

Remark When F is strictly monotone in the sense that $\langle F(w) - F(z), w - z\rangle > 0$ whenever $w \neq z$, the solution is unique: if u and u' are solutions, taking $w := u'$

in (9.68) and then writing (9.68) with u' in place of u and taking $w = u$ we obtain the two inequalities

$$\langle F(u), u' - u\rangle \geq \langle f, u' - u\rangle \quad \text{and } \langle F(u'), u - u'\rangle \geq \langle f, u - u'\rangle,$$

hence $\langle F(u) - F(u'), u' - u\rangle \geq 0$ and $u = u'$. □

When X is a Hilbert space and F is Lipschitzian and *strongly monotone*, a simple proof can be given by using the contraction fixed point theorem.

Theorem 9.33 (Stampacchia) *Suppose X is a Hilbert space and F is Lipschitzian and strongly monotone in the sense that there exists some $c > 0$ such that for all v, $w \in C$ one has $\langle F(v) - F(w), v - w\rangle \geq c\,\|v - w\|^2$. Then (9.68) has a unique solution $u \in C$.*

In particular, when F is the restriction to C of a positive definite continuous linear map $A : X \to X$, (9.68) has a unique solution $u \in C$.

Proof Let ℓ be the Lipschitz rate of F and let $c > 0$ be as in the monotonicity assumption. We pick $t > 0$ in such a way that $0 < 1 - 2ct + \ell^2 t^2 < 1$ and we set $k := (1 - 2ct + \ell^2 t^2)^{1/2}$. We observe that (9.68) can be rewritten

$$\langle tf - tF(u) + u - u, w - u\rangle \leq 0 \qquad \forall w \in C$$

or $u = p_C(tf - tF(u) + u)$, where p_C is the projection map from X to C. Let $g : C \to C$ be given by $g(v) := p_C(tf - tF(v) + v)$ for $v \in C$. Since p_C is nonexpansive, for v, $w \in C$ we have

$$\|g(v) - g(w)\| \leq \|v - w - t(F(v) - F(w))\|,$$

hence

$$\begin{aligned}\|g(v) - g(w)\|^2 &\leq \|v - w\|^2 - 2t\langle v - w \mid F(v) - F(w)\rangle + t^2\,\|F(v) - F(w)\|^2 \\ &\leq (1 - 2ct + \ell^2 t^2)\,\|v - w\|^2 = k^2\,\|v - w\|^2\end{aligned}$$

by our choice of $k \in]0, 1[$. The Contraction Theorem ensures that there exists some $u \in C$ such that $g(u) = u$. That means that u is a solution to (9.68). Uniqueness stems from uniqueness of the solution of the equation $g(u) = u$. □

Remark The Contraction Theorem ensures that the solution u is the limit of the sequence (u_n) obtained by the following algorithm in which t is chosen in such a way that $1 - 2ct + \ell^2 t^2 \in]0, 1[$:

$$u_{n+1} = p_C(u_n - t(f - Fu_n)).$$

In the general case, a preliminary result plays a key role.

Lemma 9.7 (Minty) *Under the assumptions of Theorem 9.32 the variational inequality (9.68) is equivalent to finding $u \in C$ such that*

$$\langle F(w), w - u \rangle \geq \langle f, w - u \rangle \qquad \forall w \in C. \tag{9.69}$$

Proof Let $u \in C$ be a solution to (9.68). Then by the monotonicity of F we have

$$\langle F(w), w-u \rangle = \langle F(w)-F(u), w-u \rangle + \langle F(u), w-u \rangle \geq \langle F(u), w-u \rangle \geq \langle f, w-u \rangle$$

for all $w \in C$ so that u is a solution to (9.69). Conversely, if u is a solution to (9.69), for all $v \in C$ and $t \in]0, 1[$, taking $w := u + t(v - u) \in C$ in (9.69) and simplifying, we get

$$\langle F(u + t(v - u)), v - u \rangle \geq \langle f, v - u \rangle.$$

Using weak* radial continuity of F we get $\langle F(u), v - u \rangle \geq \langle f, v - u \rangle$ for all $v \in C$. □

A first step in the proof of Theorem 9.32 is given in the next lemma.

Lemma 9.8 *If X is finite dimensional and if C is a nonempty compact convex subset of X, then, for all $f \in X^*$, the variational inequality (9.68) has a solution u.*

Proof Using a scalar product $(\cdot \mid \cdot)$ on X and identifying X and X^* we see that the continuous map $x \mapsto p_C(x - F(x) + f)$ from C into C has a fixed point u by Brouwer's Theorem. The characterization of $p_C(u - F(u) + f)$ yields

$$\langle (u - F(u) + f) - u, w - u \rangle \leq 0 \quad \forall w \in C,$$

so that u is a solution to (9.68). □

Lemma 9.9 *If X is finite dimensional, if C is a nonempty closed convex subset of X, and if F is monotone, continuous and C-coercive, then for all $f \in X^*$ the variational inequality (9.68) has a solution.*

Proof Changing F into $F - f$, which is still monotone and C-coercive, we may suppose $f = 0$. For $r \in \mathbb{R}_+$ large enough we set $C_r := C \cap rB_X$ and we denote by $u_r \in C_r$ a solution to the variational inequality

$$\langle F(u_r), v - u_r \rangle \geq 0 \qquad \forall v \in C_r. \tag{9.70}$$

Let $\overline{v} \in C$ be as in (9.67) and let us show that for $r \geq \|\overline{v}\|$ large enough we have $\|u_r\| < r$. If, on the contrary, we have $\|u_r\| = r$ we get

$$\begin{aligned}\langle F(u_r), \overline{v} - u_r \rangle &= \langle F(\overline{v}), \overline{v} - u_r \rangle - \langle F(\overline{v}) - F(u_r), \overline{v} - u_r \rangle \\ &\leq \|\overline{v} - u_r\| \left(\|F(\overline{v})\|_* - \langle F(\overline{v}) - F(u_r), \overline{v} - u_r \rangle / \|\overline{v} - u_r\| \right) < 0\end{aligned}$$

by the C-coercivity assumption, a contradiction with (9.70) since $\overline{v} \in C_r$. Thus $\|u_r\| < r$ for r large enough and for all $w \in C$, taking $t \in]0, 1[$ small enough we have $v := u_r + t(w - u_r) \in C_r$, hence

$$t\langle F(u_r), w - u_r\rangle = \langle F(u_r), v - u_r\rangle \geq 0$$

and u_r is a solution to (9.68) with $f = 0$. □

Proof of Theorem 9.32 This uses a method known as the *Ritz-Galerkin method*. We first consider the case C is bounded, and without loss of generality we suppose $f = 0$. For $v \in C$ let

$$C(v) := \{w \in C : \langle F(v), v - w\rangle \geq 0\}.$$

Since C is weakly compact and $C(v)$ is weakly closed in C, $C(v)$ is weakly compact. We want to find some u in $C(v)$ for all $v \in C$, i.e. we want to prove that $\cap_{v \in C} C(v)$ is nonempty. If, on the contrary, this intersection is empty, then by compactness there exists a finite family $v_1, \ldots, v_n$ of elements of C such that $C(v_1) \cap \ldots \cap C(v_n) = \varnothing$. Let Y be the linear space spanned by $v_1, \ldots, v_n$ and let $j : Y \to X$ be the canonical injection. Its transpose $j^\intercal$ is the restriction map $r : X^* \to Y^*$ given by $r(x^*) = x^* \mid_Y$. The map $F_Y := r \circ F \circ j : Y \to Y^*$ is still monotone and radially continuous. Thus, the preceding lemmas yield some $u \in C \cap Y$ such that $\langle F_Y(v), v - u\rangle \geq 0$ for all $v \in C \cap Y$. Considering u as an element of X, this relation implies that $u \in C(v_1) \cap \ldots \cap C(v_n)$, a contradiction. Thus $\cap_{v \in C} C(v) \neq \varnothing$ and any element u of this intersection is a solution.

When C is unbounded, we use the C-coercivity assumption as in the preceding lemma to prove the existence of a solution. □

Corollary 9.13 *If X is a reflexive Banach space X and $F : X \to X^*$ is a monotone scalarly coercive map that is radially weak* continuous then F is surjective.*

Proof Taking $C := X$, for any $f \in X^*$ we can find $u \in X$ such that $\langle F(u) - f, u - w\rangle \geq 0$ for all $w \in X$. Thus $F(u) = f$. □

Exercise If in Stampacchia's theorem the map $A := F$ is linear, continuous, and *symmetric* in the sense that $\langle Av, w\rangle = \langle Aw, v\rangle$ for all $v, w \in X$, show that the solution u of (9.68) is the minimizer on C of the convex function $h : v \mapsto \frac{1}{2}\langle Av, v\rangle - \langle f, v\rangle$.

Exercise (Penalty Method) Assume X, C, f are as in Theorem 9.32, with $0 \in C$. Suppose $j : X \to \mathbb{R}$ is a convex function of class C^1, $F := J := Dj$ and $P : X \to X^*$ is a continuous monotone operator such that $P(w) = 0$ is equivalent to $w \in C$. Show that for every $r > 0$ the equation

$$F(u_r) + \frac{1}{r}P(u_r) = 0$$

has a solution $u_r \in C$ and that for a sequence $(r_n) \to 0_+$ the sequence (u_{r_n}) converges weakly to a solution u of (9.68). [See [231, Thm 9.3.7].]

Application Let Ω be an open subset of $\mathbb{R}^d$, let $p \in]2, \infty[$ and let $F : W^1_{p,0}(\Omega) \to (W^1_{p,0}(\Omega))^*$ be given by

$$\langle F(u), v \rangle := \int_\Omega \|\nabla u\|^{p-2} \langle \nabla u \mid \nabla v \rangle \qquad u, v \in W^1_{p,0}(\Omega).$$

Using the fact that the function $j : x \mapsto (1/p)\,\|x\|^p$ is convex, so that its subdifferential $\partial j : x \mapsto \|x\|^{p-2} x$ is monotone, one can see that F is monotone. The nonlinear map F plays an important role as an example of a partial differential equation.

Additional Reading

[19, 24, 26, 27, 32, 35, 37, 50, 53, 69, 70, 96, 116, 123, 131, 141, 153, 155, 171, 175, 180, 188, 192, 195, 202, 203, 211, 214–216, 230, 238, 242, 248–250, 252, 253, 259–261, 266–269]

Chapter 10
Evolution Problems

Fleet the time carelessly, as they did in the golden world.

W. Shakespeare, *As you like it.*

I have measured out my life with coffee spoons.

T.S. Eliot.

Abstract In this chapter, problems involving time are considered. Those expressed by means of ordinary differential equations are the simplest ones. In contrast to problems involving partial derivatives, they do not require the functions spaces introduced in the preceding chapter. But for parabolic problems and hyperbolic problems, Sobolev spaces are again crucial tools. The notion of a semigroup forms a natural and unifying framework for such problems. Notions of dissipativity and monotonicity again show their usefulness.

We devote this chapter to problems involving time. The equations describing the evolution problem being studied may be ordinary differential equations or partial differential equations. As in [2] we gather these two types of problem together in order to underline the analogies and the differences between them; on the other hand, we do not consider integral equations nor delay differential equations or equations with deviated arguments.

We give a particular attention to two equations (or rather systems). The first one is the *heat equation* that rules the propagation of heat in a medium:

$$\frac{\partial u}{\partial t}(x,t) - \Delta u(x,t) = 0, \quad u(x,0) = g(x) \quad (x,t) \in \Omega \times \mathbb{P}$$

where g is the temperature of the medium at time $t = 0$, which is supposed to be known, and $u(x,t)$ represents the temperature of point x in the open subset Ω of $\mathbb{R}^d$ at time $t \in \mathbb{P} :=]0,\infty[$. Here the Laplacian Δ is taken with respect to the space variable x. This equation is a typical example of a parabolic second-order equation. Other examples are the Fokker-Planck equation and the Komolgorov equation for the probabilistic study of diffusion processes.

J.-P. Penot, *Analysis*, Universitext, DOI 10.1007/978-3-319-32411-1_10

The second type of equation we study is the *wave equation*

$$\frac{\partial^2 u}{\partial t^2}(x,t) - \Delta u(x,t) = 0, \quad u(x,0) = g(x), \; \frac{\partial u}{\partial t}(x,0) = h(x) \quad (x,t) \in \Omega \times \mathbb{P}$$

that is so often studied in geophysics for petrol exploration. Under the form of the Schrödinger equation it governs the wave-like behavior of matter in quantum mechanics. This equation is a higher dimensional version of the simpler *vibrating string equation*

$$\frac{\partial^2 u}{\partial t^2}(x,t) - \frac{\partial^2 u}{\partial x^2}(x,t) = 0, \quad u(x,0) = g(x), \; \frac{\partial u}{\partial t}(x,0) = h(x) \quad (x,t) \in T \times \mathbb{P}$$

giving the shape of a vibrating string over an interval $T := [0, \theta[$ of $\mathbb{R}$; it was solved by J. Le Rond d'Alembert in the eighteenth century. Such equations are typical examples of hyperbolic equations.

We just describe elementary aspects of the approach to such equations, referring to specialized monographs for a complete study. We discard the many nonlinear evolution problems involving partial differential equations besides the ones involving dissipativity.

10.1 Ordinary Differential Equations

Whole books have been devoted to ordinary differential equations; see f.i. [11, 79, 91, 93, 152, 154, 194, 200, 244]. Thus, in this section we only present the main facts about this field, which has been very active during the last three centuries and which still experiences important discoveries. The understanding of turbulence and chaos are examples of these new insights. We also discard stability questions that are so important for mechanical phenomena. We concentrate our study on first-order differential equations since higher order equations can be reduced to first-order equations by introducing auxiliary unknown maps. In the case of the second-order differential equation

$$x''(t) = f(t, x(t), x'(t)), \qquad x(t_0) = x_0, \; x'(t_0) = y_0$$

we substitute it with the differential equation

$$(x'(t), y'(t)) = (y(t), f(t, x(t), y(t))), \qquad (x(t_0), y(t_0)) = (x_0, y_0)$$

and we observe that the so-called right-hand side of the latter has the same regularity as the right-hand side of the former.

10.1.1 Separation of Variables

In some cases it is possible to get an explicit solution to a scalar differential equation of the form

$$x'(t) = f(x(t)), \qquad x(t_0) = x_0. \tag{10.1}$$

In the traditional formalism one rewrites this equation in the form

$$\frac{dx}{f(x)} = dt \tag{10.2}$$

and if one knows a primitive g of $1/f$, one gets $g(x) - g(x_0) = t - t_0$. If, moreover, g has an inverse h on some interval T, one obtains $x(t) = h(t - t_0 + g(x_0))$. Relation (10.2) could be given a rigorous meaning by using the formalism of differential forms. We rather assume that a solution x of equation (10.1) has an inverse y on some interval containing t_0. Then we have

$$y'(x(t)) = \frac{1}{x'(t)} = \frac{1}{f(x(t))}.$$

Thus, if we can compute a primitive g of $1/f$, we may take $y(x) = g(x) - g(x_0) + t_0$ and then get $x(\cdot)$ by taking the inverse of $y(\cdot)$.

The following simple example shows that in general one cannot expect to get a solution defined everywhere.

Example Let us consider equation (10.1) with f given by $f(x) = x^2$, $t_0 := 0$. If $x_0 = 0$, then $x(\cdot) = 0$ is a solution to (10.1). If $x_0 \neq 0$, for t close to 0, one has $x(t) \neq 0$ and $g : x \mapsto -1/x + 1/x_0$ is the primitive of $1/f$ that takes the value 0 for $x = x_0$. The relation $-1/x(t) + 1/x_0 = t$ yields $x(t) = x_0(1 - x_0 t)^{-1}$. We observe that although f is of class C^∞ on $\mathbb{R}$, the solution is just defined on the interval $]-\infty, 1/x_0[$ if $x_0 > 0$ and on the interval $]1/x_0, \infty[$ if $x_0 < 0$. □

The classical method of separation of variables for ordinary differential equations can be extended to partial differential equations. As an example, let us consider the *heat equation* on some bounded open subset Ω of $\mathbb{R}^d$:

$$\frac{\partial u}{\partial t}(x,t) - \Delta u(x,t) = 0 \qquad (x,t) \in \Omega \times \mathbb{P} \tag{10.3}$$

$$u(x,t) = 0 \qquad (x,t) \in \partial\Omega \times \mathbb{P} \tag{10.4}$$

$$u(x,0) = g(x) \qquad x \in \Omega \tag{10.5}$$

where $g : \Omega \to \mathbb{R}$ is given and the Laplacian Δ bears on the space variable x as above. We look for a solution of the form

$$u(x,t) := v(t)w(x) \qquad (x,t) \in \Omega \times \mathbb{R}_+. \tag{10.6}$$

Computing $\frac{\partial u}{\partial t}$ and Δu we get

$$v'(t)w(x) - v(t)\Delta w(x) = 0.$$

Assuming that w is an eigenfunction of Δ, i.e. that for some $\lambda \in \mathbb{R}$ we have

$$\Delta w(x) = \lambda w(x),$$

taking a solution v to the differential equation

$$v'(t) = \lambda v(t),$$

so that $v(t) = ce^{\lambda t}$ for some $c \in \mathbb{R}$, we obtain that the function u given by (10.6) solves (10.3). Assuming g can be represented as the sum of a series

$$g = \sum_{n=0}^{\infty} c_n w_n$$

with $c_n \in \mathbb{R}$, w_n being an eigenfunction of Δ with associated eigenvalue λ_n, we can expect u given by

$$u(x,t) = \sum_{n=0}^{\infty} c_n e^{\lambda_n t} w_n$$

is a solution of the heat equation. Of course, the meaning of this series has to be made precise, as does the one for the initial value g.

As a variant, let us consider the *porous medium equation* (see [130])

$$\frac{\partial u}{\partial t}(x,t) - \Delta u^{\gamma}(x,t) = 0 \qquad (x,t) \in \Omega \times \mathbb{P}$$

where $\gamma > 1$ is a fixed constant and Ω is an open subset of $\mathbb{R}^d$. Again, we look for a solution u given by relation (10.6), so that for some constant c we must have

$$\frac{v'(t)}{v^{\gamma}(t)} = c = \frac{\Delta w^{\gamma}(x)}{w(x)} \qquad (x,t) \in \Omega \times \mathbb{P},\ v(t) \neq 0, w(x) \neq 0.$$

The differential equation $v'(t) = cv^{\gamma}(t)$ yields $v(t) = ((1-\gamma)ct + b)^{1/(1-\gamma)}$ for some $b \in \mathbb{R}_+$. Looking for a solution w of the equation $\Delta w^{\gamma}(x) = cw(x)$ in the form $w(x) := a\,\|x\|^{\beta}$ with $a, \beta \in \mathbb{P}$, we see that we must have

$$0 = cw(x) - \Delta w^{\gamma}(x) = ac\,\|x\|^{\beta} - a^{\gamma}\beta\gamma(d + \beta\gamma - 2)\,\|x\|^{\beta\gamma-2},$$

hence $\beta = \beta\gamma - 2$, $\beta = 2/(\gamma - 1)$ and $ac = a^{\gamma}\beta\gamma(d + \beta\gamma - 2)$ or $c = a^{\gamma-1}\beta\gamma(d + \beta\gamma - 2)$.

Another example of the method of separation of variables occurs with the *Hamilton-Jacobi equation*

$$\frac{\partial u}{\partial t}(x,t) + H(\nabla u(x,t)) = 0 \qquad (x,t) \in \Omega \times \mathbb{R}_+, \tag{10.7}$$

where the gradient ∇ bears on the space variable x and $H : \mathbb{R}^d \to \mathbb{R}$, the Hamiltonian, is given. This time we decompose u in an additive manner by taking

$$u(x,t) = v(t) + w(x).$$

Equation (10.7) then becomes

$$v'(t) + H(\nabla w(x)) = 0 \qquad (x,t) \in \Omega \times \mathbb{R}_+.$$

Such a relation requires that there exists some constant $c \in \mathbb{R}$ such that $v'(t) = c$, $H(\nabla w(x)) = -c$ for all $(x,t) \in \Omega \times \mathbb{R}_+$. In particular, when the initial data is $u(x,0) := \langle a \mid x \rangle + b$ for some $a \in \mathbb{R}^d$, $b \in \mathbb{R}$, we get that for $c := -H(a)$, we have

$$u(x,t) = \langle a \mid x \rangle + ct + b.$$

10.1.2 Existence Results

We already proved the existence of a local solution to the differential equation

$$x'(t) = f(t, x(t)) \qquad x(0) = x_0$$

when the right-hand side f is Lipschitzian in the state variable x. In this subsection and the next one we prove such a result by using a different method and we consider the dependence on the initial condition and on parameters. We also present a more general existence result involving partial derivatives that proves to be useful in differential geometry and in the consumer theory of mathematical economics.

Theorem 10.1 (Frobenius) *Let G be an open subset of the product $X \times Y$ of two Banach spaces and let $f : G \to L(X, Y)$ be a map of class C^1. For all $(x_0, y_0) \in G$*

the differential equation

$$Dw_y(x) = f(x, w_y(x)) \qquad w_y(x_0) = y \tag{10.8}$$

has a local solution w_y of class C^1 defined on some neighborhood U of x_0 whenever y belongs to some neighborhood V of y_0 if and only if f satisfies the relation

$$D_1 f(x,y).u.v + D_2 f(x,y).(f(x,y).u).v = D_1 f(x,y).v.u + D_2 f(x,y).(f(x,y).v).u \tag{10.9}$$

for all $(x, y) \in G$, $(u, v) \in X^2$. In this case, the map $w : (x, y) \mapsto w_y(x)$ is of class C^1 from $U \times V$ into Y.

When $X := \mathbb{R}^d$, Theorem 10.1 can be formulated in terms of partial derivatives; we encourage the reader to write down such a formulation.

We observe that when $X := \mathbb{R}$ equation (10.8) reduces to an ordinary differential equation since we identify $L(\mathbb{R}, Y)$ with Y via $\ell \mapsto \ell.1$ and $Dw_y(x)$ with $w'_y(x) := Dw_y(x)(1)$. In such a case, relation (10.9) is automatically satisfied since $D_1 f(x,y).u.v = uvD_1 f(x,y).1.1$ and $D_2 f(x,y).(f(x,y).u).v = uvD_2 f(x,y).(f(x,y).1).1$ with similar relations obtained by interchanging u and v. These remarks lead to the next statement in which the notation is slightly changed.

Theorem 10.2 *Let E be a Banach space, let G be an open subset of $\mathbb{R} \times E$ and let $f : G \to E$ be a map of class C^k, $k \geq 1$. Then, for all $(t_0, x_0) \in G$ there exist an open interval T containing t_0, an open neighborhood V of x_0, and a map $w : T \times V \to E$ of class C^k such that $(t, w(t, x)) \in G$ for all $(t, x) \in T \times V$ and*

$$\frac{d}{dt} w(\cdot, x)(t) = f(t, w(t, x)) \qquad \forall (t, x) \in T \times V, \tag{10.10}$$

$$w(t_0, x) = x \qquad \forall x \in V. \tag{10.11}$$

Proof of Theorem 10.1 Let us first show that condition (10.9) is necessary in order that the assertion about the existence of a local solution be satisfied. Equation (10.8) shows that Dw_y is of class C^1, hence that the local solution w_y is of class C^2. In such a case, Schwarz' Theorem ensures that $D^2 w_y(x)$ is a symmetric bilinear map. Differentiating $x \mapsto f(x, w_y(x))$, we deduce (10.9) from this symmetry property.

Now let us show that relation (10.9) is sufficient to obtain the existence of a local solution. Without loss of generality we suppose that $(x_0, y_0) = (0, 0)$ and that $G := X_0 \times Y_0$, with $X_0 = B(0, r)$, $Y_0 = B(0, r)$. We denote by T the interval $[0, 1]$ and by $C(T, Y)$ the space of continuous maps from T to Y endowed with the norm $\|\cdot\|_\infty : z \mapsto \sup_{t \in T} \|z(t)\|$, by $C^1_0(T, Y)$ the space of maps z of class C^1 from T to Y satisfying $z(0) = 0$ endowed with the norm $\|\cdot\|_{1,\infty} : z \mapsto \sup_{t \in T} \|z'(t)\|$. By Corollary 5.14 both $C(T, Y)$ and $C^1_0(T, Y)$ are Banach spaces and the derivation $D : z \mapsto z'$ is an isometry from $Z := C^1_0(T, Y)$ onto $C(T, Y)$.

Let $V_0 := B(0, r/2) \subset Y$ and let

$$Z_0 := \{z \in Z := C_0^1(T, Y) : z(T) \subset V_0\},$$

so that for all $y \in V_0$ and $z \in Z_0$ we have $y + z(t) \in Y_0$ for all $t \in T$. Let us define a map $g : X_0 \times V_0 \times Z_0 \to C(T, Y)$ by

$$g(x, y, z)(t) := z'(t) - f(tx, y + z(t)).x \qquad (x, y, z) \in X_0 \times V_0 \times Z_0, \ t \in T.$$

As in Sect. 5.7.1, it can be shown that g is of class C^1 and since for all $t \in T$ the evaluation map $c \mapsto c(t)$ from $C(T, Y)$ to Y is linear and continuous, setting $\overline{g}(x, y, z, t) := g(x, y, z)(t)$, we see that the partial derivatives $D_i g$ of g are characterized by

$$(D_1 g(x, y, z).u)(t) = D_1 \overline{g}(x, y, z, t).u \tag{10.12}$$

$$(D_2 g(x, y, z).v)(t) = D_2 \overline{g}(x, y, z, t).v \tag{10.13}$$

$$(D_3 g(x, y, z).w)(t) = D_3 \overline{g}(x, y, z, t).w \tag{10.14}$$

for all $(x, y, z, t) \in X_0 \times Y_0 \times Z_0 \times T$ and all $(u, v, w) \in X \times Y \times Z$, so that in particular,

$$(D_3 g(x, y, z).w)(t) = w'(t) - D_2 f(tx, y + z(t)).w(t).x$$

and $D_3 g(0, 0, 0).w = w'$. Thus $D_3 g(0, 0, 0)$ is the isomorphism $D : C_0^1(T, Y) \to C(T, Y)$ given by $Dw := w'$. The implicit function theorem yields open balls U, V centered at 0 in X_0 and Y_0 respectively and a map $h : U \times V \to Z_0$ of class C^1 such that $h(0, 0) = 0$ and $g(x, y, h(x, y)) = 0$ for all $(x, y) \in U \times V$. Without loss of generality, we assume that for all $(x, y) \in U \times V$ the linear map $D_3 g(x, y, h(x, y))$ is invertible. Taking the partial derivative with respect to y in the relation $g(0, y, h(0, y)) = 0$ and using the fact that $D_2 g(0, y, z) = 0$ for all $(y, z) \in V \times Z_0$, we obtain $D_3 g(0, y, h(0, y)) \circ D_2 h(0, y) = 0$, hence $D_2 h(0, y) = 0$ and $h(0, y) = h(0, 0) = 0$ for all $y \in V$. Setting for $(x, y) \in U \times V$, $t \in T$

$$h_{x,y}(t) := h(x, y)(t), \quad w_{x,y}(t) := y + h_{x,y}(t), \quad w_y(x) := w_{x,y}(1),$$

we see that $w_y(0) = y$ for all $y \in V$ and, by definition of $h(x, y)$, we have

$$w'_{x,y}(t) = h'_{x,y}(t) = f(tx, y + h_{x,y}(t)).x. \tag{10.15}$$

It remains to show that

$$Dw_y(x).u = f(x, w_y(x)).u \qquad \forall (x, y) \in U \times V, \ \ u \in X.$$

For $t \in T$, $(x, y) \in U \times V$, $u \in X$, let us set

$$j(t) := j_{x,y,u}(t) := tf(tx, w_{x,y}(t)).u.$$

The derivative $j'(t)$ of j at t is given by

$$f(tx, w_{x,y}(t)).u + tD_1f(tx, w_{x,y}(t)).x.u + tD_2f(tx, w_{x,y}(t)).f(tx, w_{x,y}(t)).x.u$$

or, in view of assumption (10.9), by

$$f(tx, w_{x,y}(t)).u + tD_1f(tx, w_{x,y}(t)).u.x + tD_2f(tx, w_{x,y}(t)).f(tx, w_{x,y}(t)).u.x.$$

On the other hand, taking the derivative with respect to x in the relation $g(x, y, h(x, y)) = 0$, we get for all $(x, y) \in U \times V$, $u \in X$, $t \in T$

$$D_3g(x, y, h(x, y))(D_1h(x, y).u) = -D_1g(x, y, h(x, y)).u.$$

Using relations (10.12) and (10.14), this equality means that

$$D_3\overline{g}(x, y, h_{x,y}(t), t).(D_1h(x, y).u)(t) = f(tx, w_{x,y}(t)).u + tD_1f(tx, w_{x,y}(t)).u.x.$$

Plugging the second expression of $j'(t)$ in the computation of D_3g yields

$$\begin{aligned} D_3\overline{g}(x, y, h(x, y)(t), t).j(t) &= j'(t) - D_2f(tx, y + h_{x,y}(t)).j(t).x \\ &= f(tx, w_{x,y}(t)).u + tD_1f(tx, w_{x,y}(t)).u.x. \end{aligned}$$

By the injectivity of $D_3\overline{g}(x, y, h(x, y)(t), t)$ we obtain $(D_1h(x, y).u)(t) = j(t)$ for all $t \in T$, and in particular, for $t := 1$. With relation (10.15) we get $Dw_y.u = (D_1h(x, y).u)(1) = j(1) = f(x, w_y(x)).u$, so that $Dw_y(x) = f(x, w_y(x))$. □

Theorem 10.2 includes a dependence statement upon the initial position x or x_0, and a similar result is obtained in Theorem 10.1. This dependence can be extended to the dependence upon the initial condition (t_0, x_0) and even upon a parameter p.

Theorem 10.3 *Let E, P be Banach spaces, let G be an open subset of $\mathbb{R} \times E \times P$ and let $f : G \to E$ be a map of class C^k, $k \geq 1$. Then, for all $(t_0, x_0, p_0) \in G$ there exist an open interval T containing t_0, open neighborhoods U, V of p_0 and x_0 respectively, and a map $w : T^2 \times V \times U \to E$ of class C^k such that $(t, w(t, s, x, p), p) \in G$ for all $(t, s, x, p) \in T^2 \times V \times U$ and*

$$\frac{dw}{dt}(\cdot, s, x, p)(t) = f(t, w(t, s, x, p), p) \qquad \forall (t, s, x, p) \in T^2 \times V \times U, \tag{10.16}$$

$$w(s, s, x, p) = x \qquad \forall (s, x, p) \in T \times V \times U. \tag{10.17}$$

Proof This follows from the proof of Theorem 10.1 and the extension of the implicit function theorem to the case when the equation depends on a parameter. □

One can also deduce this result from the study of the differential equation

$$(x'(t), p'(t)) = (f(t, x(t), p(t)), 0).$$

Let us also observe that equations (10.10)–(10.11) can be reduced to the case of an autonomous differential equation

$$(s'(t), x'(t)) = (1, f(s(t), x(t)))$$

whose solution is of the form $(s(t), x(t)) = (t + s_0, x(t))$.

In finite dimensions, a local existence result can be given under a continuity assumption, but it does not assert local uniqueness. We admit it. The proof relies on the Ascoli-Arzela Theorem and the Brouwer's Fixed Point Theorem.

Theorem 10.4 (Arzela, Peano) *Let E be a finite dimensional normed space, let G be an open subset of $\mathbb{R} \times E$ and let $f : G \to E$ be a continuous map. Then, for any $(t_0, x_0) \in G$ there exist an interval T of $\mathbb{R}$ containing t_0 and a map $x : T \to E$ such that $x(t_0) = x_0$, $(t, x(t)) \in G$, and $x'(t) = f(t, x(t))$ for all $t \in T$.*

A local existence and uniqueness result can be given following the lines of Theorem 2.14 under the assumption that the right-hand side f is locally Lipschitzian. Its conclusion is similar to that of Theorem 10.2, but with less regularity. We return to such a question in the next subsection.

Exercises

1. Verify that the solution of the equation $x'(t) = x(t)^3$ taking the value $x_0 \in E := \mathbb{R}$ for $t = t_0$ is given by $x(t) = x_0(1 - 2x_0^2(t - t_0))^{-1/2}$ for $t \in]-\infty, t_0 + x_0^{-2}/2[$ and $x_0 \neq 0$ and $x(t) = 0$ if $x_0 = 0$. Draw the integral curves in the phase space $\mathbb{R} \times E$.
2. Verify that the solution of the equation $x'(t) = x(t)^{2/3}$ taking the value $x_0 \in E := \mathbb{R}$ for $t = t_0$ is given by $x(t) = (x_0^{1/3} + (t - t_0)/3)^3$ for $t \in]t_0 - 3x_0^{1/3}, +\infty[$.
3. Let $E := \mathbb{R}$ and let $f : E \to E$ be given by $f(e) = 2e^{1/2}$ for $e \in \mathbb{R}_+$, $f(e) = -2|e|^{1/2}$ for $e \in \mathbb{R}\backslash\mathbb{R}_+$. Verify that the equation $w'(t) = f(w(t))$ has two solutions on $t \in]t_0, +\infty[$ taking the value $x_0 := 0$ for $t = t_0$, given by $x(t) = (t - t_0)^2$, $y(t) = -x(t)$. Note that f is not Lipschitzian near 0.
4. Let E be a Banach space and let T be an open interval of $\mathbb{R}$. Suppose that for some continuous function $k : T \to \mathbb{R}_+$ a map $f : T \times E \to E$ satisfies $\|f(t, e_1) - f(t, e_2)\| \leq k(t) \|e_1 - e_2\|$ for $(t, e_1, e_2) \in T \times E^2$. Prove that for any $(t_0, x_0) \in T \times E$ there exists a unique solution $x(\cdot)$ on T of the equations $x'(\cdot) = f(\cdot, x(\cdot))$, $x(t_0) = x_0$.

5. Let $E := c_0$ be the space of sequences $x := (x_n)$ of real numbers satisfying $\lim_n x_n = 0$. Suppose E is endowed with the norm $x \to \|x\| = \sup_n |x_n|$. Let $f : E \to E$ be given by $f(x) = (y_n)$ with $y_n := |x_n|^{1/2} + (n+1)^{-1}$. Verify that f is continuous but that there is no solution to the equation $x'(t) = f(x(t))$ on some open interval containing 0 satisfying $x(0) = 0$. Note that E is infinite dimensional.

10.1.3 Uniqueness and Globalization of Solutions

A globalization of Theorem 10.1 can be devised, but we restrict our attention to the more classical globalization of Theorem 10.2. We retain its notation and we assume its conclusion, namely local existence rather than smoothness assumptions. Given $(t_0, x_0) \in G$, we define an order relation on the set of pairs (T, u) where T is an open interval of $\mathbb{R}$ containing t_0 and $u : T \to E$ is a map of class C^1 such that $u(t_0) = x_0$, $(t, u(t)) \in G$ and $u'(t) = f(t, u(t))$ for all $t \in T$: we write $(T_1, u_1) \preceq (T_2, u_2)$ if $T_1 \subset T_2$ and $u_2 \mid T_1 = u_1$. It is easy to see that these conditions define an order and that this order is inductive: if $((T_i, u_i))_{i\in I}$ is a totally ordered family, the pair (T, u) given by $T := \cup_{i\in I} T_i$, $u \mid T_i = u_i$ is a majorant of the family $((T_i, u_i))_{i\in I}$. Thus, by Zorn's Lemma, there exists a maximal solution of the equation $x'(t) = f(t, x(t))$ for all $t \in T$, $x(t_0) = x_0$. Note that this existence result does not assume uniqueness.

Example Let $E := \mathbb{R}, f(e) = 3e^{2/3}$ for $e \in E$, and let $x_0 = 0$. Then the functions $t \mapsto x(t) = (t - t_0)^3$ and $y = 0$ are two distinct maximal solutions defined on $\mathbb{R}$.

When local existence and local uniqueness are ensured, one can obtain existence and uniqueness of maximal solutions without using Zorn's Lemma. In fact one has a largest solution (T_0, u_0) for the order defined above. It is obtained by taking for T_0 the union of the open intervals T containing t_0 on which a solution u exists and by setting $u_0(t) = u(t)$ for $t \in T$. This defines u_0 unambiguously since if u and v are solutions on intervals S and T containing t_0 respectively one has $u(t) = v(t)$ for $t \in S \cap T$ as a connectedness argument shows along with local uniqueness.

The domain of this maximal (or rather maximum) solution is as large as expected in the particular case of the next statement.

Proposition 10.1 *Let $f : \mathbb{R} \times E \to E$ be continuous and such that for any compact interval T of $\mathbb{R}$ there exists some $c_T \in \mathbb{R}_+$ such that $\|f(t, e_1) - f(t, e_2)\| \le c_T \|e_1 - e_2\|$ for all $t \in T$, $e_1, e_2 \in E$. Then for every $(t_0, x_0) \in \mathbb{R} \times E$ the maximal solution of the equation $w'(t) = f(t, w(t))$ satisfying $w(t_0) = x_0$ is defined on the whole of $\mathbb{R}$.*

In particular, this conclusion holds whenever for some continuous maps $A : \mathbb{R} \to L(E) := L(E, E)$, $b : \mathbb{R} \to E$, one has $f(t, e) = A(t).e - b(t)$ for all $(t, e) \in \mathbb{R} \times E$.

Proof This follows from Theorem 2.14 asserting that $w(\cdot)$ is uniquely defined on any bounded interval of $\mathbb{R}$. □

The following lemma is useful when seeking estimates for solutions to evolution problems.

Lemma 10.1 (Gronwall) *Let a, $b : T := [0, \tau] \to \mathbb{R}$ be two continuous (or regulated) nonnegative functions. If $u : T \to \mathbb{R}$ is a continuous (or regulated) nonegative function satisfying the inequality*

$$u(t) \leq \int_0^t a(s)u(s)ds + b(t) \qquad t \in T \tag{10.18}$$

then, setting $A(t) := \int_0^t a(s)ds$, u is bounded above by

$$u(t) \leq e^{A(t)} \int_0^t e^{-A(s)}a(s)b(s)ds + b(t) \qquad t \in T. \tag{10.19}$$

In particular, if $a(\cdot)$ is a constant function with value $\alpha \in \mathbb{P} :=]0, \infty[$ and if $b(t) := \kappa t + \lambda$ for $t \in T$, with κ, $\lambda \in \mathbb{R}_+$, then one has

$$u(t) \leq (\lambda + \frac{\kappa}{\alpha})e^{\alpha t} - \frac{\kappa}{\alpha} \qquad t \in T.$$

Proof We just prove the estimate (10.19), the special case being obtained by an integration by parts. We set

$$y(t) := \int_0^t a(s)u(s)ds, \qquad z(t) := e^{-A(t)}y(t).$$

On T (or the complement of a countable subset of T) relation (10.18) implies that

$$y'(t) - a(t)y(t) \leq a(t)b(t),$$

so that

$$z'(t) \leq e^{-A(t)}a(t)b(t)$$

and $z(t) \leq \int_0^t e^{-A(s)}a(s)b(s)ds$ since $z(0) = 0$. Thus

$$y(t) = e^{A(t)}z(t) \leq e^{A(t)} \int_0^t e^{-A(s)}a(s)b(s)ds$$

and (10.19) follows from the fact that $u(t) \leq y(t) + b(t)$. □

With the help of Gronwall's Lemma we shall obtain an estimate of the distance between two approximate solutions starting from two different points.

Proposition 10.2 *Let c, δ, ε, η be positive numbers, let E be a Banach space, let G be an open subset of $\mathbb{R} \times E$, and let $f, g : G \to E$ be continuous and such* *that $\|f(t,e) - g(t,e)\| \leq \delta$ for all $(t,e) \in G$. Suppose that $\|f(t,e) - f(t,e')\| \leq c\,\|e - e'\|$ for all $(t,e) \in G$ and $(t,e') \in G$. If $t_0 \in T$, an interval of $\mathbb{R}$, and if x and y are two maps from T to E such that $(t,x(t)) \in G$, $(t,y(t)) \in G$, $\|x'(t) - f(t,x(t))\| \leq \varepsilon$, $\|y'(t) - g(t,y(t))\| \leq \eta$ for all $t \in T$, setting $x_0 := x(t_0)$, $y_0 := y(t_0)$, one has the estimate*

$$\|x(t) - y(t)\| \leq \|x_0 - y_0\|\, e^{c|t-t_0|} + \frac{\varepsilon + \eta + \delta}{c}(e^{c|t-t_0|} - 1).$$

Proof Changing t into $t - t_0$ we may assume that $t_0 = 0$. Our assumptions ensure that for all $t \in T$, $e, e' \in E$ we have

$$\left\|f(t,e) - g(t,e')\right\| \leq \left\|f(t,e) - f(t,e')\right\| + \left\|f(t,e') - g(t,e')\right\| \leq c\left\|e - e'\right\| + \delta$$

hence

$$\begin{aligned}\left\|x'(t) - y'(t)\right\| &\leq \left\|x'(t) - f(t,x(t))\right\| + c\,\|x(t) - y(t)\| + \delta + \left\|g(t,y(t)) - y'(t)\right\| \\ &\leq \varepsilon + c\,\|x(t) - y(t)\| + \delta + \eta.\end{aligned}$$

Setting $u(t) := \|x(t) - y(t)\|$ for $t \in T$, $t \geq 0$, this relation implies

$$\begin{aligned}u(t) - \|x_0 - y_0\| \leq \|(x(t) - y(t)) - (x_0 - y_0)\| &= \left\|\int_0^t (x'(s) - y'(s))ds\right\| \\ &\leq \int_0^t \left\|x'(s) - y'(s)\right\| ds \leq \int_0^t cu(s)ds + (\varepsilon + \delta + \eta)t,\end{aligned}$$

hence the result, by Gronwall's Lemma. The case $t \leq 0$ is obtained by changing t into $-t$. □

Corollary 10.1 *Let E be a Banach space, let G be an open subset of $\mathbb{R} \times E$ and let $f : G \to E$ be continuous and such that for all $(t_0, e_0) \in G$ there exist $c \in \mathbb{R}_+$ and a neighborhood G_0 of (t_0, e_0) in G such that $\|f(t,e) - f(t,e')\| \leq c\,\|e - e'\|$ for all $(t,e), (t,e') \in G_0$. Let $x(\cdot)$ and $y(\cdot)$ be two solutions of the equation $w'(t) = f(t, w(t))$ on an open interval T. Then, for any t_0, $t \in T$ one has*

$$\|x(t) - y(t)\| \leq e^{c|t-t_0|}\,\|x(t_0) - y(t_0)\|.$$

In particular, if $x(\cdot)$ and $y(\cdot)$ coincide at some $t_0 \in T$, then they coincide all over T. Moreover, for all $(t_0, e_0) \in G$ there exists a largest open interval $T_{(t_0,x_0)}$ on which there exists a solution.

Of course here $w(\cdot)$ is a solution on T if $(t, w(t)) \in G$ and if $w'(t) = f(t, w(t))$ for all $t \in T$.

Proof The local estimate is obtained by taking $\varepsilon = \eta = \delta = 0$ in the preceding proposition. It yields local uniqueness. A connectedness argument entails global uniqueness. □

Note that the assumption on f is satisfied when the partial derivative $D_2 f$ exists and is continuous. This follows from the Mean Value Theorem.

Proposition 10.3 *Let G be an open subset of $\mathbb{R} \times E$ and let $f : G \to E$ be as in the preceding corollary. Let D_f be the set of $(t, t_0, x_0) \in \mathbb{R} \times G$ such that t belongs to the largest interval $T_{(t_0,x_0)}$ containing t_0 on which a solution $x(\cdot) := x(\cdot, t_0, x_0)$ of the equation $x'(t) = f(t, x(t))$, $x(t_0) = x_0$ exists. Then D_f is open. Moreover, the flow of f, i.e. the map $(t, t_0, x_0) \mapsto x(t, t_0, x_0)$, is continuous and of class C^k iff f is of class C^k. Furthermore, for all $(t_1, t_0, x_0) \in D_f$, $(t, t_1, x_1) \in D_f$ with $x_1 := x(t_1, t_0, x_0)$, one has $T_{(t_0,x_0)} = T_{(t_1,x_1)}$ and $x(\cdot, t_0, x_0) = x(\cdot, t_1, x_1)$ on this interval.*

Proof We have seen local existence and uniqueness of solutions. Thus D_f coincides with the set of $(t, t_0, x_0) \in \mathbb{R} \times G$ such that there exist a neighborhood U of x_0 and an open interval T containing t and t_0 on which a solution w of the equation $w'(t) = f(t, w(t))$, $w(t_0) = u$ exists for all $t \in T$. This yields openness of D_f and the existence and uniqueness of the largest solution $x(\cdot, t_0, x_0)$ of the equation $x'(t) = f(t, x(t))$, $x(t_0) = x_0$. The regularity of $(t, t_0, x_0) \mapsto x(t, t_0, x_0)$, being a local property, ensues. Given $(t_1, t_0, x_0) \in D_f$, $(t, t_1, x_1) \in D_f$ with $x_1 := x(t_1, t_0, x_0)$ we note that $x(\cdot, t_0, x_0)$ and $x(\cdot, t_1, x_1)$ are two solutions of $x'(\cdot) = f(\cdot, x(\cdot))$ that coincide at t_1. Thus, $T_{(t_0,x_0)}$ is a subset of the largest interval of existence $T_{(t_1,x_1)}$ of the solution issued from x_1 at t_1 and for $t \in T_{(t_0,x_0)}$ one has $x(t, t_0, x_0) = x(t, t_1, x_1)$; in particular $x(t_0, t_1, x_1) = x_0$. Thus, the roles of (t_0, x_0) and (t_1, x_1) can be interchanged and $T_{(t_0,x_0)} = T_{(t_1,x_1)}$ with $x(\cdot, t_0, x_0) = x(\cdot, t_1, x_1)$ on this interval. □

The following result explains why the maximal interval of existence of the solution may differ from $\mathbb{R}$.

Proposition 10.4 *Let f, G be as in the preceding proposition and for $(t_0, x_0) \in G$ let $T_{(t_0,x_0)} :=]\alpha, \omega[$ be the largest open interval on which a solution of the equation $x'(\cdot, t_0, x_0) = f(t, x(\cdot, t_0, x_0))$, $x(t_0, t_0, x_0) = x_0$ is defined. Then any limit point $\overline{x}$ of $x(\cdot, t_0, x_0)$ as $t \to \omega$ is such that $(\omega, \overline{x})$ belongs to the boundary of G and a similar statement is valid for α instead of ω.*

When $G := \mathbb{R} \times E$ and $\omega < \infty$, $x(\cdot, t_0, x_0))$ has no limit point as $t \to \omega$ and if, moreover, E is finite dimensional, $x(\cdot, t_0, x_0))$ is unbounded.

Proof For the first assertion we may assume $\omega < \infty$. Suppose on the contrary that there exists some sequence $(t_n) \to \omega$ such that $(x(t_n)) \to \overline{x}$ for some $\overline{x} \in E$ satisfying $(\omega, \overline{x}) \in G$. Let $\theta \in]t_0, \omega[$ and $\tau > \omega$ and let U be an open neighborhood of $\overline{x}$ such that for all $(s, u) \in]\theta, \tau[\times U$ there exists a unique solution of the equation $x'(t, u) = f(t, x(t, u))$, $x(s, u) = u$. We pick $n \in \mathbb{N}$ large enough such that $t_n > \theta$ and $x(t_n) \in U$. Setting $y(t) = x(t)$ for $t \in]\alpha, \theta]$ and $y(t) = x(t, x(t_n))$ for $t \in]\theta, \tau[$, we get a solution of the equation $x'(t, x_0) = f(t, x(t, x_0))$, $x(t_0, x_0) = x_0$ whose domain is larger than the domain of $x(\cdot, x_0)$. This contradicts the maximality of the interval $]\alpha, \omega[$.

The second assertion stems from the emptiness of the boundary of G when $G := \mathbb{R} \times E$ and, when E is finite dimensional, from the local compactness of E. □

The flow of an autonomous differential equation (also called a *vector field*) enjoys a striking property.

Proposition 10.5 *Let U be an open subset of a Banach space E and let $f : U \to E$ be a map of class C^k with $k \in \mathbb{N}\backslash\{0\}$. For $x_0 \in U$ let $T(x_0)$ be the largest open interval of $\mathbb{R}$ containing 0 on which the maximal solution $x(\cdot, x_0)$ of $x'(\cdot, x_0) = f(x(\cdot, x_0))$, $x(0, x_0) = x_0$ is defined. Then, for $s \in T(x_0)$ and $t \in T(x(s, x_0))$ one has $s + t \in T(x_0)$ and $x(s + t, x_0) = x(t, x(s, x_0))$.*

In particular, if the solutions are defined on all of $\mathbb{R}$, the family of maps $(\varphi_t)_{t\in\mathbb{R}} := (x(t, \cdot))_{t\in\mathbb{R}} : U \to U$ is a group of diffeomorphisms of U in the sense that $\varphi_{t+s} = \varphi_t \circ \varphi_s$ for all $s, t \in \mathbb{R}$.

Proof The derivative at t of $x(s+\cdot, x_0)$ is $f(x(s+\cdot, x_0))$ and $x(s+\cdot, x_0)(0) = x(s, x_0)$, so that, by uniqueness, $x(s + \cdot, x_0)(t) = x(t, x(s, x_0))$.

When $x(\cdot, x_0)$ is defined on $\mathbb{R}$, setting $\varphi_t = x(t, \cdot)$, the relation $\varphi_{t+s} = \varphi_t \circ \varphi_s$ follows from the preceding. Since $\varphi_0 = x(0, \cdot) = I_U(\cdot)$, the identity map, and since $\varphi_{-t} = \varphi_t^{-1}$, we see that φ_t is a diffeomorphism and $(\varphi_t)_{t\in\mathbb{R}}$ is a group of transformations of U. □

Exercises

1. **(Hadamard-Levy Theorem)** Let $f : \mathbb{R}^d \to \mathbb{R}^d$ be of class C^2 such that $f(0) = 0$ and that for some $c > 0$ and all $x \in \mathbb{R}^d$, $Df(x)$ is an isomorphism satisfying $\|Df(x)^{-1}\| \leq c$. Given $x \in \mathbb{R}^d$ show that the solution w of the differential equation $\frac{\partial w}{\partial t}(t, x) = [Df(w(t, x))]^{-1}.x$ with $w(0, x) = 0$ is defined on $\mathbb{R} \times \mathbb{R}^d$. Prove that $f(w(1, x)) = x$ by showing that $f(w(t, x)) - tx$ does not depend on t. Conclude that f is a diffeomorphism from $\mathbb{R}^d$ onto $\mathbb{R}^d$.
2. **(Arnold)** Let E be a Euclidean space and let $U : E \to \mathbb{R}$ be a smooth potential with nonnegative values. Consider the *Newton equation*

$$q''(t) = -\nabla U(q(t)).$$

Setting $p(\cdot) := q'(\cdot)$, show that H given by $H(p, q) := U(q) + \frac{1}{2}\|p\|^2$ is constant along a solution $(p(\cdot), q(\cdot))$ of the equation

$$(p'(t), q'(t)) = (-\nabla U(q(t)), p(t))$$

issued from $(p_0, q_0) \in E^2$ for $t_0 = 0$. Deduce from this fact that $\|q'(t)\|^2 \leq 2H(p_0, q_0)$ and that $\|q(t) - q_0\| \leq (2H(p_0, q_0))^{1/2} |t|$. Conclude that the solution $q(\cdot)$ of the Newton equation is defined on the whole of $\mathbb{R}$.

3. Prove the same conclusion as in the preceding exercise in the case when there exists some $c \in \mathbb{R}_+$ such that $U(q) \geq -c \|q\|^2$ for all $q \in E$.
4. Verify that for $E := \mathbb{R}$, $U(q) := -q^4/2$ the solution q of the Newton equation is given by $q(t) = (t-1)^{-1}$ and cannot be extended after $t = 1$.
5. Let E_0 be an open subset of a Banach space E and let $f : E_0 \to E$ be locally Lipschitzian or of class C^1. Let $h : E_0 \to \mathbb{R}$ be such that $h^{-1}(\{r\})$ is compact for all $r \in \mathbb{R}$. Suppose that for any solution $x(\cdot)$ of the equation $x'(\cdot) = f(x(\cdot))$ the function $h(x(\cdot))$ is constant. Prove that the maximal solution $x(\cdot)$ of this equation satisfying $x(0) = x_0$ is defined on the whole of $\mathbb{R}$.
6. Applying the conclusion of Exercise 5 in the case $E := E_0 := \mathbb{R}^3, f(x,y,z) := (y-z, z-x, x-y)$ for $(x,y,z) \in E$, $h(x,y,z) := x^2+y^2+z^2$, verify that the maximal solutions of the equation $w'(t) = f(w(t))$ are defined on the whole of $\mathbb{R}$.
7. Let $E := \mathbb{R}$, $E_0 := \mathbb{P}, f : \mathbb{P} \times E \to E$ be given by $f(t,e) := t^{-2} \sin t^{-1}$. Find the limit points of its solution $x(t) := \cos t^{-1}$ on $]0, \infty[$ as $t \to 0_+$.

10.1.4 The Exponential Map

The usual exponential map $\exp : t \mapsto e^t$ on $\mathbb{R}$ can be extended to the space $L(X) := L(X,X)$ of continuous linear operators on a Banach space X by means of the formula

$$e^A := \sum_{n=0}^{\infty} \frac{1}{n!} A^n \qquad A \in L(X),$$

where $A^0 = I_X$ and A^n is defined inductively by $A^{n+1} = A \circ A^n$. The series is absolutely convergent since $\|A^n\| \leq \|A\|^n$ for all $n \in \mathbb{N}$. Considering partial sums and passing to the limit, we get

$$\left\|e^A\right\| \leq e^{\|A\|}. \tag{10.20}$$

The next lemma will be used to show the following classical result.

Lemma 10.2 *Let B, $C : X \to X$ be two continuous linear operators on a Banach space X. If B and C commute, i.e. $B \circ C = C \circ B$, then one has $e^{B+C} = e^B \circ e^C = e^C \circ e^B$.*

Proof Since B and C commute, one has

$$(B+C)^k = \sum_{j=0}^{k} \frac{k!}{(k-j)!j!} B^{k-j} \circ C^j$$

hence

$$\sum_{k=0}^{n}\frac{1}{k!}(B+C)^k=\sum_{k=0}^{n}\sum_{j=0}^{k}\frac{1}{j!(k-j)!}B^{k-j}\circ C^j=\sum_{i+j\leq n}\frac{1}{i!j!}B^i\circ C^j.$$

Thus

$$(\sum_{i=0}^{n}\frac{1}{i!}B^i)\circ(\sum_{j=0}^{n}\frac{1}{j!}C^j)-\sum_{k=0}^{n}\frac{1}{k!}(B+C)^k=\sum_{i,j\leq n,\ i+j>n}\frac{1}{i!j!}B^i\circ C^j.$$

Setting $b:=\|B\|$, $c:=\|C\|$, the norm of the right-hand side is bounded above by

$$\sum_{i,j\leq n,\ i+j>n}\frac{1}{i!j!}b^ic^j=(\sum_{i=0}^{n}\frac{1}{i!}b^i)(\sum_{j=0}^{n}\frac{1}{j!}c^j)-\sum_{k=0}^{n}\frac{1}{k!}(b+c)^k$$

and the limit as $n\to\infty$ of this expression is $e^be^c-e^{b+c}=0$. Passing to the limit in the left-hand side of the preceding relation, we get $e^B\circ e^C-e^{B+C}=0$. □

Theorem 10.5 *For a Banach space* X, $A\in L(X)$, *and any* $u_0\in X$, $u(t):=e^{tA}u_0$ *is the solution to the equation*

$$u'(t)=Au(t),\qquad u(0)=u_0. \tag{10.21}$$

Proof Given $r\in\mathbb{R}$ and $s\in\mathbb{R}\backslash\{0\}$, we note that for $u(t)=e^{tA}u_0$ we have

$$\frac{1}{s}(u(r+s)-u(r))-Au(r)=\left[\frac{e^{sA}-I}{s}-A\right]u(r)$$

with

$$\frac{e^{sA}-I}{s}-A=\sum_{n=2}^{\infty}\frac{1}{n!}s^{n-1}A^n,$$

and, for $a:=\|A\|$

$$\left\|\frac{e^{sA}-I}{s}-A\right\|\leq\sum_{n=2}^{\infty}\frac{1}{n!}|s|^{n-1}\|A\|^n=\frac{e^{|s|a}-1}{|s|}-a\to 0\text{ as }s\to 0.$$

That shows that $u'(r)$ exists and equals $Au(r)$ for all $r\in\mathbb{R}$. Since $e^{0A}=I$, u is a solution to equation (10.21). Uniqueness yields the conclusion. □

Since $e^{sA} \circ e^{tA} = e^{(s+t)A}$, we get a one-parameter group of transformations of X. We shall see an important generalization of this property. The following estimate will be used in the case $k = 0$, which is slightly simpler than the general case.

Lemma 10.3 *Let B and C be two continuous linear operators on a Banach space X, let $u(t) = e^{tB}$, $v(t) = e^{tC}$ and let $k \in \mathbb{R}_+$ be such that $\|e^{tB}\| \le e^{kt}$, $\|e^{tC}\| \le e^{kt}$ for all $t \in \mathbb{R}_+$. If $B \circ C = C \circ B$ then*

$$\|u(t)x - v(t)x\| \le te^{kt}\,\|Bx - Cx\| \qquad \forall t \in \mathbb{R}_+,\ x \in X.$$

Proof Since $B \circ u(t) = u(t) \circ B$, $B \circ v(t) = v(t) \circ B$ for all $t \in \mathbb{R}_+$ by the expansion of exp, an easy computation shows that

$$\frac{d}{ds}[u(st)v(t-st)x] = tu(st)v(t-st)(Bx - Cx)$$

for all $t \in \mathbb{R}_+$, $x \in X$. It follows that for all $t \in \mathbb{R}_+$, $x \in X$ we have

$$\begin{aligned}\|u(t)x - v(t)x\| &= \left\|\int_0^1 tu(st)v(t-st)(Bx - Cx)ds\right\| \\ &\le t\,\|Bx - Cx\| \int_0^1 e^{kst}e^{k(t-st)}ds = te^{kt}\,\|Bx - Cx\|\,.\end{aligned}$$

□

Exercises

1. (**Trotter**) Given a Banach space X and $A, B \in L(X)$, show that $e^{t(A+B)} = \lim_{n\to\infty}(e^{(t/n)A} \circ e^{(t/n)B})^n$ for all $t \in \mathbb{R}_+$.
2. Let X be a Banach space and let $F : \mathbb{R}_+ \to L(X)$ be such that $F(0) = I$, $\|F(t)\| \le 1$ for all $t \in \mathbb{R}_+$ and such that the right derivative $A := F'(0)$ of F at 0 exists in $L(X)$. Show that $e^{tA} = \lim_{n\to\infty}(F(t/n))^n$ for all $t \in \mathbb{R}_+$.
3. (**Fibonacci**) Let $A \in L(\mathbb{R}^2)$ be given by $A(e_1) := e_2$, $A(e_2) = e_1 + e_2$. Show that there exist sequences (a_n), (b_n) in $\mathbb{R}$ such that $A^n = a_nA + b_nI$.

   Given $a, b \in \mathbb{R}$, compute u_n, where (u_n) is the sequence defined by $u_0 := a$, $u_1 := b$, $u_{n+1} = u_{n-1} + u_n$. Taking $a = 1$, $b = 2$ prove that this sequence converges to the *golden ratio* $\phi := (1/2)(1 + \sqrt{5})$. Since antiquity, many philosophers, mathematicians and artists have been intrigued by this number; during the Renaissance it was called the divine proportion. It has been used in many monuments, from the Great Pyramid and the Parthenon to the United Nations building. Note that if two numbers r, s are such that $s = r\phi$ then $r + s = s\phi$.

4. Show that the map $A \mapsto e^A$ from $L(\mathbb{R}^2)$ into $GL(\mathbb{R}^2) := \mathrm{Iso}(\mathbb{R}^2)$ is not onto. [Hint: verify that $B \in GL(\mathbb{R}^2)$ defined by $B(e_1) := -2e_1$, $B(e_2) := -e_2$ is not in the image of this map.]
5. Find the solution $u(t) := (x(t), y(t), z(t))$ to the system $u'(t) = Au(t)$, $u(0) = u_0$ given by

$$\begin{aligned} x'(t) &= -y(t) + z(t) \\ y'(t) &= z(t) \\ z'(t) &= -x(t) + z(t) \end{aligned}$$

by computing e^{tA}, where A is the matrix of this system. [Hint: verify that the characteristic polynomial of A is $\lambda^3 - \lambda^2 + \lambda - 1$, so that $A^3 = A^2 - A + I$.]
6. Consider the system

$$\begin{aligned} x'(t) &= y(t) + z(t) & x(0) &= x_0 \\ y'(t) &= x(t) & y(0) &= y_0 \\ z'(t) &= x(t) + y(t) + z(t) & z(0) &= z_0. \end{aligned}$$

Verify that for some $c \in \mathbb{R}$ one has $z(t) = x(t) + y(t) + c$ for $t \in \mathbb{R}$. Verify that the solution to the system

$$\begin{aligned} x'(t) &= x(t) + 2y(t) + c & x(0) &= x_0 \\ y'(t) &= x(t) & y(0) &= y_0 \end{aligned}$$

satisfies $x(t) = ae^{-t} + be^{2t}$, $y(t) = -ae^{-t} + (b/2)e^{2t} - c/2$. Deduce from this the solution to the first system and the expression of e^{tA} where A is the matrix of this system. From this expression compute A^n.

10.1.5 The Laplace Transform

Whereras the Fourier transform is adapted to functions defined on all of $\mathbb{R}$ or $\mathbb{R}^d$, the Laplace transform is suited to functions defined on $\mathbb{R}_+ := [0, \infty[$ or $\mathbb{P} :=]0, \infty[$. We deal with functions with values in $\mathbb{R}$ or $\mathbb{C}$ but an extension to functions with values in a complex Banach space could be considered. This transform enables us to change an evolution equation into another equation, a polynomial equation if the evolution equation is a linear differential equation of order n. It is defined as follows.

Definition 10.1 Given a measurable function $f : \mathbb{P} \to \mathbb{C}$, its Laplace transform is the function $f^{\#}$ on $\mathrm{dom} f^{\#} := \{z \in \mathbb{C} : t \mapsto |e^{-tz} f(t)| \in L_1(\mathbb{R}_+)\}$ given by

$$f^{\#}(z) := \int_0^{\infty} e^{-tz} f(t) dt \qquad z \in \mathrm{dom} f^{\#}.$$

One says that f is *Laplace transformable* if $\mathrm{dom} f^{\#}$ is nonempty.

Let us observe that if $z := r + is \in \mathrm{dom} f^{\#}$ with $r, s \in \mathbb{R}$, then for all $w := p + iq$ with $p \geq r$ one has $w \in \mathrm{dom} f^{\#}$ since $\left|e^{-(p+iq)t} f(t)\right| \leq \left|e^{-(r+is)t} f(t)\right|$. Thus, for $\sigma_f := \inf\{\mathrm{Re}\, z : t \mapsto e^{-tz} f(t) \in L_1(\mathbb{R}_+)\}$, one has $\mathrm{dom} f^{\#} =]\sigma_f, \infty[\times \mathbb{R}$. If f is locally integrable and of *exponential order* in the sense that there exist a, b, c in $\mathbb{R}$ such that $|f(t)| \leq be^{at}$ for $t \in [c, \infty[$, then one has $]a, \infty[+ i\mathbb{R} \subset \mathrm{dom} f^{\#}$. It can be shown that the function $f^{\#}$ is analytic on the half-space $]\sigma_f, \infty[\times \mathbb{R}$ and such that $f^{\#}(p + iq) \to 0$ as $p \to \infty$. We leave the proof of the following propositions to the reader as exercises (see also [229]).

Proposition 10.6 *Let $f : \mathbb{P} \to \mathbb{R}$ be Laplace transformable and for $c \in \mathbb{R}$ let g be given by $g(t) := e^{ct} f(t)$. Then, for $s > c + \sigma_f$ one has $g^{\#}(s) = f^{\#}(s - c)$.*

Proposition 10.7 *Let f, $g : \mathbb{P} \to \mathbb{R}$ be Laplace transformable. Then, for the convolution $f * g$ given by $(f * g)(t) := \int_0^t f(r) g(t - r) dr$, $\mathrm{dom} f^{\#} \cap \mathrm{dom}\, g^{\#} \subset \mathrm{dom}\, (f * g)^{\#}$ and for $s \in \mathrm{dom} f^{\#} \cap \mathrm{dom}\, g^{\#}$ one has*

$$(f * g)^{\#}(s) = f^{\#}(s).g^{\#}(s).$$

Proposition 10.8 *Let $f : \mathbb{P} \to \mathbb{R}$ be continuous and such that $\sigma_f = 0$ and $f^{\#}(z) = 0$ for all $z \in \mathbb{P}$. Then $f = 0$.*

Since the Laplace transform is linear, this last result shows that the map $f \mapsto f^{\#}$ is injective on the space $C_b(\mathbb{P})$ of bounded continuous functions on $\mathbb{P}$. Moreover, if f and g are two regulated functions of exponential order on $\mathbb{R}_+$ and if $f^{\#} = g^{\#}$, then $f = g$ at any point of continuity of f and g.

We shall not determine the image of the Laplace transform; we just quote the next result from [229, Thm 5.42].

Proposition 10.9 *Let a, $b \in \mathbb{P}$, $h \in L_1(\mathbb{R}, \mathbb{R}_+)$ and let $g :]a, \infty[+ i\mathbb{R} \to \mathbb{C}$ be such that $|g(z) - b/z| \leq h(y)$ for $z := x + iy$ with $x > a$. Then there exists some continuous function f on $\mathbb{P}$ such that $f(t) \leq be^{at}$ for all $t \in \mathbb{P}$ and $f^{\#} = g$.*

The following result can used to solve linear differential equations.

Proposition 10.10 *Let $f : \mathbb{P} \to \mathbb{R}$ be of class C^1, of exponential order and such that f' is Laplace transformable. Then, for $a > \sigma_{f'}$, $a > 0$ one has*

$$(f')^{\#}(s) = sf^{\#}(s) - f(0_+) \qquad s \in [a, +\infty[.$$

Proof An integration by parts yields

$$(f')^{\#}(s) = \lim_{\varepsilon\to 0_+} ([e^{-st}f(t)]_{\varepsilon}^{1/\varepsilon} + s\int_{\varepsilon}^{1/\varepsilon} e^{-st}f(t)dt) = -f(0_+) + sf^{\#}(s).$$

□

Example Given a_0, $a_1 \in \mathbb{R}$, let us consider the second-order differential equation $f'' + a_1 f' + a_0 f = 0$. Since $(f'')^{\#}(s) = s^2 f^{\#}(s) - sf(0_+) - f'(0_+)$, one can show that

$$f^{\#}(s) = \frac{sf(0_+) + f'(0_+) + a_1 f(0_+)}{s^2 + a_1 s + a_0}$$

from which one can get f.

In particular, when f is of class C^n and when $f^{(k)}(0_+) := \lim_{r\to 0_+} f^{(k)}(r)$ exists for $k \in \mathbb{N}_n \cup \{0\}$ we see that if f is a solution to a linear differential equation of order n, its Laplace transform satisfies a polynomial equation of degree n.

Such results explain the interest of using the Laplace transform for ordinary differential equations, the differentiation being transformed into a multiplication by s along with the subtraction of the initial value. Moreover, tables containing the Laplace transforms of commonly occurring functions are available in several books, f.i. [140, 229] and on //mathworld.wolfram.com/LaplaceTransform.html. As an example of the use of the Laplace transform, let us consider the *heat equation* (in which the Laplacian Δ bears on the space variable x in an open subset Ω of $\mathbb{R}^d$)

$$\frac{\partial v}{\partial t}(x,t) - \Delta v(x,t) = 0 \qquad (x,t) \in \Omega \times \mathbb{P}$$

with initial condition $v(x,0) = f(x)$. Performing a Laplace transform with respect to t :

$$v^{\#}(x,r) := v_x^{\#}(r) := \int_0^{\infty} e^{-rt} v(x,t)dt,$$

we see that (under appropriate assumptions) the heat equation implies that

$$\Delta v^{\#}(x,r) = \int_0^{\infty} e^{-rt} \Delta v(x,t)dt = \int_0^{\infty} e^{-rt} \frac{\partial v}{\partial t}(x,t)dt$$
$$= r\int_0^{\infty} e^{-rt} v(x,t)dt + [e^{-rt}v(x,t)]_{t=0}^{t=\infty} = rv^{\#}(x,r) - f(x).$$

Thus, the function u_r on Ω given by $u_r(x) := v^{\#}(x,r)$ satisfies the stationary equation

$$\Delta u_r - ru_r = -f.$$

Therefore, the rich knowledge about the solutions to this stationary equation can be transferred to the solution of the heat equation.

Exercises

1. For f given by $f(t) := e^{ct}$ verify that $f^{\#}(z) = 1/(z-c)$ and that $(\cosh)^{\#}(z) = z/(z^2-1)$.
2. For $f : \mathbb{R} \to \mathbb{R}$ null on $\mathbb{R}_-$ and $a \in \mathbb{R}_+$ let f_a be given by $f_a(t) := f(t-a)$. Show that $(f_a)^{\#}(z) = e^{-az}f^{\#}(z)$ when both sides are defined.
3. Let $f : \mathbb{P} \to \mathbb{R}$ be locally integrable and Laplace transformable and let g be the primitive of f satisfying $g(0) = 0$. Show that $g^{\#}(s) = f^{\#}(s)/s$ for $s \in \mathbb{P}$.
4. Show by induction that if f is of class C^n on $\mathbb{P}$ and if $f^{(k)}$ is Laplace transformable for all $k \in \mathbb{N}_n$ then $(f^{(n)})^{\#}(s) = s^n f^{\#}(s) - s^{n-1}f'(0_+) - \cdots - f^{(n-1)}(0_+)$.
5. Using the Laplace transform, find the solution y of the differential equation

$$y''(t) + y(t) = t \qquad t \in \mathbb{R}_+$$

   satisfying the initial conditions $y(0) = 1$, $y'(0) = 2$. [Hint: verify that $y(t) = t + \cos t + \sin t$ by showing that $y^{\#}(s) = s(s^2+1)^{-1} + 2(s^2+1) + s^{-2} - (s^2+1)^{-1}$.]
6. Let $f : \mathbb{P} \to \mathbb{R}$ be Laplace transformable and such that $f(0_+) := \lim_{t\to 0_+} f(t)$ (resp. $f(\infty) := \lim_{t\to\infty} f(t)$) exists. Show that $sf^{\#}(s) \to f(0_+)$ as $s \to \infty$ (resp. $sf^{\#}(s) \to f(\infty)$ as $s \to 0_+$).

10.2 Semigroups

The mathematical complexity of (nonlinear) semigroups corresponds to the complexity of time-dependent processes in nature. One may observe turbulence, shock waves, and explosions. Mathematically, these phenomena are reflected by instability, strange attractors, and blowing up effects that are outside the scope of this book. Thus, we essentially limit our study to central results in the case of linear semigroups.

When A is a densely defined unbounded linear operator on a Banach space X, it is not obvious how to generalize the definition and the properties of $e^{tA} := \exp tA$ considered in the preceding section for $A \in L(X)$. In this section we first study the consequences of a property weaker than the group property $e^{(s+t)A} = e^{sA} \circ e^{tA}$ for $s, t \in \mathbb{R}$. We show that under an assumption on the resolvent of A the solvability of (10.21) with $u_0 \in D(A)$ can be established. After a subsection devoted to a general class of multimaps, we close the section with some views on the nonlinear multivalued case. We start with a study of continuous linear semigroups. In this section, for linear maps $F, G : X \to X$ and $x \in X$ we often write Fx instead of $F(x)$ and FG instead of $F \circ G$.

10.2.1 Continuous Linear Semigroups and Their Generators

We adopt the notion of a *continuous semigroup* of operators in the following sense as an appropriate generalization of the notion of a group of operators.

Definition 10.2 A family $S(\cdot) := (S(t))_{t\in\mathbb{R}_+}$ of maps $S(t) : X \to X$ is called a *semigroup* if it satisfies the conditions

(SG0) $S(0) = I_X$;

(SG) for all $s, t \in \mathbb{R}_+$ $S(s) \circ S(t) = S(s + t)$.

If, moreover, the following condition holds, then $(S(t))_{t\in\mathbb{R}_+}$ is said to be a (strongly) *continuous semigroup*:

(SGC) the map $(t, x) \mapsto S(t)x$ is continuous on $\mathbb{R}_+ \times X$.

The following weakening of condition (SGC) is often adopted:

(SGC$_0$) for all $x \in X$ the map $S(\cdot)x$ is continuous at 0 from $\mathbb{R}_+$ into X.

Lemma 10.4 *When X is a Banach space and the maps $S(t)$ are linear and continuous, conditions (SGC) and (SGC$_0$) are equivalent. Moreover, when the semigroup $S(\cdot)$ satisfies these conditions there exist $\omega \in \mathbb{R}$ and $c \in \mathbb{R}_+$ such that*

$$\|S(t)\| \leq ce^{\omega t} \qquad \forall t \in \mathbb{R}_+. \tag{10.22}$$

If $c = 1$ one says that $(S(t))_{t\in\mathbb{R}_+}$ is a *ω-contraction semigroup*. If, moreover, $\omega = 0$, one says that $(S(t))_{t\in\mathbb{R}_+}$ is a *contraction semigroup*, or more correctly, a *continuous nonexpansive semigroup*. Since in this subsection and the next one we only consider semigroups of linear operators, we often omit any mention to linearity. Moreover, the composition of two operators A, B is often written AB rather than $A\circ B$.

Proof Let us first show that condition (SGC$_0$) implies that there exists some $\delta > 0$ and $c \in \mathbb{R}_+$ such that $\|S(t)\| \leq c$ for all $t \in [0, \delta]$. If that assertion were false, we could find a sequence $(t_n) \to 0_+$ such that $\|S(t_n)\| > n$ for all $n \in \mathbb{N}$. Then, by the uniform boundedness theorem there would exist some $x \in X$ such that

$$\sup_n \|S(t_n)x\| = \infty,$$

contradicting the assumption that $(S(t_n)x) \to x$.

Let $\delta > 0$, $c \in \mathbb{R}_+$ be such that $\|S(t)\| \leq c$ for all $t \in [0, \delta]$. Given $t \in \mathbb{R}_+$ let $s \in [0, 1[$ and $n \in \mathbb{N}$ be such that $t/\delta = n + s$. Then, for $\omega := (1/\delta)\log c$ we have

$$\|S(t)\| \leq \|S(s\delta)\| . \|S(\delta)\|^n \leq c.c^n = ce^{\omega n\delta} \leq ce^{\omega t}.$$

Given $x \in X$, the continuity from the right of the map $t \mapsto S(t)x$ stems from the relation $S(t+s)x - S(t)x = S(t)(S(s)x - x)$ for $s \in \mathbb{R}_+$. The continuity from the left of this map is a consequence in the inequality

$$\|S(t-s)x - S(t)x\| \leq \|S(t-s)\| . \|x - S(s)x\| \leq ce^{\omega t} \|x - S(s)x\| .$$

Given $(\bar{t}, \bar{x}) \in \mathbb{R}_+ \times X$ and a sequence $((t_n, x_n)) \to (\bar{t}, \bar{x})$, we have

$$\|S(t_n)x_n - S(\bar{t})\bar{x})\| \leq \|S(t_n)(x_n - \bar{x})\| + \|S(t_n)\bar{x} - S(\bar{t})\bar{x}\| \underset{n\to\infty}{\to} 0$$

since $(\|S(t_n)\|)_n$ is bounded. Thus (SGC) holds. □

Lemma 10.5 *Let $(S(t))_{t\in\mathbb{R}_+}$ be a semigroup of continuous linear operators on a Banach space X and let $x \in X$, $u(t) := S(t)x$. Then $u(\cdot)$ is right differentiable on $\mathbb{R}_+$ if and only if $u(\cdot)$ is right differentiable at* 0. *Moreover, for $t \in \mathbb{R}_+$ the right derivative u'_r of $u := u(\cdot)$ satisfies $u'_r(t) = S(t)u'_r(0)$.*

If condition (SGC) is satisfied, then $u(\cdot)$ is differentiable on $\mathbb{P} :=]0, \infty[$ whenever $u'_r(0)$ exists.

Proof For the first assertion it suffices to prove that $u(\cdot)$ is right differentiable at $t \in \mathbb{P}$ when $u'_r(0)$ exists. Since $S(t)$ is linear and continuous we have

$$\frac{1}{r}(S(t+r)x - S(t)x) = S(t)(\frac{S(r)x - x}{r}) \underset{r\to 0_+}{\to} S(t)u'_r(0).$$

When (SGC) holds, to prove that $u(\cdot)$ has $S(t)u'_r(0)$ as left derivative at t we write

$$\frac{1}{-r}(S(t-r)x - S(t)x) = S(t-r)(\frac{1}{r}(S(r)x - x)) \qquad r \in]0, t[$$

and we see from (SGC) that the right-hand side converges to $S(t)u'_r(0)$ as $r \to 0_+$. □

The preceding lemma incites us to consider the set $D(A)$ of $x \in X$ such that $u(\cdot) := S(\cdot)x$ is right differentiable at 0 and the operator $A : x \mapsto u'_r(0) = (S(\cdot)x)'_r(0)$. The latter is called the *infinitesimal generator* (or just the *generator*) of the semigroup $S(\cdot)$:

$$D(A) := \{x \in X : \exists v \in X : \frac{S(t)x - x}{t} \underset{t\to 0_+}{\to} v\},$$

$$Ax := \lim_{t\to 0_+} \frac{u(t)x - x}{t} \text{ if } x \in D(A).$$

Proposition 10.11 *Let $(S(t))_{t\in\mathbb{R}_+}$ be a continuous semigroup of linear operators on a Banach space X and let A be its generator. Then the domain $D(A)$ of A is a linear subspace, A is linear; for all $u_0 \in D(A)$ and all $t \in \mathbb{R}_+$ one has $u(t) := S(t)u_0 \in D(A)$ and $u : t \mapsto S(t)u_0$ is continuously differentiable with $u'(t) = Au(t)$ for all $t \in \mathbb{R}_+$. Moreover, $Au(t) = S(t)Au_0$ and $S(t)u_0 = u_0 + \int_0^t S(r)Au_0 dr$.*

Proof The linearity of A is obvious. From the lemma we know that if $u_0 \in D(A)$ then, for all $t \in \mathbb{P}$,

$$\frac{S(s)S(t)u_0 - S(t)u_0}{s} = \frac{S(t)(S(s)u_0 - u_0)}{s} \underset{s\to 0}{\to} S(t)u_r'(0).$$

This means that $u(t) \in D(A)$, that $u(\cdot) := S(\cdot)u_0$ is differentiable at t, and that $Au(t) = S(t)Au_0$. Since $S(\cdot)Au_0$ is continuous, $u(\cdot) := S(\cdot)u_0$ is continuously differentiable on $\mathbb{R}_+$ and $u(t) - u_0 = \int_0^t u'(r)dr = \int_0^t S(r)Au_0 dr$ since $t \mapsto S(t)Au_0$ is continuous. □

Theorem 10.6 *Let A be the generator of a continuous semigroup $(S(t))_{t\in\mathbb{R}_+}$ of linear operators on a Banach space X. Then $D(A)$ is a dense linear subspace of X and A is closed in the sense that its graph is a closed subset of $X \times X$.*

Moreover, $S(\cdot)$ is determined by A in the sense that if $T(\cdot)$ is another continuous semigroup with generator A, then $T(\cdot) = S(\cdot)$.

Proof Given $x \in X$ and $t > 0$, let $y_t := \int_0^t S(s)xds$, $x_t := t^{-1}y_t$. Since the map $s \mapsto S(s)x$ is continuous, we have $(x_t) \to x$ as $t \to 0_+$. To prove the first assertion, let us show that $x_t \in D(A)$ for all $t > 0$, or, equivalently (since $D(A)$ is a linear subspace) that $y_t \in D(A)$. Now for $r \in]0, t[$ we have

$$\begin{aligned}
r^{-1}(S(r)y_t - y_t) &:= r^{-1}[S(r)\int_0^t S(s)xds - \int_0^t S(s)xds] \\
&= r^{-1}\int_0^t (S(r+s)x - S(s)x)ds = r^{-1}\int_r^{r+t} S(s)xds - r^{-1}\int_0^t S(s)xds \\
&= r^{-1}\int_t^{t+r} S(s)xds - r^{-1}\int_0^r S(s)xds \underset{r\to 0_+}{\to} S(t)x - x.
\end{aligned}$$

This shows that $y_t \in D(A)$ and $Ay_t = S(t)x - x$.

In order to prove that the graph $G(A)$ of A is closed, we consider the limit (x, y) of a sequence $((x_n, y_n))$ in $G(A)$. Since $x_n \in D(A)$, the preceding proposition ensures that

$$S(r)x_n - x_n = \int_0^r \frac{d}{dt}(S(t)x_n)dt = \int_0^r S(t)Ax_n dt \qquad \forall r > 0.$$

Since $(Ax_n)_n \to y$ and, by Lemma 10.4, there exists positive constants c, ω such that $\|S(t)Ax_n\| \leq ce^{\omega t}(\|y\| + 1)$ for n large enough, passing to the limit on n, we get

$$S(r)x - x = \int_0^r S(t)ydt \qquad \forall r > 0.$$

Thus, $r^{-1}(S(r)x - x) = r^{-1}\int_0^r S(t)ydt \to y$ as $r \to 0_+$. By definition of $D(A)$ this means that $x \in D(A)$ and $y = Ax$.

Let us prove the last assertion. For r fixed in $\mathbb{R}_+$ and $t \in [0, r]$, let us set $Q(t) := T(r-t)S(t)$. Then, for $x \in D(A)$, since $y := S(t)x \in D(A)$ by Proposition 10.11, $Q(\cdot)x$ is differentiable with

$$(Q(\cdot)x)'(t) = -AT(r-t)S(t)x + T(r-t)AS(t)x = 0$$

since $AT(r-t)y = T(r-t)Ay$. Observing that $Q(0)x = T(r)x$, $Q(r)x = S(r)x$, and $Q(r)x = Q(0)x$, we conclude that $S(r)x = T(r)x$. Since $D(A)$ is dense in X and both $S(r)$ and $T(r)$ are linear and continuous, we get $S(r) = T(r)$. □

It is natural to wonder whether any densely defined closed linear operator A on X is the generator of a continuous semigroup. The answer is negative as the counterexample of Exercise 8 shows. Under an additional assumption we shall get a positive answer in the next subsection.

Exercises

1. Show that for a continuous linear semigroup $S(\cdot)$ with generator A on a Banach space X the following assertions are equivalent:

   (a) A is everywhere defined and continuous;
   (b) the domain $D(A)$ of A is X;
   (c) the domain $D(A)$ is closed;
   (d) the map $S(\cdot)$ is continuous from $\mathbb{R}_+$ into $L(X) := L(X, X)$.

2. Show that a linear semigroup $(S_t)_{t\geq 0}$ on X is continuous if and only if it is weakly continuous in the sense that for all $x \in X$, $x^* \in X^*$ the function $t \mapsto \langle x^*, S(t)x\rangle$ is continuous on $\mathbb{R}_+$.
3. Prove that the translation semigroup $(T(t))_{t\geq 0}$ defined by $T(t)(f)(s) := f(s+t)$ for $s, t \in \mathbb{R}_+, f \in C_0(\mathbb{R}_+) := \{f \in C(\mathbb{R}_+) : \lim_{s\to\infty} f(s) = 0\}$ is a continuous semigroup when $C_0(\mathbb{R}_+)$ is endowed with the sup norm $\|\cdot\|_\infty$.
4. Consider the same question when $C_0(\mathbb{R}_+)$ is replaced with the space $C_{ub}(\mathbb{R}_+)$ of $f \in C(\mathbb{R}_+)$ that are bounded and uniformly continuous, $C_{ub}(\mathbb{R}_+)$ being endowed with $\|\cdot\|_\infty$.
5. Consider the same question when $C_0(\mathbb{R}_+)$ is replaced with the space $C_0^1(\mathbb{R}_+)$ of $f \in C_0(\mathbb{R}_+)$ that are of class C^1, with $f' \in C_0(\mathbb{R}_+)$, $C_0^1(\mathbb{R}_+)$ being endowed with the norm $f \mapsto \|f\|_\infty + \|f'\|_\infty$.
6. Let $c \in \mathbb{P}$, $\omega \in \mathbb{R}$, and let $(S(t))_{t\in\mathbb{R}_+}$ be a continuous semigroup of linear operators on a Banach space X satisfying $\|S(t)\| \leq ce^{\omega t}$ for all $t \in \mathbb{R}_+$. Setting $\|x\|' := \sup\{e^{-\omega t}\|S(t)x\| : t \in \mathbb{R}_+\}$, show that $\|\cdot\|'$ is a norm on X satisfying $\|\cdot\| \leq \|\cdot\|' \leq c\|\cdot\|$. Verify that $S(\cdot)$ is a continuous semigroup on X satisfying $\|S(t)x\|' \leq e^{\omega t}\|x\|'$ for all $x \in X$.

7. Extending operators on a real Banach space X to its complexified space, prove that a closed, densely defined linear operator A generates a continuous nonexpansive semigroup if and only if for all $\lambda \in \mathbb{C}$ satisfying $\operatorname{Re}\lambda > 0$ one has $\lambda \in \rho(A)$ and $\left\|(\lambda I - A)^{-1}\right\| \leq 1/\operatorname{Re}\lambda$. [See [113, Thm 3.5] and the next subsection.]
8. Verify the following counterexample showing that a closed linear operator A with dense domain in a Banach space X and whose spectrum is contained in some interval $]-\infty,\omega]$ is not necessarily the generator of a continuous semigroup. Take for X the space of continuous functions $f : \mathbb{R}_+ \to \mathbb{R}$ that are continuously differentiable on $[0,1]$ and satisfy $\lim_{r\to\infty} f(r) = 0$ and endow X with the norm $f \mapsto \|f\| := \sup_{r\in\mathbb{R}_+} |f(r)| + \sup_{r\in[0,1]} |f'(r)|$. Consider the operator $A : D(A) \to X$ defined by $D(A) := \{f \in X \cap C^1(\mathbb{R}_+) : f' \in X\}$, $Af := f'$. Show that $D(A)$ is dense, that A is closed and that the resolvent of A contains $\mathbb{P}$ but that if A generates a continuous semigroup $(S(t))_{t\geq 0}$ one must have $(S(t)f)(r) = f(r+t)$ for all $f \in X$, $r, t \in \mathbb{R}_+$, contradicting the fact that $S(t)f \in X$ for all $f \in X$.

10.2.2 Characterization of Generators of Continuous Semigroups

Recall that the *resolvent set* $\rho(A)$ of a closed (linear) operator A is the set of $\lambda \in \mathbb{R}$ such that $\lambda I - A$ is a bijection from $D(A)$ onto X, I standing for the identity map I_X of X. Then, the *resolvent operator*

$$R_\lambda := R_\lambda^A := (\lambda I - A)^{-1}$$

is a continuous linear operator, as seen in Proposition 3.38. The following proposition shows that R_λ^A can be considered as the *Laplace transform* of the semigroup generated by A. The estimate it provides is the first step in the characterization of generators of continuous semigroups we have in view.

Proposition 10.12 *If R_λ is the resolvent of the generator A of a continuous semigroup $(S(t))_{t\in\mathbb{R}_+}$ and if $\omega \in \mathbb{R}$ and $c \in \mathbb{R}_+$ are such that $\|S(t)\| \leq ce^{\omega t}$ for all $t \in \mathbb{R}_+$ as in Lemma 10.4, then $]\omega,\infty[\subset \rho(A)$ and for all $\lambda \in]\omega,\infty[$ one has $\|R_\lambda\| \leq c/(\lambda-\omega)$ and*

$$R_\lambda x = \int_0^\infty e^{-\lambda r} S(r)x dr.$$

In particular, if A is the generator of a nonexpansive continuous semigroup, then one has $\mathbb{P} \subset \rho(A)$ and for all $\lambda \in \mathbb{P}$ one has $\|R_\lambda\| \leq 1/\lambda$.

Proof By our assumption, for $\lambda > \omega$ we have $\left\|e^{-\lambda r}S(r)x\right\| \leq ce^{-(\lambda-\omega)r}\|x\|$, so that the integral $T_\lambda(x) := \int_0^\infty e^{-\lambda r}S(r)xdr$ is well defined and $\|T_\lambda(x)\| \leq c\|x\|/(\lambda-\omega)$.

Setting $\widetilde{S}(t) := c^{-1}e^{-\lambda t}S(t)$, we reduce the assertion of the statement to the case $\omega < 0$, $\lambda = 0$ and $c = 1$, the generator of $(\widetilde{S}(t))_{t\geq 0}$ being $\widetilde{A} := c^{-1}(A - \lambda I_X)$, so that $R_\lambda = -(c\widetilde{A})^{-1}$. Then, for $x \in X$ and $t \in \mathbb{P}$, for $\widetilde{T}_0 x := \int_0^\infty \widetilde{S}(r)xdr = c^{-1}T_\lambda x$ we have

$$\begin{aligned}\frac{\widetilde{S}(t) - I_X}{t}\widetilde{T}_0 x &= \frac{1}{t}\int_0^\infty (\widetilde{S}(t) - I_X)(\widetilde{S}(r)x)dr \\ &= \frac{1}{t}\int_0^\infty \widetilde{S}(t+r)xdr - \frac{1}{t}\int_0^\infty \widetilde{S}(r)xdr \\ &= \frac{1}{t}\int_t^\infty \widetilde{S}(s)xds - \frac{1}{t}\int_0^\infty \widetilde{S}(r)xdr = -\frac{1}{t}\int_0^t \widetilde{S}(s)xds.\end{aligned}$$

Since the right-hand side converges to $-x$ as $t \to 0_+$ we get that $\widetilde{T}_0 x \in D(\widetilde{A}) = D(A)$ and $\widetilde{A}(\widetilde{T}_0 x) = -x$.

For $x \in X$ we have

$$\widetilde{T}_0 x = \lim_{t\to\infty}\int_0^t \widetilde{S}(r)xdr$$

and similarly, for $x \in D(\widetilde{A})$, using Lemma 3.21 and the fact that $\widetilde{A}$ has a closed graph by Theorem 10.6,

$$\widetilde{T}_0\widetilde{A}x = \lim_{t\to\infty}\int_0^t \widetilde{S}(r)\widetilde{A}xdr = \lim_{t\to\infty}\int_0^t \widetilde{A}\widetilde{S}(r)xdr = \lim_{t\to\infty}\widetilde{A}\int_0^t \widetilde{S}(r)xdr.$$

Since $\widetilde{A}$ is a closed linear operator, it follows that $\widetilde{T}_0\widetilde{A}x = \widetilde{A}\widetilde{T}_0 x = -x$. These relations prove that $\widetilde{T}_0 = -\widetilde{A}^{-1}$, so that $0 \in \rho(\widetilde{A})$ and $\lambda \in \rho(A)$, $R_\lambda = -c\widetilde{A}^{-1} = c\widetilde{T}_0 = T_\lambda$. □

In the sequel, given $\omega \in \mathbb{R}_+$, we denote by $\mathcal{D}_\omega(X)$ the set of closed linear operators A whose domains are dense in X and that satisfy $]\omega, \infty[\subset \rho(A)$, the resolvent set of A, and $\|R_\lambda\| \leq 1/(\lambda - \omega)$ for all $\lambda > \omega$, where $R_\lambda := (\lambda I - A)^{-1}$ is the resolvent operator of A. This class of operators will be studied later on and generalized in the next subsection. We introduce

$$\omega(A) := \inf\{\omega \in \mathbb{R}_+ : A \in \mathcal{D}_\omega(X)\} \qquad \text{for } A \in \mathcal{D}(X) := \bigcup_{\omega\in\mathbb{R}_+} \mathcal{D}_\omega(X).$$

Given $A \in \mathcal{D}_\omega(X)$ and $r \in]0, 1/\omega[$ we set

$$P_r(:= P_r^A) := \frac{1}{r}R_{\frac{1}{r}} = (I - rA)^{-1}, \qquad A_r := \frac{1}{r}(P_r - I).$$

Note that since the graph of A is closed, the graphs of $I - rA$ and $(\lambda I - A)^{-1}$ are closed, as is easily seen, so that P_r is a continuous linear operator from X into $D(A)$. Moreover, since $(I - rA)(I - rA)^{-1} = I$, and hence $(I - rA)^{-1} - I = rA(I - rA)^{-1}$, one sees that

$$A_r x = AP_r x \qquad \forall x \in X,\ \forall r \in]0, 1/\omega[.$$

Let us give some properties of the map A_r called the *Yosida approximation* of A. We start with a study of the *proximal operator* P_r of A.

In Proposition 10.14 below we use the fact that $AP_r x = P_r Ax$ for $x \in D(A)$ (by Proposition 3.38 (a)), so that

$$A_r x = AP_r x = P_r Ax. \tag{10.23}$$

Proposition 10.13 *Given $\omega \in \mathbb{R}_+$, $r \in]0, 1/\omega[$, with $1/\omega := \infty$ if $\omega = 0$, and $A \in \mathcal{D}_\omega(X)$ the following properties hold:*

(a) $\|P_r\| \leq (1 - \omega r)^{-1}$;
(b) $\|P_r x - x\| \leq r(1 - r\omega)^{-1} \|Ax\|$ *for all $x \in D(A)$;*
(c) $\lim_{r\to 0_+} P_r x = x$ *for all $x \in X$.*

Proof

(a) By definition of $\mathcal{D}_\omega(X)$ we have $\|P_r\| = \frac{1}{r}\left\|R_{\frac{1}{r}}\right\| \leq (1 - \omega r)^{-1}$.
(b) For $x \in D(A)$ we have

$$\begin{aligned}\|P_r x - x\| &= r^{-1}\left\|R_{\frac{1}{r}} x - R_{\frac{1}{r}}(I - rA)x\right\| \\ &\leq r^{-1}\left\|R_{\frac{1}{r}}\right\| \|x - (I - rA)x\| \leq r(1 - r\omega)^{-1}\|Ax\|.\end{aligned}$$

(c) ensues for $x \in D(A)$. Since $\|P_r\| \leq 2$ for $r \in]0, 1/(2\omega)[$ and since $D(A)$ is dense in X, given $x \in X$ and $\varepsilon > 0$, picking $w \in D(A)$ with $\|w - x\| \leq \varepsilon/6$, the inequalities

$$\|P_r x - x\| \leq \|P_r x - P_r w\| + \|P_r w - w\| + \|w - x\| \leq 3\|w - x\| + \frac{r}{1 - r\omega}\|Aw\|$$

yield $\|P_r x - x\| < \varepsilon$ provided r is small enough. □

Proposition 10.14 *Given $\omega \in \mathbb{R}_+$, $r \in]0, 1/\omega[$, and $A \in \mathcal{D}_\omega(X)$ the following properties hold:*

(a) A_r is linear and continuous and $\|A_r\| \leq r^{-1}(1 - \omega r)^{-1} + r^{-1}$;
(b) $\|A_r x\| \leq (1 - r\omega)^{-1} \|Ax\|$ *for all $x \in D(A)$;*
(c) $\lim_{r\to 0_+} A_r x = Ax$ *for all $x \in D(A)$.*

Proof

(a) Since $A_r = \frac{1}{r}(P_r - I)$, this stems from Proposition 10.13(a).
(b) Since $A_r x = \frac{1}{r}(P_r x - x)$, this follows from Proposition 10.13(b).
(c) For $x \in D(A)$, by relation (10.23) and Proposition 10.13 (c), we have $\|A_r x - Ax\| = \|P_r Ax - Ax\| \to 0$ as $r \to 0_+$. □

Exercise For $\omega \in \mathbb{R}_+$ and $A \in \mathcal{D}_\omega(X)$ show that $A_r \in \mathcal{D}_{\omega(1-r\omega)^{-1}}(X)$ for $r \in]0, 1/\omega[$.

We are ready to present the main result of this section. Its proof will be presented by starting with the special case of nonexpansive continuous semigroups, which is technically simpler than the general case (it corresponds to the case $\omega = 0$).

Theorem 10.7 (Hille-Yosida) *If A is a linear operator with domain $D(A)$ in a Banach space X and if $\omega \in \mathbb{R}_+$, the following assertions are equivalent:*

(a) A is closed, densely defined, the resolvent set $\rho(A)$ of A contains the interval $]\omega, \infty[$ and for all $\lambda > \omega$ one has $\left\|(\lambda I - A)^{-1}\right\| \le (\lambda - \omega)^{-1}$;
(b) A is the infinitesimal generator of a continuous semigroup $(S(t))_{t \in \mathbb{R}_+}$ of linear continuous operators such that $\|S(t)\| \le e^{\omega t}$ for all $t \in \mathbb{R}_+$.

Corollary 10.2 *For a linear operator A with domain $D(A)$ in a Banach space X, the following assertions are equivalent:*

(a) A is closed, densely defined, $\mathbb{P} \subset \rho(A)$ and $\left\|(\lambda I - A)^{-1}\right\| \le \lambda^{-1}$ for all $\lambda \in \mathbb{P}$;
(b) A is the generator of a nonexpansive continuous semigroup.

Proof of Corollary 10.2 In Theorem 10.6 and Proposition 10.12 we have seen the implication (b)⇒(a), so that it remains to show (a)⇒(b). For $r \in \mathbb{P}$ let

$$S_r(t) := e^{tA_r} \qquad t \in \mathbb{R}_+.$$

Since $A_r = r^{-1}P_r - r^{-1}I$, we have

$$e^{tA_r} = e^{-t/r} e^{(t/r)P_r}$$

and since $\|P_r\| \le 1$ and $\left\|e^{(t/r)P_r}\right\| \le e^{\|(t/r)P_r\|} \le e^{t/r}$ we get

$$\|S_r(t)\| = \left\|e^{tA_r}\right\| \le e^{-t/r} e^{t/r} = 1.$$

Thus, for $r, s \in \mathbb{P}$, using the relation $A_r A_s = A_s A_r$, Lemma 10.3 yields

$$\|S_r(t)x - S_s(t)x\| \le t \|A_r x - A_s x\| \qquad x \in X,\ t \in \mathbb{R}_+. \tag{10.24}$$

For $x \in D(A)$ since $(A_{r(n)}x) \to Ax$ for any sequence $(r(n)) \to 0_+$, we see that $(S_{r(n)}(\cdot)x)$ is a Cauchy sequence with respect to the norm of uniform convergence on every compact interval of $\mathbb{R}_+$. Thus it converges uniformly to a limit $S(\cdot)x$ on each such interval and the limit is independent of the choice of the sequence $(r(n))$. Since $D(A)$ is dense in X, taking $x \in X$ and a sequence (x_n) in $D(A)$ with limit x and using the estimates $\|S_r(t)\| \leq 1$, $\|S_s(t)\| \leq 1$ yielding the inequality

$$\|S_r(t)(x) - S_s(t)(x)\| \leq \|x - x_n\| + \|S_r(t)(x_n) - S_s(t)(x_n)\| + \|x_n - x\|$$

we see that this convergence also holds for $x \in X$, uniformly on every compact interval of $\mathbb{R}_+$. Thus $t \mapsto S(t)x$ is continuous. Passing to the limit as $r \to 0_+$ in the relations $\|S_r(t)(x)\| \leq \|x\|$ and $(S_r(t) \circ S_r(t'))(x) = S_r(t+t')(x)$, we get that $(S(t))_{t\geq 0}$ is a semigroup of nonexpansive linear maps. Lemma 10.4 shows that the semigroup $(S(t))_{t\geq 0}$ is continuous.

It remains to show that A coincides with the generator B of $(S(t))_{t\geq 0}$. We first note that for every $x \in D(A)$ and every $\tau > 0$ the maps $u_r : t \mapsto S_r(t)(x)$ and their derivatives $u'_r : t \mapsto S_r(t)(Ax)$ converge uniformly on $[0, \tau]$ to $u : t \mapsto S(t)(x)$ and $t \mapsto S(t)(Ax)$ respectively as $r \to 0_+$. This shows that u is of class C^1 on $\mathbb{R}_+$ and that $u'(t) = S(t)(Ax)$. Thus $D(A) \subset D(B)$ and $A = B$ on $D(A)$.

Now, by Proposition 10.12, given $\lambda \in \mathbb{P}$ we have $\lambda \in \rho(B)$, so that $\lambda I - B$ is a bijection from $D(B)$ onto X. Since we also have $\lambda \in \rho(A)$ by assumption, $\lambda I - A$ also is a bijection from $D(A)$ onto X. Since $A = B$ on $D(A)$, we see that $\lambda I - B$ is a bijection from $D(A) \subset D(B)$ onto X. Thus $D(A) = D(B)$ and $A = B$. □

Proof of Theorem 10.7 Again, we only have to show (a)⇒(b). Setting $B := A - \omega I$, for $\lambda > 0$ we have that $B - \lambda I = A - (\lambda + \omega)I$ is a bijection of $D(B) = D(A)$ onto X, so that $\mathbb{P} \subset \rho(B) = \rho(A) - \omega$ and $\left\|(\lambda I - B)^{-1}\right\| = \left\|(\lambda + \omega)I - A)^{-1}\right\| \leq ((\lambda + \omega) - \omega)^{-1} = \lambda^{-1}$. Thus B is the generator of a nonexpansive continuous semigroup $(T(\cdot))$. Setting $S(t) := e^{\omega t}T(t)$ for $t \in \mathbb{R}_+$, we see that $\|S(t)\| \leq e^{\omega t}$ for $t \in \mathbb{R}_+$, that $S(\cdot)$ is a continuous semigroup and that for all $x \in D(A) = D(B)$ we have

$$\frac{d}{dt}\Big|_{t=0_+} S(t)x = \omega x + \frac{d}{dt}\Big|_{t=0_+} T(t)x = \omega x + Bx = Ax.$$

Thus $S(\cdot)$ is generated by A. □

Let us identify the class $\mathcal{D}_0$ in terms of the class of (single-valued) linear dissipative operators in the sense of the next definition. In the next subsection we will enlarge this class to nonlinear (multivalued) operators.

Definition 10.3 A linear map $A : D(A) \to X$ with domain $D(A)$ in a Banach space X is said to be *dissipative* if for any $x \in D(A)$ there exists some $x^* \in J(x) := \{x^* \in X^* : \|x^*\| = \|x\|, \langle x^*, x\rangle = \|x\|^2\}$ such that

$$\langle Ax, x^*\rangle \leq 0. \tag{10.25}$$

A linear map $A : D(A) \to X$ is said to be *accretive* if $-A$ is dissipative.

Given $\omega \in \mathbb{R}$ one says that $A : D(A) \to X$ is *ω-dissipative* (resp. *ω-accretive)* if $A - \omega I$ is dissipative (resp. $A + \omega I$ is accretive).

Kato's lemma (Lemma 6.17) shows the following characterization using a semi-scalar product $[\cdot, \cdot]$ and the semi-inner product $\langle \cdot | \cdot \rangle_+$ defined by $\langle x|y\rangle_+ := \lim_{t \to 0_+} \frac{1}{2t}(\|x + ty\|^2 - \|x\|^2)$.

Proposition 10.15 *A linear map $A : D(A) \to X$ is dissipative if and only if one of the following assertions holds:*

(a) $\|x\| \leq \|x - rAx\|$ *for all $x \in D(A)$, $r > 0$;*
(b) for some semi-scalar product $[\cdot, \cdot]$ one has $[x, Ax] \leq 0$ for all $x \in D(A)$;
(c) $\langle x| - Ax\rangle_+ \geq 0$ *for all $x \in D(A)$.*

The first characterization shows that for all $\lambda > 0$ the map $\lambda I - A : D(A) \to X$ is injective and its inverse $R_\lambda := (\lambda I - A)^{-1} : R(\lambda I - A) \to D(A)$ is continuous with norm at most $r := 1/\lambda$: given $z \in R(\lambda I - A) := (\lambda I - A)(X)$, for $x \in D(A)$ such that $z = \lambda x - Ax$, one has $\|x\| \leq \lambda^{-1} \|z\|$. In particular, if $z = 0$ then $x = 0$.

Proposition 10.16 *Let A be a dissipative linear operator. Then A is closed if and only if for some (hence all) $\lambda > 0$ the set $R(\lambda I - A)$ is closed. In particular any linear dissipative operator such that $R(\lambda I - A) = X$ for some $\lambda > 0$ is closed.*

Proof Clearly, A is closed if and only if $\lambda I - A$ is closed. In turn this is equivalent to $R_\lambda := (\lambda I - A)^{-1}$ being closed. Since $(I, R_\lambda) : D(R_\lambda) \to G(R_\lambda)$, i.e. $z \mapsto (z, R_\lambda(z))$ is a continuous bijection whose inverse is the restriction to $G(R_\lambda)$ of the canonical projection, this is equivalent to $D(R_\lambda) = R(\lambda I - A)$ being complete or closed. □

We conclude from the next theorem that the class $\mathcal{D}_0$ coincides with the set of densely defined dissipative operators A satisfying $R(I - A) = X$.

Theorem 10.8 (Lumer-Phillips) *Let A be a densely defined linear operator on a Banach space X. Then A is the generator of a nonexpansive continuous semigroup if and only if A is dissipative and $R(I - A) = X$.*

Proof Suppose A generates a nonexpansive continuous semigroup $(S(t))_{t \geq 0}$. Then, for any semi-scalar product $[\cdot, \cdot]$ on X and $x \in D(A)$ we have $[x, Ax] \leq 0$ since

$$[x, S(t)x - x] = [x, S(t)x] - [x, x] \leq \|x\| . \|S(t)x\| - \|x\|^2 \leq 0,$$

so that $[x, Ax] = [x, \lim_{t \to 0_+} (1/t)(S(t)x - x)] \leq 0$, $[x, \cdot]$ being a continuous linear form. Thus A is dissipative. Moreover, $R(I - A) = D((I - A)^{-1}) = X$ since A is the generator of a nonexpansive semigroup so that $\mathbb{P} \subset \rho(A)$ by Corollary 10.2.

Conversely, let us suppose A is dissipative and $R(I - A) = X$. Then, by the preceding proposition, A is closed. Note that for $\lambda > 0$, $z \in X$, and $x := R_\lambda z$ we have $\lambda x - Ax = z$,

$$\lambda \|x\|^2 \leq \lambda[x, x] - [x, Ax] = [x, \lambda x - Ax] \leq \|x\| . \|z\| ,$$

hence $\|R_\lambda z\| \leq (1/\lambda)\|z\|$. In particular $\lambda := 1 \in \rho(A)$ and $\|R_1\| \leq 1$. Moreover, if λ is such that $|\lambda - 1| < 1$, by Proposition 3.17 we have $\lambda \in \rho(A)$ and for $\mu \in \mathbb{R}$ satisfying $|\mu - \lambda| < \lambda \leq 1/\|R_\lambda\|$ we have $\mu \in \rho(A)$. Repeating this argument, we see that $\mathbb{P} \subset \rho(A)$ and $\|R_\lambda\| \leq 1/\lambda$. Then Corollary 10.2 shows that A is the generator of a nonexpansive continuous semigroup. □

Corollary 10.3 *Let A be a densely defined linear dissipative operator on a smooth Banach space X such that $R(I - A) = X$. Then, for all $u_0 \in X$ there exists a $u \in C^1(\mathbb{P}, X) \cap C(\mathbb{R}_+, X)$ such that $u(0) = u_0$, $u(\mathbb{P}) \subset D(A)$, and $u'(t) = Au(t)$ for all $t \in \mathbb{P}$.*

Moreover, $\|u(\cdot)\|$ is nonincreasing. In particular, one has $\|u(t)\| \leq \|u_0\|$ for all $t \in \mathbb{R}_+$.

Proof The first assertion follows from the Lumer-Phillips Theorem and Proposition 10.11.

Since

$$\frac{d}{dt}\|u(t)\|^2 = 2\langle J(u(t)), u'(t)\rangle = -2\langle u(t) \mid -Au(t)\rangle_+ \leq 0,$$

$\|u(\cdot)\|^2$ is nonincreasing. □

Corollary 10.4 *Let A be a densely defined linear operator on a Hilbert space X such that $B := A - \omega I$ is maximally dissipative (i.e. $-B$ is maximally monotone) for some $\omega \in \mathbb{R}$. Then A generates a ω-contraction semigroup $(S(t))_{t\geq 0}$ in X.*

Proof By maximal monotonicity (Proposition 9.15) $-B$ is closed (see also Corollary 10.6 below for a direct proof). Since for all $x \in D(A)$ we have $\langle(\omega I - A)x \mid x\rangle \geq 0$, given $\lambda \in]\omega, \infty[$, we have $R(\lambda I - A) = R((\lambda - \omega)I - B) = X$ by Theorem 9.28, so that $]\omega, \infty[\subset \rho(A)$. Given $y \in X$ and $x := R_\lambda y$ we have

$$\|x\| . \|y\| \geq \langle y \mid x\rangle = \langle(\lambda I - A)x \mid x\rangle \geq (\lambda - \omega)\|x\|^2$$

hence $(\lambda - \omega)\|x\| \leq \|y\|$. Thus $\left\|(\lambda I - A)^{-1}\right\| \leq (\lambda - \omega)^{-1}$ and the Hille-Yosida Theorem yields the conclusion. □

The present section culminates in the following generalization of the Hille-Yosida Theorem.

Theorem 10.9 (Feller-Miyadera-Phillips) *Given $\omega \in \mathbb{R}_+$ and some $c \in [1, \infty[$, a closed linear operator A with dense domain $D(A)$ in a Banach space X is the infinitesimal generator of a continuous semigroup $(S(t))_{t\in\mathbb{R}_+}$ of linear continuous operators such that $\|S(t)\| \leq ce^{\omega t}$ for all $t \in \mathbb{R}_+$ if and only if the resolvent set $\rho(A)$ of A contains the interval $]\omega, \infty[$ and*

$$\|(\lambda I_X - A)^{-n}\| \leq c(\lambda - \omega)^{-n} \qquad \forall\lambda > \omega, \quad \forall n \in \mathbb{N}.$$

Proof The necessary assertion has been proved in Proposition 10.12 for $n = 1$. For $n = 0$ it is obvious. For $n \geq 2$ in $\mathbb{N}$ and $R_\lambda := (\lambda I_X - A)^{-1}$ we use the relations

$$R_\lambda^n = \frac{(-1)^{n-1}}{(n-1)!}\frac{d^{n-1}}{d\lambda^{n-1}}R_\lambda, \quad R_\lambda x = \int_0^\infty e^{-\lambda r}S(r)x dr$$

obtained in Propositions 3.38 and 10.12 respectively to get by induction and differentiation under the integral symbol an integral representation of R_λ^n:

$$R_\lambda^n = \frac{(-1)^{n-1}}{(n-1)!}\int_0^\infty t^{n-1}e^{-\lambda t}S(t)dt.$$

Then, an integration by parts yields the estimate

$$\left\|R_\lambda^n\right\| \leq \frac{c}{(n-1)!}\int_0^\infty t^{n-1}e^{(\omega-\lambda)t}dt = \frac{c}{(\lambda-\omega)^n}.$$

Let us prove the sufficiency assertion. Again, passing from A to $B := A - \omega I_X$, we may suppose $\omega = 0$, so that $\mathbb{P} \subset \rho(A)$ and $\|(\lambda I_X - A)^{-n}\| \leq c\lambda^{-n}$ for all $\lambda \in \mathbb{P}$ and all $n \in \mathbb{N}$. For every $\mu \in \mathbb{P}$ let us define a new norm $\|\cdot\|_\mu$ on X by setting

$$\|x\|_\mu := \sup_{n\in\mathbb{N}} \|\mu^n(\mu I_X - A)^{-n}x\|.$$

These norms have the following properties:

(a) $\|\cdot\| \leq \|\cdot\|_\mu \leq c\,\|\cdot\|$ for all $\mu \in \mathbb{P}$.

(b) $\|\cdot\|_\lambda \leq \|\cdot\|_\mu$ for $\mu > \lambda > 0$.

(c) $\left\|(\mu I_X - A)^{-1}\right\|_\mu \leq 1/\mu$ for all $\mu \in \mathbb{P}$.

(d) $\|(\lambda I_X - A)^{-n}\|_\mu \leq \lambda^{-n}$ for all $n \in \mathbb{N}$ and all $\lambda,\ \mu \in \mathbb{P}$ satisfying $\lambda \leq \mu$.

We leave the proofs to the reader except for (d). By induction, it suffices to prove the case $n = 1$. Proposition 3.38 ensures that $R_\lambda - R_\mu = (\mu - \lambda)R_\lambda R_\mu$, hence

$$y := R_\lambda x = R_\mu x + (\mu - \lambda)R_\mu R_\lambda x = R_\mu x + (\mu - \lambda)R_\mu y.$$

This relation implies, using (c), that

$$\|y\|_\mu \leq \frac{1}{\mu}\|x\|_\mu + \frac{(\mu-\lambda)}{\mu}\|y\|_\mu$$

hence $\lambda\,\|y\|_\mu \leq \|x\|_\mu$.

In view of these properties, one can define still another norm by

$$\|x\|' := \sup_{\mu>0}\|x\|_\mu$$

which satisfies $\|\cdot\| \leq \|\cdot\|' \leq c\,\|\cdot\|$ and, for all $\lambda \in \mathbb{P}$,

$$\left\|(\lambda I_X - A)^{-1}\right\|' \leq 1/\lambda.$$

Then one can apply Corollary 10.2 on the space $(X, \|\cdot\|')$ which shows that A generates a nonexpansive continuous semigroup $(S(\cdot))$. Applying the relations $\|\cdot\| \leq \|\cdot\|' \leq c\,\|\cdot\|$, we get that $\|S(t)\| \leq c$ for all $t \in \mathbb{R}_+$. □

Exercises

1. Prove the properties of the norm $\|\cdot\|_\mu$.
2. Show that a dissipative (linear) operator A on a Banach space is closable if $R(A) \subset \mathrm{cl}(D(A))$. Show that in such a case the closure $\overline{A}$ of A is again dissipative and satisfies $R(\lambda I - \overline{A}) = \mathrm{cl}(R(\lambda I - A))$ for all $\lambda \in \mathbb{P}$.
3. Let A be a dissipative (linear) operator on a Banach space X. Prove that if both A and its transpose $A^{\intercal}$ are dissipative, then the closure of A generates a nonexpansive semigroup on X.

*10.2.3 * Dissipative and Accretive Multimaps*

Although we essentially focus our attention on linear single-valued dissipative operators, in the present subsection we make a detour through nonlinear, multivalued maps. A dissipativity assumption generalizing the one we considered gives rise to interesting properties that justify such a detour. It clearly generalizes the class of dissipative operators. Again J denotes the duality map of the Banach space X.

The class of multimaps we introduce plays a key role for evolution problems.

Definition 10.4 A multimap (or multivalued map) $M : X \rightrightarrows X$ on a Banach space X is said to be *dissipative* if for any $(x_1, y_1), (x_2, y_2) \in M$ there exists some $x^* \in J(x_1 - x_2)$ such that

$$\langle y_1 - y_2, x^* \rangle \leq 0. \tag{10.26}$$

A multimap $M : X \rightrightarrows X$ is said to be *accretive* if $-M$ is dissipative.

Given $\omega \in \mathbb{R}_+$ one says that $M : X \rightrightarrows X$ is *ω-dissipative* (resp. *ω-accretive*) if $M - \omega I$ is dissipative (resp. $\omega I - M$ is accretive).

The following characterization is similar to the one obtained for the case of a linear single-valued operator: it stems from Kato's lemma (Lemma 6.17).

Proposition 10.17 *A multimap $M : X \rightrightarrows X$ is dissipative if and only if one of the following assertions holds:*

(a) $\|x_1 - x_2\| \leq \|(x_1 - x_2) - r(y_1 - y_2)\|$ *for all* $(x_1, y_1), (x_2, y_2) \in M$, $r \in \mathbb{R}_+$;
(b) there exists a semi-scalar product $[\cdot, \cdot]$ *on* X *such that* $[x_1 - x_2, y_1 - y_2] \leq 0$ *for all* $(x_1, y_1), (x_2, y_2) \in M$;
(c) $[x_1 - x_2, y_2 - y_1]_+ \geq 0$ *for all* $(x_1, y_1), (x_2, y_2) \in M$.

The first characterization shows that for all $\lambda > 0$ the map $\lambda I_X - M : D(M) \to X$ is injective and $R_\lambda^M := (\lambda I_X - M)^{-1}$ is Lipschitzian with rate $1/\lambda$ on $R(\lambda I_X - M)$: given $z_1, z_2 \in R(\lambda I_X - M) := (\lambda I_X - M)(X)$, for $x_1, x_2 \in D(M)$ such that $z_i \in \lambda x_i - Mx_i$ for $i = 1, 2$, setting $r := \lambda^{-1}$, $y_i := \lambda x_i - z_i \in Mx_i$, one has $\|x_1 - x_2\| \leq \lambda^{-1} \|z_1 - z_2\|$. In particular, if $z_1 = z_2$ then $x_1 = x_2$. If M is single-valued and linear, M is dissipative if and only if for all $x \in D(M)$, $r \in \mathbb{R}_+$ one has

$$\|x\| \leq \|x - rMx\|,$$

or equivalently if for all $x \in D(M)$, $\lambda \in \mathbb{P}$ one has $\|x\| \leq (1/\lambda) \|(\lambda I - M)(x)\|$ or $\mathbb{P} \subset \rho(M)$ and $\|R_\lambda^M y\| \leq (1/\lambda) \|y\|$ for all $y \in R(\lambda I - M)$, $\lambda \in \mathbb{P}$. Thus the class of closed linear dissipative operators with dense domains is $\mathcal{D}_0(X)$.

Proposition 10.18 *A nonlinear (multivalued) operator* $M : D(M) \rightrightarrows X$ *on a Hilbert space* X *(identified with its dual) is monotone if and only if* M *is accretive. Thus, in a Hilbert space the definitions of dissipativity given in Definitions 9.7 and 10.4 coincide.*

Proof Given $u, w \in D(M)$, $v \in Mu$, $z \in Mw$ and $r > 0$ one has

$$\|(u + rv) - (w + rz)\|^2 - \|u - w\|^2 = 2r\langle v - z \mid u - w\rangle + r^2 \|v - z\|^2.$$

If M is monotone, the right-hand side is nonnegative, hence M is accretive. Conversely, if M is accretive, for all $r > 0$ the left-hand side is nonnegative, so that one must have $\langle v - z \mid u - w\rangle \geq 0$. □

The *resolvent operator* associated with an ω-dissipative multimap M is defined by $R_\lambda := R_\lambda^M := (\lambda I - M)^{-1}$ for $\lambda > \omega$, where I stands for I_X. In the sequel, passing from λ to $r := 1/\lambda$, and assuming M is ω-dissipative, we generalize previous definitions in Sect. 10.2.2 by setting (with $1/\omega = \infty$ if $\omega = 0$)

$$P_r(x) := (I - rM)^{-1}(x) \qquad x \in R(I - rM) \qquad r \in]0, 1/\omega[$$

$$M_r(x) := r^{-1}(P_r(x) - x) \qquad x \in R(I - rM) \qquad r \in]0, 1/\omega[.$$

The map P_r called the *proximal operator* of M is single-valued and one has

$$P_r = R_{\frac{1}{r}}(\frac{1}{r}\cdot) \qquad r \in]0, 1/\omega[.$$

The map $M_r := r^{-1}(P_r - I)$ is called the *Yosida operator* of M. It can be seen as an approximation of M. We have seen that the closed graph theorem ensures that $P_r : X \to X$ is linear and continuous when M is single-valued, linear and ω-dissipative with $r \in]0, 1/\omega[$. In the general case we have the following properties. The reader may assume $\omega = 0$ to get a simpler view, since in the sequel we focus on this case.

Proposition 10.19 *Let $M : X \rightrightarrows X$ be ω-dissipative. Then for all $r \in]0, 1/\omega[$ the (single-valued) maps M_r and P_r satisfy the following properties. In particular, if M is dissipative, M_r is dissipative and P_r is nonexpansive.*

(a) $\|P_r x - P_r x'\| \leq (1 - \omega r)^{-1} \|x - x'\|$ for all $x, x' \in R(I - rM)$.
(b) M_r is Lipschitzian with rate $(2 - \omega r)/r(1 - \omega r)$ on $R(I - rM)$.
(c) $M_r x \in MP_r x$ for all $x \in R(I - rM)$.
(d) $(1 - \omega r) \|M_r x\| \leq |Mx| := \inf\{\|y\| : y \in Mx\}$ if $x \in D(M) \cap R(I - rM)$.
(e) $\lim_{r \to 0_+} P_r x = x$ for all $x \in D(M) \cap_{r \in]0,1/\omega[} R(I - rM)$.
(f) If M is ω-dissipative and such that $D(M) \subset R(I - rM)$ for all $r \in]0, 1/\omega[$, then for all $x \in \mathrm{cl}(D(M))$ one has $(P_r x) \to x$ as $r \to 0_+$.

Proof In the sequel we take $r \in]0, \omega^{-1}[$ or $r \in \mathbb{P}$ if M is dissipative.

(a) Since $M - \omega I$ is dissipative, for $(x_1, y_1), (x_2, y_2) \in M$, one has

$$\|x_1 - x_2\| \leq \|(1 + \omega r)(x_1 - x_2) - r(y_1 - y_2)\|,$$
$$(1 - \omega r) \|x_1 - x_2\| \leq \|(x_1 - ry_1) - (x_2 - ry_2)\|,$$

so that $P_r := (I - rM)^{-1} : R(I - rM) \to X$ is single-valued and $(1 - \omega r)^{-1}$-Lipschitzian on $R(I - rM)$.

(b) Then the map $M_r := r^{-1}(P_r - I)$ is Lipschitzian on $R(I - rM)$ with rate $r^{-1}((1 - \omega r)^{-1} + 1) = r^{-1}(2 - \omega r)(1 - \omega r)^{-1}$. If M is dissipative, P_r being nonexpansive, M_r is dissipative since for $x, x' \in X$, $z^* \in J(x - x')$ one has $\langle x - x', z^* \rangle = \|x - x'\|^2 = \|z^*\|^2$, hence

$$\begin{aligned}\langle (P_r - I)x - (P_r - I)x', z^* \rangle &= \langle P_r x - P_r x', z^* \rangle - \langle x - x', z^* \rangle \\ &\leq \|x - x'\| . \|z^*\| - \|x - x'\|^2 \leq 0.\end{aligned}$$

(c) Given $x \in R(I - rM)$, $w := P_r x$, we have $w - x \in rMw$, hence

$$M_r x = r^{-1}(w - x) \in Mw = MP_r x.$$

(d) For $x \in D(M) \cap R(I - rM)$, $y \in Mx$, we have $x = P_r(x - ry)$, hence $M_r x = r^{-1}(P_r(x) - P_r(x - ry))$. Since P_r is $(1 - \omega r)^{-1}$-Lipschitzian, we get $\|M_r x\| \leq r^{-1}(1 - \omega r)^{-1} \|x - (x - ry)\| = (1 - \omega r)^{-1} \|y\|$. Passing to the infimum over $y \in Mx$, we get the announced inequality.

(e) For $x \in D(M) \cap_{r\in]0,1/\omega[} R(I - rM)$ we have

$$\|P_r x - x\| = r \|M_r x\| \leq \frac{r}{1 - \omega r} |Mx|,$$

so that $(P_r x) \to x$ when $r \to 0_+$.

(f) Suppose M is ω-dissipative and such that $D(M) \subset R(I - rM)$ for all $r \in]0, 1/\omega[$. Given $x \in \mathrm{cl}(D(M))$ and $\varepsilon > 0$, we pick $x' \in D(M) = D(M) \cap_{r\in]0,1/\omega[} R(I - rM)$ such that $\|x - x'\| < \varepsilon/4$ and we take $\delta \in]0, 1/2\omega[$ such that $\|P_r x' - x'\| < \varepsilon/4$ for all $r \in]0, \delta]$. Then, since P_r is 2-Lipschitzian we have $\|P_r x - P_r x'\| \leq 2 \|x - x'\| < \varepsilon/2$ and

$$\|P_r x - x\| \leq \left\|P_r x - P_r x'\right\| + \left\|P_r x' - x'\right\| + \left\|x' - x\right\| < \varepsilon,$$

so that $(P_r x) \to x$ when $r \to 0_+$. □

A multimap $M : X \rightrightarrows X$ is said to be *hyperdissipative* (or *m-dissipative*) if it is dissipative and if $R(I_X - M) = X$. In view of the next result, for such a multimap, for all $r \in \mathbb{P}$ the approximate map M_r and the proximal map P_r are defined on the whole of X and the last assertions of Proposition 10.19 can be simplified.

Proposition 10.20 *A dissipative multimap $M : D(M) \rightrightarrows X$ on a Banach space is hyperdissipative if and only if for all (or, equivalently for some) $r > 0$ one has $R(I_X - rM) = X$.*

Proof Let M be hyperdissipative. Given an arbitrary $r \in \mathbb{P}$ and $y \in X$ we have to prove that the equation $y \in x - rMx$ has a solution x. We rewrite it as $\frac{1}{r}y + (1 - \frac{1}{r})x \in (I - M)(x)$ or

$$x = P_1\left(\frac{1}{r}y + (1 - \frac{1}{r})x\right).$$

For $r \in]\frac{1}{2}, \infty[$ the map $x \mapsto P_1(\frac{1}{r}y + (1 - \frac{1}{r})x)$ is a contraction, so that this equation has a unique solution. Taking $s \in]\frac{1}{2}, 1[$ we see that sM is m-dissipative. Repeating the preceding argument with M changed into sM we see that for $r \in]\frac{s}{2}, \infty[$ one has $R(I - rM) = X$ and iterating this procedure we get that $R(I - rM) = X$ for all $r > 0$.

If the dissipative multimap M satisfies $R(I - sM) = X$ for some $s \in \mathbb{P}$, then for all $t \in \mathbb{P}$ the hyperdissipative multimap sM satisfies $R(I - tsM) = X$. Given $r \in \mathbb{P}$, taking $t := r/s$ we get that $R(I - rM) = X$. □

Corollary 10.5 *Any hyperdissipative multimap is maximally dissipative in the sense that any dissipative multimap whose graph contains the graph of M coincides with M.*

Proof If $(x, y) \in X \times X$ is such that

$$\|x - u\| \leq \|(x - ry) - (u - rv)\| \qquad \forall (u, v) \in M,\ \forall r > 0,$$

using the relation $R(I - rM) = X$, choosing $(u, v) \in M$ such that $u - rv = x - ry$, we see that $x = u$ and $y = v$, so that $(x, y) \in M$. □

Corollary 10.6

(a) Any maximally dissipative multimap M is closed in the sense that its graph is closed.
(b) Moreover, for any $x, y \in X$, any sequences $(r_n) \to 0_+$, $(x_n) \to x$ satisfying $x_n \in D(M_{r_n})$ for all n and $(M_{r_n}x_n) \to y$, one has $(x, y) \in M$.
(c) If J is single-valued and continuous, then M is demi-closed, i.e. sequentially closed in $X \times X_w$ and if $x, y \in X$ are such that for some sequences $(r_n) \to 0_+$, $(x_n) \to x$ with $x_n \in D(M_{r_n})$ for all n and $(M_{r_n}x_n) \to y$ weakly then one has $(x, y) \in M$.

Proof

(a) Given a sequence $((x_n, y_n))$ in (the graph of) M converging to (x, y), by dissipativity we have

$$\|x_n - u\| \leq \|(x_n - ry_n) - (u - rv)\| \qquad \forall (u, v) \in M,\ r > 0. \qquad (10.27)$$

Taking limits we get

$$\|x - u\| \leq \|(x - ry) - (u - rv)\| \qquad \forall (u, v) \in M,\ r > 0.$$

By maximality of M, we obtain $(x, y) \in M$.
(b) Now suppose $(r_n) \to 0_+$, $(x_n) \to x$ with $x_n \in D(M_{r_n})$ for all n, and $(M_{r_n}x_n) \to y$. Since $(P_{r_n}x_n) = (x_n + r_nM_{r_n}x_n) \to x$ and since $M_{r_n}x_n \in MP_{r_n}x_n$ by Proposition 10.19, we get $(x, y) \in M$ by the closedness of M.
(c) From now on we assume J is single-valued and continuous. We denote weak convergence by $\rightharpoonup$. Let $(x_n) \to x$, $(y_n) \rightharpoonup y$ with $(x_n, y_n) \in M$ for all n. For all $(u, v) \in M$, passing to the limit in the relation

$$\langle y_n - v, J(x_n - u)\rangle \leq 0$$

we obtain $\langle y - v, J(x - u)\rangle \leq 0$. Then $M \cup \{(x, y)\}$ is dissipative, so that $(x, y) \in M$ by maximality. The proof of the second assertion is similar: setting $(w_n) := (P_{r_n}x_n) = (x_n) - (r_nM_{r_n}(x_n)) \to x$ since $(r_ny_n) \to 0$, $y_n := M_{r_n}x_n \in MP_{r_n}x_n = Mw_n$ (by Proposition 10.19(c)) we have $(x, y) \in M$. □

If M is maximally dissipative and if X^* is strictly convex, so that J is single-valued, then for all $x \in X$ the set Mx is given by

$$Mx = \{y \in X : \langle y - v, J(x - u)\rangle \leq 0\ \forall (u, v) \in M\}.$$

Thus, it is closed and convex. If, moreover, X is reflexive and strictly convex, there exists in $Mx := M(x)$ a unique element of minimum norm. We denote it by M^0x or $M^0(x)$. In the sequel, if X is a reflexive space, using Asplund's Theorem (Theorem 6.23) we endow it with a norm that is strictly convex along with its dual norm.

Proposition 10.21 *Suppose X is a reflexive Banach space whose duality map is single-valued and continuous. Let $M : X \rightrightarrows X$ be hyperdissipative. Then, for all $x \in D(M)$, $(M_r(x)) \to M^0(x)$ weakly as $r \to 0_+$.*

If the norm of X satisfies the Kadec-Klee Property, for all $x \in D(M)$ one has $(M_r(w)) \to M^0(x)$ as $(r, w) \to (0_+, x)$ with $(\|w - x\| / r) \to 0$. In particular, $(M_r(x)) \to M^0(x)$ as $r \to 0_+$.

Proof Given $x \in D(M)$, $r > 0$, Proposition 10.19 (d) ensures that $\|M_r x\| \leq |Mx| := \|M^0 x\|$. Now, for any sequence $(r_n) \to 0_+$, the preceding corollary yields $(M_{r_n}x) \rightharpoonup M^0x$ since $(M_{r_n}x)$ has a weak limit point y that belongs to Mx and since the norm of y is not greater than $\|M^0x\|$, we get $y = M^0x$ and the first assertion stems from the uniqueness of this limit. The Kadec-Klee Property ensures that $(M_{r_n}x) \to M^0x$.

Since M_r is Lipschitzian with rate $2/r$, the last assertion stems from the relations $\|M_r w - M^0 x\| \leq \|M_r w - M_r x\| + \|M_r x - M^0 x\| \leq (2/r) \|w - x\| + \|M_r x - M^0 x\|$.
□

If X and X^* are uniformly convex one can show that $\mathrm{cl}(D(M))$ is convex (see [92, Prop. 13.2]).

In the case when M is hyperdissipative one has the following result concerning the differential inclusion

$$u'(t) \in M(u(t)) \qquad u(0) = x_0 \in D(M). \tag{10.28}$$

Here u is said to be a *solution* if for all $t \in \mathbb{R}_+$ $u(t) \in D(M)$, u is continuous, weakly right differentiable at t, and if its weak right derivative (simply denoted by u') satisfies relation (10.28).

Theorem 10.10 (Kōmura-Kato) *Let X be a uniformly convex Banach space whose dual X^* is also uniformly convex. Let $M : D \rightrightarrows X$ be hyperdissipative. Then the differential inclusion (10.28) has a unique solution u. Moreover, u is Lipschitzian, its right derivative u' is continuous from the right, and $\|u'(\cdot)\|$ is nonincreasing, $u'(t)$ being the element $M^0(u(t))$ of least norm in $M(u(t))$ for all $t \in \mathbb{R}_+$.*

For all $t \in \mathbb{R}_+$ the map $x_0 \mapsto S_t(x_0) := u(t)$ is nonexpansive and $(S_t)_{t \geq 0}$ is a semigroup of (nonlinear) nonexpansive maps.

If M is single-valued, the right derivative of u exists in the strong topology.

The property $u'(t) = M^0(u(t))$ is surprising: so to speak, solutions are lazy as they choose the speed that minimizes the norm!

Proof in the case M is single-valued. See [23, 93, 194] for the general case.

We first prove uniqueness. Let u and v be two solutions, $w := u - v$. Setting $j(\cdot) := (1/2) \|\cdot\|^2$, $f(\cdot) := j(w(\cdot))$, let us verify that f is right differentiable, with $f'(t) = \langle u'(t) - v'(t), J(u(t) - v(t)) \rangle \leq 0$. Since X^* is uniformly convex, j is differentiable on X, so that, setting $w_t(s) := (1/s)(w(t+s) - w(t)) = w'(t) + z_t(s)$ with $z_t(s) \to 0$ weakly as $s \to 0_+$, and taking a remainder $r_t(\cdot) := \varepsilon_t(\cdot) \|\cdot\| : X \to \mathbb{R}$ such that $j(w(t) + x) - j(w(t)) = \langle x, J(w(t)) \rangle + \varepsilon_t(x) \|x\|$ we get

$$j(w(t+s)) - j(w(t)) = s\langle w_t(s), J(w(t)) \rangle + s\varepsilon_t(sw_t(s)) \|w_t(s)\|,$$

$$\frac{1}{s}(j(w(t+s)) - j(w(t))) \to \langle w'(t), J(w(t)) \rangle \text{ as } s \to 0_+.$$

Thus f is right differentiable, with

$$f'(t) = \langle w'(t), J(w(t)) \rangle = \langle u'(t) - v'(t), J(u(t) - v(t)) \rangle \leq 0$$

since M is dissipative and $u'(t) \in M(u(t))$, $v'(t) \in M(v(t))$. Since $f(0) = 0$ and f is continuous with $f(\cdot) \geq 0$, by the Mean Value Theorem we get $f(\cdot) = 0$ and $u = v$. The preceding calculation also shows that if u and v are the solutions with initial conditions x_0 and y_0 respectively, then for all $t \in \mathbb{R}_+$ one has $\|u(t) - v(t)\| \leq \|x_0 - y_0\|$. The fact that $S_t(x_0) := u(t)$ defines a semigroup stems from uniqueness.

Next we show that if u is a solution to (10.28) and if the right derivative exists in the strong topology then the function $g(\cdot) := \|u'(\cdot)\|$ is nonincreasing. Since for a fixed $s > 0$ the map $t \mapsto v(t) := u(t+s)$ satisfies $v'(t) \in M(v(t))$, setting again $f(t) := j(u(t) - v(t))$, for $0 \leq r \leq t$ we get

$$\|u(t+s) - u(t)\| \leq \|u(r+s) - u(r)\|.$$

Dividing by s and taking limits as $s \to 0_+$, it follows that $\|u'(t)\| \leq \|u'(r)\|$.

Let us turn to existence. Since for $r > 0$ the approximate map M_r is single-valued and Lipschitzian with rate $2/r$ on $R(I - rM) = X$, the approximate equation

$$u_r'(t) = M_r(u_r(t)), \qquad u_r(0) = x_0$$

has a solution on $\mathbb{R}_+$. It is natural to wonder whether $(u_r)_r$ converges as $r \to 0_+$. Before considering this question we note that since M_r is dissipative and u_r is of class C^1, the preceding argument and Proposition 10.19 (d) show that

$$\left\|u_r'(t)\right\| \leq \left\|u_r'(0)\right\| = \|M_r(x_0)\| \leq c := \left\|M^0(x_0)\right\|. \tag{10.29}$$

Thus, combining (10.29) with the Mean Value Theorem, we have

$$\|u_r(t) - u_r(s)\| \leq c\,|t - s| \qquad \forall r \in \mathbb{P},\ \forall s, t \in \mathbb{R}_+. \tag{10.30}$$

In particular, for every $\tau \in \mathbb{P}$, and every $r, t \in [0, \tau]$, we have $\|u_r(t) - x_0\| \leq c\tau$. Since $u_r'(t) = M_r(u_r(t))$ and $M_r = r^{-1}(P_r - I)$, setting $v_r(t) := P_r(u_r(t))$, we have $ru_r'(t) = v_r(t) - u_r(t)$ so that relation (10.29) implies

$$\|u_r(t) - v_r(t)\| \leq cr. \tag{10.31}$$

Since $u_r(t) \in B[x_0, c\tau]$, we see that $v_r(t) \in B[x_0, c\tau + cr]$. Let α be the modulus of uniform continuity of J on $B[x_0, 2c\tau]$. Then for $r, s, t \in [0, \tau]$ we have

$$\|J(u_r(t) - u_s(t)) - J(v_r(t) - v_s(t))\| \leq \alpha(cr + cs). \tag{10.32}$$

Since u_r is the solution of $u_r' = M_r u_r = M v_r$, by dissipativity of M we have

$$\langle u_r'(t) - u_s'(t), J(v_r(t) - v_s(t))\rangle \leq 0.$$

Using this relation and estimate (10.32), we get for the right derivative of $f(\cdot) := \frac{1}{2}\|u_r(\cdot) - u_s(\cdot)\|^2$

$$\begin{aligned} f'(t) &= \langle u_r'(t) - u_s'(t), J(u_r(t) - u_s(t))\rangle \\ &\leq \langle u_r'(t) - u_s'(t), J(v_r(t) - v_s(t))\rangle + \alpha(cr + cs)\left\|u_r'(t) - u_s'(t)\right\| \\ &\leq 2c\alpha(cr + cs). \end{aligned}$$

Thus $\|u_r(t) - u_s(t)\|^2 \leq 4c\tau\alpha(cr + cs)$ for $r,\ s,\ t \in [0, \tau]$ and, by the Cauchy criterion, (u_r) converges to some map u as $r \to 0_+$, uniformly on compact intervals. Moreover, by (10.30), the limit u is Lipschitzian with rate c on $\mathbb{R}_+$.

For any $r \in \mathbb{P}$, $t \in \mathbb{R}_+$, $\|u_r'(t)\| = \|M_r(u_r(t))\|$ is bounded by c in view of (10.29) and since X is reflexive, given sequences $(r_n) \to 0_+$, $(t_n) \to t$ in $\mathbb{R}_+$, we can find a subsequence $((r_{k(n)}, t_{k(n)}))$ of $((r_n, t_n))$ such that $(u_{r_{k(n)}}'(t_{k(n)}))$ weakly converges to some $y(t)$. By Corollary 10.6 we have $u(t) \in D(M)$ and $y(t) = M(u(t))$. Thus, by uniqueness, $(u_r'(s))$ weakly converges to $M(u(t))$ as $(r, s) \to (0_+, t)$. In particular, $(u_r'(t)) \rightharpoonup M(u(t))$ and $M(u(\cdot))$ is weakly continuous.

For all $x^* \in X^*$, passing to the limit as $r \to 0_+$ in the relation

$$\langle x^*, u_r(t) - x_0\rangle = \int_0^t \langle x^*, u_r'(s)\rangle ds$$

we get $u(t) = x_0 + \int_0^t y(s)ds$ by the Dominated Convergence Theorem, so that u is weakly differentiable with derivative $u'(t) := y(t) = M(u(t))$.

Proposition 10.21 ensures that $(M_r(x_0))_r \to M(x_0)$ as $r \to 0_+$. Moreover, by (10.29) and by weak lower semicontinuity of the norm, for all $t \in \mathbb{R}_+$ we get

$$\|M(u(t))\| = \|y(t)\| \leq \liminf_{r\to 0_+} \|M_r(u_r(t))\| \leq c = \|M(u(0))\|.$$

For all $s \in \mathbb{R}_+$, since $v : t \mapsto u(s + t)$ is the solution of $v'(t) = M(v(t))$, $v(0) = u(s)$, we get

$$\|M(u(s + t))\| = \|M(v(t))\| \leq \|M(v(0))\| = \|M(u(s))\|,$$

i.e. $\|M(u(\cdot))\|$ is nonincreasing. It is also right continuous: given a sequence $(t_n) \to t_+$, since $(M(u(t_n))) \rightharpoonup M(u(t))$, the weak lower semicontinuity of the norm yields

$$\|M(u(t))\| \leq \liminf_n \|M(u(t_n))\| \leq \limsup_n \|M(u(t_n))\| \leq \|M(u(t))\|$$

and these relations are equalities. The Kadec-Klee Property ensures that $(M(u(t_n)))$ converges to $M(u(t))$. Thus $M(u(\cdot))$ itself is right continuous on $\mathbb{R}_+$. The relation $u(t) = x_0 + \int_0^t M(u(s))ds$ then entails that

$$\frac{1}{s}(u(t+s) - u(t)) \to M(u(t))$$

as $s \to 0_+$. Thus u is right differentiable for the strong topology. □

The semigroup $(S_t)_{t\geq 0}$ we consider has a regularizing effect in the sense that S_t can be extended to a nonexpansive map from $\mathrm{cl}(D(A))$ into $D(A)$; see [51, Thm 3.3]. Let us quote a result of interest in this direction. It bears some analogy with the classical gradient method for the numerical minimization of a differentiable convex function.

Theorem 10.11 *Let $f : X \to \mathbb{R}_\infty$ be a closed proper convex function on a Hilbert space X and let $M := -\partial f : X \rightrightarrows X$, X^* being identified with X via the Riesz isomorphism. Then the semigroup $(S_t)_{t\geq 0}$ generated by M can be extended to a semigroup of maps S_t: $\mathrm{cl}(D(M)) \to D(M)$ for all $t \in \mathbb{P}$.*

Moreover, for all $\theta > 0$ the map $u : t \mapsto S_t(x_0)$ is Lipschitzian and right differentiable on $[\theta, \infty[$ and $f \circ u$ is convex, nonincreasing and Lipschitzian on $[\theta, \infty[$ with $(f \circ u)'(t) = -\|u'(t)\|^2$.

Several approximation methods are available for solving equation (10.21); see [71, 88, 144]. Let us end this section by quoting some results dealing with a numerical approach called the *implicit time discretization* scheme. This terminology can be explained as follows. Given $\theta > 0$ one divides the interval $[0, \theta]$ into n intervals of equal length $\theta_n := \theta/n$ and, starting with $u_{0,n} := x_0$, one obtains $u_{k,n}$ inductively on k by solving the inclusions

$$\frac{u_{k+1,n} - u_{k,n}}{\theta_n} \in Mu_{k+1,n}$$

or, when M is single-valued, $(I - \theta_n M)u_{k+1,n} = u_{k,n}$, so that

$$u_{k,n} = (I - \frac{\theta}{n}M)^{-k}x_0.$$

For $t \in [k\theta/n, (k+1)\theta/n[$ one sets $u_n(t) := u_{k,n}$ or a natural convex combination of $u_{k,n}$ and $u_{k+1,n}$. One says that $u_n(\cdot)$ is an *approximate solution* to the differential inclusion

$$u'(t) \in M(u(t)) \qquad u(0) = x_0 \in D(M). \tag{10.33}$$

One says that a continuous map $u : \mathbb{R}_+ \to X$ is a *mild solution* of the preceding equation if there is a sequence (u_n) of approximate solutions that converges uniformly to u on any compact interval.

Theorem 10.12 (Crandall-Liggett) *Let $M : X \rightrightarrows X$ be a ω-dissipative multimap such that, for some $\theta > 0$,*

$$\mathrm{cl}(D(M)) \subset R(I - rM) \qquad \forall r \in [0, \theta].$$

Then (10.33) has a unique mild solution u. Moreover, it is given by

$$u(t) = \lim_{n\to\infty} (I - \frac{t}{n}M)^{-n}x_0 \qquad t > 0,$$

the convergence being uniform on compact intervals.

One can also consider the case when M depends on t, f.i. when M is replaced with $M + f(t)$, with $f \in L_1(\mathbb{R}_+, X)$. See [25, 173].

On the other hand, the proof of Corollary 10.2 shows that under its assumptions one has the convergence result

$$S(t)x = \lim_{n\to\infty} \exp(tA(I - \frac{1}{n}A)^{-1}x).$$

Exercises

1*. **(Trotter-Kato)** Let S, S_n ($n \in \mathbb{N}$) be continuous semigroups on a Banach space X satisfying for some $c > 0$, $\omega \in \mathbb{R}$ the estimates $\|S(t)\| \le ce^{\omega t}$, $\|S_n(t)\| \le ce^{\omega t}$ for $t \in \mathbb{R}_+$, $n \in \mathbb{N}$. Let A_n (resp. A) be the generator of S_n (resp. S). Prove that the following assertions are equivalent:

(a) for all $x \in D(A)$ there exists an $x_n \in D(A_n)$ such that $(x_n)_n \to x$ and $(A_n x_n)_n \to Ax$;
(b) $(R_\lambda^{A_n})_n \to R_\lambda^A$ for some $\lambda > \omega$, $R_\lambda^{A_n}$ (resp. R_λ^A) being the resolvent of A_n (resp. A);
(c) $(R_\lambda^{A_n})_n \to R_\lambda^A$ for all $\lambda > \omega$;
(d) $(S_n(t))_n \to S(t)$ uniformly on compact intervals.

2*. Let A be the generator of a continuous semigroup $S(\cdot)$ of linear operators satisfying for some $c, \omega \in \mathbb{R}_+$ the relation $\|S(t)\| \le ce^{\omega t}$. Show that for all $t \in \mathbb{R}_+$, $x \in X$ one has $S(t)x = \lim_{n\to\infty}(I_X - (t/n)A)^{-n}x$, the limit being uniform on every compact interval of $\mathbb{R}_+$.

3*. Let $M : X \rightrightarrows X$ be a maximally dissipative multimap on a Hilbert space X. Prove that the solution u to the problem $u'(t) \in M(u(t))$, $u(0) = u_0$, where u_0 is a given element of $D(M)$, belongs to $C^1(\mathbb{R}_+, X)$ and $C(\mathbb{R}_+, D(M))$.

10.3 Parabolic Problems: The Heat Equation

In this section and the following one we consider evolution problems involving second-order partial differential equations. Following the classical classification of partial differential equations into elliptic, parabolic and hyperbolic equations, for which we refer to [83], we separate our study of evolution equations into two distinct sections. In both sections, Ω denotes a bounded open subset of class C^∞ of $\mathbb{R}^d$, with boundary Γ, and T is the interval $T :=]0,\tau[$, with $\tau \in]0,\infty]$. Given essentially bounded measurable functions $a_{i,j} : \Omega \times T \to \mathbb{R}$, $a_i : \Omega \times T \to \mathbb{R}$, $a_0 : \Omega \times T \to \mathbb{R}$, we denote by L the operator defined by

$$(Lu)(x,t) := \sum_{i,j=1}^{d} D_i(a_{i,j}(x,t)D_j u(x,t)) + \sum_{i=1}^{d} a_i(x,t)D_i u(x,t) + a_0(x,t)u(x,t).$$

A simple example is the Laplacian operator $L := \Delta$ in the space variables x_i. In such a case, the general parabolic equation

$$\frac{\partial u}{\partial t}(x,t) - (Lu)(x,t) = f(x,t) \qquad (x,t) \in \Omega \times T, \tag{10.34}$$

$$u(x,t) = 0 \qquad (x,t) \in \Gamma \times T \tag{10.35}$$

$$u(x,0) = g(x) \qquad x \in \Omega, \tag{10.36}$$

in which $f \in L_2(\Omega \times T)$, $g \in H_0^1(\Omega) \cap H^2(\Omega)$ are given, turns out to be the *heat equation* when $f = 0$. It describes the distribution of the temperature in a medium Ω over time $t \in T$ with initial temperature g. The boundary condition (10.35) can be replaced with a Neuman condition

$$\frac{\partial u}{\partial n}(x,t) = 0 \qquad (x,t) \in \Gamma \times T,$$

but we limit our study to the Dirichlet condition. Many other diffusion phenomena can be described by *parabolic equations*, u measuring the concentration of a chemical, for example.

Definition 10.5 The partial differential operator $\frac{\partial}{\partial t} - L$ is said to be uniformly parabolic if there exists a constant $c_E > 0$ such that

$$\sum_{i,j=1}^{d} a_{i,j}(x,t)v^i v^j \geq c_E \|v\|^2 \qquad \forall (x,t) \in \Omega \times T,\ v := (v^1,\ldots,v^d) \in \mathbb{R}^d.$$

Thus, for all $t \in T$ the operator L is elliptic; but the above definition requires a lower bound for the ellipticity constant that is valid for the whole interval T. In the sequel, for the sake of simplicity, we suppose that the coefficients of L are smooth and do not depend on t. The Galerkin method allows one to get rid of this last restriction (see [117, Section 7.1] for instance). Here we want to easily derive an existence result from the Hille-Yosida Theorem.

We consider the operator $A : D(A) \to H := L_2(\Omega)$ with domain $D(A) := H_0^1(\Omega) \cap H^2(\Omega) \subset L_2(\Omega)$ given by $A(v)(x) := (Lv)(x)$ for $x \in \Omega$ and we look for a continuous map $u : T \cup \{0\} \to H$ satisfying $u(T) \subset D(A)$ and

$$\begin{aligned} u'(t) &= Au(t) \qquad t \in T \\ u(0) &= g, \end{aligned}$$

and we set $u(x, t) := u(t)(x)$ for $(x, t) \in \Omega \times (T \cup \{0\})$.

We introduce the bilinear form b associated with L as in (9.17) given by

$$b(u, v) := \int_\Omega \sum_{i,j=1}^{d} a_{ij} D_i u D_j v + \int_\Omega \sum_{i=1}^{d} a_i u D_i v + \int_\Omega a_0 uv \qquad u, v \in H_0^1(\Omega).$$

By Gårding's inequality (9.21) there exist constants $c > 0$, $\omega \in \mathbb{R}$ such that

$$b(u, u) + \omega \|u\|_0^2 \geq c \|u\|_1^2 \qquad \forall u \in H_0^1(\Omega), \tag{10.37}$$

where $\|\cdot\|_0$ (resp. $\|\cdot\|_1$) is the norm of $L_2(\Omega)$ (resp. $H_0^1(\Omega)$).

Proposition 10.22 *The operator A generates an ω-contraction semigroup $(S(t))_{t \geq 0}$ in $L_2(\Omega)$.*

Proof In order to apply Theorem 10.8 or rather Corollary 10.3, we verify that $A - \omega I$ is a linear maximally dissipative operator or that $\omega I - A$ is a maximally monotone operator on $L_2(\Omega)$. By Green's formula and (10.37)

$$\langle (\omega I - A)u \mid u \rangle = b(u, u) + \omega \|u\|_0^2 \geq 0$$

we see that the linear operator $\omega I - A$ is a monotone operator on $L_2(\Omega)$ endowed with its usual scalar product $\langle \cdot \mid \cdot \rangle$. Moreover, by Theorem 9.28, it is maximally monotone since for $\lambda > \omega$ and every $f \in L_2(\Omega)$ the equation

$$\lambda u - Au = f$$

has a solution $u \in D(A) := H_0^1(\Omega) \cap H^2(\Omega)$ by Theorem 9.18, so that for $\mu := \lambda - \omega > 0$ one has $R(\mu I + (\omega I - A)) = R(\lambda I - A) = L_2(\Omega)$. □

Given $f \in L_2(T, L_2(\Omega))$, $g \in L_2(\Omega)$, we say that a function $u \in L_2(T, H_0^1(\Omega))$ whose derivative u' belongs to $L_2(T, H^{-1}(\Omega))$ is a *weak solution* to the system (10.34)–(10.36) if $u(0) = g$ and

$$\langle u'(t), v \rangle + b(u, v) = \langle f(t), v \rangle \qquad \forall v \in H_0^1(\Omega).$$

It can be shown (see [117, Thm 3 p. 287]) that a function $u \in L_2(T, H_0^1(\Omega))$ whose derivative u' belongs to $L_2(T, H^{-1}(\Omega))$ is in fact in $C(T \cup \{0\}, L_2(\Omega))$, so that $u(0)$ is well defined.

Galerkin's method and regularity estimates enable one to obtain the following result (see [117, Section 7.1]). When $f = 0$ and $g \in H_0^1(\Omega) \cap H^2(\Omega)$ its second assertion is a consequence in the preceding proposition.

Theorem 10.13 *Assume $f \in L_2(\Omega \times T)$ and $g \in L_2(\Omega)$. Then there exists a unique weak solution u to the system (10.34)–(10.36).*

If $g \in H_0^1(\Omega)$, then the weak solution u is in $L_2(T, H^2(\Omega)) \cap L_\infty(T, H_0^1(\Omega))$. Moreover, for some $c > 0$ one has the estimate

$$\|u\|_{L_2(T,H^2(\Omega))} + \|u\|_{L_\infty(T,H_0^1(\Omega))} + \|u'\|_{L_2(\Omega\times T)} \le c\,\|f\|_{L_2(\Omega\times T)} + c\,\|g\|_{H_0^1(\Omega)}\,.$$

Assuming more regularity on f and g, one can get more regular solutions.

Exercise (Heat Kernel) For $x \in \mathbb{R}^d$, $t \in \mathbb{P}$ let

$$k_t(x) := \frac{1}{(4\pi t)^{d/2}} e^{-\|x\|^2/4t}.$$

Let $X := C_0(\mathbb{R}^d)$ be the space of continuous functions $w : \mathbb{R}^d \to \mathbb{R}$ such that $w(x) \to 0$ as $\|x\| \to \infty$, with the norm given by $\|w\| := \sup_x |w(x)|$. For $t \in \mathbb{P}$ and $w \in X$ let

$$S_t(w)(x) := (k_t * w)(x) := \int_{\mathbb{R}^d} k_t(x-y)w(y)d\lambda_d(y) \qquad x \in \mathbb{R}^d.$$

Prove that $S_t(w) \in X$ for all $t \in \mathbb{P}$ and $w \in X$ and that $\|S_t(w)\| \le \|w\|$. Set $S_0(w) := w$ for all $w \in X$.

Prove the **Chapman-Komolgorov equation**: for all s, $t \in \mathbb{P}$ one has

$$k_{s+t}(x) = S_t(k_s)(x) := \int_{\mathbb{R}^d} k_t(x-y)k_s(y)d\lambda_d(y) \qquad x \in \mathbb{R}^d.$$

From this deduce that $(S_t)_{t\ge 0}$ is a continuous semigroup on X. [See: [246, p. 162–169].]

It can be proved that for $g \in C_0(\mathbb{R}^d)$ the function $u : (x,t) \mapsto S_t(g)(x)$ is the solution of the heat equation

$$\frac{\partial u}{\partial t}(x,t) - \Delta u(x,t) = 0, \qquad u(x,0) = g(x) \quad x \in \mathbb{R}^d,\ t \in \mathbb{P}.$$

10.4 Hyperbolic Problems: The Wave Equation

Retaining the notation and the assumptions of the preceding subsection, let us consider now the *second-order hyperbolic problem*

$$\frac{\partial^2 u}{\partial t^2}(x,t) - (Lu)(x,t) = f(x,t) \qquad (x,t) \in \Omega \times T, \tag{10.38}$$

$$u(x,t) = 0 \qquad (x,t) \in \Gamma \times T, \tag{10.39}$$

$$u(x,0) = g(x), \quad \frac{\partial u}{\partial t}(x,0) = h(x) \qquad x \in \Omega, \tag{10.40}$$

where $g \in H_0^1(\Omega)$, $h \in L_2(\Omega)$. We now suppose $a_{i,j} = a_{j,i}$ for $i,j \in \mathbb{N}_d$. Following a scheme used for second-order ordinary differential equations, we recast this problem as a first-order system in (u, v) where $v := \frac{\partial u}{\partial t}$:

$$\left(\frac{\partial u}{\partial t}(x,t), \frac{\partial v}{\partial t}(x,t)\right) - (v(x,t), (Lu)(x,t)) = (0, f(x,t)) \qquad (x,t) \in \Omega \times T,$$

$$(u(x,t), v(x,t)) = (0,0) \qquad (x,t) \in \Gamma \times T$$

$$(u(x,0), v(x,0)) = (g(x), h(x)) \qquad x \in \Omega.$$

Note that we have added the condition $v(x,t) = 0$ for $(x,t) \in \Gamma \times T$; but this condition is a consequence in (10.39) obtained by differentiating u with respect to t. Again, by Gårding's Theorem (Theorem (9.13)) there exist constants $c > 0$, $\omega \in \mathbb{R}$ such that inequality (10.37) is satisfied, b being the bilinear form of Theorem 9.13 associated with L as above. We take

$$X := H_0^1(\Omega) \times L_2(\Omega)$$

endowed with the norm $\|\cdot\|_X$ associated with the scalar product given by

$$\langle (u,v) \mid (x,z) \rangle_X := b(u,x) + 2\omega \langle u \mid x \rangle_0 + \langle v \mid z \rangle_0 \qquad (u,v),\ (x,z) \in X,$$

where $\langle \cdot \mid \cdot \rangle_0$ is the scalar product in $L_2(\Omega)$. By (9.19) with $m = 1$, the norm $\|\cdot\|_X$ is equivalent to the product norm of $\|\cdot\|_1 = \|\cdot\|_{H_0^1(\Omega)}$ with $\|\cdot\|_0 = \|\cdot\|_{L_2(\Omega)}$ given by

$$\|(u,v)\| := (\|u\|_1^2 + \|v\|_0^2)^{1/2} \qquad (u,v) \in H_0^1(\Omega) \times L_2(\Omega).$$

We consider the operator A with domain $D(A) := (H^2(\Omega) \cap H_0^1(\Omega)) \times H_0^1(\Omega)$ and values in X defined by

$$A(u,v) := (v, Lu) \qquad (u,v) \in D(A).$$

Theorem 10.14 *The operator A generates an ω-contraction semigroup on $X := H_0^1(\Omega) \times L_2(\Omega)$. If the coefficients a_i of L are null, the operator A generates a nonexpansive semigroup on X.*

Proof Let us verify the assumptions of Corollary 10.4. Without loss of generality, we assume that $\omega \geq 1$. First, for $(u, v) \in D(A)$, by inequality (9.19) we see that $\omega I - A$ is a monotone operator:

$$\begin{aligned}&\langle(\omega I - A)(u, v) \mid (u, v)\rangle_X \\ &\quad = \omega(b(u,u) + 2\omega \|u\|_0^2 + \|v\|_0^2) - b(v,u) - 2\omega\langle v \mid u\rangle_0 - \langle v \mid Lu\rangle_0 \\ &\quad \geq \omega(b(u,u) + \omega \|u\|_0^2) + \omega^2 \|u\|_0^2 + \|v\|_0^2 - 2\omega\langle v \mid u\rangle_0 \geq 0.\end{aligned}$$

Now let us show that for $\lambda > \omega^{1/2}$ and for every $(f, g) \in X$ the equation

$$\lambda(u, v) - A(u, v) = (f, g) \qquad (u, v) \in X \tag{10.41}$$

has a solution. This equation amounts to the system

$$\begin{aligned}\lambda u - v &= f \qquad u \in H^2(\Omega) \cap H_0^1(\Omega), \quad v \in H_0^1(\Omega) \\ \lambda v - Lu &= g \qquad u \in H^2(\Omega) \cap H_0^1(\Omega), \quad v \in H_0^1(\Omega).\end{aligned}$$

These two equations imply that $v = \lambda u - f$ and

$$\lambda^2 u - Lu = \lambda f + g.$$

Since $\lambda^2 > \omega$, Theorem 9.18 ensures that this equation has a unique solution $u \in H^2(\Omega) \cap H_0^1(\Omega)$. Then $(u, \lambda u - f) \in X$ is a solution of equation (10.41): $A - \lambda I$ is surjective and $\omega I - A$ is maximally dissipative.

If the coefficients a_i of L are null we can take $\omega = 0$ in the preceding and apply Corollary 10.4. □

Given $f \in L_2(T, L_2(\Omega))$, $g \in H_0^1(\Omega)$, we say that a function $u \in L_2(T, H_0^1(\Omega))$ whose derivatives u' and u'' belong to $L_2(T, L_2(\Omega))$ and $L_2(T, H^{-1}(\Omega))$ respectively is a *weak solution* to the system (10.34)–(10.36) if $u(0) = g$, $u'(0) = h$ and

$$\langle u'(t), v\rangle + b(u, v) = \langle f(t), v\rangle \qquad \forall v \in H_0^1(\Omega).$$

Again, $u(0)$ and $u'(0)$ are well defined and, using Galerkin's method and regularity estimates, one can establish the following result (see [117, Section 7.2]).

Theorem 10.15 *Assume $f \in L_2(T, L_2(\Omega))$, $g \in H_0^1(\Omega)$, $h \in L_2(\Omega)$. Then, the weak solution u of the system (10.38)–(10.40) exists and is unique. Moreover, for* *some $c > 0$ one has the estimate*

$$\|u\|_{L_\infty(T,H_0^1(\Omega))} + \|u'\|_{L_\infty(T,L_2(\Omega))} \leq c\,\|f\|_{L_2(T,L_2(\Omega))} + c\,\|g\|_{H_0^1(\Omega)} + c\,\|h\|_{L_2(\Omega)}\,.$$

If in addition f has a weak derivative f' in $L_2(T,L_2(\Omega))$, $g \in H^2(\Omega)$, $h \in H_0^1(\Omega)$, then $u \in L_\infty(T,H^2(\Omega))$, $u' \in L_\infty(T,H_0^1(\Omega))$, $u'' \in L_\infty(T,L_2(\Omega))$, $u''' \in L_2(T,H^{-1}(\Omega))$, and their norms are bounded above by a multiple of the sum of the norms of f, f', g, h in their respective spaces.

Exercises

1. Given $c \in \mathbb{R}$ and two functions f, g of class C^2 on $\mathbb{R}$, verify that the function u given by

$$u(x,t) := f(x-ct) + g(x+ct) \qquad (x,t) \in \mathbb{R}^2$$

is a solution of the equation

$$\frac{\partial^2 u}{\partial t^2}(x,t) - c^2\frac{\partial^2 u}{\partial x^2}(x,t) = 0 \qquad (x,t) \in \mathbb{R}^2. \tag{10.42}$$

2. Let $C_c^2(\mathbb{R}^2)$ be the space of functions of class C^2 with compact support on $\mathbb{R}^2$. Show that any solution u of class C^2 of equation (10.42) is a *weak solution* of this equation in the sense that for any $\varphi \in C_c^2(\mathbb{R}^2)$ the following relation holds:

$$\int_{\mathbb{R}^2}\Big(\frac{\partial^2 u}{\partial t^2} - c^2\frac{\partial^2 u}{\partial x^2}\Big)(x,t)\varphi(x,t)dxdt = \int_{\mathbb{R}^2}\Big(\frac{\partial^2 \varphi}{\partial t^2} - c^2\frac{\partial^2 \varphi}{\partial x^2}\Big)(x,t)u(x,t)dxdt.$$

3. Prove the converse of the assertion of the preceding exercise: if a function u of class C^2 is a weak solution of equation (10.42) then u is a (classical) solution of this equation.
4. Show that if (f_n) and (g_n) are two sequences of functions of class C^2 on $\mathbb{R}$ that converge in $\mathcal{L}_2(\mathbb{R})$ to $f, g \in \mathcal{L}_2(\mathbb{R})$ respectively, then u given by $u(x,t) := f(x-ct) + g(x+ct)$ for $(x,t) \in \mathbb{R}^2$ is a weak solution to (10.42).
5. Given u_0, $u_1 \in C^2(\mathbb{R})$, verify that the function u given by

$$u(x,t) := \frac{1}{2}(u_0(x+ct) + u_0(x-ct)) + \frac{1}{2c}\int_{x-ct}^{x+ct} u_1(s)ds$$

is a solution of equation (10.42) satisfying the initial conditions

$$u(x,0) = u_0(x) \qquad \frac{\partial u}{\partial t}(x,0) = u_1(x) \qquad x \in \mathbb{R}.$$

Appendix: The Brouwer's Fixed Point Theorem

This appendix is devoted to a proof of Brouwer's Theorem. This simple but powerful result plays an important role in analysis and the proof we give uses techniques that are scattered throughout the book. Thus, it can serve as a conclusion. Moreover, we consider some striking related properties.

This result has received numerous and very different proofs, the first ones being due to H. Poincaré (1886), J. Hadamard (1910), and L.E.J. Brouwer (1912). Whereas arguments from algebraic topology are natural for such a result, nice analytical proofs have been devised by J. Milnor (1978) and C.A. Rogers (1980); see [197], [222]. The proof we present is inspired by the latter. Throughout we denote by B the closed unit ball of $\mathbb{R}^d$ and by S its unit sphere, $\mathbb{R}^d$ being endowed with its Euclidean norm and its scalar product denoted by $x.y$ for $x, y \in \mathbb{R}^d$ rather than $\langle x \mid y \rangle$ for the sake of brevity.

Theorem 10.16 (Brouwer) *Any continuous map $f : B \to B$ has a fixed point: there exists some $z \in B$ such that $f(z) = z$.*

We denote by (F) the assertion of this theorem and we consider some other assertions:

(Z) For any continuous map $v : B \to \mathbb{R}^d$ such that $v(x).x \leq 0$ for all $x \in S$ there exists some $z \in B$ such that $v(z) = 0$;

(R) There exist no continuous map $r : B \to S$ such that $r(x) = x$ for all $x \in S$.

Assertion (R) is often called the *Retraction Theorem* (a map $r : B \to S$ such that $r(x) = x$ for all $x \in S$ being called a *retraction* of B onto S) and assertion (Z) is sometimes called the *Hairy Ball Theorem*. In assertion (Z) v can be considered as a vector field on B since for all $x \in B$ one has $v(x) \in T(B, x)$, the tangent cone to the convex subset B (which is also a smooth manifold with boundary). Recall that the definition yields $T(B, x) = \mathbb{R}^d$ for $x \in U := \text{int}B = B \backslash S$ and $T(B, x) = \{w \in \mathbb{R}^d : w.x \leq 0\}$ for $x \in S$.

The relationships between the preceding assertions are remarkable.

J.-P. Penot, *Analysis*, Universitext, DOI 10.1007/978-3-319-32411-1

Proposition 10.23 *The assertions (F), (Z), (R) are equivalent.*

Proof (F)$\Longrightarrow$(Z) Suppose (Z) does not hold: there exists some continuous $v : B \to \mathbb{R}^d$ such that $v(x) \neq 0$ for all $x \in B$ and $v(x).x \leq 0$ for all $x \in S$. Then, setting

$$w(x) := \frac{v(x)}{\|v(x)\|}$$

one gets a continuous map $w : B \to S$ and by (F) there exists some $z \in B$ such that $w(z) = z$. Then one has $z \in S$ and one gets the contradiction

$$1 = \|z\|^2 = w(z).z = \frac{v(z).z}{\|v(z)\|} \leq 0.$$

(Z)$\Longrightarrow$(R) Suppose there exists a continuous map $r : B \to S$ such that $r(x) = x$ for all $x \in S$. Setting $v(x) := x - 2r(x)$ one defines a continuous vector field v satisfying $v(x).x = -\|x\|^2 = -1$ for all $x \in S$ and $v(x).r(x) = x.r(x) - 2 \leq -1$ for $x \in B$ by the Cauchy-Schwarz inequality. Thus v cannot have a zero in B, contradicting (Z).

(R)$\Longrightarrow$(F) Suppose there exists a continuous map $f : B \to B$ such that $f(x) \neq x$ for all $x \in B$. Set $g(x) := x - f(x)$, $h_x(t) := \|x + tg(x)\|^2 - 1$ for $x \in B$, $t \in \mathbb{R}$. For all $x \in B$ one has $h_x(0) = \|x\|^2 - 1 \leq 0$ and $\lim_{t\to\infty} h_x(t) = \infty$, so that there exists some $t_x \in \mathbb{R}_+$ such that $h_x(t_x) = 0$ (Fig. A.1).

In fact, t_x is the unique nonnegative root of the quadratic function $h_x : t \mapsto t^2 \|g(x)\|^2 + 2tg(x).x + \|x\|^2 - 1$, hence

$$t_x = \|g(x)\|^{-2} \left((g(x).x)^2 + \|g(x)\|^2 (1 - \|x\|^2)\right)^{1/2} - \|g(x)\|^{-2} g(x).x.$$

Since by compactness of B there exists a $c > 0$ such that $\|g(x)\| \geq c$ for all $x \in B$, the function $x \mapsto t_x$ is continuous, and $r : x \mapsto x + t_x g(x)$ is continuous too. By construction, one has $r(B) \subset S$. For $x \in S$ one has $h_x(0) = 0$, so that $t_x = 0$; this also follows from the expression of t_x: $g(x).x = \|x\|^2 - f(x).x \geq 0$, $((g(x).x)^2)^{1/2}$

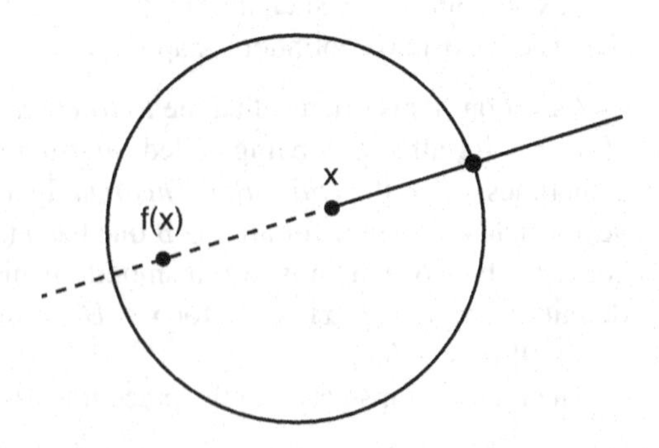

Fig. A.1 Intersecting the sphere with the half-line $x + \mathbb{R}_+(x - f(x))$

$= g(x).x$ and $t_x = 0$, so that $r(x) = x$. Assertion (R) is thus denied. Thus (F) must hold. □

Remark The implication (Z)$\Longrightarrow$(F) is immediate: given a continuous map $f : B \to B$, setting $v(x) := f(x) - x$ one has $v(x).x \leq 0$ for $x \in S$ by the Cauchy-Schwarz inequality, so that (Z) implies that there exists some $z \in B$ such that $f(z) - z = 0$. □

We prove Brouwer Theorem, or rather assertion (R), in two steps. The first one is a weakened version of (R).

Lemma 10.6 *There is no retraction of class C^1 from B onto S.*

Proof Suppose $p : B \to S$ is a retraction of class C^1 from B onto S. For $t \in T := [0, 1]$ let $p_t = (1 - t)I + tp$, where I is the identity map on B. Since $p - I$ is of class C^1 and B is compact, there exists some $c \geq 1$ such that $p - I$ is Lipschitzian with rate c on B. Then, for $t \in [0, 1/c[$ the map p_t is injective and its inverse is Lipschitzian since for $x, y \in B$ one has

$$\|p_t(x) - p_t(y)\| \geq \|x - y\| - t\,\|(p - I)(x) - (p - I)(y)\| \geq (1 - ct)\,\|x - y\|,$$

so that $x = y$ when $p_t(x) = p_t(y)$. Moreover, since $(t, x) \mapsto \det(Dp_t(x))$ is a continuous function on $T \times B$ and is 1 on $\{0\} \times B$, there exists some $\varepsilon > 0$ such that $\det(Dp_t(x)) > 0$ for $(t, x) \in [0, \varepsilon] \times B$. The inverse function theorem ensures that for $t \in [0, \varepsilon]$ the image $p_t(U)$ of the open unit ball $U := B \backslash S$ is an open subset of B, hence is contained in U. Since $p_t(S) = S$ and $U \cap S = \varnothing$, we have $U \backslash p_t(U) = U \backslash p_t(B)$, so that $U \backslash p_t(U)$ is open, $p_t(B)$ being compact. Since U is connected and is the union of the disjoint open subsets $p_t(U)$ and $U \backslash p_t(U)$, with $p_t(U)$ nonempty, we have $U = p_t(U)$.

Let

$$f(t) := \int_U \det Dp_t d\lambda_d \qquad t \in T.$$

The change of variable theorem ensures that for $t \in [0, \varepsilon]$, $f(t)$ is the measure of the open set $p_t(U) = U$. Since $f(\cdot)$ is a polynomial, it is constant on T, with value $f(0) = \lambda_d(U) > 0$. However, since $\|p(x)\|^2 = 1$ for $x \in B$ we have $Dp(x).p(x) = 0$ for $x \in B$; since $p(x)$ is nonzero, this relation shows that $Dp(x)$ is not an isomorphism, hence that $\det Dp(x) = 0$. Thus $f(1) = 0$, a contradiction. □

The second step of the proof is given by the next lemma.

Lemma 10.7 *If there were a continuous retraction from B onto S, then there would exist a retraction of class C^1 from B onto S.*

Proof Suppose there exists a continuous retraction q from B onto S. Approximating the components of q by functions of class C^∞, we would obtain a sequence (q_n) of maps of class C^∞ on B such that $(\|q_n - q\|_\infty) \to 0$, where $\|\cdot\|_\infty$ is the norm of uniform convergence. Let $h : \mathbb{R} \to [0, 1]$ be a bump function of class C^∞ satisfying $h(0) = 1$, $h(r) = 0$ for $r \in \mathbb{R} \backslash [-1, 1]$ and let $h_n : \mathbb{R}^d \to \mathbb{R}$ be given by

$h_n(x) := h(n\|x\|^2 - n)$, so that $h_n(x) = 1$ for all $x \in S$ and $(h_n(x)) \to 0$ for $x \in \mathbb{R}^d \backslash S$. Let $p_n : B \to \mathbb{R}^d$ be given by

$$p_n(x) := h_n(x)x + (1 - h_n(x))q_n(x) \qquad x \in B.$$

Since $h_n(x) \in [0, 1]$, we have $r_n(x) := \|p_n(x)\| \leq \max(\|q_n(x)\|, 1)$. Let us show that there exist some $c > 0$ and $m \in \mathbb{N}$ such that $r_n(x) \geq c$ for all $x \in B$ and $n \geq m$. Otherwise we could find an infinite subset N of $\mathbb{N}$ and $\overline{x} \in B$, $t \in [0, 1]$ such that $(r_n(x_n))_{n \in N} \to 0$, $(x_n)_{n \in N} \to \overline{x}$, $(h_n(x_n)) \to t$, hence $t\overline{x} + (1-t)q(\overline{x}) = 0$. The case $\overline{x} \in S$ would be excluded since then $q(\overline{x}) = \overline{x}$ whereas the case $\overline{x} \in U$ would be excluded too since then we would have $h_n(x_n) = 0$ for $n \in N$ large enough, hence $q(\overline{x}) = 0$, a contradiction with $q(\overline{x}) \in S$. We conclude that, for $n \geq m$, (p_n/r_n) is a map of class C^∞ from B to S satisfying $p_n(x)/r_n(x) = x$ for all $x \in S$ since then $h_n(x) = 1$ and $p_n(x) = x$, $r_n(x) = 1$. □

Additional Reading

[11, 72, 76, 79, 85, 91, 93, 114, 147, 152, 154, 159, 161, 173, 180, 182, 188, 194, 200, 204, 230, 244, 246, 249, 261, 262, 266]

References

1. R.A. Adams, Sobolev Spaces, Academic Press, 1975.
2. R.P. Agarval and D. O'Regan, Ordinary and Partial Differential Equations With Special Functions, Fourier Series, and Boundary Value Problems, Universitext, Springer, New York, 2009.
3. S. Agmon, Lectures on Elliptic Boundary Value Problems, Van Nostrand, Princeton, 1965.
4. A.A. Agrachev, Geometry of optimal control problems and Hamiltonian systems, in Nonlinear and Optimal Control Theory, A.A. Agrachev et al. (eds) Lecture Notes in Mathematics #1932, 1–59, (2000).
5. N.I. Akhiezer, The Calculus of Variations, Harwood, Chur, 1988.
6. N. Akhiezer and I. Glazman, Theory of Linear Operators in Hilbert Spaces, Pitman, 1980.
7. A. Ambrosetti and G. Prodi, A Primer of Nonlinear Analysis, Cambridge University Press, London, 1993.
8. A. Ambrosetti and D. Arcoya Alvarez, An Introduction to Nonlinear Functional Analysis and Elliptic Problems, Progress in Nonlinear Differential Equations and Their Applications #82, Birkhäuser, Boston, 2011.
9. H. Amann, Analysis II, Birkhäuser, Basel, 1999, 2008.
10. T.M. Apostol, Mathematical Analysis, Addison-Wesley, Reading, Massachusetts, 1974.
11. V.I. Arnold, Ordinary Differential Equations, Universitext, Springer, Berlin, 1992.
12. J.-P. Aubin and I. Ekeland, Applied Nonlinear Analysis, Wiley Interscience, New York, 1984.
13. J.-P. Aubin and H. Frankowska, Set-Valued Analysis, Birkhäuser, Boston, 1990.
14. D. Azé, Eléments d'Analyse Convexe et Variationnelle, Ellipse, Paris, 1997.
15. D. Azé and J.-B. Hiriart-Urruty, Analyse Variationnelle et Optimisation, Cepadues, Toulouse, 2010.
16. D. Azé, G. Constans, J.-B. Hiriart-Urruty, Calcul Différentiel et Equations Différentielles, Dunod, Paris, 2002.
17. D. Azé and J.-P. Penot, Uniformly convex and uniformly smooth convex functions, Ann. Fac. Sciences Toulouse, 4 no 4, 705–730 (1995).
18. G. Bachman, L. Narici, E. Beckenstein, Fourier and Wavelet Analysis, Springer-Verlag, New York, 2000.
19. M. Badiale and E. Serra, Semilinear Elliptic Equations for Beginners, Existence Results via the Variational Approach, Universitext, Springer, 2011.
20. H. Bahouri, J.-Y. Chemin and R. Danchin, Fourier Analysis and Nonlinear Partial Differential Equations, Grundlehren der mathematischen Wissenschaften #343, Springer-Verlag, Berlin, 2011.
21. I.J. Bakelman, Convex Analysis and Nonlinear Geometric Elliptic Equations, Springer-Verlag, Berlin, 1994.

J.-P. Penot, *Analysis*, Universitext, DOI 10.1007/978-3-319-32411-1

22. A. Balakrishnan, Applied Functional Analysis, Applications of Mathematics 3, Springer, New York, 1976.
23. V. Barbu, Nonlinear Semigroups and Differential Equations in Banach Spaces, Noordhoff, Leyden, 1978.
24. V. Barbu, Partial Differential Equations and Boundary Value Problems, Kluwer, Dordrecht, 1998.
25. V. Barbu, Nonlinear Differential Equations of Monotone Types in Banach Spaces, Springer Monographs in Mathematics, Springer, New York, 2010.
26. V. Barbu and G. Da Prato, Hamilton-Jacobi Equations in Hilbert Spaces, Research Notes in Mathematics #86, Pitman, Boston, 1983.
27. M. Bardi and I. Capuzzo-Dolcetta, Optimal Control and Viscosity Solutions to Hamilton-Jacobi-Bellman Equations, Birkhäuser, Basel, 1997, 2008.
28. T. Bascelli et al., Fermat, Leibniz, Euler, and the gang: the true history of the concept of limit and shadow, Notices of the American Mathematical Society, 61, no 8, 848–864 (2014).
29. H.H. Bauschke and P.-L. Combettes, Convex Analysis and Monotone Operator Theory in Hilbert Spaces, CMS Books in Mathematics, Springer, 2011.
30. R. Beals, Advanced Mathematical Analysis, Graduate Texts in Mathematics #12, Springer, New York, 1973.
31. J.J. Benedetto-M.W. Frazier, Wavelets, Mathematics and Applications, CRC Press, Boca Raton, 1994.
32. P. Bérard, Spectral Geometry: Direct and Inverse Problems, Lecture Notes in Mathematics 1207, Springer, 1986.
33. S.K. Berberian, Lectures in Functional Analysis and Operator Theory, Graduate Texts in Mathematics #15, Springer, New York, 1974.
34. S.K. Berberian, A First Course in Real Analysis, Springer, New York, 1994.
35. S.K. Berberian and P. R. Halmos, Lectures in Functional Analysis and Operator Theory, Graduate Texts in Mathematics, Springer, New York, 1974, 2014.
36. M. Berger, Geometry of the Spectrum, in: Differential Geometry (S.S. Chern and R. Osserman eds.), pp. 129–152, Proc. Sympos. Pure Math. Vol 27, Part 2, American Mathematical Society, Providence, 1975.
37. L. Bers, F. John, M. Schechter, Partial Differential Equations, American Mathematical Society, Providence, Rhode Island, 1979.
38. P. Billingsley, Probability and Measure, Anniversary Edition, Wiley, Hoboken, New Jersey, 1979, 1986, 1995, 2012.
39. G. Birkhoff, A Source Book in Classical Analysis, Harvard University Press, Cambridge, MA, 1973.
40. G.A. Bliss, Lectures on the Calculus of Variations, The University of Chicago Press, Chicago, 1946, 1968.
41. V.I. Bogachev, Measure Theory I, Springer, Berlin, 2007.
42. V.I. Bogachev, Measure Theory II, Springer, Berlin, 2007.
43. J.M. Borwein, Maximal monotonicity via convex analysis, J. Convex Anal. 13, 561–586 (2006).
44. J.M. Borwein and A.S. Lewis, Convex Analysis and Nonlinear Optimization. Theory and Examples, Canadian Mathematical Society, Books in Mathematics, Springer-Verlag, New York, 2000.
45. J.M. Borwein and J.D. Vanderwerff, Convex Functions: Constructions, Characterizations and Counterexamples, Cambridge University Press, Cambridge, 2010.
46. J.M. Borwein and J. Vanderwerff, Fréchet-Legendre functions and reflexive Banach spaces, J. Convex Anal. 17 no 3&4, 915–924 (2010).
47. J.M. Borwein and Q.J. Zhu, Techniques of Variational Analysis, CMS Books in Mathematics, Springer, 2005
48. J.M. Borwein and Q.J. Zhu, Variational methods in convex analysis, J. Global Optim. 35 no 2, 197–213 (2006).

49. N. Bourbaki, Elements d'Histoire des Mathématiques, Masson, Paris, 1984, Springer, Berlin, 2007.
50. A. Bressan, Lecture Notes on Functional Analysis With Applications to Linear Partial Differential Equations, Graduate Studies in Mathematics #143, American Mathematical Society, Providence, Rhode Island, 2013.
51. H. Brézis, Opérateurs Maximaux Monotones et Semi-groupes de Contractions dans les Espaces de Hilbert, Mathematics Studies #5, North Holland, Amsterdam, 1973.
52. H. Brezis, Functional Analysis, Sobolev Spaces and Partial Differential Equations, Universitext, Springer, 2011; adapted from "Analyse Fonctionnelle", Masson, Paris, 1983.
53. H. Brezis, J.M. Coron and L. Nirenberg, Free vibrations for a nonlinear wave equation and a theorem of Rabinowicz, Comm. Pure and Appl. Math. 33, 667–689 (1980).
54. M. Briane and G. Pagès, Analyse. Théorie de l'Intégration. Convolution et Transformée de Fourier, Vuibert, Paris 2012.
55. Brook and Chacon, Continuity and compactness of measures, Adv. in Math. 37, 16–26 (1980).
56. A. Browder, Mathematical Analysis. An Introduction, Undergraduate Texts in Mathematics, Springer-Verlag, New York, 1996.
57. F.E. Browder, Existence theorems for nonlinear partial differential equations, in "Global Analysis", Proceedings of Symposia in Pure Mathematics 16, American Mathemetical Society, Providence, 1970.
58. F.E. Browder, Nonlinear Operators and Nonlinear Equations of Evolution in Banach Spaces, Proceedings of Symposia in Pure Mathematics 18, Part 2, American Mathematical Society, Providence, 1976.
59. A. Brown and C. Pearcy, An Introduction to Analysis, Graduate Texts in Mathematics #154, Springer-Verlag, New York, 1995.
60. A.M. Bruckner, Differentiation of Real Functions, Springer-Verlag, Berlin, 1978, American Mathematical Society, Providence, R.I., 1994.
61. G. Buttazzo, M. Giaquinta, and S. Hildebrandt, One-Dimensional Variational Problems, Oxford University Press, Oxford, 1998
62. E. Cancès, C. Le Bris, Y. Maday, Méthodes Mathématiques en Chimie Quantique. Une introduction, Mathématiques et Applications #53, Springer-Verlag, Berlin, 2006.
63. C. Canuto and A. Tabacco, Mathematical Analysis II, Universitext, Springer-Verlag Italia, Milan, 2010.
64. M. Capinski and E. Kopp, Measure, Integral and Probability, Springer Undergraduate Mathematics Series, Springer-Verlag, London 1999.
65. L. Carleson, On convergence and growth of partial sums of Fourier series, Acta Math. 116, 135–157 (1966).
66. H. Cartan, Differential Calculus, Mifflin Co, Boston, 1971, translated from "Cours de Calcul Différentiel", Hermann, Paris, 1971.
67. J. Cerdà, Linear Functional Analysis, Graduate Studies in Mathematics #116, American Mathematical Society, Providence, 2010.
68. K.-C. Chang, Methods in Nonlinear Analysis, Springer Monographs in Mathematics, Springer-Verlag, Berlin, 2005.
69. I. Chavel, Eigenvalues in Riemannian Geometry, Academic Press, 1994.
70. Y.-Z. Chen and L.C. Wu, Second Order Elliptic Equations and Elliptic Systems, Translation of Mathematical Monographs #174, American Mathematical Society, Providence, 1998.
71. W. Cheney, Analysis for Applied Mathematics, Graduate Texts in Mathematics #208, Springer-Verlag, New York, 2001.
72. P. Cherrier and A. Milani, Linear and Quasi-linear Evolution Equations in Hilbert Spaces, Graduate Studies in Mathematics #135, American Mathematical Society, Providence, 2012.
73. Ch. Chidume, Geometric Properties of Banach Spaces and Nonlinear Iterations, Lecture Notes in Mathematics #1965, Springer-Verlag, London 2009.
74. M. Chipot, Elliptic Equations: An Introductory Course, Birkhäuser Advanced Texts, Birkhäuser, Basel 2009.

75. G. Choquet, Topology, Academic Press 1966, translated from: Cours d'Analyse, Tome II, Topologie, Masson, Paris 1969.
76. A. Cialdea and V. Maz'ya, Semi-bounded Differential Operators, Contractive Semigroups and Beyond, Birkhäuser, Basel, 2014.
77. Ph. Ciarlet, Linear and Nonlinear Functional Analysis With Applications, SIAM, Philadelphia, 2013.
78. F.H. Clarke, Functional Analysis, Calculus of Variations and Optimal Control, Graduate Texts in Mathematics #264, Springer, London, 2013.
79. E. Coddington and N. Levinson, Theory of Ordinary Differential Equations, McGraw Hill, 1955.
80. D.L. Cohn, Measure Theory, Birkhäuser, Boston, 1980, 1993, 1996, 1997.
81. P. L. Combettes, J.-B. Hiriart-Urruty and M. Théra, Preface to Modern Convex Analysis, special issue of Mathematical Programming, series B 148, no 1–2 (2014).
82. J.B. Conway, A Course in Functional Analysis, Graduate Texts in Mathematics 96, Springer, New York, 1990.
83. R. Courant and D. Hilbert, Methods of Mathematical Physics, I, II, Intersience, 1962.
84. R. Courant and F. John, Introduction to Calculus and Analysis I, Classics in Mathematics, Springer-Verlag, Berlin, 1946, 1989, 1999.
85. M.G. Crandall and A. Pazy, Nonlinear semi-groups of contractions and dissipative sets, J. Funct. Anal. 3, 376–418 (1969).
86. B. Dacorogna, Direct Methods in the Calculus of Variations, Applied Mathematical Sciences #78, Springer, New York, 2008.
87. M. Danesi, Discovery in Mathematics: An Interdisciplary Perspective, Lincom Europa, Muenchen, 2013.
88. R. Dautray and J.-L. Lions, Mathematical Analysis and Numerical Methods for Science and Technology, (6 volumes), Springer, 1988, Masson, Paris, 1988.
89. B. Davies, Integral Transforms and Their Applications, Texts in Applied Mathematics #41, Springer, 1978, 1985, 2002.
90. S.R. Deans, The Radon Transform and Some of its Applications, Wiley, New York, 1973.
91. K. Deimling, Ordinary Differential Equations in Banach Spaces, Lecture Notes in Mathematics, #596, Springer-Verlag, Berlin, 1977.
92. K. Deimling, Nonlinear Functional Analysis, Springer-Verlag, Berlin, 1985. 574
93. K. Deimling, Multivalued Differential Equations, De Gruyter Series in Nonlinear Analysis and Applications 1, De Gruyter, Berlin, 1992.
94. F. Demengel and G. Demengel, Functional Spaces for the Theory of Elliptic Partial Differential Equations, Universitext, Springer, London, 2012.
95. R. Deville, G. Godefroy and V. Zizler, Smoothness and Renormings in Banach Spaces, Pitman Monographs 64, Longman, 1993.
96. E. DiBenetto, Partial Differential Equations, Birkhäuser, Basel, 2009.
97. J. Diestel, Geometry of Banach Spaces: Selected Topics, Springer, New York, 1975.
98. J. Diestel, Sequences and Series in Banach Spaces, Graduate Texts in Mathematics vol. 92, Springer, New York, 1984.
99. J. Diestel and J.J. Uhl Jr., Vector Measures, Mathematical Surveys #15, American Mathematical Society, Providence, R.I., 1977.
100. J. Dieudonné, Treatise on Analysis, (8 volumes) Academic Press, New York and London, 1960, 1969.
101. J. Dieudonné, Panorama des Mathématiques Pures. Le Choix Bourbachique, Gauthier-Villars, Bordas, Paris, 1977.
102. A.L. Dontchev and R.T. Rockafellar, Implicit Functions and Solutions Mappings. A View from Variational Analysis, Springer Monographs in Mathematics, Springer Dordrecht, 2009.
103. P. Drábek and J. Milota, Methods of Nonlinear Analysis. Applications to Differential Equations, Birhäuser Advanced Texts, Birhäuser, Basel, 2007.
104. R.M. Dudley and R. Norvaiša, Concrete Functional Calculus, Springer Monographs in Mathematics, Springer, New York, 2011.

105. P. Dugac, Histoire de l'Analyse, Vuibert, Paris, 2003.
106. N. Dunford and J.T. Schwartz, Linear Operators, I, General Theory, Pure and Applied Mathematics vol. 7, Wiley Interscience, New York, 1958, 1967.
107. Dunkl, Ch. F., Xu, Y., Orthogonal Polynomials of Several Variables, Encyclopedia of Mathematics and its Applications 81, Cambridge University Press, Cambridge, 2001.
108. W.F. Eberlein, Weak compactness in Banach spaces, I, Proc. Nat. Acad. Sci. U.S.A. 33, 51–53, 1947.
109. R. Edwards, Functional Analysis, Holt-Rinehart-Winston, 1965.
110. C.H. Edwards Jr., The Historical Development of the Calculus, Springer, 1979.
111. Y. Eidelman, V. Milman and A.Tsolomitis, Functional Analysis. An Introduction, Graduate Studies on Mathematics #66, American Mathematical Society, Providence, Rhode Island, 2004.
112. I. Ekeland and R. Temam, Convex Analysis and Variational Problems, Classics in Applied Mathematics #28, Society for Industrial and Applied Mathematics, Philadelphia, PA, 1999, translated from the French: Analyse Convexe et Problèmes Variationnels, Dunod, Paris, 1974.
113. K.-J. Engel and R. Nagel, One-parameter Semigroups for Linear Evolution Equations, Graduate Texts in Mathematics #194, Springer, New York, 2000.
114. K.-J. Engel and R. Nagel, A Short Course on Operator Semigroups, Universitext, Springer, New York, 2006.
115. B. Epstein, Introduction to Lebesgue Integration and Infinite Dimensional Problems, Saunders, Philadelphia, 1970.
116. G. Eskin, Lectures on Linear Partial Differential Equations, Graduate Studies in Mathematics #123, American Mathematical Society, Providence, Rhode Island, 2011.
117. L.C. Evans, Partial Differential Equations, Graduate Studies in Mathematics 19, American Mathematical Society, Providence, Rhode Island, 1998, 2002, 2008, 2010 .
118. L.C. Evans and R.F. Gariepy, Measure Theory and Fine Properties of Functions, Studies in Advanced Mathematics, CRC Press, Boca Raton FL, 1992.
119. M. Fabian, P. Habala, P. Hájek, J. Pelant, V. Montesinos, and V. Zizler, Functional Analysis and Infinite Dimensional Geometry, CMS Books # 8, Springer, New York 2001.
120. M. Fabian, P. Habala, P. Hájek, V. Montesinos, and V. Zizler, Banach Space Theory, The Basis for Linear and Nonlinear Analysis, CMS Books in Mathematics, Springer, New York, 2011.
121. T. Figiel, On the moduli of convexity and smoothness, Studia Math 56(2), 121–155 (1976).
122. T.G. Foeman, The Mathematics of Medical Imagery, Undergraduate Texts in Mathematics and Technology, Springer, 2010.
123. G.B. Folland, Introduction to Partial Differential Equations, Princeton University Press, 1976.
124. I. Fonseca and G. Leoni, Modern Methods in the Calculus of Variations: L^p Spaces, Springer Monographs in Mathematics, Springer, New York, 2007.
125. J. Foran, Fundamentals of Real Analysis, Marcel Dekker, New York, 1991.
126. D.H. Fremlin, Measure Theory, Volume 1, The Irreducible Minimum, Torres Fremlin, Colchester, 2000, 2001, 2004.
127. D.H. Fremlin, Measure Theory, Volume 2, Broad Foundations, Torres Fremlin, Colchester, 2001.
128. A. Friedman, Partial Differential Equations, Holt, Rinehart and Winston, New York, 1969.
129. B. Friedman, Lectures on Applications-Oriented Mathematics, Wiley Classics Library, John Wiley, New York, 1964, 1991.
130. G. Gagneux and M. Madaune-Tort, Analyse mathématique de modèles nonlinéaires de l'ingénierie pétrolière, Mathematics and Applications no 22, Springer, Berlin, 1996.
131. P. Garabedian, Partial Differential Equations, Wiley, 1964.
132. R.F. Gariepy and W.P. Ziemer, Modern Real Analysis, PWS Publishing company, Boston, 1994.
133. M.I. Garrido and J. Jaramillo, Homomorphisms on function lattices, Monatsh. Math. 141, 127–146, (2004)
134. I.M. Gelfand and S.V. Fomin, Calculus of Variations, Prentice Hall, Englewood Cliffs, N.J., 1963.

135. N. Ghossoub and D. Preiss, A general mountain pass principle for locating and classifying critical points, Ann. Inst. H. Poincaré, 6, 321–330, (1989).
136. M. Giaquinta, Multiple Integrals in The Calculus of Variations and Nonlinear Elliptic Systems, Princeton University Press, 1983.
137. M. Giaquinta and S. Hildebrandt, Calculus of Variations I: the Lagrangian Formalism, Grundlehren der mathematischen Wissenschaften #310, Springer-Verlag, Berlin, 1996.
138. M. Giaquinta and S. Hildebrandt, Calculus of Variations II: the Hamiltonian Formalism, Grundlehren der mathematischen Wissenschaften #311, Springer-Verlag, Berlin, 1996.
139. D. Gilbarg and N. Trudinger, Elliptic Partial Differential Equations of Second Order, Springer, Classics in Mathematics, 1977, 1998, 2001.
140. J.-M. Gilsinger and M. El Jai, Eléments d'Analyse Fonctionnelle. Fondements et Applications aux Sciences de l'Ingénieur, Presses polytechniques romandes, Lausanne, 2010.
141. O. Giraud and K. Thas, Hearing shapes of drums-mathematical and physical aspects of isospectrality, Mathematical Physics 82, 2213–2245, 2010.
142. E. Giusti, Direct Methods in the Calculus of Variations, World Scientific, Singapore, 2003.
143. S. Givant, Duality Theories for Boolean Algebras with Operators, Springer Monographs in Mathematics, Cham, 2014.
144. R. Glowinski, Numerical Methods for Nonlinear Variational Problems, Computational Physics Series, Springer Verlag, Heidelberg, 1984.
145. R. Glowinski, J.-L. Lions and R. Trémolières, Numerical Analysis of Variational inequalities, North Holland, Amsterdam, 1981.
146. R. Godement, Analysis I, II, Universitext, Springer-Verlag, Berlin, 2005.
147. J. Goldstein, Semigroups of Operators and Applications, Oxford University Press, Oxford, 1985.
148. H.H. Goldstine, A History of the Calculus of Variations, Springer-Verlag, New York, 1980.
149. L. Grafakos, Classical and Modern Fourier Analysis, Graduate Texts in Mathematics #249, Springer, New York, 2008.
150. P.R. Halmos, Measure Theory, Van Nostrand, Princeton, 1950, Graduate Texts in Mathematics # 18, Springer, New York, 1974.
151. Qing Han, A Basic Course in Partial Differential Equations, Graduate Studies in Mathematics #120, American Mathematical Society, Providence 2011.
152. A. Haraux, Nonlinear Evolution Equations_Global Behavior of Solutions, Lecture Notes in Mathematics #841, Springer-Verlag, Berlin, 1981.
153. D. Haroske and H. Triebel, Distributions, Sobolev Spaces, Elliptic Equations, European Mathematical Society, Zürich, 2008.
154. Ph. Hartman, Ordinary Differential Equations, Wiley, New York, 1964.
155. H. Hattori, Partial Differential Equations, Methods, Applications and Theories, World Scientific, Singapore, 2013.
156. B. Hauchecorne and D. Suratteau, Des Mathématiciens de A à Z, Ellipses, Paris, 2008.
157. S. Helgason, The Radon Transform, Birkhäuser, Basel, 1980.
158. E. Hernandez, G. Weiss, A First Course on Wavelets, Studies in Advanced Mathematics, CRC Press, Boca Raton, Florida, 1996.
159. E. Hewitt and K. Stromberg, Real and Abstract Analysis, Springer, New York (1965).
160. E. Hille, Methods in Classical and Functional Analysis, Addison-Wesley, 1972.
161. E. Hille and R.S. Phillips, Functional Analysis and Semi-groups, American Mathemetical Society, Providence, R.I., 1974.
162. J.-B. Hiriart-Urruty and C. Lemaréchal, Fundamentals of Convex Analysis, Springer, Berlin, 2001.
163. F. Hirsch and G. Lacombe, Elements of Functional Analysis, Graduate Texts in Mathematics #192, Springer, New York, 1999.
164. R.B. Holmes, A Course on Optimization and Best Approximation. Lecture Notes in Mathematics #257, Springer-Verlag, New York, 1972.
165. R.B. Holmes, Geometric Functional Analysis and its Applications, Graduate Texts in Mathematics #24, Springer-Verlag, New York 1975.

166. A.D. Ioffe and V.M. Tikhomirov, Theory of Extremal Problems, Studies in Mathematics and Its Applications, North Holland, Amsterdam, 1979, translated from the Russian, Nauka, Moscow, 1974.
167. M. Ivanov and N. Zlateva, A (hopefully) new proof of maximality of the subdifferential operator of a convex function, preprint, February 2015.
168. J. Jacod and P. Protter, Probability Essentials, Universitext, Springer-Verlag, Berlin, 2000, 2003, 2004.
169. J. Jost, Postmodern Analysis, Universitext, Springer, New York, 1997, 2002, 2005.
170. J. Jost and X. Li-Jost, Calculus of Variations, Cambridge studies in advanced mathematics #64, Cambridge University Press, Cambridge, 1998.
171. M. Kac, Can one hear the shape of a drum?, American Math. Monthly 73, pp. 1–23 (1966).
172. T. Kato, Perturbation Theory for Linear Operators, Grundlehren der mathematischen Wissenschaften #132, Springer-Verlag, Berlin, 1966, 1976.
173. T. Kato, Accretive Operators and Nonlinear Evolution Equations in Banach Spaces, Proceedings of Symposia in Pure Mathematics 18, Part 2, pp. 138–161, American Mathemetical Society, Providence, 1976.
174. J. Kelley, General Topology, Van Nostrand, New York, 1955.
175. D. Kinderlehrer and G. Stampacchia, An Introduction to Variational Inequalities and Their Applications, Academic Press, New York, 1980.
176. A. Komolgorov and S. Fomin, Introductory Real Analysis, Prentice-Hall, 1970.
177. T.W. Körner, A Companion to Analysis. A Second First and First Second Course in Analysis, Graduate Studies in Mathematics #62, American Mathematical Society, Providence, Rhode Island, 2004.
178. A.M. Krall, Applied Analysis, D.Reidel, Dordrecht, Holland, 1986.
179. M.A. Krasnoselskii, P.P. Zabreiko, E.I. Pustylnik, P.E. Sobolevskii, Integral operators in spaces of summable functions, Noordhoff, Groningen, 1976.
180. N.V. Krylov, Lectures on Elliptic and Parabolic Equations in Sobolev Spaces, Graduate Studies in Mathematics #96, American Mathematical Society, Providence, Rhode Island, 2008.
181. D. Labate, G. Weiss, E. Wilson, Wavelets, Notices Amer. Math. Soc. 60, no1, 66–76 (2013).
182. S. Lang, Analysis II, Addison-Wesley, Reading, Massachusetts, 1969.
183. S. Lang, Real and Functional Analysis, Springer-Verlag, New York, 1993.
184. P. Lax, S. Burstein and A. Lax, Calculus With Applications and Computing, I, II, Undergraduate Texts in Mathematics, Springer-Verlag, New York, 1976, 1984.
185. E. H. Lieb and M. Loss, Analysis, Graduate Studies in Mathematics #14, American Mathematical Society, Providence, 1997, 2001.
186. J. Lindenstrauss, A short proof of Liapounoff's convexity theorem, J. Math. Mech. 15, 1966, 971–972.
187. J. Lindenstrauss and L. Tzafriri, Classical Banach Spaces, (two volumes), Springer, 1973, 1979
188. J-L. Lions, Quelques Méthodes de Résolution des Problèmes aux Limites Non-Linéaires, Dunod-Gauthier-Villars, Paris, 1969.
189. D. Luenberger, Optimization by Vector Spaces Methods, Wiley, New York, 1969.
190. B. Makarov and A. Podkorytov, Real Analysis: Measures, Integrals and Applications, Universitext, Springer-Verlag, London, 2013.
191. D.P. Maki and M. Thompson, Mathematical Models and Applications, Prentice Hall, Englewood Cliffs, N.J., 1973.
192. P.A. Markowich, Applied Partial Differential Equation. A Visual Approach, Springer, Berlin, 2007.
193. C.-M. Marle, Mesures et Probabilités, Hermann, Paris, 1974.
194. R.H. Martin Jr., Nonlinear Operators and Differential Equations in Banach Spaces, Wiley, New York, 1976.
195. V.G. Maz'ja, Sobolev Spaces, With Applications to Elliptic Partial Differential Equations, Grundlehren der mathematischen Wissenschaften #342, Springer-Verlag, Berlin, 1985, 2011.

196. R.E. Megginson, An Introduction to Banach Space Theory, Graduate Texts in Mathematics #183, Springer, New York, 1998.
197. J. Milnor, Analytic proofs of the "Hairy ball theorem" and the Brouwer fixed point theorem, Amer. Math. Monthly 85, 521–524, 1978.
198. J.–J. Moreau, Fonctionnelles Convexes, Collège de France, Paris, 1966–1967.
199. C.B. Morrey, Multiple Integrals in the Calculus of Variations, Springer-Verlag, New York, 1966.
200. D. Motreanu, N. Pavel, Tangency, Flow Invariance for Differential Equations, and Optimization Problems, Dekker, 1999.
201. F. Natterer, The Mathematics of Computerized Tomography, Teubner, Stuttgart, Wiley, Chichester, 1986, 1989.
202. J. Nečas, Direct Methods in the Theory of Elliptic Equations, Springer Monographs in Mathematics, Springer-Verlag, Berlin, 2012, translated from "Les Méthodes Directes en Théorie des Equations Elliptiques", Academia, Praha and Masson, Paris, 1967.
203. R. Osserman, Isoperimetric inequalities and eigenvalues of the Laplacian, in Proc. International Congress of Mathematicians, Helsinki, 1978 and Bull. Amer. Math. Soc. 84, pp.1182–1238, 1978.
204. A. Pazy, Semigroups of Linear Operators and Applications to Partial Differential Equations, Springer, 1983.
205. J.-P. Penot, Subdifferential calculus without qualification assumptions, J. Convex Anal. 3 (2), 1–13, (1996).
206. J.-P. Penot, Well-behavior, well-posedness and nonsmooth analysis, Pliska Stud. Math. Bulgar. 12, 141–190 (1998).
207. J.-P. Penot, The relevance of convex analysis for the study of monotonicity, Nonlin. Anal. 58, 855–871 (2004).
208. J.-P. Penot, Calculus Without Derivatives, Graduate Texts in Mathematics #266, Springer, New York, 2013.
209. R.R. Phelps, Convex Functions, Monotone Operators, and Differentiability, Lecture Notes in Mathematics #1364, Springer, Berlin, 1988.
210. E.R. Pinch, Optimal Control and the Calculus of Variations, Oxford, 1993.
211. Y. Pinchover and J. Rubinstein, An Introduction to Partial Differential Equations, Cambridge University Press, 2005.
212. M.A. Pinsky, Introduction to Fourier Analysis and Wavelets, Graduate Studies in Mathematics #102, American Mathematical Society, Providence, 2002.
213. M.A. Pons, Real Analysis for the Undergraduate. With an Introduction to Functional Analysis, Springer, New York, 2014.
214. P. Pucci and J. Serrin, The Maximum Principle, Birkhäuser, Basel, 2007.
215. J. Rauch, Partial Differential Equations, Graduate Texts in Mathematics #128, Springer, New York, 1991.
216. M. Renardy and R.C. Rogers, An Introduction to Partial Differential Equations, Springer Texts in Applied Mathematics #13, Springer, New York, 1993, 2004.
217. F. Riesz and B.Sz.-Nagy, Functional Analysis, Dover Publications, 1990, translated from Leçons d'Analyse Fonctionnelle, Académie des Sciences de Hongrie, 1954.
218. R.T. Rockafellar, Characterization of the subdifferentials of convex functions, Pacific J. Math. 17 no 3, 497–510, (1966).
219. R.T. Rockafellar, On the maximal monotonicity of subdifferential mappings, Pacific J. Math. 33 no 1, 209–216, (1966).
220. R.T. Rockafellar, Convex Analysis, Princeton University Press, Princeton, 1970.
221. R.T. Rockafellar and R. J.-B. Wets, Variational Analysis, Grundlehren der mathematischen Wissenschaften #317, Springer-Verlag, Berlin, 1998.
222. C.A. Rogers, A less strange version of Milnor's proof of Brouwer's fixed point theorem, Amer. Math. Monthly 87, 525–527 (1980).
223. H. Royden and P. Fitzpatrick, Real Analysis, Pearson, Boston, 2010.
224. W. Rudin, Real and Complex Analysis, McGraw-Hill, 1966, 1974, 1987.

225. W. Rudin, Functional Analysis, McGraw-Hill, New York, 1973, 1991.
226. S. Saks, Theory of the integral, Monografje Matematyczne, Warsaw, 1937, Dover, New York, 1947.
227. H.H. Schaefer and M.P. Wolff, Topological Vector Spaces, Graduate Texts in Mathematics #3, Springer, New York 1966, 1999.
228. M. Schechter, Principles of Functional Analysis, Graduate studies in Mathematics #36, American Mathematical Society, Providence, 2002.
229. G.E. Shilov, Elementary Functional Analysis, Dover New York, 1974.
230. R.E. Showalter, Monotone Operators in Banach Spaces and Nonlinear Partial Differential Equations, Mathematical Surveys and Monographs #49, American Mathematical Society, Providence, 1997.
231. A.H. Siddiqi, Applied Functional Analysis. Numerical Methods, Wavelet Methods, and Image Processing, Marcel Dekker, New York, 2004.
232. S. Simons, From Hahn-Banach to Monotonicity, Lecture Notes in Mathematics #1693, Springer, New York, 2008.
233. S. Simons and C. Zalinescu, A new proof for Rockafellar's characterization of maximal monotone operators, Proc. Amer. Math. Soc. 132, 2969–2972 (2004).
234. S. Simons and C. Zalinescu, Fenchel duality, Fitzpatrick functions and maximal monotonicity, J. Nonlin. Convex Anal. 6, 1–22, (2005).
235. I.M. Singer, Eigenvalues of the Laplacian and invariants of manifolds, Proc. International Congress of Mathematicians, Vancouver, 1974.
236. L. Sirovich, Introduction to Applied Mathematics, Texts in Applied Mathematics #1, Springer-Verlag, New York 1988.
237. M. Spivak, Calculus on Manifolds, Benjamin, New York, 1965.
238. G. Stampacchia, Equations elliptiques à coefficients discontinus, Presses de l'Université de Montréal, Montréal, 1966.
239. E.M. Stein and R. Shakarchi, Fourier Analysis, an Introduction, Princeton University Press, Princeton, 2003.
240. E.M. Stein and R. Shakarchi, Real Analysis, Princeton University Press, Princeton, 2005.
241. G. Strang, Introduction to Applied Mathematics, Wellesley-Cambridge, Cambridge, Mass., 1986.
242. W. Strauss, Partial Differential Equations: An Introduction, Wiley, New York,1972.
243. M. Struwe, Variational Methods: Applications to Nonlinear Partial Differential Equations and Hamiltonian Systems, Ergebnisse der Mathematik und der Grenzgebedte #34, Springer, Berlin, 2008.
244. J. Szarski, Differential Inequalities, Monografie Matematyczne #43, Panstwowe Wydawnictwo Naukowe, Warsaw, 1965.
245. P. Szekeres, A Course in Modern Mathematical Physics. Groups, Hilbert Spaces and Differential Geometry, Cambridge University Press, Cambridge, 2004.
246. K. Taira, Semigroups, Boundary Value Problems and Markov Processes, Springer Monographs in Mathematics, Springer, Berlin, 2004, 2014.
247. T. Tao, An Introduction to Measure Theory, American Mathematical Society, Providence, 2011.
248. M. Taylor, Partial Differential Equations, vol. I–III, Springer, 1966.
249. L. Tartar, An Introduction to Sobolev Spaces and Interpolation Spaces, Lecture Notes in Mathematics, Springer, Berlin Heidelberg, 2007.
250. R. Teman, Navier-Stokes Equations, North-Holland, Amsterdam, 1966.
251. L. Thibault, Sequential convex subdifferential calculus and sequential Lagrange multipliers, SIAM J. Contr. Optim. 35 no 4, 1434–1444 (1997).
252. H. Triebel, Theory of Function Spaces, (3 volumes) Birhäuser, 1983, 1992, 2006.
253. F. Tröltzsch, Optimal Control of Partial Differential Equations, Graduate Studies in Mathematics #112, American Mathematical Society, Providence, 2010.
254. M.M. Vainberg, Variational Method and Method of Monotone Operators in the Theory of Nonlinear Equations, Wiley, New York, 1973.

255. M. Väth, Topological Analysis. From the basics to the triple degree for Fredholm inclusions, De Gruyter, Berlin, 2012.
256. J.-P. Vial, Strong convexity of sets and functions, Math. Oper. Res. 8, 231–259 (1983).
257. A. A. Vretblad, Fourier Analysis and its Applications, Graduate Texts in Mathematics #223, Springer-Verlag, New York, 2003.
258. J.S. Walker, Fast Fourier Transform, CRC Press, Boca Raton, 1996.
259. H. Weinberger, A First Course on Partial Differential Equations, Blaisell, 1965.
260. J. Wloka, Partial Differential Equations, Cambridge University Press, 1987.
261. A. Yagi, Abstract Parabolic Evolution Equations and their Applications, Springer Monographs in Mathematics, Springer-Verlag, Berlin, 2010.
262. K. Yosida, Functional Analysis, Grundlehren der mathematischen Wissenschaften #123, Springer-Verlag, Berlin, 1965, 1980, 1995.
263. L.C. Young, Lectures on the Calculus of Variations and Optimal Control Theory, Saunders, Philadelphia, 1969.
264. C. Zălinescu, On uniformly convex functions, J. Math. Anal. Appl. 95, 344–374 (1983).
265. C. Zălinescu, Convex Analysis in General Vector Spaces, World Scientific, Singapore, 2002.
266. E. Zeidler, Nonlinear Functional Analysis and its Applications, (4 volumes) Springer, New York, 1994, 1990.
267. A.H. Zemanian, Distribution Theory and Transform Analysis, An Introduction to Generalized Functions, with Applications, Dover, New York, 1965.
268. W.P. Ziemer, Weakly Differentiable Functions, Graduate Texts in Mathematics #120, Springer-Verlag, New York, 1989.
269. N. Zlateva, Integrability through infimal regularization, C.R. Acad. bulgare des Sciences 68 (5), 551–560, (2015).
270. V.A. Zorich, Mathematical Analysis II, Universitext, Springer-Verlag, Berlin, 2004.

Index

J.-P. Penot, *Analysis*, Universitext, DOI 10.1007/978-3-319-32411-1

Printed in the United States
By Bookmasters